International Conference on Advancements in Nuclear Instrumentation Measurement Methods and Their Applications (ANIMMA 2017)

EPJ Web of Conferences Volume 170 (2018)

Liege, Belgium
19 - 23 June 2017

Editor:

A. Lyoussi

ISBN: 978-1-5108-5575-5

Printed from e-media with permission by:

Curran Associates, Inc.
57 Morehouse Lane
Red Hook, NY 12571

Some format issues inherent in the e-media version may also appear in this print version.

Printed by Curran Associates, Inc. (2018)

For additional information, please contact EDP Sciences – Web of Conferences at the address below.

EDP Sciences – Web of Conferences
17, Avenue du Hoggar
Parc d'Activité de Courtabœuf
BP 112
F-91944 Les Ulis Cedex A
France

Phone: +33 (0) 1 69 18 75 75
Fax: +33 (0) 1 69 28 84 91

contact-edps@webofconferences.org

Additional copies of this publication are available from:

Curran Associates, Inc.
57 Morehouse Lane
Red Hook, NY 12571 USA
Phone: 845-758-0400
Fax: 845-758-2633
Email: curran@proceedings.com
Web: www.proceedings.com

TABLE OF CONTENTS

edp sciences EPJ Web of Conferences

Statement of Peer review

In submitting conference proceedings to EPJ Web of Conferences, the editors of the proceedings certify to the Publisher that they adhere to the **Policy on Publishing Integrity** of the journal in order to safeguard good scientific practice in publishing.

1. All articles have been subjected to peer review administered by the proceedings editors.
2. Reviews have been conducted by expert referees, who have been requested to provide unbiased and constructive comments aimed, whenever possible, at improving the work.
3. Proceedings editors have taken all reasonable steps to ensure the quality of the materials they publish and their decision to accept or reject a paper for publication has been based only on the merits of the work and the relevance to the journal.

Title, date and place of the conference

ANIMMA 2017
June 19-23, 2017
LIEGE, Belgium

Proceedings editor(s):

Prof. Dr. Abdallah Lyoussi

Date and editor's signature

December, 11th 2017

https://doi.org/10.1051/epjconf/201817001001

Upgrade of the ATLAS Hadronic Tile Calorimeter for the High Luminosity LHC

Ignacio Asensi Tortajada
on behalf of the ATLAS Tile Calorimeter System
University of the Witwatersrand
Universidad de Valencia
Email: ignacio.asensi@cern.ch

***Abstract*—The Large Hadron Collider (LHC) has envisaged a series of upgrades towards a High Luminosity LHC (HL-LHC) delivering five times the LHC nominal instantaneous luminosity. The ATLAS Phase II upgrade, in 2024, will accommodate the upgrade of the detector and data acquisition system for the HL-LHC. The Tile Calorimeter (TileCal) will undergo a major replacement of its on- and off-detector electronics. In the new architecture, all signals will be digitized and then transferred directly to the off-detector electronics, where the signals will be reconstructed, stored, and sent to the first level of trigger at the rate of 40 MHz. This will provide better precision of the calorimeter signals used by the trigger system and will allow the development of more complex trigger algorithms. Changes to the electronics will also contribute to the reliability and redundancy of the system.**

Three different front-end options are presently being investigated for the upgrade, two of them based on ASICs, and a final solution will be chosen after extensive laboratory and test beam studies that are in progress. A hybrid demonstrator module is being developed using the new electronics while conserving compatibility with the current system. The status of the developments will be presented, including results from the several tests with particle beams.

1. Introduction

The Large Hadron Collider (LHC) is a proton-proton collider with 14 TeV center of mass energy and design luminosity of $10^{34}\ cm^{-1}s^{-1}$. A series of upgrades have been planned which will have an increased average luminosity 5-7 times larger than the nominal Run-2 value. The ATLAS detector [1] is one of the two general purpose particle detectors at the LHC, and it consists of multiple sub-detectors which are designed to detect interesting particle physics events in 40 million bunch-crossings per second. The Tile Calorimeter is the central hadronic calorimeter in the central region of ATLAS (Fig. 1 (top)). Its purpose is to measure energies and directions of hadrons, τ-jets and leptons and contribute to the measurement of the missing transverse energy. It's comprised of two Extended Barrels and a central Long Barrel divided in two sections, each divided into 64 slices. They are made out of alternating thin steel plates and scintillating tiles (Fig. 1 (bottom)). Wavelength shifting fibers coupled to the tiles collect the light produced in the scintillators and are read out by photomultiplier tubes (PMTs).

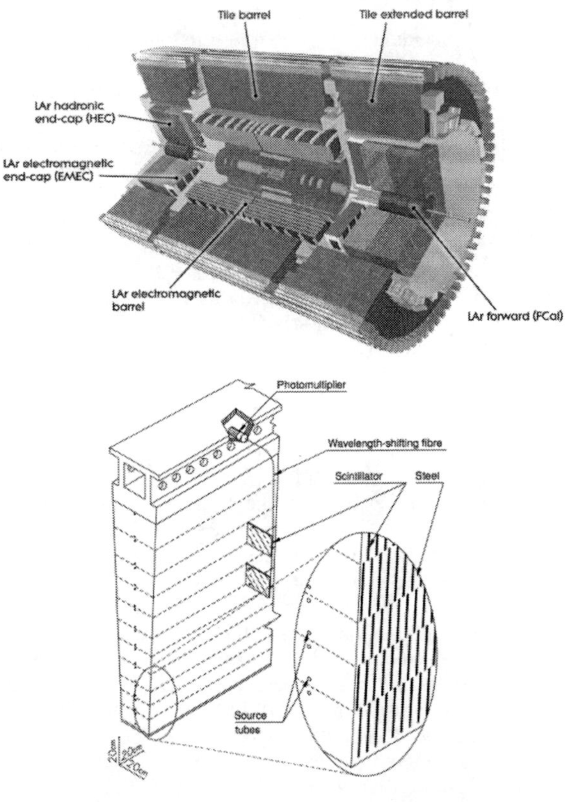

Figure 1. Artistic representation of the ATLAS calorimeters (top) and a Tile Calorimeter module slice (bottom).

2. Upgrade of the Tile Calorimeter

The Phase-II upgrade of the ATLAS Tile Calorimeter will culminate in 2024 for the HL-LHC. The detector components (absorber, scintillating tiles, fibers, and PMTs) are

https://doi.org/10.1051/epjconf/201817001001

largely in good shape and do not have to be replaced. However, the readout electronics are being completely redesigned and will provide data to the trigger at 40 MHz instead of at 100 kHz, with full cell energy precision instead of analog sums in projective towers.

A demonstrator project was defined in order to assess the long term performance of the upgraded system. The Demonstrator super-drawer is composed of four independent mini-drawers (Fig. 2 (top)), each of them equipped with 12 front-end boards of one of the three different options, shown in Fig. 3, one MainBoard for the corresponding front-end option, one Daughter Board that handles data processing and communications with the back-end electronics, one high voltage regulation board from the two options available, one adder base board plus 3 adder cards for the summation of the analog trigger signals to send to the Trigger, and one new Low Voltage Power Supply (LVPS). A fully functional demonstrator is integrated into the current ATLAS Timing, Trigger and Control (TTC) and Data Adquisition system (DAQ). The new low voltage system is designed with a high degree of redundancy, and supply voltages are generated using point-of-load regulators fed from 10V LV supplies. Active dividers in the base of each PMT block increase the linearity of the gain with High Voltage (HV). The chassis that houses the PMT blocks has been redesigned including a cable carrier for easier access during maintenance periods. The single design of the Daughter Board is equipped with a GBTx chip [2] that performs clock synchronization and recovery. The Daughter Board controls and configures the mini-drawers, and performs data serialization and duplicate transfer of the data out to the back-end electronics through the two dual QSFP+ links.

Due to the increase in radiation damage, the front-end pipeline memories will be shifted to the off-detector Tile PreProcessor (PPr) electronics, which will contribute to the serviceability of the system. A functional prototype of the Tile PPr [3] has been designed and produced by the University of the Witwatersrand in collaboration with the Instituto de Fisica Corpuscular in Valencia. A picture of the Tile PPr prototype is shown in Fig. 2 (bottom). Each PPr reads out one mini-drawer through one 10 Gbit Quad Small Form-factor Pluggable (QSFP) link, and serves a complete super-module. Data read-out and control is implemented over Ethernet using the IPBus protocol [4]. Furthermore it is equipped with a SFP module in order to send data to the existing Read-Out Drivers (RODs) resulting in a transparent integration into the data flow.

Each of the three front-end options processes the PMT signals, performs slow integration for Cesium calibration, and provide a Charge Injection System (CIS) for calibration purposes. The redesigned 3-in-1 cards provide pulse shaping with discrete components, bi-gain amplification (1x, 32x), and an analog trigger output compatible with the current Level 1 trigger system. Contrary to the QIE [5] that is a charge conveyor in an ASIC with no pulse shaping for the signal digitization. The third option is an ADC in an ASIC design called FATALIC [6], which has active pulse shaping and tri-gain amplification (1x, 8x, 64x) for the signal digitization. A picture of each of the front-end options is shown in Fig. 3.

Figure 2. Picture of the top side of a mini-drawer (top) and the Tile PPr prototype motherboard (bottom).

Figure 3. Pictures of the three front-end readout options: 3-in-1 (left), FATALIC (middle), QIE (right).

3. Testbeam campaigns

The three test-beam campaigns took place in 2015 and 2016 with the aim of testing the performance of the upgrade electronics with very good results. Several modules were equipped with these electronics and exposed to different beam particles (electrons, muons and hadrons) during these campaigns. The primary objective was to assess the status of the demonstrator based on the front-end baseline option, the modified 3-in-1. Attention was put also on the two alternative front-end options, QIE and FATALIC.

Calibration of the redesigned 3-in-1 board shows a strong performance for the upgrade system (Fig. 4). Achieved linearity is 0.3% and equivalent energy resolution is less than 0.1%. Performance results of the FATALIC (Fig. 5) and QIE (Fig. 6) are also very promising, with very little noise. Particle response will be measured again in the two test-beam campaigns scheduled for June and September of 2017

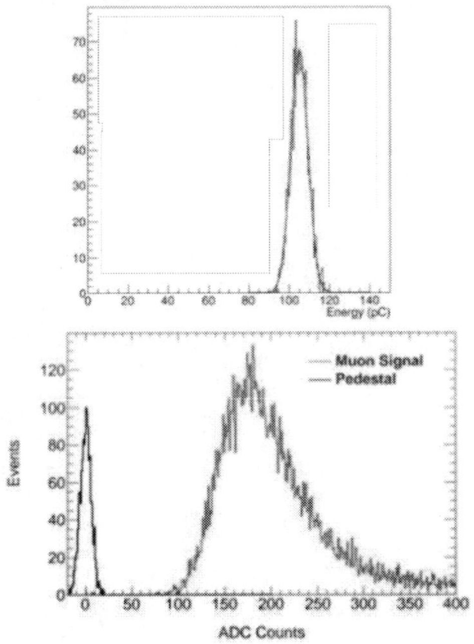

Figure 4. 3in1: Energy distribution for 100 GeV electrons (top), and muons signal response (bottom).

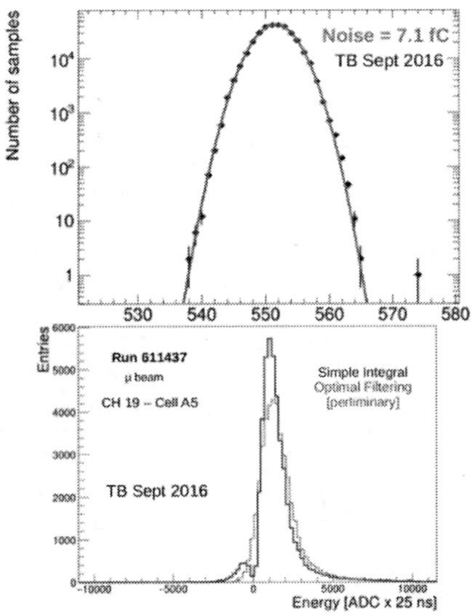

Figure 5. FATALIC: Electron signal response (top), and muons signal response (bottom).

4. Conclusions

The upgrade Phase-II program of the Tile calorimeter is progressing well. Three test-beam campaigns have taken place up to now to evaluate the performance of the three

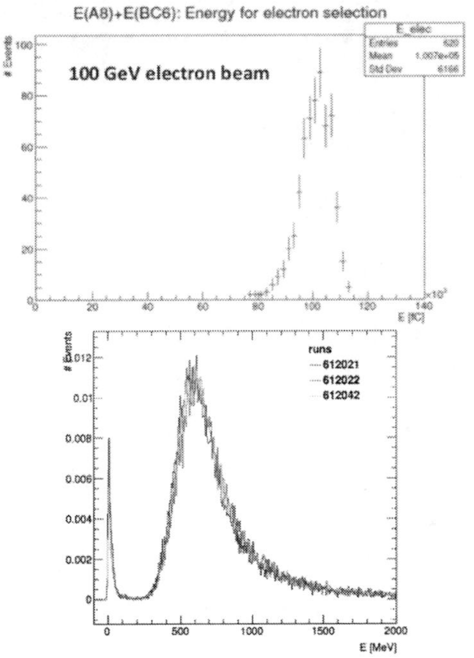

Figure 6. QIE: Energy distribution for 100 GeV electrons (top), and muons signal response (bottom).

different front-end options looking forward for a drop-down selection in 2017. The reliability and stability of the system has been visibly improved with respect to the 2015 test-beam. The Demonstrator analog trigger signals output has been integrated successfully into the TDAQ infrastructure for the commissioning of the hybrid Demonstrator. The FE and HV option final selection process has been defined based on extensive comparison of the performances and other criteria. The Tile calorimeter system plans to complete the R&D stage and take a decision by the end of summer.

References

[1] ATLAS Collaboration, "The ATLAS Experiment at the CERN Large Hadron Collider," *JINST*, vol. 3, no. S08003, p. 437, 2008. [Online]. Available: http://inspirehep.net/record/796888

[2] P. Moreira et al., "The GBT-SerDes ASIC prototype," *JINST*, vol. 5, no. 11, 2010, c11022. [Online]. Available: http://iopscience.iop.org/article/10.1088/1748-0221/5/11/C11022

[3] F. Carrio et al., "The sROD module for the ATLAS Tile Calorimeter Phase-II Upgrade Demonstrator," *JINST*, vol. 9, no. 02, p. C02019, 2014.

[4] C. Ghabrous Larrea et al., "IPbus: a flexible Ethernet-based control system for xTCA hardware," *JINST*, vol. 10, no. 02, p. C02019, 2015. [Online]. Available: http://iopscience.iop.org/article/10.1088/1748-0221/10/02/C02019

[5] A. Baumbaugh et al., "QIE10: a new front-end custom integrated circuit for high-rate experiments," *JINST*, vol. 9, no. 01, p. C01062, 2014. [Online]. Available: http://iopscience.iop.org/article/10.1088/1748-0221/9/01/C01062

[6] F. Carrio, "Upgrade for the ATLAS Tile Calorimeter Readout Electronics at the High Luminosity LHC," *NIM A*, vol. 718, pp. 69 – 71, 2013. [Online]. Available: http://www.sciencedirect.com/science/article/pii/S0168900212014052

Time-dependent characterisation of stability performance of EJ-309 detector systems

Vytautas Astromskas[1,*] and Malcolm Joyce[1,*]
[1]Engineering Department, Lancaster University, Bailrigg, Lancaster, LA1 4YW, United Kingdom

I. Introduction

Multi-element scintillator systems are increasingly used in a broad range of applications such as neutron spectrometry [1], nuclear security [2], nuclear reaction [3] positron emission tomography [4] and all detector are required to operate as identical to each other as possible. However, typically, there is a significant difference in the performance of the scintillator material, the photomultiplier tubes and/or the quality of the coupling which leads to variance in the detectors performances. This is especially important in material discrimination applications in mixed radiation fields, where a non-linear or non-matching linearity with regards to energy response can lead to false identification or discrimination of the material. Normally, the detectors are calibrated using laboratory-based radioactive gamma sources to account for the difference in energy linearity. However, this method ignores the differences in gain which is important for multiplicity assays. Finally, the knowledge of stability of the system is vital for real-world applications. The focus of this study is to investigate the normalisation of the performance of a multi-element scintillator detector system as well as the stability with respect to time and temperature.

II. Methods

Seven EJ-309 liquid scintillators coupled to photomultiplier tube (PMT) were tested in terms of the uniformity of their response to gamma radiation sources. The scintillator detectors were obtained from Scionix, Netherlands. The liquid scintillator was housed in a 3 mm thick aluminium chamber with the dimensions of 10x10x10 cm. The PMTs were connected to a Hybrid Instruments single-channel mixed field analyzer (MFA) digitizer system for signal processing and data readout. Laboratory radioactive sources, namely, Cesium-137, Sodium-22 and Cobalt-60 were used to evaluate the uniformity of the detectors' response in terms of energy linearity and efficiency.

For all experiments, a radioactive source was placed at the centre on top of the vertically held detector. For energy response and efficiency measurements, the spectrum was collected for 10 minutes in the case of Cs-137, and for 30 minutes for Na-22 and Co-60 due to different activity levels of the sources. For stability measurement, the spectrum was continuously collected for 3 hours using Cs-137 radioactive source. The detectors were energy calibrated by adjusting the applied high voltage in such way that the Cs-137 peak would be centred at the channel 200 on the MFA software. The same applied HV was, then, kept for Na-22 and Co-60 spectra in order to identify the deviations in energy linearity. The applied high voltages ranged from -1576 V to -1705 V.

* V.Astromskas and M.Joyce are with Lancaster University, Bailrigg, Lancaster, LA1 4YW, U.K (e-mail: v.astromskas@lancaster.ac.uk; m.joyce@lancaster.ac.uk)

III. Results and Discussion

The energy response of all seven detectors is shown in figure 1. The peak positions were extracted from the Gaussian fit of each individual peak. The extent of the error bars is contained within the data point. The figure indicates that the detectors have a non-uniform response with respect to each other as well as that the greater non-uniformity is observed in the high energy channels. This could be due to non-linear response in terms of energy of the PMTs or due to the higher energy peak's channel being further away from the calibration channel. The effect will be investigated and identified in future work.

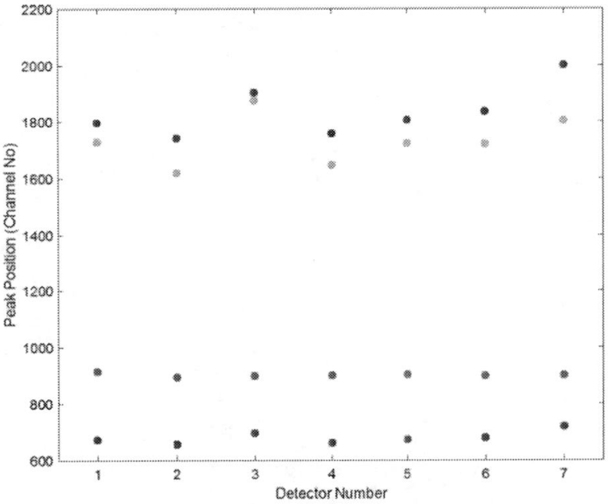

Fig. 1. The response in terms of energy for each of the seven detectors..

The linearity of the energy response is shown in figure 2. The peak positions and the error bars were extracted from the Gaussian fit. The extracted R-Square values extracted from the linear fit ranged from 0.871 to 0.927, thus, further supporting the further investigation in the energy response of the detectors is required in order to ensure uniform response in multi-detector arrangements.

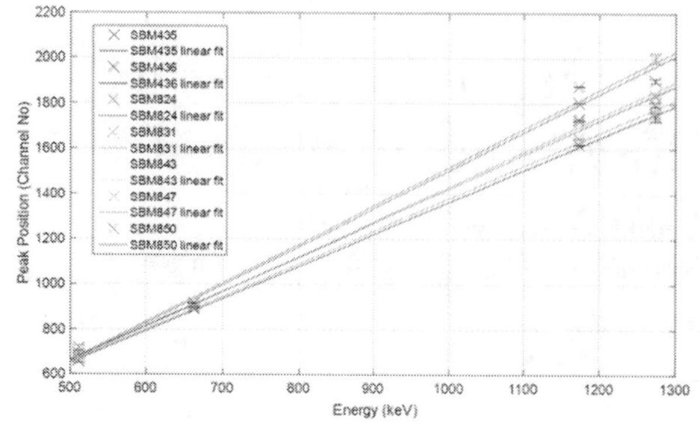

Fig. 2. The energy linearity of each of the detectors. The peak positions were fitted with linear fit.

Furthermore, the efficiency of the detectors was investigated and the results are displayed in figure 3. The errors bars are a combination of the statistical error of the total number of counts collected and the uncertainty in the fit. The greater error bars appear fig.3.(c) and fig.3.(d) due to lower activities of the sources. Despite all detectors being calibrated using the Cs-137 source, the difference in efficiencies is observed in fig.3.(b) as well. The efficiencies varied by up to 24 %. Interestingly, the significant difference in efficiencies was observed in detectors 1 and 6 despite the fact both of these detectors had applied high voltage set to -1700 V. This indicates that the difference in efficiency is not caused by applied high voltage.

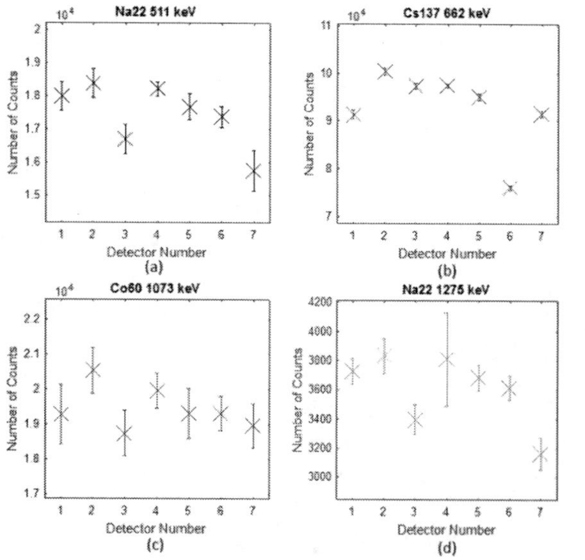

Fig. 3. The efficiency of the seven detectors for: (a) Na-22 511 keV peak; (b) Cs-137 662 keV peak; (c) Co-60 1073 keV peak; (d) Na-22 1275 keV peak.

Finally, the stability of a detector over 3 hour period is shown in figure 4. The data shown is incremented in 10 minute steps. The error bars were extracted from the statistical error of the total number of counts collected. The raw data was smoothed using a 10-point filter in order to qualitatively evaluate the stability of the detector. The figure indicates two behaviours. Firstly, that the detector requires a warm-up time of approximately 60 minutes in order to achieve a stable response. Secondly, the detector's operation is stable for the remainder of the experiment. In the future, the stability will be evaluated of all the detectors as well as the dependency of the warm-up time in terms of temperature as this is vital information for applying these EJ-309 liquid scintillator detectors in real-world applications.

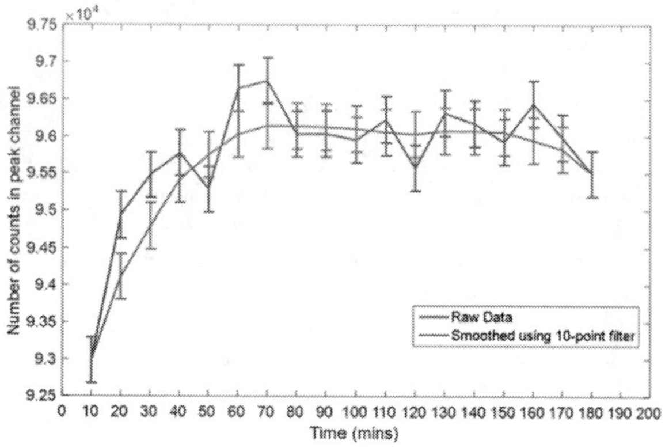

Fig. 4. The number of counts collected over 3 hour period using the Cs-137 radioactive source.

IV. Summary

In this paper, the stabilization in terms of energy response, time and temperature of multiple EJ-309 liquid scintillator detectors coupled to PMTs is going to be discussed. The results will include a detailed study of the energy response, warm-up time, as well as, temperature effect on the stability of multiple detectors. The final outcome of this research will suggest a methodology for stabilizing a multi-element scintillator system in terms of the aforementioned factors.

References

[1] R. Machrafi, A. L. Miller, and N. Khan, "New approach to neutron spectrometry with multi element scintillator," *Radiation Measurements*, vol. 80, pp. 10–16, Sept. 2015.

[2] G. Popescu, S. Herman, S. Glover, and H. Spitz, "Compton background suppression with a multi-element scintillation detector using high speed data acquisition and digital signal processing," *Journal of Radioanalytical and Nuclear Chemistry*, vol. 307, pp. 1949–1955, Mar. 2016.

[3] M. Febbraro, F. D. Becchetti, R. O. Torres-Isea, M. Ojaruega, A. M. Howard, J. J. Kolata, A. Roberts, and A. N. Villano, "Neutron Spectroscopy Without Time-of-Flight Measurement: A DSP-Based Deuterated Scintillator Array," *IEEE Transactions on Nuclear Science*, vol. 60, pp. 890–896, Apr. 2013.

[4] J. Y. Yeom, R. Vinke, V. C. Spanoudaki, K. J. Hong, and C. S. Levin, "Readout Electronics and Data Acquisition of a Positron Emission Tomography Time-of-Flight Detector Module With Waveform Digitizer," *IEEE Transactions on Nuclear Science*, vol. 60, pp. 3735–3741, Oct. 2013.

ATLAS Tile Calorimeter calibration and monitoring systems

Arely Cortés-González, on behalf of the ATLAS Collaboration
European Organization for Nuclear Research (CERN)
C/O CERN 1211 Geneve 23, CH
Email: arelycg@cern.ch
Telephone: (+41) 22 76 73797

***Abstract*—The ATLAS Tile Calorimeter is the central section of the hadronic calorimeter of the ATLAS experiment and provides important information for reconstruction of hadrons, jets, hadronic decays of tau leptons and missing transverse energy. This sampling calorimeter uses steel plates as absorber and scintillating tiles as active medium. The light produced by the passage of charged particles is transmitted by wavelength shifting fibres to photomultiplier tubes, located in the outer part of the calorimeter. Neutral particles may also produce a signal after interacting with the material and producing charged particles. The readout is segmented into about 5000 cells, each of them being read out by two photomultipliers in parallel. To calibrate and monitor the stability and performance of each part of the readout chain during the data taking, a set of calibration systems is used. This comprises Cesium radioactive sources, Laser, charge injection elements and an integrator based readout system. Information from all systems allows to monitor and equalise the calorimeter response at each stage of the signal production, from scintillation light to digitisation. Calibration runs are monitored from a data quality perspective and used as a cross-check for physics runs. The data quality efficiency achieved during 2016 was 98.9%. These calibration and stability of the calorimeter reported here show that the TileCal performance is within the design requirements and has given essential contribution to reconstructed objects and physics results.**

I. Introduction

The Tile Calorimeter (TileCal) is the central section of the hadronic calorimeter of the ATLAS experiment [1]. Tile Calorimeter [2] covers the $|\eta| < 1.7$ region of the detector. It consists of one barrel and two extended barrel sections and surrounds the Liquid Argon (LAr) barrel electromagnetic and endcap hadronic calorimeters, as seen in Figure 1. The Long Barrel (LB) and Extended Barrel (EB) sections roughly correspond to $|\eta| < 1.0$ and $|\eta| > 1.0$, respectively. The crack region between the TileCal LB and EB and the LAr electromagnetic barrel and hadronic end-cap sections are covered with special cells with a reduced size.

The TileCal provides important information for reconstruction of hadrons, jets, hadronic decays of tau leptons and missing transverse energy. This sampling calorimeter uses iron plates as absorber and scintillating tiles as active medium. The calorimeter readout is segmented into about 5000 cells (longitudinally and transversally), each of them being read by two photomultipliers (PMTs). Tile cells and electronics are organised into 4 partitions, LBA and LBC for the A side and C side of the barrel region, and separate EBA and EBC partitions in the extended barrel region. A side corresponds to positive z direction, along the beam line. Each partition is divided into 64 symmetric ϕ slices (modules), with 45 instrumented channels in LB modules and 32 channels in EB modules. Both LB and EB have up to three layers: the A-layer being the closest to the beam axis, followed by the B(C) and D-layers. The η and radial structure of the TileCal cells is shown in Figure 2a. The cells marked in yellow are the E-cells. They cover the gap/crack region with $1.0 < |\eta| < 1.6$. They are partly closer to the beam axis than the A-layer cells, thus they are characterised by a larger exposure to the irradiation and by the large amount of light they produce and that is read by the associated PMTs.

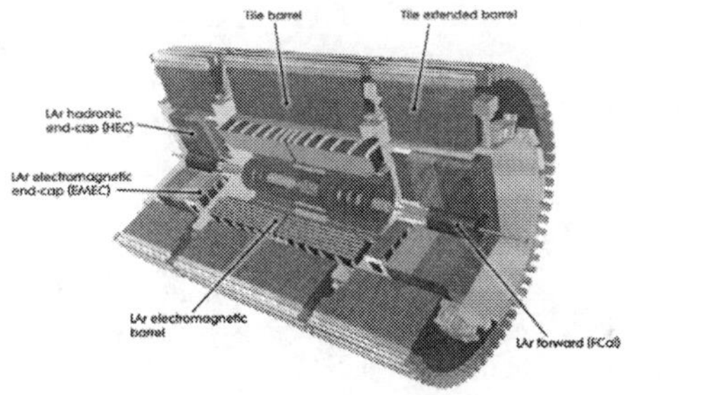

Fig. 1. A cut-away diagram of the ATLAS inner detector and calorimeter. The Tile Calorimeter consists of one barrel and two extended barrel sections [3].

The TileCal has three key elements: the scintillators, the PMTs and the read-out electronics. Figure 2b shows a schematic of the assembly of these components. The light is produced in scintillating tiles and converted into electric currents by the PMTs. Their signal is shaped and amplified with two gains. The sampling and the digitisation is realised by ADCs. Trigger signals are formed by an analogue sum of input signals and sent to the calorimeter trigger system, which also considers input from other calorimeters. During collisions, if an event is selected by the trigger system, the digitised signals are collected and processed by a Read-Out Driver. In parallel to this, integrators measure the integrated current from the PMTs.

https://doi.org/10.1051/epjconf/201817001003

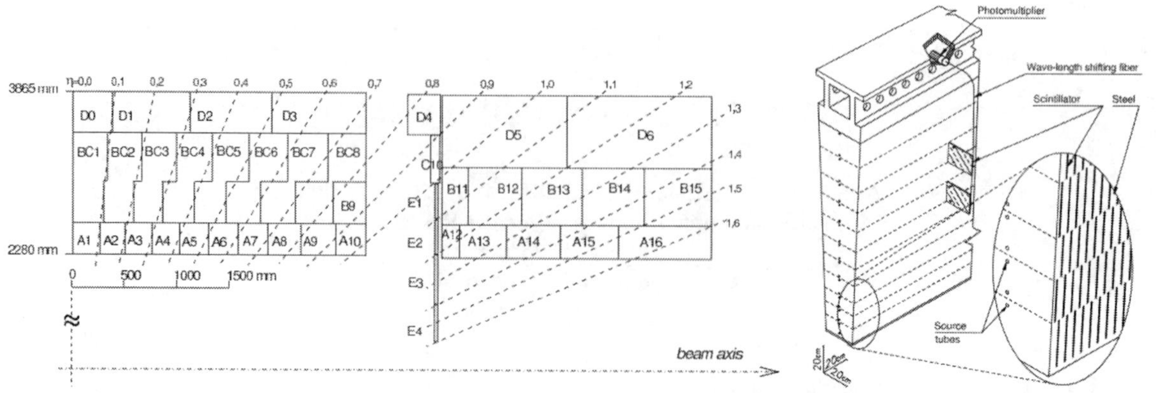

(a) (b)

Fig. 2. (a) TileCal cell and scintillator structure, including the so-called "E-cells" (E1-E4) installed in the gap/crack region, which are highlighted in yellow. (b) Schemetic of the mechanical assembly and the optical readout of the TileCal. The various components of the optical readout, namely the tiles, the fibres and the PMTs, are shown [3].

II. Calibration Systems

At each level of the signal development inside the TileCal there is a dedicated calibration system to monitor the behaviour of the different detector components. Figure 3 shows the different calibration systems along with the paths followed by the signals from different sources. Fast readout is used for physics analyses of collision data. The reconstructed energy of each TileCal channel, E(GeV), is derived from the raw response, A(ADC) as follows:

$$E(\mathrm{GeV}) = A(\mathrm{ADC}) \cdot C_{\mathrm{ADC}\rightarrow\mathrm{pC}} \cdot C_{\mathrm{pC}\rightarrow\mathrm{GeV}} \cdot C_{\mathrm{Cesium}} \cdot C_{\mathrm{Laser}} \quad (1)$$

The factors can evolve in time because of variations of PMTs high-voltage, PMTs stress induced by high light flux or optics ageing. The calibration systems are used to monitor the stability of these elements and provide per channel calibration. While $C_{\mathrm{pC}\rightarrow\mathrm{GeV}}$ was fixed during dedicated test beam campaigns, the remaining calibration constants are provided by individual systems:

- A moveable Cesium radioactive γ-source to calibrate the optic components and the PMT gains.
- Calibrations of digital gains and linearities with the charge injection system.
- Laser system to monitor the PMTs and the electronic components.
- Monitoring of beam conditions and TileCal optics with the integrator system minimum bias (MB).

A. Cesium Calibration System

The Cesium calibration system is based on three movable radioactive sources using a hydraulic control. The ^{137}Cs γ-sources move inside the calorimeter body, emitting 0.662 MeV photons to illuminate the scintillators. The signal generated by the ^{137}Cs source is collected through a special readout that integrates the analog PMT signals. This system is used to calibrate the optical components of the calorimeter as well as the PMTs. The channel response to the energy deposits is used to equalise the response of all the cells and maintain global response of the calorimeter at the electromagnetic scale. Deviation of measured Cesium signals from expected values, corrected for the Cesium decay curve, are interpreted as gain variations and translated into calibration constants (C_{Cesium}). From Run 1 to Run 2 of the LHC, a new water storage system, lower pressure in the hydraulics, and more precise water level metering during the scans have been achieved to improve stability and safety of the operation. The deviation of C_{Cesium} calibration constant from 1 for A-cells (in %) over the period between 2009 and 2016 is shown in Figure 4. The precision of the system is at the order of 0.3% with Cesium calibration scans spaced by one to three months up to 2015. During 2016 the frequency was reduced, and scans were taken only at the beginning and end of the proton-proton collisions period. The frequency of the Cesium calibration can be insufficient to track fast drifts of the PMT responses. The Laser system is used between two Cesium scans to correct for this.

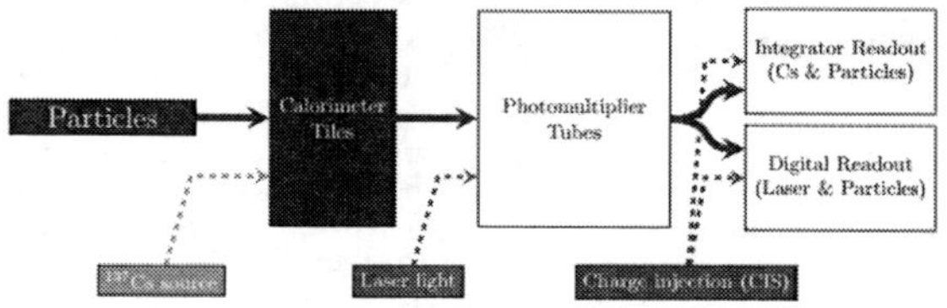

Fig. 3. Flow diagram of the readout signal path of the different TileCal calibration tools. The paths are partially overlapping, allowing for cross-checks and an easier identification of component failures [3].

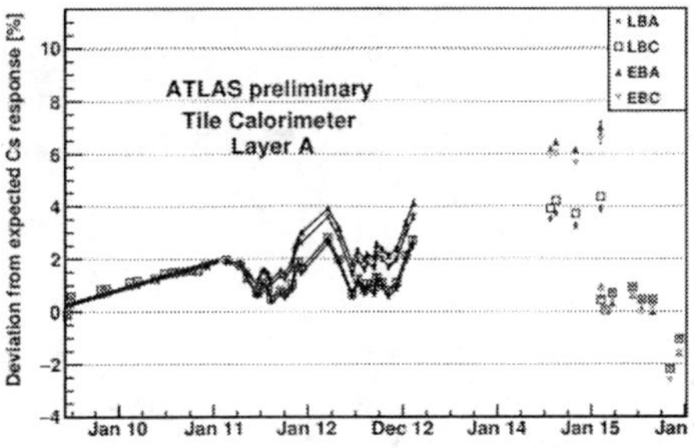

Fig. 4. Variation of Tile Calorimeter response over period of 6 years measured in Cesium calibration runs averaged over all cells in a given partition. The variation is shown for cells in the inner most layer of the calorimeter [3].

B. Charge Injection Calibration System

A Charge Injection System (CIS) is used to monitor the electronics and extract the conversions factor from ADC counts to pC ($C_{\mathrm{ADC}\to\mathrm{pC}}$). The CIS simulates physics signals in TileCal channels by injecting a known charge into the ADC and measuring the electronic response. It provides a quantitative relationship between the analog physical signals from the TileCal PMTs and the electronic response of TileCal read-out channels and corrects for non-linearity in the analog signal processing. The corresponding calibration runs are taken from daily to weekly, so $C_{\mathrm{ADC}\to\mathrm{pC}}$ constants can be regularly produced and applied to data. The overall stability of the calibration factor is at the level of 0.02% as shown in Figure 5 (for low gain) and usually less than 1% of the channels exhibit large fluctuations. CIS calibration constants are updated if the deviation from previous measured value is above the CIS systematic error (0.7%).

C. Laser Calibration System

The gain stability of each PMT is measured using a Laser calibration system [4] that sends a controlled amount of light onto the photocathode of each PMT in the absence of collisions. Deviations of any channel response with respect to its nominal response (at the time of the latest Cesium calibration) is then translated into a calibration constant: C_{Laser}. The Laser calibration runs are usually taken twice a week to monitor the individual PMT gain variations between the Cesium scans. The typical precision on the gain variation during Run 1 was better than 0.5% per channel. During the LHC Long Shutdown from 2012 to 2015, a new Laser II system was developed to correct shortcomings in electronics and light monitoring of the first system, which in turn resulted in an improved resolution. The new system has been used since the beginning of the Run 2 in 2015 to monitor the channel gain deviations. Figure 6 shows the mean gain variation per cell type between end of May and end of October 2016, covering most of the 2016 proton-proton collisions data taking period. Laser pulses are also sent during empty bunch crossings of the LHC, with a frequency of 4 Hz (1 Hz) during Run 2 (Run 1), and the timing of TileCal response is measured and used for channel per channel time calibration.

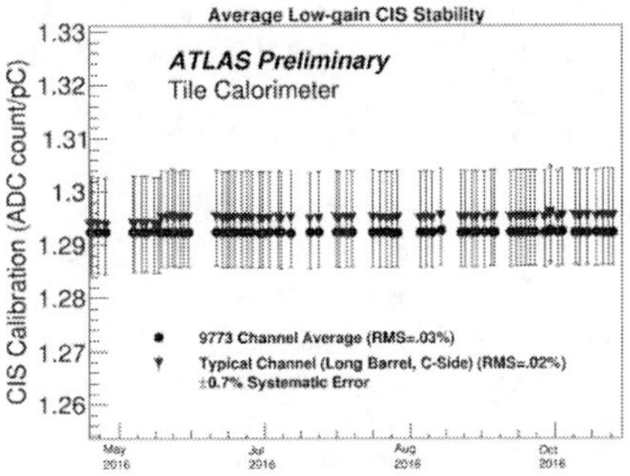

Fig. 5. Detector-wide CIS calibration constant averages of all the low gain channels for each CIS calibration run from 20 April 2016 to 28 November 2016, plotted as black circles. Channels that are unresponsive or have fluctuating CIS constants are not included in the average. The CIS constants from a typical channel (LBC19, channel 13) are additionally plotted as blue triangles for comparison. The RMS values on the plot are indicative of the fluctuation present in calibrations. In addition, there is a 0.7% systematic uncertainty present in individual calibrations, represented by the blue error bars. High gain channels present similar behaviour [3].

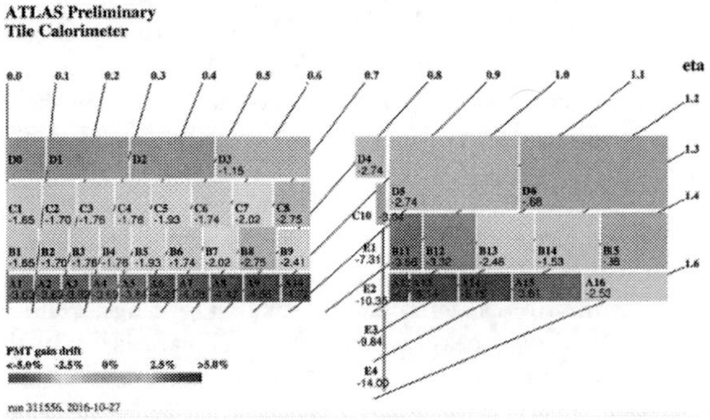

Fig. 6. Mean gain variation (%) in the TileCal PMTs that read the signal deposited in each channel, as a function of eta and radius, between May 24, 2016 and October 27, 2016. The gain in each PMT is measured using a Laser calibration system. The mean gain variation of the ~10000 TileCal channels is computed cell by cell. A total of 64 modules in phi are used for each cell while known pathological channels were excluded. The observed down-drift mostly affects cells at inner radius, that are the cells with higher current [3].

D. Minimum Bias Currents

High energy proton-proton collisions are dominated by soft parton interactions, also called Minimum Bias (MB) events. The integrator readout is the core of the MB system. It integrates the PMT signal over a large time (~ 10 ms), providing a slow control readout. The currents are recorded by all channels of the TileCal. The integrator signal is dominated by MB events. The fraction of signal coming from physics data events (e.g. containing high p_T objects) is extremely negligible. Each collision run is divided into lumiblocks (LBs) where a LB is the time interval with a fixed duration where the instantaneous luminosity is about constant. Integrator currents are recorded for each LB unit. The currents are averaged over about 20-25 integrator measurements per LB. Both the Minimum Bias and the Cesium systems measure the signal coming from scintillators and the variation in PMT response over time. Thus, they are expected to give similar results. The MB system can then be used to monitor the stability of the channels and to provide an independent cross-check of the Cesium calibration. The response of the TileCal to signals induced by the MB interactions scale with instantaneous luminosity. Because of this, the system is used to monitor and provide an independent measurement of the ATLAS instantaneous luminosity. The MB response variation measurement is possible by using a reference cell which is known to be stable in time, and which provides a luminosity measurement. The reference cell chosen for the 2016 data studies is cell D6 (outermost layer, covering the region $1.1 < |\eta| < 1.3$) which provides a measurable signal while being relatively protected from irradiation. This reference cell may in fact have a non-zero PMT gain drift, but given that this is well measured with the Laser system, it still serves as a good reference. The comparison between MB and Cesium and the Laser system can give insights on origins of channels drifts (those not related to PMT gain drifts). Figure 7 shows the response variation measured by the MB and the Laser system for cells in the innermost layer of the Extended Barrel, covering the region $(1.2 < |\eta| < 1.3)$ as a function of time. The integrated luminosity evolution during the same period is also shown in the plot. The variations observed by the MB and Cesium systems are sensitive to PMT gain drift and scintillator irradiation (ageing). The Laser system, however measures only the PMT gain drifts in time. Thus, the difference between MB, or Cesium, and the Laser measurements can be interpreted as a loss of efficiency of the scintillators by irradiation damage. The cell probed in Figure 7, A13, is the most irradiated cell in TileCal, excluding the E-cells.

The comparison of the three calibration systems for the response variation of representative cells has been performed in Run 1 and Run 2 data periods. As mentioned above, the Laser points in Figure 7 show the PMT gain drift, where drops can be observed during collision phase and a rise during machine development phases where no particle collisions take place. Minimum Bias system measures in addition any effect of irradiation of the scintillators when the response variation is higher than the PMT gain drift. While this PMT gain drifts recovers during machine stops, the irradiation effect on the scintillators does not. It is corrected for by adjusting the calibration constants.

Both the integrated luminosity and centre of mass energy $\sqrt{s}$ have an impact on the overall collected charge, Q. The charge Q is determined with the help of the luminosity coefficient (exploiting the linear dependance of the current

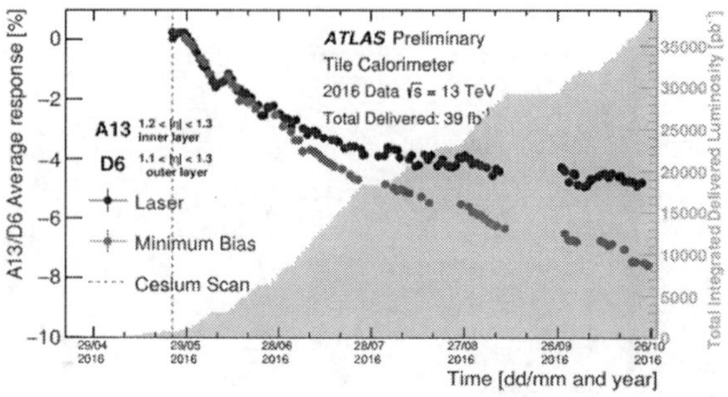

Fig. 7. The variation of the response to Minimum Bias and Laser for cells in the inner layer of the Extended Barrel, covering the region $(1.2 < |\eta| < 1.3)$, as a function of time. The response variation is derived with respect to a reference cell D6 ($1.1 < |\eta| < 1.3$, in the outermost layer of the calorimeter). Minimum Bias and Laser data cover the period from the end of May to the end of October 2016. The response variation versus time measured by the Minimum Bias system has been normalised to the response variation measured by the Laser system on May 24th (first point shown in the plot, right after the Cesium scan). The variations observed by the Minimum Bias are sensitive to PMT gain drift and scintillator irradiation. The difference between minimum bias and Laser is interpreted as an effect of the scintillators irradiation [3].

with respect to instantaneous luminosity), k:

$$Q(t) = k \cdot \int_{t_0}^{t} L(t)\mathrm{d}t,$$

where t is the given time during collisions, and t_0 is the reference start time. The irradiation effect in scintillators will depend on the collected charge and therefore depends on both the $\sqrt{s}$ and the integrated luminosity. Furthermore, geometry will also play a role, as a larger cell will experience less irradiation effect than a smaller cell with the same collected charge. Figure 8 shows the dependence of the irradiation in cells belonging to the A13 and E-cells in the EB as a function of the collected charge for 2016 data. An irradiation effect of up to 5% is reported there. The maximum effect is observed for the E3 and E4 cells. A total irradiation effect of at most 3% for the highest exposed cell has been observed for the 2012 data taking period [5]. For 2015 the largest irradiation effect was below 1% [6]. The total integrated luminosity in 2015 was ~ 5 times lower than the one in 2012. The center of mass energy $\sqrt{s}$ was 8 TeV in 2012 and 13 TeV in 2015 and 2016.

Because of the reduced frequency of the Cesium scans, and the increased integrated charge measured in the system, since the start of 2016, this system is used to correct for the scintillator ageing effect between two Cesium calibrations. This is reflected in the C_{Cesium} constant in Equation 1.

III. Data Quality Monitoring

TileCal monitoring includes identifying and masking problematic channels correcting for timing jumps, monitoring data corruption or other hardware issues. Immediately after

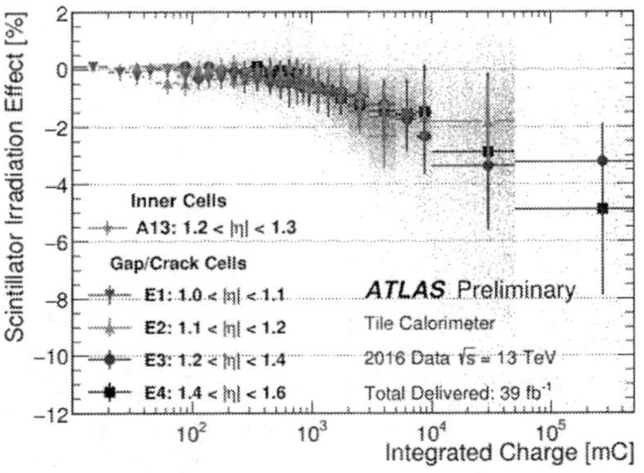

Fig. 8. The relative response variation measured by the Minimum Bias system, after the subtraction of the PMT gain drift component measured by the Laser system, as a function of the integrated charge collected in the cells in the gap/crack region in the Extended Barrel, covering the region $(1.0 < |\eta| < 1.6)$ and in the cells in the inner layer of the Extended Barrel, covering the region $(1.2 < |\eta| < 1.3)$, during the period between the end of May to the end of October 2016. The average per cell type is also shown. The integrated luminosity is the total delivered in the whole proton-proton collisions period [3].

a physics run ends, a subset of data (*express stream*) is processed and reviewed. There is a short delay before the full dataset processing starts, during which any problems identified in the express stream may be corrected with or accounted for in conditions updates. Changes to TileCal timing within collision runs are monitored by Laser calibration events in the empty bunch crossings of physics runs, used to apply timing offset corrections to data. Tile uses a specialised web interface to generate an initial report for a data quality shifter to finish, based on the results of automated data quality monitoring tests. Based on these reviews corresponding actions may be taken. If a problem cannot be corrected, it is entered into the ATLAS Defect database. If the problem is considered intolerable, then the affected data are removed from the ATLAS Good Run List and are not used in physics analyses.

Figure 9 shows the fraction of cells and channels in the detector that have been masked from December 2010. Problematic channels can be recovered by detector interventions during the maintenance periods (hatched areas in the Figure). Figure 10 shows a (η, ϕ) representation of the number of TileCal masked cells as of June 2017. Tile achieved 99.6% data quality efficiency in 2012, 100% in 2015 and 98.9% in 2016, with $\sim 1\%$ of cells masked by the end of October, 2016.

IV. Conclusion

The calibration systems of TileCal are being successfully operated since Run 1 of the LHC. These calibration and monitoring systems have been adapted between Run 1 and Run 2, to adjust to the new LHC conditions. Their

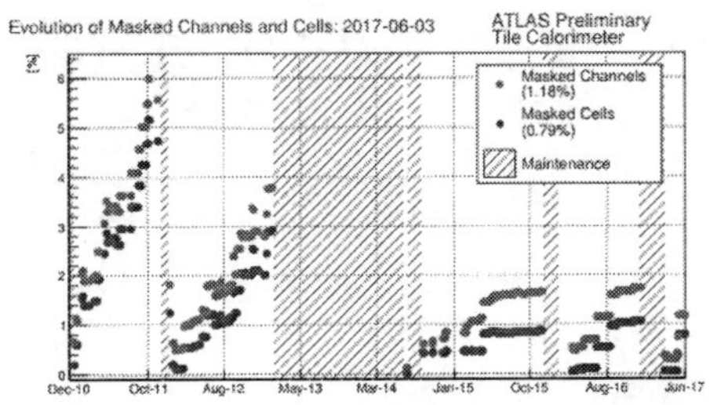

Fig. 9. Percentage of all cells and channels in the detector that are masked as a function of time starting from December 2010. The recent detector status corresponding to the beginning of June 2017 is noted. The hatched area represents the maintenance period of the detector [3].

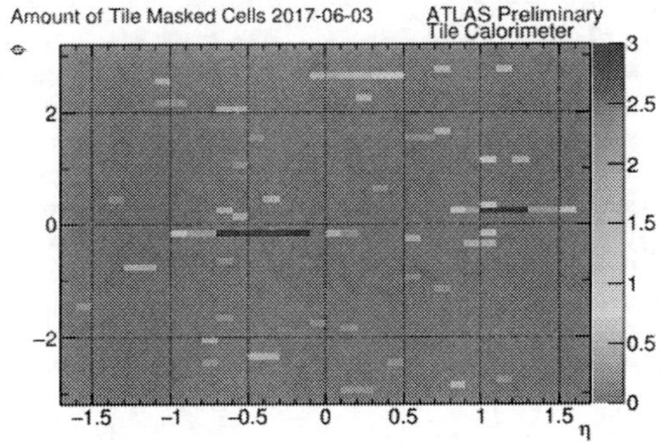

Fig. 10. (η, ϕ) representation of the number of Tile masked cells. The binning is chosen to be 0.1×0.1, corresponding to the calorimeter tower definition. Cells in Tile Calorimeter can have one or two readout channels. The bin colour corresponding to integer numbers is the integrated number of cells in the calorimeter tower. Whereas bin colour corresponding to non-integer numbers represents the fraction of masked ADCs in the tower [3].

performances have been reported here. All calibration systems are used to monitor and, when appropriate, correct for instabilities affecting the channels response. These instabilities are understood to be induced by high instantaneous luminosity, optical effects or detector ageing. The monitoring system is able to identify the fraction of pathological channels and have this information propagated to the physics analyses in ATLAS.

References

[1] ATLAS Collaboration, *The ATLAS Experiment at the CERN Large Hadron Collider*, Journal of Instrumentation 3 S08003, 2008.
[2] ATLAS Collaboration, *Readiness of the ATLAS Tile Calorimeter for LHC collisions*, Eur. Phys. J. C70: 1193-1236, 2010.
[3] ATLAS Collaboration, *https://twiki.cern.ch/twiki/bin/view/AtlasPublic/ApprovedPlotsTile*
[4] ATLAS Tile Calorimeter system, *The Laser calibration of the Atlas Tile Calorimeter during the LHC run 1*, JINST 11 T10005, 2016.
[5] Fracchia, S. *Irradiation effect on the response of the scintillators in the ATLAS Tile Calorimeter*, Tech. 116 Rep. ATL-TILECAL-PROC-2014-013 CERN Geneva, 2014. https://cds.cern.ch/record/1710463
[6] Fischer, C. *Study of TileCal Scintillator Irradiation using the Minimum Bias Integrators*, ATL-TILECAL-PROC-2016-006 CERN, Geneva, 2016. https://cds.cern.ch/record/2160267

Neutron detection devices with ^{6}LiF converter layers

Paolo Finocchiaro, Luigi Cosentino, Sergio Lo Meo, Ralf Nolte, Desiree Radeck

***Abstract*— The demand for new thermal neutron detectors as an alternative to ^{3}He tubes in research, industrial, safety and homeland security applications, is growing. These needs have triggered research and development activities about new generations of thermal neutron detectors, characterized by reasonable efficiency and gamma rejection comparable to ^{3}He tubes. In this paper we show the state of art of a promising low-cost technique, based on commercial solid state silicon detectors coupled with thin neutron converter layers of ^{6}LiF deposited onto carbon fiber substrates. Several configurations were studied with the GEANT4 simulation code, and then calibrated at the PTB Thermal Neutron Calibration Facility. The results show that the measured detection efficiency is well reproduced by the simulations, therefore validating the simulation tool in view of new designs. These neutron detectors have also been tested at neutron beam facilities like ISIS (Rutherford Appleton Laboratory, UK) and n_TOF (CERN) where a few samples are already in operation for beam flux and 2D profile measurements. Forthcoming applications are foreseen for the online monitoring of spent nuclear fuel casks in interim storage sites.**

***Index Terms*—Thermal neutron detectors, ^{3}He replacement, ^{6}LiF neutron converter, silicon detectors.**

I. Introduction

The lack and the increasing cost of ^{3}He have triggered in the last years a worldwide R&D program investigating new techniques for neutron detection. For many applications a realistic alternative is needed to ^{3}He-based neutron detectors which so far have been the most widely used systems, as they are almost insensitive to radiation other than thermal neutrons [1],[2],[3].

Several developments involving neutron detection are currently being pursued in the fields of homeland security, nuclear safeguards, nuclear decommissioning and radwaste management. Two possible applications are worth to be mentioned, namely the development of neutron sensitive panels to be placed around nuclear material in a $\approx 4\pi$ solid angle coverage for coincidence neutron counting applications [4], and the deployment of arrays of small neutron detectors for the online monitoring of spent nuclear fuel storage sites [5],[6].

This work was supported by the Istituto Nazionale di Fisica Nucleare
P. Finocchiaro and L. Cosentino are with INFN Laboratori Nazionali del Sud, Catania, Italy (e-mail: finocchiaro@lns.infn.it, cosentino@lns.infn.it).
S. Lo Meo is with ENEA Bologna, Italy (e-mail: sergio.lomeo@enea.it).
R. Nolte and D. Radeck are with PTB Braunschweig, Germany (e-mail: ralf.nolte@ptb.de, desiree.radeck@ptb.de).

In a previous paper [7] it was shown that the use of a fully depleted silicon detector, in combination with a ^{6}LiF neutron converter film, can be successfully exploited to detect thermal neutrons with a reasonable efficiency, as also suggested by other authors [8],[9]. The reliability of this technique, along with a characterization in terms of response, efficiency and gamma sensitivity, was also assessed by means of GEANT4 simulations [10]. The neutron conversion mechanism is based on the well known reaction

$$^{6}Li + n \rightarrow {}^{3}H\ (2.73\ MeV) + \alpha\ (2.05\ MeV) \tag{1}$$

which is the only possible decay channel following the neutron capture in ^{6}Li, and is free of gamma rays. Its cross section at thermal neutron energy is 940 b, and it scales with 1/v up to ≈200 keV with a back-to-back isotropic emission of the reaction products. The energy spectrum measured by the silicon detector in such a configuration has a characteristic shape, and allows to discriminate the capture reaction products from the low-energy background basically due to gamma rays.

This technique is indeed well established [11],[12], several applications are already in use, like for instance at the n-TOF spallation neutron beam facility [13],[14]. In this paper we report on the calibration of a few samples of this solid state neutron detector in a thermal neutron field, and the results are compared to the respective GEANT4 simulations.

II. Experimental setup

A. The thermal neutron field

For the calibration of the devices we used the thermal neutron field at the Physikalisch-Technische Bundesanstalt (PTB). The Thermal Neutron Calibration Facility at PTB ([15],[16]) consists of sixteen ^{241}Am-Be sources that are mounted inside a graphite block whose dimensions are 150 cm (height), 150 cm (width), and 180 cm (depth). The facility is sketched in Fig. 1 and shown in Fig. 2. The reference position is at 30 cm from the front surface of the moderator exit window and 75 cm above the floor. The neutron and photon fields at the reference position were characterized by means of measurements and Monte Carlo simulations [16]. The neutron field is highly thermalized, 98.4 % of neutrons have energies below the cadmium cut-off energy with a thermal neutron flux at the reference position of 68.3±1.9 neutrons/cm^{2}/s and a uniform field size of at least 10×10 cm^{2}. Optionally, a Cadmium plate can be installed in front of the moderator exit window that cuts the thermal neutron contribution below the

 https://doi.org/10.1051/epjconf/201817001004

cadmium threshold. This way, a pure thermal field can be obtained by applying the difference method. However, the high-energy ($E\gamma \geq 5MeV$) gamma ray flux increases by about 2÷3 orders of magnitude when the Cadmium plate is inserted. Fig. 3 shows the distribution of the neutron energies at the reference position, with and without the Cadmium plate.

B. The thermal neutron detectors

The thermal neutron detectors we discuss in this paper were named SiLiF, because they feature silicon detectors (3 cm x 3 cm x 300 μm) and ^{6}LiF converters. Two samples were calibrated at PTB (Fig. 4):

- A SiLiF1.6 made of a 1.6 μm ^{6}LiF layer (thin converter) deposited onto a carbon fiber plate (a), coupled to a 300 μm thick silicon detector (b), and enclosed in a thin aluminum box (d).
- A SiLiF64 made of a double sandwich of four 16 μm ^{6}LiF layers (thick converters) deposited onto carbon fiber plates (c), and enclosed in a thin aluminum box (d).

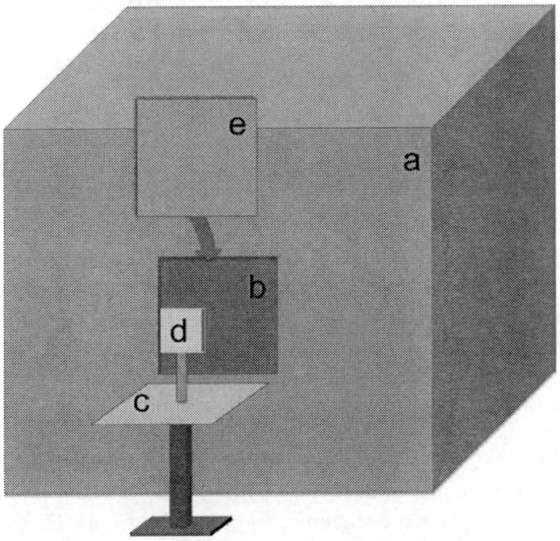

Fig. 1. Sketch of the PTB Thermal Neutron Calibration Facility. (a) The moderator/shield. (b) The neutron exit window. (c) The detector support plate. (d) Detector. (e) Optional Cadmium plate to be installed on the output window.

Fig. 2. A picture of the PTB Thermal Neutron Calibration Facility.

In Fig. 5 we show one of the calibrated detectors during the measurement in front of the Thermal Neutron Calibration Facility. The light beams from the laser alignment system are visible. The detectors were biased at 30 V, so that the silicon diodes were fully depleted. This was especially important for the SiLiF64, as the neutron converters were installed on both faces of each silicon diode and the full depletion regime is mandatory to get the same response from the front and the back sides. The operating criterium is quite simple: we decide a neutron discrimination energy threshold and declare as neutrons all the counts above that threshold. Below the threshold there will be other neutron events mixed with gamma ray events and background noise. Of course the higher the threshold the cleaner the neutron signal will be, but at the same time the lower the detection efficiency will be. Therefore one has to find a trade off between purity and efficiency.

III. Simulation procedure

The detectors were simulated by means of the well known GEANT4 toolkit ([17],[18]), including the aluminum box, the carbon fiber substrate, the ^{6}LiF neutron converter, and the silicon diode. The supporting printed circuit board, the cable and the connector were not included in the simulation. $2x10^6$ neutrons were generated for each case, with 25.3 meV kinetic energy, uniformly and perpendicularly irradiating the detector area.

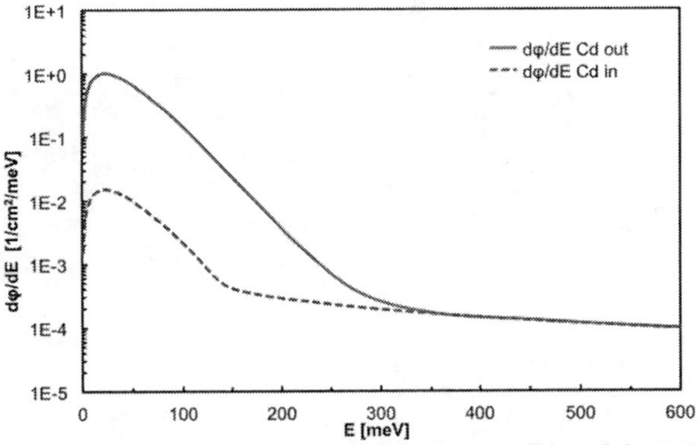

Fig. 3. Neutron energy distribution at the reference position of the PTB Thermal Neutron Calibration Facility, with and without the Cadmium plate.

IV. Measurement results

The triton and the alpha particle emitted from the reaction (1) following a neutron capture have well defined energies, which are degraded after crossing the residual converter layer thickness and the thin air layer between the converter face and the silicon detector. Therefore, the thinner the converter the sharper is the spectrum shape measured by the silicon detector. Indeed, by looking at Fig. 6, where we reported the energy spectrum (counting rate ε normalized to the unit flux) as a function of the energy, one can see that using a 1.6 μm ^{6}LiF converter the triton and alpha contributions to the spectrum are well identified. The alpha and triton endpoint energies are easily spotted on the spectrum, and this allows a perfect energy calibration of the detector, quite useful to set the neutron discrimination threshold with high precision. The agreement with the simulated spectrum is remarkable, considering that the experimental data were normalized to the nominal flux and therefore both spectra are in absolute units.

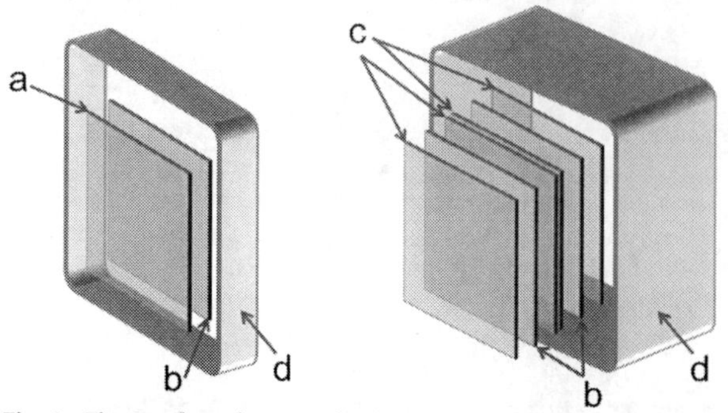

Fig. 4. The two thermal neutron detector samples calibrated at PTB. Left: SiLiF1.6 made of a 1.6 μm ^{6}LiF layer (thin converter) deposited onto a carbon fiber plate (a), coupled to a 300 μm thick silicon detector (b), and enclosed in a thin aluminum box (d). Right: SiLiF64 made of a double sandwich of four 16 μm ^{6}LiF layers (thick converters) deposited onto carbon fiber plates (c), and enclosed in a thin aluminum box (d).

Fig. 5. One of the calibrated detectors during the measurement in front of the Thermal Neutron Calibration Facility sketched in Fig. 1. The light beams from the laser alignment system are visible.

Unfortunately with such a thin converter the neutron detection efficiency is of the order of 0.5%, quite low to allow a realistic employment of this detector. However, due to its spectral features and precision of calibration, it can be quite useful as a reference for the absolute calibration of other detectors. In Fig. 7 we show the energy spectrum as measured with the SiLiF64 detector, along with the simulation result. The agreement between simulation and measurement is remarkable also in this case, thus validating both the simulation tool and the detector behavior. In Fig. 8 we reported the energy spectra as measured by the SiLiF64 with and without the Cadmium plate installed on the exit window. The spectra were normalized to the nominal unit flux respectively with and without Cadmium plate. One can immediately see that on the one hand the neutron contribution decreases by about two orders of magnitude, whereas the gamma contribution is relevant at least up to 1÷1.5 MeV.

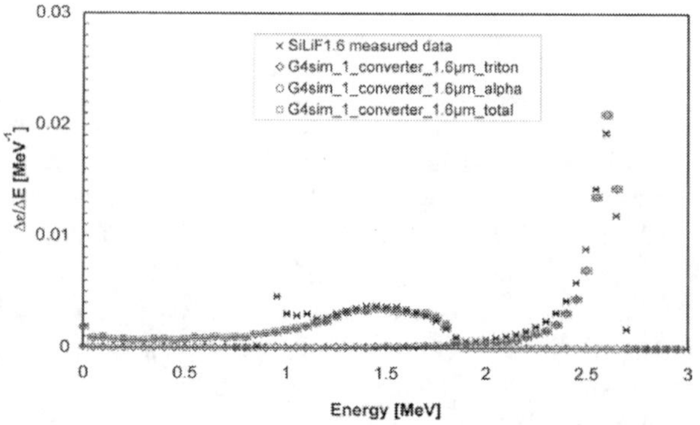

Fig. 6. Energy spectrum (counting rate ε normalized to the unit flux) measured with the SiLiF1.6 detector and compared with the simulated one. Also reported are the separate simulated contributions from alphas and tritons.

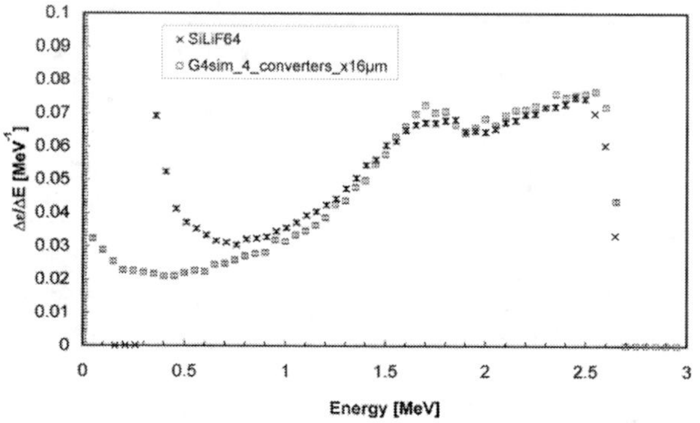

Fig. 7. Energy spectrum (counting rate ε normalized to the unit flux) measured with the SiLiF64 detector and compared with the simulated one.

Indeed, in Fig. 9 we reported the ratio between the two spectra of Fig. 8, and such a ratio shows rather clearly that in order to have a good γ/n discrimination the neutron discrimination threshold value should be chosen at 1.5 MeV. However, lower threshold values can be safely employed in applications where there are no high energy gamma rays (as a reference, the γ/n contamination probabilty from ^{60}Co gamma rays when setting a 1.5 MeV discrimination threshold is $\leq 10^{-12}$ [10]).

TABLE I
SIMULATED AND MEASURED DETECTION EFFICIENCY

detector	simulated efficiency	measured efficiency
SiLiF1.6	0.48 %	0.50 ± 0.02 (syst) %
SiLiF64	8.25 %	7.95 ± 0.35 (syst) %

In Table I we listed the simulated and measured efficiency for the two detectors.The measured data have a low statistical uncertainty, whereas the systematic one is more relevant, especially on the SiLiF64 detector, because of the energy calibration. Indeed, while the calibration of the SiLiF1.6 can be very precise due to the presence of the two very well defined endpoints for alphas and tritons, this is not the case for a thicker converter layer, as can be easily seen by comparing Fig. 6 and Fig. 7.

Therefore the systematic uncertainty was estimated by

assuming a ±10 keV uncertainty in the alpha endpoint of the SiLiF1.6 and a ±50 keV in the alpha endpoint of SiLiF64. The results look satisfactory with a quite good agreement between data and simulations. This gives us confidence on the reliability of the simulation tool and of the detectors in view of new designs.

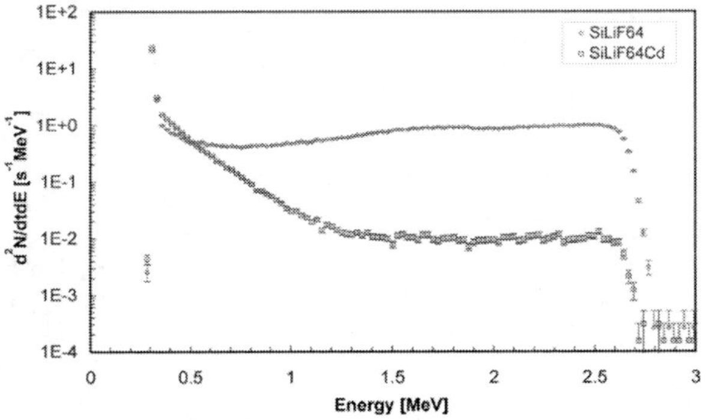

Fig. 8. Energy spectrum measured with the SiLiF64 detector, with and without the Cadmium plate.

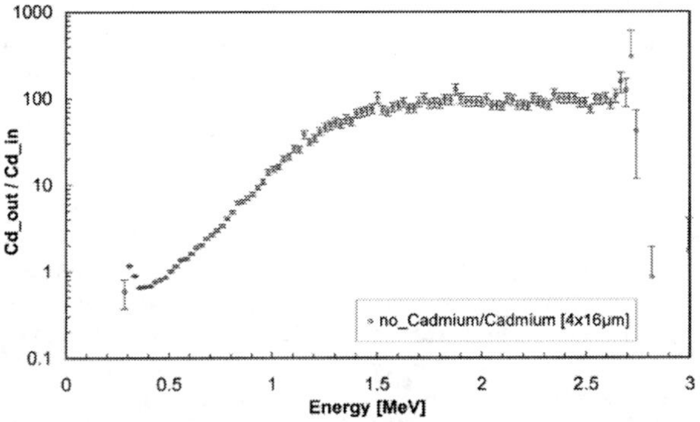

Fig. 9. Ratio between the two spectra of Fig. 8, that justifies the choice of 1.5 MeV as neutron discrimination threshold value.

V. Perspectives

These neutron detectors have also been tested at neutron beam facilities like ISIS (Rutherford Appleton Laboratory, UK), and n_TOF (CERN) where a few samples are already in operation for beam flux and 2D profile measurements. Forthcoming applications are foreseen for the online monitoring of spent nuclear fuel casks in interim storage sites, and to this aim a new set of simulations for several different configurations has already been started, along with additional calibrations using the same facility at PTB.

VI. Conclusion

In order to cope with the need of new thermal neutron detectors as an alternative to ^{3}He tubes in research, industrial, safety and homeland security applications, we have developed, tested and characterized solid state silicon detectors coupled with ^{6}LiF neutron converters. Even though their detection efficiency is only around 8%, the gamma rejection performance and the rather low cost as compared to ^{3}He tubes, make these detectors quite interesting for several applications, especially those with continuous monitoring purposes.

Acknowledgment

We thank Sebastian Fässer for his invaluable support during the calibration measurements at PTB. We are strongly indebted with Carmelo Marchetta and Antonio Massara for the preparation of the ^{6}LiF converters. A heartly felt acknowledgment is due to Marco Ripani for his constant encouragement and support within the framework of the INFN-E Strategic Project.

References

[1] D.Henzlova et al., "Current Status of ^{3}He Alternative Technologies for Nuclear Safeguards", NNSA USDOE and EURATOM, LA-UR-15-21201.

[2] R. T. Kouzes, "The ^{3}He supply problem", Tech. Rep. PNNL-18388, Pacific Northwest National Laboratory (Richland, WA, 2009).

[3] R.T. Kouzes et al., "Neutron detection alternatives to ^{3}He for national security applications", Nucl. Instrum. Methods Phys. Res. A623 (2010) 1035.

[4] N. Ensslin, in "Passive Nondestructive Assay of Nuclear Materials", NUREG/CR-5550 LA-UR-90-732, 457 (1991)

[5] P. Finocchiaro, "Radioactive Waste: A System for Online Monitoring and Data Availability", Nucl. Phys. News v.24, n.3, (2014) 34

[6] P.Finocchiaro, "DMNR: a new concept for real-time online monitoring of short and medium term radioactive waste", in *Radioactive Waste: Sources, Types and Management*, Nova Science Publishers, (2011)

[7] A. Pappalardo et al., "Characterization of the silicon+^{6}LiF thermal neutron detection technique", Nucl. Instrum. Methods Phys. Res. A810 (2016) 6.

[8] D. S. McGregor et al., "Design considerations for thin film coated semiconductor thermal neutron detectors—I: basics regarding alpha particle emitting neutron reactive films", Nucl. Instrum. Methods Phys. Res. A500 (2003) 272.

[9] M. Barbagallo et al., "Thermal neutron detection using a silicon pad detector and ^{6}LiF removable converters", Rev. Sci. Instrum. 84, (2013) 033503.

[10] S. Lo Meo et al., Nucl. Instr. Meth. A, in press, https://doi.org/10.1016/j.nima.2017.04.029

[11] L.Cosentino et al., "Silicon detectors for monitoring neutron beams in n-TOF beamlines", Rev. Sci. Instrum. 86 (2015) 073509.

[12] A.Pappalardo, C. Vasi, P. Finocchiaro, "Direct comparison between solid state Silicon+6LiF and 3He gas tube neutron detectors", Results in Physics 6 (2016) 12.

[13] F.Gunsing et al., "Status and outlook of the neutron time-of-flight facility n_TOF at CERN", Nucl. Instr. Meth. B261 (2007) 925.

[14] E. Chiaveri et al., "Proposal for n_TOF Experimental Area 2", CERN-INTC-2012-029 / INTC-O-015 09/

[15] M. Luszik-Bhadra, M. Reginatto, H.Wershofen, B.Wiegel, and A. Zimbal, "New PTB Thermal Neutron Calibration Facility: first results", Radiation Protection Dosimetry (2014), Vol. 161, No. 1–4, pp. 352–356.

[16] M. Luszik-Bhadra, D. Radeck, M. Reginatto, H. Wershofen, M. Zboril, and A. Zimbal, "The PTB Thermal Neutron Calibration Facility", PTB Report, Physikalisch-Technische Bundesanstalt, Braunschweig, to be published.

[17] S.Agostinelli et al., "Geant4 - a simulation toolkit", Nucl. Instrum. Methods Phys. Res. A506 (2003) 250.

[18] S. Lo Meo et al., "GEANT4 simulations of the n TOF spallation source and their benchmarking", Eur. Phys. Jou. A51 (2015) 160.

https://doi.org/10.1051/epjconf/201817001005

RESPONSE OF CsI:Pb SCINTILLATOR CRYSTAL TO NEUTRON RADIATION

Maria da Conceição Costa Pereira, Tufic Madi Filho, José Roberto Berretta, José Patrício Náhuel Cárdenas and Antonio Carlos Iglesias Rodrigues

Abstract - The helium-3 world crisis requires a development of new methods of neutron detection to replace commonly used ^{3}He proportional counters. In the past decades, great effort was made to developed efficient and fast scintillators to detect radiation. The inorganic scintillator may be an alternative. Inorganic scintillators with much higher density should be selected for optimal neutron detection efficiency taking into consideration the relevant reactions leading to light emission. These detectors should, then, be carefully characterized both experimentally and by means of advanced simulation code. Ideally, the detector should have the capability to separate neutron and gamma induced events either by amplitude or through pulse shape differences. As neutron sources also generate gamma radiation, which can interfere with the measurement, it is necessary that the detector be able to discriminate the presence of such radiation. Considerable progress has been achieved to develop new inorganic scintillators, in particular increasing the light output and decreasing the decay time by optimized doping. Crystals may be found to suit neutron detection. In this report, we will present the results of the study of lead doped cesium iodide crystals (CsI:Pb) grown in our laboratory, using the vertical Bridgman technique. The concentration of the lead doping element (Pb) was studied in the range $5\text{x}10^{-4}$ M to 10^{-2} M . The crystals grown were subjected to annealing (heat treatment). In this procedure, vacuum of 10^{-6} mbar and continuous temperature of 350°C, for 24 hours, were employed. In response to neutron radiation, an AmBe source with energy range of 1 MeV to 12 MeV was used. The activity of the AmBe source was 1Ci Am. The fluency was 2.6 x 10^6 neutrons/second. The operating voltage of the photomultiplier tube was 1700 V; the accumulation time in the counting process was 600 s and 1800 s. The scintillator crystals used were cut with dimensions of 20 mm diameter and 10 mm height.

Key words: neutron detection, scintillator, crystals

All Authors: Nuclear and Energy Research Institute, IPEN–CNEN/SP
Instituto de Pesquisas Energéticas e Nucleares – IPEN-CNEN/SP
Av. Prof. Lineu Prestes 2242 Cidade Universitária CEP: 05508-000 – São Paulo-SP - Brazil
Maria da Conceição Costa Pereira - macoper@ipen.br correspondent author
Tufic Madi Filho - tmfilho@ipen.br
José Roberto Berretta - jrretta@ipen.br
José Patrício Náhuel Cárdenas - ahiru@ipen.br
Antonio Carlos Iglesias Rodrigues - acirodri@ipen.br

I. Introduction

The disappearing inventory and minute natural abundance of ^{3}He gas necessitate the adoption of new technologies for the detection of neutrons. The exclusive source of ^{3}He on Earth is derived from the tritium stockpile, which decays to ^{3}He at a rate of 5.5% per year. Despite the low ^{3}He supply, and uncertain production rate in the future, this medium remains by far, the most attractive for many applications. The desired world deployment of the large area monitors to intercept the smuggling of nuclear materials, alone, could consume the entire ^{3}He supply, limiting the prospects of nuclear science and other applications that rely very heavily on ^{3}He-based detectors. Clearly, alternate neutron detection technologies have to be developed. The inorganic scintillator crystal may be an alternative.[1]

Scintillators are materials that convert the energy of ionizing radiation into a flash of light. Scintillation material can be gaseous, liquid, glass-like, organic (plastics), or inorganic. In each case the material should be transparent to its own scintillation light. Inorganic wide band gap ionic crystals are the most widely used scintillators for detection of X-rays, gamma rays, and thermal neutrons. The scintillation method is widely used for detection of various kinds of ionizing radiation. The modern physics of scintillators is a broad field of fundamental and applied research dealing with a large variety materials and applications.[2] Every year, the scintillation method finds new applications in rapidly developing fields of high energy physics, nuclear medicine, geophysics, the monitoring of environmental, in devices and systems for security inspection.

There are many different luminescence species and scintillation mechanisms possible in inorganic materials. The luminescence may be intrinsic to material and involve electron-hole recombination; free, self trapped, and defect-trapped exciton luminescence; constituent transition group per post transition group ion fluorescence; core-valence band transitions; or charge transfer transitions within a molecular complex. Or it may be extrinsic, such as luminescence associated with impurities or defects and additive dopant ions. In the role of an activator, the dopant ion may be the luminescence species or may promote luminescence as in the case of defect bound exciton emission. [3,2]

Scintillators based on cesium iodide are the leading, among materials available for solid-state detectors. Cesium iodide based scintillators have high luminescence yield accompanied by high energy resolution and radiation stability. Various impurities have benn incorporated into alkali halides and their effects on optical properties and radiation detection have been investigated. Ions such as Tl+, Br^-, Li^+, Pb^{2+} has been the theme of pursuit in our laboratory, in the last years. In continuation, a systematic investigation has been carried out in Pb^{2+} doped cesium iodide crystal to neutron response and the results are discussed in this article.

II. Material and Methods

Lead doped cesium iodine crystals (CsI:Pb) were grown in our laboratory, using the vertical Bridgman technique. [4,5,6] The concentration of the lead doping was in the range 10^{-2}M to $5x10^{-4}$M. The crystal grown were subjected to annealing (heat treatment). In this procedure, vacuum of 10^{-6} mbar and continuous temperature of 350°C, for 24 hours, were employed. Others details on the detector fabrication and characteristics may be found in our previous publication [7].

In the study of response to neutron radiation, the crystals were directly coupled to the photomultiplier tube using silicone grease viscosity of 0.5 McStokes, as optical interface. This ensured uniform refractive index across the contact surface between the crystal and photomultiplier tube. Sides of the crystals, which were not in contact with the photo-sensor, were covered with several layers of polytetrafluoroethylene (PTFE) tape to ensure good reflection of light. The electronic modules used for the processing of signals from the photomultiplier tube were constituted by conventional ORTEC electronics. The detection efficiency of the scintillator crystal was measured in two different positions: in the first (position I), the AmBe source was positioned at a distance of 100 mm from the photomultiplier tube. In the second (position II), the AmBe source was positioned at a distance of 100 mm from the photomultiplier tube, using 70 mm paraffin block as the interface. A foil of Cd was placed around the detector and the photomultiplier tube was used to avoid the scattered neutron contribution, as shown in Fig.1.

The fluency was 2.6 x 10^6 neutrons/second. The operating voltage of the photomultiplier tube was 1700 V; the accumulation time in the counting process was 600 s and 1800 s. The scintillator crystals used were cut with dimensions of 20 mm diameter and 10 mm height.

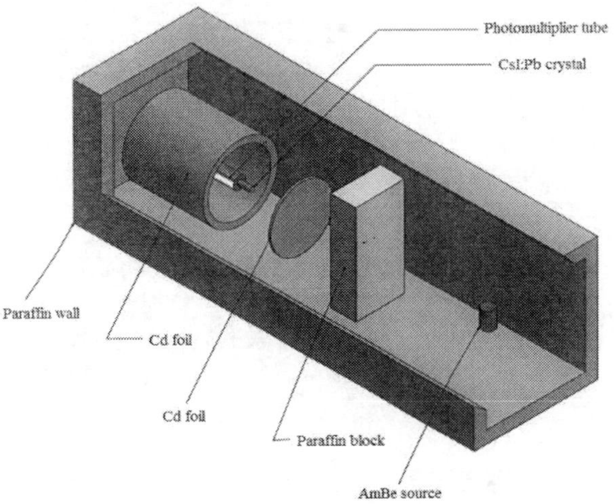

Fig.1 - Exploded schematic representation of the array used for measurements, (position II).

III. Results

Figure 2 shows the CsI: Pb crystal grown by the vertical Bridgman technique in our laboratory.

(a) (b)

Fig. 2 - CsI:Pb crystal with molar fraction 10^{-2} grown by the vertical Bridgman technique. (a) before the heat treatment; (b) after heat treatment and polishing.

The curve shown in Fig. 3 illustrates the experimental results for the neutron radiation from an AmBe source using the CsI:Pb 10^{-2} M scintillator crystal. In order to verify the the response to fast and thermal neutrons, a block of 70 mm of paraffin was used. A Cd foil was placed around the detector and the photomultiplier tube, to avoid the scattered neutrons contribution.

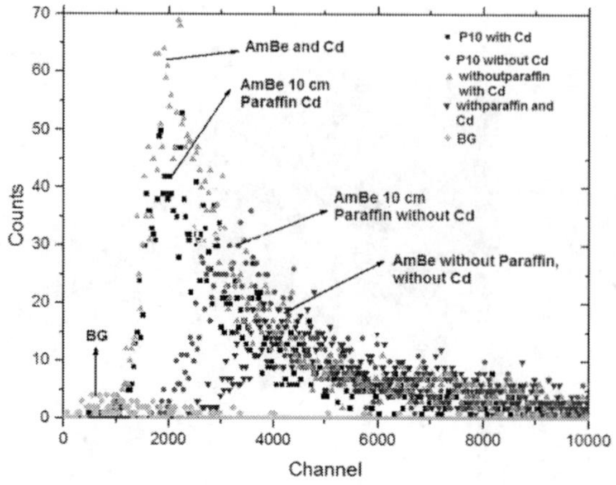

Fig. 3 - Results of the measurements using CsI:Pb crystal 10^{-2} M and neutron source, with paraffin, without paraffin, with Cd foil and without Cd foil.

Fig. 4 shows the experimental results using CsI:Pb with concentration 10^{-2} M, 10^{-3}M and 5×10^{-4} M, without paraffin and without Cd foil.

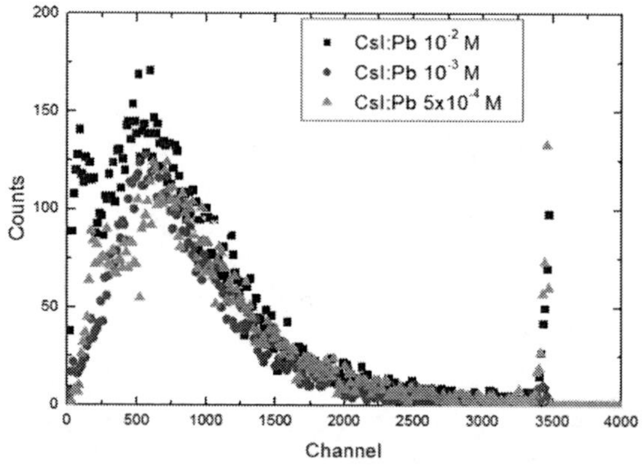

Fig. 4 - Results of the measurements using CsI:Pb crystal and neutron source, without paraffin, with Cd foil.

III. Conclusion

The addition of the lead ions (Pb^{2+}) to the CsI matrix resulted in crystals with promising results, when excited with neutron radiation. The crystal showed sensitive to fast and thermal neutron.

In the dopant concentration range (Pb) studied, the best neutron detection efficiency was obtained for the molar fraction 10^{-2}, without the use of paraffin between the source and the crystal and using cadmium.

In the experiment with the crystals at various concentrations of the dopant, without the paraffin block between the detector and the neutron source, a slight increase was observed in the counts as a function of the increase in the concentration of dopant Pb.

References

[1] A. J. PEURRUNG, Recent developments in neutron detection. Nuclear Instruments & Methods in Physics Research. A 443, pp. 400-415, 2000.

[2] G. F. KNOLL, Radiation Detection and Measurement, John Wiley & Sons, New York, NY, USA, 4th edition 2010.

[3] S. E. DERENZO, M. J. WEBER, E. B. COURCHESNE and M. K.KLINTENBERG. The quest for ideal inorganic scintillator. Nuclear Instruments & Methods in Physics Research. A 505, pp. 111-117, 2003.

[4] P. W. BRIDGMAN, Proc. Amer. Acad. Arts Sci.,60, pp. 303-383, 1925

[5] M. C. C. PEREIRA, T. M. FILHO, J. P. N. CARDENAS. Inorganic Scintillation Crystals for Neutron Detection. IEEE Transactions on Nuclear Science, v. 63, pp. 1699-1702, 2016.

[6] M. C. C. PEREIRA, T. M. FILHO, Scintillation Characteristics of CsI Crystal Doped Br under Gamma and Alpha Particles Excitation. Materials Sciences and Applications (Print), v. 05, pp. 368-377, 2014.

[7] M. C. C. PEREIRA, T. M. FILHO, M. M. HAMADA. The effect of Pb $^{2+}$ dopant in the crystal of CsI and its application as scintillation detector: a study of alpha particles. Radiation Effects and Defects in Solids (Print), v. 167, pp. 921-928, 2012

Maria da Conceição Costa Pereira: Graduated in Industrial Chemistry from the Escola Superior de Química Oswaldo Cruz, São Paulo, Brazil in 1985. Master in Nuclear Technology from the Universidade de São Paulo in 1997. PhD in Nuclear Technology from the Universidade de São Paulo in 2006. Currently is senior Researcher of the Instituto de Pesquisas Energéticas e Nucleares - Comissão Nacional de Energia Nuclear (IPEN/CNEN-SP). Has experience in Chemistry and Nuclear Engineering, acting on the following topics: radiation detectors, luminescence decay time, growth scintillators crystals. Working in the development of new materials and their applications in radiation detectors. Address to the curriculum http://lattes.cnpq.br/0977986453736808

Tufic Madi Filho: Graduated in Physics from the Pontifícia Universidade Católica de São Paulo (1976). Master in Nuclear Technology from the Universidade de São Paulo (1982) and PhD in Nuclear Technology from the Universidade de São Paulo (1999). Currently is senior Researcher of the Comissão Nacional de Energia Nuclear (National Commission of Nuclear Energy and Professor of the graduate course: "Theoretical Fundaments and Practices of the Instrumentation used in Nuclear Data Acquisition". Has experience in Physics, with emphasis on nuclear physics. Works on the following topics: nuclear reactor; alpha, beta and neutron detection, gamma spectrometry. Address to the curriculum http://lattes.cnpq.br/3251192790149021.

Professor of the course: **Dr. José Patrício Náhuel Cardenas,** Doctor of Science in Nuclear Technology, University of Sao Paulo, Brazil. Works in the development of neutrons experiments using IPEN laboratories and facilities, on a postgraduate program to support students in the use of instrumentation applied to nuclear research.

A Low-Power and In Situ Annealing Technique for the Recovery of Active Devices After Proton Irradiation

Laurent A. Francis, Amor Sedki, Nicolas André, Valéria Kilchytska, Pierre Gérard, Zeeshan Ali, Florin Udrea, and Denis Flandre.

Abstract— **In this paper, we study the recovery of on-membrane semiconductor components, such as N-type Field-Effect Transistors (FETs) available in two different channel widths and a Complementary Metal-Oxide-Semiconductor (CMOS) inverter, after the exposure to high dose of proton radiation. Due to the ionizing effect, the electrical characteristics of the components established remarkable shifts, where the threshold voltages showed an average shift of -480 mV and -280 mV respectively for 6 µm and 24 µm N-channel transistors, likewise the inversion point of the inverter showed an important shift of -690 mV. The recovery concept is based mainly on a micro-hotplate, fabricated with backside MEMS micromachining structure and a Silicon-On-Insulator (SOI) technology, ensuring rapid, low power and *in situ* annealing technique, this method proved its reliability in recent works. Annealing the N-channel transistors and the inverter for 16 min with a temperature of the heater up to 385 °C, guaranteed a partial recovery of the semiconductor based components with a maximum power consumption of 66 mW.**

Keywords— **Field Effect Transistors (FETs), In situ annealing, Micro-hotplates, Proton radiation, Radiation mitigation.**

I. Introduction

The Silicon-On-Insulator (SOI) technology is one favored hardening method for circuits exposed to natural radiations, in particular for single event effects (SEE) [1], in addition to the tremendous advantages and great resistivity at high temperatures up to 300 °C, making it a very suitable technology in harsh environments [2]. However, this technology still remains sensitive to total ionizing doses (TID) [3], where increased densities of both oxide-trapped charges and interface traps lead to massive degradations in transistor's fundamental parameters (threshold voltage, carriers mobility, leakage currents, etc.). The recovery of transistors after their irradiation can be guaranteed by the neutralization of positive trapped charges helped by thermal annealing or a tunneling effect under external electrical field. The thermal annealing process consist in the exposure of radiated devices to high temperature between 300 and 400 °C for about one hour [4], otherwise by long term stability at room temperature [5]. The latter approach is efficient but constrained by the post-CMOS thermal budget, limited by e.g. the thermally activated dopant diffusion or the degradation of the metal layers, a wide variety of reliability issues related to the different coefficients of thermal expansion for System-in-Package (SiP) technology, and by the bulky equipments and their high power consumption needed to achieve the thermal treatment. Recent works improved a revolutionary success by improving an *in situ* low power total recovery of semiconductor based devices after high gamma dose irradiation [6]. In this work, our interest is rather focusing on the proton radiation, by studying the electrical characteristics of SOI electronic components embedded in a 5 µm-thick micro-hotplate membrane and analyzing the degradations that appeared after proton irradiation. As well we are investigating the recovery of the components by thermal annealing up to 385 °C. To do so we used a special MEMS structure based on micro-hotplates, which cluster the studied electronic components and the micro-heater on a single membrane, for better power conservation and integration efficiency, where the temperature is monitored by an integrated thermo-diode as demonstrated in [7].

The first part of this paper, describes the used device's structure, specify its fabrication technology, demonstrate the used characterization technique and details the stabilization method, including the description of thermo-diode's calibration methodology. The second section, reports the applied referential characterization of the components, present the proton irradiation process of the device and study the related degradations of the semiconductor based components. The third and last section discusses the applied thermal annealing process, as well, evaluates the partial recovery of the electrical characteristics of the MOSFET based components and the CMOS inverter.

L.A. Francis, S. Amor, N. André, V. Kilchytska, P. Gérard, and D. Flandre are with the Institute of Information and Communication Technologies, Electronics, and Applied Mathematics, Université catholique de Louvain, B-1348 Louvain-la-Neuve, Belgium (e-mail: {laurent.francis}, {nicolas.andre},{ valeria.kilchytska }, {pierre.gerard}, {denis.flandre}@uclouvain.be, and sedkiamor@gmail.com). L.A. Francis is corresponding author.

F. Udrea is with the Department of Engineering, Electrical Engineering Division, University of Cambridge, 9 JJ Thomson Avenue, CB3 0FA, Cambridge, United Kingdom (e-mail: florin.udrea@ucam.uk).

S.Z. Ali and F. Udrea are with Cambridge CMOS Sensors Ltd, Deanland House, Cowley Road, Cambridge, CB4 0DL, United Kingdom, (e-mail: zeeshan.ali@ccmoss.com).

II. Experimental Procedure

A. Device's technology and description

The used technology to fabricate the MEMS structure is 0.1 µm Partially Depleted (PD) SOI (XI.10 from X-FAB, Erfurt, Germany). The device is based mainly on a micro-hotplate, obtained by a post-CMOS Deep Reactive Ion Etching (DRIE) of the 400 µm thick silicon layer. The backside etching stops when reaching the buried 1 µm thick SiO_2 layer to form a circular membrane with a diameter of 600 µm. On the front side the membrane is centered by a Joule heater with 200 µm diameter, made by a 0.5 µm thick tungsten layer. Two N-channel transistors, one CMOS inverter and one PIN diode are placed on the membrane around the micro-heater (see Figure. 1), whereas the membrane and the electronic component's temperature is monitored by the PIN diode. The benefit of the membrane consists mainly in its low thermal dissipation, which allow the micro-hotplates to reach high temperatures with good linearity, in addition to its high resistivity to the ionizing radiations making this technique very suitable in harsh environments [7].

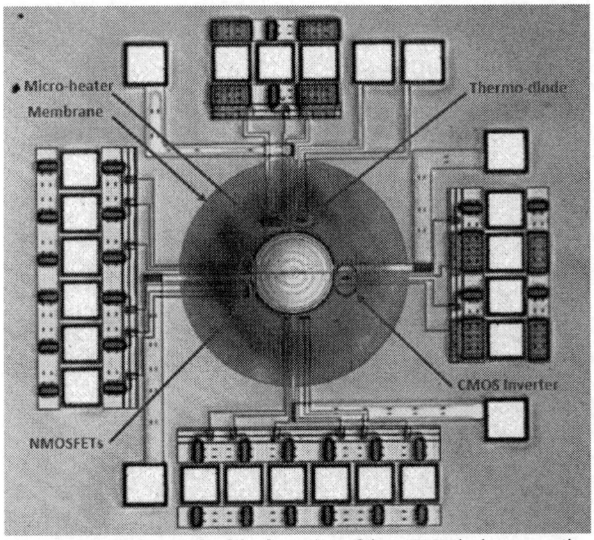

Fig. 1. Optical micrograph of the front view of the MEMS device, presenting the location of the studied on-membrane embedded components.

To avoid electrical parasites during measurements the device was packaged in a Dual In Line (DIL) ceramic package where the metallic pads were electrically connected to the 24 pins by wire-bonding, in respect to a pre-fixed order. After preparing the device, it is inserted in a Zero Insertion Force (ZIF) support, then connected to the triaxial ports of a test fixture box. The device is connected to a Semiconductor Parameter Analyzer (HP4145) to be controlled by IC-CAP software through an adjustable interface.

After the backside micromachining, the electrical characteristics of the MOSFET based components showed a significant instability in the I-V curves, this returns mainly to the creation of parasitic charges during the engraving of the silicon substrate to create the membrane. Thus, a pre-stabilization process is considered mandatory to stabilize the characteristics of the MOSFETs before starting the characterization part. In order to neutralize the trapped charges, the process consists in annealing the MOSFETs by 4 successive cycles of 30 s each, where the micro-hotplate's temperature is gradually increased up to 265 °C by supplying the micro-heater with the corresponding electrical power. Subsequently, the on-membrane semiconductor components showed as well a reliable characteristics after high temperature annealing up to 335 °C. A deep study thermal characteristics of the different devices is available in [8].

B. Temperature monitoring

Using the diode as temperature sensor, it reports the temperature of the micro-hotplate at the MOSFETs site. This technique is based on the dependency of PIN diodes characteristics to the operating temperature biased by a stable current source (here 65 µA), the diode voltage dependency is demonstrated by Equation (1).

$$V_d = V_0 - d(T - T_0) \tag{1}$$

with V_0 [V] the initial voltage at room temperature, d [V/°C] the resolution coefficient of the diode, T [°C] the operating temperature and T_0 [°C] the room temperature. Increasing the temperature of device progressively up to 300 °C by a controlled hot chuck helped to extract the drop voltage of the diode, given by 1.2 mV/°C. After this first calibration of the diode, the temperature monitoring is possible by supervising V_d variation.

III. Fast proton irradiation and effects

After applying the stabilization process, a first referential characterization is made by logging the I-V characteristics of the MOSFETs in linear regime (V_d = 60 mV), as well the Voltage Transfer Characteristics (VTC) of the CMOS inverter at room temperature. After that, the packaged device was removed from the test fixture box, installed in a test plate, where all the contacts are leaved in a floating condition, then fixed in front of the high flux fast protons beam facility of the Cyclotron Resource Centre (Université catholique de Louvain, Louvain-la-Neuve, Belgium), pending the performing of a high proton bombardment with a fluence of 10^{14} p^+/cm^2. After the irradiation process the device is reinstalled in the test fixture box for a post-irradiation characterization. The subthreshold curves of the 6 µm wide MOSFET, measured in linear regime with V_{ds}= 60 mV, before and after irradiation, are presented in Figure 2, where we clearly notice the wide negative shift of the post-radiation curve, in addition to the tremendous increase in the off-state leakage current. The I-V characteristics of the 24 µm wide channel MOSFETs, showed a similar behavior after irradiation (see Figure 3), confirming the negative shift of the threshold voltages due to the positive trapped charges. In the other hand the VTC characteristics of the inverter showed as well a severe negative transferal, seriously changing its specific noise margin. The threshold voltage of the MOSFETs is extracted by extrapolation method in linear regime (see Table I). The irradiation dose was sufficient to change the parameters of MOS components, by degrading the threshold voltages of N-channel transistors with 27% and 16% for, respectively, 6 and 24 µm-wide channels, while the input voltage corresponding to the inversion point of the inverter was also degraded by 40% with respect to its initial value (see Figure 4).

IV. THERMAL ANNEALING

The thermal annealing was applied after the irradiation of the devices, where the protocol consisted in increasing the temperature of the micro-hotplate by successive thermal upshots from room temperature up to 385 °C with a total period of 16 min. The thermal upshots were performed automatically with IC-CAP software by supplying the tungsten micro-heater with increasing voltage steps of 0.25V to reach a maximum voltage of 2 V and requiring a maximum power consumption of 66 mW. Each step of annealing is escorted by monitoring the corresponding temperature of the micro-hotplate.

After this annealing process, we clearly notice that the drain current versus gate voltage characteristics showed a considerable recovery, the subthreshold slope of the 6 µm wide NMOSFET presented in Figure 2, is significantly decreased after the annealing and tends to establish its initial position, accordingly the off-state current is decreased, which proves the efficiency in neutralizing the trapped induced positive charges.

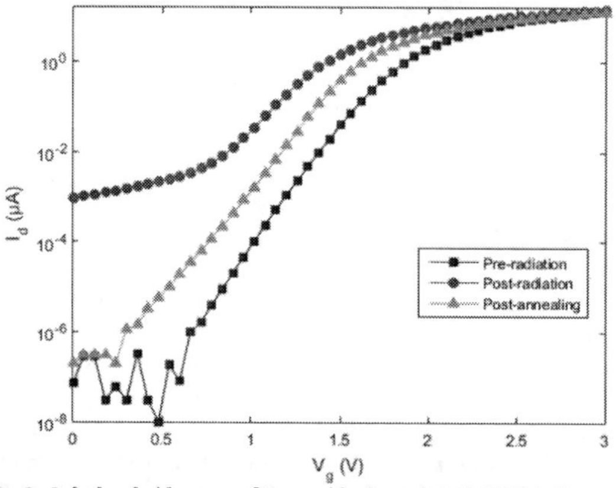

Fig. 2. Sub-threshold curves of 6 µm wide channel nMOSFET in linear regime with V_{ds}= 60mV, presenting its behavior before, after irradiation and after the annealing process.

The 24 µm wide channel NMOSFET showed a similar behavior after the annealing process, where the I-V characteristics are significantly shifted towards the initial position (see Figure. 3). The extracted values of threshold voltage presented in Table. 1 shows the considerable recovery obtained by the annealing process and prove the reliability of this technique for the different dimensions of MOSFETs.

TABLE I
THRESHOLD VOLTAGE VARIATION

	6 µm wide	24 µm wide
V_{Th} [V] Pre-radiation	1.73	1.66
V_{Th} [V] Post-radiation	1.25	1.38
V_{Th} [V] Post-annealing	1.46	1.56

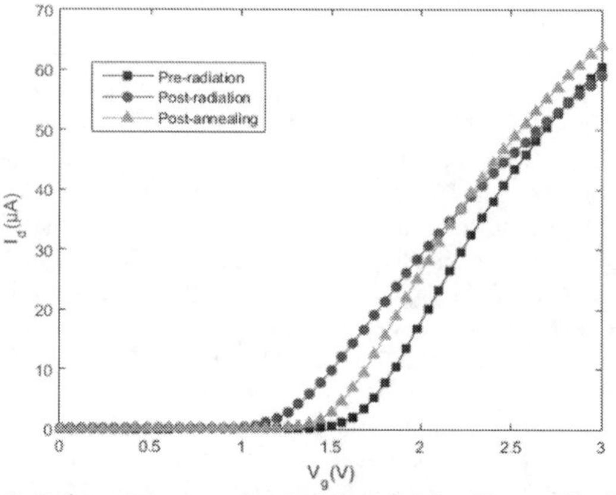

Fig. 3. Drain current versus gate voltage characteristics of 24 µm wide channel N-MOS transistor in linear regime with V_{ds} = 60 mV, showing the change in behavior before, after proton radiation and after thermal annealing.

After the annealing process, the percentage of the threshold voltage degradation was reduced to 15% and 6% respectively for the 6 µm and 24 µm wide channel transistors, while the inversion point of the single stage inverter was moved back to 24% of its initial value. Hence, the recovery of the components reached about its half.

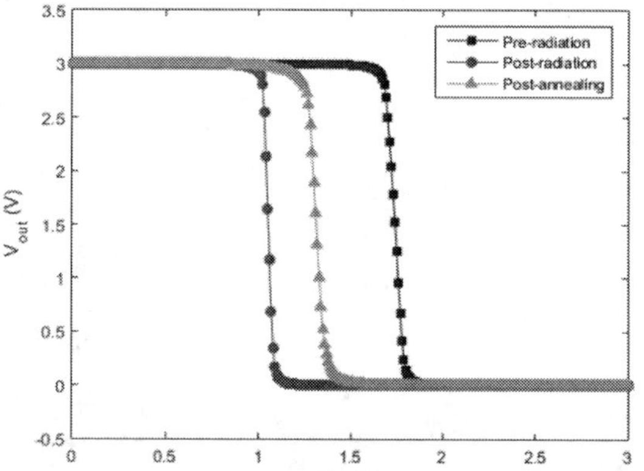

Fig. 4. Voltage transfer characteristics of the CMOS inverter presented before and after radiation and after the thermal annealing.

V. CONCLUSION

The 10^{14} p^+/cm^2 dose of fast protons irradiation was detrimental to the electrical characteristics of the on-membrane NMOSFETs and the CMOS inverter. Thanks to the low power micro-hotplates of the special MEMS structure, a high temperature annealing up to 385 °C was sufficient to perform a partial recovery of the threshold voltage and the off-state leakage currents of the NMOSFETs as well for the inversion point of the inverter. A total recovery of components is expected by increasing the annealing temperature.

https://doi.org/10.1051/epjconf/201817001006

References

[1]Manet P, Falmagne S, Garnier J, Berger G, Legat J-D, editors. SEE evaluation of a low-power 1μm-SOI 80C51 for extremely harsh environments. Radiation and Its Effects on Components and Systems (RADECS), 2009 European Conference on. IEEE. 2009.
[2]Rue B, André N, Olbrechts B, Raskin J, Flandre D, editors. High Temperature SOI CMOS Low Power circuits and micro systems for MEMS co-integrated interfaces, temperature sensing and power management applications. Caneus Workshop. 2009.
[3]Rezzak N, Zhang EX, Alles ML, Schrimpf RD, Hughes H, editors. Total-ionizing-dose radiation response of partially-depleted SOI devices. SOI Conference (SOI), 2010 IEEE International. IEEE. 2010.
[4]Gwyn C. Model for Radiation-Induced Charge Trapping and Annealing in the Oxide Layer of MOS Devices. Journal of Applied Physics.40(12):4886-92. 1969.
[5]Pejovic MM, Pejovic MM, Jaksic AB. Contribution of fixed oxide traps to sensitivity of pMOS dosimeters during gamma ray irradiation and annealing at room and elevated temperature. Sensors and Actuators A: Physical.174:85-90. 2012.
[6]Amor S, André N, Kilchytska V, Tounsi F, Mezghani B, Gérard P, et al. In-situ thermal annealing of on-membrane silicon-on-insulator semiconductor-based devices after high gamma dose irradiation. Nanotechnology.28(18):184001. 2017.
[7]Francis LA, André N, Gérard P, Ali SZ, Udrea F, Flandre D, editors. A low-power and in situ annealing mitigation technique for fast neutrons irradiation of integrated temperature sensing diodes. Advancements in Nuclear Instrumentation Measurement Methods and their Applications (ANIMMA), 2015 4th International Conference on. IEEE. 2015.
[8]Amor S, André N, Gérard P, Ali S, Udrea F, Tounsi F, et al. Reliable characteristics and stabilization of on-membrane SOI MOSFET-based components heated up to 335° C. Semiconductor Science and Technology.32(1):014001. 2016.

EPJ Web of Conferences **170**, 01007 (2018)
ANIMMA 2017 https://doi.org/10.1051/epjconf/201817001007

The PADME Detector

Paola Gianotti*
on behalf of the PADME collaboration[1]
*Laboratori Nazionali di Frascati dell'INFN
Via E. Fermi 40
I-00044 Frascati (Italy)
Email: paola.gianotti@lnf.infn.it

Abstract—**The PADME experiment, by using the positron beam of the Frascati laboratory, aims at searching for signals of a dark photon, A′ . It will evaluate the final state missing mass of the process $e^+ e^- \to A' \gamma$ by knowing the beam energy and measuring the four-momentum of the ordinary recoil photon. The precise determination of this quantity, and the capability to reject background events, are the key points for the success of the experiment. In this paper a description of each component of the PADME detector is given.**

I. Introduction

The Positron Annihilation into Dark Matter Experiment (PADME) [1] aims to search for the production of a dark photon in the process:

$$e^+e^- \to A'\gamma \tag{1}$$

To measure such a reaction, at the Frascati National Laboratory (LNF) of INFN, it is under construction a small scale detector composed of the following parts:

- a diamond active target station, capables to measure the average position and the intensity of the beam; it also includes a beam monitor device, made of ultra-thin silicon pixel detectors, to study beam characteristics in terms of intensity and divergence;
- a finely segmented, high resolution electromagnetic calorimeter, to measure four-momenta and/or veto final state photons;
- a dipole magnet, to deflect the primary positron beam out of the spectrometer and the calorimeter and to allow momentum analysis;
- a spectrometer, to measure the momenta of the charged particles in the energy range 50-400 MeV;
- a vacuum chamber, to minimize the unwanted interactions of primary and secondary particles;
- a further silicon pixel detector to monitor the not interacting positrons.

[1]G. Chiodini, P. Creti, F. Oliva (INFN Lecce), A.P. Caricato, M. Martino, G. Maruccio, A. Monteduro, V. Scherini, S.Spagnolo (INFN Lecce e Dip. di Matematica e Fisica, Università del Salento), P. Albicocco, R. Bedogni, B. Buonomo, F. Bossi, R. De Sangro, G. Finocchiaro, L.G. Foggetta, A. Ghigo, P. Gianotti, M. Palutan, G. Piperno, I. Sarra, B. Sciascia, T. Spadaro, E. Spiriti (INFN Laboratori Nazionali di Frascati), L. Tsankov, (University of Sofia "St. Kl. Ohridski), G. Georgiev, V. Kozhuharov (University of Sofia "St. Kl. Ohridski" and INFN Laboratori Nazionali di Frascati), F. Ameli, F. Ferrarotto, E. Leonardi, P. Valente (INFN Roma1), S. Fiore (INFN Roma1 e ENEA), G.C.Organtini, M.Raggi (INFN Roma1 e Dip. di Fisica, Sapienza" Università di Roma), C. Taruggi (INFN Laboratori Nazionali di Frascati e Dip. di Fisica, Università di Roma Tor Vergata")

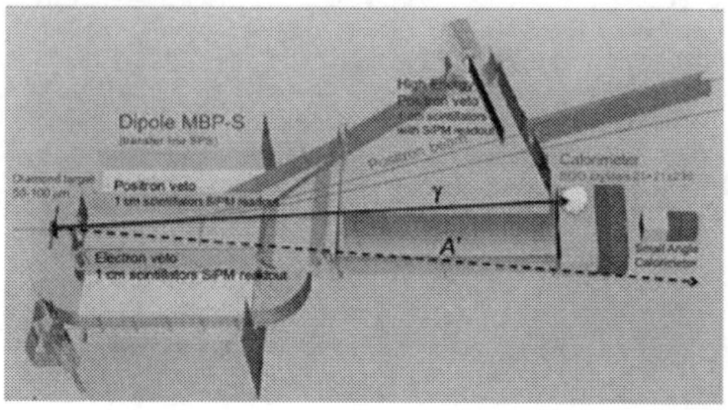

Fig. 1. Layout of the PADME experimental setup (see text for more details).

An overview of the PADME experimental setup can be seen in fig. 1.

The 550 MeV positron beam, provided by the LINAC of the LNF, crosses the diamond target and if it does not interact, it is bent by the dipole in between the end of the spectrometer and the calorimeter. If any kind of interaction causes the positron to lose more than 50 MeV of energy, the magnet bends it into the spectrometer acceptance, where a segmented detector provides a veto signal. In case reaction (1) occurs, the accompanying ordinary photon is detected by the electromagnetic calorimeter regardless of the A′ decay products. A single kinematic variable characterizing the process, the missing mass, is computed using the formula:

$$M^2_{miss} = (\overrightarrow{P}_{beam} + \overrightarrow{P}_{e^-} - \overrightarrow{P}_{\gamma})^2 \tag{2}$$

with the target electrons assumed to be at rest ($\overrightarrow{P}_{e^-} = \overrightarrow{0}$) while for the beam a nominal momentum of 550 MeV along the z axis is expected. The distribution of this variable should peak at $M^2_{A'}$ if an A′ is produced, at zero for the concurrent $e^+ e^- \to \gamma\gamma$ process, and should be smooth for the remaining background. Figure 2 shows the results of Montecarlo simulations performed to study the background. The red histogram shows all different background contributions, while the blue one is what remains after the following cuts have been applied:

- a single photon detected by the main e.m. calorimeter;
- no clusters in the small angle calorimeter;

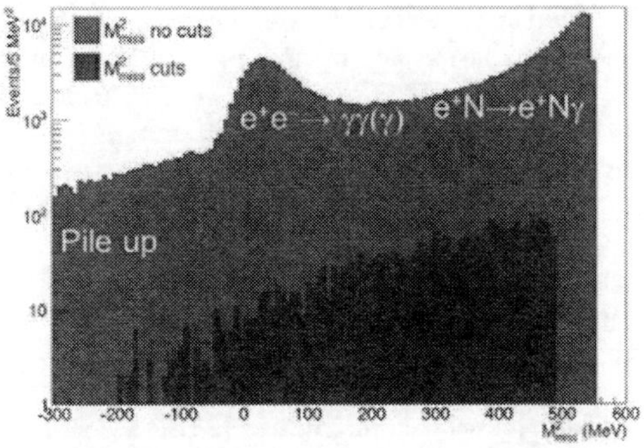

Fig. 2. Montecarlo simulated squared missing mass for background events. In red with no selection cuts, in blue after all cuts are applied. If a dark photon A′ is produced it should manifest as a pick in this distribution at the proper mass.

- no signals in the veto counters.

In this way, a sensitivity better than 10^{-3} can be reached on the coupling ε of the dark photon in a mass range up to $\approx$23 MeV/c^2 [2].

In order to perform at best this measurement, each element of the experimental setup has specific requirements that are stringent and sometimes challenging. In the following sections, it is given a description of the chosen technical solutions implemented to accomplish the experiment needs. Results of the test beams, performed to characterize each sub-component, are also illustrated.

II. The target station

To enhance the dark photon signal, proportional to the annihilation cross-section ($\propto$ Z), with respect to the Bremsstrahlung background ($\propto$ Z^2), a low Z material is needed. Carbon is the right compromise. In addition, the target should be thin enough to reduce the number of pile-up events to a level manageable by the high-energy resolution electromagnetic calorimeter. The knowledge of the beam spot hitting the target can improve the missing mass resolution, thus increasing the statistical significance of the signal. The above requirements can be satisfied by an active target made of detector-grade polycrystalline CVD diamond with strip electrodes on both sides and connected to adequate front-end electronics. Two active target prototypes (2×2 cm^2 area and 50 and 100 μm thickness) have been built and characterized with high multiplicity electron bunches at the Frascati Beam Test Facility (BTF) [3]. The readout strips (1 mm pitch in orthogonal directions on the two sides of the film) were made in the first case with in-house laser graphitization, in the second case with Cr-Au thermal evaporation by the supplier. The prototype with graphitic electrodes was fully characterized with two kind of off-board electronics, showing a good position resolution ($\sim$ 0.2 mm) on both views and timing performance ($\sim$ 2 ns) meeting the requirement of PADME. The prototype with Cr-Au electrodes was readout with the

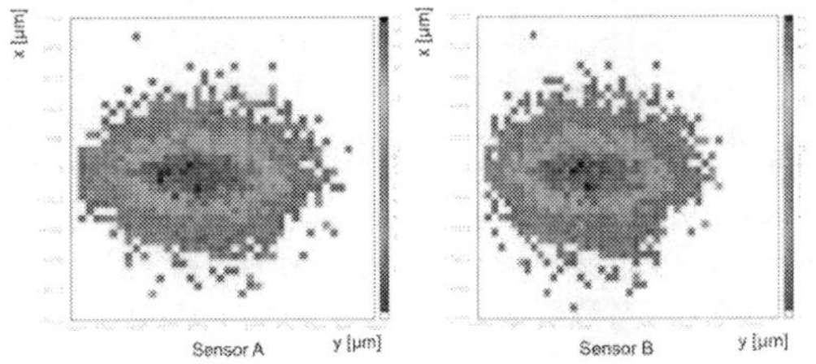

Fig. 3. Beam spot obtained with the MIMOSA sensors at the Frascati BTF.

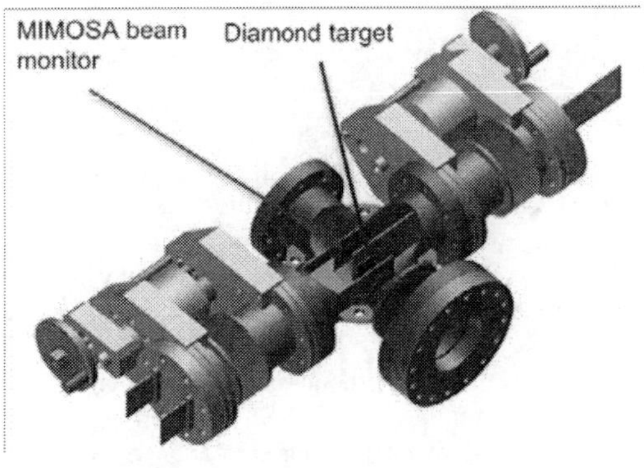

Fig. 4. CAD drawing of the target station: it houses either the thin diamond target and a beam monitoring system, based on monolithic silicon pixel detectors. Both components are remotely movable via a step motor.

on-board final electronics showing very good signal-to-noise ratio, but full characterization was impossible due to a poor quality of the contacts likely due to a not careful deposition [4].

To have a more accurate beam monitor capability, in the target region will be installed a system based on two planes of silicon pixel detectors placed up and down stream the diamond target. Each plane will consist of two MIMOSA 28 Ultimate chips, developed for the upgrade of the STAR vertex detector [5]. These devices integrate a Monolithic Active Pixel Sensor (MAPS) with a fast binary readout. Each sensor consists of a matrix of 928×960 pixels of 20.7 μm with a thickness of 50 μm. For the STAR experiment the chips, that dissipate 150 mW/cm^2, operate in air without cooling. For PADME, the detectors should be placed in vacuum and a modified PCB is under development at the LNF electronic workshop. A first prototype of this new monitoring system has been tested on the beam in April 2017. The picture of the BTF electron beam spot, obtained with the MIMOSA detectors, is visible in fig 3. Figure 4 shows a CAD drawing of the target region. Both the target and the beam monitor will be connected to step motors to allow their insertion in and out from the beam line.

III. The dipole magnet

The positron beam of 10^3/10^4 particles/pulse (depending mainly from beam pulse duration and the pile-up rate) will pass almost undisturbed through the PADME thin Carbon target. Therefore, a magnetic field of a moderate strength ($\sim$ 0.5 T) is necessary to sweep it out from the acceptance of the e.m. calorimeter. A magnet with the requested characteristics has been found at CERN within the spare dipoles of the SPS transport line (MBP-S), and it has been shipped to LNF at the end of 2015. During 2016, the dipole vertical gap has been increased to 230 mm, to better suite the experiment needs, and a campaign of measurements to map the field has been performed.

Figure 5 shows a picture of the magnet and the plots of the field map measurements. The magnet will house a vacuum vessel of proper shape hosting a segmented charged particle veto system

IV. The veto tracking system

The Bremsstrahlung is the process with the highest cross section, when the positron beam interact with the target. This leads to the production of low energy photons emitted mostly at small angles. To cut out these events, the main e.m. calorimeter has been designed with a central hole, with an aperture of $\sim 1^o$, and a fast, Small Angle Calorimeter (SAC), is placed behind to generate a corresponding veto signal. Furthermore, the magnetic field will deflect the charged tracks proportionally to their momentum. Strips of plastic scintillators ($10\times10\times180$ mm^3), readout by SiPMs, will cover both sides of the dipole gap, with the long side aligned to the magnetic field (y axis), while others will be located in the forward region outside the magnet (see fig. 1). A rough measurement of the momentum of the particles will be obtained from the position of the hit ($\sim$%). A prototype of these veto detectors is shown in Fig. 6 taken during a test, performed in April 2017, with an electron beam of 500 MeV. It consists of 16 plastic scintillator strips white painted, to better diffuse the scintillating light, readout by SiPMs Hamamatsu S13360 [8] mounted on a PCB developed by Frascati electronic service providing also signal amplification and shaping. The analog signals are then processed by CAEN V1742 high-speed digitizers [9], based on the DRS4 chip, set to a sampling speed of 5 GS/s (200 ps/sample). The light can be readout directly from the scintillator coupled to the SiPM, or through a WLS fiber placed in a groove along the bar in contact with the SiPM. During the test beam, the different possibilities have been tested. Preliminary results seem to indicate a better performance of the readout performed via the WLS fiber. An average time resolution of about 600 ps has been measured, while the detection efficiency is >0.998 for all slabs.

V. The electromagnetic calorimeter

The most important detector of the PADME setup is the electromagnetic calorimeter that should evaluate with high efficiency the energy, direction and timing of the single photons emitted from the target. The requirements for this detector are:

- high acceptance for ordinary photons;
- energy resolution $<2\ \%/\sqrt{E}$ GeV;
- time resolution $\sim$ 0.5 ns;
- angular resolution $\sim$ 1÷2 mrad.

To cut out the high number of Bremsstrahlung photons produced in the interaction with the target, the main calorimeter (ECAL) will have a central hole, and behind this, a second, small size and faster calorimeter (SAC), will be placed. The ECAL module, placed 3 m downstream the target, is a tight matrix of 616 crystals ($21\times21\times230$ mm^3) arranged in a cylinder of external radius 30 cm, with a squared central hole of 10 cm (see fig. 7). This arrangement allow a 66% acceptance for a 550 MeV positron beam. The choice of the scintillating material for the ECAL has been driven by different aspects: high density, small radiation length and Molire radius (due to the calorimeter small dimensions), high light output (due to the low energy range), cost. Excluding the cost requirement, both LYSO and BGO would have been suitable, with a slight preference for the LYSO since it has a shorter decay time that would have mitigated the effect of event pile-up caused by the beam structure. Nevertheless, the BGO crystals that were used by the L3 experiment at LEP [6], that are property of INFN, were available at the only cost of cutting them at the proper size. Therefore, after having checked that these crystals could be reconditioned and re-shaped, they have been adopted.

For the light collection, early tests, aimed at evaluating the best readout technology, showed that avalanche photodiodes (APDs), even with a relatively large area of 10×10 mm^2, have a gain that is insufficient to perform a high resolution energy measurement in the PADME energy rage (from a few to a few hundred MeV). The readout system will therefore be based on 19 mm diameter photomultipliers. On July 2016, XP1912 HZC Photonics tubes [7] have been tested at the BTF, on a calorimeter prototype made of 25 BGO crystals obtained by machining the L3 recovered ones. Figure 8 shows two units of the ECAL detector. The 25 channels of the prototype were readout by a CAEN V1742 high-speed digitizer [9] set to a sampling speed of 1 GS/s (1 ns/sample). An important feature of the BTF facility is the possibility to vary the beam intensity from 10^9 particles down to a single particle per pulse. Electrons of two different energies, 250 and 450 MeV, were fired on the ECAL prototype in order to evaluate the energy resolution. The results are shown in fig. 9 [10].

To reject backgrounds with multiple photons in the final state, namely $e^+e^- \rightarrow \gamma\gamma$ and $e^+e^- \rightarrow \gamma\gamma\gamma$, 50 cm downstream the hole of the main ECAL a fast Small Angle Calorimeter (SAC) is foreseen. The requirements for this detector are:

- good energy resolution and efficiency for photon in the 100 MeV energy range;
- capability to sustain high rates;
- time resolution of about 200 ps;
- radiation hardness up to 1 Gy per 10^{13} positrons on target.

Among inorganic crystals commonly used for e.m. calorimeters, only Barium Fluouride (BaF_2), having a fast component of the decay time of scintillation, could match the requirements. On the other hand, since the light yield in this detector is not an issue, the choice of the material can also orients

(a) Picture of the PADME dipole. The used coordinate system is also indicated.

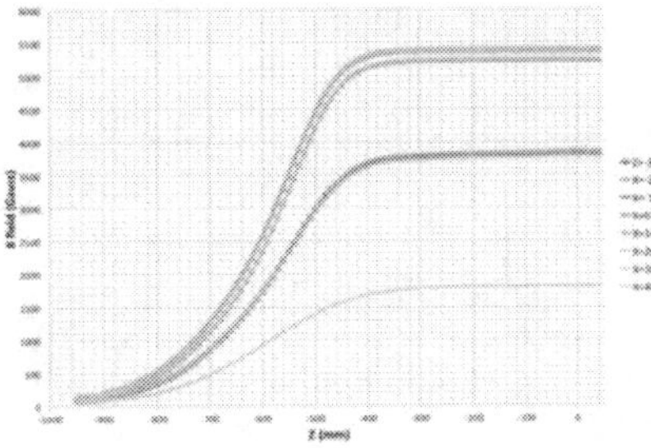

(b) Longitudinal scan of the magnetic field at different transverse positions.

Fig. 5. The PADME dipole magnet.

Fig. 6. Prototype of the veto detector ready for a beam test. The electronic cards at the bottom house SiPMs that collect the light generated in the plastic scintillators.

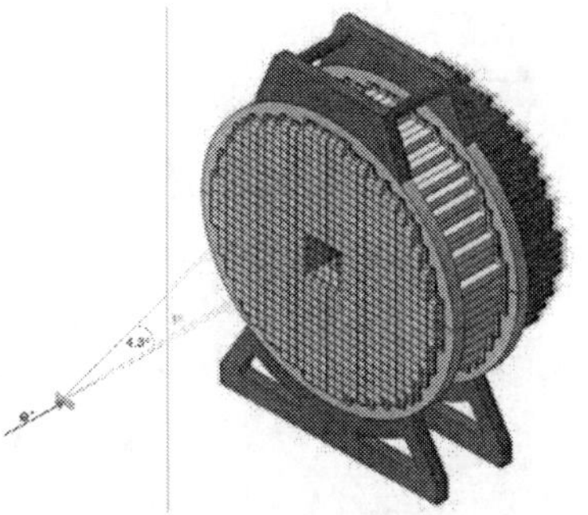

Fig. 7. Layout of the PADME main e.m. calorimeter (ECAL). it consists of an array of 616 BGO crystals.

Fig. 8. Two units of the PADME ECAL detector. They consist of a BGO crystal (10×10×230 mm^3) glued to a 19 mm diameter photomultiplier.

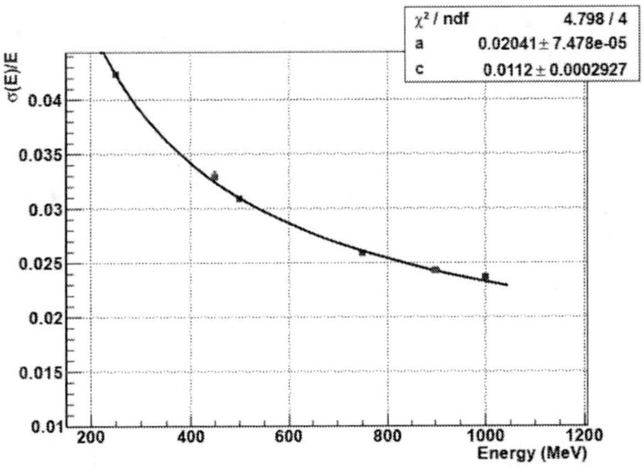

Fig. 9. Energy resolution measured with a prototype of the PADME ECAL at the Frascati BTF. Red and blue points refer to two different beam energies of 450 and 250 MeV, respectively. The line is the best fit curve whose parameters are quoted in the inset.

TABLE I
MAIN CHARACTERISTICS OF THE CRYSTALS UNDER STUDY FOR THE PADME SMALL ANGLE CALORIMETER.

	PbF_2	SF57
Density [g/cm^3]	7.77	5.51
X_0 [cm]	0.93	1.54
Moliere radius [cm]	2.12	2.61
Interaction length (λ) [cm]	22.1	20.6
λ/X0	23.65	13.3
Refraction index	1.8	1.8

toward Cherenkov emitting crystals. Lead glasses (Schott SF57) used by the OPAL experiment at CERN, are available at Frascati. Another Cherenkov counter that can match even better the needs of the experiment is Lead Flouoride (PbF_2). It owns a higher density, a better λ/X_0 ratio that imply more compact showers; better transparency down to $\sim$ 250 nm, and a factor 10 more radiation hardness with respect to SF57. Table I shows the main characteristics of both SF57 and PbF_2 materials.

In order to have a good timing response, for the photosensors, fast photomultipliers are mandatory. With a SF57 crystal, read out with Hamamatsu R9880-U100 PM [8], exposed in November 2016 to the BTF electron beam, a RMS of 600 ps in the time resolution was achieved. In April 2017, two samples of PbF_2, used by the $g-2$ collaboration at Fermilab to build their e.m. calorimeter ($30\times30\times150$ mm^3), have been shipped to Frascati. They have been coupled to the same PMs Hamamatsu and exposed to the BTF beam, together with the SF57 crystal, to compare their behaviours. The data analysis is still ongoing, but the preliminary results indicate that PbF_2 coupled with a PM with a larger, 1 inch diameter photocathode (*i.e.* R13478), could reach the experiment requirements. Further tests will be then performed during summer.

The final layout of the calorimeter will consist in a squared matrix of crystals whose number will depend from the chosen material. The two possible alternatives are: a 7×7 array of $20\times20\times200$ mm^3 SF57 crystals, or a 5×5 array of $30\times30\times150$ mm^3 PbF_2 ones.

VI. Beam monitor

As already mentioned, a big fraction of the incoming positrons will cross the diamond target without interacting, and the dipole magnet will deflect them out of the calorimeters acceptance. By placing a silicon pixel detector immediately out of the vacuum vessel, it would be possible to precisely monitor online the beam characteristics. A device, based on the Timepix3 chip [11], can be the perfect solution.

The Timepix chip family are conceived for timing measurements with the added functionality of measuring time-over-threshold, these features are perfect for particle tracking applications where timing and spatial resolution are necessary. The device can be operated in different mode allowing to acquire charge and time, time only, event counting and integral charge. The time resolution is $\sim$ 1.5 ns and it can stands particle rates up to 40 Mhits/cm^2/s.

Many tracking detectors based on the Timepix3 chip have been developed for particle physics apparatuses. PADME need to build a Timepix array covering an area of 10×3 cm^2. At present, the collaboration is considering the offers of NIKHEF and Diamond laboratories which both can provide detectors that can be easily adapted to the experiment setup.

VII. Timeline and Conclusion

After the approval at the end of 2015, an international collaboration has been formed to perform the PADME experiment. At present, it consists of about 50 people from 8 different institutions: INFN Lecce, Salento University, INFN LNF, University and INFN Roma, University of Sofia, MTA Atomki from Hungary, William and Mary College of Williamsburg (VA), Cornell University of Ithaca (NY) last two both from USA. The collaboration with Cornell is also looking to the future. In facts, at this Institute there are plans to realize an extracted positron beam line, from the CESR storage ring, of 5.3 GeV on which, an upgraded version of the PADME experiment could be installed. This will extend the region of possible mass of the dark photon search.

At present, the design of the PADME detector elements has been completed in almost all aspects, and the procurement of the main components has already started in 2016. The different sub-detectors are under construction and the installation on the beam line is foreseen for winter 2017. After a commissioning of few weeks in spring 2018, the real data taking of the experiment could start.

In order to achieve 10^{13} positrons on target in a six months data taking period, the Frascati LINAC has to be slightly modified and optimized to increase the positron pulse duration from the present value of 40 ns to 250 ns. During 2016 a new LINAC pulsing system has been installed, and dedicated machine development runs have been performed to stretch the pulse duration. Further adjustments are foreseen in 2017 to reduce the intensity variations and to keep the beam energy spread below 2%.

Acknowledgment

The authors would like to thanks MAECI that partially support this activity throughout the PGR-226 grant.

References

[1] M. Raggi and V. Kozhuarov, Adv. High Energy Phys. 2014, 959802 (2014).
[2] The PADME collaboration, "The PADME experiment Technical Proposal",
[3] P. Valente *et al.*, arXiv:1603.05651.
[4] G. Chiodini *et al.*, JINST 12, C02036 (2017).
[5] I. Valin *et al.*, JINST 7, C01102 (2012).
[6] B. Adeva *et al.*, Nucl. Instrum. Meth. A 289, 35 (1990).
[7] http://hzcphotonics.com.
[8] http://www.hamamatsu.com
[9] CAEN Mod. 1742, Technical Information Manual, rev6, 06 February 2016.
[10] M. Raggi *et al.*, Nucl. Instrum. Meth. A 862, 31 (2017).
[11] T. Poikela *et al.*, JINST 9, C05013 (2014).

EPJ Web of Conferences **170**, 01008 (2018) https://doi.org/10.1051/epjconf/201817001008
ANIMMA 2017

Advances in Monte-Carlo code TRIPOLI-4®'s treatment of the electromagnetic cascade

Davide Mancusi*, Alice Bonin*, François-Xavier Hugot* and Fadhel Malouch*
*Den-Service d'études des réacteurs et de mathématiques appliquées (SERMA),
CEA, Université Paris-Saclay, F-91191, Gif-sur-Yvette, France
Email: davide.mancusi@cea.fr

***Abstract*—TRIPOLI-4® is a Monte-Carlo particle-transport code developed at CEA-Saclay (France) that is employed in the domains of nuclear-reactor physics, criticality-safety, shielding/radiation protection and nuclear instrumentation. The goal of this paper is to report on current developments, validation and verification made in TRIPOLI-4 in the electron/positron/photon sector. The new capabilities and improvements concern refinements to the electron transport algorithm, the introduction of a charge-deposition score, the new thick-target bremsstrahlung option, the upgrade of the bremsstrahlung model and the improvement of electron angular straggling at low energy. The importance of each of the developments above is illustrated by comparisons with calculations performed with other codes and with experimental data.**

I. Introduction

An accurate description of the electromagnetic cascade is an essential prerequisite for the simulation of the response of a wide range of particle detectors, including in-core and ex-core instrumentation in nuclear reactors. For such applications, one possible simulation tool is provided by the TRIPOLI-4® code [1], a Monte-Carlo particle-transport code developed at CEA-Saclay (France) that is employed in the domains of nuclear-reactor physics, criticality-safety, shielding/radiation protection and nuclear instrumentation. The latest TRIPOLI-4 version (v10) was released in December 2015 and will soon be available to OECD member countries through the NEA Data Bank. This extended abstract reports on some recent developments in TRIPOLI-4 for the simulation of nuclear instrumentation (Section II). Section III compares the predictions of the development version of TRIPOLI-4 with those of TRIPOLI-4 v10 and of other transport codes; additionally, we compare the calculation results against selected experimental data. Conclusions and future plans are discussed in Section IV.

II. Description of some recent developments

We now proceed to describe some of the new developments in detail.

A. Bremsstrahlung model

The production of bremsstrahlung photons by electrons and positrons in TRIPOLI-4 v10 proceeds according to the specifications of the model described in Berger and Seltzer [2] and summarised in Peneliau [3]. However, the same authors presented a newer model for energy-differential cross sections in Seltzer and Berger [4]. Contrary to the 1970 model, the new model is essentially based on the interpolation of tabulated values. The table values are derived from numerical phase-shift calculations for electron energies lower than 2 MeV, and from the analytical high-energy theory above 50 MeV; a numerical interpolation scheme was used to cover the intermediate-energy region. Seltzer and Berger's 1986 model may be considered as the state of the art of the field. It is currently implemented in Geant4 [5], [6] and MCNP6 [7].

We introduced Seltzer and Berger's 1986 model and tables as the new default model for bremsstrahlung in the development version of TRIPOLI-4. A new TRIPOLI-4 keyword was added to control the choice of the bremsstrahlung model.

B. Electron angular straggling at low energy

In TRIPOLI-4, electron angular straggling is described by the Goudsmit-Saunderson model [8]. The Goudsmit-Saunderson formalism can in principle be used to condense elastic cross sections from any model/table. In TRIPOLI-4, the cross sections are extracted from different sources depending on the electron energy. Above 256 keV, TRIPOLI-4 uses the Mott cross section [9] with Molière's screening correction [10], [11], as calculated by Feshbach [12] and Sherman [13]. At 256 keV and below, TRIPOLI-4 makes use of data from the Evaluated Electron Data Library (EEDL) [14], which provides integrated and angle-differential large-angle ($\cos\theta < 1-10^{-6}$) elastic cross sections. Due to a bug in the processing of the cross sections, angular straggling was overestimated below 256 keV. The correction, which was applied to the development version of the code, results in more forward-peaked angular distributions.

C. Charge-deposition score

Self-powered neutron and gamma detectors (SPND and SPGD), often referred to as *collectrons*, are passive devices for in-core measurement of neutron and/or gamma flux. The basic mechanism underlying collectrons is that matter ionized by neutrons or gammas will in general exhibit some degree of spontaneous electrical polarization, which can be detected using appropriate instrumentation. The resulting current, which is ultimately due to electron and positron transport, increases as the radiation field becomes more intense.

In order to characterize the response of a collectron with a Monte-Carlo simulation, it is necessary to calculate the charge-

https://doi.org/10.1051/epjconf/201817001008

deposition distribution in the instrument. This capability was added to TRIPOLI-4 in version 10.

D. Thick-target bremsstrahlung mode

The full simulation of the electromagnetic cascade, including electron and positron transport, is very CPU intensive when compared to a pure photon simulation. However, pure photon transport does not take into account bremsstrahlung photons emitted by secondary electrons; when the photon energy is sufficiently high (say above 1 MeV), pure photon transport will underestimate the secondary photon flux. In cases where a rough estimation of the secondary bremsstrahlung flux is sufficient, it is advantageous to consider a simplified calculation mode (*thick-target bremsstrahlung*, TTB) where electron production and transport are directly replaced by the emission of secondary bremsstrahlung photons; this effectively amounts to replacing secondary electron production with a photon-photon scattering vertex. Since the additional secondary photons are emitted from the starting point of the suppressed electron track, the approximation is generally understood to be less severe in thick volumes, hence its name. A similar approximation exists in MCNP6 [7]. Contrary to TRIPOLI-4, MCNP6's TTB approximation is activated by default in pure photon calculations.

The TRIPOLI-4 TTB implementation proceeds along the lines indicated by Ridoux, Kitsos, Diop, *et al.* [15] and Riz [16]. The most notable difference with respect to the MCNP6 implementation concerns the fact that TRIPOLI-4 accounts for the angular straggling of the condensed electrons; therefore, the bremsstrahlung photons are not necessarily emitted around the initial electron direction. Additionally, the TRIPOLI-4 TTB implementation also allows the use of electron sources; in this case, only tertiary electrons generated by secondary photons are replaced with the corresponding bremsstrahlung photons.

III. Verification and validation

As an example of validation of the new developments for low-energy electrons, we present the comparison of TRIPOLI-4 calculation with an electron transmission/reflection experiment. Seltzer and Berger [17] characterised transmission, reflection and absorption of 100 keV to 400 keV electrons through aluminium and titanium foils of various thicknesses. They report measured values of the number transmission (T_N) and reflection coefficients (R_N), which respectively represent the average number of transmitted (reflected) electrons per incident electron. They also measured energy transmission (T_E) and reflection coefficients (R_E), which respectively represent the fraction of the beam energy that is transmitted through (reflected by) the foil. By difference, one can also define the energy absorption coefficient $\phi_A = 1 - T_E - R_E$. Finally, they also measured the angular distributions of the transmitted electron beams, which however we will not discuss in the present work.

In this work, we limit our comparisons to a subset of the measured data. We concentrate on transmission and reflection

Table I
Transmission, reflection and absorption coefficients for 100 keV electrons through 25.4 µm-thick aluminium and titanium foils, expressed in percentage points. The measured values are taken from Seltzer and Berger [17]. Calculations were performed with TRIPOLI-4 v10 and TRIPOLI-4 (development version); the Geant4 values are taken from Kadri, Ivanchenko, Gharbi, *et al.* [18]. The statistical uncertainties on the TRIPOLI-4 values are smaller than the last reported significant digit. See text for the definition of the coefficients.

		exp (%)	TRIPOLI-4 v10 (%)	TRIPOLI-4 dev (%)	Geant4 (%)
	T_N	60.2 ± 6.0	18.2	65.4	60.7
	R_N	12.4 ± 0.1	64.7	9.0	12.6
Al	T_E	39.3 ± 3.9	6.7	42.4	40.3
	R_E	7.1 ± 0.7	43.2	4.5	7.2
	ϕ_A	53.6 ± 5.4	50.2	53.1	53.2
	T_N	10.5 ± 1.1	11.7	10.1	8.2
	R_N	22.4 ± 2.2	67.8	17.8	18.6
Ti	T_E	5.0 ± 0.5	3.2	4.0	3.9
	R_E	14.2 ± 1.4	47.6	9.7	11.8
	ϕ_A	80.8 ± 8.1	49.2	86.4	84.7

of 100 keV electrons through 25.4 µm-thick aluminium and titanium targets. Table I compares the experimental values with the results of calculations performed with TRIPOLI-4 v10 and with the development version of TRIPOLI-4. Geant4 calculation results from the literature [18] are also included for comparison.

The first observation is that the TRIPOLI-4 v10 results show little sensitivity to the target. This is probably connected to the underestimation of the mean free path for elastic scattering (Sec. II-B): indeed, an overestimated elastic cross section leads to excess multiple scattering. If electrons scatter too much and at too large angles, electron transport becomes akin to diffusion; under these conditions, one would expect the transmission and reflection coefficients to be essentially dominated by the geometry, which is the same in both cases (same thickness). Comparison with the experimental data makes it quite clear that this picture is largely erroneous.

The recent TRIPOLI-4 developments bring a substantial improvement. The salient features the experimental results are correctly reproduced, at least qualitatively. As a general trend, TRIPOLI-4 seems to slightly underestimate reflection and overestimate transmission. The Geant4 results are in better agreement with the experiment. These facts may probably be explained by considering that Geant4 singles out large-angle elastic scattering as a separate process[1]. Backward elastic-scattering events, especially those that take place at small penetration depths, are expected to yield a sizeable contribution to reflection. It is clear that the effect of these collisions cannot be accurately modelled in the framework

[1]Geant4's electron-transport algorithm falls in class II of Berger *et al.*'s ([19]) classification.

of an angular-straggling model, which by definition aims at capturing the mean effect of a large number of soft collisions.

IV. Conclusions

We have briefly presented a few recent TRIPOLI-4 developments concerning the treatment of the electromagnetic cascade, namely the implementation of a new model for the energy-differential bremsstrahlung cross section and a more careful treatment of the condensation of electron elastic cross sections into angular-straggling distributions. The impact of these developments is under study by comparing TRIPOLI-4 calculation results with experimental data and with the results of other transport codes. We have also presented the new charge-deposition score, which is useful for the simulation of self-powered neutron and gamma detectors, and the new thick-target-bremsstrahlung calculation mode.

Acknowledgment

TRIPOLI-4® is a registered trademark of CEA. Partial financial support from EDF (Électricité de France) is gratefully acknowledged.

References

[1] E. Brun, F. Damian, C. M. Diop, E. Dumonteil, F.-X. Hugot, C. Jouanne, Y. K. Lee, F. Malvagi, A. Mazzolo, O. Petit, J. Trama, T. Visonneau and A. Zoia, "TRIPOLI-4®, CEA, EDF and AREVA reference monte carlo code", *Ann. Nucl. Energy*, 2014, in press. DOI: 10.1016/j.anucene.2014.07.053.

[2] M. J. Berger and S. M. Seltzer, "Bremsstrahlung and photoneutrons from thick tungsten and tantalum targets", *Phys. Rev. C*, vol. 2, pp. 621–631, 2 Aug. 1970. DOI: 10.1103/PhysRevC.2.621.

[3] Y. Peneliau, "Tests de validation de l'implémentation de la cascade électromagnétique dans le code de transport Monte Carlo TRIPOLI-4", CEA-Saclay, France, Rapport DM2S SERMA/LEPP/RT/02-3186/A, Oct. 2002.

[4] S. M. Seltzer and M. J. Berger, "Bremsstrahlung energy spectra from electrons with kinetic energy 1 keV–10 GeV incident on screened nuclei and orbital electrons of neutral atoms with $Z = 1$–100", *Atom. Data Nucl. Data*, vol. 35, no. 3, pp. 345–418, 1986. DOI: 10.1016/0092-640X(86)90014-8.

[5] J. Allison, K. Amako, J. Apostolakis, *et al.*, "Recent developments in Geant4", *Nucl. Instrum. Meth. A*, vol. 835, pp. 186–225, 2016. DOI: 10.1016/j.nima.2016.06.125.

[6] *Geant4 Physics Reference Manual*, version 10.2, Geant4 collaboration, Dec. 2015. [Online]. Available: http://cern.ch/geant4/UserDocumentation/UsersGuides/PhysicsReferenceManual/fo/PhysicsReferenceManual.pdf.

[7] T. Goorley, M. James, T. Booth, *et al.*, "Initial MCNP6 release overview", *Nucl. Technol.*, vol. 180, no. 3, pp. 298–315, Dec. 2012.

[8] S. Goudsmit and J. L. Saunderson, "Multiple scattering of electrons", *Phys. Rev.*, vol. 57, pp. 24–29, 1 Jan. 1940. DOI: 10.1103/PhysRev.57.24.

[9] N. F. Mott, "The scattering of fast electrons by atomic nuclei", *P. R. Soc. Ac*, vol. 124, no. 794, pp. 425–442, 1929. DOI: 10.1098/rspa.1929.0127.

[10] G. Molière, "Theorie der Streuung schneller geladener Teilchen i: Einzelstreuung am abgeschirmten Coulomb-Feld", *Z. Naturforsch.*, vol. 2, no. a, pp. 133–145, 1947.

[11] ——, "Theorie der Streuung schneller geladener Teilchen ii: Mehrfach- und Vielfachstreuung", *Z. Naturforsch.*, vol. 3, no. a, pp. 78–97, 1948.

[12] H. Feshbach, "The Coulomb scattering of relativistic electrons and positrons by nuclei", *Phys. Rev.*, vol. 88, pp. 295–297, 2 Oct. 1952. DOI: 10.1103/PhysRev.88.295.

[13] N. Sherman, "Coulomb scattering of relativistic electrons by point nuclei", *Phys. Rev.*, vol. 103, pp. 1601–1607, 6 Sep. 1956. DOI: 10.1103/PhysRev.103.1601.

[14] S. T. Perkins, D. E. Cullen and S. M. Seltzer, "Tables and graphs of electron-interaction cross sections from 10 eV to 100 GeV derived from the LLNL Evaluated Electron Data Library (EEDL), $Z = 1$–100", Lawrence Livermore National Laboratory, Tech. Rep. UCRL-50400, Nov. 1991.

[15] P. Ridoux, S. Kitsos, C. M. Diop, A. Assad and J. C. Nimal, "Improvement of gamma-ray S_n transport calculations including coherent and incoherent scatterings and secondary sources of bremsstrahlung and fluorescence: Determination of gamma-ray buildup factors", *Nucl. Sci. Eng.*, vol. 123, no. 2, pp. 215–227, 1996.

[16] D. Riz, "Calculation and use of multigroup cross sections including electron-photon cascade for a 3D Monte Carlo neutron-gamma transport code. Comparisons with MCNP-4B", in *Proceedings to the 2000 ANS International Topical Meeting on Advances in Reactor Physics and Mathematics and Computation into the Next Millennium (PHYSOR-2000)*, Pittsburgh, PA, USA, May 2000.

[17] S. M. Seltzer and M. J. Berger, "Transmission and reflection of electrons by foils", *Nucl. Instrum. Meth.*, vol. 119, pp. 157–176, 1974. DOI: 10.1016/0029-554X(74)90747-2.

[18] O. Kadri, V. Ivanchenko, F. Gharbi and A. Trabelsi, "Incorporation of the Goudsmit–Saunderson electron transport theory in the Geant4 Monte Carlo code", *Nucl. Instrum. Meth. B*, vol. 267, no. 23–24, pp. 3624–3632, 2009. DOI: 10.1016/j.nimb.2009.09.015.

[19] M. J. Berger *et al.*, "Monte Carlo calculation of the penetration and diffusion of fast charged particles", *Meth. Comp. Phys.*, vol. 1, pp. 135–215, 1963.

EPJ Web of Conferences **170**, 01009 (2018)
ANIMMA 2017
https://doi.org/10.1051/epjconf/201817001009

GEM detector performance with innovative micro-TPC readout in high magnetic field

I. Garzia(1)(2), M. Alexeev(3), A. Amoroso(3)(4), R. Baldini Ferroli(5)(6), M. Bertani(5), D. Bettoni(1), F. Bianchi(3)(4), A. Calcaterra(5), N. Canale(2), M. Capodiferro(5)(7), V. Cassariti(1), S. Cerioni(5), J.Y. Chai(3)(6)(8) S. Chiozzi(1), G. Cibinetto(1), F. Cossio(3)(8), A. Cotta Ramusino(1), F. De Mori(3)(4), M. Destefanis(3)(4), J. Dong(5), F. Evangelisti(1), R. Farinelli(1)(2), L. Fava(3), G. Felici(5), E. Fioravanti(1), M. Gatta(5), M. Greco(3)(4), L. Lavezzi(3)(6), C.Y. Leng(3),(6),(8), H. Li(3)(6), M. Maggiora(3)(4), R. Malaguti(1), S. Marcello(3)(4), M. Melchiorri(1), G. Mezzadri(1)(2), M. Mignone(3), G. Morello(5), S. Pacetti(9)(10), P. Patteri(5), J. Pellegrino(3)(4), A. Pelosi(5)(7), A. Rivetti(3), M. D. Rolo(3), M. Savrié(1)(2), M. Scodeggio(1)(2), E. Soldani(5), S. Sosio(3)(4), S. Spataro(3)(4), E. Tskhadadze(5)(11), S. Verma(2), R. Wheadon(3), L. Yan(3)

(1)INFN - Sezione di Ferrara - via G. Saragat 1, 44122 Ferrara, Italy
(2)Università di Ferrara, Dipartimento di Fisica e Scienze della Terra - via G. Saragat 1, 44122 Ferrara, Italy
(3)INFN - Sezione di Torino - via P. Giuria 1, 10125 Torino, Italy
(4)Università di Torino - via P. Giuria 1, 10125 Torino, Italy
(5)INFN - Laboratori Nazionale di Frascati - via E. Fermi 40, 00044 Frascati (Roma), Italy
(6)Institute of High Energy Physics, Chinese Academy of Sciences, 19B YuquanLu, Beijing, 100049, China
(7)INFN - Sezione di Roma, c/o Università la Sapienza, p.le Aldo Moro 2, 00185 Roma, Italy
(8)Politecnico di Torino, Dipartimento di Elettronica e Telecomunicazioni, C.so Duca degli Abruzzi 24, 10129 Torino, Italy
(9)INFN - Sezione di Perugia, via A. Pascoli 14, 06123 Perugia, Italy
(10)Università di Perugia, Dipartimento di Fisica e Geologia, via A. Pascoli 14, 06123 Perugia, Italy
(11)Joint Institute for Nuclear Research (JINR), Joliot-Curie 6, Dubna, Moscow region, 141980, Russia

***Abstract*—Gas detector development is one of the pillars of the research in fundamental physics. Since several years, a new concept of detectors, called Micro Pattern Gas Detector (MPGD), allowed to overcome several problems related to other types of commonly used detectors, like drift chamber and micro strips detectors, reducing the rate of discharges and providing better radiation tolerance. Among the most used MPGDs are the Gas Electron Multipliers (GEMs). Invented by Sauli in 1997, nowadays GEMs have become an important reality for particle detectors in high energy physics. Commonly deployed as fast timing detectors and triggers, their fast response, high rate capability and high radiation hardness make them also suitable as tracking detectors. The readout scheme is one of the most important features in tracking technology. Analog readout based on the calculation of the center of gravity technique allows to overcome the limit imposed by digital pads, whose spatial resolution is limited by the pitch dimensions. However, the presence of high external magnetic fields can distort the electronic cloud and affect the performance. The development of the micro-TPC reconstruction method brings GEM detectors into a new prospective, improving significantly the spatial resolution in presence of high magnetic fields. This innovative technique allows to reconstruct the 3-dimensional particle position, as Time Projection Chamber, but within a drift gap of a few millimeters. In these report, the charge centroid and micro-TPC methods are described in details. We discuss the results of several test beams performed with planar chambers in magnetic field. These results are one of the first developments of micro-TPC technique for GEM detectors, which allows to reach unprecedented performance in a high magnetic field of 1 T.**

I. Introduction

The Gas Electron Multiplier (GEM) is a newly developed gaseous electron multiplier invented in 1997 by Fabio Sauli [1]. It consists of a thin insulating polymeric foil (~ 50 μm of kapton) metal coated on both sides ($3-5$ μm of copper), and pierced by a high density regular matrix of holes with a typical diameter of about $50/70$ μm [2] realized with a photolithographic method. The application of proper potential differences between cathode, two GEM electrode sides and anode generates a very intense dipole field in the holes. A charged track passing through the GEM detector ionizes the gas. The released electrons drift into the holes following the field lines, undergo avalanche multiplication and are collected on the anode. With a single GEM foil a gain up to 10^3 can be achieved with moderate voltages. Better performance is obtained by means of three GEM (triple-GEM) foils between the cathode and the anode instead of one: gain values up to 10^5 can be achieved working at lower voltages while reducing the discharge probability. Unless otherwise specified, all the results reported here are related to the triple-GEM configuration.

II. Track reconstruction methods

GEM detectors can fulfil a large variety of requests thanks to their properties, such as the robustness, the possibility to build large area detectors, its relatively easy construction method with limited cost, the low aging effects and high reconstruction

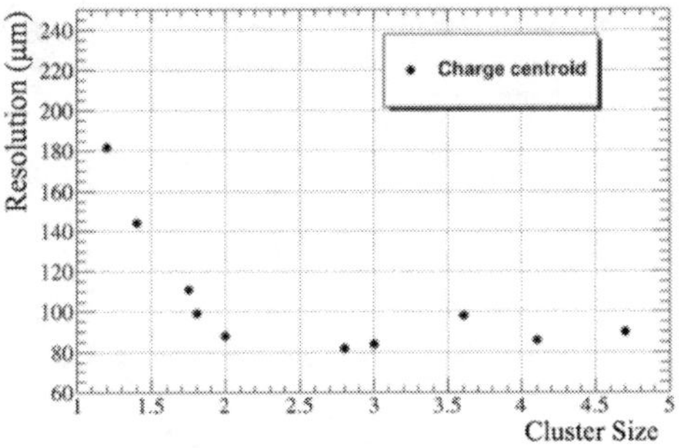

Fig. 1. Resolution from charge centroid method vs. cluster size for orthogonal tracks and without magnetic field. A planar triple-GEM with 5 mm drift gap filled with $ArCO_2$ is used for this study.

efficiency, etc. However, the spatial resolution in presence of high magnetic fields is not optimal and new tracking reconstruction methods should be investigated in order to push the spatial resolution below 150 μm.

A. *Charge centroid method*

In the case of a Gaussian charge distribution on the anode, the position x of a charged track relative to the anode can be determined by means of the weighted average of the fired strip positions x_i:

$$x = \frac{\sum_i x_i q_i}{\sum_i q_i}, \quad (1)$$

where the weights q_i are defined by the charge collected by each fired strip. Figure 1 shows the spatial resolutions as a function of the average cluster size (or number of fired strips) measured under the condition of zero magnetic field and orthogonal tracks. A resolution of 80−90 μm is achieved when the average cluster size is larger than two; for lower cluster size values, the performance of digital readout is limited by the strip pitch.

III. MICRO-TPC METHOD

When the charge distribution is not Gaussian, i.e. in presence of magnetic field or inclined tracks, the charge centroid method fails and a new reconstruction method should be investigated in order to keep a resolution of the order of 100 μm: the micro-TPC (μ − TPC) method [3]. The drift gap (gap between the cathode and the first GEM foil) can be considered as a tiny time projection chamber: the time information related to each fired strip allows to perform a local track reconstruction in the few-mm drift gap. In Figure 2 is reported the drift time distribution for an event in which eight strips are hit and for which the charge distribution on the anode is no more Gaussian. Knowing the coordinate x_i of each strip, the position of each primary ionization z_i perpendicular to the strip plane is obtained from time measurement using the electron drift velocity from GARFIELD simulation. The (x_i, z_i) points are then fitted with a linear function $z = ax + b$ and the position is extracted in the middle of the gap in order to minimize the errors [3] as

$$x = \frac{gap/2 - b}{a}. \quad (2)$$

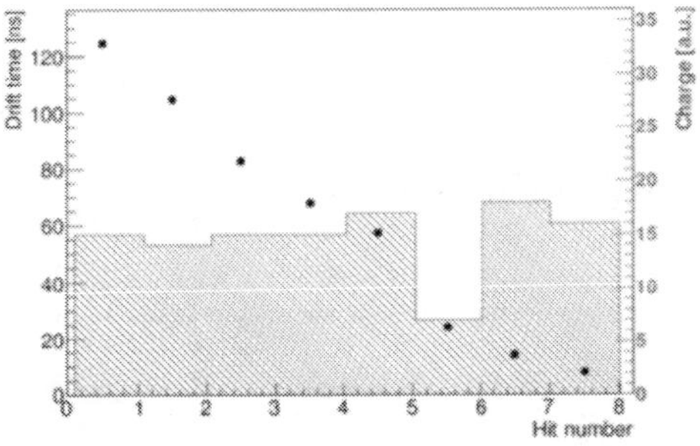

Fig. 2. Drift time distribution as a function of hit strips (points). The dashed histogram represents the charge distribution, which is no more Gaussian for this particular event.

IV. BEAM TEST RESULTS

Planar chamber prototypes (see Fig. 3) have been tested on the H4 line of the SPS at CERN [4]. The chambers are 10×10 cm^2 triple GEMs, with a bi-dimensional XY view readout and a drift gap of 3 mm or 5 mm. Two different gas mixtures are used: $Ar{:}CO_2$ (70/30) and $Ar{:}i\text{-}C_4H_{10}$ (90/10). The main goal of the beam test measurements is to study the performance as a function of several parameters, such as high voltage, field setting, gas mixture, drift gap, etc., in order to set the optimal working point needed to achieve the best efficiency and spatial resolution.

Fig. 3. Instrumented 10 × 10 cm^2 planar chamber used during beam test data taking.

All the results reported below are obtained using a chamber with 5 mm drift gap.

 https://doi.org/10.1051/epjconf/201817001009

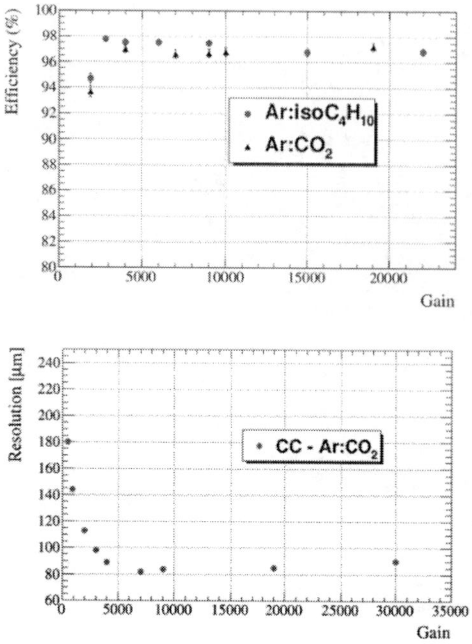

Fig. 4. Efficiency (upper plot) and resolution (lower plot) as a function of gain measured using the charge centroid method (CC) for orthogonal tracks without magnetic field.

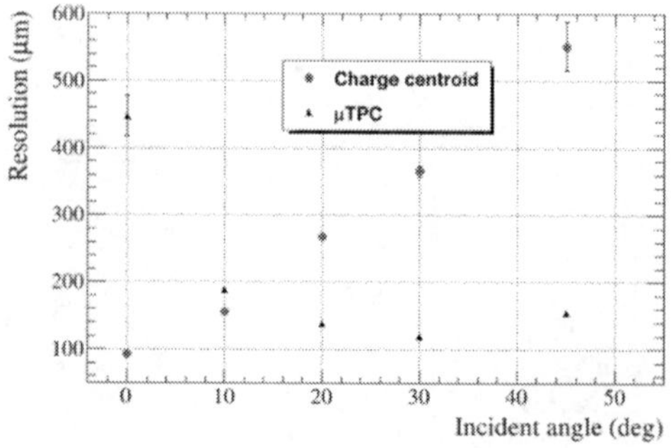

(a) B = 0 and inclined tracks.

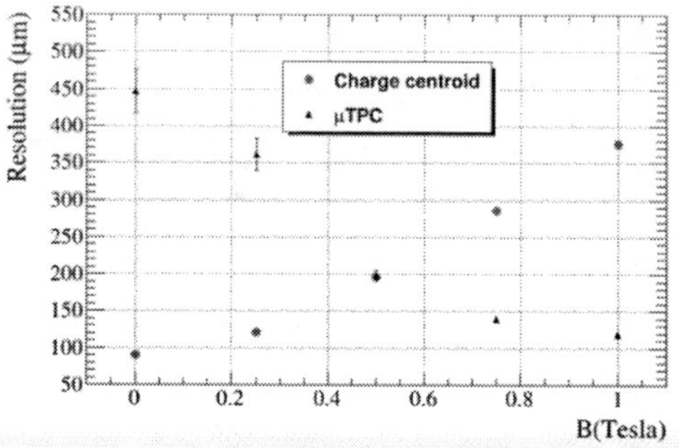

(b) B $\neq$ 0 and orthogonal tracks.

Fig. 5. Resolution measured using charge centroid and $\mu-$ TPC methods for two different scenarios: (a) inclined tracks with B = 0, and (b) B $\neq$ 0 and orthogonal tracks. These results are obtained with a gas mixture of Ar:i-C_4H_{10} and 5 mm drift gap. Points and triangles refer to charge centroid and $\mu-$ TPC results, respectively.

A. Orthogonal tracks without magnetic field

The simplest case is for events with orthogonal tracks and no magnetic field, since the charge distribution on the anode is Gaussian and the charge centroid method works properly. Only the diffusion process and the avalanche multiplication apply. Charge centroid method is very accurate: we can reach a reconstruction efficiency of $97\% - 98\%$ on the single X/Y view and a spatial resolution < 100 μm for a wide range of gain, as shown in Figure 4.

B. Inclined tracks or/and magnetic field

In most cases, charged tracks are not perpendicular to the GEM planes and in some high energy physics experiments they can be immersed in a magnetic field.

We can identify three different scenarios:

1) *B* $= 0$ *and inclined tracks:* cluster size increases for increasing incident angles and the charge distribution on the anode becomes far from a Gaussian shape. The resolution measured with the charge centroid method worsens almost linearly with the angle, while the $\mu-$ TPC method becomes more and more efficient, as shown in Figure 5a.
2) *B* $\neq 0$ *and orthogonal tracks:* the effect of the Lorentz force on the electrons must be taken into account. The magnetic field shifts the electrons avalanche with respect to the perpendicular path and enlarges the charge distribution on the anode, which leads to the deviation from a Gaussian shape. As shown in Fig. 5b, also in this case the charge centroid method deteriorates almost linearly for increasing magnetic field values, while the $\mu-$ TPC method reaches the best resolution value of about 120 μm for B = 1 T. The parameter that describes the avalanche shift is the Lorentz angle, defined as the angle between the drift direction of the electrons under the influence of the magnetic field and the direction of the electric field and it must be used to correct the reconstructed track trajectory.
3) *B* $\neq 0$ *and inclined tracks:* this is the most interesting and awkward case. Two different effects should be taken into account in order to fully understand the beam test results. As depicted in Figure 6a, a focusing and defocusing effect appears, depending on the fact that the incident track and the Lorentz angles are concordant or discordant, respectively. The measured resolution for several incident angles in a magnetic field of 1 T is shown in Figure 6b. The charge centroid and $\mu-$ TPC methods are complementary and their combination can provide a flat spatial resolution of 130 μm, which is the best resolution result obtained for a triple-GEM in a magnetic field of 1 T.

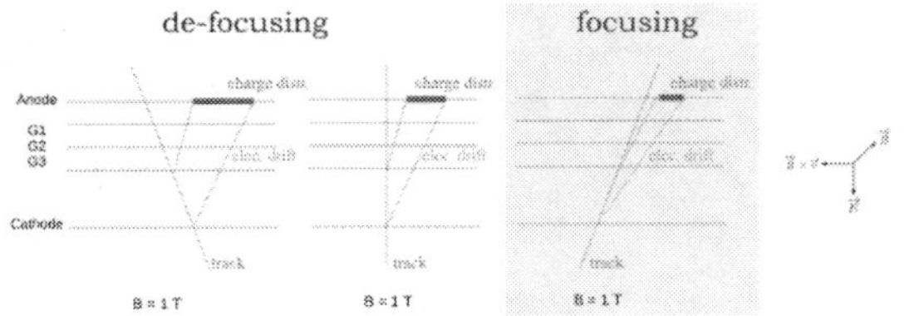

(a) De-focusing (white) and focusing (gray) effects.

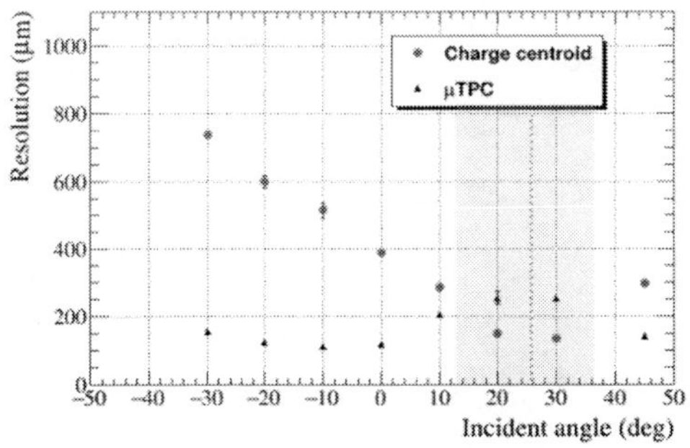

(b) Resolution vs. incident track angle for B = 1.

Fig. 6. (a) De-focusing and focusing effects for different configurations of incident angle. A de-focusing effect occurs when the Lorentz angle and the incident angle of the track are discordant, while a focusing effect (light gray highlighted area) when they are concordant. (b) Resolution vs. incident angle measured from a triple-GEM (Ar:i-C_4H_{10} gas mixture and 5 mm drift gap) located in a magnetic field of B = 1 T. Focusing region is highlighted by the light gray area and the Lorentz angle ($\sim$ 26 deg) determined by the applied fields and the gas mixture is represented by the dashed orange line. Points and triangles refer to charge centroid and $\mu -$ TPC results, respectively.

V. $\mu -$ TPC OPTIMIZATION

All the studies performed during the beam test are used to optimize the $\mu -$ TPC performance: Ar:i-C_4H_{10} gas mixture provides higher electrical stability, an higher number of primary electrons and proper diffusion properties, while the 5 mm conversion gap gives the best resolution in magnetic field for inclined tracks. Another parameter that can be used to optimize the $\mu -$ TPC algorithm is the drift field.

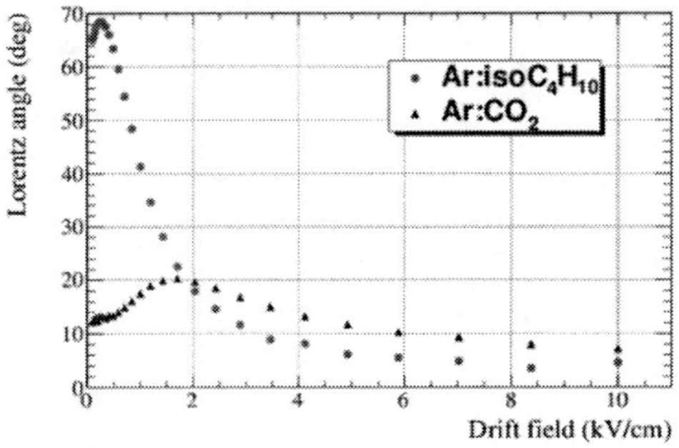

Fig. 7. GARFIELD simulation of Lorentz angle as a function of the drift field. For a gas mixture of Ar:i-C_4H_{10} (90/10) and a drift field of 1.5 kV/cm the Lorentz angle is about 26 deg, in agreement with the behavior shown in Figure 6b.

The results shown in this report are obtained with the following electrical setting within the electrodes: 1.5 kV/cm in the drift gap, 3 kV/cm between the GEM foils, and 5 kV/cm between the last GEM foil and the anode.

From GARFIELD simulation, we can study the behavior of the Lorentz angle for different drift fields and, as shown in Figure 7, we observed a very strong dependence for drift field lower than 3 kV/cm. Thus, the working region can be properly chosen by setting the drift field in such a way that the focusing region highlighted in Figure 6b moves to the left or to the right depending on the experiment requirements.

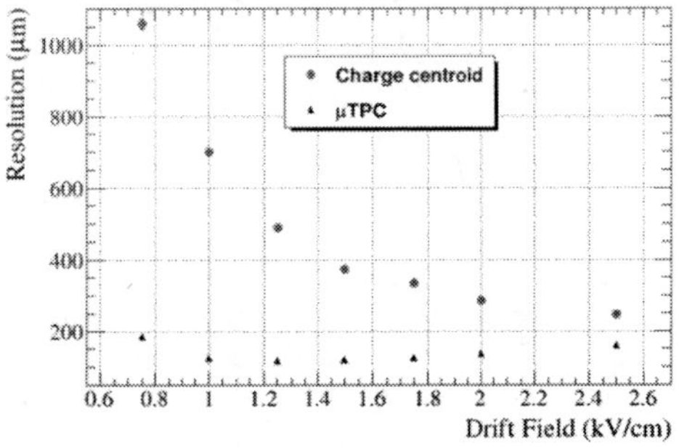

Fig. 8. Comparison of resolutions measured using charge centroid and $\mu -$ TPC method as a function of the drift field. These results are obtained using a gas mixture of Ar:i-C_4H_{10} (90/10) and 5 mm drift gap. Points and triangles refer to charge centroid and $\mu -$ TPC results, respectively.

Figure 8 shows the resolution measured using charge centroid and $\mu -$ TPC methods for several configurations of drift fields up to 2.5 kV/cm. For the charge centroid, we observe a worsening of the resolution for decreasing fields which is related to the increasing of the Lorentz angle as clearly visible in Figure 7, which in turn produces a larger smearing of the charge distribution at the anode. On the other hand, the behavior of $\mu -$ TPC is smoother and reaches the minimum resolution values around 1.5 kV/cm. The optimized resolution after the combination of the two methods is obtained for a drift field lying between 1.5 kV/cm and 2 kV/cm. In this range, we can keep the resolution stable in the full working region.

VI. APPLICATIONS

The GEM technology has a large field of applications due to its robustness, workability and low cost with respect to other technologies. In addition, it can be used to cover large area and it can be shaped to the required form. The development of the $\mu -$ TPC reconstruction method allows to obtain spatial resolutions values in high magnetic field never reached so far. This enlarges the field of applications making GEM detector suitable to work in high energy physics experiments in presence of intense field. In this perspective, the Cylindrical Gas Electron Multiplier Inner Tracker (CGEM-IT) project has been developed, which consists in the design and construction of a cylindrical GEM detector as inner tracker for the BESIII experiment [5]. The CGEM-IT is made up of three layers of cylindrical triple-GEM. The first cylindrical triple-GEM belongs to KLOE-2 experiment [6], from which we inherited the construction procedure. In order to satisfy the BESIII requirements on spatial resolution and material budget, several improvements were adopted. Rohacell 31, instead of Honeycomb, is used as support for anode and cathode and the permanglass rings located only outside the active area allow to minimize the material budget. CGEM-IT will be the first cylindrical GEM detector with analog readout inside a magnetic field, with simultaneous measurements of deposited charge and time of signal arrival. A dedicated ASIC has being developed: the Torino Integrated GEM Electronics for Readout (TIGER) [7], while a jagged strips anode configuration is adopted to reduce the inter-strips capacitance [8]. The CGEM-IT project is funded by the European Commission in the call H2020-MSCA-RISE-2014, and the detector will be installed during the Summer 2018.

VII. Conclusion

Triple-GEM planar chamber have been studied in detail under several working setting (HV, drift fields) and configurations (two different gas mixtures and 3/5 mm drift gaps). A stable spatial resolution as a function of track incident angle and/or in high magnetic field can be achieved only if $\mu-$TPC method is adopted and combined with charge centroid. A flat distribution of the order of 130 μm is expected. Additional beam test data taking and studies are planned in order to finalize and improve the combined μ − TPC - charge centroid code to reconstruct the position.

References

[1] F. Sauli, *Nucl. Instr. and Meth. A* **386**, 531 (1997).
[2] J. Benlloch *et al.*, *Nucl. Instr. and Meth. A* **419**, 410 (1998).
[3] M. Iodice, *Journal of Instrumentation* **9**, C01017 (2014).
[4] http://sba.web.cern.ch/sba/BeamsAndAreas/resultbeam.asp?beamline=H4.
[5] M. Ablikim *et al.*, *Nucl. Instr. and Meth. A* **614**, 345 (2010).
[6] A. Balla *et al.*, *Journal of Instrumentation* **9**, C01014 (2014).
[7] M. Da Rocha Rolo *et al.*, arXiv:1706.02267 (2017).
[8] I. Garzia *et al.*, *PoS TIPP2014* 292 (2014).

Development of Large-Area Charged Particle Detector with Inorganic Scintillator Plates and Wavelength Shifting Fibers

Takahiro MIZUNO, Yusaku EMOTO, Kento FUJIHARA, Hiroshi ITO, Hideyuki KAWAI, Shota KIMURA, and Atsushi KOBAYASHI

Email: takahiro.hqa125@gmail.com

Abstract– **A Large-area charged particle detector for high energy physics experiments has been developed. This detector includes inorganic scintillator plates with thicknesses of 0.5 mm and wavelength-shifting fibers (WLSFs) of 0.2 mm in diameters. The size of effective area is 1 m by 1 m. The WLSFs are connected to both plate surfaces optically, and on the top and the bottom, fibers are attached along to x and y-axis direction, respectively. The best WLSF to obtain large number of photoelectrons is determined, which enables us to detect charged particles with thinner scintillation crystals. This means an improvement of position resolution of this detector. The number of photoelectrons obtained from a new type scintillation crystal, which is more reasonable than the conventional ones are also measured. We also estimated its position resolution. This detector enables us to detect charged particles with higher position resolution and lower cost than conventional scintillation detectors and gas chambers.**

I. INTRODUCTION

When charged particles pass thorough a scintillation crystal, scintillation light is released isotropically and the light with smaller incident angle than the critical angle goes outside the crystal. The area of the light emission region on the crystal depends on the location of the light emitting point and the thickness of the crystal. As shown in Fig. 1, the emission region gets smaller as the crystal thinner. To curb this spread of the light, inorganic crystals with high densities are introduced.

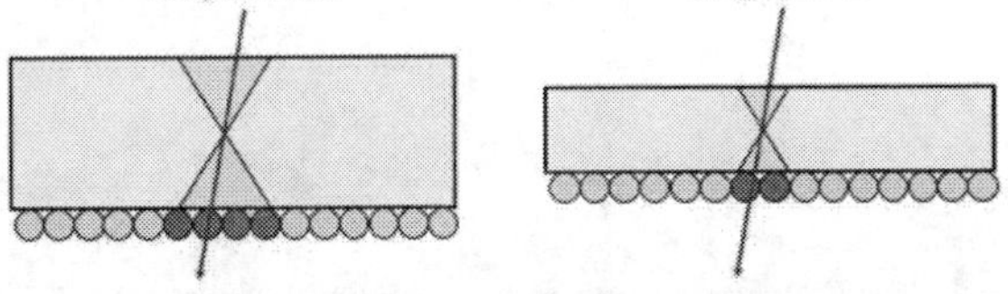

Fig. 1. Using a high density and thin inorganic crystal enables us to detect charged particles with high position resolution. When the crystal is thin, the distance to the boundary surface is short, and the spread of the light is smaller.

As a scintillation crystal, La-GPS(Ce) scintillator is used. It has chemical composition of $(La_{0.75}Ga_{0.24}Ce_{0.01})_2Si_2O_7$. It emits 30,000 to 40,000 photons/MeV. The emission wavelength is 390 to 410 nm. Its density is 5.3 g/cm^3[1]. Wavelength-shifting fiber (WLSF) is a kind of optical fiber. It absorbs the light of the particular wavelength and emits longer wavelength light. WLSF has approximately 1 m of attenuation length. Fibers are arranged on the crystals and lead the light to the photodetectors. As shown in Fig. 2, charged particles are detected with several layers of WLSFs on the crystal. They are arranged both longitudinally and laterally, so the light-emitting point is determined. One of the WLSF layers is composed of 5000 fibers of 0.2 mm in diameter.

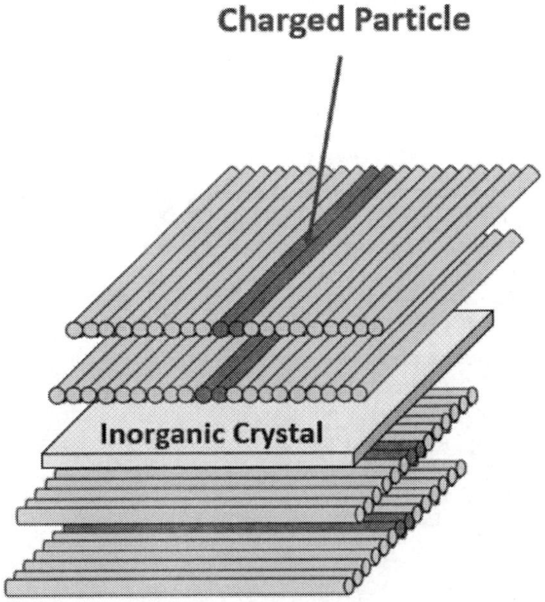

Fig. 2. The detection part is composed of several layers of WLSF on the inorganic scintillation crystal. They are arranged longitudinally and laterally.

We measured the number of photoelectrons with two kinds of fibers B-3 (300) MJ and Y-11 (300) MJ, which are manufactured by Kuraray Co. Ltd. to determine which kind of WLSF is the best to read out the light. We also measured the number of photoelectrons with high-growth-rate La-GPS crystal. High-growth-rate scintillation crystal is a scintillator which is crystalized at the higher rate than the conventional one. It is much more reasonable than the conventional one while it contains micro bubble and is not transparent completely.

II. Experimental

The number of photoelectrons was measured by using the La-GPS scintillation crystals. As a WLSF, B-3 (300) MJ and Y-11 (300) MJ are used. Fig. 3. shows the La-GPS crystal and the WLSFs used in this experiment.

Fig. 3. The La-GPS crystal (left) with the thickness of 0.50 mm is used as a scintillator. The Y-11 (300) MJ WLSFs (center) and the B-3 (300) MJ WLSFs (right) are 0.2 mm in diameter. They are arranged in sheet-like.

La-GPS crystal is manufactured by C&A Co., Ltd., which has the thickness of 0.50 mm and yield 30000 to 40000 photons/MeV light output. Peak wavelength of emission light is 390 to 410 nm[1]. B-3 (300) MJ and Y-11 (300) MJ WLSFs are made by Kuraray Co. Ltd., which are 0.2 mm in diameters. B-3 (300) MJ Fiber has emission peak of 450 nm and absorption peak of 351 nm[2]. Y-11 (300) MJ Fiber has emission peak of 476 nm and absorption peak of 430 nm[2]. Photomultiplier tubes of R9880U-210 series manufactured by Hamamatsu Photonics K.K. are used as the photodetectors. They have ultra bialkali photocathodes of 8 mm in diameter. They have spectral response in the wavelength range of 230 nm to 700 nm and the peak wavelength of response is 400 nm[3].

The number of photoelectrons was measured by changing the number of layers of WLSFs. Fig. 4. shows the setup of the experiment. The β-ray from the ^{90}Sr radiation source was detected by the 0.50 mm La-GPS crystal. The crystal was covered with the WLSF sheet. WLSF sheets with 1 to 6 layers of B-3(300) MJ and WLSF sheets with 1 to 6 layers of Y-11 (300) MJ are prepared. A PMT was attached to the end of the sheet. The detection was triggered by three PMTs. Two bundles of 5 scintillation fibers of 1.0 mm in diameters were placed orthogonally immediately above the detection part and each bundle was attached to a PMT. Another scintillation crystal was placed immediately below and attached to another PMT.

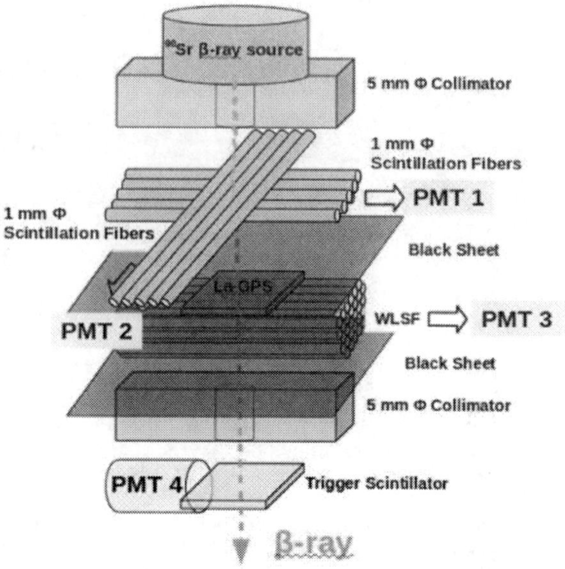

Fig. 4. The setup of the experiment is shown above. The detection is triggered by three PMTs.

The number of photoelectrons using high-growth-rate La-GPS with the thickness of 0.55 mm was measured in the same set-up. Taking the results of former experiment into consideration, we experimented using WLSF sheets with 1 to 6 layers of B-3 (300) MJ. Fig. 5. shows the high-growth-rate La-GPS crystal used in this experiment.

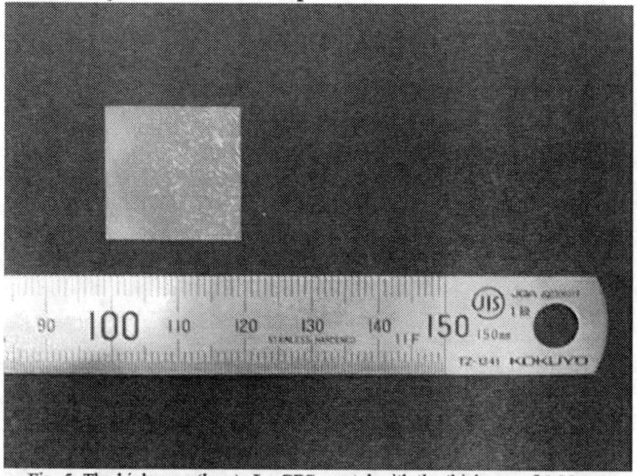

Fig. 5. The high-growth-rate La-GPS crystal with the thickness of 0.55 mm used in the experiment is shown above. It contains micro bubble and is not transparent completely. However, it is more reasonable than conventional ones.

III. Results

The results for Y-11 (300) MJ WLSF are shown in Fig. 6. and Fig. 7. The number of photoelectrons is a value on one side reading.

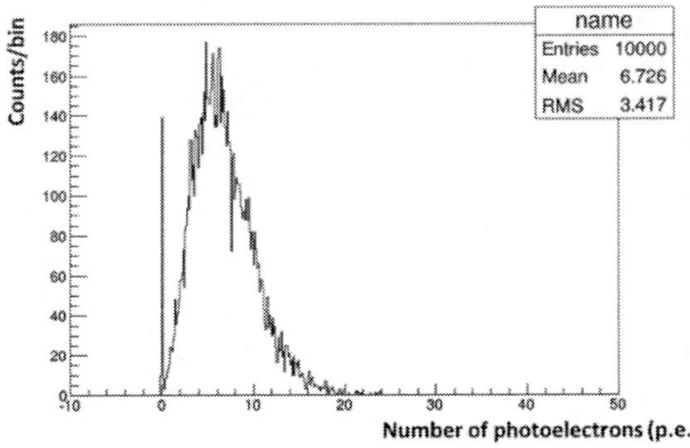

Fig. 6. The distribution of the numbers of photoelectrons detected with 0.50 mm thick La-GPS crystal, 6 layers Y-11 (300) MJ WLSF on one-side reading is shown above. 6.7 photoelectrons are detected in average.

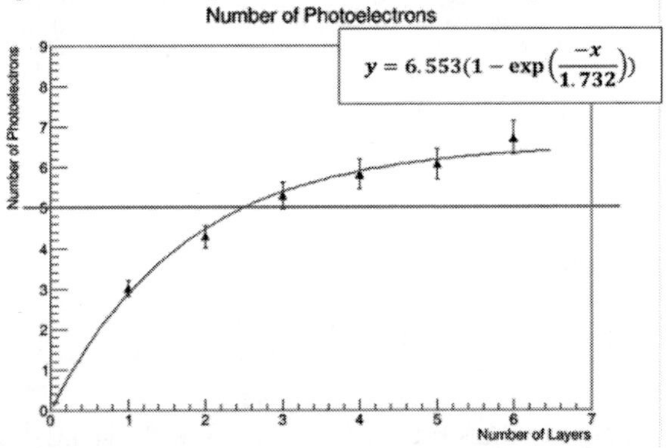

Fig. 7. Averaged numbers of photoelectrons detected on one-side reading for 0.50 mm La-GPS crystal by changing the number of Y-11 (300) MJ WLSF layers is shown above.

According to Fig. 7., the numbers of photoelectrons are gradually approaching a specific value. The experimental data could be approximated to a function y=6.553(1-e-x/1.732), and this means the number of photoelectrons is saturated to 6.553 as increasing the number of layers. Energy deposit in 1.0 mm crystal is approximately 1 MeV, thus approximately 13 photoelectrons per 1 MeV can be collected. Considering the number of photoelectrons is doubled with both sides reading, 3 WLSF layers are enough for detecting 10 photoelectrons, which corresponds to of 99% or more detection efficiency.

The results for B-3 (300) MJ WLSF are shown in Fig. 8. and Fig. 9. The number of photoelectrons is a value on one side reading.

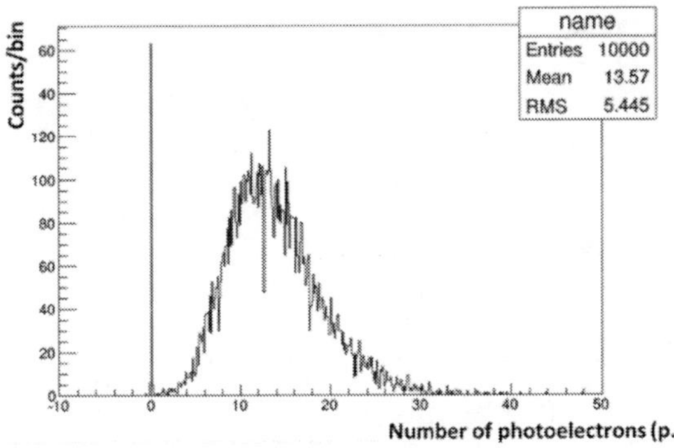

Fig. 8. The distribution of the numbers of photoelectrons detected with 0.50 mm thick La-GPS crystal, 6 layers B-3 (300) MJ WLSF on one-side reading is shown above. 13.6 photoelectrons are detected in average.

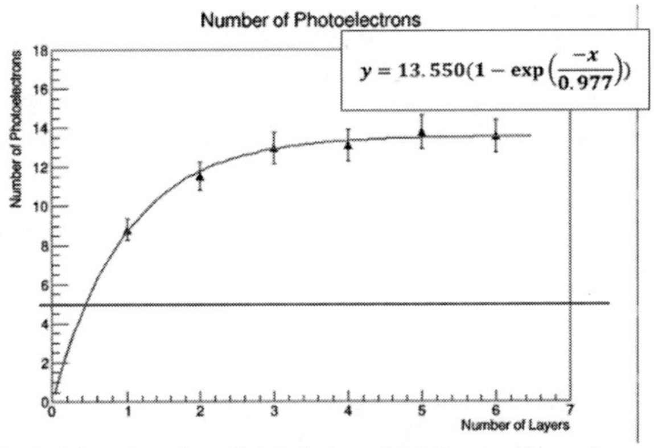

Fig. 9. Averaged numbers of photoelectrons detected on one-side reading for 0.50 mm La-GPS crystal by changing the number of B-3 (300) MJ WLSF layers is shown above.

According to Fig. 9., the numbers of photoelectrons are gradually approaching a specific value. The experimental data could be approximated to a function y=13.550(1-e-x/0.977), and this means the number of photoelectrons is saturated to 13.550 as increasing the number of layers. Energy deposit in 1.0 mm crystal is approximately 1 MeV, thus approximately 27 photoelectrons per 1 MeV can be collected. Considering the number of photoelectrons is doubled with both sides reading, 1 WLSF layer is enough for detecting 10 photoelectrons, which corresponds to of 99% or more detection efficiency.

The result for B-3 (300) MJ WLSF and 0.55 mm thick high-growth-rate La-GPS is shown in Fig. 10. The number of photoelectrons is a value on one side reading.

https://doi.org/10.1051/epjconf/201817001010

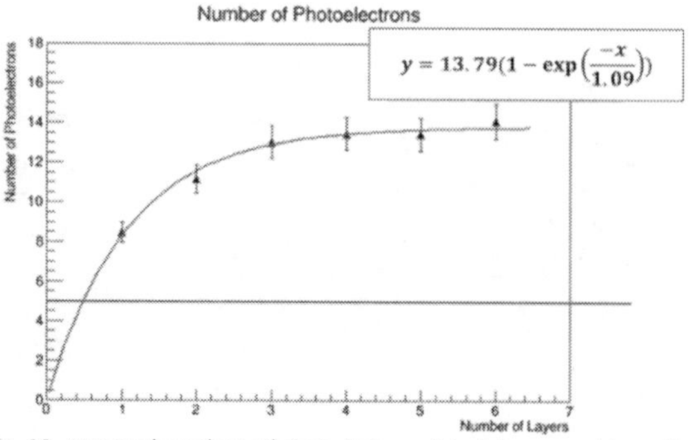

Fig. 10. Averaged numbers of photoelectrons detected on one-side reading for 0.55 mm thick high-growth-rate La-GPS crystal by changing the number of B-3 (300) MJ WLSF layers is shown above.

According to Fig. 10., the numbers of photoelectrons are gradually approaching a specific value. The experimental data could be approximated to a function y=13.79(1-e-x/1.09), and this means the number of photoelectrons is saturated to 13.79 as increasing the number of layers. It corresponds to 12.54 photoelectrons for 0.50 mm high-growth-rate La-GPS. Energy deposit in 1.0 mm crystal is approximately 1 MeV, thus approximately 25 photoelectrons per 1 MeV can be collected. Considering the number of photoelectrons is doubled with both sides reading, 1 WLSF layer is enough for detecting 10 photoelectrons, which corresponds to of 99% or more detection efficiency. The number of photoelectrons when using high-growth-rate La-GPS is approximately equal to the value when using conventional one.

IV. Estimation of the Position Resolution

The position resolution is estimated using 0.5 mm thick La-GPS crystal and 1 layer B-3 (300) MJ WLSF sheet which is composed of 10 WLSFs. The setup of the experiment is shown in Fig. 11.

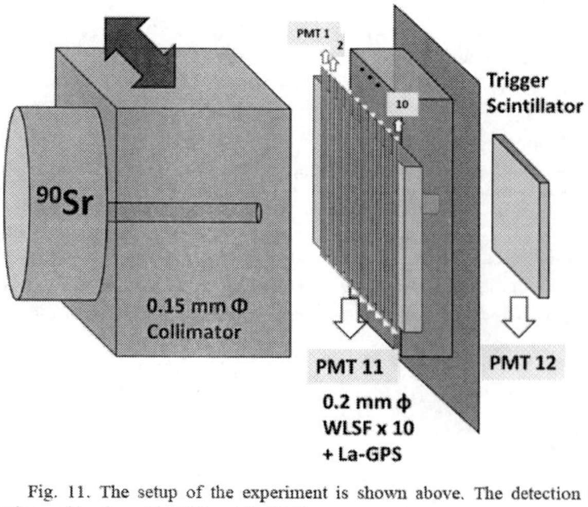

Fig. 11. The setup of the experiment is shown above. The detection is triggered by three PMT 11 and PMT 12.

The number of the photoelectrons is measured by reading out from one end of each WLSF. The position where the charged particle has passed thorough is determined by using the center of gravity calculation with the number of photoelectrons emitted from each WLSF. All 10 WLSFs are bundled and all photoelectrons emitted from fibers are read out at another end of WLSF sheet. The detection is triggered by PMT 11 and PMT 12.

Fig. 12. shows the one result of measurements of position resolution. The sigma of the distribution of reconstruction points is 0.326 mm.

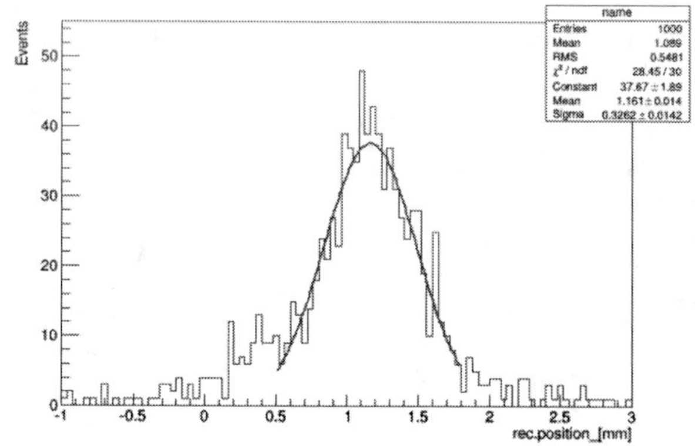

Fig. 12. The result of center of gravity calculation is shown above. The sigma of the distribution of reconstruction points is 0.326 mm.

As a result of 6 times measurements, the position resolution is 0.403 ± 0.082 mm in sigma.

V. Discussion

We propose the large-area charged particle detector with inorganic scintillator plates and wavelength shifting fibers. It is confirmed that photons from La-GPS can be read out with WLSF B-3 (300) MJ and Y-11 (300) MJ. When using 0.5 mm thickness La-GPS and B-3 (300) MJ, the more photoelectrons can be obtained and 1 WLSF layer is enough for detecting 10 photoelectrons with both sides reading. The same result is obtained when using high-growth-rate La-GPS. This is better result than 3 layers for 1.0 mm crystal[4]. Its position resolution is 0.403 ± 0.082 mm in sigma.

If using 2 layers of WLSF sheets for detection in each direction, only 300 read-out circuits are necessary for each direction; 5000 fibers in each sheet is divided into 50 bundles and each fiber is assigned a number 1 to 100 in each bundle. The fiber assigned to the same number is attached to one PMT. In the first layer, fired bundle is determined with 50 x 2 (both-sides reading) PMTs and in the second layer, the fired fiber is determined with 100 x 2 (both-sides reading) PMTs.

By using these inorganic crystals and WLSFs, "high position resolution" and "lower cost" detector can be made. In addition, photo detectors are suitable for measurements at high event rate. This detector may take the place of other conventional detectors.

Acknowledgment

We are grateful to Assoc. Prof. K. Kamada and Prof. A. Yoshikawa, Tohoku University for providing scintillation crystals.

References

[1] C&A Corporation, Product Information. Available: www.c-and-a.jp/GPS.html.
[2] Kuraray, Wavelength Shifting Fibers Available: kuraraypsf.jp/psf/ws.html.
[3] Hamamatsu Photonics K.K., Photomultiplier tube R9880U-210 Available: https://www.hamamatsu.com/jp/en/R9880U-210.html
[4] T. Mizuno et al. "Development of Large-Area Charged Particle Detectors with High Position Resolution and Low Cost", 2006 IEEE Nuclear Science Symposium Conference.

Neutron/gamma discrimination enhancement: plastic scintillators high dose irradiation and recovery time

Eva Montbarbon[1,2,3,*], Matthieu Hamel[1], Dominique Tromson[1], Marie-Noëlle Amiot[1], Camille Frangville[1], Romuald Woo[1], Robert B. Pansu[2]

[1]*CEA, LIST, Laboratoire Capteurs et Architectures Electroniques, CEA Saclay, 91191 Gif-sur-Yvette Cedex, France*
[2]*ENS Cachan, 61 avenue du président Wilson, 94230 Cachan, France*
[3]*Ecole doctorale INTERFACES, Paris-Saclay University, 91400 Saclay, France*

In the context of high-energy physics experiments, particle accelerators create a significant dose of radiations, up to 10^6 Gy for example at the Large Hadron Collider (LHC, Cern). Control detectors, such as scintillators, are affected by these high irradiation levels. The literature mentions that the light yield of organic scintillators dramatically drops when strongly irradiated. However, the tested scintillators can recover some of their light output with time. To the best of our knowledge, only the luminescent properties of plastic scintillators were analyzed under high radiation fluxes. But, plastic scintillators are also able to discriminate fast neutrons from gamma rays. In this work, we characterized neutron/gamma discriminating as well as emissive properties of plastic scintillators after high dose irradiations.

Two identical lab-made plastic scintillators containing a polystyrene-based matrix and two fluorophores were analyzed before and after high dose irradiation. Irradiation was performed using a Gamma-Cell 220 Excel with twelve ^{60}Co sources and a 53.7 Gy/min dose rate at isocenter. These two scintillators were irradiated to reach a cumulative dose equal to 10^4 Gy.

In order to measure their neutron/gamma discrimination ability, scintillators were coupled to a Hamamatsu R7724-100 photomultiplier tube and placed in front of a ^{252}Cf source (Activity 580 kBq). The anode signal fed a digitizer. Scintillation pulses were then recorded and post-processed. A charge-comparison method was implemented and FoM was evaluated. Offline treatment allowed the estimation of fast and slow decay times of neutron pulses, as well as their relative intensities. In parallel, characterization of radioluminescence properties of irradiated and non-irradiated samples was performed.

Results indicate a strong modification in the neutron/gamma discrimination capability before and after high dose irradiations and a redshift of the radioluminescence spectrum. In fact, a 10^3 Gy irradiation increases the FoM of both sensors by a factor 2.5 at least. Further, as recovery time passes after a 10^4 Gy cumulative dose, FoM improves for scintillators and is multiplied by 6 compared to the value at the zero dose. Thanks to evaluated decay times and relative intensities, we could infer that a strong irradiation does not produce a higher yield of triplet excited states but it impacts the slow decay time of the tested scintillator. This leads to a better neutron/gamma discrimination than at the zero dose. These observations lead us to believe that the intrinsic nature of the plastic material is modified under high dose irradiation.

In this paper we present the current iteration of this ongoing work. The neutron/gamma discrimination properties are presented for both plastic scintillators and first characterization results are discussed. We show that material modification under high dose irradiation manages to an improvement in neutron/gamma discrimination opening the field to applications of very high dose measurements.

 https://doi.org/10.1051/epjconf/201817001011

All observations we received thanks to these strong irradiation experiments will be detailed at the conference. The experiments planned to highlight the intrinsic modifications of scintillators will also be described.

 https://doi.org/10.1051/epjconf/201817001012

Triply Differential Ionization Cross Sections of atomic and molecular targets by single electron impact

Lena Mouawad*, Ziad El Bitar
Université de Strasbourg, CNRS
IPHC UMR 7178, F-67000
Strasbourg, France
Email*: lena.mouawad@iphc.cnrs.fr

Lena Mouawad, Ahmad Osman, Mohamad Khalil
AZM research center, EDST
Lebanese University
Tripoli, Lebanon

Paul Antoine Hervieux
Institut de Physique et Chimie des Matériaux de Strasbourg
23 rue du Loess, BP 43, 67034
Strasbourg, France

Claude Dal Cappello
Laboratoire de Physique Moléculaire et des Collisions
UMR CNRS 7565, Université de Lorraine
1 Boulevard Arago, 57078 Metz Cedex 3, France

***Abstract*—We present numerically calculated triply differential cross sections for the simple ionization of two targets, atomic hydrogen and ammonia, in the First Born Approximation framework with one Coulomb wave describing the ejected electron. The partial wave series of the wavefunctions centered around the same origin were used and the intial state molecular wavefunction was derived using information provided by Gaussian03 software for ammonia. The theoretical results are calculated in the geometrical settings and kinematical conditions of previous experiments and are compared to published experimental and theoretical data. The results show good agreement with previous studies based on the same model for hydrogen. The general shape of the triply differential cross sections obtained for ammonia are in acceptable agreement with experimental data in the binary region but not in the recoil region where our simple framework fails to reconstruct the recoil peaks.**

I. INTRODUCTION

The study of the ionization of atoms and molecules by electron impact is necessary to reveal information about the structure of the target and improve the understanding of particle interactions in many domains. The dynamics of this process is best described by the triply differential cross section (TDCS) which involves all the kinematic parameters of ionization providing the ionization probability over the ejected electron angle distribution. Determining this quantity presents many challenges both experimentally and theoretically because it requires an accurate description of the evolution of all the electrons' states that are involved as well as their mutual interactions before and after collision. Modeling these interactions becomes even more complex in the case of polyatomic molecular targets.

Extensive research has been done to provide TDCSs for a variety of atomic and molecular targets exploring different theoretical models. At high incident and ejected electron energies, the nucleus's role can be neglected and the interaction can be approximated by a simple Coulomb attraction between the ejected electron and the nucleus. Under these kinematics, the use of the first Born approximation (FBA) is justified and slight improvements are provided by other much more sophisticated models that require heavy calculations. To further simplify the calculations, many groups have used partial wave expansion which facilitates the translation of the analytical calculations to parallel numerical codes with the least computing time possible [1]–[3].

Accurately describing the molecular wave function is very important and presents many challenges for multicenter complex molecules. Defining the wave functions around a single center reduces the complexity of this task [4]. These wave functions have been expressed as linear combinations of atomic orbitals written in terms of Slater-type functions in some studies [5], [6] and in terms of Gaussian-type functions in others [3], [7]. Although Slater-type functions provide a more accurate desription of the atomic orbitals, they are difficult to express on distinct atoms of bigger molecules and hence the use of Gaussian-type orbitals is preferrable to describe molecular wave functions. Another challenge that is faced when dealing with molecular targets is defining the molecule's orientation before the collision. This cannot be done in experiments, therefore an average orientation of the molecule should be considered in theoretical calculations for appropriate comparision with experimental data. This can be done either by using orientation averaged molecular orbitals (OAMO) [8] or by calculating TDCSs for each possible orientation and then determining the average TDCS [9]. In this work, we use single center Gaussian-type molecular wave functions developed from the GAUSSIAN 03 [10] program in the 6-31G basis set.

The hydrogen atom represents the simplest case study that allows the validation of developed analytical and numerical models before moving on to molecular targets which intial state wave functions are more complex to determine. *Ehrhardt et al.* performed an experiment in which they measured the TDCSs of atomic hydrogen at incident energy of 250 eV, ejected energy of 5 eV and a detection angle of the scattered electron equal to 3 degrees and 8 degrees [11]. The experimental data showed a binary peak near the direct momentum transfer direction and a recoil peak near its backward direction. In 2012 *Sahlaoui and Bouamoud* [6] compared analytical TDCSs calculated within the FBA framework using Slater-type functions to these experimental data. Their work aimed at proposing an improvement of the FBA to model post-collision interactions by multiplying the TDCS equation with the Gamow factor [12]. We consider the results of these studies and compare them to our TDCSs for the ionization of the hydrogen atom. We also present calculated TDCSs for the ionization of ammonia and compare them to the experimental data of *El Mir et al.* [13] measured at an incident energy of 500 eV, ejected energy of 74 eV and with scattered angle of -6 degrees. These data show the same trend observed for the hydrogen atom TDCSs. In the work of *El Mir et al.* these data were compared to analytical cross sections calculated within the FBA using the one coulomb wave model for the ejected electron with Slater-type wave functions developed according to the approach proposed by Moccia [14]. They also presented theoretical results using a distorted wave to describe the ejected electron and results calculated with the Brauner-Briggs-Klar model (BBK) [15]. The FBA results also underestimated the recoil peak but predicted a better binary to recoil amplitude ratio than the other models for the $3a_1$ and $1e_1$ orbitals. We present numerical TDCSs calculated also in the FBA-1CW framework but with Gaussian-type wave functions and writing all the wave functions in partial wave series and compare our results to the experimental and analytical data of *El Mir et al.* [13].

We first begin by presenting our theoretical framework, then we provide our results for both atomic hydrogen and ammonia. Our results show very good agreement with published analytical data therebye validating our program.

II. Theoretical framework

The triply differential cross section $\sigma^{(3)}$ is given by:

$$\sigma^{(3)} = \frac{k_f k_e}{k_i} |f_{B1}|^2 \tag{1}$$

f_{B1} being the transition amplitude in the first born approximation from the state Ψ_i to the state Ψ_f:

$$f_{B1} = \frac{-1}{2\pi} < \Psi_f | V | \Psi_i > \tag{2}$$

The interaction potential V is the coulomb interaction between the incident electron and the target noting $\overrightarrow{R}$ as the position vector of the heaviest nucleus of the target with respect to the incident electron. In the case of the hydrogen atom, the interaction potential is simply written as:

$$V = -\frac{1}{\overrightarrow{R}} + \frac{1}{|\overrightarrow{R} - \overrightarrow{r_1}|} \tag{3}$$

$\overrightarrow{r_1}$ being the position vector of the bound electron e_1 with respect to the Hydrogen atom.

On the other hand, the potential interaction between the incident electron and the NH_3 molecule is given by:

$$V = -\frac{7}{\overrightarrow{R}} - \frac{1}{|\overrightarrow{R} - \overrightarrow{R_1}|} - \frac{1}{|\overrightarrow{R} - \overrightarrow{R_2}|} - \frac{1}{|\overrightarrow{R} - \overrightarrow{R_3}|} + \sum_{i=1}^{10} \frac{1}{|\overrightarrow{R} - \overrightarrow{r_i}|} \tag{4}$$

$\overrightarrow{R}$ is the position vector of the Nitrogen atom with respect to the incident electron, $\overrightarrow{R_1}$, $\overrightarrow{R_2}$ and $\overrightarrow{R_3}$ are the position vectors of the three Hydrogen atoms with respect to the Nitrogen atom, and $\overrightarrow{r_i}$ is the position vector of the ith bound electron with respect to the Nitrogen atom.

In the First Born Approximation the incident and scattered electrons are both described by plane waves, $e^{i\overrightarrow{K_i}\overrightarrow{R}}$ and $e^{i\overrightarrow{K_f}\overrightarrow{R}}$ respectively, while the ejected electron is described by a Coulomb wave $\mathscr{F}^{(-)}_{\boldsymbol{K_e}}(\overrightarrow{r})$.

Let's consider the molecular target case of ammonia. Before collision, the target electron is bound to a molecular orbital i and is described by the molecular orbital wavefunction $\psi_i(\boldsymbol{r})$. The method used to develop the molecular orbital wavefunctions from Gaussian03 was presented in details elsewhere [16], [17] so it will not be discussed here. In summary, the contracted Cartesian Gaussian type orbitals (CGTO), used as basis functions in Gaussian, are written as linear combinations of primitive gaussian type orbitals which are then written as function of spherical type orbitals. The obtained form for the wavefunction is:

$$\psi_i(x,y,z) = \sum_{\kappa=1}^{n_c} \sum_{j=1}^{n_p} \sum_{n,l,m} a_{ik} d_{\kappa j} A(ijk, nlm) \phi_b(\hat{r}) \tag{5}$$

A single center partial wave expansion of the molecular wavefunction is then performed thereby reducing the complexity of studying a multicenter molecule and that by separating the radial and angular components of the spherical gaussian functions. This was already done in details by *Kaufmann and Baumeister* [4] where a detailed form of the spherical gaussian functions is given. This leads to a single-center molecular orbital wavefunction of the following form:

$$\psi_i(x,y,z) = \sum_{\lambda,m_\lambda} \tilde{R}^i_{\lambda,m_\lambda}(r) Y_{\lambda,m_\lambda}(\Omega_{\vec{r}}) \quad (6)$$

$\vec{r}$ being the position vector with respect to the origin.

For appropriate comparision with experimental data, the molecular orientation considered in Gaussian is transformed to the laboratry frame by using the rotation operator $\mathscr{R}^\lambda_{\mu m_\lambda}(\alpha\beta\gamma)$ according to the following:

$$\psi_i(x,y,z) = \sum_{\lambda,m_\lambda} \tilde{R}^i_{\lambda,m_\lambda}(r) \sum_{\mu=-\lambda}^{\mu=+\lambda} \mathscr{R}^\lambda_{\mu m_\lambda}(\alpha\beta\gamma) Y_{\lambda\mu}(\hat{r}) \quad (7)$$

The initial state wavefunction is the product of the incident electron's plane wave $e^{i\vec{k_i}\vec{R}}$ with the intial state molecular wavefunction ψ_{is} describing the 10 bound electrons of the ammonia molecule:

$$\Psi_i = e^{i\vec{k_i}\vec{R}} \quad \psi_{is}(\vec{r_1},\vec{r_2},\vec{r_3},..,\vec{r_{10}}) \quad (8)$$

With $\psi_{is}(\vec{r_1},\vec{r_2},\vec{r_3},..,\vec{r_{10}}) = \psi(\vec{r_1})\psi(\vec{r_2})\psi(\vec{r_3}) \ \ldots \ \psi(\vec{r_{10}})$

After collision, the final state wavefunction is the product of the scattered electron's plane wavefunction $e^{i\vec{k_f}\vec{R}}$ with the ejected electron's Coulomb wavefunction $\mathscr{F}^{(-)}_{\vec{k_e}}(\vec{r_1})$ and the molecular wavefunctions of the other 9 bound electrons. We suppose that the electron e_1 is the ejected electron, with $\vec{r_1}$ being the position vector of e_1 with respect to the Nitrogen nucleus, and that all other electrons are passive electrons with frozen wavefunctions from the initial to the final state:

$$\Psi_f = e^{i\vec{k_f}\vec{R}} \quad \psi_{fs}(\vec{r_1},\vec{r_2},\vec{r_3},..,\vec{r_{10}}) \quad (9)$$

With $\psi_{fs}(\vec{r_1},\vec{r_2},..,\vec{r_{10}}) = \mathscr{F}^{(-)}_{\vec{k_e}}(\vec{r_1}) \ \psi(\vec{r_2})\psi(\vec{r_3}) \ \ldots \ \psi(\vec{r_{10}})$

After replacing 4, 8 and 9 in equation 2, and assuming a small momentum transfer from the incident electron to the target, the transition amplitude expression simplifies to that obtained with the simple potential of the Hydrogen atom:

$$f_{B1} = \frac{-2}{q^2} < \mathscr{F}^{(-)}_{\vec{k_e}}(\vec{r_1}) \mid e^{i\vec{q}\vec{r_1}} - 1 \mid \psi(\vec{r_1}) > \quad (10)$$

$\vec{q} = \vec{k_i} - \vec{k_f}$ is the transfer momentum from the incident electron to the target.

The plane, Coulomb and target's wavefunctions are written in partial wave expansion as follows:

$$\mathscr{F}^{(-)*}_{\vec{k_e}}(\vec{r_1}) = \frac{1}{(2\pi)^{\frac{3}{2}}} \sum_{l_e,m_e} (4\pi)(-i)^{l_e} e^{i\delta_{l_e}} \frac{F_{l_e}(k_e r_1)}{k_e r_1} Y_{l_e,m_e}(\hat{k_e}) Y^*_{l_e,m_e}(\hat{r_1}) \quad (11)$$

$$e^{i\vec{q}\vec{r}} = \sum_{l,m} (4\pi) i^l j_l(qr) Y^*_{l,m}(\hat{q}) Y_{l,m}(\hat{r}) \quad (12)$$

$$\psi(\vec{r_1}) = \sum_{\lambda,m_\lambda} \tilde{R}^i_{\lambda,m_\lambda}(r_1) \sum_{\mu=-\lambda}^{\mu=+\lambda} \mathscr{R}^\lambda_{\mu m_\lambda}(\alpha\beta\gamma) Y_{\lambda\mu}(\hat{r_1}) \quad (13)$$

After replacing in 1, the triply differential cross section becomes:

$$\sigma^{(3)} = \frac{k_f k_e}{k_i} \frac{2^5}{q^4} \Big| \sum_{\lambda,m_\lambda} \sum_{\mu} \sum_{l,m} \sum_{l_e,m_e} (-1)^{m_e} i^{l-l_e} e^{i\delta_{l_e}} Y^*_{l,m}(\hat{K}) Y_{l_e,m_e}(\hat{k_e}) \mathscr{R}^\lambda_{\mu m_\lambda}(\alpha\beta\gamma) \mathcal{R}^i_{le,l,\lambda,m_\lambda} \mathcal{A}^{\lambda,\mu}_{l_e,m_e,l} \Big|^2 \quad (14)$$

With $\mathcal{R}^i_{le,l,\lambda,m_\lambda} = \int r_1^2 \mathrm{d}r_1 \tilde{R}^i_{\lambda m\lambda}(r_1) \frac{F_{l_e}(k_e r_1)}{k_e r_1} (j_l(qr) - \delta_{l,0})$
and $\mathcal{A}^{\lambda,\mu}_{l_e,m_e,l} = \left(\hat{l_e}\hat{l}\hat{\lambda}\right)^{\frac{1}{2}} \begin{pmatrix} l_e & l & \lambda \\ 0 & 0 & 0 \end{pmatrix} \begin{pmatrix} l_e & l & \lambda \\ -m_e & m_e-\mu & \mu \end{pmatrix}$

The rotation operator $\mathscr{R}^\lambda_{\mu m_\lambda}(\alpha\beta\gamma)$ is represented by a rotation matrix $D^\lambda_{\mu m_\lambda}(\alpha\beta\gamma)$ such that:

$$\mathscr{R}^\lambda_{\mu m_\lambda}(\alpha\beta\gamma) = e^{-i\alpha m} D^\lambda_{\mu m_\lambda}(\alpha\beta\gamma) e^{-i\gamma mu} \quad (15)$$

The elements of the rotation matrix $D^\lambda_{\mu m_\lambda}(\alpha\beta\gamma)$ form a complete set of orthogonal functions of the Euler angles:

$$\int_0^{2\pi} \mathrm{d}\alpha \int_0^{\pi} sin(\beta)\mathrm{d}\beta \int_0^{2\pi} D^\lambda_{\mu m_\lambda}(\alpha\beta\gamma) D^{\lambda'*}_{\mu' m'_{\lambda'}}(\alpha\beta\gamma)\mathrm{d}\gamma = 8\pi^2 \frac{1}{\hat{\lambda}} \delta_{\lambda\lambda'} \delta_{\mu\mu'} \delta_{m_\lambda m'_{\lambda'}} \quad (16)$$

$$\delta_{\lambda\lambda'}, \delta_{\mu\mu'}, \delta_{m_\lambda m'_{\lambda'}} \neq 0 \ for \ \lambda = \lambda', \mu = \mu', m_\lambda = m'_{\lambda'}$$

Integrating 14 using 15 and 16 leads to the following final expression for the TDCS:

$$\sigma^{(3)} = \frac{k_f k_e}{k_i} \frac{2^5}{q^4} \sum_{\lambda,m_\lambda} \sum_{\mu} \frac{1}{\hat{\lambda}} |S^i_{\lambda,m_\lambda,\mu}|^2 \quad (17)$$

$$With \ S^i_{\lambda,m_\lambda,\mu} = \sum_{l_e,m_e} \sum_{l} (-1)^\mu i^{(l-l_e)} e^{i\delta_{l_e}} Y_{l_e,m_e}(\hat{k_e}) Y_{l,\mu-m_e}(\hat{q}) \mathcal{R}^i_{le,l,\lambda,m_\lambda} \mathcal{A}^{\lambda,\mu}_{l_e,m_e,l} \quad (18)$$

III. Results

We first consider the simplest case of atomic hydrogen and compute the triply differential ionisation cross sections in the same conditions as the ones used in the work of *Sahlaoui and Bouamoud* [6] $E_i = 250$ eV, $E_e = 5$ eV, $\theta_f = 5$ degrees.

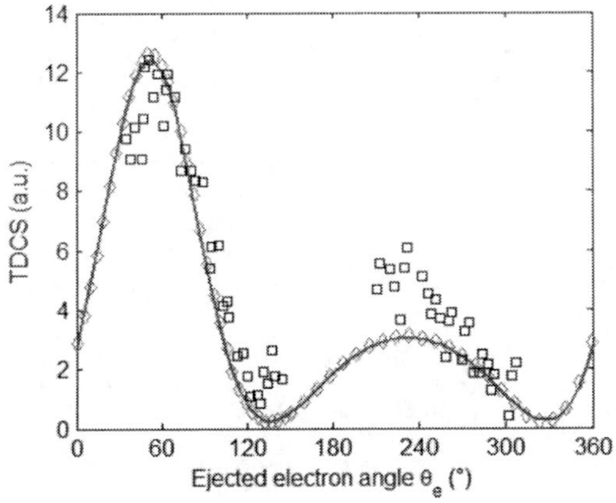

Fig. 1: Triply differential ionization cross sections of the ionization of atomic hydrogen. The considered kinematical conditions are: $E_i = 250$ eV, $E_e = 5$ eV and $\theta_f = 5$ degrees. Black squares: experimental data [11], blue triangles: analytical data of *Sahlaoui and Bouamoud* [6], red solid line: our results.

Figure 1 shows that we numerically obtain exactly what was analytically found by *Sahlaoui and Bouamoud* with the same FBA-1CW model. The binary and recoil peak positions as well as the peak widths are in good agreement with the experimental data of *Ehrhardt et al.* [11] but the relative amplitude of the peaks is not well predicted by the FBA since the recoil peak's amplitude is underestimated.

The same shape of the TDCS as function of the ejected electron angle is found in the ammonia case under the kinematical conditions considered in the work of *El Mir et al.* [13]. In figure 2, the experimental data of *El Mir et al.* are represented by the solid black squares, their 1CW analytical results are given by the blue diamonds. Our numerical results are represented by the dashed green line in figure 2 for molecular wave functions calculated using Gaussian03 and by the red solid line for molecular wave functions calculated using the approach proposed by Moccia.

Good agreement is found between the analytical triply differential cross sections of *El Mir et al.* and our data calculated using Slater-type wave functions futher validating our program. The amplitude of the binary peak is a little different with the 6-31 Gaussian basis set than using the Moccia approach to develop the molecular wave function but in general the TDCSs have the same shape using both Slater and Gaussian type wave functions. In comparision with the experimental data, the width of the binary peaks and the amplitude of the recoil peaks are underestimated. This proves the need for a more detailed description of the interactions taking place in order to have a better estimation of the experimental data.

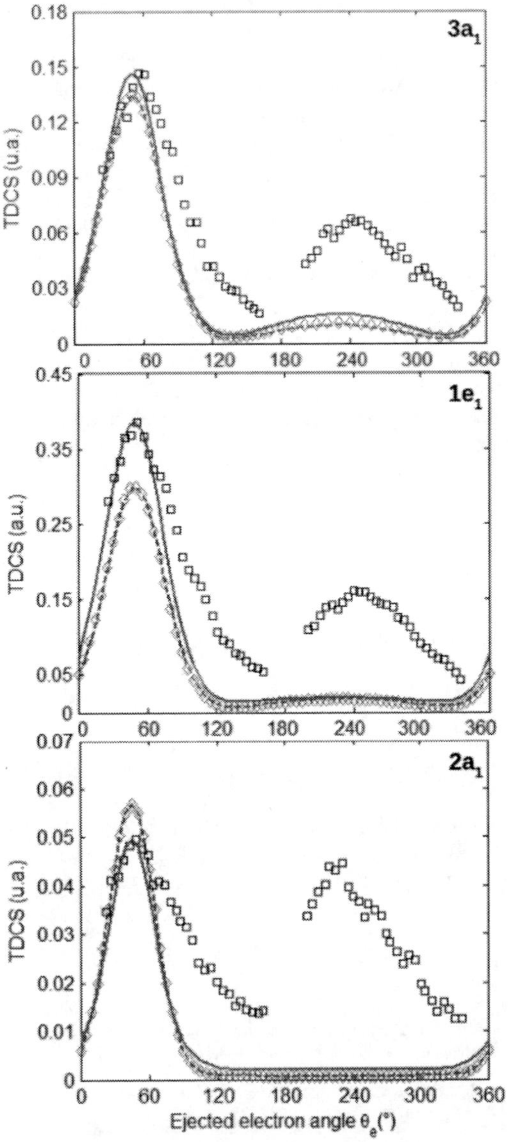

Fig. 2: Triply differential ionization cross sections of the ionization of the three outer valence orbitals of ammonia. The considered kinematical conditions are: $E_i = 500$ eV, $E_e = 74$ eV and $\theta_f = -6$ degrees. Black squares: experimental data [13], blue triangles: analytical FBA-1CW data given in the same work, green dashed line: our results using Slater-type molecular wave functions, red solid line: our results using Gaussian-type molecular wave functions.

IV. Conclusion

We present a theoretical model using partial wave expansion and based on the First Born approximation to calculate the triply differential single electron impact ionization cross sections of hydrogen and ammonia. Partial wave expansion is an important tool that facilitates programming the theoretical model and using a computing grid to reduce computing time which is of particular interest for the study of big molecules. Using Gaussian 03 to develop single-center molecular wave functions makes the corresponding program user-friendly with the ability to easily change the parameters for different molecules and kinematical conditions. As a start, we demonstrate the validity of our program by comparing the results with previously published data for atomic hydrogen and present the triply differential cross sections for the ionization of ammonia based on this formalism which are in good agreement with published data. The same discrepancies between experimental and theoretical FBA data that were noted in a previous study were observed in our work in the case of the ammonia, especially regarding the inability of FBA-1CW model to accurately describe the collision process under the considered kinematics when an electron is ejected in the backwards direction. This work is a first step to validate the functions used in our program as well as the transition from Gaussian03 output to the program where the molecular wave function is also calculated. The next step would be to use the program to determine cross sections for more complex molecules such as the DNA bases and develop the collision description into a more complex description of the interactions between the projectile and the target.

Acknowledgment

This work is co-funded by IN2P3-CNRS and IRD, France under the ARTS research grant program. It is also supported by the Projet International de Coopération Scientifique (Pics 6509) between the French CNRS and the Lebanese CNRS, Lebanese University and Saint Joseph University.

References

[1] C.-Y. Lin, C. McCurdy, and T. Rescigno, "Complex kohn approach to molecular ionization by high-energy electrons: Application to h 2 o," *Physical Review A*, vol. 89, no. 1, p. 012703, 2014.

[2] C. Champion, "Electron impact ionization of liquid and gaseous water: a single-center partial-wave approach," *Physics in medicine and biology*, vol. 55, no. 1, p. 11, 2009.

[3] I. Tóth, R. I. Campeanu, and L. Nagy, "Ionization of nh3 and ch4 by electron impact," *The European Physical Journal D*, vol. 69, no. 1, p. 2, 2015.

[4] K. Kaufmann and W. Baumeister, "Single-centre expansion of gaussian basis functions and the angular decomposition of their overlap integrals," *Journal of Physics B: Atomic, Molecular and Optical Physics*, vol. 22, no. 1, p. 1, 1989.

[5] S. Houamer, M. Chinoune, and C. Dal Cappello, "Theoretical study of (e, 2e) process of atomic and molecular targets," *The European Physical Journal D*, vol. 71, no. 1, p. 17, 2017.

[6] M. Sahlaoui and M. Bouamoud, "Electron impact single ionization of the water molecule in the second born approximation," *Journal of Physics B: Atomic, Molecular and Optical Physics*, vol. 45, no. 8, p. 085201, 2012.

[7] M. Sahlaoui and M. Bouamoud, "Cross sections for electron-impact ionization of water molecules," *Canadian Journal of Physics*, vol. 89, no. 6, pp. 723–727, 2011.

[8] J. Gao, J. Peacher, and D. H. Madison, "An elementary method for calculating orientation-averaged fully differential electron-impact ionization cross sections for molecules," *The Journal of chemical physics*, vol. 123, no. 20, p. 204302, 2005.

[9] E. Ali, K. Nixon, A. Murray, C. Ning, J. Colgan, and D. Madison, "Comparison of experimental and theoretical electron-impact-ionization triple-differential cross sections for ethane," *Phys. Rev. A*, vol. 92, p. 042711, Oct 2015.

[10] M. Frisch, G. Trucks, H. Schlegel, G. Scuseria, M. Robb, J. Cheeseman, G. Scalmani, V. Barone, B. Mennucci, G. Petersson, *et al.*, "Gaussian 09, revision b. 01. wallingford, ct: Gaussian," *Inc.,(2*, 2004.

[11] H. Ehrhardt, K. Jung, G. Knoth, and P. Schlemmer, "Differential cross sections of direct single electron impact ionization," *Zeitschrift für Physik D Atoms, Molecules and Clusters*, vol. 1, no. 1, pp. 3–32, 1986.

[12] A. S. Kheifets, A. Naja, E. M. S. Casagrande, and A. Lahmam-Bennani, "Dwba-g calculations of electron impact ionization of noble gas atoms," *Journal of Physics B: Atomic, Molecular and Optical Physics*, vol. 41, no. 14, p. 145201, 2008.

[13] R. El Mir, E. S. Casagrande, A. Naja, C. Dal Cappello, S. Houamer, and F. El Omar, "Triple differential cross sections for the ionization of the valence states of nh3 by electron impact," *Journal of Physics B: Atomic, Molecular and Optical Physics*, vol. 48, no. 17, p. 175202, 2015.

[14] R. Moccia, "One-center basis set scf mo's. ii. nh3, nh4+, ph3, ph4+," *The Journal of Chemical Physics*, vol. 40, no. 8, pp. 2176–2185, 1964.

[15] A. Lahmam-Bennani, A. Naja, E. S. Casagrande, N. Okumus, C. Dal Cappello, I. Charpentier, and S. Houamer, "Dynamics of electron impact ionization of the outer and inner valence (1t2 and 2a1) molecular orbitals of ch4 at intermediate and large ion recoil momentum," *Journal of Physics B: Atomic, Molecular and Optical Physics*, vol. 42, no. 16, p. 165201, 2009.

[16] C. Champion, C. Dal Cappello, S. Houamer, and A. Mansouri, "Single ionization of the water molecule by electron impact: Angular distributions at low incident energy," *Physical Review A*, vol. 73, no. 1, p. 012717, 2006.

[17] H. D. Hafied, *Etude théorique de l'ionisation par impact électronique des molécules d'eau en phase gazeuse et liquide*. PhD thesis, Metz, 2007.

FIPPS: Fission Product Prompt Gamma-Ray Spectrometer, a New High Efficiency HPGe Array for Nuclear Spectroscopy

P. Mutti[1,*] , A. Blanc[1], M. Jentschel[1], U. Köster[1], C. Michelagnoli[1], E. Ruiz-Martinez[1]

[1] ILL, Institut Laue-Langevin, BP 156, 6 rue Jules Horowitz, 38042 Grenoble Cedex 9, France

[*] Corresponding author – mutti@ill.eu

Within the EXILL campaign a large and efficient cluster of Ge-detectors was installed around a very well collimated neutron beam. This has allowed carrying out rather complete spectroscopic studies close to the line of stability using the (n, gamma) reaction. Neutron rich isotopes were produced by neutron-induced fission on both ^{235}U and ^{241}Pu samples and prompt gamma spectroscopy was carried out. Isotope identification was based on triple gamma coincidences a partially known level scheme and the use of triple coincidences. Because of the high multiplicity coincidence, the sensitivity in the case of weak production yields was substantially reduced.

The EXILL campaign demonstrated the potential physical output from (n, γ) and (n, f) reactions and the interest of the nuclear structure community in this type of experiments. Despite the undoubted success of EXILL we learned, as well, that there was space for further improvements on the experimental setup. One of the major problem arising from the analysis of the (n, f) data is the lack of a clear identification of the fission events. This information is very important if one wants to study the less populated nuclei in the fission fragment distribution as well as the kinematic of the fission process itself and therefore determine the mass and the kinetic energy of the fission fragments. The possibility of an additional mass information would improve significantly the capability of the germanium detectors array to identify new γ-rays. From all these considerations, we decided to propose the construction of a new ILL instrument combining an EXILL-like germanium array with the mass resolving power of a gas filled magnet. FIPPS, FIssion Product Prompt gamma-ray Spectrometer, is now part of the ILL Endurance program phase one, the next ILL's instruments upgrade plan which started this year and will run until 2019. FIPPS initial setup will be very similar to the one of EXILL. 8 high efficiency germanium clover detectors have been purchased this year to form the central ring of the spherical detectors array. The remaining space around the target will be filled with additional detectors depending on the specific setup required for the experiments. Therefore, not only more segmented germanium detectors are foreseen to complete the setup but also BEGe (broad energy germanium) that could significantly improve the resolution at very low energy to identify very weak low energy transitions. From the FATIMA collaboration, a set of $LaBr_3(Ce)$ scintillators for picosecond timing studies is also expected to be installed as well as a set of neutron detectors for the study of the fission process. FIPPS has been located at the end position of H22 thermal neutron guide. The construction of the instrument and its commissioning was completed mid-December last year and FIPPS is now officially operational for users' experiments since the first reactor cycle in 2017. In the last proposals' round of November 2016, assigning beam time only for the first reactor cycle 2017, 8 experiments have been submitted for FIPPS for a total of 108 days of beam time, meaning more than a factor of 2 overbooking; a clear indication of the interest of the nuclear physics community. For fission studies measurements, FIPPS will use a scintillating sample, where the fissioning isotope (e.g. ^{235}U) will be dissolved inside scintillator material. This trigger on fission events will clear the gamma spectrum from the huge beta background, reducing, in such a way, the need for high multiplicity coincidence. In a later stage, the HPGe array will be coupled with a gas filled magnet and auxiliary detectors. The presence of the gas filled magnet will dramatically improve the fission tagging allowing the identification of fission products directly. Moreover, the gas filled magnet will be coupled with an ionization chamber to measure the total kinetic energy of the fission fragments allowing detailed fission yield studies. The presentation will briefly discuss the physics case focusing on the experimental results obtained from the first measurement campaign. Further options like a low energy detector cluster, a plunger, a conversion electron spectrometer or neutron detectors and potentially possible experiments will be also discussed.

EPJ Web of Conferences 170, 01014 (2018) https://doi.org/10.1051/epjconf/201817001014
ANIMMA 2017

Induced Charge Fluctuations in Semiconductor Detectors with a Cylindrical Geometry

Victor V. Samedov

Abstract— **Now, compound semiconductors are very appealing for hard X-ray room-temperature detectors for medical and astrophysical applications. Despite the attractive properties of compound semiconductors, such as high atomic number, high density, wide band gap, low chemical reactivity and long-term stability, poor hole and electron mobility-lifetime products degrade the energy resolution of these detectors. The main objective of the present study is in development of a mathematical model of the process of the charge induction in a cylindrical geometry with accounting for the charge carrier trapping. The formulae for the moments of the distribution function of the induced charge and the formulae for the mean amplitude and the variance of the signal at the output of the semiconductor detector with a cylindrical geometry were derived. It was shown that the power series expansions of the detector amplitude and the variance in terms of the inverse bias voltage allow determining the Fano factor, electron mobility lifetime product, and the nonuniformity level of the trap density of the semiconductor material.**

Index Terms— **charge induction, charge carriers trapping, compound semiconductors, cylindrical geometry, energy resolution.**

I. Introduction

Now, compound semiconductors are very appealing for hard X-ray room-temperature detectors for medical and astrophysical applications. Despite the attractive properties of compound semiconductors, such as high atomic number, high density, wide band gap, low chemical reactivity and long-term stability, poor hole and electron mobility-lifetime products degrade the energy resolution of these detectors. The process of trapping in imperfections introduced during semiconductor crystal growth, such as impurity atoms, vacancies, and structural irregularities has a pronounced effect on the charge carrier transport and leads to the peak broadening. The main goal of the theory of radiation detectors is in putting forward mathematical models that adequately describe the processes occurring in the transformation of the energy of impinging radiation into the output detector signal. The energy resolution of a semiconductor detector depends on the fluctuations in the process of charge carriers generation, in the process of charge induction on the detector electrodes, caused by charge carriers trapping, on the fluctuations of the trapping centers concentration in the detector volume, on the fluctuations in the gain of amplifier and on the electronic noise. The main objective of the present study is the development of a mathematical model of the process of the charge induction in a cylindrical geometry with accounting for the charge carrier trapping. In this work, theoretical consideration of the stochastic process of the charge induction in cylindrical geometry with accounting for the charge carrier trapping was considered.

As in compound semiconductor detectors the electron mobility greatly exceeds the hole mobility, significant improvement in the energy resolution can be achieved by discarding the hole with the poorer transport properties. One way is in using special detector geometries to eliminate the contribution of holes. Two detector geometries that offer a marked increase in field strength near the collecting electrode are hemispherical and coaxial geometries. Theoretical consideration of the stochastic process of charge induction in a hemispherical geometry with accounting for electron trapping was considered in [1]. The main objective of the present study is in development of a mathematical model of the process of charge induction in a cylindrical geometry with accounting for the charge carrier trapping. The general formulae for the moments of the distribution function of the induced charge on the electrodes of a detector with a cylindrical geometry are applicable to all coaxial X-ray detectors. For example, they are applicable to the Reverse-Electrode Germanium detectors (REGe) with the ion-implanted boron p-type outer electrode, and the diffused lithium n-type inner electrode.

II. Moments of the Induced Charge Distribution Function in Coaxial Detectors

In coaxial X-ray detectors, the charge on the detector's electrodes is induced mainly in the very small high electric field region near the inner electrode. Thus, to eliminate the contribution of holes, the zero potential must be applied to the outer electrode and the potential V to the inner electrode, and X-rays must enter the detector from the outer electrode.

For low-energy X-rays the dominant interaction with the matter is a photoelectric absorption. If the length of the active region of the detector is much greater than the mean free path length of X-rays in semiconductor material, then we can assume that all X-rays interact with the detector close to the outer electrode. As the thickness of a detector is usually much greater than the size of the volume of the photoelectron energy conversion into the energy of electron-hole pairs, then we can

Manuscript received May 31, 2017.
Victor V. Samedov is with the National Research Nuclear University MEPhI (Moscow Engineering Physics Institute), 31, Kashirskoye Shosse, 115409, Moscow, Russian Federation (telephone: 007-495-323-9038, e-mail: v-samedov@yandex.ru).

consider that all electron-hole pairs are produced at the point of the X-quantum interaction, i.e. close to the outer electrode, and only electrons determine the detector signal.

The coaxial geometry on the one hand eliminates the contribution of holes, but on the other hand increases the fluctuations of the induced charge on the detector electrodes. These fluctuations are determined by the process of electron trapping in imperfections, impurity atoms, vacancies, and structural irregularities. As an electron spends most of the time in the low electric field region, the process of electron trapping has a pronounced effect on the process of charge induction on the detector electrodes. For the best performance of the coaxial detector the correct ratio of the outer to inner radii and the operating voltage must be chosen according to the electron mobility-lifetime product.

Let us consider a coaxial detector with the zero electric potential applied to the outer electrode of the radius R_2, and the electric potential V applied to the inner electrode of the radius R_1. In [2], the general formulae for the moments of the distribution function of the relative charge induced on the detector electrodes were derived

$$\langle q^n \rangle = n\int_0^1 dq \cdot q^{n-1}\exp(-F(q)) \quad (1)$$

The most important are the first two ones

$$\langle q \rangle = \int_0^1 dq\exp(-F(q)), \quad (2)$$

$$\langle q^2 \rangle = 2\int_0^1 dq \cdot q\exp(-F(q)), \quad (3)$$

that determine the mean value and the variance

$$\sigma_q^2 = \langle q^2 \rangle - \langle q \rangle^2 = 2\int_0^1 dq \cdot q\exp(-F(q)) - \left(\int_0^1 dq\exp(-F(q))\right)^2 \quad (4)$$

of the relative induced charge on the detector electrodes.

In the above formulae, according to the Ramo's theorem [3], the relative induced charge on the detector electrodes is

$$q(r) = \frac{\varphi(r)-\varphi(R_2)}{V} = \frac{\varphi(r)}{V}, \quad (5)$$

where $\varphi(r)-\varphi(R_2)$ is the electric potential difference between the electron absorption point at the radius r from the center and the creation point, which for low energy X-rays coincides with the radius of the outer electrode R_2.

The dimensionless function $F(q)$ is the analogue of the optical path length, and is equal to the ratio of the time to induce the relative induced charge on the detector electrodes q to the mean electron trapping time, usually referred to as the electron lifetime τ_e.

The function $F(q)$ has the form of the line integral along the electric field line

$$F(q) = \int_0^{s(q)} ds/\mu_e(s)\tau_e(s)E(s) \quad (6)$$

where $E(s)$ is the magnitude of the electric field strength, $\mu_e(s)$ is the electron mobility, and $\tau_e(s)$ the electron lifetime at the distance s from the electron creation point; $s(q)$ is the inverse function of the relative induced charge on the detector electrodes (5).

The electric potential and the magnitude of the electric field strength in a coaxial detector at the radius r from the center are

$$\varphi(r) = V\frac{\ln(R_2/r)}{\ln(R_2/R_1)}, \quad (7)$$

$$E(r) = E_r(r) = \frac{V}{r\ln(R_2/R_1)}. \quad (8)$$

For our case, the relative induced charge on the detector electrodes (5) is determined by the equation

$$q(r) = \frac{\ln(R_2/r)}{\ln(R_2/R_1)}, \quad (9)$$

and the inverse function of $q(r)$ is

$$r(q) = R_2\exp(-pq). \quad (10)$$

In the formula (10), the dimensionless parameter

$$p = \ln(R_2/R_1) \quad (11)$$

is the first main parameter that characterizes the coaxial geometry of a given detector.

After integration along the electric field line, the function $F(q)$ has the form

$$F(q) = \frac{1}{2}p[1-\exp(-2pq)]\xi, \quad (12)$$

where

$$\xi = \frac{R_2^2}{\mu_e\tau_e V} \quad (13)$$

is the second main dimensionless parameter that includes the physical characteristics of the coaxial detector and varies with the applied bias voltage V.

The average value of the relative induced charge on the electrodes of a coaxial detector can be received by expanding integrand in (2) in power series of the parameter ξ

$$\langle q \rangle = \int_0^1 dq\exp(-F(q)) = \sum_{n=0}^{\infty}\frac{1}{n!}\left(-\frac{p\xi}{2}\right)^n\int_0^1 dq[1-\exp(-2pq)]^n \quad (14)$$

After term-by-term integration of the binomial expansion for power n and rearranging terms, the average value of the relative induced charge on the electrodes of a coaxial detector in power series of the parameter ξ has the form

$$\langle q\rangle = \sum_{n=0}^{\infty} a_n \xi^n , \quad (15)$$

where the expansion coefficients are determined by the formula

$$a_n = \frac{1}{n!}\left(-\frac{p\xi}{2}\right)^n \left\{1 - \frac{1}{2p}\left[\begin{matrix}\Psi(n+1)+\gamma \\ -e^{-2p}\,{}_3F_2\left(1,1,1-n;2,2;e^{-2p}\right)\end{matrix}\right]\right\} . \quad (16)$$

In the formula (16) ${}_rF_s\left(a_1,\ldots a_r;b_1,\ldots b_s;x\right)$ is the generalized hypergeometric series [4], $\Psi(x)$ is the digamma function, and $\gamma = 0.57721\,56649$ is the Euler's constant [5].

The first several coefficients of expansion are

$$a_0 = 1, \quad (17)$$

$$a_1 = -\frac{1}{4}(2p-1+e^{-2p}), \quad (18)$$

$$a_2 = \frac{1}{32} p(4p-3+4e^{-2p}-e^{-4p}), \quad (19)$$

$$a_3 = -\frac{1}{576} p^2(12p-11+18e^{-2p}-9e^{-4p}+2e^{-6p}) . \quad (20)$$

In a similar way, the second moment of the distribution function of the relative induced charge on the electrodes of a coaxial detector has the form

$$\langle q^2(\tau_e^{-1})\rangle = 2\int_0^1 dq\cdot q\exp(-F(q)) = \sum_0^{\infty} b_n \xi^n , \quad (21)$$

where the expansion coefficients are determined by the formula

$$b_n = \frac{1}{n!}\left(-\frac{p\xi}{2}\right)^n \left\{\begin{matrix} 1 - \frac{n}{2p^2}\cdot {}_4F_3\left(1,1,1,1-n;2,2,2;1\right) + \frac{n}{2p^2} \\ \times e^{-2p}\left[\begin{matrix} 2p\cdot{}_3F_2\left(1,1,1-n;2,2;e^{-2p}\right) \\ +{}_4F_3\left(1,1,1,1-n;2,2,2;e^{-2p}\right)\end{matrix}\right]\end{matrix}\right\} , \quad (22)$$

where ${}_rF_s\left(a_1,\ldots a_r;b_1,\ldots b_s;x\right)$ is the generalized hypergeometric series [4].

The first several coefficients of expansion are

$$b_0 = 1, \quad (23)$$

$$b_1 = -\frac{1}{4p}(2p^2-1+e^{-2p}(2p+1)), \quad (24)$$

$$b_2 = \frac{1}{64}(8p^2-7+e^{-2p}8(2p+1)-e^{-4p}(4p+1)), \quad (25)$$

$$b_3 = -\frac{p}{3456}(72p^2-85+e^{-2p}108(2p+1) \\ -e^{-4p}27(4p+1)+e^{-6p}4(6p+1)) . \quad (26)$$

In the limit $R_2 \to \infty$ and $R_1 \to \infty$, providing that $R_2 - R_1 = D$, the formulae (15) and (21) reduce to the formulae for the first two moments of the distribution function of the induced charge in a planar detector with uniform electric field [6].

III. THE BIAS DEPENDENCE OF THE VARIANCE OF COAXIAL DETECTORS

In [7], the general formulae for the mean value and the variance of the semiconductor detector signal were derived. In the case, when all electron-hole pairs are produced at the point of the soft X-quantum interaction close to the outer electrode, and the detector signal is determined only by electrons, the dependence of the mean value and the variance of the semiconductor detector signal on the bias voltage V have forms

$$\langle Q(V)\rangle = \frac{E_0}{\varepsilon}\langle q(V)\rangle\langle g\rangle , \quad (27)$$

$$\sigma_Q^2(V) = \frac{E_0}{\varepsilon}\left(\begin{matrix} F\langle q(V)\rangle^2 + \langle q^2(V)\rangle \\ -\langle q(V)\rangle^2 + \langle q(V)\rangle\eta_g^2 \end{matrix}\right)\langle g\rangle^2 + \sigma_{\text{noise}}^2 , \quad (28)$$

where E_0 is the energy of incident monoenergetic particles, ε is the effective energy to produce one electron-hole pair, F is the Fano factor, $\langle g\rangle$ and $\eta_g^2 = \sigma_g^2/\langle g\rangle^2$ are the mean value and the relative variance of the electronic amplifier gain, σ_{noise}^2 is the electronic noise at the output of spectrometer. In the formulae (27) and (28)

$$\langle q(V)\rangle = \int d\tau_e^{-1}\rho(\tau_e^{-1})\langle q(\tau_e^{-1},V)\rangle , \quad (29)$$

$$\langle q^2(V)\rangle = \int d\tau_e^{-1}\rho(\tau_e^{-1})\langle q^2(\tau_e^{-1},V)\rangle , \quad (30)$$

are the first and the second moments of the distribution function of the relative charge induced by one electron on the detector electrodes averaged over the distribution of the inverse electron drift time τ_e^{-1} [8].

It was shown in [7], that for quasi-uniform semiconductor material with spatial fluctuations in the trap density, the inverse charge carrier drift time is the normal distribution $\rho(\tau_e^{-1})$ with the mean value $\langle\tau_e^{-1}\rangle$, the variance $\sigma_{\tau_e^{-1}}^2$, and the relative variance $\eta_{\tau_e^{-1}}^2 = \sigma_{\tau_e^{-1}}^2/\langle\tau_e^{-1}\rangle^2$ equals to the relative variance of the trap density $\eta_{N_t}^2 = \sigma_{N_t}^2/\langle N_t\rangle^2$.

As $\tau_e^{-1} = \langle\tau_e^{-1}\rangle + \Delta\tau_e^{-1}$, after introducing the new parameter

$$\chi = \xi\frac{V}{\tau_e^{-1}} = \frac{R_2^2}{\mu_e} , \quad (31)$$

averaged over the Gaussian distribution $\rho(\tau_e^{-1})$ the first and the second moments of the distribution function have forms

https://doi.org/10.1051/epjconf/201817001014

$$\langle q(V)\rangle = \int d\tau_e^{-1}\rho(\tau_e^{-1})\langle q(\tau_e^{-1},V)\rangle = \int d\tau_e^{-1}\rho(\tau_e^{-1})\sum_0^\infty a_n\xi^n$$

$$= \sum_0^\infty a_n\left(\frac{\chi}{V}\langle\tau_e^{-1}\rangle\right)^n\cdot\int d\Delta(\tau_e^{-1})\rho(\Delta(\tau_e^{-1}))\left(1+\frac{\Delta(\tau_e^{-1})}{\langle\tau_e^{-1}\rangle}\right)^n$$

$$= \sum_0^\infty a_n\left(\frac{\chi}{V}\langle\tau_e^{-1}\rangle\right)^n\int d\Delta(\tau_e^{-1})\rho(\Delta(\tau_e^{-1}))\cdot\left(1+n\frac{\Delta(\tau_e^{-1})}{\langle\tau_e^{-1}\rangle}\right)$$

$$= \sum_0^\infty a_n\left(\frac{\chi}{V}\langle\tau_e^{-1}\rangle\right)^n$$

, (32)

$$\langle q^2(V)\rangle = \int d\tau_e^{-1}\rho(\tau_e^{-1})\sum_0^\infty b_n\xi^n$$

$$= \sum_0^\infty b_n\left(\frac{\chi}{V}\langle\tau_e^{-1}\rangle\right)^n\cdot\int d\Delta(\tau_e^{-1})\rho(\Delta(\tau_e^{-1}))\left(1+\frac{\Delta(\tau_e^{-1})}{\langle\tau_e^{-1}\rangle}\right)^n$$

$$= \sum_0^\infty b_n\left(\frac{\chi}{V}\langle\tau_e^{-1}\rangle\right)^n\int d\Delta(\tau_e^{-1})\rho(\Delta(\tau_e^{-1})) \quad ,$$

$$\times\left(1+n\frac{\Delta(\tau_e^{-1})}{\langle\tau_e^{-1}\rangle}+\frac{n(n-1)}{2}\frac{(\Delta(\tau_e^{-1}))^2}{\langle\tau_e^{-1}\rangle^2}\right)$$

$$= \sum_0^\infty b_n\left(\frac{\chi}{V}\langle\tau_e^{-1}\rangle\right)^n\left(1+\frac{n(n-1)}{2}\eta_{\tau_e^{-1}}^2\right)$$

(33)

With these results, and after rearranging in powers of

$$\gamma = \chi\langle\tau_e^{-1}\rangle \quad (34)$$

for the mean value of the signal of coaxial detector we have

$$\langle Q(V)\rangle = \frac{E_0}{\varepsilon}\sum_0^\infty a_n\left(\frac{\gamma}{V}\right)^n\langle g\rangle. \quad (35)$$

$$\sigma_Q^2(V) = \frac{E_0}{\varepsilon}\sum_0^\infty c_n\left(\frac{\gamma}{V}\right)^n\langle g\rangle^2+\sigma_{\text{noise}}^2. \quad (36)$$

where

$$c_n = b_n\left(1+\frac{n(n-1)}{2}\eta_{\tau_e^{-1}}^2\right)+(F-1)\cdot\sum_{k=0}^n a_k a_{n-k}+a_n\eta_g^2,$$

(37)

and coefficients a_n and b_n are given by (16) and (22).

The first several coefficients of c_n are

$$c_0 = F+\eta_g^2, \quad (38)$$

$$c_k = \alpha_k\left[d_k+d_k^F F+d_k^{\tau_e^{-1}}\eta_{\tau_e^{-1}}^2+d_k^g\eta_g^2\right],\quad k\ge 1. \quad (39)$$

$$k=1,\ \alpha_1=\frac{1}{4},\ d_1=2(p-1)+\frac{1}{p}\left(1-e^{-2p}\right),$$

$$d_1^F=-2\left(2p-1+e^{-2p}\right),\ d_1^{\tau_e^{-1}}=0,\ d_1^g=\left(2p-1+e^{-2p}\right).$$

(40)

$$k=2,\ \alpha_2=-\frac{1}{64},$$

$$d_2=24p^2-28p+11+e^{-2p}16(p-1)+e^{-4p}5,$$

$$d_2^F=-\left(32p^2-28p+4+e^{-2p}8(4p-1)-e^{-4p}4(p-1)\right),$$

$$d_2^{\tau_e^{-1}}=-\left(8p^2-7+e^{-2p}8(2p+1)-e^{-4p}(4p+1)\right),$$

$$d_2^g=-2p\left(4p-3+e^{-2p}4-e^{-4p}\right). \quad (41)$$

$$k=3,\ \alpha_3=\frac{1}{3456},$$

$$d_3=\alpha\begin{pmatrix}504p^2-672p+247+e^{-2p}2(324p-243)\\ -e^{-4p}27(4p-11)\end{pmatrix},$$

$$d_3^F=-6p\begin{pmatrix}96p^2-112p+27+e^{-2p}(144p-63)\\ -e^{-4p}9(4p-5)+e^{-6p}(4p-9)\end{pmatrix},$$

$$d_3^{\tau_e^{-1}}=-3p\begin{pmatrix}72p^2-85+e^{-2p}108(2p+1)\\ -e^{-4p}27(4p+1)+e^{-6p}4(6p+1)\end{pmatrix},$$

$$d_3^g=-6p^2\left(12p-11+e^{-2p}18-e^{-4p}9+e^{-6p}2\right). \quad (42)$$

In all above formulae only the first order of small parameters $F<<1$, $\eta_{\tau_e^{-1}}^2=\sigma_{\tau_e^{-1}}^2/\langle\tau_e^{-1}\rangle^2<<1$ and $\eta_g^2=\sigma_{g_e}^2/\langle g\rangle^2<<1$ are retained.

The formulae (38) - (45) in the limit $R_2\to\infty$ and $R_1\to\infty$, providing that $R_2-R_1=D$, reduce to the formulae for the planar detector with uniform electric field [8].

The coefficients of expansions (35) and (36) allow determining characteristics of a coaxial detector from experimental data.

IV. Conclusion

The exact mathematical description of the processes in a semiconductor coaxial detector at registration of low energy X-rays gives the correct formulae for the mean and the variance of the output signal of a detector. It was shown that the power series expansions of the detector amplitude and the variance in terms of the inverse bias voltage allow determining the characteristics of a coaxial detector. These formulae are useful for analysis of the influence of different factors on the energy resolution of semiconductor coaxial detectors.

https://doi.org/10.1051/epjconf/201817001014

REFERENCES

[1] V.V. Samedov, "Induced charge fluctuations in hemispherical semiconductor detectors", *X-Ray Spectrometry*, 2017; to be published.
[2] V.V. Samedov, "Fluctuations in the processes of charge induction in ionization-type detectors", *X-Ray Spectrometry*, vol. 44, pp. 183–185, 2015.
[3] H. Spieler, *Semiconductor detector systems*, Oxford University Press, 2005.
[4] G. Gasper, *Basic hypergeometric series*, Cambridge University Press, 2004.
[5] *Handbook of Mathematical Functions*, Eds. M. Abramowitz and I. A. Stegun, National Bureau of Standards Applied Mathematics Series 55, Washington, D.C., 1972.
[6] V.V. Samedov, "The Basics of Experimental Determination of the Fano Factor in Intrinsic Semiconductors", *IEEE Catalog Number* CFP1324I-CDR, DOI 978-1-4799-1047-2/13/$31.00, 2013
[7] V.V. Samedov, "Theoretical approach to the energy resolution of semiconductor detectors", *Conference Record of the 2016 IEEE Nuclear Science Symposium and Medical Imaging Conference*, Paper ID R09-40, 2016.
[8] V.V. Samedov, "Theoretical consideration of the energy resolution in planar HPGE detectors for low energy X-rays", *ANIMMA 2015*, Article number 7465586, DOI: 0.1109/ANIMMA.2015.7465586, 2015.

A Complete Readout Chain of the ATLAS Tile Calorimeter for the HL-LHC: from FATALIC Front-End Electronics to Signal Reconstruction

Sergey Senkin, on behalf of the ATLAS Tile Calorimeter System
Laboratoire de Physique de Clermont-Ferrand
Campus Universitaire des Cézeaux
4 Avenue Blaise Pascal
63178 Aubière cedex, France
Email: sergey.senkin@cern.ch

***Abstract*—The ATLAS Collaboration has started a vast programme of upgrades in the context of high-luminosity LHC (HL-LHC) foreseen in 2024. We present here one of the front-end readout options, an ASIC called FATALIC, proposed for the high-luminosity phase LHC upgrade of the ATLAS Tile Calorimeter. Based on a 130 nm CMOS technology, FATALIC performs the complete signal processing, including amplification, shaping and digitisation. We describe the full characterisation of FATALIC and also the Optimal Filtering signal reconstruction method adapted to fully exploit the FATALIC three-range layout. Additionally we present the resolution performance of the whole chain measured using the charge injection system designed for calibration. Finally we discuss the results of the signal reconstruction used on real data collected during a preliminary beam test at CERN.**

I. Introduction

ATLAS [1] is one of two general-purpose experiments at the Large Hadron Collider (LHC) at CERN. It is designed to realise the full discovery potential and the vast range of physics opportunities provided by the LHC. The Tile Calorimeter (TileCal) [2] is the hadronic calorimeter covering the central region of the ATLAS detector, as can be seen on Fig. 1. It is essential for the measurement of hadrons, jets, tau leptons and missing transverse energy.

The upgrade of the LHC to the high-luminosity phase (HL-LHC) [3] foreseen in 2024 will provide invaluable opportunities for the search for new physics beyond the Standard Model, as well as the detailed studies of the electroweak symmetry breaking mechanism and precise measurements of the properties of recently discovered Higgs boson. The ATLAS Collaboration has started an extensive programme of HL-LHC upgrades for every sub-detector, including TileCal. The current readout electronics of TileCal must be upgraded to comply with the new specifications aiming for the future operating conditions.

The ASIC described in this document, named Front-end ATlAs tiLe Integrated Circuit (FATALIC) [4], has been developed to fulfil the requirements of the HL-LHC upgrade.

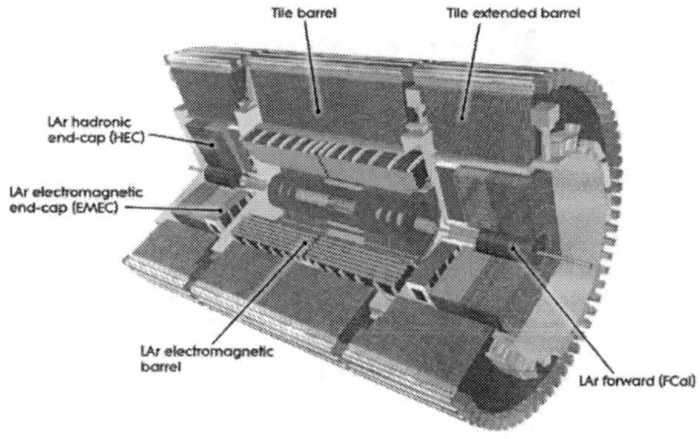

Fig. 1. A cut-away view of the ATLAS calorimeters. The TileCal comprises a central barrel region with pseudorapidity range of $|\eta| < 1.7$, and two extended barrel regions providing coverage up to $|\eta| < 2.4$.

It is one of the front-end readout options proposed for the upgrade of TileCal, based on a 130 nm CMOS technology and designed for the complete processing of the signal delivered by each Photomultiplier-Tube (PMT), including amplification, shaping and digitisation.

The document is structured as follows. The overview of the FATALIC layout as well as the main progress stages of its development and the performance measurements are given in Section II. The signal reconstruction aspects are described in Section III. Subsection III-A gives an overview of the Optimal Filtering algorithm which is used to reconstruct the amplitude and time of the signal is described. The specific details of application of this algorithm to FATALIC pulses are described in subsection III-B. Finally, the results of the signal reconstruction used on real data collected during a preliminary beam test at CERN are shown in Section IV, and a short conclusion is given in Section V.

II. FATALIC LAYOUT

The front-end readout system designed for the HL-LHC upgrade has to abide by stringent requirements, an overview of which is given in Table I. In order to comply with them and particularly with the large dynamic range requirement of 25 fC to 1.2 nC, a three-gain layout has been adopted for FATALIC.

TABLE I
HL-LHC REQUIREMENTS FOR THE FRONT-END READOUT

Technology	0.13 μm CMOS GlobalFoundries
Number of channel per ASIC	1
Polarity	negative
Dynamic range in charge	Q = 25 fC – 1.2 nC
Dynamic range in current peak	Q = 1.25μA – 60 mA
Rise time of the current peak	t_r = 4 ns
Fall time of the current peak	t_f = 36 ns
Noise (RMS)	< 12 fC
Power	as low as possible
Power supply	1.6 V
Output	word of 12 bits at 40 MHz

The overall layout of the latest ASIC iteration, FATALIC5, is shown in Fig. 2. The first stage of FATALIC is a current conveyor which splits the input signal into three ranges with gain ratios of 1, 8 and 64. Each current conveyor output is followed by a shaper and a dedicated pipelined 12-bit Analogue-to-Digital Converter (ADC) operating at 40 MHz. As a result, the analogue signal from the physics events is digitised at the ADC output, which in case for FATALIC is integrated in the ASIC itself. Due to bandwidth limitations, only two gains are forwarded to the output. Auto-selection of the data to be transmitted is performed between the high and the low gains, while the medium gain data are always included. This auto-selection is performed digitally inside the chip based on the saturation of the most sensitive channel. Fig. 3 shows an example of an analogue signal at the FATALIC shaper output (using a digital scope) and a digitised event at the ADC output, both shown for a muon event.

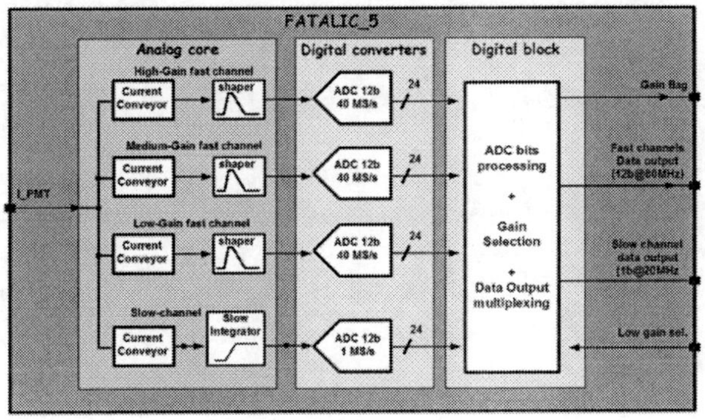

Fig. 2. Block diagram of the latest design of FATALIC ASIC.

The dynamic performance of the whole chain is measured in terms of resolution using the dedicated charge injection system designed for calibration. Fig. 4 shows the linearity results for the three gains of the previous FATALIC ASIC, FATALIC4b. It can be seen that the non-linearity is contained within 2% for the high and medium gains over their whole range. However, saturation is observed for the low gain starting at around 600 pC.

This drawback has been addressed in FATALIC5. The improvement of the linearity, particularly for the low gain, is achieved at the input stage (the current conveyor) by reducing the impact of the high input currents (about 60 mA for a charge of 1.2 nC) on the Grid-Source voltage variations of the input transistor. Use of larger transistors allows these variations to be reduced by almost one order of magnitude. Fig. 5 gives the linearity results for the latest FATALIC5 design, obtained using simulation. The non-linearity is now better than 1% over the whole dynamic range for all gains.

Another improvement implemented in FATALIC5 is the addition of the slow channel (Fig. 2), necessary for the measurement of low current. This measurement is needed for the absolute detector calibration with radioactive caesium source, which produces a known but low signal. While the three previously described channels have integration times of 25 ns (the LHC clock period), the slow channel has a 100 μs time constant inside the integrator. For this channel, the digitisation is performed over 1.2 μs by the ADC, and the integration time is 10 ms. Larger integration times are accessible by digital means, when requested. The fast and slow channels are always DC coupled, and the current is split by allocating 20% to the fast channel, and 80% to the slow one. The optimisation of this ratio is obtained using simulation. The FATALIC5 design has been recently sent to foundry, and its first performance tests are planned for July 2017.

III. SIGNAL RECONSTRUCTION

A. *Optimal Filtering in TileCal*

Optimal Filtering (OF) is a method of reconstructing the amplitude of an analogue signal from its digital samples. It is currently used in TileCal as a main reconstruction method, and is also envisaged to be used in the future with one of the front-end readout options implemented. The method is well-documented in the designated ATLAS note [5], and here we describe the main principle behind Optimal Filtering.

The digitised samples, such as those at the output of FATALIC, can be expressed as:

$$S_i = p + Ag(t_i + \tau) + n_i \tag{1}$$

where p is the signal pedestal, A is the true amplitude, $g(t)$ is the normalized reference pulse shape (taken from the shaper circuit, which provides a stable and well-defined pulse shape), τ is the phase between the expected and measured pulse times, and n_i is the noise term, which is typically modelled by a Gaussian distribution if only electronic noise is present. In order to make the phase τ an output parameter of the

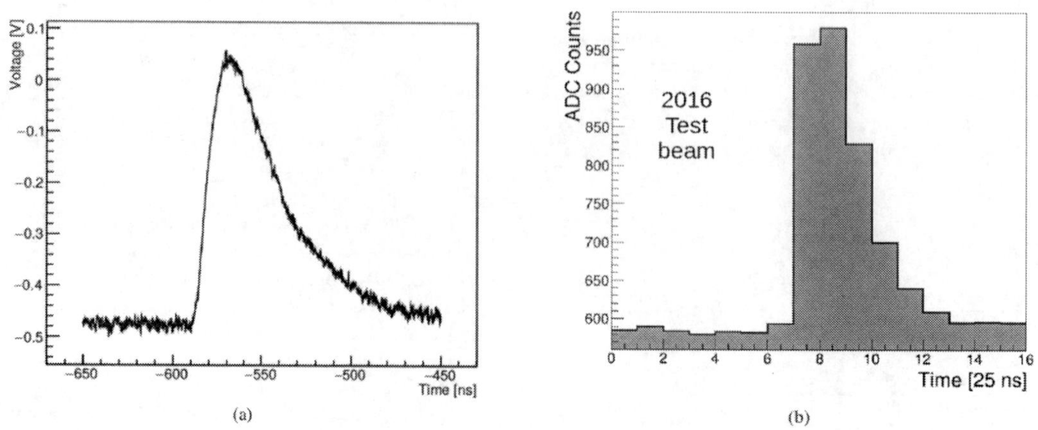

Fig. 3. (a) Analogue signal (shaper output) of a muon event. (b) Digital signal (ADC output) of a muon event through the whole chain.

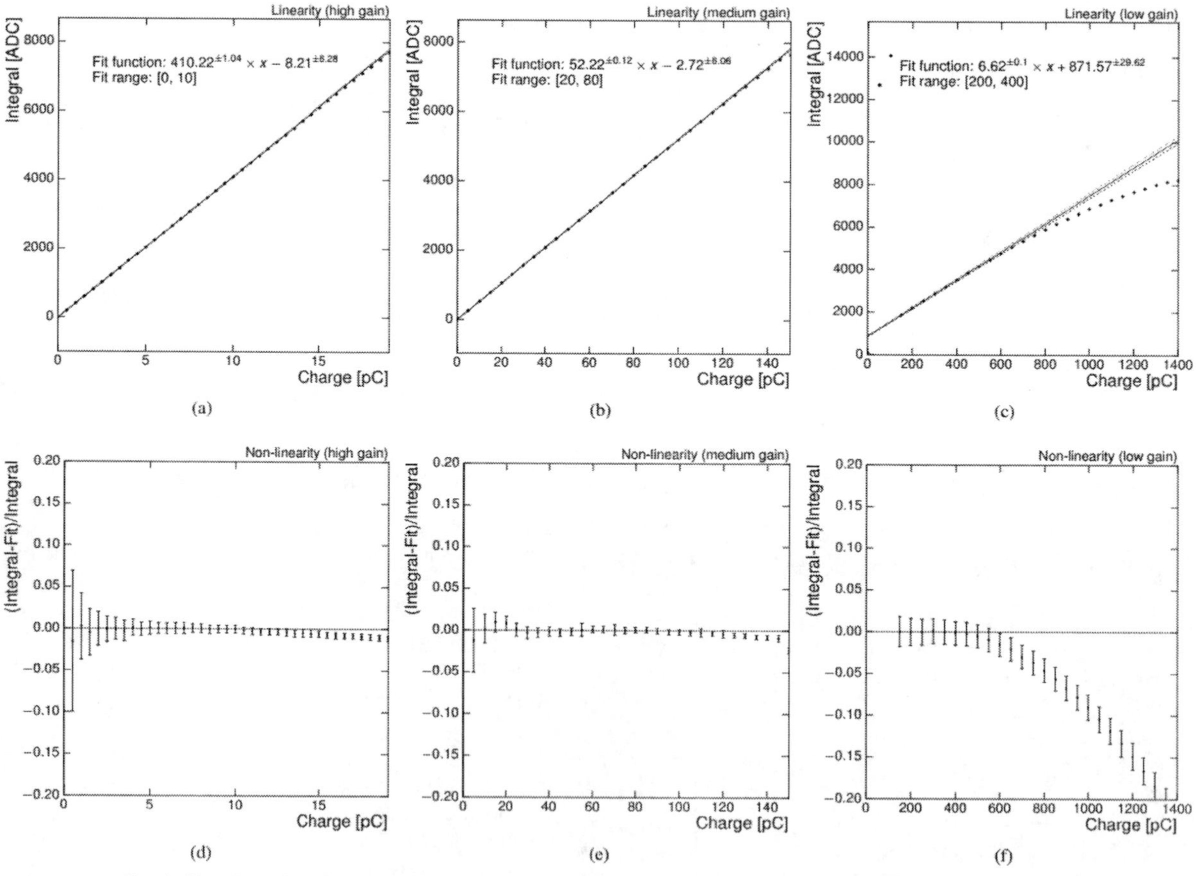

Fig. 4. Linearity and non-linearity of the previous design (FATALIC4b) for (a, d) high, (b, e) medium and (c, f) low gains.

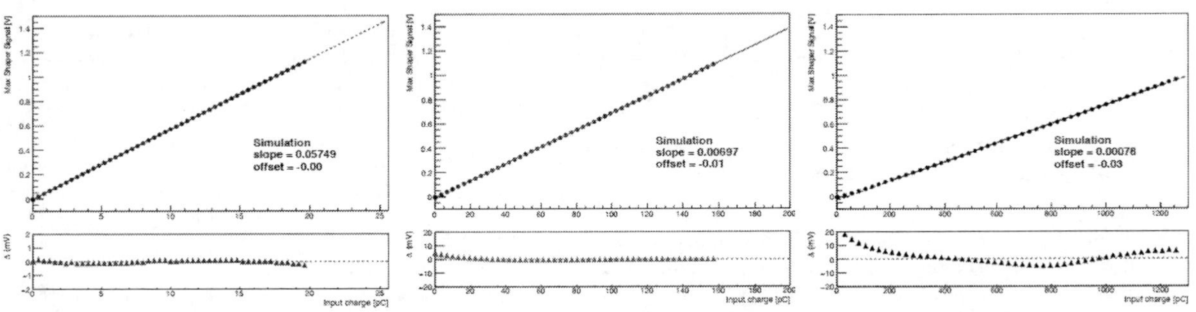

(a) (b) (c)

Fig. 5. Linearity of the latest design (FATALIC5) for (a) high, (b) medium and (c) low gains (simulation results). $\Delta(\mathrm{mV})$ in the bottom plots indicates the absolute difference between the shaper signal and the linear fit. 1 mV corresponds to 0.1% relative difference with respect to the whole dynamic range.

algorithm, the dependence of S_i on τ can be linearised via Taylor's expansion at first order:

$$S_i \simeq p + Ag(t_i) - A\tau g'(t_i) + n_i = p + Ag_i - A\tau g'_i + n_i \quad (2)$$

Three sets of free parameters a_i, b_i and c_i are defined, called OF weights, such as:

$$A = \langle \sum_{i=1}^{N} a_i S_i \rangle = \sum_{i=1}^{N} a_i \langle S_i \rangle \quad (3)$$

where N is the number of samples.

The OF procedure aims at minimizing the variance of the linear combinations of these parameters subject to a set of constraints. These constraints lead to a set of linear equations [5], which prompt the derivation of the OF weights and hence the amplitude, pulse time and pedestal can be calculated.

B. Optimal Filtering application to FATALIC

1) Analytical pulse shape: In order to validate the principle of the Optimal Filtering reconstruction method, a study was performed with the reference pulse shape taken from the analytic function of the following form:

$$V(t) = V_{\text{norm}} \times \frac{Erf(1 - e^{-\frac{t - t_{\text{up}}}{\tau_{\text{up}}}}) + 1}{2} \times e^{-\frac{t - t_{\text{down}}}{\tau_{\text{down}}}} + V_0 \quad (4)$$

where V_{norm} is the normalisation parameter, Erf is the Gaussian error function [6], t_{up} and τ_{up} are the pulse rise time parameters, t_{down} and τ_{down} are the pulse fall time parameters, and V_0 is the pedestal. This function approximates the pulse shape measured with scope (Fig. 3, a), and is represented in Fig. 6 where it is compared with the scope output. The shape is clearly asymmetrical, making it different from the pulses of other front-end options (e.g. 3in1 [7]).

2) Construction of pulses: Before the OF method is tested in real conditions, it is important to validate its performance using known sampled pulses. In this study, to produce a sampled digitised pulse, the input reference pulse shape (Fig. 6) is normalised and sampled every 25 ns which is the LHC clock period. Then the normalised pulse is multiplied by the energy

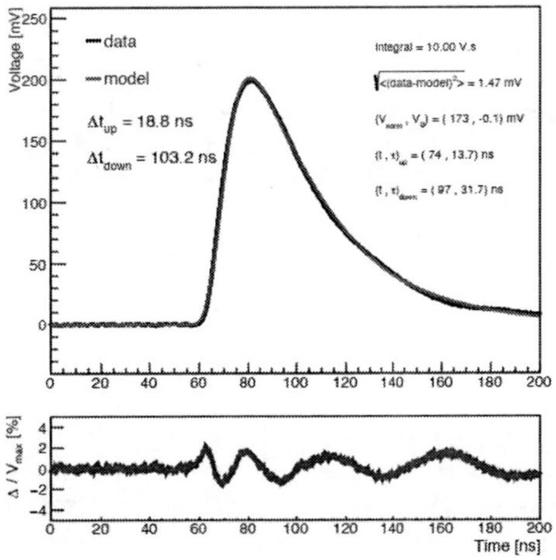

Fig. 6. Analytical pulse shape ("model") compared with the scope output ("data").

factor and added with a fixed typical pedestal value. The energy factor is obtained from a known conversion rate of the input charge to the energy deposited in a TileCal cell, which is taken from a uniform distribution. The noise factor is then added for each sample independently, uncorrelated between samples in the first approximation. The noise factors for each sample are considered as independent Gaussian distributions, RMS of which are later on referred to as noise levels.

As shown above, FATALIC layout comprises three different gains covering three dynamic ranges. While the medium gain is always present, high or low gain is selected depending on the saturation of either gain. For the purpose of validating the OF method with the analytical pulse shape whilst also simulating the gain switch behaviour, an effective digitised pulse is constructed by using the medium and alternative gain based on the following criteria:

- if both high and medium gains are saturated ($s_{\mathrm{HG}} > 4095$ ADC, $s_{\mathrm{MG}} > 4095$ ADC):
  - $s_{\mathrm{eff}} = 64 \times s_{\mathrm{LG}}$ ADC
- if only high gain is saturated, and medium gain is not ($s_{\mathrm{HG}} > 4095$ ADC, $s_{\mathrm{MG}} < 4095$ ADC):
  - $s_{\mathrm{eff}} = 8 \times s_{\mathrm{MG}}$ ADC
- if neither high nor medium gain are saturated ($s_{\mathrm{HG}} <$ 4095 ADC, $s_{\mathrm{MG}} < 4095$ ADC):
  - $s_{\mathrm{eff}} = s_{\mathrm{HG}}$ ADC

Here s_{HG}, s_{MG}, s_{LG} correspond to the signal output yields of the high, medium and low gains, and s_{eff} is the resulting signal of an effective digitised pulse encoded in 18 bits. Fig. 7 shows the digitised pulses for the medium gain, alternative gain, and the effective 18-bit digitised pulse.

3) Performance results: The main performance criteria presented here is the pulse energy resolution, defined as follows:

$$\frac{\Delta E}{E} = \frac{E_{\mathrm{OF}} - E_{\mathrm{true}}}{E_{\mathrm{true}}}, \tag{5}$$

where E_{true} is the true energy, and E_{OF} is the energy reconstructed by the Optimal Filtering method.

The energy resolution plots for a typical LHC energy range from 100 MeV to 1.5 TeV are shown in Fig. 8 for two different scenarios: with and without the electronic noise. For the latter case, the typical noise levels are considered: 3.5 ADC ($\sim$8.4 fC) for the high gain, 1.5 ADC ($\sim$29 fC) for the medium gain, and 1.5 ADC ($\sim$230 fC) for the low gain. It can be seen that the energy resolution is kept within percentage level across four orders of magnitude of the signal energy.

IV. Results with Real Data

For the final validation of the Optimal Filtering method and the whole chain of the FATALIC readout system, the signal reconstruction performance was tested with real data collected during a preliminary beam test at CERN [8]. In Fig. 9 (a) the OF method is compared with the flat estimation method, which is the basic energy reconstruction corresponding to the simple summation of digitised samples. Clearly, the Optimal Filtering reconstruction provides a significant improvement of energy resolution compared to the simple summation approach. Fig. 9 (b) shows the two-dimensional plot of the energy reconstructed using the OF method for muon events recorded by two different channels of a single cell, and the expected correlation is observed accordingly.

V. Conclusion

We have presented an overview and performance of the FATALIC front-end readout option of the TileCal High-Luminosity LHC upgrade foreseen in 2024. The characterisation of the full chain was shown, as well as the signal reconstruction was validated. The Optimal Filtering signal reconstruction method was introduced, its adaptation to fully exploit the FATALIC three-range layout was presented and its performance results were given by using both analytical pulse shapes and the real data from the beam test at CERN.

References

[1] ATLAS Collaboration, "The ATLAS Experiment at the CERN Large Hadron Collider," *JINST*, vol. 3, p. S08003, 2008.

[2] *ATLAS tile calorimeter: Technical Design Report*, 1996, no. CERN-LHCC-96-042. [Online]. Available: https://cds.cern.ch/record/331062

[3] L. Rossi and O. Brüning, "High Luminosity Large Hadron Collider A description for the European Strategy Preparatory Group," Tech. Rep. CERN-ATS-2012-236, 2012. [Online]. Available: https://cds.cern.ch/record/1471000

[4] L. Royer, "FATALIC: A Dedicated Front-End ASIC for the ATLAS TileCal Upgrade," Oct 2015. [Online]. Available: https://cds.cern.ch/record/2057101

[5] E. Fullana, J. Castelo, V. Castillo, C. Cuenca, A. Ferrer, E. Higón, C. Iglesias, A. Munar, J. Poveda, A. Ruiz-Martinez, B. Salvachúa, C. Solans, R. Teuscher, and J. Valls, "Optimal Filtering in the ATLAS Hadronic Tile Calorimeter," Tech. Rep. CERN-ATL-TILECAL-2005-001, 2005. [Online]. Available: https://cds.cern.ch/record/816152

[6] E. W. Weisstein, "Erf. From MathWorld—A Wolfram Web Resource," last visited on 24/5/2017. [Online]. Available: http://mathworld.wolfram.com/Erf.html

[7] K. Anderson *et al.*, "Design of the front-end analog electronics for the ATLAS tile calorimeter," *Nucl. Instrum. Meth.*, vol. A551, pp. 469–476, 2005.

[8] M. Volpi, L. Fiorini, and I. Korolkov, "Time inter-calibration of hadronic calorimeter of ATLAS with first beam data." Tech. Rep. ATL-TILECAL-INT-2009-003, Nov 2009. [Online]. Available: https://cds.cern.ch/record/1223049

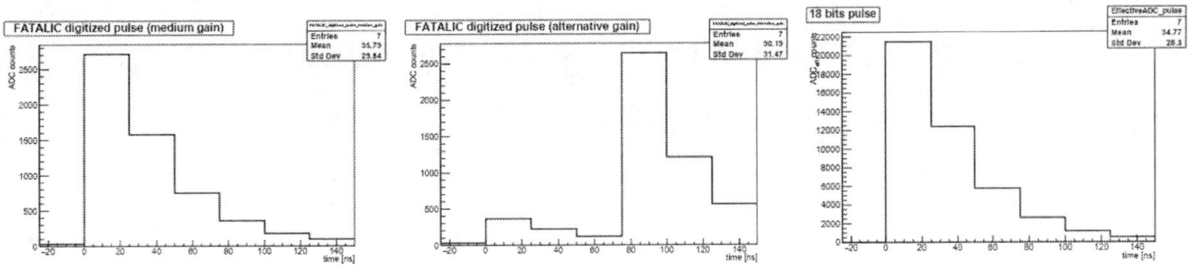

(a) (b) (c)

Fig. 7. FATALIC digitised pulse for (a) medium gain, (b) alternative gain, and (c) effective 18-bit digitised pulse.

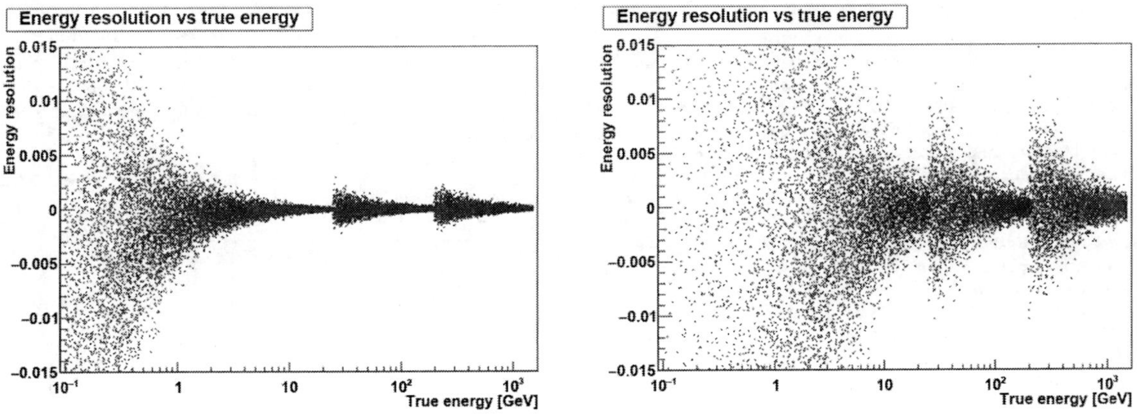

(a) (b)

Fig. 8. FATALIC energy resolution with respect to the true pulse energy using analytical pulse shape: (a) without electronic noise, (b) with typical noise.

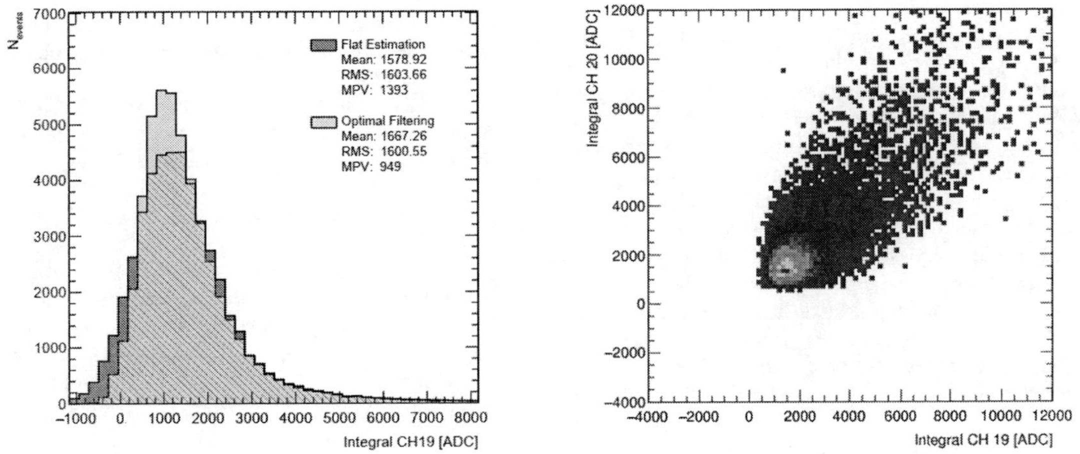

(a) (b)

Fig. 9. Energy reconstructed using the real test beam data: (a) comparison of the OF method with the simple flat estimation method, (b) energy reconstructed using the OF method for muon events recorded by two different channels of a single cell.

The Use of 3D Printing in the Development of Gaseous Radiation Detectors

Sam Fargher, Chris Steer, and Lee Thompson

Abstract—**Fused Deposition Modelling has been used to produce a small, single wire, Iarocci-style drift tube to demonstrate the feasibility of using the Additive Manufacturing technique to produce cheap detectors, quickly. Recent technological developments have extended the scope of Additive Manufacturing, or 3D printing, to the possibility of fabricating Gaseous Radiation Detectors, such as Single Wire Proportional Counters and Time Projection Chambers. 3D printing could allow for the production of customisable, modular detectors; that can be easily created and replaced and the possibility of printing detectors on-site in remote locations and even for outreach within schools.**

The 3D printed drift tube was printed using Polylactic acid to produce a gas volume in the shape of an inverted triangular prism; base length of 28 mm, height 24.25 mm and tube length 145 mm. A stainless steel anode wire was placed in the centre of the tube, mid-print. P5 gas (95% Argon, 5% Methane) was used as the drift gas and a circuit was built to capacitively decouple signals from the high voltage. The signal rate and average pulse height of cosmic ray muons were measured over a range of bias voltages to characterise and prove correct operation of the printed detector.

Index Terms—**3D Printing, Additive Manufacturing, Gaseous radiation Detectors, Drift Tubes.**

I. Introduction

ADDITIVE Manufacturing (AM), or 3D printing, is a technique used to fabricate 3D models from CAD files by fusing together successive layers of material. AM was developed in the 1980's as a method of producing basic prototypes, however the available processes were expensive and slow. Recent years have seen new technologies and developments occur that have extended the scope of AM. Improved speed and accuracy of AM processes have enabled fast production of high quality, functional products; and an increase in the range of printable materials, such as flexible and electrically conductive thermoplastics, has allowed more complex designs and geometries to be fabricated. Commercialisation has also allowed Rapid Prototyping (the iterative process of trial and error to produce a working design) to become a cost-effective, accessible and convenient production method.

A. Motivations

In 2013 M. Hohlmann [1] posed a "Grand Challenge" to the HEP instrumentation community to use AM to produce particle detectors. In this white paper, Hohlmann shows the current AM capabilities still do not meet the requirements to 3D print a fully-functional MicroPattern Gas Chambers, including a Gas Electron Multiplier, that have the spatial resolution needed for applications in HEP experiments. However, current AM techniques do have the necessary performance to allow Rapid Prototyping to be used in the R& D phase of building such Gaseous Radiation Detectors (GRDs) and in the production of GRDs of a simpler design, such as Single Wire Proportional Counters (SWPCs) and Time Projection Chambers (TPCs).

	Current capability	Performance goal
Printing resolution in x-y	~ 75 μm	~ 1 μm
Layer thickness in z	~ 50 μm	~ 1 μm
Print speed	10 cm/s	> 100 cm/s
Materials	Either polymers or metals	Polymer-metal composites
Object size	50 cm × 50 cm × 25 cm	200 cm × 100 cm × 10 cm

Fig. 1. A table comparing commercial AM capabilities to the performance needed to 3D print complete gaseous radiation detectors such as MicroPattern Gas Chambers (containing GEMs), for high spatial resolution detectors required in HEP experiments [1].

Sam Fargher and Lee Thompson were with the Department of Physics and Astronomy, University of Sheffield, UK.
Chris Steer was with AWE plc, Aldermaston, UK.
Manuscript received July 15, 2017

II. 3D Printing Wired Gaseous Radiation Detectors

Fused Deposition Modelling (FDM) is the main AM technique available to to the authors. This process works by extruding thin lines of molten thermoplastic, in the x-y plane, onto a base plate to produce a "slice" of a model. The baseplate then moves downwards in the z-axis and another "slice" is printed onto the previous, thermally fusing as the plastic is printed. This process is repeated until a 3D model is produced.

An Iarocci-style drift tube was chosen to be the first type of GRD to be printed as this is of the simplest design. Iarocci tubes are often cylindrical, a few cm in diameter, with an anode wire of diameter $\sim 100\ \mu m$, and a resistive cathode (typically graphite) coating on the inside. This internal resistive cathode allows for an external pick-up electrode to be used, allowing for easier detector construction [2]. The initial challenges of 3D printing a the drift tube using FDM are producing a hollow enclosure, as the drift tube 'roof' would have to be printed over empty space; introducing a taut anode wire in the correct position, which will remain taut; and ensuring the enclosure is airtight.

Acrylonitrile butadiene styrene (ABS) and Polylactic acid (PLA) were investigated as possible thermoplastics to print the enclosure, with a variety of different drift tube enclosure designs. Two printers were used; the Makerbot Replicator 2X for the ABS and the re:3D Gigabot (Fig. 2) for the PLA. Each

Fig. 2. Picture of the re:3D Gigabot at The University of Sheffield. The Gigabot is a large scale commercial FDM printer, with a possible print volume of dimensions 590 x 600 x 600 mm^3.

Fig. 3. Cross-section of the OpenSCAD drift tube design with gas inlet. Open sections at bottom left and right are connected to the main, central triangular prism to allow gas to flow within but these sections do not contribute to any signals.

drift tube was designed using OpenSCAD to produce an .stl file which was then converted, via slicing software, into printable .x3g or .GCODE files, depending on the printer used. The .x3g file format is a binary file used by the Makerbot and the .GCODE file format is a Computer Numerical Control (CNC) file used by the Gigabot. The .GCODE files have a significant advantage over the .x3g as GCODE is a list of commands followed by the printer; for 3D printing each successive layer of the print has its own section of code, allowing the possibility of editing the .GCODE before printing, to customise certain layers. From the tests, ABS was found to warp too easily and provide poor overhang formation for the majority of the enclosure designs, however PLA produced stable prints and excellent overhang formation for all tested designs and an inverse triangular prism drift tube was selected to be continued with. These results, along with the advantages that come with using .GCODE, meant using PLA with the Gigabot henceforth.

Having found PLA to be the best thermoplastic to use, a new drift tube was created using OpenSCAD. This drift tube had a base length of 28 mm, height 24.25 mm and tube length 145 mm and included gas inlet valves (see Fig. 3 for cross-section).

A frame was also designed, and 3D printed with the Gigabot, to hold an anode wire at high tension. The height of the frame was designed to be half the height of the drift tube, allowing the wire to run through the centre of the drift tube. A stainless steel wire, ($r = 50\mu m$), was secured to the frame using Araldite while using a 300g load to supply the correct tension. Although stainless steel is not an ideal metal for a drift tube anode wire, it was the easiest available wire for these tests.

Four pegs, each (10 x 10 x 5) mm^3, were added to the drift tube model on OpenSCAD, which corresponded to the inner corners of the wire frame (see Fig. 4). These pegs ensure the wire frame can only be placed in the correct position to leave the anode wire running central down the tube.

The drift tube .GCODE was generated, usingSimplify3D slicing software, and subsequently edited to pause the process mid-print, at the height of the wire frame; lower the baseplate; and wait for 120 seconds, allowing enough time for the frame holding the anode wire to be introduced. Once in place, the 3D printing process continued, depositing molten plastic over the anode wire, securing it in place. The drift tube walls were printed with a 20% infill, however this amount of plastic was not sufficient to grip the wire and maintain the tension required for operation once the wire was cut free from the frame. Instead of increasing the infill of the entire drift tube, the .GCODE corresponding to the 5 printed layers, above and below the anode wire, were replaced with the .GCODE generated with 100% infill, which gripped the anode wire sufficiently.

Once printed, the anode wire was cut free from the frame, leaving a hollow drift tube with a central anode wire. Each end of the anode wire was soldered to an SHV connector (with one end connected via a 1MΩ resistor) fixed to 3D printed end caps to provide a barrier to the bare wire, which would be connected to a high voltage (see Fig. 5). The drift tube was then coated with copper shielding spray to provide RF protection and act as an external cathode (grounded) as an internal cathode has not yet been introduced into the drift tube design.

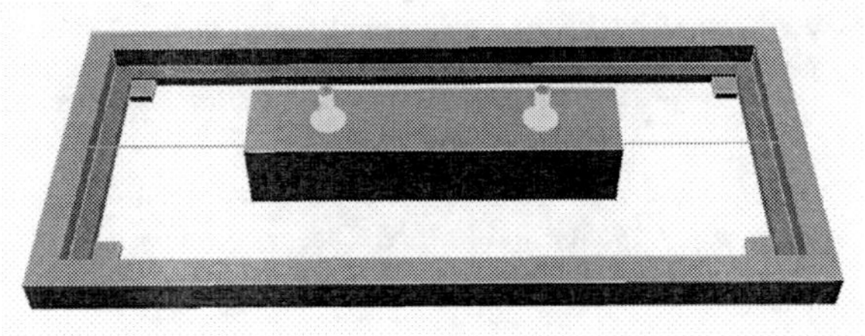

Fig. 4. OpenSCAD design of drift tube, wire rig and pegs, to allow the wire (red) to be correctly placed within the drift tube. Although all designs are collected here, the drift tube and pegs are one design and the wire rig is another design to be printed separately. The wire (red) is not part of either model but has been added to show position.

A simple circuit was built to take the output of the drift tube and capacitively decouple any signals from the bias voltage before being passed to an oscilloscope. An airtight metal box was modified to take the gas output of the drift tube and monitor the oxygen level, using a SGX oxygen sensor, as a measure of contaminants in the drift gas. The oxygen monitor box, circuit enclosure and circuit diagram can be seen in Fig. 5.

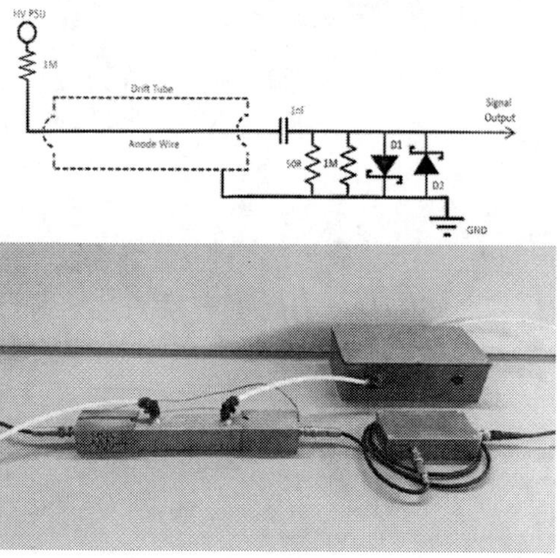

Fig. 5. UPPER: Circuit diagram of signal decoupling circuit. The 50 Ω resistor was included to better match the impedance of the signal output system. Components D1 and D2 are Schottky Diodes included as a precaution, to limit the signal amplitude and protect the oscilloscope electronics. LOWER: Picture of completed drift tube connected to signal decoupling circuit (small metal box) and oxygen sensor (large metal box).

III. CHARACTERISATION

To characterise the drift tube, the average pulse height and signal rate was determined for a range of applied bias voltages. P5 gas (95% Argon 5% Methane) was flowed through the drift tube at a constant flow rate of 0.2 litres/min and the oxygen level monitored and maintained below 0.0025%. The temperature of the gas was maintained at 299.0 ± 1.5 K. A range of bias voltages were applied to the anode wire via a V6533P 6 Channel VME Programmable HV Power Supply and the raw signals, decoupled from the bias voltage, were passed onto a Teledyne LeCroy WaveRunner Oscilloscope to record the signal pulse heights and signal rate. Cosmic ray muons were used to produce signals in the drift tube and a minimum of a few thousand signals were recorded for each of the applied bias voltages.

Fig. 6 shows the variation of signal rate against the bias voltage applied to the anode wire. These results were obtained from a relatively low statistics data run, however, the graph does show that the signal rate generally increases with bias voltage; as expected with GRDs. It should be noted that although the drift tube has been operated at high voltages, uncommon for most GRDs, these voltages are typical for Iarocci tube operation.

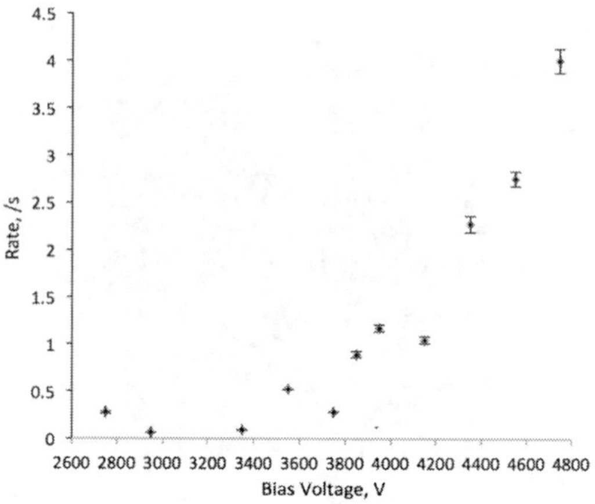

Fig. 6. Graph of average signal rate against applied bias voltage and errors.

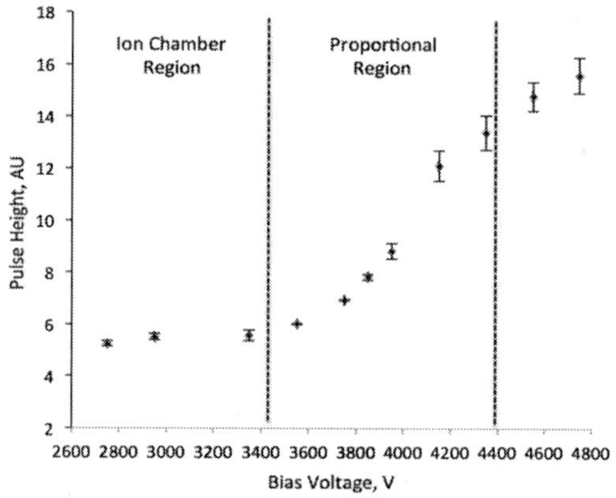

Fig. 7. Graph of pulse heights against applied bias voltage and with errors. The assumed regions of operation are labelled. The pulse heights are in arbitrary units as the raw signals are used, and a integrator and shaper need to be used to fix appropriate units to values.

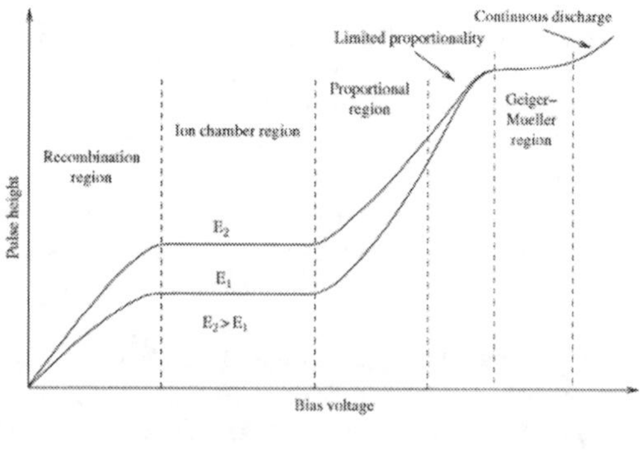

Fig. 8. Graph depicting the typical variation in average pulse height, with bias voltage, found for GRDs. The operation regions are also labelled [3].

Fig. 7 shows the variation in average pulse height against the bias voltage applied to the 3D printed drift tube anode wire, and the typical variation of average pulse height, with bias voltage, expected for GRDs are shown in Fig. 8. Currently, the range of bias voltages that the drift tube has been operated with is not sufficient to provide a clear correlation between the data and the expected variation, however, the general shape of the measured curve demonstrates the expected behaviour at lower voltages and the assumed regions of operation have been labelled on the graph, i.e. Ion Chamber Region and Proportional Region. As noted in the figure caption, only the raw signals have been used and the next stage is to add a pulse integrator and shaper to provide better signals for analysis.

IV. Conclusion

A simple Iarocci-style drift tube was 3D printed, proving the concept that simple GRDs can be 3D printed at current AM capabilities. The characterisation of the drift tube is underway and the results appear to show the drift tube is operating correctly; giving reason to continue investigating AM as a means of producing GRDs. Upon completion of characterising the current 3D printed drift tube, the next stage of this project is to produce a method of introducing an internal cathode into the drift tube, to allow readout. If this is successful, it is hoped that the project will move forward to developing a method of printing more complex GRDs, such as Time Projection Chambers.

References

[1] M. Hohlmann, "Printing out Particle Detectors with 3D-Printers ? a Potentially Transformational Advance for HEP Instrumentation", in APS-DPF Snowmass Summer Study, The Division of Particles and Fields of the American Physical Society, 2013.

[2] E. Iarocci, "Plastic streamer tubes and their applications in high energy physics", Nuclear Instruments and Methods in Physics Research, vol. 217, no. 1-2, pp. 30-42, 1983.

[3] S. Ahmed, "Physics And Engineering Of Radiation Detection", [S.l.]: Elsevier, 2007.

https://doi.org/10.1051/epjconf/201817001017

The role of 14 MeV neutrons in light element nucleosynthesis

V. Valkovic, D. Sudac, and J. Obhodas

***Abstract*—The 14 MeV neutron induced nuclear reactions which might contribute to better understanding of so-called "cosmological lithium problem" have been discussed. The BBN theory predicted ^{7}Li overestimated value could be lowered by ^{2}n induced reactions on Li-isotopes and/or ^{7}Be destruction via resonance in d+^{7}Be→^{9}B process.**

Proposals for ^{6}Li+^{2}n and ^{7}Li+^{2}n reactions cross section evaluations and measurements are discussed in some details.

We discuss also a preliminary measurements of n-n coincidences from the ^{10}B(n,2n)^{9}B reaction with two neutron detectors placed outside the cone of 14 MeV tagged neutron beam. From the time-of-flight measurements and known geometry we deduce the neutron energies and calculate the ^{9}B missing mass spectrum.

***Index Terms*—astrophysics, astrochemistry, boron, gamma-rays, neutrons, nuclear physics, scintillation counters.**

I. Introduction

THE 14 MeV neutrons are produced in the ^{2}H+^{3}H→^{4}He+n nuclear reaction, (Q=17.590 MeV). In spite of large cross sections, both for their production and their interactions with light nuclei, their role in light element nucleosynthesis has not been studied in great many details.

By now it is accepted that the Big Bang nucleosynthesis, BBN, starts with the p-p chain with ^{1}H + ^{1}H → ^{2}H + e$^+$ + ν_e nuclear reaction. According to the present model all neutrons end up bound in the most stable light element ^{4}He. Heavier nuclei do not form in any significant quantity both because of the absence of stable nuclei with mass number 5 or 8. As a result Solar system shows very low relative abundance of Li, Be and B with respect to other elements.

One of the most important unresolved problem in nuclear astrophysics is so-called "cosmological lithium problem". It refers to the large discrepancy between the abundance of primordial ^{7}Li predicted by the standard BBN theory and the value inferred from the so called "Spite plateau" in halo stars. In fact, the predictions of the BBN theory reproduce successfully the observations of all primordial abundances except for ^{7}Li, which is overestimated by more than a factor of three.

V. Valkovic was with the Institute Ruder Boskovic, Zagreb, Croatia until his retirement in 2004 (e-mail: valkovic@ irb.hr).

D. Sudac is with the Institute Ruder Boskovic, Zagreb , Croatia (e-mail: dsudac@irb.hr).

Jasmina Obhodas is with the Institute Ruder Boskovic, Zagreb, Croatia (e-mail: jobhodas@irb.hr).

Among the possible resolutions to this discrepancy are (i) ^{7}Li depletion in the atmosphere of stars, (ii) systematic errors originating from the choice of stellar parameters – most notably the surface temperature and (iii) systematic errors in the nuclear cross sections used in the nucleosynthesis calculations.

We propose Li-isotopes destruction by the following series of nuclear reactions:

(i) ^{6}Li+^{2}n→^{8}Li→^{8}Be+e+$\bar{\upsilon}$ followed by ^{8}Be→2α.

(ii) ^{7}Li+^{2}n→^{9}Li→^{9}Be+e+$\bar{\upsilon}$ followed by ^{9}Be→n+2α

Where the ^{2}n symbol represent neutron-neutron final state interaction (fsi) as observed in the n+d→p+n+n reaction for E_n = 14 MeV bombarding energy. (Nuclear reactions ^{1}H+^{2}H→^{3}H+e$^+$+ν_e and/or ^{2}H+^{2}H→^{3}H+^{1}H can than lead to 14 MeV neutron producing reaction ^{2}H+^{3}H→^{4}He+n). During BBN, in the period of transition from quark-gluon plasma to baryonic matter, probability of ^{2}n formation [n-n fsi or n-n bound state, if existed at that time] was increased. This would have led to Li distraction via reactions (i) and (ii).

Since the primordial ^{7}Li is mainly produced by β-decay of ^{7}Be ($t_{½}$=53.2 days) the abundance of ^{7}Li is essentially determined by the production and destruction of ^{7}Be. The neutron induced reactions can also play a role in the destruction of ^{7}Be, in particular the ^{7}Be(n,α)^{4}He reaction in the energy range of interest for BBN, in particular between 20 and 100 keV. Recent activity in solving the "lithium problem" in big bang nucleosynthesis has focused on the role that putative resonances may play in resonance-enhanced destruction of ^{7}Li. Particular attention has been paid to the reactions involving ^{9}B compound nuclear system d+^{7}Be→^{9}B.

II. The ^{2}N-induced nuclear reactions

The neutron-neutron scattering length of a_{nn} = -21.7 F deduced from the measured proton spectrum from the n+d→p+n+n reaction at E_n = 14 MeV bombarding energy (as shown in Fig.1, from ref. [1]) allows the two neutrons to interact with target nucleus as a single projectile. Dominant contribution to "f" is coming from the -1/a_{nn} term in the effective range expansion. In the repeated measurement [2] the pure nuclear value for the neutron-neutron $^1S_0$ scattering length is found to be a_{nn}=-22.5±1 fm. Scattering amplitude is a sum of partial wave amplitudes which are defined by scattering phase shifts. Let us remind that scattering length characterizes effective size of the target.

The support for this approach comes from the observed dineutron decay of halo nuclei and structure and decay correlations of the neutron systems beyond dripline. Nuclei studied by Kohley et al. [3]: ^{16}Be, ^{13}Li, ^{10}He and ^{26}O. The correlations of the 3-body decay for the ^{16}Be and ^{13}Li were extracted and demonstrated a strong enhancement between two neutrons. ^{16}Be and ^{13}Li exhibit direct two-neutron decay since a sequential decay is energetically forbidden. Two-neutron radioactivity was observed in the case of ^{26}O ground state.

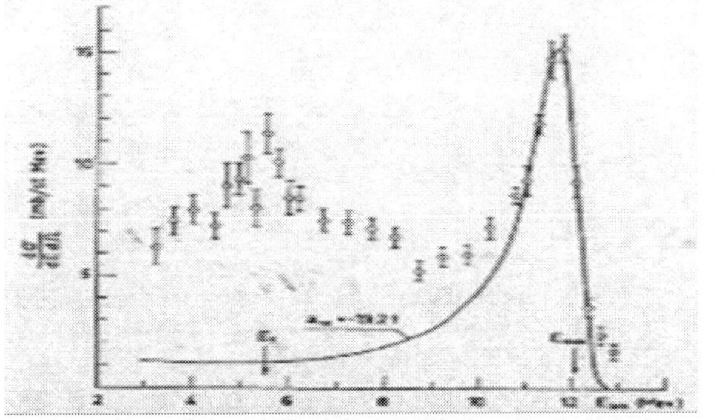

Fig 1: The experimental data for D(n,p)2n at 14.4 MeV at 4° in the laboratory system compared with the calculations for a_{nn} = -19.2 fm after smearing and normalization [1]. This corresponds to the case when two neutrons are moving together with zero relative energy backwards in the center-of-mass system. E1 indicates the proton energy in the case where a neutron and proton are going together with zero relative energy in the forward direction in the center-of mass system.

First observation of ground state dineutron decay: ^{16}Be. In experiment [4] ^{16}Be ground state was populated via single proton knockout reaction from ^{17}B. ^{16}Be is bound with respect to the emission of one neutron and unbound to two-neutron emission. The dineutron character of the decay is evident by a small emission angle between the two neutrons. The two-neutron separation energy of ^{16}Be was measured to be 1.35 MeV – agreement with shell model calculations. ^{16}Be → ^{14}Be$_{g.s.}$ + 2n. From the time reversal considerations dineutron induced reactions should be feasible.

A. Experiment #1

The neutron-neutron scattering length of a_{nn} = -21.7 F deduced from the measured proton spectrum allows the two neutrons to interact with target nucleus as a single projectile.
Reactions of interest:
A step which might be of interest is

$$^{1}H + ^{2}H \rightarrow ^{3}H + e^{+} + \nu_e$$

or

$$^{2}H + ^{2}H \rightarrow ^{3}H + ^{1}H \quad (Q=4.03\ MeV)$$

which can lead to one or more neutron producing reactions

$$^{2}H + ^{2}H \rightarrow ^{3}He + n \quad (Q=2.45\ MeV)$$

$$^{2}H + ^{3}H \rightarrow ^{4}He\ (3.5\ MeV) + n\ (14.1\ MeV) \quad (Q=17.590\ MeV)$$

$$^{3}H + ^{3}H \rightarrow ^{4}He + n + n \quad (Q=11.3\ MeV)$$

Reaction ^{2}H+^{3}H→^{4}He+n has peak in the total cross section (σ_T = 5 b) at approximately deuteron energy of 100 KeV (T=1.16x10^{9} °K), while reaction ^{2}H+^{2}H→^{3}He+n total cross section is peaked at around deuteron energy of 2.0 MeV. During BBN, in the period of transition from quark-gluon plasma to baryonic matter, probability of 2n formation [n-n fsi or n-n bound state (?)] was increased. This would have led to Li distraction via reactions shown in Fig.2.

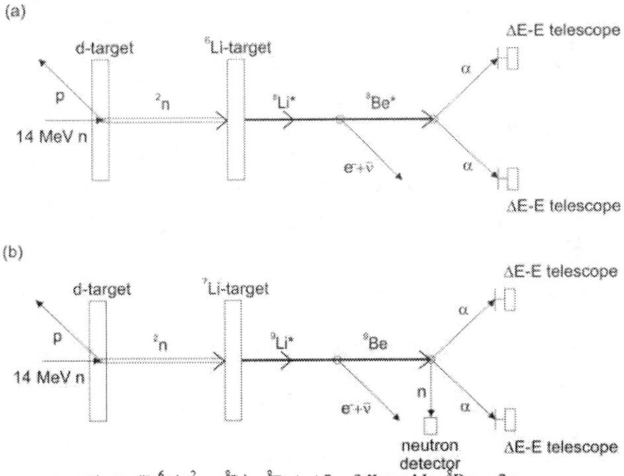

Fig.2: (i) ^{6}Li+^{2}n→^{8}Li→^{8}Be+e+ῡ followed by ^{8}Be →2α.
(ii) ^{7}Li+^{2}n→^{9}Li→^{9}Be+e+ῡ followed by ^{9}Be → n+2α.

B. Other Reactions

In real experiment one needs to bring ^{2}H and Li targets / nuclei as close as possible. One possibility is using lithium hydrides ^{6}Li^{2}H and ^{7}Li^{2}H. Lithium hydride has chemical formula LiH and the crystal structure fcc (NaCl type) as shown in Fig. 3.

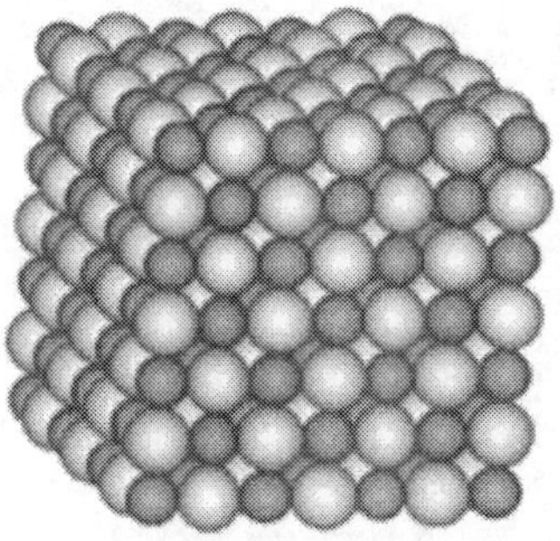

Fig.3: The crystal structure of lithium hydride with lattice constant a=0.40834 nm..

Another possible target is ^{9}Be as indicated in Fig.4.

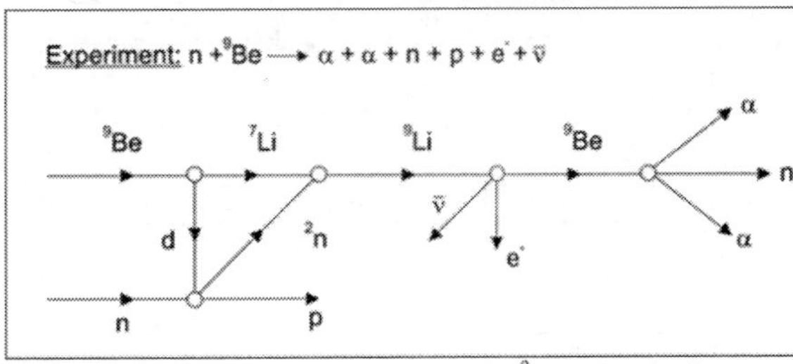

Fig. 4: Incoming neutron interacting with d in ^{9}Be nucleus leading to ^{7}Li+^{2}n interaction and sequence of decays as indicated.

III. ^{10}B(N,2N)^{9}B REACTION

Since the primordial ^{7}Li is mainly produced by β-decay of ^{7}Be (t½=53.2 days) the abundance of ^{7}Li is essentially determined by the production and destruction of ^{7}Be. The neutron induced reactions can also play a role in the destruction of ^{7}Be, in particular the ^{7}Be(n,α)^{4}He reaction in the energy range of interest for BBN, in particular between 20 and 100 keV. Recent activity in solving the "lithium problem" in Big bang nucleosynthesis has focused on the role that putative resonances may play in resonance-enhanced destruction of ^{7}Li. Particular attention has been paid to the reactions involving ^{9}B compound nuclear system d+^{7}Be→^{9}B.

The concept of mirror nuclei is well established and mirror pairs like ^{7}Li-^{7}Be, ^{13}C-^{13}N, ^{15}N-^{15}O, ^{17}O-^{17}F and ^{19}F-^{19}Ne are known to have nearly identical energy level schemes. The properties of the mass-9 system, in which the ^{9}B partner is particle unbound, even in the ground state, have been difficult to determine; the ground state is unbound to breakup into p+^{8}Be by 186 keV. There has been a large theoretical and experimental effort directed towards predicting and observing the low-lying states of ^{9}B, especially the first excited ½$^{+}$ state (see Table I).

The unbound ½$^{+}$ state at 1.68 MeV in the mirror ^{9}Be has been known for many years and yet the existence and properties of the state in ^{9}B are not clear. The state is hard to define because it is difficult to excite and very broad. Furthermore, it exists amongst much more intensely populated peaks with large widths [1-2]. Different nuclear reactions like ^{9}Be(^{6}Li,^{6}He)^{9}B, ^{9}Be(^{3}He,t)^{9}B etc., were used to investigate the first-excited ½$^{+}$ state of ^{9}B, but nevertheless the ½$^{+}$ state remains elusive.

TABLE I
EXPERIMENTAL DATA AND THEORETICAL CALCULATIONS OF 1ST EXCITED STATE IN ^{9}B

Reference	Reaction	First excited state in ^{9}B	
		E_x (MeV)	Γ (MeV)
Marion and Levin	^{9}Be(p,n)^{9}B	1.4	1
Saji (1960)	^{9}Be(p,n)^{9}B	1.4	1
Spencer, Floyd and Young (1960)	^{10}B(^{3}He,α)^{9}B	Did not find	Did not find
Symons and Treacy (1962)	^{12}C(p,α)^{9}B	1.7±0.2	1
Bauer, Anderson and Wong (1964)	^{9}Be(p,n)^{9}B	Did not find	Did not find
Teranishi and Furubayashi (1964)	^{9}Be(p,n)^{9}B	1.7	-
Farrow and Hay (1964)	^{10}Be(p,d)^{9}B	Did not find	Did not find
Islam and Tracy (1965)	^{12}C(p,α)^{9}B	1.50±0.05	broad
Slobodrian et al. (1967)	^{9}Be(p,n)^{9}B	1.4	-
Kroepfl and Browne (1967)	^{9}Be(^{3}He,t)^{9}B	1.5	0.7
Anderson et al. (1970)	^{9}Be(p,n)^{9}B	1.4	0.0
Gul et al. (1970)	^{7}Li(^{3}He,n)^{9}B	Did not find	Did not find
Chou et al. (1978)	^{9}Be(p,n)^{9}B	Did not find	Did not find
Byrd et al. (1983)	^{9}Be(p,n)^{9}B	Did not find	Did not find
Sherr and Bertsch	calculation	0.9	1.4
Kadija et al. (1987)	^{9}Be(^{3}He,t)^{9}B	1.16±0.5	1.0±0.2
Burlein et al. (1988)	^{9}Be(^{6}Li,^{6}He)^{9}B	1.32±0.08	0.86±0.26
Arena et al. (1988)	^{10}B(^{3}He,α)^{9}B	1.8±0.2	0.9±0.3
Catford et al. (1992)	^{9}Be(^{6}Li,^{6}He)^{9}B	Did not find	Did not find
Efros and Bang (1999)	calculation	1.1	1.5
Akimune et al. (2001)	^{10}B(^{3}He,α)^{9}B	1.8±0.22	0.6±0.3
Scholl et al. (2011)	^{9}Be(^{3}He,t)^{9}B	1.85±0.13	0.7±0.27
Baldwin et al. (2012)	^{6}Li(^{6}Li,t)^{9}B	0.8-1.0	1.5
Karki (2013)	^{9}Be(p,n)^{9}B	Did not find	Did not find
Fortune and Sherr (2013)	^{9}Be(^{6}Li,^{6}He)^{9}B	1.27(1.31)	1.38

Table II shows the first few excitation levels of ^{9}B. A very broad the first excited level at 1.5 MeV (?) lies between two intensive peaks at 0 MeV and 2.345 MeV.

TABLE II
THE FIRST FEW EXCITED STATE IN ^{9}B

E_{level} (keV)	J^{π}	$T_{1/2}$
0	3/2-	0.54 keV
1500 ?		≈ 1.2 MeV
2345	5/2-	81 keV
2751	5/2+	614 keV

Our approach to the problem consists using the experimental setup sketched in Fig.5. The 14 MeV neutron beam is produced in the ^{3}H(d,n)^{4}He nuclear reaction. The n-n coincidences from the ^{10}B(n,2n)^{9}B reaction are measured by two neutron detectors placed outside the cone of 14 MeV tagged neutron beam. From the time-of-flight measurements (n-detector start, α-detector stop) and known geometry one can deduce the neutron energies and calculate the ^{9}B missing mass spectrum. This could lead to much more precise information about the low lying states of ^{9}B.

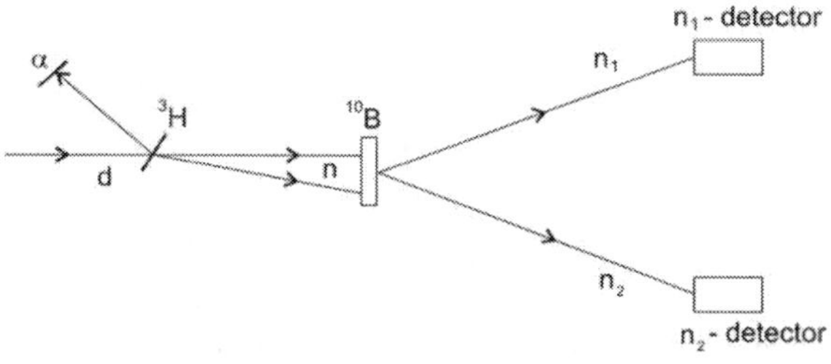

Fig.5: Sketch of experimental set-up for the measurement of ^{10}B(n,2n)^{9}B reaction.

A. Experimental set-up

The used experimental set-up is shown on Fig. 6. Two plastic scintillator 10cmx10cmx10 cm were used as neutron detectors, equipped with adequate photomultiplier tubes (PMT). API120 neutron generator (NG) produced by ThermoFisher was used as a neutron source. NG is equipped with the YAP:Ce scintillator which serves as an alpha detector. YAP:Ce is connected to the Hamamatsu R1450 PMT. Boron target was boron carbide, CB4, 106 g in mass, volume Φ55 mm x 50 mm, in plastic bottle. NG was tilted for 8° so that tagged neutron cone axis goes along the thick dashed line. Boron target center of mass was placed 22.5 cm from the NG tritium target inside the NG. Distance between the boron target center and neutron detector face was 50 cm. Exact position of the tagged neutron cone axis was found by 3''x3'' Ne213 neutron detector which was put in front of NG at two different distances from the tritium target. Ne213 was moved horizontally and vertically. For each position tagged neutrons were counted and normalized to the same number of emitted alpha particles. Point at which maximum occurred was taken as a point on tagged neutron cone axis. Neutron detectors were placed outside the tagged neutron cone. Neutron beam intensity was kept fixed during the measurements and was around 5 x 10^6 n/s.

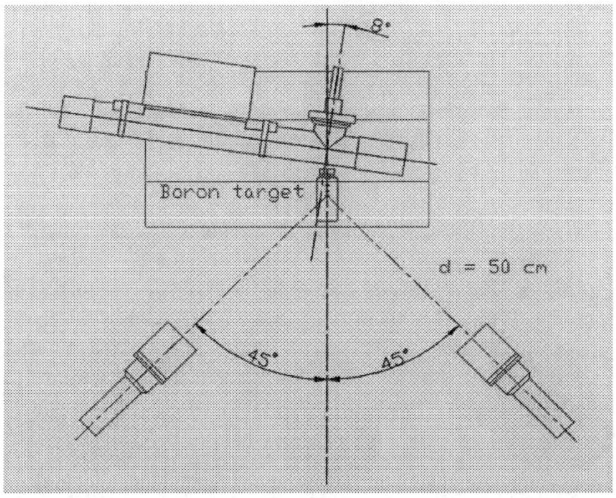

Fig. 6: Experimental set-up. As a neutron detectors two plastics scintillators 10 cm x 10 cm x 10 cm equipped with the available PMT (3"-12 stages 3M-P-E1-X-N) were used. Neutron detectors are called left and right from NG point of view.

Electronic set-up is shown on a Fig. 7. Fast output from the neutron detectors goes through Constant Fraction

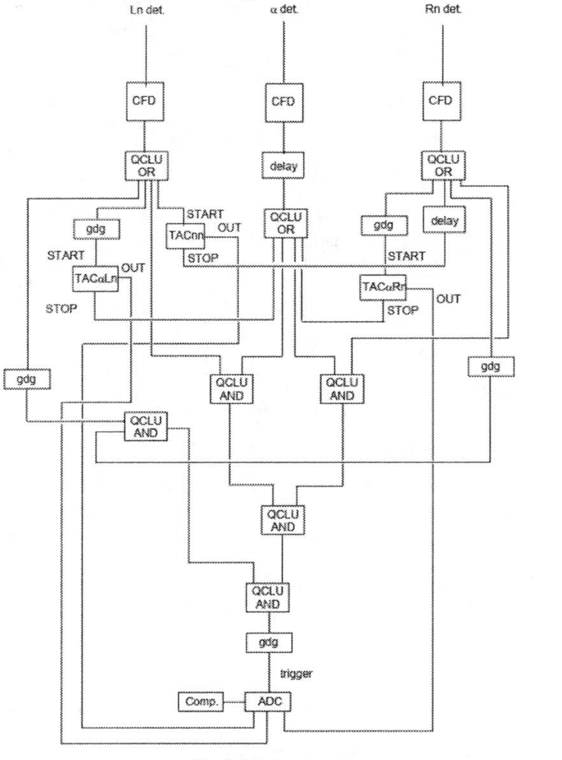

Fig. 7: Electronic set-up.

Discriminators (CFD) and splitting units to the START of the time-to-amplitude converters ($TAC_{\alpha Ln}$/ $TAC_{\alpha Rn}$), while STOP signal to the $TAC_{\alpha Ln}$/ $TAC_{\alpha Rn}$ comes from the alpha detector through associated CFD, splitting unit and delays. Left detector starts the TAC_{nn}, while right detector stops it. Outputs from all detectors were fed also to the Quad Coincidence Logic Units (QCLU) working in AND mode. There are three AND units for αLn, αRn and nn coincidences, one AND unit for simultaneous αLnRn coincidences and additional one for insurance. Output from the last AND unit serve as a trigger for Analog-to Digital Converter (ADC) which accepts outputs from all three TACs.

B. Experimental results

In total, sixty hours of measurements were collected during the period of two weeks or ten days. Each day set of two measurements were done, one measurement lasting for three hours. Fig.8 shows the time spectra $TAC_{\alpha Ln}$ (upper) and $TAC_{\alpha Rn}$ (lower), while Fig. 9 shows the TAC_{nn} spectrum, respectively.

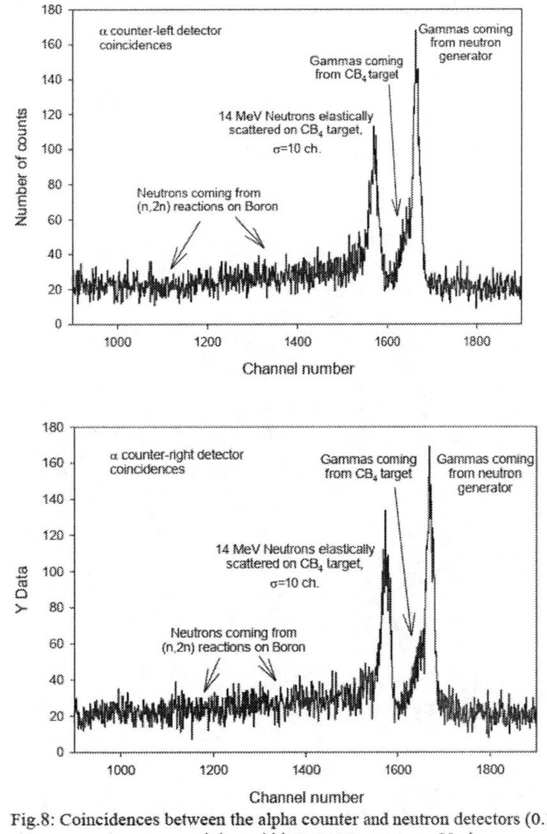

Fig.8: Coincidences between the alpha counter and neutron detectors (0.11 ns/ ch). Scattered neutron peak has width FWHM = 2.34 σ = 23 ch.

In both $TAC_{\alpha Ln}$ and $TAC_{\alpha Rn}$ spectra two peaks are clearly visible, one coming from the gamma rays induced in NG, and another one belonging to the 14 MeV neutrons elastically scattered on CB_4 target (absent in target off configuration).

https://doi.org/10.1051/epjconf/201817001017

Time distance between this two peak correspond to the time difference between the time needed by gamma ray to travel the distance between the NG and neutron detector, and the time needed for 14 MeV neutron to travel the distance between the NG and the CB_4 target plus the time needed for 5.07 MeV neutron to travel the distance between the target and neutron detector. The second peak, the one belonging to the 5.07 MeV neutrons, is used for TACs calibration.

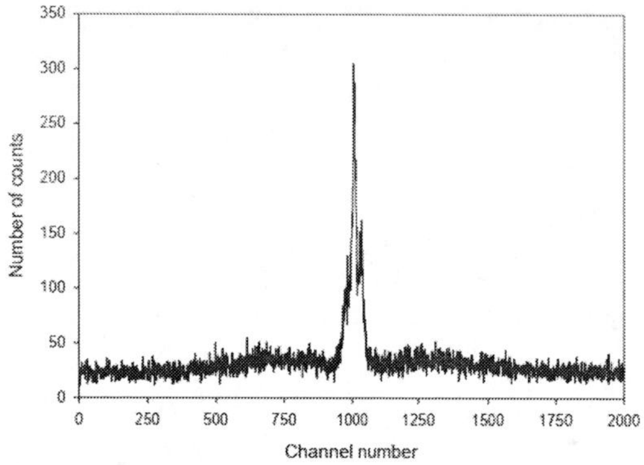

Fig. 9: Coincidences between two neutron detectors. Signal from the left detector is a START signal for time-to-amplitude converter.

Velocities, v_{Rn} and v_{Ln}, of the two outgoing neutrons are measured by t-o-f measurements using the associated alpha particle method and by knowing the position of the elasticaly scattered 14 MeV neutrons in t-o-f spectrum. Expresion for excitation energy E in dependence on the output neutrons velocities v_{Rn}, v_{Ln} is shown on Fig.10.

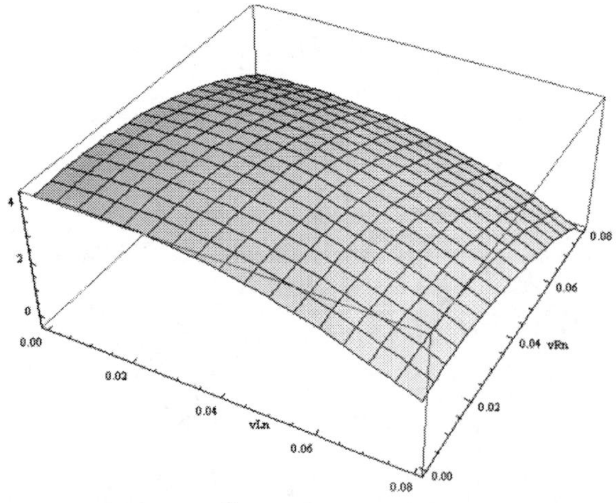

Fig. 10: Excitation energy of ^9B in dependence on neutron velocities.

The number of events for $^{10}B(n,2n)^9B^*$ reaction was found by the following way:

By convention v_{Rn} was chosen to be the velocity of neutron coming to the right neutron detector. For a given ch_{Rn}, v_{Rn} was calculated by following equation.

$$v_{Rn} = \frac{\frac{d}{30}}{a_{Rn}\left(-ch_{Rn}+ch_{oRn}\right)+\frac{d}{5.07}} \quad , \quad v_{Ln} = \frac{\frac{d}{30}}{a_{Ln}\left(-ch_{Ln}+ch_{oLn}\right)+\frac{d}{5.07}}$$

The v_{Ln} velocity was calculated from corresponding channel number ch_{Ln}, see above.

The position of the above event in TAC_{nn} spectrum is calculated by formula

$$ch_{nn} = ch_{nn0} + \frac{1}{30\, a_{nn}}\left(\frac{d}{v_{Ln}} - \frac{d}{v_{Rn}}\right)$$

where ch_{nn0} correspond to simultaneously events recorded by neutron detectors. For $ch_{nn} > ch_{nn0}$ signal came to the left detector was earlier in time than signal came to the right one. For $ch_{nn} < ch_{nn0}$ the opposite is true. Only those events recorded in the channel windows ($ch_{Rn}\pm 30$) and ($ch_{Ln}\pm 30$) and ($ch_{nn}\pm 30$) are recorded as a true one.

Sixty channels windows were chosen as a compromise solution. In average around eight events were recorded during the 3 hours measurement which is not far from the theoretically estimated value. For narrower windows accepted events drops down, while broader windows includes more false events.

Fig.11 shows the excitation levels of ^9B found by procedure described above. Through the experimental data a curve was fitted described by eq. (6). Expected two peaks are found at 0 MeV (ground level) and 2 MeV (second excited level). The fitting parameters are peaks width (b and b1) and assumed flat background (y0). The second excited level has underestimated energy by 345 keV because the poor statistics and because the calibration problems described in the next chapter. Nevertheless, the peaks width is found to be b=(0.525 ± 0.065) MeV for ground level and b1=(0.37 ± 0.04) MeV for second excited level, what should be compared with theoretical prediction 0.56 MeV and 0.4 MeV, respectively. Correlation factor was found to be R = 0.5318.

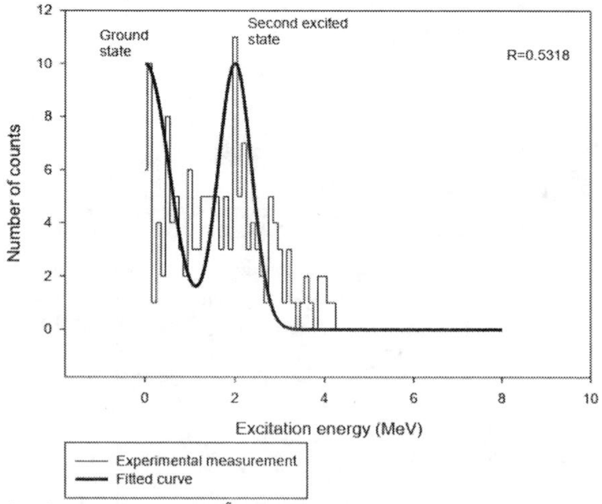

Fig. 11: Excitation levels of ^9B, fitted curve described below.

$$f(x) = y0 + 10\, e^{-0.5\frac{(x)^2}{b^2}} + 10\, e^{-0.5\frac{(x-2)^2}{b1^2}}$$

Fitting parameters: b = (0.525 ± 0.065) MeV
b1 = (0.37 ± 0.04) MeV
y0 = 2 x10^{-9} ± 0.3

Fig. 12 shows the same experimental results, but this time instead of two peaks, three peaks are fitted (centered at energies 0 MeV, x0 MeV and 2 MeV, eq. 7). Fitting parameters are position of the first excited level (x0), first excited level width (b2) and peak high (a). Since correlation factor is somewhat larger (R = 0.5707), our data support existence of a peak between the ground level and second excited level, although data is insufficient for concluding statement.

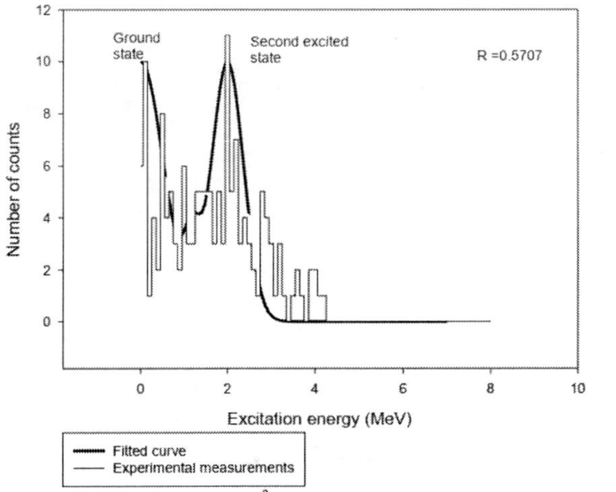

Fig.12: Excitation level of ^{9}B, fitted curve described below. R is correlation parameter.

$$f(x) = a\, e^{-0.5\frac{(x-x0)^2}{b2}} + 10\, e^{-0.5\frac{(x)^2}{0.525^2}} + 10\, e^{-0.5\frac{(x-2)^2}{0.37^2}}$$

Fitting parameters: b2 = 0.2 ± 0.1
x0 = 1.2 ± 0.1
a = 2.5 ± 1.4

By calculating the area under the curve (7) and comparing it with the areas under the fitted curves on fig. 12, it is possible to roughly estimate the cross section for ^{10}B(n,2n)^{9}B* nuclear reaction and particularly, cross section for excitation to first excitation level. Total cross section was found to be σ~10 mb which should be compared with 26 mb from www.nndc.bnl.gov. It should be noted that our rough estimation ignore the neutron detector efficiency dependence on neutron/gamma energy and all neutrons/gammas not detected because its signals did not pass the CFD. Total cross section for excitation to 1.2 MeV (first excitation level) is found to be 0.7 mb.

IV. Conclusion

According to the results presented above we can conclude that it is possible to check for the existence of the first excitation level of 9B nuclei and to measure its energy as well as width, by using the 10B(n,2n)9B* nuclear reaction. In order to do it properly distance between the CB4 target and neutron detectors should be at least one meter. To compensate the lack of solid angle, pure 10B isotope should be used as well as several pairs of 10 cm x 10 cm x 10 cm neutron detectors (or equivalent). TACs calibration should be done at least once per day (every six hours) in full TACs range. Our preliminary results support the existence of the first excitation level at (1.2 ± 0.1) MeV, and with FWHM = 2.35 (0.2± 0.1) MeV = (0.47 ± 0.235) MeV which is in agreement whit some of data presented in table 1. Total cross section for excitation to 1.2 MeV (first excitation level) is estimated to be 0.7 mb.

Acknowledgment

We would like to thank to dr. K. Nad from.Rudjer Boskovic Institute, Zagreb, Croatia and mr. J. Jaegly from CEA Cadarache, France for assisting in ^{10}B(n,2n)^{9}B reaction data taking process.

References

[1] K.Ilakovac, L.G.Kuo, M.Petravic and I.Slaus, Attemp to Determine the n-n Scattering Lenght from the Reaction D(n,p)2n, *Phys.Rev.* 124(1961)1923

[2] M. Cerineo, K. Ilakovac, I. Šlaus, P. Tomaš and V. Valković: Charge Dependence of Nuclear Forces and the Break-up of Deuterons and Tritons, *Phys. Rev.* 133 (1964) B948-955

[3] Z.Kohley *et al.* 2014, "Structure and decay correlations of two-neutron systems beyond the dripline", Journal of Physics, Conference Series 569, 012033.

[4] A.Spyrou *et al.* 2012, "First Observation of Ground State Dineutron Decay: ^{16}Be", *Phys.Rev.Lett.* 108, 102501.

[5] T.D.Baldwin at al., „First excited 12+ state in ^{9}B", *Phys. Rev.* C 86 034330 (2012)

[6] Akimune, H et al., „Evidence for a 3.8 MeV state in ^{9}B", *Phys. Rev.* C. 64:1,2001.

[7] Anderson, J. D. et al., „Fast neutron spectroscopy of the Reaction ^{9}Be(p,n)^{9}B at 20 MeV", *Phys. Rev.* C. 2: 319, 1970.

[8] Arena, N. et al. „Energy and Width Measurement of the 9B First Excited State Observed by the ^{10}B(^{3}He,α)^{9}Be and ^{10}Be(^{3}He,α)^{9}B(α)^{5}Li Reactions", Europhys. Lett., 6: 517, 1988.

[9] Baldwin, T. D. et al., "First excited 12+ state in ^{9}B", Phys. Rey. C. 86, 0343301, 2012.

[10] Bauer, R. W., Anderson, J. D. and Wong, C. „ A search for an excited state of ^{9}B near 1.7 MeV", Nucl. Phys. 56: 117, 1964.

[11] Burlein, M. et al., „Energies and widths of states in ^{9}B", Phys. Rey. C. 38: 2078, 1988.

[12] Byrd, R. C. et al., „ Cross-section Measurement and Lane model analysis for the ^{9}Be(p,n)^{9}B Reaction", Nucl. Phys. A. 399:94, 1983.

[13] Catford, W. N. et al., „The missing 1/2+ mirror state in ^{9}B". Nucl. Phys. A. 550: 517, 1992.

[14] Chou, J. C. et al., „Investigation of 9B and 9Be Levels in a Kinematically Complete ^{9}Be(p,pn)^{8}Be Experiment", J. of the Phys. Soc. Japan. 44:1-6, 1978.

[15] Crain, I. K. „Possible direct causal relation between geomagnetic reversals and biological extinctions". Bull. Geol. Soc. Am. 82:2603-2606, 1971.

[16] Efros, V. D., Bang, J. M. „The first excited states of ^{9}Be and ^{9}B". Eur. Phys. J. A 4: 33-39, 1999.

[17] Farrow, E. F. and H. J. Hay, H. J. „Search for a low lying 1/2+ level in ^{9}B", Phys.Lett. 11:50, 1964.

[18] Fortune, H. T. and Sherr, R. „Update on ^{9}B (½+)". Nuclear Physics A 898:78-83, 2013.

[19] Gul, K., Armitage, B. H. and Hooton, B. W. „The ^{18}O(^{3}He,n)^{20}Ne and ^{7}Li(^{3}He,n)^{9}B Phys". Rey. A. 153:390, 1970.

[20] Islam, M. M. and Treacy, P. B. „The first excited state of ^{9}B". Nucl. Phys. 70: 236-240. „^{9}Be(^{3}He,t) spectra", Phys. Rey. C. 36: 1269, 1965.

[21] Karki, A. A Study of 9B Spectroscopy via the ^{9}Be(p,n)^{9}B Reaction using the Neutron Time-of-Flight Technique. Thesis. Ohio University, Athens, OH, USA, 2013.

[22] Kroepfl, J. J. and Browne, C. P. „Energy level structure of ^{9}Be and ^{9}B", Nucl. Phys. A. 108: 289, 1967.

[23] Marion, J. B. and Levin, J. S. „Investigation of the ^{9}Be(p,n)^{9}B and

^{9}Be(p,αγ)^{6}Li Reactions". Phys. Rev. 115: 144, 1959.
[24] Paris, M. W., Hale, G., Hayes, A. and G. Jungman G. „R-matrix analysis of reactions in the 9B compound system". arXiv:1304.3153, Nuclear Data Sheets 120:184-187, 2013.
[25] Saji, Y. „Energy Spectrum and Angular Distributions of neutrons from the reaction ^{9}Be(p,n)^{9}B at 8 to 14 MeV of proton energies", J. Phys. Soc. Japan. 15:367, 1960.
[26] Scholl, C. et al., „High resolution study of the ^{9}Be(3He,t)^{9}B reaction up to the ^{9}B triton threshold", Phys. Rev. C. 84, 0143081, 2011.
[27] Sherr, R. and Bertsch, G. „Coulomb energy systematics and the missing J^{π} = ½+ state in ^{9}B". Phys. Rev C32: 1809-1816, 1985.
[28] Slobodrian, R. J. et al., „High-Resolution fast-neutron spectroscopy of the reaction ^{9}Be(p,n)^{9}B at 20 MeV", Phys. Rey. Lett. 19:595, 1967.
[29] Spencer, R. R., Floyd, G. C. and Young, T. E. „Energy levels of ^{9}Be and ^{9}B", Nucl. Phys. C. 21: 310, 1960.
[30] Symons, G. D. and Treacy, P. B. „Evidence for 1.7 MeV level in ^{9}B", Phys. Lett. 2:175, 1962.

EPJ Web of Conferences **170**, 01018 (2018) https://doi.org/10.1051/epjconf/201817001018
ANIMMA 2017

Counting neutrons with a commercial S-CMOS camera

VAN ESCH Patrick – Institut Laue Langevin Mail: vanesch@ill.fr)
MUTTI Paolo - Institut Laue Langevin Mail: mutti@ill.fr)
RUIZ-MARTINEZ Emilio - Institut Laue Langevin Mail: ruizmartinez@ill.fr)
ABAD GARCIA Estefania - ESS Bilbao Mail: eabad@essbilbao.org)
MOSCONI Marita - ESS Bilbao Mail: mmosconi@essbilbao.org)
ORTEGA Jon - ESS Bilbao Mail: jortega@essbilbao.org)

Abstract

It is possible to detect individual flashes from thermal neutron impacts in a ZnS scintillator using a CMOS camera looking at the scintillator screen, and off line image processing. Some preliminary results indicated that the efficiency of recognition could be improved by optimizing the light collection and the image processing. We will report on this ongoing work which is a result from the collaboration between ESS Bilbao and the ILL. The main progress to be reported is situated on the level of the on-line treatment of the imaging data. If this technology is to work on a genuine scientific instrument, it is necessary that all the processing happens on line, to avoid the accumulation of large amounts of image data to be analyzed off line. An FPGA-based real-time full-deca mode VME-compatible CameraLink board has been developed at the SCI of the ILL, which is able to manage the data flow from the camera and convert it in a reasonable "neutron impact" data flow like from a usual neutron counting detector. The main challenge of the endeavor is the optical light collection from the scintillator. While the light yield of a ZnS scintillator is a priori rather important, the amount of light collected with a photographic objective is small. Different scintillators and different light collection techniques have been experimented with and results will be shown for different setups improving upon the light recuperation on the camera sensor. Improvements on the algorithm side will also be presented. The algorithms have to be at the same time efficient in their recognition of neutron signals, in their rejection of noise signals (internal and external to the camera) but also have to be simple enough to be easily implemented in the FPGA. The path from the idea of detecting individual neutron impacts with a CMOS camera to a practical working instrument detector is challenging, and in this paper we will give an overview of the part of the road that has already been walked.

Overview

We use the commercial S-CMOS camera HAMAMATSU ORCA-Flash4.0 V2 in our experiments, which is a 2048 x 2048 pixel camera with a sensor of 13.2 mm x 13.2 mm that can run at 100 fps and has a single pixel readout noise of 1.6 electrons rms. The readout can be done in USB-3 mode (up to 30 fps), or in Camera Link mode.

The fundamental idea is that such a camera can be sensitive to individual flashes of less than ~100 photo-electrons if these are sufficiently concentrated in space (pixels) and time (frames), and as such, allow for single-neutron impact measurement on scintillator screens, which would transform a scintillator/camera setup from an integrating detector into a single-neutron counting detector in those areas of the image where the neutron intensity is weak. This would tremendously increase the signal-to-noise ratio for weak neutron signals, while preserving the linearity of response that integrating detectors have at high intensity. This would also allow for a very high dynamic range in the images taken. The signal-to-noise ratio would also be improved due to neutron-gamma discrimination, something an integrating detector cannot do very well.

It is indeed well-known that individual neutron counting detectors, using their non-linear neutron detection processing (for instance, a simple threshold crossing), have signal-to-noise ratios that are orders of magnitude better than integrating detectors with linear processing when the count rates are low, because all the noise is essentially reduced to zero. However, it is also known that individual counting detectors hit dead time limitations on the high side of the counting scale which limits their dynamical range, while integrating detectors remain linear up to very high count rates. We set out to try to combine the best of both worlds.

However, the difficulty resides in obtaining a sufficiently clear signal per individual neutron so that the efficiency is high enough and the error rate is low enough in order to be able to talk about reliable neutron detection. This will depend on one hand on the physics of the devices used, and on the other hand, on the specific detection algorithms used. Ideally, one would like to build such a system using commercially available components, but it is possible that the performance of such a system is not sufficient to obtain a working system. We will report on the status of this ongoing research which is a collaborative effort between the ILL and ESS-Bilbao.

Earlier reported work showed that a relatively simple algorithm had some potential to discriminate neutron impacts, but at that point, we had no idea of any efficiency of the process involved. In other words, we reported already that some individual neutrons could be discriminated from non-neutron situations. In the mean time, we have studied several setups in better controlled circumstances, and we will report on the results of these tests. The results show that we have low, but non-zero efficiencies, and that we have low, but statistically significant, signal-to-noise ratios as it stands, but that we obtain enough light to be able in principle to obtain better performances than the ones obtained and reported as of today.

We also show progress on integrating the data processing on-line. Indeed, although the results reported have all been obtained with tedious high volume off-line processing, if ever such a system is to function on-line, we want to avoid to have to store individual 8 MB images at 100 fps. We want the neutron-recognition algorithm to run on-line. If the algorithm is simple enough, we want to do this with an FPGA-based solution.

Camera results

A quick estimation gives an idea of what the challenge is about: if we have a per-pixel rms noise level of 1.6 electrons, and the light of an impact is spread over a square of 8x8 pixels, and we integrate the amount of photo-electrons over this square, the expected rms noise level is 12.8 electrons, if the readout noise is statistically independent from pixel to pixel, which it is experimentally observed to be. If we require a 6 sigma threshold to detect a light flash, this light flash would need about 77 electrons to be deposited during one frame, in this square. The quantum efficiency of the sensor is such, that this corresponds to about 100 photons to reach the sensor. If the screen is at 200 mm from the camera, and the lens is a 50 mm focal lens with a theoretical F number of about 1, we will capture theoretically about 1/250 of the emitted light with perfect optics ; if we take 50% optical efficiency, this would bring us to the need of emitting about 50 000 photons from the scintillator screen. Even though this napkin estimation is rather rough, it gives an idea of what is needed.

The only known class of scintillators that can yield sufficient photons are ZnS based scintillators, which are known to emit up to 170 000 photons in a Li-6 conversion event. So, a priori, such events should potentially be individually detectable by our system. This is what a priori justifies our attempts at realizing such a system. The principal difficulty is not to waste this light from its point of generation to the camera sensor.

We have tested essentially two classes of ZnS based scintillators: the "green" ones and the "blue" ones ; with different layer thickness, from different manufacturers, and different dopant/scintillator concentrations.

We have tested a commercial objective with a focal length of 50 mm and an f-number of 0.95, as well as a Canon objective which was a special-purpose objective made for an X-ray medical radiography machine with focal length also 50 mm and an f-number of 0.7.

Integral measurements of photo-electron yield per converted neutron give encouraging numbers. With a "black" He-3 neutron proportional counter, we estimated the neutron flux of a collimated and attenuated neutron beam ; we measured the fraction of the beam that was absorbed by the scintillator ; and we measured the integrated amount of photo-electrons in a sum of images of the beam with the camera. By dividing the integrated amount of photo-electrons by the absorbed fraction of the neutron flux during the exposition time, we could estimate the number of photo-electrons per absorbed (and potentially converted) neutron in the scintillator. The "blue" type scintillator with the Canon (f-number 0.7) lens gives consistent estimates of 350 photo-electrons per neutron and some 270 photo-electrons with the commercial (f-number 0.95) lens. The "green" type scintillator with the Canon lens gives about 430 photo-electrons while the commercial lens gives us 320 photo-electrons.

With these numbers, we see not only that there is a reasonable agreement between quick order-of-magnitude estimations of the expected number of photo-electrons on the camera, and the

 https://doi.org/10.1051/epjconf/201817001018

actual measured values, but we also see that there is good hope that our proposition will actually work.

However, when we run a relatively simple threshold-crossing algorithm to find individual impacts, the results are at the same time disappointing and encouraging. The disappointing part is that if we take the best combination, which is the green scintillator with the Canon lens, and we look at short exposure images, we have a meager detection efficiency of single impacts of at best 6% of the converted neutrons, with a signal-to-noise ratio that is not better than a factor of 10. That said, this signal-to-noise ratio does indicate a statistically significant signal (we are not recognizing random events), but the very low efficiency and the bad signal-to-noise ratio indicates us that our algorithm is not sophisticated enough to separate noise from light pulses.

So this is essentially the puzzling state in which our research is at this point: we know that we have enough photo-electrons for the principle to work ; but if we apply a simple threshold algorithm, even though the results are statistically significant, they are of such mediocre quality that we have no working system.

Board development

The structure of the acquisition system at the ILL consists of a general control software system, NOMAD, which is developed in-house, that communicates with a VME master through a first point-to-point optical link. A VME slave board, the IoxOS 1210 board (consisting of a Virtex 6 FPGA and 2 CPU units) connects to a second optical fibre token ring, on which all "data sources" are connected. Most detectors at the ILL being neutron counting detectors, they send blocks of time-stamped neutron impacts to the data acquisition board, that can transmit them in list-mode to NOMAD directly, or that can histogram them locally in various acquisition modes (like time-of-flight, kinetic mode etc...). At regular times, NOMAD reads out the status of the histograms through the VME link.

A VME slave board, the IoxOS 1200, housing a Virtex 6 Xilinx FPGA, has been complemented by an in-house board that has CameraLink receivers and an optical fibre interface: our GG_CameraLink board. The hardware has reached a prototype level that is functioning sufficiently well in order to have started implementing a part of the algorithm: the frame grabber, decoding the CameraLink signals that come from the Hamamatsu camera, which has a slightly specific version of the CameraLink protocol.

The idea is to integrate the GG-CameraLink board in this optical token ring, by having it look like a normal neutron-counting detector on one hand, if our on-line algorithm of detecting individual neutrons works well, but integrating also the possibility to send full integrating images to the acquisition board, be it at relatively low frame rates.

Correlated Production and Analog Transport of Fission Neutrons and Photons using Fission Models FREYA, FIFRELIN and the Monte Carlo Code TRIPOLI-4®.

Jérôme M. Verbeke
Lawrence Livermore National Laboratory
Nuclear and Chemical Sciences Division
P.O. Box 808
Livermore, California 94551–0808
U.S.A.
Email: Verbeke2@llnl.gov

Odile Petit
Commissariat à l'Energie Atomique
et aux Energies Alternatives
Université Paris-Saclay
F-91191 Gif-sur-Yvette, cedex
France
Email: Odile.Petit@cea.fr

Olivier Litaize & Abdelhazize Chebbouhi
Commissariat à l'Energie Atomique
et aux Energies Alternatives
F-13108 Saint-Paul-lès-Durance
France
Email: Olivier.Litaize@cea.fr
Abdelhazize.Chebbouhi@cea.fr

***Abstract*—Fission modeling in general-purpose Monte Carlo transport codes often relies on average nuclear data provided by international evaluation libraries. As such, only average fission multiplicities are available and correlations between fission neutrons and photons are missing. Whereas uncorrelated fission physics is usually sufficient for standard reactor core and radiation shielding calculations, correlated fission secondaries are required for specialized nuclear instrumentation and detector modeling. For coincidence counting detector optimization for instance, precise simulation of fission neutrons and photons that remain correlated in time from birth to detection is essential. New developments were recently integrated into the Monte Carlo transport code TRIPOLI-4 to model fission physics more precisely, the purpose being to access event-by-event fission events from two different fission models: FREYA and FIFRELIN. TRIPOLI-4 simulations can now be performed, either by connecting via an API to the LLNL fission library including FREYA, or by reading external fission event data files produced by FIFRELIN beforehand. These new capabilities enable us to easily compare results from Monte Carlo transport calculations using the two fission models in a nuclear instrumentation application. In the first part of this paper, broad underlying principles of the two fission models are recalled. We then present experimental measurements of neutron angular correlations for ^{252}Cf(sf) and ^{240}Pu(sf). The correlations were measured for several neutron kinetic energy thresholds. In the latter part of the paper, simulation results are compared to experimental data. Spontaneous fissions in ^{252}Cf and ^{240}Pu are modeled by FREYA or FIFRELIN. Emitted neutrons and photons are subsequently transported to an array of scintillators by TRIPOLI-4 in analog mode to preserve their correlations. Angular correlations between fission neutrons obtained independently from these TRIPOLI-4 simulations, using either FREYA or FIFRELIN, are compared to experimental results. For ^{240}Pu(sf), the measured correlations were used to tune the model parameters.**

I. Introduction

Spontaneous fission is characterized by the emission of bursts of neutrons. If multiplying fissile materials surround these spontaneous fissions, the bursts are amplified to form fission chains. This unique fission chain signature has been used for many decades to detect and authenticate nuclear materials. Most nonproliferation and verification neutron measurement systems rely on detecting thermalized neutrons. Typically ^{3}He tubes record the arrival times of thermalized neutrons. Unfortunately, the neutron properties at the time of emission are obscured by down-scattering. Scintillators, on the other hand, directly detect unmoderated fission neutrons because inelastic scattering of neutrons on hydrogen results in the emission of a recoil proton, ionizing the scintillation material, enabling detection on a nanosecond time scale. When fast neutron detectors are used, the neutron-neutron correlation from fission can be used to characterize fissile samples.

With fast detectors, characterization of nuclear materials becomes possible by measurements of the fission neutron angular anisotropy. In spontaneous and low energy fission, almost all of the neutron emission comes from the fully accelerated fission fragments whose back-to-back motion is imprinted on the neutron directions in the laboratory frame. Thus small angle correlations are expected from neutrons emitted from the same fragment, whereas large angle correlations arise from opposite fragments. Anisotropy measurements can provide valuable information for identifying the composition of materials.

General-purpose Monte Carlo codes are available for modeling neutron transport. They have traditionally employed "average fission models" for modeling fission, characterized by uncorrelated secondary particle emission. This approximation is sufficient for the calculation of average quantities such as flux, energy deposition and multiplication. However, correlations are important for modeling anisotropy of fission neutrons, because measurement of anisotropy relies on measuring time-correlated neutrons.

Based on physics models and experimental data, the event-by-event fission generators FREYA and FIFRELIN are able to predict a host of correlation observables, including correlations in neutron multiplicity, energy, and angles, and the energy sharing between neutrons and photons. The combination of the Monte Carlo radiation transport code TRIPOLI-4.10 with these fission generators enables users to directly model fission event-by-event and transport fission secondaries through complex detector geometries while keeping them fully correlated from birth to detection.

The paper is organized as follows. We first describe the fission event generators FREYA, FIFRELIN and the Monte Carlo radiation transport code TRIPOLI-4. We then present experimental measurements of neutron-neutron angular correlations between ^{252}Cf(sf) and ^{240}Pu(sf) fission neutrons. Next, we compare these correlations to those obtained by simulations and discuss the differences between the two fission models. Model parameters are adjusted to best match the experimental data.

II. FISSION EVENT GENERATORS

A. FREYA

FREYA [1]–[9] requires the fission fragment mass distribution Y(A) and the total kinetic energy TKE(A) for the particular excitation considered (FREYA can simulate both neutron-induced fission and spontaneous fission). The yields Y(A) are either taken to be the measured fragment yields directly [3] or a five-Gaussian fit to the data which makes it possible to parameterize the energy dependence of the yields [4] in the case of neutron-induced fission. In order to generate an event, FREYA first selects the mass split based on the provided Y(A). The fragment charges are then sampled from normal distributions. The linear and angular momenta of the two fragments and their internal excitations are subsequently sampled, as explained below. After their formation, the fully accelerated fragments de-excite first by neutron evaporation and then by photon emission. For a given split of nucleus A_0 into light and heavy fragments, L and H respectively, the Q-value is given by $Q = M_0c^2 - M_Lc^2 - M_Hc^2$. For a given total fragment kinetic energy TKE, the total excitation energy available for rotational and statistical excitation of the two fragments is then $E^*_{sc} = Q - \overline{TKE}$ and the corresponding 'scission temperature' T_{sc} is obtained from $E^*_{sc} = (A_0/e_0)T^2_{sc}$ where e_0 is the scale of the level density parameter. The inclusion of angular momentum in FREYA was described in Ref. [6]. In addition to any overall rigid rotation of the dinuclear configuration prior to scission, which imparts a mean angular momenta to the two fragments, the fragments also acquire fluctuations around those mean values from the wriggling and bending modes (consisting of rotations in the same or opposite sense around an axis perpendicular to the dinuclear axis) but ignores tilting and twisting modes (in which the fragments rotate around the dinuclear axis). The degree of these fluctuations is governed by the adjustable parameter c_S: $T_S = c_ST_{sc}$. The fluctuations vanish for $c_S = 0$ so the fragments emerge with the angular momenta dictated by the overall rigid rotation of the scission configuration (which is usually very small for induced fission - and entirely absent for spontaneous fission). The default value, $c_S = 1$, leads to $\bar{S}_L \sim 6.2\hbar$ and $S_H \sim 7.6\hbar$ for ^{252}Cf(sf), see Ref. [6] for details.

After accounting for the total rotational energy of the two fragments, E_{rot}, one is left with a total of $E_{stat} = E^*_{sc} - E_{rot}$ for statistical fragment excitation that is distributed between the two fragments.

First a preliminary partition, $E_{stat} = \acute{E}^*_L + \acute{E}^*_H$, is made according to the heat capacities of the two fragments which are assumed to be proportional to the corresponding Fermi-gas level density parameters, i.e. $\acute{E}_L/\acute{E}_H = a_L/a_H$, where the level density parameter a can depend on the pairing energy and shell corrections of the fragment. If the shell corrections are negligible, or the available energy is large, then $a_i = A_i/e_0$, i.e. a_i is simply proportional to the fragment mass number A_i, and this renormalization is immaterial. A default value of e_0 = 10/MeV is typically used. If the two fragments are in mutual thermal equilibrium, $T_L = T_H$, the total excitation energy will, on average, be partitioned as above. But because the observed neutron multiplicities suggest that the light fragments tend to be disproportionately excited, the average excitations are modified in favor of the light fragment,

$$\overline{E}^*_L = x\acute{E}^*_L,\ \overline{E}^*_H = E_{stat} - \overline{E}^*_L, \quad (1)$$

where the adjustable model parameter x is expected be larger than unity. It was found that $x = 1.27$ leads to reasonable agreement with $\nu(A)$ for ^{252}Cf(sf) [3].

After the mean fragment excitation energies have been assigned, FREYA considers the effect of thermal fluctuations. In Weisskopf's statistical model of the nucleus, which describes the excited nucleus as a degenerate Fermi gas, the mean excitation of a fragment is related to its temperature T_i by $\overline{E}^*_i = \tilde{a}_iT^2_i$ [10]–[12] and the associated variance in the excitation is $\sigma^2_{E_i} = -\partial^2 ln\rho_i(E_i)/\partial E^2_i = 2\overline{E}^*_iT_i$. Therefore, for each of the two fragments, an energy fluctuation δE^*_i is sampled from a normal distribution of variance $2c^2\overline{E}^*_iT_i$. The fragment excitations are adjusted accordingly with

$$E^*_i = \overline{E}^*_i + \delta E^*_i,\ i = L, H. \quad (2)$$

Energy conservation is accounted for by making a compensating opposing fluctuation in the total kinetic energy,

$$\mathrm{TKE} = \overline{\mathrm{TKE}} - \delta E^*_L - \delta E^*_H. \quad (3)$$

The average TKE, $\overline{\mathrm{TKE}}$, has been adjusted by dTKE to reproduce the average neutron multiplicity, $\bar{\nu}$. The factor c multiplying the variance explores the effect of the truncation of the normal distribution at the maximum available excitation. Its value affects the neutron multiplicity distribution $P(\nu)$. A default value, $c = 1.0$, had previously been assumed.

The evaporated neutrons are assumed to be isotropic in the frame of the emitting nucleus, apart from a very slight flattening due to the nuclear rotation. Their energy is sampled from a black-body energy spectrum, $dN_n/dE_n \sim E_n\exp(-E_n/T_{\max})$, where $T_{\max}$ is the maximum possible temperature in the daughter nucleus (corresponding to emission of a very soft neutron).

FREYA generally assumes that neutron evaporation continues until the nuclear excitation energy is below the threshold $S_n + Q_{min}$, where S_n is the neutron separation energy and Q_{min} is small. The default value is $Q_{min} = 0.01$ MeV, so that neutron evaporation continues as long as energetically possible. After neutron evaporation has ceased, the excited product nucleus will undergo sequential photon emission to rid itself of the remaining excitation energy. First, statistical photons are emitted isotropically with an energy distribution sampled from a black-body spectrum modulated by a giant-dipole resonance form factor. Each emission reduces the magnitude of the angular momentum by $dS = 1\hbar$. The cascade is continued until the statistical excitation is below a specified threshold, ϵ_{min}, which can be detector dependent. However, when the photon decay cascade leads to an excitation below any of the tabulated RIPL-3 [13] levels, FREYA switches to a discrete cascade based on the RIPL-3 data. The discrete cascade is continued until the half-life t_ℓ exceeds a specified value, t_{max}, which should be adjusted to reflect the detector response time. If the photon decay for a particular fragment is not in the RIPL-3 tables and the statistical excitation is below ϵ_{min}, the remaining de-excitation occurs by collective photon emission that each reduce the angular momentum by $2\hbar$.

B. FIFRELIN

A FIFRELIN calculation consists on one hand in sampling the initial state of a pair of fission fragments emerging from a binary fission event (mass A, nuclear charge Z, excitation energy E, spin J and parity π) and on the other hand in de-exciting both fragments [14]–[17]. In this work, mass and kinetic energy are sampled from experimental data [18], [19] while nuclear charge is estimated from unchanged charge density hypothesis, corrected by polarization function and proton/neutron even-odd factors. Total excitation energy is derived from energy conservation since kinetic energy has already been sampled from experimental data.

A rotational energy (collective mode of excitation) is calculated from a simple rotating liquid drop model involving a nuclear moment of inertia and a total angular momentum which is sampled following Bethe's work [20]: $P(J) = (J + 1/2)/\sigma^2 \exp(-(J+1/2)^2/2\sigma^2)$. For primary fission fragments (before de-excitation), the spin cut-off parameter σ is a free parameter (one for light fragment group σ_L and another one for heavy group σ_H). A third free parameter is the fraction of rigid body moment of inertia involved in the calculation of the rotational energy but it is generally fixed to one.

Intrinsic excitation energy (obtained after subtracting rotational energy from total excitation energy) is shared between fragments using an $E = aT^2$ relation involving the level density parameter a (following Ignatyuk's prescription) and the temperature T of the fragment. This system of two equations (one equation per fragment) is solved by iteration since the level density parameter is energy dependent and the excitation energy is the unknown quantity.

For a given fragmentation, the temperature of the light fragment T_L is linked to the temperature of its heavy partner T_H by a mass dependent temperature ratio law $R_T(A) = T_L/T_H$. This law involves two additional free parameters $R_T^{\rm min}$, $R_T^{\rm max}$.

Finally five free parameters have been described to simulate the entry zone in energy/spin of the primary fission fragments. These parameters will be chosen to reproduce a target fission observable such as average prompt neutron multiplicity $\overline{\nu}$ (3.76 for ^{252}Cf).

The de-excitation process in an energy, spin, parity set can be solved through the so-called Hauser-Feshbach formalism accounting for neuton/gamma competition. In FIFRELIN, it has been implemented in a specific way using the notion of Nuclear Realization developed in [21] for gamma emission and extended here to neutron/gamma coupled emission [15], [16]. A nuclear realization is a nuclear level scheme (energy E, spin J, parity π of a set of levels written $[EJ\pi]$) plus the associated partial widths $\Gamma_p([EJ\pi]_i \to [EJ\pi]_f, \alpha)$, describing transitions between initial level i and final level f. $[EJ\pi]$ can be a unique level or a set of levels inside an energy bin. Parameter p stands for 'particle' and can be a gamma, neutron or conversion electron. Parameter α is related to quantum numbers (XL for multipolarity-type of gammas and (l, j) for orbital and total angular momenta of neutrons). The algorithm also accounts for Porter-Thomas fluctuations of partial widths during the cascade.

Inside a nuclear level scheme the excitation energy range is decomposed in three parts:

- from the ground state to a cut-off energy $E_{cut-off}$ provided by RIPL-3 database, the level scheme is supposed to be complete and $[E, J, \pi]$ set as well as experimental transition intensities are accounted for (strictly speaking spins are not always known and the code then completes this lack of information by using spin distribution models),
- from $E_{cut-off}$ to E_{bin} the scheme is completed by theoretical discrete levels sampled from level density models (several models are available in FIFRELIN as well as for photon strength functions involved in the calculation of partial radiative widths).
- from E_{bin} to the maximum excitation energy available the level scheme is simply decomposed on energy bins (roughly $\delta E = 10$ keV width). Again theoretical level density models are used to calculate the average number of levels $N([E, J, \pi]_i)$ inside bin i.

Finally a standard Monte Carlo procedure is used to sample the different transitions of a cascade that can correspond to neutron, gamma or conversion electron (since internal conversion coefficients are accounted for at low energy).

The simulation of a given cascade is finished when a β^-, $\beta^- n$ or ground-state level is reached. In order to compare with fission experiments, a time coincidence window (between fission fragment and gammas) is considered by accounting for half-lives of nuclear levels provided by RIPL-3 database. The emission occurs regarding the probability associated to the half-life of the initial level. A cascade ends when the cumulative time from the beginning of the cascade exceeds the coincidence time-window.

III. Radiation transport Monte Carlo code

TRIPOLI-4 is a continuous-energy Monte Carlo transport code developed in CEA Saclay, France [22]. It simulates neutron, photon, positron and electron transport. It is dedicated to shielding, reactor physics with or without depletion, criticality safety and nuclear instrumentation. Particle histories are simulated in a three-dimensional geometry, either surface-based and/or combinatorial, or produced by the ROOT software. International nuclear data libraries are used in the ENDF/B format to directly access continuous-energy cross sections, emission spectra and anisotropy data. Various variance reduction techniques are also available, and specifically a built-in module based on importance sampling. The TRIPOLI-4 code can be run in parallel on multi-core machines as well as on heterogeneous networks of stations or massive parallel machines.

Previous work has been performed in order to use either FREYA or FIFRELIN fission models, instead of nuclear evaluated data from international libraries, when sampling neutrons emitted by neutron induced fission in TRIPOLI-4. First, the LLNL fission Library was coupled to TRIPOLI-4 to model fission on an event-by-event basis through a direct call to FREYA at each fission [23]. Typically, this led to an increase in simulation time of the order of 15% only. Then, FIFRELIN fission model and TRIPOLI-4 have been coupled through external files [24]: the standalone FIFRELIN code has to be run beforehand to provide files gathering data for many independent fission events. For both fission models, TRIPOLI-4 produces neutrons from these fission events and simulates their transport up to the detectors. Usual Monte Carlo average tallies, such as fluxes or reaction rates, can then be estimated. For nuclear instrumentation applications, it is useful to store TRIPOLI-4 generated particle tracks into ROOT trees, which can easily be post-processed afterwards by imposing cuts on detector variables, e.g. time and position.

Analog transport mode is also available in TRIPOLI-4 for neutrons and photons [25]. It deactivates implicit capture and produces as many individual secondaries as required by the multiplicity of the reaction, for example in the case of neutron scattering with multiple outgoing neutrons.

For this paper, the production of photons in TRIPOLI-4 through the FREYA and FIFRELIN fission models has been added. The principle is the same as for neutrons: photons are produced via the fission model and TRIPOLI-4 transports them. The case of spontaneous fission has also been addressed: for FREYA through an external source linked to TRIPOLI-4 which performs itself the calls to FREYA; for FIFRELIN through a special kind of source feature reading neutrons and photons data from the FIFRELIN fission event files.

IV. Neutron-neutron angular correlations

A. *Experiments*

Figure 1 shows the geometrical configuration of the detector array used to measure the neutron-neutron correlations. The array consists of seventy-seven scintillators. Each detector in the array is cylindrical in shape, 10.16 cm diameter by 7.62 cm deep and filled with EJ-301 scintillating material [26]. Thirteen detectors sit over a cavity formed by an octagonal array underneath. Each arm of the octagon is a vertical tower made of eight scintillators. The measurement cavity is also octagonal, with 60 cm between the faces of opposite towers, and stands 50 cm tall. The tightly-packed system has 2π solid angle coverage, resulting in an overall geometric efficiency of 50%. Each of the 77 scintillators is individually read out by a photomultiplier tube (PMT).

Fig. 1. Photograph of the 77 liquid scintillator array on low mass floor.

All detectors were calibrated in energy using a ^{137}Cs source. Neutron-photon discrimination was achieved using the pulse shape discrimination method, i.e. by simultaneously measuring the charges in the peak and the tail of the PMT pulse. The detectors were synchronized using Compton scattering between neighbors.

The purpose of our measurements is to answer the question "are neutrons from spontaneous fission more likely to be emitted back-to-back, at 90° or in the same direction?" Correlations between neutrons arise from fissions simultaneously emitting multiple neutrons. In our experimental method, a pair is tagged as correlated when two neutrons are detected within 40 ns of each other, in which case they are assumed to have originated from the same spontaneous fission. Two detected neutrons are uncorrelated if the second one comes after a delay of at least 100 μs. A time window opens after that delay to count uncorrelated neutrons. The length of this time window depends on the neutron source strength. It is 1 ms for our ^{252}Cf(sf) source and 100 ms for the weaker ^{240}Pu(sf) source. The ratio of correlated to uncorrelated event rates is proportional to the probability of detecting two neutrons from the same spontaneous fission. Since we know the locations of all detectors with respect to the source, we can easily derive this probability as a function of the angular separation between the two neutrons.

The angular correlation between fission neutrons is measured for several neutron kinetic energy cutoffs. Neutron time of flight is applied to filter neutrons in kinetic energy. A photon from spontaneous fission is used to trigger the time-of-flight measurement window and a neutron is employed to close it. This time of flight method produces correlation distributions between neutrons that are detector independent. More details on the experimental setup are available in Ref. [27].

Measuring a 230 μCi ^{252}Cf source in the center of the detection system for 30 minutes, we obtained the two-neutron angular correlations shown as full circles in Fig. 2. In this paper, data always refers to the experimental measurements. The different colors identify the neutron kinetic energy cutoffs applied to the neutrons.

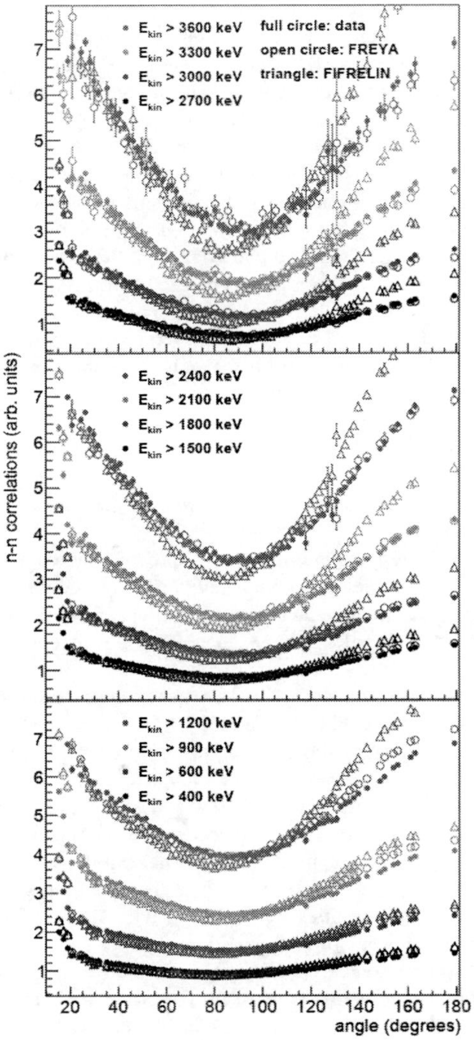

Fig. 2. The cross talk corrected two-neutron angular correlation for ^{252}Cf(sf) as a function of the angular separation for several neutron kinetic energy thresholds.

While not shown here, there is an overall good agreement between our data and other data sets [28], [29], which is encouraging in the light of the fact that our approach is completely independent, including different detectors, experimental setups, and analyses.

All distributions presented in this paper are corrected for neutron cross talk. Cross talk occurs when the same neutron is detected by two or more detectors. It affects primarily the pairs of detectors at the smallest separation angles, because these detectors are neighbors and thus tightly coupled, and they have the least material between them to prevent a neutron from scattering from one detector to the other. There is no reliable experimental analysis that can isolate counts due to cross talk on an event-by-event basis, so we resorted to Monte Carlo simulations using FREYA to remove integral cross talk counts from the experimental coincidences [27]. The simulated cross talk contribution is shown in Fig. 3 as a function of the detector separation angle, as seen from the source. In the simulations, it is important to account for the scintillation light-dependent neutron detection efficiency of the detectors.

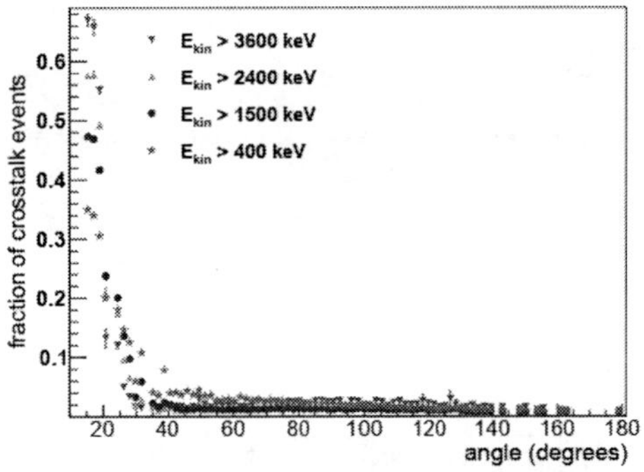

Fig. 3. Fraction of coincidences attributed to neutron cross talk as a function of the detector separation angle and for several neutron kinetic energy thresholds.

Figure 2 shows that for angles smaller than 25°, the neutron cross talk correction appears to undercorrect the correlations, leading to a small rise. Since neutron cross talk correction is estimated by simulations, this rise can be attributed to either inaccurate modeling of either the scintillation material or the scintillator geometry. It could also be attributed to an underprediction of the zero degree correlation in the FREYA simulation of the neutron cross talk.

We then proceeded to measure a 4.5 gm sample of ^{240}Pu (98% pure) of intensity 4,590 neutrons/s. The data analyzed were collected by placing the ^{240}Pu source in the center of the detection system for 23 hours. The two-neutron angular correlations are shown as full circles in Fig. 4.

B. Simulations

Using TRIPOLI-4 in combination with the fission event generators FREYA and FIFRELIN, we simulated neutron-neutron angular correlations.

The first simulations are shown for the ^{252}Cf(sf) source in Fig. 2, where the open circles and triangles are the FREYA and FIFRELIN calculations, respectively. For FREYA and

 https://doi.org/10.1051/epjconf/201817001019

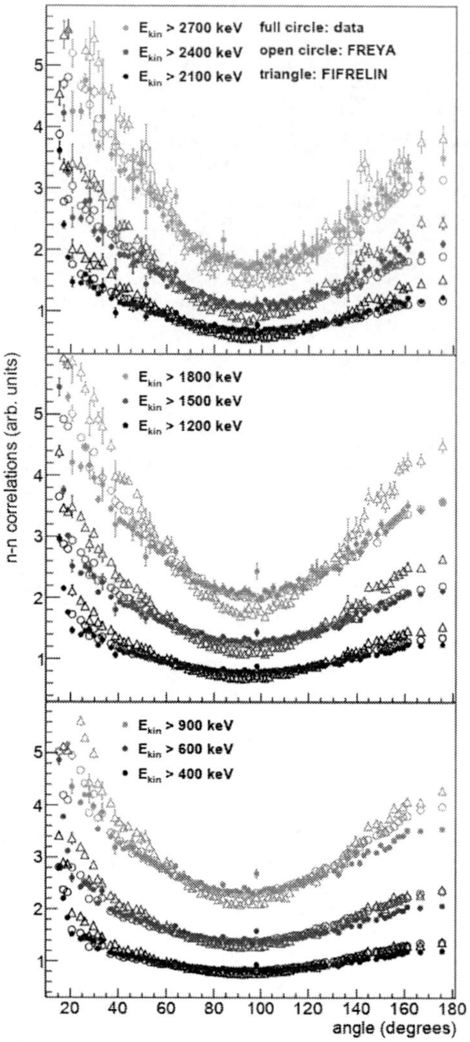

Fig. 4. The cross talk corrected two-neutron angular correlation for ^{240}Pu(sf) as a function of the angular separation and for several neutron kinetic energy thresholds.

FIFRELIN, the numbers of fissions simulated were equivalent to 40 and 56 seconds of data taking, respectively.

The neutron-neutron correlation arises because, while the neutrons are emitted isotropically in the rest frame of the fragment, the boost to the laboratory frame means that the neutrons will preferentially follow the fragments. Thus if one neutron is emitted from each fragment, they will be found at 180° apart while, if both are emitted from the same fragment, the angular separation is 0°. The 0° correlation includes two parts: both neutrons emitted from the light fragment and both emitted from the heavy fragment. Since the light fragment is higher velocity to conserve momentum, the correlation from two-neutron emission from the light fragment at 0° is larger. The three contributions combine to give peaks at 0° and 180° with a dip at 90°.

The light fission fragment also emits neutrons with larger kinetic energy on average, because it receives a disproportionately higher excitation energy than the heavy fragment. Therefore, increasing the neutron energy threshold increases the average energy of the neutrons that remain to form the correlation. Thus the higher the neutron energy threshold, the larger the bias toward emission from the light fragment and the higher the 0° correlation relative to the back-to-back correlation at 180°. In addition, the higher energy cutoff preferentially selects neutrons that are emitted in the direction of the boost rather than those opposite the boost direction which enhances the 0° correlation as the energy increases.

The simulations for ^{240}Pu(sf) are shown next in Figure 4 with the same open symbols. The numbers of fissions simulated were equivalent to 6h and 45min of data taking for FREYA and FIFRELIN, respectively.

We note qualitative differences in the ^{252}Cf(sf) and ^{240}Pu(sf) correlations due to the different average neutron multiplicities. Since the average neutron multiplicity of ^{252}Cf(sf) is ~ 3.76, each fragment can emit more than one neutron and any two emitted neutrons can be used to form the correlation function. On the other hand, the average neutron multiplicity of ^{240}Pu(sf) is ~ 2.1 so that, on average, the neutron-neutron correlation is formed from the only neutrons emitted during the fission. In addition, the average neutron energy of neutrons emitted from ^{252}Cf(sf) is higher than those from ^{240}Pu(sf) so that increasing the energy threshold is more likely to result in two peaks of equal strength for ^{252}Cf(sf) than for ^{240}Pu(sf). These characteristics can be observed in the data as well as in the simulations.

While the simulations do not reproduce the data perfectly, both fission event generators qualitatively reproduce the shapes of the experimental data. Without them the distributions would be flat. For angles smaller than 25°, we generally observe deviations between measurements and simulations, likely due to the reasons stated earlier, i.e. insufficient details for cross talk modeling in the simulation, etc.

Let us discuss the FREYA results. FREYA includes several physics-motivated model parameters. In particular, the parameter x describing how the excitation energy is partitioned between the light and heavy fission fragments. If the two fragments are in mutual thermal equilibrium, the total excitation energy will, on average, be partitioned according to the heat capacities of the fragments. However, observed neutron multiplicities suggest that the light fragment tends to be disproportionately excited. Therefore the average excitation energy is modified in favor of the light fragment. For ^{252}Cf(sf), $x = 1.27$, which means that the excitation energy of the light fragment is $\sim 30\%$ higher than that of the light fragment. For ^{240}Pu(sf), a default value of $x = 1.2$ was assumed previously since this was close to the value obtained for ^{239}Pu($n_{\rm th}$,f). We note, however, that calculations in Refs. [5], [27], showed that the neutron-neutron angular correlations are sensitive to the value of x. Therefore, we could use our data to determine x

for ^{240}Pu(sf).

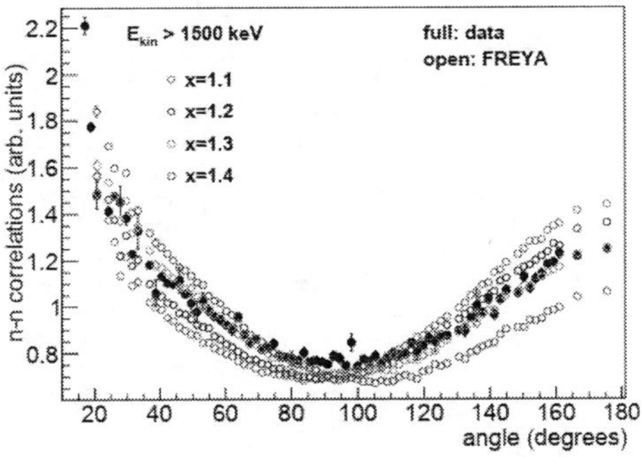

Fig. 5. The cross talk corrected two-neutron angular correlation for ^{240}Pu(sf) as a function of the angular separation and for a neutron kinetic energy threshold of 1500 keV. FREYA calculations with x = 1.1, 1.2, 1.3 and 1.4 are also shown.

To try to determine the value of x that agrees best with our ^{240}Pu(sf) data, we compared the data to four different x values between 1.1 and 1.4. The results of the FREYA/TRIPOLI-4 simulations are shown in Fig. 5. Table I lists the values of the neutron multiplicities $\bar{\nu}$, $\bar{\nu}_L$ and $\bar{\nu}_H$ for the light and heavy fission fragments. Increasing x for FREYA effectively shifts and tilts the correlation from approximately equal intensity at 0° and 180° with $x = 1.1$ to a significantly higher correlation at 0° for $x = 1.4$. This behavior can be explained by the characteristics of neutron evaporation, as stated earlier. The higher the value of x, the more neutrons the light fragment emits. This neutron yield asymmetry leads to a larger correlation at 0°. The neutron correlation measurements determined that $x = 1.3$ gives the best match between simulations and data, a value slightly greater than the original one.

TABLE I
AVERAGE NEUTRON MULTIPLICITIES (LIGHT AND HEAVY FISSION FRAGMENTS AND TOTAL) FROM ^{240}PU(SF) AND ^{252}CF(SF). PARAMETERS IN 3rd COLUMN IS x FOR FREYA AND R_T^{MIN}-R_T^{MAX}-σ_L-σ_H FOR FIFRELIN.

isotope	model	parameters	$\bar{\nu}_L$	$\bar{\nu}_H$	$\bar{\nu}$
^{240}Pu(sf)	FREYA	1.1	1.241	0.993	2.23
		1.2	1.365	0.870	2.23
		1.3	1.487	0.748	2.23
		1.4	1.608	0.610	2.22
	FIFRELIN	0.35-1.40-8.5-8.0	1.335	0.807	2.14
		0.5-1.50-8.0-7.5	1.427	0.688	2.11
		0.6-1.70-8.0-7.5	1.552	0.556	2.11
^{252}Cf(sf)	FREYA	1.27	2.18	1.58	3.76
	FIFRELIN	0.35-1.40-11-11	2.059	1.675	3.73

Figures 2 and 4 show that for energies below 1200 keV, FREYA overpredicts the 180° correlations. The agreement improves at higher energies. We have adjusted x to the measured ^{240}Pu(sf) neutron-neutron angular correlation data assuming it is single-valued, an assumption common to all isotopes in FREYA. However, comparison with ^{252}Cf(sf) data on the neutron multiplicity as a function of fragment mass suggests that x should be mass dependent. Adjusting x as a function of fragment mass may improve the overall comparison of the angular correlation data with the simulations.

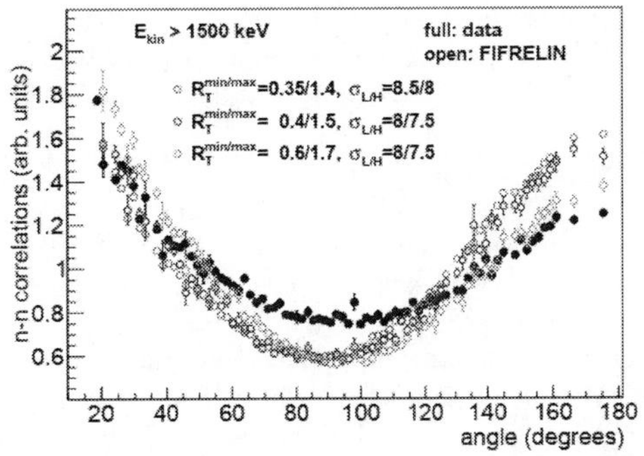

Fig. 6. The cross talk corrected two-neutron angular correlation for ^{240}Pu(sf) as a function of the angular separation and for a neutron kinetic energy threshold of 1500 keV. FIFRELIN calculations with different sets of parameters are also shown.

Compared to FREYA's calculation scheme mentioned above, FIFRELIN does not adjust its free parameters to reproduce the neutron-neutron correlation for ^{252}Cf(sf). The objective of FIFRELIN is to predict fission observables once a fission target observable is achieved. Generally the average prompt neutron multiplicity $\bar{\nu}$ is obtained by adjusting free parameters (temperature ratio law and spin cut-off for primary spin distributions). In the case of ^{252}Cf(sf) the pair ($\bar{\nu}_L$,$\bar{\nu}_H$) is the target observable and consequently the excitation energy sharing between fragments is more accurately estimated. For instance these average multiplicities are respectively 2.06 and 1.68 for light and heavy fragment groups (compared to reference 2.05 and 1.70). The mean neutron energy spectrum is also well reproduced ($<E>$=2.12 MeV in perfect agreement with Mannhart's standard reference [30]). Nevertheless the spectrum shape is not perfect and improvements are still necessary. We have shown that it can be greatly improved [31] by adding 2% of scission neutrons (emitted at scission before acceleration of fragments). For ^{240}Pu(sf) we have no information on the pair ($\bar{\nu}_L$,$\bar{\nu}_H$). In such case, FIFRELIN normally defaults to adjusting its four model parameters to reproduce the single $\bar{\nu} = 2.14$ value. This is not ideal because of compensation that may arise between light and heavy fragments groups. In this work, in addition to $\bar{\nu}$, angular correlation data can be used to fix the set of model parameters. The effect is shown in Fig. 6. The last set of parameters in Table I was selected to achieve an overall good agreement with measured correlations (Fig.4).

V. CONCLUSION

The event-by-event fission generators FREYA and FIFRELIN have been integrated into and coupled with the Monte Carlo code TRIPOLI-4. The combination of a physics-based fission event generator and an established radiation transport code leads to new capabilities: the simulation of correlations that conventional neutron Monte Carlo codes cannot predict. Users can now directly model fissions event-by-event and transport fission secondaries through complex detector geometries while keeping them fully correlated from production to detection.

Two-neutron angular correlation distributions were measured for ^{252}Cf(sf) and ^{240}Pu(sf) using an array of liquid scintillators. The shape of the correlation distributions depends on how the excitation energy is partitioned between the two fission fragments. Experimental data suggest that the lighter fragment is disproportionately excited for spontaneous fission and thermal neutron induced fissions. With the codes TRIPOLI-4 along with FREYA and FIFRELIN, we were able to reproduce the overall shape of the experimentally-measured neutron-neutron angular correlations. Moreover, for ^{240}Pu(sf), the measured asymmetry was used as a means to adjust the FREYA model parameter x, which controls the energy partition between the fragments. It is important to note that this parameter was so far inaccessible in other measurements. Indeed, there are no existing data available to quantify this energy partition for ^{240}Pu(sf). Model parameters in FIFRELIN are not directly linked to the excitation energy sharing but temperature and spin distributions before neutron emission. It is thus more difficult to tune on neutron-neutron correlations. Nevertheless without any information on ($\bar{\nu}_L$,$\bar{\nu}_H$) the measured neutron-neutron correlations are very useful since they are a signature of the excitation energy sharing and thus of the neutron multiplicity per fragment group. In addition this data can be measured for several neutron kinetic energy thresholds, which enables us to detect potential problems in the simulation of the prompt fission neutron spectrum.

ACKNOWLEDGMENT

This work was performed under the auspices of the U.S. Department of Energy by Lawrence Livermore National Laboratory under Contract DE-AC52-07NA27344. TRIPOLI-4® is a registered trademark of CEA. The authors wish to thank Electricité de France (EDF) for partial financial support.

REFERENCES

[1] J. Randrup and R. Vogt, "Calculation of fission observables through event-by-event simulation," *Phys. Rev. C*, vol. 80, p. 024601, 2009.

[2] R. Vogt, J. Randrup, J. Pruet, and W. Younes, "Event-by-event study of prompt neutrons from ^{239}Pu(n,f)," *Phys. Rev. C*, vol. 80, p. 044611, 2009.

[3] R. Vogt and J. Randrup, "Event-by-event study of neutron observables in spontaneous and thermal fission," *Phys. Rev. C*, vol. 84, p. 044621, 2011.

[4] R. Vogt, J. Randrup, D. A. Brown, M. A. Descalle, and W. E. Ormand, "Event-by-event evaluation of the prompt fission neutron spectrum from ^{239}Pu(n,f)," *Phys. Rev. C*, vol. 85, p. 024608, 2012.

[5] R. Vogt and J. Randrup, "Event-by-event study of photon observables in spontaneous and thermal fission," *Phys. Rev. C*, vol. 87, p. 044602, 2013.

[6] J. Randrup and R. Vogt, "Refined treatment of angular momentum in the event-by-event fission model FREYA," *Phys. Rev. C*, vol. 89, p. 044601, 2014.

[7] R. Vogt and J. Randrup, "Neutron angular correlations in spontaneous and neutron-induced fission," *Phys. Rev. C*, vol. 90, p. 064623, 2014.

[8] J. M. Verbeke, J. Randrup, and R. Vogt, "Fission reaction yield algorithm FREYA for event-by-event simulation of fission," *Comp. Phys. Comm.*, vol. 191, pp. 178–202, 2015.

[9] ——, "Fission reaction yield algorithm FREYA 2.0.2 user manual," *LLNL-SM-705798, Lawrence Livermore National Laboratory*, submitted to Comp. Phys. Comm.

[10] V. F. Weisskopf, "Statistics and nuclear reactions," *Phys. Rev.*, vol. 52, p. 295, 1937.

[11] J. Terrell, "Fission neutron spectra and nuclear temperatures," *Phys. Rev.*, vol. 113, p. 527, 1959.

[12] D. Madland and J. Nix, "New calculation of prompt fission neutron spectra and average prompt neutron multiplicities," *Nucl. Sci. Eng.*, vol. 81, pp. 213–271, 1982.

[13] R. Capote and al., "RIPL – reference input parameter library for calculation of nuclear reactions and nuclear data evaluations," *Nucl. Data Sheets*, vol. 110, pp. 3107–3214, 2009.

[14] O. Litaize and O. Serot, "Investigation of phenomenological models for the Monte Carlo simulation of the prompt fission neutron and γ emission," *Phys. Rev. C*, vol. 82, no. 5, p. 054616, Nov. 2010. [Online]. Available: http://link.aps.org/doi/10.1103/PhysRevC.82.054616

[15] D. Regnier, O. Litaize, and O. Serot, "Preliminary results of a full Hauser-Feshbach simulation of the prompt neutron and gamma emission from fission fragments," *Phys. Procedia*, vol. 47, pp. 47 – 52, 2013. [Online]. Available: http://www.sciencedirect.com/science/article/pii/S1875389213004379

[16] ——, "An improved numerical method to compute neutron/gamma deexcitation cascades starting from a high spin state," *Comp. Phys. Comm.*, vol. 201, pp. 19 – 28, 2016. [Online]. Available: http://www.sciencedirect.com/science/article/pii/S0010465515004452

[17] O. Litaize, O. Serot, and L. Berge, "Fission modelling with FIFRELIN," *Eur. Phys. J. A*, vol. 51, no. 12, p. 177, 2015.

[18] N. Varapai, F. J. Hambsch, S. Oberstedt, O. Serot, G. Barreau, N. Kornilov, and S. Zeinalov, "Development of a Digital Technique for the Determination of Fission Fragments and Emitted Prompt Neutron Characteristics," *AIP Conf. Proc.*, vol. 798, pp. 369–372, 2005.

[19] L. Demattè, C. Wagemans, R. Barthélémy, P. D'hondt, and A. Deruytter, "Fragments' mass and energy characteristics in the spontaneous fission of ^{236}Pu, ^{238}Pu, ^{240}Pu, ^{242}Pu, and ^{244}Pu," *Nucl. Phys. A*, vol. 617, pp. 331–346, 1997.

[20] H. A. Bethe, "An attempt to calculate the number of energy levels of a heavy nucleus," *Phys. Rev.*, vol. 50, pp. 332–341, Aug 1936. [Online]. Available: https://link.aps.org/doi/10.1103/PhysRev.50.332

[21] F. Becvar, "Simulation of γ cascades in complex nuclei with emphasis on assessment of uncertainties of cascade-related quantities," *Nucl. Instrum. Meth. Phys, Res. A*, vol. 417, pp. 434 – 449, 1998. [Online]. Available: http://www.sciencedirect.com/science/article/pii/S0168900298007876

[22] TRIPOLI 4® Project Team, "TRIPOLI-4®, CEA, EDF and AREVA reference Monte Carlo code," *Ann. Nucl. Energy*, vol. 82, pp. 151–160, 2015.

[23] J. M. Verbeke and O. Petit, "Stochastic analog neutron transport with TRIPOLI-4 and FREYA: Bayesian uncertainty quantification for neutron multiplicity counting," *Nucl. Sci. Eng.*, vol. 183, pp. 214–228, 2016.

[24] O. Petit, C. Jouanne, O. Litaize, O. Serot, A. Chebboubi, and Y. Pénéliau, "FIFRELIN-TRIPOLI-4 coupling for Monte Carlo simulations with a fission model. Application to shielding calculations," *to be published in Proc. ICRS-13 & RPSD-2016 (EPJ - Web of Conference)*.

[25] O. Petit and E. Dumonteil, "Analog neutron transport for nuclear instrumentation applications with the Monte Carlo code TRIPOLI-4," *Nucl. Technol.*, vol. 192, p. 259, 2015.

[26] (2016). [Online]. Available: http://www.eljentechnology.com

[27] J. M. Verbeke, L. F. Nakae, and R. Vogt, "Neutron-neutron angular correlations in spontaneous fission of ^{252}Cf and ^{240}Pu," *Phys. Rev. C*, submitted for publication.

[28] A. M. Gagarski, I. S. Guseva, V. E. Sokolov, G. V. Val'ski, G. A. Petrov, D. O. Krinitsin, D. V. Nikolaev, T. A. Zavarukhina, and V. I. Petrovaal, "Neutron-neutron angular correlations in spontaneous fission of ^{252}Cf," *Bull. Russ. Acad. Sci.: Phys.*, vol. 72, p. 773, 2008.

[29] S. A. Pozzi, B. Wieger, A. Enqvist, S. Clarke, M. Flaska, M. Marcath, E. Larsen, R. C. Haight, and E. Padovani, "Correlated neutron emissions from ^{252}Cf," *Nucl. Sci. Eng.*, vol. 178, pp. 250–260, 2014.

[30] W. Mannhart, "Evaluation of the ^{252}Cf fission neutron spectrum between 0 Mev and 20 Mev," *Proc. Advisory Group Mtg. Neutron Sources, Leningrad, USSR, 1986 (IAEA-TECDOC-410), Vienna*, 1987.

[31] O. Serot, O. Litaize, and A. Chebboubi, "Influence of scission neutrons on the prompt fission neutron spectrum calculations," *Proc. Int. Conf. ND-2016, 11-16 September, 2016, Brugges, Belgium*, 2016.

Piezoelectric Materials Under Natural and Man-Made Radiation: The Potential for Direct Radiation Detection

Megan Wart, Evan Simpson, and Marek Flaska

Abstract—Radiation detection systems used for monitoring long term waste storage need to be compact, rugged, and have low or no power requirements. By using piezoelectric materials it may be possible to create a reliable self-powered radiation detection system. To determine the feasibility of this approach, the electrical signal response of the piezoelectric materials to radiation must be characterized. To do so, an experimental geometry has been designed and a neutron source has been chosen as described in this paper, which will be used to irradiate a uranium foil for producing fission fragments. These future experiments will be aimed at finding the threshold of exposure of lead zirconate titanate (PZT) plates needed to produce and electrical signal. Based on the proposed experimental geometry the thermal neutron beam-line at the Breazeale Reactor at The Pennsylvania State University will be used as the neutron source. The uranium foil and neutron source will be able to supply a maximum flux of 1.5e5 fission fragments/second*cm^2 to each of the PZT plates.

Index Terms—radiation detection, fission fragments, lead zirconate titanate, piezoelectric materials

1 Introduction

RADIATION detection is very important for maintaining the safety of nuclear facilities. Radiation detection systems are used to monitor radiation levels for personnel safety, as well as to ensure the integrity of nuclear materials storage, such as dry cask waste storage at reactor facilities, and interim or long term waste storage facilities. For monitoring waste storage it is useful to have a compact and rugged detection system that has low or no power requirements, since the waste containers are packed close together and are often exposed to the elements.

In an effort to create a self-powered detection system, the use of piezoelectric materials as a direct radiation sensing medium has been studied. Piezoelectric materials are able to convert between mechanical and electrical energy due to electric dipoles within the crystal structure of the material. When mechanically stressed, piezoelectric materials produce an electrical signal. Conversely, during the application of an electrical voltage the piezoelectric material will its change shape [1]. This behavior makes these materials useful for applications such as sensors, actuators, energy harvesting and generation, micro/nano-electronic-mechanical systems, and piezoelectric igniters.

In previous work it has been shown that lead zirconate titanate (PZT) is sufficiently sensitive to produce an electrical signal when bombarded with high energy heavy ions. PZT samples were exposed directly to a 400 MeV/n xenon beam, resulting in a visible electrical signal [2]. It was also shown that the PZT samples could detect acoustic waves in liquid created by high energy xenon ions.

• *M. Wart, E. Simpson and M. Flaska are with the Department of Mechanical and Nuclear Engineering, The Pennsylvania State University, State College, PA, 16802.*
E-mail: mxf5309@psu.edu

Manuscript received July 15, 2017.

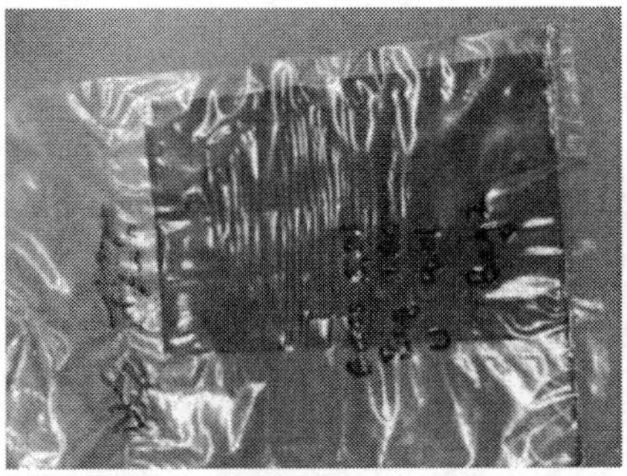

Fig. 1: 93.19% enriched metallic uranium foil.

Based on these reported behaviors, it may be possible to use piezoelectric materials to detect lower energy radiation through the force applied on the piezoelectric materials by radiation. To test this assumption, an experimental set up to irradiate PZT plates with fission fragments was designed and evaluated for radiation safety and feasibility. Fission fragments were chosen as the initial experimental radiation source due to their high energy and large mass, which will results in large amounts of energy being deposited in a short distance. To create a source of fission fragments a metallic uranium foil, shown in Figure 1, enriched to 93.19%, will be irradiated with a neutron source.

PZT was chosen as the first piezoelectric material to be tested since it is inexpensive and commonly available. A frame, shown in Figures 2 and 3 was then constructed to support the uranium foil, which is very fragile, and the PZT plates. The foil will be positioned between two sheets of mylar in order to provide contamination control and placed on top of an aluminum plate. A nylon plate was used for the

Fig. 2: Aluminum and nylon frame used to hold uranium foil and PZT plates.

Fig. 3: Cross section view of the frame. PZT is indicated by black, Aluminum is indicated by gray and nylon is indicated by white.

other half of the frame. The nylon plate has two sets of wells and channels carved into it to hold two PZT plates and the wiring.

Two possible options of neutron source are available and have been evaluated for feasibility with the proposed experimental geometry: a 1 Curie plutonium-beryllium (PuBe) source and the thermal neutron beam-line at the Breazeale Reactor at Penn State. The PuBe source has a strength of 1.4e6 neutrons/second and, for the purpose of these evaluations, it was assumed to be an isotropic point source. The thermal neutron beam-line has a flux of 3.7e7 neutrons/cm^2*second and a 7 inch diameter [3]. It was assumed that the beam-line neutrons are mono-energetic with an energy of .025 eV. Both neutron sources can be placed at various distances from the uranium foil. For the models discussed in this paper the neutron source was always placed 1 cm from the nylon plate.

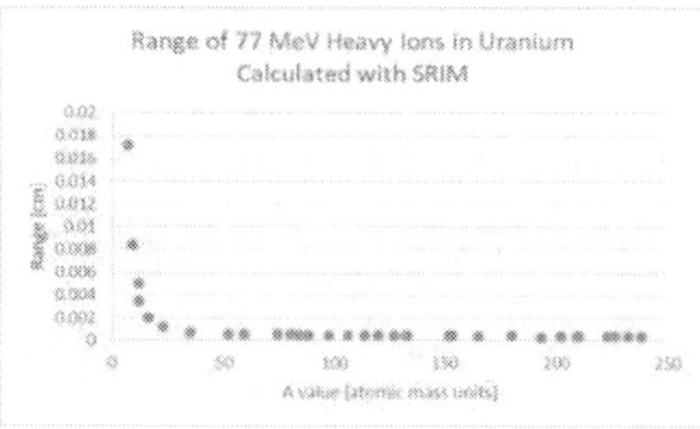

Fig. 4: Ranges for 77 MeV heavy ions as calculated with SRIM.

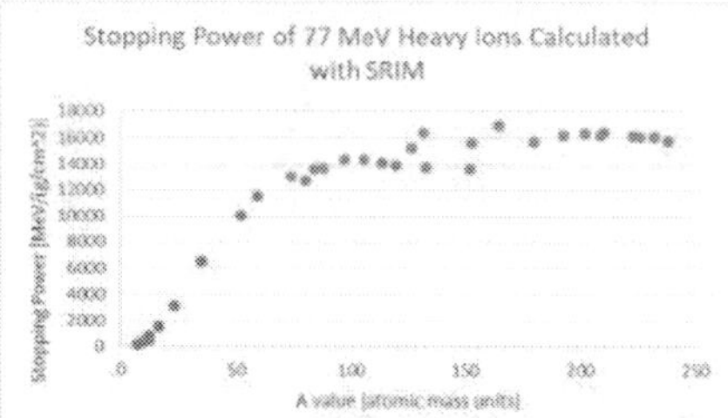

Fig. 5: Stopping powers for 77 MeV heavy ions as calculated with SRIM.

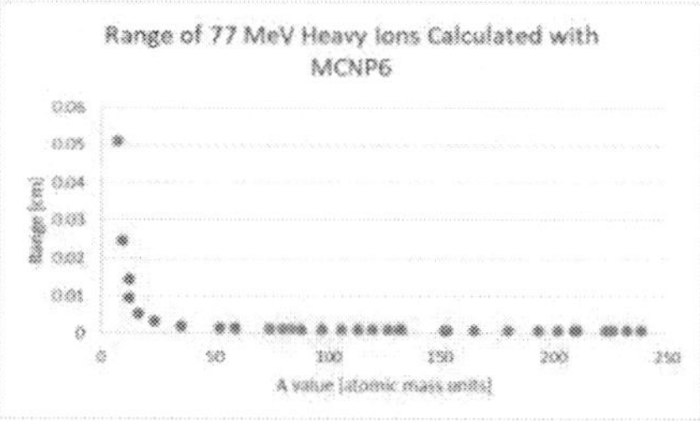

Fig. 6: Ranges for 77 MeV heavy Ions as calculated with MCNP6.

2 Calculation of Ranges of Fission Fragments in PZT

The first step in determining the fission fragment exposure of the PZT plates in the geometry was to determine how far the fission fragments created in the uranium foil would be able to travel. This distance controls what percentage of the fission fragments are able to escape the foil and interact with the PZT plates. To find the range of the fission fragments, models using both Stopping Range of Ions in Matter (SRIM) and Monte Carlo N-Particle Transport 6 code (MCNP6) were evaluated [4], [5]. For these models it was assumed that the uranium foil was 100% uranium-235.

SRIM is an open source software that uses quantum mechanics to calculate ion transport data. For the SRIM range calculations, an infinitely thick plate of uranium-235 was modeled. This plate was bombarded with a variety of mono-energetic heavy ions, each run using a different heavy ion, and the resulting ranges were tabulated. For these models the heavy-ions were all given an energy of 77 MeV, since 77 MeV is the average energy of a fission fragment created by fissioning a uranium-235 nucleus. The stopping power, which describes the energy lost per unit path length of the radiation in a particular material, was also calculated with these SRIM models and tabulated. The plots of the resulting ranges and stopping power are shown in Figures 4 and 5.

For comparison the range and stopping power of the same selection of 77 MeV heavy ions were then calculated using MCNP6. MCNP6 is a Monte Carlo method based radiation transport code. It uses a combination of the Janni proton range data tables and SPAR FORTRAN code with a density correction factor to calculate the geometry independent ranges and stopping powers for heavy ions in different target materials [6], [7]. The plots of the resulting ranges and stopping powers are shown in Figures 6 and 7.

The resulting ranges from both the SRIM and MCNP6 models were very similar for 77 MeV ions in uranium-235 and both followed the same trend of the range rapidly decreasing initially, then trailing off as the atomic mass value of the ions increased. The lightest ion had a range of .51

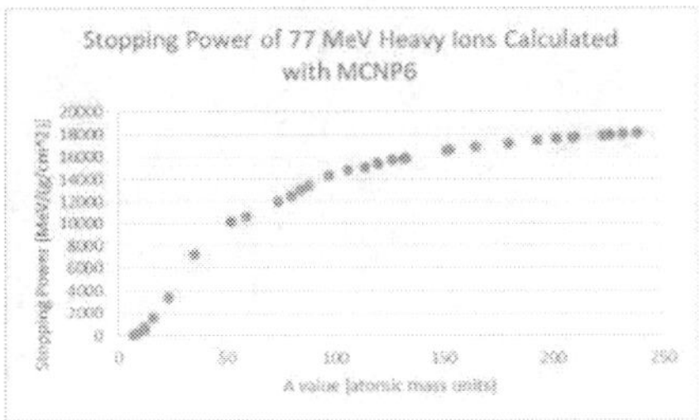

Fig. 7: Stopping powers for 77 MeV heavy ions as calculated with MCNP6.

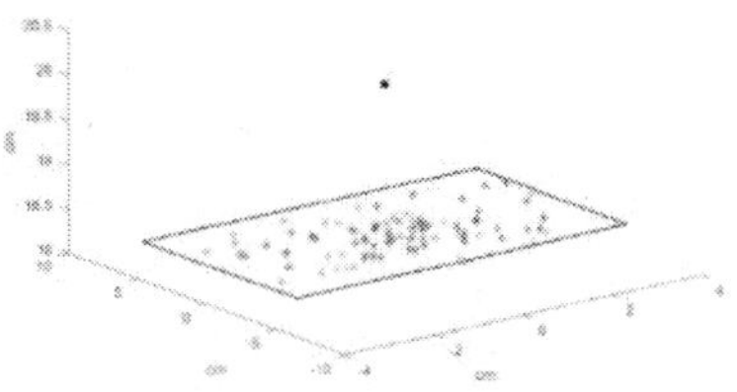

Fig. 8: Fission fragments born in the uranium foil. The PuBe source is indicated by the black dot.

mm and the heaviest had a range of 3.8 μm, which indicates that a portion of the fission fragments born in the foil will be able to escape the foil and interact with the PZT plates. The ranges calculated by MCNP6 were in general lower than those calculated using SRIM. The stopping powers calculated using SRIM show artifacts that are likely due to interpolations for points where there is no data available. The MCNP6 stopping power results were smooth and were typically slightly lower than those calculated with SRIM.

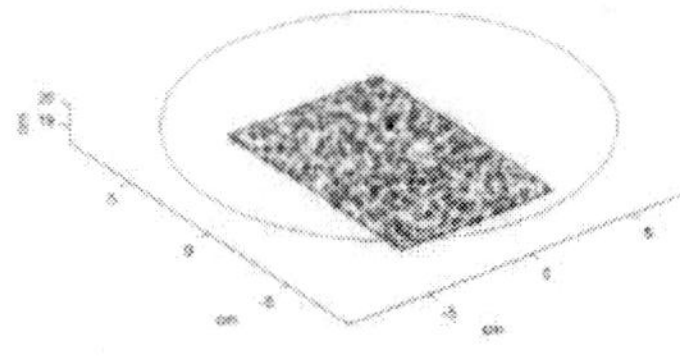

Fig. 9: Fission fragments born in the uranium foil. The thermal neutron beam-line is centered on the black dot and its perimeter is represented by the blue circle.

3 Source Selection: Fission Rate, Fission Fragment Production, and Fission Fragment Flux Calculations

In order to chose which neutron source is the better option, the fission fragment flux that the PZT plates will be exposed to, was calculated. First, the fission rates were calculated using a model of the experimental geometry in MCNP6 and the given source strengths. For the PuBe source, the fission rate in the uranium foil for the given geometry was determined to be 7.4e2 fissions/second. For the thermal neutron beam-line, the fission rate was calculated to be 2.2e8 fissions/second in the uranium foil. To determine the fission fragment production rate the fission rate was doubled, since fissions typically produce two fission fragments. This resulted in fission fragment production rates of 1.5e3 fission fragments/second for the PuBe source and 4.4e8 fission fragments/second for the thermal neutron beam-line.

Next, the average range of the average mass fission fragments was determined using SRIM, so that the volume of uranium foil contributing fission fragments to the fission fragment flux, which the PZT plates will interact with, could be determined. The average energy of a fission fragment from a uranium-235 fission was already determined to be 77 MeV and the average mass was calculated to be 117 atomic mass units. This gives a range of 4.69e-4 cm. Since the PZT plates are 45 mm by 45 mm, this gives a contributing foil volume of .0095 cm^3 for each of the PZT plates.

The fission rates were then normalized by volume. This resulted in a volumetric fission rate of 1.1e3 fissions/second*cm^3 for the PuBe source and a volumetric fission rate of 3.24e8 fissions/second*cm^3.

The fission fragment flux into each foil was then calculated assuming that each fission could only contribute one fission fragment, due to the directionality of the fission fragment trajectories. Using the volume normalized fission rates and the average range, the fission fragment flux into each PZT plate was determined to be .511 fission fragments/second*cm^2 for the PuBe source and 1.5e5 fission fragments/second*cm^2 for the thermal neutron beam-line.

To be able to understand the 2D distribution of the fissions and fission fragment production in the foil, MCNP6 Particle Track (PTRAC) outputs were used to create visualizations of the fission fragments being born in the foil. The PTRAC files were filtered and analyzed using Matlab scripts, in order to determine the location of the births of all the heavy ions in the foil. These were recorded and plotted, as seen in Figures 8 and 9, for both of the neutron sources. The PuBe source model in MCNP6 used 967,864,929 starting neutrons and the thermal neutron beam-line model used 1,248,130,865 starting neutrons. It can be seen in these results that the PuBe source created in a clustering of fission in the center due to it being an isotropic point source, while the thermal neutron beam-line gave a relatively uniform distribution over the surface of the foil. The PTRAC file was then analyzed to create visualizations of the heavy ion particle tracks that were born in the foil and entered the PZT. These visualizations are shown in Figures 10 and 11. Once again the PuBe source results in a higher density of fission fragments towards the center while the thermal neutron beam-line results in an relatively even distribution over the foil area.

4 PZT Sample Preparations

For the first round of experiments using the proposed experimental geometry, PZT plates were purchased. In order to prepare them for the experiments, custom contacts were created and technique for wiring the samples was developed. The custom contacts were patterned to provide differing

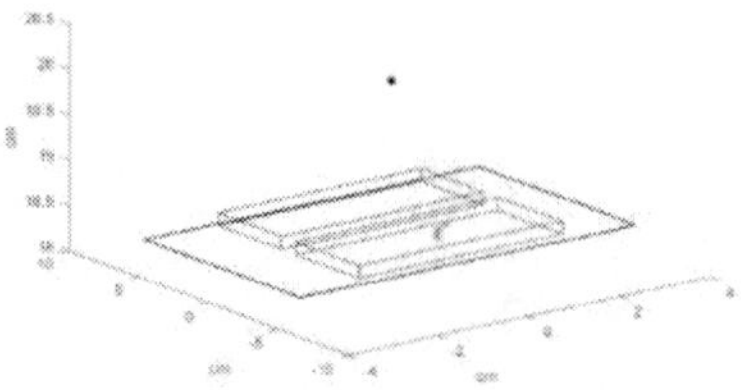

Fig. 10: Fission fragment tracks leaving the foil and entering the PZT plates. The PuBe source is centered on the black dot.

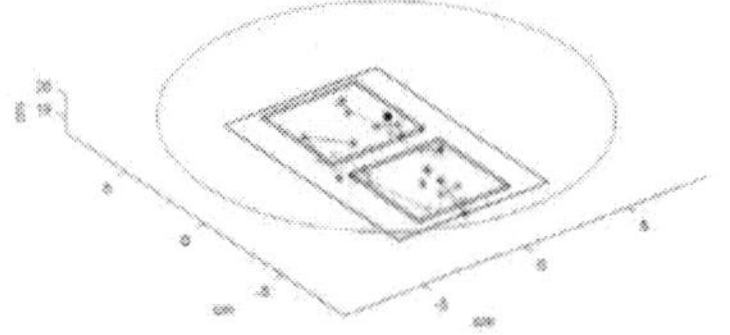

Fig. 11: Fission fragment tracks leaving the foil and entering the PZT plates. The thermal neutron beam-line is centered on the black dot and its perimeter is represented by the blue circle.

amounts of exposed PZT. This exposed PZT is important since it allows the fission fragments to deposit more energy directly into the PZT, instead of having to pass through the silver contact. A variety of comb shaped patterns were created with different comb tooth widths to give a range of electrode area coverages. Table 1 shows the various areas and comb widths for the contacts. For piezoelectric materials it is important to keep the material under its Curie temperature: the temperature threshold where the crystalline structure goes through a phase change and it looses its poling. To ensure the PZT plates remained under their Curie temperature of 320 degrees Celsius all preparations occurred at room temperature and required no heating.

For the contact preparations, the original fired-on silver electrodes were removed by polishing with Al_20_3. A mask was then created with Kapton tape on the surface to protect the exposed PZT surface area when a new patterned silver contact was sputtered on. The PZT plate was covered on all exposed surfaces with Kapton tape. Then the mask was cut into the Kapton tape using a lazer cutter and the excess Kapton tape was removed. A masked PZT plate is shown in Figure 12. The sputtered silver was put on with a thickness of 100 nm. An example of the final custom contacts are shown in Figure 13. The wiring for each PZT plate was attached to a ring terminal which was connected to the PZT plate using a silver conductive epoxy. A wired sample is shown in Figure 14.

Fig. 12: Kapton tape masking on PZT plates after cutting on the left, and post removal of excess Kapton tape on the right.

TABLE 1: Custom electrode areas and comb teeth widths.

width of non-electrode bands [mm]	width of electrode bands [mm]	percent area non-electrode	percent area electrode
2	2	49	51
2	4	40	60
2	8	30	70
2	10	25	75
2	2	51	49
4	2	60	40
8	2	70	30
10	2	75	25

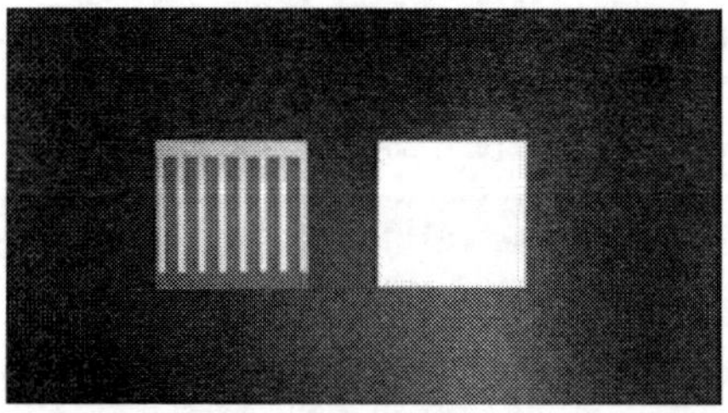

Fig. 13: PZT plate with solid commercial electrode on the right and a custom sputtered electrode on the left. The custom electrode coves 40% of the surface area of the face of the plate.

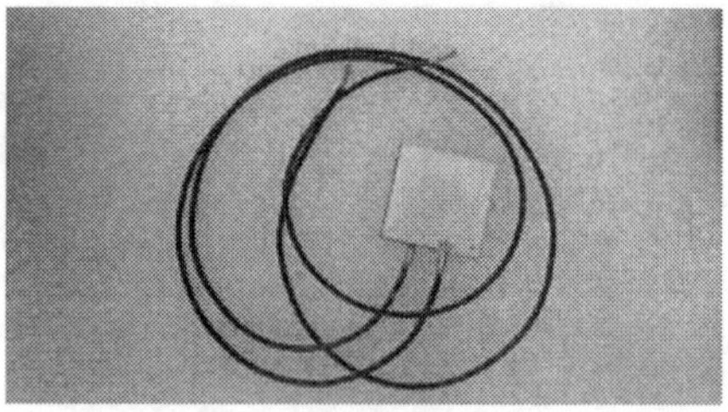

Fig. 14: PZT plate with commercial electrodes, epoxied ring terminals and wiring.

5 Conclusions and Future Work

For the proposed experimental geometry including the frame, uranium foil, PZT plates, and neutron source, the thermal neutron beam-line at the Breazeale Reactor will provide the larger fission fragment flux entering the PZT plates. This larger fission fragment flux will provide a greater force

https://doi.org/10.1051/epjconf/201817001020

on the PZT plates, which will result in the production of a larger electrical signal. This fission fragment flux will be a maximum of 1.5e5 fission fragments/second*cm^2.

The next step in this project will be to implement the experimental geometry evaluated in this paper. This will begin with PZT plates that have been prepared and will use a decreasing distance from the neutron source to control the fission fragment flux. In the future, other types of radiation, instead of fission fragments, will be tested including alpha particles and neutrons. Other piezoelectric materials will also be considered. PZT is very common but is also considered relatively resistant to radiation [8] therefore custom-made, highly sensitive piezoelectric foils will also be tested.

References

[1] A. Arnau and D. Soares, "Fundamentals of piezoelectricity," *Piezoelectric Transducers and Applications*, pp. 1–38, 2008.
[2] S. Takechi, T. Miyachi, M. Fujii, N. Hasebe, K. Mori, H. Shibata, T. Murakami, Y. Uchihori, and N. Okada, "Radiation detector based on piezoelectric lead zirconate titanate material," *Nuclear Instruments & Methods in Physics Research Section a-Accelerators Spectrometers Detectors and Associated Equipment*, vol. 586, pp. 309–313, 2008.
[3] J. J. Duderstadt and L. J. Hamilton, "Nuclear Reactor Analysis," pp. 138–142, 1976.
[4] J. F. Ziegler, "SRIM-The Stopping and Range of Ions in Matter," 2013. [Online]. Available: www.srim.org
[5] "MCNP6-a general Monte Carlo N-Particle Transport Code, version 6.1," Los Alamos, New Mexico, 2013.
[6] J. F. Janni, "Energy loss, range, path length, time-of-flight, straggling, multiple scattering, and nuclear interaction probability." *Atomic Data and Nuclear Data Tables*, vol. 27, no. 2-3, pp. 147–339, 1982.
[7] T. W. Armstrong and K. C. Chandler, "Stopping powers and ranges for muons, charged pions, protons, and heavy ions," *Nuclear Instruments and Methods*, vol. 113, no. 2, pp. 313–314, 1973.
[8] A. Holmes-Siedle and L. Adams, *The Handbook of Radiation Effects*, 2nd ed. Oxford University Press, 2002.

https://doi.org/10.1051/epjconf/201817002001

Intrinsic Fabry-Perot Sensors for Magnetic Field Detection

Christian Broadway, Frédéric Descamps, Damien Kinet, Christophe Caucheteur, and Patrice Mégret

***Abstract*—Within the context of ensuring stable nuclear fusion, it is important to monitor and control a number of parameters including the magnetic field associated with plasma circulation. Optical fibre sensing techniques have seen a surge in promulgation and research advances in recent years, due to their immunity to electromagnetic radiation and compact dimensions. Prior work has shown that fibre Bragg gratings are one method of recovering the induced magnetic field, with the main point of interest being their use as distributed point sensors. However, Bragg grating inscription leads to the creation of linear birefringence that increases detector noise and could obscure a given signal. We have hypothesised that by using an intrinsic Fabry-Perot cavity comprised of two identical Bragg gratings, we could obtain a more accurate detector with the removal of photo-induced birefringence in the detection region. We present a proof of concept optical fibre sensor based on an intrinsic Fabry-Perot cavity that shows spectrally visible amplitude modulation. Finally, we demonstrate faster data processing that allows real time monitoring of a given scenario.**

***Index Terms*—Fiber Optic Current Sensors, Fiber Bragg grating sensors.**

I. Introduction

Many applications exist for magnetic field sensors in the modern world, each with differing requirements for an eventual sensor. Three such examples are geophysical imaging, magnetic resonance imaging and field measurements within a TOKAMAK environment.

Magnetic sensing can be achieved by a range of optical and non-optical means. These include magneto-electric sensors[1], magneto-strictive devices [2], opto-fluidics [2], phased shifted Bragg structures [3], polarization based optical sensors [4], [5], [6] and optical birefringence analysis [7], [8]. Each methodology has specific advantages and disadvantages that can indicate specific applications.

Optical fibre sensors share an immunity to electrical interference and typically offer a passive sensor configuration. The wide variety of sensing configurations, materials, geometry and a typically small size can deliver sensitivity inversely proportional to its own dimensions has led to widespread research into applications from acoustics [9] to humidity [10].

One detection configuration is the inscription of periodic modulations within the core of a given optical fibre, known as a Bragg grating. These gratings cause a narrowband reflection around a central wavelength known as the Bragg wavelength.

C. Broadway, F. Descamps, D. Kinet, C. Caucheteur, and P. Mégret are with the Electromagnetism and Telecommunications Department of the University of Mons, 31 Boulevard Dolez, 7000 Mons, Belgium (email: christian.broadway@umons.ac.be, damien.kinet@umons.ac.be, christophe.caucheteur@umons.ac.be, patrice.megret@umons.ac.be)

Bragg gratings are most commonly known for sensing temperature and strain [11] measurements but are also used for a wide variety of parameters including hydrogen detection [12] and acousto-optics [13]. Bragg gratings act as point detectors, isolating the reflected response from any perturbation along the fibre length. This has led to distributed sensing, where cascaded Bragg gratings can be used in applications such as civil engineering [14], providing a location for a given stress along with a given magnitude.

Working to develop point detectors for magnetic fields, our group has undertaken theoretical and practical work to apply Bragg grating technology in this domain [15], [16]. We have previously demonstrated a detector with good linearity that distinguishes between inherent birefringence in the optical fibre and externally induced birefringence, proving the concept and showing the potential for distributed magnetic field sensing. However, the inscription of Bragg gratings causes a significant localised increase in the inherent birefringence of an optical fibre within the grating region. This has rendered mandatory complex mathematical treatments to extract the induced field strength, presenting a blockage to real time analysis.

In this work, we present an intrinsic Fabry-Perot cavity composed of two identical Bragg gratings, inscribed within the fibre core and separated from each other by a given distance. This cavity is sensitive to perturbations between the gratings where no additional birefringence has been induced from the inscription process. We compare this cavity to a standard Bragg grating, demonstrating faster data extrapolation that allows for real time magnetic field monitoring.

II. Theoretical Principles

The state of light polarization is the shape drawn by the tip of the electric field vector as a function of time at a given location along the fibre. For a monochromatic wave propagating in the z direction of a Cartesian coordinate system (x, y, z), the general solution of the wave equation can be stated as:

$$E_x = E_{0x} \cos(\omega t + \varphi_x) \tag{1}$$

$$E_y = E_{0y} \cos(\omega t + \varphi_y) \tag{2}$$

where $E_{0x(y)}$ are real numbers representing the amplitudes of the x and y components respectively with $\varphi_{x(y)}$ representing the given phase. Analysis of these two relations shows that the polarisation state is (1) linear when $\Delta\varphi = \varphi_y - \varphi_x = 0$ or π, (2) circular when $E_{0x} = E_{0y}$ and $\Delta\varphi = \pi/2$ or $3\pi/2$, and (3) elliptical in the general case.

https://doi.org/10.1051/epjconf/201817002001

Equations (1) and (2) can be written compactly using the Jones vector, a 2 by 1 complex vector:

$$\mathbf{V}(\omega) = \begin{bmatrix} E_{0x}\, e^{j\varphi_x} \\ E_{0y}\, e^{j\varphi_y} \end{bmatrix} \tag{3}$$

Guiding properties of perfectly circular singlemode fibres leads to the propagation of two orthogonal degenerate modes LP_{01x} and LP_{01y}, referred to as polarization modes or eigenmodes [17]. These modes propagate at the same velocity and experience the same effective refractive index ($n_x = n_y$). However, optical fibres are always slightly elliptical and this imperfection removes the mode degeneracy, i.e. n_x is no longer equal to n_y. The fibre therefore presents linear internal birefringence defined as:

$$B_l = |n_x - n_y| \tag{4}$$

Intrinsic birefringence in standard fibres typically corresponds to a refractive index difference in the region of 10^{-7}. However, linear birefringence can also be externally introduced by mechanical perturbations such as stress.

For chiral materials, eigenmodes are the left-handed and right-handed circularly polarised fields, characterised by effective refractive indices n_L and n_R. In a similar way to linear birefringence, circular birefringence can be defined as:

$$B_c = |n_L - n_R| \tag{5}$$

Under normal conditions, optical fibres do not present intrinsic circular birefringence, but twisting or magnetic fields can induce such birefringence.

When optical fibres are subjected to a longitudinal magnetic field B, an induced circular birefringence is created, this being known as the Faraday effect [18]. If polarised light is injected at the input of a fibre with length l, the output polarisation direction is rotated by angle α:

$$\alpha = V B\, l \tag{6}$$

where V is the Verdet constant. A magnetic field can be retrieved by either examining the angle of rotation or the induced circular birefringence that causes the rotation.

One method to extract the induced birefringence is to work with the fundamental fibre parameters, such as polarization dependent loss (PDL) and polarization mode dispersion (PMD). The PDL is the ratio of the maximum transmitted power to the minimum transmitted power over all polarization states, while the PMD (also called differential group delay (DGD)) is the maximum difference in group delay over all polarization states.

Propagation within an optical fibre can be described by the Jones' formalism that relates $\mathbf{V}_{\text{out}}$ with $\mathbf{V}_{\text{in}}$ Jones vectors through a 2 by 2 complex transfer matrix:

$$\mathbf{V}_{\text{out}}(\omega) = \begin{bmatrix} A(\omega) & B(\omega) \\ C(\omega) & D(\omega) \end{bmatrix} \mathbf{V}_{\text{in}}(\omega) \tag{7}$$

It can be shown that PDL [19] and PMD [20] can be extracted from the measurement of the 4 coefficients $A(\omega)$, $B(\omega)$, $C(\omega)$ and $D(\omega)$of the transfer matrix.

We affirm that a given magnetic field will induce additional birefringence in an optical fibre and modify the aforementioned fibre properties, directly correlating changes in these values to the magnetic field. One method of measuring a magnetic field can be to observe these values.

III. Experimental setup

Fibre Bragg grating (FBG) based optical sensors can be used to measure magnetic fields as described in II. These FBGs are a periodic modulation of the fibres refractive index, creating a passband in the reflected spectrum from the fibre and a stopband in the transmitted spectrum. While FBGs are widely used in different sensing applications, the photo-inscription process that causes the refractive index modulation also causes an increase in the region of one order of magnitude in linear fibre birefringence. This obscures the induced birefringence and has previously required complex data processing, preventing real time monitoring [8].

To overcome this limitation, we have defined and inscribed an intrinsic Fabry-Perot cavity formed from two identical uniform FBGs of 2 mm long and separated by a cavity with a length of 6.83 mm as shown in figure 1.

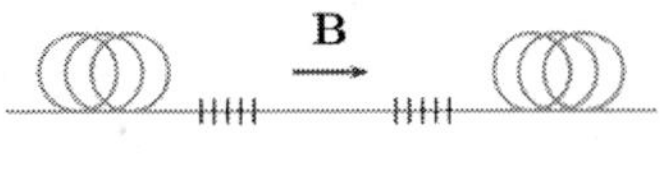

Fig. 1. Fabry-Perot cavity composed of two identical FBG with a cavity length of 6.83 mm.

The Fabry-Perot cavity is placed inside a cylindrical electromagnet with an air gap of 6 mm (cf. figure 2). The cavity length of 6.83 mm was defined so as to be slightly larger than the air gap to isolate the FBGs from the magnetic field and only expose the cavity between electromagnet plates. The electromagnet generates a magnetic field when stimulated by a current source and a optical vector analyser from LUNA is used to retrieve the spectra, PDL and PMD from the transfer matrix (7).

Fig. 2. Experimental setup: an electromagnet is used to generate a magnetic field between the two FBGs of figure 1.

We have characterized the electromagnet response by inducing a given current and measuring the induced magnetic field and change in internal temperature, the origin of which being electromagnet heating from the induced current. We observe in figure 3 a near linear relationship between current and magnetic field and a variation of temperature of less than 1 °C.

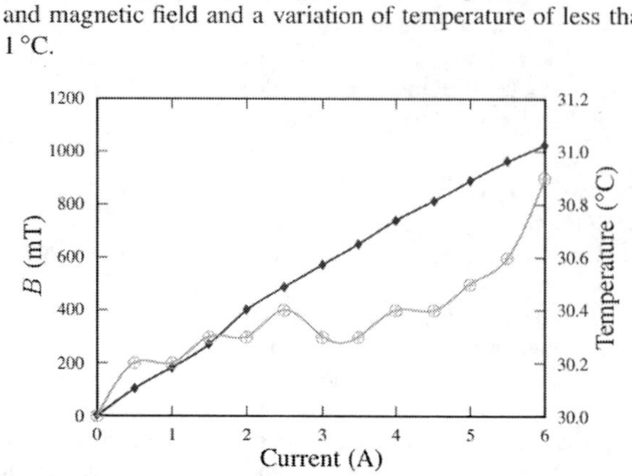

Fig. 3. Induced magnetic field strength (blue) and internal electromagnet temperature (orange) vs. current.

The Fabry-Perot cavity effect generates an interferometric peak pattern within the envelope of a uniform FBG spectrum (as shown in figure 4). There is a red-shift of the interference

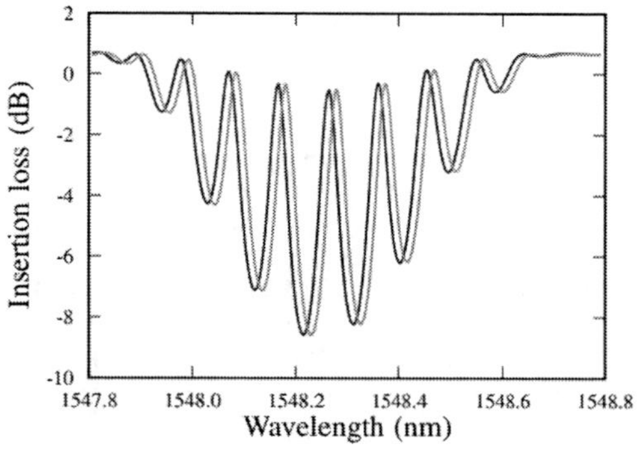

Fig. 4. Fabry-Perot spectral profile for currents of 0 A (blue) to 6 A (orange).

pattern between the two curves, which can be attributed to the associated temperature change, given that Bragg wavelength shifts in the region of 10 pm/°C have been well reported.

IV. Results

Using the LUNA polarimeter, we obtain key propagation parameters along the fibre transmission line. As opposed to examining Mueller matrix coefficients as previously done [8], we take the fundamental properties of the fibre such as polarization dependent loss (PDL) and polarisation mode dispersion (PMD), analysing the results to determine a correlation to the induced magnetic field.

A. PDL results

As explained in section II, the PDL is based on the amplitude component of the cavity response. The induced modulation caused by the magnetic field can be observed in

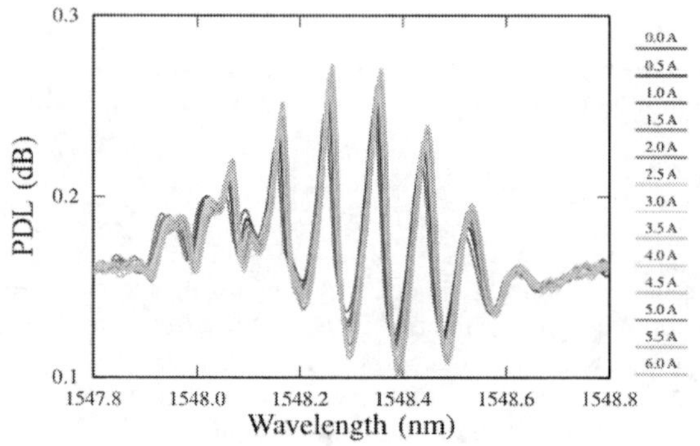

Fig. 5. PDL vs. wavelength per step current.

figure 5, showing the PDL spectrum change per 0.5 ampere step.

Due to the inhomogeneous nature of the profile, we identify the strongest central peak for data processing, giving figure 6. The closer look afforded shows a distinctive wavelength shift along with a noticeable amplitude shift. As mentioned previously, a temperature change is responsible for the wavelength shift and originates in electromagnet heating.

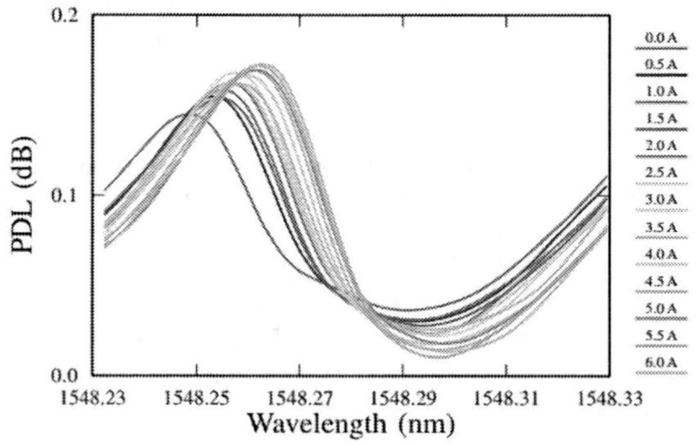

Fig. 6. Selected peak from figure 5.

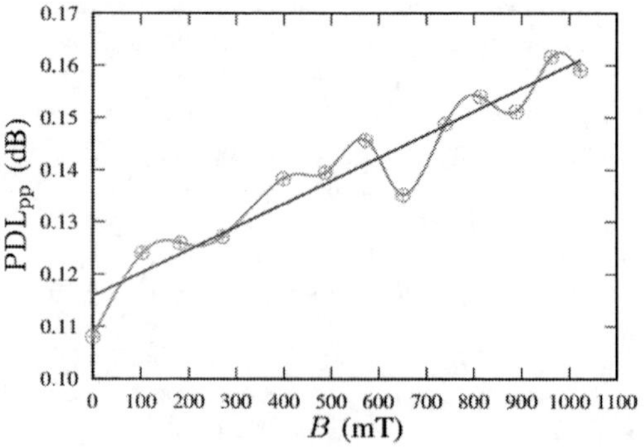

Fig. 7. Peak to peak PDL (orange) and least square fit (green) vs. the induced magnetic field.

From this, we calculate the peak to peak values for each ampere step, converting amperes to tesla as per figure 3, yielding figure 7. We have displayed a least square fit, which can be expressed as:

$$\mathrm{PDL[dB]} = 4.412\,77 \times 10^{-5}\ B[\mathrm{mT}] + 0.115\,796 \quad (8)$$

We can observe an increase of 0.044 dB/T, where the precision of the LUNA is ±0.03 dB. Due to the narrow difference between these figures, it is difficult to confirm the trend with a good degree of precision, except to say that the trend is positive and conforms to expectations founded on our previous work.

B. PMD results

We have shown in section II that the PMD is based on the phase component of the cavity response. The induced modulation caused by the magnetic field can be observed in figure 8, showing the PMD spectrum change per 0.5 ampere step.

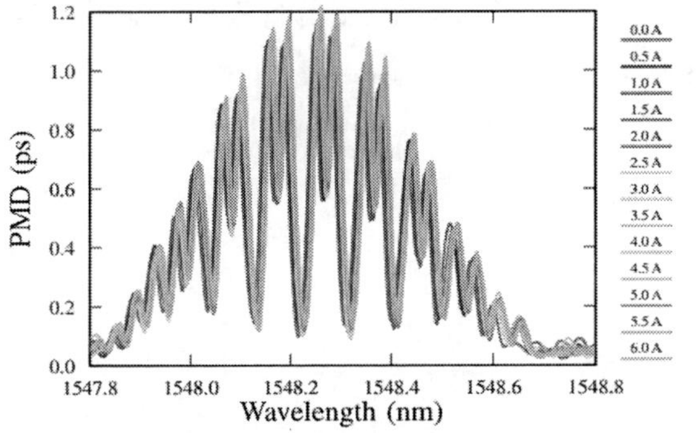

Fig. 8. PMD vs. wavelength per step current.

As with the PDL, we select the strongest and most homogeneous peak for data processing, seen in figure 9. Applying the same peak to peak value measurement, we correlate it to the induced magnetic field in figure 10.

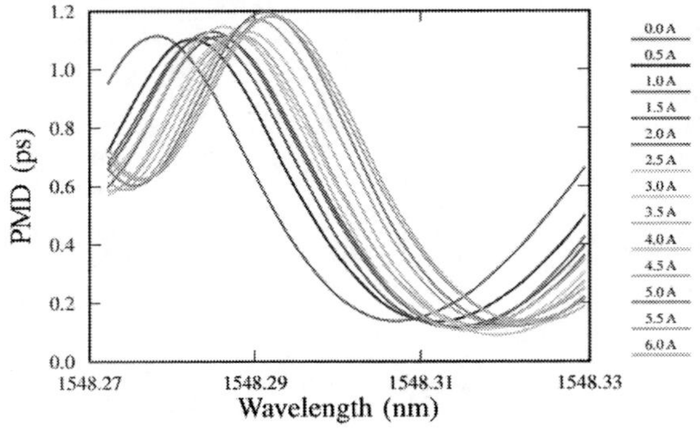

Fig. 9. Selected peak from figure 8.

A least square fit is again shown from which we give a derivation of the sensitivity:

$$\mathrm{PMD[ps]} = 8.154 \times 10^{-5}\ B[\mathrm{mT}] + 0.971\,917 \quad (9)$$

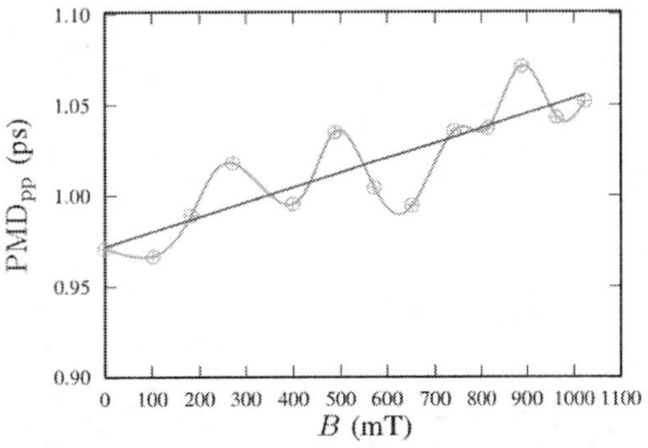

Fig. 10. Peak to peak PMD (orange) and least square fit (green) vs. induced magnetic field.

We can observe a sensitivity of 0.081 ps/T for a precision of ±0.05 ps. While the trend of the response is positive, it is clear that the magnitude of the total change is of the same order as the precision. As with the PDL, we can see that the trend matches that observed in our previous work except that now the actual change is visible in the PMD spectrum without the need for further data processing. Additionally, a given peak of the PMD spectrum has a width of 60 pm, only yielding 24 points of actual data. This lack of sample density severely compromises the precision of the results. It is clear that were further studies to be based on the PMD, it would be better to have a polarimeter with a higher wavelength resolution.

V. Conclusion

We have presented an intrinsic Fabry-Perot cavity that can detect an induced magnetic field. The cavity exhibits no photo-induced birefringence and acts as a point detector, with the possibility for distributed sensing. We have demonstrated the possibility of real time monitoring by reducing the complexity of our data processing, made possible through the removed photo-induced birefringence. We have explored both phase and amplitude based detection and have demonstrated an amplitude component sensitivity of 0.044 dB/T and a phase component sensitivity of 0.081 ps/T.

We have successfully demonstrated the concept, but are conscious of the additional work to make this concept a more attractive one. Improvements include testing photonic crystal and larger diameter fibres, where greater fundamental sensitivity may be achievable. Changes to FBG inscription and an optimization of cavity parameters should deliver a more homogeneous spectrum that would permit data redundancy and an improved sensitivity. Finally, consideration of detector capabilities would yield more data points in the target region, giving a clearer response around peak values that could provide a more homogeneous progression.

Acknowledgment

C. Broadway is supported by H2020 European project Modern 2020, C. Caucheteur is supported by the F.R.S.-FNRS and P. Mégret is supported by the Belgian Science Policy IAP7/35.

REFERENCES

[1] T. Nan, Y. Hui, M. Rinaldi, and N. Sun, "Self-biased 215 MHz magnetoelectric NEMS resonator for ultra-sensitive DC magnetic field detection," *Scientific reports*, vol. 3, p. 1985, 2013. [Online]. Available: https://www.nature.com/articles/srep01985

[2] H. V. Thakur, S. M. Nalawade, S. Gupta, R. Kitture, and S. N. Kale, "Photonic crystal fiber injected with Fe_3O_4 nanofluid for magnetic field detection," *Applied Physics Letters*, vol. 99, p. 161101, 2011. [Online]. Available: http://aip.scitation.org/doi/abs/10.1063/1.3651490

[3] P. Orr, P. Niewczas, M. Stevenson, and J. Canning, "Compound phase-shifted fiber Bragg structures as intrinsic magnetic field sensors," *IEEE Journal of Lightwave Technology*, vol. 28, no. 18, pp. 2667–2673, 2010.

[4] Y. Su, B. Zhang, Y. Zhu, and Y. Li, "Sensing circular birefringence by polarization-dependent parameters in fiber Bragg gratings and the influence of linear birefringence," *Optical Fiber Technology*, vol. 18, pp. 51–57, 2012.

[5] H. Peng, Y. Su, and Y. Li, "Evolution of polarization properties in circular birefringent fiber Bragg gratings and application for magnetic field sensing," *Optical Fiber Technology*, vol. 18, pp. 177–182, 2012.

[6] L. I. Chong-Zhen and W. U. Bao-Jian, "Theory and experiment on polarization-dependent loss of magnetooptical fiber Bragg grating system with a polarization controller," *Microwave and Optical Technology Letters*, vol. 53, no. 10, pp. 2224–2228, 2011.

[7] S. Bette, C. Caucheteur, M. Wuilpart, and P. Mégret, "Theoretical and experimental study of differential group delay and polarization dependent loss of Bragg gratings written in birefringent fiber," *Opt. Communication*, vol. 269, pp. 331–337, 2007.

[8] F. Descamps, D. Kinet, S. Bette, and C. Caucheteu, "Magnetic field sensing using standard uniform FBGs," *Optics Express*, vol. 24, no. 23, pp. 26 152–26 160, 2016.

[9] C. Broadway, D. Gallego, A. Pospori, M. Zubel, D. J. Webb, K. Sugden, G. Carpintero, and H. Lamela, "A compact polymer optical fibre ultrasound detector," in *Proc. SPIE 9708, Photons Plus Ultrasound: Imaging and Sensing*, A. A. Oraevsky and L. V. Wang, Eds., vol. 9708. SPIE, 2016, pp. 970 813–970 813–7. [Online]. Available: http://dx.doi.org/10.1117/12.2212420

[10] W. Zhang and D. J. Webb, "Humidity responsivity of poly(methyl methacrylate)- based optical fiber Bragg grating sensors," *Optics Letters*, vol. 39, no. 10, pp. 3026–3029, 2014. [Online]. Available: http://dx.doi.org/10.1364/OL.39.003026

[11] C. Caucheteur, F. Lhommé, K. Chah, M. Blondel, and P. Mégret, "Simultaneous strain and temperature sensor based on the numerical reconstruction of polarization maintaining fiber Bragg gratings," *Optics and Lasers in Engineering*, vol. 44, no. 5, pp. 411–422, 2006.

[12] C. Caucheteur, M. Debliquy, D. Lahem, and P. Mégret, "Hybrid fiber gratings coated with a catalytic sensitive layer for hydrogen sensing in air," *Optics Express*, vol. 16, no. 21, pp. 16 854–16 859, 2008.

[13] C. A. F. Marques, L. Bilro, L. Kahn, R. A. Oliveira, D. J. Webb, and R. N. Nogueira, "Acousto-optic effect in microstructured polymer fiber bragg gratings: Simulation and experimental overview," *Journal of Lightwave Technology*, vol. 31, no. 10, pp. 1551–1558, 2013.

[14] D. Kinet, P. Mégret, W. Goossen, L. Qiu, D. Heider, and C. Caucheteur, "Fiber Bragg grating sensors toward structural health monitoring in composite materials: Challenges and solutions," *Sensors*, vol. 14, no. 4, pp. 7394–7419, 2014.

[15] F. Descamps, M. Aerssens, A. Gusarov, P. Mégret, V. Massaut, and M. Wuilpart, "Simulation of vibration-induced effect on plasma current measurement using a fiber optic current sensor," *Optics Express*, vol. 22, no. 12, pp. 14 666–14 680, 2014.

[16] M. Wuilpart, M. Aerssens, A. Gusarov, P. Moreau, and P. Mégret, "Plasma current measurement in thermonuclear fusion reactors using a photon-counting POTDR," *IEEE Photonics Technology Letters*, vol. 29, no. 6, pp. 547–550, 2017.

[17] M. Wuilpart and M. Tur, "Polarization effects in optical fibers," in *Advanced Fiber Optics: Concepts and Technology*, L. Thévenaz, Ed. EPFL Press, 2011, ch. 2, pp. 29–86.

[18] J. L. Cruz, M. V. Andres, and M. A. Hernandez, "Faraday effect in standard optical fibers: dispersion of the effective Verdet constant," *Appl. Opt.*, vol. 35, no. 6, pp. 922–927, 1996.

[19] B. L. Heffner, "Deterministic, analytically complete measurement of polarization-dependent transmission through optical devices," *IEEE Photonics Tech. Lett.*, vol. 4, pp. 451–454, 1992.

[20] ——, "Automated measurement of polarization mode dispersion using jones matrix eigenanalysis," *IEEE Photonics Tech. Lett.*, vol. 4, pp. 1066–1069, 1992.

Monitoring and Hardware Management for Critical Fusion Plasma Instrumentation

Paulo F. Carvalho, Bruno Santos, Miguel Correia, Álvaro M. Combo, António P. Rodrigues, Rita C. Pereira, Ana Fernandes, Nuno Cruz, Jorge Sousa, Bernardo B. Carvalho, António J.N. Batista, Carlos M.B.A. Correia and Bruno Gonçalves

Abstract — **Controlled nuclear fusion aims to obtain energy by particles collision confined inside a nuclear reactor (Tokamak). These ionized particles, heavier isotopes of hydrogen, are the main elements inside of plasma that is kept at high temperatures (millions of Celsius degrees). Due to high temperatures and magnetic confinement, plasma is exposed to several sources of instabilities which require a set of procedures by the control and data acquisition systems throughout fusion experiments processes. Control and data acquisition systems often used in nuclear fusion experiments are based on the Advanced Telecommunication Computer Architecture (AdvancedTCA®) standard introduced by the Peripheral Component Interconnect Industrial Manufacturers Group (PICMG®), to meet the demands of telecommunications that require large amount of data (TB) transportation at high transfer rates (Gb/s), to ensure high availability including features such as reliability, serviceability and redundancy. For efficient plasma control, systems are required to collect large amounts of data, process it, store for later analysis, make critical decisions in real time and provide status reports either from the experience itself or the electronic instrumentation involved. Moreover, systems should also ensure the correct handling of detected anomalies and identified faults, notify the system operator of occurred events, decisions taken to acknowledge and implemented changes. Therefore, for everything to work in compliance with specifications it is required that the instrumentation includes hardware management and monitoring mechanisms for both hardware and software. These mechanisms should check the system status by reading sensors, manage events, update inventory databases with hardware system components in use and maintenance, store collected information, update firmware and installed software modules, configure and handle alarms to detect possible system failures and prevent emergency scenarios occurrences. The goal is to ensure high availability of the system and provide safety operation, experiment security and data validation for the fusion experiment. This work aims to contribute to the joint effort of the IPFN control and data acquisition group to develop a hardware management and monitoring application for control and data acquisition instrumentation especially designed for large scale tokamaks like ITER.**

Index Terms — **Fusion, Control, Data Acquisition, Monitoring, Management, Hardware, Software**

I. Introduction

Control and data acquisition (C&DAQ) systems applied to nuclear experiments of large fusion devices play an important key role to monitor and control the plasma parameters such as temperature, pressure, density, magnetic field, shape and position.

The instrumentation activities can help to prevent fusion experiments to terminate unexpectedly and without notice. Nuclear environment factors like radiation, dust, temperature, voltage and current peaks, distance between the systems and the reactor, signals routing of sensors and actuators, expose the instrumentation to several noise sources and electronic interferences that contribute to increase the hardware failure condition.

This scenario is not desirable since these systems are required to operate in continuous regimes without human local intervention over long time periods.

Instrumentation monitoring and hardware management capabilities are desirable to check the platform health status, mitigate system errors, provide platform inventory information, ease hardware replacement, update system firmware, handle alarms and send notification messages to operators.

A software architecture for monitoring and hardware management is being considered for development taking advantages of the xTCA® [1] high reliability, serviceability and redundancy key features to reduce the platform downtime, give support to data integrity validation and prevent hardware failure.

These features adds robustness to the platforms, reduces maintenance costs and provides instrumentation safe operation.

This paper describes the methodology behind the proposed software architecture aiming high availability to xTCA® C&DAQ instrumentation especially designed for fusion experiments of large nuclear devices like ITER.

II. System Overview

The hardware components are commonly distributed through a computer network organized into clients, servers and xTCA® C&DAQ platforms as depicted in Fig. 1.

Clients communicate with the servers through TCP/IP to send commands and receive the corresponding data.

Servers connect to both clients and xTCA® C&DAQ platform controllers through IPMI over LAN, execute the received commands and respond with the requested data.

Paper submitted for review in July of 2017. IST activities received financial support from "Fundação para a Ciência e Tecnologia" through Project UID/FIS/50010/2013.

Paulo F. Carvalho, Bruno Santos, Miguel Correia, Álvaro M. Combo, António P. Rodrigues, Rita C. Pereira, Ana Fernandes, Nuno Cruz, Jorge Sousa, Bernardo B. Carvalho, António J.N. Batista and Bruno Gonçalves are with the Instituto Plasmas e Fusão Nuclear, Instituto Superior Técnico, Av. Rovisco Pais, 1049-001, Lisboa, Portugal (e-mail: pricardofc@ipfn.ist.utl.pt).

Carlos M.B.A. Correia is with LibPhysUC, Departamento de Física, Universidade de Coimbra, 3004-516 Coimbra, Portugal.

C&DAQ platform controllers perform the required tasks to acknowledge the received instructions, report back in case of error occurrences and, if not, respond with the requested data towards monitoring or the task operation result regarding hardware management.

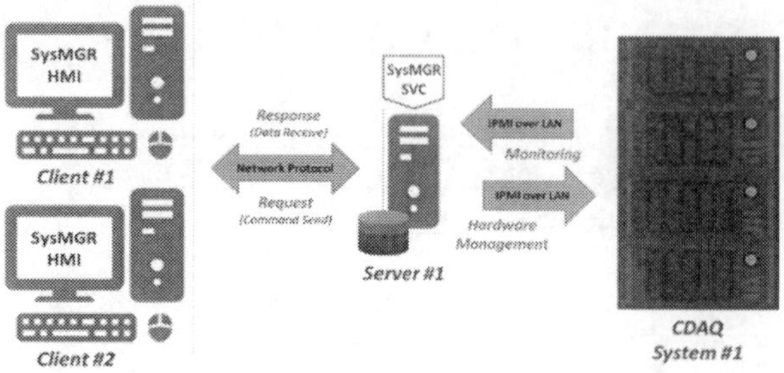

Fig. 1. System Overview.

III. Software Architecture

The software modules shall be organized according to the diagram of Fig. 2.

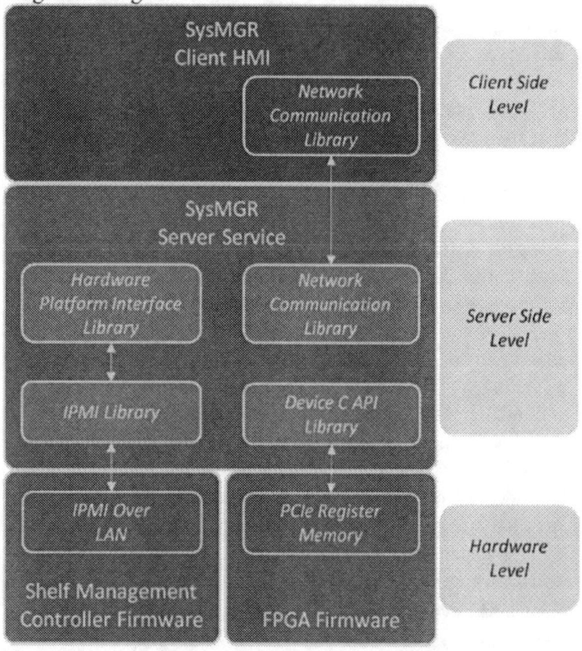

Fig. 2. Software Architecture and Communication Interfaces.

The high layer includes the System Manager (SysMGR) Human Machine Interfaces (HMI) running on the client computer side. The middle layer corresponds to the SysMGR service module running in the server computer side. It receives the requests from the SysMGR client HMI side and communicates with the platform management controllers and processor units through HPI, IPMI over LAN and Device C API libraries. The received data shall be published in the computer network and available to all connected SysMGR client HMIs. The low layer corresponds to the firmware of the platform management controllers and processor units.

This layer receives the instructions sent from the middle layer side, executes them and return the requested data or task operation result.

IV. Technology Overview

Radisys Platform Management, ScorpionWare® System Manager, FlowPilotTM, NATView and IPMIView are five currently available solutions to monitor and manage the hardware of Advanced TCA based control and data acquisition platforms. Compass System Manager and SpiderWare were also two available projects that were meantime descontinued. Common features offered by these solutions go far beyond sensor monitoring, alarm handling and some limited user, hardware and firmware management capabilities. Cross sensor data from different instrumentation, correlate information that can support data integrity and validation and enable hardware failure prediction implicates a different methodology. This should go through a centralized architecture to monitor environment and instrumentation sensors extended to include hardware failure prediction algorithms to help error mitigation and prevent malfunction scenarios before they could occur, identify possible error sources and notify the system operators in advance. Adding these capabilities to critical instrumentation exposed to several factors that increases the hardware failure condition might contribute to the high availability level required.

V. Targeting ITER-CODAC Compatibility

Since the proposed architecture aims xTCA® C&DAQ instrumentation also compatible with ITER device it is desirable to study an architecture implementation that makes usage of the ITER-CODAC framework. This can reduce the time and expertise required to implement a future software solution based on the proposed architecture. The Instituto de Plasmas e Fusão Nuclear (IPFN) has developed a Fast Plant System Controller (FPSC) prototype [2] similar to the fast controllers found in ITER. The prototype is based on the AdvancedTCA® architecture.

The platform supports control and data acquisition boards up to 12 modules and provides allocation for 2 more HUB timing and AMC carrier boards already included in the ITER Catalogue. Using all the knowledge and expertise gathered with the described prototype, to ease the integration of the SysMGR application into the ITER CODAC environment, the software modules should be developed in compliance with the Plant Control Design Handbook (PCDH) requirements and implemented with ITER best programming practices. The HMI running in the client computer side can be developed using the Control System Studio (CSS) Integrated Development Environment (IDE) distributed with the ITER CODAC package.

The software module running in the server computer side can be an Enhanced Physics and Industrial Control System (EPICS) Device Support module developed in C++ programming language based in the Nominal Device Support (NDS) architecture.

This module should include communication protocols based on IPMI over LAN and optionally PCI/PCIe Device C API to provide access to the hardware for monitoring and hardware management.

VI. IPMI Function Library

Although IPMI is broadly used to manage server PC systems the majority of them use the IPMITool to perform the required tasks. Since this tool is a command line application and not a function library it is not a simple task to use the included source files into an application. A research study found that it is easier and straightforward to use OpenIPMI. This function library is currently under development and provides interesting examples to help developers to start using it.

VII. EPICS Device Support

Using the ITER-CODAC MakeBaseApp.pl [epics] script to create an EPICS Device Support source code skeleton to start with, it possible now to include the OpenIPMI function library and use the provided IPMI examples to connect each dedicated EPICS sensor Process Variable (PV) to the corresponding IPMI read function. The device support st.cmd script was configured and prepared to read the server IP address, the username, password, authentication type and access privilege and run an initialization routine to collect and store all the system available information into an array of positions. This information will be useful to read each sensor value, store in the array and use the EPICS PV record udf field and sensor name to find the corresponding sensor value position. After this, the sensor PV is published in the computer network and can be seen by the CSS operator interface panel by pointing the widgets PV name field to the desired monitored system sensor.

VIII. Future Work

The work developed so far demonstrates that it is possible to integrate the described system architecture with the ITER CODAC framework. The next steps will be focused in the following aspects: (i); extend the described methodology to read all available system sensors and map them into the corresponding EPICS PVs; (ii) configure all sensor alarms and thresholds into EPICS PVs and provide this information to the system operator; (iii) implement task execution whenever an alarm event occurs; (iv) detect all system C&DAQ and timing boards status; (v) implement a system logger to display system messages; (vi) extend monitoring and hardware management capabilities to other xTCA® system; (vii) correlate acquired time stamped data with alarm occurrences to support data integrity and validation; (viii) demonstrate the possibility to implement hardware failure prediction/prevention algorithms; (ix) add, if required, management support to applications whenever a board is added/removed from system via PCIe hotplug and (x) ease the system firmware update procedures.

IX. Conclusion

The proposed software architecture aims to provide:

- A centralized software solution to monitor and manage de hardware of several xTCA® C&DAQ instrumentation extended with environment sensor data can eliminate the need of additional software modules and provide cross data correlation between systems;
- Information about the health status of monitored xTCA® C&DAQ platforms to check if the hardware is operating as expected;
- Monitored sensor data that allows the implementation of algorithms to predict/prevent hardware failure [3] This can prevent occurrence of unexpected scenarios and reduce hardware maintenance costs;
- Correlation between timestamped archived data with event occurrences makes possible to support data integrity and validation algorithms;
- Hardware and firmware maintenance procedures such as replacement of damaged cards via PCIe hot-plug [4][5][6] and system firmware update [7] without powering off the entire platform can contribute to reduce the instrumentation downtime and therefore increase the systems availability;
- Compatibility with ITER CODAC environment can reduce time spent and expertise required to monitor and manage xTCA® C&DAQ instrumentation.

References

[1]. PICMG, PICMG® 3.0 Revision 3.0 AdvancedTCA® Base Specification, March 24, 2008.

[2]. B. Gonçalves, *et al.*, "Engineering design of ITER prototype fast plant system", Fusion Engineering and Design, Vol. 86, Issues 6–8, Pages 457-1574, 2011.

[3]. Doug Turnbull, Neil Alldrin, "Failure Prediction-Hardware or Software", IEEE Transaction on Aerospace, Vol. 2, Issue 2, Pages 550-557, 1964.

[4]. PCI Express® Base Specification, Revision 3.0, November 10, 2010.

[5]. Paulo F. Carvalho, *et al.*, "PCI express hotplug implementation for ATCA based Instrumentation", Fusion Engineering and Design, Vol. 96-97, Pages 738-741, October 2015.

[6]. Paulo F. Carvalho, *et al.*, "PCIe Hot-Plug Event Handling Tasks using PICMG Standard Interrupt Mechanism for ATCA based Instrumentation", Fusion Engineering and Design, Vol. 96-97, Pages 738-741, May 2017.

[7]. Ana Fernandes, *et al.*, "FPGA Remote Update for Nuclear Environments", IEEE Transactions on Nuclear Science, Vol. 63, No. 3, June 2016.

Design and test of irradiation-related components in ITER radial x-ray camera

L. Hu[1], H. Cao[1], J. Zhao[1], L. Niu[1], Y. Chen[1], N. Zhou[1], J. Zhan[1], H. Yu[1], K. Chen[1], S. Li[1], X. Sheng[1], J.Shen[1], S. Qin[1], C. Feng[1], J. Ge[1], B. Zhang[1], S. Zhang[1]

[1]Institute of Plasma Physics, Chinese Academy of Sciences, Hefei 230031, China

E-mail contact of main author: lqhu@ipp.ac.cn

ITER Radial X-ray Camera is designed to measure the poloidal profile of the plasma x-ray emission with high spatial and temporal resolution and will be installed inside equatorial port 12 [1-2]. Overall view of RXC is shown in Fig.1.

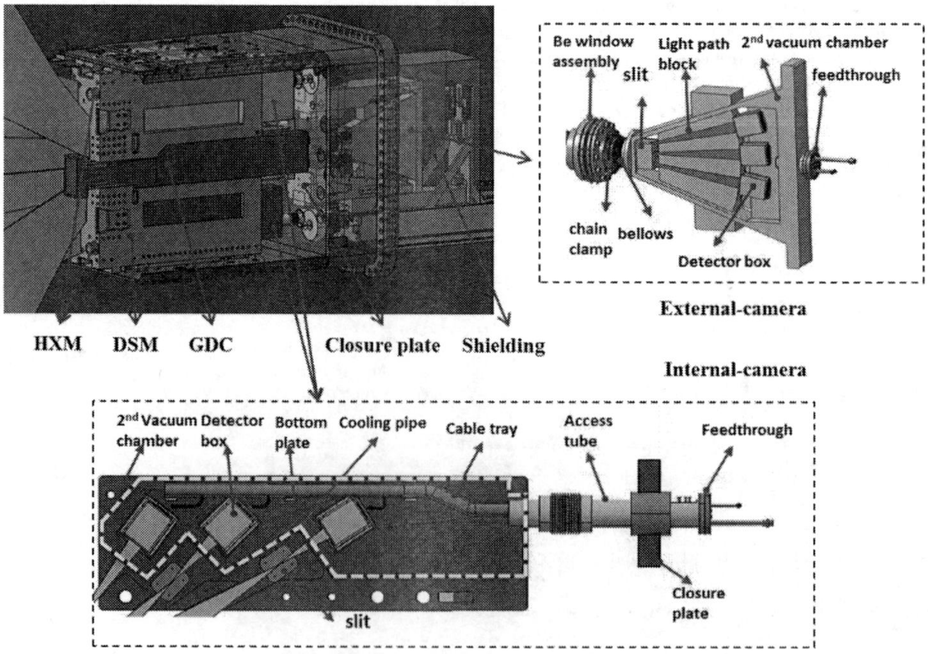

Fig.1. Overall view of RXC

As can be seen from Fig.1, RXC, Glow Discharge Cleaning (GDC) and Hard X-ray Monitor (HXM) share middle module of Equatorial 12 port plug [3]. RXC is composed of in-port and ex-port cameras which view the outer and core region respectively through vertical slots in the Diagnostics Shield Module (DSM). A concrete shielding shell enclosing the external camera is designed to reduce radiation leakage into outside environment.

Due to harsh environment of neutron and gamma radiation, nuclear radiation hardness of many components has to be considered in design and test, including silicon detector array, electronics, cabling, and also remote handling during maintenance. To protect detector against neutron damage, boron carbide blocks are envisaged to put in front of and around detectors to mitigate neutron flux onto detector. The neutron flux onto detectors was calculated. The neutron flux onto internal camera detector in DD operation is in the order of 10^7 n/cm^2/s, which means the silicon diode detector can survive DD operation without detector replacement [4]. As to external camera, it is possible the silicon detector can survive the whole DT operation with optimized shielding. To guarantee detector survival and performance, advanced detector which should be radiation-hard is being developed. Apart from neutron damage to detector, another important factor to camera design is maintenance under the condition of activation, especially during DT phase. Considering this factor, remote handling strategy has been worked out based on ITER remote handling procedure and tooling.

As for electronics put in port cell, preliminary design of highly integrated pre-amplifier and program controllable mid-amplifier has been completed, and many tests have been done to investigate the radiation hardness performance of the amplifiers together with detectors and cables, although shielding cabinet has been designed to house and enhance the survivability of electronics. In the Cf-252 neutron test with flux of ~ 2.23×10^3 n/cm^2/s and fluence greater than 1.0×10^{10} n/cm^2, there was no performance degradation in detector and pre-amplifier. In the Co-60 gamma test with experimental setup shown in Fig.2, under gamma dose rate of 0.5Gy/min and integrated dose of 200Gy, no performance change was found in pre-amplifier and mid-amplifier except for some radiation damage to the power module in the board of pre-amplifier.

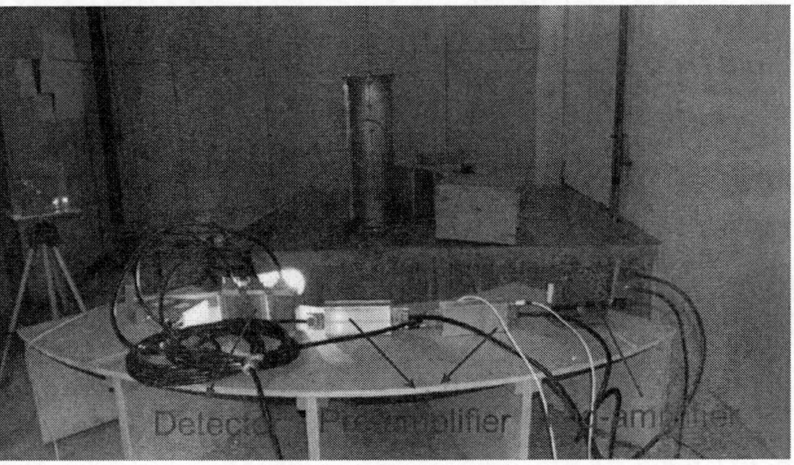

Fig.2. Experimental setup for gamma test

In subsequent accelerator neutron test using 14.9 MeV D-T neutron source where the neutron flux is ~ 5×10^7 n/cm^2/s and neutron fluence 2×10^{12} n/cm^2, no function problem was found in pre-amplifier and cable, while there were occurrences of single event effect in mid-amplifier. The detector underwent 10 percent decrease of sensitivity to visible light. The design optimization of electronics and further neutron test in reactor will be carried out in the future to improve performance of radiation hardness and qualify the design.

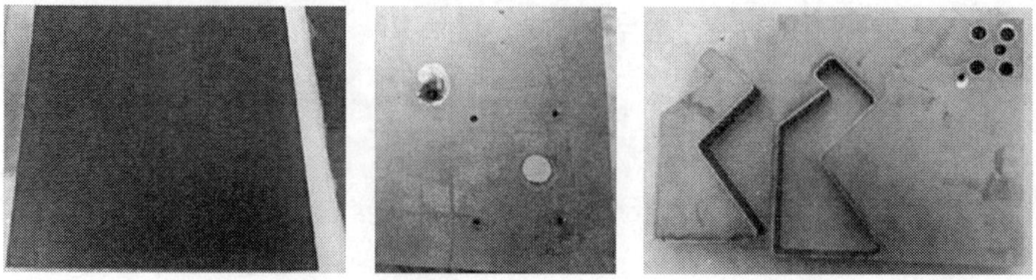

(a) (b) (c)

Fig.3 Manufacturability study of B4C plate. (a) Raw B4C plate; (b) after EDM hole processiong; (c) after wire EDM.

In addition, Manufacture study of B4C plate and tests were carried out to assess the feasibility of using B4C plate as shielding material. Fig.3 shows the raw B4C plate and those after machined. To produce hole with irregular shape inside B4C plate as shown in Fig.3 (c), the raw B4C plate was firstly drilled rough circular holes with Electrical Discharge Machining (EDM) method, and then cut with wire EDM method. The processed sample surface roughness can be as low as 1.6, making it promising to act as boundary of light of sight or to reduce assembly tolerance.

References

[1] A. E. COSTLEY, et al., Measurement Requirements and the Diagnostic System on ITER: Modifications Following the Design Review, in: 22nd IAEA Fusion Energy Conference, Geneva, Swiss, 2008.

[2] S. WANG, et al., Electromagnetic behavior on ITER radial soft x-ray camera, in: 2013 IEEE 25th Symposium on Fusion Engineering (SOFE), San Francisco, United States, 2013.

[3] G. VAYAKIS, et al., Current status of ITER diagnostics development, in:26th ITPA Spring Meeting, Pohang, Korea, 2014.

[4] B. ALPER, et al., The JET soft x-ray diagnostic systems, *Rev. Sci. Instrum.*,68 (1997) 778.

The intensive DT neutron generator of TU Dresden

Axel Klix, Toralf Döring, Alexander Domula and Kai Zuber

Abstract—**TU Dresden operates an accelerator-based intensive DT neutron generator. Experimental activities comprise investigation into material activation and decay, neutron and photon transport in matter and R&D work on radiation detectors for harsh environments. The intense DT neutron generator is capable to produce a maximum of 10^{12} n/s. The neutron source is a solid-type water-cooled tritium target based on a titanium matrix on a copper carrier. The neutron yield at a typical deuteron beam current of 1 mA is of the order of 10^{11} n/s in 4π. A pneumatic sample transport system is available for short-time irradiations and connected to wo high-purity germanium detector spectrometers for the measurement of induced activities. The overall design of the experimental hall with the neutron generator allows a flexible setup of experiments including the possibility of investigating larger structures and cooled samples or samples at high temperatures.**

***Index Terms*—DT neutron generator, fusion neutronics, neutron flux detector**

I. Introduction

TU Dresden (TUD) has been involved in experimental neutronics and nuclear physics research for several decades. A new neutron laboratory replacing the previous one at the branch campus in the city of Pirna Copitz was constructed by the Institute of Nuclear and Particle Physics (IKTP) in the early 2000 and came into full operation in 2005. It is located at the premises of the Helmholtz-Zentrum Dresden-Rossendorf. The main experimental device is an accelerator-based intense DT/DD neutron generator which is used for a broad variety of experimental work ranging from ultra-low background experiments to the development of radiation detectors for harsh environments.

II. Experimental Infrastructure

A. Neutron generator

The neutron generator is based on a commercial air-insulated accelerator system purchased from HVEE (Amersfoort, Netherlands). The high voltage supply is of Greinacher multiplier (Cockroft-Walton) type and can produce an output voltage from 30 to 300 kV with a maximum current of 15 mA. The ion source is a type 358 duoplasmatron (HVEE) which can deliver a deuteron beam of several mA (data sheet, literally). The deuterons are typically extracted with a 45 kV voltage and then analyzed by a mass filter magnet and the selected monatomic D^+ ions directed into the vertically arranged acceleration tube. While passing through a second magnet the deuteron beam is bend into the horizontal and focused onto a solid target by a triplet of quadrupole magnets. The general layout of the machine is shown in Fig. 1.

The neutron-generating targets are copper disks coated with a titanium layer. Tritium/deuterium is absorbed into the titanium layer. The tritium targets currently used were manufactured by SODERN and have an active diameter of 6 cm and are loaded with 20 Ci / 740 GBq of tritium with a tritium/titanium ratio >1.5. The deuterium targets are loaded to a similar tritium/titanium ratio. The targets are water-cooled on the external side of the tritium/deuterium target to dissipate the deuterium beam power (up to 3000 W).

The vacuum system interface for the target assemblies utilizes standard vacuum parts, it is therefore easily possible to use a different target assembly custom-made for a particular experiment for example to achieve a higher neutron flux density in a sample.

The accelerator is capable of providing the deuterium beam in DC mode as well as in pulsed mode. Pulsed mode is available with pulse widths of 10-100 μs and repetition frequencies from 1 to 1000 Hz. The accelerator is also prepared for nanosecond pulsed mode with a pulse width of 30 ns and repetition frequencies of 1, 2, and 5 MHz. A buncher is currently not installed in the beam line.

All functions of the accelerator are controlled by a PC via CAN bus interfaces. The PC is connected to the CAN bus controlers in the accelerator via optical fiber.

Primary monitoring of the neutron source is done through counting of the alpha particles associated with the DT reaction in the tritium target and the protons associated with the branch of the DD reaction resulting in p + ^{3}He. For this purpose, a silicon surface barrier detector is mounted in the beam line which is shielded from scattered accelerator deuterons by a 4 μm aluminum foil. The detector is 69.2 cm away from the tritium target and approximately 2 cm above the beam line. It is covered by an aperture with a diameter of 1.55 mm. The deuterium targets are mounted with a slightly different target assembly, in this case the distance between the silicon detector and the deuterium target is 50.0 cm. Additional monitoring is provided by a fission chamber loaded with 85 mg of ^{238}U.

Axel Klix is with the Karlsruhe Institute of Technology, Eggenstein-Leopoldshafen, D-76344 Germany (e-mail: axel.klix@kit.edu).
Toralf Döring, Alexander Domula and Kai Zuber are with TU Dresden, Dresden, D-01062 Germany.

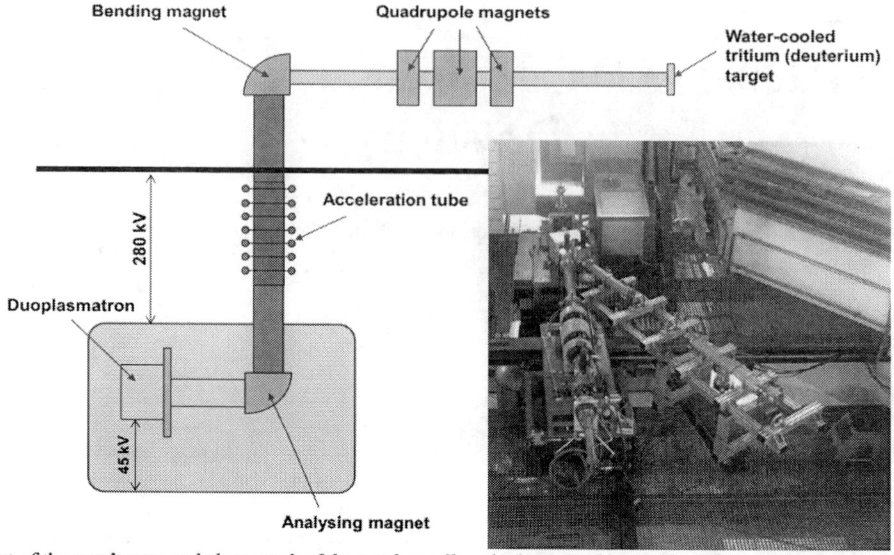

Fig. 1. Layout of the accelerator and photograph of the two beam lines in the second level. The left beam line ends in the wobble target. The right beam line is under construction.

The target of the neutron generator is at a height of 4.40 m above the floor of the experimental room and >4.00 m to the walls. The room has a height of 8.90 m, a length of 10.50 m and a width of 8.00 m. Structures in the room are kept to a minimum to reduce the influence of room-returned neutrons as much as possible.

B. Radio nuclide neutron and gamma sources

Three radio nuclide neutron sources are available for measurements and calibration:

TABLE I
AVAILABLE RADIO-NUCLIDE BASED NEUTRON SOURCES

Source	Activity*) (Bq)	Neutron yield (n/s)
AmBe	1.71×10^{11}	1.25×10^{7} **)
^{252}Cf	1.56×10^{5}	1.81×10^{4}
^{252}Cf	4.90×10^{7}	5.70×10^{6}

*) activities as of July 2017
**) using a standard conversion factor of $2.7*10^6$ n/Ci

These neutron sources are complemented by a ^{137}Cs source with an activity of 1.35×10^7 Bq and a ^{60}Co source of 6.29×10^5 Bq, both activities as of July 2017. In addition, various gamma sources with several tens of kBq are available for calibration purposes.

C. Pneumatic sample transport system

A pneumatic sample changer (PSC) was constructed in collaboration of KIT and TUD with the aim to provide a test bench for the development of a neutron activation system for the European test blanket modules (TBM) for ITER and for the study of short-living induced activities [1].

Up to thirty samples can be prepared and kept ready for irradiation in a storage. The storage can serve also for keeping samples for a certain cooling time. The PSC can transport samples to two high-purity germanium (HPGe) detectors. The irradiation end can be positioned freely to some extent, in particular it can be also used to irradiate samples for example inside a blanket assembly which is important for the development of a neutron activation system for the TBMs or simply to shape the neutron spectrum and enhance the fraction of thermal neutrons.

The PSC is controlled by a programmable logic controller (PLC). The PLC exchanges transport orders and status information with a PC via RS232 connection. Low level control of the PLC is done via a set of orders which are programmed into the PLC. This set of orders is extendable by reprogramming the PLC. High level control of the system is

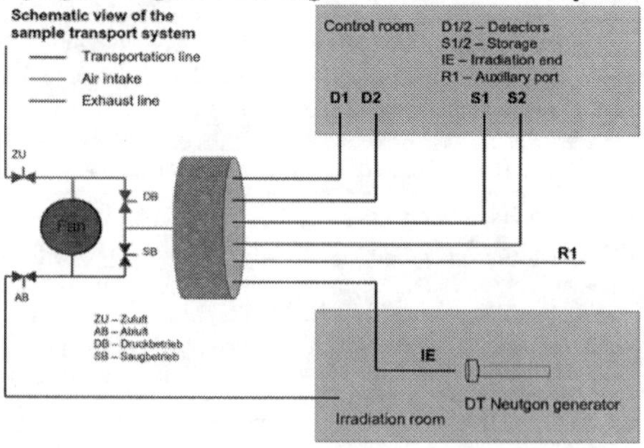

Fig. 2. Schematic view of the pneumatic sample transport system for neutron activation measurements.

provided by a LabView interface and a C# interface which were both written locally. They allow to set up a measurement program to process a set of samples.

The sample containers can hold a cylindrical sample with either 1.0 or 1.8 cm diameter. Their typical transport time from the irradiation end to one of the gamma spectrometers is approximately 11 s. This comparably long time is due to the special design of the distributor of the PSC which must ensure that activated air from the irradiation end cannot be released into occupational areas.

D. Detectors for neutrons and photons

Two HPGe detectors with an efficiency of 30% are available on-site, a horizontally oriented DSG NGC-3019 and a vertically oriented ORTEC GEM25P4-76. Several multi-channel analyzer options are available for gamma spectrometry: a) digital signal processor (DSP) Lynx (Canberra), b) DSP DT5780 (CAEN), c) analog processing with a MPA-3 system (multi-parameter acquisition system from FastComtec).

Two NE-213 liquid scintillators are available for neutron/photon spectroscopy in the MeV range. The NE-213 scintillator allows to discriminate photons and neutrons by means of pulse shape analysis. One of the detectors is of cylindrical shape with a diameter of 1.5" and a height of 1.5" and coupled to the photo multiplier unit via a 50 cm long light guide. The second NE-213 detector is also of cylindrical shape but with a diameter of 2" and a height of 2". Both detectors are made available from PTB Braunschweig through a mutual research agreement. The detectors were characterized experimentally at PTB, and response matrices for neutrons and photons are available for unfolding calculations [2]. An analog pulse shape discrimination and pulse height analysis system based on a Canberra PSD and the MPA-3 mentioned above had been used in previous experiments. Recently a DT5720 digitizer with digital pulse shape discrimination capability has been purchased from CAEN.

Two proportional counters LND 281 (LND Inc.) filled with 405.3 kPa hydrogen and a small addition of N for in-situ calibration are available for medium energy neutron spectroscopy.

A ^{3}He proportional counter tube 24NH15 (Canberra), several BF_3 counter and Li glas scintillators are available for online thermal neutron detection.

III. Experimental Program

The activity induced by fast neutrons is of concern for many application areas. IKTP plays a leading role in several ultra-low background experiments in search for the double-beta decay. Detector material activation by cosmogenic neutrons needs to be understood and supporting experiments with TUD-NG are ongoing to study for example nuclear level schemes and decay chains [3]. Similar experiments are performed to check evaluated nuclear data files utilized in shut-down dose rate calculations for fusion reactors [4].

The understanding of the neutron and photon transport in materials and structures is of importance for radiation shielding applications in areas ranging from medical to energy production but also for the design and operation of fusion reactor breeding blankets. A series of experiments with simple mock-ups of European fusion reactor tritium breeding blankets were performed with the aim to measure neutron and photon flux spectra under DT neutron irradiation inside the mock-ups and compare the measurements with simulations [5,6].

In the last years a program to develop radiation detectors for harsh environments was undertaken in collaboration with several European universities and research institutions. The work aims in particular at qualifying nuclear instrumentation for the European ITER TBM. Among others, the PSC is used to test activation probes for a neutron activation system [7], silicon carbide fast neutron detectors have been successfully tested at operating temperatures up to 500 OC under DT neutron irradiation, and self-powered neutron/photon detectors are investigated in a similar way.

Other applications of the neutron generator cover such fields as single event upset tests in microprocessors, cross section measurements of neutron reactions relevant for astrophysics and geology and others.

References

[1] A. Klix, A. Domula, U. Fischer, D. Gehre, P. Pereslavtsev, I. Rovni, "Test Facility for a neutron flux spectrometer system based on the foils activation technique for neutronics experiments with the ITER TBM," Fus. Eng. Des. 86 (2011) 2322.

[2] M. Guldbakke, H. Klein, A. Meister, U. Scheler, S. Unholzer, J. Pulpan, M. Tichy, "Response Matrices of NE213 Scintillation Detectors for Neutrons," Proc. of the Eighth ASTM-EURATOM Symposium on Reactor Dosimetry, 29 Aug.-3 Sept 1993.

[3] A. Domula

[4] A. Klix, A. Domula, R. Forrest, K. Zuber, "Measurement and Analysis of Activation Induced in Titanium with Fusion Peak Neutrons," J. Nucl. Mat. 417 (2011) 688.

[5] K. Seidel, P. Batistoni, U. Fischer, H. Freiesleben, A. Klix, D. Leichtle, E. Poenitz, S. Unholzer, R. Villari, "Measurement and analysis of neutron and gamma-ray flux spectra in a neutronics mock-up of the HCPB Test Blanket Module," Fus. Eng. Des. 82 (2007) 2212.

[6] A. Klix, P. Batistoni, R. Böttger, D. Lebrun-Grandie, U. Fischer, J. Henniger, D. Leichtle, R. Villari, „Measurement and Analysis of Neutron Flux Spectra in a Neutronics Mock-up of the HCLL Test Blanket Module," Fus. Eng. Des. 85 (2010) 1803.

[7] A. Klix, A. Domula, U. Fischer, D. Gehre, G. Kleizer, I. Rovni, "Preliminary experimental test of activation foil materials with short half-lives for neutron spectrum measurements in the ITER TBM," Fus. Sci. Techn. 64 (2013) 604

[8] D. Szalkai, R. Ferone, F. Issa, A. Klix, M. Lazar, A. Lyoussi, L. Ottaviani, P. Tüttő, and V. Vervisch, "Fast Neutron Detection With 4H-SiC Based Diode Detector up to 500 ∘ C Ambient Temperature," IEEE Trans. Nucl. Sci. 63(3) (2016) 1491

[9] Spnd Prasoon P. Raj, A. Klix, M. Angelone, U. Fischer, „Self-Powered Detectors for ITER Test Blanket Modules," IEEE Nuclear Science Symposium, Strasbourg, France, November 3, 2016

Data Acquisition System for ITER Neutron Diagnostic Divertor Neutron Flux Monitor

V. A. Fedorov, E. S. Martazov, Yu. A. Parishkin, N. A. Selyaev, [1]Yu. A. Kashchuk and [1]V. A. Vorobiev

National Research Nuclear University MEPhI, [1]Institution "Project Center ITER"

The ITER nuclear diagnostics Divertor Neutron Flux Monitor (DNFM) is designed to determine the ITER fusion power. DNFM diagnostic provides a neutron yield and fusion power measurements for both deuterium-deuterium and deuterium-tritium plasmas with the required time resolution and uncertainty.

The data acquisition system (DAQ) is a part of the DNFM diagnostic. This system is designed for signal processing of the three neutron detectors modules, each including six fission chambers (FC).

The DAQ system performs the following functions:

- FCs bias generation and signals detection;
- transformation of the measurement data into neutron flux values at the location of the detector module;
- detectors module FCs and measurement channels calibration with external neutron source;
- data transfer to the upper systems – CODAC;
- diagnostics of measurement channels and their components, including a detector modules and a communication lines.

The structure of DNFM diagnostics in ITER is presented in Figure 1. There are three distributed subsystems runs under the CODAC. The green shows the DNFM diagnostic components and the blue shows the CODAC components.

The DNFM diagnostic used three detector modules located around the ITER Vacuum Vessel through 120 degrees. Detector modules are mounted on the Divertor Cassettes under the Dome. Each DNFM detector module consists of two detector units (DU) and each detector unit consists of three fission chambers. The first DU contains the FCs with a radiator of ^{235}U and the second DU contains the FCs with a radiator of high purity ^{238}U. Fission chambers are manufactured with different sensitivities to satisfy ITER's requirements for time resolution, dynamic range and measurement uncertainty.

The DNFM Data Acquisition System consists of the three parallel independent subsystems (LPxx Subsystem) for the detector modules. Each Subsystem consists of the DAQ DNFM Control System Cubicle (CSC) and the DAQ DNFM Signal Conditioning Cubicle (SCC). The DAQ DNFM System Control Cubicle will be located in the Diagnostic Building and the DAQ DNFM Signal Conditioning Cubicle will be located in the Lower Ports (LP) in Tokamak Building.

The DNFM Data Acquisition System calculates the neutron flux at the location of the detector modules and transmits diagnostic measurements to the CODAC System every 1 ms.

Data exchange with the CODAC system is performed on four networks: PON, SDN, DAN and TCN. There are two terminals linked to the CODAC system – the physicist terminal and the engineer terminal.

The DNFM diagnostic will be calibrated in situ using neutron sources during the brake in work of the reactor before the start of the nuclear phase (work with deuterium plasma). The expected dynamic range of neutron flux measurements by the divertor monitor in the operating mode is 7 orders of magnitude.

The DNFM data acquisition system is designed to work with an FC with a nominal bias voltage of up to 300 V, an average pulse charge of about 2×10^{-13} C and a maximum output current of less than 3 mA. Each FC is connected to the system in a differential way using triaxial cables with mineral insulation (MI cables) with an impedance of 50 Ohm and a length of up to 20 m. The FCs bias voltage settings are performed remotely. Each FCs power supply is independent in terms of control and monitoring and has remote access.

The measuring channels in the Signal Conditioning Cubicle are galvanically isolated from each other and from the power supply for a voltage of at least 1 kV.

To ensure the neutron flux dynamic range of 7 orders of magnitude the fission chambers signals processing concept used in this system presupposes the presence in one control channel of three processing methods – pulse, Campbelling (fluctuation) and current.

The neutron flux measurement period in the operating mode is less than 1 ms. The settling time for the half-height level at the output of the measuring channel with a step change of the input signal in the measurement range does not exceed 1 ms.

CSC consists of a PXI system, a data acquisition system Fast Controller (industrial PC), parts of the rack status monitoring kits and linkage with the CODAC System (Hab&Cubicle health) and a switching component.

The Signal Processing Unit (SPU), components of switching and shielding of SPU from external influencing factors are placed in the SCC. The SPU consists of a pre-amplifier module, a control and preliminary processing module and a remote galvanic isolation module. The design provides for the possibility of removing the pre-amplifiers modules from the control and preliminary processing module for a distance of up to 30 meters. This will significantly reduce the amount of electronic equipment means to be shielded from hard radiation and exclude elements with high sensitivity to high-intensity magnetic fields.

In the SCC the pulses of the FCs are converted into voltage pulses, pre-processed and filtered. The analog signals of the measuring chains are converted to digital and transmitted to the CSC via fiber-optic communication lines.

In the CSC the data from the measurement chains are processed by processing and registration device. The processing and registration device consists of the PXI system based on NI PXIe-1075 and the Fast Controller and performs the following functions:

- control signals generation for the block of preliminary amplifiers;
- data collection from the normalizing converters of the measuring chains and their transformation into signals proportional to the neutron flux at the location of the detector module;
- calculation of the neutron flux at the location of the detector module in the entire of neutron flux measurement range;
- diagnostics of DAQ hardware;
- calibration of measuring channels.

The reduced count rate calculation of neutron pulses is performed for each chain and for the FC as a whole, taking into account the weight coefficients of the chains. The reduced count rate and neutron flux at the location of the detector module are also calculated. All calculated and measured data are transmitted to the CODAC system.

As a part of the development on the "Divertor Neutron Flux Monitor" diagnostic a mockup of the registration system for the DNFM detector module was produced. This mockup passed a number of laboratory and reactor tests which showed the efficiency of the presented solutions. So the non-linearity of the pulse chain was less than 1%, and the non-linearity of the current and Campbell chains did not exceed 2%. The tests also showed that the required performance in 1 ms with the current chain can be ensured only at currents of more than 100 nA.

Based on the mockup of the registration system for the DNFM detector module a three-channel system was developed and manufactured for the acceptance testing of the detector units.

Some of the solutions in part of CSC described above were adopted as basic when a use case example for ITER neutron diagnostics has been developed by ITER Organization.

Studies on Flat Sandwich-type Self-Powered Detectors for Flux Measurements in ITER Test Blanket Modules

Prasoon Raj, Maurizio Angelone, Toralf Döring, Klaus Eberhardt, Ulrich Fischer, Axel Klix, Ronald Schwengner

***Abstract*— Neutron and gamma flux measurements in designated positions in the test blanket modules (TBM) of ITER will be important tasks during ITER's campaigns. As part of the ongoing task on development of nuclear instrumentation for application in European ITER TBMs, experimental investigations on self-powered detectors (SPD) are undertaken. This paper reports the findings of neutron and photon irradiation tests performed with a test SPD in flat sandwich-like geometry. Whereas both neutrons and gammas can be detected with appropriate optimization of geometries, materials and sizes of the components, the present sandwich-like design is more sensitive to gammas than 14 MeV neutrons. Range of SPD current signals achievable under TBM conditions are predicted based on the SPD sensitivities measured in this work.**

I. Introduction

THE breeding blanket will be a critical component of a power reactor based on nuclear fusion. Breeding of tritium for the fuel cycle of reactor, heat extraction and radiation shielding will be its main functions. To test the performance of such a blanket under representative conditions, test blanket modules (TBM) based on different blanket concepts will be inserted in designated equatorial ports of ITER reactor. As part of the nuclear experiments in TBMs during ITER campaigns, responses like tritium production, nuclear heating, material activation etc. will be measured at different locations in TBMs. The results of these experiments will be compared with those obtained from calculations. ITER being a large tokamak device with physical characteristics like those of a fusion power reactor, these tests will be used for a validation of the computational tools (e.g. Monte-Carlo and activation/inventory codes) and nuclear data employed in neutronic design studies of fusion reactor components.

Manuscript received June 1, 2017. The work leading to this publication has been funded partially by Fusion for Energy under the Specific Grant Agreement F4E-FPA-395-1. This publication reflects the views only of the authors, and Fusion for Energy cannot be held responsible for any use which may be made of the information contained therein.

Authors would like to gratefully acknowledge the support of scientists and staff at the laboratories in TRIGA reactor Mainz and ELBE accelerator of Helmholtz-Zentrum Dresden-Rossendorf (HZDR).

P. Raj, U. Fischer and A. Klix are with the Karlsruhe Institute of Technology (KIT), Hermann-von-Helmholtz-Platz 1, 76344 Eggenstein-Leopoldshafen, Germany (corresponding author e-mail: prasoon.raj@kit.edu).

M. Angelone is with the E.N.E.A. C.R., I-00044 Frascati (Rome), Italy.

T. Döring is with the Neutron Laboratory of Technical University of Dresden (TUD), 01328 Dresden, Germany.

K. Eberhardt is with the Institute for Nuclear Chemistry, Johannes Gutenberg-University (JGU), 015228 Mainz, Germany.

R. Schwengner is with the Institute for Radiation Physics, Helmholtz-Zentrum Dresden-Rossendorf (HZDR), 01328 Dresden, Germany.

Online measurements of neutron and gamma fluxes will be important tasks in these experiments, since the TBMs will not be accessible during ITER campaigns. European Union (EU) has two proposed TBM systems to be installed in ITER and tested: the HCLL and the HCPB TBMs [1]. With high radiation, high temperature, and electromagnetic interferences the operating conditions for nuclear detectors in TBM are detrimental. Development of suitable nuclear instrumentation for such harsh conditions is an ongoing project. Activation foils, diamond detectors and self-powered detectors (SPD) are among the detector classes chosen for further study.

Self-powered detectors are commonly utilized as flux monitors in fission power reactors. They have compact and robust design and they do not require external bias voltage, making them highly desirable for application in regions with less accessibility like nuclear reactor cores. They are often composed of standard metals, mineral oxides and alloys, making them desirable with respect to nuclear non-proliferation and export-control guidelines. For these reasons, some regular SPDs, one commercially manufactured ad-hoc SPD [2] and several different variants in flat sandwich-like design [3] are under investigation.

II. Development of Test Detector

A. Description of a Self-Powered Detector

An SPD is a multi-layered device [4] with two electrodes separated by a mineral oxide insulation layer. There is a central *emitter* layer with high neutron or gamma interaction cross-section. Under these interactions, the material produces high energy electrons. These electrons move outwards and are stopped in the outer electrode layer called *collector*. The separation of charges forms the detector signal. The response is usually measured as a direct current (DC) signal. The rate of neutron/gamma interactions, thereby the rate of electron emission and magnitude of DC signal are proportional to the incident flux. *Sensitivity* of an SPD is defined as SPD current per unit of incident flux, expressed in units of A cm^2 s^1.

Traditionally, SPDs are made in a coaxial cylindrical geometry. A wire (or rod) of emitter is packed in a closed tube of collector with a tube or beads of mineral oxide between them for electrical insulation. It can also be designed in a flat geometry with foils of insulator and collector sandwiching a foil of emitter. A schematic of this design is shown in Fig. 1 with the current measurement connections.

The variation of SPD signal under dynamic reactor conditions has a profile in time. This is based on the half-life corresponding to the exact atomic or nuclear process which creates most of the fast electrons. The most commonly used

SPDs in fission reactor cores, known as *Delayed Self-Powered Neutron Detector (SPND)* respond mainly via (n, β^-) process. In this process, neutron absorption reaction leads to formation of a radioactive daughter nuclide in the emitter that undergoes beta minus decay. Due to the half-life of beta decay, the signal does not vary promptly with variation in incident flux. The alternative (n, γ, e^-) process involves emission of high-energy gamma in the neutron-activated emitter material and subsequent interactions through photoelectric effect, Compton scattering and pair-production routes to produce fast secondary electrons, which form the detector signal. This is a prompt process, making the so-called *Prompt SPND* a good choice for reactor power-level monitoring. A third common process, (γ, e^-) process is the principle at the heart of a *Self-Powered Gamma Detector (SPGD)*. In this, external gammas from the surrounding of the detector are detected through secondary electron emission in the emitter. Usually, a detector is optimized for high response to measure neutrons or gammas. Nevertheless, the radiation field in which an SPD is applied is mixed with both neutrons and gammas and the output signals are mixed with contributions from both.

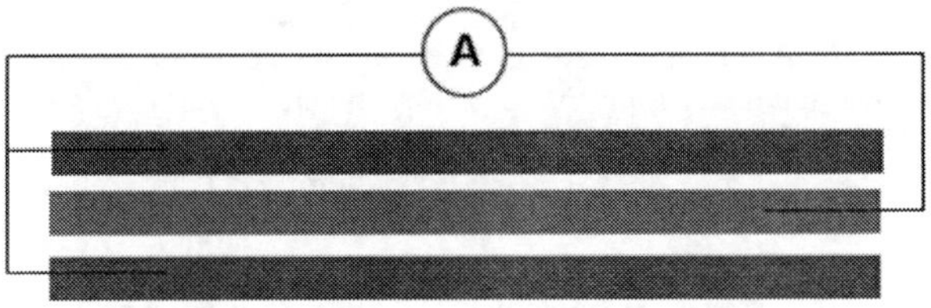

Fig. 1. Schematic of an SPD showing emitter layer in green, insulation in white and collector layers in red. An ammeter across emitter and collector (grounded) can be used to see the DC signal from the detector.

B. *Design and Construction of Flat Sandwich-like SPD*

Test blanket modules (TBM) of ITER have neutrons and gammas over a wide energy range [1]. The neutrons range from thermal energy to 14 MeV D-T neutron energy. Majority of them have intermediate energies. Total neutron flux intensity can go up to the order of 10^{14} n cm^{-2} s^{-1}. Unfortunately, no available neutron sources provide comparable field characteristics for detector tests. Gammas at positions relevant to SPDs range in energies between few keV to 30 MeV, intensities going up to the order of 10^{13} γ cm^{-2} s^{-1}. Dedicated detectors for measurement of TBM neutrons (SPND) and gammas (SPGD) are vital. To this end, theoretical and experimental investigations are underway in European laboratories [2, 3].

Based on a set of criteria, several materials were chosen for emitter of a delayed SPND for TBM [2]. These criteria included high saturation activity, short half-life of product nuclide and high beta energy. Designing a pure delayed SPND for TBM neutrons is challenging because of the relatively lower reaction cross-sections towards fast neutrons. On the other hand though, the probability of photon production is higher than that of beta emission for fast neutrons. So, prompt current due to neutron-induced gammas in detector dominates the signal. This is unlike the case of thermal neutrons, where both the processes have similar cross-sections, and the delayed current dominates for short half-life beta decays. Overall sensitivity of SPND for fast neutrons is lower. Another important difference is that possibility of threshold reactions gives a way to exclusively measure fast neutron contribution in TBM. Beryllium, vanadium, chromium and a few other materials are chosen for investigation as emitter of SPDs for TBM. For insulation alumina and for collector Inconel-600 alloy are used in these studies.

A test detector (Fig. 2) in flat geometry was designed with foils of collector and insulator sandwiching a foil of emitter. High-purity foils with dimensions 25 mm by 25 mm by 0.5 to 2 mm were used in different experiments. The model is flexible with openable parts to test different material combinations in different thicknesses. The assembled detector is packaged in a case of aluminum which also works as an electromagnetic shield. SPDs in flat geometry are of interest for application in TBMs. A foil of emitter can be encapsulated in a case made from collector material with insulation material filling the gap. One can think of a multi-layer strip-like detector or a chip-like compact detector which can be integrated with minimal invasion.

Although multiple materials have been tested and studied with this detector, results recorded only with vanadium SPD are presented in the upcoming section. Vanadium is a good reference material to study the newly implemented flat geometry, as its response is simple and well-understood for thermal neutrons. It is a candidate emitter for SPNDs [2] in TBMs, expected to provide information on thermal as well as fast neutrons. Also, it is a relatively easy material to handle in the laboratory than beryllium and chromium. The detector (V-SPD) has 1 mm thick layers of vanadium emitter and Inconel-600 collector, and 0.5 mm thick alumina insulator.

Fig. 2. Constructed flat sandwich-type SPD with an electromagnetic case made from aluminum and coaxial cable connection for signal

III. Irradiation Tests And Results

The V-SPD has been irradiated with three test sources: thermal neutrons, 14 MeV fast neutrons and bremsstrahlung photons. Small DC signals (50 fA to 500 pA) have been measured in these tests. The flat SPD currents are proportional to the incident neutron [3] and gamma flux, as required for a flux monitor. It has been established in tests with multiple material combinations and flux levels varying over orders of magnitude. At small currents, there are chances of disturbances due to various currents produced in mechanical, electromagnetic, thermal and radiation induced effects in detector and cables. The measurements have been performed with care, reducing these sources as far as possible.

A proof of the working of flat sandwich-type SPD is achieved by irradiating the V-SPD in the TRIGA Mark II reactor of the Johannes-Gutenberg University Mainz. Experimental position is near the hot end of the graphite thermal column (GTC) of the reactor with total thermal neutron flux of the order of 10^{10} n cm^{-2} s^{-1} [5]. The building up and saturation of the signal during first few minutes of irradiation and its decay after shutdown of the reactor showed characteristics of the vanadium emitter ($T_{1/2}$ = 3.75 min) like expected. Presence of a prominent prompt component in the signal shows enhanced sensitivity to gammas produced in the reactor and detector materials. Sensitivity of V-SPD calculated at saturation (Table I) is of the same order as the commonly reported sensitivities of cylindrical SPNDs in thermal reactor.

Table I
Measured saturated currents, incident particle flux intensities (measured with Au foil activation in thermal neutron field, with calibrated α-detector in fast neutron field and estimated with validated Monte-Carlo simulation in bremsstrahlung photon field) and sensitivities of V-SPD in three test scenarios

Test Field	Current 10^{-12} A	Flux 10^{9} cm^{-2} s^{-1}	Sensitivity 10^{-20} A cm^{2} s
Thermal neutrons	492	26.8	1.84
Fast Neutrons	6.48	1.56	0.42
Photons	6.20	0.06	9.60

For fast neutron irradiation, 14 MeV Neutron Generator of Technical University of Dresden (TUD-NG) is utilized. TUD-NG is an accelerator based intense D-T neutron source situated in Dresden-Rossendorf. It provides a neutron spectrum with peak around 14 MeV and flux intensity going up to 10^{10} n cm^{-2} s^{-1} near the tritium target. In fast neutron field, the detector signal rises and disappears almost instantaneously with the neutron flux. The delayed current component is negligibly small (approx. 6.3% of the total). It is a sum of contributions from ^{51}Ti and ^{52}V beta electrons, betas from aluminum case, collector and insulator materials and decay gamma from activated detector and surrounding materials.

The TUD-NG is characterized by a mixed neutron-gamma field. An estimation has been made in a Monte-Carlo model calculation of the generator, for the ratio of intensities of neutrons and gammas. At the experimental position, total gamma flux intensity is approx. 19% of the total neutron flux intensity. To obtain the neutron to gamma signal ratio for the tests at TUD-NG, a comparable test in a pure gamma field is required. Whereas radioisotope based line gamma sources are frequently used for such tests, a photon source is used in this work. At the beam dump of the ELBE [6] accelerator of Helmholtz-Zentrum Dresden-Rossendorf, there is a high intensity high-energy bremsstrahlung photon source created which is used for photo-activation experiments. The ELBE photon source has a continuous energy spectrum ending close to the energy of the electron beam. Beam parameters are selected (I_{beam} = 0.5 μA, E_{beam}=10 MeV) to have an energy spectrum of photons close to the one of gammas in TUD-NG. SPD current is proportional to the photon flux, showing again that the detector has linear and high response to photons.

The pure gamma sensitivity measured in ELBE irradiation of V-SPD (Table I) is substantially higher than the optimistic thermal neutron sensitivity assessed in case of TRIGA irradiation. On multiplying this with the calculated total flux of gammas (having energy spectrum similar to ELBE beam dump) at the experimental position in TUD-NG, one obtains that about 81% of the signal in TUD-NG is due to detection of gamma. With the rest 19% of the signal, the 14 MeV neutron sensitivity of V-SPD is determined to be approx. 4.15 x 10^{-21} A cm^{2} s^{-1} as reported in Table I. Therefore, in 14 MeV neutron field, the sandwich-type V-SPD has a higher sensitivity towards gammas than neutrons.

It is clear that flat SPDs can be applied as SPGD for gamma measurements in TBM. Signals in the range of nA to μA can be expected with the present design. This can be increased by changing emitter material to a high-Z metal, and further optimizing the geometry. On the other hand, the neutron detection is also feasible but it is tricky. Thermal neutron fluxes can be measured by separating the delayed currents due to thermal capture reaction(s). But for fast neutrons, an indirect mechanism can be used in which neutron-induced gammas are detected in a prompt SPND. Essential design improvements and further tests are required before optimized prototype SPDs for application in TBM can be realized.

IV. Conclusions

As part of the ongoing task on development of nuclear instrumentation for European test blanket modules of ITER, self-powered detectors in flat geometry have been constructed. The present sandwich-like vanadium SPD in aluminium casing is responsive to neutrons and gammas. The currents expected with this detector in TBM environment can be reliably measured. Gammas can be measured directly with such a detector, while fast neutrons need to be measured indirectly through neutron-induced prompt gammas.

References

[1] P. Calderoni, "Status of the HCLL and HCPB Test Blanket System instrumentation development," *21st Top. Meet. Technol. Fusion Energy (TOFE), 9-13 Novemb. 2014, Anaheim, CA.*

[2] M. Angelone, A. Klix, M. Pillon, P. Batistoni, U. Fischer, and A. Santagata, "Development of self-powered neutron detectors for neutron flux monitoring in HCLL and HCPB ITER-TBM," *Fusion Eng. Des.*, vol. 89, no. 9–10, pp. 2194–2198, 2014.

[3] P. Raj, M. Angelone, U. Fischer, and A. Klix, "Self-Powered Detectors for Test Blanket Modules in ITER," in *IEEE Nuclear Science Symposium, 29 Oct. - 5 Nov. 2016, Strassbourg, France*, 2016.

[4] J. W. Hilborn, "Self-Powered Neutron Detectors for Reactor Flux Monitoring," *Nucleonics*, vol. 22, pp. 69–74, 1964.

[5] B. Wortmann, "Ph.D. Thesis: Auslegung und Optimierung einer Bestrahlungseinrichtung fuer die Bor-Neutroneneinfangtherapie an autotransplantierten Organen," Technical Unversity of Dresden, 2008.

[6] R. Schwengner, R. Beyer, F. Dönau, E. Grosse, A. Hartmann, A. R. Junghans, S. Mallion, G. Rusev, K. D. Schilling, W. Schulze, and A. Wagner, "The photon-scattering facility at the superconducting electron accelerator ELBE," *Nucl. Instruments Methods Phys. Res. Sect. A Accel. Spectrometers, Detect. Assoc. Equip.*, vol. 555, no. 1–2, pp. 211–219, 2005.

Detectors for High Count Rate Measurements with a Compensation for MPPC Gain Dependence on Temperature

A. Urban, A. Brosławski, G. Bołtruczyk, M. Gosk, S. Korolczuk, D. Rybka, V. Kiptily, M. Nocente, D. Rigamonti, M. Tardocchi, I. Zychor and JET Contributors*

Multi Pixel Photon Counter (MPPC) is one of devices called silicon photomultiplier (SiPM). It is characterized by a fast response time, high gain coefficient, high photon detection efficiency resulting in good energy resolution, low voltage operation, resistance to mechanical shocks, compactness and immunity to a magnetic field. A MPPC gain is temperature dependent, so it is necessary to use a device which allows to maintain a constant value of a MPPC gain.

We report on two devices designed at the National Centre for Nuclear Research (NCBJ): FilterBox@NCBJ and MTCD@NCBJ to be used at the Joint European Torus (JET) during high count rate measurements.

The experiments at JET during D-T campaigns will be performed in harsh radiation conditions, especially at high rates up to ~1 Mcps and at relatively high gamma-ray energy about a few MeV. Devices designed to replace existing detector at JET have to fit to available space and must use cabling installed almost 20~years ago. The device should be user-friendly and easy operated by both engineers and physicists. It is worth to notice that in case of a completely new system, some new designed elements could be optimized in comparison with a presented solution.

A device for real-time temperature monitoring and MPPC gain stabilization was designed and produced at NCBJ, necessary due to a very strong voltage-temperature dependence characteristic for MPPC.

In Fig. 1, the overall scheme of electronics for the upgraded system is shown.

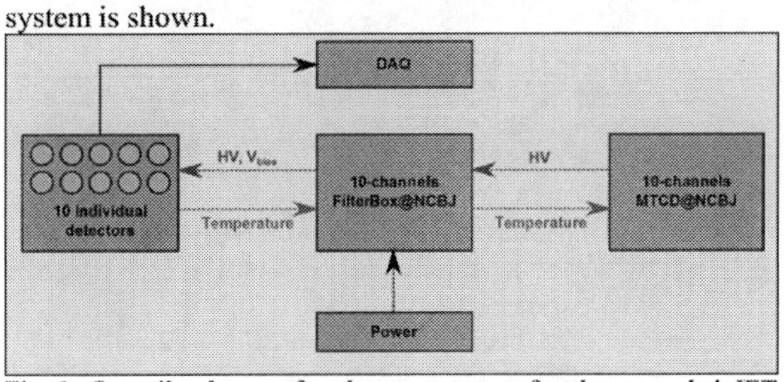

Fig. 1. Overall scheme of a detector system for the upgraded JET Gamma-ray Camera. HV denotes a bias voltage for MPPC, LV - power supply for both temperature sensor and active elements.

The most important requirements for electronics could be summarized as follows:

- setting of MPPC bias voltage,
- integrated temperature sensor,
- integrated power supply MAX1932,
- advanced power supply closed loop control.

A detector system consists of a set of individual capsules with a scintillator and a dedicated printed circuit board (PCB). On each PCB, a temperature sensor is mounted. The TSIC 506F sensor is characterized by an outstanding accuracy of ±0.1K, an excellent long-term stability and a very low current consumption of 30 μA during operation. The MPPCs, type S13361-3050NE-04 from Hamamatsu, were used.

An active system based on a transimpedance amplifier (TIA) was designed for MPPC signal read-out to obtain a signal characterized by a high output amplitude with low time-constant. TIA is mounted on the same PCB described above.

In Fig. 2 a detailed schematic of TIA is shown. All results presented in this paper were obtained with the same experimental conditons and a CAEN Desktop Digitizer DT5730 was used for data acquisition.

This scientific work was partly supported by Polish Ministry of Science and Higher Education within the framework of the scientific financial resources in the years 2015-2017 allocated for the realization of the international co-financed project. This work has been carried out within the framework of the EUROfusion Consortium and has received funding from the Euratom research and training programme 2014-2018 under grant agreement No 633053. The views and opinions expressed herein do not necessarily reflect those of the European Commission.

Arkadiusz Urban, Andrzej Broslawski, Grzegorz Boltruczyk, Marcin Gosk, Stefan Korolczuk, Dominik Rybka, and Izabella Zychor (e-mail: arkadiusz.urban@ncbj.gov.pl) are with Narodowe Centrum Badań Jądrowych (NCBJ), 05-400 Otwock, Poland.

Vasily Kiptily is with Culham Centre for Fusion Energy, Culham, United Kingdom.

Massimo Nocente and Davide Rigamonti are with Dipartimento di Fisica "G. Occhialini", Università degli Studi di Milano-Bicocca,Milano, Italy.

Massimo Nocente, Davide Rigamonti and Marco Tardocchi are with Istituto di Fisica del Plasma "P. Caldirola", CNR, Milano, Italy.

*See the author list of "Overview of the JET results in support to ITER" by X. Litaudon et al. to be published in Nuclear Fusion Special issue: overview and summary reports from the 26th Fusion Energy Conference (Kyoto, Japan, 17-22 October 2016).

 https://doi.org/10.1051/epjconf/201817002007

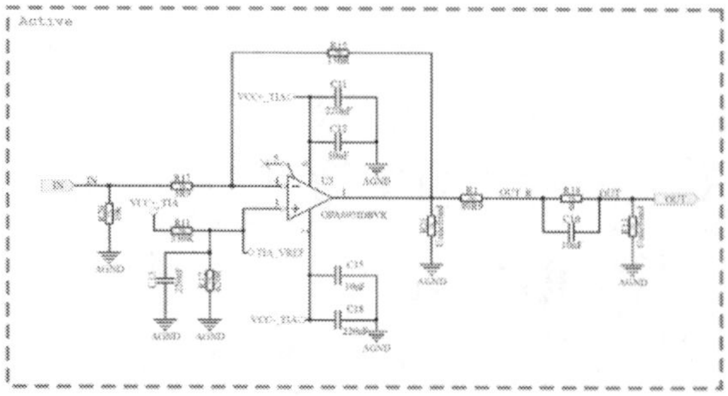

Fig. 2. Detailed schematic of TIA.

Information on temperature values is sent from a capsule to the FilterBox@NCBJ, based on a Microsemi ProAsic3 FPGA.

In Fig. 3 a photo of FilterBox@NCBJ is presented.

Fig. 3 FilterBox@NCBJ.

The main functions of the FilterBox@NCBJ are:

1. to serve for all individual capsules in each part of the Gamma-ray camera,
2. to filter a bias voltage for MPPC in a capsule,
3. to power active elements, e.g., TIA and a temperature sensor, mounted on PCB. For each capsule, a separated DC linear voltage regulator is used to minimize noise and crosstalk between detectors,
4. to provide communication with the MTCD@NCBJ using three independent communication channels based on a RS485 standard. Via RS485 actual detector temperature values are read from FILTERBOX@NCBJ.

FilterBox@NCBJ is supplied from a dedicated low voltage laboratory power supply.

The firmware for the Microsemi ProAsic3 FPGA on FilterBox@NCBJ was designed in a VHDL language. It incorporates modules for reading a temperature from 10 sensors. The power supply for each temperature sensor can be switched on/off individually. Temperature readout is done continuously with a programmed frequency and temperature values are accessible by reading dedicated registers.

The MPPC Temperature Compensation Device (MTCD@NCBJ) is using a measured dependence of a breakdown voltage on temperature to maintain a constant value of the MPPC gain. The MPPC Temperature Compensation Device (MTCD@NCBJ) with integrated power supplies comprises two main parts: one is connected with 10 adjustable MPPC bias voltage channels for each individual capsule, the other one is used to determine an optimal value of a bias voltage which guarantees a constant gain. Each channel has its own isolated converter to eliminate ground loops, followed by a low-dropout regulator to minimize a ripple on output. Connections are protected against electrostatic discharges. All functions are controlled from a personal computer.

In Fig. 4 the MTCD@NCBJ board is shown.

Fig. 4. MTCD@NCBJ with integrated power supplies.

Power supplies used in MTCD@NCBJ are characterized by an output voltage up to 80 V, filtering of the MPPC voltage and an option to limit current.

A development board MicroZed based on the Xilinx Zynq®-7000 was mounted in MTCD@NCBJ. A semiconductor device Zynq contains dual core ARM Cortex-A9 CPU with FPGA peripherals what allows to implement firmware both in HDL and high level language. A main algorithm was implemented in C language under Linux operating system to provide in an easy way a connection to MTCD@NCBJ via Ethernet or USB and investigation of an actual status of the device.

A control of 10 analog-to-digital (ADC) and 10 digital-to-analog (DAC) converters was implemented in VHDL. Modules allows to be controlled by CPU via standard AXI(AMBA) on chip interconnection.

PID (proportional–integral–derivative) regulator is applied to calculate a difference between actual and desirable voltage value. PID algorithm was implemented in C to control each high-voltage channel separately. A desirable value of a voltage is calculated based on temperature read from FilterBox@NCBJ.

In Fig. 5 a logic of MTCD@NCBJ is shown.

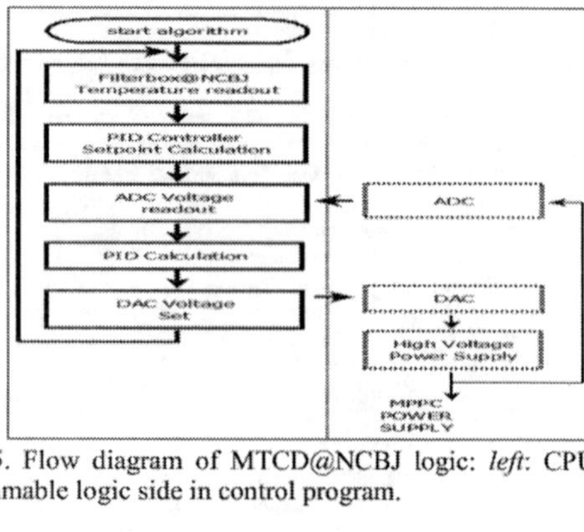

Fig. 5. Flow diagram of MTCD@NCBJ logic: *left*: CPU, *right*: programmable logic side in control program.

A protection of overvoltage was implemented in a control part of a program to guarantee that detectors will not be destroyed by applying too high voltage.

The output voltage, determined by a 20-bits digital-analog converter (DAC1220E), is delivered through the buffer to the feedback loop of the HV switching regulator (MAX1932ETC+).

Measurements detailed:

- 20 mm×15 mm cylindrical $CeBr_3$ scintillator,
- MPPC type S13361-3050NE-04 from Hamamatsu,
- active system based on a transimpedance amplifier (TIA) to obtain a signal characterized by a high output amplitude with low time-constant,
- ^{137}Cs source emitting 661.7 keV gamma linne with an activity of 400 MBq.

Preliminary results are presented in Figs. 6 and 7 as well as in Table 1.

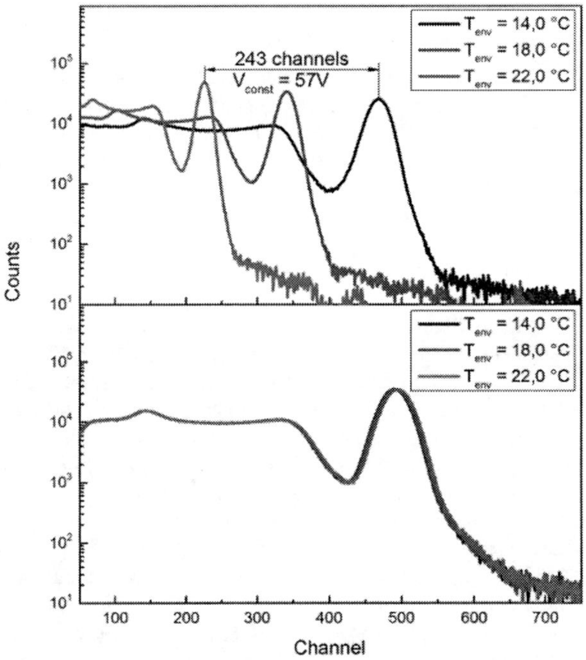

Fig. 6. Peak position as a function of MPPC temperature.
Upper: without MTCD@NCBJ.
Lower: with MTCD@NCBJ.

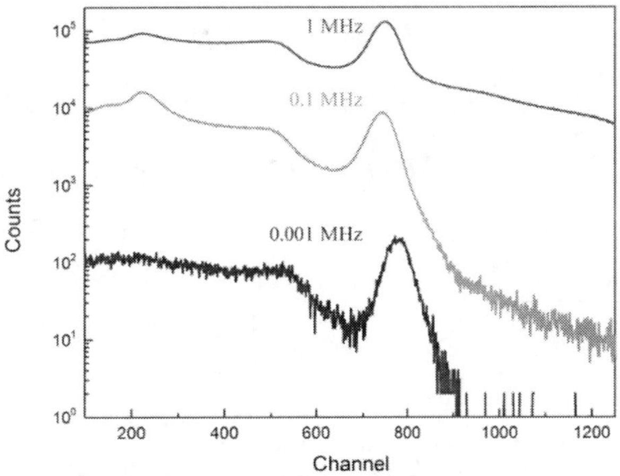

Fig. 7. Peak position as a function of count rate.

Table 1. Peak position and FWHM as a function of count rate.

Rate, Mcps	Peak position (PP), ch	Δ=|PP-PP_{av}|/PP_{av}, %	FWHM, %
1	748	2.5	8.48
0.1	743	3.3	8.46
0.001	778	1.2	7.65

REFERENCES

- I. Zychor *et al.*, "High performance detectors for upgraded gamma ray diagnostics for jet DT campaigns", *Physica Scripta* vol. 91, 064003, 2016.
- G.Boltruczyk *et al.*, "Development of MPPC+based detectors for high count rate DT campaigns at JET", *Fusion Engineering and Design*, 2017.
- A. Pellegrino, "Summary Actel ProAsic/3E Iradiation Tests", Actel Corporation, Nikhef, OT FE Architecture Mini-Review March 2013.
- S. Rezgui, "Radiation-Tolerant ProASIC3 FPGAs Radiation Effects", Actel Corporation, *Tech. Report,* April 2010.
- AVNET, "MicroZed Evaluation Kit", datasheet 2017.
- Hamamatsu Photonics K.K., "MPPC (Multi-Pixel Photon Counter) arrays," S13361-3050 series datasheet, April 2016.
- CAEN Electronic Instrumentation, Digitizer DT5730 datasheet [rev May 2016].

EPJ Web of Conferences 170, 02008 (2018) https://doi.org/10.1051/epjconf/201817002008
ANIMMA 2017

Radiation induced currents in mineral-insulated cables and in pick-up coils: model calculations and experimental verification in the BR1 reactor

Ludo Vermeeren, Willem Leysen and Benoit Brichard

Abstract — **Mineral-insulated (MI) cables and Low-Temperature Co-fired Ceramic (LTCC) magnetic pick-up coils are intended to be installed in various position in ITER. The severe ITER nuclear radiation field is expected to lead to induced currents that could perturb diagnostic measurements. In order to assess this problem and to find mitigation strategies models were developed for the calculation of neutron- and gamma-induced currents in MI cables and in LTCC coils.**

The models are based on calculations with the MCNPX code, combined with a dedicated model for the drift of electrons stopped in the insulator. The gamma induced currents can be easily calculated with a single coupled photon-electron MCNPX calculation. The prompt neutron induced currents requires only a single coupled neutron-photon-electron MCNPX run. The various delayed neutron contributions require a careful analysis of all possibly relevant neutron-induced reaction paths and a combination of different types of MCNPX calculations.

The models were applied for a specific twin-core copper MI cable, for one quad-core copper cable and for silver conductor LTCC coils (one with silver ground plates in order to reduce the currents and one without such silver ground plates). Calculations were performed for irradiation conditions (neutron and gamma spectra and fluxes) in relevant positions in ITER and in the Y3 irradiation channel of the BR1 reactor at SCK•CEN, in which an irradiation test of these four test devices was carried out afterwards.

We will present the basic elements of the models and show the results of all relevant partial currents (gamma and neutron induced, prompt and various delayed currents) in BR1-Y3 conditions. Experimental data will be shown and analysed in terms of the respective contributions. The tests were performed at reactor powers of 350 kW and 1 MW, leading to thermal neutron fluxes of 1E11 n/cm²s and 3E11 n/cm²s, respectively. The corresponding total radiation induced currents are ranging from 1 to 7 nA only, putting a challenge on the acquisition system and on the data analysis.

The detailed experimental results will be compared with the corresponding values predicted by the model. The overall agreement between the experimental data and the model predictions is fairly good, with very consistent data for the main delayed current components, while the lower amplitude delayed currents and some of the prompt contributions show some minor discrepancies.

This work was supported by F4E via the contract F4E-OFC-358-01-01-03.

Ludo Vermeeren and Willem Leysen are with the Belgian Nuclear Research Center, SCK•CEN, Boeretang 200, BE-2400 Mol, Belgium (e-mails: ludo.vermeeren@sckcen.be and willem.leysen@sckcen.be).

Benoit Brichard is with Fusion for Energy, Av. Josep Pla 2, Torres Diagonal Litoral B3, 08019 Barcelona, Spain (e-mail: benoit.brichard@f4e.europa.eu).

Index Terms—**ITER, LTCC pick-up coils, Mineral-Insulated Cables, Neutrons, Radiation effects**

I. Introduction

During the burning process in ITER, a large amount of radiation will be released mainly in the form of energetic neutrons and gamma-rays. Due to their uncharged state, neutrons and gamma-rays can easily escape the plasma and potentially damage (either permanently or temporarily) components and devices, such as ITER Diagnostic Components installed in ITER [1,2]. Magnetic Pick-Up Coils and mineral-insulated (MI) cables are intended to be installed in the divertor region and in the ITER vessel behind the blanket modules. The Magnetic Pick-Up Coils are fabricated based on the Low Temperature Co-fired Ceramic (LTCC) technology [3,4]. In view of the high ITER nuclear radiation field at the locations of these items a specific study of radiation effects on MI cables and LTCC coils was performed. In this paper we discuss the radiation induced currents between the inner conductors and the grounded parts of the devices, as these currents might perturb the proper functioning of the devices. Elaborating the approach described in [5] and [6], we will present MCNPX [7] based models for the calculation of the various contributions to the radiation induced currents and we will discuss an irradiation test in the BR1 research reactor at SCK•CEN proving the validity of the models.

II. Description of the devices under test

A. Description of the selected MI cables

One MI cable with two inner copper conductor (twin-core) and one with four inner copper conductors (quad-core) were selected for this study, both with magnesia insulator and with

https://doi.org/10.1051/epjconf/201817002008

AISI 316L sheath (containing approximately resp. 0.81% and 1.64% Mn). The twin-core MI cable has an outer diameter of 4.2 mm and a conductor diameter of 0.8 mm. The quad-core MI cable has an outer diameter of 3.6 mm and a conductor diameter of 0.44 mm. For the model calculations, the cables are considered to be infinitely long and the results are presented per unit length.

B. Description of the LTCC pick-up coils

The LTCC magnetic sensor prototypes consist of a sintered stack of ceramic layers with printed metallic circuits, which constitute a number of conductive turns connected in series in order to obtain a magnetic pick-up coil. Figure 1 illustrates the external layout of the prototypes as well as the inside part showing the wiring.

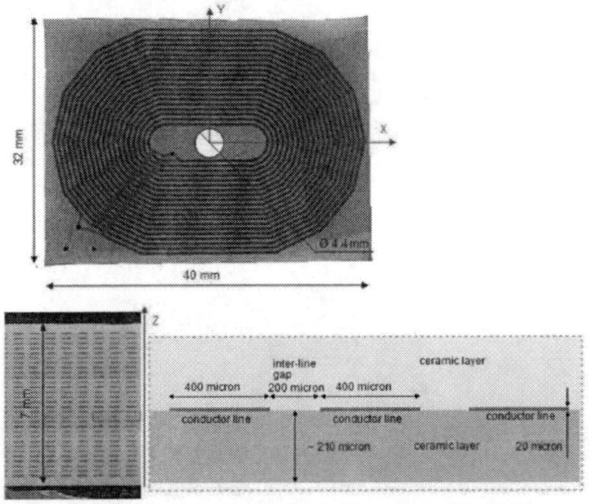

Fig. 1. details on the geometry and the winding pattern of the LTCC sensor prototypes

Two types of LTCC sensors were studied; both of them have silver as conductor material and DuPont-951 ceramic [8] as ceramic insulator. The composition of the latter material was approximated as 43 at% Al + 31 at% Si + 20 at% O + 6 at% Ca, with a density of 3.1 g/cm³. Both LTCC sensors have nominal dimensions of 40mm*32mm*7mm and are composed of a bulk DuPont-951 ceramic part, a 30 layer electrical circuit embedded in the ceramic and two electrical contacts equipped with a pair of copper wires connecting to the measuring equipment. For one of them, 50 µm thick silver ground plates (with 40% effective surface) were added at the lower and upper side of the sensor, each of them connected to a copper wire. The aim of these plates is to counterbalance the neutron induced current in the silver windings, thus reducing the net current.

III. MCNPX MODELLING

A. Basic principles

Neutron and gamma-induced interactions lead to the generation of a current between the inner conductors and the grounded parts [5]. After the impact of neutrons or gammas, several possible interactions lead to the creation of energetic electrons. A fraction of these free electrons is sufficiently energetic to cross the insulator. These electrons constitute a current between the inner conductors and the grounded parts. Conventionally, currents are taken positive if electrons cross the insulator from the core wire towards the ground. If the 'circuit' is closed by connecting a current meter, this corresponds to a positive (conventional) current through the meter from the central conductor side to the ground.

Two classes of free electron generating processes can be considered. On one hand, the production of beta rays upon decay of unstable isotopes formed after neutron capture leads to a delayed current with a response time depending on the half-life of the activation product. On the other hand, prompt processes stem from external gamma rays — or from gamma rays produced quasi-instantaneously after neutron interactions — creating free electrons by photoelectric effect, Compton scattering, or pair formation.

In principle, the following contributions to the induced current can be identified:

A. I(n,β_core): current due to neutron interactions in the central conductor and subsequent beta decay

B. I(n, β_insulator): current due to neutron interactions in the insulator and subsequent beta decay

C. I(n, β_ground): current due to neutron interactions in the (grounded) sheath and subsequent beta decay

D. I(n, β_surrounding): current due to neutron interactions in the sensor surroundings (if applicable) and subsequent beta decay

E. I(n,γ,e): current due to gammas created by neutron interactions in the sensor (and, if applicable, in the surroundings) and subsequent gamma-electron interactions

F. I(γ,e): current due to fast electrons generated in the sensor (and, if applicable, in the surroundings) by external gammas

Components E and F each require one single MCNPX calculation: starting from a neutron or gamma source at the outer surface of the sensor (with the appropriate spectrum), the deposited charge in the relevant parts of the cable is calculated directly, including all (prompt) interactions between neutrons, gammas and electrons.

On the other hand, MCNPX does not include the delayed effects related to beta decay of activation products, so these contributions have to be modelled explicitly. This requires two steps: (1) the reaction rate of each possible contribution should be calculated, and (2) an electron transport calculation should be performed for each contribution, starting from the appropriate beta source (spectrum, and, if relevant, spatial distribution); this second step yields the deposited charges in the relevant parts of the cable per source beta.

For thermal neutrons, neutron capture (the (n,γ) reaction) is the dominant reaction. But for a spectrum with a high contribution of fast neutrons like the ITER spectrum, several other reaction channels can be open, like (n,2n), (n,p) and (n,α). For all these reactions on all isotopes present in the sensor, it should be analyzed whether a beta-emitting nucleus is formed and a preliminary assessment of the possible contribution to the RIEMF current is required based on the atomic concentration of the isotope, on the cross section of the reaction, on the half-life of the activation product and on the beta end-point energy.

For the cable core wires (pure copper), both natural Cu isotopes contribute in principle through the following schemes:

- $^{63}Cu(n,\gamma)^{64}Cu(\beta+ decay)\text{-->}^{64}Ni$, $T_{1/2}$ = 12.7 h (61%)
- $^{63}Cu(n,\gamma)^{64}Cu(\beta- decay)\text{-->}^{64}Zn$, $T_{1/2}$ = 12.7 h (39%)
- $^{63}Cu(n,2n)^{62}Cu(\beta+ decay)\text{-->}^{64}Zn$, $T_{1/2}$ = 9.74 min
- $^{65}Cu(n,\gamma)^{66}Cu(\beta- decay)\text{-->}^{66}Zn$, $T_{1/2}$ = 5.12 min
- $^{65}Cu(n,2n)^{64}Cu(\beta+ decay)\text{-->}\ ^{64}Ni$, $T_{1/2}$ = 12.7 h (61%)
- $^{65}Cu(n,2n)^{64}Cu(\beta- decay)\text{-->}\ ^{64}Zn$, $T_{1/2}$ = 12.7 h (39%)
- $^{65}Cu(n,p)^{65}Ni(\beta- decay)\text{-->}^{65}Cu$, $T_{1/2}$ = 2.52 h

The cable insulators are composed of magnesium and oxygen. After analysis, the retained contributions are:

- $^{24}Mg(n,2n)^{23}Mg$ (β+ decay)-->^{23}Na, $T_{1/2}$ = 11.3 s
- $^{24}Mg(n,p)^{24}Na$ (β- decay)-->^{24}Mg, $T_{1/2}$ = 14.96 h
- $^{16}O(n,2n)^{15}O$ (β+ decay)-->^{15}N, $T_{1/2}$ = 2.03 min
- $^{16}O(n,p)^{16}N$ (β- decay)-->^{16}O, $T_{1/2}$ = 7.13 s

The cable sheaths contain mainly iron, nickel and chromium, and also about 1.5% of manganese. The relevant reactions are:

- $^{56}Fe(n,p)^{56}Mn$ (β- decay)-->^{56}Fe, $T_{1/2}$ = 2.58 h
- $^{58}Ni(n,p)^{58}Co$ (β+ decay)-->^{58}Fe, $T_{1/2}$ = 70.8 d
- $^{60}Ni(n,p)^{60}Co$ (β- decay)-->^{60}Ni, $T_{1/2}$ = 5.272 a
- $^{64}Ni(n,\gamma)^{65}Ni$ (β- decay)-->^{65}Cu, $T_{1/2}$ = 2.52 h
- $^{52}Cr(n,p)^{52}V$ (β- decay)-->^{52}Cr, $T_{1/2}$ = 3.75 min
- $^{55}Mn(n,\gamma)^{56}Mn$ (β- decay)-->^{56}Fe, $T_{1/2}$ = 2.58 h

For the LTCC windings (pure silver), both natural Ag isotopes contribute in principle through the following schemes:

- $^{107}Ag(n,\gamma)^{108}Ag(\beta- decay)\text{-->}^{108}Cd$, $T_{1/2}$ = 2.41 min
- $^{107}Ag(n,2n)^{106}Ag(\beta+ decay)\text{-->}^{106}Pd$, $T_{1/2}$ = 24 min
- $^{109}Ag(n,\gamma)^{110}Ag(\beta- decay)\text{-->}^{110}Cd$, $T_{1/2}$ = 24.6 s
- $^{109}Ag(n,2n)^{108}Ag(\beta- decay)\text{-->}\ ^{108}Cd$, $T_{1/2}$ = 2.41 min
- $^{109}Ag(n,p)^{109}Pd(\beta- decay)\text{-->}\ ^{109}Ag$, $T_{1/2}$ = 13.43 h

All these contributions are taken into account in the calculations. In principle, also the metastable states ^{106m}Ag, ^{108m}Ag and ^{110m}Ag are populated by the mentioned reactions, but these are neglected here, since:

(1) ^{106m}Ag decays via electron capture only, emitting no betas;

(2) ^{108m}Ag decays via internal transitions and electron capture, emitting no betas (moreover, it has a half-life of 418 years, so the decay rate is very low);

(3) ^{110m}Ag has three main decay paths: 2% via internal transitions (no beta), 67% via beta decay with a beta end-point energy of 0.06 MeV only (so all emitted betas will be stopped locally) and 31% via beta decay with a beta end-point energy of 0.51 MeV only (so an average beta energy of about 0.2 MeV, which is also not sufficient to contribute efficiently to the current; moreover, the half-life is long (250 days), leading anyway to low decay rates.

The mentioned schemes also apply to the contributions from the grounded silver planes (if present), but, as in that case the betas originate from the other electrode, the sign is inverted.

The LTCC insulator material is supposed to be composed of aluminum, silicon, oxygen and a small fraction of calcium. After analysis, the retained contributions are:

- $^{27}Al(n,\gamma)^{28}Al$ (β- decay)-->^{28}Si, $T_{1/2}$ = 2.25 min
- $^{27}Al(n,p)^{27}Mg$ (β- decay)-->^{27}Al, $T_{1/2}$ = 9.46 min
- $^{27}Al(n,\alpha)^{24}Na$ (β- decay)-->^{24}Mg, $T_{1/2}$ = 14.96 h
- $^{28}Si(n,p)^{28}Al$ (β- decay)-->^{28}Si, $T_{1/2}$ = 2.25 min
- $^{30}Si(n,\gamma)^{31}Si$ (β- decay)-->^{31}P, $T_{1/2}$ = 2.62 h
- $^{16}O(n,p)^{16}N$ (β- decay)-->^{16}O, $T_{1/2}$ = 7.13 s

B. Special treatment of electrons stopped in the insulator

MCNPX only calculates the particle transport down to a preset lower energy limit. More specifically, the electron transport is calculated until an energy of 1 keV at which it is supposed to remain at the same position. The subsequent drift of electrons out of the insulator under the action of the created space charge field is not calculated by MCNPX. Part of these electrons will drift back to the beta emitting cell, and another part to the other electrode; only the latter electrons will contribute to the current. So the calculation of the current requires an extra calculation step. For cylindrically symmetric configurations, formalisms have been developed to determine the fraction f of electrons drifting back to the original electrode, e.g. [9].

For the geometries considered here, the cylindrical symmetry is broken, so the simple analytical model cannot be applied anymore. Therefore, a similar procedure as in [5]. was followed: the Poisson equation was solved in an iterative way, assuming a constant charge density and zero boundary conditions. From the resulting space charge field profile, the zero-field surface was deduced and introduced as an additional surface in MCNPX to subdivide the insulator in two parts: a part close to the copper wires (in which all the deposited betas are supposed to drift towards the copper wires) and a part close to the sheath (in which all deposited betas are supposed to drift towards the sheath).

Fig. 2. Calculated space charge potential profile for the quad-core MI cable, with the deduced zero-field surface shown as a red line

To determine the boundary surface for the cables, a rectangular 2D grid is defined representing the insulator and the Poisson equation for the space charge potential is solved, assuming a homogeneous charge deposition and zero potential boundaries at the interfaces with the core wires and the sheath. This is done in Excel by iteratively calculating each cell as the average of the potentials in each of the four neighbours plus a constant value (the space charge density, which for our purpose can take any non-zero value). As an example, the resulting potential contours for the quad-core cable are shown in Figure

2, together with the deduced zero-field surface approximation: a combination of four symmetric planes defined by x,y=±0.94mm and a concentric cylinder with radius 1.1 mm.

For the planar LTCC geometry, the zero field surfaces are assumed to be in the middle of the insulator parts.

C. Calculation of prompt neutron induced contribution

A coupled neutron-photon-electron MCNPX calculation is performed, starting from a neutron source which is uniformly distributed at the outer surface of the cable (or at a cylindrical surface enveloping the LTCC), with an inward initial direction distribution proportional to the cosine of the angle with the normal to the surface, and with the relevant energy spectrum. As output we obtain the deposited charge in each cell of the geometry. MCNPX provides the results normalized per source particle. In order to normalize the results to the neutron flux, the neutron flux per source neutron is calculated by MCNPX too.

The effective charge transfer per source neutron from the central conductor to the grounded parts can be calculated as the total charge deposited in the central conductors and in the surrounding insulator parts.

The total prompt neutron induced current (for the cables per unit length) can then be calculated by dividing the effective charge transfer per source neutron by the ratio between the real flux and the flux per source neutron.

D. Calculation of prompt gamma induced contribution

This calculation follows the same procedure as for the prompt neutron contribution, only a suitable gamma source is defined instead of a neutron source and a couples photon-electron MCNPX calculation is performed.

E. Calculation of delayed neutron induced contributions

First, for each geometry and each neutron spectrum of interest, a "neutron only" MCNPX calculation is performed in order to obtain the rate densities of all reactions specified in Section III.A. The JEFF-3.2 neutron cross section library [10] was consistently used for all calculations.

Each of the considered reactions leads to the formation of a beta-emitting isotope (with a certain half-life); so, in equilibrium conditions, each reaction corresponds to one decay. The decay properties (involving beta emission) were studied in detail for each isotope involved: the relative probabilities of beta-plus decay paths (with possibly various beta end point energies), of beta-minus decay paths (with possibly various beta end point energies), and of decay paths without beta emission (mainly electron capture). The beta spectra were calculated as the sum of the corresponding Fermi beta distribution functions.

In order to obtain the induced current, one needs to calculate the probability that an emitted beta effectively contributes to the current, i.e. the probability that it crosses the insulator. This calculation is done by an "electron only" MCNPX calculation. After defining the problem geometry in MCNPX, an electron source is defined with the appropriate beta spectral distribution and homogeneously distributed within the appropriate cells (and with random initial direction). As output from MCNPX we obtain the charge deposition in each cell (normalized per source electron). The effective charge transfer per source electron from central conductors to ground can be calculated as the total charge deposited in the central conductors and in the surrounding insulator parts.

Each (equilibrium) delayed neutron-induced current contribution is then calculated as the product of the MCNPX reaction rate density, the volume of the cell in which the interactions take place, the effective charge transfer per beta (with the correct sign, depending on whether positrons or electrons are the charge carriers), the number of betas per decay, and the ratio between the real flux and the flux (per source neutron per second) as outputted by MCNPX.

IV. MCNP RESULTS

A. Neutron and gamma spectra

Calculations were performed for typical ITER conditions, as well as for the conditions in the Y3 channel of the BR1 reactor at SCK•CEN, in which validation tests were performed later on. Figs. 3 and 4 show the BR1 neutron and gamma spectral data used for the calculations (total neutron flux 5.998 10^{11} n/(cm²s), total gamma flux 7.9 10^{10} γ/(cm²s)). The total neutron and gamma fluxes at the location of the MI cables in ITER was assumed to be 2.28 10^{13} n/(cm²s) and 2.1 10^{13} γ/(cm²s)); the fluxes at the LTCC coil location were taken as 1.41 10^{13} n/(cm²s) and 0.72 10^{13} γ/(cm²s)).

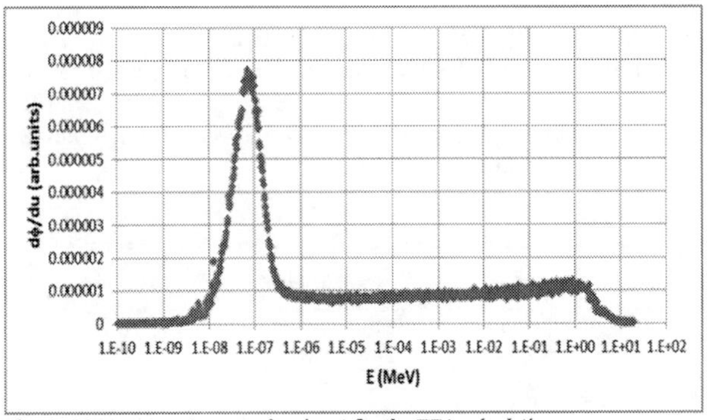

Fig. 3. Neutron spectrum used as input for the BR1 calculations

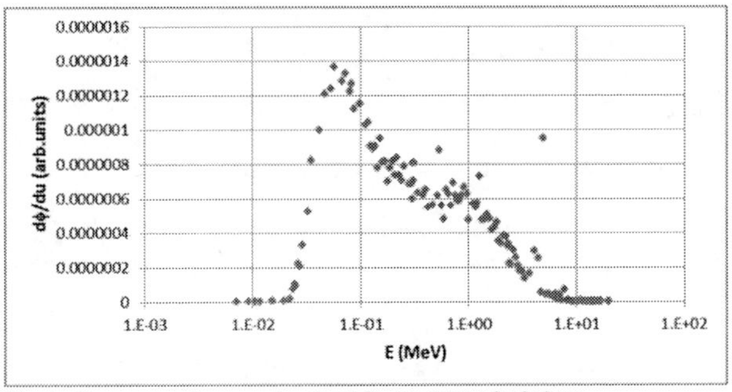

Fig. 4. Gamma spectrum used as input for the BR1 calculations

B. *MI cables*

Table I shows the calculated absolute current contributions of the various processes in the twin-core and the quad-core MI cable (in nA per m cable length, sum of all wires) for the BR1 case. Similar calculations were performed for the ITER case.

TABLE I

Process	2-core	4-core
$^{63}Cu(n,\gamma)^{64}Cu$	0.0015 nA	0.017 nA
$^{65}Cu(n,\gamma)^{66}Cu$	0.54 nA	0.44 nA
$^{64}Ni(n,\gamma)^{65}Ni$	-0.001 nA	-0.001 nA
$^{55}Mn(n,\gamma)^{56}Mn$	-0.132 nA	-0.223 nA
Prompt neutron	0.78 nA	0.47 nA
Prompt gamma	0.05 nA	0.08 nA
Total	1.12 nA	0.80 nA

The prompt neutron contribution appears to be dominant. Amongst the delayed contributions, the $^{65}Cu(n,\gamma)^{66}Cu$ reaction is the dominant reaction, leading to a main response time of 5.1 min / (ln(2)) = 7.36 min. The contribution by the $^{55}Mn(n,\gamma)^{56}Mn$ reaction is also rather large and negative; its response time is much larger: 2.58 h / ln(2)) = 4.04 h. The results for the ITER case show non-negligible contributions from the $^{63}Cu(n,2n)^{62}Cu$ reaction (response time 9.74 min / ln(2)) = 14.05 min) and the $^{16}O(n,p)^{16}N$ reaction (response time 7.13 s / ln(2)) = 10.5 s); the prompt gamma contribution is also very important in that case.

C. *LTCC coils*

Table II summarizes the calculated absolute current contributions of the various processes for the LTCC coils without ground plate and with silver ground plates (20μm effective thickness) in the BR1 conditions. Similar calculations were performed for the ITER case.

TABLE II

Process	No ground plate	With ground plate
from windings		
$^{107}Ag(n,\gamma)^{108}Ag$	0.5 nA	0.5 nA
$^{109}Ag(n,\gamma)^{110}Ag$	3.6 nA	3.2 nA
from ground plate		
$^{107}Ag(n,\gamma)^{108}Ag$	-	-0.44 nA
$^{109}Ag(n,\gamma)^{110}Ag$	-	-1.9 nA
from insulator		
$^{27}Al(n,\gamma)^{28}Al$	0.4 nA	0.2 nA
$^{30}Si(n,\gamma)^{31}Si$	0.003 nA	0.001 nA
Prompt neutron	2.0 nA	2.0 nA
Prompt gamma	0.2 nA	0.2 nA
Total	6.7 nA	3.8 nA

In the BR1 case, the dominant contributions are the prompt neutron current and the current due to beta decay of ^{110}Ag, formed after neutron capture by ^{109}Ag. In the ITER case, the gamma induced current becomes dominant and additional delayed components due to (n,p), (n,α) and (n,2n) reactions are also contributing to some extent.

When adding two silver ground plates (each with effective thickness 20 μm), an extra current with opposite sign is generated, due to activation and decay of silver nuclei in these ground plates. The resulting total current is about a factor 2 lower than in the geometry without ground plates.

V. Experimental test in the BR1 reactor

The MIC and LTCC coils were irradiated in the Y3 channel of the BR1 reactor. BR1 is an air cooled graphite moderated thermal reactor with a nominal thermal power of 700 kW; for short periods the power can be raised to 1 MW. The thermal and epithermal neutron flux at the center of channel Y3 were determined by activation dosimetry, using an Al/Co and an Al/Ag foil, placed near the LTCC coils. For the reference BR1 power of 700 kW, the conventional thermal neutron flux was found to be 1.98 10^{11} n/(cm²s) and the epithermal flux per unit lethargy 1.5 10^{10} n/(cm²s). Taking into account the neutron spectrum used for the calculations, the measured thermal neutron flux corresponds to a total neutron flux of 4.5 10^{11} n/(cm²s).

For the LTCC coil calculations, a reference total flux of 5.998 10^{11} n/(cm²s) was assumed; so the model results have to be normalized by a factor 4.5/5.998 to get the results for the real conditions at 700 kW.

The ends of the MI cables were also positioned near the center of the Y3 channel, but they extend over half of the reactor axis, through the reflector and the shielding to the side of the reactor. In order to obtain the integrated flux over the complete cable length, the known axial neutron flux profile was used. For a maximum total neutron flux of 4.5 10^{11} n/(cm²s), it leads to a length-integrated total flux of 8.62 10^{11} n/(cm²s)m. Hence, the MCNP model results from should be renormalized by a factor 8.62/5.998 before comparison with the experimental data at 700 kW.

Several test runs at varying conditions were performed with continuous monitoring of the currents and the reactor power. For the first two test runs, the reactor power was ramped up to 350 kW and scrammed after about 30 minutes. In the first case, the currents of the MI cables were measured individually for each core wire, while in the second, all cores wires of each cable were physically connected and the global currents were measured. Similarly two more test runs were performed with the reactor power rising to 350 kW, after 5 minutes stabilization followed by an increase up to 1000 kW, stabilization for 60 minutes, power reduction to 350 kW and scram. The decay of the signals was monitored continuously for about 16 hours.

All recorded MI cable currents were fitted with a least squares procedure as the sum of a constant offset, a prompt signal (proportional to the reactor power), a delayed component due to activation of ^{65}Cu and subsequent decay of ^{66}Cu and a delayed component due to activation of ^{55}Mn (four free parameters). Similarly the LTCC coil current were analyzed as the sum of an offset, a prompt signa and delayed components due to activation of ^{107}Ag and ^{109}Ag.

Figure 5 shows an example of the data recorded for an MI cable: the current from one of the quad-core cable wires during the 1000 kW test run. The fitted partial contributions are included in the graph as well as a plot of the pointwise differences between the total fit curve values and the

 https://doi.org/10.1051/epjconf/201817002008

experimental data. Taking into account the limited number of free parameters (4) and the uncertainty on the experimental data (including the reactor power data) the agreement is very satisfactory.

Figure 6 shows similar data for the LTCC coil without ground plates and figure 7 for the LTCC coil with ground planes. For the latter, an insulation resistance problem resulted in a strong, slowly varying background. Hence the fitting procedure could not be applied as such; the data could only be analyzed in first order as the sum of a linearly varying background and a total neutron induced component (considered to be prompt).

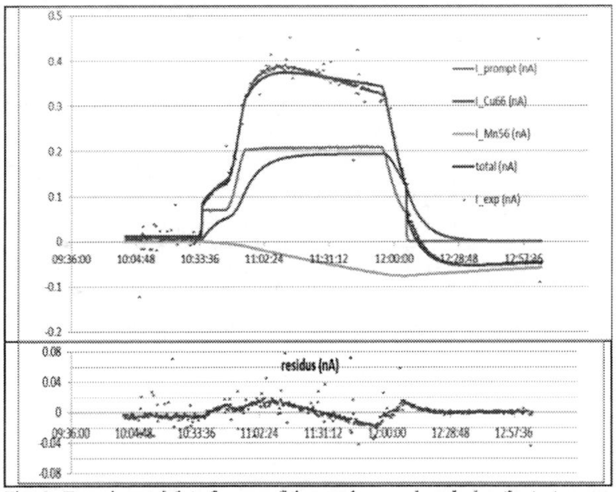

Fig. 5. Experimental data for one of the quad-core wires during the test run up to 1000 kW, including the curves with the various signal contributions, resulting from the fitting procedure; residue plot.

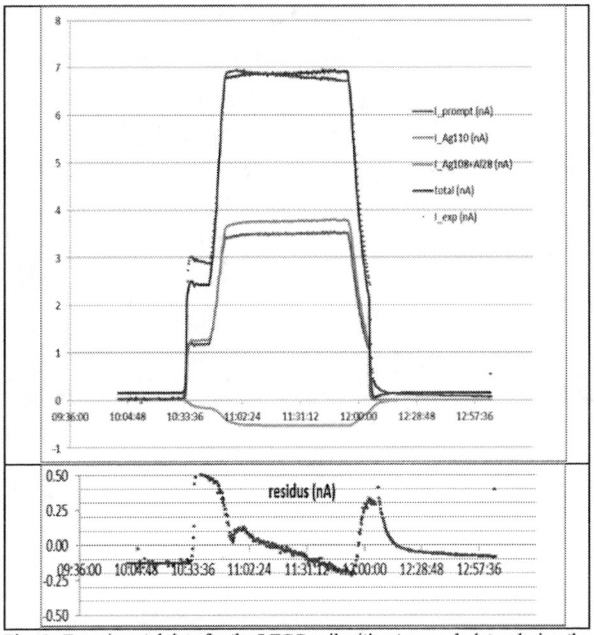

Fig. 6. Experimental data for the LTCC coil without ground plates during the test run up to 1000 kW, including the curves with the various signal contributions, resulting from the fitting procedure; residue plot.

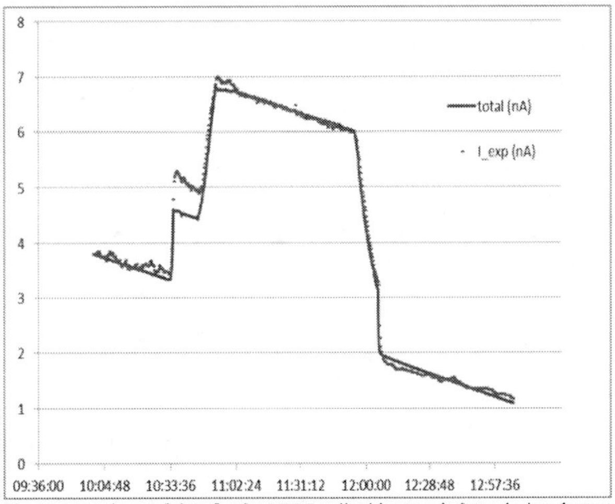

Fig. 7. Experimental data for the LTCC coil with ground plates during the test run up to 1000 kW, including the result of a manual fit assuming a linearly decreasing background.

VI. VALIDATION OF THE MCNPX BASED MODELS

Tables III to V summarize the fit results for both MI cables and both LTCCs from an overall analysis and from the analyses of the four runs separately (always normalized to a reactor power of 700 kW). The averages and the standard deviations over the five data sets are also given and the corresponding predicted current contributions from the model are included.

For the twin-core MI cable, the observed prompt current amounts to 55% of the predicted current. Part of this discrepancy can be ascribed to uncertainties in the geometry, the composition of the materials and the neutron/gamma spectra, and also to the fact that surrounding materials in the irradiation setup are not taken into account in the model. However, it is also known [5,11] that especially prompt gamma-induced currents are difficult to model accurately and that significant uncertainties are associated with these particular model results. The observed current contribution due to ^{66}Cu agrees with the model result within about two standard deviations (13% difference) which is considered to be acceptable taken into account all uncertainties. The uncertainty on the observed current contribution due to ^{56}Mn is larger, but it is perfectly consistent with the model prediction (within one standard deviation).

For the quad-core MI cable, the observed prompt current differs from the predicted current by only 30%, which is considered good in view of the modelling uncertainties. The observed current contribution due to ^{66}Cu agrees perfectly with the theoretical result. The experimental current contribution due to ^{56}Mn is larger than the model prediction - possibly since the real MI cable sheath thickness seems to be somewhat larger than the nominal value used for the modelling; still the data agree within 1.5 standard deviations.

TABLE III: ANALYSIS OF TWIN-CORE CABLE DATA

I(nA) @ 700 kW	prompt	^{66}Cu	^{56}Mn
350 kW – A*	0.643	0.938	-0.017
350 kW – B**	0.612	0.972	-0.441
1000 kW – A*	0.664	0.852	-0.275
1000 kW – B**	0.640	0.841	-0.233
overall	0.639	0.863	-0.258
Average+std.dev.	0.64±0.02	0.89±0.05	-0.24±0.14
model	1.17	0.77	-0.19

* sum of the current recorded for the individual wires
** current recorded with joined core wires

TABLE IV: ANALYSIS OF QUAD-CORE CABLE DATA

I(nA) @ 700 kW	prompt	^{66}Cu	^{56}Mn
350 kW – A*	0.502	0.731	-0.261
350 kW – B**	0.466	0.730	-0.560
1000 kW – A*	0.581	0.528	-0.749
1000 kW – B**	0.552	0.464	-0.715
overall	0.531	0.537	-0.767
Average+std.dev.	0.53±0.04	0.60±0.11	-0.61±0.19
model	0.76	0.63	-0.32

* sum of the current recorded for the individual wires
** current recorded with joined core wires

TABLE V: ANALYSIS OF LTCC COIL DATA

I(nA) @ 700 kW	Prompt*	^{110}Ag*	^{108}Ag*	total**
350 kW – A	2.60	2.41	0.223	2.80
350 kW – B	1.99	3.06	0.134	2.94
1000 kW – A	2.41	2.59	-0.367	2.70
1000 kW – B	1.19	3.28	0.005	2.80
overall	1.52	3.31	-0.233	2.59
Average	1.94	2.93	-0.05	2.77
+ std.dev.	±0.53	±0.36	±0.22	±0.12
model	1.71	2.71	0.38	2.80

* LTCC coil without ground plates
** LTCC coil with ground plates

The LTCC coil without any silver ground planes shows a prompt current contribution in very good agreement with the calculation result. The ^{110}Ag related component is also perfectly predicted by the model. The experimental determination of the lower amplitude ^{108}Ag related component was hampered by the quality of the data; moreover, interference with a contribution from activation and decay of aluminum in the supporting structure (leading to a component with a very similar half-life) cannot be excluded. The experimental data give essentially a zero current with a significant uncertainty, while the model predicts a pure ^{108}Ag contribution with an amplitude of almost two times the standard deviation of the experimental results.

Finally, for LTCC7 only a reliable value for the total current could be deduced from the experimental data (due to the significant and varying background). The total expected current of +2.8 nA at a BR1 power of 700 kW is in perfect agreement with the experimental value. Extrapolating the model calculations, a nearly perfect annihilation of the total current could be obtained when using two silver ground plates with an effective average thickness of 50 µm.

So the overall agreement between the experimental data and the MCNPX based model predictions is fairly good, with very consistent data for the main delayed current components, and some minor discrepancies for the lower amplitude delayed currents and some of the prompt contributions.

Acknowledgment

The work leading to this publication has been funded by Fusion for Energy under the Specific Contract F4E-OFC-358-01-01-03. This publication reflects the views only of the authors, and Fusion for Energy cannot be held responsible for any use which may be made of the information contained therein.

References

[1] G. Vayakis, E.H. Hodgson, V. Voitsenya and C.I. Walker, Fusion Science and Technology, vol. 55, pp. 699-750, 2008.
[2] G. Vayakis et al., Journal of Nuclear Materials, vol. 417, pp. 780-786, 2011.
[3] G. Chitarin, R. Delogu, A. Gallo and S. Peruzzo, Fusion Engineering and Design, vol. 84, p. 593-598, 2009.
[4] Y. Imanaka, Multilayered low temperature co-fired ceramics (LTCC) technology, Springer, USA, 2005.
[5] L. Vermeeren and R. Van Nieuwenhove, Review of Scientific Instruments 75, pp. 4667-4674, 2003.
[6] S. Peruzzo et al., Fusion Engineering and Design, vol. 88, pp. 1302-1305, 2013.
[7] D.B. Pelowitz, Ed., "MCNPX Users Manual Version 2.7.0" LA-CP-11-00438 (2011).
[8] DuPontTM Green TapeTM 951 Low Temperature Ceramic System, http://www.dupont.com/content/dam/dupont/products-and-services/electronic-and-electrical-materials/documents/prodlib/951.pdf, accessed on May 18, 2017.
[9] N.P. Goldstein, IEEE Transactions on Nuclear Science, vol. NS-20, pp.549-556, 1973.
[10] JEFF-3.2 evaluated data library - Neutron data, https://www.oecd-nea.org/dbforms/data/eva/evatapes/jeff_32/, accessed on May 17,/2017
[11] L. Vermeeren and M. Wéber, Fusion Engineering and Design, vol. 82, pp. 1185-1191, 2007.

Measurement of plasma current in Tokamaks using an optical fibre reflectometry technique

Marc Wuilpart, Willem Leysen, Andrei Gusarov, Philippe Moreau, and Patrice Mégret

***Abstract*—An optical time-domain reflectometer sensitive to the polarization of light is proposed for the measurement of plasma current in the Tore Supra fusion reactor. The measurement principle relies on the Faraday effect i.e. on the generation of a circular birefringence along an optical fiber subject to an axial magnetic field. The circular birefringence induces a polarization rotation that can be mapped along the fiber thanks to an optical-time domain reflectometer followed by an linear polarizer. A proper fitting of the measurement trace then allows determining the applied plasma current. The sensor has been experimentally validated on the Tore Supra tokamak fusion reactor for a plasma current range going from 0.6 to 1.5 MA. A maximum error of 13.50% has been observed for the lowest current.**

***Index Terms*—Optical time-domain reflectometry, Faraday effect, current sensing, polarimetry.**

I. Introduction

In tokamak fusion reactors, magnetic fields are used to squeeze the plasma in the shape of a torus [1]. In practice, it is fundamental to properly monitor the intensity of the plasma current in order to evaluate the plasma stability. Nowadays, the measurement of the plasma current is performed by means of various kinds of inductive coils (Rogowski, pick-up coils) [2]. Since these sensors require a temporal integration to calculate the value of the plasma current, they are prone to drift during long plasma discharges. Moreover the radiation load induces spurious electrical effects like radiation-induced conductivity, radiation-induced electrical degradation and radiation-induced electromotive force that perturb the proper operation of electrically-based current sensors [3].

FOCS (Fiber-optic Current Sensors) [4] constitute an alternative. They are based on the monitoring of the state of polarization (SOP) rotation undergone by a lightwave propagating in an optical fiber. In the presence of a current surrounded by the fiber, the magnetic field aligned with the fiber axis has for effect to rotate the light SOP via the so-called Faraday effect. From the rotation angle, it is possible to retrieve the surrounded current. This approach presents two drawbacks in the field of tokamak fusion reactor diagnosis. First, the FOCS only measures the SOP rotation within the range $0\text{-}2\pi$. This is problematic for high plasma current measurement like in the future ITER (20 MA). Second, the presence of a lead fiber (fiber section situated between the interrogating device and the loop surrounding the vacuum vessel) deteriorates the FOCS performance [6]. To eliminate these issues, a POTDR (Polarization Optical Domain Reflectometer) can be used since a POTDR trace contains information about the SOP spatial evolution along the fibre. Such an approach was proposed in [5] where currents up to 2.5 kA were tested. The drawback of this technique is to require the full measurement of the backscattered SOP.

In this paper, we summarize the results obtained when using a simple approach (commercially available OTDR followed by a linear polarizer) for plasma current measurement. The sensor has been experimentally validated on the Tore Supra tokamak fusion reactor of the CEA (Cadarache, France) for a plasma current range going from 0.6 to 1.5 MA.

The next section of this paper described the principle of the POTDR approach. In Section III, the modeling of the sensor is described and the method to extract the plasma current is explained. Finally, section IV describes the experimental results obtained on the Tore Supra fusion reactor. Let us note that a theoretical background describing polarization effects in optical fiber and their modeling using the Jones formalism can be found in [11].

II. Measurement principle and signal analysis

The measurement scheme is depicted in figure 1. A photon-counting optical time domain reflectometer launches optical pulses into the lead and sensing fibers (the sensing fiber is the fiber loop placed around the vacuum vessel) via a linear polarizer to fix the input SOP. The role of the polarization controller is explained in section III. When a pulse propagates down the fiber, it is continuously scattered in all directions via the Rayleigh scattering process [9]. A part of this scattered signal is captured by the fiber and propagates back towards the source. The backscattered signal then passes again through the linear polarizer before reaching the OTDR detector where its power is measured as a function of time. The time scale can be converted into a distance scale so that the POTDR trace provides the backscattered power as a function of the backscattering location as described in more detail in [11]. When a plasma current circulates in the vacuum vessel, it generates a magnetic field along the sensing fiber. The component of the magnetic field parallel to the fiber axis generates a circular birefringence via the Faraday effect [7] and induces a SOP rotation. The Faraday-induced birefringence is a non-reciprocal effect so that the direction of the SOP rotation is identical in both directions. In practice, the light SOP varies along an optical fiber even if no magnetic field is applied. A fiber is indeed subject to intrinsic linear birefringence due

Marc Wuilpart and Patrice Mégret are with the Department of Electromagnetism and Telecommunications, University of Mons, Mons, Belgium e-mail: marc.wuilpart@umons.ac.be

Andrei Gusarov and Willem Leysen is with the Belgian Nuclear Research Center SCK-CEN, Mol, Belgium.

Philippe Moreau is with the CEA (IRFM), Saint-Paul-lez-Durance, France.

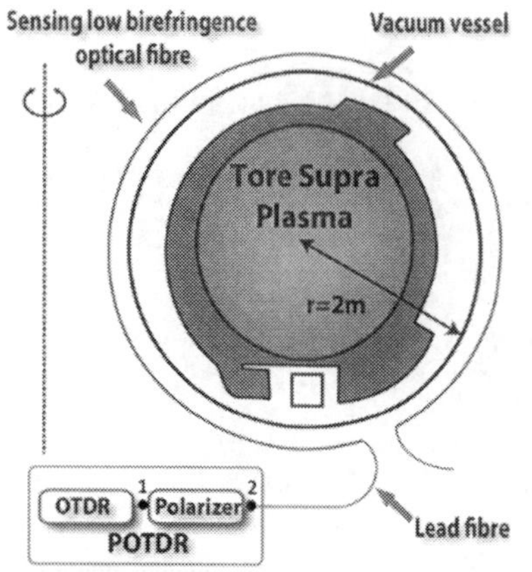

Fig. 1. Section of the vacuum vessel of the Tore Supra tokamak and P-OTDR setup for plasma current monitoring.

to a non perfect circular symmetry of the refractive index distribution in the fiber cross-section. This birefringence adds to the current-induced circular birefringence and therefore constitutes a parasitic effect for current sensing. To minimize the detrimental effect of the intrinsic birefringence, a single-mode low-birefringence fiber (beat length L_B of 480 m) is used as the sensing fiber in the proposed measurement setup.

The SOP variation along the fiber length is given by $J(z) = \mathbf{M}(z)\mathbf{M_L}J_{in}$ where $J(z)$ is the normalized Jones vector of the SOP at a distance z starting from the sensing fiber and J_{in} is the input SOP fixed by the input linear polarizer. $\mathbf{M_L}$ is the Jones matrix of the concatenation of the polarization controller with the lead fiber. Making the hypotheses that the magnetic field is constant (circular vacuum vessel with a central plasma current) and that there are neither twist-induced circular birefringence nor polarization mode coupling (the fiber polarization modes are constant) along the sensing fiber, the backscattered SOP observed at port 2 of the polarizer (see figure 1) after Rayleigh backscattering at a distance z is given by $J_B(z) = \mathbf{M_L}^T\mathbf{M}(z)\mathbf{M}(z)\mathbf{M_L}J_{in}$. As the lead fiber is only characterized by a reciprocal birefringence (no magnetic field), its backward Jones matrix is indeed the transpose of the forward matrix while the hypotheses of no mode coupling and no twist together with the non-reciprocal property of the Faraday effect result to a backward matrix equal to the forward matrix within the sensing fiber as explained in [12]. $\mathbf{M}(z)$ is given by [10]:

$$\mathbf{M}(z) = \begin{pmatrix} \alpha + j\beta\cos(2q) & -\gamma + j\beta\sin(2q) \\ \gamma + j\beta\sin(2q) & \alpha - j\beta\cos(2q) \end{pmatrix} \quad (1)$$

where

$$\alpha = \cos(\Delta z) \qquad \beta = \frac{\delta}{2}\frac{\sin(\Delta z)}{\Delta} \quad (2)$$

$$\gamma = \rho\frac{\sin(\Delta z)}{\Delta} \qquad \Delta = \sqrt{\rho^2 + \frac{\delta^2}{4}} \quad (3)$$

δ is the linear intrinsic birefringence of the fibre given by $\delta = 2\pi/L_B$ and q is the angle of its fastest axis. ρ is the circular birefringence induced by the longitudinal magnetic field B_l and is given by $\rho = VB_l$ [9], yielding:

$$\rho = \frac{V\mu_0 I_P}{2\pi r} \quad (4)$$

where r is the radius of the vacuum vessel circular section and I_P the plasma current. V is the Verdet constant equal to 0.54 rad/(m.T) for standard silica fibers and for a wavelength of 1550nm [8]. The backscattered SOP obtained at port 1 of the polarizer is given by $J_P(z) = \mathbf{M_P}J_B(z)$ where $\mathbf{M_P}$ is the polarizer Jones matrix. Considering the power level backscattered at the polarizer port 2 and the angle of the linear polarizer as the reference power and the reference angle, respectively, yields to a expression of $P_B(z)$ (normalized power of the signal measured at the OTDR detector) given by:

$$P_B(z) = r_1 + r_2\cos(4\Delta z) + r_3\sin(4\Delta z) \quad (5)$$

where r_1, r_2 and r_3 are real numbers depending on δ, ρ, q and the $\mathbf{M_L}$ matrix coefficients. A more detailed explanation about how equation (5) was obtained can be found in [12].

Based on this theoretical analysis, the proposed plasma current measurement principle consists in four steps: (1) measurement of a POTDR trace, (2) calculation of$P_B(z)$ from the POTDR trace by shifting to the linear scale and normalizing by the maximum power, (3) least-squares fitting of $P_B(z)$ allowing finding and (4) calculation of I_P via equations (3) and (4) considering that δ is known.

III. Experimental results

The proposed technique was tested on the Tore Supra experimental thermonuclear fusion reactor. As shown in figure 1, the vacuum vessel is circular with a radius $r = 2$ m. A low-birefringence fiber was placed around the vacuum vessel. The polarization controller was configured to maximize the backscattered power in the sensing fiber region. The POTDR traces were obtained at 1550 nm using a pulse duration of 100 ps, a repetition rate of 40 kHz and a peak power of 16.8 mW. A sliding averaging window of 1.89 m and a time averaging of 10 s were applied to increase the signal-to-noise ratio of the OTDR traces. After extracting the sensing from the POTDR trace, moving to the linear scale and then, normalizing by the maximum backscatter power, the fitting of the experimental curve by equation (4) is performed to extract I_P. The value of δ considered to determine I_P is $2\pi/L_B$ where $L_B = 480$ m (obtained from the low-birefringence fiber datasheet).

The blue trace on figure 2 corresponds to the normalized backscattered power in the sensing fiber distance interval (a more complete trace including the lead fiber can be found in reference [12]) and was obtained for a plasma current set-point of 1.5 MA. One can clearly observe a cosine-like behavior as expected by the theoretical analysis. The green trace corresponds to the fitting by equation (5). The results obtained for various plasma current set-points are shown in Table I. The relative error lies between 1.00 was 13.50%. The worst error was obtained for the smallest I_P value for which the ratio between the POTDR signal variation over the measurement distance range and the random noise level is the smallest, leading to a less accurate fitting.

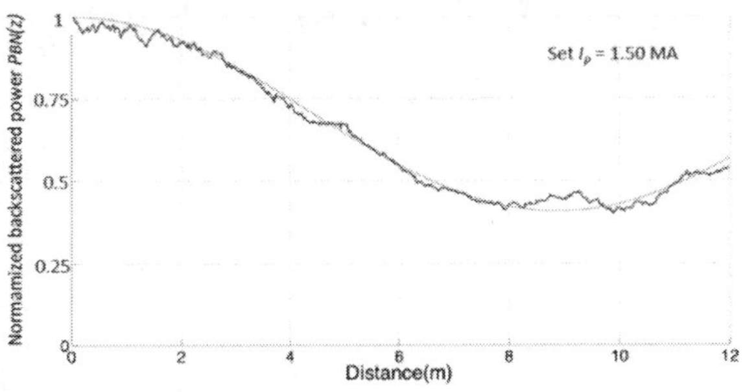

Fig. 2. Normalized POTDR trace in the sensing region.

Set I_P (MA)	Measured I_P (MA)	Relative error(%)
1.5	1.475	1.67
1.3	1.357	4.38
1.2	1.248	4.00
1.0	0.999	1.00
0.8	0.791	1.13
0.7	0.693	1.00
0.6	0.519	13.50

TABLE I
EXPERIMENTAL RESULTS

IV. CONCLUSION

We presented a fiber-optic current sensor for the measurement of plasma current in thermonuclear fusion reactors. The sensor includes a low-birefringence fiber placed around a the vacuum vessel. The interrogator consists of a POTDR unit. The plasma current can be calculated after fitting the experimental trace and removing the contribution of the intrinsic linear birefringence. The sensor has been experimentally validated on the Tore Supra fusion reactor for a plasma current range going from 0.6 MA to 1.5 MA.

REFERENCES

[1] D. van Houtte, G. Martin, A. Bécoulet, J. Bucalossi, G. Giruzzi, G.T. Hoang, Th. Loarer, B. Saoutic and the Tore Supra Team, Recent fully non-inductive operation results in tore supra with 6 min, 1 gj plasma discharges, *Nuclear Fusion*, vol. 44, no. 5, L11, April 2004.

[2] Chen Qing, Li Hong-bin, Huang Ben-xiong and Dou Qiao-qi., Rogowski sensor for plasma current, *Sensors Journal, IEEE*, vol. 9, no. 3, pp. 293-296, March 2009.

[3] R. Van Nieuwenhove and L. Vermeeren, Study of the radiation induced electromotive force effect on mineral insulated cables for magnetic diagnostics in ITER, *Fusion Engineering and Design, 22nd Symposium on Fusion Technology*, vol. 66-68, pp. 821-825, September 2003.

[4] Ph. Moreau, B. Brichard, A. Fil, Ph. Malard, P. Pastor, A. Le-Luyer, F. Samaille and V. Massaut, Test of fiber optic based current sensors on the tore supra tokamak, *Fusion Engineering and Design*, vol. 86, no. 6-8, pp. 1222-1226, October 2011.

[5] L. Palmieri, D. Sarchi and A. Galtarossa, Distributed measurement of high electric current by means of polarimetric optical fiber sensor, *Optics Express*, vol. 23, no. 9, pp. 11073-11079, April 2015.

[6] M. Aerssens, F. Descamps, A. Gusarov, P. Mégret, P. Moreau and M. Wuilpart, Influence of the optical fiber type on the performances of fiber-optics current sensor dedicated to plasma current measurement in ITER, *Applied Optics*, vol. 54, no. 19, pp. 5983-5991, July 2015.

[7] T. Yoshino, Theory for the Faraday effect in optical fiber, *J. Opt. Soc. Am. B*, vol. 22, no. 9, pp. 1856-1860, September 2005.

[8] M. V. Andres, J. L. Cruz and M. A. Hernandez, Faraday effect in standard optical fibers: dispersion of the effective verdet constant, *Applied Optics*, vol. 35, no. 6, pp. 922-927, February 1996.

[9] A. J. Rogers, Polarization-optical time domain reflectometry: a technique for the measurement of field distributions, *Applied Optics*, vol. 20, no. 6, pp. 1060-1074, March 1981.

[10] A. J. Rogers, Y. R. Zhou and V. A. Handerek, Computational Polarization-optical time domain reflectometry for measurement of the spatial distribution of PMD in optical fibres, *Proc. Optical Fibre Measurement Conference 1997*, pp. 126-129, 1997.

[11] M. Wuilpart, C. Caucheteur et P. Mgret, Measurement of magnetic field using Rayleigh backscattering in optical fibres, *Proceed. ANIMMA 2011*, 2011.

[12] M. Wuilpart, M. Aerssens, A. Gusarov, Ph. Moreau and P. Mgret, Plasma Current Measurement in Thermonuclear Fusion Reactors Using a Photon-Counting POTDR, *Photonics Technology Letters*, vol. 29, no. 6, pp. 547-550, March 2017.

Optical fiber distributed sensing based on Rayleigh scatter frequency domain reflectometry for Sodium leakage detection within SFR reactors

S. Armiroli [a],R. Cotillard [a], G. Laffont [a], M. Girard [b], D. Cambet [b], J.-P. Jeannot [b], P. Charvet [b], S. Albaladéjo [b] and S. Lusso [b].

[a]CEA, LIST, Laboratoire Capteurs et Architecture Electronique, 91191 Gif-sur-Yvette cedex, France.
[b]CEA, DEN, DTN, 13108 Saint Paul lez Durance cedex, France.

As part of the Generation IV international forum, CEA is performing research on Sodium-cooled fast neutron reactors. Liquid sodium holds many advantages such as its boiling point around 900°C, its high thermal conductivity, its ability to solidify itself after cooling. Moreover Sodium is a weak absorber of neutrons.

Because of the sodium's high chemical reactivity, its use as coolant in nuclear reactor leads to new safety issues and technical challenges such as the development of efficient leak detection system adapted to pipelines or tanks. Laboratories of the French Alternative Energies and Atomic Energy Commission (LIST-LCAE, DEN-DTN) are working together to realize tests and studies about an innovative sodium leakage detection system involving optical fiber distributed temperature and strain sensing techniques.

In this paper, we propose to illustrate how optical fiber sensors are used to monitor liquid sodium leaks at the outer surface of a large diameter pipe. Single mode optical fibers are promising candidates for instrumentation within any nuclear facilities due to their low intrusivity, their electromagnetic immunity, their capability to withstand high temperatures ranges and their reliability even when operating in harsh and radiative environments. The optical fiber sensing technology used in this paper is the Optical Frequency Domain Reflectometry. The operating principle of the instrument is a frequency analysis of the Rayleigh backscattered light along the core of the optical fiber. Collecting and processing this backscattered signal, an OFDR trace is obtained which is similar to a fingerprint of the heterogeneities (refractive index variations or distortion of the fiber's geometry) distributed along the length of the optical fiber's core. The apparition of any spatially strain or temperature change along the fiber's core will locally modify the Rayleigh backscattered signal. One can then follow in quasi real time the temperature change or strain modifications over a typically 70 meter long predefined segment of the optical fiber sensor. The sensing optical fiber then operates as a distributed sensor. The measurement resolution of this technique is typically 1°C and 10 micro-strains (similar to 0.001% of elongation) but specific temperature calibration has to be done for high temperature operating range. In this study, we are using a single mode optical fiber SMF1500[9/125], 9µm core diameter and 125µm cladding diameter similar to those used for telecommunication application. The optical fiber is protected using a metallic sheaths to avoid any damages by a direct contact with liquid sodium.

The leakage experiment is realized in the CEA\DEN\DTN laboratory. A 600 mm diameter pipe mock-up is instrumented by a 20 m longoptical fiber probe. The lagging of the mock-up is composed by several layers of thermal materials to confine the leakage on the main pipe and preserve the external environment from damages. The pipe of the mock-up can be maintained at high temperatures up to 550°C to simulate the liquid sodium temperature. The leakage is simulated by a secondary pipe placed inside the mock-up and its extremity leaves the main pipe on the top of it. The optical fiber is disposed directly on the main pipe and between the lagging's layers following a helical pattern. Any leak of Sodium is detected as a localized temperature change or "hot spot". We can then study the sodium leakage expansion around the pipe over time and estimate the seriousness of the leak thanks to alerts. In a near future, a 3D cartography will be proposed to display on real time the sodium expansion around the pipe.

Keywords: Optical Fiber Sensor, OFDR, Rayleigh backscattering, leakage detection, liquid sodium, fast reactor

 https://doi.org/10.1051/epjconf/201817003002

Characterization and localization of partial-discharge-induced pulses in fission chambers designed for sodium-cooled fast reactors

G. GALLI[1,3], H. HAMRITA[1], C. JAMMES[2], M.J. KIRKPATRICK[3], E. ODIC[3], Ph. DESSANTE[3], Ph. MOLINIE[3], B. CANTONNET[4], J-C. NAPPÉ[4]

***Abstract*—During the operation of the Superphenix and Phenix reactors, an aberrant electrical signal was detected from the fission chambers used for neutron flux monitoring. This signal, thought to be due to partial electrical discharge (PD) is similar to the signal resulting from neutron interactions, and is generated in fission chambers at temperatures above 400 °C. This paper reports work on the characterization and localization of the source of this electrical signal in a High Temperature Fission Chamber (HTFC).**

The relation between the shape of the PD signal and various parameters (nature and pressure of the chamber filling gas, electrode gap distance, and fission chamber geometry) are first described. Next, experiments designed to identify the location within the chambers where the PD are being generated are presented. After verification and refinement of the results of these localization studies, it should be possible to propose changes to the fission chamber in order to reduce or eliminate the PD signal.

***Index Terms*—Fission chamber, high temperature, neutron detector, partial discharge, sodium-cooled fast reactors, triple point effect.**

I. Introduction

THE purpose of high temperature fission chamber (HTFC) is to detect neutrons inside the core of sodium-cooled fast reactors [1]. Multiple uses are envisaged such as reactor power control and fuel cladding failure detection.

To operate in–core, the HTFC will have to withstand the high operating temperatures, up to 650°C, of the sodium-cooled fast reactors and also to operate under high irradiation, up to 10^{10} n/cm².s.

After years of study [2-5], it is now known that an electrical signal, more or less similar to the signal resulting from neutron interactions, is generated in fission chambers at temperatures above 400 °C.

This unwanted signal poses challenges, especially during reactor start-up when the PD signal count may be on the same order of magnitude as the neutron signal count [5].

In addition to problems with reactor start-up, the unwanted signal will prohibit fuel clad failure detection, which relies on a very precise neutron count in the sodium lines.

The study presented in this paper aims to characterize this PD signal compared with the neutron pulses, and to locate the problematic zone of the chamber.

A brief summary of electrical discharge theory is described in Section II. The experimental setup is described in Section III. All results on pulse shape characterization are presented in section IV. All results on PD localization and a brief overview of the methods used are presented in section V.

II. Notions of Electric Discharge Theory

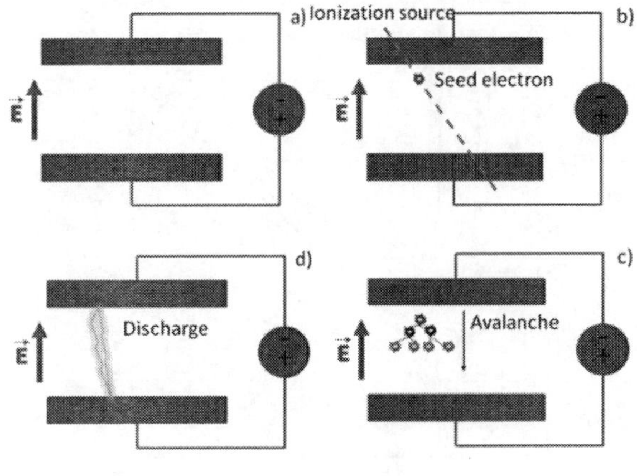

Fig. 1. Electric discharge scheme.

Electrical discharge occurs when a seed electron is accelerated by an imposed electric field, acquiring energy superior to the ionization energy of the gas atoms/molecules, leading to an electron avalanche. The schema shown in figure 1 describes the basic process. Assuming the presence of a seed electron, the conditions which will lead to an electron avalanche are

Submitted June 1, 2017. Work supported by the CEA, the French Alternative Energies and Atomic Energy Commission.

G. Galli and H. Hamrita are with the CEA, LIST, Sensors and Electronic Architectures Laboratory, Gif/Yvette, France, (Giacomo.GALLI@cea.fr). C. Jammes is with the CEA, DEN, DER, Instrumentation, Sensors and Dosimetry Laboratory, Cadarache, Saint-Paul-lez-Durance, France. M.J. Kirkpatrick, E. Odic, Ph. Dessante and Ph. Molinie are with the GeePs | Group of electrical engineering - Paris, UMR CNRS 8507, CentraleSupélec, Univ. Paris-Sud, Université Paris-Saclay, Sorbonne Universités, UPMC Univ Paris 06, 3 & 11 rue Joliot-Curie, Plateau de Moulon 91192 Gif-sur-Yvette CEDEX, France. B. Cantonnet and J-C. Nappé are with PHOTONIS France, Nuclear Instrumentation, Brive-la-Gaillarde, France.

described by Paschen's Law (eq. 1). Sources of seed electrons include cosmic rays, photo-ionization, etc.

$$V = \frac{B \cdot P \cdot d}{\ln(A \cdot P \cdot d) - \ln\left(\ln\left(1 + \frac{1}{\gamma}\right)\right)} \quad (1)$$

Where P and d are the filling gas pressure and the inter-electrode distance, γ is the secondary emission coefficient (the yield of electrons from positive ion impact on the cathode - normally on the order of 10^{-4}-10^{-2}) and A and B are coefficients which depend on the gas nature and are tabulated, for the main gases, in Table 1.

TABLE 1
PASCHEN LAW EMPIRICAL COEFFICIENTS

	A ($Torr^{-1}cm^{-1}$)	B ($V\ Torr^{-1}cm^{-1}$)
H_2	5	130
N_2	12	342
Air	15	365
He	3	34
Ne	4	100
Ar	14	180
Xe	26	350

Figure 2 shows Paschen curves for different gases. The abscissa is the product of the pressure and the electrode separation distance, and the ordinate the voltage above which a discharge should occur. The breakdown voltage passes by a minimum value on the order of a few hundred volts, depending on gas type.

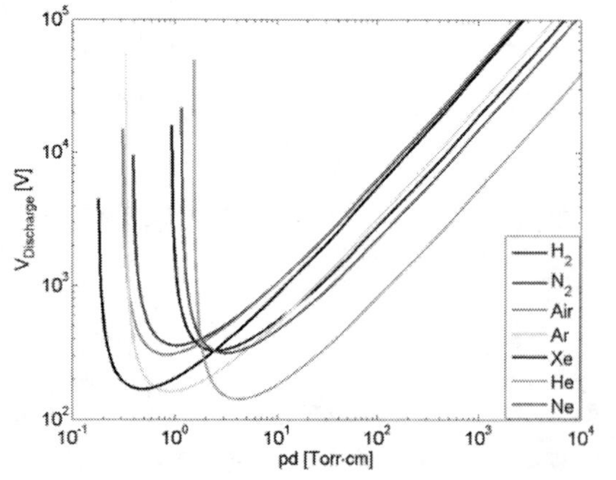

Fig. 2. Paschen curve for different gasses.

Paschen's Law is derived for a case of infinite planar electrodes, in other words, homogeneous applied electric field. This is of course different from the case of a fission chamber, especially from the point of view that, in a fission chamber, the electrodes are held in place and electrically insulated from each other by ceramic. The presence of the insulating ceramic material, which may become electrically charged, and whose dielectric properties are dependent on the temperature, is likely an important aspect of this type of system. In addition, the complex geometry of the system in combination with the presence of the ceramic insulator produces local electric field intensification.

III. EXPERIMENTAL SETUP

Figure 3 shows both the full and partial fission chambers used in this work; all these chambers were manufactured by Photonis using the same processes. A complete fission chamber consists of an active zone with cylindrical electrodes covered in fissile material, and a zone which serves to assure the electrical connection between the active zone and the cable which conducts signals to an amplifier. This means that there are two electrical connections in a complete chamber. The first two fission chambers, whose parameters are shown in table 1, both have the geometry shown in figure 3 (HTFC 1A and 1B), but are filled with argon to two different pressures, while the third fission chamber listed in the table is smaller (figure 2 (HTFC 2)). The last two chambers listed in Table 2 are partial chambers, having no active zone but containing either both or only one of the aforementioned electrical connectors as shown in figure 3 (HTFC 3 and 4).

All of the chambers are filled with pure Argon and the complete chambers were initially tested at room temperature and in the presence of neutron flux to ensure their proper operation under standard conditions [3].

After these tests, the fission chambers were heated to different temperatures and polarized in order to observe the appearance of partial discharges.

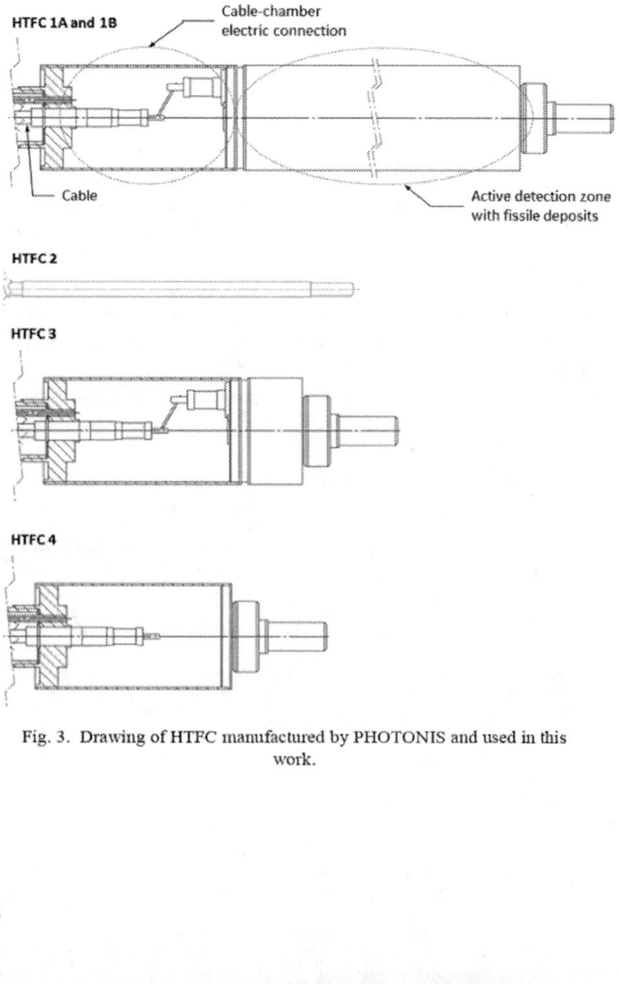

Fig. 3. Drawing of HTFC manufactured by PHOTONIS and used in this work.

TABLE 2
HTFCS PARAMETERS

Name	*HTFC 1A*	*HTFC 1B*	*HTFC 2*	*HTFC 3*	*HTFC 4*
Diameter	48 mm	48 mm	7 mm	48 mm	48 mm
Electrodes distance	1.5 mm	1.5 mm	0.5 mm	--	--
Bias Voltage	400V	400V	200V	400V	400V
Gas pressure	1.5 bar	3.5 bar	9 bar	3.5 bar	3.5 bar
N. electrodes	3	3	2	--	--
N. fissile deposits	4	4	2	--	--

IV. Pulse shape characterization

The characterization of the electrical signals produced by the chambers was done for a large number of pulses (~500-10000) due to the stochastic nature of the underlying process. A given experimental condition is therefore presented here using the average pulse waveform and a corresponding histogram of the pulse Full Width at Half Maximum (FWHM).

A. Neutrons

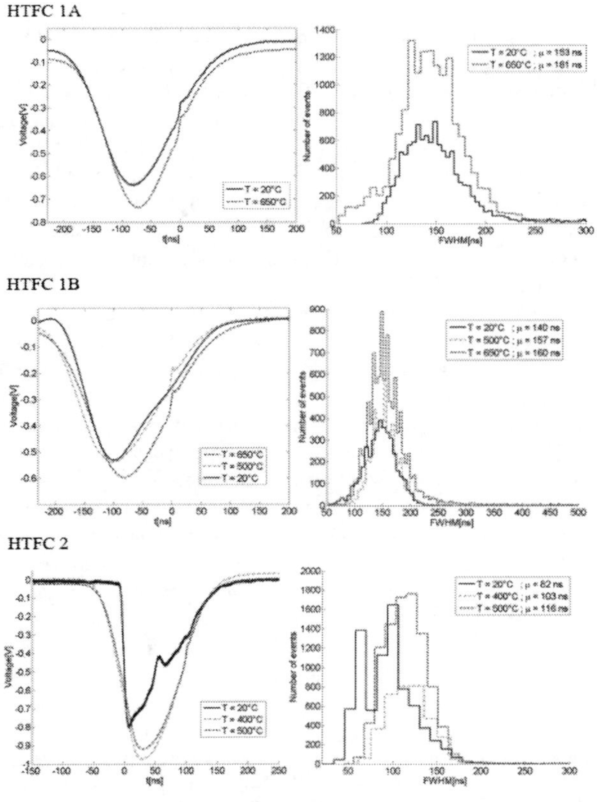

Fig. 4. (Left side) Average neutrons pulses and (Right side) histogram of Full Width Half Maximum.

Neutron detection signal characterization, as function of temperature, obtained from the three complete fission chambers (two chambers with geometry 1 from figure 3, and one with geometry 2) is shown in figure 4. In these experiments, the three fission chambers were polarized to their nominal bias voltage, and subjected to neutron irradiation. As seen in the figure 4, the resulting pulses have a Gaussian shape, with the exception of HTFC 2 at ambient temperature, with very similar rise and fall times; As expected, an increase in temperature causes an enlargement of the pulse FWHM. The data in figure 4 also shows that, for the smaller fission chamber (HTFC 2), the FWHM value is significantly reduced compared to the FWHM value observed for the HTFC 1A and 1B. The gas pressure does not seem to significantly affect the signal shape for the three fission chambers.

B. Partial discharges

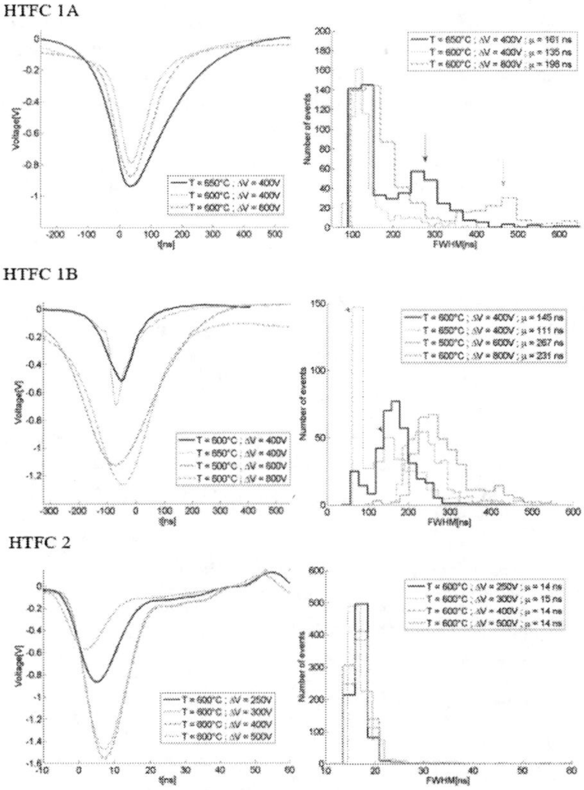

Fig. 5. (Left side) Average PD pulses and (Right side) histogram of Full Width Half Maximum.

PD pulse characterization as function of temperature, obtained from three different fission chambers, is shown in figures 5. It can be noted again here that partial discharges are only observed for cases where fission chambers are heated above about 400°C.

The PD pulses, like neutron pulses, have a Gaussian shape with very similar rise and fall times, perhaps as a consequence of the amplifier used for signal detection. .

Stable trends in the value of the FWHM when varying the voltage or temperature are not observed as was shown for the case of neutron pulses.

However, when the diameter of the fission chamber including its electrical connection section is reduced, the FWHM tends to decrease accordingly.

Figure 5 also shows that while only one kind of PD seems to be present in HTFC 2, in HTFC 1A and 1B when the product temperature and voltage is increased beyond a certain limit, the PD pulses seem to move towards higher values of FWHM, which are indicated by arrows.

At last, HTFC 2 shows that, the PD pulses seem not to be affected by the rise of voltage and they have a small standard deviation in FWHM histogram, while HTFC 1A and 1B show that, the PD pulses seem to be affected by the rise of voltage or temperature, which increases their standard deviation in FWHM histogram.

Figure 6 shows the PD count as a function of time, acquired through the PING time-stamping acquisition system, developed at the French Atomic Commission (CEA) [6]. The larger fission chambers (HTFC 1A and 1B) have been tested twice: the first test (solid line) was made in a conventional laboratory while the second tests (dashed line) were made in the underground and shielded SAPHIR facility [7] which is located at CEA in Saclay (France).

As seen in the figure 6, when the tests are performed in Saphir facility, both fission chambers have a very low PD count, and when the tests are performed in laboratory the PD count is significantly higher. The cause of very low PD count when the fission chambers were tested in the Saphir facility is then perhaps reduced ambient radiation of the Saphir facility compared to the standard laboratory.

Therefore if this reasoning is right in the experiences performed in Saphir facility the electric germ generation frequency is very small compared to the experiences performed in the standard laboratory and in this way is possible to explain the differences which we observe in figure 6.

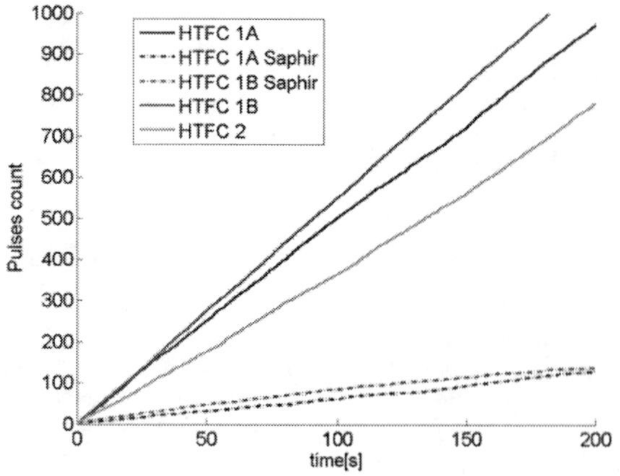

Fig. 6. Count of PD pulses as a function of time at 650°C and 400V.

All of this information suggests that the chamber geometry plays a key role in the PD pulses and that there are perhaps multiple locations within the chambers where PD generation occurs.

V. Neutron-Discharge shape comparison

In this section the pulse characterizations discussed in subsections A and B of part III are compared for each fission chamber.

Comparison of figures 4 and 5 shows that, in HTFC 2, neutron and PD pulses are simple to discriminate, using FWHM information, and that it is possible that they are located in different parts of the fission chamber.

Comparison of figures 4 and 5 shows that, in HTFC 1A and 1B, contrary to HTFC 2, neutron and PD pulses are very hard to discriminate because the voltage amplitude is quite similar and FWHM histogram cover the same range of values in the two cases.

VI. Discharge localization: Methods and Results

To perform the discharges localization, we used two different methods: the first involves the use of a Phase Resolved Pattern Diagrams (PRPD), and the second is based on the comparison between the PD observed in real neutron detectors (HTFC 1B) as opposed to the PD observed in the HTFC 3 and 4, which were constructed specifically to test for electrical discharges outside of the active zone of the real fission chambers.

A. PRPD method

To perform this experiment, the HTFC 2 was placed in a tubular furnace at 650°C without any neutron source, and connected to a 50 Hz AC voltage supply. An Omicron partial discharge analyzer was then used to create the pattern diagrams, as shown in figure 7. In the PRPD, partial discharge events are displayed as points with a total charge value represented in the ordinate and the timing with respect to applied voltage in abscissa. For multiple similar discharges, a color code is used for the display, which creates the patterns seen in figure 7.

In this way it is possible to characterize different PD phenomena according to the patterns obtained or according to the way in which the patterns vary by modifying the voltage or frequency.

In figure 7, PDs were characterized with increasing applied AC voltage and it is observed that an increase in voltage leads to a left-hand motion of the resulting pattern.

Thanks to these results it is proposed that an insulator is involved in the phenomenon [8], and that the PD are, most likely, located in the chamber-cable electric connection zone, where most of the insulators are located, and not in the active part of the detector. This observation is supported by a separate experiment in which HTFC 1A and 1B were partially heated: either the active zone was heated, both active zone and cable connection zones were heated, or the cable only was heated. Partial discharges were only observed in these experiments for the case when the cable-chamber electrical connection was heated. The fact that a heated insulator is involved in the formation of partial discharges is interesting. It is possible that increased charge transport on the surface of the insulators is responsible for the formation of partial discharges through a mechanism which may involve the metal-insulator-gas triple point. Further work will investigate this mechanism.

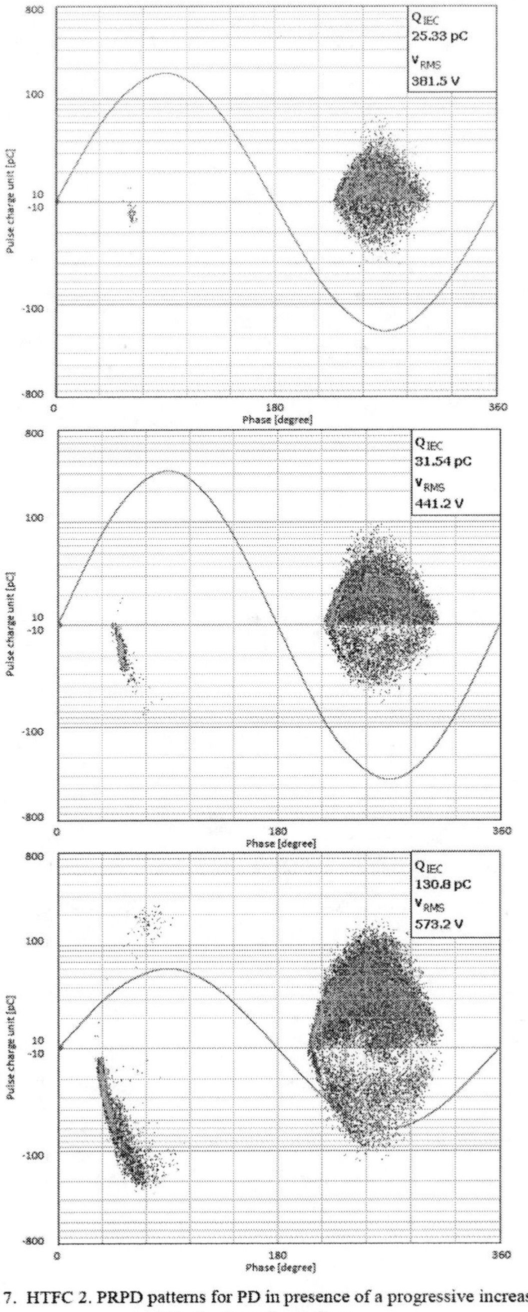

Fig. 7. HTFC 2. PRPD patterns for PD in presence of a progressive increase of 50 Hz alternative voltage.

B. Chamber-Cable connection tests

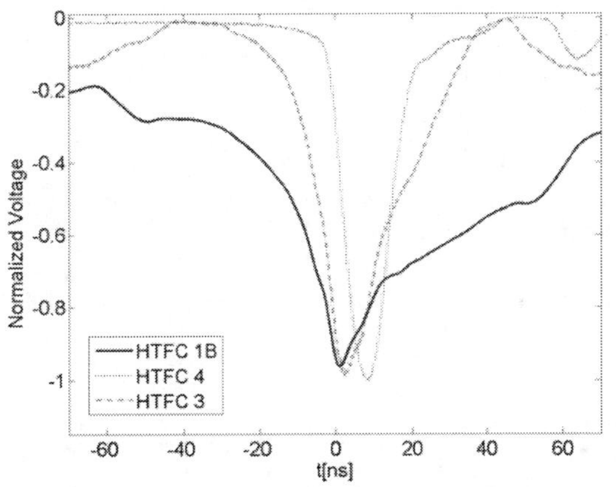

Fig. 8. Normalized average pulses of HTFC 1B (blue), HTFC 3 (green) and HTFC 4 (red).

In order to further investigate the location of the partial discharge observed in the fission chambers at high temperature, the two partial chambers shown in figure 3 (HTFC 3 and 4) were constructed by PHOTONIS and tested for partial discharge activity. Neither of the two partial chambers contain the active zone with cylindrical electrodes covered in fissile material. The first partial chamber (HTFC 3) contains both the chamber-cable electrical connection and the inner electrical connector which connects to the active zone in complete chambers. The second partial chamber contains only the external electrical connector. Otherwise, both of the partial chambers are similar to the first complete chamber shown in figure 3 (same diameter, materials, and geometry).

Both the partial chambers and the complete chamber were tested at the same temperature and the average pulse shape and pulse count were compared.

Figure 8 shows the normalized average pulse waveform for the three cases at 600°C and 400V.

The FWHM of the signal pulses appears to rise with increasing complexity of the chambers.

Figure 9 shows the PD count as a function of time, acquired through the PING time-stamping acquisition system, developed at the French Atomic Commission (CEA) [6], where the slope of the curves gives us an idea of the number of PD per second over a period of time. The partial chamber with only one of the two electrical connectors has a very low PD count, the full chamber has a high PD count, and the partial chamber with both connectors has an intermediate PD count. These results therefore indicate that the problematic zone is the junction between the active zone of the complete chamber to the intermediate zone linking the active zone with the cable, and not the junction between the intermediate zone and the cable. It is not clear at present why the PD count changes for different acquisitions made through the same fission chamber (the two blue curves in the figure 9), but this may be due to the statistical character of the partial discharge phenomenon. Further work will investigate the mechanisms leading to these results and seek to define design modifications to prevent PDs in fission chambers operating at high temperature.

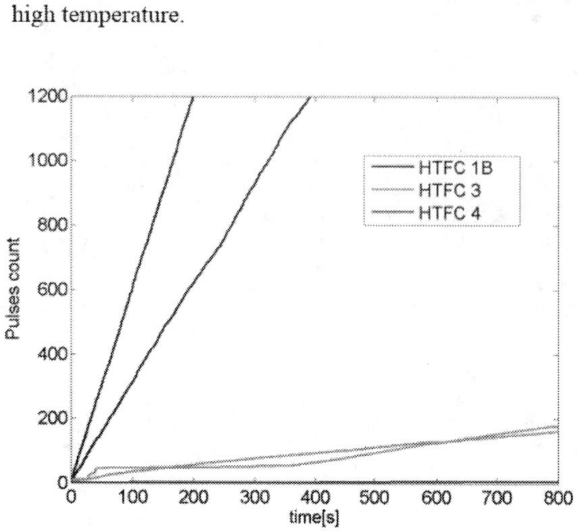

Fig. 9. Count of PD pulses as a function of time.

VII. Conclusion

Results presented in this paper demonstrate firstly that it is difficult to discriminate between neutron detection signals and signals from partial discharges using the analysis FWHM for the case of the larger fission chambers (HTFC 1A and 1B) but may be feasible for the smaller fission chamber design (HTFC 2). The filling gas pressure was found not to significantly affect the PD pulse shape, while some broadening of the pulse signals was observed with increasing temperature. Change the environment of experiences can change completely the magnitude of PD count to the same fission chamber. Tests using an AC power supply and partial discharge analyzer indicate that the mechanism leading to partial discharges in the chambers involves an insulator.

In addition, the results on PD pulse localization show that the majority of the PDs are located near the electrical connection linking the active zone of the chamber with the intermediate zone, and therefore likely involve the alumina insulator in this connector.

Further tests will be performed to better understand the problematic zone of the chambers and the specific mechanism responsible for these discharges. A better understanding of the mechanism should then lead to the definition of design changes for the elimination of the problem of partial discharges in high temperature fission chambers.

References

[1] C. Jammes, P. Filliatre, B. Geslot, T. Domenech, and S. Normand, "Assessment of the high temperature fission chamber technology for the French fast reactor program", *IEEE Trans. Nucl. Sci.*, vol. 59, 2012.

[2] J. P. Trapp, S. Haan, L. Martin, J. L. Perrin and M. Tixier, "High Temperature Fission Chambers: State-of-the-Art", *In-Core Instrumentation and Reactor Core Assessment. Proceedings of a Specialist Meeting*, 1997.

[3] H. Hamrita, C. Jammes, G. Galli and F. Laine, "Rejection of partial-discharge-induced pulses in fission chambers designed for sodium-cooled fast reactors", *Nuclear Inst. and Methods in Physics Research, A*, to be published. DOI: 10.1016/j.nima.2016.11.055.

[4] J. F. Miller, Fission chambers, *Wiley Encyclopedia of Electrical and Electronics Engineering*, to be published DOI: 10.1002/047134608X.W5203.pub2.

[5] C. Jammes *et al.*,"Progress in the development of the neutron flux monitoring system of the French GEN-IV SFR: simulations and experimental validations", *Proceeding in ANIMMA*, 2015.

[6] S. Normand, V. Kondrasovs, G. Corre, et al., "PING for nuclear measurements: First results", *IEEE Transactions on Nuclear Science*, vol. 59, 2012.

[7] F. Carrel, et al "Characterization of Old Nuclear Waste Packages Coupling Photon Activation Analysis and Complementary Non-Destructive Techniques", *IEEE Transactions on Nuclear Science*, vol. 61, 2014.

[8] C. Hudon and M. Belec, "Partial Discharge Signal Interpretation for Generator Diagnostics", *IEEE Trans. Dielectr. Electr. Insul.*, Vol. 12, 2005.

Suggestion of Design Evaluation Plan based on Star Life Cycle to introduce the Information Minimalism Concept of KOREA Nuclear Plant

Gwi-sook Jang [a*], Seung-min Lee[b], Gee-yong Park[a]
[a]I&C and HFE Research Division, Korea Atomic Energy Research Institute, 989-111, Daedeok-daero, Yuseong-gu, Daejeon 305-353, Republic of Korea
[b]Maritime Reactor Development Center, Korea Atomic Energy Research Institute, 989-111, Daedeok-daero, Yuseong-gu, Daejeon 305-353, Republic of Korea

***Abstract*—The design of Korea Nuclear Power Plant (NPP) main control rooms (MCR) has been changed to be fully digitalized. Five or six display devices are assigned to each operator in NPP MCR to provide the information of safety parameter and plant status, and various control functions by connecting computerized control devices. Under this circumstance, the distributed displays can induce a dispersion of the operators' attention and increase the workload while conducting monitoring and control tasks efficiently. In addition, to support human operators to reduce their workload and increase the performance, the concepts of the ecological interface design (EID) and the operator-centered design were applied to the design HMI display. However these designs are applied to a limited set of screens and did not differ largely from the traditional HMI design in that the layout of the information is somewhere similar to P&IDs.**

In this paper, we propose a design evaluation plan based on star life cycle to introduce the information minimalism concept for designing an HMI display.

***Index Terms*—Monitoring System, Human Factors Engineering, Human-Machine Interface**

I. Introduction

The traditional HMI (Human Machine Interface) design approach for a Korea NPP (Nuclear Power Plant) is a heavily mimic-based hard-paneled interface based on P&ID (Piping and Instrumentation Diagram). Recently, the design of the Korea Nuclear Power Plant (NPP) main control rooms (MCR) was changed to be fully digitalized, as shown in Fig.1. Five or six display devices are assigned to each operator in an NPP MCR to provide the information of safety parameter and plant status, and various control functions by connecting computerized control devices.

This was expected to minimize the operator's movement and enable the operator to operate even when seated, greatly enhancing the operator's macro performance. On the other hand, because the information based on the existing P&IDs was implemented as a hierarchical structure on the VDU (Visual Display Unit) with the display space constraint, it required additional tasks such as information navigation in addition to primary tasks such as monitoring, situation assessment, planning, and implementation.

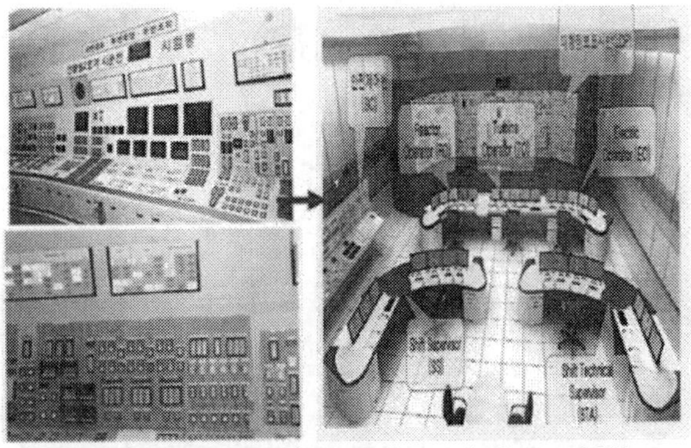

Fig. 1. Status of a main control room in Korean NPP

According to NUREG/CR-6633, as the interface is distributed over many displays limited in size, operators will have to navigate through them to access the information they are looking for. The display shown on each screen is chosen by the operator, e.g., a mimic-based display, trends, and alarm systems. While this flexibility offers some advantages, studies have shown that operators often get lost, experiencing a hard time managing screens and finding the information they are particularly looking for and thus reducing the operator performance [1,2].

In practice, the information display of the digital based control room of a Korean NPP is operated with a hierarchy structure of system-oriented, component-oriented, and function-oriented based on existing P&IDs. The operator obtains the monitoring and controller information necessary for operation through the information navigation by hierarchically distributed information over five or more display pages to perform a particular operation mode. As a result of the intervention of these additional tasks, the reaction time is delayed, and it is difficult to maintain the cognitive context, and can cause a cognitive error under a situation such as an emergency.

Therefore, it is necessary that an information structure and visualization shall be designed to minimize the cognitive burden of recognition, memory, and judgment on the relationship between complex systems when an unexpected event occurs. The information structure and visualization design is based on a careful analysis and understanding of the work domain of the operator.

P&ID based layouts tend to require more display space. Overuse of mimic layouts can result in visual clutter. P&ID based layouts of controls are usually less easily operated configurations than other array conventions. Thus, a display method that minimizes the P&ID based display is emerging.

Although they are from different domains (e.g. process controls and nuclear power generation controls, etc.), common topics are raised issues and/or concerns of an HMI for plant operators. These common issues are summarized and translated as the following design aspects [3]:

- Minimalism: To support operator understanding and situation awareness, utilizing graphical representation of plant assets can be beneficial. However, unnecessary details of the graphics shall be avoided, and reduced to visual noise and clutter.
- Color palette: A color palette has been defined for the HMI to reduce operator's cognitive workload and make abnormal conditions more apparent.
- Visual coding: A visual coding refers to any visual treatments to indicate or imply certain context around data.
- Screen hierarchy: Categorizing the screens systematically can improve the navigation efficiency, and, moreover, describes more details of the process and/or assets.
- Navigation: Having multiple forms of navigation improves the operator's speed of screen transition and awareness of screen locations.
- Information layout: As operators form their mental model of the operation (processes and assets) based on the location of the physical equipment and assets in the plants, utilizing the P&ID for process overview to a certain extent can be useful.
- Contextual data: Data can transform into useful information when provided with a proper context, representing the context of the data, (e.g., targets, operating ranges, alarm limits, process relationship) supports faster operator understanding and situation awareness.

The above concept of an HMI design was estabilshed as shown in Fig.2.

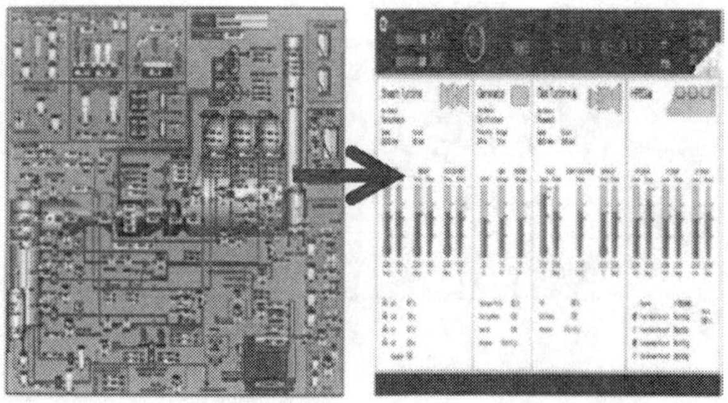

Fig.2. An example of resulting plant overview screen

However, the HMI design that minimize the P&ID as based on the above example has not yet proven its effectiveness.

To solve the above problem, the designer has to decide how to reduce the amount of information to be displayed on the screen so that the user can work easily. In addition, it is important to provide a means for the operator to easily move the information display space and naturally make a cognitive transition in the display design with a large amount of information networks.

Thus, we propose a design evaluation plan based on the star life cycle to introduce the information minimalism concept for designing an HMI display. These days a minimalism concept and the star life cycle methodology is emerging in the HMI design approach.

Minimalism is used to support the operator understanding and situation awareness, and utilizing a graphics representation of plant assets can be beneficial. However, unnecessary details of the graphics shall be avoided to reduce visual noise and clutter.

This paper proposes a monitoring and controller display method for each operation mode based on log analyses of operator actions as an implementation method of minimalism. This method minimizes the existing information navigation by providing the monitoring and control means necessary for the operator to perform a specific operation mode on one or two display pages.

The method provides the monitoring and controller information for each operation mode of NPP to operator based on the understanding of the work domain of the operator. And the method complements the monitoring information and the control means for each operation mode through analysis of the operation behavior log obtained during operation to improve the efficiency of the display for each operation mode.

The star life cycle methodology, suggested by Hartson and Hix, is an evaluation-centered iterative usability engineering process. The star lifecycle does not specify any ordering of activities. In fact, the activities are highly interconnected: you can move from any activity to any other, provided you first go through the evaluation activity. This reflects the findings of the empirical studies.

An evaluation is central to this model, and where an activity is completed, its results must be evaluated. It is expected that the suggested evaluation plan will be valid to estimate an applicability of various HMI design-related theories, and the HMI design-related theoretical features evaluated by the suggested evaluation will be reflected in the new monitoring and control system design.

II. Information Minimalism Concept

Fig.3 shows the monitoring and controller display method for each operation mode based on log analyses of the operator actions, and the detailed descriptions of each module are as follows.

A. Monitoring and controller DB for each plant mode

The monitoring and controller database for each operation mode is composed of the initial analysis information and complementary information based on the operation log analysis during operation. The initial analysis information means the basic information of the monitoring and the controller to be provided to the operator for each operation mode.

This was confirmed through the task analysis of the operator for an NPP. The complementary information is the condition evaluation and the action data for each the operation mode based on the operation behavior log analysis.

In addition, the initial analysis information and complementary information is stored in the database in the form of ECA rules (on Event if Conditions then Action). Basic display information for each operation mode and complementary information based on operation behavior analysis are added / deleted / modified to the shared memory through the rule interface, which is an operator interface module.

The rules interface is an interface for system users and designers to define, manage, delete, and modify the display rules.

B. Monitoring and controller management module for each operation mode

The management module is a main module that drives the monitoring and controller database to provide the information indicator with the monitoring and the control information necessary for the task in each operation mode. This module consists of a display condition evaluator for each operation mode and a display actor for each operation mode.

If the operator selects a specific operation mode, the management module provides necessary information for the corresponding monitoring and controller means to the operator. In addition, this module is responsible for running the operator's demands.

The display condition evaluator for each operation mode is connected to the event and rule triggering to sequentially drive the display rule. Then, the condition of this rule is evaluated, and if the condition is satisfied, the action of the rule is executed.

C. Operation behavior log analysis and complementary module

To complement the monitoring and control information required by the operator according to the operation mode, this module logs and stores the object and control means monitored by the operator according to the operation mode. Then this module preprocesses and performs statistics and correlation analysis on periodically stored log information.

The analysis result is prepared based on the condition evaluation and action data for displaying necessary information for each operation mode. The complementary module generates only the data to be supplemented during online operation and reflects the complementary data to the system after confirming the operator according to the administrative procedure.

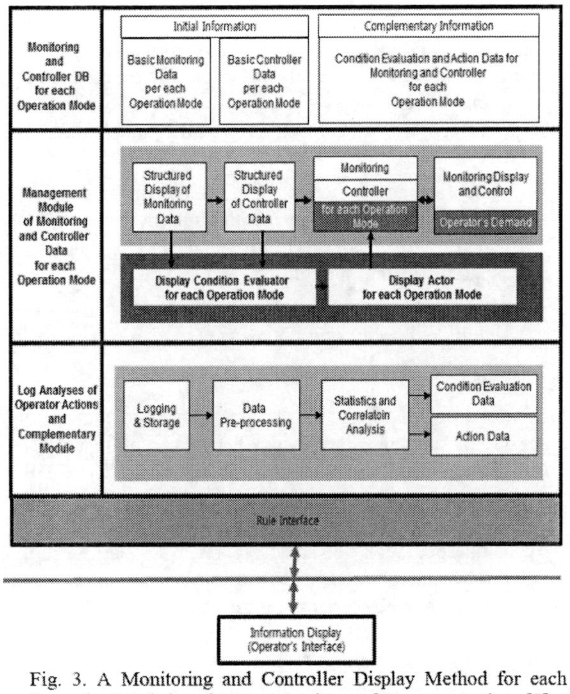

Fig. 3. A Monitoring and Controller Display Method for each Operation Mode based on Log Analyses of Operator Actions [5]

III. Operation Sequence

1) As shown in Fig. 3, the "monitoring and controller DB for

each operation mode" is a database composed of ECA rules. The monitoring and controller information to be provided to the operator for each operation mode are confirmed by the operator's task analysis at the beginning of the monitoring system design are stored in advance as ECA (on event if condition then action) rules.

2) When the operator selects the start operation mode button after the monitoring system starts to operate, the monitoring system activates the monitoring information and controller manager module for each operating mode based on the "initial analysis information ECA DB". The monitoring information and control means necessary for performing the specific driving task are provided in a screen of 1 to 2 pages. In particular, the management module comprises a condition evaluator for determining the display condition for each selected operation mode, and an activity unit for activating the display method for each operation mode that should be provided to the information display when the condition is satisfied.

3) In addition, all operations of the operator are logged and stored until the operator completes the related tasks after selecting the specific operation mode start button.

4) If operator select the specific operation mode completion button, logging and saving of the operator's action are stopped.

5) When the operation history of each operation mode is accumulated enough to analyze the statistical correlation at the time of the periodic test of the monitoring system, if the operator selects the log analysis button, the preprocessing for analyzing the stored log is performed and the following analysis is performed. And the result is provided to an information display.

- The number and frequency of monitoring and controllers are detected by the operator through screen navigation in addition to the screen provided previously for each specific operation mode
- Correlation between specific operating modes and additional monitoring and control periods with high frequency

6) Based on the results of the log analysis of the operator's action, the supplementary items (display condition and activity information for each operation mode) of the ECA DB are determined through surveys of the operators.

7) If the operation log analysis result is determined, the condition evaluation and activity information for each operation mode determined by the tool provided by the rule interface are input to the "ECA DB based on log analysis" according to the administrative procedure.

IV. Design Evaluation Plan based on Star Life Cylce

Making the user interface design and development process iterative is a way to ensure that users can get involved in the design, and that different kinds of knowledge and expertise related to the computer system can be brought into play as needed. This paper shall use an approach adapted from a model proposed by Hix and Hartson (1993). This is known as the star life cycle for obvious reasons, as you can see from its appearance in Fig 4.

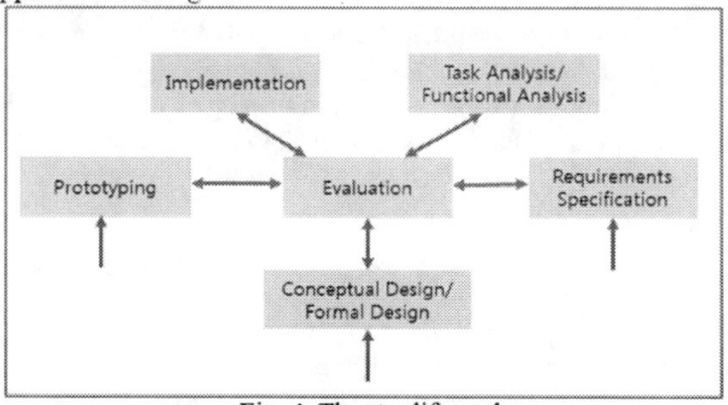

Fig. 4. The star life cycle

The central point of the star is an evaluation, which is viewed as being relevant at all stages in the life cycle and not just at the end of the product development as the classic life cycle tends to suggest. The evaluation is concerned with gathering data about the usability of a design or product by a specified group of users for a particular activity within a specified environment or work context. Users should be involved in every part of the user interface design and development life cycle [4].

- Early in the design process when the requirements are being specified
- During prototyping, to test designs and options
- During training/after delivery of the system

This paper proposes design evaluation plan based on stat life cycle. Monitoring system design for NPP consists of obtaining the design input data, setting the driving concept, developing a functional requirement, concept design, detailed design, testing, and evaluation (See Fig.5).

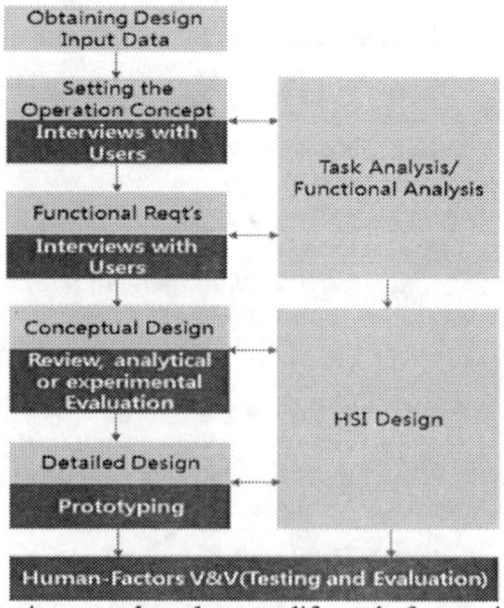

Fig. 5. Design steps based on star life cycle for monitoring system

 https://doi.org/10.1051/epjconf/201817003003

The monitoring system design work is carried out and integrated by the human-factors design team through the concept, basic and detailed design stages by receiving the results of the human-factors planning and analysis tasks. The design procedure and method of the nuclear plant monitoring system are as follows.

A. Obtaining design input data

Obtain the following input data for monitoring system design.

- Operator job requirements
- System requirements
- Regulatory requirements

B. Setting the operation concept

Set up the driving concept according to the structure and role of the nuclear plant operation group. The concept of driving includes the following contents.

- Verify the relationship between operator and plant automation
- Provide high-level description for operators to perform their duties using surveillance systems
- Mentioning of mutual relations among operators

C. Development of function requirement for monitoring system

The functional requirements of the monitoring system are developed to represent the operation concept, operator functions and duties, and operator requirements for safety. It also includes various kinds of detailed functional requirements for monitoring system such as alarm, indication, and control.

D. Conceptual design of surveillance system

The monitoring system performs a concept design according to the pre-established functional requirements for monitoring system. To meet the design requirements in this process, various design plans are examined considering the following items.

- Latest monitoring implementation technology
- Technological feasibility
- Human performance aspect
- Tradeoffs between monitoring technologies

The conceptual designs examined are selected through operational experience and a literature review, analytical or an experimental evaluation. In addition, the monitoring system performance requirements are identified for the selected conceptual designs.

E. Detailed design of monitoring system

In the detailed design stage of the monitoring system, detailed design requirements (hereinafter referred to as 'style guides') for specific monitoring systems are developed and applied to the features, layout and environmental design of the monitoring system screen and control systematically. The style guide is divided into high-level general guidelines for monitoring the system design and detailed design guidelines that apply specifically to specific plants. Detailed design guidelines that apply specifically to specific plants are developed to meet the baseline level of general guidelines. The design criteria for the detailed design of the monitoring system are based on the contents of the plant concept design, the standard used in the field of nuclear power plants, the industry standard or guidelines, and the general characteristics and practices of the monitoring system required by the plant operator.

The design criteria for the main detailed monitoring system design are as follows.

- Designed to support primary task requirements, such as plant monitoring and control, while minimizing the secondary workload required to operate monitoring system facilities
- Minimizes the possibility of human error and maximizes the detection capability
- Consider communication between operators, collaboration and workload, information feedback, on-site environment, inspection, test and maintenance task, and safety importance.
- Determine the monitoring screen layout considering roles of individual operators and various deployment strategies
- Designed to support the operator's task performance in all operation and operating environment conditions
- Designed to take into account situations in which the operator's ability to perform tasks may be degraded.
- Be able to support the work of inspection, maintenance, and testing, and to ensure that these tasks do not interfere with plant monitoring and control

F. Testing and evaluation for monitoring system

Testing and evaluation of monitoring system design is repeated throughout the entire development process of the monitoring system design. The test and evaluation are repeated in the human-factors verification and verification process, and the method is as follows.

- Trade-off evaluation: derive discrepancies through a style-guide-based, checklist-based evaluation and design alternatives to discrepancies. Analyze conflicts between initial design (plan) and alternative design (plan), and confirm design (plan) considering safety and human-factors efficiency. If there are trade-offs in terms of operator performance for several designs, they are determined by the Analytic Hierarchy Process (AHP) method. The AHP method is applied to a large number of decision-making fields because it is convenient to convert the complex criteria into a simplified hierarchical structure and compares them to each other to rank the comparison objects. Therefore, the criteria for evaluation and 2 to 3 candidates for the design are ranked as follows in each design through each step of AHP, and objective results are obtained.

- ✓ Operator's task requirements
- ✓ Operator performance and restrictions
- ✓ HSI performance requirements
- ✓ Inspection and testing requirements
- ✓ Maintenance requirements

• Performance-based testing: A test that evaluates the performance of the operator after simulating the monitoring system design (draft). The monitoring system is evaluated in accordance with the conformance standards of the performance as follows.

- ✓ Purpose of test: Evaluate the feasibility of the monitoring system design (draft) according to the conformity criterion of the degree of performance as a method to evaluate the performance of the operator after simulating the design of a supervisory system.

- ✓ General method of test: The design evaluation of the monitoring system first defines the scope of the test environment, development schedule, development restrictions, and development related tests. This test defines verification test items and test limitations, selection and training of testee, selection of test scenarios, defining test scales and measurement methods, conducting preliminary tests, collection and analysis of test data, and analysis of the test results.

- ✓ Requirements for test equipment: The requirements for the test equipment describe the structure of the test equipment, the functions of the equipment, the development and operation plans of the test equipment, and the requirements for the test plan. They also describe the hardware and software requirements of the test equipment, the development environment requirements, and the associated requirements and development limitations of developing the test equipment.

- ✓ Selection of performance evaluation method: the performance of the monitoring system design is evaluated by the accuracy and appropriateness of the information delivery time (operator response time) and the information transmitted to the operator. In addition, the acceptance criterion for performance evaluation is that the improved monitoring system design is 'allowed' if the supportability of the power plant situation to the improved supervisory system design is better than the level of task supportability in the interface environment of the existing main control room, and is evaluated as worthy. As the measurement variables of the performance evaluation test, the recognition time, recognition accuracy, objective situation recognition using SACRI (Situation Awareness Control Room Inventory), and subjective situation recognition using SART (Situational Awareness Rating Technique) are utilized.

- ✓ Selection and training of testee: Construct testee group using experts who have experience of operating a NPP. Experts belonging to the human-factors design team or those with a high degree of experience or proficiency are excluded in order to reduce the bias of the tester, which may affect the test results. It produces training materials on the plant design and operation developed by experts such as the fluid, core, main machine and safety analysis field in the ergonomics design team. Prior to securing a plant evaluation test room, classroom education was conducted and practical training was conducted when the evaluation test was completed.

- ✓ The test design and data analysis methodology: The performance evaluation is based on a randomized factorial design. Verification of differences in the perceived performance time, perceptual accuracy, and SART Rating Scales was performed using the Paired Samples t-test. In addition, the verification of the difference of situation recognition through SACRI is verified by comparing the sensitivity (d ') and the criterion (response criterion (β)) of the operator under test conditions A and B. The sensitivity and judgment criteria are derived based on the SDT (Signal Detection Theory), and the statistical significance is obtained through normalization of the data collected through a SACRI evaluation.

V. Results and Future Plans

A traditional HMI design approach of an NPP was heavily based on the P&IDs. A clear trend in Korea NPPs is to modernize the control rooms, moving from panel-based traditional control rooms to computerized solutions. A significant amount of work was done on the process control HMI design using an ecological interface design and the operator centered interface. These designs are applied to a limited set of screens and did not differ largely from the traditional HMI design in that the layout the information is somewhere similar to the P&IDs.

The digital-based nuclear plant monitoring system has a problem in that there is a large amount of information to be

https://doi.org/10.1051/epjconf/201817003003

displayed in a limited display space and that the capacity of the operator to process the visual information is limited. In order to solve the above problems, the latest digital technology and display technology have been introduced.

In this paper, we proposed a design evaluation plan based on star life cycle to introduce the information minimalism concept for designing HMI display. The information minimalism concept is the monitoring and controller display method for each operation mode based on log analyses of operator actions.

This method provides the monitoring information and control means necessary for the operator according to operation mode to perform a specific operation. This method can reduce the information searching process time and the transition between the display pages. So, this method can be provided as an alternative to overcome the problems of P&IDs based hierarchical information provision.

This method will be validated by establishing a validation facility in the future. In addition, this concept is added to the concept of operator support with the existing display configuration and navigation as it is, and it extends its application range while actually utilizing it.

References

[1] O'Hara, J., & Brown, W. (2002), "The effects of interface management tasks on crew performance and safety in complex, computer-based systems", NUREG/CR-6633. Washington, D.C.:U.S. NRC.

[2] A.O.Braseth, C. Nihlwing, Lessons Learned from Halden Project Research on Human System, Nuclear Engineering and Technology, Vol.41 No.3, April, 2009.

[3] S.Y Kim, J.V.Wormer, Human-Machine Interface Design for Power Generation Plant Operators

[4] D.Stone, C.Jarrett, "User Interface Design and Evaluation", Morgan Kaufmann, 2005, pp. 19-20.

[5] Gwi-sook Jang, Seung-min Lee, Gee-yong Park, "A Monitoring and Controller Display Method for each Operation Mode based on Log Analyses of Operator Actions", Transactions of the Korean Nuclear Society Spring Meeting Jeju, Korea, May 18-19, 2017

Gwi-sook. Jang was born in Daegu, Republic of Korea in 1967. She received her M.S. degree in computer science from the Kyungpook National University, Daegue, Korea in 1992. She received her Ph.D. in computer science from the Chungnam National University, Daejeon, Korea in 2010. She has worked for the Korea Atomic Energy Research Institute since February 1992. She has designed the monitoring system for a Small Modular Reactor (SMR) and a research reactor as principal researcher. Dr. Jang is a member of the Korean Nuclear Society

Seung M. Lee was born in Seoul, Republic of Korea in 1984. She received a B.S. degree in physics from Chung-Ang University, Seoul, in 2008 and an M.S. degree and Ph.D. in nuclear engineering from the Korea Advanced Institute of Science and Technology (KAIST), Daejeon, in 2010 and 2015. From 2015 to 2016, she was a Postdoctoral Researcher with the Mechanical Technique Laboratory in at KAIST. Since 2016, she has been a Senior Researcher with the Maritime Reactor Development Division, Korea Atomic Energy Research Institute (KAERI). Her research interests include autonomous operating system, information processing and cognitive load degradation for operators in nuclear power plants, and design of a Small Modular Reactor (SMR) control room and monitoring display.

Dr. Lee was a recipient of the International Symposium on Symbiotic Nuclear Powers System for the 21st Century (ISSNP2013) Best Student Award in 2013.

Gee-Yong Park received a B.S. degree in nuclear engineering from Hanyang University, Seoul, Korea, in 1990 and an M.S. degree and Ph.D in nuclear and quantum engineering from the Korea Advanced Institute of Science and Technology (KAIST), Daejeon, Korea, in 1992 and 1996, respectively. From 1996, he has been being involved at the Korea Atomic Energy Research Institute (KAERI) and is currently developing of instrumentation and control (I&C) systems. His research interest includes the development of diagnosis and control methodologies for nuclear power plants, safety analysis of safety-critical software, and reliability analysis of digital systems. Dr. Park is a member of the Korean Nuclear Society.

Radiation-Tolerance Assessment of a Redundant Wireless Device

Q. Huang, *Member, IEEE* and J. Jiang, *Fellow, IEEE*

***Abstract*— This paper presents a method to evaluate radiation-tolerance without physical tests for a commercial off-the-shelf (COTS)-based monitoring device for high level radiation fields, such as those found in post-accident conditions in a nuclear power plant (NPP). This paper specifically describes the analysis of radiation environment in a severe accident, radiation damages in electronics, and the redundant solution used to prolong the life of the system, as well as the evaluation method for radiation protection and the analysis method of system reliability. As a case study, a wireless monitoring device with redundant and diversified channels is evaluated by using the developed method. The study results and system assessment data show that, under the given radiation condition, performance of the redundant device is more reliable and more robust than those non-redundant devices. The developed redundant wireless monitoring device is therefore able to apply in those conditions (up to *10 M Rad (Si)*) during a severe accident in a NPP.**

***Index Terms*— COTS, radiation-tolerance assessment, rad-hardened by design**

I. Introduction

SEMICONDUCTOR-BASED electronic systems are highly susceptible to high energy particles present in strong radiation fields, such as those in aeronautics and space sectors, nuclear power plants, and the nuclear warfare. Taken Fukushima Daiichi nuclear disaster as an example, one of the lessons learnt is the level of difficulty to obtain environmental conditions inside the plant due to the presence of large amount of radioactive substances, which can cause damages to semi-conductor based devices and systems. One approach to levitate such a problem is to employ radiation hardened (rad-hardened) devices in such systems. Preliminary analysis indicates, however, that this approach is prohibitively expensive. Furthermore, many new data processing algorithms and communication technologies require devices to support high speed, large storage, and low power consumption. Rad-hardened devices may not even meet those advanced system requirements. Another approach is to rely on regular commercial off-the-shelf (COTS) devices, but to utilize rad-hardened techniques to reduce the vulnerability of radiation effects and to prolong the life of the system during the mission of the deployment. One approach is to take advantage of the triplication of important circuits and subsystems to detect and to correct radiation induced errors. Needless to say, the assessment of the radiation-tolerance is a critical step in such design.

The authors are with the Department of Electrical and Computer Engineering, The University of Western Ontario, London, Ontario, N6A 5B9, Canada (e-mail: jjiang@eng.uwo.ca).

In general, the performance of rad-hardened systems can be evaluated in two ways: (1) physical tests: which uses external perturbation sources (natural and accelerated particle radiation, laser beam, pin forcing, etc.) to create a similar radiation environment to evaluate the performance of the design. This approach is very precise but could be excessively complicated and expensive; and (2) simulation with analysis: which uses logic relationships of the circuits and systems to access internal elements and insert the effect of a radiation induced fault according to the fault model. However, a limitation of this approach is that it is difficult to assess the radiation-tolerance of the whole system [1]. To address these issues, a new approach by combining the radiation protection and the rad-hardened design is developed. To demonstrate this method, a wireless monitoring device is developed for high level radiation fields during a serve accident in nuclear power plants.

The reminder of the paper is organized as follows. Post-accident environments, radiation damages on electronics, as well as potential solutions are described in detail in Section II with the problem statement. Subsequently, the evaluation of radiation protection scheme is discussed in Section III. The radiation-tolerance assessment of a redundant architecture is discussed in Section IV. Finally, using a wireless monitoring device as an example, the implementation and the assessment of its radiation-tolerance are presented in Section V.

II. Analysis of Radiation Damages

A. Post-Accident Environments in NPPs and Radiation Effects on Electronics

In the event of a nuclear accident, a significant amount of radiation can be released, which may include alpha (α), beta (β) particles, gamma (γ) rays, *x*-rays, and neutron particles [2]. Taken Fukushima Daiichi disaster as an example, in February 2017, the level of radiation particles from melted fuel has been estimated up to 530 *Sv/h* inside the containment of No.2 reactor [3]. This level of radiation is so high that can cause severe damages to electronic devices if deployed there.

In general, radiation particles lose their energy through either non-ionization processes (displacement damage) and/or ionization processes when they pass through semiconductor devices. Radiation effects on electronics can generally be classified into three categories:

(1) Displacement damage (DD), which is caused by long-term non-ionizing effects, when incident particles have enough energy to knock an atom free from its normal lattice structure in the semiconductor and onto an interstitial site. It will change properties of the device and lead to permanent damage [4];

(2) Total ionizing dose (TID), which refers to the total amount of energy deposited by the particles passing through semiconductor materials. It is an important consideration when semiconductors devices are exposed to a strong radiation environment [5]; and

(3) Single event effect (SEE), which is primarily caused by single particle ionization and/or secondary particle formation. According to the failure mechanism of semiconductor materials, a single event effect can further be divided into two types: (1) non-destructive effects, which can be recovered by performing a system reset, or data re-initialization, including single event transient (SET), single event upset (SEU), and single event functional interrupt (SEFI); and (2) destructive effects, which are terminal and cannot be recovered, including single event latchup (SEL), single event snapback (SES), single event burnout (SEB), and single event gate rupture (SEGR) [6].

This work mainly focuses on the damages due to ionizing radiation. Displacement damages are therefore not considered. Considering short-term radioactive release (less than *24* hours) in each stage of a nuclear accident [7], it is assumed that the post-accident monitoring devices have to survive the first 24 hours of the accident. Taking the highest level of radiation in the Fukushima accident (*530 Sv/h*) as the radiation rate, the total radiation dose after the first *24* hours can be obtained as follows:

$$R = 530\,Sv/h \times 24\,h = 12720\ Sv = 1272.0\ K\ Rad\ (Si).$$

Total radiation dose (*10 M Rad (Si)*) are considered as the limit in this work.

B. A Potential Solution

1) Radiation Protection to Mitigate Damages of Total Dose

As shown in Fig.1, most semiconductor devices will experience radiation damages when the total radiation dose is more than *20 K Rad (Si) (1 Gy = 100 Rad (Si))* [8]. Electronic systems made with regular COTS components will definitely be damaged if it is directly exposed to a radiation environment whose total dose is *10 M* Rad (Si). Radiation protection is an effective solution to mitigate radiation damages for the total dose. The reliability and the lifespan of electronic systems are determined by the type of radiation it is exposed to, the radiation resistance of its semiconductors, and the properties of the shielding materials used [9]. In this work, radiation protection is used where ever necessary to reduce the total dose to less than *20 K* Rad (Si).

2) Radiation Protection to Mitigate Damages of Total Dose

On the other hand, radiation shielding is not effectively against single event effects [10]. Since system recovery and timely power off reset can be effective against radiation damages, in this work, a radiation-tolerant architecture, shown in Fig.2, is developed to prolong both device and system life through independent built-in redundant channels, on-line fault detection, real-time preventive remedial actions, and rapid power off/on recovery, as well as reduction of common-mode damage modes. The architecture consists of an active triple modular redundancy (TMR) core with spare units for replacement of the failed active one. Both measurements and self-diagnosis functions are accomplished in the TMR core.

Each channel can be divided into three layers: the input layer, the decision layer, and the output layer. Self-diagnostic functions, such as fault detection, fault diagnosis, prognostic assessment, and reconfiguration suggestions, are accomplished in the decision layer. The decision layer has two types of buses: an internal bus, which is used to exchange information with other channels; and an *IO* bus, which is used for the master selection. All buses are independent, so other channels

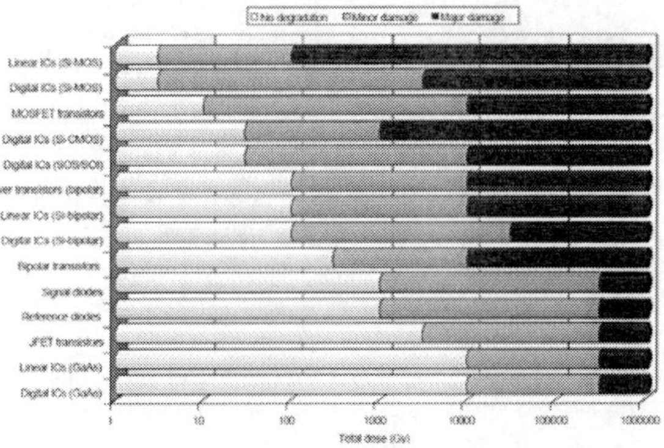

Fig. 1. Radiation tolerance by a family of COTS components [8]

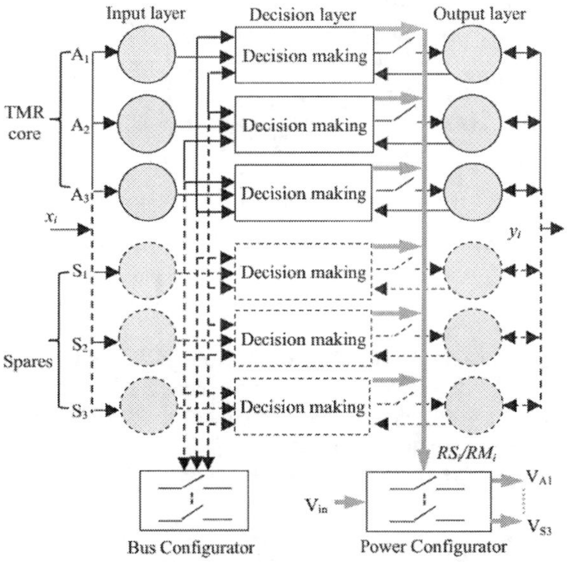

Fig. 2. The proposed rad-hardening architecture

will work normally when one channel fails or has faults. More specifically, a "configurator" is designed to guarantee that the system only ever has three channels working simultaneously and is built with passive devices, such as resistance, capacitance, and non-electronic relays, etc., which are not sensitive to ionizing radiation. The configurator therefore can withstand a high level radiation.

The operating principle of the proposed system works as follows: when one channel has faults or fails to operate, which will be detected by the function of self-diagnosis and/or the function external-diagnosis; then, decision-making units in other channels will generate the reconfiguration suggestions to cut power supply in a timely manner and its space channel will power-on to form a new TMR core. A number of radiation tolerant strategies are involved in protecting the system when one channel encounters radiation damages. These are summarized in Tab. I.

C. *Problem Statement*

The objective of this paper is to:

(1) To evaluate the effectiveness of the proposed radiation protection architecture; and

(2) To establish reliability assessment model against radiation damages.

III. Evaluation of Radiation Protection

A. *Notations*

u	the attenuation coefficient of shielding material
u_l	the linear attenuation coefficient of shielding material
p	the density of shielding material
d	the shielding material thickness
I	the intensity after passing the shielding thickness d
I_0	the original intensity
B	the build-up factor, is greater than 1
E	the energy of the gamma radiation

B. *Attenuation of Gamma Radiation*

When gamma radiation passes through a material under conditions of a 'good' geometry, as shown in Fig.3, a straight-line relationship between the logarithm of the intensity and the thickness of the shielding can be illustrated as follows [11]:

$$I = I_0 e^{-ud}. \quad (1)$$

The linear attenuation coefficient (u) is the probability per unit thickness that particles interact with the material. This value is dependent upon the atomic number Z of the material and its density (p). This relation can also be described through the use of the linear attenuation coefficient as follows [12]:

$$I = I_0 e^{-(u_l/p)(pd)}. \quad (2)$$

C. *Build-up Factor*

However, under conditions of a 'poor' geometry, as shown in Fig.4, Eqn. (1) underestimates the required shielding thickness. The shielding thickness can be estimated by the use of a build-up factor (B), which is defined as the ratio of the intensity of the radiation at any point in a beam to the intensity of the primary radiation only at that point; it can be a function of the total attenuation coefficient, the thickness of shielding material, and the energy of the gamma radiation [10-13]. Eqn.2 can, therefore, be estimated by Eqn.3 [11].

$$I = B(ud, E) \cdot I_0 \cdot e^{-(u_l/p)(pd)}. \quad (3)$$

Build-up factors have been calculated for different levels of gamma energies and for various shielding materials, which can be found in "American National Standard for Gamma-Ray Attenuation Coefficients and Buildup Factor for Engineering Materials", ANSI/ANS-6.4.3-1991 [14].

TABLE I
Summary of Radiation-Tolerant Strategies

Tolerant strategies	Description
Hardware selection	To investigate radiation hardness of different semiconductor technologies and devices; and to select COTS components with high radiation resistance.
Diversity	Enforcing differences, such as different semiconductor technologies but functionally equivalent components, different software and algorithms for the same function, etc., are applied to achieve the diversity.
Multi-processor cooperation	Redundancy communication buses and *IO* checking bus; when one channel has non-responsive failure or wrong messages, it will be reset or power off.
Self-test technology	Self-test techniques, such as monitoring voltage and current, CPU self-checking, etc.; the recovery strategies will be triggered when they are out of range.
Watchdog timer	Combined with hardware watchdog and software watchdog.
Modular software design	Many modules with different functions, error detection and correction are used for the recovery strategies.
Prognostic assessment	Accurately prediction of the circuit performance according to present measurement and the analysis of devices responses, then make a decision to the reconfigurator.
Fault detection and diagnosis	Online analysis of radiation responses of devices and circuits to timely diagnose the radiation fault for fast recovery.

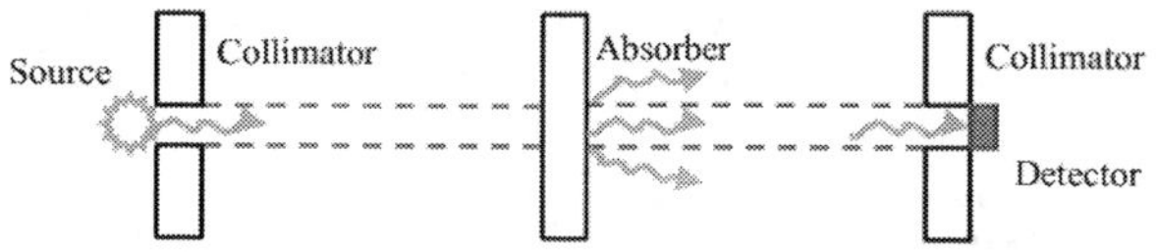

Fig. 3. Measurement of the attenuation of gamma radiation under conditions of 'good' geometry [11]

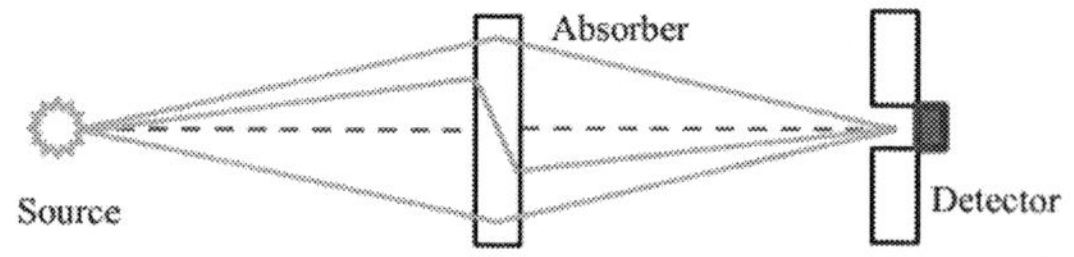

Fig. 4. Gamma radiation attenuation under conditions of broad beam geometry [11].

D. Calculation of Shielding Thickness

As previous discussion, with the given radiation source, the given radiation dose rate, and the known shielding material, based on Eqn.1 to Eqn.3, the required shielding thickness under a poor geometry can be calculated as follows:

$$d = \ln\left(B \cdot {}^{I_0}\!/_{I} \right) / u. \tag{4}$$

According to this equation, various shielding materials can be selected and compared, then the designed shielding thickness can be also evaluated to achieve the design objective, which is to reduce the total dose to less than *20 K* Rad (Si).

IV. Reliability Assessment

A. Attenuation of Gamma Radiation

$R(t)$	reliability of an item at time t
$R_{ij}(t)$	reliability of the jth layer in the channel i at time t
$R_{C_i}(t)$	reliability of the channel i at time t
$R_S(t)$	reliability of the proposed system at time t
$R'(t)$	reliability of an item at time t under total radiation dose D_t
$R'_{ij}(t)$	reliability of the jth layer in the channel i at time t under total radiation dose D_t
$R'_{C_i}(t)$	reliability of the channel i at time t under total radiation dose D_t
$R'_S(t)$	reliability of the proposed system at time t under total radiation dose D_t
m	the number of redundant channels in the system
λ	the failure rate of an item
λ_{ij}	the failure rate of the jth layer in channel i
λ_{ij_k}	the failure rate of kth component of the jth layer in channel i
Δ	radiation degradation factor under total radiation dose D_t
P_0	the value of a characteristic parameter before exposure
P_t	the value of the characteristic parameter after a total radiation dose D_t
P_f	the value of the characteristic parameter at failure of total radiation dose D_f

B. Assessment Model

The reliability function $R(t)$ represents the probability that an item (component, subsystem, or system) will perform the designed functions over a given time interval $[0, t]$ under specific operating environment and conditions [15]. Conventional analysis methods for system reliability are dependent on probabilistic approaches, which incorporate all failure events as random events. These methods are based on two fundamental assumptions: (1) binary state assumptions, where the system can only be in either of the two states (fully functioning or fully failed); and (2) probability assumptions, where the system failure behavior is fully characterized by the probability measures [16]. However, there are many uncertainties when a device or system works in a harsh environment, which may include the strong levels of radiation, extremely high temperature, and high humidity, etc.; those uncertainties challenge the assumptions in the "conventional" reliability analysis of the components and increase the probability of failure of the item (component, subsystem, and/or system) [17]. Therefore, the analysis employed by the conventional methods may not represent a realistic situation in a harsh environment. Radiation effects should be considered in the reliability analysis.

This work first establishes the assessment model for non-radiation conditions through the failure rate, in order to obtain the reliability of the proposed architecture under radiation conditions. The failure rate $\lambda(t)$ of an item expresses the "possibility to failure" of the item after time t has passed [15]. It is estimated from the mean number of failures per unit time, which can be expressed by failure in time (*FIT*) as follows:

$$1\,FIT = 10^{-9} \quad failure/hour.$$

The reliability $R(t)$ of the item can then be determined from the failure rate $\lambda(t)$ with the consideration of $R(0) = 1$ as follows [15]:

$$R(t) = e^{-\int_0^t \lambda(\tau) d_\tau}. \tag{5}$$

Assuming that the failure rate is independent of time (t), then $\lambda(t) = \lambda$, Eqn.5 can be simplified to [15]

$$R(t) = e^{-\lambda t}. \tag{6}$$

Considering that the proposed architecture consists of an input layer, a decision layer, and an output layer; and the reliability of diversified channels are all different, the reliability of the jth layer in the channel i, which consists of n_l components, can be evaluated using the formula.

$$R_{ij}(t) = e^{-\lambda_{ij} t} = \prod_{k=1}^{n_l} e^{-\lambda_k t} = e^{-\sum_{k=1}^{n_l} \lambda_{ij_k} t} \quad (i = 1, ..., m;\ j = 1, 2, 3). \tag{7}$$

The reliability of the channel i can be described as follows:

$$R_{C_i}(t) = e^{-\lambda_{C_i} t} = \prod_{j=1}^{3} e^{-\lambda_{ij} t} = e^{-\sum_{j=1}^{3} \lambda_{ij} t} \quad (i = 1, 2, 3). \tag{8}$$

Under non-radiation conditions, the reliability model of the proposed architecture can then be described as follows:

$$R_S(t)=\left(1-\prod_{i=1}^{m}\left(1-R_{i1}(t)\right)\right)\times\left(1-\prod_{i=1}^{m}\left(1-\left(R_{i2}(t)\times R_{i3}(t)\right)\right)\right). \tag{9}$$

Specifically, as previously discussed, cases of all three channels simultaneously encountering failures are not considered in this work, common cause failure (CCF) is therefore not considered.

C. Radiation Degradation Factor

To take radiation effects in consideration in system reliability analysis, a new method for electronic systems has been developed in [17], [18]. This method uses radiation degradation factors(Δ), instead of the usual failure rate data of an item in the reliability model, as input to describe the radiation response of this item under a total radiation dose D_t, which will lie in the interval $[0,1]$ and can be defined as follows:

$$\Delta=\min\left\{\left|(P_0-P_t)/(P_0-P_f)\right|,\ 1\right\}. \tag{10}$$

A detailed description of the radiation degradation factor can be found in [18]:

$$\Delta=\begin{cases}(P_0-P_t)/(P_0-P_f) & for\ P_0\ge P_t>P_f\ or\ P_0\le P_t<P_f\\ 0 & for\ P_t>P_0>P_f\ or\ P_t<P_0<P_f\\ 1 & for\ P_0>P_f>P_t\ or\ P_0<P_f<P_t\end{cases}. \tag{11}$$

Previous research [17], [18] has derived the parameter values of radiation degradation from real radiation test data; with radiation degradation function which are used to describe how the material and/or components change their properties under given radiation conditions. Radiation degradation functions are separated into three categories in [18], as shown in Fig.5:

1) Piece-wise linear radiation degradation function, with logarithmic dose values and linear parameter values;

2) Linear radiation degradation function in the entire range of exposure; and

3) Constant radiation degradation function. The value 1 is up to D_f, and the value 0 is assumed to fail abruptly at the threshold dose.

Due to the fact that semiconductor components may have a number (n_p) of critical parameters, in this study, the radiation degradation factor is defined as the mean value of those degradation factors of all critical parameters, which can be described as follows:

$$\Delta=\frac{\sum_{i=1}^{n_p}\min\left\{\left|(P_{i_0}-P_{i_t})/(P_{i_0}-P_{i_f})\right|,\ 1\right\}}{n_p}. \tag{12}$$

The reliability ($R'(t)$) of the item under total radiation dose

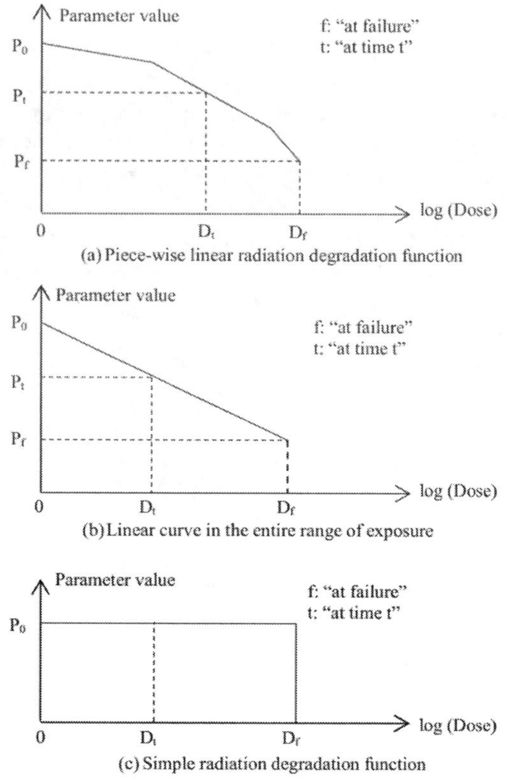

(a) Piece-wise linear radiation degradation function

(b) Linear curve in the entire range of exposure

(c) Simple radiation degradation function

Fig. 5. Radiation degradation function [18].

D_t can then be described in Eqn.13.

$$R'(t)=(1-\Delta)\cdot R(t)=(1-\Delta)\cdot e^{-\lambda t}. \tag{13}$$

The reliability of the jth layer in the channel i under the total radiation dose D_t can be evaluated in Eqn.14.

$$R'_{ij}(t)=\prod_{k=1}^{n_l}(1-\Delta_k)e^{-\lambda_k t}\quad(i=1,\ldots,m;\ j=1,2,3). \tag{14}$$

Assuming that the channel i consists of n_c components, the reliability of the channel i under total radiation dose D_t can be described in Eqn.15.

$$R'_{C_i}(t)=\prod_{k=1}^{n_c}(1-\Delta_k)e^{-\lambda_k t}\quad(i=1,2,3). \tag{15}$$

Based on Eqn.5 and Eqn.15, the reliability model of the redundant architecture mentioned in Fig.2 under total radiation dose D_t can be derived as Eqn.16.

$$R'_S(t)=\left(1-\prod_{i=1}^{m}\left(1-R'_{i1}(t)\right)\right)\times\left(1-\prod_{i=1}^{m}\left(1-\left(R'_{i2}(t)\times R'_{i3}(t)\right)\right)\right). \tag{16}$$

V. A Case Study

A. Development and Evaluation of Radiation Protection

1) Radiation Shielding

To enhance the radiation tolerance, the proposed radiation-tolerant architecture in Fig.2 is further protected by a structure of radiation shielding as illustrated in Fig.6 to increase the radiation tolerance while to avoid common-mode damage. This design consists of three protection layers, with the second and third layers (Fig.6 (b)) inserted into the first protective layer (Fig.6 (a)). The third layer is smaller than the second, and each layer is composed of different materials, which are determined by the particular semiconductor devices used in the electronic system. The layout of the proposed architecture is also illustrated in Fig.6 (c).

The detailed parameters of radiation shieling are listed as follows:

- The material of the first shielding is copper, the size is *26cm*26cm*20cm*, and its thickness is *1cm*;
- The material of the second shielding is lead, the size is *24cm*24cm*18cm*, and its thickness is *6cm*; and
- The material of the third shielding is tungsten, the diameter is *4cm*, and its thickness is *1cm*.

2) The Evaluation of Radiation Shielding

As illustrated in Fig.6, with different shielding materials for different layers, shielding thickness, radiation angle, and radiation locations, all have different effects on the radiation exposure experienced by electronic systems in the six identified areas, as shown in Fig.6(a) $(A_1 - A_3, S_1 - S_3)$. Using Co-60 as a radiation source, the capability of the protections can be summarized as follows:

- Under the condition with dose rate *2700 Sv/h*, at the given *24h*, the highest total dose in six areas is about *100 K Rad (Si)*;
- Under the condition with dose rate *1350 Sv/h*, at the given *24h*, the highest total dose in six areas is about *50 K Rad (Si)*;
- Under the condition with dose rate *530 Sv/h*, at the given *24h*, the highest total dose in six areas is about *24 K Rad (Si)*;
- Under the condition with dose rate *70 Sv/h*, at the given *24h*, the highest total dose in six areas is about *2.6 K Rad (Si)*.

B. Design and Assessment of Redundant Wireless Monitoring Devices

1) Implementation of Monitoring Devices

As previously mentioned, TMR core of the developed radiation-tolerant architecture consists of an input layer, a decision layer, and an output layer. More specifically, the input layer consists of input sources, a source encoder, and a channel encoder; the decision layer consists of the decision-making unit; and the output layer consists of a digital modulator and transceiver. For simplity, only one temperature sensor is considered herein.

Component selection is an important step in the design phase of COTS-based rad-hardened systems. Radiation effects are different on different devices, circuits and systems, depending on their unique materials, structure, manufacturing

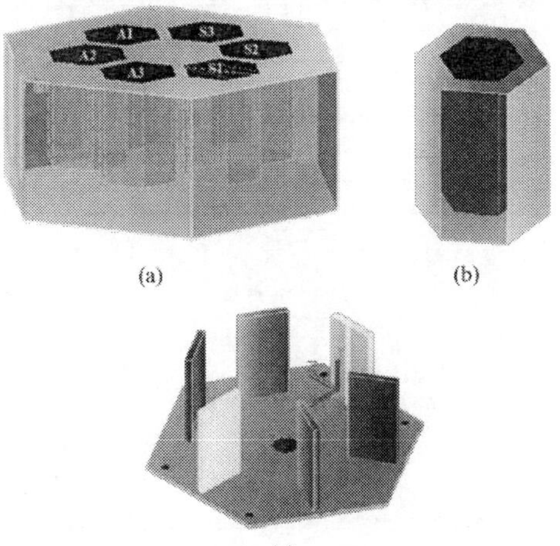

Fig. 6 . (a), (b) The proposed structure of radiation protection. (c) The mounting of electronic circuits in the proposed radiation-tolerant architecture.

technologies, and applications. In this work, referring to radiation test data from the NASA Goddard Space Flight Center [19], all wireless channels and their spares are implemented with diversity hardware, whose detailed implementation and related information are listed in Table II. Specifically, those components such as resistance, capacitance, etc., are not listed due to their robustness to radiation.

2) The Reliability Assessment of Monitoring Device

In the assessment process for system reliability in Eqn.16, the reliability model considers only the factor of total radiation dose. The difficulty of the assessment lies in determining the degradation factors for semiconductor devices. Most of radiation degradation factors under different radiation doses in this study come from NASA Goddard Space Flight Center radiation test data [19], which are available from on-line resources; others are derived from the existing literature [20-22].

In this work, the failure rates of semiconductor components listed in Column (7) of Table II come from the online resources of their manufacturers, while those radiation degradation factors listed in Column (8) to (11) of Table II. Specifically, P_f of some components are not easy to obtain from the NASA database and literature, and are instead derived from specification limits of electronic parameters.

According to Eqn.5 to Eqn.16, under four different total radiation doses (*10 K Rad (Si), 20 K Rad (Si), 50 K Rad (Si), 100 K Rad (Si)*), the system reliabilities of the developed wireless monitoring node are shown in Fig.7. It shows that system reliability decreases significantly when total radiation dose increases. Moreover, according to the compassion result for the reliabilities of different channels (*A1/S1, A2/S2, A3/S3*) and redundant systems (*FT (m=3), FT (m = 6)*), also shown in Fig.7, the reliability of the developed redundant architecture is much improved than those of non-redundant channels under the given radiation conditions.

TABLE II
THE IMPLEMENTATION OF RADIATION-TOLERANT WIRELESS MONITORING DEVICE

Channel	Function	Type	Q.	Tech.	Manu.	FIT	R.D.F.	Δ_{10K}	Δ_{20K}	Δ_{50K}	Δ_{100K}	Radiation effects
A1&S1	*Source encoder*	NPN BJT	1	Bipolar	Semicoa	2.45	A	0.194	0.3201	0.4267	0.4591	TID
		Voltage reference	1	Bipolar	TI	3.30	A	0.0774	0.1010	0.2104	0.3432	SEU, SEL, TID
		OP amp	3	Bipolar	National Semi.	1.85	A	0.0208	0.0365	0.0383	0.0365	SEL, SET, TID
	Channel encoder	Voltage reference	1	Bipolar	TI	3.30	A	0.0642	0.1099	0.5158	0.5786	SEU, SEL, TID
		AD	1	Bipolar	Analog Devices	0.20	A	0.0178	0.0486	0.0633	0.0649	SEU, SEL, SEFI, TID
	Decision making& Digital modulator	E^2PROM	1	CMOS	Atmel	2.2	A	0.0023	0.0244	0.1341	0.1326	SEU, SEL, SEFI, TID
		Micro controller	1	CMOS	Microchip	3.3	A	0.0187	0.0465	0.1001	0.1179	SEE, TID
	Transceiver	Voltage reference	1	Bipolar	Linear	3.30	C	0.0000	0.0000	0.0000	0.0000	SEU, SEL, TID
		Diode	1	Bipolar	Toshiba	3.30	A	0.0000	0.0577	0.0145	0.0769	SEU, SEL, TID
		433MHz RF	1	Bipolar	RFMD	1.90	A	0.0395	0.0745	0.1503	0.1810	SEE, TID
A2&S2	*Source encoder*	Voltage reference	1	BiCMOS	TI	3.30	A	0.1510	0.0181	0.0087	0.0094	SEU, SEL, TID
		OP amp	3	CMOS	Analog Devices	0.28	A	0.0409	0.0770	0.2989	0.2168	SEL, SET, TID
	Channel encoder	OP amp	1	CMOS	Analog Devices	0.28	A	0.2377	0.3964	0.6620	0.6537	SEL, SET, TID
		Voltage reference	1	CMOS	TI	3.30	A	0.1408	0.3371	0.3204	0.3846	SEU, SEL, TID
		AD	1	BiCMOS	Analog Devices	0.25	A	0.1735	0.1503	0.2741	0.3345	SEE, TID
	Decision making& Digital modulator	Micro controller	1	CMOS	Atmel	5.60	A	0.0638	0.0654	0.0985	0.1190	SEU, SET, SEL,SEFI,TID
		Logic gate	1	CMOS	TI	0.50	A	0.133	0.0244	0.1850	0.2432	SEU, SEL, TID
	Transceiver	433MHz RF	1	CMOS	Freescale Semi.	2.0	A	0.1026	0.1336	0.2310	0.2451	SEE, TID
A3&S3	*Source encoder*	Voltage reference	1	HSCMOS	Allegro	3.3	A	0.1408	0.3371	0.3204	0.3846	SEU, SEL, TID
		OP amp	3	BiFET	TI	0.2	A	0.0689	0.1551	0.3673	0.5151	SEL, SET, TID
	Channel encoder	Voltage reference	1	TTL	Analog Devices	3.3	A	0.0039	0.0216	0.0223	0.0644	SEU, SEL, TID
		AD	1	LC^2MOS	Analog Devices	1.6	A	0.0181	0.0229	0.0246	0.0209	SEE, TID
		OP amp	2	Hybrid	MOTOR OLA	0.2	A	0.1360	0.0764	0.1757	0.2717	SEL, SET, TID
	Decision making& Digital modulator	Micro controller	1	TTL	SILICON	2.26	A	0.0109	0.0134	0.0149	0.0168	TID, SEU, SET
		Logic gate	1	TTL	TI	3.3	A	0.0469	0.0494	0.0480	0.0724	SEU, SEL, TID
	Transceiver	433MHz RF	1	TTL	SILICON	1.90	A	0.0479	0.0781	0.1108	0.1567	SEE, TID
		Voltage reference	1	BiMOS	TI	3.3	A	0.0055	0.0269	0.0238	0.0646	SEU, SEL, TID

R.D.F. (radiation degradation function) – A (piece-wise linear radiation degradation function), B(linear curve in the entire range of exposure), C (simple radiation degradation function)

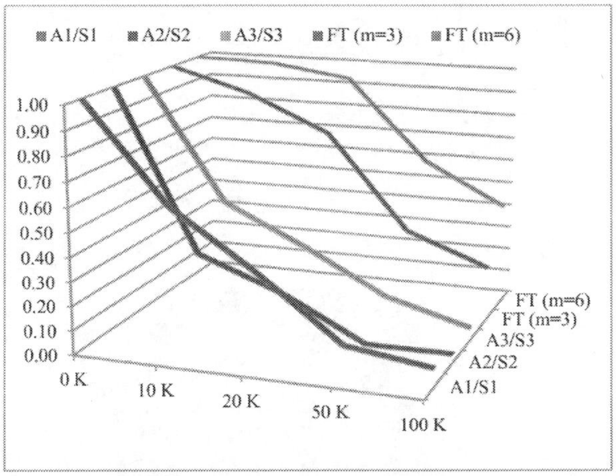

Fig. 7. The compassion of the reliability of the developed redundant system and no-redundant channels under the given total dose (0, 10 K Rad (Si), 20 K Rad (Si), 50 K Rad (Si), 100 K Rad (Si)).

In a summary, radiation assessment of the developed wireless device for various dose rates under the given time (*24h*) is illustrated in Fig.8. It shows that the reliability of the developed wireless device under dose rate *530Sv/h* at *24 h* is about *89.6%*. The developed wireless device can therefore work in those high level radiation fields and meets the design objective.

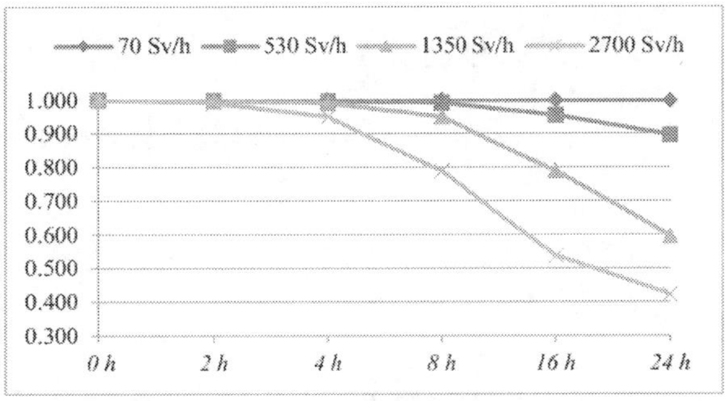

Fig. 8. Radiation assessment of the developed wireless device under various dose rates.

VI. Conclusion

In this work, the assessment of radiation-tolerance of a wireless monitoring device is given for high level radiaiton fields, which is developed by using radiation protection and radundant techniques. The study results show that total dose can be effectively decreased. In addition, the radiation assessment shows, under given radiation conditions (*10 K Rad (Si), 20 K Rad (Si), 50 K Rad (Si), 100 K Rad (Si)*), the reliability of the developed system (*99.0%, 94.2%, 59.6%, 42.3%*) can be much improved than those of non-redundant channels (*60.3%, 37.5%, 11.9%, 7.2%; 33.6%, 20.7%, 3.2%, 3.4%; and 48.5%, 32.5%, 14.3%, 5.1%*).

According to assessment studies, the developed device can work in high level radiation fields whose total dose is up to *1 M Rad (Si)*; and it may be available used in those conditions with total dose *10 M Rad (Si)*. It therefore provides an effective and economical solution to obtain up-to-date information during a severe accident in a nuclear power plant.

Acknowledgment

This work is financially supported by the University Network of Excellence in Nuclear Engineering (UNENE) and the Natural Sciences and Engineering Research Council of Canada (NSERC).

References

[1] H. M. Quinn, D. A. Black, W. H. Robinson, and S. P. Buchner, "Fault simulation and emulation tools to augment radiation-hardness assurance testing," IEEE Transactions on Nuclear Science, vol. 60, pp. 2119-2142, 2013.

[2] R., Sharp, M. Decreton, "Radiation tolerance of components and materials in nuclear robot applications," Reliability Engineering and System Safety, vol. 53, pp. 291-299, 1996.

[3] https://www.theguardian.com/environment/2017/feb/03/fukushima-daiichi-radiation-levels-highest-since-2011-meltdown

[4] J.R. Srour, C.J. Marshall, and P.W. Marshall, "Review of displacement damage effects in silicon devices," IEEE Transactions on Nuclear Science, vol. 50, pp. 653-670, 2003.

[5] D.M. Fleetwood, "Total ionizing dose effects in MOS and low-dose-rate-sensitive linear-bipolar devices" IEEE Transactions on Nuclear Science, vol. 60, pp. 1706-1730, 2013.

[6] V. Ferlet-Cavrois, L.W. Massengill, and P. Gouker, "Single event transients in digital CMOS – a review," IEEE Transactions on Nuclear Science, vol. 59, pp. 1767-1790, 2013.

[7] Policy Elements Post-accident management in the event of nuclear accident," the Steering Committee for the Management of the Post-Accident Phase of a Nuclear Accident (CODIRPA), 2012.

[8] G. C. Messenger and M. S. Ash, "The effects of radiation electronic systems," Van Nostrand Reinhold, New York, 2nd edition, 1991.

[9] J. K, Shultis, R. E. Faw, "Radiation shielding technology," Health Physics, vol. 88, pp. 587-612, 2005.

[10] K.A. LaBel and M.M. Gates, "Single-event-effect mitigation from a system perspective," IEEE Transaction on Nuclear Science, vol. 43, pp. 654-660, 1996.

[11] https://www.nucleonica.com/

[12] E. Yilmaz, H. Baltas, E. Kiris, i. Ustabas, U. Cevik, and A. M. El-Khayatt, "Gamma ray and netron shielding properties of some concrete materials," Annals of Nuclear Energy, vol. 38, pp. 9, 2011.

[13] C. Suteau and M. Chiron, "An iterative method for calculating Gammy-ray build-up factors in multi-layer shields," Radiation Protection Dosimetry, vol. 116, pp. 489-492, 2005.

[14] American National Standard for Gamma-Ray Attenuation Coefficients and Buildup Factors for Engineering Materials", ANSI/ANS-6.4.3-1991, 1991.

[15] Y. Song and B. Wang, "Survey on reliability of power electronic systems," IEEE Transactions on Power Electronics, vol. 28, pp. 591-604, 2013.

[16] V. Ravi, P. J. Reddy, and H. J. Zimmermann, "Fuzzy global optimization of complex system reliability," IEEE Transactions on Fuzzy Systems, vol. 8, pp. 241-248, 2000.

[17] K. Lauridsen, P. Christensen, and H. E. Kongso, "Assessment of the reliability of robotic systems for use in radiation environments," Reliability Engineering and System Safety, vol. 53, pp. 265-276, 1996.

[18] K. Lauridsen, M. Decreton, C. C. Seifert, and R. E. Sharp, "Reliability and radiation tolerance of robots for nuclear applications," The final report of CEC/TELEMAN project, No. FI2T-CT90-0011, ENTOREL, 1996.

[19] https://radhome.gsfc.nasa.gov/radhome/tid.htm

[20] G. C. Messenger and M. S. Ash, "The effects of radiation electronic systems," Van Nostrand Reinhold, New York, 2nd edition, 1991.

[21] L. P. Houssay, "Robotics and radiation hardening in the nuclear industry," Master thesis, University of Florida, 2000.

[22] R. D. Kulkarni, and V. Agarwal, "Reliability analysis of a modern power supply under nucler radiation effects," Power Electronics and Drive Systems, PEDS 2003.

Study of Lamb Waves for Non-Destructive Testing Behind Screens

P. Kauffmann, M.-A. Ploix, J.-F. Chaix, C. Gueudré, G. Corneloup and F. Baqué

Abstract—**The inspection and control of sodium-cooled fast reactors (SFR) is a major issue for the nuclear industry. Ultrasonic solutions are under study because of the opacity of liquid sodium. In this paper, the use of leaky Lamb waves is considered for non-destructive testing (NDT) on parallel and immersed structures assimilated as plates. The first phase of our approach involved studying the propagation properties of leaky Lamb waves. Equations that model the propagation of Lamb waves in an immersed plate were solved numerically. The phase velocity can be experimentally measured using a two dimensional Fourier transform. The group velocity can be experimentally measured using a short-time Fourier transform technique. Attenuation of leaky Lamb waves is mostly due to the re-emission of energy into the surrounding fluid, and it can be measured by these two techniques.**

Index Terms—**Group velocity, Leaky Lamb waves, Non-Destructive Testing (NDT), Phase velocity, Re-emission, Sodium Fast Reactor (SFR), Ultrasound.**

I. Introduction

The sodium-cooled fast reactor (SFR) technology has been chosen for the French 4th generation of nuclear power plants. With liquid sodium as coolant, the improvement of in-service inspection and repair (ISI&R) techniques has been identified as a major issue for the ASTRID project (Advanced Sodium Technological Reactor for Industrial Demonstration) [1], [2]. Ultrasonic solutions are being studied in order to inspect the main vessel and different parts of ASTRID. Fig. 1 shows different inspection strategies with ultrasonic transducers inside and outside the main vessel. Immersion of specific ultrasonic transducers in liquid sodium is under study [3], [4] (1st strategy in Fig. 1). However, inspections using a transducer located outside the main vessel

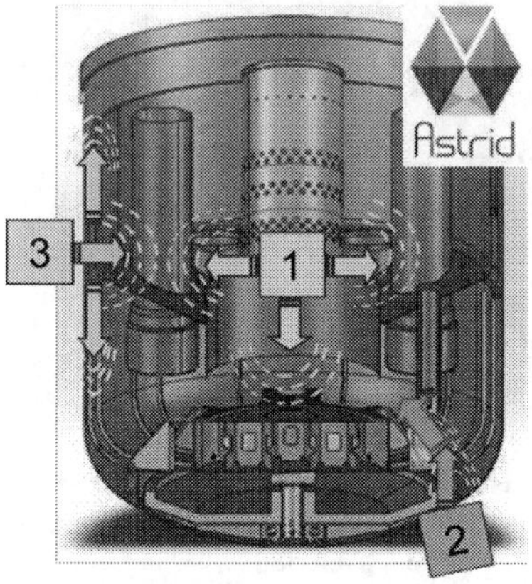

Fig. 1. Different ultrasonic inspections strategies for ASTRID: inspection with a transducer immersed in sodium (1st strategy), with a transducer in the inter-vessel gap using immersed structures as a wave guide (2nd strategy), and through the main vessel (3rd strategy)

(in the inter-vessel gap filled with gas) have shown promising potential [5], [6]. The second strategy indicated in Fig. 1 is also under study [7]. We will focus on the third strategy indicated in Fig. 1.

The objective is to not only inspect the main vessel, but also the structures behind it. This includes the *redan* (the component separating the hot and the cold pool), the internal baffles, sodium pump, and the heat exchanger skirts. The diameter of the main vessel and internal baffles is quite large (about 16 m) and justify to approximate locally their shape by two parallel and immersed plates (Fig. 2). The challenge is then to propagate ultrasonic waves from one plate to another and to control them.

Mechanical waves that propagate in an elastic plate are called Lamb waves [8]. Their use for NDT is widely studied in the case of plates in vacuum (or in air) [9], [10]. However, the configuration of several parallel and immersed plates is quite innovative. References are rare, even if an artificial notch in the second and third plate has been observed experimentally under certain conditions [5], [6]. We can also cite [11]: it has been shown that for a system of two semi-immersed plates, Lamb waves propagate from the first plate to the second.

As a first step, this paper describes the propagation and re-emission of Lamb waves in the case of an immersed plate. Equations used to calculate the phase velocity, group velocity and attenuation are specified. Experiments in water with two different data processing techniques (2D Fourier transform and short-time Fourier transform) are briefly presented.

Manuscript submitted June 1, 2017. This work was supported by the CEA/ Cadarache, France.

P. Kauffmann, M.-A. Ploix, J.-F. Chaix, C. Gueudré, and G. Corneloup are with Aix Marseille Univ, CNRS, Centrale Marseille, LMA, Laboratoire de Mécanique et d'Acoustique, Equipe Ondes et Imagerie, site d'Aix en Provence, Marseille, France (e-mail: pierre.kauffmann@cea.fr).

P. Kauffmann and F. Baqué are with the CEA Cadarache DEN/DTN/STCP/LIET (Laboratoire d'Instrumentation et d'Essais Technologiques), Saint Paul Lez Durance, France (e-mail: francois.baque@cea.fr).

Finally, the experimental results are discussed.

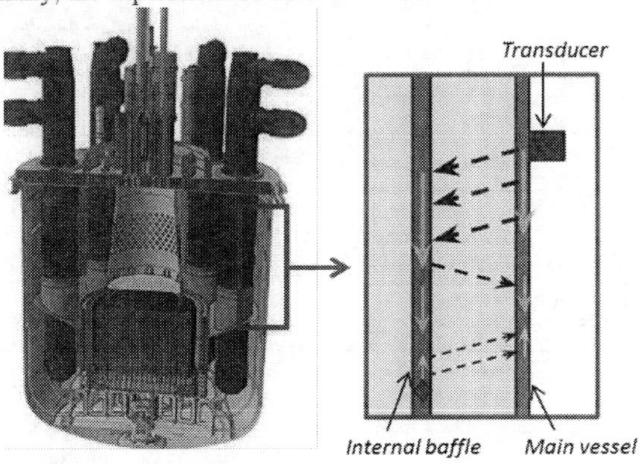

Fig. 2. Cross-section of the ASTRID reactor block with location of the area to be inspected (red box) and a schematic view of this area with the expected ultrasonic path

II. Propagation and attenuation of Lamb waves

Two types of Lamb modes can be differentiated: symmetric and antisymmetric, which are designated with the letters S and A respectively. Symmetric (and antisymmetric respectively)
For an elastic plate of thickness h end density ρ immersed in a fluid of density ρ_0, propagation equations can be found in [12]. These equations can be solved numerically, and it allows us to determine the phase velocity c_p (Fig. 3) and attenuation k_2 (Fig. 4) of the Lamb modes.

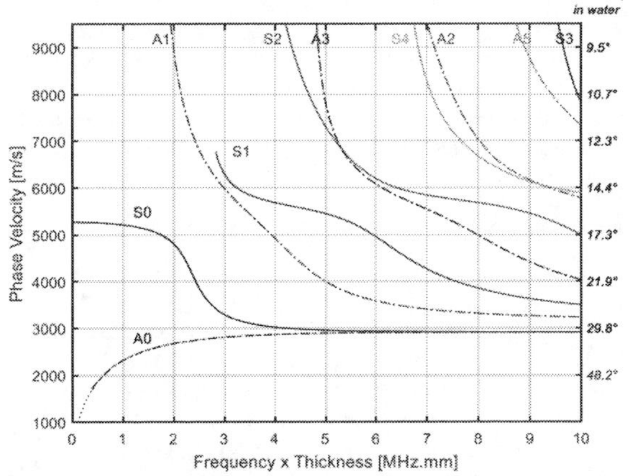

Fig. 3. Phase velocity of Lamb modes for a stainless steel plate ($h = 7.8\ mm$, $\rho = 7\,950\ kg/m^3$, $c_L = 5\,750\ m/s$ and $c_T = 3\,150\ m/s$) immersed in water. Results for a plate in vacuum are also plotted but are too similar to be distinguished from the immersed case

Fig. 3 shows that the phase velocity of the Lamb modes is strongly dependent on the frequency. This dispersive propagation has two major consequences. First of all, short time wave packets will spread in space and time as they propagate. Therefore, the pulse duration increases with propagation, and the maximum amplitude measured decreases because of this spreading. Secondly, the velocity of the wave packet is no longer the phase velocity, but the group velocity.

The group velocity can be calculated from the phase velocity [13].

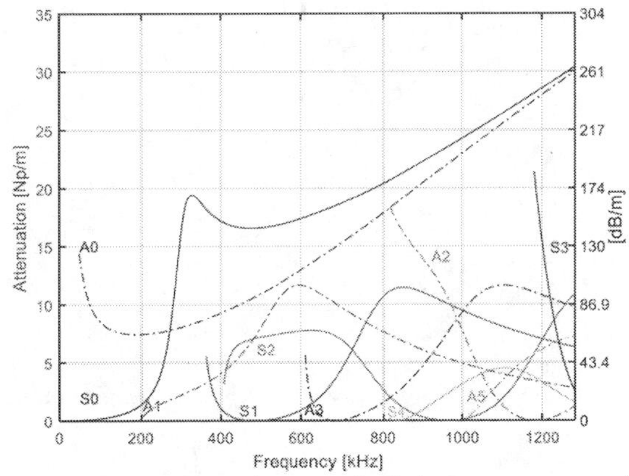

Fig. 4. Attenuation coefficient k_2 computed with the same parameters as in Fig. 3. The x-axis now represents the frequency and no longer the product frequency thickness because attenuation is strongly dependent on the thickness. The results given are for a plate that was 7.8 mm thick.

For more information, the effects of dispersive propagation on a pulse are studied in detail in reference [14].

Phase velocity, group velocity and leaky attenuation are key characteristics to performing NDT behind screens. We will present the experimental set-up that we used to measure these characteristics in the following section.

III. Experimentation in water and data processing

Since experiments in liquid sodium are complex and liquid sodium is quite similar to water from an acoustic viewpoint, experimental measurements and observations were performed in water at room temperature.
As shown in Fig. 5, we considered a single plate immersed in water. A transducer was used to send a pulse at a given central frequency and a chosen incident angle θ_i. The incident angle was chosen on the basis of the Snell-Descartes law (1) in order to excite the phase velocity of a chosen Lamb mode in the plate. Some of the incident angles in water are shown on the right scale in Fig. 3.

$$\sin(\theta_i) = \frac{c_w}{c_p} \tag{1}$$

with c_w the speed of sound in water, c_p the phase velocity of the chosen Lamb mode, and θ_i the incident angle to the normal of the plate.
However, this technique may excite more than one Lamb mode if their respective phase velocities are close enough regardless the spreading angle of the beam [15].

In order to separate the contribution of each mode, we used a 2D Fourier transform method [16]. Wave propagation was no longer integrated into the {time; position} domain, but into the {frequency; wave number} domain.
Moreover, the attenuation was measured by reproducing the

2D Fourier transform twice in two different propagation zones. This technique is presented in [17], but is applied now to measure the leaky attenuation and not anymore the attenuation of composite material.

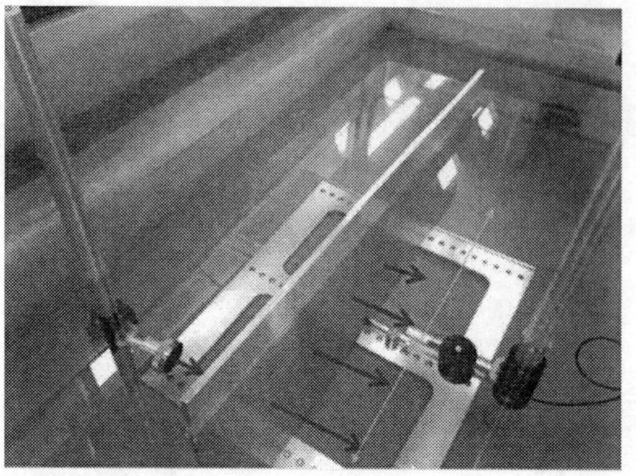

Fig. 5. Experimental bench: a transducer sends bulk waves at a chosen incident angle θ_i onto a plate 7.8 mm thick, immersed in water. A needle hydrophone scans the transmitted field on a line parallel to the plate. Bolds arrows symbolize the waves in water (incident and re-emitted waves). The thin arrow symbolizes hydrophone displacement

In the literature, group velocity are generally measured by measuring the speed of the envelope [14], [18]. In the case of an immersed plate, however, attenuation may be too strong and multi-modal propagation is likely to append, preventing the measurement of velocities on the raw data. This is why we chose to use a time frequency technique to separate the contribution of each mode in the frequency range. This technique is detailed in [19] but is applied now to the case of an immersed plate and not anymore to a plate in air.

In addition to the ultrasonic transducer used in Fig. 5, two other transducers with different central frequencies were used around 110 kHz (i.e. 0.9 MHz.mm) and 250 kHz (i.e. 1.95 MHz.mm). For each transducer, the experimentation was conducted for two or three different incidence angles in order to study the various Lamb modes.

IV. Conclusion

We have shown that techniques used in the literature to measure the propagation of Lamb waves can be applied for an immersed plate. Then the study of the leaky attenuation becomes possible. This will enable us to understand and predict the behavior of an ultrasonic wave in a set of several parallel and immersed plates. It is a primordial step to performing NDT behind screens.

Further study will focus on the propagation of Lamb waves in a set of parallel plates, including finite-element simulations and experimentation in water. The reflection on a defect will also be studied.

References

[1] F. Jadot, F. Baqué, J. P. Jeannot, G. de Dinechin, J. M. Augem, et J. Sibilo, « ASTRID sodium cooled fast reactor: Program for improving in service inspection and repair », in *2011 2nd International Conference on Advancements in Nuclear Instrumentation, Measurement Methods and their Applications (ANIMMA)*, 2011, p. 1-8.

[2] F. Baqué, F. Jadot, R. Marlier, J. F. Saillant, et V. Delalande, « In Service Inspection and Repair of the sodium cooled ASTRID reactor prototype », in *Proceedings of ICAPP'15*, Nice (France), 2015.

[3] F. Baqué, F. Reverdy, J.-M. Augem, et J. Sibilo, « Development of Tools, Instrumentation and Codes for Improving Periodic Examination and Repair of SFRs », *Sci. Technol. Nucl. Install.*, n° 718034, p. 19, 2012.

[4] F. Baqué, F. Jadot, R. Marlier, J. F. Saillant, et V. Delalande, « In Service Inspection and Repair of Sodium cooled ASTRID Prototype », in *2015 4th International Conference on Advancements in Nuclear Instrumentation Measurement Methods and their Applications (ANIMMA)*, 2015.

[5] G. Corneloup, M.-A. Ploix, J.-F. Chaix, I. Lillamand, et F. Baqué, « Potential of Ultrasounds for NDT of a structure located behind parallel immersed plates », in *Review of progress in Quantitative NonDestructive Evaluation: volume 30A; volume 30B*, 2011, vol. 1335, p. 1695-1700.

[6] F. Baqué, K. Paumel, G. Corneloup, M. A. Ploix, et J. M. Augem, « Non destructive examination of immersed structures within liquid sodium », in *2011 2nd International Conference on Advancements in Nuclear Instrumentation, Measurement Methods and their Applications (ANIMMA)*, 2011, p. 1-5.

[7] Q. Elie, F. L. Bourdais, K. Jezzine, et V. Baronian, « Study of guided wave transmission through complex junction in sodium cooled reactor », in *2015 4th International Conference on Advancements in Nuclear Instrumentation Measurement Methods and their Applications (ANIMMA)*, 2015, p. 1-5.

[8] H. Lamb, « On Waves in an Elastic Plate », *Proc. R. Soc. Lond. Math. Phys. Eng. Sci.*, vol. 93, n° 648, p. 114-128, mars 1917.

[9] I. A. Viktorov, *Rayleigh and Lamb Waves: Physical Theory and Applications*, Softcover reprint of the original 1st ed. 1967 edition. Springer, 2013.

[10] P. D. Wilcox, M. J. S. Lowe, et P. Cawley, « Mode and transducer selection for long range lamb wave inspection », *J. Intell. Mater. Syst. Struct.*, vol. 12, n° 8, p. 553-565, 2001.

[11] M. Schmitt *et al.*, « Detection of coatings within liquid-filled tubes and containers by mode conversion of leaky Lamb waves », *J. Sens. Sens. Syst.*, vol. 2, n° 1, p. 73-84, mai 2013.

[12] L. G. Merkulov, « Damping of normal modes in a plate immersed in a liquid », *Sov Phys Acoust*, vol. 10, n° 2, p. 169-173, 1964.

[13] J. L. Rose, *Ultrasonic Waves in Solid Media*. Cambridge University Press, 1999.

[14] P. Wilcox, M. Lowe, et P. Cawley, « The effect of dispersion on long-range inspection using ultrasonic guided waves », *NDT E Int.*, vol. 34, n° 1, p. 1-9, janv. 2001.

[15] T. D. K. Ngoc et W. G. Mayer, « A General Description of Ultrasonic Nonspecular Reflection and Transmission Effects for Layered Media », *IEEE Trans. Sonics Ultrason.*, vol. 27, n° 5, p. 229-235, sept. 1980.

[16] D. Alleyne et P. Cawley, « A two-dimensional Fourier transform method for the measurement of propagating multimode signals », *J. Acoust. Soc. Am.*, vol. 89, n° 3, p. 1159-1168, mars 1991.

[17] M. Castaings, C. Bacon, B. Hosten, et M. V. Predoi, « Finite element predictions for the dynamic response of thermo-viscoelastic material structures », *J. Acoust. Soc. Am.*, vol. 115, n° 3, p. 1125-1133, mars 2004.

[18] X. Luis Dean-Ben, C. Trillo, A. F. Doval, et J. L. Fernandez, « Phase and group velocity measurement of ultrasonic guided wavetrains in plates by pulsed TV holography », *J. Acoust. Soc. Am.*, vol. 127, n° 4, p. 2210-2219, avr. 2010.

[19] M. Niethammer, L. J. Jacobs, J. Qu, et J. Jarzynski, « Time-frequency representation of Lamb waves using the reassigned spectrogram », *J. Acoust. Soc. Am.*, vol. 107, n° 5, p. L19-L24, mai 2000.

Wave propagation simulation in the upper core of sodium-cooled fast reactors using a spectral-element method for heterogeneous media

Masaru Nagaso, Dimitri Komatitsch, Joseph Moysan and Christian Lhuillier

***Abstract*—ASTRID project, French sodium cooled nuclear reactor of 4th generation, is under development at the moment by Alternative Energies and Atomic Energy Commission (CEA). In this project, development of monitoring techniques for a nuclear reactor during operation are identified as a measure issue for enlarging the plant safety. Use of ultrasonic measurement techniques (e.g. thermometry, visualization of internal objects) are regarded as powerful inspection tools of sodium cooled fast reactors (SFR) including ASTRID due to opacity of liquid sodium. In side of a sodium cooling circuit, heterogeneity of medium occurs because of complex flow state especially in its operation and then the effects of this heterogeneity on an acoustic propagation is not negligible. Thus, it is necessary to carry out verification experiments for developments of component technologies, while such kind of experiments using liquid sodium may be relatively large-scale experiments. This is why numerical simulation methods are essential for preceding real experiments or filling up the limited number of experimental results. Though various numerical methods have been applied for a wave propagation in liquid sodium, we still do not have a method for verifying on three-dimensional heterogeneity. Moreover, in side of a reactor core being a complex acousto-elastic coupled region, it has also been difficult to simulate such problems with conventional methods. The objective of this study is to solve these 2 points by applying three-dimensional spectral element method. In this paper, our initial results on three-dimensional simulation study on heterogeneous medium (the first point) are shown. For heterogeneity of liquid sodium to be considered, four-dimensional temperature field (three spatial and one temporal dimension) calculated by computational fluid dynamics (CFD) with Large-Eddy Simulation was applied instead of using conventional method (i.e. Gaussian Random field). This three-dimensional numerical experiment yields that we could verify the effects of heterogeneity of propagation medium on waves in Liquid sodium.**

***Index Terms*—Acoustics, Acoustic simulation, Finite element analysis, Spectral-element method,**

I. Introduction

THE 4th generation nuclear reactors, which use liquid metals for medium of their coolant circuits, is under development as an international project [1,2,3]. Inspection and repairing systems for inside of a reactor core are regarded as important issues for enlarging its safety. Based on this viewpoint, improvement of measurement tools' accuracy, sensitivity, frequency of measurements and reliability are required [8,6]. Ultrasonic measurement systems have been studied and developed among LIET CEA Cadarache and LMA CNRS as possible methods for liquid sodium with no transparency [5].

It is known that strong flow fields and heterogeneities of temperature and density of gas bubbles caused by flows occurs above core sub-assemblies during operation of reactor from studies [9, 25, 26]. Also the effects of these heterogeneities of gas density and temperature gradient on a wave propagation have been researched [20, 7]. By these studies, it was revealed that the effects of those heterogeneities are not negligible as for estimating wave propagations inside of reactor core. Because of the use of liquid metals that is a characteristic 4th generation of nuclear reactors, possible scales and the number of times are limited. Thus numerical methods for predicting this phenomenon is demanded.

There exist several studies on the numerical methods for a wave propagation in heterogeneous medium. D. Fiorina simulated the fluctuation of sound celerity caused by heterogeneity of a medium by applying Gaussian random field (GRF). He concluded that it is possible to estimate the amount of possible effects of medium fluctuation on a wave propagation only for limited propagation distance and strength of fluctuation, which was confirmed for two dimensional acoustic problems with a point and plane source and for isotropic fluctuation condition (i.e. no strong flow) [10, 18].

This method which applying GRF is recognized as a method which required relatively fewer computational cost than methods which use the CFD. B. Lü improved this method and applied for three dimensional problems by applying the fluctuation for the time-of-flight and signal amplitudes calculated by a ray (tracing) method beforehand [19]. B. Iooss expanded this application of GRF for anisotropic fluctuation. He proposed to used anisotropic GRF for the fluctuation in two-dimensional flow fields. At the same time, he concluded that the fluctuation pattern generated by Gaussian method is far

M. Nagaso and C. Lhuiller are with DEN/DTN/STCP/LIET, CEA Cadarache, 13108 Saint Paul Lez Durance, France (e-mail: masaru.nagaso@cea.fr; christian.lhuillier@cea.fr).

D. Komatitsch is with LMA, CNRS UPR 7051, Université d'Aix-Marseille, Centrale Marseille, 13453 Marseille cedex 13, France (e-mail : komatitsch@lma.cnrs-mrs.fr).

J. Moysan is with Université d'Aix-Marseille, LMA UPR 7051 CNRS, 13625 Aix-en-Provence, France (e-mail : joseph.moysan@univ-amu.fr).

from the real situation because this is a stochastic and idealized method [13].

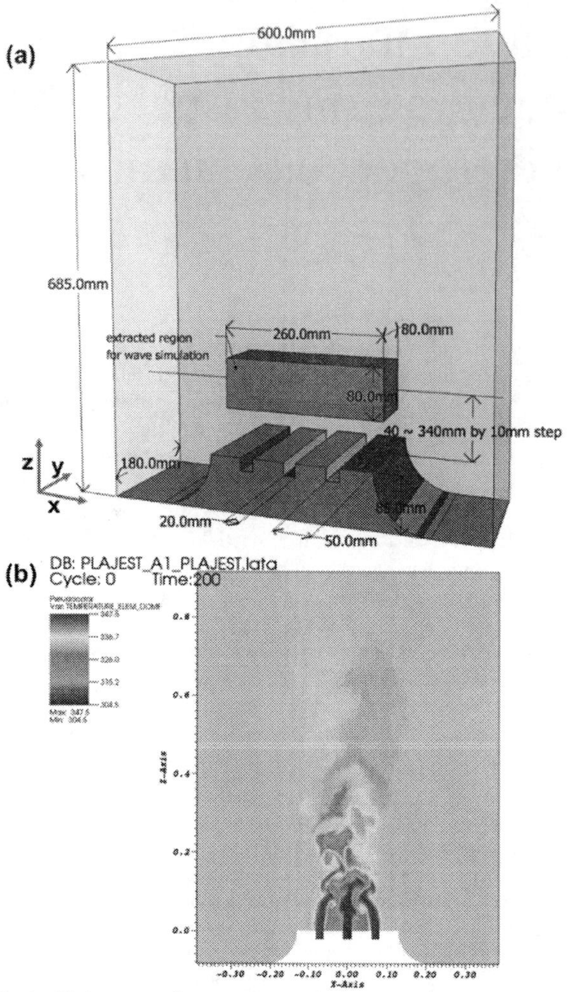

Fig. 1. (a) Geometry of PLAJEST CFD simulation. 3 sodium jets outflow from the gaps with 20 mm width. The region for wave propagation simulation is extracted (purple box). Multiple simulations are carried out with changing the z coordinate of extracted region from 40 mm to 340 mm above the outlet at the intervals of 10 mm. (b) A snapshot of CFD result at time 200.0 seconds, x-y cross-sectional plane at y = 0.09 meter. The temperature pattern just above the outlets are simple and stable while it becomes more complex and unstable when z coordinate goes the farther.

Additionally, only the ray (tracing) method [23] or the pencil method [11] have been used as numerical methods for a wave propagation. There are limitations on application of ray tracing techniques for heterogeneous medium. There are limitations on application of ray tracing techniques for heterogeneous medium.

The objective of this study is to simulate the effect of heterogeneity on three-dimensional wave propagation by applying the spectral element method, which is one of higher order finite element methods. As the method to simulate the fluctuation of medium, we applied an CFD simulation results, which was calculated by CEA STMF, then we studied the effects of realistic fluctuated fields on the three-dimensional acoustic field from a circular piston plate.

In the section 2, the model target of our numerical study is explained. In the section 3, the numerical tools which we used for the wave simulation, the details of CFD and interpolation of medium properties are described. Then we show our first results about the application of three-dimensional wave simulation with CFD.

II. Numerical methods for a wave propagation

In our study, we simulate a wave propagation with three dimensional spectral-element method (SEM) for the discretization scheme in spatial dimension. We used a numerical code called SPECFEM. SPECFEM is a generic name of numerical codes, belonging CIG (Computational Infrastructure for Geodynamics), for a wave propagation simulation which uses the SEM. Initially this was developed by Dimitri Komatitsch and Jean-Pierre Vilotte at Institut de Physique du Globe (IPGP) in Paris, France from 1995 to 1997 and then by Dimitri Komatitsch and Jeroen Tromp at Harvard University and Caltech, USA for a seismic wave simulation. [16, 17, 30]

On SPECFEM, continuous Galarkin typed SEM with higher (4th)-order Gauss-Lobatto-Legendre is implemented. The higher order basis function brings better accuracy for surface-wave, acoustic-elastic coupling problems than the other types of finite-element method using lower-order polynomials. The time domain is discretized by Newmark-beta method.

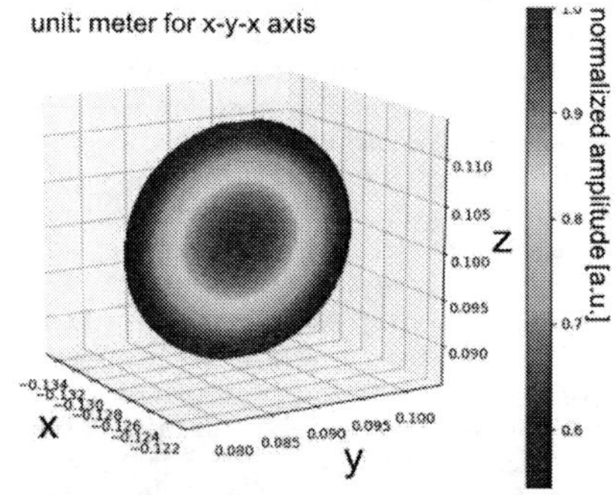

Fig. 2. The circular source plane applied for the simulations. The color shows the maximum amplitudes of the emissions at each point normalized by the maximum amplitude at the center of the circle. The plane source is composed of monopole point sources on a circular plane with a diameter of 0.0254 meter. The intervals of each point sources are same as the element size of SPECFEM mesh (i.e. approximately 8.0×10^{-4} meter). Each source point emits a 1MHz Ricker wavelet. The maximum amplitude of each emission is multiplied by hamming window function having a maximum point at the center of the circular plane.

III. Modeling of the propagation medium

A. PLAJEST experiment / CFD simulation

We selected an experimental and numerical study called PLAJEST as the modeling target of our study since PLAJEST is also targeting the same object of the same condition i.e. the upper-core region of SFR in operation. This study program is collaboratively worked between Japanese Atomic Energy

Agency (JAEA), the U.S. Department of Energy (DOE) and CEA. The configuration of this experiment is shown in Figure 1. The main purpose of PLAJEST is to observe heat conduction at a liquid-solid boundary in a sodium cooling circuit because the thermal fluctuations lead to high frequency thermal fatigue and crack in adjoining structures of SFR.

After the experiment by JAEA [14] [15], thermo-fluid analysis has been carried out by CEA/STMF [4]. STMF used a CFD tool called TrioCFD (known as Trio_U by 2015) for their calculation and LES was employed as a turbulent model. Tetrahedral element with 4 nodes was selected for TrioCFD computation model. Total number of element reached 5,582,706 and characteristic mesh length was set to 1.40 mm. Initial 200 seconds of their calculation is excluded as duration for stabilization then 200 ~ 210 seconds with 0.1 milli-seconds of time step is available as candidates for our wave simulation. 3 jets of sodium exist in this setup. Sodium with lower temperature (304.5 °C) is emitted from central jet and with higher temperature (347.5 °C) from two outer jets. The average flow velocity is 0.51 m/s for every jet. The simulated temperature field at time = 200.787 seconds is also shown in figure 1 (b). Their simulation results achieved good accordance with the experiment results by JAEA in terms of Spectral power density and standard deviation of temperature values. [4]

B. Wave simulation in realistic fluctuated medium

For the modeling of sodium state, we took few hypothesis:

- The density of gas bubble is constant everywhere in whole simulated region.
- The effect of flow on a wave propagation is vanishingly small, so called frozen fluid.

Thus, the heterogeneity of the temperature field is taken into account for our study.

As can be seen in figure 1 (b), the temperature field is spatially (and also temporally) stable in the region of just above the jets' outlet. We call this field as a static temperature field. The pattern of temperature field becomes more complex and temporal fluctuation occurs as the region goes farther from the outlets. We call this field as a fluctuation field. Our former study [21] applied the hypothesis that the temperature field in the region near the outlet may be described by superimposing a static field and fluctuation field. The static field was provided as very simple profiles including parabolic shapes. The fluctuation fields were generated by GRF. However, as described above, GRF method is only validated for the field without strong flows and we are still not sure about to what extent GRF is applicable for the region including flows. It is expected that GRF is almost perfectly applicable for the simulation region being sufficiently far from the outlets and the difference of acoustic simulation results between with GRF and with CFD will be increased when the region becomes closer the outlet.

Hence one of the objectives of this study is to examine the applicability of the GRF approximation for the medium fluctuation by comparing the results from GRF-applied wave simulation and CFD-applied simulation and changing the distance from the sodium outlets.

In this paper, as a previous step, we apply the temperature field calculated by CFD to our three-dimensional wave simulation, then observe the modified acoustic fields and relation between the distance from the outlets.

We use the Sobolev's empirical equations [24] for the relationships between temperature and wave celerity in sodium:

$$C_p = 2723.0 - 0.531\ T_{kelvin} \tag{1}$$

where C_p is the celerity of ultrasonic waves in meters per second and T_{kelvin} is sodium temperature in Kelvin degrees.

Density is also temperature dependent:

$$\rho = 1014.0 - 0.235\ T_{kelvin} \tag{2}$$

C. Configuration of a three-dimensional wave simulation

In order to decrease the total computational costs, the target domain for our acoustic simulation is partially extracted from the entire region calculated in the CFD side and other domain which is acoustically not interested is excluded from our simulations. We extracted the rectangular parallelepiped region drawn as the purple box in figure 1. The boundary lengthes for y. z axis (perpendicular to the direction of propagation) are defined to become about three times of the diameter of the source plane. The dimension for x axis is taken enough for acoustic waves' passing through three sodium jets completely. Herewith, we observe the modification of acoustic fields before and after passing through each jet.

We carried out multiple simulations by changing the z coordinate of the simulation region from z = 0.040 meter (just above the jets) to z = 0.340 meter at 0.010 meter interval (z origin is set as the same level of the edges of outlets, which z value means the central coordinate of the simulated box). X and y coordinates are kept in all simulations.

As the source of all simulations, we used circular plane source, which is composed of monopole point sources on a circular plane with a diameter of 0.0254 meter. The intervals of each point sources are same as the element size of SPECFEM mesh (i.e. approximately 8.0×10^{-4} meter). Each source point emits a 1MHz Ricker wavelet. The maximum amplitude of each emission is multiplied by hamming window function having a maximum point at the center of the circular plane. Figure 2 shows this source plane. The color shows the maximum amplitudes of the emissions at each point normalized by the maximum amplitude at the center of the circle.

The Finite-Element (FE) Mesh for SPECFEM3D is generated by FE meshing software CUBIT (developed by Sandia National Laboratories, United States). Both 1st and 2nd-order hexahedral FE may be imported to SPECFEM. We selected 1st-order elements for our simulation because the targeted model does not have any curved geometry. It is necessary to convert the data format while the mesh data is exported from CUBIT to SPECFEM3D. In order to speed-up this conversion process, we used

IOSS (IO Systems) library included the FE analysis supporting software SEACAS (also developed by Sandia National Laboratories, United States). After being imported into the SPECFEM, the GLL points are inserted into each element. Determination of an element size is done following 2 conditions, which are CFL condition (3) and the number of elements par one wave length,

$$C_p \frac{\Delta t}{\Delta x_{gll}} \leq \alpha \quad (3)$$

where Δt is the interval of each time increment. Δx_{gll} is the minimum interval of GLL points.

We selected the averaged Courant number $\alpha = 0.4$ and the wave celerity $C_p = 2416.268$ m/s (in sodium with lowest temperature value 274.5°C during the CFD simulation) for the calculation of a mesh size and time step length. In the equation 3, dx is not the mesh size itself, but it is the interval of GLL points. We decided the average mesh size $\Delta x = 8.05 \times 10^{-4}$meter, time step interval $\Delta t = 2.3 \times 10^{-8}$ second. For the number of time step updates, we set 5000 steps for having enough temporal duration for waves' passing through entire simulation domain. The mesh model used our simulations has 3,250,000 elements and 3,722,406 nodes and total number of computation nodes including GLL points became 215,320,764.

D. Interpolation of temperature field from CFD to wave simulation

The element type of TrioCFD is a tetrahedral FE with the average element size $\Delta x = 1.400 \times 10^{-3}$ meter. The temperature value is defined on the central point of each element, while the element type of SPECFEM3D is a hexahedral with average element size $\Delta x = 8.050 \times 10^{-4}$. The temperature values need to be defined on the corner nodes and every internal GLL point. Thus the CFD temperature data needs to be interpolated for importing to SPECFEM3D. For this process, we used a software library so called MEDCoupling, which is one of the libraries included in a scientific numerical method supporting tool SALOME. SALOME is the open source software developed by CEA. By using MEDCoupling, interpolation of physical fields between several types of FE or nodes is possible. In our application, the temperature field of TrioCFD is at first interpolated onto the corner nodes of SPECFEM3D elements using conservative interpolation. After being imported into SPECFEM3D, all temperature values on corner nodes are interpolated on the GLL points using tri-linear interpolation.

Figure 3 shows the interpolated temperature field on SPECFEM3D mesh. For visualization purpose, the mesh size is 10 times larger than the mesh actually used.

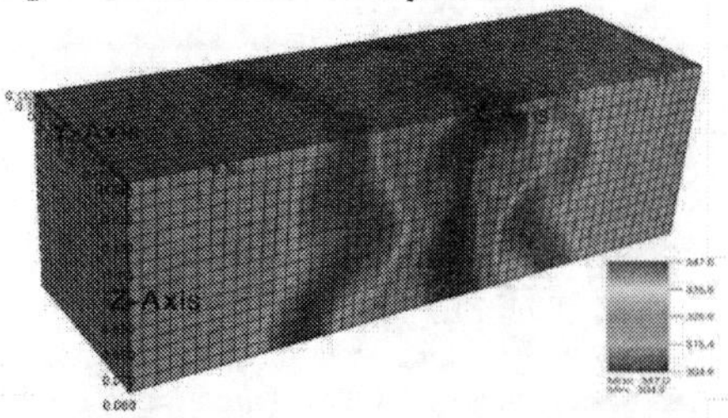

Fig. 3. the interpolated temperature field on SPECFEM3D mesh. For visualization purpose, the mesh size is 10 times larger than the mesh actually used.

E. Computational resources

In this study, all of simulations were performed on 2 super computers, CURIE (CEA TGCC) and OCCIGEN (CINES). By the helps of those advanced computer systems and gigantic capacities dedicated for scientific calculations, the computation domain was divided into 256 parts then the parallelized calculations were carried out. The average duration for 1 acoustic simulation takes about 15 ~ 20 minutes excluding mesh generation and interpolation of temperature fields.

IV. Results

Figure 4 shows the positions of the plane source (blue) and receiver surfaces (orange and red). For the visualization purpose, acoustic fields calculated here are indicated as cross-section surfaces. The acoustic fields and the effects from the temperature heterogeneity were observed for each simulation on x-y, y-z receiver planes and y-z receiver planes positioned at x = -0.105, -0.080, -0.035, 0.000, 0.035, 0.080, 0.105 meter. The y-z plane at x = 0.000 meter is the middle point of the simulation domain. The others are placed at front and back of each sodium jet.

Figure 5 shows part of acoustic fields obtained from all of the simulations. Here an acoustic field refers the maximum pressure values at each spatial point. In figure 5, acoustic fields on x-y, y-z and y-z plane (only x = 0.035, 0.105 meter) are indicated for the simulations whose middle z coordinate value of simulation domains are z = 0.04, 0.12, 0.24, 0.34 meter. Additionally, the top row is the results of homogeneous case. In the homogeneous case, the temperature of medium was set as 333.167°C which is the ambient temperature value of all heterogeneity cases. From this figure, it is found that the acoustic field of z=0.040, which is close to the outlets, and homogeneous case are quite similar. It is because that the temperature boundary at just above the outlets is almost parallel to the wave front. Also at z = 0.340 meter, the mixing of the hot and cold jets matured enough and the magnitude of temperature difference becomes small; therefore, the wave less effected by the heterogeneity of this altitude. However, at z = 0.120 and 0.240 meter, it can be seen that the acoustic fields are bent by strong heterogeneity and the transmission loss towards propagation distance becomes greater.

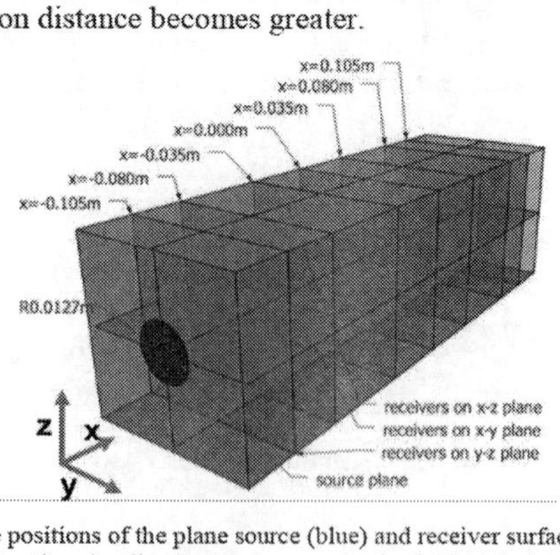

Fig. 4. The positions of the plane source (blue) and receiver surfaces (orange and red). For the visualization purpose, acoustic fields calculated here are indicated as cross-section surfaces.

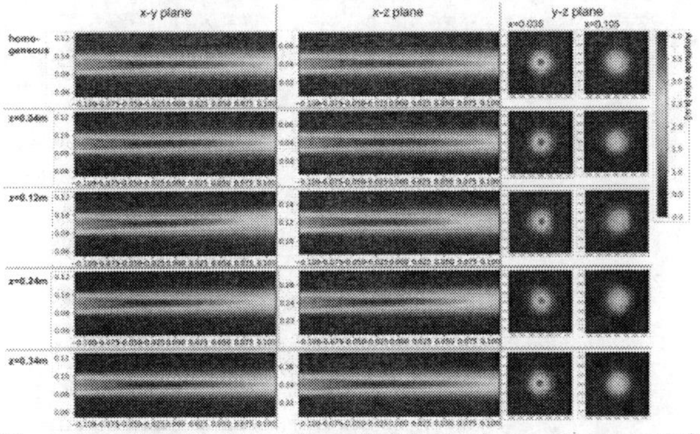

Fig. 5. Acoustic fields obtained from all of the simulations. Here an acoustic field refers the maximum pressure values at each spatial point. Acoustic fields on x-y, y-z and y-z plane (only x = 0.035, 0.105 meter) are indicated for the simulations whose middle z coordinate value of simulation domains are z = 0.04, 0.12, 0.24, 0.34 meter. Additionally, the top row is the results of homogeneous case. In the homogeneous case, the temperature of medium was set as 333.167°C which is the ambient temperature value of all heterogeneity cases.

In figure 6, values of acoustic fields on z axis of each y-z receiver planes are shown. Red lines represent the results of homogeneous cases and blue lines represent all of the results from heterogeneous cases at each z position of simulation domains. Amplitude values are normalized by the maximum amplitude (i.e. pressure) among all simulations. Also, the values on horizontal axis are normalized by the diameter of the source plane. The larger x values become, i.e. the farther the y-z receiver plane goes, the more dispersive among the amplitude curves become.

How the maximum amplitude of pressure recorded on each y-z receiver plane changes depending the positional changes for z direction of simulation domains is indicated in figure 7. At planes x=-0.105, -0.080 meter, positioning near to the plane source, maximum amplitudes take approximately constant values regardless of distances from outlets. In contrast, at the farthest and second farthest y-z planes at x = 0.105, 0.080 meter, it was found that the amplitude value of simulation domain z = 0.120 meter is approximately 10 % smaller than the other values. At the same time the second peak may be recognized at z = 0.240 meter.

It is considered that the effect of temperature heterogeneity becomes stronger under 2 conditions, that are:

- Mixing of hot and cold flow are grown enough so that the shape of temperature boundary is curved and not parallel to the emitted wave front.
- Temperature difference is kept large.

At the altitude z = 0.120 meter, it is about the distance from outlets where the temperature boundary starts to be changed while the temperature difference is relatively still high compared to the region above.

On the other hand, it seems that, at z = 0.240 meter, the temperature difference there is already small and even smaller that z = 0.230 meter. This might be caused by the temperature boundary shape became a key factor of the total effects rather than the temperature difference.

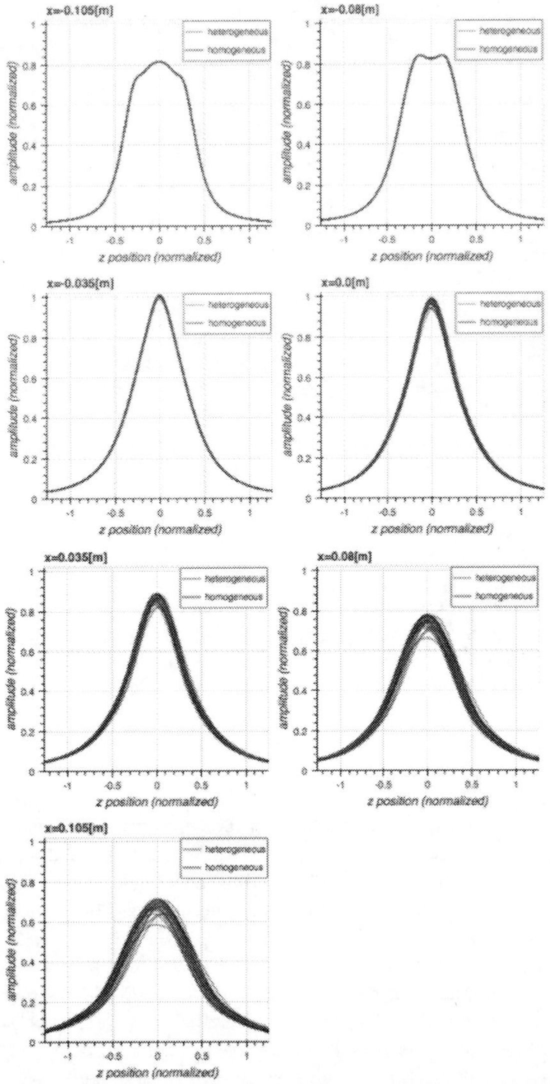

Fig. 6. Values of acoustic fields on z axis of each y-z receiver planes are shown. Red lines represent the results of homogeneous cases and blue lines represent all of the results from heterogeneous cases at each z position of simulation domains. Amplitude values are normalized by the maximum amplitude (i.e. pressure) among all simulations. The values on horizontal axis are normalized by the diameter of the source plane.

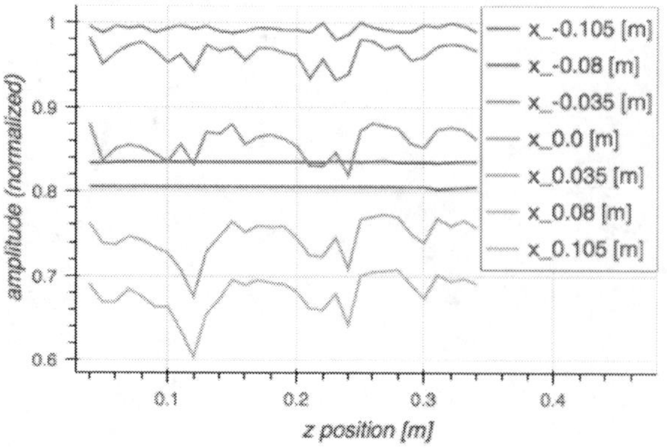

Fig. 7. The maximum amplitude of pressure recorded on each y-z receiver plane changes depending the positional changes for z direction of simulation domains. The minimum peak may be found at z = 0.12 meter on the lines x = 0.08, 0.105 meter. The amplitude value of simulation domain z = 0.120 meter is approximately 10 % smaller than the other values. The second peak is also found at z = 0.24 meter.

V. Conclusion

This study observed the effects of three-dimensional heterogeneity of temperature field on a wave propagation under a high temperature liquid sodium environment by numerical simulation using spectral element method.

The heterogeneous temperature field were imported from CFD Large-eddy simulation which was carried out based on an experiment and validated its result by comparing with the experimental result.

As a result, it was found that there is a certain relationship between a strength of effects of temperature heterogeneity on acoustic fields and a distance from the outlets of sodium jets, i.e. the state of mixing flows. More concretely, the amount of effects on maximum acoustic pressure can vary up to about 10 % at a certain altitude.

Accordingly, we may assume that the magnitude of temperature difference in a medium and the shape of temperature boundary are the key factor which decide the effect of medium heterogeneity on a wave propagation.

One single time step was examined for this study. We will continue this observation for other time shots in order to obtain an index concerning a relation between the strength of effects from medium heterogeneity and the distance from the outlets.

Subsequently, by comparing with a wave propagation with a heterogeneity generated by GRF, we will verify the availability of isotropic assumption on heterogeneity which was taken in GRF for the medium where a sodium flow exists.

Acknowledgment

References

[1] *Use of fast reactors for actinide transmutation*, 1st ed. Vienna, Austria: International Atomic Energy Agency, 1993.

[2] H. Abderrahim, P. Kupschus, E. Malambu, P. Benoit, K. Van Tichelen, B. Arien, F. Vermeersch, P. D'hondt, Y. Jongen, S. Ternier and D. Vandeplassche, "MYRRHA: A multipurpose accelerator driven system for research & development", *Nuclear Instruments and Methods in Physics Research Section A: Accelerators, Spectrometers, Detectors and Associated Equipment*, vol. 463, no. 3, pp. 487-494, 2001.

[3] T. Abram and S. Ion, "Generation-IV nuclear power: A review of the state of the science", *Energy Policy*, vol. 36, no. 12, pp. 4323-4330, 2008.

[4] P. E. Angeli, "Large-Eddy Simulation of thermal striping in WAJECO and PLAJEST experiments with TrioCFD", In Proc. NURETH-16, Chicago, USA, 2015.

[5] F. C. A. Baqué, F. Reverdy, J.-M. Augem, and J. Sibilo, "Development of Tools, Instrumentation and Codes for Improving Periodic Examination and Repair of SFRs," *Science and Technology of Nuclear Installations*, vol. 2012, pp. 1–19, 2012.

[6] J.L. Berton and G. Loyer, "Continuous monitoring of the position of two subassemblies heads of Phénix at 350 MWth power and 550 C temperature", in Conf. SMORN VII, Avignon, France, June. 19–23, 1995.

[7] M. Cavaro, C. Payan, J. Moysan and F. Baqué, "Microbubble cloud characterization by nonlinear frequency mixing", *The Journal of the Acoustical Society of America*, vol. 129, no. 5, pp. EL179-EL183, 2011.

[8] J. Jeannot, G. Rodriguez, C. Jammes, B. Bernardin, J. Portier, F. Jadot, S. Maire, D. Verrier, F. Loisy and G. Prele, "R&D program for core instrumentation improvements devoted for French Sodium fast reactors", *ANIMMA-2*, Ghent, Belgium, 2011.

[9] D. Tenchine, "Some thermal hydraulic challenges in sodium cooled fast reactors", *Nuclear Engineering and Design*, vol. 240, no. 5, pp. 1195-1217, 2010.

[10] D. Fiorina, "Application de la méthode de sommation de faisceaux gaussiens à l'étude de la propagation ultrasonore en milieu turbulent", Ph.D. disseration, École Centrale de Lyon, France, 1999.

[11] N.Gengembre, "Pencil method for ultrasonic beam computation", In *Proc. the 5th World Congress on Ultrasonics, Paris, 2003*.

[12] T. Hughes, *finite element method : linear static and dynamic finite element analysis*, 1st ed. 2000.

[13] B. Iooss, C. Lhuillier and H. Jeanneau, "Numerical simulation of transit-time ultrasonic flowmeters: uncertainties due to flow profile and fluid turbulence", *Ultrasonics*, vol. 40, no. 9, pp. 1009-1015, 2002.

[14] N. Kimura, H. Miyakoshi, H. Ogawa, H. Kamide, Y. Miyake and K. Nagasawa, "Study on convective mixing phenomena in parallel triple-jet along wall Comparison of temperature fluctuation characteristics between sodium and water", in Proc. *Nureth-11, Avignion*, France, 2005.

[15] N. Kimura, H. Kamide, P. Emonot and K. Nagasawa, "Study on thermal striping phenomena in triple-parallel jet-Transfer characteristics of temperature fluctuation in sodium and water based on conjugated numerical simulation", in Proc. NUTHOS-7, Seoul, Korea, Oct. 2008.

[16] D. Komatitsch, "Fluid–solid coupling on a cluster of GPU graphics cards for seismic wave propagation", *Comptes Rendus Mécanique*, vol. 339, no. 2-3, pp. 125-135, 2011.

[17] S. Lee, D. Komatitsch, B. Huang and J. Tromp, "Effects of Topography on Seismic-Wave Propagation: An Example from Northern Taiwan", *Bulletin of the Seismological Society of America*, vol. 99, no. 1, pp. 314-325, 2009.

[18] C. Lhuillier, D. Fiorina and D. Juvé, "Simulation of Ultrasound Propagation in a Thermally Turbulent Fluid Using Gaussian Beam Summation and Fourier Modes Superposition Techniques," in Proc. 16th ICA, Seattle, USA, 20–26 June, 1998.

[19] B. Lü, M. Darmon and C. Potel, "Stochastic simulation of the high-frequency wave propagation in a random medium", *Journal of Applied Physics*, vol. 112, no. 5, p. 054902, 2012.

[20] N. Massacret, J. Moysan, M. Ploix, J. Jeannot and G. Corneloup, "Modelling of ultrasonic propagation in turbulent liquid sodium with temperature gradient", *Journal of Applied Physics*, vol. 115, no. 20, p. 204905, 2014.

[21] M. Nagaso, J. Moysan, S. Benjeddou, N. Massacret, M. Ploix, D. Komatitsch and C. Lhuillier, "Ultrasonic thermometry simulation in a random

fluctuating medium: Evidence of the acoustic signature of a one-percent temperature difference", *Ultrasonics*, vol. 68, pp. 61-70, 2016.
[22] N. Rawlinson, H. Jurg and S. Malcolm, "Seismic ray tracing and wavefront tracking in laterally heterogeneous media." *Advances in Geophysics, vol.* 49, pp.203-273, 2008
[23] G. Spencer and M. Murty, "General Ray-Tracing Procedure", *Journal of the Optical Society of America*, vol. 52, no. 6, p. 672, 1962.
[24] V. Sobolev, "Database of Thermophysical Properties of Liquid Metal Coolants for GEN-IV" , SCK-CEN, Mol, Belgium, Scientific Rep. BLG-1069, Nov. 2010.
[25] D. Tenchine, V. Barthel, U. Bieder, F. Ducros, G. Fauchet, C. Fournier, B. Mathieu, F. Perdu, P. Quemere and S. Vandroux, "Status of TRIO_U code for sodium cooled fast reactors", *Nuclear Engineering and Design*, vol. *242*, pp. 307-315, 2012.
[26] D. Tenchine, C. Fournier and Y. Dolias, "Gas entrainment issues in sodium cooled fast reactors", *Nuclear Engineering and Design*, vol. *270*, pp. 302-311, 2014.
[27] O. Witte, M. Roth and G. Müller, "Ray tracing in random media", *Geophysical Journal International*, vol. 124, no. 1, pp. 159-169, 1996.

 https://doi.org/10.1051/epjconf/201817004001

Characterization of gamma field in the JSI TRIGA reactor

Klemen Ambrožič, Vladimir Radulović, Luka Snoj
"Jožef Stefan" Institute, Jamova cesta 39, SI-1000 Ljubljana, Slovenia

Adrien Gruel, Mael Le Guillou, Patrick Blaise
CEA, DEN, DER, SPEx, Experimental Programs Laboratory
Cadarache, F-13108 St. Paul-Lez-Durance, France

Christophe Destouches, Loïc Barbot,
CEA, DEN, DER, SPEx, Instrumentation Sensors and Dosimetry Laboratory,
Cadarache, F-13108 St. Paul-Lez-Durance, France

***Abstract*—Research reactors such as the "Jožef Stefan" Institute TRIGA reactor have primarily been designed for experimentation and sample irradiation with neutrons. However recent developments in incorporating additional instrumentation for nuclear power plant support and with novel high flux material testing reactor designs, γ field characterization has become of great interest for the characterization of the changes in operational parameters of electronic devices and for the evaluation of γ heating of MTR's structural materials in a representative reactor γ spectrum.**

In this paper, we present ongoing work on γ field characterization both experimentally, by performing γ field measurements, and by simulations, using Monte Carlo particle transport codes in conjunction with R2S methodology for delayed γ field characterization.

***Index Terms*—Gamma-rays, Nuclear electronics, Radiation hardening (electronics), Reactor instrumentation, Scintillators.**

I. Introduction

The "Jožef Stefan" Institute (JSI) TRIGA reactors is a pool type, 250 kW reactor with U Zr-H fuel elements arranged in an annular core and equipped with diverse irradiation facilities, which have during its 50 years of operation [1] been used primarily for neutron experimentation such as neutron radiography [2], neutron activation analysis [3], bulk material testing [4], [5], [6] and electronic component radiation hardness studies [7], [8], etc.

The neutron field in irradiation facilities has been thoroughly computationally characterized [9] and experimentally validated [10]. Experiments have also been included in the International Reactor Physics Experimental Evaluation Project (IRPhEP) handbook [11], and the reactor has become a reference center for neutron irradiation of detectors for the ATLAS experiment in CERN [12].

Recent increased interest in development of radiation tolerant electronic devices for operation inside Nuclear Power Plant (NPP) containment buildings [13], particle accelerators [14], aerospace applications [15], [16] and for evaluation of structural material γ heating in Material testing reactors (MTR) [17], [18] has highlighted the need for utilization of reactor irradiation positions for pure γ irradiation.

γ-rays inside a nuclear reactor can generally be divided into two groups by their mode of creation: prompt γ-rays, which are produced during reactor operation and are emitted promptly after nucleus interaction, such as fission [19], scattering [20] or radiative capture [21]. On the other hand, delayed γ-rays are emitted as a consequence of radioactive decay of fission and activation products. To date, evaluations suggest that delayed γ field contributes more then 30 % to total γ flux during reactor operation [22], [23], and is the only major source of γ rays after reactor shutdown.

Characterizing the γ field and distinguishing the time dependent (delayed) and independent (prompt) contributions presents several challenges that we aim to overcome. The characterization can roughly be divided into two parts: computational characterization and modeling and experimental validation. Novel techniques for delayed γ modeling coupling steady state Monte Carlo particle transport codes with radioactive inventory calculations codes such as the Rigorous-2-Step (R2S) [24] methodology are utilized and implemented, and numerous detector types are used for experimental validation, measuring both integral and differential quantities of interest.

The established framework will serve for routine qualification of the γ field in the JSI TRIGA irradiation facilities γ field and sample responses to radiation. The method could also be applied later on for qualification of activated material prior to decommissioning of nuclear reactors.

II. JSI TRIGA REACTOR

The JSI TRIGA reactor is controlled by 4 control rods and is equipped with numerous irradiation ports, both in-core, placed in the fuel element positions with the exception of the Triangular channel (TriC) which occupies 3 fuel element positions (Fig. 3), and ex-core, with 40 irradiation positions located inside the graphite reflector, which can be rotated into desired position by a rotary carousel. Some of the in-core and carousel irradiation positions are equipped with a fast pneumatic transfer system for sample irradiation. There are also 3 larger horizontal irradiation positions with inner radius of 15 cm. Two of them extending radially outwards from the reactor core, Radial Piercing Port (RPP) piercing the graphite reflector and Radial Beam Port (RBP) extending outwards from the graphite reflector. The Tangential Channel (TanC) horizontal irradiation position passes the reactor core tangentially. The in-core irradiation

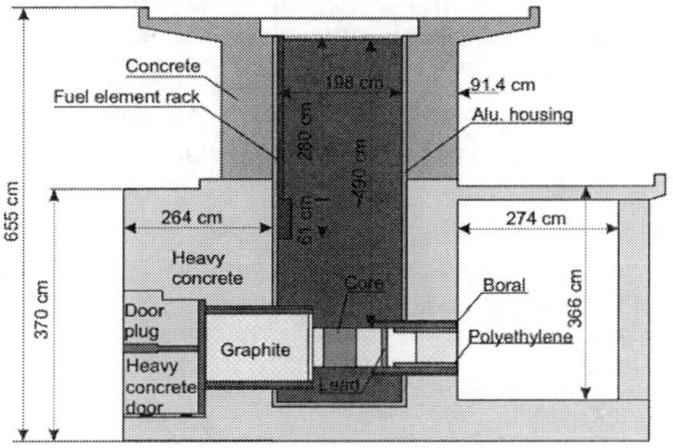

Fig. 1. JSI TRIGA reactor side-cut view schematic.

positions have graphite or aluminum spacers at the bottom, which define the sample height in relation to the fuel elements to fuel element center-height. Samples are usually lowered into positions inside aluminum or polyethylene capsules.

Fuel elements can also be extracted from the reactor

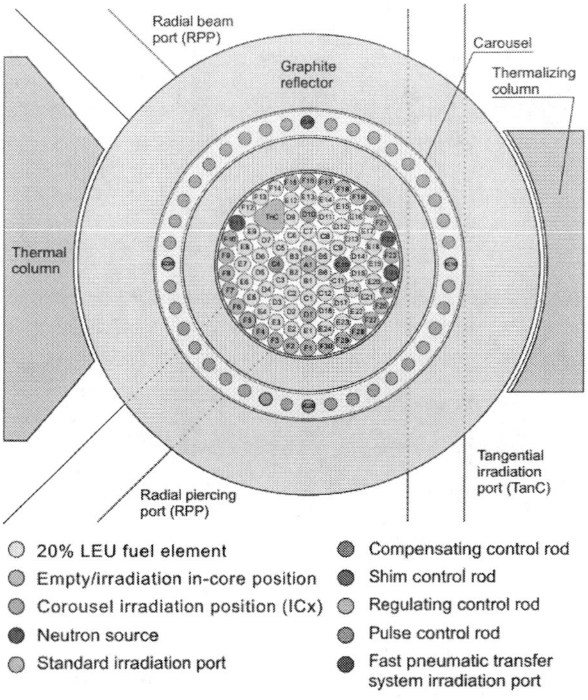

Fig. 2. JSI TRIGA reactor core schematic with core configuration no. 189.

core and relocated to the fuel element rack located at the edge of the reactor rank (Fig.1), which can hold up to 6 fuel elements. Since fuel elements are highly radioactive after reactor operation, they can be used as a viable pure γ-ray source. For pure γ irradiation with such an arrangement, an additional irradiation facility in form or an aluminum box with inner dimensions of $30\,\mathrm{cm}\times20\,\mathrm{cm}\times20\,\mathrm{cm}$, submerged to close proximity of the fuel elements, with a cable guide tube leading to the reactor pool surface has been constructed especially for γ irradiation of larger electronic components and assemblies (Fig. 4).

The existing irradiation positions were primarily used for neutron irradiation of samples, and have only recently been utilized for γ irradiation after reactor shutdown. Although γ dose-rates after reactor shutdown exceed $1\times10^{4}\,\mathrm{Gy\,h^{-1}}$, the γ spectrum is time dependent and difficult to measure with a spectrometer due to the high dose-rate, further investigation in characterizing the γ

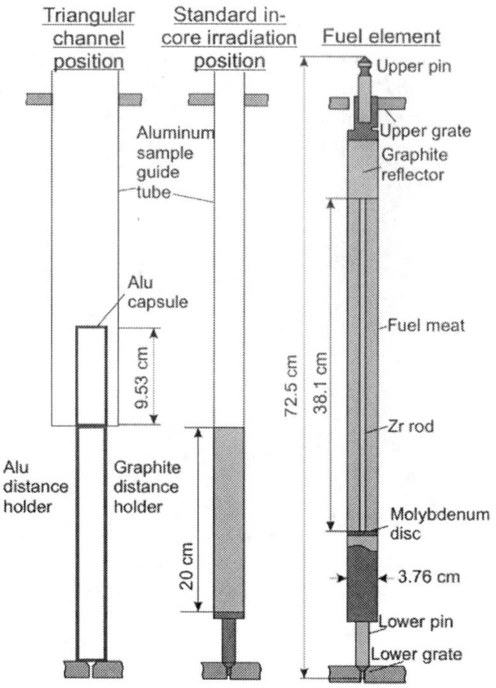

Fig. 3. JSI TRIGA fuel elements and in-core irradiation position schematics.

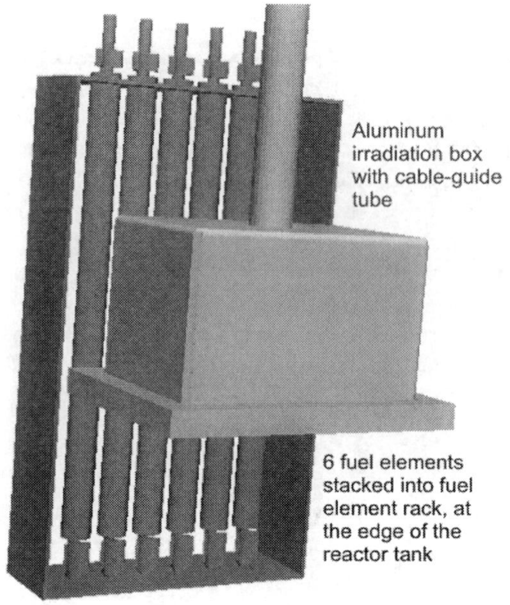

Fig. 4. Schematic of 6 fuel elements stacked in a rack, with Al. irradiation box at close proximity.

field is required both computationally and by use of different detectors.

III. Gamma field measurements

The measurements of the γ field inside a strong mixed neutron-γ field presents several difficulties such as discrimination between neutron and γ contributions, and further, discriminating between prompt and delayed γ field, where the latter is also dependent on reactor operational history. In case of the JSI TRIGA, where the reactor power ranges from $10\,\mathrm{W}$ to $2.5\times10^5\,\mathrm{W}$, and taking into account background radiation, which is in order of few $\mathrm{Gy\,h^{-1}}$, detectors with measurement ranges spanning at least 6 orders of magnitude must be used. Ionization and fission chambers, radiation sensing field effect transistors (RadFETs) [25] and thermoluminescent dosimeters (TLD) using novel readout and annealing techniques [26] have proven to be prime candidates for γ field measurements inside a reactor.

Discrimination between neutron and γ contributions is usually performed by using similar detectors, which have different sensitivities to different particle types, as is the case with fission [27] and ionization chambers [28], [29], as well as for TLD detectors, where some can be loaded with natural lithium and others with enriched ^{7}Li. However, qualifying the RadFETs and their temperature dependence [30] is still in progress, and solutions such as different oxide layer thickness and encapsulation of detectors with a neutron filter such as Cd, or by increasing γ production by applying a thin layer of B on the RadFET surface [31] are under investigation.

Radiation detectors measure radiation by charge collection (IC) or trapping (TLD, OSLD), which is released by radiation interaction with matter, so charged particle equilibrium (CPE) must be established in order for the detector to measure the correct quantity [32]. The measured quantity must therefore previously be qualified in a stable γ source, such as ^{137}Cs, ^{60}Co, X-ray and others, preferably high energy sources. CPE is achieved by encapsulating the sensitive part of the detector with an equilibrator part of sufficient thickness [33]. The material of the equilibrator dictates the detector response function, since the actual dose on the equilibrator material is measured. γ ray energies inside a reactor reaching up to $10\,\mathrm{MeV}$ and further must also be considered. Although the average γ energy is in order of an MeV and is comparable with ^{60}Co spectrum, pair production becomes the dominant interaction process at higher energies instead

https://doi.org/10.1051/epjconf/201817004001

of Compton effect [32], and therefore larger equilibrator thicknesses are required.

While these detectors can be used for a range of reactor powers, they still only measure integral responses, and give no information on incident γ energy. To this end a novel, miniature, scintillator crystal based spectrometer is under development. Currently, optimizations on scintillation crystal size and readout techniques are being performed, in order not to oversaturate the crystal and read piled-up pulses, but still have a crystal large enough, for γ-rays to deposit all of their energy inside the scintillation crystal. Techniques on resolving pile-up pulses [34] somewhat relax these constraints.

A joint JSI and French Atomic Energy and Alternative Energies Commission (CEA) γ measurement campaign has already been performed at the JSI TRIGA reactor, where all of the above mentioned detectors (or development versions for detectors under development) were utilized for γ-ray measurements. A compact core configuration has been established in order to have a homogeneous neutron field, and reactor operation was planned beforehand, gradually increasing the reactor power from zero power, with extracted neutron source, to establish baseline background signal, to full power in a series of runs, increasing the power exponentially. Special detector holders were constructed for positioning the detectors in centers of their respective irradiation positions, in order to minimize experimental uncertainty. Two fission chambers were also inserted symmetrically to the reactor core center for additional reactor power reading and determining the flux redistribution due to control rod positions. One of the ionization chambers was positioned in a fixed position, while others were used for axial and radial γ profile measurements. A RadFET detector and scintillation spectrometer were also utilized, and were lowered in the carousel positions, in order not to oversaturate them or to minimize impact on the detector. During the operations, several sets of TLDs in aluminum equilibrators provided by the CEA were enclosed inside a polyethylene capsule and lowered into 3 different positions: in-core Central channel (A1) and F26 and carousel IC40 position at different reactor powers.

On-line measurements during reactor transients such as power increase (Fig.5) and reactor scram (Fig.6) are of special interest, since delayed γ flux time dependency can directly be extracted, by appropriate IC and FC detector signal weighting and subtraction.

These experiments will be repeated again, in a similar fashion, aiming to replicate the conditions of the previous experimental campaign, although using another set of TLD detectors without equilibrators, provided by Institute of Nuclear Physics, Polish Academy of Sciences (IJF-PAN), different type of RadFET detectors and an improved version of the scintillation spectrometer, with fission and ionization cells being the same. The aim is, to reproduce these measurements several times, with a large number of detectors in order to minimize measurement uncertainty. The obtained experimental results will serve for validation of the delayed γ computational scheme.

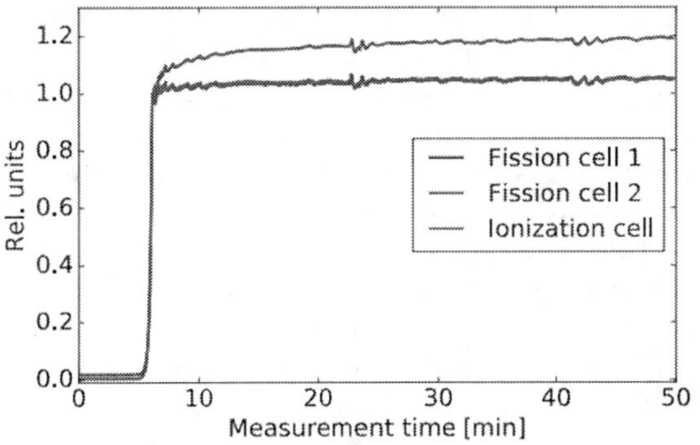

Fig. 5. Fission and ionization cell signal, normalized to value after reactor start-up to 4 kW. Fission chamber signals overlap, while ionization chamber signal increases with time due to neutron activation and delayed γ-rays.

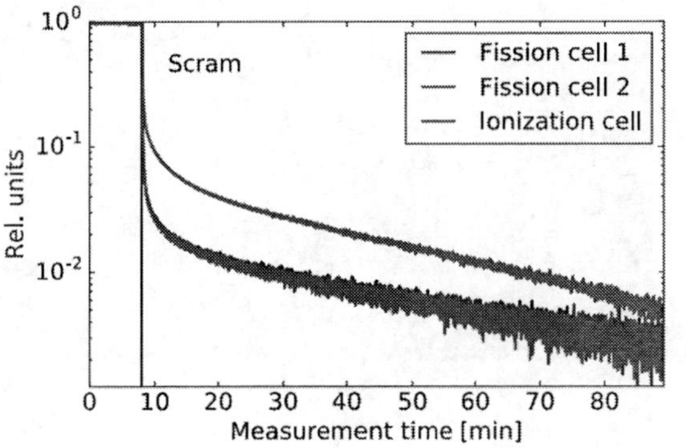

Fig. 6. Fission and ionization cell signal, after scram from steady-state 200 W, showing the delayed γ emitters decaying, and contributing to ionization chamber signal. Online measurements for long periods after a scram can provide valuable information about delayed γ, even for longer-lived activation products.

IV. Computational Modeling

Computational modeling of the γ field inside a nuclear reactor needs to be able to couple transport of neutrons and γ-rays for kerma approximation. We used the Monte Carlo particle transport code MCNP v6.1 (MCNP) [35] and ENDF/B-VII.0 [36] for prompt γ irradiation. The reactor computational model is based on an experimental benchmark computational model [37], with added details, such as fuel holding pins, additional irradiation positions, reactor instrumentation to most accurately reflect the real state of the reactor. Since MCNP only enables steady state eigenvalue calculations, only the prompt part of the γ field is calculated directly by MCNP. The γ flux and anbient dose equivalents, derived by the use of ICRP-21 [38] and ANSI-ANS6.1.1 ambient dose equivalent factors were calculated throughout the reactor. In the present work, core loading pattern 189 (Fig.2) was simulated, with transient and safety control rod fully withdrawn, regulating rod at position 382 and shim rod in position 350 and TriC inserted into the core. Results in Fig. 7 shows only prompt γ contribution, with measurements suggesting an increase $\approx$30 %, once the delayed γ are accounted for. For accurate estimation of the delayed γ contribution, an R2S methodology Python [39] script was developed, coupling MCNP and FISPACT-II [40] radioactive inventory calculation code, which has only been applied to fusion applications [41], [42] and has to our knowledge never been applied to a fission reactor before. The model geometry is divided by a 3D mesh, where the neutron flux and spectra and homogenized material compositions are calculated in each mesh voxel. Inventory calculations are performed using these data and reactor operation parameters, such as reactor power, operation duration and cool-down to calculate radioactive isotope inventory at each time of interest. Delayed particle spectra are calculated for each voxel of the mesh, and secondary particle transport calculations performed with MCNP and a source routine developed at KIT [43].

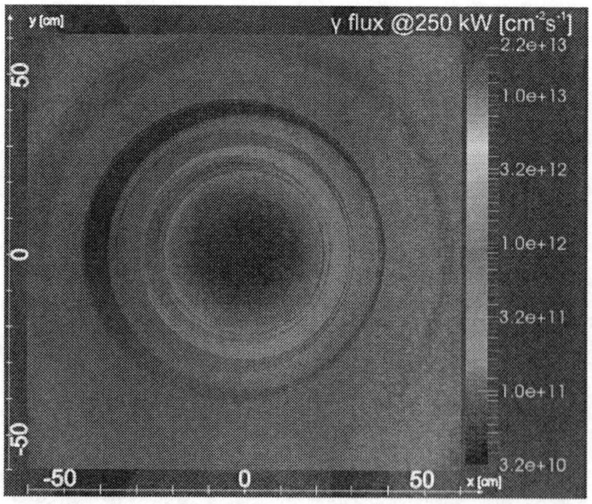

Fig. 7. Prompt γ flux in the JSI TRIGA core at fuel element center height @ 250 kW.

To accurately reproduce the measurements obtained experimentally, one must consider renormalization of computational results to the reactor power [44], as well as flux redistribution [45] due to control rod insertion and its effects on reactor power readout instrumentation. In addition, secondary particles other then γ-rays should also be transported, as they may produce additional γ-rays due to their high energy. Special care must also be taken for energy and charge deposition inside the detectors and electron and positron transport must be considered as well, since the detector size is usually smaller then the characteristic dimension required to achieve CPE.

Simulating different particle types, especially charged particles, required in order to accurately reproduce detector responses and additional delayed γ generation, is however computationally very intensive, and special care should be taken in order to optimize accuracy vs. computational time.

V. Conclusions

With the development of novel, high flux reactors and the need for radiation tolerant electronics for nuclear and aerospace technology, γ field effects studies on reactor component heating and changes in operational parameters of electronic components and assemblies, γ field characterization of JSI TRIGA irradiation facilities will enable further studies and development in that area. With a thoroughly characterized γ field, sample characteristics changes will be more accurately qualified.

An R2S methodology code, coupling MCNP and FISPACT-II codes for delayed γ field calculations, has been developed an with addition of prompt γ field calculations performed by the MCNP code, a generalized framework for reactor γ field calculations will be established. In addition, results from experimental γ measurement campaigns at the JSI TRIGA reactor will serve for code validation and improvement, as well as further development of novel detector types, such as before mentioned γ spectrometer. Due to the novelty of R2S methodology use in the irradiation facilities of a

fission reactor, special care should be taken to obtain experimental results of benchmark quality, on which other R2S methodology codes could also be tested and verified.
The established computational framework will not only serve for qualification of sample responses under γ irradiation, but could also serve for qualification of radioactive inventory in reactor structural materials, which will become of great interest after reactor decommissioning and long-term storage, leading to waste storage space optimization and minimizing the associated cost.

REFERENCES

[1] B. Smodiš and L. Snoj, "Utilization and Application of the Slovenian TRIGA Reactor," "Jožef Stefan" Institute, Ljubljana, Slovenia, Tech. Rep., 2011. [Online]. Available: http://www-pub.iaea.org/MTCD/Publications/PDF/P1575_CD_web/datasets/papers/A29\%20Smodis.pdf

[2] J. Rant, Z. Milič, J. Istenič, T. Knific, I. Lengar, and A. Rant, "Neutron radiography examination of objects belonging to the cultural heritage," *Applied Radiation and Isotopes*, vol. 64, no. 1, pp. 7 – 12, 2006. [Online]. Available: http://www.sciencedirect.com/science/article/pii/S096980430500206X

[3] A. R. Byrne, "Neutron activation analysis using TRIGA," "Jožef Stefan" Institute, Ljubljana, Slovenia, Italy, Tech. Rep., 1972, iNIS-IT–09N0001.

[4] E. Huseynov, A. Garibov, and R. Mehdiyeva, "Effect of neutron flux, temperature and frequency on the permittivity of nanocrystalline silica," *International Journal of Modern Physics B*, vol. 28, no. 30, p. 1450213, 2014.

[5] ——, "Influence of neutron irradiation and temperature on the electric conductivity of SiO2 nanoparticles," *Journal of Electrostatics*, vol. 74, pp. 73 – 78, 2015. [Online]. Available: http://www.sciencedirect.com/science/article/pii/S0304388615000029

[6] ——, "TEM and SEM study of nano SiO2 particles exposed to influence of neutron flux," *Journal of Materials Research and Technology*, vol. 5, no. 3, pp. 213 – 218, 2016. [Online]. Available: http://www.sciencedirect.com/science/article/pii/S2238785415001106

[7] I. Mandić, V. Cindro, A. Gorišek, G. Kramberger, and M. Mikuž, "Online integrating radiation monitoring system for the ATLAS detector at the Large Hadron Collider," *IEEE Transactions on Nuclear Science*, vol. 54, no. 4, pp. 1143–1150, Aug 2007.

[8] G. Kramberger, V. Cindro, I. Mandić, M. Mikuž, and M. Zavrtanik, "Determination of detrapping times in semiconductor detectors," *Journal of Instrumentation*, vol. 7, no. 04, p. P04006, 2012.

[9] L. Snoj, G. Žerovnik, and A. Trkov, "Computational analysis of irradiation facilities at the JSI TRIGA reactor," *Applied Radiation and Isotopes*, vol. 70, no. 3, pp. 483 – 488, 2012. [Online]. Available: http://www.sciencedirect.com/science/article/pii/S0969804311005963

[10] L. Barbot, C. Domergue, S. Braud, C. Destouches, J. F. Villard, L. Snoj, Štancar Ž., V. Radulović, and A. Trkov, "Neutron field characterization of irradiation locations applied to the slovenian TRIGA reactor," in *Advancements in Nuclear Instrumentation Measurement Methods and their Applications (ANIMMA), 2013 3rd International Conference on*, June 2013, pp. 1–5.

[11] E. Sartori and J. Blair Briggs, "IRPhEP-handbook, International Handbook of Evaluated Reactor Physics Benchmark Experiments," Nuclear Energy Agency of the OECD (NEA), Mar 2008.

[12] L. Snoj and B. Smodiš, "45 Years of TRIGA Mark II in Slovenia," in *Proc. International Conference Nuclear Energy for New Europe 2011*, 2011, article no. 401.

[13] A. F. Fernandez, A. I. Gusarov, B. Brichard, S. Bodart, K. Lammens, F. Berghmans, M. Decreton, P. Megret, M. Blondel, and A. Delchambre, "Temperature monitoring of nuclear reactor cores with multiplexed fiber bragg grating sensors," *Optical Engineering*, vol. 41, no. 6, pp. 1246–1254, 2002. [Online]. Available: http://dx.doi.org/10.1117/1.1475739

[14] A. Serrano, J. Andrés, N. P. Barradas, F. M. Cardeira, J. Casas-Cubillos, A. C. Fernandes, F. Peláez, P. Gomes, I. C. Gonçalves, A. Hernández Cachero *et al.*, "Radiation tests on commercial instrumentation amplifiers, analog switches & DAC's," in *Proceedings of the 8th Conference on Electronics for LHC Experiments.* CERN Publishing Service, 2001, pp. 117–121.

[15] R. C. Lacoe, J. V. Osborn, R. Koga, S. Brown, and D. C. Mayer, "Application of hardness-by-design methodology to radiation-tolerant ASIC technologies," *IEEE Transactions on Nuclear Science*, vol. 47, no. 6, pp. 2334–2341, Dec 2000.

[16] S. Duzellier, "Radiation effects on electronic devices in space," *Aerospace Science and Technology*, vol. 9, no. 1, pp. 93 – 99, 2005. [Online]. Available: http://www.sciencedirect.com/science/article/pii/S1270963804001129

[17] H. Amharrak, J. D. Salvo, A. Lyoussi, A. Roche, M. Masson-Fauchier, J. C. Bosq, and M. Carette, "Analysis and recent advances in gamma heating measurements in MINERVE facility by using TLD and OSLD techniques," in *2011 2nd International Conference on Advancements in Nuclear Instrumentation, Measurement Methods and their Applications*, June 2011, pp. 1–9.

[18] M. Lemaire, C. Vaglio-Gaudard, A. Lyoussi, and C. Reynard-Carette, "For a better estimation of gamma heating in experimental reactors and devices: Stakes and work plan from calculation methods to nuclear data," in *2013 3rd International Conference on Advancements in Nuclear Instrumentation, Measurement Methods and their Applications (ANIMMA)*, June 2013, pp. 1–6.

[19] F. Pleasonton, R. L. Ferguson, and H. Schmitt, "Prompt gamma rays emitted in the thermal-neutron-induced fission of u 235," *Physical Review C*, vol. 6, no. 3, p. 1023, 1972.

[20] R. B. Day, "Gamma rays from neutron inelastic scattering," *Phys. Rev.*, vol. 102, pp. 767–787, May 1956. [Online]. Available: https://link.aps.org/doi/10.1103/PhysRev.102.767

[21] M. Lone, R. Leavitt, and D. Harrison, "Prompt gamma rays from thermal-neutron capture," *Atomic Data and Nuclear Data Tables*, vol. 26, no. 6, pp. 511 – 559, 1981. [Online]. Available: http://www.sciencedirect.com/science/article/pii/0092640X8190005X

[22] D. Fourmentel, V. Radulović, L. Barbot, J. F. Villard, G. Žerovnik, L. Snoj, M. Tarchalski, K. Pytel, and F. Malouch, "Delayed Gamma Measurements in Different Nuclear Research Reactors Bringing Out the Importance of Their Contribution in Gamma Flux Calculations," *IEEE Transactions on Nuclear Science*, vol. 63, no. 6, pp. 2875–2879, Dec 2016.

[23] V. Radulović, D. Fourmentel, L. Barbot, J.-F. Villard, T. Kaiba, Ž. Gašper, and L. Snoj, "Measurements of miniature ionization chamber currents in the JSI TRIGA Mark II reactor demonstrate the importance of the delayed contribution to the photon field in nuclear reactors," *Nuclear Instruments and Methods in Physics*

Research Section A: Accelerators, Spectrometers, Detectors and Associated Equipment, vol. 804, pp. 149–154, 2015.

[24] A. Davis, "Radiation Shielding of Fusion Systems," Ph.D. dissertation, University of Birmingham, 2010.

[25] F. Ravotti, "Development and characterization of radiation monitoring sensors for high energy physics experiments of the CERN LHC accelerator," Ph.D. dissertation, L'UNIVERSITE MOTPELLIER II, November 2006.

[26] B. Obryk, P. Batistoni, S. Conroy, B. D. Syme, S. Popovichev, I. E. Stamatelatos, T. Vasilopoulou, and P. Bilski, "Thermoluminescence measurements of neutron streaming through JET Torus Hall ducts," *Fusion Eng. Des.*, vol. 89, no. 9-10, pp. 2235–2240, 2013.

[27] S. Normand, P. Delacour, B. Lescop, L. Laribiere, T. Nguyen, C. Blandin, and J.-C. Trama, "A new small fission chamber for in core and wide range neutron flux measurement," CEA, Tech. Rep., 2006.

[28] U. Chica, M. Anguiano, M. Vilches, and A. M. Lallena, "Quality indexes based on water measurements for low and medium energy X-ray beams: A theoretical study with PENELOPE," *Medical Physics*, vol. 41, no. 1, pp. 012 101–n/a, 2014, 012101. [Online]. Available: http://dx.doi.org/10.1118/1.4836556

[29] D. Fourmentel, P. Filliatre, J. Villard, A. Lyoussi, C. Reynard-Carette, and H. Carcreff, "Measurement of photon flux with a miniature gas ionization chamber in a material testing reactor," *Nuclear Instruments and Methods in Physics Research Section A: Accelerators, Spectrometers, Detectors and Associated Equipment*, vol. 724, pp. 76 – 82, 2013. [Online]. Available: http://www.sciencedirect.com/science/article/pii/S0168900213005214

[30] J. Hofman, A. Holmes-Siedle, R. Sharp, and J. Haze, "A method for in-situ, total ionising dose measurement of temperature coefficients of semiconductor device parameters," *IEEE Transactions on Nuclear Science*, vol. 62, no. 6, pp. 2525–2531, Dec 2015.

[31] J. K. Shultis and D. S. McGregor, "Efficiencies of coated and perforated semiconductor neutron detectors," *IEEE Transactions on Nuclear Science*, vol. 53, no. 3, pp. 1659–1665, June 2006.

[32] F. H. Attix, *Introduction to radiological physics and radiation dosimetry*. John Wiley & Sons, 2008.

[33] H. Amharrak, J. D. Salvo, A. Lyoussi, P. Blaise, M. Carette, A. Roche, M. Masson-Fauchier, A. Pepino, and C. Reynard-Carette, "Development and optimization of nuclear heating measurement techniques in zero power experimental reactors," *IEEE Transactions on Nuclear Science*, vol. 61, no. 5, pp. 2515–2526, Oct 2014.

[34] M. Miklavec, B. Löher, D. Savran, R. Novak, S. Širca, and M. Vencelj, "Pile-up correction techniques for real-time dosimetry in photon radiotherapy," in *Nuclear Science Symposium and Medical Imaging Conference (NSS/MIC), 2012 IEEE*. IEEE, 2012, pp. 3880–3882.

[35] J. T. Goorley, M. R. James, T. E. Booth, F. B. Brown, J. S. Bull, L. J. Cox, J. W. J. Durkee, J. S. Elson, M. L. Fensin, R. A. I. Forster, and et al., *Initial MCNP6 Release Overview - MCNP6 version 1.0*. Los Alamos National Laboratory (LANL), Jun 2013, Report number: LANL Report LA-UR-13-22934. [Online]. Available: http://www.osti.gov/scitech/servlets/purl/1086758

[36] M. Chadwick, P. Obložinskỳ, M. Herman, N. Greene, R. McKnight, D. Smith, P. Young, R. MacFarlane, G. Hale, S. Frankle *et al.*, "ENDF/B-VII. 0: Next Generation Evaluated Nuclear Data Library for Nuclear Science and Technology," *Nuclear data sheets*, vol. 107, no. 12, pp. 2931–3060, 2006.

[37] R. Jeraj and M. Ravnik, "Reactor U(20)-Zirconium Hydride Fuel Rods In Water With Graphite Reflector, International Handbook of Evaluated Criticality Safety Benchmark Experiments," NEA/NSC/DOC/(95) 03/III, Tech. Rep., 1999.

[38] J. Hale, "Data for Protection Against Ionizing Radiation from External Sources: Supplement to ICRP Publication 15. ICRP Publication 21. By The International Commission on Radiological Protection. paper, pp.100, Elmsford, N.Y., Pergamon Press, 1973." *Radiology*, vol. 111, no. 3, pp. 716–716, jun 1974. [Online]. Available: http://dx.doi.org/10.1148/111.3.716b

[39] G. van Rossum and F. L. Drake, *The Python Language Reference Manual*. Network Theory Ltd., 2011.

[40] J.-C. Sublet C, J. W. Eastwood, J. G. Morgan, M. Fleming, and M. R. Gilbert, *The FISPACT-II User Manual*, issue 7 ed., UK Atomic Energy Authority, Culham Science Centre, Abingdon, Oxfordshire, OX14 3DBr, 9 2015.

[41] P. Sauvan, J. P. Cataln, F. Ogando, R. Jurez, and J. Sanz, "Development of the R2SUNED Code System for Shutdown Dose Rate Calculations," *IEEE Transactions on Nuclear Science*, vol. 63, no. 1, pp. 375–384, Feb 2016.

[42] P. Batistoni, M. Angelone, L. Petrizzi, and M. Pillon, "Benchmark experiment for the validation of shut down activation and dose rate in a fusion device," *Journal of Nuclear Science and Technology*, vol. 39, no. sup2, pp. 974–977, 2002.

[43] D. Leichtle and E. Relson. (2013) Source rutine for cartesian steady state gamma source. Karsruhe Institue of Technology. [Online]. Available: https://github.com/svalinn/r2s-act/blob/master/mcnp_source/source_gamma.F90

[44] G. Žerovnik, M. Podvratnik, and L. Snoj, "On normalization of fluxes and reaction rates in MCNP criticality calculations," *Annals of Nuclear Energy*, vol. 63, pp. 126 – 128, 2014. [Online]. Available: http://www.sciencedirect.com/science/article/pii/S0306454913004003

[45] Ž. Štancar, L. Snoj, and L. Barbot, "Analysis of the TRIGA Mark II Research Reactor Ex-Core Detector Response," in *2015 24th International Conference Nuclear Energy for New Europe*. Društvo jedrskih strokovnjakov, September 2015, pp. 31–32.

EPJ Web of Conferences **170**, 04002 (2018) https://doi.org/10.1051/epjconf/201817004002
ANIMMA 2017

Last Improvements of the CALMOS Calorimeter Dedicated to Thermal Neutron Flux and Nuclear Heating Measurements inside the OSIRIS Reactor

H. Carcreff, L. Salmon, V. Lepeltier, J.M. Guyot E. Bouard

***Abstract*—Nuclear heating inside an MTR reactor needs to be known in order to design and to run irradiation experiments which have to fulfill target temperature constraints. To improve the nuclear heating knowledge, an innovative calorimetric system CALMOS has been studied, manufactured and tested for the 70MWth OSIRIS reactor operated by CEA. This device is based on a mobile calorimetric probe which can be inserted in any in-core experimental location and can be moved axially from the bottom of the core to 1000 mm above the core mid-plane. Obtained results and advantages brought by the first CALMOS-1 equipment have been already presented. However, some difficulties appeared with this first version. A thermal limitation in cells did not allow to monitor nuclear heating up to the 70 MW nominal power, and some significant discrepancies were observed at high heating rates between results deduced from the calibration and those obtained by the "zero method". Taking this feedback into account, the new CALMOS-2 calorimeter has been designed both for extending the heating range up to 13W.g^{-1} and for improving the "zero method" measurement thanks to the implementation of a 4-wires technique. In addition, the new calorimeter has been designed as a real operational measurement system, well suited to characterize and to follow the radiation field evolution throughout the reactor cycle. To meet this requirement, a programmable system associated with a specific software allows automatic complete cell mobility in the core, the data acquisition and the measurements processing. This paper presents the analysis of results collected during the 2015 comprehensive measurement campaign. The 4-wires technique was tested up to around a 4 W.g^{-1} heating level and allowed to quantify discrepancies between "zero" and calibration methods. Thermal neutron flux and nuclear heating measurements from CALMOS-1 and CALMOS-2 are compared. Thermal neutron flux distributions, obtained with the Self-Power Neutron Detector suited to the CALMOS-2 calorimetric probe, are compared with those obtained with current devices. This campaign allowed to highlight advantages brought by the human machine interface automation, which deeply refined the profiles definition. Finally, the decay of the reactor residual power after shutdown could be performed after shutdown, demonstrating the ability of such type of calorimeter to follow the heating level whatever the thermohydraulic conditions, forced or natural convection regimes.**

***Index Terms*— Nuclear heating, calorimetry, in-core measurements, OSIRIS reactor, calorimetric cell, sample**

I. Introduction

Nuclear heating inside an MTR reactor has to be known in order to predict sample temperatures reached during an irradiation experiment and check that it fulfills experimental requirements, but also for safety reasons, to demonstrate that temperatures will not exceed material limits. This determination is usually obtained through calorimetry [1, 2]. Homemade calorimeters, working in permanent mode, have been used for many years for the qualification of irradiation locations in the pool type OSIRIS reactor, which were of five stage calorimeters [1], piled-up along the core height. Therefore, they provided only five measurement points to fit the heating profile. These calorimeters were reliable but had inherent drawbacks, coming essentially from the static nature of this technology. The main goal of the CALMOS R&D program was to transpose the current in-core static measurement system to a mobile one, in order to plot the nuclear heating distribution as finely as required. This innovative calorimetric system was studied and developed at the Saclay research center. The so-called CALMOS-1 device, equipped with a displacement system suited to in-core measurements was tested in 2014. Results obtained during the cell development, modelling, mock-ups irradiation, and first results obtained with the complete system up to 10 W.g^{-1} heating level have been already detailed and presented [3, 4]. However, major problems came along while operating CALMOS-1. Indeed, because of safety rules, a thermal limitation does not allow to perform measurements at nominal power in any core locations. Secondly, some significant discrepancies are

This paragraph of the first footnote will contain the date on which you submitted your paper for review. It will also contain support information, including sponsor and financial support acknowledgment. For example, "This work was supported in part by the U.S. Department of Commerce under Grant BS123456."

Hubert .Carcreff is with the CEA/DEN/DANS/DM2S/SERMA, 91191, FRANCE (e-mail: Hubert.Carcreffr@ cea.fr).

Laurent.Salmon is with the CEA/DEN/DANS/DM2S/STMF, 91191, FRANCE (e-mail: Laurent.Salmon@ cea.fr).

V.Lepeltier is with the CEA/DRF/PSAC/USPS/SPRE/SRR, 91191, FRANCE (e-mail: Valerie.Lepeltier@ cea.fr).

Jean-Marie Guyot is with the CEA/DEN/DDCC/UADS/SEROS, 91191, FRANCE (e-mail: Jean-Marie.Guyot@ cea.fr).

Eric Bouard is with the CEA/DEN/DDCC/UADS, Gif sur Yvette, 91191, FRANCE FRANCE (e-mail: Eric.Bouard@cea.fr)

observed at high heating rates between results deduced from the preliminary calibration and those obtained by the "zero method" which is thought to be an absolute measurement. The CALMOS-1 design is not suited for investigating further on. In addition, the displacement system is not equipped with any automation functionalities and was operated manually, hence requiring the presence of an operator. Based on this feedback, a new CALMOS-2 calorimeter has been designed with significant improvements brought to the calorimetric cell and on the displacement system. The radiation field characterization was made in all in-core locations and for many core configurations, measuring nuclear heating and thermal neutron flux up to respectively 12 $W.g^{-1}$ and 2.5 10^{14} $n.cm^{-2}.s^{-1}$.

II. CALMOS-2 Calorimetric Cell Design

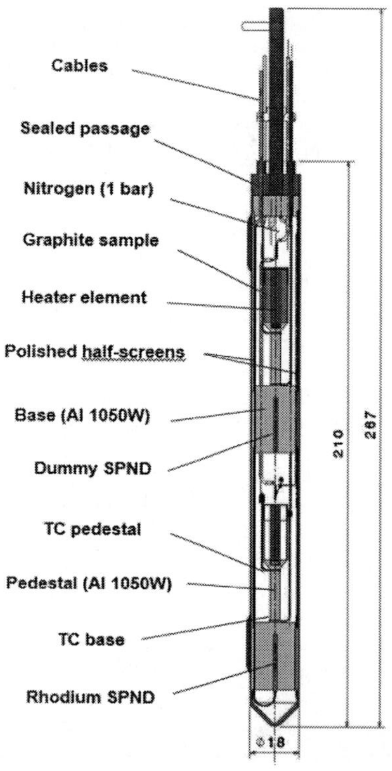

Fig 1 – Cross section of the CALMOS-2 calorimetric probe (dimensions in mm)

The global design is similar to CALMOS-1 (Fig.1). The key point of this design is still to stack up two cells (one empty and one filled with the graphite sample) coaxially inside the same external sleeve and then to move the whole probe with the help of the displacement system. Structural modifications brought to the first version are discussed hereafter. It is thus possible to obtain heating rate data at any level in the core and therefore to determine accurately the heating profile inside and above the core (which was not possible with previous static calorimeters). Description, working modes and advantages of such calorimeter have already been presented and discussed [3].

III. Reminder of the Probe Calibration and the Zero Method

Prior to any measurement, a calibration is performed under non irradiating conditions and natural convection cooling. Assuming that the calorimeter has a perfect linear response, we define a K calibration coefficient in $W.g^{-1}.°C^{-1}$ [3]. If the sample and reference cells are in a same irradiation field, the total heating rate E_T ($W.g^{-1}$) in graphite is given by:

$$E_T = \Delta\Delta T.\ K.\ K_c(E) \qquad (1)$$

Where $\Delta\Delta T$ is defined as $\Delta\Delta T = \Delta T$ sample – ΔT reference. ΔT refers to the temperatures difference between the pedestal and base thermocouples of each cell (Fig.1) [3]. $K_C(E)$ is a correction coefficient taking into account inherent nonlinearities due to heat leakages by radiation, conduction in gas and the aluminum conductivity dependence with temperature [5]. In CALMOS-2, Kc(E) is evaluated both by using the zero method (up to around 4 $W.g^{-1}$) and with the help of the CAST3M Finite Element Model [4, 6]. Once the probe is calibrated, the heating rate measurement in the radiation field is performed in two steps [3]. Starting from a given static position in the radiation field, temperatures of the sample cell are recorded. Then, taking advantage of the moving system, the whole calorimeter is shifted (upward or downward) and temperatures of the reference cell are recorded when this latter is located at the same position as the sample was. When cells are at the same level, $\Delta\Delta T$ is measured and heating is deduced using (1). The step by step displacement of the calorimeter, allows plotting the whole profile. An alternative measurement can be performed by the "zero method". Such procedure aims to match with an absolute measurement [3, 4]. The two cells being at the same altitude in the core, the nuclear energy deposit inside the sample can be evaluated by adjusting an electrical power applied to the reference cell heater (simulating the energy deposit by radiation) so as to equalize ΔT sample and ΔT reference (see Fig.7a). In that case the heating rate E_T inside the sample is given by:

$$E_T = \frac{W_e}{M_E}.K_0 \qquad (2)$$

Where We is the dissipated electric power (W) in the reference cell and M_E the graphite sample mass (g). As there are inherent small differences of thermal transfer capabilities between the two cells (slopes in $°C.W^{-1}$), we add a K_0 correction factor [3, 4].

IV. Major Improvements of CALMOS-2

A - Zero Method by a 4-wires Measurement Technique

A first major improvement in CALMOS-2 avoids any assumption about the resistance value of the heater element embedded in the reference cell. In balance process (zero method), the nuclear energy deposit in the sample is evaluated by adjusting the electrical power applied inside the reference cell heater element so as to equalize ΔT sample and ΔT reference. Therefore, calculations of the W_E energy deposit assume that the effective resistance r of the heater element is well known, and more importantly, that r does not change with ageing, temperature or irradiation conditions. To suppress completely this assumption and to measure more accurately the energy deposit by joule effect, the reference cell is equipped with a four wires element heater, two for the intensity and two for the voltage measurements (Fig. 2 and Fig.3a).

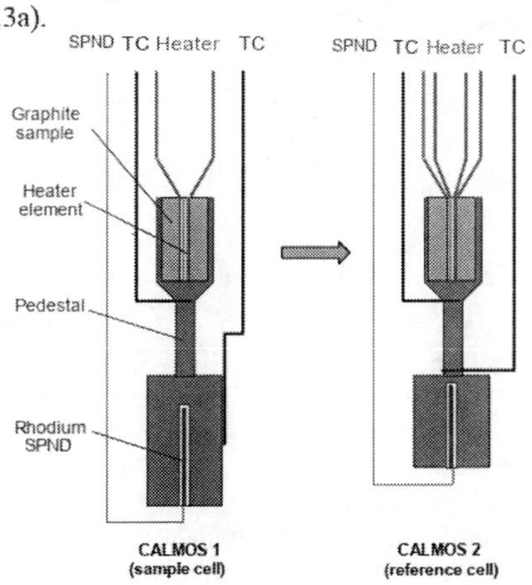

Fig.2 – Evolution between both versions of CALMOS (Heater element and thermocouples location)

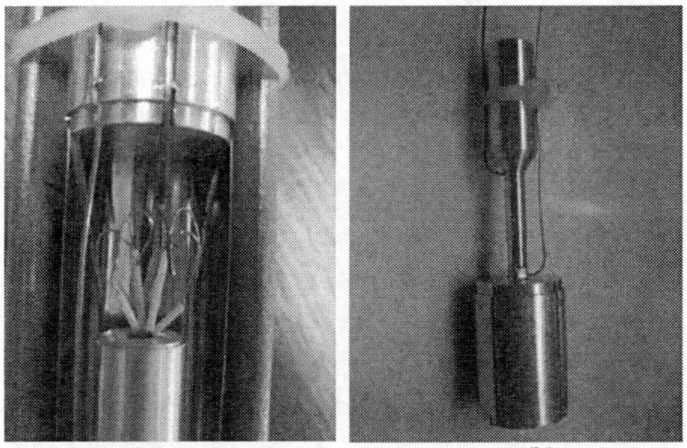

(a) (b)

Fig. 3 – (a) 4-wires resistance implementation in lower cell, (b) New position of thermocouples on the pedestal

Therefore, the applied electric power We during the zero method is now evaluated by We = U I (instead of We = r I^2). However, such implementation led to one of the major difficulty in manufacturing, because the whole wiring has to be inserted inside an overall 17mm diameter with each wire insulated from each other (Fig. 3a). In addition, the heater is only 2.8mm in diameter.

B - Thermocouples Location

The second improvement relates to thermocouples (TC) location. The difference between the two designs is illustrated in Fig. 2 and Fig.3b. While the TC on top of pedestal is unchanged, the TC which was inserted on the base surface (CALMOS-1) is moved to the pedestal bottom. The objective is:

- First, looking for the balance between the two temperatures difference by replacing nuclear energy deposit in the sample cell by joule effect in the reference one, assumes that cells are identical. However, they don't have exactly the same transfer capability (slope in °C.W^{-1}). The lower the difference is, the lower the K_0 influence in equation (2) is. The best way to tend K_0 to unity is limiting the thermal resistance between the two measurement points to the only aluminum cylinder (pedestal, Fig. 3b). The effect is significant. The slopes difference in CALMOS-1, measured during the calibration phase [3, 4], was 2.2%, whereas in CALMOS-2, the new TCs location leads to an only 0.3% difference. The K_0 correction coefficient is now close to 0.99 against 0.955 for CALMOS-1,
- Secondly, instead of having a thermocouple placed in an intermediate position between the external base and the sheath internal surface (with a mechanical clearance), the thermocouple is now embedded inside the pedestal itself, leading to a more accurate modelling assessment. There is almost no difference anymore between calculated and measured slopes during calibration.

C - Volume of Each Cell Base and Rhodium SPND location

The height of the base is reduced from 35 to 25mm (Fig. 2a), involving no change on cell slopes (°C.W^{-1}), but the modification reduces absolute temperatures. Indeed, all the energy deposit inside the base plays the role of an offset for absolute temperatures. Lower this volume is, lower the deposit inside is and lower are the temperatures in the pedestal at equal heating level [5]. First, the margin vs boiling point (117°C in OSIRIS conditions) at the envelope external surface is easier to keep. In addition, it is easier to meet the required margin with respect to the melting point of aluminum (650°C for AW1050 alloy [7]) at any point of the cell. Measurements carried out in the 64North location showed, under same conditions, a "pedestal temperature" in CALMOS-2 (420°C at 70MW) around 20°C lower than CALMOS-1. The new design offers wider margins with regard to the aluminum melting temperature.

The new rhodium SPND (10mm in emitter length), specially designed for the R&D CALMOS program, previously located in central axis of the sample cell base is now moved downward to the reference cell (Fig.1). Such change allows to perform measurements in the lower part of the core. The lowest accessible altitude is now -297mm instead of -191mm/core mid-plane for CALMOS-1.

V. Finite Element Model

The thermal design of the CALMOS-2 probe was performed using the thermohydraulic CAST3M finite element code. The model was adapted from the one used for CALMOS-1. Both models neglect nitrogen convection inside cell cavities. If this effect is significant, it is assumed leading same effects between the reference and the measurement cells. It also neglects heat loss by radiation, which should remain low even at high temperatures because of the presence of polished screens (stainless steel) surrounding the cells, offering a very low emissivity coefficient. Taking into account this effect in the model would require to evaluate correctly the emissivity coefficient of screens internal surface, which is very difficult.

Finally we can notice that for both models, material properties are set to depend on temperatures. Therefore, the initial gap between the base and external tube, in the order of magnitude of a few tens of micrometers, depends also on temperature.

Differences between CALMOS-2 and CALMOS-1 designs are:

- Geometrical evolutions (height of each cell base, total length of the probe, gas height above the element heater),
- The change of thermocouples location, which are located on both extremities of the pedestal in CALMOS-2,
- The Aluminum (AW 1050 alloy) thermal conductivity dependence with temperature in CALMOS-1 model was set so as to find good agreements between calculated and measured temperatures in many experiments. But the CALMOS-2 modelling takes into account the real conductivity, which has been measured by a specialized lab ("Laboratoire National de Métrologie et d'Essai") on specimens withdrawn from the material used for the calorimetric probe manufacturing.

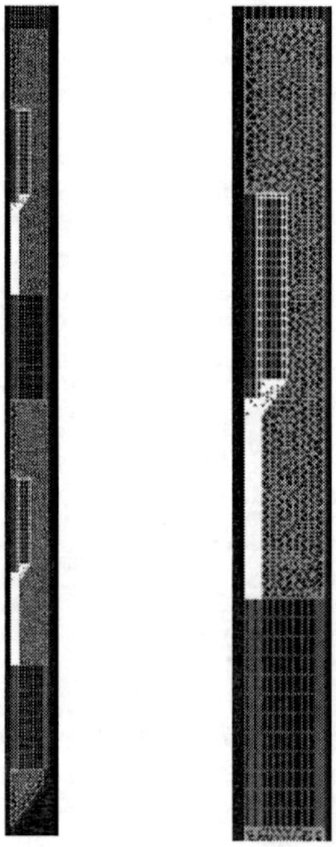

(a) (b)

Fig.4 - CALMOS2 meshing of the probe (a), zoom on the reference cell (b)

Temperatures measured during the calibration phase of CALMOS-2 were compared to calculated ones. Observed discrepancies between calculated and measured cell slopes (close to 7.32 $°C.W^{-1}$ for each one) do not exceed 0.15%. This excellent C/M agreement comes essentially from the new TCs location, allowing a more relevant modelling of temperatures measurement points. The mechanical clearance between the sheath and the external surface of the base does not play any role anymore in the measured temperature difference.

Comparisons were also made on the zero method experiments and shows results which are in good agreement. However, that led us to emphasize a particular phenomenon: even if ΔT between both cells are equal after applying the Joule effect in the reference cell, absolute temperatures of the sample cell are systematically a little higher than those of the reference one. This effect had been previously observed on CALMOS-1, and could be confirmed in CALMOS-2 thanks to the 4-wires technique. However, the discrepancy between cells remains limited, around 5°C at 4W/g for a ΔT total difference of around 100°C. There is no effect on final measurement uncertainties, but this difference has to be considered in order to perform the measurement, particularly for high heating levels. On the other hand, this phenomenon is perfectly reproduced by modelling. In order to understand this systematic discrepancy, the model was modified by implementing graphite in both cells (both cells are therefore identical). The difference still remains. Therefore that seems to come from a small thermic environment dissymmetry between cells. Further investigations showed that it is created by nitrogen temperature differences. Indeed, nitrogen in contact with the lower surface of the base of the sample cell is heated up by the reference cell, whereas nitrogen of this latter is cooled by the low part of the external tube. At $1W.g^{-1}$, $4W.g^{-1}$ and $12W.g^{-1}$, differences are respectively: 2.2; 5 and 6°C. Therefore, the higher the nuclear heating is, the higher is the temperature difference. This difference should have very limited effect on heat loss difference between both cells and on the measurement uncertainty. That does not seem to explain why the nonlinearity coefficient Kc seems to follow a parabolic curve with nuclear heating.

VI. Generalities

OSIRIS is a pool type light water reactor with an open core. The core is a compact unit, with an horizontal section of 60cm x 70cm and a height of 70cm. The core housing contains a rack of 56 cells and is loaded with 38 fuel elements and up to 7 beryllium elements (Fig. 5) [3].

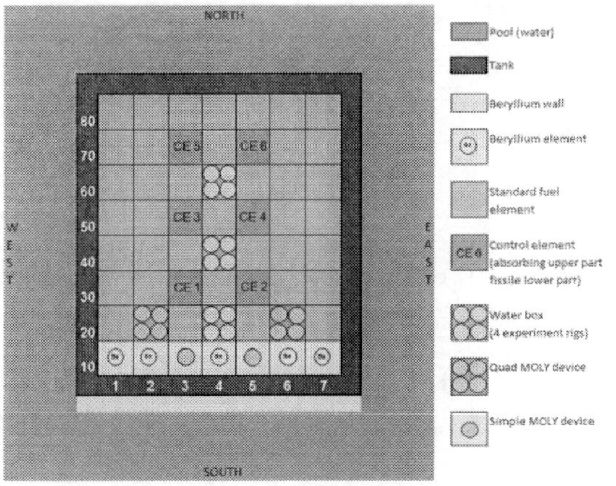

Fig. 5 – Cross section of OSIRIS reactor

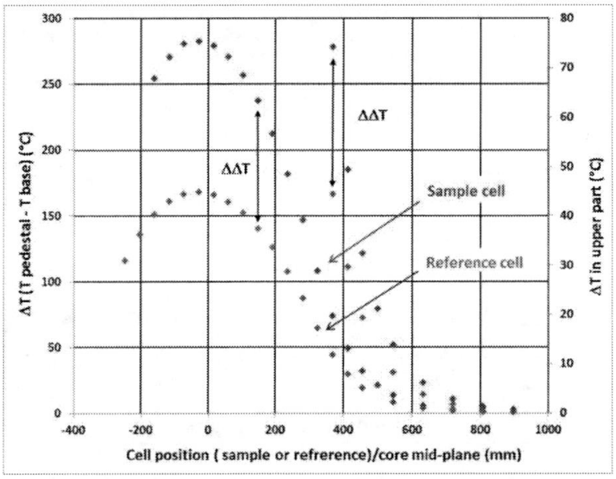

Fig. 6 – Raw temperatures recorded in 64NE location

Two locations (22, 26) are used for radioisotopes production for medical application. The remaining locations (24, 44 and 64) are dedicated to experiments, and equipped with water boxes which can contain up to four experiment rigs, with a 37mm maximum internal diameter. The CALMOS device is suited to in-core measurements in the 24, 44 and 64 locations. Nuclear heating at mid-plane varies from around 4 to 12 W.g^{-1} (in graphite), depending on the selected location. Fig. 6 shows an example of signal acquisition at 69.7MWth in the 64NorthEast location with CALMOS-2. On the graph are plotted ΔT sample, ΔT reference and ΔΔT difference, from which the nuclear heating is deduced. We can see that altitudes lower than -160 mm cannot be reach with the sample cell (red points), whereas those higher than +808m are not accessible with the reference cell (blue points) due to the probe structure (Fig. 1). Zoomed scale (on right) is dedicated to the upper part.

VII. Use of the Zero Method in Upper Part of the Core

The zero method can be considered as an absolute heating measurement and has to confirm, at least for low heating levels, results obtained with preliminary calibration. To study the linearity loss, an intensive use of this procedure was made during the measurement campaign (Fig. 7). However, it could only be performed in upper part where heating levels remain under around 4 W.g^{-1}, due to the maximum intensity current (2 A) applicable to heater wires. Fig. 8 gathers 21 heating measurements obtained both by calibration and zero methods on the 0 - 4 W.g^{-1} range. All measurements are obtained by the new 4-wires technique, measuring together voltage (U) and current (I) applied to the reference heater. Fig. 8 shows that all results are within 2% discrepancy up to around 2.5 W.g^{-1}. Accounting for uncertainties, we can consider that both methods are very close to each other up to that point. Beyond that level, the zero method provides values slightly higher than those obtained by calibration. That leads to apply to the calibration a Kc factor close to 1.04 at 3.5 W.g^{-1} and 1.05 at 4 W.g^{-1}.

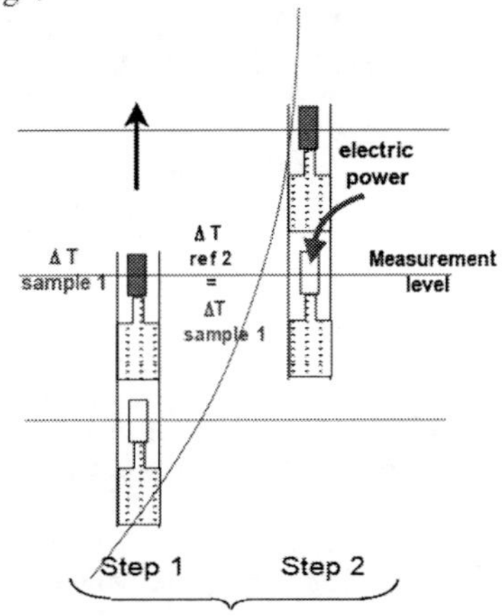

Fig.7 – Zero method procedure (step 1: recording of the ΔT sample at the measurement level, step 2: moving up the probe and applying an energy deposit in the element heater so as to equalize temperatures difference between both cells)

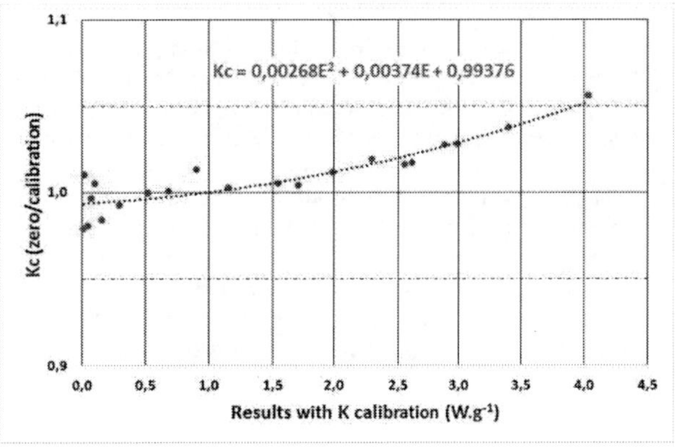

Fig. 8 – Results obtained by the zero and the calibration methods in upper part of the core up to 4 W. g^{-1}. Measurement of the Kc correction factor.

VIII. Uncertainties on Nuclear Heating Measurements

The following evaluation is restricted to measurements deduced from calibration with relation (1) and up to 4 W.g^{-1}, thanks to the 4-wires technique allowing a Kc experimental evaluation. As a first approach, the K coefficient, the ΔΔT difference and the Kc coefficient can be assumed as independent:

- The uncertainty associated with K, depending on cell slopes determination (°C.W^{-1}), is 3.8% at one standard deviation (1 σ) and independent of the heating level,
- ΔΔT value comes from the measurement of four temperatures. The uncertainty of each TC (calibration certificate) and those coming from the measurement device, lead to uncertainties decreasing with ΔΔT rise: 12% for 5°C, 5.5% for 11°C and 1.9% for 42°C ΔΔT (1 σ),
- Kc was evaluated both by the zero method and the CAST3M Finite Element Model [4, 6]. The only experimental process is kept in this evaluation. Despite of the low measurements number, points fit well with the curve in Fig. 8. From an analytical determination we assess a Kc uncertainty to 2% (1 σ).

Table 1 – Uncertainties on heating measurements deduced from calibration

Heating (W.g^{-1} in graphite)	0.5	1	2	4
σ E_T/E_T at 1 σ (%)	12.7	7.0	5.4	4.7

Therefore, if the measurement is made through the preliminary calibration, associated uncertainties are those of Table 1, as a function of the heating level. If the heating measurement is performed through the zero method, which is the reference, the assessed uncertainty is 4% (1σ), accounting for the main parameters (U, I, sample mass) and the representability of We in applied powers.

IX. Comparison of Heating and Thermal Flux Profiles

For each scan, CALMOS-2 allows measuring separately the nuclear heating, the thermal flux or both at the same time. One notes that, the rhodium has a response both in thermal and epithermal ranges. As it moves inside in-core and ex-core areas, the neutron spectrum changes and has to be calculated along the total stroke to be able to transpose the current to a neutron flux. Therefore, the epithermal flux/thermal neutron flux ratio has been calculated by the TRIPOLI-4® Monte Carlo code for all accessible altitudes accounting the cell geometry, rods position and the actual core loading [8]. The associated uncertainty to the conventional thermal flux by this rhodium SPND is 8% (1σ) [9].

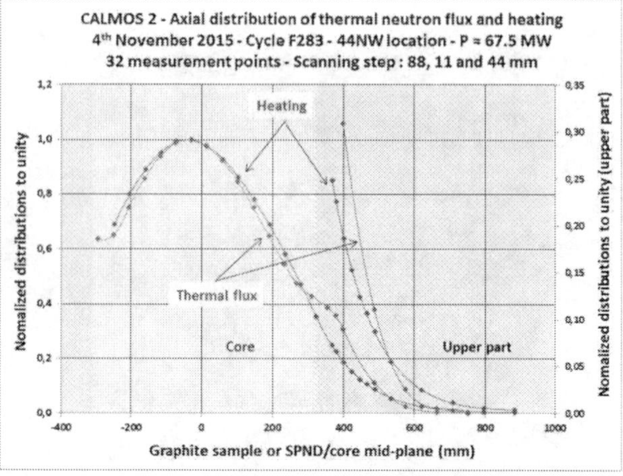

Fig.9 – Simultaneous messurement of heating and thermal neutron flux in the 44North-East location. Values are normalized to unity.

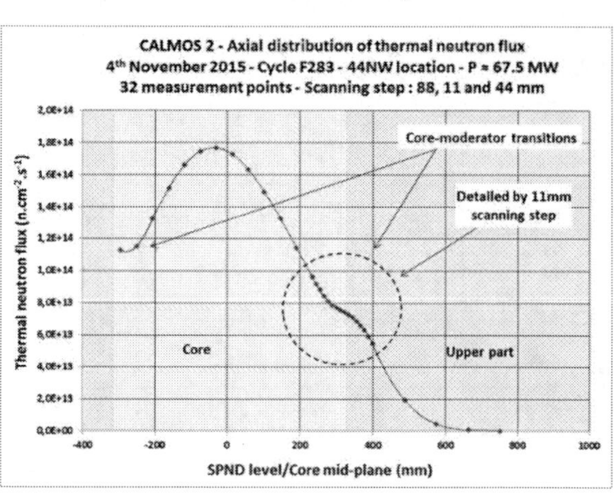

Fig.10 – Measurement of the thermal neutron flux in the 44 North-East location. Detail of the thermal flux distribution at the core-moderator transition.

Fig. 9 shows an example with 32 simultaneous measurement points, heating and thermal flux, performed in the 44NorthWest location during the F283 cycle, which allows observing behaviors of both quantities along the scanning height. The Figure shows both distributions normalized to unity. Fig. 10 relates to the only thermal neutron flux in absolute values. That shows the possibilities offered by the HMI automatic displacement system allowing to refine deeply the profiles. Two significant phenomena could be well highlighted and measured:

- As the probe moves away from the core, the heating attenuation is quite different from the thermal flux one (see right scale on Fig. 9 zooming measurements in upper part). Thermal flux decreases faster than heating, around a ratio 2 each 40mm against 80mm for heating,
- The slope break in thermal flux evolution at the moderator-core transition (+320mm/mid-plane), is well emphasized (Fig. 10) with a scanning step reduced to 11mm in this area, only possible by an automatic displacement system.

X. Checking of the Consistency between both Prototypes

CALMOS-2 and CALMOS-1 were loaded side by side (37mm distance) in the two north positions of the 64 core location during the F284 cycle. Two distributions of thermal flux and nuclear heating were obtained by each device, but operated separately so as to avoid any influence on each other. Fig.11 shows the heating comparison whereas Fig.12 relates to the thermal flux.

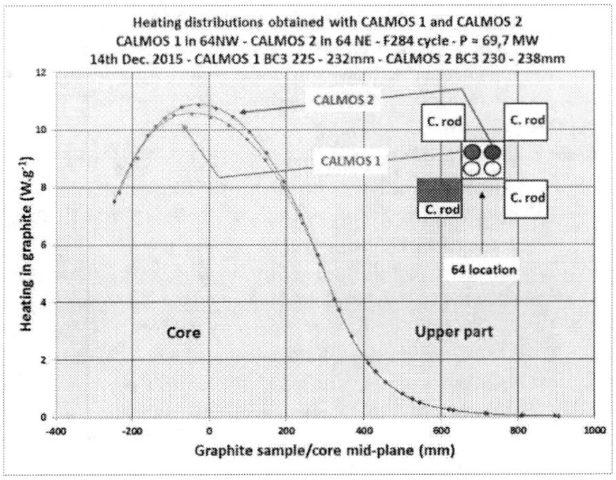

Fig. 11 – Distributions of nuclear heating obtained with both prototypes located in 64 North location

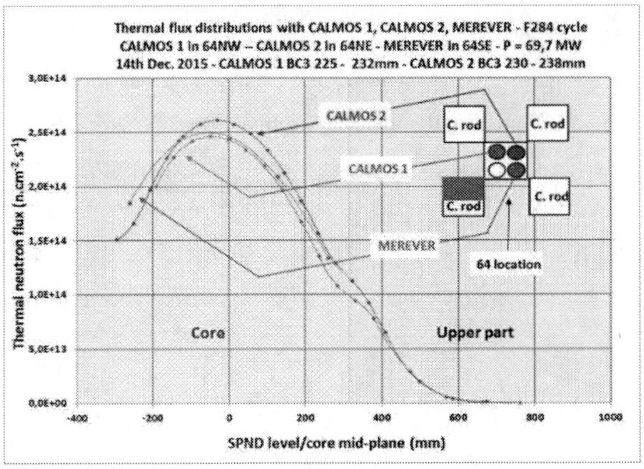

Fig. 12 - Distributions of the thermal neutron flux obtained in 64 location with CALMOS-1 (blue) CALMOS-2 (red) and the MEREVER device (green) equipped with standard rhodium SPNDs

During the scans, made at the 70MWth power, among the 4 control rods surrounding the location only BC3 (in brown) in the south part is in the first third of its stroke. Such configuration leads approximately, to same irradiation conditions in north part of the 64 location. Results show that:

- Both heating profiles are almost identical in Fig. 11. The only significant difference appears in the mid-plane region in which the CALMOS-2 signal (red curve) is around 3% higher,
- Thermal neutron flux distributions in Fig.12 seem to confirm the heating behavior. Indeed, at mid-plane, measurements by CALMOS-2 (red curve) are also higher around 6% than CALMOS-1 (blue curve). However, conversely to heating recording, we note a thermal flux in CALMOS-1 remaining lower up to the fissile-moderator transition in upper part,
- To reinforce this comparison, the current static device called MEREVER was placed also in 64SE location (Fig. 12). It is instrumented with 5 rhodium standard SPNDs (50mm emitter length) inside an aluminum rod and allows to plot the thermal flux profile with 5 in-core measurements covering +/-260mm/mid-plane. In comparison with those obtained with CALMOS devices, we see also that profiles are similar, except for low altitudes under the mid-plane (green curve), demonstrating that the thermal neutron flux is influenced by the rods position, closer to the MEREVER device.

Such results show that, both devices give coherent responses in spite of significant geometry differences (K calibration coefficients with a 16.6% discrepancy). In addition, the global consistency in thermal flux measurement found between CALMOS and MEREVER devices, shows that the 10mm short emitter SPND, specially developed for the CALMOS program, seems qualified. To complete the study, two scanning were performed in this configuration with both CALMOS devices moving downward simultaneously in the core so as to evaluate their mutual influence. Results show that the influence remains lower than 3% for heating measurements for in-core region, which is acceptable.

XI. Recording of the Residual Power Decay

The CALMOS-2 calorimeter has been used to follow the residual power decay after shutdown at the end of the cycle. Fig. 13 shows an example of a 10 hours recording performed after shutdown at the end of the F283 cycle. After an 8 hours elapsed time, the residual heating level is around 150 mW.g^{-1}. At this time, the reactor procedure requires to stop the primary pumps, involving a change of cooling conditions for the calorimeter, switching from the forced convection to the natural convection regime.

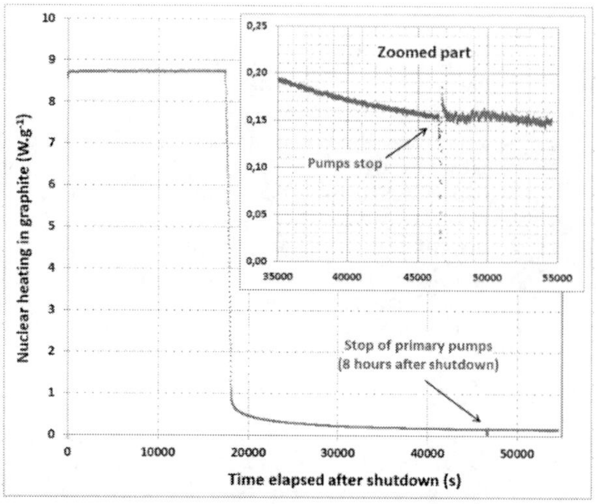

Fig. 13 – Decay of the residual power after the end of the F283 OSIRIS cycle. Recording during 10 hours after shutdown. The signal behaviour before and after the change of cooling regime appears in the zoomed part (upper right corner).

The zoomed part in Fig. 13 shows that the signal is disturbed at the time of pumps stop and the noise becomes stronger in natural convection. However, the decay is still coherent after the event, and the signal tends to find again the same slope of decay, demonstrating that the calorimeter remains reliable in both cooling regimes. That allows to use such type of calorimeter to follow the residual power independently of the OSIRIS thermohydraulic conditions.

XII. Conclusion

The comprehensive 2015 measurement campaign in the OSIRIS reactor with the CALMOS-2 calorimeter allowed to highlight different possibilities offered by this new device. The automatic cell mobility allows to obtain very well defined profiles, both for nuclear heating and thermal neutron flux measurements. The scanning steps number and the total measurement time can be adjusted to core conditions and to target regions, while keeping relevant measurements with regard to the control rods movement. Implementation of the 4-wires technique in the reference cell allows to be more confident in the zero method. It enables assessing the probe non-linearity and to reevaluate associated uncertainties up to 4 $W.g^{-1}$. However, they could not be entirely evaluated up to the nominal power. A complete study, including the model adjustment of the cell and all the data processing during the measurement campaign, is ongoing. The intensive use of CALMOS-2 in the core, cumulating a total 70 hours automatic scanning time, has demonstrated its reliability for in-core conditions for the calorimeter itself and the enhanced displacement system. In addition, it has been demonstrated the ability of such device to measure the evolution of the residual power after shutdown whatever the thermohydraulic conditions, forced or natural convection regimes. The device meets better safety requirements and becomes an equipment well suited to the surveillance and qualification of an MTR, preparing the transfer of this technique to the incoming JHR reactor in France.

Acknowledgments

Authors wish to thank the DER/SPEx department of CEA/Cadarache for fruitful discussions along this R&D program, the DRSN/SEROS and DM2S/SERMA departments of CEA/Saclay for their help during the measurement campaign and neutron calculations required for results processing.

References

[1] [1] IAEA, International Intercomparison of Calorimeters, Report N° 128, 1970.

[2] M.S. Kim, S.Y. Hwang, H.S. Jung, K.H. Lee, "Measurements of Nuclear Heating Rate and Neutron Flux in HANARO CN Hole for Designing the moderator cell of neutron source." Proceedings of IGORR, (2005).

[3] H. Carcreff, L. Salmon, and C. Courtaux, "First In-Core Measurement Results Obtained with the Innovative Mobile Calorimeter CALMOS inside the OSIRIS Material Testing Reactor", *IEEE Transactions on Nuclear Science*, Vol.61, N°.4, Aug 2014.

[4] H. Carcreff and coll, "First In-Core Simultaneous Measurements of Nuclear Heating and Thermal Neutron Flux obtained with the Innovative Mobile Calorimeter CALMOS inside the OSIRIS Reactor" *IEEE Transactions on Nuclear Science*, Vol. 63, N°.5, Oct 2016.

[5] J. Brun and coll, "Numerical and Experimental Calibration of a Calorimetric Sample Cell Dedicated to Nuclear Heating Measurements", *IEEE Transactions on Nuclear Science*, Vol.59, N°.6, Dec 2012.

[6] "CAST3M code", http://www-cast3m.cea.fr/

[7] "Alliages d'aluminium", http://www.euralliage.com

[8] F. Chevallier, "Calculs d'indice épithermique dans le dispositif CALMOS-2 pour l'interprétation du signal collectron dans le réacteur OSIRIS". Forthcoming CEA report.

[9] D. Fourmentel and coll. "Comparison of Thermal Neutron Flux Measured by ^{235}U Fission Chamber and Rhodium Self-Powered Neutron Detector in MTR". *IEEE Transactions on Nuclear Science*, Vol.61, N°.4, Aug 2014.

The CABRI fast neutron Hodoscope: Renovation, qualification program and first results following the experimental reactor restart

V. Chevalier[1], S. Mirotta[1], J. Guillot[1], B. Biard[1]

[1]*IRSN, PSN-RES/SEREX/L2EP, Cadarache, F-13108 Saint-Paul-Lez-Durance, France*

***Abstract*— The CABRI experimental pulse reactor, located at the Cadarache nuclear research center, southern France, is devoted to the study of Reactivity Initiated Accidents (RIA). For the purpose of the CABRI International Program (CIP), managed and funded by IRSN, in the framework of an OECD/NEA agreement, a huge renovation of the facility has been conducted since 2003. The Cabri Water Loop was then installed to ensure prototypical Pressurized Water Reactor (PWR) conditions for testing irradiated fuel rods.**

The hodoscope installed in the CABRI reactor is a unique online fuel motion monitoring system, operated by IRSN and dedicated to the measurement of the fast neutrons emitted by the tested rod during the power pulse. It is one of the distinctive features of the CABRI reactor facility, which is operated by CEA. The system is able to determine the fuel motion, if any, with a time resolution of 1 ms and a spatial resolution of 3 mm. The hodoscope equipment has been upgraded as well during the CABRI facility renovation. This paper presents the main outcomes achieved with the hodoscope since October 2015, date of the first criticality of the CABRI reactor in its new Cabri Water Loop configuration. Results obtained during reactor commissioning phase functioning, either in steady-state mode (at low and high power, up to 23 MW) or in transient mode (start-up, possibly beyond 20 GW), are discussed.

***Index Terms*— CABRI, Hodoscope, pulse mode, RIA, fast neutron detection system.**

I. Nomenclature

ϕ_{core}	Neutronic flux in the center of the core
$\sigma_{f,i}$	Microscopic fission x-section for the isotope i
$N_{i,core}$	Atomic density of fissile material of the core
$Q_{f,i}$	Q-value of the fission for the isotope i
$N_{i,testrod}$	Atomic density of fissile material of the test rod
n_m	Measured counting rate
n_t	True counting rate
n_{rod}	Measured counting rate due to test rod neutrons
n_{sc}	Measured counting rate due to scattered neutrons
n_{bg}	Measured counting rate due to CABRI core neutrons
τ	Dead time of detection chain
ϑ	Dead time type, $0 \leq \vartheta \leq 1$

II. Introduction

For enhancing safety of nuclear power plants (NPP), the Institut de Radioprotection et de Sûreté Nucléaire (IRSN) carries out some experimental programs in order to improve the understanding of the fuel behavior under severe accident conditions. One of them is the CABRI International Program, managed and funded by IRSN under an OECD/NEA agreement, launched to study Reactivity Initiated Accidents (RIA) in representative PWR conditions.

For this purpose, the renovation of the CABRI facility has been conducted by CEA, its operator. This pool type reactor is designed to submit a test rod placed in the center of the core to a RIA. The old experimental sodium loop, crossing axially the entire core, has been replaced by a pressurized water loop, so as to reproduce the thermal hydraulic conditions of a Pressurized Water Reactor (PWR), for testing the cladding of a fuel rod under prototypical conditions of RIA in PWRs.

The power transients are generated by the unique reactivity injection system of CABRI [1]. ^{3}He is previously introduced inside 96 tubes (so-called "transient rods", visible in Figure 6) located in 4 banks among the CABRI fuel rods. The very fast depressurization of this strong neutron absorber into a discharge tank, through two flow channels (low and high flow rates), is then finely adjusted to trigger the desired power pulse.

The CABRI facility is equipped with two nondestructive measurement systems operated by IRSN teams:

- The IRIS facility, for performing X-ray radiography and tomography imaging before and after a power transient thanks to a linear electron accelerator, as well as quantitative gamma scanning analyses;
- The Hodoscope, an online fuel motion measurement system, which aims at analyzing the fuel motion deduced from the detection of fast neutrons emitted by the tested rod, in real time (with a time step of 1ms) during the transient.

In section III of this article, a brief description of the hodoscope measuring system will be given. In section 0 comparison of results performed in the old sodium loop and in the new water loop will be discussed. Furthermore, in section V a comparison of dosimetry measurements will be analyzed, and in section VI neutron fluxes measured by the hodoscope will be compared to those obtained by Monte-Carlo simulations. In section VII dead time analysis of the different detectors will be investigated as well. Finally, in VIII current studies will be presented, and conclusion is given in section IX.

III. General description of the system

The CABRI reactor has been equipped with the hodoscope since 1977. This system measures in real time the behavior of the test rod in the center of the CABRI core during the transient. The hodoscope is able to quantify the amount of fuel ejected in the milliseconds following failure, time-dependent axial fuel mass distributions and to follow the fuel clusters after failure. In addition, the initial state of the test rod is precisely known thanks to the hodoscope measurement of the fissile length and of the axial power profile.

The hodoscope centerpiece is a 3 m steel collimator (in green in Figure 1) of 3 columns and 51 rows, so 153 channels. Behind each one of these line-of-sights, a ^{237}Np fission chamber detector (FC) and a proton recoil proportional detector (PR) measure the fast neutrons coming from the core and test rod. Two different technologies of detectors are used in CABRI in order to follow all the experiments from low power (~50 kW) up to 20 GW. Proton recoil counters are better suited for low power (up to ~1 GW which means 2.5 million impulsions per second), thanks to a higher efficiency and no discrimination of the γ noise, and ^{237}Np fission chambers are used for higher power. The Proton Recoil detectors are operated at 2500 Volts and an energy threshold at 400 keV. The ^{237}Np fission chamber detector (FC), operated at 600 V, have low gamma sensitivity and an energy threshold of 650 keV. The signal acquisition rate may be adjusted between 1 ms and 10 s according to the count rate.

The hodoscope collimator has three different degrees of freedom, for adjusting its position in front of the test rod. The distance from the collimator nose to the core axis may be adjusted manually in translation, while two Direct Current Motors are used to move the collimator in verticality and rotation. Each one of the 306 detectors is linked by a μ-metal cable to its electronic chain placed 30 m far from the detector. These electronics have been renewed taking benefit from the renovation period [2].

In order to keep the noise/signal ratio as low as possible, the CABRI driver core is traversed radially by the irradiation channel. In that way, none of the fuel rods of the core are in direct view of the hodoscope detectors. Nevertheless, the noise to signal ratio remains low and requires specific data treatment algorithms.

The collimator was installed in order to pixelize the test rod placed in the center of the reactor. Each detector looks at a slice of the test rod. In steady state conditions, the signals measured by all detectors are stable, whereas in transient conditions (i.e. a RIA pulse) they may vary according to a fuel displacement, a fuel densification (yielding to an increasing signal), or a fuel dilution or ejection (yielding to a decreasing signal).

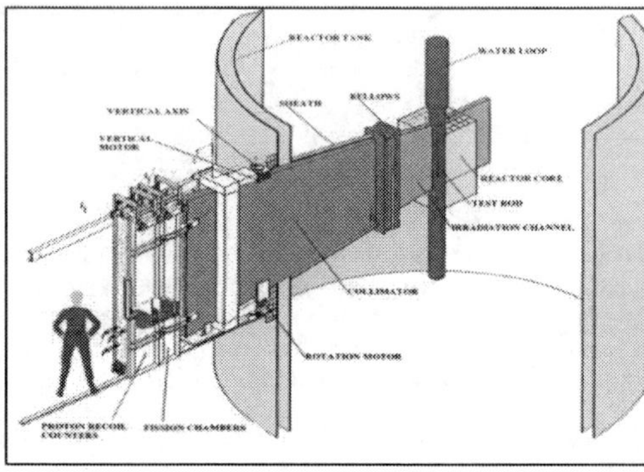

Figure 1: Hodoscope detector system

IV. Comparison of new Water Loop with Sodium Loop

The CABRI restart campaign, started in October 2015, could be split into four main kinds of experiments:

1. Low power experiments – part 1, performed to measure neutronic parameters of the reactor. In this part the hodoscope was not used because the sensitivities of PR and FC are too low to measure signals when core power is limited to 10 kW;
2. Low power experiments – part 2, in this second part reactor power has reached 80 kW; at this power level the hodoscope has given first results;
3. High power experiments in steady state conditions, these experiments has taken the core power up to 23 MW, in order to calibrate the core detectors [3];
4. High power experiments where perfomed to reproduce power pulses up to 21 GW.

Results obtained during the CABRI restart campaign (phase 3 and 4) are compared to those obtained in the past, in order to observe potential modification in detector response. This comparison allows to determine whether the hodoscope counting chain was damaged or not during the facility renovation. Indeed, some major works were conducted during the renovation of the facility which led to a 12 year shut off period for the hodoscope, including seismic reinforcement in the immediate proximity of the detectors and the collimator (kept in place during the renovation) and of the electronic chains which had to be moved away and back.

In this campaign the device positioned in the center of the CABRI core inside the Water Loop (WL) was equipped with a dummy lead rod which does not influence the neutronic profile in the center of the core. In order to observe the differences due to the new Water Loop, we have compared these results to those obtained in the last startup campaign of CIP02 (S9CIP02) experiments on 2002. S9CIP02 experiment was chosen because of its experimental conditions, which were comparable to that of a test realized in the new commission campaign. In both cases, given the absence of a real fuel rod in the test cell, the hodoscope recordings are then only background signal coming from the core fuel and not "useful" neutrons from a test rod.

In these experiments, the same depressurization parameters, ^{3}He initial pressure (hence the same control rod position) and initial power were set.

Next sub section will focus on the axial power profiles measured by the hodoscope in steady state conditions (during a 100 kW power plateau) with both FC and PR detectors and in transient conditions (during power peak transient) just for FC.

A. Power profiles at 100 kW steady state conditions

Power profiles may be compared in steady state conditions with ^{3}He in the transient rod, before triggering the pulse.

Uncertainties in this case are much higher because of the low counting rate at 100 kW. Such acquisitions are performed at this low power level for determining the initial and final state of the test rod without heating it too much. The acquisition rate is then tuned to its minimum value (1 ms) to follow the power transient.

An average over three columns (for FC in the same row) has been calculated, taking into account neither a dead time correction (at this low counting rate the detectors are in their linear range) nor a sensitivity coefficient (would be required only for a quantitative analysis).

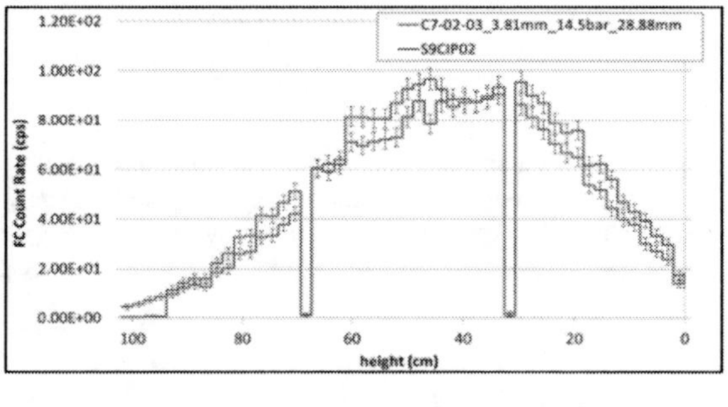

Figure 2: FC axial power profiles in Steady State Condition

The comparison is presented in Figure 2 for FC and in Figure 3 for PR. Axial height is counted from the bottom of the CABRI core fuel rods. Two detector chains have a low counting rate because two line-of-sights are filled with stainless steel in order to strengthen the collimator. The FC comparison is consistent, with differences inside a 2σ confidence interval. No degradation of the measurement system can be observed. This result makes feel confident about the global response of the system in steady state condition at low power.

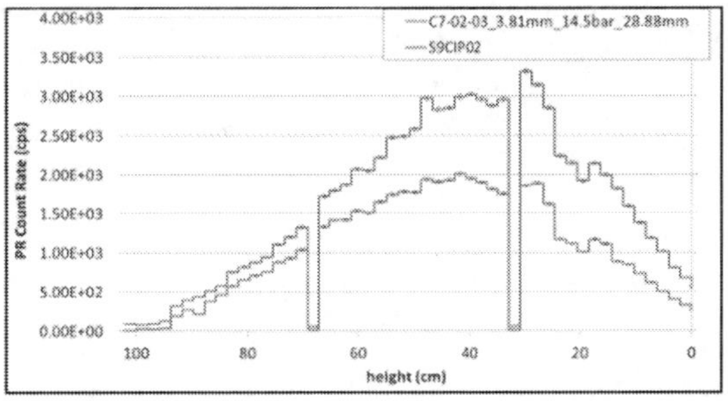

Figure 3: PR axial profiles in Steady State Condition

For PR detectors, a decreased measured signal is observed. This is due to the refilling of the PR detectors with CH_4 (all PR detectors are interconnected), increasing the gas circuit pressure to 1.23 bar instead of the past 1.05 bar. Decreasing the measured counting rate allows to reach a higher power before saturation of PR detectors.

B. Power Profile in transient conditions

It is now possible to compare axial power profiles during power transient where the count rate is higher, hence the uncertainties are lower. Profiles presented in Figure 4 are calculated just at the maximum of the power transient, and show a normalized axial power profile obtained in the last experimental campaign compared again to S9CIP02.

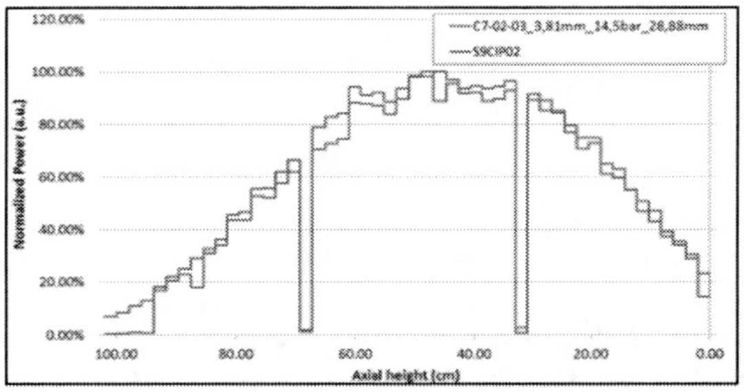

Figure 4: Normalized FC axial power profiles in Transient Condition (Max Power of 5 GW)

Uncertainties are less than 0.5% and are not appreciable in the figure.

As for the previous comparison, results shown in Figure 4 are neither evaluated with a dead time correction nor a sensitivity correction, which are not necessary for this qualitative comparison.

At this step, the general functioning of the hodoscope system is demonstrated. Indeed, during the power steady states and the pulses, it was observed:

- The correct operation of all measuring channels (detectors, associated electronics) whatever the power level and the acquisition rate,
- The correct operation of the remote acquisition module,

- The proper functioning of real-time tracking of all channels on the new acquisition station and monitoring PC,
- The correct operation of the software and numerical tools dedicated to preparation, acquisition and analysis.

V. COMPARISON WITH DOSIMETRY MEASUREMENT

The CABRI commissioning campaign included several experiences with dosimetry measurements [4]. Wire dosimeters were placed at several known elevations in the center of the test cell to measure the axial power profile of the CABRI core. These results were compared to those obtained with the hodoscope PR detectors in order to determine their qualitative response. On Figure 5 are shown the axial profile measured with the PR detectors of the hodoscope and the dosimeter counting rates. The power plateau for this experience was set at 30 kW and the measurements are consistent.

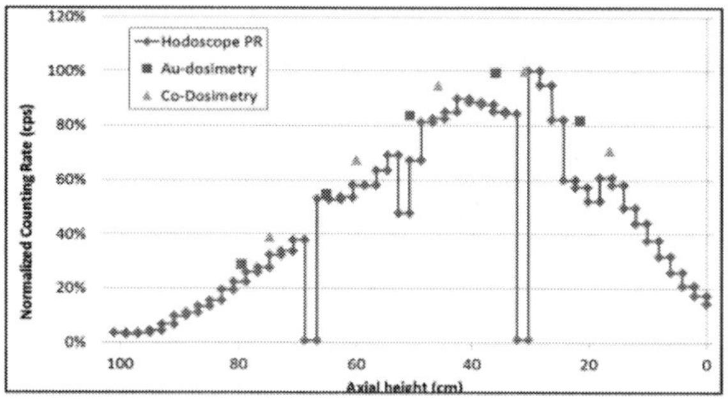

Figure 5: Hodoscope power profile vs. dosimetry in Steady State condition

An important aspect to observe in this graph is that hodoscope PR detectors measure fast neutron flux coming from the core, whereas dosimeters measure thermal and epithermal fluxes of the core. Consistency between these curves is due to a constant ratio between thermal and fast flux during steady state conditions. This assumption is also verified in transient conditions.

VI. COMPARISON WITH NUMERICAL SIMULATION

In order to solve the Boltzmann Equation, all deterministic code needs to do some approximation. For CABRI reactor, deterministic codes do not give very satisfying answers. Therefore, Monte Carlo code have been used to determine neutronic parameters (i.e. β_{eff}, λ, K_{eff}, etc.), such as the MORET code developed by IRSN [5]. The MORET model of CABRI core is shown on Figure 6.

In this section, results obtained with MORET 5.B.1, using JEFF3.1 libraries, are compared to experimental results.

Monte Carlo codes give results in steady state conditions, so a comparison has been carried out with experimental results obtained during a 23 MW power plateau.

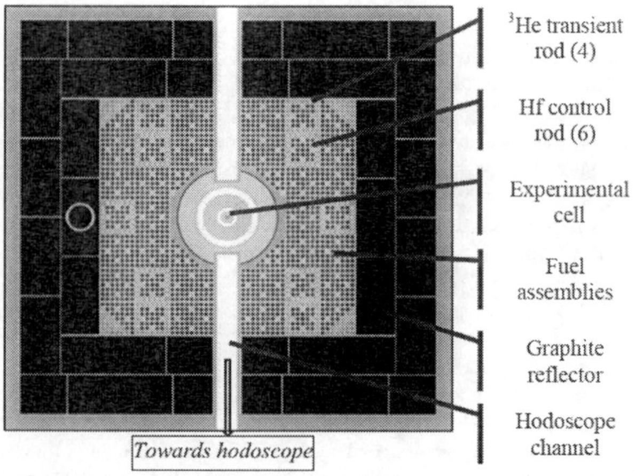

Figure 6: MORET model of the CABRI core

For this comparison, dead time correction (as detailed in section VII) and sensitivity coefficient estimated in 1994 were used in order to obtain physical results.

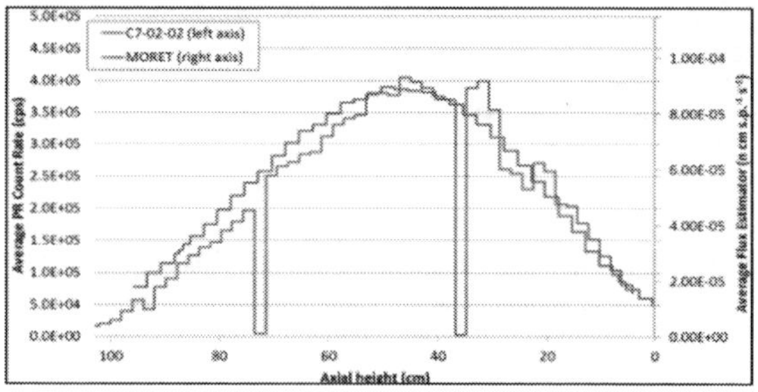

Figure 7: MORET 5 compared to PR profile in Steady State Condition

As we can see in Figure 7, Monte Carlo predictions (right axis) are consistent with experimental measurements (left axis). This result makes feel confident with calculated results, and especially the coupling factor calculation expressed in Eq.(1).

$$CF = \frac{P_{core}}{P_{testrod}} = \frac{\sum_i \phi_{core} \cdot \sigma_{f,i} \cdot N_{i,core} \cdot Q_{f,i}}{\sum_i \phi_{core} \cdot \sigma_{f,i} \cdot N_{i,testrod} \cdot Q_{f,i}} \quad (1)$$

This coupling factor estimation is of crucial importance for the experiment, since it allows the estimation of the energy that will be deposited into the test rod during a power transient, based on the core power measurement.

VII. DEAD TIME CHARACTERIZATION

The analysis of the first commissioning tests also allows the characterization of the dead time of detectors. The PRs as well as the FCs are all operated in pulse mode. This is possible because of the important distance between the core and the

detectors. As shown in Figure 8, for the FC, when the counting rate arises up to 4 million impulses per second, the detector's response becomes nonlinear. In order to solve this problem, a dead time correction must be applied.
This correction is applied just to FC detectors because PR detectors become useless at high power.
Till today, for dead time correction of FC detectors Eq. (2) was used regardless the counting rate. This correction is in great accord with the true counting rate when dead time losses are smaller than ~15%. Beyond this limit this correction is not efficient. This is because Eq. (2) describes the behavior of an electronic chain with a non-parallelizable dead time. [5]

$$n_m = \frac{n_{th}}{1+\tau \cdot n_{th}} \qquad \textbf{(2)}$$

In reality both FC and PR detectors present an intermediate behavior between paralyzable and nonparalyzable dead time, which is characterized by a parameter ϑ. Therefore, the more complex Eq. (3) has to be used for a good evaluation of dead time losses.

$$n_m = \frac{n_t \cdot \vartheta}{e^{n_t \cdot \tau \cdot \vartheta} + \vartheta - 1} \qquad \textbf{(3)}$$

Using this new approach, ϑ and τ have to be evaluated. In order to do that, the following methodology was used:

1) For low counting rate (i.e. 10% of losses) Eq.(2) gives us dead time of the electronic chain[1];
2) Then Eq.(3) is solved to determine theta using experimental results at high counting rate. Indeed, the true counting rate according to the measured one has to be known;
3) These two parameters (τ, ϑ) are used in Eq.(4) that is a Taylor expansion of rearranged Eq.(3) in order to find the true counting rate. Indeed, the true counting rate according to the measured one has to be known.

$$n_t = n_m \cdot \Big[1 + n_m \cdot \tau + \left(1 + \frac{\vartheta}{2}\right) \cdot (n_m \cdot \tau)^2 + \left(1 + \frac{3}{2}\vartheta + \frac{\vartheta^2}{6}\right) \cdot (n_m \cdot \tau)^3 + \cdots \Big] \qquad \textbf{(4)}$$

With this methodology all 306 dead time parameters were obtained.

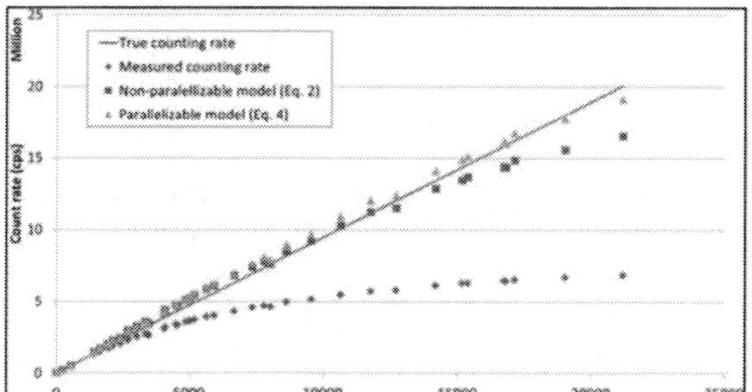

Figure 8: Comparison of dead time correction algorithms

[1] These values are consistent with those evaluated by the double oscillators methodology [6]

As we can see in Figure 8 the counting rate reevaluated after dead time correction is consistent with the true counting rate. Results found for three FCs are summed up in Table 1 for three detection chains. This analysis must be realized for each detector of the hodoscope system. This dead time correction allows the hodoscope to monitor with a good linearity even the most powerful transients of CABRI.

Table 1 – DEAD TIME CHARACTERISTIC FOR THREE FC

	Upper row (FC189)	Middle row (FC237)	Bottom row (FC278)
τ	82.6 ns	85.4 ns	87 ns
ϑ	0	0.134	0.572

VIII. PERSPECTIVES

In June 2017 a calibration campaign is planned in order to adjust the sensitivity coefficients for both FC and PR detectors. This campaign will be split in two parts:

- The Collimator Alignment campaign will aim at verifying the collimator position viewing the rod in the test cell;
- The Detector Calibration campaign will be devoted to the determination of the sensitivity coefficient for each FC and PR detector.

This campaign is necessary to obtain absolute quantitative measurements with the hodoscope system.

A. Aligning the hodoscope collimator

The Hodoscope Collimator may be motioned according to different degrees of freedom presented in Figure 9, so as, z-axis rotation and y-axis rotation allow to align the lines of sight of the hodoscope with the test rod fuel stack.
The first part of the campaign will be dedicated to the rotation alignment (c.f. z-axis rotation). In this way, it is possible to look at the test rod through the collimator. The graph presented in Figure 10 is obtained with this experiment. The second part of the campaign consists on moving the collimator under the y-axis in order to align with the vertical axis as the core. This verification will be based on the counting profiles obtained in the left, middle and right columns of the hodoscope. If the alignment is good, the fuel rod may be seen by only one column at a time. The distance between the nose of the collimator and the test rod will be verified with the measurement of the fissile length of the fuel stack of a fresh fuel rod placed into the experimental cell. Since the lines of sight are not parallel but convergent, this measured length will vary according to the distance of the collimator from the fuel rod.
This campaign will follow an iterative way: a power plateau of ~10 MW will be scrutated by the hodoscope while rotating the collimator, in front of the test rod. Once the data analyzed, a correction will be determined for the vertical motor position. A new power plateau will be performed to check this new position, etc.
When the global alignment is alright, it will be possible to determine the optimal measuring position. Indeed, by rotating

 https://doi.org/10.1051/epjconf/201817004003

the collimator, the operators may choose which column does see the rod and which one does not.

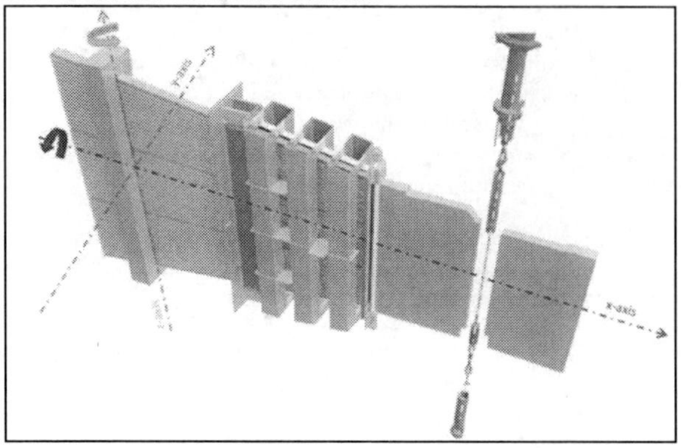

Figure 9: Hodoscope collimator movements

When a detector is in front of the test rod, its counting rate will be given by:

$$n_m = n_{rod} + n_{sc} + n_{bg} \qquad \textbf{(5)}$$

When the collimator is shifted away from the test rod axis, hodoscope detectors will count just background and scattered neutrons coming from the CABRI core, as shown by Eq.(6).

$$n_m = n_{sc} + n_{bg} \qquad \textbf{(6)}$$

Figure 10 shows the response of one row when the collimator is rotated in front of the test rod. When the collimator is in front of the test rod a maximum will be obtained. When it is far from the test rod just background and scattered neutrons are measured.

The standard position of the hodoscope, named "split" position, consists in placing the collimator between left and central row. By this way, with left and central row we are able to measure counting rate expressed in Eq.(5), whereas with the right column it is possible to measure neutrons coming from the core only (c.f.Eq.(6)).

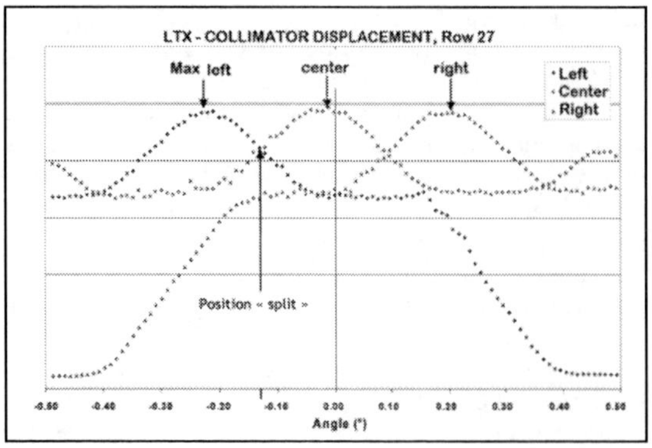

Figure 10: Measured counting rate for left, center and right detectors of a common row during collimator rotation

The split position is preferred to a more classical position when just one column looks at the test rod because in this way signal variations are seen by two columns simultaneously reducing uncertainties.

This campaign should be performed each time a test device will be loaded in the experimental cell with a new test rod.

B. Detectors Calibration Campaign

After the collimator alignment, the calibration campaign will be performed to adjust the detectors efficiency. In order to evaluate these efficiencies (different for each detector), the experiment consists first in irradiating a fresh UO_2 fuel rod in CABRI for a couple of hours (duration limited by the core cooling system capacities) at a constant power while monitoring fast neutrons emitted from this fuel rod by the hodoscope system. After the irradiation, the axial profile of the non-volatile fission products created during the irradiation of the test rod will be measured on the IRIS facility by γ-spectroscopy analysis.

Although the direct measurement with the hodoscope is based on fast neutrons while the subsequent IRIS counting is based on gamma measurement, this comparison is relevant assuming both particles are proportionally generated by the thermal fissions of UO_2. Each fission produces indeed $\bar{\nu}$ neutrons, which are directly detected by the hodoscope, some prompt γ and fission products that will be detected in IRIS. γ emitted by the radioactive decay of fission products will give us the amount and the location of the fission products. In this way it is possible to build the axial profile of CABRI core. Efficiency coefficients will finally be obtained by comparison of this profile with those measured with the hodoscope system.

IX. Conclusion

After a long standby and renovation period, the restart campaign of the CABRI reactor has allowed to verify the good operability of the hodoscope detectors, their electronic chains and the command and control system, and verifying its operating domain. It was also useful to obtain the detector response at different core power levels, up to 23 MW in steady-state mode and up to more than 20 GW in transient mode. Furthermore, with these experiments the dead time parameters of each detector were characterized. The hodoscope is now ready to be aligned and calibrated thanks to a dedicated campaign that will be carried out in order to adjust the sensitivity coefficients for both FC and PR detectors.

References

[1] B. Duc, B. Biard, P. Debias, L. Pantera, J.-P. Hudelot and F. Rodiac, "Renovation, improvement and experimental validation of the Helium-3 transient rods system for the reactivity injection in the CABRI reactor," IGORR 2014 Conference, Bariloche, Argentina, November 2014.

[2] S.Mirotta, J.Guillot, V.Chevalier and B.Biard, "Qualification and characterization of electronics of the fast neutrons Hodoscope detectors using neutrons from CABRI core," In ANIMMA2017 Conference Proceedings,

https://doi.org/10.1051/epjconf/201817004003

Liege, June 2017.

[3] J. Lecerf, Y. Garnier, J. Girard, C. Domergue, L. Gaubert and C. Manenc, "Study of the linearity of CABRI experimental ionization chambers during RIA transients," in ANIMMA 2017 Conference Proceedings, Liege, June 2017.

[4] J. Lecerf, Y. Garnier, J. Hudelot, J. Girard, C. Domergue, L. Gaubert, M. Maillot, C. Manenc and S. Mirotta, "Analysis of the dosimetry campaign performed during the CABRI commissioning tests," In ANIMMA2017 Conference Proceedings, Liege, June 2017.

[5] G. F. Knoll, "Radiation Detection and Measurement," John Wiley and Sons, 3rd edition, 1999.

[6] J. Gostely et E. Carnal, «La methode des deux oscillateurs pour la mesure du temps mort dans l'instrumentation nucléaire,» Nuclear Instruments and Methods, Vol 150, Iss 3, Pgs 459-464 , 15 April 1978.

[7] B. Cochet, A. Jinaphanh, L. Heulers and O. Jacquet, "Capabilities overview of the MORET5 Monte Carlo code," Annal of Nuclear Energy, Vol 82, Pgs 74–84, August 2015.

Test of Fibre Bragg Gratings samples under High Fast Neutrons Fluence

G. Cheymol, L. Remy, A. Gusarov, D. Kinet, P. Mégret, G. Laffont, T. Blanchet, A. Morana, E. Marin and S. Girard *Senior Member, IEEE*

Abstract– **Optical fibre sensors (OFS) are worthy of interest for measurements in nuclear reactor thanks to their unique features, particularly compact size and remote multi-point sensing for some of them. But besides non negligible constraints associated with the high temperature environment of the experiments of interest, it is well known that the performances of OFS can be severely affected by high level of radiations. The Radiation Induced Attenuation (RIA) in the fibre is probably most known effect, which can be to some extent circumvented by using rad hard fibres to limit the dynamic loss. However, when the fast neutron fluence reaches 10^{18} to 10^{19} n/cm², the density and index variations associated to structural changes may deteriorate drastically the performances of OFS even if they are based on rad hard fibres, by causing direct errors in the measurements of temperature and/or strain changes. The aim of the present study is to access the effect of nuclear radiations on the Fabry Perot (FP) and of Fibre Bragg Grating (FBG) sensors through the comparison of measurements made on these OFS - or part of them - before and after irradiation [1].**

In the context of development of OFS for high irradiation environment and especially for Material Testing Reactors (MTRs), Sake 2 experiment consists in an irradiation campaign at high level of gamma and neutron fluxes conducted on samples of fibre optics – bare or functionalised with FBG. The irradiation was performed at two levels of fast neutron fluence: 1 and 3.10^{19} n/cm² (E>1MeV), at 250°± 25°C, in the SCK•CEN BR2 reactor (Mol Belgium). An irradiation capsule was designed to allow irradiation at the specified temperature without active control. The neutron fluence was measured with activation dosimeters and the results were compared with MCPN computations. Investigation of bare samples gives information on the density changes, while for the FBGs both density and refractive index perturbation are involved. Some results for bare fibres were reported recently. In this paper, we will focus on the measurements made on FBGs that have been manufactured by different laboratories on SMF 28 fibers: CEA, University of St-Etienne and University of Mons. Tested gratings have been written using various conditions (type of fibre, of laser, writing wavelength, power density, post writing thermal annealing,...), leading to various behaviours after Sake 2 irradiation. Bragg wavelength and reflectivity have been measured before and after irradiation thanks to a special mounting at the same temperature. It appears that a change in the shape after irradiation of the Bragg peak disturb the retrieval of the Bragg wavelength.

The measurements show that for nearly all gratings the Bragg peak remains visible after the irradiation, and that Radiation Induced Bragg Wavelength Shifts (RI-BWSs) vary from few pm (equivalent to an error of less than 1°C for a temperature sensor) to nearly 1 nm (equivalent to 100°C) depending of the FBG types. High RI-BWSs could indeed be expected when considering the huge refractive index variation and compaction of the bare fibre samples that have been measured by other techniques. Post writing thermal annealing is confirmed as a key parameter in order to obtain a more radiation tolerant FBG. Our results show that specific annealing regimes allow making FGBs suitable to perform temperature measurements in a MTR experiment.

Index terms– **Fiber Bragg Gratting, optical fiber sensors, radiation effect, optical fibres, neutron, gamma, compaction.**

References

[1] L. Remy, G. Cheymol, A. Gusarov, A. Morana, E. Marin and S. Girard, "Compaction in optical fibres and fibre Bragg gratings under nuclear reactor high neutron and gamma fluence" IEEE Transactions on nuclear science, Vol. 63, N°4, August 2016.

Manuscript received xx. This work was supported by French Nuclear Atomic Agency, Direction of Nuclear Energy, Nuclear Instrumentation project and SCK.CEN, Belgian Nuclear Research Center.

G. Cheymol, L. Remy, are with Den – Service d'Etudes Analytiques et de Réactivité des Surfaces (SEARS), CEA, Université Paris-Saclay, F-91191, Gif sur Yvette, France (33(0)1.69.08.62.71; e-mail: guy.cheymol@cea.fr, e-mail: laurent.remy@cea.fr).

A. Gusarov is with SCK•CEN - Belgian Nuclear Research center, Boeretang 200 2400 Mol Belgium (e-mail: andrei.goussarov@sckcen.be).

D. Kinet and P. Mégret are with the Electromagnetism and Telecommunications Department of the University of Mons, 31 Boulevard Dolez, 7000 Mons, Belgium (e-mail: damien.kinet@umons.ac.be).

G. Laffont and T. Blanchet are with the Technological Research Division (TRD) of the French Alternative Energies and Atomic Energy Commission. F-91191, Gif sur Yvette, France (e-mail : guillaume.laffont@cea.fr; thomas.blanchet@cea.fr)

T. Blanchet, A. Morana, E. Marin, and S. Girard are with Université de Saint-Etienne, Laboratoire Hubert Curien, UMR CNRS 5516, 18, rue du Pr. Lauras, F-42000 Saint-Etienne, France (e-mails : adriana.morana@univ-st-etienne.fr ; emmanuel.marin@univ-st-etienne.fr; sylvain.girard@univ-st-etienne.fr).

https://doi.org/10.1051/epjconf/201817004005

Assessment of the ^{3}He pressure inside the CABRI transient rods - Development of a surrogate model based on measurements and complementary CFD calculations

Olivier Clamens, Johann Lecerf, Jean-Pascal Hudelot, Bertrand Duc, Thierry Cadiou, Patrick Blaise, and Bruno Biard

***Abstract*—CABRI is an experimental pulse reactor, funded by the French Nuclear Safety and Radioprotection Institute (IRSN) and operated by CEA at the Cadarache research center. It is designed to study fuel behavior under RIA conditions. In order to produce the power transients, reactivity is injected by depressurization of a neutron absorber (^{3}He) situated in transient rods inside the reactor core. The shapes of power transients depend on the total amount of reactivity injected and on the injection speed. The injected reactivity can be calculated by conversion of the ^{3}He gas density into units of reactivity. So, it is of upmost importance to properly master gas density evolution in transient rods during a power transient. The ^{3}He depressurization was studied by CFD calculations and completed with measurements using pressure transducers. The CFD calculations show that the density evolution is slower than the pressure drop. Surrogate models were built based on CFD calculations and validated against preliminary tests in the CABRI transient system. Studies also show that it is harder to predict the depressurization during the power transients because of neutron/^{3}He capture reactions that induce a gas heating. This phenomenon can be studied by a multiphysics approach based on reaction rate calculation thanks to Monte Carlo code and study the resulting heating effect with the validated CFD simulation.**

***Index Terms*—CABRI, ^{3}He depressurization, CFD, TOP effect.**

I. Introduction

CABRI is an experimental pulse reactor operated by CEA (Commissariat à l'Énergie Atomique et aux Énergies Alternatives) at the Cadarache research center. Since 1978, the experimental programs have been aiming at studying the fuel behavior under Reactivity Initiated Accident (RIA) conditions. In order to study PWR high burn up fuel and new cladding materials behavior under such transients, the facility was modified to accept a pressurized water loop in its central part able to reproduce thermal-hydraulics characteristics representative of PWR nominal operating conditions (155 bar, 300°C). This project, which began in 2003 and supported first commissionning power tests from October 2015 to March 2017, was driven within a broader scope including both an overall facility refurbishment and a complete safety review. The global modifications were conducted by CEA. The experiments take place in the framework of the OECD/NEA CIP (CABRI International Program) Project led by IRSN (Institut de Radioprotection et de Sûreté Nucléaire), which financially supports the refurbishment as well as the operational costs of the facility. The CIP program will investigate several fresh and burnt UOx and MOX LWR fuel samples with new cladding materials under RIA conditions, with a foreseen completion by the end of 2023.

Power transients are generated by a dedicated so-called transient rods system [1] allowing the very fast depressurization of ^{3}He tubes positioned inside the CABRI core. This paper focuses on the study of the ^{3}He depressurization of CABRI transient rods. The main objective is to properly reproduce the ^{3}He density evolution in the transient rods situated in the CABRI core, from experimental data provided by pressure transducers situated in the valve and piping system far from the core. The first part of this paper consists of a brief description of the transient rods system. In a second part, the CFD (Computational Fluid Dynamics) approach is addressed. This approach allows the evaluation of state parameters in the entire system and not only at the pressure transducers locations. The last part of the paper deals with the explanation of the TOP effect that affects the depressurization during the transient over power.

O. Clamens, J. Lecerf, J-P. Hudelot, B. Duc DEN CAD/DER/SRES CEA Cadarache, Bt 721. 13108 St Paul Lez Durance, France e-mail: olivier.clamens@cea.fr.
T. Cadiou, DEN CAD/DER/SESI CEA Cadarache, Bt 1222. 13108 St Paul Lez Durance, France.
P. Blaise, DEN CAD/DER/SPEX CEA Cadarache, Bt 238. 13108 St Paul Lez Durance, France.
B. Biard, IRSN/PSN-RES/SEREX Cadarache, BP3 13115 Saint-Paul-Lez-Durance Cedex, France.

II. CABRI reactivity injection system

CABRI is a pool-type reactor, with a core made of 1487 stainless steel cladded 6 wo% enriched UO_2 fuel rods. The reactor is able to reach a 25 MW power level in steady state conditions. The reactivity is controlled via a system of 6 bundles made of 23 hafnium control and safety rods. The key feature of the CABRI reactor is its unique reactivity injection system [1]. This device allows the very fast depressurization of the ^3He (strong neutron absorber) into a discharge tank. The ^3He is previously introduced inside 96 tubes (so called "transient rods") located in 4 banks among the CABRI fuel rods (see Fig. 1a).

The CABRI transient rods system is made of the following main components (see Fig. 1b):

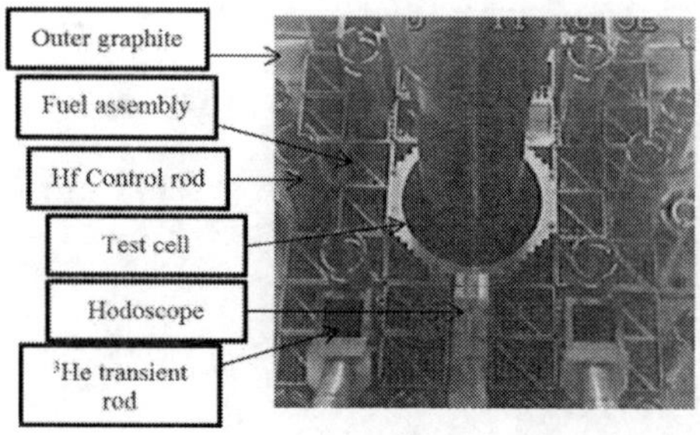

(a) Radial view of the CABRI core

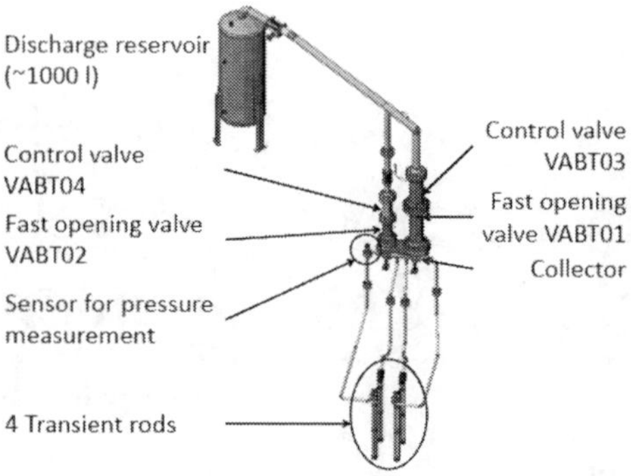

(b) Main components of the CABRI transient rods system

Fig. 1. CABRI transient rods system

- 4 fuel assemblies (7x7 pins) equipped on their periphery with 24 tubes instead of 24 fuel rods. The 4 transient assemblies are pressurized to the target pressure (15 *bar* maximum) by the use of a compressor which pumps the ^{3}He from its storage tank via a devoted circuit.
- From the top of this collector, two flow channels (low and high flow rates) lead to a 1000 *l* discharge tank set under vacuum before operation. Both channels are equipped with a fast-opening valve (respectively with small and large diameters) followed by a controlled valve.
- A specific control device that triggers the different orders of the experimental sequence as for the opening time of the two fast-opening valves and the shutdown of the reactor control rods.
- Two different pressure transducers measuring the ^{3}He pressure at the inlet of the collector. For design reasons, the ^3He pressure cannot be measured directly in the transient rods.

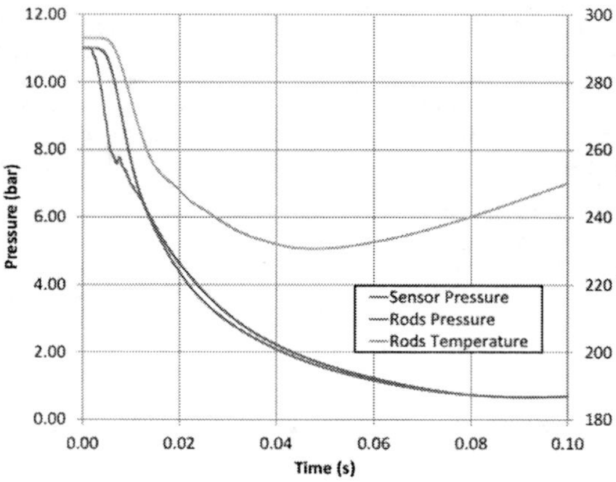

Fig. 2. Helium pressure and temperature evolution during a depressurization according to CFD calculation

III. CFD simulation of ^{3}He depressurization in transient rods

The CFD modeling, unlike an analytical approach, can precisely handle complex geometries. The ^{3}He pressure evolution during the depressurization will then be calculated in the entire transient rods system, and not only at the pressure transducers location. The complete validation of the CFD Simulation is described in [2].

The ^{3}He gas depressurization induces a temperature drop in transient rods (see Fig. 2). Assuming an ideal gas, the gas quantity "n" is defined as in (1).

$$n = \frac{PV}{RT} \tag{1}$$

A good state parameter that can be linked to the injection of reactivity is the ^{3}He density, that is proportional to the number of atoms (2).

$$d_{^3He} = \frac{nM}{V} = \frac{PM}{RT} \tag{2}$$

The density evolution inside the transient rods is slower than the pressure "P" evolution (see Fig. 3), as temperature "T" varies in about the same proportions. That induces a slower calculated evolution of the reactivity injection.

Several types of surrogate models were built based on validated CFD simulations. The CEA's URANIE uncertainty platform [3] was used for the surrogate models construction.

In fact, transient rods depressurization is a little different when core power evolves. This little difference can have big effects on power transients. This effect appears when the gas pressure and the core power are both relatively high. It is named "TOP effect" as "Transient Over Power effect". We can observe it on the pressure curves measured during power transients (Fig. 4).

IV. The TOP effect

The TOP effect comes from ^{3}He heating during power transients. As power increases, the thermal neutron flux also increases. So, the neutron absorption by ^{3}He intensifies. This reaction produces two charged particles : proton and tritium. One part of their energies is deposited in the ^{3}He gas by ionization before reaching the metallic wall of the transient

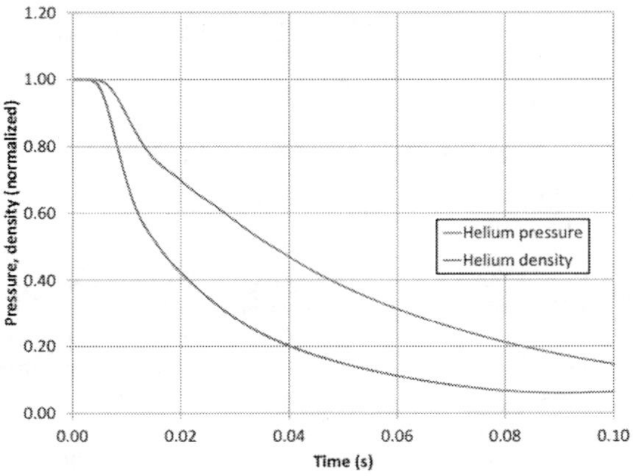

Fig. 3. Helium pressure and density evolution during a depressurization according to CFD calculation

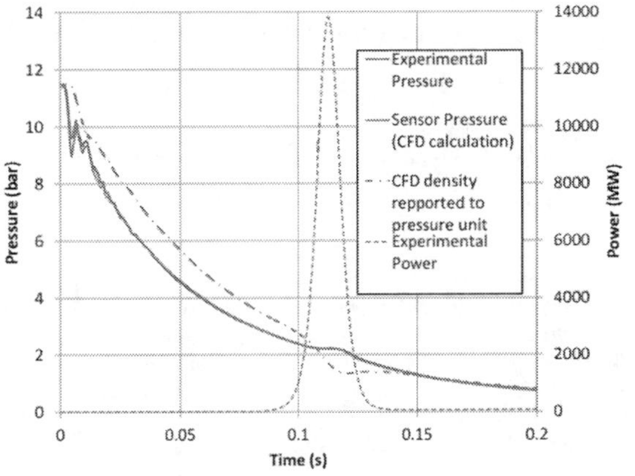

Fig. 5. CFD calculation of simple depressurization by high flow rate channel (VABT03 aperture $\sim$ 40 %) including TOP effect

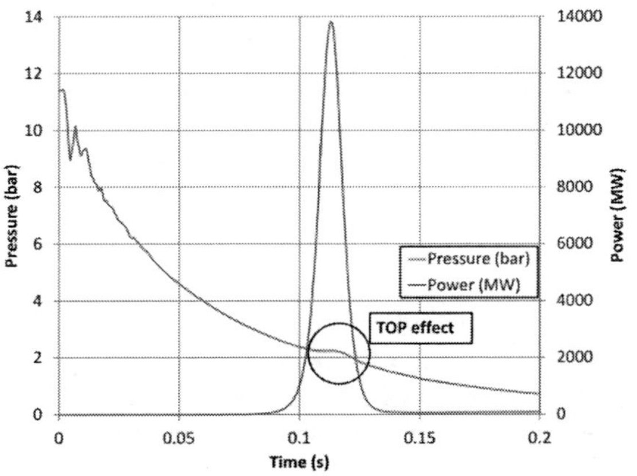

Fig. 4. TOP effect visualization on a recent CABRI power transient (2017)

rods. Denser is the gas, higher the probability of ionization is and more important the deposited energy is. The direct effect of this energy deposit is that the gas temperature increases. A temperature increase is equivalent to a pressure increase. The pressure gap between rods and flow channels implies a faster depressurization of helium from the transient rods. This finally implies a rise of the reactivity injection speed. The TOP effect increases the maximum power and the energy deposit. For relatively slow transients (at least 20 ms FWHM), it can represent more than the half of the maximum power and at least 30 % of the energy deposit. Thus the TOP effect has to be taken into account in order to have an accurate predictive tool.

We simulated a depressurization by entering the experimental power and a function linking power, helium density and helium function, in the existing simulation. Results of this simulation are plotted on Fig. 5. We can observe that density is decreasing faster at the moment of the peak, resulting in an increase of the injected reactivity.

V. Conclusions

The present study points out differences between measured pressure and ^{3}He density in transient rods. It shows that the gas density evolution is slower than pressure evolution because of temperature changes in the rods. Surrogate models were developed in order to replace old models based on analytical solution of the problem (with simplification of the geometry). The study demonstrates that the ^{3}He density evolution is different if core power is boosted due to gas heating by neutron/^{3}He interactions. This effect, named TOP effect, affects density evolution by increasing depressurization speed during the transients. It explains some difficulties in the CABRI power transients prediction. Surrogate models are in development in order to be used in future power transients calculations.

Appendix A
Nomenclature

Name	Definition
n	Amount of substance of the gas (in moles)
P	Pressure
V	Volume
R	Gas constant (8.314 $J.K^{-1}.mol^{-1}$)
T	Absolute temperature
$d_{^3He}$	Mass density of the gas
M	Molar mass of the gas

References

[1] B. Duc, B. Biard, P. Debias, L. Pantera, J.-P. Hudelot, and F. Rodiac, "Renovation, improvement and experimental validation of the Helium-3 transient rods system for the reactivity injection in the CABRI reactor," in *International Group On Research Reactors*, 2014, Bariloche, Argentina, November 17 - 21.

[2] O. Clamens, J. Lecerf, B. Duc, J.-P. Hudelot, T. Cadiou, and B. Biard, "Assesment of the CABRI transients power shape by using CFD and point kinetic codes." in *Proc. Int. Conf. PHYSOR2016*, Sun Valley, USA, 2016, pp. 1747–1758.

[3] F. Gaudier, "URANIE: The CEA/DEN Uncertainty and Sensitivity platform," *Procedia - Social and Behavioral Sciences*, vol. 2, no. 6, pp. 7660–7661, Jan. 2010. [Online]. Available: http://www.sciencedirect.com/science/article/pii/S1877042810013078

Design of a delayed neutron detection system for the ASTRID Sodium-cooled fast reactor

Romain Coulon, Emmanuel Rohée, Jonathan Dumazert, Sara Garti, Philippe Filliatre, Christian Jammes

Abstract – **Delayed neutron detection systems are key apparati in Sodium cooled Fast Reactors to prevent safety accidents. Based on the feedbacks from Phenix and Superphenix, DND systems are studied by means of the Monte-Carlo particles transport code MCNP6. Promising designs are proposed and investigated as a result of the simulation study.**

***Index Terms*— Fast reactor; Sodium; Fuel failure; Fission products; Neutron detection.**

I. Introduction

IN 4TH generation reactors, the sodium coolant has to be kept free from any fuel contamination. The objective of DND systems in the safety policy of the reactor is to detect the release of fission products emitting delayed neutrons (FPDN) as bromine 87 or iodine 137. In-vessel fission chambers have been studied to efficiently detect the loss of FPDN but ex-vessel DND system is still a favorite, as immune to any interference from core-neutrons [1]. In this ex-vessel configuration, remaining interferences can arise from:

- original pollution of the cladding by fuel,
- photoneutrons produced by (γ,n) photonuclear reactions on deuterium nuclei when polyethylene is implemented as a moderator.

A solution for photoneutron interference removal has been developed in which graphite is implemented instead of polyethylene for neutron cooling. Previous work has proven detection limit improvements in the graphite configuration compared to older DND system design, and therefore a safety improvement for SFR is expected [2]. This new study shows the tuning of DND system designs where dimensions are optimized and detector technologies are compared (helium-3 and boron-10 proportional counters).

II. METHOD

To carry out the study, Monte-Carlo particle transport simulations have been performed using MCNP6 code. A digital model of the system has been designed allowing the simulation of an intrinsic counter behavior by estimation of deposited energy distributions from light ions ($^{3}He(n,p)$ and $^{10}B(n,a)$ reactions), which is a new feature from MCNP6 code [3]. The response of simulated detectors is compared to the results of experimental tests, after an energy calibration.

The following specifications have been considered. The system has a cylindrical shape with a diameter equal to 70 cm and composed from the center by:

- A continuous circulation of sodium into a volume,
- A lead shied allowing limiting gamma dose rate from ^{24}Na on counters (pile-up),
- A graphite or polyethylene moderator maximizing capture probabilities in ^{3}He or ^{10}B nuclei,
- Helium 3 or boron 10 counters located into the moderator.

Fig. 1 shows a schematic view of an ex-vessel DND system. The primary sodium coolant is pumped from the hot pool, at the outlet of each assembly through the control plug or in different locations around the core. It circulates through a volume located at the outside of the vessel. This volume is thermally insulated and surrounded by a gamma rays shielding reducing photoneutron activity and gamma ray pile-up on the detectors. Neutron detectors are enclosed in a moderator.

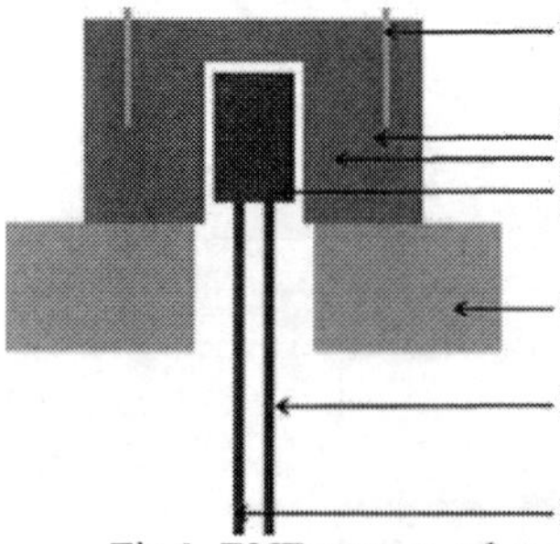

Fig.1. DND system schematic view.

III. Results

The thickness required to ensure the operability of neutron detectors against gamma ray pile-up is firstly calculated. The gamma dose rate behind the shield is estimated for different values of lead thickness. A flux point tally (f5) is used and transformed in dose rate thanks to a transfer function provided by ICRP87 [4]. Fig. 2 shows the dose rate as a function of the lead thickness considering a

R. Coulon, E. Rohée, J. Dumazert, and S. Garti are with the CEA, LIST, F-91191 Gif-sur-Yvette, France. (e-mail: romain.coulon@cea.fr ; phone: +33169088491).
P. Filliatre and C. Jammes are with the CEA, DEN, F-13108 Saint-Paul-les-Durance, France. (e-mail: christian.jammes@cea.fr ; phone: +33442253926).

^{24}Na activity equal to 310 GBq.kg^{-1} in a volume of 2.8 liters.

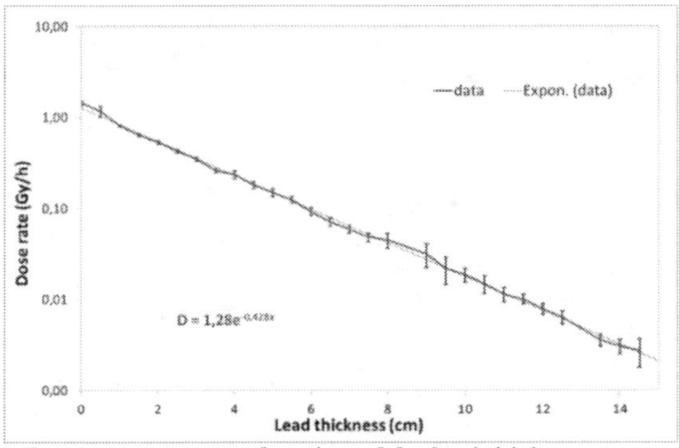

Fig.2. Dose rate as a function of the lead thickness.

According to this result and dose rate limits which are 10 mGy.h^{-1} for ^{3}He counters and 100 mGy.h^{-1} for ^{10}B counters, the obtained minimal thickness are therefore 12 cm and 8 cm for respectively ^{3}He and ^{10}B counters.

Two types of moderator are considered in this work: polyethylene and graphite. It has been observed on Phenix and Superphenix that a significant level of photoneutron signal is induced by (γ,n) reactions on deuterium isotopes naturally contained in polyethylene [5]. This background signal changes as a function of the reactor operating parameters, notably neutron power. Graphite has been proposed in previous work in order to suppress this interference signal [2]. Indeed, the photoneutron reaction thresholds for carbon isotopes lies above 5 MeV, far from the 2.74 MeV energy of the highest gamma ray in ^{24}Na spectral emission (see Fig. 3).

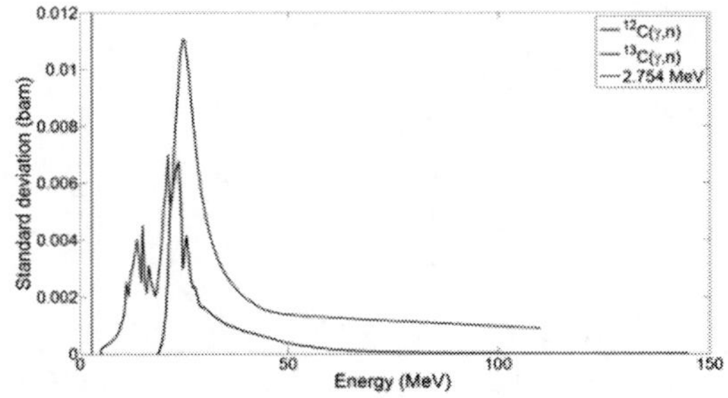

Fig.3. Photoneutron cross-section of carbon isotopes (ENDF-VII data base).

The neutron spectrum used in calculations is one of ^{87}Br (Fig. 4). Neutrons are emitted homogeneously into the sodium volume, and the reaction rate normalized to one emitted neutron is calculated by MCNP6 (F4 +SD4+ FM4 tally). The estimated count rate is obtained by weighting this result with fission product activity, fission product emission rate, and a factor taking into account the losses in detector (self-absorption into the deposit or deposition below the discrimination threshold).

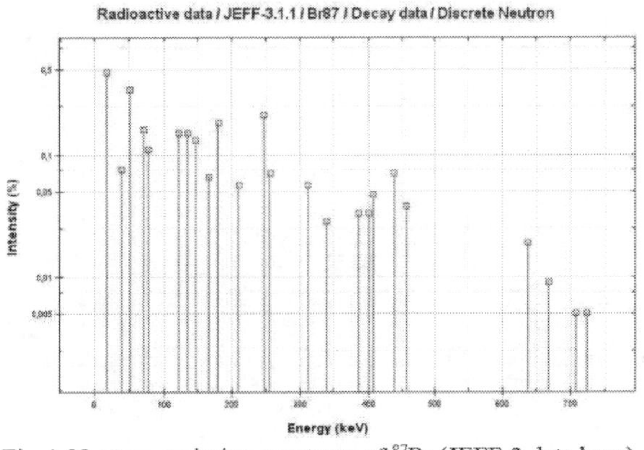

Fig.4. Neutron emission spectrum of ^{87}Br (JEFF-3 data base).

Two models of detectors have been retained for the study: ^{3}He proportional counter 65NH45 (MIRION) and ^{10}B proportional counter CPNB48 (PHOTONIS). The position of the detectors has been varied in order to find the configuration that maximizes neutron sensitivity. Fig. 5 shows the neutron sensitivity as a function of the position of counters into the moderator. The use of graphite moderator instead of polyethylene induces a sensitivity reduction of a factor approximatively equal to 3. It can be also concluded that ^{3}He counter provides sensitivity approximatively the twice higher than the ^{10}B counter.

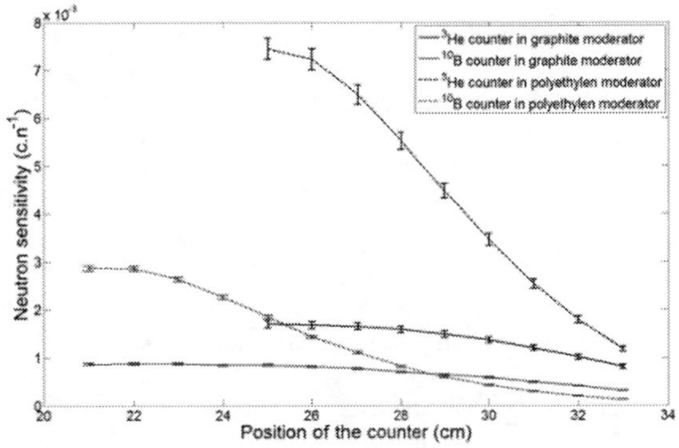

Fig.5. Neutron sensitivity as a function of counter locations and for different configurations.

Photoneutron noise is evaluated using photonuclear data LA150U implemented in MCNP6 [6]. The isotopic abundance of deuterium in polyethylene is equal to 2.145*10^{-3} %mass. In order to accelerate the calculation, this abundance is artificially increased and estimations are a posteriori corrected. A photoneutron signal equal to 0.5 cps in the ^{3}He counter in polyethylene and 23.5 cps in the case of ^{10}B counter in polyethylene has been estimated.

Table 1 presents the results obtained by simulation for each configuration:

- ^{3}He counter in graphite moderator (^{3}He/C),

- ^{10}B counter in graphite moderator ($^{10}B/C$),
- ^{3}He counter in polyethylene moderator ($^{3}He/CH_2$),
- ^{10}B counter in polyethylene moderator ($^{10}B/CH_2$),

Table 1. Results obtained by simulation.

	$^{3}He/C$	$^{10}B/C$	$^{3}He/CH_2$	$^{10}B/CH_2$
Lead thickness	12 cm	8 cm	12 cm	8 cm
Location in moderator*	2.25 cm	4.25 cm	1.25 cm	2.25 cm
Neutron sensitivity	$1.68.10^{-3}$ $c.n^{-1}$	$8.83.10^{-4}$ $c.n^{-1}$	$7.44.10^{-3}$ $c.n^{-1}$	$2.86.10^{-3}$ $c.n^{-1}$
Photoneutron signal	0 $c.s^{-1}$	0 $c.s^{-1}$	0.49 $c.s^{-1}$	4.7 $c.s^{-1}$

* from the lead surface to the detector axis

While photoneutrons can be suppressed, neutrons induced by the pollution will constitute a remaining background which will also change as a function of core parameters. Some algorithms are studied to better manage this background baseline [7-10].

IV. CONCLUSION

The DND system is a crucial part of the safety system of a sodium cooled fast reactor. A simulation study has been conducted in order to propose some designs allowing detecting efficiently and with reliability cladding failures.

Two types of detectors technology: ^{3}He and ^{10}B proportional counters and two types of moderator: graphite and polyethylene have been tested. The use of graphite suppresses the photoneutron signal while reducing the detection sensitivity by a factor three.

References

[1] P. Filliatre, C. Jammes, N. Chapoutier, *et al.*, "In vessel detection of delayed neutron emitters from clad failure in sodium cooled nuclear reactors: An estimation of the signal", *Nuclear Instruments and Methods* A, vol. 744, pp. 1-10, 2014.

[2] E. Rohée, R. Coulon, C. Jammes, P. Filliatre, *et al.*, "Delayed Neutron Detection with graphite moderator for clad failure detection in Sodium-Cooled Fast Reactors", *Annals of Nuclear Energy*, vol. 92, pp. 440-446, 2016.

[4] "MCNP6 User's Manual Version 1.0". Technical report LANL LA-CP-13-00634, chap. 3, pp. 71-72, 2011.

[5] ICRP, 1987. Data for Use in Protection against External Radiation. ICRP Publication 51. Ann. ICRP 17 (2-3).

[5] C. Berlin, "Evolution of FFD-FFL instrumentation on French fast reactors", *In proceedings of IAEA Fuel Failure Detection and Location in LMFBRs Meeting*, pp. 36-44, 1982.

[6] M.C. White, R.C. Little, M.B. Chadwick, "Photonuclear Physics in MCNP(X)". *Proceedings of the ANS meeting on Nuclear Applications of Accelerator Technology,* 1999.

[7] R. Coulon, V. Kondrasovs, J. Dumazert, *et al.*, "Estimation of Nuclear Counting by a Nonlinear Filter Based on a Hypothesis Test and a Double Exponential Smoothing" *IEEE Transactions on Nuclear Science*, vol. 63, no. 5, pp. 2671-2676, 2016.

[8] R. Coulon, V. Kondrasovs, J. Dumazert, *et al.*, "Estimation of Nuclear Counting by a Nonlinear Filter Based on a Hypothesis Test and a Double Exponential Smoothing" *IEEE Transactions on Nuclear Science*, vol. 63, no. 5, pp. 2671-2676, 2016.

[9] P. Filliatre, C. Jammes, J.-P. Jeannot, F. Jadot, "In vessel detection of delayed neutron emitters from clad failure in sodium cooled nuclear reactors: Information treatment", *Annals of Nuclear Energy*, vol. 65, pp. 385-389, 2014.

[10] O.I. Albutova, D.A. Lukyanov, "Investigation of dependence of BN-600 reactor sector fuel cladding leak detection system responses on the operation parameters", *Nuclear Energy and Technology*, vol.1, pp. 248-252, 2015.

Fission Chambers Detector Unit Mockup Testing at Research Reactors

V. P. Alferov, V. A. Fedorov, A. V. Kudryavtsev, E. S. Martazov, N. A. Selyaev, [1]A. V. Batyunin, [1]Yu. A. Kashchuk and [1]V. A. Vorobiev
National Research Nuclear University MEPhI, [1]Institution "Project Center ITER"

The required range of neutron radiation intensity monitoring of the International Experimental Thermonuclear Reactor (ITER) is from 10^{14} to 10^{21} n/s. One of the systems of neutron diagnostics, called the Divertor Neutron Flux Monitor (DNFM), contains as primary converters fission chambers (FCs) placed in the ITER Vacuum Vessel beneath the Divertor Dom. During the operation the FCs will be exposed to the intense neutron radiation and electromagnetic fields. The expected neutron flux at the location of the DNFM FC is 10^6 - 10^{13} n·cm^{-2}·s^{-1}.

A distinctive feature of the neutron diagnostic channel for ITER is the desired time resolution of 1 ms and a measuring error of 10%. Such a requirement can be realized only at an FC signal registration rate of at least $1 \cdot 10^5$ s^{-1}. Therefore, the pulse mode of processing the FC signals is very limited and it will probably be used only for calibration and diagnostics of the equipment. The working monitoring range of the channel with one FC will be less than 7 orders of magnitude. This explains the necessity of using several FCs with different sensitivity in the DNFM. The hardware of the signal processing equipment will be the same for all FCs.

The wide-range channel equipment circuitry development for DNFM recording signals is based on the experience of the MEPhI in the development, implementation and operation of wide-range neutron flux monitoring equipment for control and protection systems (CPS) for various types of nuclear reactors. An important factor in the design of such equipment is the use of the fluctuation (Campbell) mode of fission chamber operation and signal processing equipment. The use of the fluctuation mode made it possible to radically solve the problem of the neutron flux monitoring in CPS channels in the 10-order range using a single FC. The equipment operates in the pulse, fluctuation and current modes.

The work which related to the testing of the DNFM detector unit (DU) mockup by the influence of intensive fluxes of neutron and gamma radiation and evaluation of technical decisions of the neutron flux monitoring channel was performed with use of the research reactors. The fission chamber Signals Processing Unit (SPU) was used. This device based on microprocessors with wide application of high integration chips, including complex programmable logic devices (CPLD).Taking into account the requirements for ITER neutron diagnostics and the conditions for DU testing, the SPU circuitry was adjusted and the software of microprocessors and CPLDs was modified to control the temperature of the DU. In addition, a test complex including hardware and software for controlling equipment, processing and recording measurement data is assembled for a detailed research of the DU characteristics.

The test complex includes the remote thermo assembly and the control and signal processing equipment. Thermo assembly includes an electric heater (EH) and a thermocouple (TC). Control and signal processing equipment consist of heater power supply (PS), DU signals processing unit (SPU), industrial personal computer (IPC), providing recording of signals and control of the SPU operation modes and automatic test unit (ATU) for recording the shape of pulses and the pulse height amplitude spectrum of fission chambers.

The DNFM detector unit mockup on the base of a fission chamber KNT 25 with a communication line was tested at the first stage of the tests. The mass of uranium in the chamber is equal to 1 g and the electrodes coating density is 1 mg/cm^2. The communication line was made of a double coaxial cable with fiberglass insulation with length of more than 3 m. The fission chamber with the communication line was placed in a sealed enclosure filled with an inert gases mixture with a pressure of up to $5 \cdot 10^5$ Pa. At the second stage the mockup of a similar design based on the KNT-30 fission chamber with a uranium mass of 100 mg was tested. For testing DU at temperatures up to 350 °C it was mounted in a thermo assembly with an electric heater and thermocouple.

The SPU provides receiving of the pulse-current signal from the DU and signals from the temperature sensors, amplification, digital processing and transmission in digital code of the measuring and diagnostic information to the IPC. Also the SPU controls the power supply of the electric heater EH by commands from the IPC.

The measurement methods and SPU software were developed for carrying out laboratory and reactor tests of the equipment. They provide:

- parameters setting, acquisition and archiving of information coming from the SPU;
- data processing and heater control of the thermal assembly with DU;
- real time display of data on all signals and parameters (controlled and calculated) on the screen in digital and graphical form;
- presentation, editing and printing of digital and graphical information of measurement results and processing.

The SPU provides DU fission chamber bias by a DC high-voltage power supply in the range from 0 to 500 V which controlled by commands from the IPC and provides the threshold voltage generation for the FC analog processing module.

The FC analog processing module consist of three measuring chains - pulse, fluctuation and current. The DU fission chamber signal enters to the differential input of the FC analog processing module which provides:

- amplification and preliminary conversion of FC pulses, fluctuation and average current;
- noise and current pulses discrimination caused by gamma radiation;
- FC current pulses transformation to standard form voltage pulses and calculation by CPLD of the counting rate (pulse chain);
- logarithmic transformation and pre-filtering of the average and fluctuation FC currents.

The experiments to determine the DNFM detector unit mockup and the wide-range channel for monitoring the neutron flux characteristics were performed at the research reactors of the IRT MEPhI and VVR-c. The detector unit mockup was studied under DU temperature range from 20°C to 350°C with the neutron flux range from 10^3 to 10^{12} $n \cdot cm^{-2} \cdot s^{-1}$.

The tests confirmed the ability to control the change in reactor power using the SPU fluctuation chains before and after the neutron fluence of $4.2 \cdot 10^{15}$ $n \cdot cm^{-2}$ in the range of four decades with the FC background current up to $1 \cdot 10^{-4}$ A. The measuring chains operating ranges in term of the equivalent count rate are determined as a result of processing the measurement data for pulse chain up to $1 \cdot 10^6$ s^{-1}, for fluctuation chain from $4 \cdot 10^5$ to $4 \cdot 10^9$ s^{-1} and for current chain from $2 \cdot 10^8$ to $1 \cdot 10^{10}$ s^{-1}.

The lower boundaries of the operating ranges for the fluctuation and current chains can be reduced several times by taking into account their background signals. Integral nonlinearity of all measuring chains in operating ranges does not exceed ± 2%.

Satisfaction with speed requirements (1 ms measurement cycle) leads to a significant increase in statistical noise in measurements results for the pulse and fluctuation chains. We can estimate its contribution to the statistical error at different power levels and refine the measurement chains ranges boundaries using the ratio of the rms deviation (σ_F) to the average value of the signal (F) as a noise parameter. Therefore, the lower limit of the pulse chain must be increased to $4 \cdot 10^5$ s^{-1} and for the fluctuation chain up to $2 \cdot 10^6$ s^{-1} for a relative statistical noise of less than 5%.

High temperature ultrasonic sensor for fission gas characterization in MTR harsh environment

O. Gatsa, P. Combette, E. Rozenkrantz, D. Fourmentel, C. Destouches, and J.Y. Ferrandis

Abstract— **In the contemporary world, the measurements in hostile environment is one of the predominant necessity for automotive, aerospace, metallurgy and nuclear plant. The measurement of different parameters in experimental reactors is an important point in nuclear power strategy. In the near past, IES (Institut d'Électronique et des Systèmes) on collaboration with CEA (Commissariat à l'Energie Atomique et aux Energies Alternatives) have developed the first ultrasonic sensor for the application of gas quantity determination that has been tested in a Materials Testing Reactor (MTR). Modern requirements state to labor with the materials that possess stability on its parameters around 350 °C in operation temperature. Previous work on PZT components elaboration by screen printing method established the new basis in thick film fabrication and characterization in our laboratory. Our trials on Bismuth Titanate ceramics showed the difficulties related to high electrical conductivity of fabricated samples that postponed further research on this material. Among piezoceramics, the requirements on finding an alternative solution on ceramics that might be easily polarized and fabricated by screen printing approach were resolved by the fabrication of thick film from Sodium Bismuth Titanate (NBT) piezoelectric powder. This material exhibits high Curie temperature, relatively good piezoelectric and coupling coefficients, and it stands to be a good solution for the anticipated application.**

In this paper, we present NBT thick film fabrication by screen printing, characterization of piezoelectric, dielectric properties and material parameters studies in dependence of temperature. Relatively high resistivity in the range of $1 \cdot 10^{13}$ Ohm·cm for fabricated thick film is explained by Aurivillius structure in which *a*- and *b*- layers form perovskite structure between oxides of *c*- layer. Main results of this study are presented and discussed in terms of feasibility for an application to a new sensor device operating at high temperature level (400° C). Piezoelectric parameters enhancement and loss reduction at elevated temperatures are envisaged to be optimized. Further sensor development and test in MTR are expected to be realized in the near future.

Index Terms— **acoustic measurements, high temperature transducer, nuclear instrumentation, screen printing, sodium bismuth titanate.**

This work, presented at the International Conference on Advancements in Nuclear Instrumentation Measurement Methods and their Applications, June 19-23, 2017

O. Gatsa, P. Combette, E. Rosenkrantz, and J. Y. Ferrandis, are with the CNRS-University Montpellier 2, Institute of Electronics and Systems, UMR 5214, F-34095 Montpellier, France (e-mail: oleksandr.gatsa@ies.univ-montp2.fr).

D. Fourmentel and C. Destouches are with the Reactor Studies Department, French Nuclear Energy and Alternative Energies Commission, CEA Cadarache, F-13108 Saint Paul-Lez-Durance, France (e-mail: damien.fourmentel@cea.fr).

I. INTRODUCTION

NUCLEAR fuel rods lifetime optimization will increase the combustion rate of nuclear fuel, hence reducing the volume of waste to be recycled. In the case of previous studies, the measurements of fission gas composition in REMORA-3 experiment [1] via in-pile control of fuel rods interior environment have been carried out. The gas mixtures (Helium and fission gases) were successfully monitored during the experiment up to operation temperature of 200°C. The measurement principle was based on data processing of the registered echograms. REMORA-3 experiment was a first step in the field of gas composition monitoring under harsh environment conditions in a nuclear plant. It opens new possibilities in the field of qualitative measurements by an acoustic approach.

Currently, we work on a new ultrasonic sensor for an experimental material testing reactor Jules Horowitz with a temperature constraint of 400°C. The difficulties with brazing and ensuring of a parallelism between an active element and stainless steel plate led us to work on another sensor configuration concept which is based on the screen-printing. Manufacturing of the ceramic material by this method ensures high flexibility on sensor fabrication: direct deposition can be performed on different structure types; the deposition area shape of material can be easily chosen.

II. EXPERIMENTAL PROCEDURE

In this paper, the studies of a manufacture process of high temperature piezoelectric material (Curie temperature of 650°C) are presented. Three types of a Sodium Bismuth Titanate (NBT) samples, chemical formula Na0.5Bi4.5Ti4O15, were manufactured by the screen printing method. Two types of samples were in a form of a detached disk. One type of sample was directly deposited on an alumina substrate. The diameters of the samples were about 0.5 and 1 cm. The thickness was 100 µm for detached disks and 40 µs for on alumina NBT. Before processing with piezoelectric characterization, the samples were polarized in the silicon oil under an electric field of 100 kV/cm.

The samples were tested regarding their morphological, chemical, resistivity, piezoelectric, dielectric, ferroelectric properties. By studying the evolution of dielectric properties as a function of temperature we were able to determine the optimal conditions for a high temperature sensor manufacture. The first device of an NBT ceramic material on alumina plate was tested on its acoustic properties. The measurements of the speed of sound in liquid substances were carried out.

III. RESULTS AND DISCUSSIONS

By analyzing the difference in the manufacture process such as grain size of a raw material and sintering conditions, we were able to study samples morphologies. The grain size of powder determines the morphology of ceramics. By increasing sintering time for one type of samples, we were able to process with more dense ceramic structure.

The importance on a chemical composition of a raw NBT material was shown to be the main factor in establishing the material properties. In a case of materials for high temperature applications, we are looking for a ceramic possesses high resistivity. High resistivity means endurance to a voltage and temperature shock thus reliability of sensor operation. From its measurement two types of fabricated samples have a relatively good resistivity.

In order to confirm resistivity values, dielectric properties were measured at room temperature and compared with the state of the art. They are in good agreement with previously reported NBT bulk ceramics [2] for the samples with relatively high resistivity of more than $1 \cdot 10^{13}$ Ohm·cm.

Piezoelectric charge coefficient has been found to be higher for the samples with a low dielectric loss (high resistivity), approaching the state of the art of 16 pC/N for bulk NBT [3].

The dielectric properties monitoring as a function of temperature show that the best figure evolution (relative stability of dielectric permittivity and dielectric loss up to 400°C) was obtained for the sample fabricated from a small grain size powder, sintered for 15 minutes. Furthermore, it was a material containing lower quantity of sodium. In an NBT, sodium is substitute bismuth in the A-site of perovskite in Aurivillius structure and is an important element in reducing of the conductivity (this assumption is based on a high conductivity of a Bismuth Titanate ceramic which is a ceramic of the same family not containing sodium [4]). However, in a pure state, sodium is a good electrical conductor [5]. Apparently, for the samples with a low dielectric stability on temperature sodium is in an increased quantity, which explains that loss increases dramatically with a temperature.

Based on material properties evolution as a function of temperature we choose an optimal process on material fabrication. The active NBT material together with screen printed electrodes were deposit on an alumina substrate. The device was joined with a centimetric acoustic cavity filled by water or ethanol. The acoustic signal was generated and acquired by pulser/receiver and oscilloscope. The measurements under room temperature conditions ensure device performance. The speed of sound for water and ethanol were in the good agreement with the state of the art [6, 7]. In the case of measurement of gas the signal processing as reported in [1] and [8] will be applied.

IV. CONCLUSION

This study proves the feasibility of acoustic sensor manufacture by screen printing method for high temperature applications. Measurements of dielectric properties as a function of temperature demonstrate the importance of the samples manufacture conditions. Their variation, for on alumina sample, ensures its applicability in the desirable temperature range from 300 to 400 °C. We suggest following the same process of fabrication for subsequent sensors.

The piezoelectric coefficient of samples was in the range of theoretical value for NBT bulk ceramic. Acoustic test for water and ethanol show a high efficiency in the measurement of the speed of sound.

The investigation of sensor properties as a function of temperature for a long-time scale period is the next step in sensor characterization. The experiment with a gas composition measurement are planned to be carried out in the near future.

REFERENCES

[1] Rosenkrantz, E., Ferrandis, J. Y., Augereau, F., Lambert, T., Fourmentel, D., & Tiratay, X. (2013). An Innovative Acoustic Sensor for In-Pile Fission Gas Composition Measurements. *IEEE Transactions on Nuclear Science*, *60*(2), 1346-1353.

[2] Wang, C. M., Zhao, L., Liu, Y., Withers, R. L., Zhang, S., & Wang, Q. (2016). The temperature-dependent piezoelectric and electromechanical properties of cobalt-modified sodium bismuth titanate. *Ceramics International*, *42*(3), 4268-4273.

[3] Subbarao E. A family of ferroelectric bismuth compounds. Journal of Physics and Chemistry of Solids. 1962;23(6):665–676.

[4] Shulman, Holly S., et al. "Microstructure, electrical conductivity, and piezoelectric properties of bismuth titanate." *Journal of the American Ceramic Society* 79.12 (1996): 3124-3128.

[5] Weast, Robert C., and M. J. Astle. "CRC." Handbook of Chemistry and Physics. Boca Raton, Florida: Chemical Rubber Company Publishing. pp. E 110 (1984).

[6] Del Grosso, V. A., and C. W. Mader. "Speed of sound in pure water." *the Journal of the Acoustical Society of America* 52.5B (1972): 1442-1446.

[7] Wilson, Wayne, and David Bradley. "Speed of sound in four primary alcohols as a function of temperature and pressure." *The Journal of the Acoustical Society of America* 36.2 (1964): 333-337.

[8] Rosenkrantz, E. (2007). Conception et tests d'un capteur ultrasonore dédié à la mesure de la pression et de la composition des gaz de fission dans les crayons combustibles (Doctoral dissertation, Univ. Montpellier 2).

EPJ Web of Conferences **170**, 04009 (2018) https://doi.org/10.1051/epjconf/201817004009
ANIMMA 2017

Innovative hybrid pile oscillator technique in the Minerve reactor: open loop vs. closed loop

Benoit Geslot, Adrien Gruel, Stéphane Bréaud, Pierre Leconte, and Patrick Blaise

***Abstract*—Pile oscillator techniques are powerful methods to measure small reactivity worth of isotopes of interest for nuclear data improvement. This kind of experiments has long been implemented in the Mineve experimental reactor, operated by CEA Cadarache. A hybrid technique, mixing reactivity worth estimation and measurement of small changes around test samples is presented here. It was made possible after the development of high sensitivity miniature fission chambers introduced next to the irradiation channel. A test campaign, called MAESTRO-SL, took place in 2015. Its objective was to assess the feasibility of the hybrid method and investigate the possibility to separate mixed neutron effects, such as fission/capture or scattering/capture. Experimental results are presented and discussed in this paper, which focus on comparing two measurements setups, one using a power control system (closed loop) and another one where the power is free to drift (open loop). First, it is demonstrated that open loop is equivalent to closed loop. Uncertainty management and methods reproducibility are discussed. Second, results show that measuring the flux depression around oscillated samples provides valuable information regarding partial neutron cross sections. The technique is found to be very sensitive to the capture cross section at the expense of scattering, making it very useful to measure small capture effects of highly scattering samples.**

***Index Terms*—Neutron cross section measurement; hybrid pile oscillator; minerve experimental reactor.**

I. Introduction

PILE oscillator techniques are powerful methods to measure small reactivity worth of isotopes of interest for nuclear data improvement [1-3]. The principle is to create, in a well-known neutron flux, a weak and localized perturbation, the characteristics of which (amplitude, phase, energy spectrum) can be connected to some physical parameters (i.e. cross sections). Pile oscillator techniques are called "open loop" if the reactor power is free to drift during the measurement. On the contrary, it is called closed loop when a power control system is activated. Such pile oscillator experiments have long been held in the Minerve facility, operated by CEA at Cadarache [4]. A closed loop reactivity oscillator method is implemented based on measuring the global reactivity perturbation associated to samples motion in the core center. The technique was recently extended by introducing CEA-made high sensitivity miniature fission chambers (MFCs) close to the irradiation channel in order to measure local perturbations of the neutron flux. New detectors and associated acquisition systems were developed by CEA for that purpose [5].

A measurement campaign, called MAESTRO-SL (in which SL stands for "local signal"), took place in 2015 in the Minerve reactor, in a core configuration close to the one used for the MAESTRO oscillation campaign [6-7]. Its objective was to assess the feasibility of measuring local perturbation and, then, to investigate on unfolding capture and scattering cross sections. Closed loop experiments were first performed to provide reference results. Then, open loop measurements were done on a subset of samples.

B. Geslot is with CEA DEN/Cad/DER/SPEx/LPE, Cadarache, Bat.238, F-13108, Saint Paul Lez Durance (e-mail: benoit.geslot@cea.fr).

A. Gruel and P. Blaise are with CEA DEN/Cad/DER/SPEx/LPE, Cadarache, Bat.238, F-13108, Saint Paul Lez Durance.

S. Bréaud is with CEA DEN/Cad/DER/SPEx/LDCI, Cadarache, Bat.238, F-13108, Saint Paul Lez Durance.

P. Leconte is with CEA DEN/Cad/DER/SPRC/LEPh, Cadarache, Bat.238, F-13108, Saint Paul Lez Durance.

II. Experimental setup

Minerve is a pool type reactor operated at low power (100 W maximum). At the center of the core, an irradiation channel is used to introduce various test samples in the core. Samples are held in an oscillator device used to perform pile oscillator experiments. Their reactivity worth (in the range of a few pcm) are obtained and compared against reference materials (e.g. gold, boron, lithium).

In the MAESTRO-SL configuration, four detector locations (8.5 mm in diameter) were accommodated close to the oscillation channel (13.4 mm in diameter) in a guiding device made of aluminum alloy. Four MFCs loaded with enriched uranium deposits (Fig. 1) were used to monitor the thermal flux in the central channel vicinity.

A boron chamber, operated in current mode, was used to monitor the reactor power during experiments and to provide a signal proportional to the global perturbation introduced by oscillated samples. This detector was located in the core reflector, far enough from the core center so that its signal is proportional to the reactor global fission rate.

The power control system (PCS) is based on a rotary cadmium control rod (the so-called "pilot rod") located in the reflector, next to the driver zone. Reactivity is ajusted by changing the angle between rotor and stator, which varies the active cadmium surface. A retroaction loop fed by the boron chamber signal is used to stabilize reactor power.

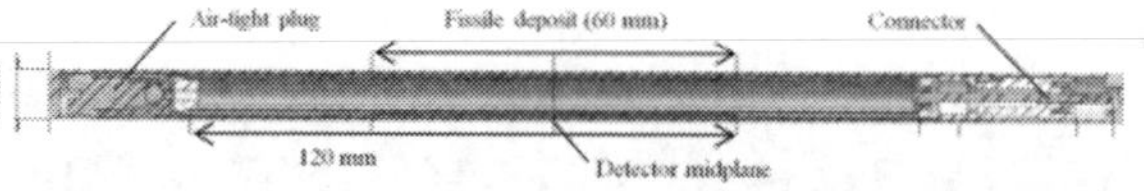

Fig. 1. Drawing of the CF8R-lg fission chamber (dimensions are in mm).

Oscillated samples were made of a column of material of interest (MOI), sometimes diluted into a matrix such as water, and enclosed in a cladding sheath (aluminum, Zircaloy-4 or both). Most samples contained scattering materials (e.g. D2O, Be) or neutron absorbing materials (e.g. acid boric solutions, gold wires). Two samples contained fissile material (UO_2 pellets at 0.25 % and 3 % enrichment in ^{235}U).

III. Description of experiments

In CLO experiments, the global signal, denoted as $S_G(t)$, was issued by the angular sensor that monitored the pilot rod motion. Thanks to the reactivity retro-action loop, all detectors recorded pseudo-square waveforms very similar in shape to the one of the sample motion in and out of the core (trapezoidal), as shown on Fig. 2. Both local and global signals were first rescaled by dividing by their average. Their amplitudes are respectively denoted ΔL and ΔG.

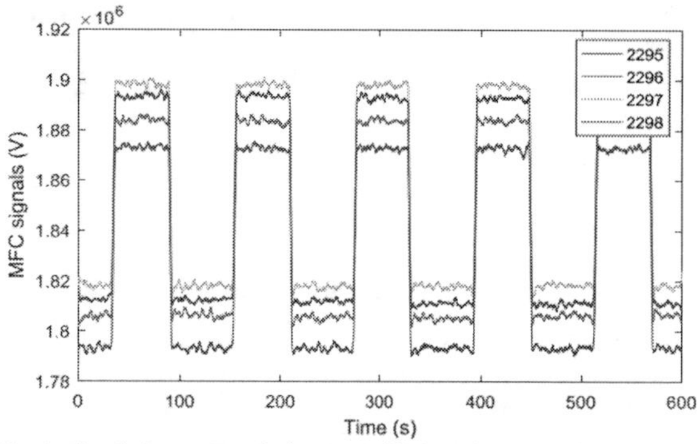

Fig. 2. Signals from MFCs during CL oscillation of D_2O sample.

In OLO experiments, reactor power was free to evolve along with the oscillator motion. Global signal S_G was taken from a boron chamber located in the core reflector. During an oscillation, S_G exhibited increasing and decreasing transients driven by delayed neutron precursors.

As for CLO experiments, local signal S_L was obtained from local MFCs counting rate. In this case, S_L is a mix of a global component driven by the reactor power and a local component proportional to thermal flux changes (in amplitude and spectrum) around the sample.

OLO data processing required first to correct for slow power drifts during experiments. Then, signals were normalized to their average. This way, it is straightforward to derive the local perturbation by subtracting the global component to the local signal, as follows:

$$L(t) = \frac{S_L(t)}{\bar{S}_L} - \frac{S_G(t)}{\bar{S}_G}, \tag{1}$$

As illustrated on Fig. 3, $L(t)$ is a square signal, that can be compared to CLO signal. Amplitude of $L(t)$ is denoted ΔL.

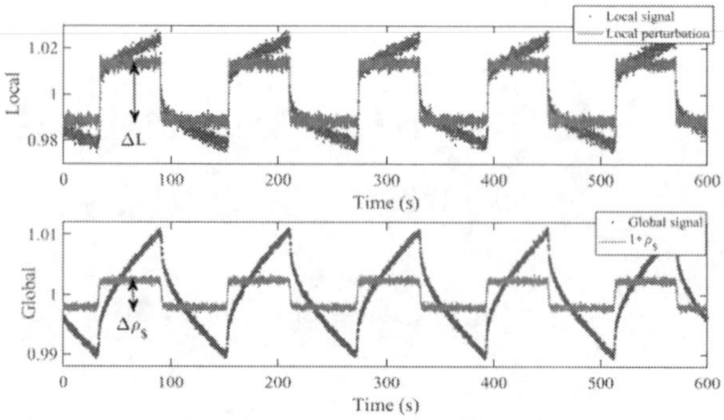

Fig. 3. Local and global signals recorded during the OLO of D_2O sample.

Analysis of $S_G(t)$ can be done by Fourier Series analysis. Because it is a non-square signal, information is contained in the phase as well as in the amplitude of the signal harmonics. Let $TF[S_G](f)$ be the Fourier transform of signal $S(t)$, f_0 be the oscillation frequency and T the oscillation duration. The modulus $|S_G|_n$ of the nth harmonics is expressed as:

$$|S_G|_n = \frac{1}{T}\left|TF[S_G](n \cdot f_0)\right|, \tag{2}$$

IV. Results and discussion

In order to compare CL and OL results, it was found necessary to rescale OLO results. As expected, global signals had different units because they came from different detectors. A linear fit was applied then to obtain a 2.65 coefficient between angular captor and boron chamber signals (R^2 coefficient was 0.99995). Unexpectedly, a systematic sensitivity shift between OLO and CLO was observed on inlocal signals. This was a slight difference (2.5 %), that probably came from a bias introduced by the normalization step, due to the fact that signals are not exactly symmetrical. A correction was made by applying linear fit to correct OLO data ($R^2 = 0.99989$). After rescaling, OLO and CLO data were found to be in very good agreement.

In the past, local oscillation experiments were performed to separate effects from different cross sections. With adequate calibration procedure, it was possible to measure at the same time fission and radiative capture cross sections in fissile materials [8]. Another interesting experiment was the measurement of very small capture cross sections in strongly scattering materials [9]. Here, independent measurements of both local and global perturbations is achieved. Such hybrid experiment should provide additional information on sample cross sections.

On Fig. 4, results from the whole CLO campaign are plotted in a 2D space using ΔL and ΔG as horizontal and vertical coordinates. 121 oscillation results are represented as dots whereas circles feature averaged results per sample.

A clear linear behavior can be seen for samples of the same type. Straight lines drawn on the graph indicate directions corresponding to neutron effects such as radiative capture in gold (green line), scattering (blue line) and neutron production (red line).

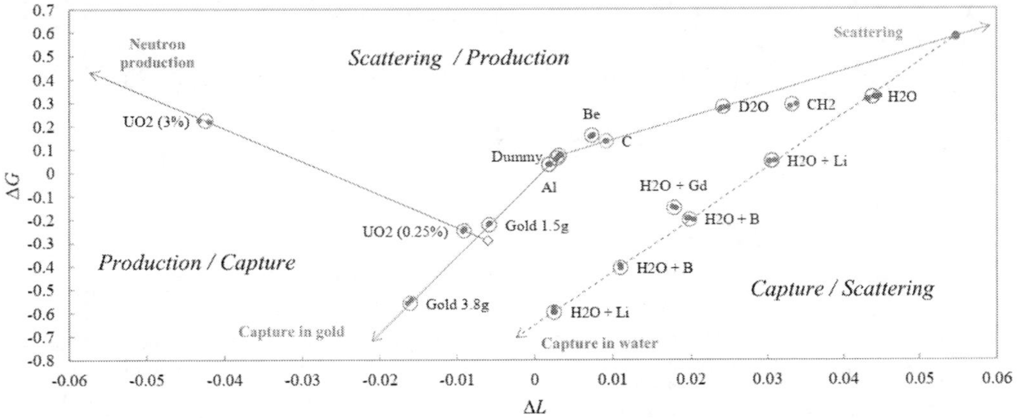

Fig. 4. CLO results in (ΔG,ΔL) plane. Dots are measurements and circles are averaged results per sample.

Samples coordinates on the graph are related to the macroscopic cross sections of their components (cladding, matrix and MOI). Let a vector ***DT*** be defined by a dummy sample D and a test sample T. Its norm $\|\boldsymbol{DT}\|$ is a quantity proportional to the test sample MOI macroscopic cross sections. The proportionality coefficient is complex to determine as it is strongly dependent on the neutron flux shape and spectrum in the sample's vicinity. It is still possible to decompose vector ***DT*** into components inside a coordinate space given by two directed lines. Those lines can be graded using reference samples of known composition. Samples components could then be expressed in physical units (mass of reference material or cross sections in barns).

The plane (ΔG,ΔL) can thus be divided into separate areas corresponding to mixed neutron effects, such as production/capture area (for fissile sample) or capture/scattering area (for water based samples). Notice that capture and neutron production lines are nearly perpendicular, which is in favor of an optimum separation of the two effects. Separation is less favorable for capture and scattering.

V. Conclusion and outlooks

Pile oscillator techniques are powerful methods to measure small reactivity worth of isotopes of interest for nuclear data improvement. Recently, hybrid pile oscillation experiments (i.e. mixing local and global signals) were carried out in the Minerve facility using newly designed CEA-made high sensitivity miniature fission chambers.

The coupled analysis of samples reactivity effect (global signal) and thermal flux depression surrounding the samples (local signal) allows separating effects into components: fission and capture for fissile samples, scattering and capture for other samples.

Experiments demonstrated the feasibility of measuring in Minerve thermal flux changes as low as 0.1 %. Because local flux variations are small and rapidly fading away off samples, a major drawback of the technique is to be rather sensitive to small geometry differences among samples. Adapted sample design and careful experimental setup should mitigate the impact on results quality.

Experimental results show clearly that separating fissile and capture effects is optimum in Minerve setup. A fair separation of scattering and capture effect, although less favorable, achievable.

A comprehensive analysis of the whole results using newly developed capabilities in the TRIPOLI-4 Monte Carlo code should start soon.

Acknowlegements

This work was conducted by CEA/Cadarache in the framework of the I3P collaboration regrouping CEA, EdF and AREVA.

References

[1] W. K. Foell, "Small-sample reactivity measure-ments in nuclear reactors". American Nuclear Society, Hinsdale, IL, USA, 1972.

[2] J. I. Hoover *et al.*, "Measurement of Neutron Absorption Cross Sections with a Pile Oscillator," *Phys. Rev.*, vol. 74, no. 8, pp. 864–870, Oct. 1948, 10.1103/PhysRev.74.864

[3] A. L. Dekker and H. R. Kleijn, "The local pile oscillator as a device for measuring epithermal neutron absorption," *Nucl. Instr. Meth.*, vol. 55, pp. 87–92, Jan. 1967.

[4] M. Antony, *et al.*, "Oscillation experiments techniques in CEA MINERVE experimental reactor," 1st ANIMMA Int. Conference, Marseille, June, 2009.

[5] B. Geslot, *et al.*, "A hybrid pile oscillator experiment in the Minerve reactor". Annals of Nuclear Energy, 108, pp. 268–276, 2017.

[6] P. Leconte, *et al.*, "MAESTRO: An ambitious experimental programme for the improvement of nuclear data of structural, detection, moderating and absorbing materials - First results for ^{nat}V, ^{55}Mn, ^{59}Co and ^{103}Rh," in ANIMMA 2013 Int. Conf., Marseille, France, 2013.

[7] P. Leconte, *et al.*, "Thermal neutron activation experiments on Ag, In, Cs, Eu, V, Mo, Zn, Sn and Zr in the MINERVE facility," in WONDER 2015, Aix-en-Provence, 2015, vol. 111.

[8] B. Laponche, M. Brunet, and Y. Bouedo, "Oscillation of Fissile Materials in Thermal-Neutron Critical Assemblies: The Equivalent Sample Method," *Nucl. Sci. Eng.*, vol 48, no. 3, Jan. 1972.

[9] A. Ertaud, V. Raievski, and D. Breton, "Méthode d'oscillation dans la pile, appliquée à la comparaison d'échantillons de graphite," *J. Phys. Radium*, vol. 12, no. 8, pp. 784–788, 1951.

Tests and foreseen developments of fibered-OSLD gamma heating measurements in low-power reactors

A. Gruel, M. Le Guillou, P. Blaise, C. Destouches, S. Magne

Abstract— **In this paper are presented test measurements of a fibered-OSLD system performed during a dedicated experimental phase in EOLE zero-power reactor. The measurement setup consists of an OSLD crystal connected onto the extremity of an optical fiber and a laser stimulation system, manufactured by the CEA/LIST in Saclay. The OSL sensor is remotely stimulated via an optical fiber using a diode-pumped solid-state laser. The OSL light is collected and guided back along the same fiber to a photomultiplier tube. Results obtained using this system are compared to usual gamma heating measurement protocol using OSLD pellets. The presence of induced radio-luminescence in the OSLD during the irradiation was also observed and could be used to monitor the gamma flux.**

The feasibility of remote measurements is achieved, whereas further developments could be conducted to improve this technique since the readout procedure still requires to withdraw the OSLD off the gamma flux (hence from the core) on account of the dose rate (around a few Gy.h^{-1}), and the readout time remains quite long for on-line applications. Several improvements are foreseen, and will be tested in the forthcoming years.

Index Terms— **nuclear heating, optical fibers, optically stimulated luminescent dosimeters, remote readout**

I. Introduction

THE accurate determination of nuclear heating in materials is a major issue for the development of nuclear technology and in particular the design of new core concepts, such as the Jules Horowitz material testing reactor [1].

In material testing reactors the neutron and gamma flux levels allow to measure directly the nuclear heating by calorimetry based techniques for instance. In mock-ups or zero-power reactors (ZPRs), such as EOLE and MINERVE reactors at Cadarache [2], the low operating level – typically from some watts to some kilowatts – requires the use of different methods such as dosimetry. In the past few years, several experimental programs [3][4] were devoted to the improvement of gamma heating measurements using thermo-luminescent and optically stimulated luminescent dosimeters (TLDs and OSLDs, respectively). However, the main drawback of nuclear heating measurement in low-power reactors remains the off-line gamma heating estimation, given that both TLD and OSLD stimulations (thermal and optical respectively) allowing the readout of the absorbed doses can only be performed after irradiation and out of core. To circumvent this disadvantage, an effort is being made on the development of *in situ* and on-line measurement of the gamma heating. One of the considered solutions is the use of optical fibered systems.

In the first section are presented the main principles of gamma heating measurements in low-power reactors, and especially the system that was used for. In the second section are presented the results of test measurements performed in the EOLE reactor.

Submitted on 01/06/2017

A. Gruel, M. Le Guillou, P. Blaise and C. Destouches are with CEA, DEN, DER, Experimental Physics Service, F-13108 Saint Paul lez Durance, France (e-mail: adrien.gruel@cea.fr).

S. Magne is with CEA, LIST, Laboratoire Capteurs Architectures Électroniques, F-91191 Gif-sur-Yvette, France.

II. Gamma Heating Measurements Using Optical Fibers

A. Overview and general principles

Nuclear heating in material arises from the local deposition of energy carried by neutrons, prompt photons issued from fissions, radiative captures and inelastic neutron scattering, and delayed photons emitted by fissions and activation product decays. This energy is transferred to the electrons through neutral particle interactions, and finally deposited in the material.

Among the different existing techniques, two are particularly suitable for gamma heating measurements in ZPR: thermo-luminescent dosimetry (TLD) and optically stimulated luminescent dosimetry (OSLD). Both techniques are based on the luminescence emitted by the recombination of trapped electrons after a thermal or optical stimulation of the detector. Usually TLDs and OSLDs are small pellets of luminescent material with high photon sensitivity stacked in metal pillboxes. The main drawback is that the readout process is performed offline, out of core.

In order to allow a remote and/or on-line measurement of the gamma heating in the core, different types of fibered setups are also based on the same physical principles, the optical fiber carrying both the light or thermal stimulation for the readout procedure and the TL or OSL signal emitted consequently. A comprehensive review of gamma dosimetry techniques can be found in [5].

B. Description of the fibered-OSLD system

A complete description of the system can be found in [6][7], its main principles are summed up hereafter. An overview of the system can be seen Fig. 1. The OSL reader includes a 16-channel optical fiber switch actuated by a stepper motor, an optical case and power/USB electronics. The OSL reader is linked to a laptop handled through dedicated software. An OSL fiber sensor consists of an optical fiber cable (armoured with iron thread) with an SMA connector at one end and a steel case at the other. The steel case is rigidly fixed onto the cable extremity. It protects the OSL crystal (Al_2O_3:C crystals, ∅ 1 mm, 1 mm long) affixed near the end of a silica fiber and provides a stable coupling of light. The sensors exhibit circular symmetry along the fiber/crystal axis.

After irradiation, each OSL sensor is remotely stimulated using a continuous wave (CW) diode-pumped solid-state laser (@532 nm, 200 mW initially) and bleached for a next use. The reader works in the CW mode (i.e. CW-OSL). The OSL resulting from the stimulation of the crystal, emitted at 410 nm, is then collected by the optical fiber, separated by a dichroic beam-splitter, filtered and eventually detected by a photomultiplier in photon-counting mode.

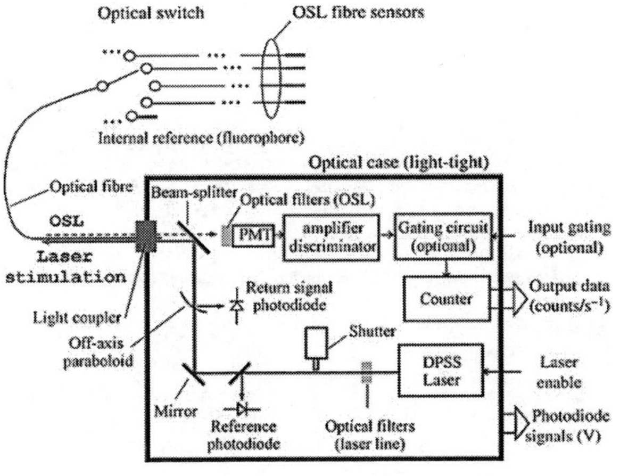

Fig. 1. Schematic of the fibered-OSL reader of the CEA/LIST [6].

III. Test measurements in the EOLE reactor

A. Absolute dose calibration

The first step in gamma dose measurements is to characterize the detector response to the absorbed gamma dose. Detectors were calibrated in a pure gamma field at a distance of 80 cm of a ^{60}Co source, whose β^- decay leads to the emission of two gamma rays at 1.17 and 1.33 MeV. The three sensors show the same behavior with significantly different sensitivities, and a second order polynomial was adjusted to the data (Fig. 2).

For comparison measurements in EOLE, OSLD pellets (alumina encapsulated in polyester, from Landauer [8]) stacked in stainless-steel pillboxes were also calibrated in a pure gamma field, up to an integrated dose of 2 Gy. Their responses are linear.

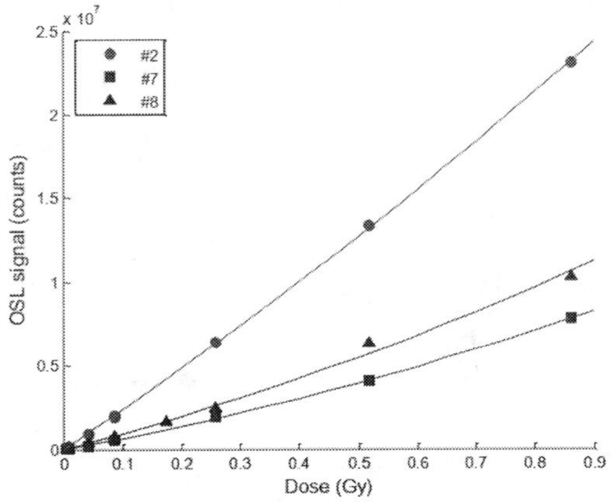

Fig. 2. OSL signal versus dose for the three fibered-OSL sensors (markers) and calibration curves (lines).

B. Experimental setup

This system was tested in the EOLE reactor. In order to reach better statistics, measurements were carried out with the most sensitive fibered OSL sensor (#2). The reader was located on the top of the EOLE core, the fiber going inside the core through a plastic tube connected at the upper end of an aluminum overclad.

The OSLD crystal connected to the fiber was set slightly above the core mid-plane, maintained in position by an aluminum shim. Below this sensor, two Landauer nanoDots OLSD pellets in a stainless-steel pillbox were used for comparison purpose, as long as other ones in an equivalent position, but without the fiber atop of them (Fig.3).

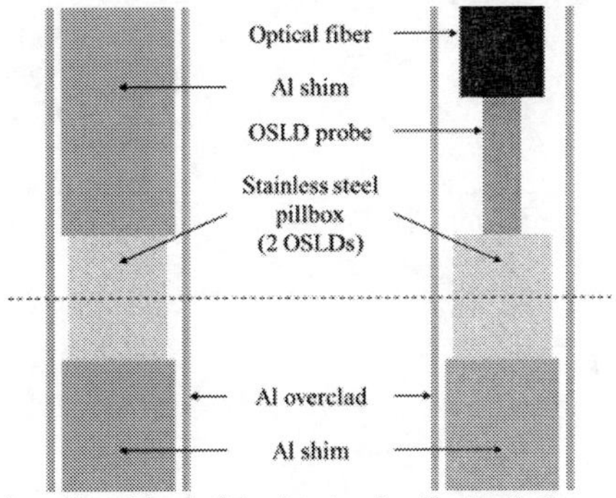

Fig. 3. Schematic axial cut of the detectors localization in the core (dashed-line: mid-core plane). Inside the overclad is air, outside is air (background measurement) or borated water (irradiation).

C. Results and discussion

The results for the comparison between the fibered-OSLD and the OSLD pellets are shown in Table I. Measurements carried out during irradiations show a very good agreement, similar to their respective uncertainties. The discrepancy between the OSLD pellets and the fibered-OSLD in the same channel is 3.5%, while the difference between the fibered-OSLD and the OSLD pellets in the equivalent position is 0.1%. However, the fibered-OSLD was only calibrated up to 0.9 Gy, so the absolute dose value at 1.2 Gy could slightly be biased. Complementary calibration measurements to higher doses must be carried out to obtain a more accurate value.

TABLE I
COMPARISON BETWEEN OSLD PELLETS AND THE FIBERED-OSLD SYSTEM (UNCERTAINTIES AT 1 STANDARD DEVIATION).

Detector	Irradiation
fibered-OSLD	1197 ± 33 mGy
OSLD pellets in same channel as fibered-OSLD	1241 ± 11 mGy
OSLD pellets in equivalent position	1196 ± 13 mGy

Another interesting feature of the fibered-OSL system is the presence of a radio-luminescence (RL) signal during the irradiations. However, in order to use this signal for dose or dose rate measurements, it would require a calibration of the RL response of the sensors with respect to dose and to dose rate [9].

Nevertheless, some limitations appeared when using this system in EOLE. The main one was that the read-out time of the fibered sensor, from the start of the laser stimulation to the bleaching of the OSL crystal, is similar to the irradiation duration. Another improvement could be to enable the readout of the OSL during irradiation, which requires the increase of the laser power.

IV. Conclusion and outlooks

Remote gamma heating measurements were performed in the EOLE reactor using a fibered setup. Standard OSLD pellets were also used for comparison purpose during the irradiations in the core.

The two measurement techniques are in good agreement (within a few percent), and the reproducibility of the measurements of the fibered-OSLD was measured around 3%.

Further development could improve the use of this remote system in ZPRs, such as the decrease of the readout time down to a few seconds, the characterization of the behavior of the detector when the readout is performed during irradiation, and the calibration and the use of the RL signal observed.

Acknowledgments

The authors wish to acknowledge the operator team of the EOLE reactor, P.-G. Allinei and P. Chaussonnet (CEA/DEN/DTN/SMTA/LMN) for their participations to the calibration campaign, as well as L. De Padua and N. Makaryan (CEA/DEN/D2S/SPR/LRIR).

References

[1] M. Lemaire *et al.*, Experimental validation of photon-heating calculation for the Jules Horowitz Reactor, *Nucl. Instr. Meth. A* (2015), Vol. 780, pp. 68-80DOI: 10.1016/j.nima.2015.01.054
[2] J.-C. Bosq *et al.*, The Use of EOLE and MINERVE Critical Facilities for the Generation 3 Light Water Reactors Studies, *RGN* (2011), Vol. 5, pp. 63-69. DOI: 10.1051/rgn/20115063
[3] D. Blanchet, Ph.D. thesis, Université Blaise Pascal, France, 2006
[4] H. Amharrak, Ph.D. thesis, Aix-Marseille Université, France, 2012
[5] M. Le Guillou *et al.*, State of the art on nuclear heating measurement methods and expected improvements in zero power research reactors, *EPJ Nuclear Sci. Technol.* (2017), Vol. 3, No. 11, DOI: 10.1051/epjn/2017002
[6] S. Magne *et al.*, Multichannel dosemeter and Al2O3:C optically stimulated luminescence fibre sensors for use in radiation therapy: evaluation with electron beams, *Radiat. Prot. Dosim.* (2008), Vol. 131, No. 1, pp. 93-99. DOI: 10.1093/rpd/ncn226
[7] S. Magne *et al.*, Multichannel Dosimeter and α-Al_2O_3:C Optically Stimulated Luminescence (OSL) Fiber Sensors for Use in Radiation Therapy—Evaluation With Photon Beams, *IEEE Trans. Nucl. Sci.* (2011), Vol. 58, No. 2, pp. 386-394, DOI: 10.1109/TNS.2011.2114369
[8] C. A. Perks *et al.*, Medical dosimetry using Optically Stimulated Luminescence dots and microStar readers, in *12th Int. Congr. Int. Radiation Protection Assoc.*, Buenos Aires, Argentina, 2008
[9] S. Magne *et al.*, Fiber-coupled, time-gated Al_2O_3:C radioluminescence dosimetry technique and algorithm for radiation therapy with LINACs, *IEEE Trans. Nucl. Sci.* (2013), Vol. 60, No. 4, pp. 2998-3007, DOI: 10.1109/TNS.2013.2263640

https://doi.org/10.1051/epjconf/201817004011

Measurement station for interim inspections of Lightbridge metallic fuel rods at the Halden Boiling Water Reactor

C. Hartmann, A. Totemeier, S. Holcombe, J. Liverud, M. Limi, J.E. Hansen, E. Navestad

***Abstract*— Lightbridge Corporation has developed a new Uranium-Zirconium based metallic fuel. The fuel rods are manufactured via a co-extrusion process, and are characterized by their multi-lobed (cruciform-shaped) cross section. The fuel rods are also helically-twisted in the axial direction. Two experimental fuel assemblies, each containing four Lightbridge fuel rods, are scheduled to be irradiated in the Halden Boiling Water Reactor (HBWR) starting in 2018. In addition to on-line monitoring of fuel rod elongation and critical assembly conditions (e.g. power, flow rates, coolant temperatures, etc.) during the irradiation, several key parameters of the fuel will be measured out-of-core during interim inspections. An inspection measurement station for use in the irradiated fuel handling compartment at the HBWR has therefore been developed for this purpose. The multi-lobed cladding cross section combined with the spiral shape of the Lightbridge metallic fuel rods requires a high-precision guiding system to ensure good position repeatability combined with low-friction guiding. The measurement station is equipped with a combination of instruments and equipment supplied from third-party vendors and instruments and equipment developed at Institute for Energy Technology (IFE). Two sets of floating linear voltage differential transformer (LVDT) pairs are used to measure swelling and diameter changes between the lobes and the valleys over the length of the fuel rods. Eddy current probes are used to measure the thickness of oxide layers in the valleys and on the lobe tips and also to detect possible surface cracks/pores. The measurement station also accommodates gamma scans. Additionally, an eddy-current probe has been developed at IFE specifically to detect potential gaps or discontinuities in the bonding layer between the metallic fuel and the Zirconium alloy cladding. Potential gaps in the bonding layer will be hidden behind a 0.5-1.0 mm thick cladding wall. It has therefore been necessary to perform a careful design study of the probe geometry. For this, finite element analysis (FEA) has been performed in combination with practical validation tests on representative fuel dummies with machined flaws to find the probe geometry that best detects a hidden flaw. Tests performed thus far show that gaps down to 25 µm thickness can be detected with good repeatability and good discrimination from lift-off signals.**

The work presented here was performed at IFE's facilities in Halden in collaboration with Lightbridge Corporation. It is part of the irradiation test program developed by Lightbridge to demonstrate the performance of the Zr-U metallic fuel. The fuel irradiation and interim inspections will be performed in an isolated loop of the Halden Boiling Water Reactor.

C. Hartmann, S. Holcombe, J. Liverud, M. Limi, J.E. Hansen and E. Navestad are with Institute for Energy Technology, P.O. Box 173, NO-1751 Halden, Norway (e-mail: christian.hartmann@ife.no).

A. Totemeier is with Lightbridge Corporation, 11710 Plaza America Dr., Ste. 2000, Reston, VA 20190 USA (e-mail: atotemeier@ltbridge.com).

Index Terms—**Lightbridge, metallic fuel, NDT, interim inspection, eddy-current probe**

I. Introduction

LIGHTBRIDGE Corporation is a U.S. company that has developed a design and fabrication process for a novel nuclear fuel [1], consisting of helically twisted metallic Uranium-Zirconium alloy fuel rods. Lightbridge's fuel technology can improve plant safety, significantly enhance the nuclear power industry's economics and increase power output of existing and new light water nuclear reactors. In order to verify the performance of the Lightbridge fuel and acquire experimental data to support implementation of this fuel type into commercial nuclear power reactors, test irradiations of Lightbridge metallic fuel rods will be carried out in the Halden Boiling Water Reactor (HBWR), a research reactor located in Halden, Norway.

In addition to monitoring characteristic fuel performance parameters in real time during the test irradiation [2], interim inspections of the test fuel rods will be performed between operating cycles. Since the metallic, helically twisted Lightbridge fuel rods differ from the standard cylindrical Uranium Oxide fuel, new inspection equipment was needed in order to perform the interim inspections. For several inspection techniques it was sufficient to design a system with fixtures capable of holding the helically twisted fuel rods; however, some inspections required development of new measurement instrumentation. This paper describes some key features of the new interim inspection equipment and development of new instrumentation.

II. Lightbridge fuel

The fuel rods are manufactured through a co-extrusion process, and are characterized by their cruciform-shaped cross section. The fuel rods are also helically twisted in the axial direction. A rendering of the Lightbridge fuel rod design is shown in Fig. 1 with cross-sections showing the main dimensions and the definitions of the lobe and valley of the fuel rod.

 https://doi.org/10.1051/epjconf/201817004011

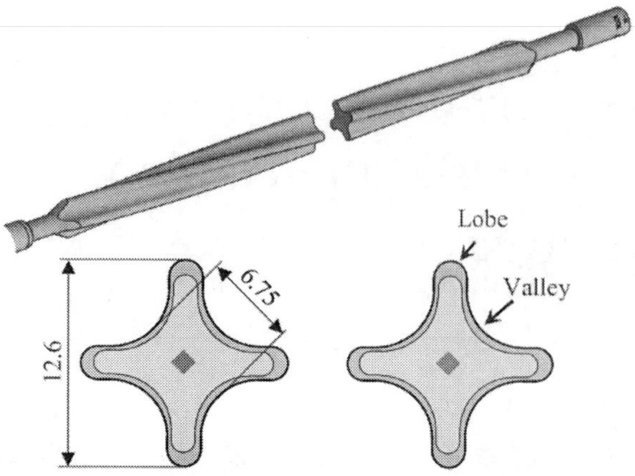

Fig. 1. Lightbridge metallic fuel rod with cross sections showing the characteristic, cruciform design. The Zirconium-alloy cladding is metallically bonded to the Uranium-Zirconium fuel core. The central displacer (diamond in the center) consists of Zirconium or a Zirconium alloy containing burnable neutron absorbers.

Lightbridge fuel rods differ from conventional fuel currently used in commercial nuclear power reactors. Conventional fuel rods consist of cylindrical Zirconium-alloy tubes filled with cylindrical ceramic Uranium Dioxide fuel pellets, where there is a gas-filled gap between the fuel pellet outer surface and the inner surface of the Zirconium-alloy cladding tubes. The Lightbridge fuel is cruciform-shaped instead of cylindrical, metallic instead of ceramic, and the cladding is metallically bonded to the fuel instead of incorporating the presence of a gas-filled gap. These features mean that thermal performance of the Lightbridge fuel is superior to conventional fuel rods, and enable operating the fuel at higher power density while at the same time increasing safety margins.

III. Inspection measurement station

A conceptual model of the interim inspection equipment is shown from different angles in Fig. 2. The interim inspection equipment is designed to be placed into a fuel handling compartment. The upper end of the Lightbridge fuel rod is held in a chuck fixed to the rotational motor. The lower end of the Lightbridge fuel rod passes through the measurement station. The rotational motor fixture is moved vertically such that the Lightbridge fuel rod moves through the measurement station. The rotational motor is used to read out the rotational position of the rod, and can be used to rotate the rod into position for alignment with the guide wheels in the measurement station.

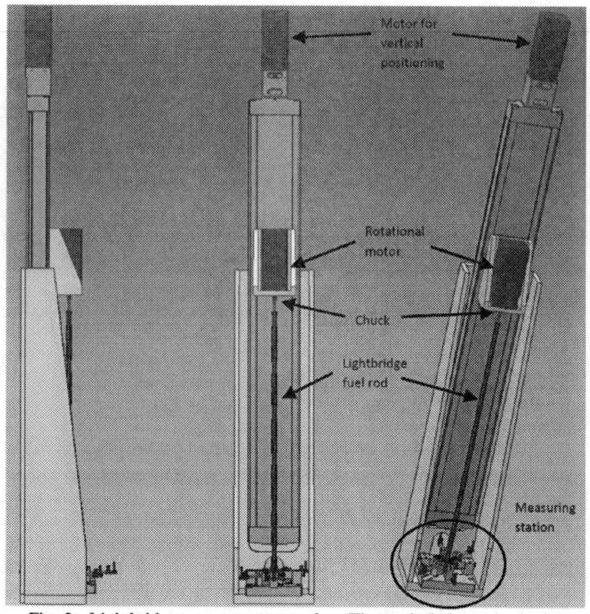

Fig. 2. Lightbridge measurement station. The encircled area in the lower right corner is shown in detail in Fig. 3.

A conceptual model of the measurement station is shown in Fig. 3. Two spring-loaded guide wheels, placed 180^0 apart, guide the Lightbridge fuel rod through the measurement station and maintain the position of the rod relative to the measurement probes. Different kinds of probes can be fixed into the measurement station where the probes are placed in order to perform measurements on the fuel rod lobe tips and in the valleys between the lobes. The types of probes which may be installed include eddy-current probes for measuring oxide thickness, LVDTs for measuring diameter and swelling, and an eddy-current probe for investigating fuel-clad metallurgical bond integrity.

IV. Fuel growth

Lightbridge metallic fuel rods are expected to swell in the radial direction, where swelling is expected to preferentially occur in the valleys between the lobes.

A. Measurement principle

Radial fuel growth is planned to be investigated by measuring the distance between opposing valleys and by measuring the circumscribed diameter (i.e. the distance between opposing lobe tips) using opposing mechanical probes, each coupled to LVDTs. The LVDTs will be placed into the measurement head as shown in Fig. 3.

https://doi.org/10.1051/epjconf/201817004011

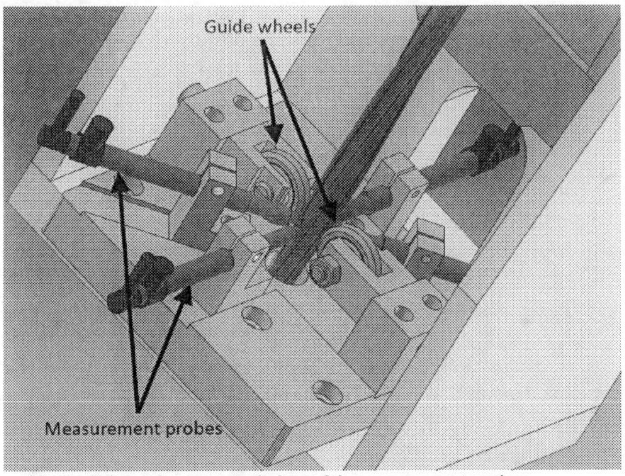

Fig. 3. Schematic conceptual view of the measurement station.

B. Mock-up tests

Testing of the fuel growth measurement precision was performed using a setup similar to that shown in Fig. 4.

Fig. 4. Photo of a mock-up of the measurement station used for testing the design.

The distance between opposing probes was measured on a dummy Lightbridge fuel rod at 6 axial locations between the valleys and at 6 axial locations between the lobe tips. The measurements were repeated 10 times, where each axial location was measured once before moving to the next axial location. The standard deviations of the measured thicknesses were calculated for the set of 10 measurements acquired for each of the 6 axial positions. The maximum and minimum standard deviations are shown in Table I along with the maximum and minimum absolute magnitudes of the deviations within each set of measurements.

TABLE I
STANDARD DEVIATIONS AND ABSOLUTE DEVIATIONS

	Standard deviation of the set of measurements performed at each axial position of the dummy rod		Absolute deviation of the set of measurements performed at each axial position of the dummy rod	
	Maximum from all positions	Minimum from all positions	Maximum from all positions	Minimum from all positions
Lobe tips	0.6 μm (1σ)	0.3 μm (1σ)	±1.0 μm	±0.5 μm
Valleys	0.8 μm (1σ)	0.4 μm (1σ)	±1.0 μm	±0.5 μm

V. OXIDE THICKNESS

The oxide thickness on the fuel cladding will be measured at each interim inspection, using IFE's eddy-current system, based on an Elotest N300 instrument and eddy-current probes from GE.

A. Measurement principle

A picture of the oxide thickness measurement probe mounted in the measurement head is shown in Fig. 5.

Fig. 5. Eddy-current probe for oxide thickness measurements mounted in mock-up test station together with a fuel rod dummy.

Prior to the execution of the oxide thickness measurements, the probe and electronics will be calibrated on an un-oxidized reference sample utilizing calibration foils of known thickness.

B. Mock-up tests

A dummy rod for testing of oxide thickness measurements was manufactured by IFE. It consisted of four segments, which were assembled into a single dummy rod. Two of the segments, assembled on the upper and lower ends of the dummy rod, were un-oxidized and two segments assembled in the middle of the dummy rod, each had different oxide thicknesses. The assembled dummy rod is shown in Fig. 6.

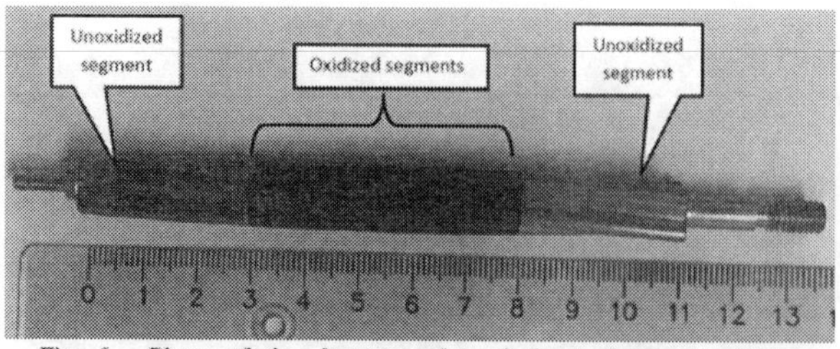

Fig. 6. Photo of the dummy rod used for testing oxide thickness measurements in the mock-up of the measurement station.

The two oxidized segments were oxidized in steam and the resulting oxide thickness was assessed using Light Optical Microscopy (LOM) on a 5 mm long segment cut from the end of each oxidized segment. The oxidation parameters and the resulting oxide thicknesses are shown in Table II.

TABLE II
OXIDATION PARAMETERS AND RESULTING OXIDE THICKNESS

	Oxidation parameters		Oxide thickness (µm)
	Time (minutes)	Temperature (°C)	
Sample 1	30	750	5.5
Sample 2	30	900	13.0

Oxide thickness eddy-current measurements were performed on the lobe and in the valley of the dummy test rod. The results from these tests are summarized in Table III.

TABLE III
SUMMARY OF OXIDE THICKNESS MEASUREMENTS

	13 µm oxide thickness segment		5.5 µm oxide thickness segment	
	Lobe	Valley	Lobe	Valley
Minimum (µm)	10.0	14.6	3.0	3.7
Maximum (µm)	13.0	16.6	4.6	5.3
Radial variation (µm)	< ±0.8	< ±0.6	< ±0.5	< ±0.5

VI. SURFACE CRACKS

It is planned to use a third-party eddy-current probe for detection of surface cracks. The development and qualification of a surface crack-detection system is pending.

VII. GAMMA SCANNING

Gamma rays are emitted by radioactive nuclides present in irradiated fuel and the content and distribution of these nuclides corresponds to operating conditions of the fuel. Measuring the distribution of the gamma-ray activity can therefore be used to assess various fuel performance characteristics, such as power and/or burnup distribution, or migration of volatile nuclides within the fuel.

Lightbridge fuel rods will be characterized by measuring the axial distribution of selected gamma-rays. The fuel rods will be placed into the interim inspection equipment in a fuel handling compartment, a collimated High Purity Germanium (HPGe) detector is inserted through an opening in the compartment wall, and the fuel rod will be moved axially past the collimator opening. The guide-wheel fixture in the interim inspection equipment will hold the fuel rod lobes as the fuel is moved axially, thereby rotating the fuel rod with the helical twist such that the fuel geometry relative to the detector will be the same for all axial positions, i.e. the axial gamma scanning will follow the twist of the fuel rods. In this way, radial as well as axial information will be acquired.

VIII. FUEL-CLADDING DE-BONDING

When manufactured, a bond is formed between the fuel and cladding during the co-extrusion process. This bond provides good thermal conductivity between fuel and cladding. The bond is expected to be maintained during irradiation, but if any gaps between cladding and fuel should be formed, it is of interest to be able to detect these flaws during interim inspections and inspect them thoroughly during post-irradiation examinations (PIE). See Fig. 7.

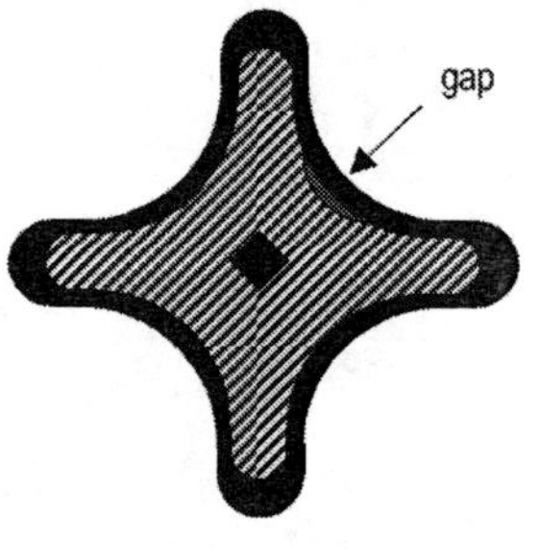

Fig. 7. Cross-section of fuel rod with gap indicated. The gap is shown in red.

Since the metallic fuel is electrically conducting, it should be possible to detect the presence of a buried gap with eddy-current measurements, although the fuel has quite low electrical conductivity relative to the cladding material, as can be seen in Table IV.

TABLE IV
ELECTRICAL CONDUCTIVITIES OF LIGHTBRIDGE FUEL AND CLADDING MATERIALS

Part	Material	Electrical conductivity (at 20 °C)
Fuel	Uranium-Zirconium	σ_{fuel} = ~ 0.63 MS/m
Cladding	Zirconium alloy	$\sigma_{cladding}$= ~1.35 MS/m

The materials are non-ferromagnetic.

Third-party eddy-current probes were tested on a representative sample, but the tested probes did not produce a reliable response to the gap buried beneath the cladding surface. A simple single-coil eddy-current probe for absolute mode operation [3] has therefore been designed for this purpose. The design process and test results are described in the subsequent sections. The current version of the probe is limited to performing gap-detection scans in the valleys of the fuel.

A. Principle of eddy-current gap detection

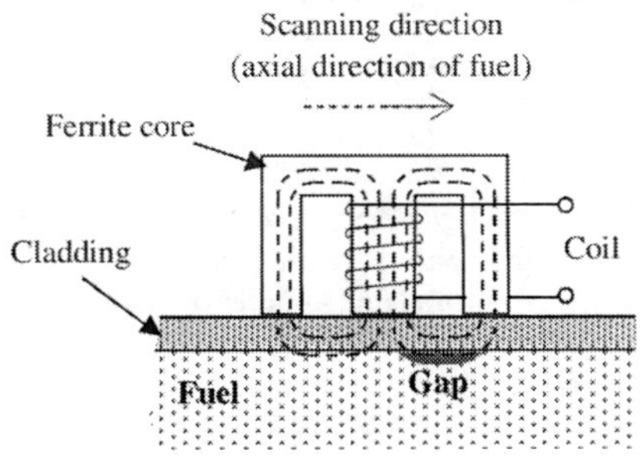

Fig. 8. Schematic illustration of a single-coil eddy-current probe scanning over a gap between fuel and cladding.

The eddy-current probe consists of a ferrite core wound with a copper winding. The winding is operated at high frequency (250 kHz), and sets up a magnetic field as illustrated in Fig. 8. Since the ferrite core is open in the section facing the fuel rod surface, the magnetic field-lines fringe into the surface of the fuel rod.

The high-frequency magnetic field causes eddy currents to be induced in the parts of the fuel rod that coincide with the path of the magnetic field. Since the eddy-currents generate losses and attenuate the magnetic field, the probe impedance will be dependent on the geometry and composition of the fuel rod. Given that a significant part of the magnetic field penetrates beyond the fuel cladding and into the fuel, the occurrence of a gap can be detected as a change in the probe's impedance.

The terminals of the probe's coil are connected to an Elotest N300 electronics unit. For the simulations performed, a simplified representation of the impedance bridge circuit has been used. See Fig. 9.

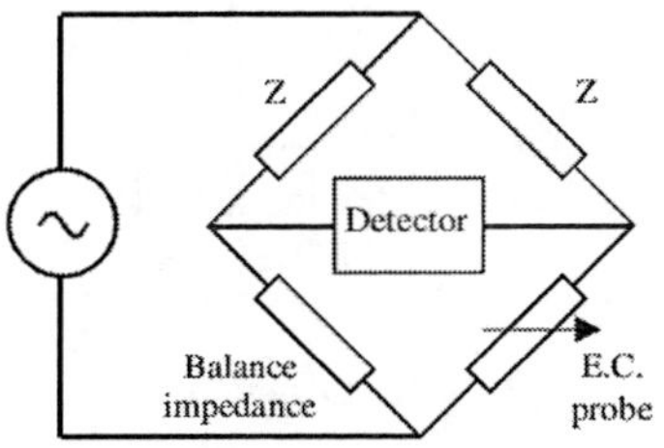

Fig. 9. Simplified impedance bridge circuit used for simulations. The balance impedance is a replicate of the eddy-current probe, but it is held at static conditions over a test sample.

The detector outputs a signal in the impedance plane with an in-phase (I) and quadrature (Q) component.

The bridge is balanced [3] at the beginning of a scan, so that all subsequent changes in the eddy-current probe's impedance during the scans shift the balance of the bridge away from zero output.

The phase of the impedance plane representation can be shifted, and gain can be applied individually to the I and Q-axes. As will be seen in subsequent paragraphs, the phase is shifted so that the gap-response is aligned with the Q-axis. This allows for a high gain to be separately applied to the gap-response.

B. Probe design

The design process has been done iteratively, with finite element method (FEM) simulations in combination with mock-up tests at various frequencies.

Fig. 10 shows a screenshot from a 2D FEM study that was performed on a C-core geometry to investigate the magnetic field's ability to penetrate through the cladding layer. The slot opening *a* was varied, and the impedance change in response to the gap was calculated. The simulations were performed at 250 kHz.

The results from these calculations indicate that a slot opening in the range 1 – 3 mm produces the highest response to a gap. This range has been used as a guideline in choosing probe-core geometries.

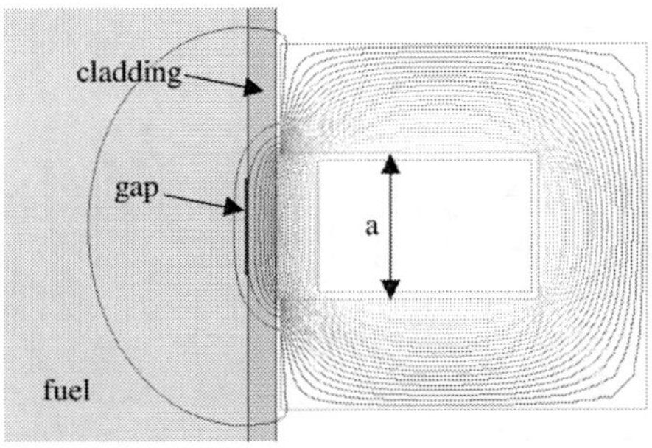

Fig. 10. Screenshot from a 2D FEM-study investigating the magnetic field penetration as a function of core slot-opening *a*. The simulations were performed at 250 kHz.

Several probe geometries have been simulated and tested as part of the design process. All designs have been based on commercially available ferrite cores. When iterating on the design, the various probes' sensitivity to the presence of gaps have been central in the evaluation of the performance of each probe design. The probe-cores also have to be sufficiently small to allow for smooth guidance along the surface of the helically twisted fuel rods. In practice, there was also a lower limit on core sizes due to the limitations set by the available equipment for winding and preparing the cores for mock-up tests.

Initially, C and E-cores in ferrite were assumed to be good candidates for use in the gap-detection probe. Although designs based on these core-types proved to be sensitive to the presence of gaps, they also have a large stray-field around the open part of the coil, which tended to give an unwanted response to surrounding structures. This interfered with the gap detection.

After iterating on the design a core from TDK was chosen for the probe. It is an RM 4-core made out of N87 ferrite. Although it is not a fully closed cup-core [4], it was the best performing core among the tested geometries. The core has been modified so that it mates closely with the curved surface of the fuel cladding. Fig. 11 shows the core before and after modification.

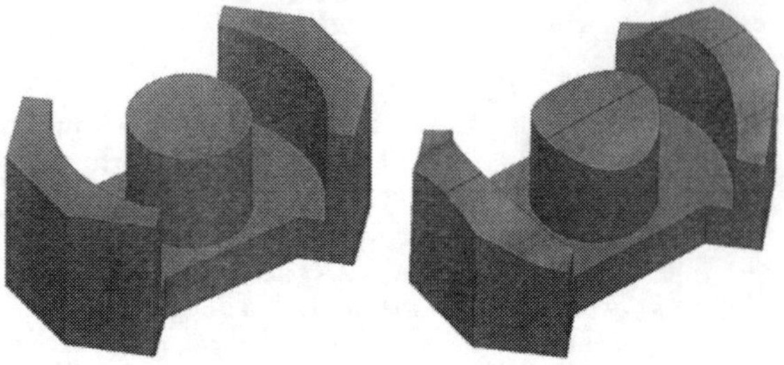

Fig. 11. Core before (left) and after (right) modification.

The main dimensions of the core are given in Fig. 12.

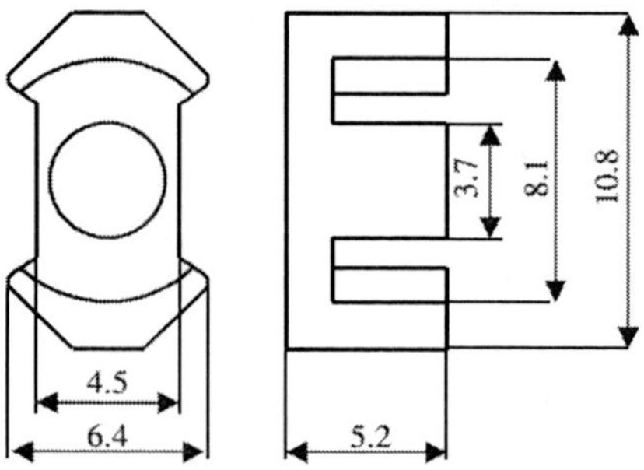

Fig. 12. Main dimensions of core. All dimensions are given in millimeters.

The core is wound with 40 turns of 0.15 mm copper wire. This wire dimension is sufficiently small for operation up to approximately 750 kHz.

Only the simulations and tests performed on the latest version of the probe are reported here.

C. FEA of gap detection probe

All FEM simulations have been performed in ANSYS Electromagnetics Suite. The probe has been modeled in 3D in eddy-current mode at 250 kHz. The model has been set up to calculate probe resistance and reactance.

Matlab has been used to simulate the measurement bridge's response to the changes in resistance and reactance in response to the presence of gaps. See Fig. 9.

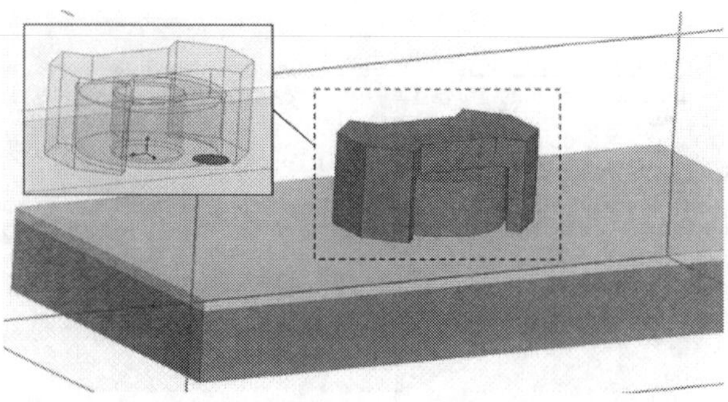

Fig. 13. FEM-model of probe. Picture inset at top left corner shows the gap in purple, located under the cladding layer.

The FEM simulation model is shown in Fig. 13. To reduce model complexity, the cladding and fuel have been modeled as flat. This simplification was necessary to preserve sufficient model resolution at acceptable computation times.

1) Gap simulations

In the inset in the top left corner of Fig. 13, the probe, cladding and fuel are rendered semi-transparent so that the gap geometry is visible in purple. The position of this gap geometry has been parameterized to enable multiple calculations to be performed with incremental changes in gap position.

Three different gap geometries have been simulated. These are shown in Fig. 14.

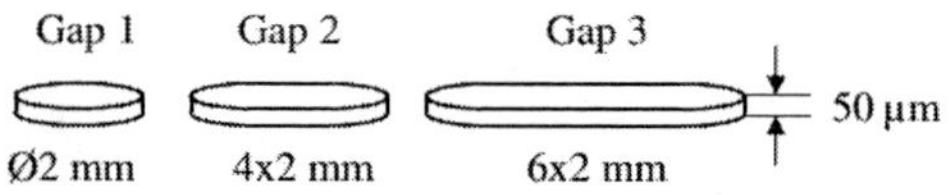

Fig. 14. Simulated gap geometries.

All three gaps are 50 μm thick and 2 mm wide, whereas the length of the gaps varies from 2 to 6 mm. The gaps are located under a cladding layer of thickness 0.61 mm.

2) Lift-off simulations

In addition to simulating the response to gap-sweeps, lift-off simulations have been performed. These simulations show the response to the probe being removed from the surface of the fuel cladding as illustrated in Fig. 15.

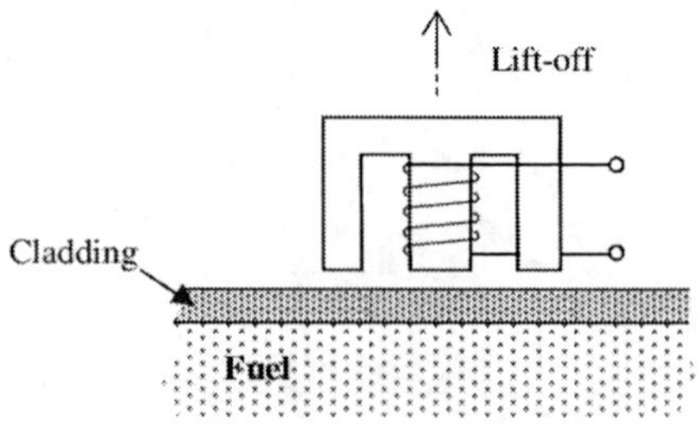

Fig. 15. Lift-off. The probe is removed from the surface of the fuel cladding

The lift-off response was simulated to check the ability to distinguish between signal changes originating from the presence of a gap, and signal changes originating from improper surface contact, etc.

The lift-off simulations have been performed without the presence of any gaps. The lift-off was simulated from 0 – 1 mm.

D. Simulation results

The following three figures show the simulation results for each gap including lift-off. The results are presented in the impedance plane. The scale is identical in all three figures. The plane has been shifted so that the gap response is aligned with the Q-axis.

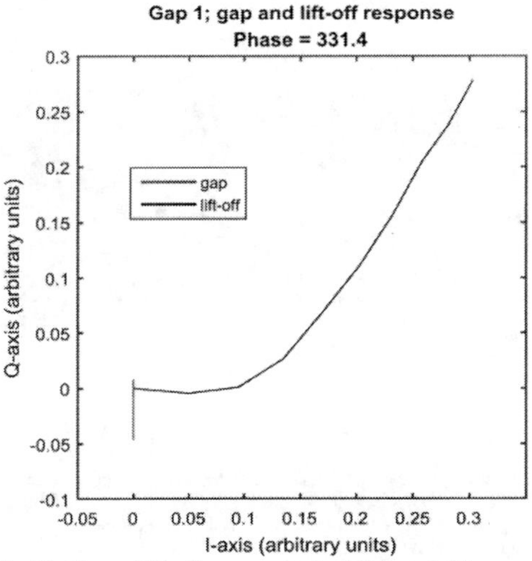

Fig. 16. Gap and lift-off response to simulated gap 1. The gap response is plotted in red, while the lift-off response is plotted in black.

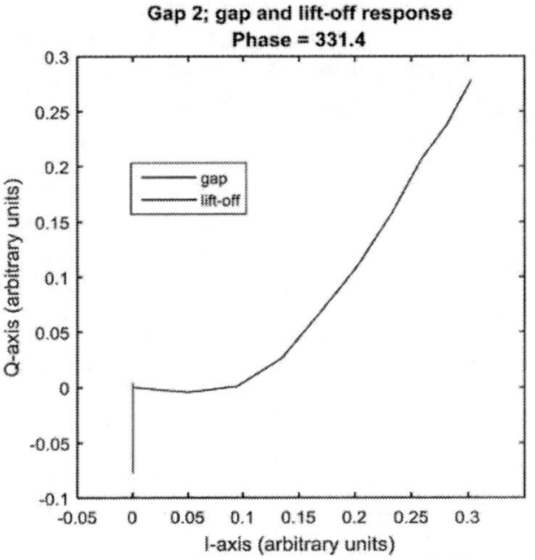

Fig. 17. Gap and lift-off response to simulated gap 2. The gap response is plotted in red, while the lift-off response is plotted in black.

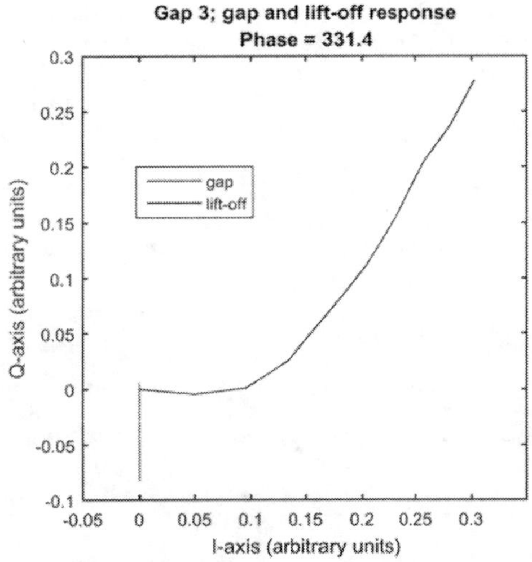

Fig. 18. Gap and lift-off response to simulated gap 3. The gap response is plotted in red, while the lift-off response is plotted in black.

The simulations results show good discrimination between gap and lift-off response. The gap-response amplitude also increases relative to the lift-off signal for increasing gap-sizes, as would be expected. Note that the underlying lift-off simulation is identical for all three figures.

In Fig. 19, the gap (Q-axis) responses have been plotted as a function of probe position. The three responses have been arranged in the same plot. All signals are in the same scale on both axes. The probe is symmetrical around the center leg, so the simulated half-wave response from each gap has been duplicated and mirrored around the center position to replicate a full sweep.

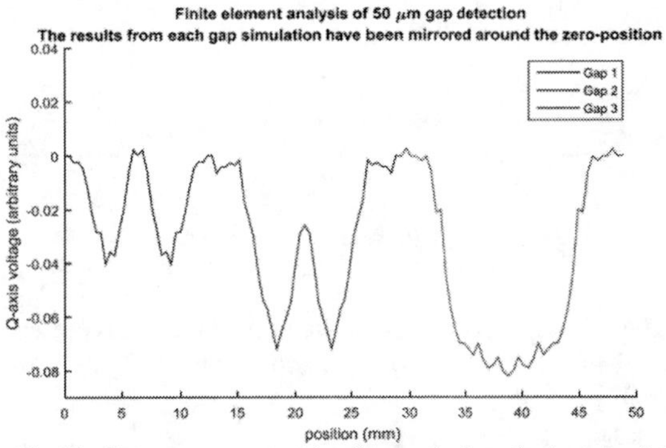

Fig. 19. Gap response vs. position. The results from the three simulations have been arranged in sequence. The three signal responses are in the same scale.

The simulated results show that the gap response amplitude increases with increasing gap lengths. The weakening in the gap response signal at the center position is also less pronounced for increasing gap lengths. A transition from a two-crested to a single-crested response can be observed when the gap-length reaches 6 mm.

https://doi.org/10.1051/epjconf/201817004011

E. Mock-up tests

A dummy fuel rod has been specifically designed and manufactured for testing of the eddy current probe. It consists of an inner part representing the fuel and an outer part representing the cladding. The parts have been produced by electrical discharge machining (EDM). On the surface of the part representing the fuel, several small gaps of different areas and depths have been milled. These gaps represent zones of de-bonding. The two dummy rod parts are shown prior to final assembly in Fig. 20. The inset picture shows the assembled dummy rod.

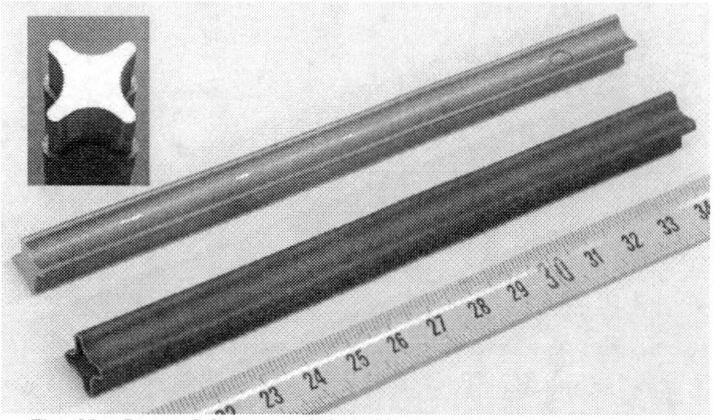

Fig. 20. Parts of dummy rod for eddy-current gap detection tests. The upper part represents the fuel, with small gaps machined onto the surface. The gaps shown are oval, whereas the opposite side is equipped with rectangular gaps. The lower part represents the cladding. The inset shows the assembled dummy rod. The dummy rod is not helically shaped.

An overview of the gap geometries on the two sides of the dummy rod is shown in Fig. 21.

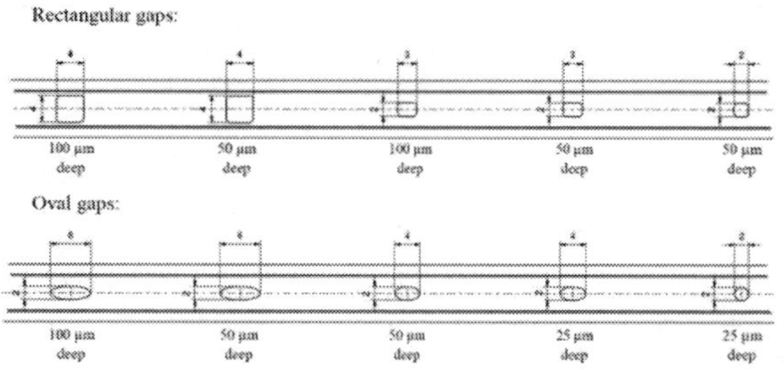

Fig. 21. Gap geometry with dimensions on the two opposite sides of the dummy rod. The same gap-geometry is reproduced in Figs. 25 and 26.

The materials for the parts have been chosen to represent the electrical conductivity of the actual fuel rod as close as possible. Table V summarizes the parameters of the selected materials.

TABLE V
ELECTRICAL CONDUCTIVITY OF DUMMY ROD MATERIALS

Part	Material	Electrical conductivity (at 20 °C)
Fuel	Titanium Grade 5	$\sigma_{Ti\text{-}5} = 0.562$ MS/m
Cladding	Zy-4	$\sigma_{Zy\text{-}4} = 1.35$ MS/m

The gap-detection probe is shown integrated with the measurement probe holder in Fig. 22. The wound core is integrated with a spring-loaded guiding system to enable steady positioning to the cladding surface while the rod moves in the axial direction.

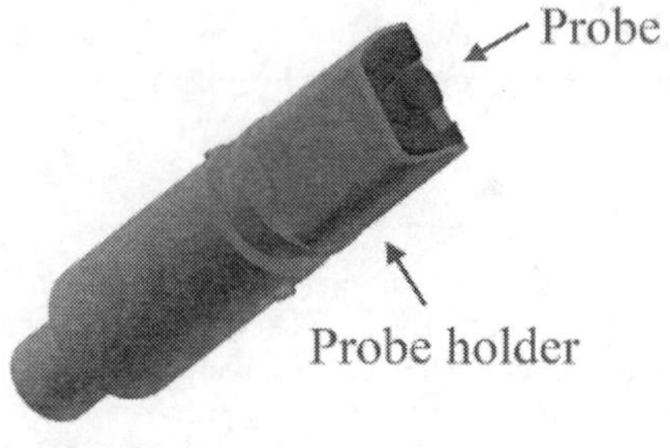

Fig. 22. Rendering of gap-detection probe mounted in probe-holder.

In Fig. 23, the probe and probe-holder is shown mounted in the mock-up measurement station.

Fig. 23. Close-up of dummy fuel rod and gap-detection probe in measurement station mock-up.

F. Test results

Fig. 24 shows the response to the 50 µm deep 4 x 4 mm rectangular gap combined with a lift-off signal. The good separation between gap and lift-off responses allow for heavy gain to be applied to the gap (Q-axis) signal.

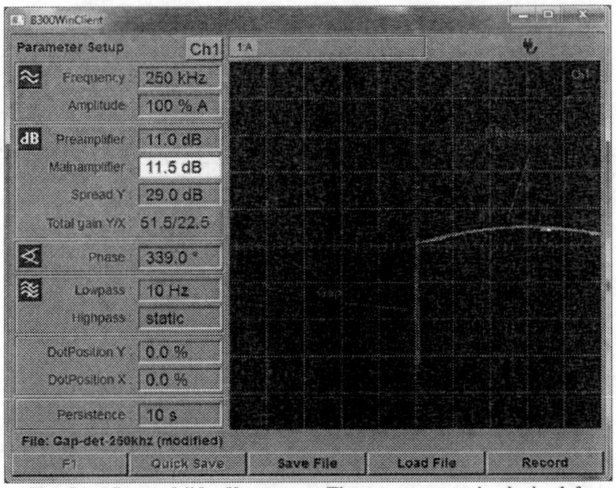

Fig. 24. Gap and lift-off response. The gap-response is obtained from scanning a 50 μm deep 4 x 4 mm rectangular gap.

The gap/lift-off discrimination was similarly good for the other gaps.

The separation in the gap and lift-off response (including phase) is similar to the simulated results in Figs. 16–18.

The two next figures show the gap (Q-axis) response as a function of dummy rod position. Since it proved difficult to obtain sufficiently good contact between the fuel and cladding parts of the dummy rod, it was necessary to shorten the length of the cladding part with a few centimeters. This significantly improved the contact between the parts, but the outermost gaps on each side were affected by this modification, as can be seen in the plots.

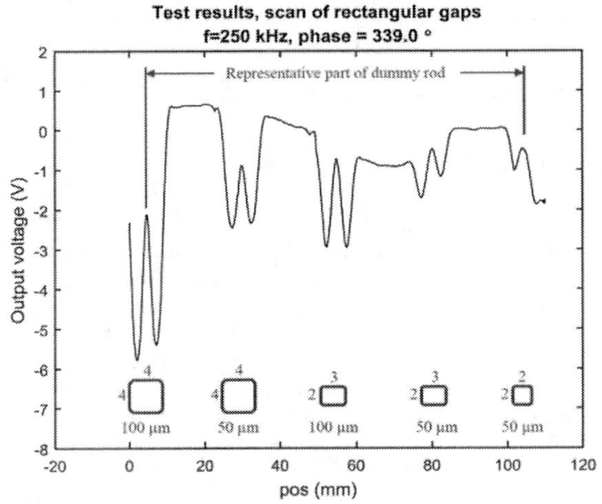

Fig. 25. Test results from gap-detection scan on the side of the dummy rod with rectangular gaps. The dimensions (in millimeters) and depth of each gap is included in the figure.

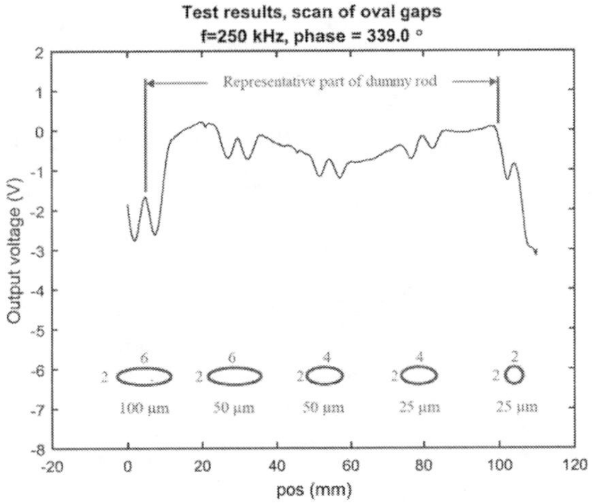

Fig. 26. Test results from gap-detection scan on the side of the dummy rod with oval gaps. The dimensions (in millimeters) and depth of each gap is included in the figure.

The test results show that the crack detection probe is able to produce a distinctive response to each of the gaps. The signal drift between the gaps is believed to be a response to a small opening between the fuel and cladding part of the dummy rod, since it was very difficult to obtain a perfect fit between the two.

The gap-detection probe is generally more sensitive to the rectangular gaps, but also gives a distinct response to oval gaps with thickness as low as 25 μm. Also note that all scans were performed at the same phase.

There is correspondence between simulated and experimental results, but the transition from a double-crested to a single-crested response is not observed in the practical tests, as predicted in the simulated gap-response rightmost in Fig. 19. It is still expected that such a response will be observed for longer gap lengths.

Acknowledgment

The authors wish to thank the personnel at the instrument workshop in the NEW-department at IFE for their know-how and dedicated effort in manufacturing and assembling the test equipment.

References

[1] J. Malone, A. Totemeier, N. Shapiro, S. Vaidyanathan, "Lightbridge Corporation's advanced metallic fuel for light water reactors," *Nuclear Technology.*, vol. 180, pp. 437-442, Dec. 2012.

[2] M.A. McGrath, P. Bennett, H. Thoresen, "Utilization of the Halden Reactor for supporting development of fuel with enhanced accident tolerance," presented at *Top Fuel* Boise, ID, USA, Sept. 11-15, 2016.

[3] T. Stepinski, "Essential parameters in eddy current inspection," Uppsala University, Uppsala, Sweden, SKI report 00:30, May 2000.

[4] J. Blitz, *Electrical and magnetic methods of non-destructive testing*, 2nd ed., London, UK: Chapman & Hall, 1997, pp. 139-141.

Calculations to Support On-line Neutron Spectrum Adjustment by Measurements with Miniature Fission Chambers in the JSI TRIGA Reactor

Tanja Kaiba, Vladimir Radulović, Gašper Žerovnik and Luka Snoj
Reactor Physics Department
Jožef Stefan Institute
Jamova cesta 19, SI-1000 Ljubljana
e-mail: tanja.kaiba@ijs.si

Damien Fourmentel, Loïc Barbot and Christophe Destouches
CEA, DEN, DER,
Instrumentation, Sensors and Dosimetry Laboratory
Cadarache, F-13108 St-Paul-Lez-Durance
e-mail: damien.fourmentel@cea.fr

***Abstract*—Preliminary calculations were performed with the aim to establish optimal experimental conditions for the measurement campaign within the collaboration between the Jožef Stefan Institute (JSI) and Commissariat à l'Énergie Atomique et aux Énergies Alternatives (CEA Cadarache). The goal of the project is to additionally characterize the neutron spectrum inside the JSI TRIGA reactor core with focus on the measurement epi-thermal and fast part of the spectrum. Measurements will be performed with fission chambers containing different fissile materials (^{235}U, ^{237}Np and ^{242}Pu) covered with thermal neutron filters (Cd and Gd). The changes in the detected signal and neutron flux spectrum with and without transmission filter were studied. Additional effort was put into evaluation of the effect of the filter geometry (e.g. opening on the top end of the filter) on the detector signal. After the analysis of the scoping calculations it was concluded to position the experiment in the outside core ring inside one of the empty fuel element positions.**

1. Introduction

The power distribution in a nuclear reactor can be monitored by neutron detectors, the signal of which depends on the neutron spectrum. The neutron spectrum is divided into the: thermal, epi-thermal and fast part. The thermal part of the neutron spectrum is approximately following Maxwellian shape around thermal neutron energy (0.0253 eV). The epi-thermal part lies between thermal and fast part and follows approximatelly $1/E$ dependance, it is followed by the fast part with fission spectrum. To be able to efficiently define and optimize the design of the nuclear instrumentation detectors, it is of great importance to know the neutron spectrum at different positions inside the reactor core as accurately as possible. Within the bilateral collaboration project between the Jožef Stefan Institute (JSI) and the Commissariat à l'Énergie Atomique et aux Énergies Alternatives (CEA Cadarache), an extensive experimental campaign is planned at the JSI TRIGA reactor [1] with the aim to establish an on-line neutron spectrum adjustment method. In the campaign, fission rates will be measured using miniature fission chambers [2] developed at the CEA Cadarache with different fissile deposits, unfiltered and covered with different neutron filters. It was decided to use ^{235}U, ^{242}Pu and ^{237}Np fission chambers in combination with Cd and Gd transmission filters [3], [4]. Fission chambers containing different fissile deposits have different energy sensitivity profiles due to different energy dependences of the fission cross-section. The transmission filters act as thermal neutron absorbers and shift the sensitivities to the epi-thermal and fast parts of the neutron spectrum. From the measured fission rates, the magnitudes of the thermal, epi-thermal and fast components of the neutron spectra will be deduced. In the experimental campaign we will further characterize the neutron fields in the JSI TRIGA reactor, and validate the computational model against measurements. This paper presents the prior support calculations, performed to establish the experimental limitations. The effect of different filter geometries (dimensions, material and presence of openings) on the neutron spectra was evaluated.

2. Computational model

Neutron transport calculations using the Monte Carlo neutron transport code MCNP6 [5] were performed with the ENDF/B-VII.1 [6] incident neutron cross section library. A full 3D JSI TRIGA reactor model used in calculations is based on the criticality benchmark model [7] which is thoroughly described in the International Handbook of Evaluated Criticality Safety Benchmark Experiments [8]. Our computational model has been expanded, verified and validated by many experiments for different calculations: the effective multiplication factor k_{eff} [9], power peaking factors [10], reactor kinetic parameters [11], flux and reaction rate distributions [12], [13], [14], neutron spectra and other reactor parameter calculations. In Figure 1 the visual presentation of the JSI TRIGA Mark II reactor model is shown.

Reaction rates obtained from MCNP calculations were normalized taking into account normalization factor [15]:

$$C = \frac{P\overline{\nu}}{k_{eff}\overline{w}_f}, \tag{1}$$

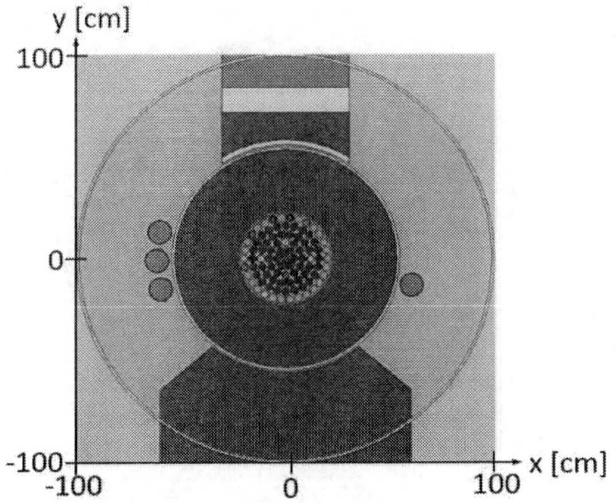

Figure 1. Visual presentation of the JSI TRIGA Mark II MCNP model developed at the JSI Reactor Physics Department.

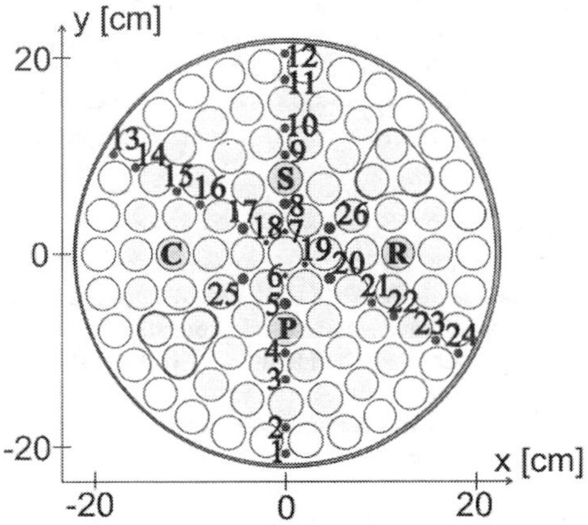

Figure 2. Schematic view of the JSI TRIGA core configuration with fuel elements in light yellow and control rods in dark yellow. Measuring positions are denoted with red dots and numbers.

where P represents reactor power, k_{eff} stands for effective multiplication factor obtained for individual MCNP simulation, $\bar{\nu}$ represents calculated average number of neutrons produced per fission and $\bar{w}_f$ is average energy released per fission contributing to the measurable thermal power (neutrino energy is not taken into account). The $\bar{\nu}$ and $\bar{w}_f$ are constants for individual core configuration and were calculated using JSI TRIGA MCNP model with ENDF/B-VII.1 nuclear data libraries [6] to be: $\bar{\nu}$ = 2.439 and $\bar{w}_f$ = 193.7 MeV. Normalization was performed for full reactor power 250 kW.

3. Preliminary design

The schematic view of the JSI TRIGA reactor core and measuring positions inside the core (red dots) is presented in Figure 2. For all calculations the safety (S) and pulse (P) control rods were completely withdrawn, while the compensating (C) and regulating (R) control rods were in approximately the same position - half withdrawn. In all calculations R control rod was positioned at step 504 and the C control rod at step 582, where step 200 indicates a fully withdrawn control rod and step 900 a fully inserted control rod.

In the JSI TRIGA reactor the neutron spectrum is highly thermalized, with approximately 92 % of fissions caused by thermal neutrons, with energies below 0.625 eV. The calculated neutron spectra in different measuring positions are presented in Figure 3. We can observe that the total neutron flux is the highest closer to the center of the core (MP5) and decreases towards the edge of the core (MP1). In general the ratio between the thermal and fission part of the neutron spectra increases towards the edge of the core. The detector response depends on the magnitude of the neutron flux and on the shape of the neutron spectrum. If the effect of the neutron spectrum change on the detector signal is studied, the effect of the neutron flux change has to be filtered out to achieve accurate evaluation. This can be performed by relative comparison of measurements and calculations performed with and without thermal neutron filters.

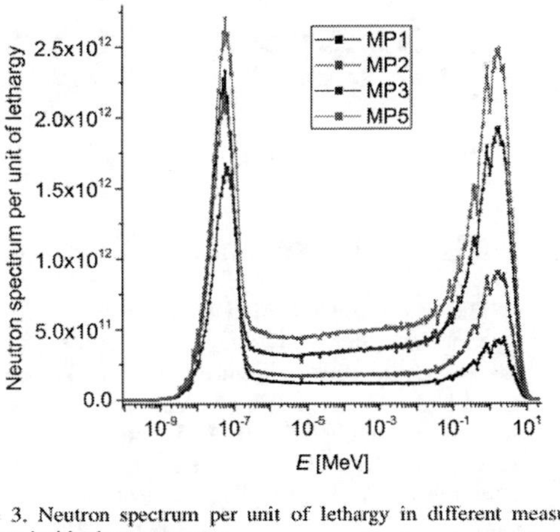

Figure 3. Neutron spectrum per unit of lethargy in different measuring positions inside the reactor core.

In previous research the reactor core has already been characterized with ^{235}U fission chambers, which are highly sensitive to the thermal part of the neutron spectrum [16]. To evaluate the fast and epi-thermal part of the neutron spectrum, fission chambers with fissile coatings with large fission cross section for epi-thermal and fast neutron energies, have to be used. Three different fission chamber coating materials were evaluated to cover different energy regions: ^{235}U, ^{237}Np and ^{242}Pu. To minimize the signal due to the thermal neutrons, the fission chambers will be covered with thermal neutron transmission filters. Chosen filter materials are: Cd and Gd.

For preliminary calculations transmission filters were modelled as hollow cylinders, without explicitly modelled fission chambers inside, due to the assumed negligible effect on the relative fission rate distribution. The effect of the filter length and configuration on the neutron spectrum inside the filter was studied. To evaluate the effect of the filter length on the detected signal, calculations with single and double filter length were compared. It was assumed that even a small opening inside the filter can result in a high contribution to the detected signal by the thermal part of the neutron spectrum. Therefore configurations with both endings opened and closed were analysed. Different filter configurations are schematically presented in Figure 4. The chosen thicknesses of transmission filters were: 1.0 mm for Cd and 250 μm for Gd with length of 5.5 cm (single) and 11.0 cm (double).

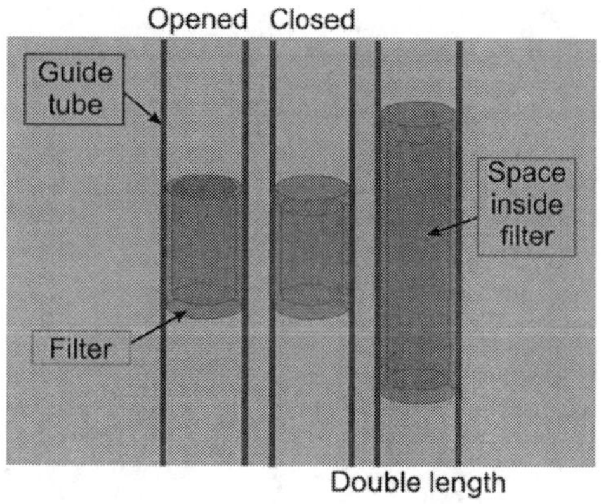

Figure 4. Schematic view of different filter configurations.

4. Results of preliminary calculations

The neutron spectrum in MP1 (see Figure 2) for different transmission filter materials and different filter configurations was analysed. In Figure 5 calculated neutron spectra per unit of lethargy are presented through the entire energy range. It can be observed that both filter materials efficiently decrease the thermal neutron part. In Figure 6 a closer view of the thermal and epi-thermal part of the neutron spectrum is presented. The effect of the transmission filter geometry on the neutron spectrum is presented in Figure 7 for the Cd and in Figure 8 for the Gd filter. It can be observed that the filter length does not affect the neutron spectrum significantly, however the difference is noticeable when it is opened. If the transmission filter is opened, the thermal part of the neutron spectrum becomes non-negligible, especially in case of the Gd filter.

For better understanding of the effect of the transmission filters on the detector response in MP1, the total fission rate (R) over all energy groups for ^{235}U, ^{237}Np and ^{242}Pu was calculated and compared. Results for the fission rates calculated through the entire space inside the filter are

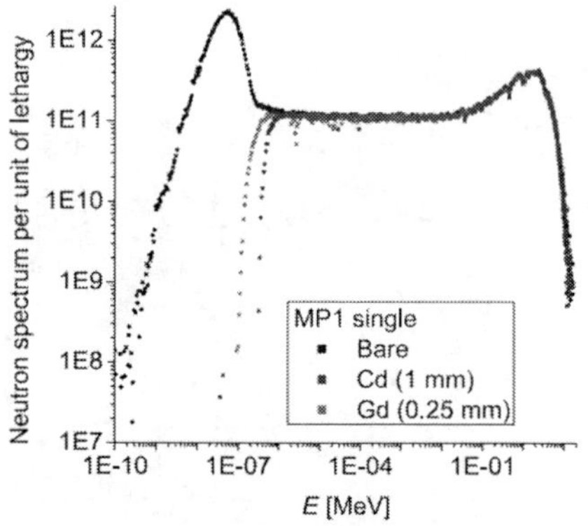

Figure 5. Neutron spectrum per unit of lethargy in MP1 for different thermal neutron filters.

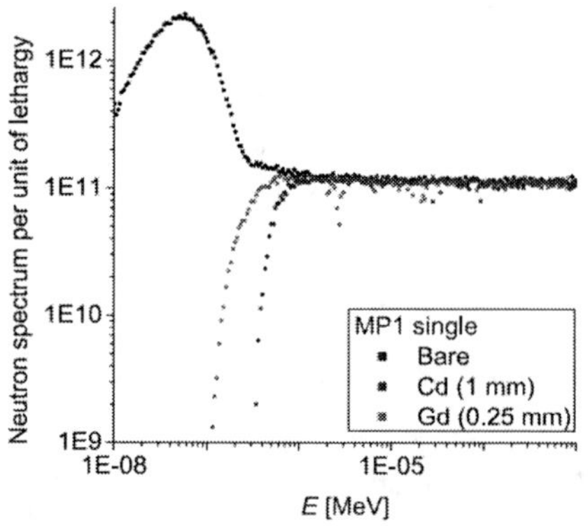

Figure 6. Neutron spectrum per unit of lethargy in MP1 for different thermal neutron filters with closer look at the thermal and epi-thermal part.

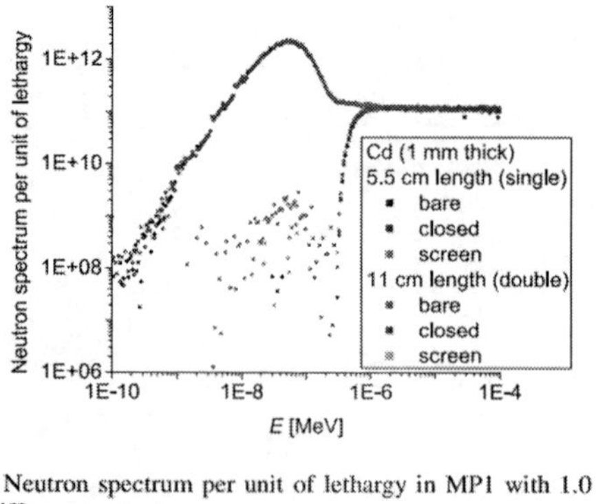

Figure 7. Neutron spectrum per unit of lethargy in MP1 with 1.0 mm Cd filter of different geometry.

reported in Tables 1 and 2. As expected, the thermal neutron filter has the largest effect on the fission rate in case of ^{235}U. The difference between the open and closed configuration of the filter is expected. Due to the opening inside the filter more thermal neutrons come to the fission chamber

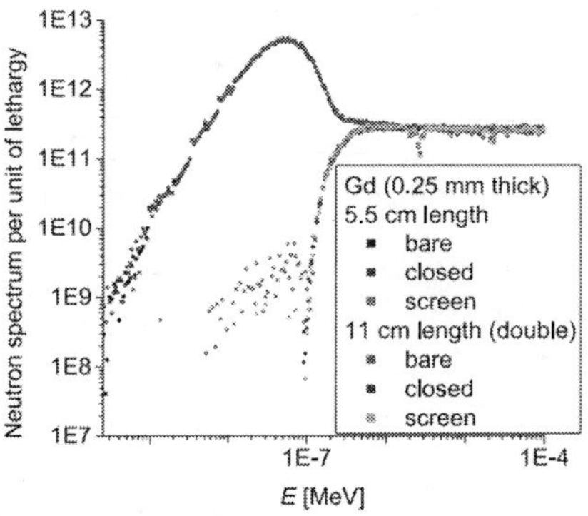

Figure 8. Neutron spectrum per unit of lethargy in MP1 with 250 μm Gd filter of different geometries.

and contribute to the detector signal. For this preliminary comparison, the fission rate through the entire space inside the filter was considered. However, if the fission chambers are positioned at the bottom of the filter a lower thermal neutron contribution to the detected signal is expected. In the comparisons of the fission rates inside single and double length filters, the calculated fission rates were divided by the bare fission rates in order to eliminate the effect of spatial position.

TABLE 1. COMPARISON OF FISSION RATES IN ACTIVE PART OF THE FISSION CHAMBER FOR CD FILTER.

Fissile material	Filter/Bare$^{(a)}$ (single, closed)	Single/Double$^{(a)}$ (closed)	Open/Closed$^{(a)}$ (single)
^{235}U	0.01490 $\pm$ 0.00004	1.0095 $\pm$ 0.0032	1.5925 $\pm$ 0.0069
^{237}Np	0.9181 $\pm$ 0.0018	1.0194 $\pm$ 0.0018	1.0007 $\pm$ 0.0020
^{242}Pu	0.9130 $\pm$ 0.0019	1.0078 $\pm$ 0.0026	1.0024 $\pm$ 0.0021
$^{(a)}$ Reported uncertainties are 1 σ statistical uncertainties.			

TABLE 2. COMPARISON OF FISSION RATES IN ACTIVE PART OF THE FISSION CHAMBER FOR GD FILTER.

Fissile material	Filter/Bare$^{(a)}$ (single, closed)	Single/Double$^{(a)}$ (closed)	Open/Closed$^{(a)}$ (single)
^{235}U	0.02078 $\pm$ 0.00005	1.0078 $\pm$ 0.0024	1.4289 $\pm$ 0.0050
^{237}Np	0.9314 $\pm$ 0.0018	1.0092 $\pm$ 0.0024	1.0007 $\pm$ 0.0020
^{242}Pu	0.9271 $\pm$ 0.0018	1.0102 $\pm$ 0.0025	0.9994 $\pm$ 0.0020
$^{(a)}$ Reported uncertainties are 1 σ statistical uncertainties.			

5. Optimization of irradiation position

After comparing the multiplication factor (k_{eff}) for different configurations, it was confirmed that transmission filters have a high impact on the reactivity. A single Cd (1.0 mm thick) or Gd (0.6 mm thick) filter positioned in the Central Channel (center of the reactor core) has approximately (-500$\pm$ 4) pcm negative reactivity. If the filter is placed in the outside core ring, in one of the empty fuel element positions, its negative reactivity amounts to approximately (-30 $\pm$ 1) pcm. The difference in negative reactivities of the Cd and Gd filters is within the statistical uncertainty of the calculations. Therefore, it was concluded that the best option is to place the experiment in one of the empty fuel element positions in the outer ring of the reactor core (F ring).

6. Conclusion

Preliminary calculations in support of the experimental campaign, using fission chambers containing different fissile material and covered with thermal neutron transmission filters, were performed. It was determined that the filter geometry can have a large effect on the detector signal, e.g. opening inside the filter can significantly increase the contribution of thermal neutrons (especially in the case of ^{235}U fission chamber). Due to the high negative reactivity of the spectrum filters it was decided to place the experiment in the outside core ring in one of the empty fuel element positions. After analysing preliminary calculations, improved computational design can be made including explicitly modelled fission chambers inside the filters. The experimental campaign is scheduled in autumn 2017. The measured magnitudes of the fission rates of different materials in combination with different transmission filters will be compared to the calculated results. The obtained results will serve as a basis for future research, in which the effect of the control rod movement on the neutron spectrum distribution will be studied.

Acknowledgments

The research is funded by the bilateral project "Experimental on-line neutron spectra adjustment method using various fission chambers with adequate fission fissile isotopes" between the Ministry of education, science and sport of the Republic of Slovenia and Commissariat à l'énergie atomique et aux énergies alternatives (CEA) under contract number BI-FR/CEA/15-17-005.

References

[1] Snoj, L., Smodiš, B., 2011. 45 Years of TRIGA Mark II in Slovenia. In: Proceedings of the International Conference Nuclear Energy for New Europe 2011, Bovec, Slovenia, Paper no. 401.

[2] Barbot, L., Domergue, C., Breaud, S., Destouches, C., Villard, J.-F., Snoj, L., Štancar, Ž., Radulović, V., Trkov, A, 2013. Neutron field characterization of irradiation locations applied to the slovenian TRIGA reactor. In: Proceedings of the International Conference Advancements in Nuclear Instrumentation Measurement Methods and their Applications (ANIMMA), 2013, Marseille, France. doi: 10.1109/ANIMMA.2013.6727954.

[3] Filliatre, P., Oriol, L., Jammes, C., Vermeeren, L., 2008. Reasons why Plutonium 242 is the best fission chamber deposit to monitor the fast component of a high neutron flux. Nuclear Instruments And Methods in Physics Research - Section A, 593 (3), pp. 510-518.

[4] Filliatre, L., Oriol, L., Jammes, C., Vermeeren, L., 2009. Joint estimation of the fast and thermal components of a high neutron flux with a two one-line detector system. Nuclear Instruments and Methods in Physics Research - Section A, 603 (3), pp. 415-420.

https://doi.org/10.1051/epjconf/201817004012

[5] Goorley, T., et al., 2012. Initial MCNP 6 Release Overview. LA-UR-11-07082, Los Alamos National Laboratory, also Nuclear Technology, 180, pp. 298-315 (Dec 2012).

[6] Chadwick, M.B., et al., 2011. ENDF/B-VII.1 Nuclear Data for Science and Technology: Cross Sections, Covariances, Fission Product Yields and Decay Data. LA-UR 11-05121, Los Alamos National Laboratory, also: Nuclear Data Sheets, 112 Issue 12, pg. 2887-2996 (Dec 2011).

[7] Jeraj, R., Ravnik, M., 1999. TRIGA Mark II Reactor: U(20) – Zirconium Hydride Fuel Rods in Water with Graphite Reflector, IEU-COMP-THERM-003, International Handbook of Evaluated Critical Safety Benchmark Experiments, Organization for Economic Cooperation and Development – Nuclear Energy Agency, NEA/NSC/SOC(95)03, Paris, 1999.

[8] ICSBEP, 2009. International Handbook of Evaluated Critical Safety Benchmark Experiments, Organization for Economic Cooperation and Development - Nuclear Energy Agency, NEA/NSC/DOC(95)03, Paris, published on DVD, ISBN 978-92-64-99054-8.

[9] Ravnik, M., Jeraj, R., 2003. Research reactor benchmarks. Nucl. Sci. Eng. 145, 145-152.

[10] Snoj, L., Ravnik, M., 2008. Power peakings in mixed TRIGA cores. Nucl. Eng. Des. 238 (9), 2473-2479.

[11] Snoj, L., Kavčič, A., Žerovnik, G., Ravnik, M., 2010. Calculation of Kinetic Parameters for Mixed TRIGA Cores with Monte Carlo. An. Nucl. Ene. 37 (2), 223-229. doi:10.1016/j.anucene.2009.10.020.

[12] Snoj, L., Trkov., A., Jaćimović, R., Rogan, P. Žerovnik, G., Ravnik, M., 2011. Analysis of neutron flux distribution for the validation of computational methods for the optimization of research reactor utilization. Appl. Radiat. Isot. 69, 136-141. doi: 10.1016/j.apradiso.2010.08.019.

[13] Žerovnik, G., Kaiba, T., Radulović, V., Jazbec, A., Rupnik, S., Barbot, L., Fourmentel, D., Snoj, L., 2015. Validation of the neutron and gamma fields in the JSI TRIGA reactor using in-core fission and ionization chambers. Appl. Radiat. Isot. 96, 27-35. doi:10.1016/j.apradiso.2014.10.026.

[14] Štancar, Ž., Snoj, L., Barbot, L., Domergue, C., 2015. Evaluation of Fission Rate Profile Experimental Benchmark at the JSI TRIGA Mark II Reactor Using Monte Carlo Method. In: Proceedings of the Joint International Conference on mathematics and Computation (M&C), Supercomputing in Nuclear Applications (SNA) and the Monte Carlo (MC) Method, 19-23 April 2015, Tennessee, USA.

[15] Žerovnik, G., Podvratnik, M., Snoj, L., 2014. On normalization of fluxes and reaction rates in MCNP criticality calculations. An. Nucl. Ene. 63, 126-128. doi:10.1016/j.anucene.2013.07.045

[16] Kaiba, T., et al., 2017. Evaluation of neutron flux and fission rate distributions inside the JSI TRIGA Mark II reactor using multiple in-core fission chambers, submitted to Annals of Nuclear Energy.

Self-Powered Neutron Detector Calibration Using a Large Vertical Irradiation Hole of HANARO

Myong-Seop Kim, Byung-Gun Park and Gi-Doo Kang

***Abstract*—A calibration technology of the self-powered neutron detectors (SPNDs) using a large vertical irradiation hole of HANARO is developed. The 40 Rh-SPNDs are installed on the polycarbonate plastic support, and the gold wires with the same length as the effective length of the rhodium emitter of the SPND are also installed to measure the neutron flux on the SPND. They are irradiated at a low reactor power, and the SPND current is measured using the pico-ammeter. The external gamma-rays which affect the SPND current response are analyzed using the Monte Carlo simulation for various irradiation conditions in HANARO. It is confirmed that the effect of the external gamma-rays to the SPND current is dependent on the reactor characteristics, and that it is affected by materials around the detector. The current signals due to the external gamma-rays can be either positive or negative, in that the net flow of the current may be either in the same or the opposite direction as the neutron-induced current by the rhodium emitter. From the above procedure, the effective calibration methodology of multiple SPNDs using the large hole of HANARO is developed. It could be useful for the calibration experiment of the neutron detectors in the research reactors.**

***Index Terms*—HANARO, self-powered neutron detector (SPND), neutron irradiation, research reactor, Monte-Carlo simulation, calibration method**

I. Introduction

SELF-Powered Neutron Detectors (SPNDs) have been widely used for monitoring the neutron flux in reactors as well as in irradiation facilities. In its simplest form, the detector operates on the basis of directly measuring the beta decay current following neutron capture [1]. Because the SPND doesn't need to use the external bias supply applied to the detector, it is used not only for a typical neutron detector in the nuclear power plants but also for real-time neuron flux monitoring in the research reactors [2].

For the neutron flux monitoring using the SPND, sensitivity that is an intrinsic characteristic of the detector is required. In general, the sensitivity which is provided by the manufacturer is used. However, when developing the SPND or using the SPND for research purposes, it is necessary to calibrate the sensitivity accurately. The sensitivity of the SPND is calibrated by irradiating the detector and measuring the current signals and the neutron flux in the reactors. A conventional approach for the SPND calibration is the irradiation one or several SPNDs in the reactor core and measurement the sensitivity [3], [4], [5], [6], [7]. In order to calibrate multiple SPNDs simultaneously, a large irradiation hole and precise calibration method for multiple SPNDs are required.

HANARO, a 30 MW research reactor in KAERI, is a good tool for a determination of the sensitivity of the SPND. In order to estimate the sensitivity of the SPND precisely, the reactor should be operated at low-power, and in this situation, influence of the external γ-rays is not insignificant. In this paper, contribution of the external γ-rays was analyzed using the Monte Carlo simulation for various irradiation conditions in HANARO of the rhodium-based SPND.

II. Sensitivity of SPND

General SPND consists of three parts of emitter, insulator, and collector. The emitter captures neutrons, and emits electrons that reach the external sheath after going through the insulator polarizing the device. The electrons can be collected using a current amplifier connecting the emitter to the sheath. The electric current of the SPND depends on the neutron fluxes at the detector location, and characteristics of the detector, which is called detector sensitivity. The sensitivity of SPND is generally defined as

$$S = \frac{I}{\phi} = \frac{I_\beta + I_\gamma + I_{cable}}{\phi} \qquad (1)$$

where ϕ is the neutron flux in the vicinity of the SPND location, I is the current signal of the SPND, I_β is the current signal by beta-decay after the (n,γ) reaction, I_γ is the current signal by γ-ray interaction, and I_{calbe} is the current signal caused from the connecting cable, and can be considered negligible for standard SPND by compensated lead wire. I_β is proportional to the neutron flux, whereas I_γ is dependent on the reactor characteristics and affected by materials around the detector. The gamma response of the SPND owing to photoelectric or Compton effects produce electrons with enough energy to pass from the collector to the emitter or vice versa. Probability of both effects increases with atomic number and mass of electrodes materials [1]. Consequently, the current signals can

The authors are with the Neutron Utilization Research Division, Korea Atomic Energy Research Institute, Daejeon 34057, Korea (email: mskim@kaeri.re.kr; bgpark@kaeri.re.kr; dooki@kaeri.re.kr)

be either positive or negative, in that the net flow of the current may be either in the same or the opposite direction as the neutron-induced current.

III. Reference Measurement

A. Preparation of Experiment

For the calibration of SPND sensitivity using a large vertical irradiation hole of HANARO, an irradiation rig and an irradiation basket were designed and manufactured. The irradiation rig is composed of a body made of Aluminum (Al) -6061, and a flexible extension hose to guide the signal cables from the irradiation hole to the electronics on the reactor pool. The irradiation basket is inserted into the irradiation rig with multiple SPNDs. Polycarbonate was used for the irradiation basket to minimize the delayed γ-rays from neutron activation. On the outer side of the cylindrical shape irradiation basket, there are 40 U-grooves to mount the multiple SPNDs. In addition, grooves of 0.1 mm are provided between the grooves on which the SPNDs are mounted too install neutron monitors.

Sensitivity calibration is performed for the SPNDs with rhodium (Rh) emitter. The specifications of the SPND used at the sensitivity calibration are summarized in Table 1.

B. Measurement of SPND Sensitivity at Various Positions in the Large Vertical Hole

SPND sensitivities at various positions in the large vertical hole were measured and calibrated. The 40 Rh-SPNDs were installed on the irradiation basket and irradiated at a low reactor power about 1 kW. Fig. 1 shows the schematic of the multiple SPND calibration configurations in the large vertical hole of HANARO. During the irradiation of the SPNDs, all of the SPND current signals were collected for 16670 s using the Keithley 6517A picoammeter. In the measurement, a ground connection box was used to reduce the charging effect of the SPNDs. In addition a shielding box of copper was used to remove the noise which can be easily captured by long metallic SPND sheath.

The neutron flux during the irradiation was determined by activation of gold (Au) wires with the same length as the effective length of the Rh emitter. The Au wire has decayed about 10 days after the end of irradiation, and the γ-ray

TABLE I
SPND SPECIFICATIONS

Parameters	Figures
Materials	
Emitter	Rh
Insulator	Al_2O_3
Collector	Inconel-600
Core wire	Inconel-600
Dimensions	
Diameter of core wire	0.25 mm
Diameter of emitter	0.46 mm
Outer diameter of insulator	1.02 mm
Outer diameter of collector	1.56 mm
Effective length of emitter	400 mm
Cable length	35.8 m

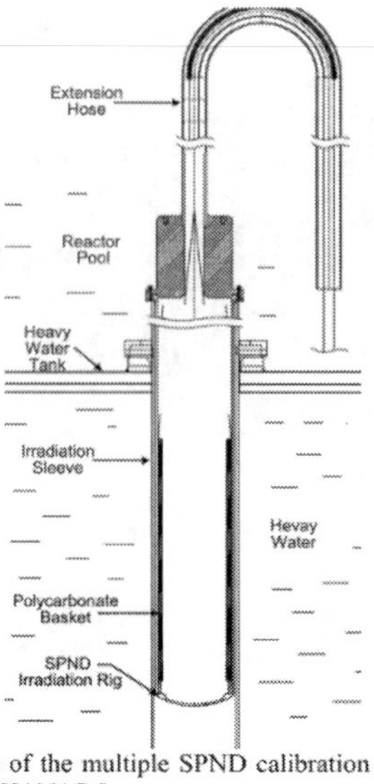

Fig. 1. Schematic of the multiple SPND calibration configuration in the large vertical hole of HANARO.

spectrum from the irradiated Au wire was measured for 1 hours of live time by using an HPGe detector. In the measured spectrum, peak area of 411.8 keV γ-ray from the Au-198 was analyzed and used to determine the saturation activity of the Au wire. The neutron flux at the position of the SPND was determined by averaging the neutron fluxes determined by the Au wires on both sides of the SPND.

Fig. 2 shows the measured neutron flux (Westcott) and determined sensitivity of SPND at each irradiation position. As shown in the Fig. 2, the maximum neutron flux was recorded at position 35, which was the nearest position to the reactor core. The measured sensitivities of SPNDs were (3.8 ~ 5.4) × 10^{-20} A/n/cm^2-s, and the standard deviation was 0.468. Because all SPNDs have the same specification, theoretically they should have the same sensitivity regardless of the irradiation position. However, the measured sensitivity is apparently dependent on the irradiation position. The sensitivity is relatively high at position where the neutron flux is low, and low at position where the neutron flux is high.

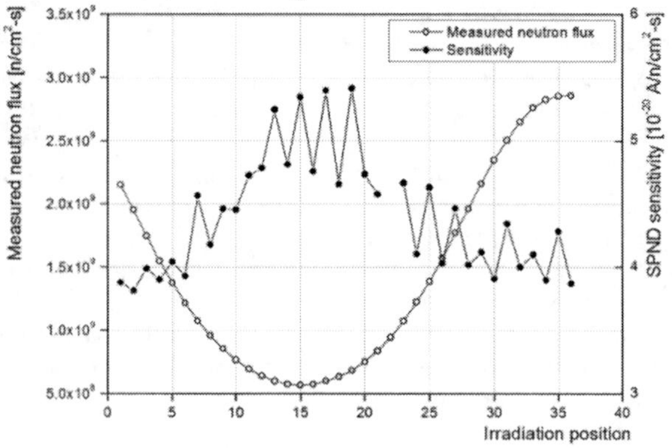

Fig. 2. Measured neutron flux and sensitivity of SPND, vs. the irradiation position.

IV. Evaluation of SPND Current Response Using Monte Carlo Simulation

In order to analyze the factors which can affect the SPND current response on the irradiation position, the Monte Carlo simulation for various irradiation conditions in HANARO was performed.

A. Modeling of the Measurement

The MCNP equilibrium core model was employed for a detailed whole core representation of the HANARO. In the core model, the isotopic density distributions of 13,104 fuel segments of HANARO core were investigated by the HANARO fuel management system, whose major components are WIMS and VENTURE codes. Based on the burnup dependent nuclide number density, power distribution and burnup distribution data from the WIMS and VENTURE codes, the MCNP model for HANARO burned-core in the middle of the 59th operation cycle was generated. The control rod position of the core model is 350 mm which is the distance from the bottom of its fully-inserted position. In the calculation, the reactor power was assumed to be 1 kW as in the measurement.

Fig. 3 shows the MCNP model for the SPND. The main components of the SPND are Rh emitter, an Inconel sheath collector, an aluminum oxide (Al_2O_3) insulator and an Inconel wire. In the SPND modeling, geometry and material composition of each component were based on the SPND that was used for the reference measurement. In the Fig. 3, R_e, R_i and R_c represent the radius of emitter, insulator and collector, respectively. Rh emitter of 40 cm length and the insulator and collector of 50 cm length were modeled for MCNP calculation.

An irradiation rig and basket for the SPND irradiation were assumed in a cylindrical shape. Top views of the models of the irradiation rig and basket are shown in Fig. 4. Outside of the irradiation rig, there is an Al sleeve that protects the surface of the heavy water reflector tank from the rig. And there is light water between the Al sleeve and the irradiation rig and between the irradiation rig and the irradiation basket. In addition, light water is filled inside the irradiation basket. In the calculation, 40 irradiation positions were located in a circle along the irradiation basket. The SPNDs are located at the center of each U-groove of the irradiation basket. The axial position of the center of the Rh emitter was assumed to be 12.5 cm from the longitudinal center of the reactor core.

B. Calculation of the Transport of Neutrons and Photons in the Measurement

In order to analyze the factors which can affect the SPND current response, the spatial distributions of the neutron flux, photon flux, and neutron capture reaction and γ-ray interaction rates in each component of the SPND were calculated using the MCNP6 code. The transport calculation was performed in criticality mode. F4 tallies and F4 tallies with multipliers for tallying the reaction rate of the MCNP code were used to calculate the flux and reaction rate, respectively.

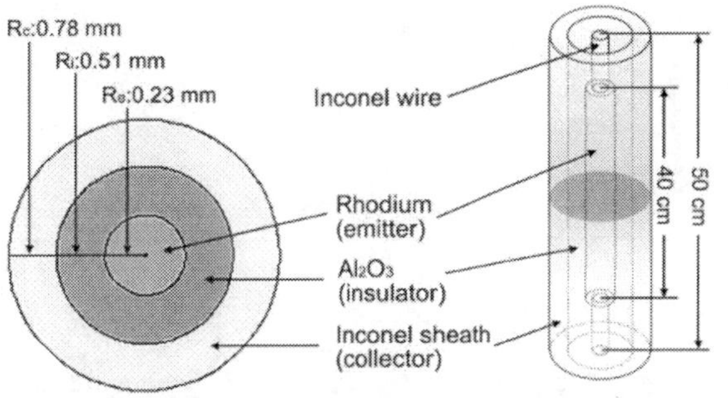

Fig. 3. Configuration of SPND calculation model (not to scale).

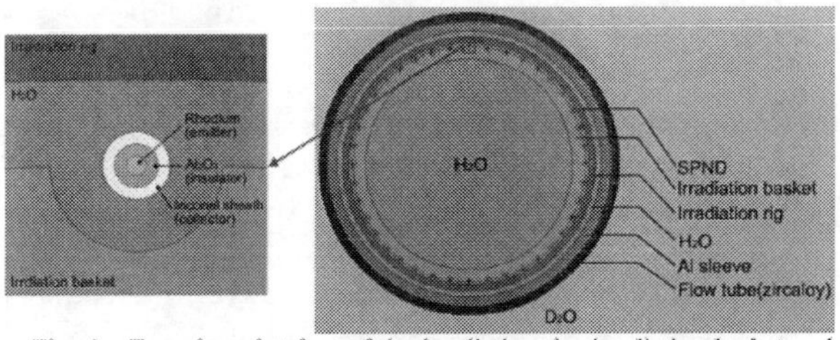

Fig. 4. Top view sketches of the irradiation rig, irradiation basket and SPND of the calculation model.

Fig. 5 shows the calculation result of the normalized neutron flux at Rh emitter of the SPNDs, which were installed at different positions (1 ~ 40) of the irradiation basket. As shown in the figure, the maximum neutron flux was recorded at position 11, which was the nearest position to the reactor core. Measured neutron flux at each position is described with the calculated one in Fig. 5. In order to compare with the calculation results, the irradiation positions of the measurement are displayed in accordance with the calculation results. The calculated spatial distribution of the neutron flux is similar to the measured one. Maximum difference of the neutron flux between the calculation and measurement is 11%. The inconsistency is caused by the neutron self-shielding effect. In the calculation, diameter of the Rh emitter was 0.46 mm which can make the self-shielding correspondingly higher than the Au wire of 0.01 mm diameter in the measurement. Moreover, the Rh emitter is surrounded by the insulator and the collector materials, and thus some of neutrons are shielded by the surrounding materials in the calculation.

Fig. 6 shows the calculated neutron and photon flux distributions at each component of the SPNDs at 40 irradiation positions. As shown in Fig. 6, the maximum neutron flux and photon flux were recorded at position 11 whereas the minimum fluxes were recorded at position 31 or 32, which was the opposite the direction to position 11. Among the four components of the SPND, neutron flux is the highest at the Inconel sheath, followed by the insulator, the emitter and the Inconel wire, regardless of the irradiation position. In contrast, the photon fluxes of the collector, the insulator and the emitter did no show a significant difference for the same irradiation position except for the Inconel wire. This is because the γ-rays from the fuels and the prompt γ-rays from neutron capture in the materials near SPND are not attenuated significantly when they incident to the Rh emitter. In consideration of the neutron

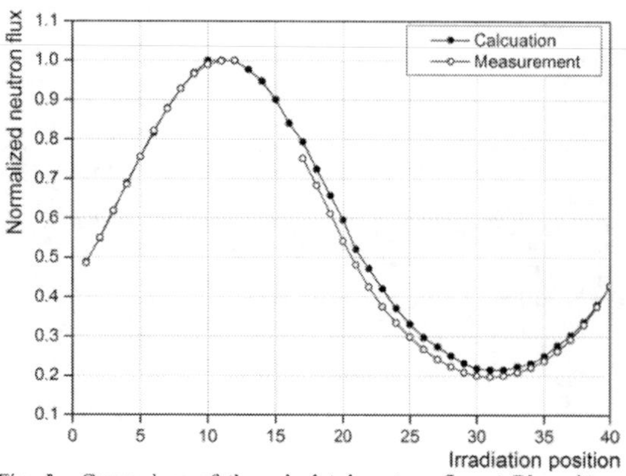

Fig. 5. Comparison of the calculated neutron flux at Rh emitter and measured one.

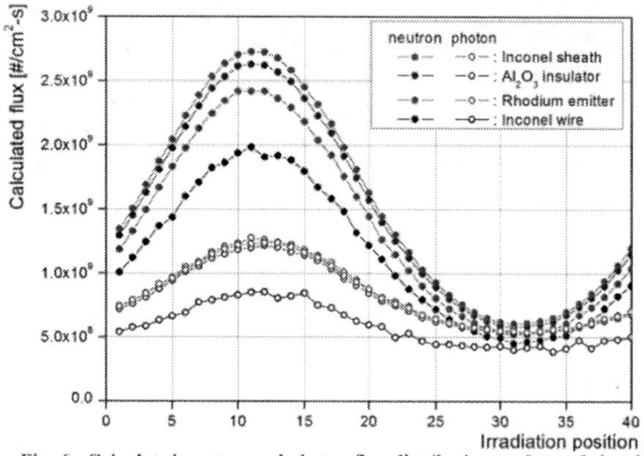

Fig. 6. Calculated neutron and photon flux distributions at Inconel sheath, Al_2O_3 insulator, Rhodium emitter and Inconel wire for 40 SPNDs

and photon fluxes as a function of the irradiation position, the ratio of the neutron flux at position 31 to position 11 for the Rh emitter ($\phi_{n,emitter31}/\phi_{n,emitter11}$) is 0.22, whereas the ratio of the neutron flux at position 31 to position 11 for the Rh emitter ($\phi_{\gamma,emitter31}/\phi_{\gamma,emitter11}$) is 0.44. This means that there is flux gradient in the large vertical hole in radial direction, and the neutron flux is decreased significantly by water in the irradiation basket from position 11 to position 31. However, the γ-rays are correspondingly less affected by water in the irradiation basket than neutrons.

Neutron capture (n,γ) reaction rate and total γ-ray (γ,tot) interaction rate distributions for 40 SPNDs are shown in Fig. 7. The ratio of the minimum position to maximum position can be estimated in the same manner as for the ratio of neutron and photon fluxes. For the Rh emitter, $RR_{n,emitter31}/RR_{n,emitter11}$ is 0.23, which is similar to the $\phi_{n,emitter31}/\phi_{n,emitter11}$. On the other hand, $RR_{\gamma,emitter31}/RR_{\gamma,emitter11}$ is 0.51, which is slightly higher than $\phi_{\gamma,emitter31}/\phi_{\gamma,emitter11}$.

C. Analysis of SPND Current Response from the (n,γ) and (γ,tot) Reaction Rates

Among the components of the SPND, the (n,γ) reaction at the Rh emitter is dominant, and insulator and Inconel wire can be

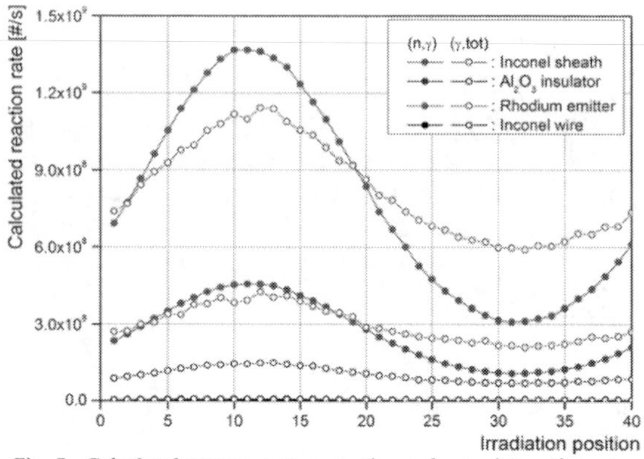

Fig. 7. Calculated neutron capture reaction and γ-ray interaction rates at Inconel sheath, Al_2O_3 insulator, Rhodium emitter and Inconel wire for 40 SPNDs

negligible. The spatial distribution of the (n,γ) reaction rate at the Rh emitter is equivalent with that of the neutron flux. Therefore, it can be regarded that the (n,γ) reaction at the Rh emitter is irrelevant to the difference of the SPND sensitivity according to the irradiation position.

The (n,γ) reaction rate at the Inconel sheath are approximately 1/3 of those at the Rh emitter regardless of the irradiation position. In order for the neutrons that are captured in the Inconel sheath to contribute to the SPND current, electrons after beta-decay should be reached the Rh emitter. In this case, negative current is produced and thereby a part of positive current by electrons from the Rh emitter is compensated. However, a diameter of the Rh emitter is less than Inconel sheath, therefore partial electrons from Inconel sheath can be reached the Rh emitter. Furthermore, current-contributing electron should have energy greater than EMIN which is the minimal energy to pass the insulator. From the Warren's model [8], the EMIN of the SPND in this study is 300 keV which is calculated by considering the diameter and thickness of the Rh emitter and Al_2O_3 insulator. Cr-55, Fe-59 and Ni-65 are main radioisotopes which are produced after (n,γ) reaction in the Inconel sheath, and energies of the released electrons are higher than 300 keV. However, abundance and (n,γ) reaction cross-sections of the parent nuclides are negligibly small. As a consequence, it is not needed to consider the (n,γ) reaction in the Inconel sheath to the current sensitivity of the SPND.

As shown in Fig. 7, (γ,tot) interaction rate is the highest at the Inconel sheath, followed by the Rh emitter, the insulator and the Inconel wire, regardless of the irradiation position. In order to contribute the SPND current, electrons from the Rh emitter and Inconel sheath should be reached the Inconel sheath and Rh emitter, respectively. The electrons from the insulator can be reached both Rh emitter and Inconel sheath. To estimate the SPND current response by the (γ,tot) interaction, it is assumed that 1) electrons produced by the (γ,tot) interaction can escape from the initial material 2) direction of the released electrons is isotropic 3) released electrons can pass the insulator. Considering the geometry of the SPND, the most electrons from the Rh emitter can be reached the Inconel sheath, and contribute the positive SPND current. In contrast, partial electrons from the Inconel sheath can be reached the Rh emitter because the diameter of the emitter is quite small. In the case of the insulator, partial electrons reach the Rh emitter, and most electrons reach the Inconel sheath.

The probability of electrons from the Inconel sheath reaching the Rh emitter was estimated by using the Monte Carlo method. In the estimation the SPND model of the MCNP calculation was used. Electrons were emitted to isotropic direction from the Inconel sheath region, and the electrons reaching the Rh emitter were counted. The probability of electron reaching was estimated to be 0.0753. This value is used as a correction factor to the (γ,tot) interaction rate at the Inconel sheath. As a result, (γ,tot) interaction rate at the Inconel sheath in Fig. 7 is decreased 7.53% of the initial value, and then the dominant contribution to the SPND current is (γ,tot) interaction rate at the Rh emitter.

D. Correction of External γ-rays to the SPND Current

Unless the contribution of the external γ-rays to the SPND current is the same for all irradiation positions, it cannot be easy to correct the external γ-rays because different correction factors should be used for each irradiation position. The external γ-rays can be classified into 1st and 2nd γ-rays. The 1st γ-rays are emitted from the fission product and activated structural materials in the reactor, and they can be produced in zero power of the reactor. Therefore the contribution of the 1st γ-rays can be corrected by subtracting the SPND current measured before operation of the reactor.

Whereas the 2nd γ-rays are depends on the reactor power. If the ratio of I_γ to the I_β in (1) is same at each position, the sensitivity of the SPND can be determined irrelevant to the irradiation position. However, as shown in Fig. 7, $RR_{\gamma,emitter11}/RR_{n,emitter11}$ is 0.29 at position 11, but $RR_{\gamma,emitter31}/RR_{n,emitter31}$ is increases 0.70 at position 31. In this case, therefore, contribution of 2nd γ-rays to the SPND positive current increases at position 31, therefore the sensitivity of SPND at position 31 can be determined higher than that at position 11. In the reference experiment, the sensitivity of SPND installed at far from the reactor core was measured 40% higher than that at the nearest position to the reactor core. That is the γ-rays can be travel longer than neutrons in the irradiation basket. Contribution of the 2nd γ-rays can be reduced by removing water in the irradiation basket, and increasing the path length of the neutrons.

E. Estimation of Corrected SPND Sensitivity

The irradiation baskets not containing light water inside the basket were considered to investigate the sensitivities of the SPNDs at each irradiation position. Using the MCNP code, the (n,γ) reaction rate and (γ,tot) interaction rate distributions of the SPND components at each irradiation position were calculated. Calculation model is same with the previous one except the water in the irradiation basket was replaced to air.

Fig. 8 shows the spatial distribution of the (n,γ) reaction rate and (γ,tot) interaction rate for 40 SPNDs installed at the irradiation basket without water. In the figure, (γ,tot) interaction rate of the Inconel sheath was corrected distribution by using the correction factor of 0.0753. In this case, neutron absorption and scattering by water disappear. The (n,γ) reaction rate at the Rh emitter at position 31 was 66.6% of that at position 11. The ratios of (γ,tot) to (n,γ) rate at the Rh emitter are the same within 2% at all positions, for example $RR_{\gamma,emitter11}/RR_{n,emitter11}$ and $RR_{\gamma,emitter31}/RR_{n,emitter31}$ are 0.23 and 0.25, respectively.

From the calculation results, sensitivity of the SPND was defined as

$$S = \frac{RR_{n,emitter} + RR_{\gamma,emitter}}{\phi_{n,emitter}} \tag{2}$$

where, S is the sensitivity, RR_n is the (n,γ) reaction rate, RR_γ is the (γ,tot) interaction rate, and ϕ is the neutron flux. The sensitivity was estimated for the SPNDs installed at the irradiation basket with water and without water conditions. Calculation results are shown in Fig. 9. As shown in Fig. 9, the spatial distribution of the sensitivity of the irradiation basket with water is more than 20% difference in the average. However the spatial distribution of the sensitivity of the irradiation basket without water is within 4% in the average for all positions.

V. CALIBRATION OF SPND SENSITIVITY

The sensitivities of multiple SPNDs were calibrated based on the calibration method determined from the MCNP calculation. In this measurement, the irradiation basket without water was used. Additionally, before raising the reactor power, the current of the SPNDs in zero reactor power was measured for 300 s. After measurement of zero power current, reactor power was increased to 1.5 kW and the current of the SPNDs were measured for 10866 s. In this experiment, the nearest irradiation position to the reactor core was 35.

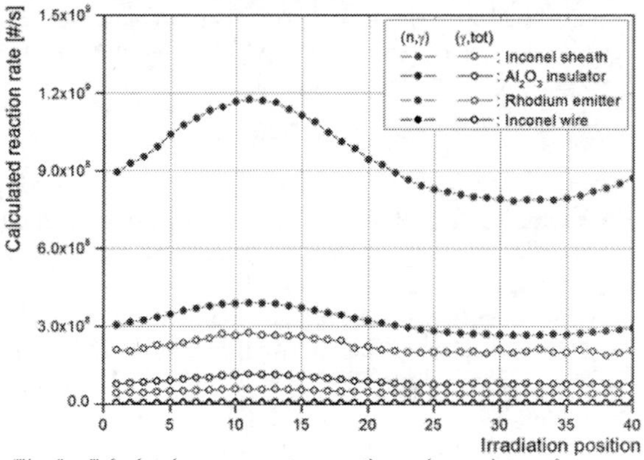

Fig. 8. Calculated neutron capture reaction and γ-ray interaction rates at each irradiation position of the irradiation basket without water.

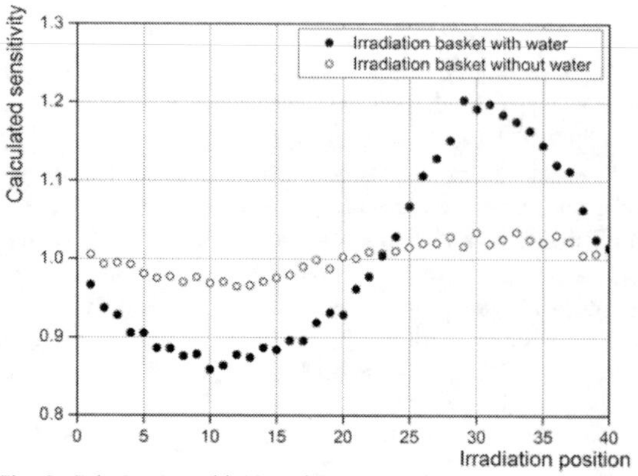

Fig. 9. Calculated sensitivities of 40 SPNDs installed at the irradiation baskets with water and without water.

Fig. 10 shows the spatial distributions of the SPND current in the zero power condition, 1.5 kW reactor power. The current in zero power subtracted data are also depicted in the figure. In the zero power measurement, only 1st γ-rays contribute to the SPND current. The measured current of the SPNDs were 9 ~ 25 pA, which depends on the irradiation position. In the 1.5 kW reactor power, neutrons, 1st γ-rays and 2nd γ-rays contribute to the SPND current together. In this condition, the current of the SPNDs were 70 ~ 130 pA. It is found that 10 ~ 20% of the SPND current is come from the 1st γ-rays. The subtracted current distribution shows more smooth shape than before subtraction.

The sensitivity of the SPND was determined by using the subtracted current and measured neutron flux. Spatial distributions of the measured sensitivity and calculated one which is shown in Fig. 9 are shown in Fig. 11. For comparison, number of irradiation position in the measurement is adjusted to the calculation. Distributions of the measured and calculated sensitivities are equivalent for irradiation positions. In the measured result, sensitivities of SPNDs installed odd number irradiation positions are larger than those installed even number irradiation positions. This is because the SPNDs at the odd position and the SPNDs at the even position were made of different manufacturers. When compared by odd positions and even positions separately, the deviation of the sensitivity to the average value is within 5%. The standard deviation of the measured sensitivities was 0.072, which is improved value than 0.468 for the reference measurement.

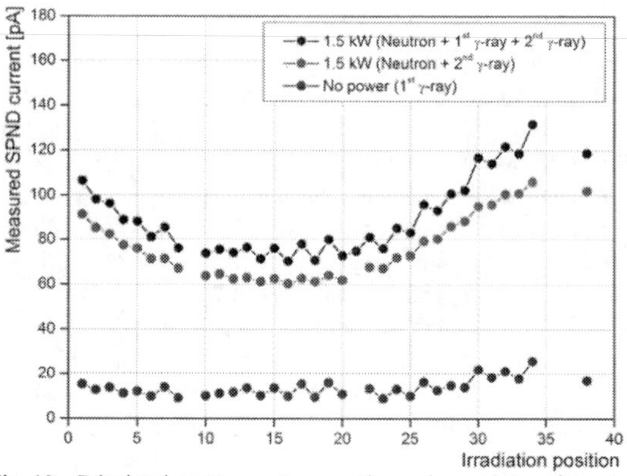

Fig. 10. Calculated neutron capture reaction and γ-ray interaction rates at each irradiation position of the irradiation basket without water.

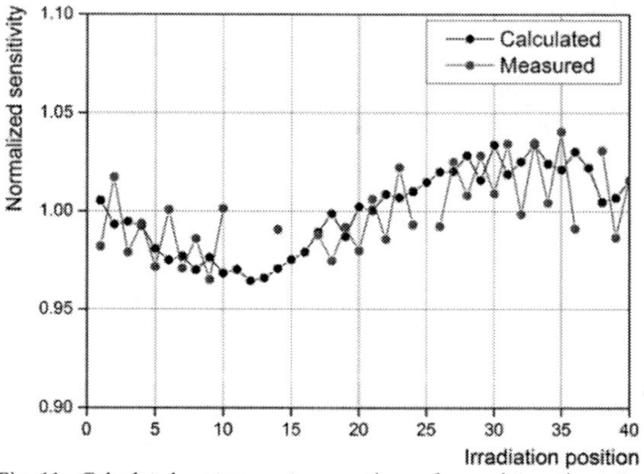

Fig. 11. Calculated neutron capture reaction and γ-ray interaction rates at each irradiation position of the irradiation basket without water.

VI. CONCLUSION

We described a method for calibrating sensitivity of multiple SPNDs installed at a large vertical hole of HANARO. Employing Monte Carlo simulation the contribution of the γ-rays to the current signal was estimated at various positions of the large vertical hole. The sensitivity variations of the multiple SPNDs were significantly improved by removing the contribution of 1st γ-rays emitted from the core, and by using the irradiation basket without water. The procedure can be useful for the calibration experiment of the neutron detectors in the research reactors.

REFERENCES

[1] G. F. Knoll, Radiation Detection and Measurement, New York: John Wiley & Sons, 2000.
[2] M. S. Kim and G. D. Kang, "Precise Control of Neutron Irradiation Fluence in the Neutron Transmutation Doping Process in HANARO Using SPND and Zirconium Foils", IEEE Trans. Nuclear Science, vol. 63, no. 3, pp. 1625-1629, 2016
[3] M. N. Baldwin and J. E. Rogers, "Calibration of the Self-Powered Neutron Flux Monitor in a Simulated PWR Lattice", IEEE Trans. Nuclear Science, vol. 16, no. 1, pp. 171-176, 1969
[4] N. P. Goldstein, "A Monte-Carlo Calculation of the Neutron Sensitivity of Self-Powered Detectors", IEEE Trans. Nuclear Science, vol. 20, no. 1, pp. 549-556, 1973
[5] L. A. Banda, "Operational Experience in Rhodium Self-Powered Detectors", IEEE Trans. Nuclear Science, vol. NS-26, no. 1, pp. 910-915, 1979
[6] P. S. Rao and S. C. Misra, "Neutron Sensitivity of Vanadium Self Powered Neutron Detectors" Nucl. Instr. and Meth. in Phys. Res. A 253, pp. 57-60, 1986
[7] M. Alex and M. D. Ghodgaonkar, "Development of an Inconel Self Powered Neutron Detector for In-Core Reactor Monitoring", Nucl. Instr. and Meth. in Phys. Res. A 574, pp. 127-132, 2007
[8] H. D. Warren, "Calculational model for self-powered neutron detector, Nucl. Sci. Eng., Vol. 48, pp. 331-342, 1972

The CANDELLE experiment for characterization of neutron sensitivity of LiF TLDs

M. Le Guillou, A. Billebaud, A. Gruel, G. Kessedjian, O. Méplan, C. Destouches, P. Blaise

***Abstract*— As part of the design studies conducted at CEA for future power and research nuclear reactors, the validation of neutron and photon calculation schemes related to nuclear heating prediction are strongly dependent on the implementation of nuclear heating measurements. Such measurements are usually performed in low-power reactors, whose core dimensions are accurately known and where irradiation conditions (power, flux and temperature) are entirely controlled. Due to the very low operating power of such reactors (of the order of 100 W), nuclear heating is assessed by using dosimetry techniques such as thermoluminescent dosimeters (TLDs). However, although they are highly sensitive to gamma radiation, such dosimeters are also, to a lesser extent, sensitive to neutrons. The neutron dose depends strongly on the TLD composition, typically contributing to 10-30% of the total measured dose in a mixed neutron/gamma field. The experimental determination of the neutron correction appears therefore to be crucial to a better interpretation of doses measured in reactor with reduced uncertainties. A promising approach based on the use of two types of LiF TLDs respectively enriched with lithium-6 and lithium-7, precalibrated both in photon and neutron fields, has been recently developed at INFN (Milan, Italy) for medical purposes. The CANDELLE experiment is dedicated to the implementation of a pure neutron field "calibration" of TLDs by using the GENEPI-2 neutron source of LPSC (Grenoble, France). Those irradiation conditions allowed providing an early assessment of the neutron components of doses measured in EOLE reactor at CEA Cadarache with 10% uncertainty at 1σ.**

***Index Terms*— nuclear heating, thermoluminescent dosimeters, neutron sensitivity**

I. Introduction

The accurate determination of nuclear heating remains a key point in the design studies of power and research reactors. The experimental validation of neutron and photon propagation calculation schemes related to nuclear heating prediction requires the implementation of nuclear heating measurements usually performed in zero-power reactors (ZPRs). Due to the very low operating power of such reactors, nuclear heating is assessed by using dosimetry techniques rather than calorimetry techniques which are usually used in power reactors like MTRs. In particular, the use of thermoluminescent dosimeters (TLDs) is remarkably well suited to gamma heating. However, although they are highly sensitive to gamma radiation, TLDs are also, to a lesser extent, sensitive to neutrons. Since the latter currently relies on literature data with relatively high uncertainties (up to 50-100%), a promising approach to experimentally determine this neutron contribution is based on the use of two types of doped lithium fluoride TLDs pre-calibrated both in gamma and neutron fields. Such approach has been recently developed for medical purposes at the national institute for nuclear physics (INFN, Milan, Italy) [1].

Within the framework of the French interdisciplinary program for nuclear, energy, environment, waste and society (NEEDS), the CANDELLE experiment is dedicated to the implementation of a pure neutron field calibration of LiF TLDs by using the GENEPI-2 neutron source of the laboratory of subatomic physics and cosmology (LPSC, Grenoble, France) [2]. The first phase of the project, presented in this paper, involved carrying out preliminary irradiations to ensure the exploitability of the TLD glow curves. In the next section are presented the main principles of gamma heating measurements performed at CEA Cadarache, as well as those of the discrimination method developed at INFN. In the following two sections are described the implementation and the results of the test irradiations carried out at LPSC, and a preliminary assessment of the neutron contributions to the total doses measured by LiF TLDs in the EOLE reactor.

Submitted on 01/06/2017

M. Le Guillou, A. Gruel, C. Destouches and P. Blaise are with CEA, DEN, DER, Experimental Physics Service, F-13108 Saint Paul lez Durance, France (e-mail: adrien.gruel@cea.fr).

A. Billebaud, G. Kessedjian and O. Méplan are with LPSC, Université Grenoble-Alpes, CNRS/IN2P3, 53 avenue des Martyrs, F-38026 Grenoble, France.

II. Gamma heating measurements using TLDs

A. *Overview and general principles*

The measurement methodology with TLDs implemented at CEA Cadarache is fully described in [3], [4] and [5]–[7], and summed up hereafter. The TLD pellets are usually made of luminescent materials such as doped calcium or lithium fluorides or alumina (CaF_2, LiF, Al_2O_3). In order to ensure the charge particle equilibrium (CPE) in the TLDs, they are encapsulated in 2 mm thick pillboxes made of the material to be investigated. The encapsulated TLDs are firstly calibrated in a pure gamma field nearby a ^{60}Co source, and then

https://doi.org/10.1051/epjconf/201817004014

irradiated in mixed gamma-neutron field at the positions of interest in reactor. The luminescent signals emitted by the irradiated TLDs, i.e. their glow curves (GCs), are read out by thermal stimulation after withdrawal of the TLDs from the core. The absorbed doses are finally deduced from the integral of the GCs, thanks to the calibration curves and the correction factors related to the presence of the pillboxes and the neutron contribution to the total measured dose.

B. Description of the gamma-neutron discrimination method

The absorbed neutron dose strongly depends on the TLD composition, typically contributing to 10-30% of the total measured dose in mixed gamma-neutron field. Researchers from INFN recently proposed a method aimed at discriminating the gamma and neutron components of the GCs of LiF TLDs irradiated in mixed field. A complete description of this method can be found in [1], and its main principles are presented below.

As shown in Fig. 1, GCs of doped lithium fluoride TLDs (LiF:Mg,Ti) exhibit two peaks whose height ratio depends on the relative contributions of gammas and neutrons to the total absorbed dose. The method relies on the deconvolution of these two contributions from the GC of a TLD-700 (^{7}Li-enriched, i.e. low neutron sensitivity) irradiated in mixed field, using the calibration GC of the same TLD irradiated in a pure gamma field, and the one of a TLD-600 (^{6}Li-enriched, i.e. high neutron sensitivity) irradiated with thermal neutrons.

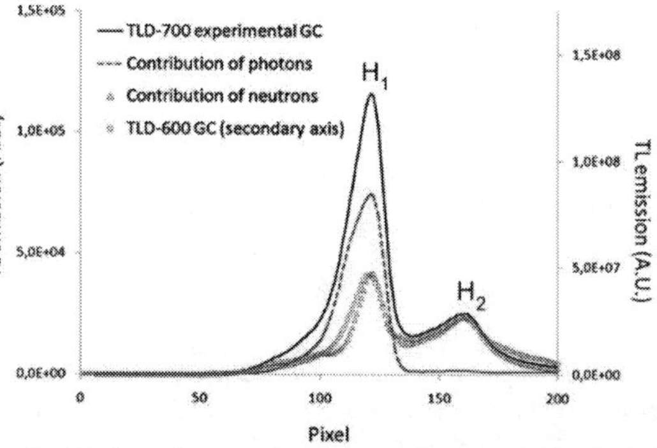

Fig. 1. Contributions of gammas (red dashes) and neutrons (pink triangles) to the GC of a TLD-700 (^{6}Li/^{7}Li ~ 0.01%) irradiated in mixed gamma-neutron field (blue line), compared with the GC of a TLD-600 (^{6}Li/^{7}Li ~ 95.6%) irradiated with thermal neutrons (green squares, secondary axis) [1].

The assumption is made that, after background subtraction, the heights H_1 and H_2 of the two peaks exhibited by the GC of a TLD-700 irradiated in mixed gamma-neutron field, i.e. in reactor, can be expressed as:

$$H_1 = D_\gamma H_1^\gamma + \phi_n H_1^n, \tag{1}$$
$$H_2 = D_\gamma H_2^\gamma + \phi_n H_2^n, \tag{2}$$

where D_γ is the absolute gamma dose, ϕ_n is the thermal neutron fluence, and H_1^γ, H_1^n, H_2^γ, H_2^n are the respective heights of the gamma (γ) and neutron (n) contributions to the first (subscript 1) and the second (subscript 2) peaks of the TLD-700 GC, normalized to dose and fluence units. The gamma dose in mixed field (corrected for the neutron contribution) is then given by the following relation:

$$D_\gamma = \frac{H_2 R_n - H_1}{H_2^\gamma R_n - H_1^\gamma} \quad \text{where} \quad R_n = \frac{H_1^n}{H_2^n}, \tag{3}$$

The R_n ratio is obtained from the GC of an uncalibrated TLD-600 irradiated with thermal neutrons, assuming that the gamma contribution for this type of TLD is usually negligible, due to the ^{6}Li enrichment. The peak heights H_1^γ and H_2^γ are deduced from the TLD-700 calibration GCs in pure gamma field and normalized to dose unit. The accuracy of the method can be checked as shown in Fig. 1, where the neutron component of the TLD-700 GC (pink triangles), obtained by subtracting the gamma dose calculated with equation (3), is compared with the GC of a TLD-600 irradiated with thermal neutron (green squares).

III. The CANDELLE experiment

A. Objectives

As part of the French interdisciplinary program NEEDS, the CANDELLE experiment (French acronym for neutron calibration of enriched-lithium based luminescent detectors) aims at evaluating the feasibility of the neutron calibration of LiF TLDs by using the nuclear energy experimental platform PEREN of LPSC.

B. Experimental setup

The irradiations were carried out with the fast neutron pulsed source provided by the GENEPI-2 electrostatic accelerator of the PEREN platform [2]. Fast neutrons were produced through the D-D nuclear reaction target at an energy of 3.1 MeV.

The Li-enriched TLD square pellets (3.2 × 3.2 mm^2, 0.9 mm thick), namely TLD-600 (^{6}LiF:Mg,Ti, 95.6% ^{6}Li) and TLD-700 (^{7}LiF:Mg,Ti, 99.9% ^{7}Li), were stacked in aluminum pillboxes. The irradiation configuration is presented in Fig. 2. The pillboxes were located at around 3 cm from the target, under which conditions the fast neutron flux at 0° was estimated to be about 5 × 10^3 n.cm^{-2}.s^{-1}.

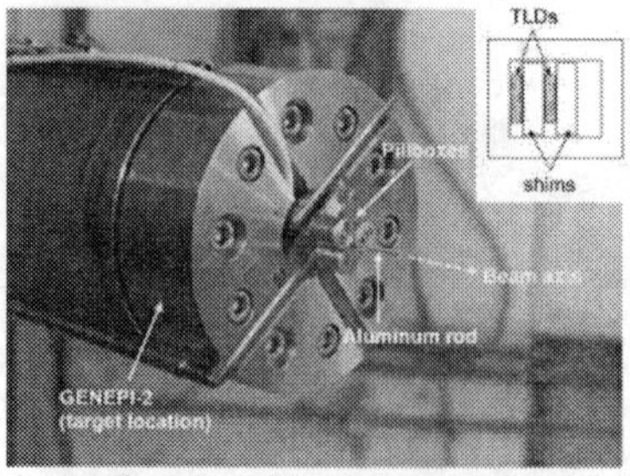

Fig. 2. Irradiation configuration of the TLDs at the PEREN platform [CNRS].

C. *Results and discussion*

The total fluency reached is around 2×10^8 n.cm^{-2}. Considering that the gamma production due to activation of the pillbox material is negligible (preliminary calculations), the averaged peak height ratios in pure fast neutron field are 4.66 ± 5% for TLD-600 (standard deviation of 3.7% on all TLD-600 measurements) and 4.97 ± 5% for TLD-700 (standard deviation of 12.2% on all TLD-700 measurements).

The peak height ratios are similar for both types of TLDs. The standard deviation for TLD-700 is higher because of the very statistics compared to the one of TLD-600 (higher neutron cross-section of ^{6}Li).

IV. Evaluation of the neutron contributions to the dose measured in mixed field

A. *Gamma dose calibration*

TLDs were calibrated in a pure gamma field at a distance of 80 cm of a ^{60}Co source, whose β− decay leads to the emission of two gamma rays at 1.17 and 1.33 MeV. TLD-600 and TLD-700 were calibrated at doses ranging from 100 mGy to 2 Gy.

The average peak heights ratios are within 45 to 50 ± 5%. The same results are observed for TLD-600. The peak height ratios are very different from those obtained in a pure neutron flux: the second peak is almost noticeable on some GCs. Among all detectors the discrepancy of the average peak heights at different doses is quite variable as it ranges from 5 to 15%.

B. *Irradiation in the EOLE reactor*

This analysis method has been tested on measurements carried out in the EOLE ZPR at Cadarache. Several locations in the core with different neutron and gamma spectra have been investigated with TLDs.

The order of magnitude of the fluency reached at the end of an irradiation at the core center (maximum flux) was around 5×10^{10} n.cm^{-2}, to be compared to the 2×10^8 n.cm^{-2} reached during the neutron calibration at the PEREN platform. Given the measurement location in the core the peak height ratios vary between 25 and 40. They indeed depend on the neutron and gamma flux shape and on the gamma/neutron ratio at the measurement location.

C. *Results and discussion*

The neutron contributions to the total doses in different locations are presented Table I. The relative uncertainty on each measurement is around 5%. In each location several TLDs measurements were made, leading to discrepancies between 5 to 10%, which is the same order of magnitude as the final uncertainty.

TABLE I
NEUTRON CONTRIBUTION IN DIFFERENT LOCATIONS OF THE CORE. THE DISCREPANCY IS THE STANDARD DEVIATION BETWEEN THE DIFFERENT TLDS MEASUREMENTS IN A SAME LOCATION.

Neutron contribution	Central core	Core periphery	Water
Average value	9.0%	8.7%	11.9%
Relative uncertainty (1σ)	5.2%	5.2%	5.1%
Relative discrepancy	10.0%	6.5%	4.9%

After the neutron calibration at the PEREN platform and the gamma calibration with a ^{60}Co source, the neutron contribution to the total dose absorbed in a mixed gamma-neutron field can be assessed with a final relative uncertainty of around 10%.

V. Conclusion and outlooks

Irradiation of TLDs in a fast neutron field at the GENEPI-2 neutron source of LPSC (Grenoble, France) allowed measuring the peak height ratios for LiF TLDs, hence providing an early assessment of the neutron components of doses measured in the mixed gamma-neutron field in the EOLE reactor at CEA Cadarache with 10% uncertainty at 1σ.

Acknowledgment

The authors wish to acknowledge S. Rey and F. Villa from the accelerator group of LPSC, L. De Padua and N. Makaryan from the radiation protection service (SPR) of CEA Cadarache, as well as the operator team of the EOLE reactor.

References

[1] G. Gambarini *et al.*, "Determination of gamma dose and thermal neutron fluence in BNCT beams from the TLD-700 glow curve shape", *Radiat. Meas.*, Vol. 45, pp. 640-642, 2010. DOI: 10.1016/j.radmeas.2010.02.006.

[2] F. Villa *et al.*, "Multipurpose applications of the accelerator-based neutron source GENEPI2". *5th International Meeting of Union for Compact Accelerator-Driven Neutron Sources (UCANS V)*, May 2015, Padova, Italy, Il Nuovo Cimento C 38, pp. 182, 2016. DOI: 10.1393/ncc/i2015-15182-2

[3] H. Amharrak, Ph.D. thesis, Aix-Marseille Université, France, 2012.

[4] M. Le Guillou *et al.*, "State of the art on nuclear heating measurement methods and expected improvements in zero power research reactors", *EPJ Nuclear Sci. Technol.*, Vol. 3, No. 11, 2017. DOI: 10.1051/epjn/2017002.

[5] H. Amharrak *et al.*, "Analysis and recent advances in gamma heating measurements in MINERVE facility by using TLD and OSLD techniques", *IEEE Trans. Nucl. Sci.*, Vol. 59, No. 4, pp. 1360-1368, 2012. DOI: 10.1109/ANIMMA.2011.6172901.

[6] H. Amharrak *et al.*, "Development and optimization of nuclear heating measurement techniques in Zero Power experimental Reactors", *IEEE Trans. Nucl. Sci.*, Vol. 61, No. 5, pp. 2515-2526, 2014. DOI: 10.1109/TNS.2014.2355917.

[7] P. Blaise *et al.*, "Nuclear heating measurement in critical facilities and experimental validation of code and libraries – An application to prompt and delayed γ nuclear data needs", *Phys. Proc.*, Vol. 59, pp. 3-16, 2014. DOI: 10.1016/j.phpro.2014.10.002.

EPJ Web of Conferences **170**, 04015 (2018)
ANIMMA 2017
https://doi.org/10.1051/epjconf/201817004015

Study of the linearity of CABRI experimental ionization chambers during RIA transients

J. Lecerf, Y. Garnier, JP. Hudelot, B. Duc, L. Pantera

johann.lecerf@cea.fr

CEA, DEN, CAD/DER/SRES/LPRE, Cadarache, F-13108 Saint-Paul-lez-Durance, France

***Abstract*– CABRI is an experimental pulse reactor operated by CEA at the Cadarache research center and funded by the French Nuclear Safety and Radioprotection Institute (IRSN).**

For the purpose of the CABRI International Program (CIP), operated and managed by IRSN under an OECD/NEA framework it has been refurbished since 2003 to be able to provide experiments in prototypical PWR conditions (155 bar, 300 °C) in order to study the fuel behavior under Reactivity Initiated Accident (RIA) conditions.

This paper first reminds the objectives of the power commissioning tests performed on the CABRI facility. The design and location of the neutron detectors monitoring the core power are also presented.

Then it focuses on the different methodologies used to calibrate the detectors and check the consistency and co-linearity of the measurements.

Finally, it presents the methods used to check the linearity of the neutron detectors up to the high power levels (~20 GW) reached during power transients. Some results obtained during the power tests campaign are also presented.

I. Introduction

CABRI is an experimental pulse reactor operated by CEA and funded by the French Nuclear Safety and Radioprotection Institute (IRSN) at the Cadarache research center. Since 1978 the experimental programs have aimed at studying the fuel behavior under Reactivity Initiated Accident (RIA) conditions. Since 2003, it has been refurbished in order to be able to provide RIA and LOCA (Loss Of Coolant Accident) experiments in prototypical PWR conditions (155 bar, 300 °C). This project is part of a broader scope including an overall facility refurbishment and a safety review. The global modification has been conducted by the CEA project team and funded by IRSN, which is operating and managing the CIP experimental program (CABRI International Program), in the framework of an OECD/NEA agreement.

During the reactor restart, commissioning tests were performed for all equipment, systems and circuits of the reactor. In particular power commissioning tests have been carried out in 2016 and 2017.

This paper focuses on the analyses of the linearity of CABRI experimental ex-core compensated ionization chambers used to measure online the power of the CABRI driver core during RIA power transients.

In the first part, the method used and the tests realized to calibrate the neutron detectors at low power will be explained.

The second part will concentrate on the method used to demonstrate the linearity of the detectors under transient conditions.

Each one of the above experimental goals will be described, as well as the target uncertainties regarding the power measurement.

II. CABRI Reactor and Commissioning Tests

A. Description of the CABRI reactor

CABRI is a pool-type reactor, with a core made of 1487 UO_2 fuel rods with a 6 % ^{235}U enrichment with stainless steel clads. The reactor is able to reach a 25 MW steady state power level. The reactivity is controlled via a system of 6 control and safety rods made-up of 23 hafnium pins for each one (see Fig. 1). The core is cooled by a forced water flow of 3215 $m^3.h^{-1}$ when the core power is higher than 100 kW and by natural convection with the pool water otherwise [1].

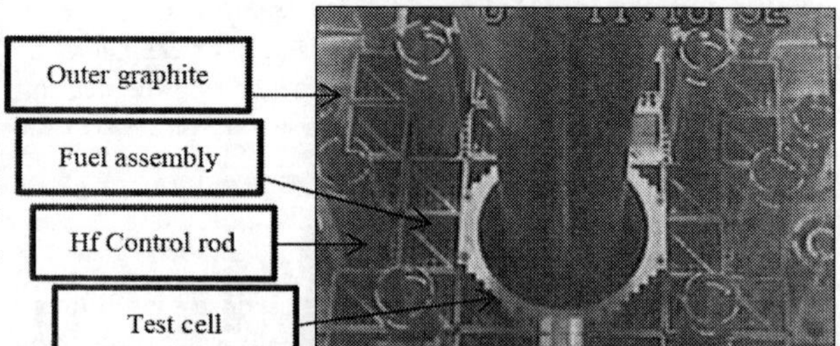

Fig. 1. CABRI core

The key feature of the CABRI reactor is its reactivity injection system [2]. This device allows the very fast depressurization of the ^{3}He (strong neutron absorber) previously introduced inside 96 tubes (so called "transient rods") located among the CABRI fuel rods. The rapid absorber depressurization translates into an equivalent reactivity injection possibly reaching 3.9 $ within a few 10 ms. The power consequently bursts from 100 kW up to ~20 GW (see Fig. 2) in a few ms and decreases just as fast due mainly to the Doppler effect. The total energy deposit in the tested rod is adjusted by dropping the control and safety rods after the power transient.

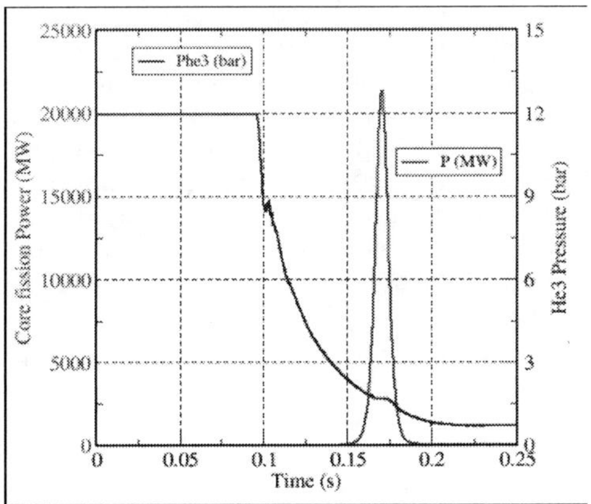

Fig. 2. Typical CABRI ^{3}He Pressure and core power shapes during a RIA transient

B. *Focus on the power commissioning tests*

The neutronics commission tests aimed at precisely characterizing the neutronics parameters of the CABRI core [3]: reactivity effects, power distributions and kinetic parameters.

After carrying out those tests at low power (< 100 kW), the CABRI facility was ready for power commissioning tests.
The main objectives of these tests were:

- to assess the thermal leakage coefficient associated with the water mixing in the upper water box area during the reactor forced cooling phase (see Fig. 3: the 6 control and safety rods and the experimental pressurized water loop are not represented in the figure),
- to calibrate the neutron detectors thanks to thermal balance measurements performed on water of the primary cooling system passing through the core. These thermal balance measurements are carried out up to the maximal steady state CABRI driver core power level (~25 MW),
- to verify the co-linearity of the experimental neutron detectors during the power increase as well as various steady state power levels,
- to study the consistency of the different experimental neutron detectors measurements and the linearity of each detector during power transients up to power levels of about 20 GW.

III. Design and location of neutron detectors

During steady states and transients conditions, the core power is measured by the mean of specific ex-core boron ionization chambers.

Experimentally, five detectors are used to monitor the core power during transients, four of them were manufactured by Reuter Stokes and the last one by Merlin Gerin (see Table I). They are operated in the current mode.

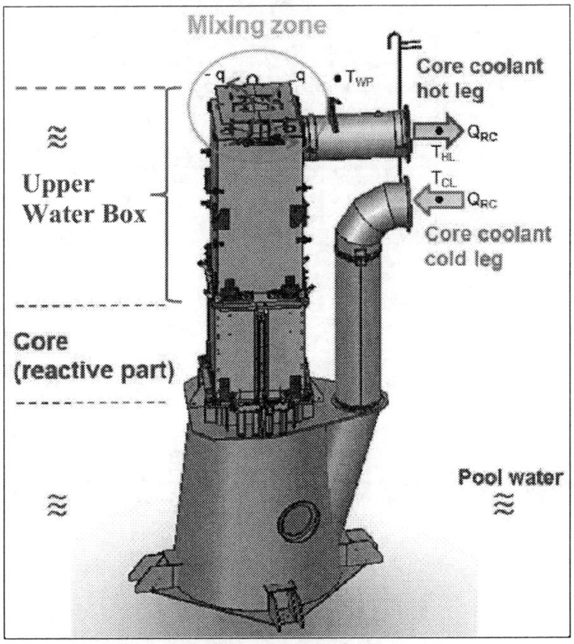

Fig. 3. Mixing zone in the upper water box area

The neutron detectors are positioned outside the core, behind a lead shield designed to reduce the gamma flux (see Fig. 4).

Thanks to increasing distances between the driver core and the detectors, the experimental signals overlap on different ranges of operation of the chambers and electronic chains which allows to measure the whole power range in the case of power transients (i.e. from few kW up to ~20 GW) (see Table I and Fig. 4).

In the Figure 4, HN1 and HN2 are the neutron detectors used to control the reactor in normal operating conditions (except for transients). They are used in the safety chain of the reactor. The response time of their electronic chain and acquisition system is too long to be able to monitor correctly a power transient.

G2.1 and G2.2, which are the detectors closest to the core, are mainly used to measure power during steady state conditions (before transients). They also follow the beginning

 https://doi.org/10.1051/epjconf/201817004015

of the transients before the overlap of the experimental detectors signals is ensured by the G3 detectors.

Note that preliminary tests on the neutron detectors were performed at low power levels in order to check their operability, to determine their high-voltage, discrimination curves and their neutron sensitivity.

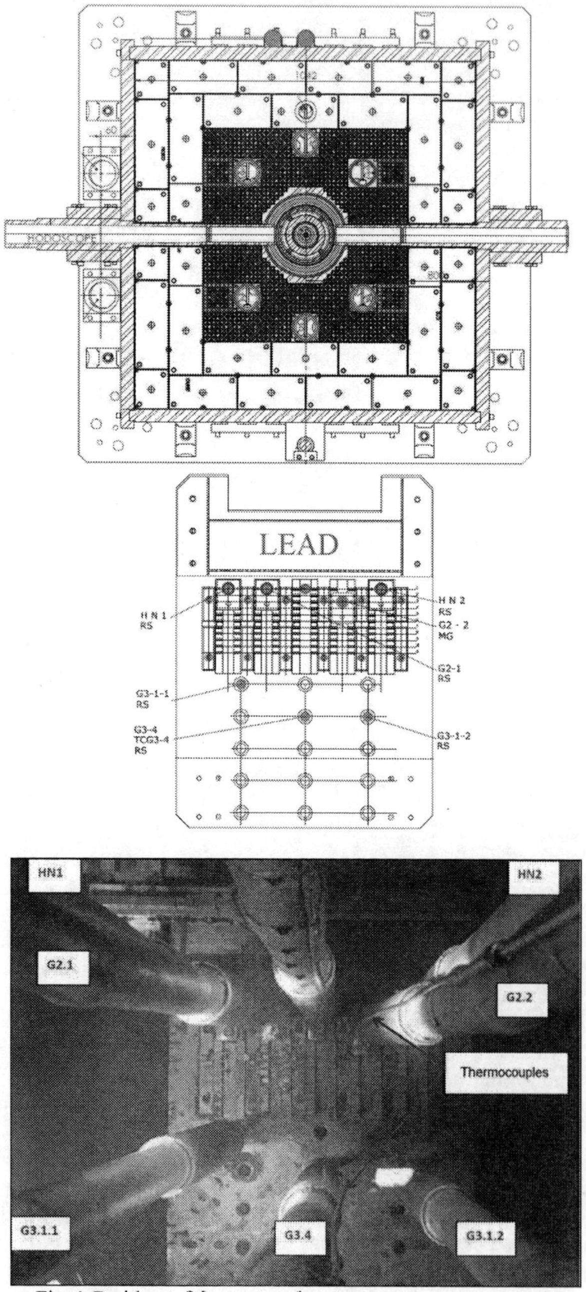

Fig. 4. Positions of the neutron detectors

TABLE I. CABRI EXPERIMENTAL BORON IONIZATION CHAMBERS

	G2.1	G2.2	G3.1.1	G3.1.2	G3.4
Supplier	Reuter Stokes	Merlin Gerin	Reuter Stokes	Reuter Stokes	Reuter Stokes
Min Power	~10 kW		~10 MW	~100 MW	
Saturation	~450 MW		> Max transient power of CABRI		

IV. CALIBRATION OF THE NEUTRON DETECTORS

The power commissioning tests aimed at calibrating the neutron detectors thanks to thermal balance measurements performed on the primary cooling system.

This calibration realized in steady state conditions (up to 25 MW) is very important since the core power during the transients is only assessed online by the neutron detectors.

The objective of the thermal balance measurements is to define the neutron detector sensitivity coefficient c which connects the power P_{ND} (W) to the detector signal S (V) according to the following equations [4] (see Fig. 3):

$$P_{ND} = c.S \tag{1}$$

$$P_{ND} = P_{TB} \tag{2}$$

$$P_{TB} = \rho.C_p.Q_{RC}.(T_{HL} - T_{CL}) + P_{TL} \tag{3}$$

$$P_{TL} = K.(T_{HL} - T_{PW}) \tag{4}$$

With

- P_{ND} core power measured by the Neutron Detector (W)
- c neutron detector sensitivity coefficient (W/V)
- S detector Signal (V)
- P_{TB} core power measured by Thermal Balance (W)
- ρ water density (kg/m^3)
- C_p water isobar specific heat (J/kg.°C^{-1})
- Q_{RC} Reactor Coolant flow rate (m^3/s)
- T_{HL} Hot Leg water temperature (°C)
- T_{CL} Cold Leg water temperature (°C)
- P_{TL} Thermal power Leak through the upper box (W)
- K thermal leakage coefficient (W/°C) (see § IV. A)
- T_{PW} Pool Water temperature (°C)

A. Experimental evaluation of the thermal leakage coefficient

The thermal leakage coefficient has been measured during primary cooling system operations with significant temperature differences between the reactor pool and the outside tanks (see Fig. 5). This measurement does not require power operating phases of the reactor, i.e. at zero power.

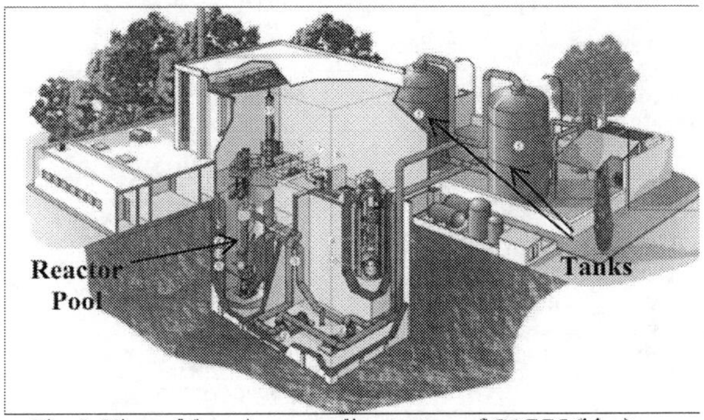

Fig. 5. View of the primary cooling system of CABRI (blue)

The mixing zone in the upper part of the water box (above the core, downstream since the cooling water circulates down to top) is illustrated on the figures 6 and 7. In this case, the temperature of the pool water was higher than the temperature of water inside the primary cooling system. The first figure was obtained from images taken with a thermal camera at the beginning of the homogenization phase (cold water in blue). The second one comes from 3D CFD calculations performed with the STAR-CCM+ code [5].

Fig. 6. Thermal camera pictures of the mixing zone in the upper part of the water box

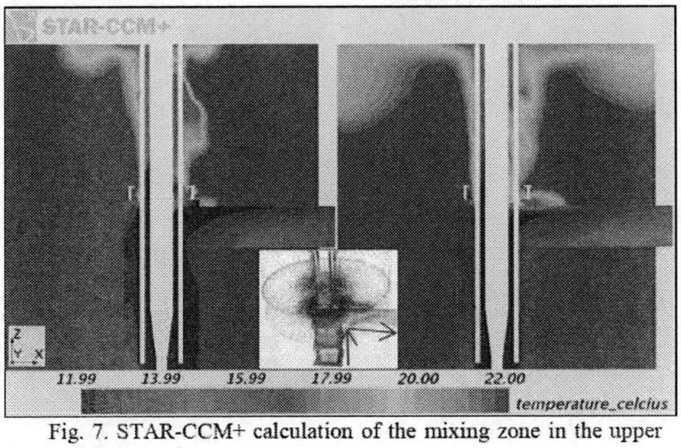

Fig. 7. STAR-CCM+ calculation of the mixing zone in the upper part of the water box

In order to assess the thermal leakage coefficient K, mobile thermocouples have been used to monitor the temperature near the mixing zone.

Measurements were made during three homogenization phases of the primary cooling system when there were significant water temperature differences between the pool and the tanks. The value obtained is about 0.3 MW/°C with an estimated uncertainty of 20 % (2σ). This high level of uncertainty does not significantly affect the global uncertainty of the thermal balance measurements of power because the thermal power leakage represents only approximately 8 % of the total power obtained by thermal balance.

B. Calibration of the neutron detectors by thermal balances and check for the consistency and co-linearity of the measurements

During the power commissioning tests, thermal balances with increasing power levels (8, 10, 12, 15, 19 and up to 23 MW) were used to calibrate the neutron detectors. The neutron detector sensitivity coefficients were readjusted whenever the power delivered by the detectors deviated from the core power measured by thermal balance.

Some tests have also been carried out with varying the initial temperature of the primary cooling system to know the neutron detectors sensitivity dependence on this parameter.

This sensitivity variation is equal to -0.30 %/°C for all neutron detectors.

The consistency and co-linearity of the neutron detectors power measurements was checked during the different power increases and steady states of power.

The figure 8 presents the results obtained during the different phases of the power increase to reach a 23 MW final steady state of power (comparisons of HN and G2 detectors). For power higher than a few kW, the signals are perfectly consistent.

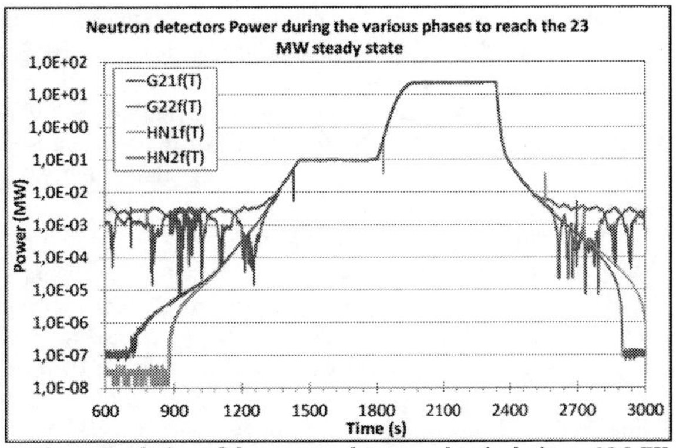

Fig. 8. Evolution of the neutron detectors signals during a 23 MW steady state power level

The three G3 neutron detectors signals are not represented on this figure because there are not representative for such low power level (see Table I). For these detectors, the sensitivity coefficients, obtained thanks to thermal balances at powers below 25 MW, are not precise and will be refined by power comparisons with G2 detectors (up to 450 MW) during power transients (see § V).

https://doi.org/10.1051/epjconf/201817004015

V. Linearity of the neutron detectors during power transients

The last objective of the power commissioning tests concerning the neutron detectors was to check their linearity during the power transients.

RIA power transients have been realized increasingly in energy, that means starting from low ^{3}He pressure and low valve aperture of the transient circuit [2].

The coherence and co-linearity of the neutron detectors power measurements was checked during the different power transients.

The G3 detector sensitivity coefficients have been refined from the power comparisons with G2 detectors during transients up to their saturation (~ 450 MW) as shown on Fig. 9.

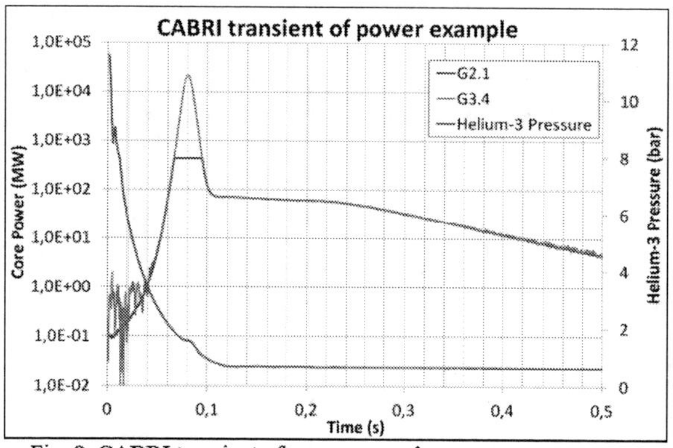

Fig. 9. CABRI transient of power example

The linearity of the neutron detectors and their capability to measure precisely the energy deposit during power transients have been checked using transient calorimetric balances.

The energy deposit in the core during the transient is evaluated by determining the energy carried away by the primary cooling system (integration of the instantaneous power) until the core power drops below the initial power before transient (see Fig. 10).

This energy is then compared to the energy integrated from the power of the neutron detectors during power transients.

This comparison is carried out for several power transients with various total energies and maximum power peaks.

The figure 11 and the table II show an example of results for the G3.4 neutron detector and conclude to its good linearity for different energy and power ranges, covering the whole range of the CABRI capabilities.

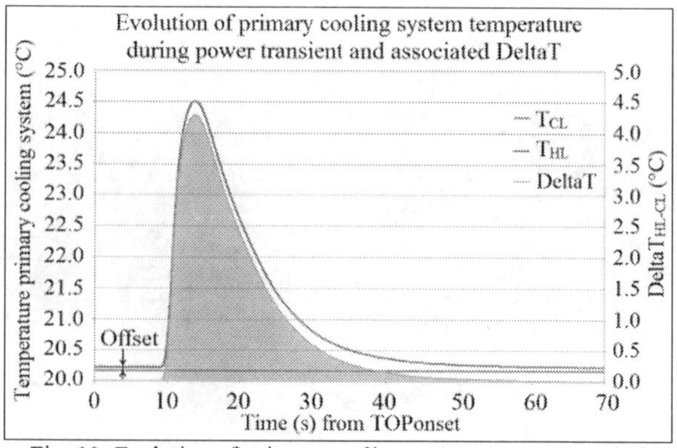

Fig. 10. Evolution of primary cooling system temperature during power transient

Table II. G3.4 Integrated power (G3.4) and calorimetric balances results on the core energy comparisons

Maximal power peak (GW)	E_CTB* (MJ)	E_G3.4** (MJ)	E_G3.4 / E_PCS
3.5	77.4	80.0	0.968
8.1	128.7	130.1	0.990
12.8	170.4	170.0	1.002
16.8	201.9	202.8	0.996
5.9	208.1	209.6	0.993
0.6	55.4	58.1	0.952
19.1	233.5	232.9	1.003
21.3	245.5	244.6	1.004

* *Calorimetric Thermal Balance energy on primary cooling system*

** *G3.4 energy (integration of the neutron detector power)*

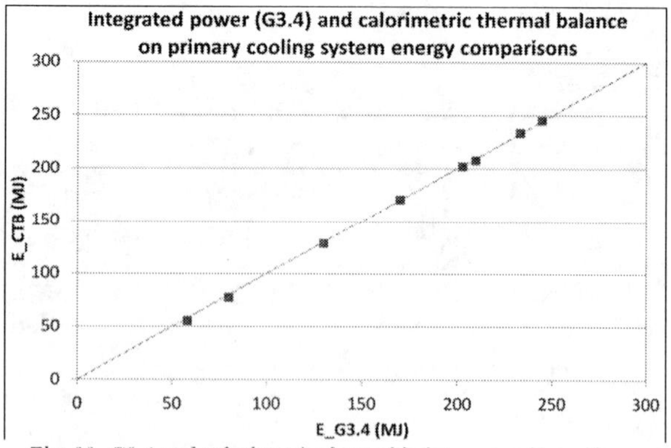

Fig. 11. G3.4 and calorimetric thermal balances results on the core energy comparisons

Given the results obtained during the power commissioning tests, an envelope uncertainty of 8 % (2σ) can be applied to the energy deposited in the core, measured from the experimental neutron detectors.

VI. Conclusion

This paper presents the results of the power tests campaign performed during the commissioning tests of the CABRI reactor.

The methods used to qualify the core power measurements (either steady state or power transients) were explained and some results of the power commissioning tests were presented.

The neutron detectors have been calibrated thanks to thermal balances performed on the primary cooling system for steady states of power up to 23 MW.

The consistency and co-linearity of the neutron detectors power measurements were checked.

Finally an important point was devoted to the study of the linearity of the detectors on typical RIA power transients considered in the CIP program.

References

[1] JP. Hudelot, Y. Garnier, J. Lecerf, M. Fournier, S. Magnetto, E. Gohier, "CABRI reactor commissioning: results and analysis of the tests on the primary cooling system and on the control and safety rods," IGORR2014 conference, 17-21 November 2014, Bariloche, Argentina

[2] B. Duc, B. Biard, P. Debias, L. Pantera, JP. Hudelot, F. Rodiac, "Renovation, improvement and experimental validation of the Helium-3 transient rods system for the reactivity injection in the CABRI reactor", IGORR2014 conference, 17-21 November 2014, Bariloche, Argentina

[3] G. Ritter, O. Guéton, F. Mellier, and D. Beretz, "Neutron Commissioning in the New CABRI Water Loop Facility," IEEE Transactions on Nuclear Science, Vol. 57, n°5, October 2010

[4] F. Jeury, L. Pantera, Y. Garnier, "CABRI experimental reactor: experimental reassessment of CABRI core power and measurement uncertainties", IGORR2013 conference, 13-18 October 2013, Daejeon, Korea

[5] CD-Adapco User Guide, STAR-CCM+ Version (11.02)

Qualification and characterization of electronics of the fast neutron Hodoscope detectors using neutrons from CABRI core

S. Mirotta, *IRSN*, J. Guillot, IRSN, V. Chevalier, IRSN, B. Biard, IRSN

Abstract— **The study of Reactivity Initiated Accidents (RIA) is important to determine up to which limits nuclear fuels can withstand such accidents without clad failure. The CABRI International Program (CIP), conducted by IRSN under an OECD/NEA agreement, has been launched to perform representative RIA Integral Effect Tests (IET) on real irradiated fuel rods in prototypical Pressurized Water Reactors (PWR) conditions. For this purpose, the CABRI experimental pulse reactor, operated by CEA in Cadarache, France, has been strongly renovated, and equipped with a pressurized water loop. The behavior of the test rod, located in that loop in the center of the driver core, is followed in real time during the power transients thanks to the hodoscope, a unique online fuel motion monitoring system, and one of the major distinctive features of CABRI. The hodoscope measures the fast neutrons emitted by the tested rod during the power pulse with a complete set of 153 Fission Chambers and 153 Proton Recoil Counters. During the CABRI facility renovation, the electronic chain of these detectors has been upgraded. In this paper, the performance of the new system is presented describing gain calibration methodology in order to get maximal Signal/Noise ratio for amplification modules, threshold tuning methodology for the discrimination modules (old and new ones), and linear detectors response limit versus different reactor powers for the whole electronic chain.**

Index Terms— **CABRI, Hodoscope, pulse mode, nuclear electronic chain, RIA accident**

S. Mirotta, IRSN/PSN-RES/SEREX/L2EP, Cadarache, Saint Paul Lez Durance 13115 FRANCE (email: salvatore.mirotta@irsn.fr),
J. Guillot, IRSN/PSN-RES/SEREX/L2EP, Cadarache, Saint Paul Lez Durance 13115 FRANCE,
V. Chevalier, IRSN/PSN-RES/SEREX/L2EP, Cadarache, Saint Paul Lez Durance 13115 FRANCE,
B. Biard, IRSN/PSN-RES/SEREX/L2EP, Cadarache, Saint Paul Lez Durance 13115 FRANCE

I. Introduction

One of the most challenging problems of our day, in nuclear industry, is to maintain Nuclear Power Plants (NPP) in safe working conditions during all their life. In order to achieve this goal, defense in depth strategy is used to develop nuclear reactor. This strategy consists in some physical barrier between de radioactive materials and the external environment. The first barrier of this methodology is realized by the fuel rod cladding. It is really important that the cladding resists to all reactor's operations (either in normal operating conditions or in severe accident conditions).

Despite all precautions to maintain mechanical conditions for the cladding, in some cases, (i.e. during severe accidents) it may fail. Among the different accidents that could occur in a NPP and could cause clad failure, one of the most important is Reactivity Initiated Accident (RIA). This kind of accident produces a high power excursion in several milliseconds (ms) which may damage cladding and the core geometry. The cladding rupture probability due to a RIA accident depends on several factors (i.e. Zirconia thickness, fuel burnup, high hydrogen content into the cladding, new cladding materials, etc.). CABRI core was built in order to determine up which limits fuel rod cladding does not fail under power excursion due to RIA accident.

Since the beginning, CABRI was equipped by a sodium loop in order to reproduce thermal hydraulic conditions of Sodium Fast Reactors (SFR). In the last period, an intensive renovation of the facility has permitted the replacement of the sodium loop by a pressurized water loop. This new circuit allows experimental studies of RIA in all its phases for Pressurized Water Reactors (PWR).

A test rodlet, refabricated from a rod irradiated in a NPP, is placed into that pressurized water loop, in the center of the core, and is submitted to a short but powerful transient, that may reach 20 GW in some ms. In CABRI, two none destructive systems were built in order to study the behavior of fuel under RIA conditions:

- A non-destructive examination facility, for performing X-ray radiography and tomography imaging before and after power transient thanks to a linear accelerator;
- An online fuel motion measurement system named hodoscope, which aims at analyzing the fuel motion deduced from the detection of fast neutrons emitted by the tested rod, in real time (with a time step of 1ms) during the transient.

These neutrons emerging from the tested rod are collimated towards a bank of detectors through 153 line-of-sights in a huge steel piece (3m length and more than 2ton weight). At the rear face of the collimator, 306 detectors are arranged by pair in a 51 row by 3 column matrix. The collimator is motioned in rotation so as to align the detectors with the test rod.

Two different technologies of detectors are used in CABRI in order to cover the whole power range of the transient, starting at low power (~100 kW) up to more than 20 GW. Proton Recoil Counters (PR) are better suited for low power measurement while ^{237}Np Fission Chambers (FC) are dedicated to high power monitoring. Both detectors measure the fast neutrons coming from the test rod and from the CABRI driver core fuel. For limiting the neutron contribution due to the CABRI core and maximizing the signal due to the tested rod, the core is divided into two symmetrical half parts by two air channels so that the collimator does not directly see the core fuel. Each detector is associated to a remote electronic chain which role is to shape signals and select interesting events.

Recently, with the collaboration of CEA LIST, these electronic chains have been renewed because of difficulties to maintain old systems. This article will describe the behavior of new amplifier and discriminator compared to those of the old electronic chain. The main goal of this renovation consists in replacing the obsolete integrated components with sustainable discrete components, while maintaining the same characteristics. Although electronics for both PR and FC detectors were created, in this article we will just discuss results obtained for FC detectors because results are similar.

In II a brief description of the electronic chain is presented, in III experimental measures with the amplifier are carried out. In IV, measurements with the discriminator are realized as well. Moreover, in V the analysis of the entire chain with CABRI neutrons is investigated to determine their real performances. Furthermore, conclusion and perspectives are discussed in section VI.

II. Detector chain

All the electronic chains of the hodoscope system are analogic. In Figure II.1, one of the 2x153 measurement chains is described. Detectors are placed behind a collimator 4 m far from the test rod. A 30 m μ-metal cable links detector with amplifier. Amplifier and discriminator are connected together, discriminator produce a TTL signal each time impulsions coming from the amplifier pass a threshold value. At the end of the chain a computer converts the TTL signals in counts which are registered in a scaler with a memory of 2^{16} bytes. As shown in the figure, the High Voltage (HV) and signals share the same cable.

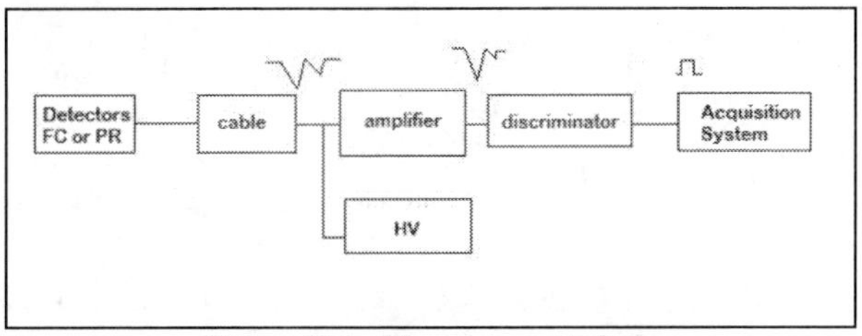

Figure II.1 - Measurement chain

All 306 detectors are used in pulse mode (detectors are placed really far from the core and the fast neutron flux is of the right magnitude for this kind of analysis). Amplifiers and discriminators are placed far (30 m) from the detectors and they are linked to them by a coaxial cable with double μ-metal shielding. These electronics were recently renewed in order to assess obsolescence of electronic components.

The new electronics was developed in collaboration with CEA LIST laboratories, and they have been tested in CABRI.

Each pair consisting of an amplifier and a discriminator is powered by 3 Direct Current (DC) voltage (12V, 5V and -5V). Each DC source supplies 10 measurement chains. Similarly, a HV polarizes 10 detectors simultaneously. All components into the electronic chain are adapted with an impedance of 50 Ω. [1]

III. Amplifier

A. Description

The amplifier is a trans-resistance amplifier, which collects current produced from the particles interacting with gas into the ionization chambers and converts this current in a pulse voltage signal.

This amplifier consists of different stages. As detector signal and HV share the same cable, the input stage comprises a circuit for decoupling and filtering the HV. Moreover, in order to avoid any damage to the following amplification stage, a protection circuit against over current is implemented as well. The amplification stage is divided into two levels made with sustainable and discrete components. In addition, a variable resistance makes it possible to adjust the gain. Finally the output circuit adjusts the impedance equal to 50 Ω.

Moreover, the amplifier has two different outputs:

- The first one, called *differentiated output*, is connected to the input of the discriminator. It provides shorts signals so as the electronic supports high count rate (The count rate can reach 6.10^6 cps for fission chambers and $2.9.10^6$ cps for proton recoil detectors without saturation).
- The second one, called *not differentiated output*, is generally plugged with 50 Ω resistance. It provides signals which shape is adapted to spectroscopy

amplifier, so that it can be used to determine the threshold value of the discriminator (interesting events corresponding to fast neutrons interaction in the detector are selected by their higher amplitude)

B. Gain analysis

The gain is defined as in Eq. 1:

$$G = \frac{V_{out,\max}}{V_{in,\max}} \quad \frac{[V]}{[V]} \qquad \textbf{Eq. 1}$$

As the prototypes characteristics are identical compared to the old electronics, the new amplifier gain was measured in order to set a similar value. It was decided to set the gain to 910 as resumed in Table III.1.

In addition, it is interesting to study the evolution of the signal/noise ratio according to the gain. As shown in Figure III.1, the value is constant for each measured gain.

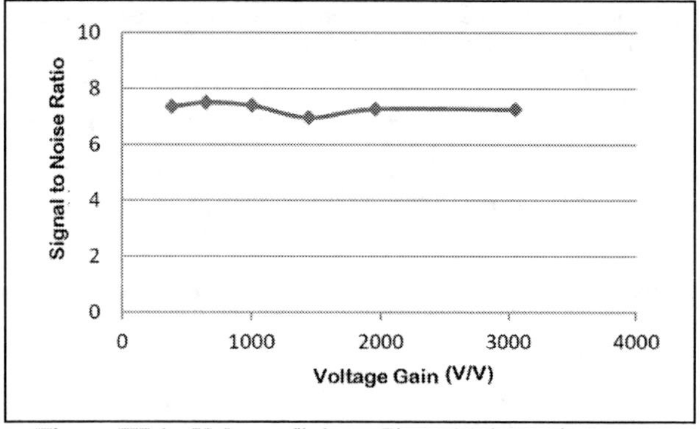

Figure III.1 - Voltage Gain vs Signal/Noise ratio for the Prototype

TABLE III.1 - MEASURED GAIN

Original (V/V)	Prototype (V/V)
1020	910

C. Amplifier Noise

In this section, the noise of the amplifier obtained during the measurement campaign is analyzed.

Figure III.2 and Figure III.3 show the noise level of the original and the new amplifier, respectively. The noise level of the new amplifier is slightly smaller than the old one. This result is probably linked to the lower gain of the prototypal amplifier compared with the original amplifier (cf. Table III.1).

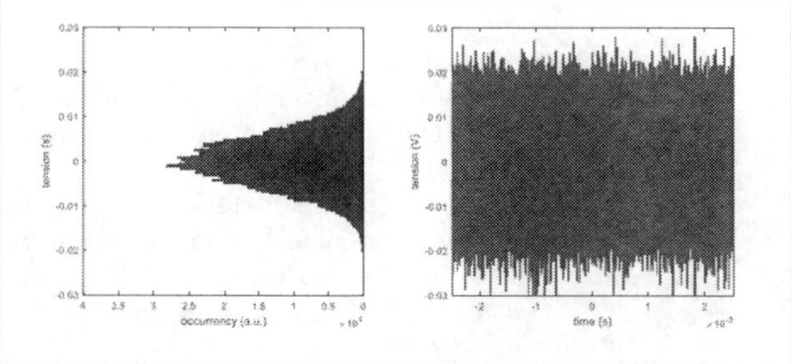

Figure III.2 - Noise level of original amplifier

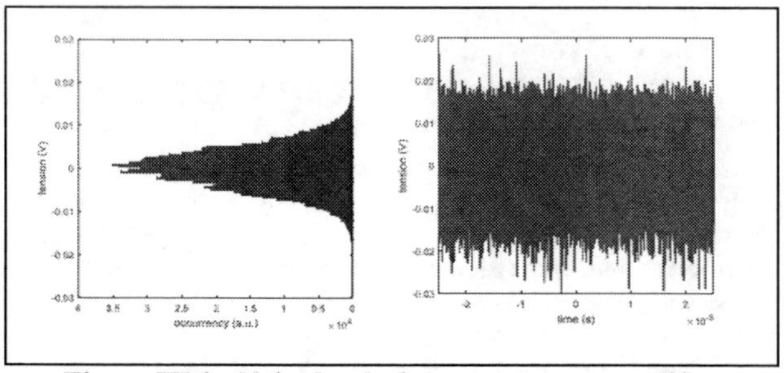

Figure III.3 - Noise level of new prototype amplifier

D. Amplifier Response

In this section the response of the amplifier tested under different conditions is discussed. First of all, the amplifier was tested using a calibrated function generator. Thereafter, its response to an Am-Be source was observed.

1) With function generator

In this part, the response of the original amplifier compared to the old amplifier is shown. The measurement was performed using a generator function. The input waveform has the characteristics listed in Table III.2.

TABLE III.2 - GENERATED IMPULSIONS

Full Weight at Half Maximum (FWHM)	40 ± 2 ns
Min Value	-3.7 ± 1 μA
Frequency	5 kHz

These impulsions were injected into the amplifier. In Figure III.4 and in Table III.3, measured characteristics of the pulses delivered by the differentiated output are shown. It is possible to see that the new prototype produce a waveform smaller than the original amplifier. This is due to different gain of two amplifiers. However, shapes of both waveforms are quite similar. All the impulsions are correctly amplified.

TABLE III.3 – WAVEFORM CHARACTERISTICS FOR DIFFERENTIATED OUTPUT

	Original	Prototype
FWHM	28 ± 2 ns	38 ± 2 ns
Min Value	-194 ± 5 mV	-170 ± 5 mV
Frequency	5 kHz	5 kHz

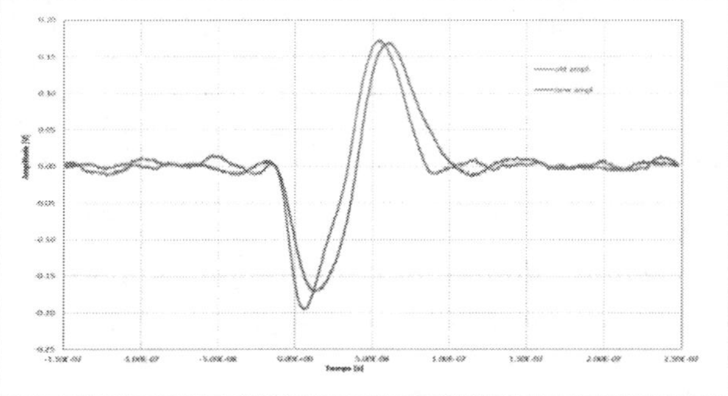

Figure III.4 - Waveform for differentiated output

2) With Am-Be Source

A test of the entire measurement chain was realized with an Am-Be source placed in front of a fission chamber. The original amplifier/discriminator and the new one have been permutated in order to observe the response of the electronic chain without any detectors influence. The source used for this test emits 4.33 10^5 n/s over 4π solid angle. This source produces very small neutron flux compared to neutronic current coming from CABRI core ($\sim 10^9$ $n \cdot cm^{-2} \cdot s^{-1}$).

Figure III.5-a shows a histogram (including 1000 impulsions) of the pulses amplitudes, measured with a digital oscilloscope, for the differentiated output of the old amplifier. Otherwise, the same distribution calculated for the new prototype is shown in Figure III.6-a.

In Figure III.5-b, the distribution of FWHM of pulses delivered by the original amplifier (differentiated output, negative part) is shown. This graph allows us to appreciate the stochastic behavior of the detector chain. This distribution could be assimilated to a Gaussian distribution centered on 21 ns with a standard deviation of 1.6 ns. In Figure III.6-b, the same analysis was carried out for the new prototype. This distribution is centered on 24 ns with a deviation of 1.3 ns.

As shown in figures, both distributions are similar. Hence it was concluded that the behavior of the new amplifier fulfills all requirements for CABRI experiments.

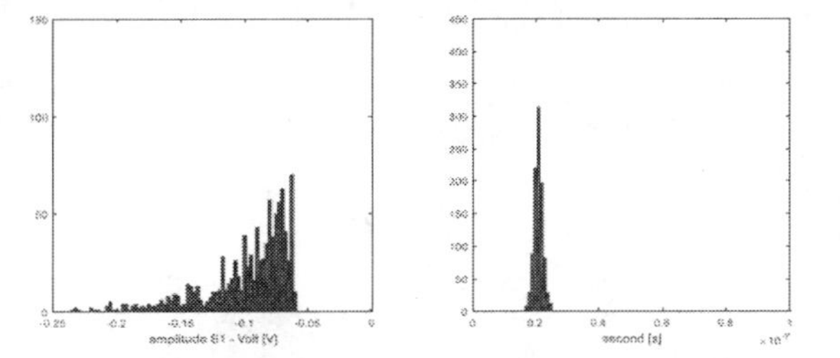

a. Minimum voltage for differentiated output *b. FWHM for impulsion mesured by differentiated output*

Figure III.5 - Characterization of original amplifier with Am-Be source

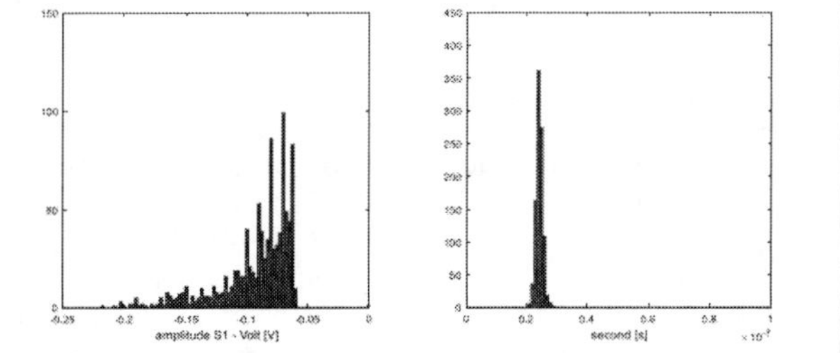

a. Minimum voltage for differentiated output *b. FWHM for impulsion mesured by differentiated output*

Figure III.6 - Characterization of new prototype amplifier with Am-Be source

Finally, the Table III.4 presents the count rates of pulses delivered by the differentiated output of both old and new amplifiers. These results are consistent with experimental conditions.

TABLE III.4 - COUNT RATE FOR AM-BE SOURCE AFTER ONE HOUR EXPOSURE

	Gain (V/V)	*Discriminator Threshold (mV)*	*Count Rate (cps)*
Original	1020	-41.8 ± 0.5	0.30 ± 0.01
Prototype	910	-33.0 ± 0.5	0.39 ± 0.01

IV. DISCRIMINATOR

A. Description

The discriminator is connected just downwards the amplifier. This component is used to create a TTL signal whenever the minimum value of the impulsion (negative part), which comes from the amplifier, is lower than a given threshold.

These TTL signals are counted by the computer.

The discriminator consists of 4 stages:

- The input stage, used to create the input voltage of the comparator, and to generate the threshold voltage as described by Eq. 2;

- The comparator stage, which compares the minimum value of the signal produced by the amplifier with a threshold;
- The shaping stage, which makes TTL signals when stimulated by the comparator stage
- The output stage, which is realized to charge the discriminator with 50 Ω impedance.

The threshold signal is constructed by two components, so that it may be changed either by means of a potentiometer or by remotely by computer. In fact, the computer produces an analogical voltage, by the means of a Digital/Analogical board, (V_{ext}) which is injected at the entrance of the discriminator and added to the potentiometer threshold (V_p, given in potentiometer turns; 10 potentiometer turns are equal to -5V). With this system, the discriminator response can be easily determined as a function of the threshold.

$$V_s = \frac{V_P}{-2} - V_{ext} \quad [mV] \qquad \textbf{Eq. 2}$$

B. TTL signal produced

When the shaped impulsions pass the threshold, a standard TTL signal is produced by the discriminator. TTL signals allow transporting easily information by the means of a normal coaxial cable to the computer. To be registered by the computer, the amplitude and the width of signals delivered by the discriminator should be greater than 1.5 V and 5 ns, respectively.

Figure IV.1 shows the shape of the two different TTL signals produced by the original discriminator and the prototype.

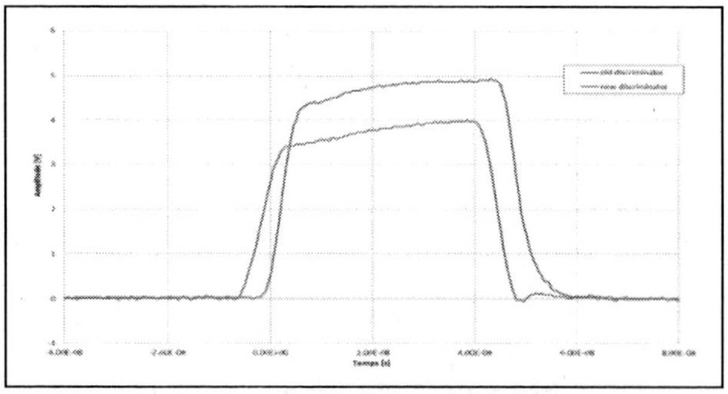

Figure IV.1 - TTL signal of the original discriminator compared to new prototype

In Table IV.1 the characteristic of the TTL signals produced from the new discriminator and the original one are presented. Characteristic of input signals are given in Table III.2.

The characteristics of the new discriminator are consistent with the original one.

TABLE IV.1 - TTL CHARACTERISTICS

	Original	*Prototype*
FWHM	46 ns	47 ns
Height	4.5 V	3.5 V
Frequency	5 kHz	5 kHz

V. ANALYSIS WITH CABRI CORE

A. Discriminator threshold determination

The determination of the threshold value for the discriminator is quite important in order to realize measurements with FC or PR.

The threshold has to be adjusted so as the discriminator delivers TTL signals whenever fast neutrons (interested events) either emitted by the CABRI core or the test rod interact in the detector. The threshold aims at suppressing pulses coming from noise. The noise can be produced either by electronic and or not interesting particles interacting in the detector (gamma, alpha for fission chamber). The discriminator threshold has to be adjusted whenever electronics or detectors characteristics change.

The electronic chain response has been determined according to threshold values (cf. Eq. 2), changing with 5 mV steps.

In Figure V.1 the response can be observed: new prototype has exactly the same behavior compared to the original one.

Furthermore this figure shows that the threshold is chosen (cf. TABLE III.4) for the two electronic chains so as to eliminate noise leading up to excessive counting. Only interesting events in the plateau zone are registered. Thresholds have not been modified.

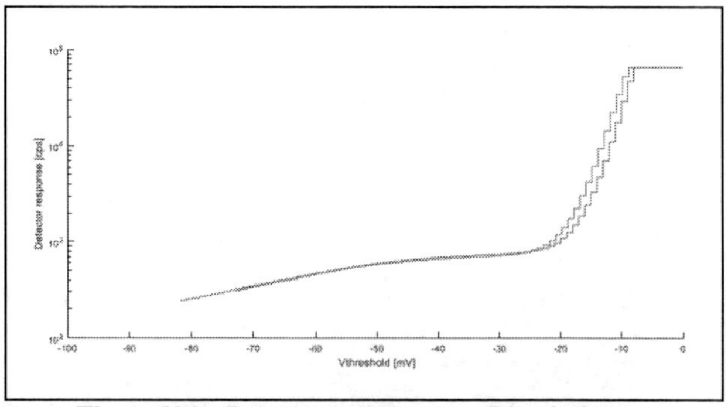

Figure V.1 - Detectors response vs. Discriminator threshold (in fuchsia the new prototype, in violet the original one)

B. Chain dead time evaluation

Another aspect of this new prototype concerns the dead time of the chain. In order to control properly the dead time of the chain, the constructor has imposed the dead time of the chain using a great dead time into the discriminator to produce the TTL signals. In this way, the dead time of the chain is constant, and does not depend anymore neither on collection time of the couples electron/ion at the detectors electrodes, nor on the propagation time of pulses throw the amplifier.

Using this concept, Eq. 3 can be written:

$$\tau_{chaine} = \tau_{det} + \tau_{amp} + \tau_{discr} \cong \tau_{discr} \quad \text{Eq. 3}$$

With τ_{det}, τ_{amp}, τ_{discr} the dead time generated by the detector, the amplifier and the discriminator, respectively.

Therefore, we can measure it without using neutron source.

A function generator with two channels is used to measure the dead time [2]. It is set so that the first channel generates, square pulse train with frequency f_1, and the second channel generates square pulse train with frequency f_2. The two pulses trains are injected at the amplifier input, and the TTL signals are viewed by means of an oscilloscope at the discriminator output. At the beginning, f_1 is equals to f_2 and the count rate amounts to f_1+f_2. Changing the frequency of the second pulse train, the second TTL signal observed at the discriminator output get close to the first, until it disappears (count rate equals to f_1). The time measured between the two pulses just before the second one disappears corresponds to the dead time. Using this methodology, dead times, presented in Table V.1, are evaluated.

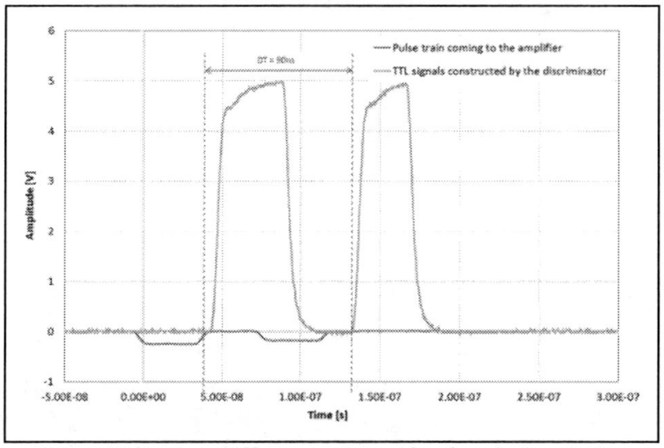

Figure V.2 - Two different pulse trains just before losing the second one

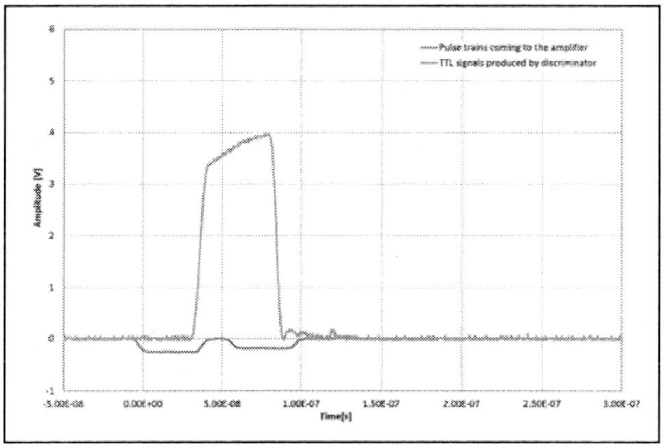

Figure V.3 – Single pulse train just after losing the second one

TABLE V.1 - DEAD TIME

Original	*New prototype*
90 ns	88 ns

These results are consistent, the new electronic works properly. These values are used in parallelizable dead time formula [3] which gives us the real count rate.

C. *Detector linearity*

In this section, the linear responses of original and prototype chains have been analyzed. Comparison is made using a boron deposition chamber placed near the core [4].

This boron detector is stimulated by thermal neutron coming from the core, both in steady state conditions (power plateau) and in transitory condition (during power peak).

Our ^{237}Np detectors are sensible to fast neutrons. In any case, the proportion of thermal neutron and fast neutron is constant in steady state condition. It is assumed that this ratio is constant also during transient of the reactor [5].

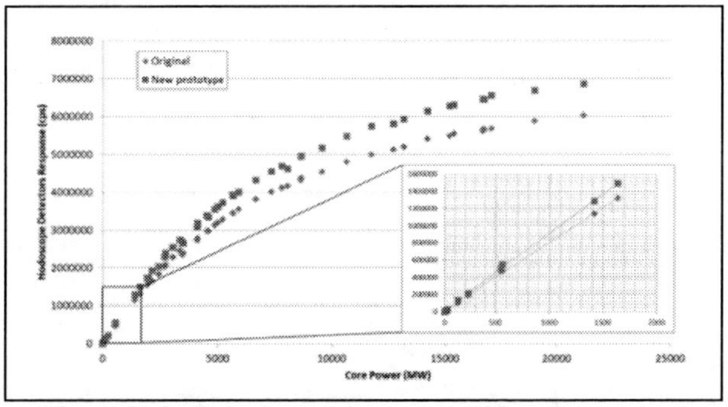

Figure V.4 - Detectors responses at different powers of CABRI reactor without dead time correction

In Figure V.4 the count rates of original and new prototype electronics chain vs core power measured by boron detectors is shown. From this figure it is possible to see that the counting chains including fission chambers have a linear behavior under 1.5 million cps which corresponds to ~1.6 GW of CABRI power. For higher power, a dead time correction has to be taken into account, due to the behavior of the electronic chain [3].

Two chains do not give the same response because they are connected to different FC. If we normalize signals to the maximum values of each set, the two curves will superpose perfectly.

VI. CONCLUSION

All tests realized in this campaign fulfill our expectation. The response of the new prototype is similar to its predecessor respecting all needed requirements for our experiments. Furthermore, this new solution will be easier to maintain thanks to the full discrete architecture of the amplifier. These new devices will be installed in CABRI Hodoscope measurement system current 2018.

In the near future, an industrialization campaign will be programmed in order to supply more devices for others measurement chains of the hodoscope.

References

[1] G. Baratte, «Ensemble de mesure pour hodoscope,» CEA LIST, Saclay.

[2] J. Gostely et E. Carnal, «La methode des deux oscillateurs pour la mesure du temps mort dans l'instrumentation nucléaire,» Nuclear instruments and methods, 1978.

[3] V. Chevalier, S. Mirotta, J. Guillot and B. Biard, "The CABRI fast neutron Hodoscope: renovation, qualification program and first resutls following the experimental reactor restart," IRSN, Cadarache, 2017.

[4] J. Lecerf, "Study of the linearity of CABRI experimental chambers during RIA transients," CEA, Cadarache, 2017.

[5] G. Augier and K. Baumung, "Quantitative fuel motion determination with the CABRI fast neutron hodoscope : evaluation methods and results," Nuclear Technology, 1990.

EPJ Web of Conferences **170**, 04017 (2018) https://doi.org/10.1051/epjconf/201817004017
ANIMMA 2017

Current Mode Neutron Noise Measurements in the Zero Power Reactor CROCUS

O. Pakari, V. Lamirand, G. Perret, L. Braun, P. Frajtag, A. Pautz

***Abstract*—The present article is an overview of developments and results regarding neutron noise measurements in current mode at the CROCUS zero power facility. Neutron noise measurements offer a non-invasive method to determine kinetic reactor parameters such as the prompt decay constant at criticality $\alpha = \beta_{eff}/\Lambda$, the effective delayed neutron fraction β_{eff}, and the mean generation time Λ for code validation efforts. At higher detection rates, i.e. above 2×10^4 cps in the used configuration at 0.1 W, the previously employed pulse charge amplification electronics with BF_3 detectors yielded erroneous results due to dead time effects. Future experimental needs call for higher sensitivity in detectors, higher detection rates or higher reactor powers, and thus a generally more versatile measurement system. We, therefore, explored detectors operated with current mode acquisition electronics to accommodate the need. We approached the matter in two ways:**

1) By using the two compensated ^{10}B-coated ionization chambers available in CROCUS as operational monitors. The compensated current signal of these chambers was extracted from core monitoring output channels.

2) By developing a new current mode amplification station to be used with other available detectors in core. Characteristics and first noise measurements of the new current system are presented. We implemented post-processing of the current signals from 1) and 2) with the APSD/CPSD method to determine α. At two critical states (0.5 and 1.5 W), using the ^{10}B ionization chambers and their CPSD estimate, the prompt decay constant was measured after 1.5 hours to be $\alpha = (156.9 \pm 4.3)$ s^{-1} (1σ). This result is within 1σ of statistical uncertainties of previous experiments and MCNPv5-1.6 predictions using the ENDF/B-7.1 library. The new system connected to a CFUL01 fission chamber using the APSD estimate at 100 mW after 33 min yielded $\alpha = (160.8 \pm 6.3)$ s^{-1}, also within 1σ agreement.

The improvements to previous neutron noise measurements include shorter measurement durations that can achieve comparable statistical uncertainties and measurements at higher detection rates.

***Index Terms* — Research reactors, neutron noise, current acquisition, kinetic parameter measurement**

Transcript received 1st June 2017

O. Pakari, L. Braun, and P. Frajtag are with the LRS, EPFL (Corresponding author: O. Pakari, email: oskari.pakari@epfl.ch). G. Perret is with the LRS, Paul Scherrer Institut (PSI). V. Lamirand and A. Pautz are both with the LRS, EPFL and NES, PSI.

I. Introduction

NEUTRON noise measurements in zero power reactors provide the means to determine kinetic parameters such as the prompt neutron decay constant α, the delayed neutron fraction β_{eff}, and the mean prompt generation time Λ in a non-invasive manner. These values then contribute to code validation and nuclear database efforts.

The kinetic parameters of the CROCUS zero power reactor core have been successfully determined using reactivity change [1] and non-invasive neutron noise measurements with the Feynman-α and auto and cross power spectral density (APSD/CPSD) methods [2, 3]. The measurement station was also successfully used in the MINERVE zero power facility [4]. The previous measurements in CROCUS were undertaken with several sub-critical and critical states with powers up to 0.1 W with four BF_3 neutron detectors in the control rod and peripheral fuel rod positions (see Fig. 2). Measurements at higher power or with more sensitive detectors could not be performed with the developed measurement station based on pulse spectrometry electronics due to pile-up in the detector and dead time effects in the amplification stage. Future experimental aspirations in CROCUS [5] call for more sensitivity, higher precision, and extended versatility in detector positioning. For this purpose, we examined the option of using current measurement electronics for neutron noise measurements in CROCUS. This paper is dedicated to describing the latest developments, beginning with an introduction to the CROCUS facility in Section II and a detailed description of the used instrumentation. In Section III, we present the measurement of the kinetic parameters using the available current signal of core monitors. The measurement setup alongside the APSD/CPSD analysis method with associated post processing details and results are discussed. In Section IV we outline the development and first noise measurements of a novel current amplifier used with Photonis CFUL01 fission chambers (1g ^{235}U). This is followed by an outlook on future developments.

II. The Research Reactor CROCUS

A. Reactor facility description

The CROCUS reactor is a two-zone, uranium-fueled, H_2O-moderated facility operated by the Laboratory for Reactor Physics and Systems Behaviour (LRS) at the Swiss Federal Institute of Technology Lausanne (EPFL) (see Fig. 1). It is a

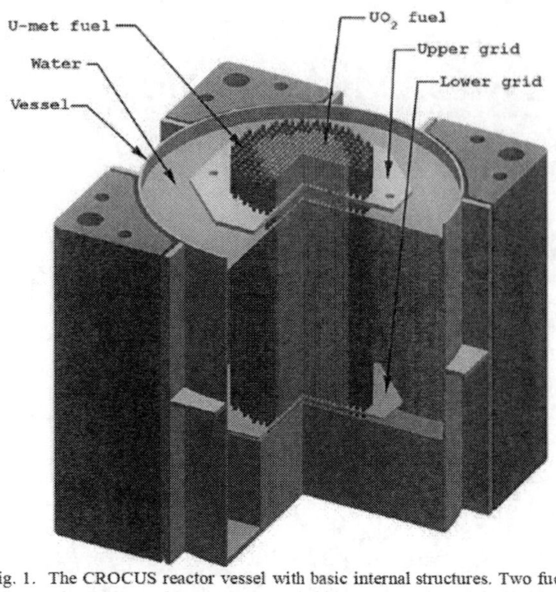

Fig. 1. The CROCUS reactor vessel with basic internal structures. Two fuel zones, an inner UO_2 and outer U_{met} zone, held in an approximately cylindrical array by two grids. Water is filled from the bottom to establish criticality.

zero power reactor, with a nominal power of up to 100 W. The core is approximately cylindrical in shape with a diameter of about 58 cm and a height of 100 cm. The core reactivity is controlled by a variation of the water level with an accuracy of 0.1 mm (equivalent to 0.4 pcm) and/or by means of two control rods containing naturally enriched boron carbide (B_4C) sintered pellets located symmetrically within the outer fuel zone. Two different kinds of fuel rods make up the reactor core of CROCUS (see Fig. 1 and 2). The central zone is loaded with 336 UO_2 fuel rods (1.806 wt.%-enriched), set in a square lattice with a pitch of 1.837 cm. The peripheral zone is loaded with up to 176 thicker, U_{met} fuel rods (0.947 wt.%-enriched) with a pitch of 2.917 cm, also in a square lattice. All fuel rods have an aluminum cladding and are maintained in a vertical position by the upper grid and lower grid plates spaced 100 cm apart. Both grid plates incorporate a cadmium layer with a thickness of 0.5 mm to limit axial thermal leakage to surrounding structures. The active fuel length starts at the top surface of the lower cadmium layer and extends to 100 cm. The core is located in an aluminum water tank, its diameter being 130 cm and thickness being 1.2 cm. Demineralized light water (H_2O) is used as moderator and reflector.

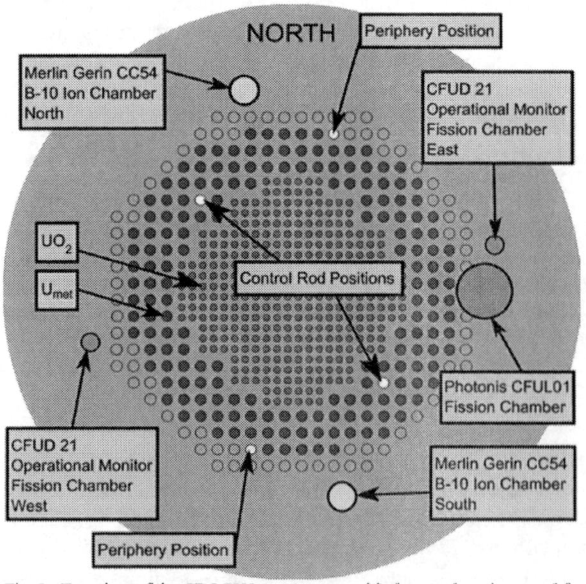

Fig. 2. Top view of the CROCUS reactor core with detector locations used for noise measurements and core monitoring.

B. Core instrumentation

CROCUS is equipped with four operational monitor detectors. Two Merlin Gerin CC54 Compensated ^{10}B coated Ionisation Chambers (CIC north & CIC south) with a sensitivity of 3×10^{-5} nA/n_{th} are used in current mode for flux control and core monitoring. The CICs have an active length of 355 mm and diameter of 49 mm. Two Photonis CFUD 21 fission chambers (FC) in the eastern and western periphery serve as the operational monitors and are operated in pulse mode. They are part of the safety channels and are calibrated with respect to reactor power with gold foil irradiation experiments, and are, thus, reactor power monitors (see Fig. 2). The readout of the fission chamber monitors is used for fission rate F_0 determination via

$$F_0 = \overline{C_{mon}}\,\kappa / E, \quad (1)$$

with $\overline{C_{mon}}$ being the average monitor count rate in cps during the experiment, E being the energy released per fission in J and κ^{-1}=2428.0±96 cps/W being the calibration factor for the western chamber. κ was determined using eight gold foils irradiated in the center of the core with even axial spacing at 15 W for 1 hour. The effective cross sections for the gold foils were calculated using MCNP5-v1.6 with the ENDF/B-7.1 library.

C. Pulse mode neutron noise measurement station

The established neutron noise measurement station consists of four BF_3 neutron detectors in control rod and peripheral fuel rod positions connected to standard gamma spectrometry preamplifiers and amplifiers (see Fig 2.). The pulses are then read out via Multi Channel Scaler cards on a computer. The system is currently limited to measurements at sub-criticality or critical states with powers below 0.1 W at 18 kcps due to pile-up in the detector and dead time effects in the amplifying stage.

D. Limitations and motivation for current mode measurement

Neutron noise measurements generally require sensitive detectors to minimize detector shape or other spatial effects [6]. The BF_3 detector has a sensitivity of 4.7×10^{-2} counts/n_{th} with an active length of 100 mm and a 7.5 mm diameter. Other available detectors such as Vniitfa ^3He ionization chambers (2.82 counts/n_{th}, 340 mm active length, 32 mm diameter) and Photonis CFUL01 fission chambers (1 count/n_{th}, 211 mm active length, 48 mm diameter) are too sensitive for in-core usage with the existing electronics, even at low powers or sub-critical states. Thus to increase our flexibility in detector choice and ultimately detector position in CROCUS, we decided to investigate the option to use current mode acquisition. The integration of charge eliminates dead time effects and extends the effective utility range in terms of power [7]. This has been successfully implemented by CEA in the zero power facilities MINERVE and EOLE using an in-house developed system [11] to measure the kinetic parameters. A cross-comparison measurement campaign between this system, SCK•CEN and PSI/EPFL stations was undertaken within the Venus-Eole-Proteus scientific cooperation and will be reported on soon.

In the following sections we present two approaches to examine the feasibility of current mode measurements for neutron noise applications in CROCUS: firstly, by directly measuring the already available current output of the CIC monitors using an advanced oscilloscope and by applying noise analysis techniques to determine the kinetic parameters. This is compared to previous results to verify the validity of the new method. Secondly, by showing the development status of a new current amplification and measurement system, to be used for various detectors, that allows for flexible detectors and detector location choice.

III. Measurement of the Kinetic Parameter α in Current Mode in CROCUS using Core Monitors

A. Acquisition of CIC signals

We used two higher reactor power configurations (0.5 W and 1.5 W) for the current mode acquisition (see Table I) to show the increased capability in terms of detection rates compared to the previous BF_3 setup. Calculations showed an order of magnitude higher detection rate in the CFUL01 fission chamber in the periphery compared to a BF_3 detector in the control rod position. The acquisition line is depicted in Fig. 3. Both CIC signals were directly connected to the 10 MΩ DC inputs of a Teledyne Lecroy Wavesurfer 10 oscilloscope [12]. At a sampling rate of 10 kHz the acquisition lasted up to 33 min per measurement. We measured three consecutive times to achieve a total measurement duration of about 1.5 hours at both reactor powers.

B. Power spectral density noise analysis

The auto or cross power spectral density (APSD/CPSD) analysis method, often referred to as the Cohn-α method [7], is

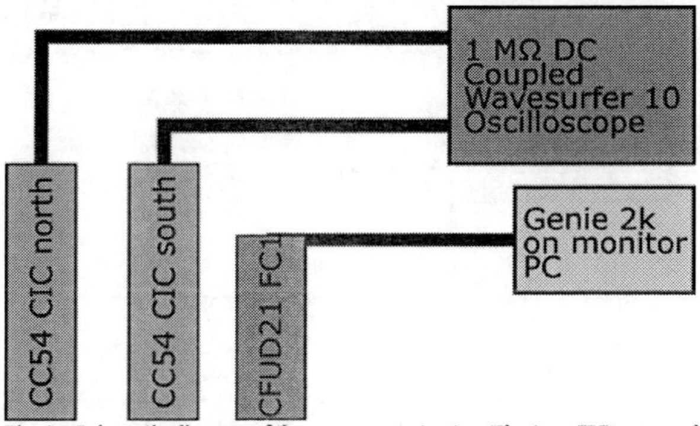

Fig. 3. Schematic diagram of the measurement setup. The two CICs are used in current mode for noise measurements whilst the calibrated CFUD21 fission chamber 1 is used for fission rate determination.

TABLE I
USED REACTOR CONFIGURATIONS FOR NEUTRON NOISE MEASUREMENTS

Power (W)	Count rate of FC 1 monitor (cps)	Mean voltage of CIC (V)	Fission rate (s^{-1})
0.5	1229	1.63	1.57E+10
1.5	3904	5.29	5.00E+10

the most common method used for current mode signals. By calculating the Fourier transform of auto- and cross-correlation functions of a detector signal, the reactor noise at criticality can be fitted by the following general expression in dependency of the angular frequency ω [4, 8, 9, 10]:

$$G_{ij}(\omega) = \frac{2S_iS_jD_v}{\Lambda^2 F_0}\frac{1}{\omega^2+\alpha^2} + S_iS_j\delta_{ij}. \quad (2)$$

with δ_{ij} being the Kronecker delta, and S_i the measured detector signal. D_v is the Diven factor [13], calculated for CROCUS in [3] to be 0.806. From a lumped parameter perspective white noise is transferred through a low-pass filter with cut-off frequency at $f_0 = \alpha/2\pi$. The plateau at frequencies much smaller than f_0 are proportional to the effective delayed neutron fraction β_{eff}.

The calculation of the one-sided APSD and CPSD was implemented in a MATLAB script using the Welch overlapped segment estimator method. A Hann window with 50% overlap was chosen for smoothing [14] and 24,000 total discrete Fourier transform points to allow for a 0.5 Hz resolution. As the oscilloscope had a sampling frequency of 10 kHz, the bandwidth of the PSD is 2500 Hz.

Code predictions for the prompt decay constant using MCNP5-v1.6 with the ENDF/B-7.1 library yielded $\alpha = (158.6 \pm 1.6)\ s^{-1}$, hence an expected cut-off frequency of $f_0 = \alpha/2\pi = 25.2$ Hz. Thus the region of interest for curve analysis was limited to 100 Hz. For this region of the APSD or CPSD a fit function of the shape

$$G_{ij}(\omega) = \frac{a}{\omega^2 + f_0^2} + c \quad (3)$$

was used, according to Equation 2. c is introduced to account for white noise introduced by imperfect electronics and uncorrelated current. For uncertainty estimation the fit data was weighted by the standard deviation, for an APSD/CPSD being

$$\sigma_{G_{ii}}(\omega) = \frac{G_{ii}}{\sqrt{N_{DFT}}};\ \sigma_{G_{ij}}(\omega) = \sqrt{\frac{G_{ij}(\omega)G_{ij}^*(\omega)}{N_{DFT}\gamma_{ij}^2(\omega)}} \quad (4)$$

with N_{DFT} being the number of discrete Fourier transforms and $\gamma_{ij}(\omega)$ being the coherence function between the two detectors,

$$\gamma_{ij}(\omega) = \frac{|G_{ij}(\omega)|}{\sqrt{G_{ii}(\omega)G_{jj}(\omega)}}, \quad (5)$$

also calculated using a magnitude squared estimate function in MATLAB.

TABLE II
RESULTS FOR THE PROMPT DECAY CONSTANT AT CRITICALITY

Power (W)	0.5	1.5
APSD north (s^{-1})	187.5 ± 5.2	189.2 ± 4.6
APSD south (s^{-1})	187.2 ± 4.8	193.2 ± 4.7
CPSD (s^{-1})	156.8 ± 3.2	160.2 ± 3.5

C. Experimental Results for CIC noise measurements

The APSD and CPSD estimates for the prompt decay constant α based on the previously described measurement and processing setup are presented in Table 2, with example CPSDs with fits and residuals displayed in Fig. 4. The listed uncertainties were calculated by fitting the APSD/CPSD data with differing upper frequency limits. The limits were chosen in 10 Hz increments between 50 and 100 Hz. The final uncertainty is the propagated uncertainty of the fit uncertainty on α over all five fits. The APSD estimates are subject to transfer function biases which are eliminated by cross-correlating. The CPSD estimate yields the most accurate value and thus the best estimate using the CICs. The prompt decay constant is found to be **(156.8 ± 3.2) s^{-1}** at 0.5 W and **(160.2 ± 3.5) s^{-1}** at 1.5 W after 1.5 hours of measurement time, respectively.

D. Comparison to previous measurements and code predictions

With the pulse mode setup two major campaigns were undertaken, the first [2] yielded $\boldsymbol{\alpha}$ **= (146.6 ± 6.3) s^{-1}** after 6 to 7 hours of measurement time using the Feynman-α method. A further refined campaign [3] used the APSD/CPSD estimate and found $\boldsymbol{\alpha}$ **= (154.4 ± 2.4) s^{-1}** after 5 hours of measurement time. The MCNP prediction of $\boldsymbol{\alpha}$ **= (158.6 ± 1.6) s^{-1}** is thus met by both the previous measurements within 2σ and 1σ, respectively. The current mode measurements both agree within 1σ with the code predictions. We conclude that current mode measurements in this detection configuration are equally viable for kinetic parameter measurements in CROCUS compared to pulse mode measurements, and similar in terms of measurement time to achieve comparable statistical uncertainties, as the measurements produce a similarly precise value compared to code predictions.

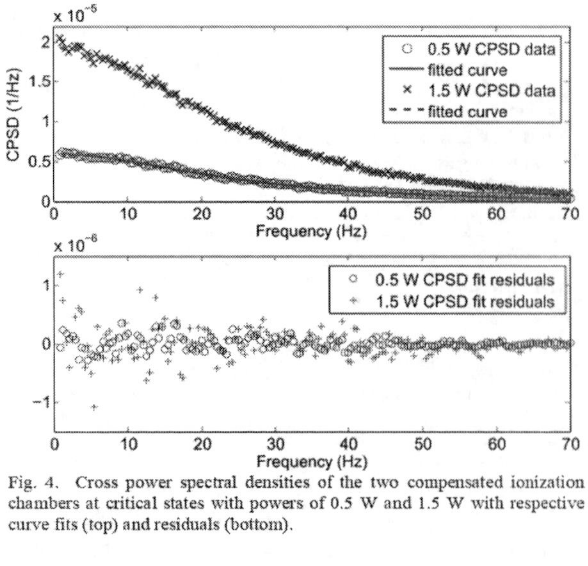

Fig. 4. Cross power spectral densities of the two compensated ionization chambers at critical states with powers of 0.5 W and 1.5 W with respective curve fits (top) and residuals (bottom).

IV. Novel current amplifier for flexible detector use

A. Design of current amplifier

The CICs offer only a limited flexibility in terms of positioning as they are fixed to the grid and are relatively large. For a comparable or higher flexibility than with the BF_3 detectors a current amplifier usable with, among others, Vniitfa ^{3}He (2.82 counts/n_{th}) and Photonis CFUL01 (1 count/n_{th}) detectors is being developed for this purpose. As the main purpose of the system is noise measurements, the set of specifications required can be summarized to be:

- Constant DC amplification in the range of nA to mA to allow for a simple acquisition system on FPGA basis.
- Constant transfer function in the region of interest for noise measurements i.e. 0 Hz to ~200 Hz.

The prototype board logic is depicted in Fig. 5. It consists of two main components:

i) A primary stage to pass through the high voltage to the detector's anode and to extract and amplify a current in the detector using a measurement resistance of 10 kΩ and an operational amplifier with a gain of 470 and a cut-off frequency of 2 kHz at -40 dB/decade. The current input range is hereby 2 pA to 20 pA. Using a Photonis CFUL01 fission chamber, this results in an effective minimum measurable current in the chamber of 1 nA. In this configuration the measurement range in terms of reactor power is 0 W to 10 W. An analog switch to augment the range in dependence of test results is foreseen for the final design.

ii) A secondary stage for active (doubled low-pass RC) filtering purposes with an effective gain of 0 and a cut-off frequency of 3.3 kHz at -25 dB/decade. Hereafter a BNC output connector gives the signal to acquisition hardware.

The transfer function of the prototype was measured and is depicted in Fig. 6. Indeed, in the range of interest for neutron noise measurements (1-100 Hz) the transfer function exhibits a plateau.

B. Current mode neutron noise measurements in CROCUS using the new amplifier

The new current amplifier was connected to a Photonis CFUL01 fission chamber as depicted in Fig. 5. set in the eastern reflector of the CROCUS core (see Fig 2.). The Plexiglas channel for the fission chamber is clamped to the grid and allows for flexible mounting in various locations in the reflector zone. We measured the output signal again using the Teledyne

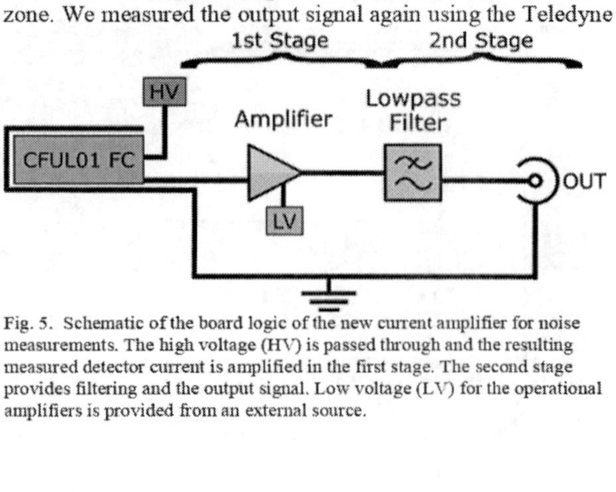

Fig. 5. Schematic of the board logic of the new current amplifier for noise measurements. The high voltage (HV) is passed through and the resulting measured detector current is amplified in the first stage. The second stage provides filtering and the output signal. Low voltage (LV) for the operational amplifiers is provided from an external source.

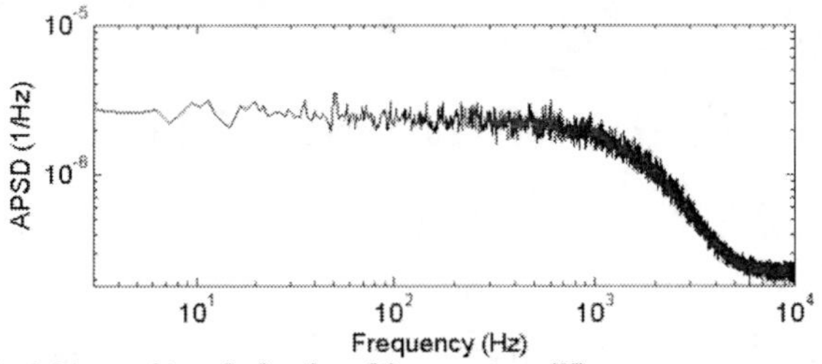

Fig. 6. Measured transfer function of the current amplifier.

Lecroy Wavesurfer 10 oscilloscope at a DC 10 MΩ input coupling. At this prototype stage, only one detector was used. A measurement time of 33 min at 100 mW was chosen as a first reference state. The signal was then treated via the MATLAB routine described in Section III B.

C. Experimental results for new current station noise measurements

The APSD was fitted following Equation 3 to determine the prompt decay constant (see Fig 7). The final value for the APSD estimate for α at 100 mW is **(160.8 ± 6.3) s^{-1}**, and is in the same agreement range as the CIC measurement. The higher statistical uncertainty is attributed to the APSD method. CPSD analyses could allow for further precision once more experimental channels and electronics have been manufactured. This first measurement proves nonetheless the capability of the new system and allows us to further pursue its development.

V. Conclusion and Outlook

Future experimental activities in the CROCUS zero power facility call for improved and more flexible neutron noise measurement methods. A previously developed system using BF_3 detectors and gamma spectroscopy electronics was limited in detection rate and sensitivity. For this purpose, the possibility of current mode acquisition was explored.

Two ^{10}B compensated ionization chambers, used as operational monitors, already output an analog current signal, which proved usable for such measurements. Using an oscilloscope and post-processing via the CPSD method the prompt neutron decay constant α, was determined to be **(156.8 ± 3.2) s^{-1}** at 0.5 W and **(160.2 ± 3.5) s^{-1}** at 1.5 W using the CICs. Additionally, the development status of a novel current amplifier for other available detectors was presented. Characteristics and first current mode noise measurements using a CFUL01 fission chamber were shown. Preliminarily measurements using one detector with the new system yielded, using APSD analysis,

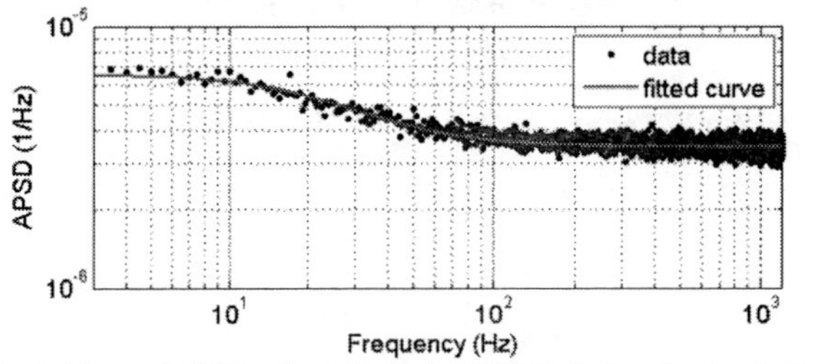

Fig. 7. Measured APSD using a Photonis CFUL01 fission chamber in the periphery of the CROCUS core connected to the new current amplifying system.

(160.8 ± 6.3) s^{-1}. All results are within 1σ agreement with Monte Carlo predictions and previous measurements.

Neutron noise measurements in current mode allowed for shorter measurement durations as well as measurements at higher detection rates and thus meet the requirements for future experimental needs. Future improvements include a full characterization and validation of the new system using more detectors examining the full range of operation. This will allow for more cross correlation measurements and the calculation of β_{eff} and Λ by solving APSD/CPSD normalization.

The final design comprising versatile electronics as well as flexible detector positioning will then be used for experiments for spatial and spectral noise theory validation within CROCUS.

Acknowledgment

The authors thank D. Godat for his assistance in operating the reactor and with the mechanical aspects of detector positioning.

References

[1] J.-M. Paratte, R. Früh, U. Kasemeyer, M.A. Kalugin, W. Timm, and R. Chawla. "A benchmark on the calculation of kinetic parameters based on reactivity effect experiments in the CROCUS reactor" Annals of Nuclear Energy, 33(8):739–748, May 2006.
[2] V. Roland, G. Perret, G. Girardin, P. Frajtag, and A. Pautz, "Neutron noise measurement in the CROCUS reactor," in Proc. IGORR, 2013.
[3] G. Perret, G. Girardin, P. Frajtag, M. Hursin, 2014. Decay constant and delayed neutron fraction measurements in CROCUS – TM-41-14-02 Rev. 1. Tech. rep., Paul Scherrer Institut, Villigen, 5232, Switzerland.
[4] G. Perret, B. Geslot, A. Gruel, P. Blaise, J. Di-Salvo, G. De Izarra, & A. Pautz. (2017). Kinetic parameter measurements in the MINERVE reactor. *IEEE Transactions on Nuclear Science*, *64*(1), 724-734.
[5] V. Lamirand, M. Hursin, P. Frajtag, A. Pautz, G. Perret & O. Pakari (2016). Future experimental programmes in the CROCUS reactor. In *Conference proceedings of RRFM/IGORR 2016* (No. EPFL-CONF-218310, pp. 284-292).
[6] J. R. Sheff, & R. W. Albrecht, (1966). The Space Dependence of Reactor Noise I-Theory. *Nuclear Science and Engineering*, *24*(3), 246-259.
[7] G. F. Knoll. Radiation detection and measurement. Wiley & Sons, 2000.
[8] C. E. Cohn, "A simplified theory of pile noise," *Nucl. Sci. Eng.*, vol. 7, no. 5, pp. 472–475, 1960.
[9] R. E. Uhrig. Random Noise Techniques in Nuclear Reactor Systems. Ronald Press, 1970.
[10] M. M. R. Williams. Random Processes in Nuclear Reactors. Pergamon Press, Oxford, 1974.
[11] G. de Izarra, C. Jammes, B. Geslot, J. Di Salvo, C. Destouches SPECTRON, a neutron noise measurement system in frequency domain Rev. Sci. Instrum., 86 (11) (2015), p. 115111 http://scitation.aip.org/content/aip/journal/rsi/86/11/10.1063/1.4935250
[12] Teledyne Lecroy Wavesurfer 10 Datasheet, Accessed 9. May 2017: http://cdn.teledynelecroy.com/files/pdf/wavesurfer10-datasheet.pdf
[13] B. Diven, H. Martin, R. Taschek, and J. Terrell, "Multiplicities of Fission Neutrons," Phys. Rev., vol. 101, no. 3, pp. 1012–1015, Feb. 1956
[14] Harris, Fredric J. "On the use of windows for harmonic analysis with the discrete Fourier transform." *Proceedings of the IEEE* 66.1 (1978): 51-83.

EPJ Web of Conferences 170, 04018 (2018)
ANIMMA 2017
https://doi.org/10.1051/epjconf/201817004018

Fabrication and Testing of a Modular Micro-Pocket Fission Detector Instrumentation System for Test Nuclear Reactors

Michael A. Reichenberger, *Member, IEEE*, Daniel M. Nichols, *Member, IEEE*, Sarah R. Stevenson, Tanner M. Swope, Caden W. Hilger, Jeremy A. Roberts, Troy C. Unruh, and Douglas S. McGregor, *Member, IEEE*

***Abstract*— Advancements in nuclear reactor core modeling and computational capability have encouraged further development of in-core neutron sensors. Measurement of the neutron-flux distribution within the reactor core provides a more complete understanding of the operating conditions in the reactor than typical ex-core sensors. Micro-Pocket Fission Detectors have been developed and tested previously but have been limited to single-node operation and have utilized highly specialized designs. The development of a widely deployable, multi-node Micro-Pocket Fission Detector assembly will enhance nuclear research capabilities. A modular, four-node Micro-Pocket Fission Detector array was designed, fabricated, and tested at Kansas State University. The array was constructed from materials that do not significantly perturb the neutron flux in the reactor core. All four sensor nodes were equally spaced axially in the array to span the fuel-region of the reactor core. The array was filled with neon gas, serving as an ionization medium in the small cavities of the Micro-Pocket Fission Detectors. The modular design of the instrument facilitates the testing and deployment of numerous sensor arrays. The unified design drastically improved device ruggedness and simplified construction from previous designs. Five 8-mm penetrations in the upper grid plate of the Kansas State University TRIGA Mk. II research nuclear reactor were utilized to deploy the array between fuel elements in the core. The Micro-Pocket Fission Detector array was coupled to an electronic support system which has been specially developed to support pulse-mode operation. The Micro-Pocket Fission Detector array composed of four sensors was used to monitor local neutron flux at a constant reactor power of 100 kWth at different axial locations simultaneously. The array was positioned at five different radial locations within the core to emulate the deployment of multiple arrays and develop a 2-dimensional measurement of neutron flux in the reactor core.**

***Index Terms*— In-Core Instrumentation, MPFD, Micro-Pocket Fission Detector, Reactor Instrumentation, TRIGA.**

I. Introduction

Research nuclear reactors commonly utilize neutron-sensitive radiation detectors located external to the reactor core to monitor reactor power [1, 2]. Advancements in nuclear reactor core modeling and computational capability have encouraged further development of in-core neutron sensors. Measurement of the neutron-flux distribution within the reactor core provides a more complete understanding of the operating conditions in the reactor than ex-core sensors. Solutions presently exist to monitor the neutron fluence experienced during reactor experiments, however information regarding minute variations in flux due to local fuel burnup, control rod motion, and experimental effects are often not captured.

Small, accurate, and robust neutron-flux sensors are an important development for the enhancement of advanced nuclear fuel testing [3]. The environment within a nuclear reactor core is not ideal for the operation of many types of radiation detectors. The high-radiation environment, including intense fields of neutrons, gamma-rays, and beta-particles quickly overwhelm most conventional radiation detectors. Even specially developed neutron detectors (such as Self-Powered Neutron Detectors (SPNDs) and sub-miniature neutron detectors) must make corrections to the measured data to compensate for the high count rate and significant burnup induced by neutron-fluxes often on the order of 10^{14} n cm^{-2} s^{-1} [2, 4, 5]. Conventional in-core sensors often utilize neutron-sensitive coatings with large neutron-absorption cross sections, whose presence can also decrease the local neutron flux near the sensor.

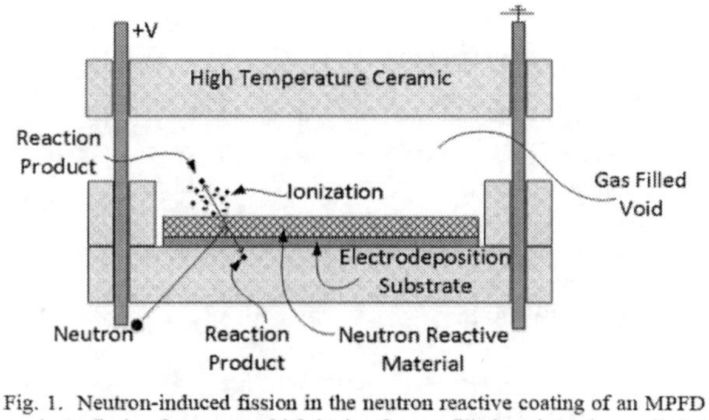

Fig. 1. Neutron-induced fission in the neutron reactive coating of an MPFD produces fission fragments which ionize the gas-filled pocket. The small size reduces ionization from non-fission fragments such as gamma-rays and beta-particles.

This work, presented at the International Conference on Advancements in Nuclear Instrumentation Measurement Methods and their Applications, June 19-23, 2017, was supported in part by the U.S. Department of Energy under Grants DE-NE0008408 and NE0008305.

M. A. Reichenberger, D. M. Nichols, S. R. Stevenson, T. M. Swope, C. W. Hilger, J. A. Roberts, and D. S. McGregor are part of the S.M.A.R.T. Laboratory at Kansas State University, Manhattan, KS 66502 USA (e-mail: mar89@ksu.edu).

T. C. Unruh is part of the HTTL at the Idaho National Laboratory, Idaho Falls, ID 83415 (e-mail: troy.unruh@inl.gov).

Micro-Pocket Fission Detectors (MPFDs), illustrated in Fig. 1, have been constructed and tested previously [6, 7]. MPFDs exhibit exceptional gamma-ray discrimination due to their small chamber size and have successfully operated in neutron fluxes as high as 10^{15} n cm^{-2} s^{-1} [7]. The one-of-a-kind instruments developed for previous experiments have successfully demonstrated the ability of MPFDs to measure neutron flux in numerous test reactor conditions [7]. In-core testing at the Kansas State University TRIGA Mk. II research reactor included steady-state reactor operations from 10 kWth to 700 kWth reactor power, positive power transients with periods of 10 sec., 20 sec., and 40 sec., negative power transients from -\$0.10 to -\$0.80 [7].

Future deployment of MPFDs at numerous nuclear reactor research facilities will benefit from the development of a more standardized MPFD instrument package. In addition to the development of an integrated electronics package (which includes high-voltage supply, pre-amplification, shaping, and discrimination), a modular sensor array has been developed. The use of this sensor array will enable the deployment of a network of MPFDs into the core of various research nuclear reactors. The data provided by an array of MPFDs deployed through the core of a research reactor can then be used to enhance material experimentation and certification, or to benchmark nuclear reactor simulations.

II. SENSOR FABRICATION

The modular MPFD array utilized a fully encapsulated design which included an electronic plug and a vacuum/fill assembly for ionization gas. The MPFD sensors were located near the end of the array which was encapsulated by an approximately 183-cm long, 0.794-cm diameter stainless steel tube. The sensors were equally spaced within the bottom portion of the tube as shown in Fig. 2. The spacing within the MPFD array was designed to yield equal spacing of sensors throughout the fuel region of the TRIGA reactor core, as illustrated in Fig. 3.

Each MPFD node was fabricated using polished alumina disks separated by machined alumina spacers. A 1.5-mm diameter electrode was evaporated onto the bottom disk and electrodeposited with natural uranium [8]. X-ray fluorescence was used to confirm the presence of uranium on each sample. Then, alpha spectrometry was used to measure the activity of each sample and to determine the mass of uranium which was deposited, summarized in Table 1.

TABLE I
URANIUM MASS FOR EACH MPFD NODE

MPFD Node	Uranium Mass (μg)
0 (bottom of array)	0.619 ± 0.014
1	0.548 ± 0.025
2	0.630 ± 0.026
3 (top of array)	0.533 ± 0.023

The array was assembled by insulating 8 30-AWG alumel wires with silica insulation between the electrical plug and the sensor region. Each MPFD sensor required a single anode and cathode wire. The MPFDs were indexed such that each ionization chamber was exposed to one anode wire and one cathode wire on opposing sides of the sensor as illustrated in Fig. 1. Therefore, the sensors did not share any common anodes nor cathodes. The assembly was inserted into the encapsulation tube and heated to approximately 100 C during three 1-hr vacuum purges each followed by a backfill with ultra-high

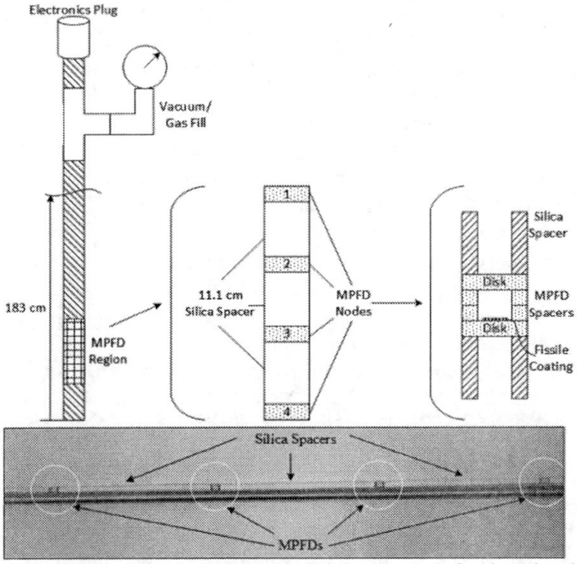

Fig. 2. The encapsulated MPFD array included an electronic feedthrough and vacuum/fill assembly. Silica spacers were used to separate four MPFD nodes at the bottom of the array by 11.1-cm, together spanning the full length of fuel in the TRIGA reactor core.

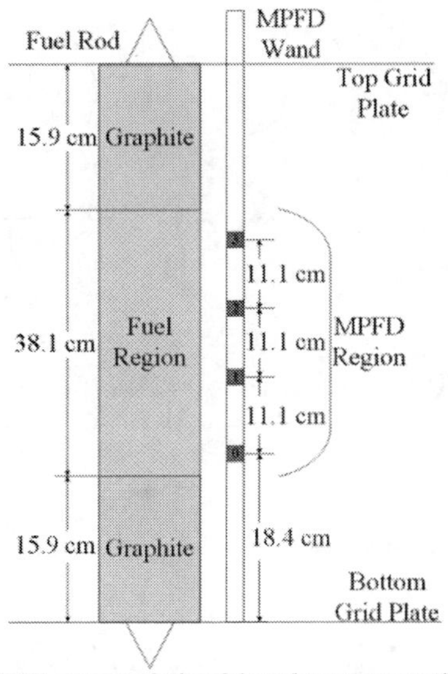

Fig. 3. The MPFD array was deployed through 8-mm penetrations in the fuel grid-plate. The distribution of sensors was centered on the same vertical axis as the reactor fuel in the TRIGA reactor core.

purity neon gas to 30 psig. Neon was of particular interest as an ionization gas because of its higher charge-carrier mobility and average ionization energy compared to argon [9]. Although not necessary for testing at the KSU TRIGA Mk. II research nuclear reactor, neon could be a useful alternative for higher flux applications.

III. Experimental Procedure

The MPFD array was first deployed into the grid-plate penetration between the B and C fuel rings at the KSU TRIGA Mk. II research nuclear reactor. The plug at the top of the MPFD array was connected to a 7.62 meter multi-conductor cable with additional tinned copper overbraid to reduce RF pickup from other electrical sources in the reactor pool. The MPFD-4 electronic support system was used to process the signal from all four MPFD sensors simultaneously and to interface with a PC. Custom LabView® software was developed to monitor the MPFD signals and to log the count rate for each channel in real-time. A 200 V bias was applied across each of the MPFD nodes and a 0.25 μs shaping time with a gain of 12 was used to shape the MPFD signals. Finally, a discrimination threshold of 800 mV was set to discriminate noise. The internal pulsar, designed to emulate a typical neutron-induced signal, registered a peak amplitude of 1.6 V.

The reactor was then brought to 1 kWth with cooling circulation pumps engaged and neutron-induced pulses were observed on all 4 MPFD channels. Then reactor power was increased to 100 kWth and sustained for at least 10 minutes during which time the count rate from the MPFDs were logged every 250 ms. The reactor was shut down and the MPFD array was moved to the C-ring flux port.

A single 10-minute measurement was conducted at 100 kWth

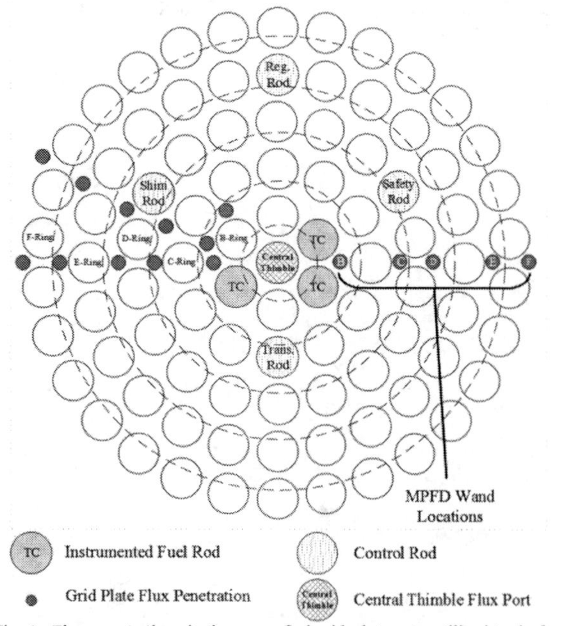

Fig. 4. Flux penetrations in the upper fuel grid-plate were utilized to deploy the MPFD array at three different radial locations in the TRIGA reactor core, outside the B-ring, D-ring, and F-ring of fuel.

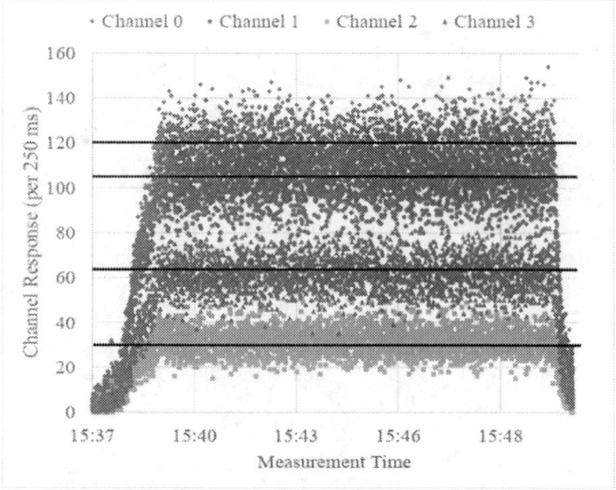

Fig. 5. Each of the four MPFDs were monitored in real-time as the reactor power was increased to 100 kWth and maintained for 10 minutes. The sensor response increased with reactor power and was stable at constant power.

reactor power for each of the 5 flux ports depicted in Fig. 4. Detector response was logged every 250 ms during each operation and integrated to determine an average count rate over the entire measurement period.

IV. Experimental Results

The real-time MPFD response rate was logged in 250 ms intervals, shown in Fig. 5. Data logging was initiated as reactor power was increasing and continued until after the 10 minute measurement interval when the reactor was shut down. An increasing count rate was observed as the reactor power approached 100 kWth. The sensor response remained stable throughout the duration of the measurement until the reactor was shut down. The real-time data was integrated over the 10-minute measurement to determine the average count rate for each sensor, summarized in Table II, and highlighted in Fig. 5. The average count rate for each channel was also divided by the mass of uranium deposited within each node to determine the mass-normalized count rate, summarized in Table II.

The average count rate for each channel was determined for each of five flux penetrations in the upper fuel grid plate of the KSU TRIGA Mk. II research nuclear reactor. The detector response for each MPFD decreased as the array was moved radially outward, shown in Fig. 6. Although reactor cooling was on for the duration of the measurements the in-core moderator temperature increased from 33 C to 36 C in the B-ring between the first and final measurement.

TABLE II
AVERAGE MPFD COUNT RATE AT 100 KWTH IN THE B-RING OF THE KSU TRIGA MK. II RESEARCH NUCLEAR REACTOR
(AVERAGED OVER 10-MINUTE MEASUREMENT)

MPFD Node	Raw Count Rate (CPS)	Mass Normalized Count Rate (CPS/μg)
0 (bottom of array)	482 ± 0.9	779 ± 17.7
1	429 ± 0.9	782 ± 35.7
2	130 ± 0.5	207 ± 8.6
3 (top of array)	257 ± 0.7	483 ± 20.9

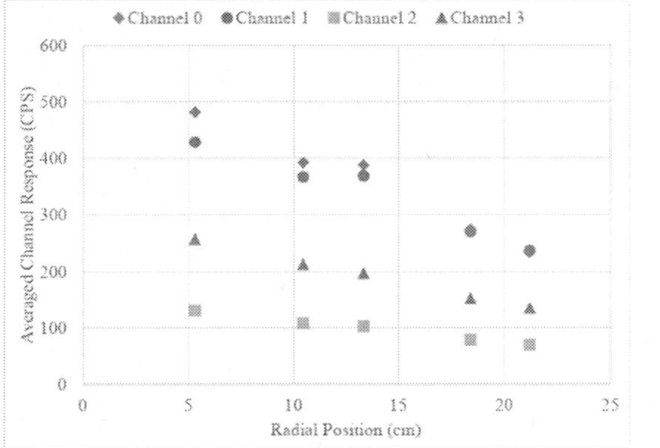

Fig. 6. The sensor response for each MPFD in the array decreased as the array was moved radially outward from the center of the reactor core.

V. Conclusions

Each of the four MPFDs responded to changes in the neutron flux in the reactor core and maintained a stable response with constant reactor power. Previous testing of MPFDs has only been conducted using ultra-high purity argon as an ionization gas. The successful use of neon as an ionizing gas for MPFDs has been demonstrated. The response of each sensor also decreased as the array was moved radially outward. However, the mass-normalized response for each sensor did not produce the expected results.

Although a different raw count rate was expected for each sensor due to the axial distribution of MPFDs in the reactor and the different amount of fissile material present in each detector, the mass normalized response was expected to be very similar for opposing nodes. Nodes 0 and 3, and nodes 1 and 2 were positioned within the array to be equidistant from the central axis of the fuel region in the core. Therefore it was expected that the mass-normalized response for the opposing sensors would reflect the relative symmetry of the neutron flux within the core. Re-assessment of the fissile mass present in each MPFD node will help to determine the source of error in determining the mass-normalized sensor response.

Work continues to develop a method of calibrating MPFDs in order to extract neutron flux magnitude from the count rate reported by the sensors. Improvements to the fabrication process have made possible the assembly and deployment of multiple MPFD arrays. Future studies will expand on the results of this preliminary work, measuring the neutron flux in the reactor core with axial and radial spacing simultaneously.

Acknowledgments

This paper presents work supported by the US Department of Energy Office of Nuclear Energy Contracts DE-NE0008408 and NE 0008305. MPFDs were fabricated at the Kansas State University Semiconductor Materials and Radiological Technologies laboratory (Manhattan, KS). Sensor testing was conducted at the Kansas State University TRIGA Mk. II research nuclear reactor facility (Manhattan, KS).

References

[1] J. J. Duderstadt and L. J. Hamilton, Nuclear Reactor Analysis, Wiley & Sons, Inc., 1976.

[2] G. F. Knoll, Radiation Detection and Measurement, Hoboken, N.J.: John Wiley, 2010.

[3] T. Unruh, J. Rempe, D. McGregor, P. Ugorowski and M. Reichenberger, "NEET Micro-Pocket Fission Detector - FY 2013 Status Report," Idaho national Laboratory, Idaho Falls, ID, 2013.

[4] G. Imel, "Performance of Hafnium and Gadolinium Self Powered Neutron Detectors in the TREAT Reactor," *Nucl. Inst. and Meth. Res. B,* no. 111, pp. 325-336, 1996.

[5] C. Blandin, S. Breaud, L. Vermeeren, M. Webber, "Development of New Sub-Miniature Fission Chambers: Modeling and Experimental Tests," *Prog. in Nucl. Energy,* vol. 43, pp. 349-355, 2003.

[6] D. S. McGregor, M. F. Ohmes, R. E. Ortiz, A. S. M. S. Ahmed, J. K. Shultis, "Micro-Pocket Fission Detectors (MPFD) for In-Core Neutron Flux Monitoring," *Nucl. Inst. and Meth.,* no. A554, pp. 494 - 499, 2005.

[7] M. A. Reichenberger, T. D. F. George, R. G. Fronk, P. B. Ugorowski, J. A. Geuther, J. A. Roberts, T. Ito, H. B. Vo-Le, S. R. Stevenson, D. M. Nichols, D. S. McGregor, "Advances in the Development and Testing of Micro-Pocket Fission Detectors (MPFDs)," in *International Conference on Research Reactors: Safe Management and Effective Utilization*, Vienna, Austria, Nov. 16-20, 2015.

[8] M. Reichenberger, T. Ito, P. B. Ugorowski, B. W. Montag, S. R. Stevenson, D. M. Nichols and D. S. McGregor, "Electrodeposition of Uranium and Thorium onto Small Platinum Electrodes," *Nucl. Instrum. and Meth.,* no. A812, pp. 12-16, 2016.

[9] N. Tsoulfanidis, "Measurement and Detection of Radiation," Taylor and Francis, Washington, 1995.

EPJ Web of Conferences 170, 04019 (2018)
ANIMMA 2017 https://doi.org/10.1051/epjconf/201817004019

Review of Nuclear Heating Measurement by Calorimetry in France and USA

C. Reynard-Carette, G. Kohse, J. Brun, M. Carette, A. Volte, A. Lyoussi

Abstract—**This paper gives a short review of sensors dedicated to measuring nuclear heating rate inside fission reactors in France and USA and especially inside Material Testing Reactors. These sensors correspond to heat flow calorimeters composed of a single calorimetric cell or of two calorimetric cells at least with a reference cell to obtain a differential calorimeter. The aim of this paper is to present the common running principle of these sensors and their own special characteristics through their design, calibration methods, and in-pile measurement techniques, and to describe multi-sensor probes including calorimeters.**

Index Terms— **Calorimetry, Nuclear Heating, Research Reactors, Fission**

I. Introduction

THE nuclear energy deposition rate in materials, often called nuclear heating rate or nuclear absorbed dose rate, corresponds to a relevant thermal parameter for fission and fusion reactors. It corresponds to the rate of energy deposited by various interactions between nuclear rays and matter in a mixed (n,γ) field. This parameter is required to choose and design accurately thermal and mechanical systems, fluid loops (with or without phase change), instrumentation and probes, but also to interpret in-pile experiments after irradiation and to control ageing of structural materials. Appropriate measurement techniques for this parameter depend on the level and range expected in reactor.

For Zero Power Reactors (ZPRs), nuclear heating quantification can be performed, for instance, by means of OSLDs (Optically Stimulated Luminescent Detectors) and TLDs (Thermo-Luminescent Detectors) [1]. Quantification with these two kinds of sensor is based on three successive steps: out-of-pile calibration, irradiation, and finally post-irradiation measurements. Consequently, these sensor types lead to an integrated result for the irradiation duration (nuclear adsorbed dose).

For Material Testing Reactors (MTRs) or other reactors such as Boiling Water Reactors (BWRs), nuclear heating rate quantification is possible thanks to other techniques, more suited to the higher level of nuclear heating occurring inside their experimental channels, and called in-pile calorimeters [2-23]. This paper focuses mainly on a review of in-pile calorimeters used in French and American reactors. First of all, the operating principle of the calorimetry technique is given. Then characteristics of each sensor are detailed (design, associated instrumentation, calibration, operating protocols and measurement methods). Then some examples of probes including calorimeters are presented.

II. In-Pile Calorimetry Principle

Calorimetry techniques are used to measure energy/heat in several fields such as material sciences, chemistry, thermodynamics but in nuclear fission field for MTRs a specific running principle is applied due to the high energy deposition rate occurring (up to 20 kW/kg in the case of the future MTR under construction in France: Jules Horowitz Reactor). The MTR calorimeters corresponds to non-adiabatic calorimeters (called heat flow calorimeters) which exchange energy continuously with their external surroundings (fluid flow in experimental channels, loops) instead of conventional adiabatic calorimeters.

One major advantage of this calorimeter principle is that it can provide online data while ensuring in-pile calorimeter structure and running integrity.

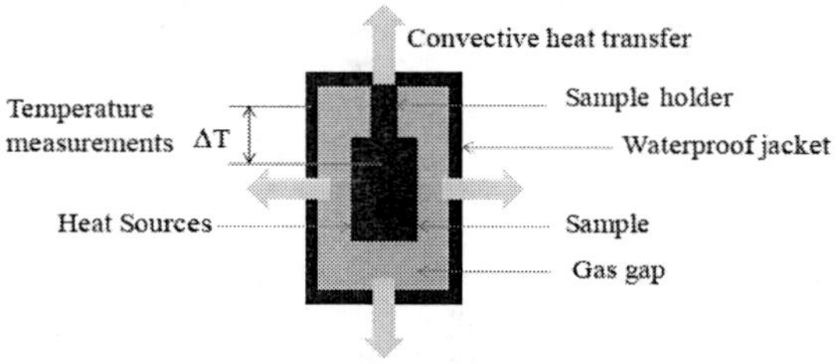

Fig. 1. General scheme of an in-pile calorimeter

A general scheme of an in-pile calorimeter is presented in Figure 1. Quantification of nuclear heating rate in a material sample by means of such sensors is based on measurements of steady temperatures or temperature differences in specific locations inside the sensor body. The specific locations are chosen in order to obtain suitable response and sensitivity, which are controlled by thermal resistances depending on the sensor design, its geometry, its materials and its gas gaps.

The sensor response can be estimated numerically or analytically by solving the heat equation by considering source terms equal to nuclear heating rates and specific boundary conditions corresponding to parietal heat exchanges (empirical correlations to impose a global heat exchange coefficient, for instance). The sensor response can be defined experimentally thanks to out-of-pile or in-pile calibrations. Two methods of calibration can be used, depending on the sensor type.

C. Reynard-Carette, J. Brun, M. Carette and A. Volte are with Aix Marseille Univ, Université de Toulon, CNRS, IM2NP, Marseille, France (e-mail: adrien.volte@im2np.fr, christelle.carette@univ-amu.fr, julie.brun@univ-amu.fr, michel.carette@univ-amu.fr). G. Kohse is with Nuclear Reactor Laboratory, Massachusetts Institute of Technology, Cambridge, USA (kohse@mit.edu). A. Lyoussi is with CEA, DEN, DER, Instrumentation Sensors and Dosimetry Laboratory, Cadarache, F-13108, Saint Paul-lez-Durance, France (e-mail: abdallah..lyoussi@cea.fr,)

III. Calorimeter Types

Different types of calorimeter are used inside fission reactors. Non-adiabatic calorimeter types can be defined by their assembly. The first category consists of single-cell calorimeters and can be found in Europe in MTRs [2-9] and in USA in BWRs [10-15]. The second class is differential calorimeters and corresponds to French sensors used in MTRs [2, 5, 16-23]. The European calorimeters are used to determine axial distributions of the nuclear heating rate in experimental channels whereas USA calorimeters are used to calibrate local power range detectors dedicated to monitoring the thermal neutron flux during BWR operation.
Schemes of several calorimeters are given in Figure 2.

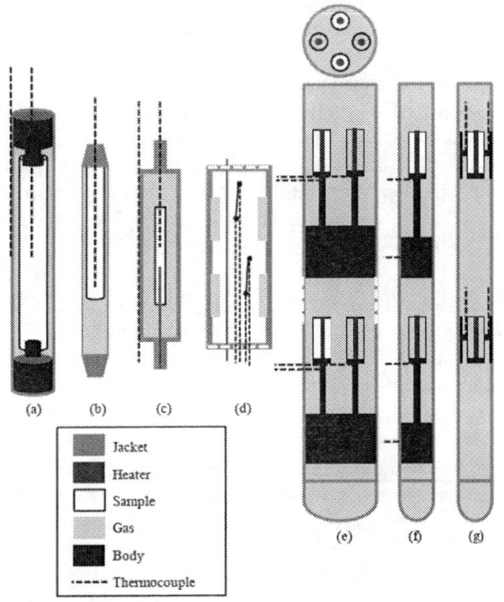

Fig. 2. Schemes of single-cell calorimeters (from (a) to (d)) and differential calorimeters (from (e) to (g)).

The first category (from (a) to (d)) contains a single calorimetric cell which is in general composed of a sample (white zone), a gas gap (grey zone), a jacket (dark grey zone), a sample holder (black zone), thermocouples (black dotted line) and sometimes a heater (red zone). The nature of the gas and the sample depends on the calorimeter. The Polish calorimeter called KAROLINA (a) has a graphite sample and is filled with helium [2-4, 21]. The gamma thermometer (b) [3-6] contains a stainless steel sample and can be filled with helium or argon according to the targeted nuclear heating level. The number of temperature sensors depends on the calorimeter too. For the gamma thermometer (b) and the USA calorimeter [10-13] (d), only a temperature difference is measured between a hot junction and a cold junction, whereas for the other single-cell calorimeters two absolute temperatures are measured by means of two thermocouples located inside the sample and on the jacket surface, respectively, and then a temperature difference is deduced. The drawing (d) represents two superposed USA calorimeters. Several USA gamma thermometers are superposed to realize fixed in-core measurements at several heights (up to 9 [10]).

The second category (from (e) to (g)) has several calorimetric cells including two kinds of cell [16-23]. This category corresponds to the only current French technique. In the past, single-cell calorimeters having the design (c) were used [7].
The first kind of calorimetric cell called sample cell or measurement cell is similar to a single cell. It contains a material sample. The second cell is used as a reference cell. Its design is the same as the sample cell but it does not contain a material sample. It is used to compensate for the nuclear heating deposition on the calorimetric cell parts distinct from the sample. Consequently, in that case the sensor is called a differential calorimeter because the nuclear heating rate inside the material sample is determined by a heat balance between the sample cell and the reference cell (heat difference thanks to a difference of temperature differences). Three differential calorimeters are presented in Figure 2. The material sample of all these sensors is made of graphite and their gas gap is filled with nitrogen. The specific characteristics of these sensors are induced by the cell number, the sensor-body material and the cell geometry.
The sensor (e) is a five-stage calorimeter [19]. Each stage corresponds to four calorimetric cells (two sample cells + two reference cells). Each cell head is instrumented by a heater and a thermocouple and is linked to a base by a cylindrical rod to produce mainly axial heat evacuation to the external surroundings. This fixed calorimeter leads to the measurement of an axial profile of the nuclear heating rate in five positions without moving the sensor and by taking account of radial flux gradients.
The sensor (f) has two superposed calorimetric cells separated by polished screens (a sample cell in the upper part and a reference cell in the lower part) [2, 5, 16-19, 22]. The cell geometry is quite similar to those of the five stage-calorimeter (head + rod + base). This calorimeter type was integrated inside different multi-sensor probes (cf. Section VI). The external diameter of this sensor is smaller than the previous one (17 mm against 32 mm) but due to the superposition of the two cells the sensor has to be moved to quantify the nuclear heating rate at a given axial position inside the experimental channel in order to avoid the influence of the axial flux gradients. For this calorimeter each cell contains a heater and two thermocouples (located in the upper part of the rod and on the external surface of the base respectively). The nature of the material for its cell structure is aluminum as for that of each five-stage calorimeter cell. However, the type of aluminum depends on the targeted nuclear heating rate (higher thermal conductivity for in-core measurements).
The last differential calorimeter called CALORRE (g) [22, 23], the newest sensor, has two superposed calorimetric cells with a new geometry. The vertical rod is replaced by a horizontal fin linked to a vertical annular fin in contact with the internal surface of the jacket. This geometry leads to a more compact calorimetric cell with radial heat transfer to evacuate the energy instead of axial heat flow. The length of CALORRE cell is reduced to the length of the head of CARMEN cell (23.1 mm against 79 mm).
Each cell contains a heater and two thermocouples located on the horizontal fin (one close to the head surface and one close to the vertical fin surface). The material structure of each cell is stainless steel instead of aluminum in order to increase the

thermal resistance and the cell sensitivity due to limitations of the dimensions of the horizontal fin.

IV. Calorimeter Calibration

The calibration is a crucial step which depends on the calorimeter type. In the case of a calorimeter with no heater such as KAROLINA calorimeter and gamma thermometers, the calibration method is based on a transient thermal operating protocol in three phases. During the first phase, the calorimeter is put inside a furnace to reach a steady hot temperature (cf. Fig. 3). Then during the second phase, the sensor is removed from the furnace to be inserted inside another set-up providing a colder surrounding (such as a fluid flow loop with a lower temperature) [2-4].

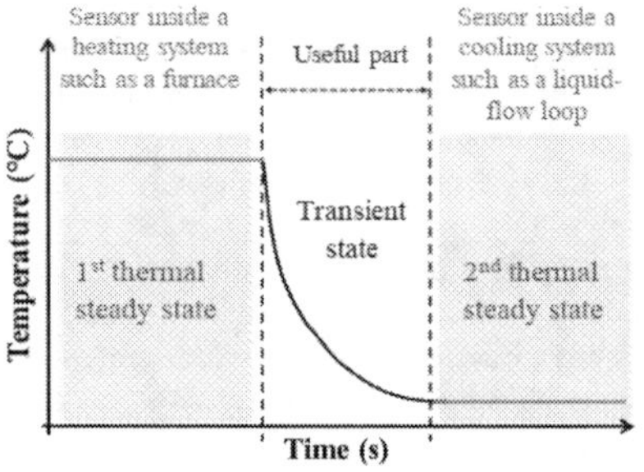

Fig. 3. Scheme of the calibration operating protocol of the calorimeters without heater.

During the last phase, the sensor remains inside the loop to reach a steady cold temperature. Thanks to this calibration operating protocol, the cooling time constant of the sensor can be determined and used to deduce the sensor sensitivity (ε in °C.g/W) by considering the thermal heat capacity of the material sample and its mass [4-5, 10]. This type of transient calibration can be realized inside the reactor by withdrawing the sensor from the experimental channel.

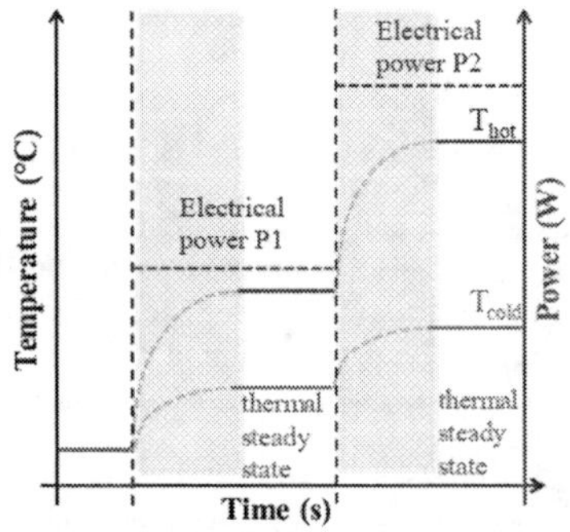

Fig. 4. Scheme of the calibration operating protocol of a calorimetric cell with a heater.

In the case of a calorimeter integrating heaters such as differential calorimeters, the calibration method is based on the analysis of successive steady thermal phases (cf. Fig. 4). In that case, different nuclear heating rates are experimentally simulated inside a localized zone of the sensor (in general inside the head center and more recently inside the rod and the base too [21]) by generating different consecutive values of a heat source by means of electrical current injections (increment by increment after reaching a steady thermal state) [2, 20-23]. Thanks to this operating protocol applied to each calorimetric cell, a curve representing the difference of the average steady temperatures versus the injected electrical power is plotted. This curve corresponds to the calibration curve of each calorimetric cell. The calorimetric cell sensitivity in °C/W can be deduced. This calibration procedure is applied under non irradiation conditions in laboratory in general but it can be used inside reactors.

V. In-Pile Measurements

During the irradiation campaigns, the nuclear heating rate can be determined by measuring steady temperature difference(s) for the two categories of calorimeters. However, the differential calorimeters allow the application of three measurement methods due to its several calorimetric cells and its heaters whereas a single-cell calorimeter is associated to one method. In the case of a single-cell calorimeter without heater, the determination of the nuclear heating value at an axial position depends on the sensor sensitivity obtained during the calibration step and on the steady temperature difference measured at this position. The nuclear heating value is defined by the ratio between ΔT and ε [10].

In the case of differential calorimeters, three in-pile measurement methods can be applied [5, 16-17] (cf. Fig. 5). For the first one the heaters inside the cells are not used. The steady temperature difference of each cell is measured consecutively when the two cells are at the same axial position. For the two other methods, an electrical current is applied inside one heater (located inside the sample cell for the addition method or located inside the reference cell for the zero method). These two methods are more time consuming and have range limitations induced by electrical current limitations.

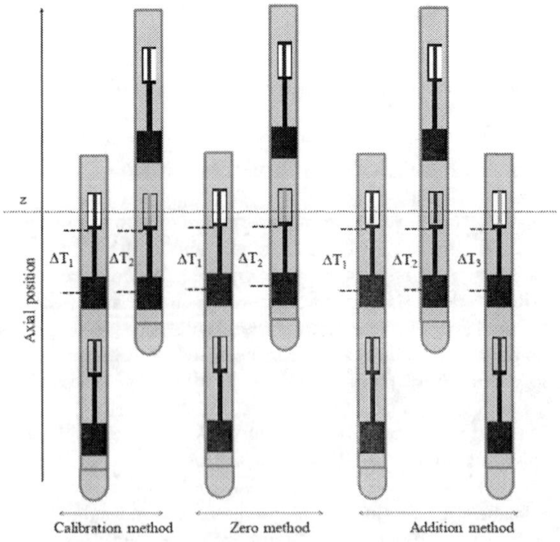

Fig. 5. Scheme of the three in-pile measurement methods for a differential calorimeter composed of two superposed calorimetric cells (passive heater in red, active heater in orange).

For the zero method, an electrical current is applied inside the reference cell heater to obtain the same steady temperature difference for the two cells at the same axial position.
For the addition method, three steps are required. The first and the second steps are identical to the calibration method steps. During the third step, the sample cell is moved to its initial position and an electrical current is applied inside its heater in order to increase its steady temperature difference.
For the three methods, the nuclear heating value at a specific axial location is deduced by considering the calibration curve coefficients of the two cells, the steady temperature difference measurements of the two cells at this axial position during the different steps and a heat balance associated with the calorimetric cells (by taking into account of the nuclear heating and the additional energy induced by the heaters in the case of the zero and addition methods) [2-5].

VI. Examples of Probes Integrating Calorimeters

Current in-pile instrumentation generally consists of multi-sensor probes where calorimeters are associated with other sensors. Three probes are described in Figure 6.

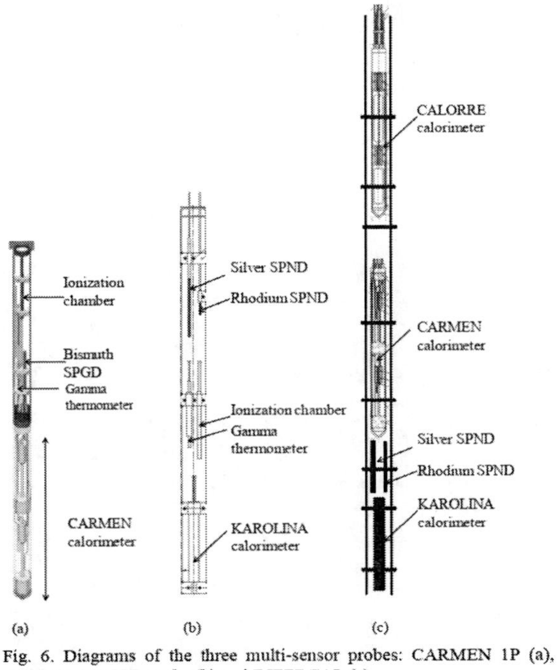

Fig. 6. Diagrams of the three multi-sensor probes: CARMEN 1P (a), GAMMA MAJOR probe (b) and INTERCAL (c).

The first probe (a), called CARMEN 1P and developed in the framework of the IN-CORE program, hosts different sensors dedicated to photon measurements and nuclear heating: a differential calorimeter, a gamma thermometer, an ionization chamber and a SPGD (bismuth) [2, 5]. The differential calorimeter used has the same global design as CALMOS type calorimeter but different size characteristics of the cell rod and base in order to have a more sensitive sensor for OSIRIS reactor periphery quantifications (up to 2 W/g). In fact, the probe called CALMOS, composed of a differential calorimeter, a SPND (rhodium) and a mechanical displacement system was designed to perform in-core measurement in OSIRIS (up to 13W/g) [16-18]. The main aim of the experiments realized with CARMEN 1P inside several experimental channels of OSIRIS periphery was to propose an advanced radiation measurement methodology by coupling this probe with a second probe called CARMEN 1N focusing on neutron detections (probe dedicated to neutron measurements including two fission chambers (242Pu and 235U), an ionization chamber and a SPND (rhodium)).
The second probe (b) designed in the framework of the GAMMA MAJOR program focuses on neutron and photon measurements by means of KAROLINA calorimeter, a gamma thermometer, an ionization chamber and two SPNDs (silver and rhodium) [2-4]. The major aim of the experiments carried out inside different experimental channels of MARIA reactor was to test the Polish calorimeter and to use experimental results to validate new numerical works.
The last probe (c), called INTERCAL and developed in the framework of the IN-CORE program [22], contains two SPNDs (silver and rhodium) and three types of calorimeters: KAROLINA calorimeter, CARMEN type calorimeter (with a rod dimension change and thus new calibration coefficients) and CALORRE calorimeter. The main aim of the experiments carried out inside one experimental channel of MARIA reactor was to study the CALORRE calorimeter for the first time under real irradiation conditions, to validate it and to compare its response with the two conventional calorimeters.

VII. Conclusions

Calorimetric measurements are crucial in fission reactors to quantify the nuclear heating rate. This relevant parameter is required to design loops, devices, irradiation experiments and conditions but it can participate to monitor other parameters too. Nuclear heating rate measurements can be realized by means of calorimeters such as single-cell calorimeters and differential calorimeters. According to their design and their assembly into a probe, these kinds of sensors can be used as a fixed technique (USA gamma thermometer, French five-stage calorimeter) or as a mobile technique (French CALMOS, CARMEN and INTERCAL probes for instance). The present tendency is the development of multi-sensor probes allowing the measurements of several parameters simultaneously and hosting calorimeters.
The choice of the calorimeter can be done by considering several criteria such as the nuclear heating rate level, the sensor size and sensitivity, the calibration operating protocol, the in-pile measurement methods, and the sensor displacement, among other factors.

REFERENCES

[1] H. Amharrak *et al.*, State of the art on nuclear heating in a mixed (n/γ) field in research reactors, Nuclear Instruments and Methods in Physics Research A, 2014, Volume 749, (2014) p. 57-67.
[2] J. Brun *et al.*, Responses of Single-Cell and Differential Calorimeters: from Out-of-Pile Calibration to Irradiation Campaigns, IEEE TNS, Vol. 63, NO. 3, June 2016, p. 1630-1639.
[3] M. Tarchalski *et al.*, Comparison Of Calculation Scheme With Multisensor Prompt And Delayed Nuclear Heating Measurements In The Maria Reactor, ENC 2016, 9-13 October 2016, Warsaw, Poland.
[4] M. Tarchalski, Nuclear heating measurements in the MARIA reactor and implementation of neutron and photon calculation scheme, thesis 2016.

[5] D. Fourmentel *et al.*, Nuclear Heating Measurements in Material Testing Reactor: a Comparison Between a Differential Calorimeter And a Gamma Thermometer, IEEE TNS, Vol. 60, Issue:1, Part:2, 2013, p. 328 – 335.
[6] B. Yu. Volkov, Qualification of Power Determination for F&M Experiments in Research Reactors, RRFM 2016, 13-17 March 2016, Berlin, Deutschland.
[7] A.W. Boyd, International Intercomparison of Calorimeters, A report of measurements made in the reactor Melusine at CEN Grenoble, March 1970.
[8] M. Varvayanni *et al.*, A point kernel model for the energy deposited on samples from gamma radiation in a research reactor core, Annals of Nuclear Energy 35 (2008) p. 2351–2356.
[9] Wang Qiu-kuan *et al.*, Advanced Measuring (Instrumentation)Methods for Nuclear Installations: A Review, Science and Technology of Nuclear Installations, Volume 2012, Article ID 672876, 9 pages.
[10] Don W. Miller et al., A review on gamma thermometer applications in nuclear reactors, The 12th International Topical Meeting on Nuclear Reactor Thermal Hydraulics (NURETH-12), Pittsburgh, Pennsylvania, USA, September 30-October 4, 2007.
[11] GE-Hitachi Nuclear Americas, LPRM Calibration oand Power Shape Monitoring, report NEDO-33197-A, Revision 3, Class I, October 2010.
[12] R. Raghavan et al., Application of the gamma thermometer as BWR fixed in-core calibration system, Proceedings of NEA Specialists' Meeting on In-core Instrument and Core Assessment, Mito-shi, Japan, October 14–17 (1996).
[13] Y. Yamauchi et al., Development of improved core monitoring system, Journal of Power and Energy Systems, Vol. 2, No. 2, 2008, pp.712-719.
[14] J. Rempe *et al.*, Status Report on Efforts to Enhance Instrumentation to Support Advanced Test Reactor Irradiations, INL Report, December 2012.
[15] J. P. Waring *et al.*, Recent Reactor Testing and Experience with Gamma Thermometers, IEEE TNS, Vol. 30, Issue: 1, Feb. 1983, p. 791 – 794.
[16] H. Carcreff *et al.*, Development, Calibration and Experimental Results Obtained with an Innovative Calorimeter (CALMOS) for Nuclear Heating Measurements, IEEE TNS, Vol. 59, N°4, p. 1369-1376, 2012.
[17] H. Carcreff *et al.*, "First In-Core Measurement Results Obtained with the Innovative Mobile Calorimeter CALMOS inside the OSIRIS Material Testing Reactor", IEEE TNS, Vol. 61, N°4, pp. 1951-1962, 2014.
[18] H. Carcreff *et al.*, First In-Core Simultaneous Measurements of Nuclear Heating and Thermal Neutron Flux Obtained With the Innovative Mobile Calorimeter CALMOS Inside the OSIRIS Reactor, IEEE Transactions on Nuclear Science, vol. 63, issue 5, pp. 2662-2670, 2016.
[19] A. Péron *et al.*, Simulation and Comparison of the Calorimeters Measuring the Nuclear Heating in the OSIRIS Reactor, with the TRIPOLI-4 Monte-Carlo Code, IEEE Transactions on Nuclear Science, Volume: 62, Issue: 3, June 2015, pp. 1218 – 1225.
[20] J. Brun *et al.*, Comparison of the thermal calibration response of two calorimetric cells dedicated to nuclear heating measurements IEEE TNS, Vol. 61, Issue: 4, Aug. 2014, p. 2254-2261, 2014.
[21] C. De Vita *et al.*, Study of the Influence of Heat Sources on the Out-of-Pile Calibration Curve of Calorimetric Cells Used for Nuclear Energy Deposition Quantification, IEEE TNS, Vol.63, Issue: 4, Aug. 2016, p. 2323 – 2330.
[22] J. Brun, *et al.*, Irradiation campaign in MARIA reactor of an innovative calorimeter. Comparison of its nuclear heating measurements with two kinds of classical sensors. ANIMMA « Advancements for Nuclear Instrumentation and Measurement Methods and their Applications », 20 - 23 June 2017, Liège, Belgium.
[23] A. Volte *et al.*, Study of the Flow Temperature and Ring Design Influence on the Response of a new Reduced-Size Calorimetric cell for Nuclear Heating Quantification. ANIMMA « Advancements for Nuclear Instrumentation and Measurement Methods and their Applications » 20 - 23 June 2017, Liège, Belgium.

On Use of Multi-Chambered Fission Detectors for In-Core, Neutron Spectroscopy

Jeremy A. Roberts

Abstract—**Presented is a short, computational study on the potential use of multichambered fission detectors for in-core, neutron spectroscopy. Motivated by the development of very small fission chambers at CEA in France and at Kansas State University in the U.S., it was assumed in this preliminary analysis that devices can be made small enough to avoid flux perturbations and that uncertainties related to measurements can be ignored. It was hypothesized that a sufficient number of chambers with unique reactants can act as a real-time, foil-activation experiment. An unfolding scheme based on maximizing (Shannon) entropy was used to produce a flux spectrum from detector signals that requires no prior information. To test the method, integral, detector responses were generated for single-isotope detectors of various Th, U, Np, Pu, Am, and Cs isotopes using a simplified, pressurized-water reactor spectrum and flux-weighted, microscopic, fission cross sections, in the WIMS-69 multigroup format. An unfolded spectrum was found from subsets of these responses that had a maximum entropy while reproducing the responses considered and summing to one (that is, they were normalized). Several nuclide subsets were studied, and, as expected, the results indicate inclusion of more nuclides leads to better spectra but with diminishing improvements, with the best-case spectrum having an average, relative, group-wise error of approximately 51%. Furthermore, spectra found from minimum-norm and Tihkonov-regularization inversion were of lower quality than the maximum entropy solutions. Finally, the addition of thermal-neutron filters (here, Cd and Gd) provided substantial improvement over unshielded responses alone. The results, as a whole, suggest that in-core, neutron spectroscopy is at least marginally feasible.**

Index Terms—**neutron spectroscopy, fission chambers, maximum entropy**

I. Introduction

IN recent years, substantial interest has grown in the development of in-core instrumentation that can provide data for validation of advanced computational models for reactor analysis. Work at Commissariat à l'énergie atomique et aux énergies alternatives (CEA) [1] to produce "sub-miniature" fission chambers and at Kansas State University (KSU) [2], [3] to produce "micro-pocket" fission chambers has led to extremely small devices that provide opportunities for multiple, in-core flux monitors that can provide a detailed map of the core power over varying operational conditions. Some work has been done to determine continuous flux shapes from the point-wise measurements [4], and ongoing research aims to provide devices capable of exhibiting linear response across many decades of power levels, e.g., as observed in pulsed reactors like the Transient Reactor Test Facility (TREAT) at Idaho National Laboratory and various university reactors, including the Training, Research, Isotopes, General Atomic (TRIGA) Mark II research reactor at KSU.

With their small size, it may be possible to use several, co-located sub-miniature or micro-pocket fission detectors to measure a neutron spectrum at a point. Even in the earliest developments at KSU, it was proposed to deploy two-chamber devices loaded with uranium and thorium, respectively, in order to distinguish between thermal and fast fluxes [2]. However, the threshold energy for thorium fission is approximately 1 MeV, too large to provide sufficient detail in the epithermal region. Here, alternative reactants, in greater number, were explored to understand how the resulting signals from multiple chambers could be used to resolve the neutron spectrum. In particular, it was hypothesized that a sufficient number of chambers with unique reactants can act as a real-time, foil-activation experiment and that, with appropriate prior modeling and uncertainties, a Bayesian framework can be used to provide improved estimates for the energy spectrum. In this preliminary effort, no such prior information was assumed, leaving the full Bayesian framework for future work. Rather, the focus was placed squarely on understanding how much information can be extracted from an unknown spectrum via a set of integral responses. The key throughout was the presumed availability of devices small enough that several can be placed in a small region and that any perturbations to the local flux (magnitude or spectrum) are negligible.

II. Methods

Shown in Fig. 1 are the microscopic fission cross sections for several nuclides of Th, U, Np, Pu, Am, and Cm. The values shown are evaluated at 300 K. The same cross sections are shown in the 69-group WIMS [6] format condensed using the simplified, pressurized-water reactor (PWR) spectrum shown in Fig. 3. Observe that the spectrum used has no fine structure and, hence, represents a very simple test case for unfolding. On a related note, the geometry of any detector of interest will be sufficiently small (by assumption) that the infinite-dilution approximation is valid. In other words, self-shielding effects due to the detectors themselves are negligible and need not be accounted for in determination of the multigroup constants and, hence, the detector response functions.

Suppose an N-chamber device were produced having a sufficient mass of each reactant shown in Fig. 2 to produce a measurable, negligible-uncertainty signal in some application

This work was presented at the International Conference on Advancements in Nuclear Instrumentation Measurement Methods and their Applications, June 19-23, 2017.

J. Roberts is with the Department of Mechanical and Nuclear Engineering, Kansas State University, Manhattan, KS, 66502 (email: jaroberts@ksu.edu).

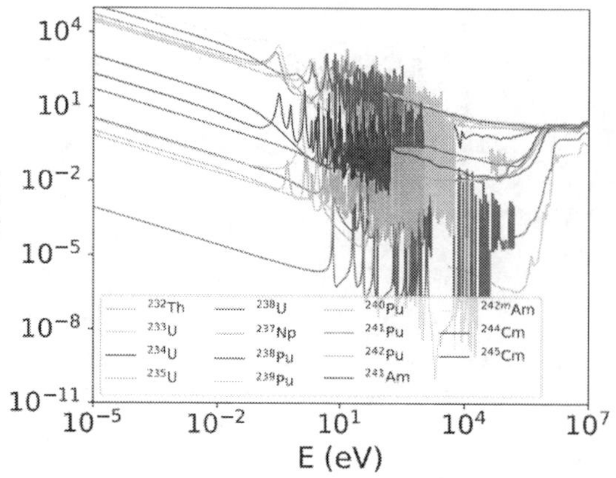

Fig. 1. Microscopic fission cross sections for several nuclides. All data was pulled directly from NNDC [5].

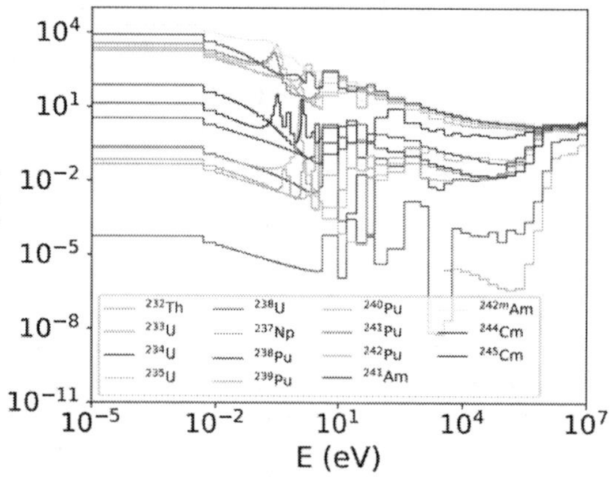

Fig. 2. Condensed microscopic, fission cross sections in the WIMS 69-group format.

reactor. Suppose further that the goal is to determine the flux spectrum of the application reactor using the 69-group WIMS energy structure. The resulting system of equations is

$$r_i = \sum_{g=1}^{69} \Sigma_{fg,i}\phi_g\,, \qquad i = 1 \ldots N\,, \tag{1}$$

or, in matrix form,

$$\mathbf{r} = \mathbf{\Sigma}_f \boldsymbol{\phi}\,, \tag{2}$$

where ϕ_g is the group-integrated flux, and it has been assumed that there are no losses due to dead time or discriminators. This is a major assumption with respect to the technology, but if the response of each device is linear, and if any losses are proportional to the signal similarly for all reactants, then the equations need only be modified by a single, multiplicative constant on the right. For real-world cases, further work must be done to include such features of the system explicitly.

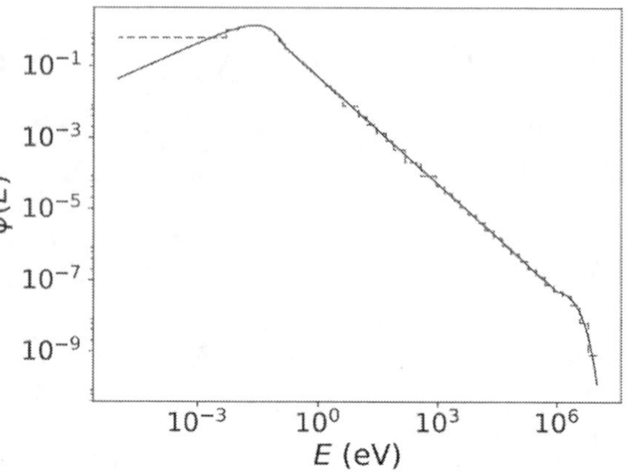

Fig. 3. Simplified, PWR-like spectrum used to weight cross sections as response functions; the multigroup equivalent is also shown.

Equation (1) is an underdetermined system, meaning that specialized techniques or additional constraints are needed. Such a problem belongs to the class of spectrum unfolding problems, itself a class of inverse problems. Many methods have been proposed to solve such problems, whether arising from Bonner-sphere spectroscopy or other applications. One simple inversion technique is to compute the minimum-norm solution, i.e.,

$$\boldsymbol{\phi} = \mathbf{\Sigma}[\mathbf{\Sigma}\mathbf{\Sigma}^T]^{-1}\mathbf{r}\,, \tag{3}$$

which yields a spectrum $\boldsymbol{\phi}$ having the smallest Euclidean norm $|\boldsymbol{\phi}|_2 = \sqrt{\boldsymbol{\phi}^T\boldsymbol{\phi}}$. Although unique, the minimum-norm solution is not guaranteed to be strictly positive.

A second, straightforward inversion technique is Tihkonov regularization [7], i.e.,

$$\boldsymbol{\phi} = [\mathbf{\Sigma}^T\mathbf{\Sigma} + \mathbf{T}^T\mathbf{T}]^{-1}\mathbf{\Sigma}\mathbf{r}\,, \tag{4}$$

where a common regularization matrix is $\mathbf{T} = a\mathbf{I}$ for a scalar a and identity matrix $\mathbf{I}$. Like the minimum-norm solution, the Tikhonov-regularization solution may also contain negative values.

For the work presented here, the primary method used is based on entropy maximization [8], in which the (Shannon) entropy of the spectrum is defined as

$$S = -\sum_{g=1}^{69} \phi_g \ln \phi_g\,. \tag{5}$$

The flux spectrum with the greatest entropy that satisfies both Eq. (1) and

$$\sum_{g=1}^{69} \phi_g = 1 \tag{6}$$

is found by solving the nonlinear, constrained optimization problem resulting from (1)–(6). Equation 6 implies that the integrated flux is known, which may not be the case in practice. However, the results below indicate that the unfolding process is not severely sensitive to the assumed, integral flux, meaning that even a crude estimate can be used. If the constraint is not employed, the results are much less favorable.

III. Results

To demonstrate the use of entropy maximization for extracting spectral information from a set of unique, fission-chamber responses, a simple test problem was studied. In particular, the PWR spectrum shown in Fig. 3 was taken to be the application spectrum, with which reference response functions (here, just the microscopic, multigroup, fission cross sections) were defined. Then, spectra were unfolded for four base cases: (1) using only ^{235}U, ^{238}U, and ^{232}Th, (2) using the case-1 nuclides in addition to ^{237}Np and ^{238}Pu, (3) using the case-2 nuclides in addition to ^{239}Pu, ^{240}Pu, ^{241}Pu, and ^{242}Pu, and (4) using all nuclides shown in Fig. 2. Furthermore, case 3 was explored with and without the use of thermal-neutron filters (^{113}Cd and ^{155}Gd), while case 4 was analyzed using minimum-norm and Thikonov-regularization inversion in addition to maximum-entropy inversion.[1]

The spectra resulting from maximum-entropy unfolding are shown in Fig. 4 and Fig. 5, with the group-wise, relative error shown in Fig. 6. The four cases led to average, relative, group-wise errors of 201%, 126%, 63%, and 51%, respectively. The rather simple algorithm (with no prior information), is able to produce reasonably close spectral shapes with as few as 5 responses. It was observed in scoping studies that inclusion of ^{238}Pu has an important effect on the unfolding process. In particular, and as can be observed in Fig. 6, the error between groups 30 and 45 (or energies from 0.78 to 2.1 eV) is reduced substantially with inclusion of ^{238}Pu and its transmutation precursor, ^{237}Np. However, errors remain in this energy range and lower energies, which indicates that additional, thermal-sensitive information (e.g., from thermal filters like Cd) may be of use.

Recall that (6) implies that the total flux is known as part of the maximum entropy unfolding process. However, the total flux may not be known, or it may be known only imprecisely. To understand how the total flux constraint impacts the unfolded spectrum, case 4 was solved using three different total fluxes: 1.0, 0.5, and 2.0. For all three cases, a flux normalized to 1.0 was used to generate the responses. In other words, for a total flux of 0.5 and 2.0, (1) and (6) used *inconsistent* fluxes, as would be the case were the total flux not known. Shown in Fig. 7 are the group-wise, relative errors resulting from the study. On the average, if the total flux is doubled or halved (relative to the flux used for generating responses), the resulting average, group-wise error is less than a factor of three. In other words, the unfolding is not extremely sensitive to the total flux normalization.

The original scope of this study intended to focus on the simplest of maximum entropy techniques. However, a comparison to simple inversion techniques is warranted. Shown in Fig. 8 are the results for case 4 based on maximum entropy,

[1]Note, all results and figures were computed using Python (3.6.0) with NumPy (1.11.3), SciPy (0.18.1), and Matplotlib (2.0.0). All code to generate the figures and this document are available at https://github.com/robertsj/fission_chamber_spectrometer.

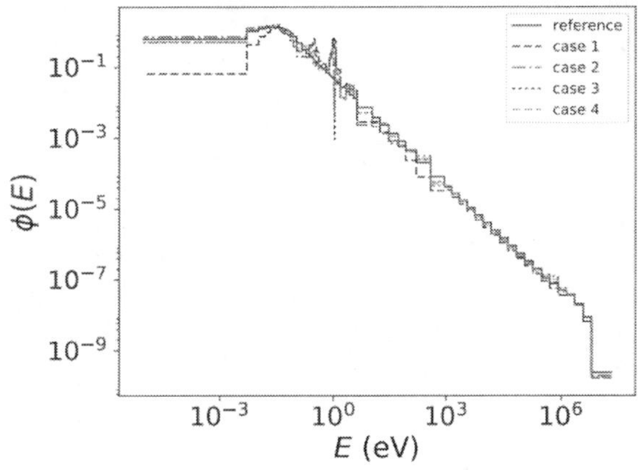

Fig. 4. Unfolded, multigroup spectra for cases 1–3 with reference spectrum for comparison.

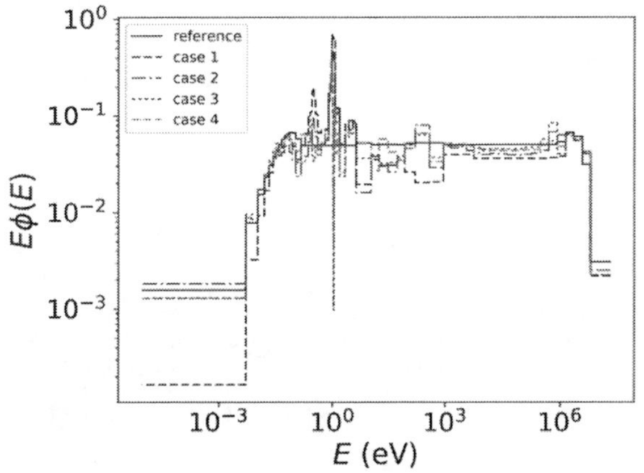

Fig. 5. Same as Fig. 4 but per unit lethargy.

minimum-norm, and Tikhonov-regularization. For the latter, two values of a were used: 0.1 and 0.01. Overall, the maximum-entropy spectrum is closer (qualitatively and quantitatively) to the reference spectrum. Both the minimum-norm and Tikhonov-regularization solutions contained negative values at energies near 1 eV. Finally, the Tikhonov-regularization solution appears to be rather insensitive to the value of a used in this particular application.

Inspired by other work presented at this conference [9], a study was performed to determine the utility of thermal-neutron filters for providing additional responses given a fixed number of fissile nuclides. In particular, case 3 was revisited with the addition of ^{113}Cd and ^{155}Gd. As a simple approximation, the attenuation of thermal neutrons due to 0.025 mm foils of each isotope was built directly into the fission response functions, the results of which are shown in Fig. 9 and Fig. 10. The filters were only applied to the nuclides with appreciable thermal

 https://doi.org/10.1051/epjconf/201817004020

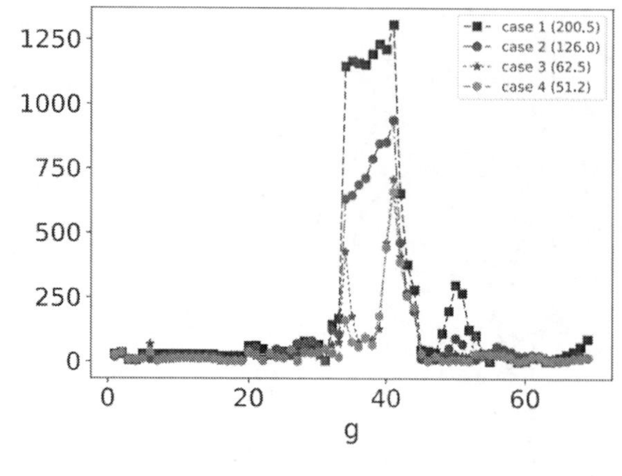

Fig. 6. Group-wise, relative errors (%). Average error shown in parentheses inside the legend.

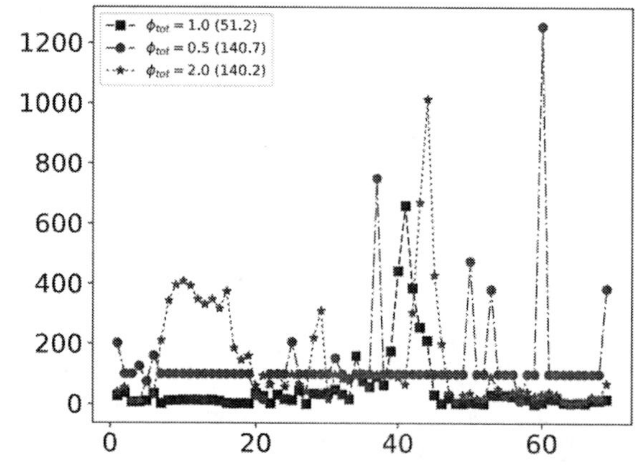

Fig. 7. Impact of total flux normalization on group-wise, relative errors (%) for case 4. Average errors shown in parentheses inside the legend.

sensitivity, i.e., ^{233}U, ^{235}U, ^{238}Pu, ^{239}Pu, and ^{241}Pu.

The results of the filter study are shown in Fig. 11, with the results near 1 eV magnified in Fig. 12. The results indicate that filtering improves the results, with use of Cd leading to a reduction from 63% to 39% in the average, group-wise, relative error. Use of Gd is less effective, reducing the error to 53%. A combination of both filters leads to nearly a reduction by two in the error. Note, although only the same number of fissile nuclides were used in this scenario, the addition of one filter doubles the number of unique responses available, while two filters triples the number of responses.

IV. Conclusion

The primary question addressed by this work has been whether or not a multichambered fission detector could be used for in-core spectroscopy. The basic conclusion to be

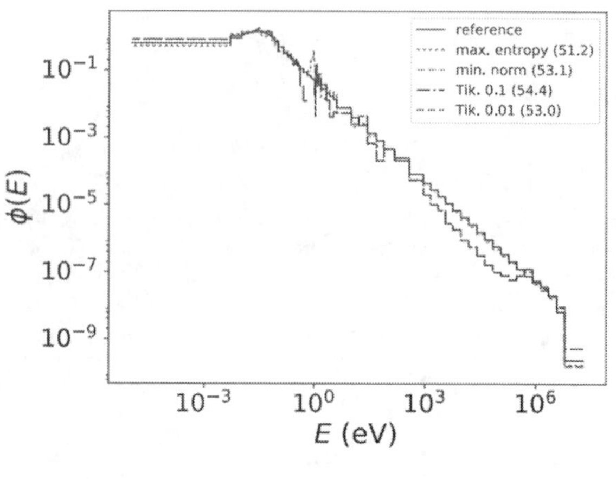

Fig. 8. Comparison of maximum-entropy, minimum-norm, and Tikhonov-regularization spectra for case 4. Average errors (%) shown in parentheses inside the legend.

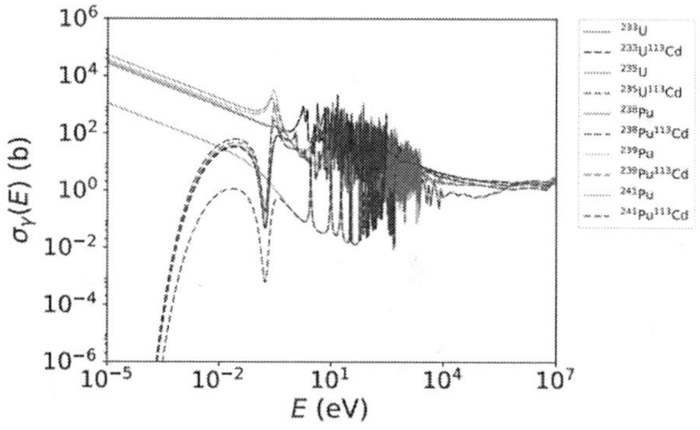

Fig. 9. Cd-filtered and unfiltered responses for thermal-sensitive nuclides of case 3.

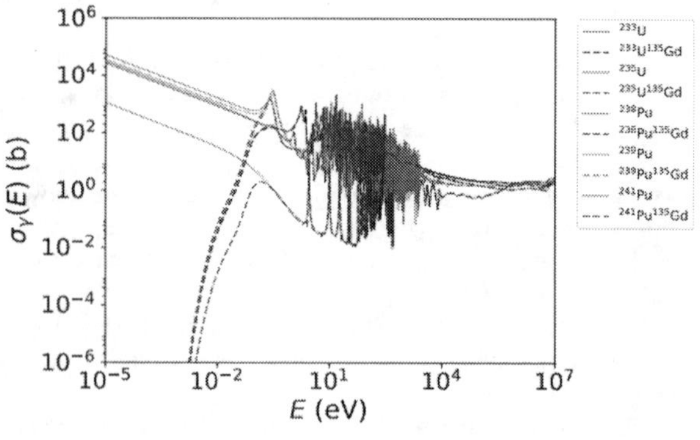

Fig. 10. Gd-filtered and unfiltered responses for thermal-sensitive nuclides of case 3.

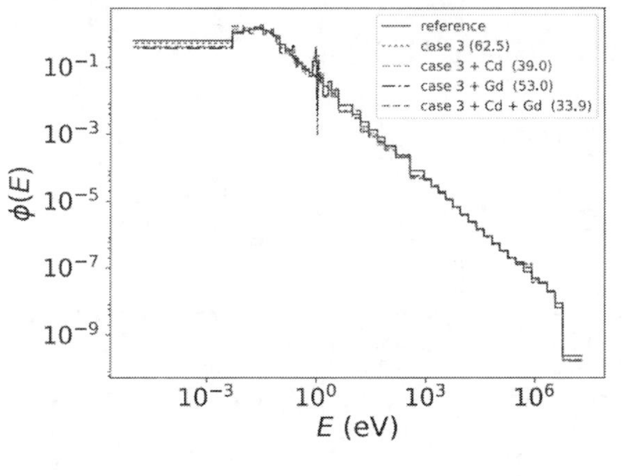

Fig. 11. Unfolded spectra for case 3 with and without filters. Average errors (%) shown in parentheses inside the legend.

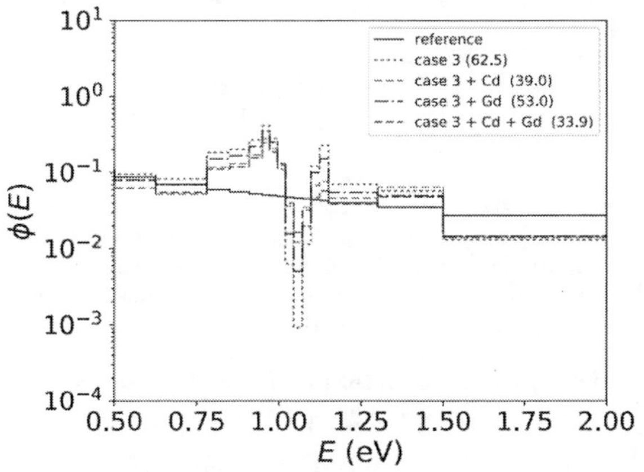

Fig. 12. Same as Fig. 11 but zoomed in on energies near 1 eV.

made is that such spectroscopy is possible, to an extent. Overall, the maximum entropy approach, when coupled to 5 or more responses (whether from unique nuclides or use of filters), can provide spectra with average, group-wise, relative errors of less than 100% assuming that the total flux is well known.

Clearly, this work is scoping in nature, and the algorithms used throughout are simple and incorporate no prior information. More complicated methods should be explored that admit uncertainties in the responses and use of prior spectral information, e.g., like those implemented in the MAXED unfolding code [10], [11]. In addition, more work should be performed to identify candidate filters to maximize the utility of small subsets of fissile (or fissionable) nuclides given that the availability of some nuclides included in this study may be limited.

Acknowledgment

The author would like to thank Prof. Amir Bahadori for some helpful discussions.

References

[1] C. Blandin, S. Breaud, L. Vermeeren, and M. Weber, "Development of New Sub-miniature Fission Chambers: Modelling and Experimental Tests," *Progress in Nuclear Energy*, vol. 43, no. 1, pp. 349–355, 2003.

[2] D. S. McGregor, M. F. Ohmes, R. E. Ortiz, A. S. Ahmed, and J. K. Shultis, "Micro-pocket Fission Detectors (MPFD) for In-Core Neutron Flux monitoring," *Nuclear Instruments and Methods in Physics Research Section A: Accelerators, Spectrometers, Detectors and Associated Equipment*, vol. 554, no. 1, pp. 494–499, 2005.

[3] M. A. Reichenberger, T. C. Unruh, P. B. Ugorowski, T. Ito, J. A. Roberts, S. R. Stevenson, D. M. Nichols, and D. S. McGregor, "Micro-Pocket Fission Detectors (MPFDs) for In-Vore Neutron Detection," *Annals of Nuclear Energy*, vol. 87, pp. 318–323, 2016.

[4] J. K. Shultis, "Determining Axial Fuel-Rod Power-Density Profiles from In-Vore Neutron Glux Measurements," *Nuclear Instruments and Methods in Physics Research Section A: Accelerators, Spectrometers, Detectors and Associated Equipment*, vol. 547, no. 2, pp. 663–678, 2005.

[5] "National Nuclear Data Center," https://www.nndc.bnl.gov/, accessed: 2017-07-03.

[6] "Methods of steady-state reactor analysis in nuclear design."

[7] J. L. Mueller and S. Siltanen, *Linear and Nonlinear Inverse Problems with Practical Applications*. SIAM, 2012.

[8] S. Itoh and T. Tsunoda, "Neutron Spectra Unfolding with Maximum Entropy and Maximum Likelihood," *Journal of Nuclear Science and Technology*, vol. 26, no. 9, pp. 833–843, 1989.

[9] T. Kaiba, V. Radulovic, D. Fourmentel, L. Barbot, C. Destouches, G. Zerovnik, and L. Snoj, "Calculations to Support to Online Neutron Spectrum Adjustment by Measurements with Miniature FissionChambers in the JSI TRIGA Reactor," in *Proc. of ANIMMA 2017*.

[10] M. Reginatto and P. Goldhagen, "MAXED, a Computer Code for Maximum Entropy Deconvolution of Multisphere Neutron Spectrometer Data." *Health Physics*, vol. 77, no. 5, pp. 579–583, 1999.

[11] M. Reginatto, P. Goldhagen, and S. Neumann, "Spectrum Unfolding, Sensitivity Analysis and Propagation of Uncertainties with the Maximum Entropy Deconvolution Code MAXED," *Nuclear Instruments and Methods in Physics Research Section A: Accelerators, Spectrometers, Detectors and Associated Equipment*, vol. 476, no. 1, pp. 242–246, 2002.

Performance assessment of imaging plates for the JHR transfer Neutron Imaging System

E. Simon, P. Guimbal

***Abstract*–The underwater Neutron Imaging System to be installed in the Jules Horowitz Reactor (JHR-NIS) is based on a transfer method using a neutron activated beta-emitter like Dysprosium. The information stored in the converter is to be offline transferred on a specific imaging system, still to be defined. Solutions are currently under investigation for the JHR-NIS in order to anticipate the disappearance of radiographic films commonly used in these applications. We report here the performance assessment of Computed Radiography imagers (Imaging Plates) performed at LLB/Orphée (CEA Saclay).**

Several imaging plate types are studied, in one hand in the configuration involving an intimate contact with an activated dysprosium foil converter: Fuji BAS-TR, Fuji UR-1 and Carestream Flex XL Blue imaging plates, and in the other hand by using a prototypal imaging plate doped with dysprosium and thus not needing any contact with a separate converter foil. The results for these imaging plates are compared with those obtained with gadolinium doped imaging plate used in direct neutron imaging (Fuji BAS-ND).

The detection performances of the different imagers are compared regarding resolution and noise.

The many advantages of using imaging plates over radiographic films (high sensitivity, linear response, high dynamic range) could palliate its lower intrinsic resolution.

***Index Terms* — Neutron Imaging, Material Testing Reactor, Indirect Neutron Imaging, Non-Destructive Examination, Jules Horowitz Reactor.**

I. Introduction

Among non-destructive examination methods commonly used in material testing reactors, neutron imaging is one of the oldest techniques. The neutron imaging system (NIS) of the Jules Horowitz Reactor (JHR) will follow the one that was used from the 1960s at the OSIRIS reactor, whose JHR will be the successor at Cadarache, France. The design of the neutron imaging bench is guided by the constraints of managing highly radiating objects. As the imaging system must be installed as close as possible to the irradiated fuel rods to be inspected, direct imaging is not an option. No direct imaging system which would be sensitive to thermal neutrons but insensitive to very high gamma-ray flux is applicable.

The underwater Neutron Imaging System to be installed in the Jules Horowitz Reactor (JHR-NIS) is based on an indirect method (transfer method) using a neutron activated beta-emitter like Dysprosium-165. The information stored in the converter foil has to be offline transferred by a specific imaging system to be defined.

Alternative solutions are currently under investigation for the JHR-NIS in order to anticipate the disappearance of historic transfer neutron radiography imagers used at MTR like specific radiographic films. The desired imaging device is the one that offers the best spatial resolution (image resolution performance), and the best detection efficiency (release time of the digital image). One of the objectives is also to improve the imaging process by working with linear response imagers and digital systems. We report here the performance assessment of Computed Radiography imagers (Imaging Plates) performed at LLB/Orphée for neutron imaging by transfer from a dysprosium neutron converter.

Several imaging plate types are studied, first in the configuration involving an intimate contact with the neutron converter: Fuji BAS-TR, Fuji UR-1 and Carestream Flex XL Blue imaging plates, second by using a prototypal imaging plate doped with dysprosium thus not needing any contact with a separate converter foil but having to be exposed to the neutron flux and cope with the associated issues. The results obtained for these imaging plates are compared with those obtained with a gadolinium doped imaging plate designed for direct neutron imaging (Fuji BAS-ND).

The detection performances of the different imagers are compared regarding resolution and noise.

II. Material and Methods

In order to work with a comparable neutron flux as the one expected in the future JHR device, the tests have been carried out at the neutron beam line G3.2 at LLB/Orphée research reactor.

A. Neutron Beamline

The characteristics of the G3.2 neutron beam line at LLB/Orphée are summarized in table I hereafter. The neutrons at G3.2 have an average energy of 3.5 meV (LH_2 source), colder than the expected thermal neutrons from JHR at the location of the imaging plane of NIS. The neutron capture cross section on ^{164}Dy is thus enhanced by a factor 2.45 regarding the configuration of future standard operation at JHR-NIS. The flux of 7.88 10^8 n/cm²/s is also higher than the one expected by a factor 15.8. The overall time scale factor to take into account in irradiation is then of the order of 38.7, leading to exposition times of dysprosium converter of a few dozen seconds to be representative of the JHR-NIS setup.

Manuscript received June 1st, 2017. E. Simon, is with CEA, DEN, Cadarache, Nuclear Measurement Laboratory, F-13108 Saint-Paul-lez-Durance, France, email : eric.simon@cea.fr

P. Guimbal is with CEA, DEN, Cadarache, Jules Horowitz Reactor Service, F-13108 Saint-Paul-lez-Durance, France, email : philippe.guimbal@cea.fr

TABLE I
G32 BEAMLINE CHARACTERISTICS

Quantity	value
Average Flux	7.88 10^8 n/cm²/s
Average energy	3.5 meV
Energy distribution	1.8 meV → 10 meV
Angular dispersion	0.4°
Area	25 × 50 mm²

B. Neutron converter

The neutron converter used is a metal dysprosium foil 100 µm thick. The foil is mounted on a 1 mm aluminum support to facilitate handling. The isotope of interest ^{164}Dy has a natural abundance of 28.18%. Neutron capture produces both ^{165}Dy and ^{165m}Dy with half-lives respectively 2.33h and 1.26 min. Only ^{165}Dy is used for imaging the activation distribution of the foil. A 15 minutes cooling is thus used after each irradiation phase to avoid interference from the metastable isotope. ^{165}Dy is a 100% beta emitter with two main beta levels : 1191.9 keV (15%) and 1286.6 keV (83%). The highest intensity emitted gamma-ray is at 94.7 keV (3.8%).
The intimate contact between the neutron converter and the imaging plate is done by depressurization.

C. Imaging Plates and Readers

To assess the performances of many imaging plate models, two different IP readers have been used, first a FLA7000 reader (Fuji) and next a HPX-1 reader (Carestream) with both a maximum readout resolution of 25 µm.
FLA7000 is however limited to only magnetic imaging plates in our setup (Fuji BAS and Protoype IP), where HPX-1 reader has no such limitations.

D. Image Quality Indicators

The spatial resolution performance assessment is carried out by using gadolinium patterns (Siemens star and bars pattern). The gadolinium bar pattern ranges from 0.5 to 20 line pairs/mm, thus allowing to directly build the experimental modulation transfer function (contrast as a function of spatial frequency). The pattern carrier is set in contact with the surface of the dysprosium foil or of the imaging plate depending on the assessed system.

III. RESULTS

A. Detection efficiency and noise

The detection efficiency is of crucial interest in view of the use of imaging plates as imagers in a transfer neutron imaging system. The response of imaging plates is assessed by forming images of the neutron beam by transfer from the dysprosium neutron converter. The signal is recorded after an irradiation of the converter during a time T_{irrad}, followed by a cooling time T_{cool} of 15 minutes to avoid any interference of the short half-lived ^{165m}Dy gamma emission and next a contact time $T_{contact}$ where the dysprosium foil is put in close contact with the imaging plate. By varying the irradiation times at Orphee neutron beam, a wide range of equivalent JHR neutron flux can be assessed, assuming typical irradiation times.
The equivalent JHR neutron flux range assessed extends from 4.3 10^6 n/cm²/s (2 s irradiation at Orphee beamline) to 4.3 10^7 n/cm²/s (20 s irradiation at Orphee) assuming an irradiation time of 15 minutes at JHR.
The detection efficiency of the different IP is assessed by measuring the signal as a function of the neutron flux by varying irradiation times and keeping a constant cooling time (15 minutes). The contact time is set to 1 hour, and the plates are kept in the dark between contact exposition and reading.
Figure 1 shows the evolution of the signal of Flex XL Blue, BAS-TR and UR-1 imaging plates read by HPX-1 reader in function of the neutron flux when using the highest sensitivity of the reading.

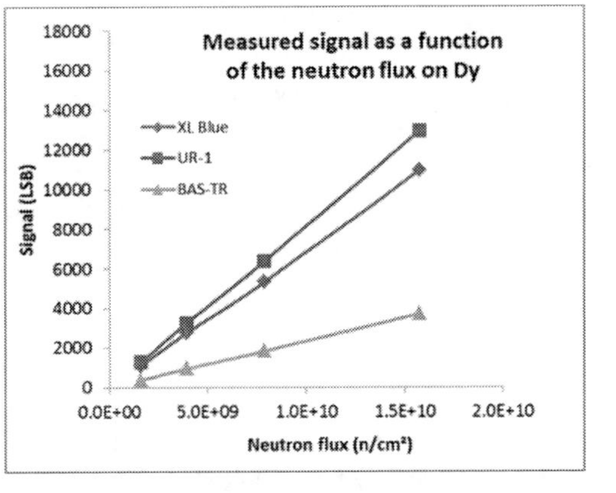

Fig. 1. Evolution of the signal measured on Flex XL Blue, BAS-TR and UR-1 imaging plates in function of the integrated neutron flux.

The dynamic range of the various imaging plate models is of the same order. All models show a good linearity within the signal range. BAS-TR IP however has a lower sensitivity of a factor close to 3 regarding HR Flex Blue and UR-1. This lower sensitivity can be related to the smaller thickness of the active phosphor layer encountered in BAS-TR (50 µm) compared to the phosphor layer of 160 µm in the other two models. This phosphor structure difference will also impact the spatial resolution performance. BAS-TR IP could thus be used up to a JHR neutron flux of 7.5 10^8 n/cm²/s with a 15 minutes dysprosium irradiation time.
The measured signal over noise ratios are as expected increasing as a function of the signal but are almost independent of the reader parameters (photomultiplier gain or laser intensity). The older models (BAS-TR and BAS-ND) show a lower value and lower increase of the S/N ratio than UR-1 or Flex XL Blue models.

B. Spatial Resolution

The intrinsic spatial resolution of imaging plates is investigated by computing their experimental modulation transfer function.
The spatial resolution achievable is defined by a limit of 10% on contrast. An image of Siemens star Image Quality Indicator is shown in figure 2. The computed Modulation Transfer

Functions of different imaging plates are shown on fig 3. The spatial resolution results are summarized in table II, using both readers set at maximum scanning resolution (25 µm).

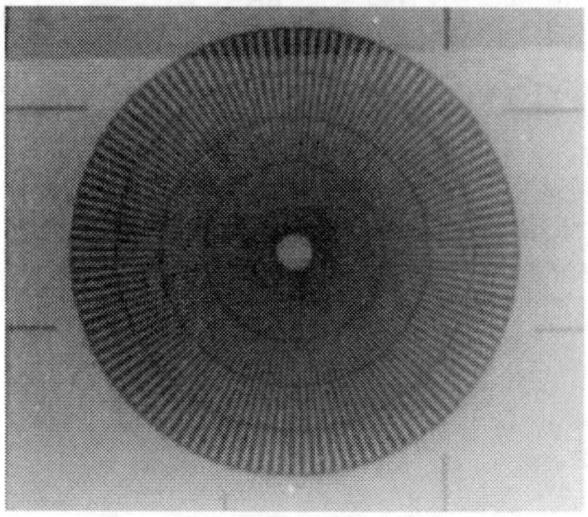

Fig. 2. IQI (Siemens star) image recorded on BAS-TR IP (HPX-1 reader).

TABLE II
SPATIAL RESOLUTION

IP Type	HPX-1 10% MTF (lp/mm)	HPX-1 Spatial resolution (µm)	FLA-7000 10% MTF (lp/mm)	FLA-7000 Spatial resolution (µm)
Fuji BAS-TR	5.4	92	5	100
Fuji-UR1	3.7	135	-	-
Carestream XL Blue	3.7	135	-	-
Prototype Dy-doped	-	-	3.8	130
Fuji BAS-ND	5.8	86	5.5	90

The intrinsic resolution measured on Fuji UR-1 and Carestream Flex XL Blue is similar (135 µm). Fuji BAS-TR imaging plate shows a better intrinsic resolution slightly below 100 µm, thanks to its thinner phosphor layer and the absence of protective layer on top of the active layer. The resolution performance of imaging plates set in contact with an activated foil cannot compete with a neutron detection imaging plate incorporating gadolinium whose resolution is measured around 90 µm. Nevertheless, the prototype IP incorporating dysprosium instead of gadolinium do not show better resolution performances than BAS-TR.

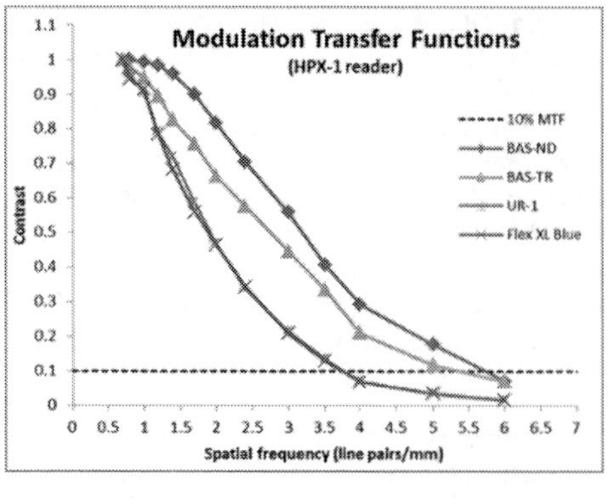

Fig. 3. Modulation Transfer Function computed from the bar pattern image quality indicator for the four imaging plate models assessed using the HPX-1 reader.

IV. CONCLUSION

All imaging plates models assessed show a good sensitivity and a good linearity over a wide signal range. The intrinsic noise appears relatively more important in BAS-TR and BAS-ND IP than in other IP but it remains acceptable even at low signal level (S/N around 50 at 100 LSB), which authorizes the use of such imaging detectors at relatively low flux. The resolution performance obtained with dysprosium doped imaging plate is slightly poorer than those achieved using a Fuji BAS-TR, Carestream HR Flex Blue or Fuji UR-1 in transfer imaging. Even if the intrinsic resolution of imaging plates appears worse than single-sided radiographic films, the impact on the final resolution of the neutron imaging system of JHR may remain low, where the geometrical blurring is the dominant resolution impactor.

Further, the many advantages of using imaging plates over radiographic films (high sensitivity, linear response, high dynamic range) could palliate its lower intrinsic resolution. Nevertheless, in order to get a quantitative response from imaging plates in neutron imaging by transfer, the light exposition between the converter exposition and readout has to be carefully managed.

Imaging plates initially conceived to perform tritium low energy beta imaging like BAS-TR, incorporating the thinnest phosphor layer with no additional protective layer, appears to be a promising on-the-shelf solution to replace a radiographic film-based imager in high resolution neutron imaging by transfer.

ACKNOWLEDGMENTS

The authors wish to thank Frederic Ott at LLB/Orphee for help and support, as well as Marc Masson at CEA/Marcoule for providing the imaging plate reader.

REFERENCES

[1] T. Suzuki et al. "Characteristics and Corrections of the Fading of Imaging Plate", *Journal of Nuclear Science and Technology*, 34:5, p 461-465 (May 1997), 10.1080/18811248.1997.9733692

[2] D. B. Pelowitz et al., MCNPX 2.7.0 extensions, LA-UR-11-02295, Los Alamos National Laboratory (2011)

[3] M.Tamaki et al. "Dy-IP characterization and its application for experimental neutron radiography tests under realistic conditions", *Nuclear Instruments and Methods in Physics Research A*, Volume 542, Issues 1–3, 21 April 2005, Pages 320–323, 10.1016/j.nima.2005.01.156

Measurements Methods for the analysis of Nuclear Reactors Thermal Hydraulic in Water Scaled Facilities

C. Spaccapaniccia, P.Planquart, J.M. Buchlin, *von Karman Institute of Fluid Dynamics Belgium*

***Abstract*—The Belgian nuclear research institute (SCK•CEN) is developing MYRRHA. MYRRHA is a flexible fast spectrum research reactor, conceived as an accelerator driven system (ADS). The configuration of the primary loop is pool-type: the primary coolant and all the primary system components (core and heat exchangers) are contained within the reactor vessel, while the secondary fluid is circulating in the heat exchangers. The primary coolant is Lead Bismuth Eutectic (LBE). The recent nuclear accident of Fukushima in 2011 changed the requirements for the design of new reactors, which should include the possibility to remove the residual decay heat through passive primary and secondary systems, i.e. natural convection (NC). After the reactor shut down, in the unlucky event of propeller failures, the primary and secondary loops should be able to remove the decay heat in passive way (Natural Convection). The present study analyses the flow and the temperature distribution in the upper plenum by applying laser imaging techniques in a laboratory scaled water model. A parametric study is proposed to study stratification mitigation strategies by varying the geometry of the buffer tank simulating the upper plenum.**

***Index Terms*—Natural Convection Loop, Particle Image Velocimetry, Laser Induced Fluorescence, Pool Type Reactors**

I. INTRODUCTION

In a Natural Circulation Loop (NCL) the flow is driven by Archimedes' forces and transports heat from a heat source to a heat sink without the aid of mechanical moving parts. Thanks to this, NCLs find applications in many industrial and domestic fields: nuclear reactor core cooling, solar heaters, electronic components cooling, house heater, etc..

In industrial or domestic heat transport systems, the fluid circulates through hydraulic elements for which pressure drops may be relevant: in these fields a *thermosyphon* loop is most often encountered. A *thermosyphon* loop is a particular case of NCL in which the heat sink is placed at higher level with respect to the heater's thermal center.

The object of this analysis is the experimental, pool type reactor MYRRHA conceived by the nuclear research institute SCK•CEN. In the pool type reactor, all the thermo-hydraulic elements of the primary loop are submerged in the primary coolant. Figure 1-a sketches the main elements of the primary loop: 1-core, 2-barrer, 3-upper plenum 4-heat exchanger, 5-propellers, 6-lower plenum. The barrel is a perforated cylinder containing the upper core structure (instrumentation and safety bars).

After a shutdown event, the core continues to produces heat by spontaneous decay of radioisotopes. The decay heat is equal to 10% of the nominal power. In the unlucky event of propellers failure, the decay heat should be removed passively by Natural Convection. Thus, the primary loop of the nuclear reactor MYRRHA operating in passive heat removal conditions can be considered an analogous NCL. The NCL under analysis is a closed, turbulent, single phase *thermosyphon* where the flow is generated by buoyancy forces. Figure 1 extrapolates and schematizes the NCL starting from a symmetry axis of the pool type reactor. The distance in vertical (y) and horizontal (x) direction between the heating and cooling elements are indicated respectively as H and L. The sketch considers the presence of the upper plenum acting like a buffer tank interposed between core and heat Exchangers (HE).

Few studies exist on NCL provided with a buffer tank. Furthermore, literature concerning this kind of systems usually involves NCLs with constant diameter where the tank's effect is limited to heat storage (Misra et al. [1]) or temperature and flow stabilization (Naveen et al. [2], Misale et al. [3]). It is therefore important to analyze the flow behavior and temperature distribution inside the buffer tank of the NCL

Paper submission: 1/06/2017
Spaccapaniccia C.*, Planquart P. * and Buchlin J.M*.
*Author for correspondence
Von Karman Institute for Fluid Dynamics,
Rhode Saint Genese, 1640, Belgium,

E-mail: chiara.spaccapaniccia@vki.ac.be
E-mail: philippe.planquart@vki.ac.be
E-mail: buchlin@vki.ac.be

outlined in Figure 1-b. In particular, given the buoyancy dominated nature of the flow, thermal stratification is expected to occur in the buffer tank. Stratification and pressure drops in the primary loop are expected to contribute negatively to the NC flow.

The objective of the study was therefore to verify the influence of the geometry of the loop and of the upper vessel on the mass flow. The methodology chosen to address the problem is a parametric study carried out in a 1:8 scaled water model (AQUARIUM) of the of the NCL of Figure 1-b. The study consisted in the simultaneous measurement of mass flow, velocity field and temperature field in different buffer tank configurations, using Thermocouples (TCs) and optical imaging techniques such as PIV (Particles Image Velocimetry) and LIF (Laser Induced Fluorescence).

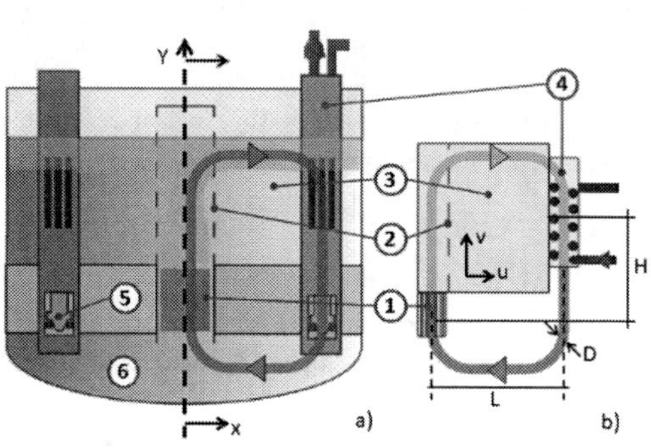

Figure 1 NCL in a pool type reactor

II. SCALING AND FACILITY DESCRIPTION

In passive heat removal conditions, a mass flow is established as a consequence of a balance between frictional pressure losses of the hydraulic loop and Archimedes forces induced by heating and cooling power (Welander [4]). The pressure head generated can be defined as: (Todreas[5]):

$$\Delta P_B = \beta\rho(T_h - T_c)gH \tag{1}$$

If the pool geometry is approximated as axial symmetric, the dominating components of velocity are v (vertical) and u (radial) (Figure 1-b). Under this assumption, and with the hypothesis of equal heat provided and removed, the equations of transport of momentum and energy describing the NCL under analysis are:

$$u\frac{\partial u}{\partial x} + v\frac{\partial u}{\partial y} = -\frac{1}{\rho_0}\frac{\partial p_x}{\partial x} + v\frac{\partial^2 u}{\partial x^2} + v\frac{\partial^2 u}{\partial y^2} \tag{2}$$

$$u\frac{\partial v}{\partial x} + v\frac{\partial v}{\partial y} = -\frac{1}{\rho_0}\frac{\partial p_y}{\partial y} + v\frac{\partial^2 v}{\partial x^2} + v\frac{\partial^2 v}{\partial y^2} - g\beta(T_h - T_c) \tag{3}$$

$$u\frac{\partial T}{\partial x} + v\frac{\partial T}{\partial y} = \alpha\frac{\partial^2 v}{\partial x^2} + \alpha\frac{\partial^2 v}{\partial y^2} \tag{4}$$

The characteristic lengths chosen for the normalization of x and y are H and L as indicated in Figure 1-b. T_h and T_c are the hottest and coldest temperature achieved in the NCL (usually at core and HE's exit). The characteristic pressure chosen is the overall pressure drop (friction and local losses) in the hydraulic loop ΔP_{ch}, therefore:

$$U\frac{\partial V}{\partial X} + V\frac{\partial V}{\partial Y} = \tag{5}$$

$$-\frac{1}{\rho_0}\frac{\Delta P_{ch}}{V_{ch}^2}\frac{\partial P_y}{\partial Y} + \frac{v}{V_{ch}H}\left(\frac{H^2}{L^2}\frac{\partial^2 V}{\partial X^2} + \frac{\partial^2 V}{\partial Y^2}\right) - \frac{Hg\beta(T_h - T_c)}{V_{ch}^2}\theta$$

Since in the NCL the buoyant pressure head should balance the pressure drops (Todreas [6]) the expression for the characteristic velocity V_{ch} can be determined by balancing the pressure term with the buoyant term of Equation 4:

$$\frac{Hg\beta(T_h - T_c)}{V_{ch}^2} = -\frac{1}{\rho_0}\frac{\Delta P_{ch}}{V_{ch}^2} = Eu \tag{6}$$

$$\Rightarrow V_{ch} = \sqrt{\frac{Hg\beta(T_h - T_c)}{Eu}}$$

where Eu is the Euler number for pressure drops.In the NCL, flow rate, pressure drops and buoyancy forces are not independent quantities, thus it is not possible to know a priory V_{ch} and ΔP_{ch} . In the rigorous non-dimensional analysis, in order to escape this problem, the characteristic velocity V_{ch} (or the characteristic pressure drops ΔP_{ch}) is chosen in order to obtain Eu=1. For the present case Eu=1 implies $V_{ch}=(Hg\beta\Delta T)^{0.5}$.

Replacing this expression of V_{ch} in the normalized equations of momentum and energy we obtain the following non-dimensional equations for the steady state:

$$U\frac{\partial U}{\partial X} + V\frac{\partial U}{\partial Y} = -\frac{\partial P}{\partial X} + \sqrt{\frac{1}{Gr_H}}\left(A_R^2\frac{\partial^2 U}{\partial X^2} + \frac{\partial^2 U}{\partial Y^2}\right) \tag{6}$$

$$U\frac{\partial V}{\partial X} + V\frac{\partial V}{\partial Y} = -\frac{\partial P}{\partial Y} + \sqrt{\frac{1}{Gr_H}}\left(A_R^2\frac{\partial^2 V}{\partial X^2} + \frac{\partial^2 V}{\partial Y^2}\right) + \theta \tag{7}$$

$$U\frac{\partial \theta}{\partial X} + V\frac{\partial \theta}{\partial Y} = \sqrt{\frac{1}{Pr \cdot Gr_H}}\left(A_R^2\frac{\partial^2 \theta}{\partial X^2} + \frac{\partial^2 \theta}{\partial Y^2}\right) \tag{8}$$

where the appearing dimensionless groups are:

$$ND_1 = A_R^2 Gr_H^{-0.5}, \quad ND_2 = A_R^2(Pr \cdot Gr_H)^{-0.5} \tag{9}$$

Respecting the similarity of both the groups ND1 and ND2 is possible only if the Pr number of the fluids is the same. However, this is not feasible when studying LBE flow with water facilities.

 https://doi.org/10.1051/epjconf/201817004022

The geometry of a nuclear reactor's primary loop is very complex, especially the geometry of the reactor's core, and it is unfeasible to reproduce every geometrical detail with a scaled model. However, the geometry of the upper plenum can be reproduced accurately. The exact matching of the Grashof number is also unfeasible with a low power water model. However, Gr_H can be relaxed until the flow is in the correct regime (turbulent natural convection for $Gr_H > 10^9$).

The resulting model (called AquariumII) and its main dimensions are shown in Figure 2. AquariumII is a 1:8, 2D slab model of the MYRRHA's NCL. The facility is approximately 1 m high and 0.8 m wide. The flow circulates in the anticlockwise direction. The main elements of the system are:

- Heater: series of flat electrical resistances (maximum total power =3 kW).
- Barrel Plate: an interchangeable perforated plate with circular holes.
- Buffer Plate: a plate with adjustable height to simulate the Heat Exchangers entrance level.
- Heat Exchanger: a copper spiral in which water cooled by an external refrigerating system is circulating.

H	0.4 m
L	0.5 m
Lv	0.39 m
Hv	0.55 m
w	0.16 m
hr	0.076 m
h	0.05 m
b	0.15 m
$Gr_H/Gr_{H,M}$	$5\ 10^{-5}$
Pr/Pr_M	$3\ 10^{2}$

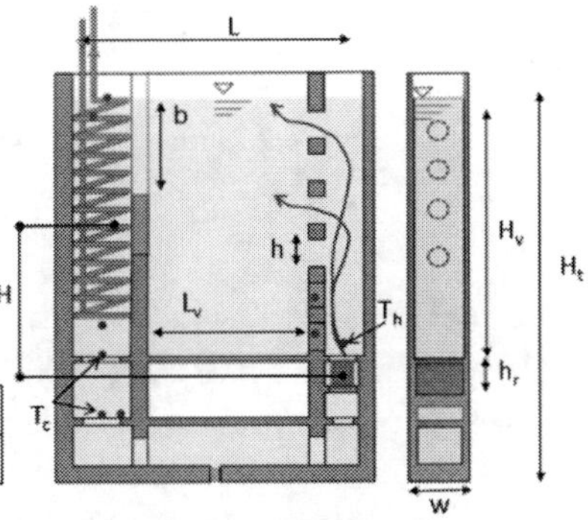

Figure 2 sketch of AQUARIUM II

The peculiarity of the model is that the position or the characteristics of each one of these elements can be changed to carry out a parametric study. The cold leg is accessible through a front window, in order to insert pressure drops elements, in case of necessity. The buffer tank is provided with four optical accesses (top, bottom, front and side) composed of glass windows.

Configurations Tested

Figure 3 shows the configurations chosen to carry out the parametric study. Configuration 1 was obtained by increasing the free area of the barrel plate through opening all the passages available. Configuration 2 was obtained by shiftiong the heat exchanger entrance towards the bottom of 15 cm. Configuration 3 was obtained by reducing the area above the heater through the immersion of a triangular full shape.

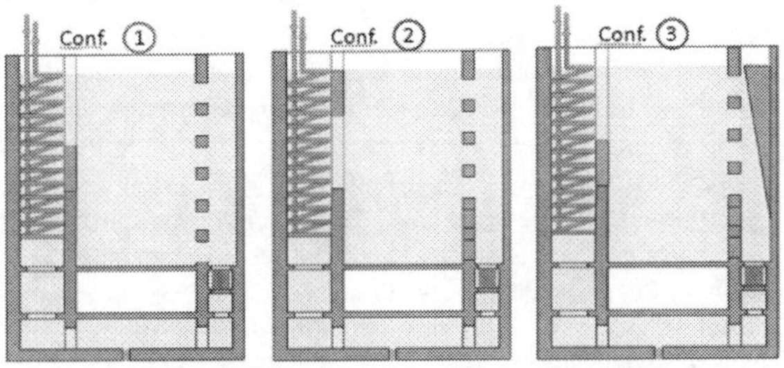

Figure 3 Parametric study configurations

III. EXPERIMENTAL SET UP AND METHODOLOGY

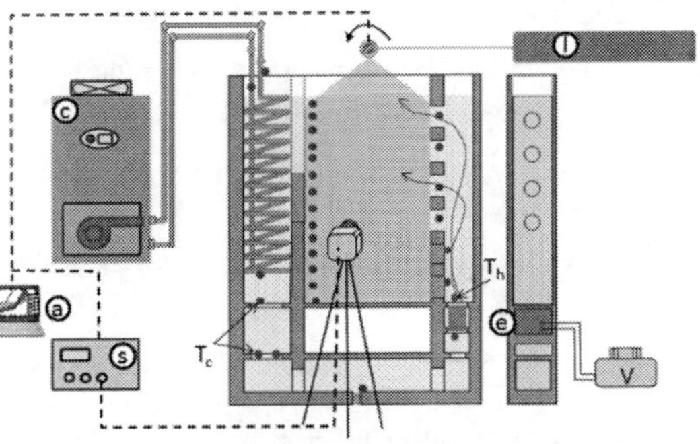

Figure 4 Experimental set up

Figure 4 shows the reference configuration of Aquarium II and the instrumentation arrangement. The TCs are indicated with red dots. Water is seeded with a proper concentration of Rhodamine B and with fluorescent tracing particles, for temperature and velocity measurements. An Argon-Ion Laser produces a continuous beam picked at 515 nm. The beam power selected is 2.1 W, and it is kept constant during the measurements by the laser remote control. Since it is necessary to measure the full tank test section (55x39 cm), a rotating mirror was used to scan the test section with the continuous laser beam. The laser beam scans a section of the plenum, while the rolling shutter CCD camera, provided with a high pass filter of 550 nm, detects the fluorescent signal for temperature measurements and the fluorescent light emitted by the particles. Each image acquisition occurs when the disk rotating together with the mirror completes a revolution: an opening in the disk activates a sensor that sends a signal to the synchronizer, which activates the camera, triggering the recording of one exposure. The image acquisition frequency is limited by the rotating mirror speed (41 Hz). The signal triggering the acquisition of the first image also activates the TCs acquisition, which records temperature values simultaneously to the image acquisition but with an independent and higher frequency (70 Hz).
The Rhodamine B concentration used for the LIF experiments is 1.6 mg/liter. The PIV tracers seeding the water are fluorescent polyethylene microspheres of the size of 40 μm with spectrum

 https://doi.org/10.1051/epjconf/201817004022

emission similar to the one of Rhodamine B: in this way a single image will contain information on both temperature (RhB signal) and velocity (particles signal). The concentration of particles is 0.025 g/liter.

Figure 5 shows the acquisition time line of the simultaneous temperature and velocity measurements performed. Before the beginning of the test, a reference temperature is recorded by means of TCs and a set of reference images is recorded by the CCD camera. The TCs measurement covers all the duration of the experiments and it is divided in 5 different recordings. Every image set includes 2400 frames, covering approximately one minute of experiment. In all the tests performed the heat exchanger and the electrical resistances are switched on simultaneously at one minute after the beginning of the experiment.

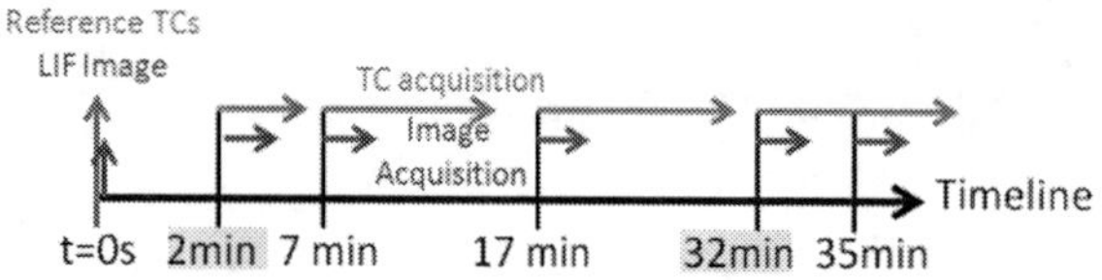

Figure 5 Data acquisition time line.

IV. DATA TREATMENT AND PROCESSING

One camera imaging techniques require a pre-processing of the images in order to separate the signals of the particles and the signal of the background.

Westerweel et al. [6] and Adrian and Westerweel [7] analysed the histogram of a synthetic PIV image. Westerweel et al. found out the background part of a PIV image can be fit by a normal distribution, while Adrian and Westerweel showed that an exponential approximation can be applied to the particles signal, when the background noise is characterized by lower intensity with respect to the particle signal. The intensity counts of an excited solution of water and RhB with a low concentration of dye are lower than the intensity counts of a fluorescent particle which contains the same dye, thanks to the light enhancement due to the Mie scattering. Therefore the threshold separating particles and temperature signal can be found on the image histogram, by approximating it with a Gaussian and an exponential distribution (Figure 6).

The approximation is done iteratively using a guess threshold point, until the point of best fitting is found. The result of the procedure is displayed in the images at the top right of Figure 6, which shows the image before and after the threshold, where the particle signal is successfully removed.

One of the major problems for a LIF image quality is the appearance of striations forming because of gradient of refractive index. Existing literature (Grafsronningenet al. [8], Coolen et al. [9]) suggests to remove geometrical elements like lines by the convolution of the two dimensional Fourier transform of the image with a proper filter. Figure 7 shows the filter used to correct the striations. Once the LIF images are normalized and corrected, a calibration must be used to convert the LIF signal into temperature values. For cases of large facilities like Aquarium II, achieving a uniform temperature in every point of the test section is possible only when the water is at ambient temperature. The calibration is therefore performed by associating the value of normalized intensity measured in small image windows placed nearby the TCs with the temperature measured by the corresponding TC. Figure 8 shows a calibration curved retrieved for five image acquisitions (corresponding to one experiment).

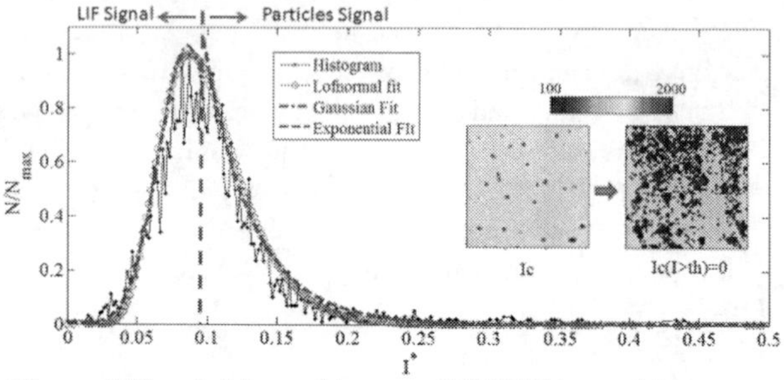

Figure 6 Threshold searching on a PIV-LIF image histogram

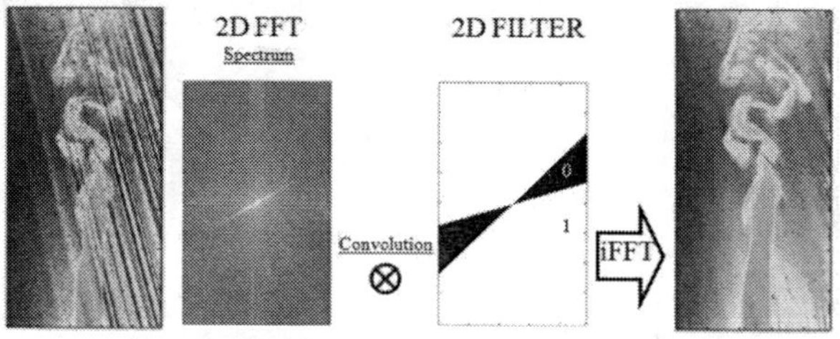

Figure 7: Striations removal logic

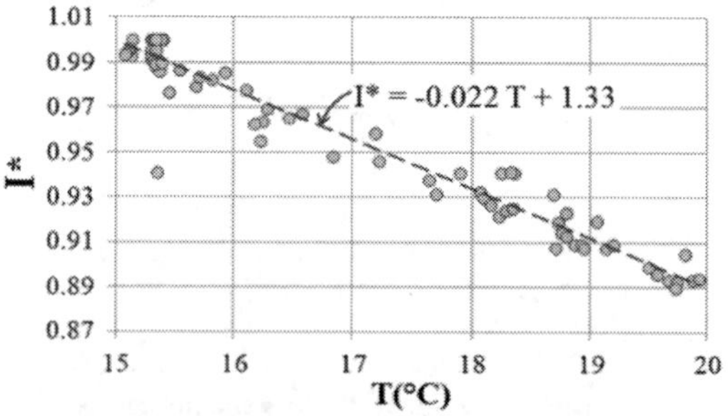

Figure 8 LIF signal-TC signal correlation curve used for calibration

Finally, the parameters of interest allowing to characterize the performance of the loop are loop mass flow and stratification number S. The loop mass flow was calculated through integration of the velocity profiles measured with PIV at every aperture of the barrel.

$$\dot{m}_{loop} = 2\pi\rho \sum_{i=1}^{N} \int_{0}^{h/2} r u(r)\, dr \qquad (10)$$

The stratification number S was derived for one line of TCs using the definition of Gandhi et al. [11] (Equation 11).

$$S = \frac{1}{J-1}\sum_{j=1}^{J-1}\frac{T_{j+1}-T_j}{dy}\frac{(J-1)dy}{T_{max}-T_{min}} \quad (11)$$

V. RESULTS

Figure 9 shows the stratification measured in the plenum of Aquarium II. The values of temperature are averaged over one minute of acquisition to match the image acquisition period. The stratification measurements performed with TCs will be discussed together, with the aid of the measured velocity and temperature fields. Considering the group formed by the first six uppermost TCs, the topmost one indicates a lower temperature with respect to the others. This is due to the room conditions: the room temperature was not higher than 15 degrees. The same group of 6 thermocouples shows an increase in temperature along the experiment. The explanation of this can be easily found when observing the streamlines, velocity and temperature fields displayed in Figure 10.

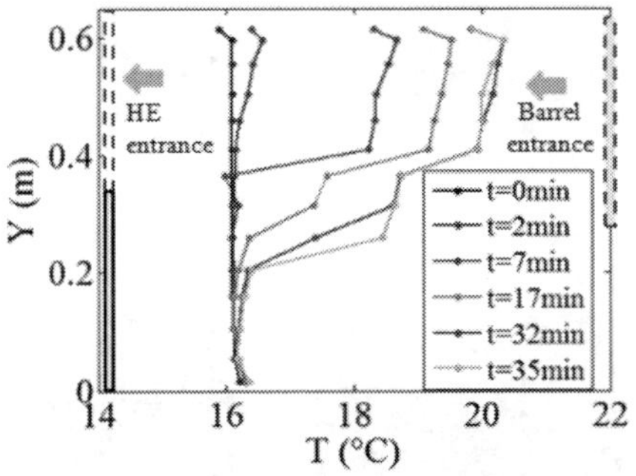

Figure 9 - evolution of temperature profiles measured at the HE entrance for the reference configuration of Aquarium II

The image on the left of Figure 10 shows the velocity magnitude field measured by laser scanning PIV and the associated streamlines. The image on the right shows the temperature field measured with LIF. The vector field is overlapped to the temperature field.

The six uppermost TCs are immersed in the hot stream coming from the barrel plate. As shown in Figure 10, a strong vortex is produced at the beginning of the test in correspondence of the heat exchanger entrance. At the beginning of the test, the hot fluid coming from the plume region and crossing the barrel plate is highly buoyant since the plenum is not stratified yet, which translates in higher flow momentum. The vortex entrains flow from the heat exchanger region, which is still at initial temperature condition. The result of this is an inclined temperature profile in the first six thermocouple of the plenum. The mass flow re-entering the plenum doesn't possess enough momentum to reach the barrel plate again and it spreads on the lower part of the plenum, opening wider and colliding against the hot flow coming from the bottommost hole of the barrel.

The vortex divides therefore in two flows: one joins the hot buoyant jet outgoing from the barrel, the other deviates toward the bottom of the plenum. The Temperature field of Figure 10 shows lower values of temperature in the lower part of the plenum. This is in accordance with the value of temperature detected by the Thermocouples.

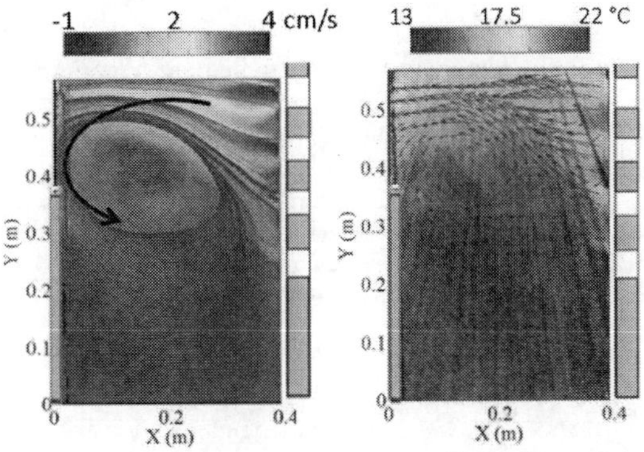

Figure 10 – velocity, streamlines and temperature field at t=2min for the reference configuration of Aquarium II

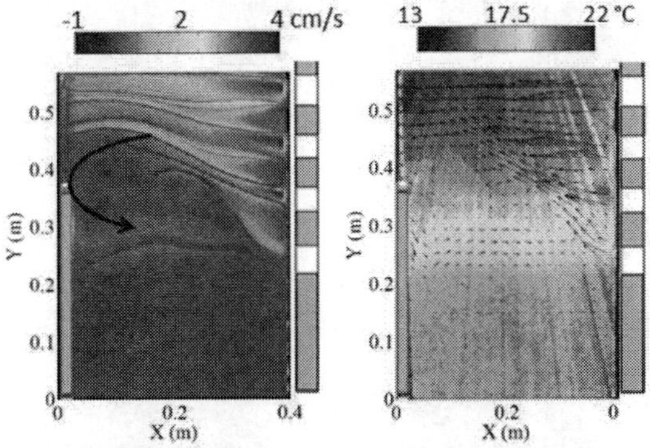

Figure 11 Reference Conf., 32 min after beginning of the test

The temperature field shows that the hot buoyant plumes outgoing from the barrel conquers space on the upper part of the plenum and force the vortex downward. This is consistent with the model of the plume in confined space derived by Baines et al. [10], for which the warm fluid occupies the upper part of the plenum and it is progressively replaced by the incoming fluid, moving towards the bottom.

Figure 11 shows velocity and temperature fields obtained for the reference configuration at 32 minutes after the beginning of the experiment. We can observe horizontal plumes outgoing from the barrel and forming a central vortex: the vortex size is greatly reduced with respect to the beginning of the test. Smaller scale vortices develop at the boundary layers of the lowest horizontal plumes, as a result of Kelvin-Helmotz instability occurring at the shear layer between two streams possessing different values of velocity. A colder mass flow re-circulating in the tank form the heat exchanger is detected by both PIV measurements (Figure 11) and TCs measurements (Figure 9). Both The temperature profile measured with the TC array and the LIF data (Figure 11) show very well the separation between the upper mixing region and the bottom, cold and almost motionless region.

https://doi.org/10.1051/epjconf/201817004022

Nevertheless the measurements involved a full transient of the NCL, the velocity and temperature results discussed and compared in the following section are the ones obtained at the end of each experiment.

Effects of increased barrel porosity, Configuration 1

In Configuration 1 the flow is recalled and entrained in the plume from the bottommost barrel apertures (Figure 12). The hot flow outgoing from the top of the barrel crosses the tank until the bottom, carrying with him the heat of the warm fluid, and re-enter partially the barrel region. The recirculation involves in this case almost all the tank, improving the thermal mixing. and leads to a more uniform temperature profiles (Figure 13, first graph) with respect to the reference configuration.

Configuration 1 shows higher values of loop mass flow with respect to other configurations tested (), indicating better performances. This is due to the communication between the upper heater warm region and the cold bottom tank region, which yields a stack effect: the difference in temperature between the tank fluid and the fluid above the heater causes a difference of pressure which contributes positively to the mass flow.

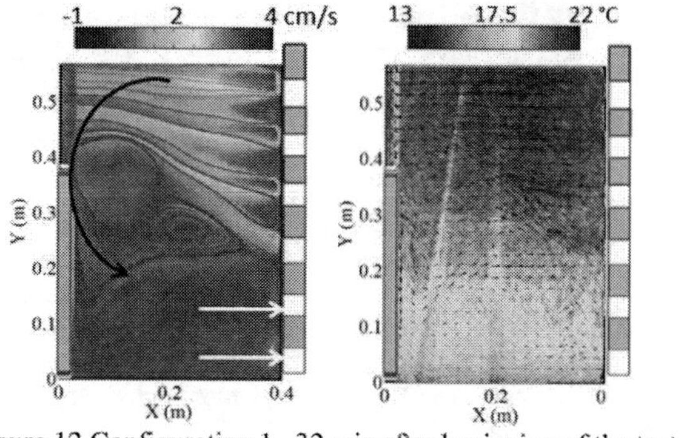

Figure 12 Configuration 1, 32 min after beginning of the test

Effects of HE window position, Conf.2

The tests performed in the reference configuration and the Conf. 2 show that the position of the central vortex is strongly influenced by the position of the heat exchanger entrance, as can be noticed when comparing the streamlines of Figure 11 and the left plot of Figure 14.

The effects of moving the heat exchanger entrance downwards are mostly negative. For instance, temperatures achieve higher values in the topmost part of the tank (Figure 13-second graph): this is due to the increased blockage effect caused by the presence of an obstacle between the upper tank and the heat exchangers.

On the other hand, the border between motionless region and the mixing region moves towards the bottom increasing the mixing zone (Figure 14-left).

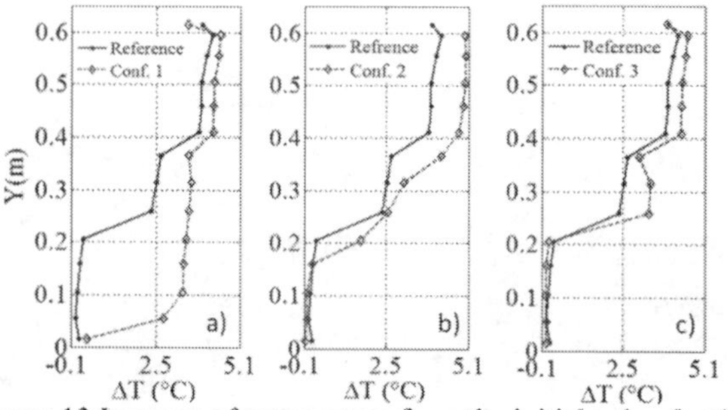

Figure 13 Increase of temperature from the initial value for the different configurations.

Effects of blockage in the barrel, Conf.3

Figure 14-right shows the velocity magnitude field obtained decreasing progressively the barrel test section by means of a slope-shaped obstacle at 32 minutes after the beginning of the tests. Comparing the field of Figure 14-right with the velocity field of Figure 11 it can be observed that the size of the plume exiting the topmost hole is reduced when the plume passage area in the plume region is reduced, due to the higher resistance encountered by the hot flow in the barrel region.

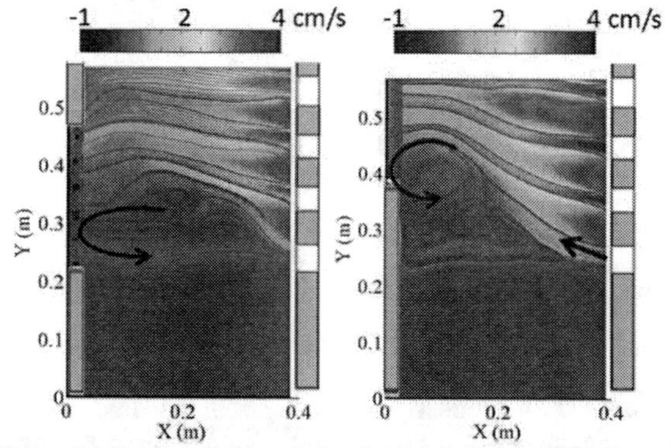

Figure 14 Conf. 2 (left) and Conf. 3 (right) velocity fields, 32 min after beginning of the test

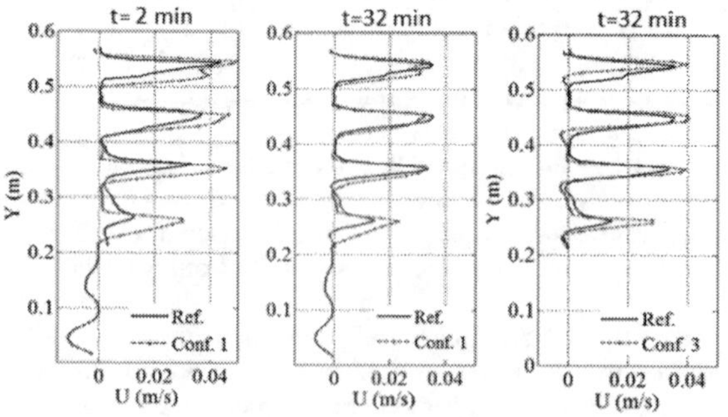

Figure 15 Velocity profiles measured at the barrel exit

At the same time, the flow outgoing from the bottommost hole becomes more important. In the reference configuration the plume outgoing from the bottommost hole merges immediately after the barrel with the buoyant plumes above, while when the passage area in the barrel is reduced (Configuration 3) this plume in particular assumes a horizontal direction and spread further in the tank.

The last graph of Figure 15 shows the velocity profiles measured at the barrel exit for the Reference geometry and Configuration 3. The effect of redistribution of the mass flow in between the holes is clearly visible in the profiles.

Summary

The experimental results obtained for the different configurations tested are summarized in Figure 16 in terms of loop mass flow and stratification number. It can be observed that the highest mass flow is achieved for Configuration 1. Configuration 1 is also the only case tested in which the mass flow increases with the time. The lowest mass flow is obtained for Configuration 3, due to the additional flow resistance introduced by decreasing the barrel area. At the end of the transient, Configuration 1 shows the lowest value of stratification number, as expected from the more uniform temperature profiles illustrated in Figure 13.

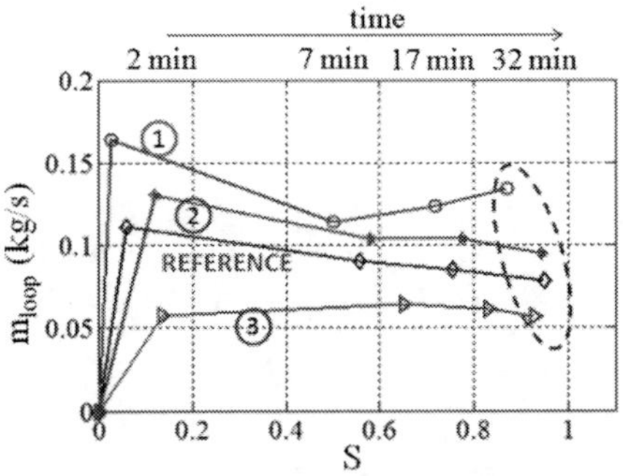

Figure 16 Loop mass flow measured for different configurations along the experiments with S

VI. CONCLUSIONS

Velocity and temperature measurements with innovative one-camera optical techniques were carried out in the buffer tank of a water NCL simulating passive decay heat removal in a liquid metal pool type reactor. It was observed that the loop mass flow partly completes the loop and partly circulates back in the buffer tank, bypassing the loop. A parametric study was carried out in order to understand the effect of the buffer tank geometry on the stratification. It was observed that the geometry of the plenum effects the temperature gradients and the value of m_{loop}. The lowest values of S and the highest values of m_{loop} were obtained with Configuration 1. This configuration allows the communication between cold upper plenum and hot barrel duct, yielding to a stack effect. The difference of temperature between these two regions causes a pressure difference which contributes positively to the mass flow. Furthermore, the communication between plenum and barrel duct allows the entrainment of flow in the plume developing right above the heater, reducing the stagnation in the plenum and increasing the mixing. As a result, the plenum temperature profile is more uniform with respect to the reference configuration. As observed thanks to the measurements carried out in Configuration 3, the progressive reduction of the barrel cross sectional area allows to distribute more uniformly the m_{loop} between the barrel apertures without affecting sensibly the temperature distribution in the tank. The flow fields measured in Configuration 2 show that size and position of the convection cell developing at the HE entrance is affected by the heat exchanger window's height and location. The measurements will serve as a base for validating CFD simulations of the NCL. The simulations will allow more extensive parametric studies, leading to the optimization of barrel bottom apertures' size and HE window position.

VII. NOMENCLATURE

x	[m]	Cartesian axis direction
y	[m]	Cartesian axis direction
r	[m]	radius
H	[m]	Thermal conductivity
L	[m]	Volumetric heat generation density
T_h	[K]	Interfacial thermal resistance
T_c	[K]	Temperature
p	[kg/ms^2]	pressure
ΔP_{ch}	[kg/ms^2]	Characteristic pressure drops
u	[m]	Radial velocity
v	[m/s]	Axial velocity
V_{ch}	[m/s]	Characteristic Velocity
β	[1/K]	Volumetric expansion coefficient
a	[m^2/s]	Thermal diffusivity
g	[m/s^2]	Gravity
ρ	[kg/ m^3]	density
ν	[m^2/s]	Kinematic viscosity
m_{loop}	[kg/ s]	Loop mass flow
Θ	[-]	T- T$_c$/ T$_h$- T$_c$

Non DImensional Numbers

$\mathcal{Gr}_{\mathcal{X}}$	[-]	NCL Grashof number
$\mathcal{Pr}$	[-]	Prandtl number
$\mathcal{A}_{\mathcal{R}}$	[-]	Aspect ratio
$\mathcal{Eu}$	[-]	Euler number ofr pressure drops
S	[-]	Stratification number

Abbreviations

LBE	Eutectic Lead Bismuth
NCL	Natural Convection Loop
PIV	Particle Image Velocimetry
LIF	Laser Induced Fluorescence
TCs	Thermocouples
HE	Heat Hexchangers

VIII. REFERENCES

[1] R. Misra. *Thermal stratification with thermosyphon effects in solar water heating systems*. Energy Conversion and Management, 1994.

[2] K. Naveen, K. N. Iyer, J. B. Doshi, and P. K. Vijayan. *Investigations on single-phase natural circulation loop dynamics. Part 2: Role of wall constitutive laws.* Progress in Nuclear Energy, 2014

[3] M. Misale, P. Garibaldi, J. C. Passos, and G. Ghisi de Bitencourt. *Experiments in a single-phase natural circulation mini-loop* Experimental Thermal and Fluid Science, 2007.

[4] P. Welander. *On the Oscillatory Behaviour of a Differentially Heated Fluid Loop*. J. Fluid Mechanic, 1967.
[5] N.E.Todreas, M.S.Kazimi, Nuclear Systems II -Elements of Thermal Hydraulic design
[6] J. Westerweel. *On velocity gradients in PIV interrogation. Experiments in Fluids*, 2008
[7] R. J. Adrian and J. Westerweel. Particle Image Velocimetry. 2010.
[8] S. Grafsrønningen, A. Jensen. *Simultaneous PIV/LIF measurements of a transitional buoyant plume above a horizontal cylinder*. International Journal of Heat and Mass Transfer, 2012.
[9] M. C. J. Coolen, R. N. Kieft, C. C. M. Rindt, and A. A. Van Steenhoven. *Application of 2-D LIF temperature measurements in water using a Nd : YAG laser*. Experiments in Fluids, 1999.
[10] W. Baines and J.S.Turner. *Turbulent Natural Convection from a source in a confined region. J.* Fluid Mechanic, 1968.
[11] M. Gandhi, J. B. Joshi, and P. K. Vijayan. *Study of two phase thermal stratification in cylindrical vessels: CFD simulations and PIV measurements*. Chemical Engineering Science, 2013.
[12] C. Spaccapaniccia. *Experimental study of natural convective internal flows*. PhD thesis, Université Libre de Bruxelles, September 2016.

Reaction Rate Benchmark Experiments with Miniature Fission Chambers at the Slovenian TRIGA Mark II Reactor

Žiga Štancar, Tanja Kaiba and Luka Snoj
Reactor Physics Department
Jožef Stefan Institute
Jamova cesta 19, SI-1000 Ljubljana
e-mail: ziga.stancar@ijs.si

Loïc Barbot, Christophe Destouches, Damien Fourmentel and Jean-François Villard
CEA, DEN, DER,
Instrumentation, Sensors and Dosimetry Laboratory
Cadarache, F-13108 St-Paul-Lez-Durance
e-mail: loic.barbot@cea.fr

Abstract*—A series of fission rate profile measurements with miniature fission chambers, developed by the Commisariat à l'énergie atomique et aux énergies alternatives, were performed at the Jožef Stefan Institute's TRIGA research reactor. Two types of fission chambers with different fissionable coating (^{235}U and ^{238}U) were used to perform axial fission rate profile measurements at various radial positions and several control rod configurations. The experimental campaign was supported by an extensive set of computations, based on a validated Monte Carlo computational model of the TRIGA reactor. The computing effort included neutron transport calculations to support the planning and design of the experiments as well as calculations to aid the evaluation of experimental and computational uncertainties and major biases. The evaluation of uncertainties was performed by employing various types of sensitivity analyses such as experimental parameter perturbation and core reaction rate gradient calculations. It has been found that the experimental uncertainty of the measurements is sufficiently low, i.e. the total relative fission rate uncertainty being approximately 5 %, in order for the experiments to serve as benchmark experiments for validation of fission rate profiles. The effect of the neutron flux redistribution due to the control rod movement was studied by performing measurements and calculations of fission rates and fission chamber responses in different axial and radial positions at different control rod configurations. It was confirmed that the control rod movement affects the position of the maximum in the axial fission rate distribution, as well as the height of the local maxima. The optimal detector position, in which the redistributions would have minimum effect on its signal, was determined.

1. Introduction

The use of advanced Monte Carlo computational codes for reactor core calculations has increased significantly in the last decade, thus generating the need for experimental verification through the use of benchmark experiments. At the Jožef Stefan Institute (JSI) Reactor Physics Department a Monte Carlo model of the TRIGA Mark II reactor has been created in MCNP - Monte Carlo N-Particle Transport Code (MCNP) [1]. The model was mainly used for criticality calculations, but was later upgraded and used also for calculation of other physical parameters of the core [2], [3], [4], such as the beta-effective and power distribution. Lately it has been thoroughly used for the computational support of experimental campaigns and for safety assessments of reactor experiments and modifications. The model has already been verified against experimental data for kinetic parameters and Au (n,γ) and Al (n,α) reaction rates in the reactor irradiation facilities [5], [6]. From 2011 a series of experiments was performed in collaboration with Commisariat à l'énergie atomique et aux énergies alternatives (CEA), in order to experimentally verify calculated fission (U-235 and U-238) rate profiles in the core between the fuel elements by using miniature fission chambers.

The benchmark experiments performed at the JSI TRIGA Mark II reactor were aimed at measuring in-core fission rate distributions. Specially designed miniature fission chambers developed by the CEA were used [7]. The fission chambers were inserted into the core of the TRIGA reactor using guide tubes, which enabled the measurement of axial fission rate profiles at multiple radial positions. A Monte Carlo computational model of the reactor and the experiment configuration was constructed, which was used to calculate the reaction rate distribution inside the core of the reactor and to evaluate the experimental and computational uncertainties together with major benchmark biases. Major uncertainty sources like fuel composition, axial and radial positions of the fission chambers, core temperature effects and control rod position were investigated an additional analyses of the effect of fuel burn-up and computational model geometric biases were performed. A comparison of the measured and calculated fission rate profiles was made and a relatively good agreement between the two sets of data was found, with average relative discrepancies being within one sigma standard uncertainty. In small research reactors such as the JSI TRIGA, the neutron flux profile inside the reactor core changes significantly in the axial and radial direction. Moreover, the effect of the neutron flux redistribution due to the control rod movements is significant. Until now the reactor ex-core power monitoring

https://doi.org/10.1051/epjconf/201817004023

system has been used, which is highly sensitive to the effects mentioned above. Therefore a new on line in-core system has been devised which would utilize multiple miniature detectors distributed throughout the core to average out the dependence of the signal on the detector and control rod position [8].

The evaluation results can serve the purpose of testing and verifying the innovative design of the absolutely calibrated miniature fission chambers, as well as help validate the values of basic reactor parameters. These are needed for the normalization of Monte Carlo neutron transport calculation results [9]. The high value of the absolute reaction rate benchmark was recognized with the inclusion of the experiment into the International Reactor Physics Experiment Evaluation Project Handbook (IRPhEP) [10].

2. Experiment description

The experiment was performed at the JSI TRIGA Mark II reactor, which is a pool type light-water reactor, with a maximum steady state power of 250 kW. The core is submerged into a 6.25 m high and 2 m wide aluminum tank filled with water and has an annular configuration, consisting of six concentric rings, where cylindrical fuel rods clad with stainless steel are positioned. The fuel material is a homogeneous mixture of 88 wt % ZrH and 12 wt % of 20 % enriched uranium. The core itself is composed of two 1.9 cm thick aluminum supporting grids in which holes of different diameters are drilled. These serve the purpose of positioning the fuel elements, control rods and additional nuclear instrumentation. In total there are 91 locations for fuel elements and additional 26 smaller holes, 9 of which were used for the insertion of the fission chambers. The reactivity in the reactor is controlled by four individually operated control rods labelled the safety (S), transient (T), compensating (C) and regulating (R) control rod. The core is surrounded by a circular graphite reflector which contains a rotary specimen rack, as can be seen in Figure 1. Five ex-core detectors are positioned on the outside of the graphite reflector, which consist of neutron and gamma detectors used for measuring the reactor thermal power. Each of the five channels has an individual power measuring interval based on the sensitivity of the detector. At steady-state reactor operation the power is generally determined with the linear channel. The nuclear detectors are periodically calibrated with the calorimetric calibration method using electrical heaters [11] with 2 % uncertainty.

Fission rate distributions in the core of the TRIGA reactor were measured in the experiment, therefore the type of the detector was chosen on the basis of several operational constraints and expected neutron flux. The detector used was a watertight miniature fission chamber (FC) connected to an integrated mineral cable, which was designed, manufactured and absolutely calibrated by CEA [7]. The detectors structure material was made of stainless steel, while the fission chambers were filled with a gaseous mixture of argon and nitrogen. They displayed a cylindrical shape with a 3 mm diameter. In the experiment two different fission chambers

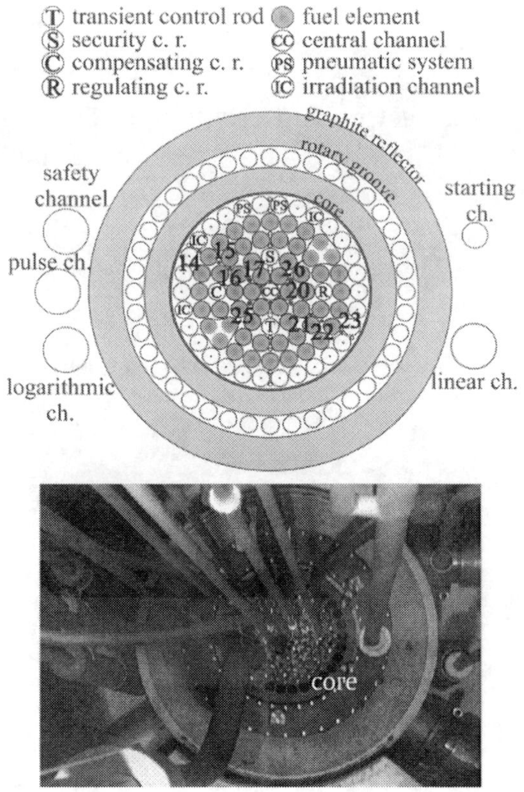

Figure 1. Top view of the reactor core, graphite reflector and ex-core nuclear detectors, with denoted measuring positions used for fission chamber insertion.

were used, which differed in the type and amount of fissionable material deposited on the walls of the active part – one with a fissionable coating composed mainly of ^{235}U (98.490 % at.) and the other of ^{238}U (98.964 % at.) [12]. The fission chambers were operate in *pulse mode* at low fission rates, i.e. when individual pulses can be separated. The measurements presented in the article have been performed at relatively low neutron flux, i.e. 10^{10} neutrons cm^{-2} s^{-1} to 10^{11} neutrons cm^{-2} s^{-1} (reactor power $\sim$ 1 kW). Initial measurements have been made to ensure a linear response of the fission chambers, the stability of the signal at steady state reactor power and to define the reactor power at which axial profiles would be recorded. It has been found that the optimal power was approximately 100 W for the ^{235}U fission chamber and 1000 W for the ^{238}U fission chamber. Measurements of fission rates were performed along the complete core height at 9 radial measurement positions, namely MP14, MP15, MP16, MP17, MP20, MP21, MP22, MP23, MP25 and MP26 denoted in Figure 1. FCs were deployed into the reactor core using a specially designed positioning system, comprising several hollow Al guide tubes, a drive mechanism and a data acquisition system.

An integrated cable connecting the chambers and the data acquisition system was used for inserting and withdrawing the FCs into and out of the reactor core. The FC position was regulated by a pneumatic drive consisting of a series of valves and pistons. The axial positioning was ensured by an incremental system which measures the FC position relative to the reference position at the end of the guide tube and a linear system for absolute position measurements through the use of a magnetized strip. Fission reaction rates were measured at 24 axial locations inside the core with increment steps of approximately 25 mm. A photograph of the positioning system is shown in Figure 2. The fission chamber positioning system was fixed on a horizontal support rod above the reactor pool with the stationed pneumatic system control panel providing the necessary pneumatic feed.

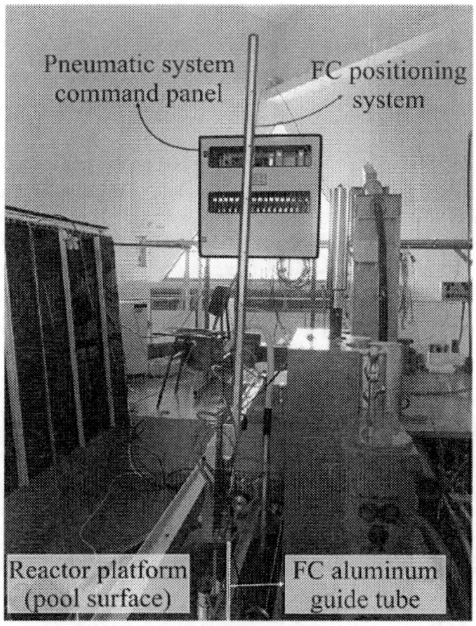

Figure 2. Photograph of the FC positioning system instalment above the pool of the TRIGA reactor.

3. TRIGA Monte Carlo computational model

The fission chamber experiment was modelled and inserted in the existing Monte Carlo model of TRIGA Mark II reactor [4], which has been validated against measurements in the past [6], [13]. The calculations were performed using the state-of-the-art Monte Carlo transport code MCNP version 6.1 [1] and ENDF/B VII.1 nuclear data library [14]. The fission reaction rates were calculated and averaged over the 4 mm high sensitive area of the fission chambers in 1 mm increments over the whole height of the reactor core. For measurement to calculation comparison the computation results had to be converted to absolute reaction rate units, i. e. counts per second. This was carried out using a normalisation factor K [9]:

$$R = K \cdot R_{mc} \ ; \ K = \frac{P\, \overline{\nu}}{w_f\, k_{eff}} \ , \qquad (1)$$

where R is the absolute reaction rate, R_{mc} is the Monte Carlo calculated reaction rate, P is the reactor thermal power, $\overline{\nu}$ is the average number on neutron released per fission, w_f is the average deposited energy per fission event and k_{eff} is the calculated effective multiplication factor of the TRIGA computational model. The values of $\overline{\nu}$ and w_f are constant in general and have values of $\overline{\nu} = 2.439$ and $w_f = 193.7$ MeV $\pm$ 0.1 MeV [15]. Values of the calculated effective multiplication factor of approximately 1.0475 were higher than those of a critical system because fuel burn-up was not taken into account in the reactor model. The schematic top and side views of the TRIGA benchmark model can be seen in Figure 3.

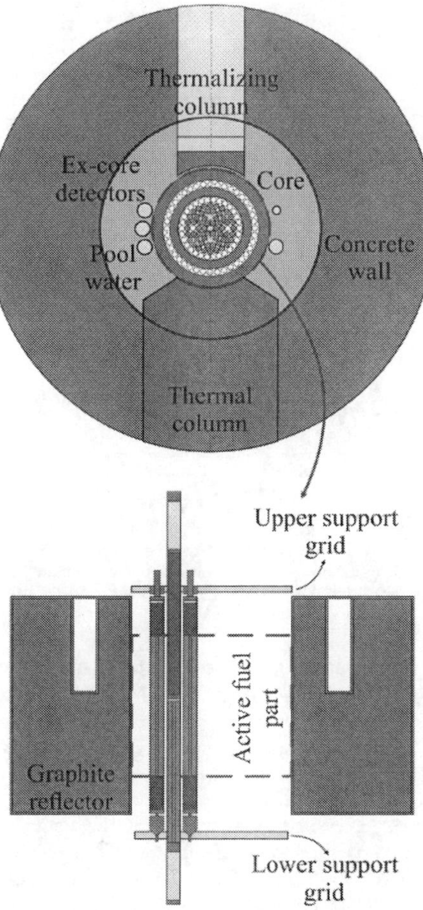

Figure 3. Schematic top and side views of the TRIGA reactor benchmark model. The reactor core within the water filled pool and the surrounding concrete shield are shown in the left-hand side. A close-up of the core is shown in the right-hand side with denoted fuel elements, control rods, support grids and graphite reflector.

4. Fission rate axial profiles

The measured fission rate profiles and the ones calculated with the TRIGA Monte Carlo computational model were compared [8], [10]. The reference point (relative 0) of the fission chambers axial position was set to coincide with the top of the upper support grid and increasing towards the bottom of the reactor core. A comparison of absolute and relative (normalized to reaction rate maximum) fission rate profiles is shown for the ^{235}U fission chamber at measuring position MP20. One can see a relatively good agreement with the discrepancies being smallest at the active part of the core. The average relative differences in this area reach a value of less than 5 %.

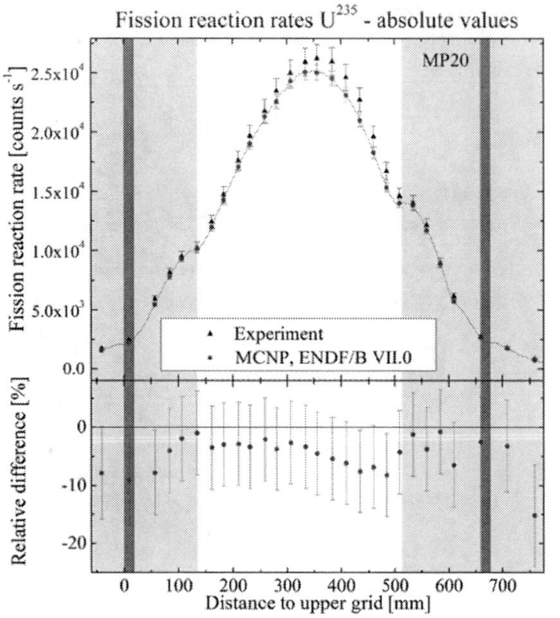

Figure 4. Comparison and relative discrepancies between the measured absolute fission rates and the ones computed with the Monte Carlo method for measuring position MP 20 – positioned near the centre of the core.

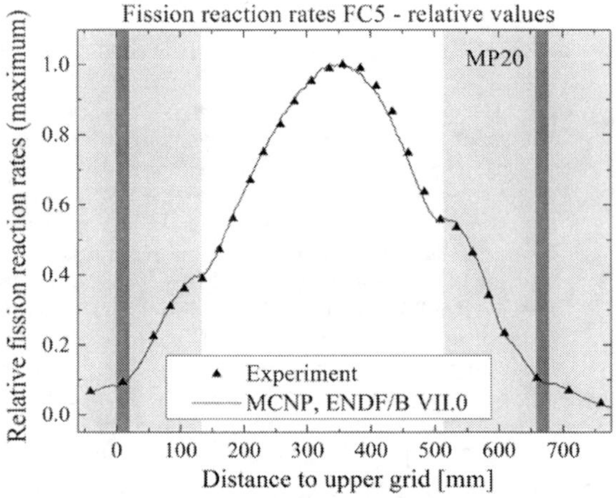

Figure 5. Comparison between the measured relative fission rates and the ones computed with the Monte Carlo method for measuring position MP 20.

5. Effect of control rod movement on the fission rate axial profiles

In small research reactors such as TRIGA, the neutron flux and fission rate inside the reactor core can change significantly in the axial and radial direction [8], [10], [16], [17], [18]. The major effect of the neutron flux redistribution is due to the control rod movement. The neutron flux redistributions affects the power readings of current ex-core power monitoring system [11]. Using multiple in-core detectors gives a possibility to average out the dependence of the detector signal on the control rod position [19].

Experiments presented in this section were performed with multiple CEA manufactured miniature FCs, containing 10 μg of 97.663 % enriched ^{235}U, positioned in MP20, MP26, MP25 and MP17, which are denoted in Figure 1. Those measuring positions were chosen due to their symmetric position inside the reactor core and the symmetric position in regard to the regulating and compensating control rods. For performing and analysing the measurements the CEA design MONACO (Multichannel Online Neutron Acquisition Campbell mOde) system [20] was used. All FCs were operating in pulse mode and were positioned inside aluminium guide tubes, which enabled vertical movement through the reactor core. Positions of miniature FCs were changed manually in 5 cm steps. The schematic view of fuel and control rods configuration inside the reactor core is presented in Figure 6, where on top in grey is the upper grid, where reference point for measurements and calculations was chosen. Due to the reference position above the reactor core all other axial positions have a negative sign in Figure 6. It should be noted that reactor was critical during the measurement with constant reactor power 100 W, therefore when the regulating control rod was withdrawn, the reactivity was compensated by insertion of the compensating control rod. The regulating and compensating control rods consist of two parts: upper part contains strong neutron absorber B_4C, while the bottom part contains fuel.

Fission rate axial profiles were measured in above mentioned measuring positions at different control rod configurations. Fission rate axial profiles were normalized to the surface area below the graph, which enabled comparison of their shapes. The reported calculation uncertainties are 1σ statistical uncertainties originating from Monte Carlo method. Relative calculation uncertainty was lower than 1 %. The reported 1σ experimental uncertainties were obtained taking into account the uncertainty due to a number of counts measured, deviations in the reactor power during measurement, uncertainty in FC design and MONACO

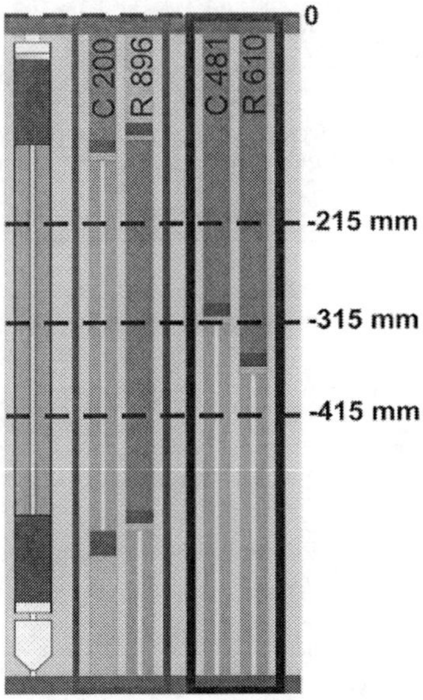

Figure 6. Schematic view of fuel and two control rods configuration inside the reactor core with a chosen reference scale. On each control rod its position is denoted in steps.

system. The total experimental uncertainty was obtained for individual FC and for individual measurement and was calculated to be ~3%. The major contribution to the experimental uncertainty is due to the uncertainty in the mass of the fissile deposit.

In Figure 7 measured and calculated fission rate axial profiles in MP20 at two different control rod configurations are presented. Difference between both axial profiles can be observed. Similar difference in axial profiles would also be observed if other measuring positions would be evaluated. The configuration of the evaluated control rod positions can be observed in Figure 6. In case of completely inserted regulating control rod (red configuration) inside the active fuel region is only absorber part of the regulating control rod in this case compensating control rod is completely withdrawn with only fuel follower within active fuel region. In this configuration the neutron flux distribution inside the core is more homogeneous in the axial direction and therefore the maximum of the axial profile is approximately in the middle of the active fuel height (red dashed line in Figure 7). When both control rods are only partially withdrawn (black configuration) the fuel follower of both control rods is in the lower part of the reactor core and absorber is inside the top part, which leads to more fissions in the lower part of the reactor core and maximum in the fission rate axial profile shifts to lower axial position as can be observed in Figure 7. Another deviation between both profiles that can be observed are also heights of the local maxima, which are consequence of the graphite inserts inside the fuel rod, which act as neutron reflector. In case of completely inserted regulating control rod (red configuration) both local maxima have approximately the same height, this is consequence of homogeneously distributed fission rate in axial direction. When both control rods are only partially withdrawn (black configuration), local maximum in the lower part is higher, because there is fuel part of the control rod and the top local maximum is lower, because there is neutron absorber.

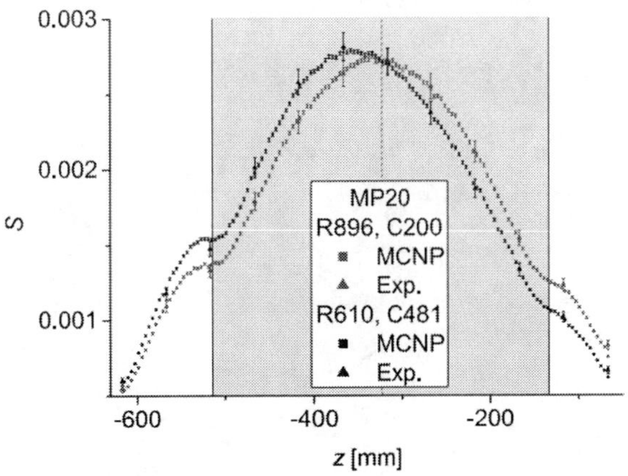

Figure 7. Comparison of fission rate axial profiles for MCNP calculations (squares) and experiment (triangles) at two different control rod positions. Red colour represents regulating control rod at step 896 and black colour regulating control rod at step 610. All distributions are normalized to the surface area below the graph. Error bars represent 1σ statistical and measurement uncertainties. Area marked in grey represents the active fuel region and red dashed line represents middle of the active fuel height.

6. Determination of optimal detector position

After successful validation of computational model the fission rate has been calculated throughout the entire reactor core at different control rod configurations [18]. Redistributions due to the control rod movement were studied by observing χ^2 distribution over the core. For evaluating the optimal detector position the χ^2 with respect to the different control rod positions as function of detector position has been calculated using the expression:

$$\chi^2(x_i, y_j, z_k) = \sum_m \left(\frac{N_{0,i,j,k} - N(x_i, y_j, z_k, m)}{\sigma_m} \right)^2, \quad (2)$$

where the sum is performed over different control rod configurations. N is a function of 3 spatial coordinates and therefore $N(x_i, y_j, z_k, m)$ represents calculated ^{235}U fission rates in fixed position inside the reactor core at m^{th} control rod configuration and σ_m represents statistical uncertainty of the individual $N(x_i, y_j, z_k, m)$. $N_{0,i,j,k}$ stands for average value of the calculated ^{235}U fission rates at fixed position and over the different control rod positions. Calculated χ^2 distribution of fission rate in two different plane views is presented in Figure 8. The χ^2 represents the deviations from the average value. In desired detector position (x_i, y_j, z_k) the deviations from the average detector signal due to the control rod movement should be minimal and therefore the calculated $\chi^2(x_i, y_j, z_k)$ should be minimized. It can be observed that the largest deviations are around control rods and other positions inside the reactor core have much lower variations. It can be concluded that the best option for positioning neutron detector would be on the reactor core central line, symmetrically to the regulating and compensating control rod and approximately in the middle of the active fuel height.

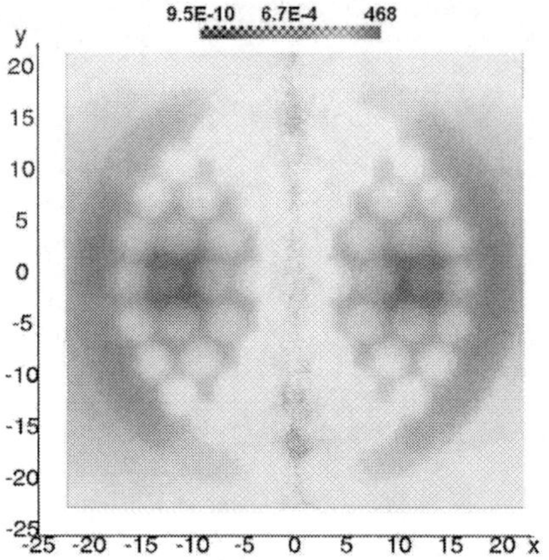

(a) xy plane

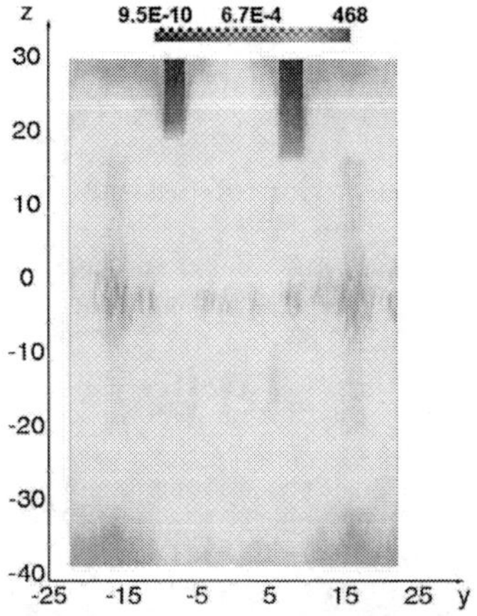

(b) yz plane

Figure 8. χ^2 distribution in different planes, approximately in the middle of the reactor core. Values of χ^2 are represented in colours ranging from low values in purple (dark) to high values in yellow (light).

7. Evaluation of experimental and computational uncertainty

The comparison between the measured profiles of absolute reaction rates and those computed by Monte Carlo method displays a relatively good agreement. The fission rate values match within 1 σ standard uncertainty at most measuring positions and for both fission chambers, which implies on a sufficient accuracy of the computational model of the TRIGA Mark II reactor. To complete the validation an evaluation of the measurement and calculational uncertainties was performed. The combined experimental and computational uncertainty assessment are presented in Table 1 and Table 2. Several major biases were also studied - this included the core temperature effect because of using cross section data at room temperature in the reference computational model of the TRIGA reactor. The latter was found to have a negligible effect on the fission rate computations due to the low thermal power of the reactor at which the measurements were performed. The largest identified bias was the fuel burnup. Namely the reference model used fresh fuel, which at a relatively low TRIGA fuel burnup ranging from 0.7 MWd to 4.8 MWd, introduced a fission rate bias of up to 3.5 %.

TABLE 1. Average experimental relative standard uncertainty of the fission chamber response.

Experimental uncertainty source	Average FC response uncertainty [%]
FC calibration	$\pm$ 3.2
Fuel composition	$\pm$ 1.4
^{234}U and ^{236}U impurities	$\pm$ 1.4
Hf impurities	$\pm$ 1.3
FC axial position	$\pm$ 2.6
FC radial position	$\pm$ 0.5
Control rod position	$\pm$ 0.3
Combined uncertainty	$\pm$ 4.8

TABLE 2. Average computational relative standard uncertainty of the fission chamber response.

Computational uncertainty source	Average FC response uncertainty [%]
Normalization factor K	$\pm$ 2.2
Nuclear data libraries	$\pm$ 1.5
MC statistical uncertainty	From $\pm$ 0.4 to $\pm$ 3
Combined uncertainty	From $\pm$ 2.7 to $\pm$ 4.0

8. Conclusion

The presented paper focused on the various reaction rate benchmark experiments that were performed using CEA developed miniature fission chambers at the TRIGA Mark II research reactor, which operates at the Jožef Stefan Institute in Slovenia. In support of the measurements a detailed Monte Carlo calculational model of the TRIGA reactor was developed, upgraded and extensively validated against experimental data. The experiments presented in the paper based on recording axial scans of fission reaction rates at multiple measuring positions in the reactor core. A complete model verification and estimation of different uncertainty sources was performed, which enabled an assessment of the combined experimental and calculational reaction rate uncertainty. The comparison between the experimentally and computationally gained data showed good agreement for both fission chamber types with the average discrepancies in the active fuel part being below 5 %. In addition the redistribution of the axial profile due to the control rod movement was also studied. It was observed that the position of control rods can affect the position of maximum as well as the heights of the local maxima. To minimize the effect of control rod movement on the detector signal optimal detector position inside the reactor core was determined to be on the reactor core central line, symmetrically in regard to the regulating and compensating control rod and approximately in the middle of the active fuel height.

Because there are very few experiments that have been conducted on the basis of absolute fission rate measurements and supported by an extensive uncertainty assessment, the experiment evaluation was included into the International Reactor Physics Experiment Evaluation Project Handbook to further contribute to the research reactor community [10].

References

[1] J. T. Goorley, M. R. James, T. E. Booth, F. B. Brown, J. S. Bull, L. J. Cox, J. W. Durkee, J. S. Elson, M. L. Fensin, R. A. Forster, J. S. Hendricks, H. G. Hughes, R. C. Johns, B. C. Kiedrowski, R. L. Martz, S. G. Mashnik, G. W. McKinney, D. B. Pelowitz, R. E. Prael, J. E. Sweezy, L. S. Waters, T. Wilcox, and A. J. Zukaitis, *Initial MCNP6 Release Overview - MCNP6 Version 1.0*, Los Alamos National Laboratory, 2014.

[2] L. Snoj and M. Ravnik, "Calculation of Power Density with MCNP in TRIGA Reactor," in *Proceedings of the International Conference Nuclear Energy for New Europe 2006*. Portorož, Slovenia, 18 - 21 September: Nuclear Society of Slovenia, 2006.

[3] L. Snoj, A. Kavčič, G. Žerovnik, and M. Ravnik, "Calculation of kinetic parameters for mixed TRIGA cores with Monte Carlo," *Annals of Nuclear Energy*, vol. 37, no. 2, pp. 223–229, feb 2010. [Online]. Available: http://linkinghub.elsevier.com/retrieve/pii/S0306454909003302

[4] L. Snoj, A. Trkov, R. Jaćimović, P. Rogan, G. Žerovnik, and M. Ravnik, "Analysis of neutron flux distribution for the validation of computational methods for the optimization of research reactor utilization," *Applied Radiation and Isotopes*, vol. 69, no. 1, pp. 136–141, jan 2011. [Online]. Available: http://linkinghub.elsevier.com/retrieve/pii/S0969804310003611

[5] R. Jeraj and M. Ravnik, "TRIGA Mark II Reactor: U(20)-Zirconium Hydride Fuel Rods in Water with Graphite Reactor, IEU-COMP-THERM-003," in *International Handbook of Evaluated Criticality Safety Benchmark Experiments*. Paris, France: NEA/NSC/DOC(95)03, OECD NEA, 2010.

[6] V. Radulović, Ž. Štancar, L. Snoj, and A. Trkov, "Validation of absolute axial neutron flux distribution calculations with MCNP with 197Au(n,γ)198Au reaction rate distribution measurements at the JSI TRIGA Mark II reactor," *Applied Radiation and Isotopes*, vol. 84, pp. 57–65, 2014.

[7] L. Barbot, C. Domergue, S. Breaud, C. Destouches, J.-F. Villard, L. Snoj, Ž. Štancar, V. Radulović, and A. Trkov, "Neutron field characterization of irradiation locations applied to the slovenian TRIGA reactor," in *2013 3rd International Conference on Advancements in Nuclear Instrumentation, Measurement Methods and their Applications (ANIMMA)*. IEEE, 2013, pp. 1–5. [Online]. Available: http://ieeexplore.ieee.org/document/6727954/

[8] T. Kaiba, G. Žerovnik, A. Jazbec, Ž. Štancar, L. Barbot, D. Fourmentel, and L. Snoj, "Validation of neutron flux redistribution factors in JSI TRIGA reactor due to control rod movements," *Applied Radiation and Isotopes*, vol. 104, 2015.

[9] G. Žerovnik, M. Podvratnik, and L. Snoj, "On normalization of fluxes and reaction rates in MCNP criticality calculations," *Annals of Nuclear Energy*, vol. 63, pp. 126–128, jan 2014. [Online]. Available: http://linkinghub.elsevier.com/retrieve/pii/S0306454913004003

[10] Ž. Štancar, L. Snoj, and L. Barbot, "Reaction Rate Distribution Experiments at the Slovenian JSI TRIGA Mark II Research Reactor, TRIGA-FUND-RESR-002," in *International Handbook of Evaluated Reactor Physics Benchmark Experiments*. Paris: NEA/NSC/DOC(2006)1, OECD NEA, 2017.

[11] Ž. Štancar and L. Snoj, "An improved thermal power calibration method at the TRIGA Mark II research reactor," *Manuscript submitted for publication*, 2017.

[12] Ž. Štancar, L. Snoj, L. Barbot, C. Destouches, D. Fourmentel, and J.-F. Villard, "Computational validation of the fission rate distribution experimental benchmark at the JSI TRIGA Mark II research reactor using the Monte Carlo method," *Manuscript submitted for publication*, 2017.

[13] L. Snoj, A. Trkov, M. Ravnik, and G. Žerovnik, "Testing of cross section libraries on zirconium benchmarks," *Annals of Nuclear Energy*, vol. 42, pp. 71–79, apr 2012. [Online]. Available: http://linkinghub.elsevier.com/retrieve/pii/S030645491100452X

[14] M. Chadwick, M. Herman, P. Obložinský, M. Dunn, Y. Danon, A. Kahler, D. Smith, B. Pritychenko, G. Arbanas, R. Arcilla, R. Brewer, D. Brown, R. Capote, A. Carlson, Y. Cho, H. Derrien, K. Guber, G. Hale, S. Hoblit, S. Holloway, T. Johnson, T. Kawano, B. Kiedrowski, H. Kim, S. Kunieda, N. Larson, L. Leal, J. Lestone, R. Little, E. McCutchan, R. MacFarlane, M. MacInnes, C. Mattoon, R. McKnight, S. Mughabghab, G. Nobre, G. Palmiotti, A. Palumbo, M. Pigni, V. Pronyaev, R. Sayer, A. Sonzogni, N. Summers, P. Talou, I. Thompson, A. Trkov, R. Vogt, S. van der Marck, A. Wallner, M. White, D. Wiarda, and P. Young, "ENDF/B-VII.1 Nuclear Data for Science and Technology: Cross Sections, Covariances, Fission Product Yields and Decay Data," *Nuclear Data Sheets*, vol. 112, no. 12, pp. 2887–2996, dec 2011. [Online]. Available: http://linkinghub.elsevier.com/retrieve/pii/S009037521100113X

[15] Ž. Štancar, *Computational analysis of power profile measurements in the TRIGA reactor (in Slovene)*. Master Thesis, 2014.

[16] Ž. Štancar, L. Snoj, and L. Barbot, "Evaluation of the Effect of Burn-Up on Neutron Flux and Reaction Rate Distributions in the TRIGA Mark II Reactor," in *Proceedings of the International Conference PHYSOR 2016 - Unifying Theory and Experiments in the 21st Century*. Sun Valley, Idaho, USA, May 1-5 2016: American Nuclear Society, 2016, pp. 1950–1959.

[17] G. Žerovnik, T. Kaiba, V. Radulović, A. Jazbec, S. Rupnik, L. Barbot, D. Fourmentel, and L. Snoj, "Validation of the neutron and gamma fields in the JSI TRIGA reactor using in-core fission and ionization chambers," *Applied Radiation and Isotopes*, vol. 96, pp. 27–35, feb 2015. [Online]. Available: http://linkinghub.elsevier.com/retrieve/pii/S0969804314003819

[18] T. Kaiba, G. Žerovnik, L. Barbot, D. Fourmentel, C. Destouches, A. Jazbec, and L. Snoj, "Evaluation of neutron flux and fission rate distributions inside the JSI TRIGA Mark II reactor using multiple in-core fission chambers," *Manuscript submitted for publication*, 2017.

[19] G. Žerovnik, L. Snoj, A. Trkov, L. Barbot, D. Fourmentel, and J.-F. Villard, "Measurements of Thermal Power at the TRIGA Mark II Reactor in Ljubljana Using Multiple Detectors," *IEEE Transactions on Nuclear Science*, vol. 61, no. 5, pp. 2527–2531, oct 2014. [Online]. Available: http://ieeexplore.ieee.org/lpdocs/epic03/wrapper.htm?arnumber=6906305

[20] M. Thevenin, L. Barbot, G. Corre, R. Woo, C. Destouches, and S. Normand, "Digital Real-Time Multiple Channel Multiple Mode Neutron Flux Estimation on FPGA-based Device," *EPJ Web of Conferences*, vol. 106, p. 05009, feb 2016. [Online]. Available: http://www.epj-conferences.org/10.1051/epjconf/201610605009

Impact of the Test Device on the Behavior of the Acoustic Emission Signals: Contribution of the Numerical Modeling to Signal Processing

Oumar Issiaka Traore, Paul Cristini, Nathalie Favretto-Cristini, Laurent Pantera and Sylvie Viguier-Pla.

***Abstract*—In a context of nuclear safety experiment monitoring with the non destructive testing method of acoustic emission, we study the impact of the test device on the interpretation of the recorded physical signals by using spectral finite element modeling. The numerical results are validated by comparison with real acoustic emission data obtained from previous experiments. The results show that several parameters can have significant impacts on acoustic wave propagation and then on the interpretation of the physical signals. The potential position of the source mechanism, the positions of the receivers and the nature of the coolant fluid have to be taken into account in the definition a pre-processing strategy of the real acoustic emission signals. In order to show the relevance of such an approach, we use the results to propose an optimization of the positions of the acoustic emission sensors in order to reduce the estimation bias of the time-delay and then improve the localization of the source mechanisms.**

***Index Terms*—Acoustic Emission, Reactivity Initiated Accident, Spectral Finite Element Modeling**

I. Introduction

Reactivity Initiated Accident (RIA) is a nuclear reactor accident that involves an unexpected and very fast increase in fission rate and reactor power due to the ejection of a control rod. The power increase may damage the fuel clad and the fuel pellets of the reactor. The French Alternative Energies and Atomic Energy Commission (CEA) operates a pool-type reactor dedicated to fuel behavior study in RIA conditions. During the experiments, several non-destructive methods are used to inspect the reactor and to get information on the behavior of the fuel clad and the fuel pellet. Among them the acoustic emission (AE) technique is of major interest. It has the advantage of being simple to adapt to nuclear-oriented purposes and allows a quasi-real-time monitoring of the experiment [1]–[5]. One goal of the AE testing is to process the AE signals in such a way that they can be associated with specific physical source mechanisms occurring in the tested structure or material. In most of the applications, the starting point of this processing is to clean as much as possible the received signal from the potential bias introduced by several parameters among which the choice and the setting of the AE sensors [6], [7], the environmental noise, the wave propagation path in the structure and the signals acquisition system.

Depending on the stochastic behavior of noise, several denoising methods exist in order to cope with the environmental noise [8]–[10]. In most of the cases, the transfer functions associated with the sensor and the acquisition system are known, a simple deconvolution is then enough to remove their impacts [11]. In the case of the wave propagation path, except in some rare cases (e.g. homogeneous media), the correction of its impact on the received signal is a complex problem. For instance, when the structure is complex such as the test device of a nuclear safety experiment reactor, the potential impacts of several parameters have to be taken into account, making difficult the estimation of the transfer function by experimental methods. Other tools like analytical computations or numerical simulations then become interesting alternatives to experimental methods. Indeed, contrary to experimental methods, they can easily integrate potential changes of the position of the source mechanism (so of the transfer function) and then the impact of these changes on the transfer function.

The main objective of this article is to use the numerical modeling in the particular case of RIA experiments in order to explore different scenarios of wave propagation in test device and compare the associated transfer functions. In addition, we also study the impact of a change of the heat transfer fluid (sodium or pressurized water) and the impact of changes of the positions of the AE sensors.

In the second section, we present and study the environment of the RIA experiment with a focus on the test device. Its geometry and configuration are studied in order to get a simplified problem. The third section is devoted to the numerical modeling of the test device by using a spectral finite element method. The obtained numerical signals are analyzed and the impacts of the position of the source mechanisms, of the position the sensors and of a change of coolant fluid are studied. In the last section, we discuss the results and analyze their impact on the experimental protocol.

II. Description of the Test Device

The experimental reactor of interest is an open pool-type research reactor composed of a driver core made of a thousand

Oumar Issiaka Traore, Paul Cristini and Nathalie Favretto-Cristini are with Aix-Marseille Univ., CNRS, Centrale Marseille, LMA, France (Email: toumarissiaka@gmail.com, cristini@lma.cnrs-mrs.fr, favretto@lma.cnrs-mrs.fr)

Laurent Pantera is with CEA, DEN, DER/SRES, Cadarache,F13108 Saint-Paul-Lez-Durance, France (Email: laurent.pantera@cea.fr)

Sylive Viguier-Pla is with Université de Perpignan via Domitia, LAMPS, 52 av. Paul Alduy, 66860 Perpignan Cedex 9, France and Université Paul Sabatier, 118 Route de Narbonne, F-31062 Toulouse Cedex 9, France of UO_2 rods and specially designed to support an injection of reactivity. The reactor includes an experimental loop specially designed to receive in the center of the driver core the instrumented test device housing the fuel rod to be tested. This test device is equipped with several sensors (including two piezoelectric AE sensors or microphones) to control the experiment and to characterize the behavior of the fuel rod during the power burst. These sensors are made of lithium niobate cristal with a pass-band of 5-400 kHz and designed to work in nuclear harsh environment up to 600 degrees Celsius. They are located at the top and the bottom of the fuel rod, respectively, and are distant of 2 m (Fig. 1). A (40 dB) preamplifier is associated with each sensor. The sampling rate is 2.5 μs.

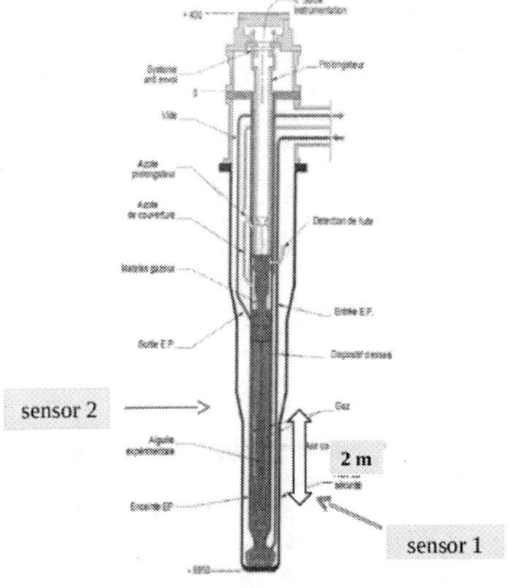

Figure 1: Illustrative picture of the complete test device with the positions of the two AE sensors

In such a context, we recorded the operating noise of the reactor during an experiment realized in 1997. The acquired signal and its spectrum are shown in Fig. 2. It can be observed a very strong component around the frequency of 180 kHz. As this signal is observed before the power burst, this 180 kHz component cannot correspond to a physical mechanism. However, it is present during all the experiment and biases the interpretation of the signals corresponding to source mechanisms. Then, it is of crucial importance to identify its origin in order to remove it in future experiments. The deterministic character of this $180kHz$ component makes us think that it can be associated with an environmental noise or with an impulse response of the test device but it is impossible to determine its origin only from data-processing. This is a typical example of a problem which can be addressed with numerical modeling. In this case, we guess that it will provide new insights that will help us to determine the origin of this strong frequency component.

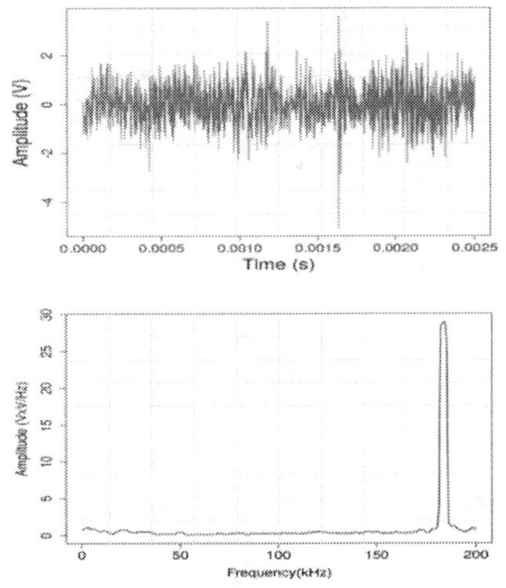

Figure 2: Example of operating noise of the reactor recorded during an experiment realized in 1997. Signal recorded at the second sensor (top) and its spectrum (bottom)

III. Modeling of wave propagation on the test device

As it is impossible to model the test device with all its complexity we chose to focus on a $3m$ long region which includes the positions of two AE sensors and the fuel rod to be tested. This limited domain is sufficient to study the wave propagation path from the potential positions of source mechanisms to the receivers. It corresponds to a stratified cylinder with eight layers of different types of media (cf. Fig. 3 and 4). Due to the presence of a Xenon layer (a neutron absorber) which has a much smaller impedance value than other layers, it is not necessary to consider wave propagation above this layer. We finally end up with a simplified model depicted in Fig. 4 whose physical characteristics are given in Table I.

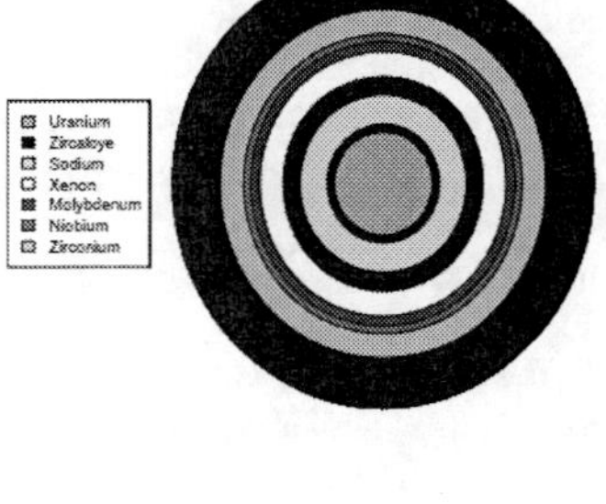

Figure 3: Radial section of the test device at $z = 0$. Illustration of the different types of media constituting the cylinder. Note that the number and the type of the layers can change depending on the position on the $z - axis$.

https://doi.org/10.1051/epjconf/201817004024

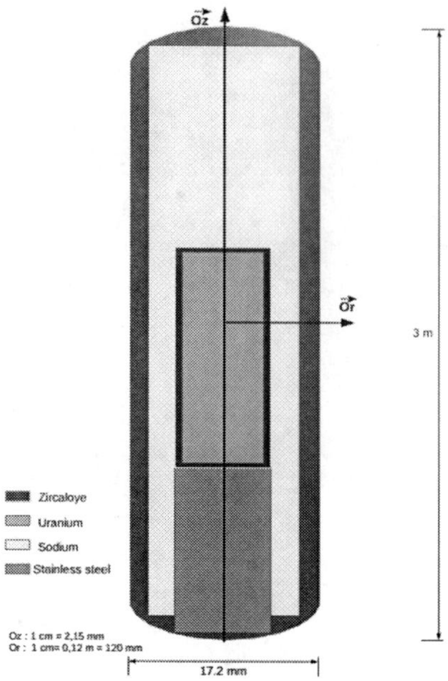

Figure 4: Longitudinal section of the simplified test device of $3m$ of length and $17.2mm$ of width.

Table I: Density and wave speeds for layers and materials associated with the simplified test device

Media	ρ (Kg/m^3)	c_l (m/s)	c_t (m/s)
Uranium	11000	3370	2020
Zircaloy	6500	4720	2360
Inox	7900	5460	3070
Sodium	968	2300	-
Water	1000	1436	-
Xenon	6	1090	-

For the sake of simplicity, we consider an axysimmetric model (with respect to the $z - axis$), which simplifies greatly the computations.

A. Brief overview on the chosen numerical tools

The spectral-element method belongs to the finite-element family of methods. It was initially developed for fluid mechanics [12] and is now commonly used for the solution of elastodynamics problems [13], [14]. It combines the accuracy of a pseudo spectral method with the flexibility of a finite-element method.
All numerical simulations were performed with the open-source SPECFEM2D software and meshes of the simplified model were performed with the GMSH software[15], [16].
One advantage of this numerical model is that it works in the time-domain. Among all the futures provide by the SPECFEM2D software, the possibility of generating source signal with a very wide-band spectrum analogous to a Dirac delta function is very attractive. It allows the generation of transfer functions from the source to the different receiver positions. It is thus possible to compare the results obtained from numerical simulations with experimental acquisitions.

B. Results Analysis

Fig. 5 shows the superposition of the normalized spectrum of the operating noise depicted in the Fig. 2 and of the transfer function of the test device with a source located on the fuel clad. We recover, with a slight shift toward high frequencies, the very strong frequency component around 180 kHz which has been observed in the experimental data (cf. Fig. 5). It can then be concluded that the origin of this frequency peak is associated with a resonance of the structure of the test device.

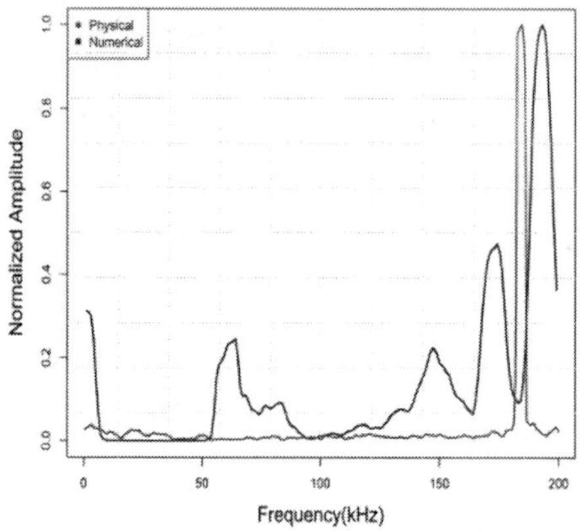

Figure 5: Superposition of the normalized spectra of the operating noise of the reactor (red) and of the transfer function of the test device associated with a source located on the fuel clad (blue)

In the following, we will use only use a Dirac type source. This allows to get the transfer functions associated with different combinations of source and sensors positions and then to analyze the impact of the position of the source mechanism (respectively of the receiver) on the physical signals. Note that the temporal signals corresponding to the spectra which will be depicted in the following have been normalized.

Fig. 6 corresponds to transfer functions associated with different source positions and one receiver position (the sensor 1). It can be seen that the transfer functions change from a position to another. The same results is observed for the second receiver or when analyzing the impact of the position of the receiver (cf. Fig.7). We then conclude that there is no global transfer function which can explain the impact of the test device on wave propagation. This result has two important practical consequences. The first one is that depending on the receiver position, one can observe changes on the spectrum for the same type of source mechanism. The second consequence is that it is not possible to use classical deconvolution methods to handle the impact of the source mechanism when its position is unknown.

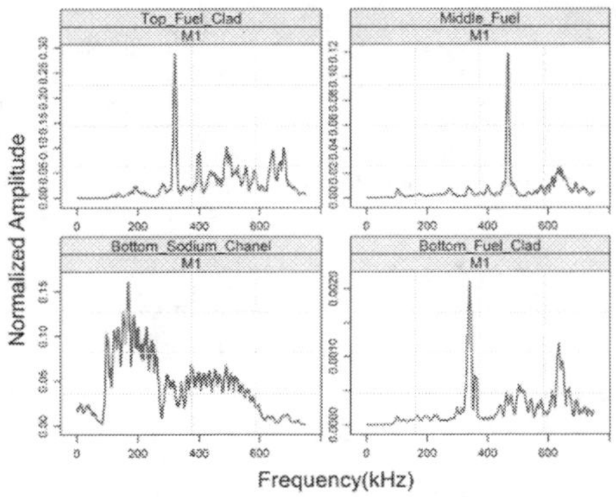

Figure 6: Illustration of the impact of the position of the source on the transfer function of the test device. Spectra of transfer functions associated with different positions of the source.

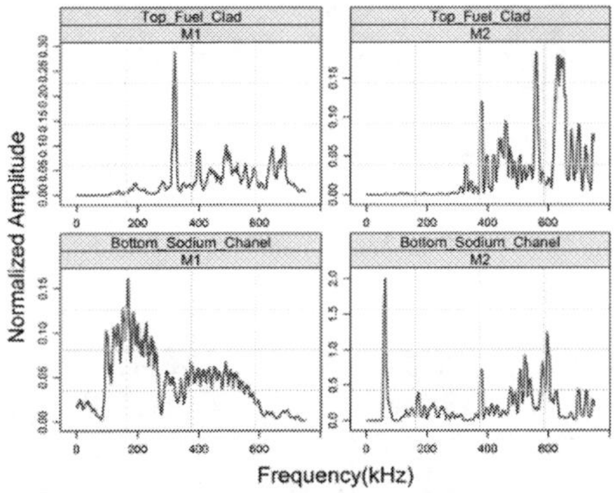

Figure 7: Illustration of the impact of the position of the receiver on the transfer function of the test device. Spectra of transfer functions associated with different positions of source and receiver.

We notice that these results are consistent from a theoretical point of view. Indeed, the test device being a stratified media, moving the position of the source mechanism (respectively of the receiver) from a layer to another leads to change the wave propagation path and then of the excitation of the resonant frequencies.

1) Detailed analysis of the transfer functions for different frequency ranges: Let us consider a source located at three different positions (bottom, middle and top) in the second layer associated with the fuel clad. Analysis of the spectra of the recorded transfer functions in the frequency range $[0, 200]$ kHz highlights two interesting results (cf. Fig. 8). First, we observe that up to $80kHz$ the transfer functions are quasi-constant. Thus, for source mechanisms situated in this layer (e.g. clad failures), there is no significant impact of the test device. Second, whatever the position of the sensor, we observe a peak of energy around $180kHz$ (more pronounced on sensor 2) associated with a resonant frequency of the test device (cf. Fig 5). This peak is not observed when sources are located in other layers indicating that the associated resonance is strongly connected to the structure of the fuel/clad couple.

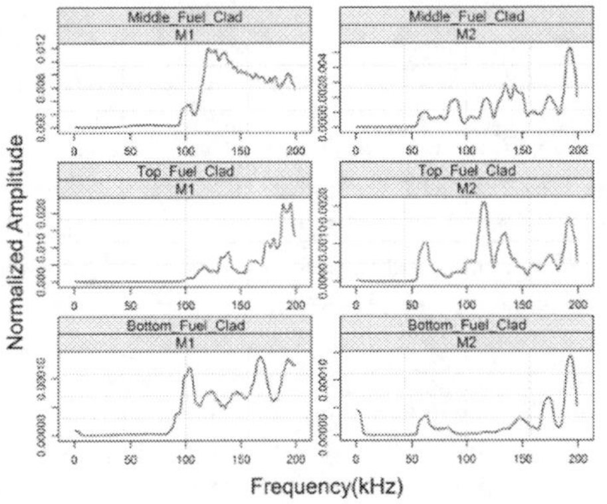

Figure 8: Focus on the spectra associated with different positions of the source in the frequency range $[0, 200]$ kHz.

2) Impact of a change of heat transfer fluid: For visualization purposes, we increase the sizes of the simplified model so that we are able to identify how the acoustic energy is distributed into the model. Three zones can be identified (cf. Fig. 9) independently of the nature of the heat transfer fluid. Nevertheless, it can be noticed that the oscillations present in zone 2 corresponding to the fluid part have a much higher frequency content with a water loop than with a sodium loop.

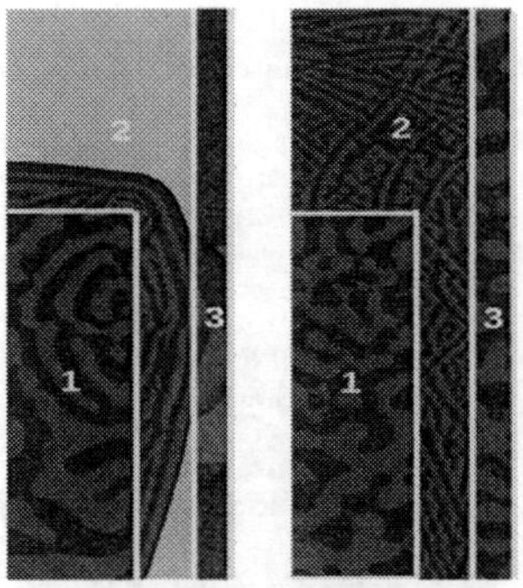

(a) Water loop

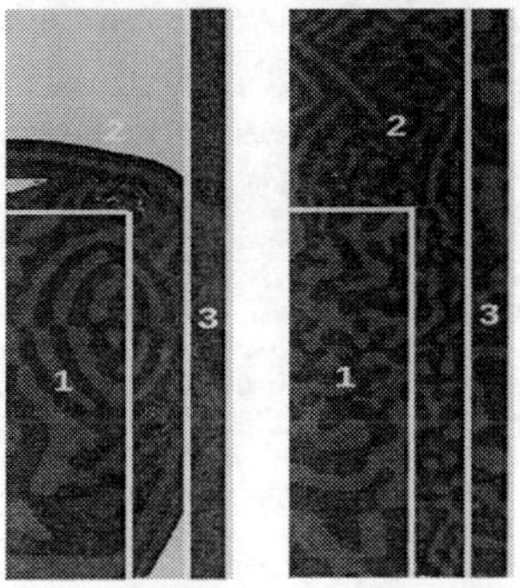

(b) Sodium loop

Figure 9: Snapshots of wave propagation with two different coolant fluids.

IV. Discussions and Conclusion

The objective of this article was to use numerical modeling to analyze the impact of the structure of the test device on recorded physical AE signals in the case of RIA experiments. A resonant frequency of the test device observed in physical signals has been identified with the help numerical computations. This resonant frequency being only connected to the structure of the test device, it can be removed safely from recorded signals, making their interpretation more effective. The results we obtained showed that there is a strong dependence of the transfer function on the source/receiver relative positions. It is thus impossible to define a global transfer function. We also conclude that it is not possible to use classical deconvolution methods when the source position is unknown. Additionally, numerical simulations suggests that frequencies up to 80 kHz have to be preferred because of the flat response of the transfer functions in this frequency band. No perturbation on the signal coming from the structure of the test device should be expected in this frequency band. A perspective of this work which is a direct consequence of the feasibility of performing numerical computations is the possibility of optimizing the receiver positions. From the previous results, we can assert that AE sensors should always be located before the Xenon layer and should be put at the same distance from the $z - axis$. This would reduce the estimation bias of the time-delay between the two AE sensors and then improve the source mechanism localization.

References

[1] SP. Ying. The use of acoustic emission for assessing the integrity of a nuclear reactor pressure vessel. *NDT International*, 12(4):175–179, 1979.

[2] P. Runow. The use of acoustic emission methods as aids to the structural integrity assessment of nuclear power plants. *International Journal of Pressure Vessels and Piping*, 21(3):157–207, 1985.

[3] Q. Ai, C. Liu, X. Chen, P. He, and Y. Wang. Acoustic emission of fatigue crack in pressure pipe under cyclic pressure. *Nuclear Engineering and Design*, 240(10):3616–3620, 2010.

[4] L. Hégron, P. Sornay, and N. Favretto-Cristini. Compaction of a bed of fragmentable particles and associated acoustic emission. *IEEE Transactions on Nuclear Science*, 61(4):2175–2181, 2014.

[5] N. Favretto-Cristini, L. Hégron, and P. Sornay. Identification of the fragmentation of brittle particles during compaction process by the acoustic emission technique. *Ultrasonics*, 67:178–189, 2016.

[6] J. Roget. *Essais non destructifs: L'émission acoustique: Mise en oueuvre et applications*. AFNOR; CETIM, 1988.

[7] G. Gautschi. *Piezoelectric sensorics: force, strain, pressure, acceleration and acoustic emission sensors, materials and amplifiers*. Springer Science & Business Media, 2002.

[8] O.I. Traoré, L. Pantera, N. Favretto-Cristini, and S. Viguier-Pla. Emission acoustique et traitement du bruit : cas de signaux expérimentaux en contexte nucléaire. In *13e Congrès Français d'Acoustique*, CFA 2016, Le Mans, France, April 2016.

[9] O.I. Traore, L. Pantera, N. Favretto-Cristini, P. Cristini, S. Viguier-Pla, and P. Vieu. Structure analysis and denoising using singular spectrum analysis: application to acoustic emission signals from nuclear safety experiments. *Measurement*, 104:78 – 88, 2017.

[10] V. Barat, Y. Borodin, and A. Kuzmin. Intelligent acoustic emission signal filtering methods. *Journal of Acoustic Emission*, 28:109–119, 2010.

[11] C. U. Grosse and M. Ohtsu. *Acoustic emission testing*. Springer Science & Business Media, 2008.

[12] Y. Maday and A.T. Patera. Spectral element methods for the incompressible navier-stokes equations. In *IN: State-of-the-art surveys on computational mechanics (A90-47176 21-64). New York, American Society of Mechanical Engineers, 1989, p. 71-143. Research supported by DARPA.*, pages 71–143, 1989.

[13] D. Komatitsch. *Spectral and spectral-element methods for the 2D and 3D elastodynamics equations in heterogeneous media*. Theses, Institut de physique du globe de paris - IPGP, May 1997.

[14] J. Tromp, D. Komattisch, and Q. Liu. Spectral-element and adjoint methods in seismology. *Communications in Computational Physics*, 3(1):1–32, 2008.

[15] C. Geuzaine and J.F. Remacle. Gmsh: A 3-d finite element mesh generator with built-in pre-and post-processing facilities. *International Journal for Numerical Methods in Engineering*, 79(11):1309–1331, 2009.

[16] C. Geuzaine and J.F Remacle. *Gmsh Reference Manual*. http://www.geuz.org/gmsh, 2.11.0 edition, November 2015.

In-Pile Qualification of the Fast-Neutron-Detection-System

D. Fourmentel, J-F. Villard, C. Destouches, B. Geslot, L. Vermeeren, M. Schyns

***Abstract*— In order to improve measurement techniques for neutron flux assessment, a unique system for online measurement of fast neutron flux has been developed and recently qualified in-pile by the French Alternative Energies and Atomic Energy Commission (CEA) in cooperation with the Belgian Nuclear Research Centre (SCK•CEN). The Fast-Neutron-Detection-System (FNDS) has been designed to monitor accurately high-energy neutrons flux (E > 1 MeV) in typical Material Testing Reactor conditions, where overall neutron flux level can be as high as 10^{15} n.cm^{-2}.s^{-1} and is generally dominated by thermal neutrons. Moreover, the neutron flux is coupled with a high gamma flux of typically a few 10^{15} γ.cm^{-2}.s^{-1}, which can be highly disturbing for the online measurement of neutron fluxes.**

The patented FNDS system is based on two detectors, including a miniature fission chamber with a special fissile material presenting an energy threshold near 1 MeV, which can be ^{242}Pu for MTR conditions. Fission chambers are operated in Campbelling mode for an efficient gamma rejection. FNDS also includes a specific software that processes measurements to compensate online the fissile material depletion and to adjust the sensitivity of the detectors, in order to produce a precise evaluation of both thermal and fast neutron flux even after long term irradiation.

FNDS has been validated through a two-step experimental program. A first set of tests was performed at BR2 reactor operated by SCK•CEN in Belgium. Then a second test was recently completed at ISIS reactor operated by CEA in France. FNDS proved its ability to measure online the fast neutron flux with an overall accuracy better than 5%.

***Index Terms*—Nuclear Measurements, Fission Chamber, Reactor Instrumentation**

I. Introduction

FAST neutron flux (E ≥ 1 MeV) is a key neutron parameter in nuclear reactors. Its measurement is particularly relevant to assess material damage under irradiation. In very low power research reactors, such as ZPRs (Zero Power Reactors), the fast neutron flux is usually measured on-line with specific fission chambers, containing for example ^{238}U coating. In more powerful reactors, such as Material Testing Reactors (MTRs) or nuclear power plants (NPPs), techniques used in ZPRs are not appropriate. This is mainly due to the rapid evolution of the fissile coating under high neutron flux that reduces the sensitivity of the sensor to the 'fast' spectrum domain (cf. II.B). This neutron parameter is therefore not accessible on-line and is generally evaluated afterward by activation dosimetry or by Monte Carlo calculations. However, in the current context of increasing requirements for in-pile measurements, particularly for experiments in MTRs, the on-line measurement of fast neutron flux has become a major objective.

In order to ensure the quality and the relevance of irradiation programs in the future Jules Horowitz Reactor (JHR), the French Alternative Energies and Atomic Energy Commission (CEA) has significantly increased its research and development effort in the field of in-pile instrumentation during the last decade. In this context and in the framework of the Joint Instrumentation Laboratory between the CEA and the Belgian Nuclear Research Centre (SCK•CEN), a measurement system dedicated to on-line measurement of fast neutron flux has been developed.

II. FNDS specifications

On-line measurements of fast neutron flux are based on the use of a fission chamber with a specific fissile coating, presenting an energy threshold near the fast domain (E > 1 MeV). If the use of such fission chamber is not a problem at very low neutron flux, it faces major difficulties under intense mixed neutron and gamma fluxes. First of all, because of the low neutron sensitivity of this type of fission chamber, the high gamma flux of MTRs leads to a high gamma contribution to the signal. Furthermore, neutron captures on the nuclei of the fissile coating create isotopes that have a high cross-section in the thermal spectrum domain. Thus the 'fast' sensitivity of the detector decreases rapidly. Isotopic impurities of the coating must also be taken into account. Figure 1 illustrates this difficulty for the use of ^{242}Pu, which has a suitable response in the fast spectrum domain [1].

We can notice that impurities and isotopes created by neutron captures induce an increasingly high sensitivity to thermal neutrons.

D. Fourmentel, J-F. Villard and C. Destouches are with the Alternative Energies and Atomic Energy Commission: CEA, DEN, DER, Instrumentation Sensors and Dosimetry Laboratory, Cadarache, 13108 Saint-Paul-Lez-Durance, France (e-mail: damien.fourmentel@cea.fr).

B. Geslot is with the Alternative Energies and Atomic Energy Commission: CEA, DEN, DER, Experimental Programs Laboratory, Cadarache, 13108 Saint-Paul-Lez-Durance, France.

L. Vermeeren and M. Schyns are with the SCK·CEN Belgian Nuclear Research Centre, Boeretang 200, B-2400 Mol, Belgium.

https://doi.org/10.1051/epjconf/201817004025

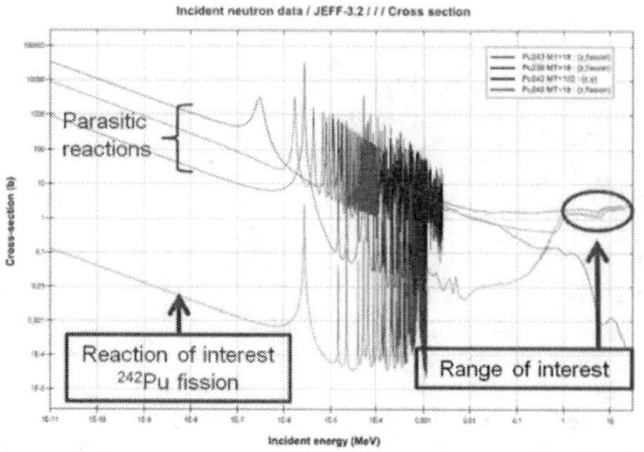

Fig. 1. Fission cross sections of a typical ^{242}Pu fissile coating showing that impurities and isotopes created by capture must be taken into account for fast neutron flux measurements

A. FNDS general description

In order to compensate difficulties mentioned above, FNDS includes an acquisition system allowing measurements with fission chambers in Campbelling mode. This mode allows a very efficient rejection of the gamma contribution [2] while offering a very wide dynamic range. FNDS also integrates signal processing software that firstly calculate in real time the evolution of the fissile coating of the fission chamber using an evolution code DARWIN [3], and that secondly deduce the corresponding evolution of the neutron sensitivity of the fast fission chamber. This software extracts in real-time the value of the fast neutron flux from the signal.

In order to allow a precise evaluation of thermal neutron fluence received by the 'fast' fission chamber, FNDS is equipped with a second detector that is sensitive to thermal neutron flux and positioned near the 'fast' fission chamber. This second detector may be a fission chamber with ^{235}U coating or a Self-Powered Neutron Detector (SPND).

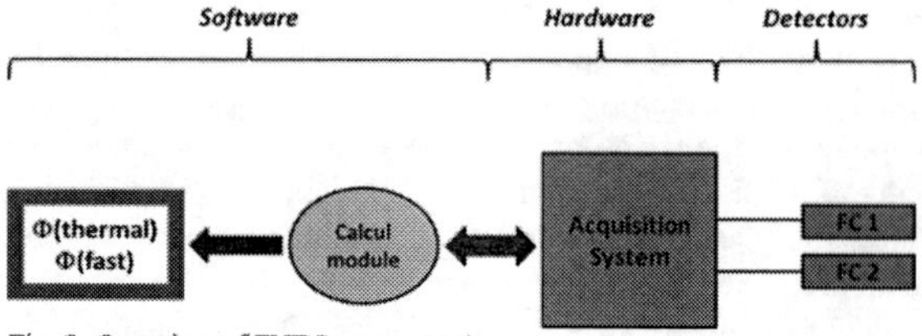

Fig. 2. Overview of FNDS components

FNDS thus consists in 2 detectors for 'fast' and 'thermal' spectrum domains, an acquisition system operating in Campbelling mode and a calculation module for on-line analysis of neutron fluxes (Figure 2). These innovations were patented for the detectors [4] and for the entire measurement system [5].

B. Detectors

An important bibliography study was carried out to establish the state of the art of neutron and photonic measurements in MTRs [6], and led us to select fission chamber sensors. It was shown that the most suitable fissile isotope for the estimation of the fast neutron flux is ^{242}Pu under the typical irradiation conditions of MTRs, such as OSIRIS (CEA Saclay – France), BR2 (Mol – Belgium) and JHR reactors [1, 7]. Figure 3 illustrates the raison why ^{242}Pu coating is the best choice compared to other fissile isotopes; it shows the stability of its fast neutron sensitivity at high fluence.

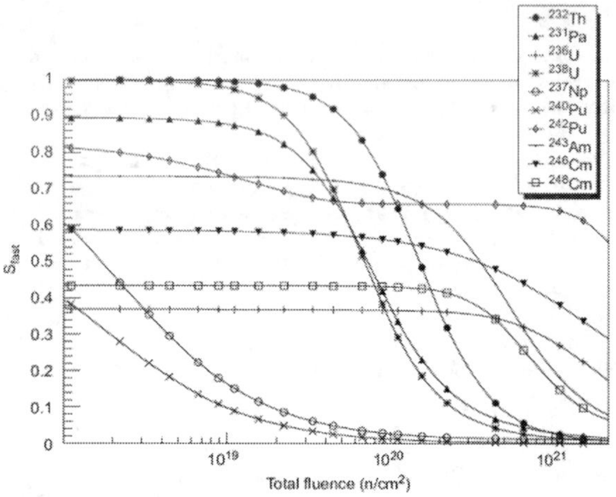

Fig. 3. Sensitivity to fast neutrons for initially pure deposits under typical BR2 neutron spectrum [3]

A typical 'fast' fission chamber selected for MTRs contains 100 µg of ^{242}Pu while the typical associated 'thermal' detector contains 10 µg of ^{235}U. Both sensors have a 3 mm outer diameter body and integrated mineral insulated cable.

These sensors have been developed and manufactured at the Fission Chamber Manufacturing Workshop located in CEA Cadarache.

C. Acquisition system

The acquisition system has 2 channels for the sensor couple. Each channel is equipped with a high-voltage power supply for polarization with adjustable voltage (typically 250 V to reach the saturation plateau of a 3 mm fission chamber). The measured low-level current signals pass through a preamplifier (manufactured by CEA Saclay), amplification stage and then are filtered and digitalized for processing with the measurement interpretation software.

There are typically 3 modes to operate fission chambers: pulse, fluctuation (or Campbelling) and current modes. In MTRs conditions, the advantage of the Campbelling mode is the radical limitation of the gamma contribution on the signal [2] compared to current mode. Indeed, MTR neutron fluxes are too high to operate such fission chambers in pulse mode.

In Campbelling mode, the measurement of the variance of low fission chamber signal with a specific bandwidth has required the development of a dedicated acquisition system. As such a system was not available from Industry; prototypes were specially designed at CEA. Studies are on-going for the industrialization of such acquisition system in a short term [7].

D. Processing software

The FNDS processing software calculates the sensitivity of each fission chamber to the thermal, epithermal and fast neutrons from specific input data (variance measurements, calibration coefficients, mass and isotopic composition of the fissile coatings, nuclear data library, etc.) and then calculates the thermal and fast neutron fluxes versus time.

Processing software includes DARWIN code [3], developed by the CEA, to compute fission cross sections of each isotope present in the fissile coating and to calculate the isotopic evolution of these coatings with fluence and time. The method to combine measurements and estimate fast neutron flux is detailed in [6, 8].

III. In-Pile qualification of FNDS

FNDS has been validated through a two-step experimental program. A first set of tests was performed at BR2 reactor through the FICTIONS-8 experiment. Two FNDS prototypes were operated in-pile during nearly 1000 hours. This experiment exhibited the consistency of the measurement of thermal to fast neutron flux ratio with Monte Carlo calculations, as well as the right compensation of fissile material depletion [9].

A second set of tests was then necessary to qualify the absolute fast neutron flux measurement process. The main requirement was to perform measurements in well-controlled conditions (neutron spectrum, flux gradient, etc.) in order to minimize external sources of uncertainty and bias. In this context, it was decided to carry out the EFICAF (Estimation of Flux in Isis by Fission Chamber) experiment in ISIS reactor (CEA Saclay, France). ISIS reactor allows stable and reproducible irradiation conditions. Its neutron flux level is also sufficient to operate fission chambers in Campbelling mode. Moreover, its flexibility allowed to carry out all the measurements that were necessary for the qualification (on-line measurements and activation dosimetry measurements without external disturbance).

Table 1 and Table 2 summarize the results of respectively thermal and fast neutron fluxes measured by FNDS and compared with values given the TRIPOLI4 Monte-Carlo code and activation dosimetry.

TABLE I
THERMAL NEUTRON FLUX EVALUATIONS AND DEVIATIONS BETWEEN FNDS AND TRIPOLI4 CALCULATIONS AND BETWEEN FNDS AND ALCO1% MEASUREMENTS

	CF U235 ($n.cm^{-2}.s^{-1}$)	(TRIPOLI/CF U235)-1	(AlCo 1%/CF U235)-1
1st measurements	9.34E+11	11.0%	7.5%
2nd measurements	9.15E+11	13.3%	9.7%

TABLE II
FAST NEUTRON FLUX EVALUATIONS AND DEVIATIONS BETWEEN FNDS AND TRIPOLI4 CALCULATIONS AND BETWEEN FNDS AND FE, NI MEASUREMENTS

	CF Pu242 ($n.cm^{-2}.s^{-1}$)	(TRIPOLI/CF Pu242)-1	(Fe-Ni/CF Pu242)-1
1st measurements	4.39E+11	7.9%	4.4%
2nd measurements	4.40E+11	7.7%	4.1%

These results confirm the good agreement for absolute thermal neutron flux using Campbelling mode operation and the corresponding calibration process [10]. The deviation between FNDS fast neutron flux measurements and dosimetry results is mainly covered by the uncertainties of the activation dosimetry measurements (typically 4% for fast neutron flux evaluation). The results of this experiment confirm the quality of the on-line fast neutron flux measurements at a few 10^{11} $n.cm^{-2}.s^{-1}$ ($E > 1$ MeV) with this type of miniature fission chambers and therefore confirm FNDS capability to measure in real-time the absolute thermal and fast neutron fluxes with an excellent accuracy.

IV. Conclusion

EFICAF experiment demonstrated the ability of FNDS to measure on-line thermal and fast neutron fluxes. Deviations between fast neutron flux measured by FNDS and activation dosimetry results are below 5%. These results confirm the excellent accuracy of the analysis method using a couple of ^{235}U and ^{242}Pu fissile coatings and special calibration in Campbelling mode.

FNDS is now operational and is assumed to be the first and unique acquisition system able to provide an on-line measurement of the fast neutron flux in MTR conditions. This system will of course be used to perform spectral neutron characterization of JHR channels, but it may also be implemented in future irradiation experiments, for a better and real-time evaluation of the fast neutron flux received by material and fuel samples.

References

[1] P. Filliatre *et al.*, "Reasons why Plutonium 242 is the best fission chamber deposit to monitor the fast component of a high neutron flux", Nuclear Instruments and Methods in Physics Research A 593 (2008) 510–518.

[2] L. Vermeeren *et al.*, "Experimental Verification of the Fission Chamber Gamma Signal Suppression by the Campbelling Mode", IEEE Transactions on Nuclear Science, vol. 58, n°2, April 2011.

[3] A.Tsilanizara *et al.*, J.Nucl.Sci.Technol.37 (2000)845.

[4] "Détecteur pour la mesure en ligne des neutrons rapides dans un réacteur" Patent FR 2925750 A1, 2009.

[5] "Dispositif de mesure en ligne d'un flux de neutrons rapides et épithermiques" Patent WO 2010043554 A1, 2010.

[6] D. Fourmentel, "Mesures neutroniques et photoniques combinées pour la caractérisation précise des canaux expérimentaux du futur réacteur d'irradiation Jules Horowitz (RJH)", PhD thesis, Aix-Marseille University, 2013.

[7] M. Thevenin *et al.*, "Digital Real-Time Multiple Channel Multiple Mode Neutron Flux Estimation on FPGA-based Device", International Symposium on Reactor Dosimetry (ISRD 2015), 18-23 May 2015, Aix-en-Provence, France.

[8] P. Filliatre, L. Oriol, C. Jammes *et al.*, "Joint estimation of the fast and thermal components of a high neutron flux with a two on-line detector system", Nuclear Instruments And Methods in Physics Research Section A, 603 (3), p. 415-420, 2009.

[9] B. Geslot, *et al.*, "New measurement system for on line in core high-energy neutron flux monitoring in materials testing reactor conditions" Review of Scientific Instruments, Vol. 82, 033504 (2011).

[10] D. Fourmentel, P. Filliatre, L. Barbot *et al.*, "Comparison of Thermal Neutron Flux Measured by 235U Fission Chamber and Rhodium Self-Powered Neutron Detector in MTR", IEEE Transactions on Nuclear Science, Vol. 61, No. 4, 2014.

EPJ Web of Conferences 170, 04026 (2018) https://doi.org/10.1051/epjconf/201817004026
ANIMMA 2017

Study of the Flow Temperature and Ring Design Influence on the Response of a New Reduced-Size Calorimetric Cell for Nuclear Heating Quantification

A. Volte, C. Reynard-Carette, J. Brun, C. De Vita, M. Carette, T. Fiorido, A. Lyoussi, D. Fourmentel, J-F. Villard and P. Guimbal

Abstract—**This paper concerns experimental studies of different designs of a new compact calorimetric cell under laboratory conditions. This kind of cell is used for the measurement of the nuclear heating rate inside Material Testing Reactors thanks to differential calorimetry. The results, obtained by applying an operating protocol corresponding to a preliminary out-of-pile calibration step, are presented for three designs. The influence of the horizontal-fin design is shown on the calibration curve and the sensor sensitivity. The influence of the external fluid flow temperature is given for the quarter design. The different responses of the calorimetric cell are explained by taken into account a 1D analytical thermal model coupling thermal conductive and radiative transfers.**

Index Terms—**Calorimetry, New design, Calibration curve, Energy deposition rate, Nuclear fission.**

I. Introduction

CONTINOUS improvements in the nuclear energy field require research works. For instance, the enhancement of nuclear fuels or a better knowledge of material damages may benefit from feedbacks from existing nuclear power reactors, from qualified numerical models, and from experimental tests and qualification programs conducted in research infrastructures such as gamma or X ray sources, electron or ion accelerators, Zero Power Reactors and Material Testing Reactors (MTRs). To perform these kinds of programs, innovative and accurate instrumentation is needed to provide more and more accurate results and consequently to better understand complex phenomena occurring under harsh nuclear conditions.

This paper deals with research works on specific sensors called calorimeters dedicated to measuring online the energy deposition rate per units of mass induced by radiation interactions with matter in experimental channels of MTRs. The nuclear heating rate is a crucial parameter required during thermal and mechanical designs of experimental devices and after during result interpretation. Nuclear heating rate measurements in MTRs are usually performed directly by using single-cell calorimeters (including gamma thermometers) or differential calorimeters [1-4]. The measurement principle of these non-adiabatic calorimeters is almost similar and is based on temperature rise measurements by means of thermocouple (absolute temperatures or temperature differences). Then steady in-pile temperatures are associated to preliminary calibration curves (obtained under laboratory conditions or inside a reactor) to determine the nuclear heating rate. A single-cell calorimeter is composed of a jacket filled with conductive gas, a sample at its center, one or two thermocouples and sometimes a heating element. A differential calorimeter is made of two identical calorimetric cells at least (one containing a sample and the other kept empty to serve as a reference for energy deposition in the calorimetric-cell structure) with each cell instrumented by two thermocouples and a heating element dedicated to the in-pile or out-of-pile calibration.

Since 2009, Aix-Marseille University and French Alternative Energies and Atomic Energy Commission have been working together to improve calorimetric measurements thanks to differential calorimeter in particular and propose innovative sensors for the future Jules Horowitz Reactor (JHR) which is under construction at CEA Cadarache. The major challenges in these studies are to design sensors able to be operated up to a nuclear heating level of 20 $W.g^{-1}$, and to reduce their size.

This paper focuses on experimental studies of three configurations of a new compact design of a differential calorimeter called CALORRE, and its tests under off-pile laboratory conditions without nuclear radiations.

The first part describes the experimental set-up previously developed [1,—3], the design of the three CALORRE calorimetric cells and the operating protocol applied to obtain the temporal responses of the sensors and to deduce their calibration curves.

A. Volte, C. Reynard-Carette, J. Brun, M. Carette, C. De Vita, and T. Fiorido are with Aix Marseille Univ, Université de Toulon, CNRS, IM2NP, Marseille, France (e-mail: adrien.volte@im2np.fr, christelle.carette@univ-amu.fr, julie.brun@univ-amu.fr, michel.carette@univ-amu.fr, cedric.devita@im2np.fr, tomas.fiorido@im2np.fr). A. Lyoussi, D. Fourmentel, and J-F. Villard are with CEA, DEN, DER, Instrumentation Sensors and Dosimetry Laboratory, Cadarache, F-13108, Saint Paul-lez-Durance, France (e-mail: abdallah..lyoussi@cea.fr, damien.fourmentel@cea.fr, jean-francoi.villard@cea.fr). P. Guimbal is with CEA, DEN, DER, Jules Horowitz Reactor Section, Cadarache, F-13108, Saint Paul-lez-Durance, France (e-mail: philippe.guimbal@cea.fr)

The second part is dedicated to the presentation of the experimental results: ~~the~~ responses of the quarter-design configuration obtained for various thermal conditions (cooling fluid temperatures from 23°C to 63°C and electrical power up to 6W) and ~~the~~ calibration curves for the three designs at constant temperature equal to 33°C. Then an analysis is performed thanks to a theoretical 1-direction thermal model based on an energy balance in the radial direction by considering thermal conductive exchanges in the horizontal fin, conductive exchanges in the gas layers located between the head and the vertical fin and the radiative exchanges between the outer surface of the head and the inner surface of the vertical fin.

The contribution of each heat transfer is discussed for all configurations to explain the different calibration curves.

II. Experimental set-up and operating protocol

A. *Experimental set-up*

The experimental set-up is composed of a thermostatic bath which contains a cylindrical jacket hosting a calorimetric cell, a power supply, a computer and an acquisition system linked to a high accuracy burster 0.1Ohm (cf. Fig.1.). The bath allows the control of the velocity for a large temperature range (from 23 to 63°C for our studies) of the cooling flow fluid (water in this case) that circulates in a parallel tube with a height of 400mm and a diameter of 99mm. So, the cooling flow that operates in a vertical channel of an experimental research reactor is simulated by the vein of the bath. The Joule's effect generates with the power supply under non irradiation conditions inside each calorimetric cell imitates the nuclear heating rate occurring in materials inside the reactor. Temperatures of the calorimetric cell, the injected electrical current and the voltage are measured by means of an Agilent acquisition system coupled with 2 K-type thermocouples and a high accuracy burster respectively.

Three calorimetric cells were successively inserted inside the outer jacket and immersed into the vein of the bath to determine their response and establish their calibration curve. Each calorimetric cell is composed of a cylindrical head containing a sample and a heating element, a horizontal fin and a vertical fin. These three concentric parts are made of stainless steel (AISI 316L).

Fig. 1. Experimental set up used to obtain laboratory calibration curves under laboratory conditions.

Two K-type thermocouples are inserted inside the horizontal fin: one close to the external surface of the head (T_{hot}) and one near the internal surface of the vertical fin at 180° (T_{cold}). For each design a different geometry of the horizontal fin was tested. The heights of the calorimetric cells and of the horizontal fin are kept constant: 23.1mm for the cell and 0.5mm for the horizontal fin.

Figure 2 gives the picture of the quarter design of the CALORRE calorimetric cell and schemes of all tested designs. The quarter design has four metal sectors with an angle equal to 22.5° and four empty sectors with an angle equal to 67.5°. The two other designs correspond to a full horizontal fin called full design and a half design with eight metal sectors and eight empty sectors with a same angle equal to 22.5° respectively.

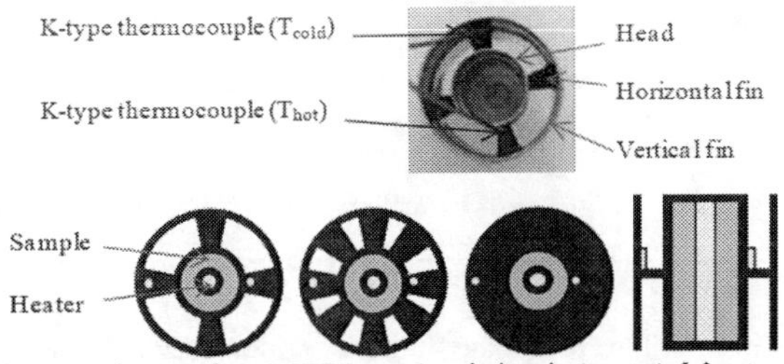

Fig. 2. Picture of CALORRE quarter design instrumented by two thermocouples and diagrams of the three designs (top-views of quarter, half and full designs and a general cross section).

B. *Operating protocol*

For each velocity and temperature, a specific electrical current is injected inside the heating element in such a way that the power range is from 0 to 6W. Temperatures and electrical power (P) are measured as a function of the time. An increment of 0.5W is carrying out at intervals of 30 minutes to have an acquisition time higher than the response time of the sensor in order to determine the mean temperature for a long steady state duration (cf. Fig.3.). For each applied electrical current the difference between the mean steady hot temperature and the mean steady cold temperature (ΔT) is calculated and plotted as a function of the injected electrical power. This kind of curve is called calibration curve (cf. Fig.4).

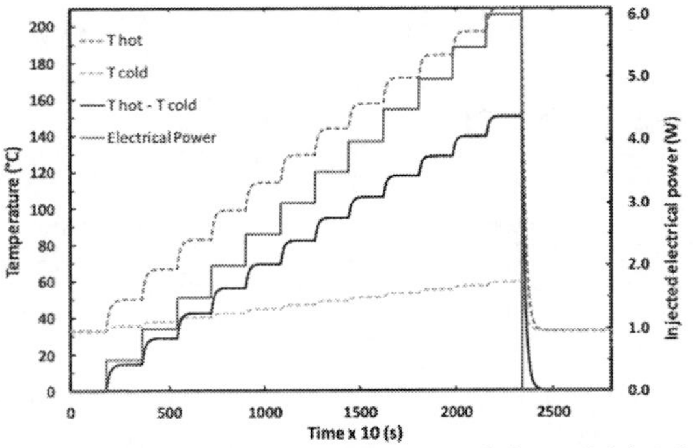

Fig. 3. Temperatures of CALORRE quarter design and injected electrical power versus time.

This protocol was followed for different temperatures (from 23°C to 63°C) and velocities (from Re=609 to Re=1607) for the quarter design.

III. EXPERIMENTAL RESULTS

Figure 4 presents the mean temperature difference versus the injected electrical power in the case of the quarter design of CALORRE for five fluid flow temperatures and a same fluid velocity corresponding to Re=1607.

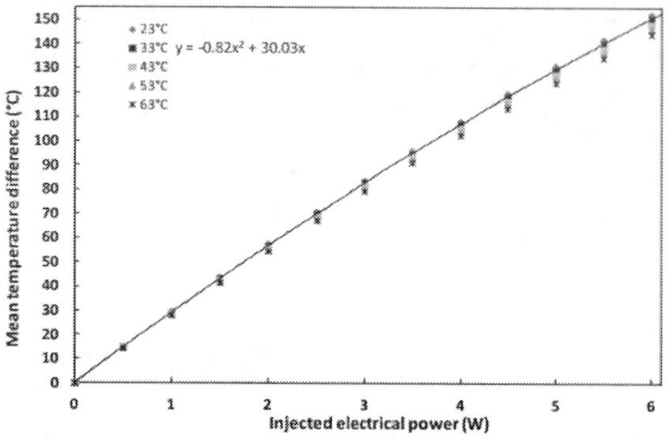

Fig. 4. CALORRE quarter design calibration curves obtained for several fluid flow temperatures.

First of all, whatever the fluid temperature the calibration curve is not linear. It corresponds to a second-order polynomial fit. Consequently, the sensitivity of this calorimetric cell decreases when the electrical power is increased. This behavior was already observed for CARMEN type calorimeter [1,3-4]. The flow temperature change leads to a variation of the calibration curve of the quarter-design cell. The mean temperature difference obtained for a same electrical power decreases with an increase of the fluid flow temperature.

Table 1 gives the calibration data obtained for the three designs of CALORRE for a same fluid temperature equal to 33°C. These three designs have a second-order polynomial calibration curve too. More the horizontal fin contains empty sectors, higher is the sensitivity and the non-linearity. The sensitivity of the quarter design is higher than CALMOS type calorimeter sensitivity [2] and close to CARMEN type calorimeter sensitivity used during the MARIA campaign [5]. This is due to the different natures of material used for the calorimetric cell structures: stainless steel for CALORRE type calorimeter instead of aluminum for CALMOS and CARMEN type calorimeters.

TABLE I
CALIBRATION DATA FOR 3 CALORRE DESIGNS AT 33°C

Sensor design	Equation of the calibration curve	Sensitivity	Sensitivity in 3W
	ΔT in °C with P in W	°C.W^{-1}	°C.W^{-1}
Full	ΔT = -0.16*P²+11.45*P	-0.32*P+11.45	10.49
Half	ΔT = -0.54*P²+22.65*P	-1.08*P+22.65	19.41
Quarter	ΔT = -0.82*P²+30.03*P	-1.64*P+30.03	25.11

The influence of the ring design and the flow temperature on the calibration curve can be explained by using a 1D thermal model. Indeed, due to the axisymmetric geometry of the calorimetric cell and by neglecting thermocouples, it is possible to apply a 1D thermal model by taking into account heat transfers only into the radial direction between the head and the vertical fin. This model is composed of three components.

The first term is due to the thermal conduction through the horizontal fin (into the metal and into the gas if this fin contains empty sectors). The power (P_1) induced by this heat exchange can be deduced from this formula:

$$P_1 = \frac{(T_{hot}-T_{cold})}{\frac{\ln\left(\frac{r_{head}}{r_{verticalfin}}\right)}{H_{horizontalfin}(n\theta\lambda_{gas}+m\varphi\lambda_{metal})}} \quad (1)$$

with r_{head} the external radius of the head (m), $r_{verticalfin}$ the internal radius of the vertical fin (m), $H_{horizontalfin}$ the height of the horizontal fin (m), n the number of empty sectors, m the number of metallic sectors, θ the angle of each empty sector (rad), φ the angle of each metallic sector (rad), λ_{gas} and λ_{metal} the thermal conductivities (W.m^{-1}.K^{-1}) of the gas gap and the metal structure of the cell respectively estimated at the mean temperature ($\frac{T_{hot}+T_{cold}}{2}$) in K.

The second term corresponds to the thermal conduction inside the gas gaps above and below the horizontal fin. The power evacuated through the gas gap (P_2) is given by this equation:

$$P_2 = \frac{2\pi\lambda_{gas}H_{gas}(T_{hot}-T_{cold})}{\ln\left(\frac{r_{head}}{r_{verticalfin}}\right)} \quad (2)$$

with H_{gas} the total height of the gas gaps above and below the horizontal fin (m).

The last term relates to thermal radiative exchanges between the head and the vertical fin. The power can be estimated by using this expression:

$$P_3 = \frac{2\pi\sigma H_{gas} r_{head}(T_{hot}{}^4-T_{cold}{}^4)}{\frac{1}{\varepsilon_{head}}+\frac{1-\varepsilon_{verticalfin}}{\varepsilon_{verticalfin}}\left(\frac{r_{head}}{r_{verticalfin}}\right)} \quad (3)$$

with σ the Stefan-Boltzmann constant (W.m^{-2}.K^{-4}), $\varepsilon_{verticalfin}$ and ε_{head} the emissivities of the internal surface of the vertical fin and the external surface of the head respectively.

Thanks to this 1D model, the contribution of each type of heat exchange can be deduced and used to explain the responses of the three designs. The results are given for an electrical power equal to 3W for all designs in Figure 5. The contribution of each thermal exchange depends on the horizontal fin design. In the case of the full design the main thermal contribution is the thermal conduction occurring into the ring. Consequently, the non-linearity of the calibration curve of the full design is mostly due to the variation of the thermal conductivity of the structure versus temperature. Then, the thermal conductive transfers through the ring are less important for the quarter design than those existing with the full design. But the thermal radiative transfers are enhanced in the case of the quarter design due to the presence of gas sectors which imply a greater thermal resistance of the horizontal fin and consequently an increasing in the sensor temperature. The new non-linear radiative component contributes also on the loss of the sensor linearity. According to the targeted nuclear heating level, this behavior should be taken into account to find a compromise to choose the suitable sensor configuration.

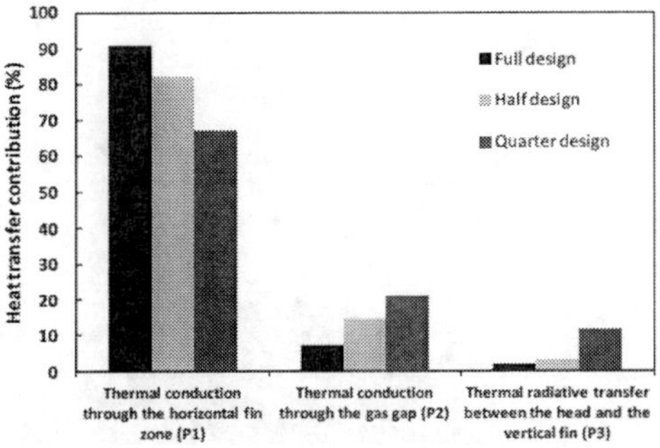

Fig. 5. Contribution of thermal conductive and radiative transfers for the three designs of CALORRE.

IV. Conclusions and Outlooks for Future Studies

A new reduced-size calorimetric cell was studied under non irradiation conditions and for different ring configurations. The experimental results in laboratory show that the sensitivity of the compact sensor depends on the fluid temperature and can be tuned by changing the design of the horizontal fin. The sensitivity of the full design is equal to 10.49°C.W^{-1} against 25.11°C.W^{-1} for the quarter design for an injected electrical power of 3W. However, the ring change modifies the sensitivity and the non-linearity of the sensor response. The 1D thermal model allows the behavior interpretation. When the thermal resistance of the horizontal ring is increased, the sensitivity is higher but the non-linearity coefficient is greater too due to the variation of the thermal conductivity of the structure material versus the temperature and the increasing in the thermal radiative transfer.

This new compact calorimetric cell could be used for a higher nuclear heating rate such as 20W.g^{-1} by choosing another structure material more conductive, and/or by reducing the size of the sensor to decrease the structure weight and consequently the undesired nuclear heating deposition on the structure leading to higher temperature field inside each cell, and/or by increasing the height of the horizontal fin. A new design will be studied by considering thermal property measurements versus temperature and by realizing 3D thermal numerical simulations with COMSOL Multiphysics with a complete differential calorimeter composed of two superposed calorimetric cells under laboratory conditions and then under irradiation conditions.

ACKNOWLEDGMENT

The IN-CORE program and MAHRI-BETHY programs are supported by FEDER, Conseil Régional PACA and Ville de Marseille.

References

[1] J. Brun, M. Tarchalski, C. Reynard-Carette, K. Pytel, A. Lyoussi, J. Jagielski, D. Fourmentel, J-F. Villard and M. Carette, " Numerical and Experimental Thermal Responses of Single-cell and Differential Calorimeters: Out-of-Pile Calibration to Irradiation Campaigns", IEEE Transactions on Nuclear Science, Vol. 63, N°3, pp. 1630-1639, 2016.

[2] H. Carcreff, L. Salmon, J. Bubendorff and V. Lepeltier, "First In-Core Simultaneous Measurements of Nuclear Heating and Thermal Neutron Flux Obtained With the Innovative Mobile Calorimeter CALMOS Inside the OSIRIS Reactor", IEEE Transactions on Nuclear Science, Vol. 63, N°5, pp. 2662-2670, 2016.

[3] J. Brun, C. Reynard-Carette, A. Lyoussi, C. De Vita, M. Carette, M. Muraglia, D. Fourmentel, P. Guimbal, and J.-F. Villard, "Comparison of the Thermal Response of Two Calorimetric Cells Dedicated to Nuclear Heating Measurements during Calibration", IEEE Transactions on Nuclear Science, Vol. 61, N°4, pp. 2254-2261, 2014.

[4] D. Fourmentel, C. Reynard-Carette, A. Lyoussi, J.F. Villard, J. Y. Malo, M. Carette, J. Brun, P. Guimbal and Y. Zerega, "Nuclear Heating Measurements in Material Testing Reactor: A Comparison Between a Differential Calorimeter and a Gamma Thermometer", IEEE Transactions on Nuclear Science, Vol. 60, N°1, pp. 328-335, 2013.

[5] J. Brun, C. Reynard-Carette, M. Carette, M. Tarchalski, K. Pytel, R. Prokopowicz, J. Jagielski, D. Fourmentel, L. Barbot, A. Lyoussi, J-F. Villard and P. Guimbal, "Irradiation Campaign in MARIA Reactor of an Innovative Calorimeter. Comparison of its Nuclear Heating Measurements with Two Kinds of Classical Sensors" presented at ANIMMA 2017, Liège, Belgium, 19-23 June 2017.

Nuclear instrumentation in VENUS-F

J. Wagemans, L. Borms, A. Kochetkov, A. Krása, C. Van Grieken, G. Vittiglio

***Abstract*—VENUS-F is a fast zero power reactor with 30 wt% U fuel and Pb/Bi as a coolant simulator. Depending on the experimental configuration, various neutron spectra (fast, epithermal, and thermal islands) are present. This paper gives a review of the nuclear instrumentation that is applied for reactor control and in a large variety of physics experiments. Activation foils and fission chambers are used to measure spatial neutron flux profiles, spectrum indices, reactivity effects (with positive period and compensation method or the MSM method) and kinetic parameters (with the Rossi-alpha method). Fission chamber calibrations are performed in the standard irradiation fields of the BR1 reactor (prompt fission neutron spectrum and Maxwellian thermal neutron spectrum).**

***Index Terms*— Activation, calibration, fast reactor, fission chamber, neutrons, reactor instrumentation**

I. Introduction

VENUS is a zero power reactor that is in operation at the Belgian Nuclear Research Centre SCK•CEN since 1964. Until 2007 VENUS was a thermal, water moderated reactor that mostly served as a mock-up for PWR and BWR reactors. Many configurations have been investigated and published which served to validate nuclear data and calculation codes.

In the frame of lead-based fast reactor and accelerator driven system research, and in particular for the MYRRHA project [1], VENUS was transformed to a mock-up of a fast reactor (VENUS-F) that can be operated in critical and sub-critical mode. For the latter, the GENEPI-3C accelerator was installed at SCK•CEN. This work was performed within the GUINEVERE program (2008-2011) [2] and the FREYA project (2011-2016) [3].

This paper starts with a brief presentation of the VENUS-F facility. The main focus is on the different types of instrumentation that is used for the reactor control and for performing the large variety of experiments. The importance of accurate detector calibration and the importance of the support of MCNP calculations is indicated.

II. The VENUS-F facility

The VENUS-F reactor is placed in an open-top stainless steel cylindrical vessel that contains a 12×12 square lattice of assemblies with a pitch of 8 cm. The lattice is surrounded with a bottom, top and radial lead reflector, see Fig. 1.

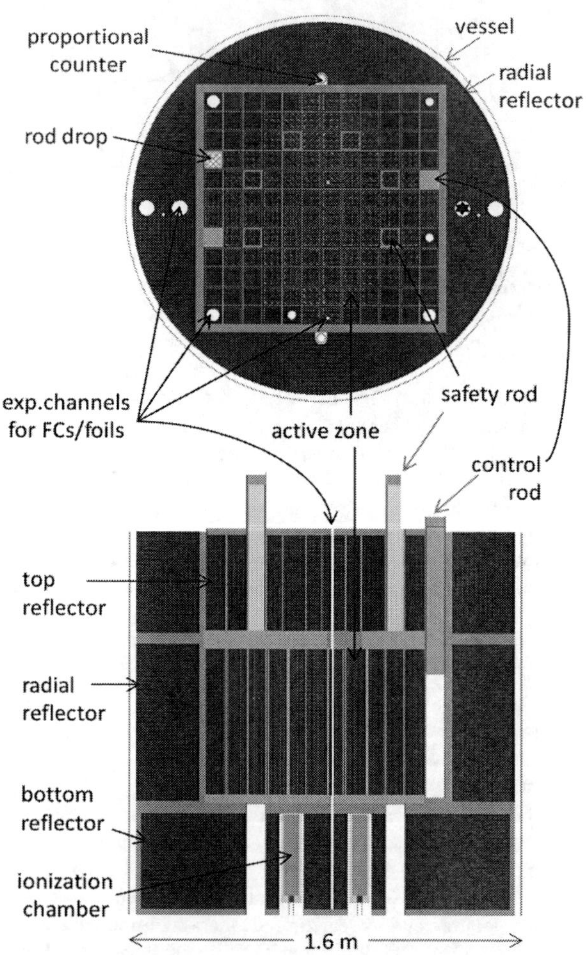

Fig. 1. Horizontal (top) and vertical (bottom) cut through the VENUS-F reactor (an example of a critical configuration). Lead – blue, uranium – red, stainless steel – green, B_4C – yellow (safety rod) and orange (control rod), white - air. An AmBe source is indicated with a star.

Each assembly can be filled with fuel, reflector material (Pb or C), B_4C (safety/control rod, rod drop) or any other material (e.g. PE acting as a thermal spectrum in-pile section). An assembly can contain a guiding tube and serve as an

Submitted on 10/07/2017. This work was supported by the following projects of the European Commission: 6th EURATOM Framework Programme Integrated Project EUROTRANS (under the Contract F16W-CT-2005-516520), the 7th FP FREYA project (No. 269665), and Horizon 2020 MYRTE project (Contract No. GA 662186).

All the authors are with the Belgian Nuclear Research Centre SCK•CEN, Boeretang 200, Mol, 2400, Belgium (e-mail of the corresponding author: jwagemans@ sckcen.be).

experimental assembly with a channel for insertion of detectors. A guiding tube penetrates through the top reflector and the entire length of an assembly. In certain positions, a guiding tube can run through the bottom reflector as well, see bottom part of Fig. 1.

The composition of a fuel assembly is variable. It consists of a 5×5 grid of rods: fuel rods (30 wt% metallic uranium) and coolant simulator rods (solid Pb or Bi). Optionally, Al_2O_3 rods (for simulating oxide fuel) and experimental channels are inserted instead of fuel or coolant rods, see Fig. 2.

Thanks to the VENUS-F versatility in both the composition of the fuel assembly as well as the assembly loading pattern, four types of a fuel assembly have been loaded at VENUS-F so far and nine core configurations have reached criticality [3]. For each critical configuration, a variety of experimental channels in fuel assemblies, reflector assemblies, in-pile sections and even in the radial reflector outside the assembly lattice have been exploited, see top part of Fig. 1.

Additionally, a large number of subcritical VENUS-F configurations with the GENEPI-3C accelerator inserted in the middle of the core has been investigated. These are published elsewhere [5]-[6].

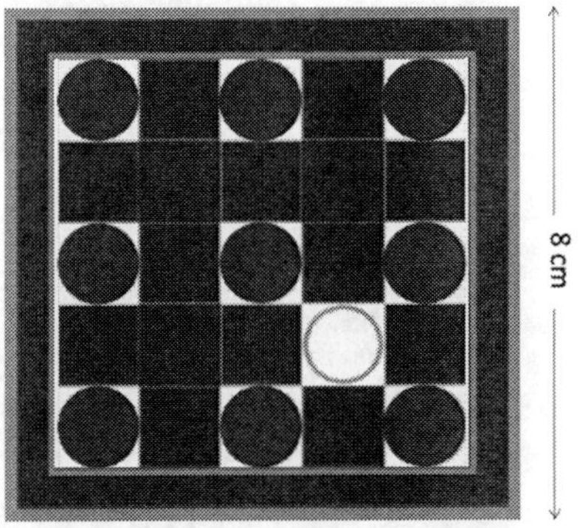

Fig. 2. Horizontal cut through an experimental fuel assembly with one rod replaced by a guiding tube. Lead – blue, stainless steel – green, uranium –red, white - air.

III. Instrumentation

VENUS-F is equipped with standard instrumentation for reactor control, which remains constantly loaded, and specific instrumentation for physics experiments, which are activation foils and fission chambers inserted in different experimental channels depending on the purpose of the experimental campaign.

A. Instrumentation for reactor control

To monitor the neutron flux during fuel loading and reactor startup, two BF_3 proportional counters are used. BF_3 gas (90 wt% enriched in ^{10}B) acts as proportional gas and neutron detection material, see Table 1. To increase the probability of neutron detection in ^{10}B, the BF_3 tubes are surrounded by a polyethylene moderator (4 or 8 mm thick cylinders). The impact of the local neutron moderation on the reactor core is insignificant because the BF_3 counters are located in the radial reflector outside the assembly lattice, see top part of Fig. 1.

As VENUS-F is a zero-power reactor, the burnup is negligible and its fuel remains fresh. The neutron flux from spontaneous fission is thus not sufficient for a reliable startup. Every VENUS-F reactor startup is driven by an external Am-Be source (intensity of 2×10^6 neutrons/s), which can be inserted in the radial reflector, see top part of Fig. 1.

For monitoring the reactor power during normal operation, four ionization chambers (PNI 1089 type supplied by Plessey) are placed in the bottom reflector, see bottom part of Fig. 1. The ionization chambers have 94 wt% enriched ^{10}B deposit and hydrogen filling gas, see Table 1. Two of them are connected to linear current chains, whereas the other two are connected to logarithmic current chains. The signals are used for example to the reactor period determination.

For monitoring the course of temperature in the core, thermocouples are connected to Pb or Bi rods in fuel assemblies. Typically, a temperature increase of 4 K is detected after 7 h of reactor operation at 100 W [7].

TABLE I
INSTRUMENTATION FOR THE VENUS-F REACTOR CONTROL

	Proportional counters	Ionization chambers
Type	12EB40 (20th Century Electronics Ltd)	PNI 1089 (Plessey)
Body	Cu	Al, electrodes coated with 94 wt% ^{10}B
Diameter	25 mm	63.5 mm
Length	210 mm	342.9 mm
Filling gas	BF_3, > 90 wt% ^{10}B	H
Active length	120 mm	193 mm
Voltage	2400 V	400-600 V
Neutron Sensitivity	3 (cts/s)/(n/cm²/s)	2.10-14 A/(n/cm²/s)

B. Fission Chambers

A large number of fission chambers is available at VENUS-F. The fission chambers vary in deposit type and mass, see Table 2. The low-mass (~ 10 µg-1 mg) fission chambers are mainly applied for measuring spatial neutron flux profiles [7] and spectrum indices [8]. The fissile deposits include ^{234}U, ^{235}U, ^{238}U, ^{237}Np, ^{239}Pu, ^{240}Pu, ^{242}Pu and ^{241}Am. The detector calibrations are performed in the standard irradiation fields of the BR1 reactor at SCK•CEN [9]. An example of a fission fragment spectrum obtained with a low-mass fission chamber is shown in Fig. 3.

The fission chambers with heavy deposits (up to 1 g ^{235}U, ^{238}U) are applied for measurements of reactivity effects (using the MSM method [10]), kinetic parameters (using the Rossi-alpha method [11]) and sub-criticality measurements [5],[6].

All fission chambers are operated in pulse mode. A Programmable High Speed Pulse Amplifier (PHSPA) electronic unit and a Labview-based data acquisition developed at SCK•CEN are used. PHSPA contains several amplifier functions (current mode pre-amplifier, amplifier), filters, discriminators, counter, and an integrated high voltage source (up to 2 kV). A complete measurement chain can be realized using only this amplifier module and a PC.

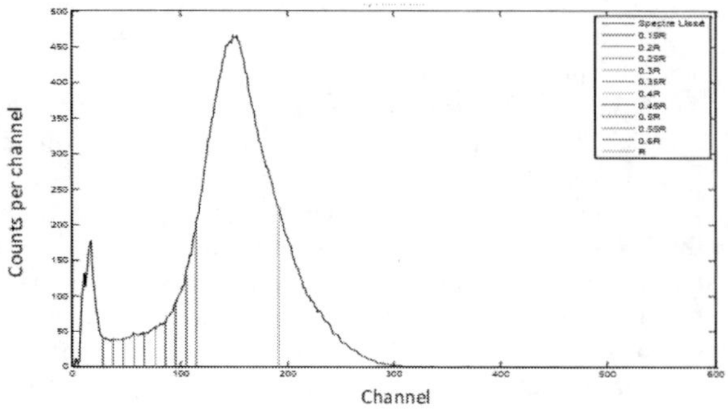

Fig. 3. Fission fragment spectrum measured with a low-mass ^{235}U fission chamber (90 μg).

TABLE II
FISSION CHAMBERS FOR VENUS-F EXPERIMENTS

	Fission chambers	
Application	Spatial fission rate distrubutions, spectrum indices	Reactivity effects, kinetic parameters, control rod calibration, sub-criticality
Manufacturer	CEA, Centronic	Photonis, Reuter-Stokes
Deposit mass [mg]	~0.01 -1	~ 1 (CFUF34), 10 s(CFUM21), 100 (RS), 1000 (CFUL01)
Deposit isotope	$^{234,235,238}U$, ^{237}Np, $^{239,240,242}Pu$, ^{241}Am	^{235}U, ^{238}U
Active length [mm]	10-25	27; 120; 127; 211
Filling gas	Ar/N	Ar/N
Body	Ti, steel	Al
Diameter [mm]	1.5; 4; 8	4.7; 26; 48
Length [mm]	23; 75	85; 165; 227; 337
Voltage [V]	300-500	700

C. *Activation foils*

Activation dosimetry is a standard technique to determine neutron flux and fluence and to obtain information on the neutron spectrum. At SCK•CEN a laboratory equipped with high-purity Germanium detectors is situated in the same building as the BR1 and VENUS-F research reactors, which allows fast gamma spectrometry measurements without the need for transport of the activated foils. For standard situations specific activities can be determined with an uncertainty (1 σ) lower than 2 %.

A set of foils with different energy responses can be used for comparison with neutronics calculations or for spectrum unfolding to obtain information on the neutron spectrum. Such stack of foils consists for example of Au, Al and In, and can be irradiated in the experimental fuel assemblies of VENUS-F.

Uranium foils are of particular interest for core characterization measurements in VENUS-F. Different types of foils varying in size, shape, thickness and enrichment are available for experiments. These foils can be used to determine absolute $^{235}U(n,f)$ and $^{238}U(n,f)$ fission rates, for example to be included in the set of foils for spectrum unfolding, to determine the $^{238}U(n,\gamma)/^{235}U(n,f)$ and $^{238}U(n,f)/^{235}U(n,f)$ spectrum indices, and to determine fission rate profiles in the reactor.

When absolute values from fission foil irradiations are required (for example for spectrum index measurements), it may be necessary to correct for fission product escape. Indeed the fragments produced in fission events occurring near the foil surface have a certain probability to escape from the foil. As a consequence the measured specific activity of such foil will be lower than the expected specific activity for a given incident neutron flux. This problem can be avoided by encapsulating the fission foil in e.g. Al during its irradiation and performing the activity measurement on the foil with its encapsulation. When encapsulation is not possible, the correction has to be determined. For metallic uranium foils of about 0.1 mm the correction is of the order of 3 %. This was experimentally determined by irradiating a uranium foil sandwiched between so-called catcher foils (Al) and measuring fission products in all foils.

References

[1] Van den Eynde *et al.*, "MYRRHA design revision 1.6 and the phased implementation plan," in *Proc. TCADS-3*, Mito, Japan, 6-9 Sep 2016, pp. 141-150.

[2] P. Baeten *et al.*, "The GUINEVERE project at the VENUS facility", in *Proc. PHYSOR-2008*, Interlaken, Switzerland, 14-19 Sep 2008.

[3] A. Kochetkov *et al.*, "The Lead-Based VENUS-F Facility: Status of the FREYA Project", *EPJ Web of Conferences* 106, 06004 (2016).

[4] A. Krása *et al.*, "The VENUS-F reactor for nuclear data validation", in *Proc. ISRD-16,* Santa Fe, USA, May 7-12, 2017.

[5] V. Bécares *et al.*, "Reactivity determination and monitoring in FREYA Project subcritical cores: Assessment and correction of spatial and energy effects," in *Proc. TCADS-3*, Mito, Japan, 6-9 Sep 2016, pp. 234-251.

[6] T. Chevret *et al.*, "Spatial effects during beam interruption experiments for ADS subcriticality monitoring using a Monte Carlo alpha-mode expansion of the neutron flux," in *Proc. TCADS-3*, Mito, Japan, 6-9 Sep 2016, pp. 252-265.

[7] A. Krása *et al.*, "Comparative study of nuclear data libraries on Monte Carlo modelling of MYRRHA mockup critical cores in the VENUS-F reactor," in *Proc. ND-2016*, Brugges, Belgium, Sep 12 – 16 2016.

[8] A. Kochetkov *et al.*, "Extensive analysis of C/E results of fission rates ratios measurements in several fast lead VENUS-F cores." in *Proc. ND-2016*, Brugges, Belgium, Sep 12 – 16 2016.

[9] J. Wagemans *et al.*, "The 235U prompt fission neutron spectrum in the BR1 reactor at SCK•CEN," *EPJ Web of Conferences* 106, 06003, 2016.

[10] J.L. Lecouey *et al.*, "Estimate of the reactivity of the VENUS-F subcritical configuration using a Monte Carlo MSM method," *Annals of Nuclear Energy* 83 (2015) 65–75.

[11] X. Doligez *et al.*, "Effective delayed neutron fraction measurement in the critical VENUS-F reactor using noise techniques," in *Proc. ANIMMA-2015*, Lisbon, Portugal, 20-24 April 2015, DOI: 10.1109/ANIMMA.2015.7465614.

Improving gross count gamma-ray logging in uranium mining with the NGRS probe

C. Carasco, B. Pérot, J.-L. Ma, H. Toubon, A. Dubille-Auchère

***Abstract*– AREVA Mines and the Nuclear Measurement Laboratory of CEA Cadarache are collaborating to improve the sensitivity and precision of uranium concentration measurement by means of gamma ray logging. The determination of uranium concentration in boreholes is performed with the Natural Gamma Ray Sonde (NGRS) based on a NaI(Tl) scintillation detector. The total gamma count rate is converted into uranium concentration using a calibration coefficient measured in concrete blocks with known uranium concentration in the AREVA Mines calibration facility located in Bessines, France. Until now, to take into account gamma attenuation in a variety of boreholes diameters, tubing materials, diameters and thicknesses, filling fluid densities and compositions, a semi-empirical formula was used to correct the calibration coefficient measured in Bessines facility. In this work, we propose to use Monte Carlo simulations to improve gamma attenuation corrections. To this purpose, the NGRS probe and the calibration measurements in the standard concrete blocks have been modeled with MCNP computer code. The calibration coefficient determined by simulation, 5.3 $s^{-1} \cdot ppm_U^{-1} \pm 10\%$, is in good agreement with the one measured in Bessines, 5.2 $s^{-1} \cdot ppm_U^{-1}$. Based on the validated MCNP model, several parametric studies have been performed. For instance, the rock density and chemical composition proved to have a limited impact on the calibration coefficient. However, gamma self-absorption in uranium leads to a nonlinear relationship between count rate and uranium concentration beyond approximately 1% of uranium weight fraction, the underestimation of the uranium content reaching more than a factor 2.5 for a 50 % uranium weight fraction. Next steps will concern parametric studies with different tubing materials, diameters and thicknesses, as well as different borehole filling fluids representative of real measurement conditions.**

***Index Terms* — Uranium mining, NaI(Tl) scintillators, MCNP**

I. Introduction

The detection of gamma rays emitted by ^{238}U and its daughter nuclei is one of the reference method for detecting uranium in borehole. One approach consists in inferring the uranium concentration in boreholes from the total count rate provided by scintillation counters such as NaI(Tl) detectors. This approach, known as gross count gamma-ray logging [1], requires a calibration coefficient which is usually measured in concrete blocks of well-defined composition with known uranium concentration. However, the measurement conditions of the calibration coefficient are not always representative of field conditions and corrections of the measured calibration coefficient are needed. This paper reports numerical simulations performed with the MCNP code in view to improve the accuracy of uranium concentration determination in gross count gamma-ray logging.

II. Validation of the MCNP numerical model

To assess uranium concentration in boreholes, a GEOVISTA Natural Gamma Ray Sonde (NGRS) [2] equipped with a $\varnothing = 2.4$ cm × H = 3.9 cm cylindrical NaI(Tl) scintillation crystal is used. The uranium concentration is linearly related to the NGRS count rate via a calibration coefficient K. In practice, K is measured using 70 cm × 70 cm × 70 cm concrete blocks of known uranium concentration, at AREVA Mines calibration facility in Bessines, France.

The calibration measurement geometry presented in Fig. 1 has been modeled with the MCNP Monte Carlo code. Being a reliable and renown photon-neutron transport code, MCNP is already used in the field of uranium logging to model measurement methods involving neutron sources [3] or measurements based on gamma-ray detection [4], this last approach being also manageable with free available codes like GEANT 4 [5]. The modeled concrete blocks have a 1.9 $g \cdot cm^{-3}$ density and an isotropic, homogeneously distributed gamma-ray source corresponding to 2906 ppm of uranium, 15.6 ppm of thorium and 3.7 % of potassium. With a 65 keV energy threshold, the calculated reference calibration coefficient is $K_{air}^{MCNP} = 5.3\ s^{-1} \cdot ppm_U^{-1}$ for a tubeless NGRS probe in the calibration block, with only air filling the borehole (no fluid), which is compatible with the measured reference calibration coefficient: $K_{air}^{meas} = 5.2\ s^{-1} \cdot ppm_U^{-1}$.

Measurements with the NGRS probe inside tubes of different thicknesses and nature have also been modeled with MCNP. Table I shows that the calculated tube attenuation corrections match the measurements.

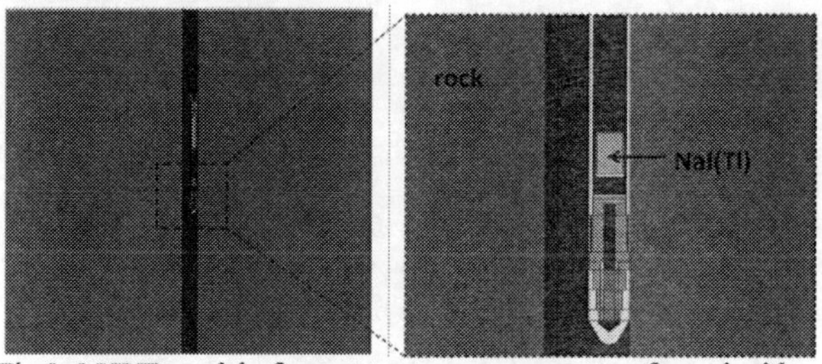

Fig.1. MCNP model of a count rate measurement performed with a NGRS probe positioned inside an air-filled borehole drilled in a 70 cm × 70 cm × 70 cm calibration concrete block.

Summary submitted on June 1st, 2017.

Cédric Carasco, Bertrand Pérot and Jean-Luc Ma are with CEA, DEN, Cadarache, Nuclear Measurement Laboratory, F-13108 Saint-Paul-lez-Durance, France.

Hervé Toubon and Anne Dubille-Auchère are with AREVA Mines, 92053 La Défense (Courbevoie), France.

TABLE I. CALCULATED AND MEASURED CALIBRATION COEFFICIENTS K^{MCNP} FOR DIFFERENT TUBES (NATURE AND THICKNESS) AND ATTENUATION CORRECTIONS RELATIVE TO THE TUBELESS CONFIGURATION

	No tube	2 mm thick Al tube	5 mm thick Al tube	5 mm thick steel tube
K_{tube}^{MCNP} ($s^{-1} \cdot ppmU^{-1}$)	5.33	5.23	5.09	3.94
$\frac{K_{air}^{MCNP}}{K_{tube}^{MCNP}}$	1	1.02	1.05	1.35
$\frac{K_{air}^{meas}}{K_{tube}^{meas}}$	1	1.02	1.06	1.37

III. Sensitivity parametric studies

A. Influence of rock composition

The impact of the rock mineralogy on the calibration factor has been studied replacing SiO_2 with $CaCO_3$, $CaSO_4$-$2H_2O$ or Na_2CO_3. As shown in Table II, rock composition has a small impact on the calibration factor. This lack of sensitivity is caused by the lesser atoms sensitivity of the mass attenuation coefficients when photons have an energy larger than 100 keV.

TABLE II. RATIO OF THE CALCULATED CALIBRATION COEFFICIENTS FOR DIFFERENT ROCK MINERTALOGY AND ENERGY THRESHOLDS.

Threshold	$CaCO_3$ d=1.6	$CaSO_4$-$2H_2O$ d=1.6	Na_2CO_3 d=1.6
65 keV	0.94	0.94	1.03
100 keV	0.96	0.96	1.02

B. Gamma self-absorption in uranium

Figure 2 shows calculation results on a wide range of uranium mass fractions, for a NGRS probe inside an iron tube filled with water, located in a 70 cm edge concrete cube similar to Bessines calibration block. The deviation from the linear fit due to self-attenuation, obtained for uranium mass fractions lower than 1 %, begins above uranium 1 %, and the self-attenuation effect reaches a factor of 2.6 for 50 % of uranium.

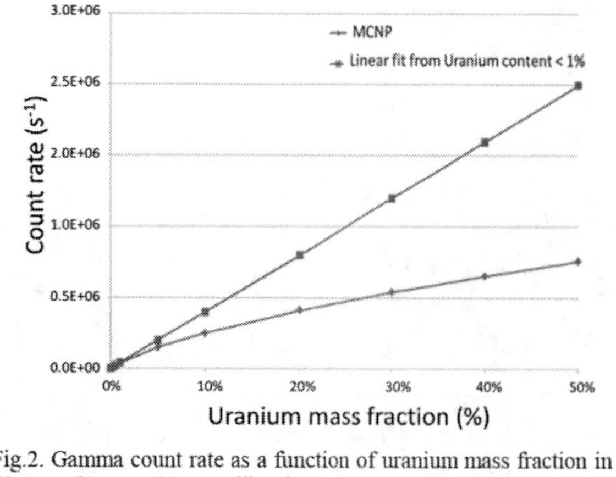

Fig.2. Gamma count rate as a function of uranium mass fraction in a 70 cm × 70 cm × 70 cm calibration concrete rock.

C. Gamma range seen by the NGRS probe

Calculations have been performed to study the range of the gamma rays detected by the NGRS probe, as shown in Fig. 3. The saturation of the signal is reached for a sphere with a radius of about 100 cm. Therefore, we have calculated the calibration coefficient in a concrete block with an edge of 200 cm, instead of 70 cm for Bessines blocks. The result is K_{∞}^{MCNP}=5.8 $s^{-1} \cdot ppm_U^{-1}$ instead of 5.3 $s^{-1} \cdot ppm_U^{-1}$, indicating that calibration coefficients measured in Bessines are underestimated by about 10 %.

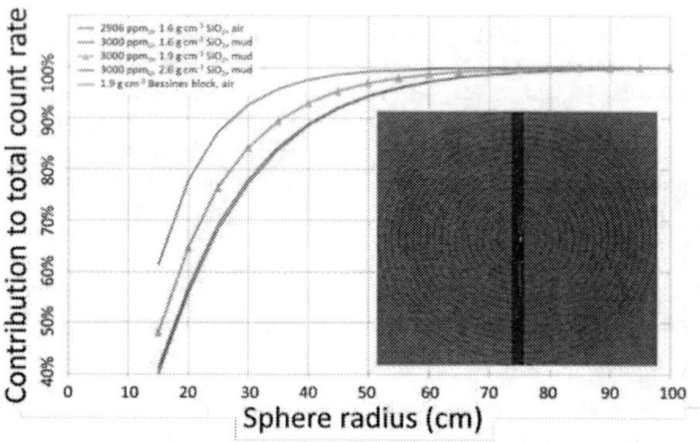

Fig.3. MCNP model and cumulated signal due to uranium present in spheres with an increasing radius for different rocks density and borehole fillings.

D.Influence of rock density

Calculations with 2906 ppm_U have been performed with a 200 cm edge SiO_2 block, the reference gamma-ray source, and a tubeless borehole filled with mud (simulated by water with a density of 1.2 $g \cdot cm^{-3}$), showing that the calibration coefficient slightly decreases from 5.44 to 5.36 $s^{-1} \cdot ppm_U^{-1}$ when the density varies from 1.2 to 2.6 $g \cdot cm^{-3}$. Since calculations keep the uranium concentration to 2906 ppm_U, increasing rock density increases the amount of uranium. The calibration coefficient is therefore poorly sensitive to rock density because the increase of gamma attenuation is compensated by a larger amount of uranium when density rises.

IV. Conclusion

The MCNP Monte Carlo simulation code has been extensively used to model uranium gamma logging with a NGRS probe. MCNP reproduces with a good accuracy the gross gamma counting measurements performed with the NaI(Tl) gamma scintillator of the probe, such as the calibration coefficient K ($s^{-1}.ppm_U^{-1}$) measured in the calibration facility of Bessines, France, which is used to convert the signal (s^{-1}) in terms of uranium weight fraction (ppm_U): the coefficient calculated with MCNP is 5.3 $s^{-1} \cdot ppm_U^{-1} \pm$ 10%, which is in good agreement with the one measured in Bessines, 5.2 $s^{-1} \cdot ppm_U^{-1}$. Sensitivity studies have been performed with MCNP, showing that the rock density and mineralogy have a limited influence on K coefficient. The uranium concentration induces a significant self-absorption effect above a uranium mass

fraction of 1 %, which leads to an underestimation of the real uranium content up to a factor of 2.6 for 50 % of uranium.
The validated MCNP model will then been used to check the validity of gamma attenuation corrections applied to the reference calibration coefficient measured in Bessines, in a tubeless borehole filled with air, in view of in situ gamma-ray measurements with different fluids filling the borehole and the tube, and with different tubes housing the NGRS probe.
These simulations are the first step of a larger study on uranium gamma logging techniques by CEA and AREVA Mines. Gross count gamma-ray measurement indeed may underestimate the quantity of uranium in case of disequilibrium in the uranium chain due to in situ leaching techniques (roll fronts). For a better assessment of the uranium content, next studies will focus on the use of high energy resolution gamma-ray spectroscopy [6].

References

[1] Practical Borehole Logging Procedures for Mineral Exploration, with Emphasis on Uranium, IAEA Technical Reports Series No. 259 (1986)
[2] http://www.geovista.co.uk/pages/gr-calip.htm
[3] R. P. Gardner , A. Sood, On the future of Monte Carlo simulation for nuclear logs, Applied Radiation and Isotopes 68 (2010) 932–935
[4] D. C. McKeon, Input preprocessor for creating mcnp models for use in developing large calculational databases, IEEE Transactions On Nuclear Science 41, 1069-1073 (1994).
[5] N. Velker, B. Banzarov, F. Inanc, Al. Vinokurov, N. Simonov, Evaluating Geant4 platform for nuclear well-logging problems, Nucl. Instr. Meth. B 297, 102–109 (2013).
[6] T. Marchais, B. Pérot, C. Carasco, P.-G. Allinei, P. Chaussonet, J-L Ma, H. Toubon, L. Pauthier, Optimization of gamma-ray spectroscopy for uranium mining, ANIMMA 2017, International Conference on Advancements in Nuclear Instrumentation Measurement Methods and their Applications, 19-23 June, Liège, Belgium.

https://doi.org/10.1051/epjconf/201817005002

Simulated Performances of a Very High Energy Tomograph for Non-Destructive Characterization of large objects

Marc Kistler, Nicolas Estre and Elsa Merle

***Abstract*— As part of its R&D activities on high-energy X-ray imaging for non-destructive characterization, the Nuclear Measurement Laboratory has started an upgrade of its imaging system currently implemented at the CEA-Cadarache center. The goals are to achieve a sub-millimeter spatial resolution and the ability to perform tomographies on very large objects (more than 100-cm standard concrete or 40-cm steel). This paper presents results on the detection part of the imaging system.**

The upgrade of the detection part needs a thorough study of the performance of two detectors: a series of CdTe semiconductor sensors and two arrays of segmented $CdWO_4$ scintillators with different pixel sizes. This study consists in a Quantum Accounting Diagram (QAD) analysis coupled with Monte-Carlo simulations. The scintillator arrays are able to detect millimeter details through 140 cm of concrete, but are limited to 120 cm for smaller ones. CdTe sensors have lower but more stable performance, with a 0.5 mm resolution for 90 cm of concrete. The choice of the detector then depends on the preferred characteristic: the spatial resolution or the use on large volumes. The combination of the features of the source and the studies on the detectors gives the expected performance of the whole equipment, in terms of signal-over-noise ratio (SNR), spatial resolution and acquisition time.

***Index Terms*—CdTe, $CdWO_4$, detective quantum efficiency (DQE), high-energy imaging, quantum accounting diagram (QAD), signal-over-noise ratio (SNR), tomography.**

I. Introduction

As PART of its R&D activities on high-energy imaging for non-destructive characterization, the Nuclear Measurement Laboratory is working on the upgrade of the tomograph implemented at the CEA-Cadarache Center. This upgrade takes into account the optimization of the detection system. That is why two different types of detectors are studied: a series of CdTe semiconductor sensors and two arrays of segmented $CdWO_4$ scintillators with different pixel sizes.

The comparison of the detectors is carried out by three indicators: the detective quantum efficiency (DQE), the spatial resolution and the signal-over-noise ratio (SNR). The DQE and the spatial resolution make it possible to compare the intrinsic performance of the detectors. The DQE can be calculated theoretically using parameters derived from a Quantum Accounting Diagram (QAD) analysis. The spatial resolution of the detector, relative to the object plane, gives the component of the tomograph spatial resolution due to the detection part. Coupled with a X-ray source (up to 20 MeV in the current studies), it allows to evaluate the spatial resolution of the complete system. The signal-over-noise ratio, on the other hand, depends on the DQE, but also on the input X-ray flux and on the acquisition time (the flux being determined by the source and the object to be imaged). The acquisition time depends in part on the type of detector used and the associated technique (horizontal scan or fixed object).

The goal is therefore to obtain a sub-millimeter spatial resolution, while keeping a significant SNR at the output of the detector, in order to obtain images with sufficient contrast and limited noise.

II. Detectors Description and QAD Analysis

The QAD analysis of a detector consists in representing its detection process as a series of cascaded stages of three types: gain stage, spread stage and additive noise stage [1]. This precise and standardized representation allows accurate comparisons and analyses of several detection systems at each stage of the detection process.

A. Multi-CdTe Detector

In the 1990s, the CEA-Leti has designed a detector based on a series of 25 CdTe sensors, separated by lead collimators and followed by an electronic acquisition chain [2]. This device still shows remarkable intrinsic performances [3].

Each sensor consists of a thin CdTe semiconductor lamella with a section of 0.8 mm x 15 mm. Its thickness of 25 mm and its high density ($\rho = 5.8$ g / cm^3) allow to capture at least 40% of the incident photons (Fig. 3). Photons deposit an amount of energy during the interaction, leading to the production of

This paragraph of the first footnote will contain the date on which you submitted your paper for review.

M. Kistler and N. Estre are with CEA Cadarache, DEN/Nuclear Measurement Laboratory, F-13108 Saint-Paul-lez-Durance, France (e-mail: marc.kistler@cea.fr).

E. Merle is with CNRS, IN2P3/Subatomic Physics and Cosmology Laboratory, F-38026 Grenoble, France.

Fig. 1. Multi-CdTe detector, with the collimators [3].

TABLE I
SUMMARY OF THE GAINS AND POISSON EXCESSES FOR THE STAGES OF THE CDTE DETECTOR

#	Description	Type of process	Symbol	Gain g_i	Poisson excess ε_{g_i}
1	X-ray interaction	Binomial gain	g_1	0.399	-0.399
2	X/Deposited energy conversion (MeV/ph-X)	Gain	g_2	1.72	-0.446
3	Energy spread in the sensor	Spread	$T_3(\omega)$	1	-1
4	Conversion in e^-/e^+ pairs	Gain	g_4	2.31×10^5	-0.85
5	Electrons collection	Binomial gain	g_5	0.78	-0.78
6	Additive electronic noise	Noise	$\sigma_a^2/\overline{N_6}$	1	-1

electron-hole pairs. The voltage created at the output of the integrator circuit is proportional to the number of collected charges. The QAD analysis is in this case composed of six stages, summarized in Table I. See reference [1] for the parameters definition.

Since the sensors are non-continuous, the complete acquisition of a tomographic section requires a translation of the object, in addition to its rotation and a precise synchronization of the acquisition with the object displacement and the firing frequency of the X-ray source. This results in an acquisition time per tomographic slice close to 30 min.

B. Segmented $CdWO_4$ Scintillator Array

The company X-Scan Imaging offers a series of linear X-ray cameras. These detectors are composed of segmented $CdWO_4$ scintillator, forming needles, coupled by optical taper to an off-axis linear matrix of photodiodes. The two models studied, XIH8804-66 and XIH8808-66, differ in the size of their needles, with a section of 0.4 mm × 0.6 mm and 0.8 mm × 1.2 mm respectively. With an ideal active detection length of 1.7 m and respectively 4224 and 2112 pixels, theses detectors allow measurements in half-field on objects with a 140 cm diameter without having to translate the object.

The detection process combines a high-density scintillator material ($\rho = 7.9$ g/cm^3) and therefore a high X-ray interaction efficiency, with small continuous pixels. The QAD analysis consists of seven stages. The values have been evaluated by simulations (parameters 1 to 5) or realistic assumptions (parameters 6 and 7).

Fig. 3 and 4 show that both detectors have an equivalent interaction efficiency higher than 40% and that each photon deposits an energy of a few MeV. Due to its small section, the $CdWO_4$ 0.4 model has the lowest energy deposition values.

Fig. 2. Segmented $CdWO_4$ scintillator [4].

TABLE II
SUMMARY OF THE GAINS AND POISSON EXCESSES FOR THE STAGES OF THE $CDWO_4$ DETECTOR

#	Description	Type of process	Symbol	Gain g_i	Poisson excess ε_{g_i}
1	X-ray interaction	Binomial gain	g_1	0.436	-0.436
2	X/Deposited energy conversion (MeV/ph-X)	Gain	g_2	0.85 / 1.55	-0.71 / -0.61
3	Energy spread in the needles	Spread	$T_3(\omega)$	1	-1
4	Conversion to optical quanta (ph/MeV)	Gain	g_4	1.50×10^4	0
5	Optical fiber efficiency	Binomial gain	g_5	0.31	-0.31
6	Quantum efficiency of the photodiodes	Binomial gain	g_6	0.48	-0.48
7	Additive electronic noise	Noise	$\sigma_a^2/\overline{N_7}$	1	-1

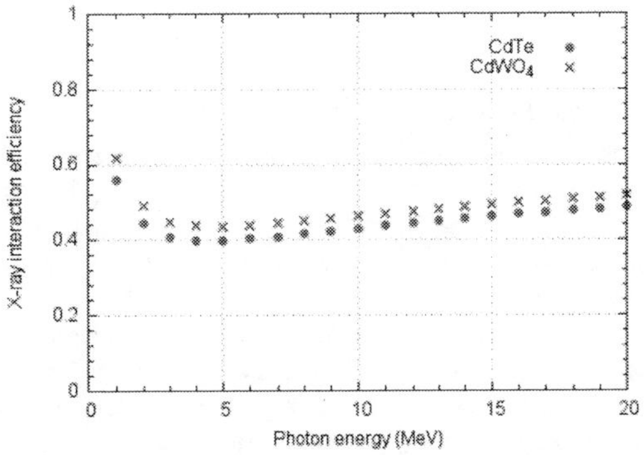

Fig. 3. X-ray interaction efficiency g_1 of each detector as a function of incident photon energy.

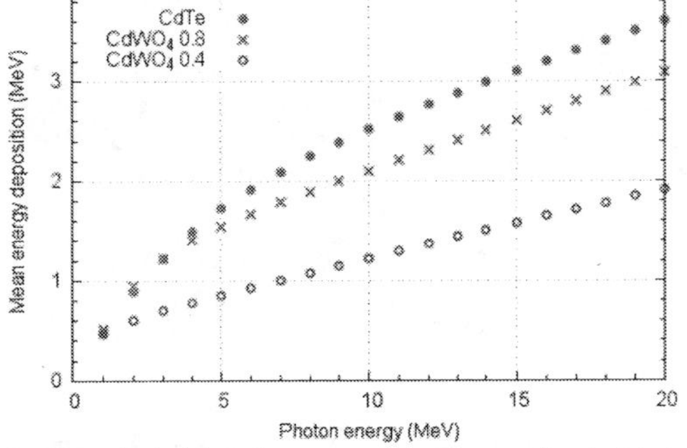

Fig. 4. Mean single interaction energy deposition g_2 of each detector as a function of incident photon energy.

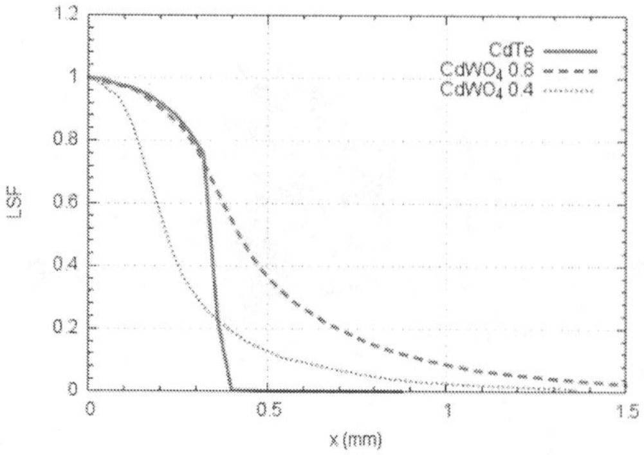

Fig. 5. Line Spread Function (LSF) of each detector.

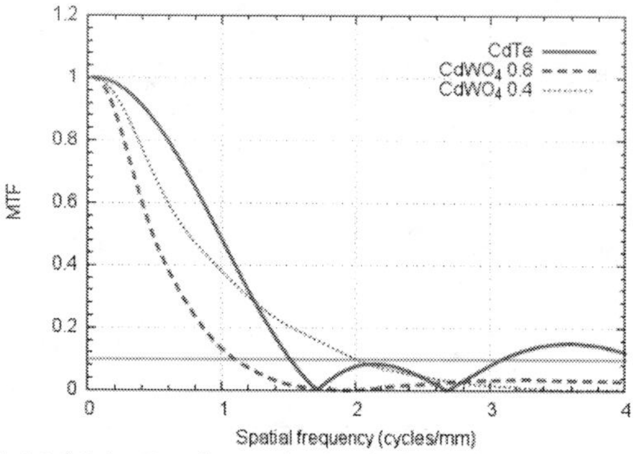

Fig. 6. Modulation Transfer Function (MTF) of each detector.

III. Results

A. Spatial Resolution and MTF

The spreading stages are materialized by the effect of signal spreading on the average number of quanta. This effect is directly characterized by the square of the Modulation Transfer Function (MTF) [1]. In order to evaluate the MTF, we first calculate the Line Spread Function (LSF), which is the one-dimension signal spread distribution. In this study, the LSF is evaluated by Monte-Carlo simulation (code MCNP6.1 [5]). The MTF is then obtained by Fourier Transform of the LSF and can be included in the QAD calculations.

The MTF also provides the resolution limit that we define as the spatial frequency at which the MTF is 10%. Beyond this frequency, it is considered that the signal is not exploitable. The LSF and MTF obtained for the two detectors are shown in Fig. 5 and 6 respectively.

The LSF of the CdTe detectors is close to a sensor width rectangular function, thanks to the lead collimation. $CdWO_4$ arrays have more complex distributions, with a fine central part and a wider peripheral part, which reflects the spreading of the signal due to the absence of collimation. The resolution limit of the small $CdWO_4$ needles remains better than the CdTe one in spite of a larger signal spreading. The large $CdWO_4$ needles (which have the same width as the CdTe sensors) have the poorest resolution.

Finally, the resolution limits are calculated in the object plane of the tomograph via the radiographic magnification factor, which is 1.6 for our setup. These values are summarized in Table III.

TABLE III
RESOLUTION LIMITS IN DETECTOR AND OBJECT PLANES FOR EACH DETECTOR

	$CdWO_4$ 0.4	CdTe	$CdWO_4$ 0.8
Detector plane	2 mm^{-1}	1.5 mm^{-1}	1 mm^{-1}
Object plane	3.2 mm^{-1}	2.4 mm^{-1}	1.6 mm^{-1}

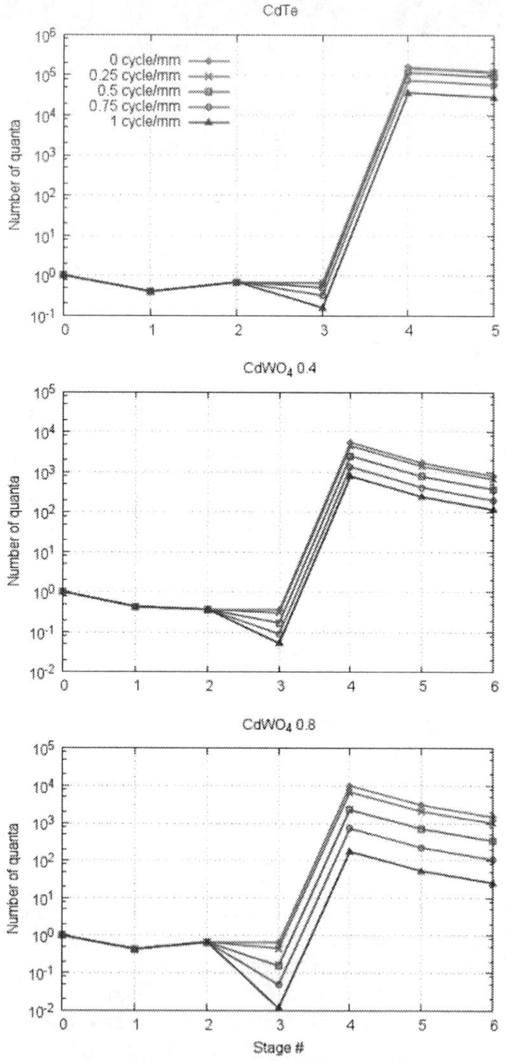

Fig. 7. Spatial-frequency dependent QAD of each detector.

B. QAD

As QAD parameters are evaluated, the diagrams can then be computed (Fig. 7). For the three detectors, they represent the evolution of the number of quanta $P_i(\omega)$ at each detection stage i and for different spatial frequencies ω:

$$P_i(\omega) = \prod_{j=1}^{i} \overline{g_j} \, |T_j(\omega)|^2 \quad (1)$$

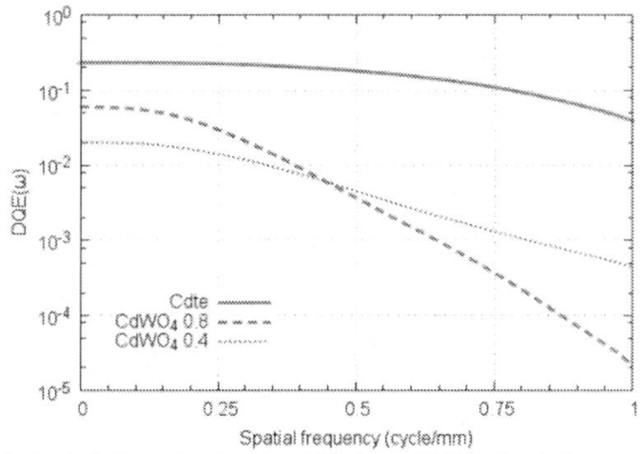

Fig. 8. Intrinsic Detective Quantum Efficiency (DQE) of each detector.

The diagrams show that the detection processes are equivalent: first a loss of signal in the first stage, then a spreading that appears through the separation of the curves according to the spatial frequencies, followed by a significant gain of several decades, and finally a small decrease due to the electrons or visible photons collecting efficiency. The larger separation of the curves for the large $CdWO_4$ suggests a higher sensitivity of its DQE to the spatial frequency.

C. *DQE*

The formulas of the DQE for each detector are obtained from the theoretical equation [6]. The three curves are plotted in Fig. 8.

$$\text{CdTe: } DQE(\omega) = \frac{g_1 g_2 |T_3(\omega)|^2}{1+(g_2+\epsilon_{g_2})|T_3(\omega)|^2+\frac{\epsilon_{g_4}}{g_4}+\frac{1+\sigma_a^2/\overline{N_6}}{g_4 g_5}} \quad (2)$$

$$\text{CdWO}_4\text{: } DQE(\omega) = \frac{g_1 g_2 |T_3(\omega)|^2}{1+(g_2+\epsilon_{g_2})|T_3(\omega)|^2+\frac{1+\sigma_a^2/\overline{N_7}}{g_4 g_5 g_6}} \quad (3)$$

CdTe sensors have the best intrinsic DQE over the entire frequency range. The $CdWO_4$ is more frequency dependent and therefore drops faster, especially the large one after 0.5 cycle/mm.

IV. Discussion

For a single incident photon, multi-CdTe detector is therefore the most efficient. But for the acquisition of a tomographic slice with an identical measurement time, it is necessary to compare the signal-over-noise ratios at the output of the detectors. Indeed, the SNR depends not only on the DQE, but also on the SNR at the detector input:

$$SNR_{out} = \sqrt{DQE} \times SNR_{in} \quad (4)$$

The input SNR depends on the photonic flux, the sensors section and the acquisition time of the signal per sensor. These last two values differ between each detector, and can counterbalance a low DQE, finally delivering a readable SNR to the detector output. Fig. 9 shows the output SNR of each detector as a function of the equivalent thickness of concrete imaged with a nominal flux of 10^6 X / mm² / pulse, for spatial frequencies of 0.5 and 1 cycle/mm.

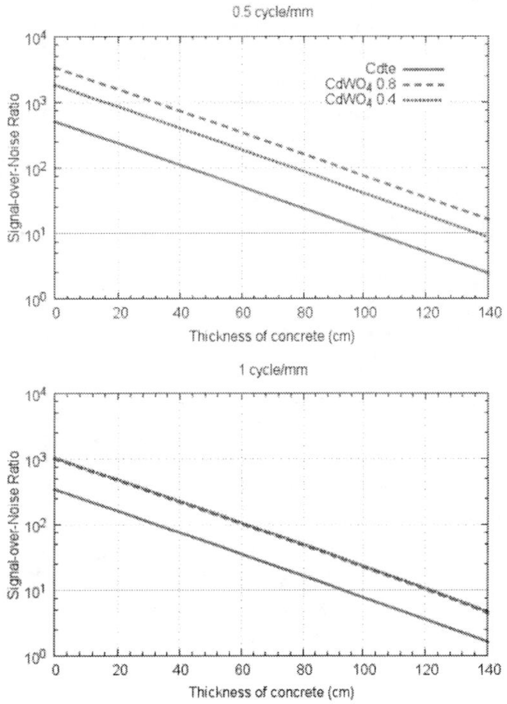

Fig. 9. Output Signal-over-Noise Ratio (SNR) of each detector as a function of the thickness of concrete, for 0.5 and 1 cycle/mm.

The readability of an image is arbitrarily fixed for a minimum SNR of 10. With the CdTe detector, it is possible in this setup to discern details up to the millimeter scale (0.5 cycle/mm) in a 100 cm thick concrete block, and sub-millimeter details through 90 cm of concrete (1 cycle/mm). For $CdWO_4$, the millimeter resolution is reached for 140 cm of concrete, with better performance for the large needles. Both models have a 500 µm resolution across 120 cm.

Finally, we should notice that the resolution limits discussed above are first approximations because the real limits will depend in the first place on the contrasts offered by the objects inspected at this level of detail. Taking into account these levels of contrasts is beyond the scope of this study and will be characterized later on and verified experimentally.

V. Conclusion

This study of two possible detectors for high energy tomography reveals notable performance, in terms of spatial resolution as well as penetration capacity. It appears that the multi-CdTe detector has the best intrinsic performance, with a relatively stable DQE, which provides a SNR that is slightly spatial frequency dependent and therefore spatial resolution dependent. For an equal acquisition time, the linear $CdWO_4$ scintillator arrays have a better SNR input, which counterbalances their lower DQE and results in a higher output SNR.

Thanks to the pixel size, the small model of $CdWO_4$ has the best spatial resolution; the large model and the multi-CdTe detector have a lower resolution. In parallel, a practical asset of the $CdWO_4$ scintillator is the large width of the detector that dispenses with the transverse displacement of the object. However, reaching a sub-millimeter resolution with a penetration capacity of 140 cm of concrete seems to be difficult.

This study by QAD analysis will be completed by an experimental verification in order to compare *in situ* the performance of the detectors.

References

[1] J.-P. Bissonnette *et al.*, "A quantum accounting and detective quantum efficiency analysis for video-based portal imaging," *Med. Phys.*, vol. 24, no. 6, pp. 815-826, Jun. 1997.

[2] F. Glasser *et al.*, "Application of cadmium telluride detectors to high energy computed tomography," *Nucl. Instrum. Methods Phys. Res. A*, vol. 322, pp. 619-622, Nov. 1992.

[3] N. Estre *et al.*, "High-energy X-ray imaging applied to nondestructive characterization of large nuclear waste drums," in *IEEE Transactions on Nuclear Science*, vol. 62, no. 6, pp. 3104-3109, Dec. 2015.

[4] X-Scan Imaging Corporation, XI880 X-ray LDA. Available: https://x-scanimaging.com/cammods/xi8800/, Accessed on: May 16, 2017.

[5] Los Alamos National Laboratory, "MCNP6 Users Manual - Code Version 6.1," Jun. 2014.

[6] I. A. Cunningham *et al.*, "A spatial-frequency dependent quantum accounting diagram and detective quantum efficiency model of signal and noise propagation in cascading imaging systems," *Med. Phys.*, vol. 21, no. 3, pp. 417-427, Mar. 1994.

Gamma-ray spectroscopy measurements and simulations for uranium mining

T. Marchais, B. Pérot, C. Carasco, P-G Allinei, P. Chaussonnet, J-L Ma, H. Toubon

***Abstract*— AREVA Mines and the Nuclear Measurement Laboratory of CEA Cadarache are collaborating to improve the sensitivity and precision of uranium concentration evaluation by means of gamma measurements. This paper reports gamma-ray spectra, recorded with a high-purity coaxial germanium detector, on standard cement blocks with increasing uranium content, and the corresponding MCNP simulations. The detailed MCNP model of the detector and experimental setup has been validated by calculation vs. experiment comparisons. An optimization of the detector MCNP model is presented in this paper, as well as a comparison of different nuclear data libraries to explain missing or exceeding peaks in the simulation. Energy shifts observed between the fluorescence X-rays produced by MCNP and atomic data are also investigated. The qualified numerical model will be used in further studies to develop new gamma spectroscopy approaches aiming at reducing acquisition times, especially for ore samples with low uranium content.**

***Index Terms*— Nuclear measurements, gamma-ray spectroscopy, germanium detector, uranium mining, MCNP6**

I. Introduction

Gamma logging for uranium exploration is currently based on total counting with Geiger Müller gas detectors or NaI(Tl) scintillators. However, the total count rate interpretation in terms of uranium concentration may be impaired in case of roll fronts, when the radioactive equilibrium of the natural ^{238}U radioactive chain is modified by differential leaching of uranium and its daughter elements (Th, Ra, Rn, etc.). Indeed, in the case of secular equilibrium, more than 95% of gamma rays emitted by uranium ores come from ^{214}Pb and ^{214}Bi isotopes, which are in the back-end of ^{238}U chain. Consequently, these latter might still produce an intense gamma signal even when uranium is not present in the ore, or in smaller quantity than in the secular equilibrium. Therefore, additional gamma measurements of core samples are performed in surface with high energy resolution, high-purity germanium (HP Ge) detectors. These last allow the detection of imbalances using the ratio of different gamma lines of the ^{238}U decay chain, such as the well-know 1001 keV of ^{234m}Pa, one of the first daughters in the beginning of the chain. However due to the low intensity of this gamma ray (0.84%), its detection in ore samples of a few hundred grams with a low uranium concentration may require measurement times of several hours. Therefore, we plan to use more intense peaks at lower energy, such as self- fluorescence X rays of uranium, to reduce acquisition times. Fluorescence is induced by the main gamma radiations emitted in the ore by ^{214}Pb (242, 295, 352 keV), ^{214}Bi (609 keV) and ^{226}Ra-^{235}U (186 keV). After Compton scattering, these gamma rays produce a high continuum of lower energy photons in the K-edge region of uranium (115.6 keV), resulting in a large photoelectric absorption rate and, subsequently, in the emission of uranium fluorescence X-rays, like the most intense 98.4 keV line. In order to study the potential of these new approaches and to establish a numerical simulation model, a measurement campaign was carried out in the radiometric calibration station of AREVA Mines in Bessines, France [1], with a coaxial HPGe detector. This station holds seven cement blocks with increasing uranium contents, up to 1% weight fraction (i.e. 10,000 ppm). The optimization of the MCNP model of the HPGe detector will be presented, as well as a comparison of different nuclear data libraries to explain missing or exceeding peaks in the simulation. Energy biases between the fluorescence X-rays produced by MCNP and atomic data will also be investigated.

Manuscript submitted on May 17, 2017. This work was supported by collaboration between CEA R&D and AREVA Mines.

Thomas Marchais, Bertrand Pérot, Cédric Carasco, Pierre-Guy Allinei, Pascal Chaussonnet and Jean-Luc Ma are with CEA, DEN, Cadarache, Nuclear Measurement Laboratory, F-13108 Saint-Paul-Lez-Durance, France.

Hervé Toubon is with AREVA Mines, F-92084 Paris La Défense Cedex, France.

II. The radiometric calibration station

AREVA Mines radiometric calibration station in Bessines aims to measure the counting rate due to gamma radiation emitted by seven independent cubic standard blocks with 70 cm edges (Fig.1). The uranium content of each block has been determined by chemical analysis (Table I). These blocks allow the calibration of radiometric probes.

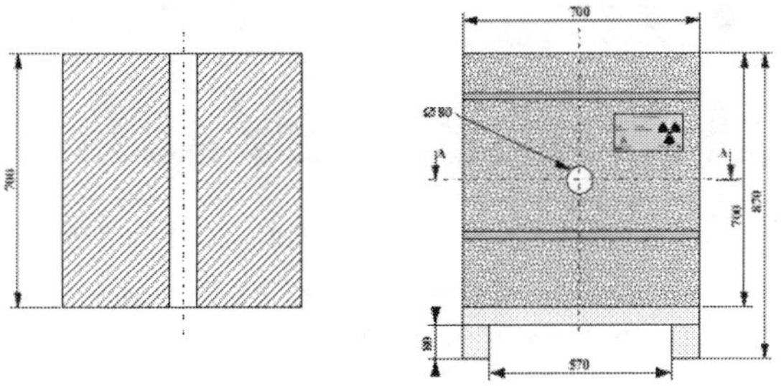

Fig. 1. Drawing of a standard block [1]

https://doi.org/10.1051/epjconf/201817005003

TABLE I. CHARACTERISTICS OF THE SEVEN BLOCKS

Block #	Uranium mass fraction from chemical analyses (ppm)	Cement density (g.cm^{-3})
B1	0	1.89
B2	500	1.92
B3	1000	1.89
B4	1900	1.91
B5	2900	1.92
B6	4800	1.93
B7	9700	1.92

The Nuclear Measurement Laboratory of CEA Cadarache has carried out a series of gamma-ray spectroscopy measurements on these standard blocks (Fig. 2). Results have confirmed the uranium content of each block, as shown further in Table II. The gamma spectra acquired during this campaign (Fig. 3) will also be used to validate the MCNP numerical model and to identify new information (X or γ peaks, Compton continuum, K-edge, etc.) that could be used to improve the assessment of uranium concentration. This campaign has been performed with a type N, coaxial HPGe detector, model GR1020, CANBERRA.

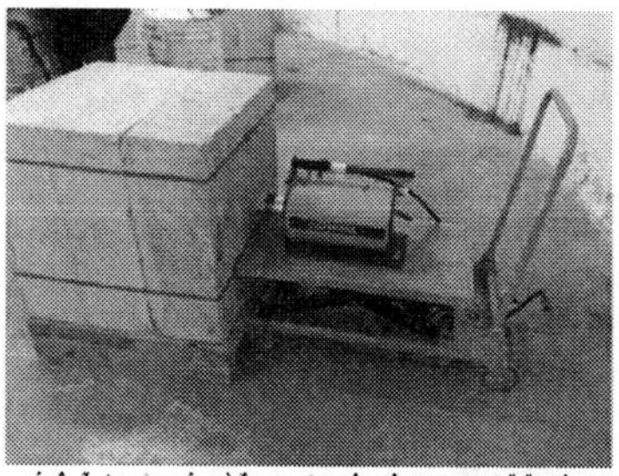

Fig. 2. Coaxial detector inside a standard cement blocks at Bessines.

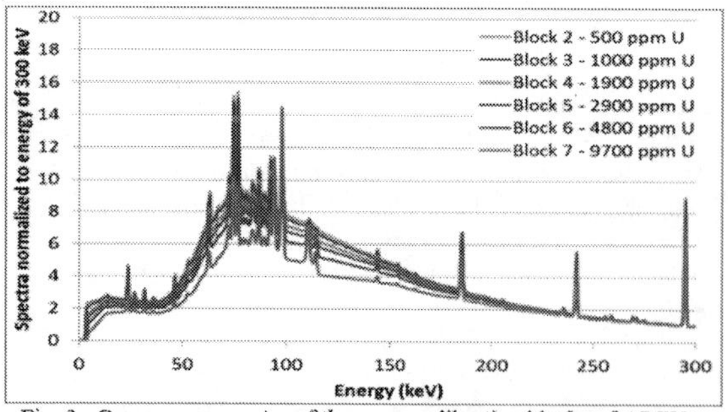

Fig. 3. Gamma-ray spectra of the seven calibration blocks of AREVA radiometric station in Bessines, France.

III. Simulation of the coaxial detector

The coaxial detector was modeled with the MCNP Monte Carlo computer code [2]. The geometric model of the detector is based on the manufacturer scheme, on X-ray radiography, and on measurements performed with a multi-energy beam coming from a highly collimated ^{152}Eu gamma source (see Fig. 4), which allowed to precisely estimate the position and active area of the crystal. The calculated and experimental detector efficiencies are reported in Fig. 5. The measured efficiency was built with calibration point sources (^{241}Am 59.5 keV line, ^{109}Cd 88.0 keV peak, and ^{152}Eu gamma-rays from 121.8 keV to 1408 keV) at a distance of 50 cm from the detector head. Note that to obtain the good agreement observed in Fig. 5, a dead layer has been implemented in the model of the germanium crystal, in the back, external edge of the germanium cylinder, to reflect a deficit of charge collection in this region. A copper electrode connecting the cooling system and the germanium crystal is also described, as well as the external aluminum cover.

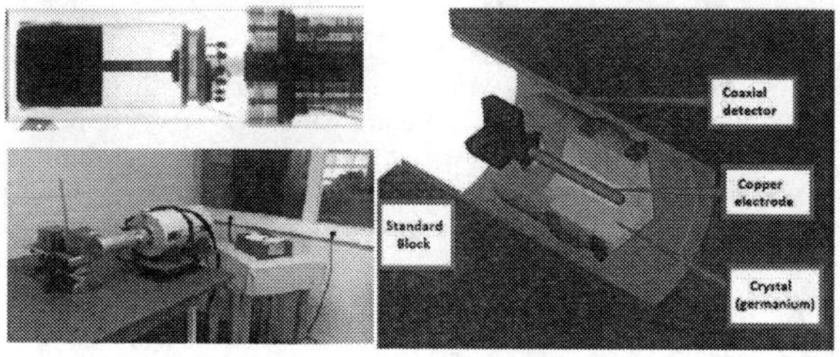

Fig. 4. X-ray radiography and the modeling of the coaxial detector.

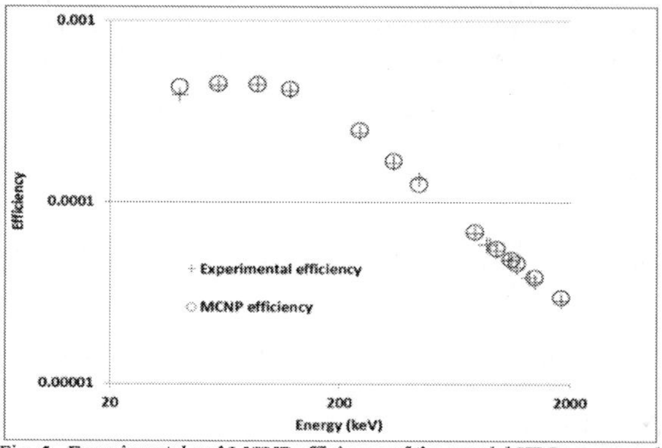

Fig. 5. Experimental and MCNP efficiency of the coaxial HPGe measured with point sources at a distance of 50 cm from the detector entrance face.

IV. Verification of the uranium content in the standard concrete blocks

The global detection efficiency, including self-absorption in the standard cement blocks, has been calculated with MCNP with a first objective to determine the uranium mass concentration of each block, using the ^{234m}Pa gamma ray at 1001 keV. Equation (1) allows converting the net area of this peak into uranium mass concentration (ppm).

$$Cm_U = \frac{(1+\eta)\times 10^6 \times M_{mol}(^{238}U)\times T_{1/2}(^{238}U)}{N_A \times \ln(2)} \times \frac{S_n(1001\ keV)}{Tc \times M_{Block} \times I_\gamma(1001\ keV)\times Eff(1001\ keV)} \quad (1)$$

With:

Cm_U : Uranium mass concentration of each block (ppm)

$S_n(1001\ keV)$: Net area of the peak at 1001 keV measured on the gamma spectrum (number of counts, dimensionless number)

Tc: Acquisition "live time" (i.e. real time corrected for dead time) of the gamma spectrum (s)

$I_\gamma(1001\ keV)$: Intensity of emission of the peak at 1001 keV of ^{234m}Pa (0.84 %) in the ^{238}U filiation

$Eff(1001\ keV)$: Detection efficiency calculated with MCNP for each block (number of count in the full-energy peak at 1001 keV per photon of 1001 keV emitted in the block)

$T_{1/2}(^{238}U)$: Radioactive period of ^{238}U (s)

N_A: Avogadro constant ($6{,}023{\cdot}10^{23}$ atoms per mole, dimensionless number)

$M_{mol}(^{238}U)$: Molar mass of ^{238}U (g)

η: Mass fraction of ^{235}U in natural uranium (0.72 %)

The mass concentration obtained by gamma-ray spectroscopy using (1) is in good agreement with chemical analysis, as shown in Table II. The reported uncertainty includes the detector calibration and modeling uncertainty (2.6 %), the statistical uncertainty on the net area of the peak at 1001 keV (between 0.9 % and 22 %, depending on uranium concentration), the uncertainty on the 1001 keV gamma-ray intensity (1.32 %), the statistical uncertainty of MCNP calculations (less than 1 %), and the uncertainty of the MCNP modeling of the concrete blocks and measurement setup. Concerning this last, the main source of uncertainty concerns concrete density, which varies from 1.89 to 1.93 g.cm^{-3} depending on the measured block. The dispersion of the efficiency values calculated on the blocks with MCNP is equal to 0.97 %. This value has been retained to be the uncertainty of the modeling of the measurement scene at 1 σ. The global uncertainty is the quadratic sum of these uncertainties.

TABLE II. URANIUM MASS CONCENTRATIONS (PPM) OF BESSINES CALIBRATION BLOCKS MEASURED BY GAMMA-RAY SPECTROSCOPY AND CHEMICAL ANALYSIS

Block	Chemical analysis	Gamma-ray spectroscopy (and 1 σ uncertainty)	Relative difference
B1	0	2.1 (± 22 %)	N/A
B2	500	459(± 3.7 %)	- 8.2 %
B3	1000	958(± 1.9 %)	- 4.2 %
B4	1900	1960(± 2.3 %)	3.4 %
B5	2900	2800(± 3.3 %)	- 3.5 %
B6	4800	4550(± 1.8 %)	- 5.2 %
B7	9700	9530(± 2.0 %)	- 1.8 %

V. SIMULATION OF THE STANDARD CONCRETE BLOCKS GAMMA SPECTRA

The simulation of the concrete blocks is based on a cement chemical composition given by AREVA Mines, including uranium concentrations (and densities) given in Table I. An isotropic gamma source uniformly distributed in the cement block is implemented in MCNP. The natural ^{238}U and ^{235}U radioactive chains are supposed in secular equilibrium. The source term retains only 14 predominant gamma emitters observed in the experimental spectra: ^{238}U, ^{234}Th, ^{234m}Pa, ^{230}Th, ^{226}Ra, ^{214}Pb, ^{214}Bi, ^{210}Pb, ^{235}U, ^{231}Th, ^{227}Th, ^{223}Ra, ^{219}Rn, and ^{211}Pb. The intensity and energy of each gamma ray are taken from the JEFF 3.1.1 library, and only the lines with energy larger than 40 keV are modeled.

Fig. 6 shows a good general agreement between simulated and experimental spectra, only small differences being observed for minor peaks. For instance, the 89.5 keV peak was initially absent from the simulated spectrum due to a missing emission line of ^{214}Pb at 89.8 keV in the JEFF 3.1.1 library (Bi $K_{\beta2}$ X-ray, intensity of 0.67% in ENSDF [3]), combined with an underestimation emission of ^{214}Bi at 89.2 keV (Po $K_{\beta3}$ X-ray, 0.085% in JEFF 3.1.1 instead of 0.116% in ENSDF [3]). After implementing ENSDF data [3], the calculated 89.5 keV peak was found in satisfactory agreement with experiment (Fig. 7).

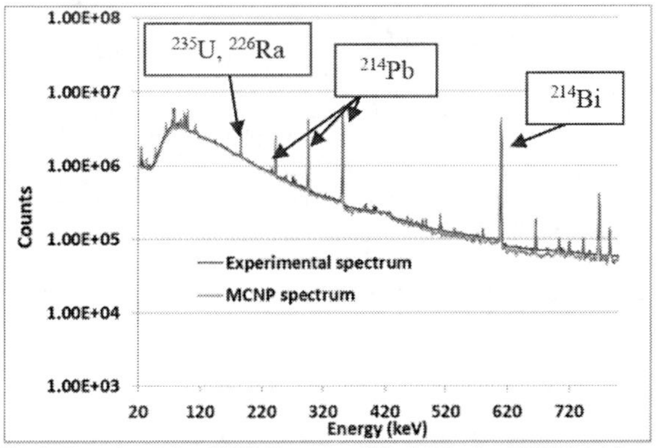

Fig. 6. Experimental and MCNP spectra of block B5 using ENSDF database.

We can also note in Fig. 7 that a peak is missing in the simulation at 110.4 keV, in the broad bump including both 110.4 and 111.3 keV uranium fluorescence X-rays ($K_{\beta3}$ and $K_{\beta1}$, respectively [4]). Indeed, MCNP manual [5] indicates that for elements with atomic number Z > 37, the X-ray emission $K_{\beta3}$ is not simulated, and $K_{\beta1}$ X-ray includes all transition intensities from layers L to K, which explains the absence of the 110.4 keV peak ($K_{\beta3}$) and a higher intensity at 111.3 keV ($K_{\beta1}$).

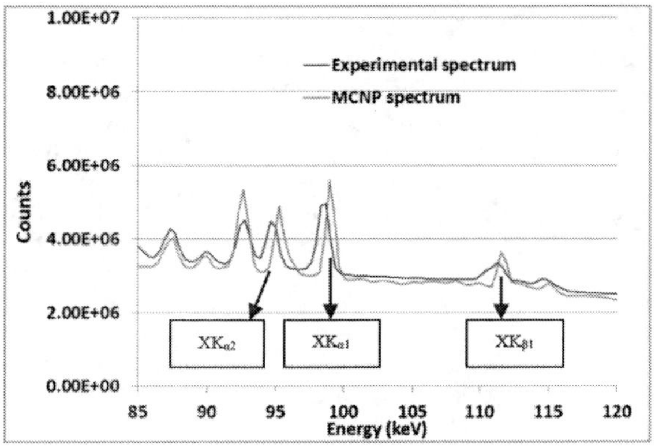

Fig. 7. Experimental and MCNP spectra of block B5 at low energy.

Fig. 7 also shows a significant energy shift between calculation and experiment concerning uranium fluorescence

X-rays. Further investigation of the MCNP fluorescence peaks energy shift has been performed for 53 materials having atomic numbers between 40 (zirconium) and 92 (uranium) irradiated by a 1 MeV photon beam. The energy shift between MCNP and the expected fluorescence X-ray energy shown in Fig. 8 as a function of Z reproduces the observation previously made in [6].

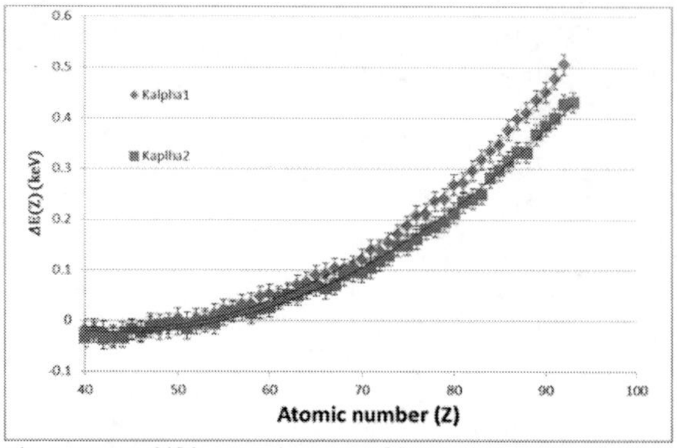

Fig. 8. Energy shift between the energy of fluorescence X-rays calculated with MCNP and the theoretical values taken from [3].

In further studies, this energy shift will be corrected when processing MCNP output data to allow a better agreement with the experimental spectrum.

VI. Conclusion and prospects

Seven calibration blocks with known uranium concentrations have been measured by high resolution gamma-ray spectroscopy at AREVA Mines calibration station, in Bessines, France, and simulated with MCNP computer code. First, a detailed model of the coaxial HPGe detector has been carried out using calibration acquisitions with standard point sources of ^{109}Cd, ^{152}Eu, and ^{241}Am. Then, the uranium mass concentrations of the blocks (in ppm), originally determined by chemical analysis, has been confirmed by gamma-ray spectroscopy with the 1001 keV gamma ray of ^{234m}Pa, in the ^{238}U radioactive chain. Finally, we obtained a very good agreement between the overall gamma spectra of the blocks calculated with MCNP and the ones measured in Bessines. Only minor discrepancies concerning small peaks have been observed and corrected, which were due to unprecise or missing nuclear data, as well as small bias in the energy of fluorescence X-rays produced by MCNP, which will be corrected in future studies.

We will now use the validated numerical model to study new gamma-ray spectroscopy methods, in view to characterize uranium faster than with the reference 1001 keV line. Indeed, its intensity (0.84 %) leads to hours of acquisition time for ore samples of a few hundred grams with uranium concentration lower than 1 % mass fraction. Among the information that is not yet used, uranium self-fluorescence X-rays in the 100 keV region (K_α and K_β transitions) can be detected in a much shorter time. They are induced by gamma radiations of the ore itself and could be used to characterize uranium up to a depth of several centimeters, hence being appropriate for small ore samples.

References

[1] http://www.areva.com/EN/operations-598/bessines-a-standardsetting-industrial-facility
[2] MCNP6TM, User's manual – Version 1.0 - LA-CP-13-00634, Rev. 0 – May 2013 – Denise B. Pelowitz, editor
[3] The Lund/LBNL Nuclear Data Search, Version 2.0, February 1999, S.Y.F. Chu, L.P. Ekström and R.B. Firestone, LBNL, Berkeley, USA, and Lund University, Sweden, http://nucleardata.nuclear.lu.se/toi/
[4] M.C. Miller, X-ray Fluorescence, Chapter 10 of Passive Nondestructive Assay of Nuclear Materials (1991) pp. 313-335, NUREGCR5550, Los Alamos National Laboratory.
[5] Morgan C. White, Los Alamos National Laboratory, Los Alamos, - "Photoatomic Data Library MCPLIB04 : A New Photoatomic Library Based On Data from ENDF/B-VI Release 8" - LA-UR-03-1019- Dec. 2002.
[6] Cook, Matthew Tyler, "Hybrid K-edge Densitometry as a Method for Materials Accountancy Measurements in Pyrochemical Reprocessing." PhD diss., University of Tennessee, 2015. http://trace.tennessee.edu/utk_graddiss/3329
[7] Gordon GILMORE, John HEMINGWAY – "Practical gamma-ray spectrometry" - 1995
[8] T. Jensen, T. Aljundi, C. Whitmore, H. Zhong, and J. N. Gray, "X-Ray, K-Edge Measurement of Uranium Concentration in Reactor Fuel Plates," Ames Lab., IA (United States), IS--5129, Nov. 1997, **http://dx.doi.org/**10.2172/671994
[9] P. Pin and B. Perot, "Characterization of Uranium in Bituminized Radioactive Waste Drums by Self-Induced X-Ray Fluorescence," *IEEE Trans. Nucl. Sci.*, vol. 61, no. 4, pp. 2131–2136, Aug. 2014.

Development of a new bench for puncturing of irradiated fuel rods in STAR hot laboratory

B. Petitprez, P. Silvestre, P. Valenza, A. Boulore *(Fuel Research Department, CEA Cadarache)*, T. David *(Sealing Laboratory Maestral, CEA Marcoule)*

Abstract— **A new device for puncturing of irradiated fuel rods in commercial power plants has been designed by Fuel Research Department of CEA Cadarache in order to provide experimental data of high precision on fuel pins with various designs. It will replace the current set-up that has been used since 1998 in hot cell 2 of STAR facility with more than 200 rod puncturing experiments. Based on this consistent experimental feedback, the heavy-duty technique of rod perforation by clad punching has been preserved for the new bench. The method of double expansion of rod gases is also retained since it allows upgrading the confidence interval of volumetric results obtained from rod puncturing. Furthermore, many evolutions have been introduced in the new design in order to improve its reliability, to make the maintenance easier by remote handling and to reduce experimental uncertainties.**

Tightness components have been studied with Sealing Laboratory Maestral at Pierrelatte so as to make them able to work under mixed pressure conditions (from vacuum at 10^{-5} mbar up to pressure at 50 bars) and to lengthen their lifetime under permanent gamma irradiation in hot cell. Bench ergonomics has been optimized to make its operating by remote handling easier and to secure the critical phases of a puncturing experiment. A high pressure gas line equipped with high precision pressure sensors out of cell can be connected to the bench in cell for calibration purposes. Uncertainty analyses using Monte Carlo calculations have been performed in order to optimize capacity of the different volumes of the apparatus according to volumetric characteristics of the rod to be punctured. At last this device is composed of independent modules which allow puncturing fuel pins out of different geometries (PWR, BWR, VVER).

After leak tests of the device and remote handling simulation in a mock-up cell, several punctures of calibrated specimens have been performed in 2016. The bench will be implemented soon in hot cell 2 of STAR facility for final qualification tests. PWR rod punctures are already planned for 2018.

Index Terms— **fuel rod, pressure, puncturing, remote-handling, experimental uncertainties**

I. OBJECTIVES OF THE NEW ROD PUNCTURING BENCH

Fission gas release from fuel under irradiation and the resulting pressure increase are key parameters influencing the behaviour of a nuclear Pressurized Water Reactor (PWR) fuel rod in nominal and accident conditions. This data is obtained experimentally at the CEA Cadarache centre by a device that has been used in hot cell 2 of the STAR facility since 1998 with more than 200 PWR rod puncturing experiments [1]. In order to broaden the CEA's experimental capability, a new device for puncturing of irradiated fuel rods from different types of commercial power plants has been designed by the CEA Fuel Research Department at Cadarache. Based on the consistent experimental feedback obtained with the current device, the double expansion puncturing method [2] has been kept for the new device. The design of this new device has been optimised to simplify operation and maintenance by remote-handling, to guarantee reliable puncturing experiments and to reduce uncertainties associated with the experimental results.

II. ASSEMBLING, OPERATION AND MAINTENANCE OF THE BENCH BY REMOTE-HANDLING

In compliance with the ALARA approach, all the phases of the device service life have to be performed in the hot cell by remote-handling. At the same time, availability of the bench has to be guaranteed to allow up to twenty rod puncturing experiments per year. Thus, a modular design has been adopted with several independent components to be assembled directly in the hot cell. As for the current machine, three main volumes – the calibration assembly (VB), the rod assembly (VM) and the expansion assembly (VD) – comprise the gas circuit of the new bench as shown in Figure 1.

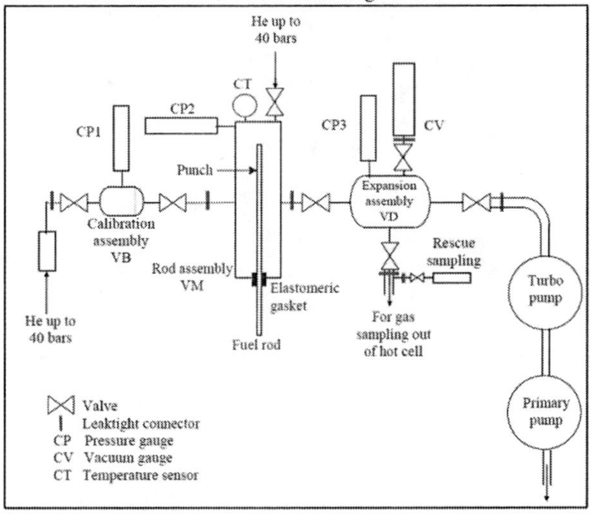

Fig. 1. Schematic diagram of the puncturing device

The three volumes can then be connected and disconnected independently thanks to gasket fittings easy to handle with remote manipulators. A guiding support receives the different volumes and helps to connect them to each other with alignment pins and fasteners. At the end of assembling, a punching column is fixed on the rod assembly and the support. Pressure gauges have also been designed to be screwed onto the modules by remote-handling through an appropriate replaceable sealing joint as represented in Figure 2.

Fig. 2. Pressure gauge seal replacement by remote-handling

Leak tightness is key parameter of the puncturing device; its leak rate should remain below 10^{-7}Pa.m^3/s for all operating conditions, i.e. from 10^{-3} Pa (secondary vacuum) to 5.10^6 Pa of helium. As metallic seals could not be used in this case, it was crucial to find an elastomeric seal material resistant enough to permanent gamma irradiation in a hot cell and suitable for our working conditions.

This is why different kinds of elastomers were tested in the MAESTRAL Laboratory [3] at the CEA Pierrelatte centre to characterise leak-tightness and mechanical behaviour after several months of gamma irradiation in hot cell. Figure 3 shows the leak-tight behaviour during a 3-phase pressure test which reproduces a puncturing experiment for two different elastomers at three stages of irradiation.

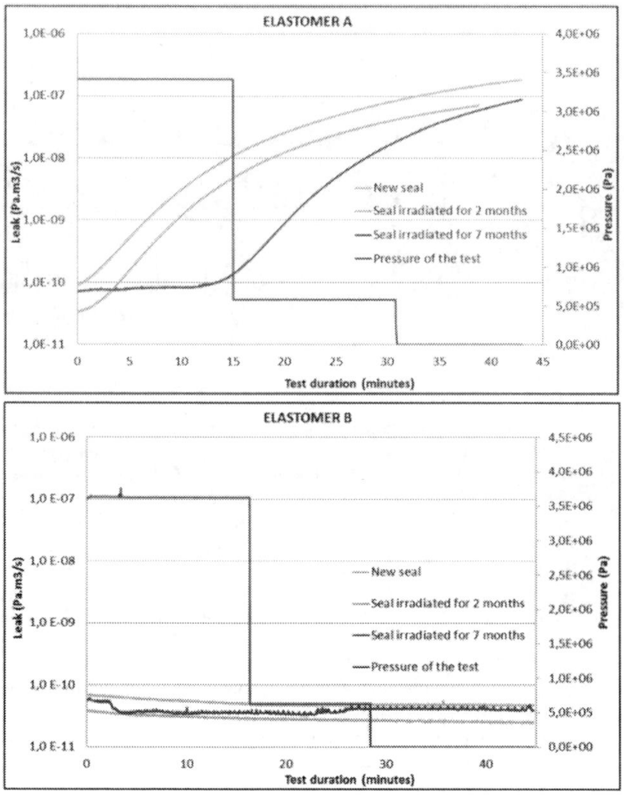

Fig. 3. Comparison of leak-tight behaviour between two different elastomers (fresh and after irradiation)

The most appropriate elastomer was chosen as the sealing material to be used for the fittings, the gauge seals and the gasket around the fuel rod. Additional tests in the MAESTRAL Laboratory made it possible to determine the optimised torque for tightening the different seals of the bench using a torque wrench adapted for remote-handling. The tests performed in the MAESTRAL Laboratory also allowed defining the replacement frequency of the device sealing components.

The bench ergonomics were also studied to simplify operations with remote manipulators and to make it easier to monitor the actions trough the shielded window. For instance, all of the device valves have been identified and designed for secure handling as shown in Figure 4.

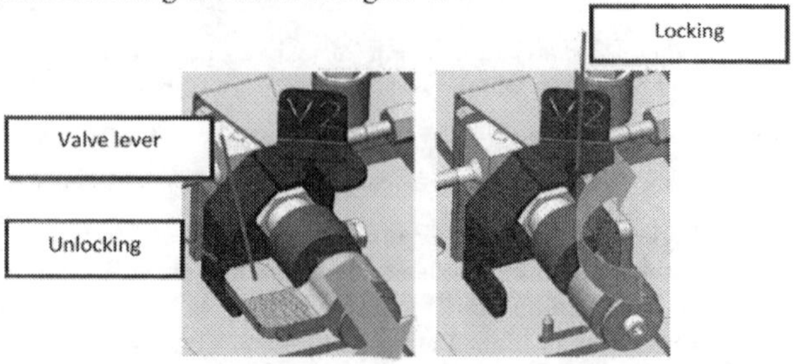

Fig. 4. Bench valves design for remote-handling

III. A MORE RELIABLE PUNCTURING BENCH FOR ALL TYPES OF FUEL ROD GEOMETRIES

As rod puncturing is a one-shot experiment, the leak-tightness of the puncturing device has to be guaranteed once the fuel rod has been placed in rod assembly for the whole experiment, from rod punching to gas sampling. This is why a helium pressurization line capable of providing up to 5.10^6 Pa of helium has been implemented in hot cell 2 of the STAR facility. Thanks to a pressure connection on the puncturing device rod assembly (see Figure 1), the line can be connected directly to the rod assembly while the fuel rod is in place ready to be punched. It is therefore possible to perform a leak test in the rod assembly at the pressure level expected after rod puncturing. This change in the experimental procedure with the new device secures the puncturing experiment.

The reliability of the puncturing experiment is also guaranteed by the double expansion method [2], which is an extension of the conventional single expansion method with an additional experimental step. Using the double expansion puncturing method with the new bench sketched in Figure 1 enables to express rod inner pressure Pc and rod free volume VL as a system of two equations with two unknowns:

$$\begin{cases} Pc * VL = P1 * (VL + VM) \\ P1 * (VL + VM) = P2 * (VL + VM + VD) \end{cases}$$

where P1 is the pressure measurement after the first expansion of rod gases in VM and P2 is the pressure measurement after the second expansion in VD.

The rod free volume is still measured by single expansion at the end of the puncturing experiment to confirm the measurement and it can be compared with the result obtained by double expansion. It can also be used as a back-up measurement. After this stage of the experimental procedure, the volumetric data of the fuel rod has been determined and confirmed, i.e. the rod free volume, the inner rod pressure at 20°C and the total gas volume in the rod at 20°C and 1 atmosphere.

It is then necessary to determine the elementary and isotopic compositions of the gas contained in the fuel rod, in particular to assess the fission gases released from fuel since the beginning of irradiation. To do this, it is necessary to collect a leak-tight sample of the rod gas that has expanded in the puncturing device circuit. The sample is then sent to the mass spectrometry laboratory for analysis. All these operations require many handling operations in hot cell and later in a glove box outside the hot cell. In order to prevent the risk of gas loss during this phase, a removable sampling phial (see Figure 5) has been added to the new device. In case of gas loss during sampling, the phial can be used as back-up for gas sampling. After decontamination, the phial can be removed from the cell so the gas content can be transferred into a secondary ampoule to be sent for mass spectrometry analysis.

Fig. 5. Back-up gas sampling system for the new puncturing bench

The modular design of the new puncturing bench allows receiving all geometries of commercial fuel rods irradiated in nuclear plants. A rod assembling VM such as that represented in Figure 6 can be designed and built on demand for any given geometrical specifications (outer diameter and plenum length of fuel rod). It can then be installed on the bench and connected to the two other modules (calibration volume VB and expansion volume VD) by means of the gasket fittings. In the same manner, a pressure gauge can be screwed into the CP2 position of the new rod assembly. At last two different positions are available on the punching column for the failing mass that allows adapting the perforation energy to the cladding characteristics of the rod to be punctured.

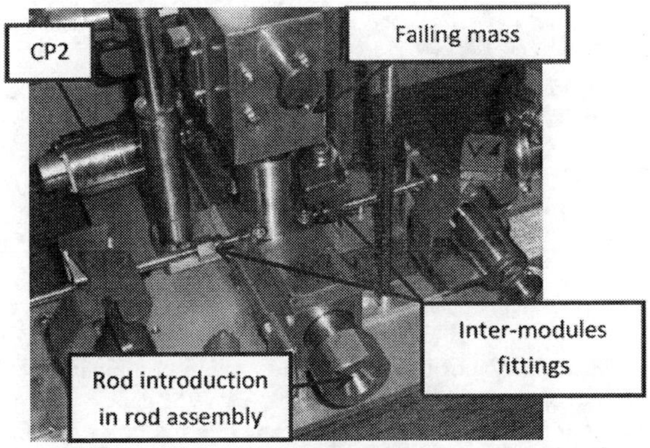

Fig. 6. Rod assembly design for the new puncturing bench

IV. PROVIDING MORE ACCURATE EXPERIMENTAL RESULTS

During any reactor cycle, the plant operator has to take into account pressure margins in fuel assemblies to maintain a certain level of flexibility for power incursions and/or burn-up increase [4]. This monitoring uses fuel performance codes that have been qualified based on the data-base of rod puncturing results. The rod pressure margin calculations also take into account experimental uncertainties on these puncturing results. This is why it is essential to reduce as much as possible the experimental uncertainties associated to the quantities of interest (rod inner pressure, total gas volume in the rod, free volume) obtained with the new bench.

An optimisation approach using Monte-Carlo calculations was adopted in the early design phase to determine the dimensions of the sampling volume VB, the rod assembly VM and the expansion volume VD. First the URANIE software developed by the CEA has been used for uncertainty quantification [5], sensitivity analysis and optimisation to implement the measurement model defined by the mathematical relations between all quantities known to be involved in a rod puncturing experiment. Then the optimisation problem consisted in identifying the input data to fix (e.g. volumetric characteristics of the rod to puncture and measuring features of the three pressure gauges) and the quantities to make vary within a given range (VB, VM, VD and P1 corresponding to the pressure measurement after the first expansion). Lastly Monte-Carlo calculations were performed with varying set combinations (P1; VB; VM; VD) and the results were filtered to minimise the standard deviation results. Figure 7 schematises the optimisation process for the given input data with all Monte-Carlo calculations represented in green lines and the resulting standard deviation histogram for rod pressure calculations (stdPc). The blue lines represent the combinations of values (P1; VB; VM; VD) that minimise the value of stdPc.

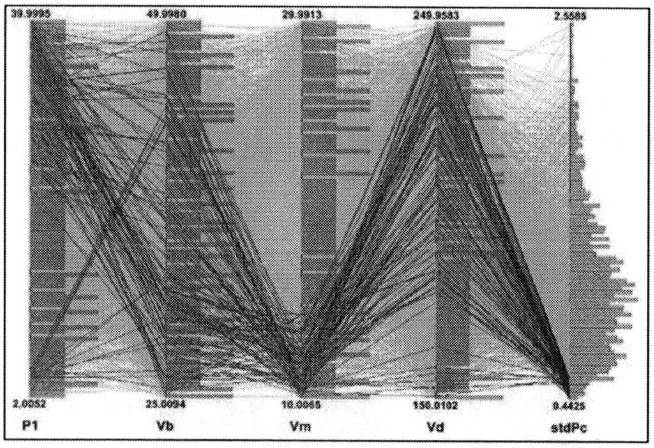

Fig. 7. Optimization process to design bench volumes (VB, VM, VD) that minimise rod pressure uncertainty (StdPc)

From a practical point of view, these combinations allow defining capacity ranges for VB, VM and VD where puncturing result uncertainties are minima for a given rod. Specifications for the new puncturing bench were finally established according to the results of these optimisation calculations.

The accuracy of the on-line pressure and temperature measurements also contributes to the resulting uncertainties related to the experimental rod puncturing results. Therefore the helium pressurisation line has been equipped with high precision pressure gauges located out of cell to take into account the instrumentation drift on the three pressure gauges implemented on the puncturing bench (CP1, CP2 and CP3). These high precision sensors measure the absolute pressure with a maximal relative error of ±0.01% and are periodically calibrated at the French metrology laboratory LNE. They can then be used as secondary measurement standards to calibrate the pressure gauges CP1, CP2 and CP3 directly in the hot cell providing an operating pressure measurement accuracy better than ±0.5% (k=2). Furthermore a new calibration procedure has been developed to determine the capacity of the puncturing bench reference volumes VB and VD with an accuracy close to ±1% (k=2). These calibrations provide thus the standard measurement uncertainties u(xi) associated with the different parameters xi that are used in the uncertainty budget of a puncturing experiment. According to uncertainty propagation methodology by second order Taylor development, the following equation illustrates the calculation used to determine the uncertainty associated with the inner rod pressure result Pc obtained at the end of rod puncturing experiment:

$$u(P_c) = \sqrt{\sum_i \left(\frac{\partial P_c}{\partial CPi}\right)^2 * u(CPi)^2 + \left(\frac{\partial P_c}{\partial VB}\right)^2 * u(VB)^2 + \left(\frac{\partial P_c}{\partial VD}\right)^2 * u(VD)^2 + \left(\frac{\partial P_c}{\partial T}\right)^2 * u(T)^2}$$

where CPi correspond to pressure measurements by CP1, CP2 and CP3 during the experiment and T is the rod gas temperature measured during the experiment.

The final uncertainty U(Pc) on the rod pressure result is : U(Pc) = k*u(Pc) with k=2 so as to guarantee a coverage interval at 95%. It is important to note that u(Pc) is the analytical expression corresponding to the standard deviation std(Pc) obtained by probabilistic calculations used for the volume bench design. So comparison of results obtained by both methods is an efficient way of validating the experimental uncertainty estimations.

V. Conclusion

Many improvements have been introduced in the design of the new rod puncturing device and the puncturing experiment procedure has been upgraded as well. As a result, the bench maintenance can be performed entirely by remote-handling within a few days and it will now take two days to complete a puncturing experiment. Furthermore the bench has been designed to puncture all types of commercial fuel rod geometries. Its reliability has been improved for evaluating the quantities of interest (rod inner pressure, total volume of gas in rod, free volume) as well as for assessing the rod gas composition. Presently, work has be done to improve the accuracy of the puncturing results so the relative uncertainties can be reduced to ±4% (k=2) for the rod inner pressure and ±5% (k=2) for the rod free volume. The new rod puncturing bench will be implemented in hot cell of STAR facility in 2018 and it should be fully qualified at the end of the year. PWR rod punctures are already planned afterwards.

VI. Acknowledgments

The authors would like to thank AREVA-NP and EDF for their financial and technical support in this study.

VII. References

[1] *Irradiated Fuel Puncturing Devices Implemented at the LECA-STAR Laboratory of CEA-Cadarache, Y. Pontillon, B. Gleizes, H. Capdevilla, P. Silvestre, P. Plume*, 45th HOTLAB conference, 2008

[2] *A new apparatus for determination of the free volume of a fuel rod using the double expansion method, L. Desgranges, M.H. Faure, A. Thouroude*, Nuclear Technology, vol. 149, January 2005

[3] http://technetics.com/engineering/rd-innovation/

[4] *Nuclear fuel safety criteria technical review (second edition)*, ISBN 978-92-64-99178-1, OECD 2012

[5] *URANIE: The CEA/DEN Uncertainty and Sensitivity platform, F. Gaudier*, Procedia Social and Behavioral Sciences 2 (2010) 7660–7661

Detection of tiny amounts of fissile materials in large-sized containers with radioactive waste

V.F. Batyaev, S.V. Skliarov
Dukhov Research Institute of Automatics (VNIIA), 127055 Moscow, Russia
bogolubov@vniia.ru

***Abstract*—The paper is devoted to non-destructive control of tiny amounts of fissile materials in large-sized containers filled with radioactive waste (RAW). The aim of this work is to model an active neutron interrogation facility for detection of fissile materials inside NZK type containers with RAW and determine the minimal detectable mass of U-235 as a function of various parameters: matrix type, nonuniformity of container filling, neutron generator parameters (flux, pulse frequency, pulse duration), measurement time. As a result the dependence of minimal detectable mass on fissile materials location inside container is shown. Nonuniformity of the thermal neutron flux inside a container is the main reason of the space-heterogeneity of minimal detectable mass inside a large-sized container. Our experiments with tiny amounts of uranium-235 (<1 g) confirm the detection of fissile materials in NZK containers by using active neutron interrogation technique.**

***Index Terms*—Fissile materials, Neutron generator, Monte-Carlo simulating**

I. Introduction

One of the urgent challenges of today's nuclear power engineering is to detect and control fissile material (FM) in various radioactive waste (RAW). The expediency of its solution lies both in ensuring long-term nuclear and radiation safety of RW storage sites, and in preventing theft of FM.

The major part (at least 99%) of RAW in Russia belong to low and very low level wastes (LLW) [1], which, as a rule, are packed into the NZK containers [2], the FM content there being estimated in the range ~ 10^{-2} to ~ 10 mg/kg of LLW [3]. The legislation of the Russian Federation obliges the certification procedure of RAW packages before their disposal, in particular, for the presence of FM. The opening of the container for such certification is not permitted.

It is clear that the existence of fission products, as well as the neutron emitting actinides in the certified RAW packages leads to increased the gamma and/or neutron background around the containers and the deterioration of FM control. Nevertheless, at the first stage of designing the required certifier, our theoretical calculations were made under the assumption that only uranium isotopes and no fission products and other actinides were present in the NZK. The expediency of this assumption is supported by the fact that part of the nuclear fuel cycle facilities produce RAW containing uranium isotopes only.

Previously, we carried out a calculation work [4] comparing non-destructive methods for controlling small amounts of FM in large-sized containers with radioactive waste, in order to select the method and determine the ways to implement it. It was shown that the use of the passive gamma-neutron control method is difficult due to the small amount of FM in containers and the high absorbing capability of both the container itself and the filler matrix. Executed evaluations of FM responses to the effect of neutron (2.5 MeV and 14 MeV) and gamma sources (10 MeV) on the container show the advantage of active neutron methods for controlling large containers in front of the photonuclear method in terms of the absolute values of response radiation ensuring control of these small amounts of FM in the NZK.

This work consists in modeling the active neutron control system for FM in the NZK containing LLW and determining the minimum detectable mass of FM depending on various factors (measurement duration, number of neutron detectors, type of matrix and uneven distribution of FM in the NZK), and also in improving the proposed method.

II. Modeling of active neutron control system for detection of FM in NZK containing LLW

The first step is to determine the sensitivity of the neutron-radiation method, depending on the position of uranium-235 in various zones of the NZK. For this purpose, a mathematical model of the container and its contents was prepared for computational modeling of neutrons transport and interaction.

The container is a hollow concrete cube with external dimensions of 165cm x 165cm x 134cm, a wall thickness of 11cm, a weight of 3.5 tons, which is filled with ~ 3.2 tons of concrete matrix containing LLW.

As the material of the walls, reinforced concrete with a density of 2.5 g/cm3 was selected, while FM was placed in the concrete matrix of the container. The container matrix was split into 25 zones (Figure 1), with FM (uranium-235) sequentially located in the middle of each of the zones of the NZK over the height of the zone. As neutron detectors, SNM-18 3-He-counters were simulated, surrounded by polyethylene and a cadmium shield. The source of neutrons was chosen as a pulsed point isotropic one with 14 MeV energy. An example of the obtained time distributions is shown in Fig. 2, which shows that the first exponent corresponds to the detection of initial fast neutrons and its decay constant is determined by the thickness inner moderator inside the cadmium shield of the counter. The second exponent corresponds to the thermalized neutrons (background) and fast neutrons of FM fission. The decay constant of the second exponential is determined by the walls of the NZK and its filling.

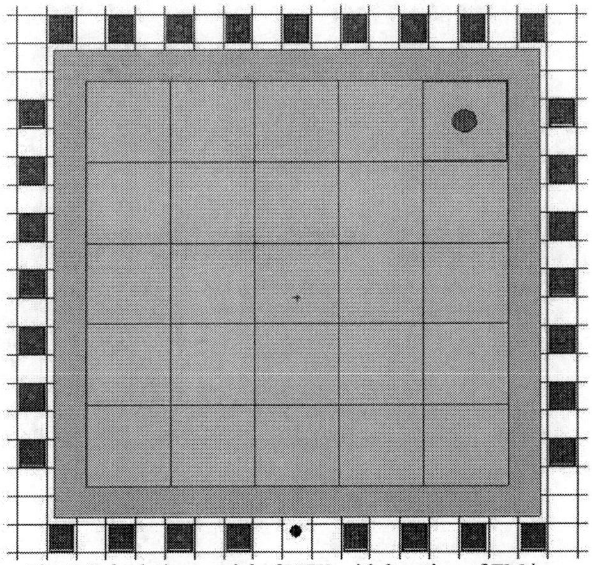

Fig. 1 Calculation model of NZK with location of FM in one of container zones. The dot shows the position of the neutron source (14 MeV neutron generator ING-07T produced at VNIIA).

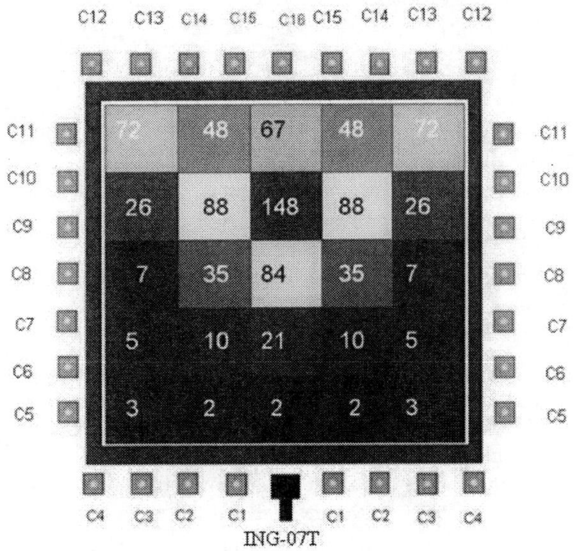

Fig. 3 – Simulated minimum detectable mass of ^{235}U (mg) depending on its location in container.

The simulation results allowed to determine the minimum detectable mass of FM as a function of its position in the NZK container as well as its dependence on the number of neutron counters used (Figures 3 and 4), provided that their backgrounds are the same. The wide range of the minimum detectable mass of uranium-235 (from 2 to 84 mg) shown in Figures 3 and 4 is due to the non-homogeneity of the thermal neutron field inside the container, which in practice can be partially neutralized by rotating the container during the measurement.

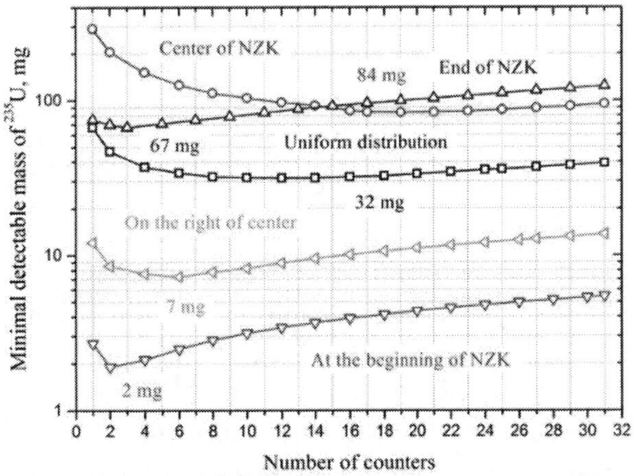

Fig. 4 - Simulated minimum detectable mass of ^{235}U (mg) depending on FM location in container and number of counters used in analysis.

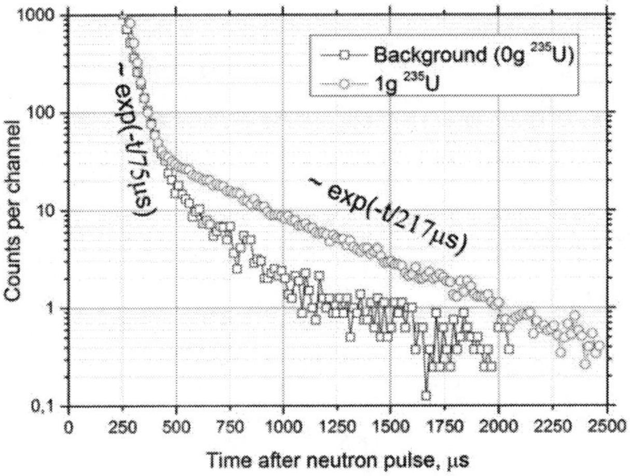

Fig. 2 - Time distributions of counting rate of neutron counter closest to ING after neutron pulse. The flux of ING-07T is 5×10^8 n/s, measurement duration is 8 minutes.

III. External shielding of neutron counters

The above simulation studies were made without taking into account the materials surrounding the system (floor, walls, ceiling). Therefore, the second stage was to determine their effect primarily on the background sensitivity of neutron counters. For this purpose, calculations were made of the time spectra of background neutrons reflected from surrounding walls, located 3 m from the system (Fig. 5).

Fig. 5 shows that in the time range of interest of 500-2,000 µs after the neutron pulse, the neutron detectors will be exposed to reflected neutrons with energy from thermal to several dozen eV. Neutrons with energy below ~0.5 eV can be absorbed by

 https://doi.org/10.1051/epjconf/201817005005

the cadmium shield of the counters, while neutrons with high energies (epithermal) can be detected, thus making an unnecessary contribution to the background.

To minimize this effect, it was proposed to cover the neutron counters with a neutron-absorbing external shield. The calculated time distribution of the count rate of a neutron detector with and without a shield in the form of borated polyethylene are shown in Fig. 6. In addition, we carried out the corresponding experiments, which confirmed the calculated results.

Application of these shields should ensure that the minimum detectable mass of FM estimated in the previous paragraph is preserved.

IV. Conclusion

The simulated and experimental results of detecting of small amounts of FM in large-sized NZK containers are presented. Executed estimates of the minimum detectable mass, depending on the position of FM in the NZK, show a strong scatter of its magnitude (from units to dozens of mg), which is due to the inhomogeneity of the thermal neutron field inside the NZK-MR.

The conducted experiments on detection of FM behind concrete shielding showed an unambiguous possibility of detecting small amounts of FM (dozens of mg) in the NZK.

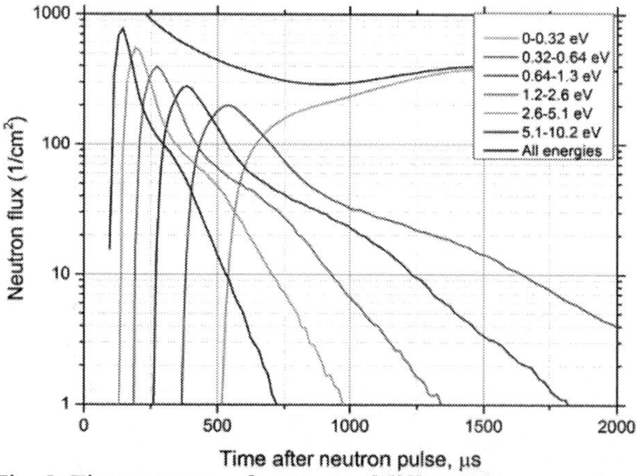

Fig. 5. Time spectrum of neutrons of different energy groups reflected from concrete wall 3 m apart from 3-He counters

It was demonstrated that the background level of the reflected neutrons can make a decisive contribution into the value of the minimum detectable mass of FM, which requires appropriate shielding of neutron detectors from epithermal neutrons.

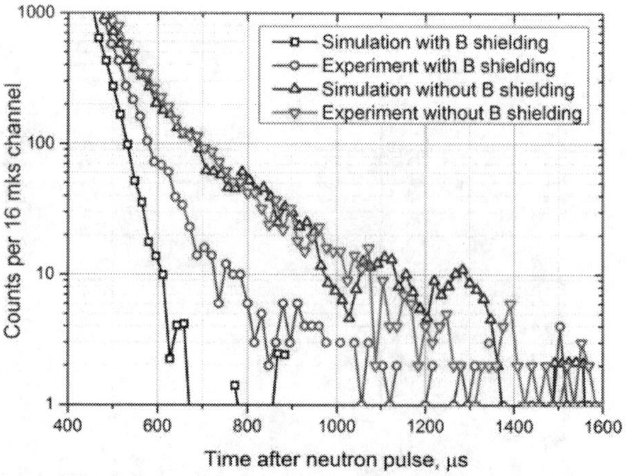

Fig. 6. Computational modeling of neutron detectors shielding with 8% borated polyethylene of 3 cm thickness.

References

[1] www.norao.ru/waste/where-is/
[2] NZK-MR concrete container, http://www.radon.ru/eng/activities/radwaste_management/conditioning/
[3] V.F. Batyaev, S.V. Sklyarov, Potential of active neutron interrogation to control fissile materials in "closed" fuel cycle, IAEA conference, Wien 2015
[4] V.F. Batyaev, S.V. Sklyarov, Comparative Analysis of Non-destructive Methods to Control Fissile Materials in Large-size Containers. International Conference Nuclear Data 2016, Bruges, Belgium, 2016

A Photofission Delayed γ-ray Spectra Calculation Tool for the Conception of a Nuclear Material Characterization Facility

D. BERNARD*, O. SEROT*, E. SIMON*, L. BOUCHER* and S. PLUMERI+

Abstract—**The photon interrogation analysis is a non-destructive technique allowing to identify and quantify fissile materials in nuclear waste packages. This paper details an automatic procedure which has been developed to simulate the delayed γ-ray spectra for several actinide photofissions. This calculation tool will be helpful for the fine conception (collimation, shielding, noise background optimizations, etc.) and for the on-line analysis of such a facility.**

Index Terms—**Photonic Interrogation, Nuclear Material Characterization, Photofission Yields, Delayed γ, Decay Data.**

I. Introduction

THIS paper proposes a new tool to help the conception of a new facility in order to detect and distinguish actinides in a waste package. The technique used is based on the analysis of delayed photons emitted after photon-burst-induced fissions. Because of the difference on fission yields, the delayed photons spectra are a signature of different actinides, if any.

The first section will present the overall studied facility for the photon beam inducing fissions and the delayed gamma detection set-up. The second part will be devoted to the specific developed tool, called DEGAS (Delayed Gamma Spectrum), to obtain γ spectra (including fission yields and decay data libraries, Bateman depletion equations solving - involving numerical issues and the verification of the whole procedure). At the end, we will propose an automated tool to select specific γ-ray lines ratios to detect a specific actinide in the waste package.

II. The Examination Facility

This facility, under design phase, has to detect the amount of actinides and to distinguish uranium from plutonium species in nuclear waste packages. Some of these waste packages are made of concrete in a big cylindrical container of more than 800 L. Despite of the photon attenuation through concrete, the photon interrogation is one of the most ambitious ways to interrogate the waste package. The photon pulsed beam is generated by an electron convertor made of tungsten or other metal (copper, silicon,...) allowing the Bremsstrahlung emission. The LINAC maximum energy is supposed to be flexible between 15 to 25 MeV. During this irradiation phase (roughly 2 to 3 hours), the waste package is axially rotating to make uniform the irradiation dose. The beam is then stopped and after a reasonable decay time (1 to 5 minutes), High Purity Germanium detectors are deployed all around the package to permit the photon spectrum acquisition for a large duration (around 2 to 3 hours). This photon spectrum contains delayed γ-rays and X-rays emitted by the fission products decay. Its analysis then gives the quantitative information on the fissile mass present in the waste container.

The next sections will be devoted to the developed tools to perform the fine conception of this facility.

III. Bateman Equations

The irradiation phase and the cooling/acquisition phase will be described sequentially.

A. Irradiation Phase

1) Statement of the problem

γ-induced fission of actinides leads to around 1000 independent fission products for which we have to calculate the atomic number densities $\boldsymbol{N_i(t)}$, at a given time $\boldsymbol{t}$, by solving the coupled Bateman equations [1] :

$$\frac{dN_i}{dt} = -\lambda_i N_i + \sum_{j\neq i} B_{j\to i}\lambda_j N_j + \int Y_{f\to i}(E_\gamma) N_f \sigma_{(\gamma,f)}(E_\gamma)\phi(E_\gamma, r, \omega t)dE_\gamma \qquad (1)$$

The first two terms are associated to decay probabilities ($\boldsymbol{\lambda}$) and branching ratios ($\boldsymbol{B}$) for the disappearance and creation of the $\boldsymbol{i}$ fission product (based on the specific JEFF-3.1.1/Decay Data Library [2]). The last term involves the production rate by folding (with respect to the incident photon energy $\boldsymbol{E_\gamma}$) the photofission cross section ($\boldsymbol{\sigma}$) of the fissile actinide f times its

Paper submitted to ANIMMA conference for proceedings and to IEEE transactions in June 2017. This work was supported by the CEA/DEN and ANDRA/R&D collaborative project DISN/ANDRA/CHK/A-CARAC-01-12.
*CEA/DEN/DER&DTN Cadarache, F-13108 Saint-Paul-Lez-Durance, France, corresponding authors: david.bernard@cea.fr, olivier.serot@cea.fr, eric.simon@cea.fr, lionel.boucher@cea.fr.
+ANDRA, 1-7 rue Jean Monnet, F-92290 Châtenay-Malabry, France, corresponding author : stephane.plumeri@andra.fr.

 https://doi.org/10.1051/epjconf/201817006001

independent fission yield ($\boldsymbol{Y}$) to the incident photon flux $\boldsymbol{\phi}$. The $\boldsymbol{N_i}$ number density is calculated at the distance $\boldsymbol{r}$ (location of the fission event) from the center of the cylindrical waste package. The latter is rotating during this irradiation phase with a period $T = 2\pi/\omega$ close to 2 to 3 minutes.

In order to solve equation (1), we assume first the decoupling in energy, space and time of the photon incident flux $\boldsymbol{\phi(E_\gamma, r, \omega t) = \varphi(E_\gamma) \times \psi(r) \times T(t)}$. In addition, since the rotation period of the waste package is much lower than the irradiation time, the time dependent production term is reduced to the averaged value. By this way, equation (1) is transformed into the following equation:

$$\frac{dN_i}{dt} \cong -\lambda_i N_i + \sum_{j \neq i} B_{j \to i} \lambda_j N_j + \widetilde{Y}_{f \to i} \times \overline{\tau_f}(r) \times m_f \quad (2)$$

Where:

- $\widetilde{Y}_{f \to i}$ stands for the energy-averaged independent isotopic fission product yield $\boldsymbol{i}$ for an actinide $\boldsymbol{f}$,
- $\overline{\tau_f}(r)$ its specific fission rate (relative to the mass $\boldsymbol{m_f}$ that we look for).

For a given fissile isotope and by identification of equation (1) and equation (2), the effective yield $\widetilde{Y}_{f \to i}$ is defined as:

$$\widetilde{Y}_{f \to i} = \frac{\int_0^{E_\gamma^{end}} Y_{f \to i}(E_\gamma)[\sigma_{(\gamma,f)}(E_\gamma)\varphi(E_\gamma)]dE_\gamma}{\int_0^{E_\gamma^{end}} [\sigma_{(\gamma,f)}(E_\gamma)\varphi(E_\gamma)]dE_\gamma} \quad (3)$$

Where:

- The incident γ-spectrum $\boldsymbol{\varphi(E_\gamma)}$ is calculated with MCNPX [3] and depends on the electron/photon conversion target and on the maximum energy of the LINAC (see figure 1).

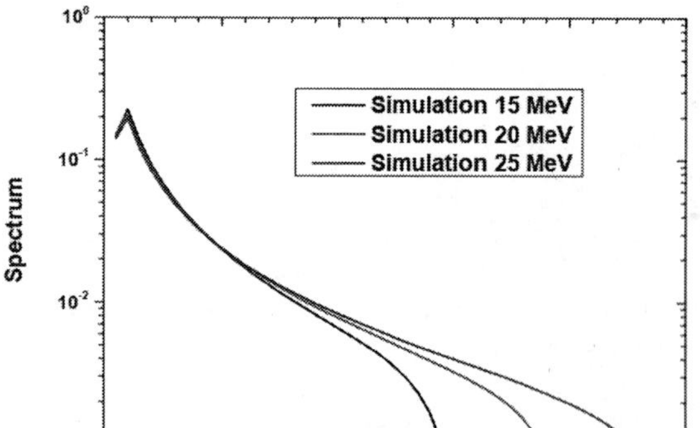

Fig. 1. Incident photon spectra in the direction of the electron beam. Calculations done with MCNPX (target conversion is W).

The maximum spectrum energy $\boldsymbol{E_\gamma^{end}}$ will be fixed to 15 MeV for the whole paper but can be modified by the user as an input data of the developed tool DEGAS.

- The photofission cross sections $\boldsymbol{\sigma_{(\gamma,f)}(E_\gamma)}$ are taken from the ENDF/B-VII.1 library [4] for major actinides (^{235,238}U, ^{238,239,240,241}Pu, ^{241}Am) and from the TENDL-2015 library [5] for ^{242}Pu and ^{242m}Am:

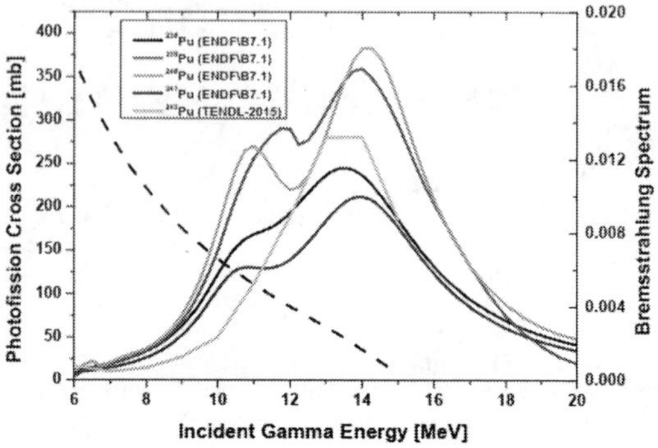

Fig. 2. Evaluated photofission cross sections (left axis) with incident γ-spectrum (dashed line, right axis).

- The independent fission product yield for a given fissioning nucleus at a given $\boldsymbol{E_\gamma}$ incident energy, $\boldsymbol{Y_{f \to i}(E_\gamma)}$, is determined from a Monte Carlo code called GEF (acronym for GEneral description of Fission observables) [6]. This code first calculates fission fragment yields before prompt neutron emission, from the concept of fission modes developed by Brosa [7]. These fission modes correspond to the various valleys observed in the deformation potential energy surface of the fissioning nucleus. When the nucleus starts its deformation from the ground state up to the scission point, it follows one of those valleys with a given probability. The depth and width of each mode are determined in order to deduce the main properties of the fission fragments (yield, deformation at scission, kinetic energy…). In a second step, the total excitation energy available at scission is calculated and shared between the two nascent fragments. Finally, thanks to a statistical treatment, the emission of prompt neutrons and prompt gammas are simulated to predict their properties (energy spectrum, multiplicity ...). In this way, the GEF code is able to calculate almost all fission observables, in particular the isotopic and isomeric yields after prompt neutron evaporation which we are needed for. It should also be noted that GEF can simulate any type of fission: spontaneous fission, neutron-induced fission, photofission, etc. Nevertheless, in the case of photofission, the current version of the code does not allow taking into account the second and third chance of fission which should not significantly affects the present results. Examples of effective independent mass yields obtained with a 15 MeV LINAC and a W conversion target are shown in figure 3. One of the main fission characteristics is clearly visible: the stabilization of the mass yield in the heavy mass region (around the mass 140) and consequently a shift of the mass yield in the light mass region from one fissioning nucleus to another, which ensures the conservation of the fissioning nucleus mass (removing the prompt neutron multiplicity).

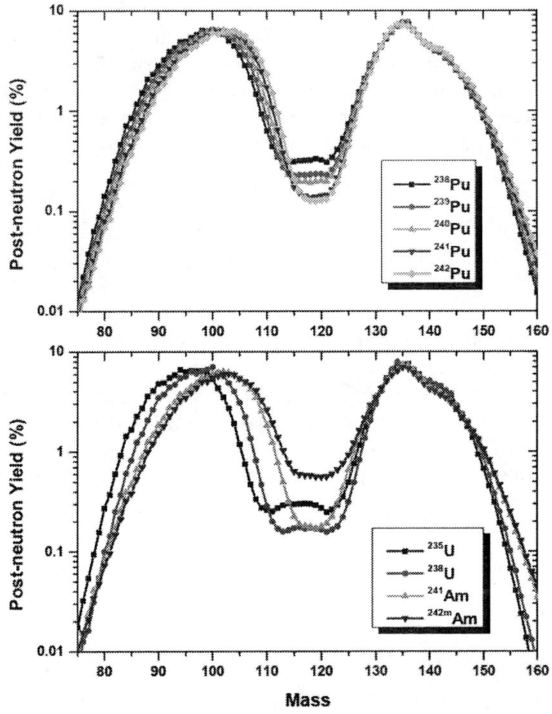

Fig. 3. Independent effective fission (15MeV+W) mass yields obtained from the GEF code (version 2015 V2.2) [6].

The average excitation energy available for a given fissioning nucleus can be easily calculated:

$$< E *>= \frac{\int_0^{E_\gamma^{end}} E_\gamma \sigma_{(\gamma,f)}(E_\gamma)\varphi(E_\gamma)dE_\gamma}{\int_0^{E_\gamma^{end}} [\sigma_{(\gamma,f)}(E_\gamma)\varphi(E_\gamma)]dE_\gamma} \quad (4)$$

These energies are reported in Tab. 1 for all fissioning nuclei investigated in the present work.

TABLE I
SURVEY OF THE AVERAGE EXCITATION ENERGY AVAILABLE (IN MEV) FOR THE FISSIONING NUCLEI, CONSIDERING AN INCIDENT γ- SPECTRUM GENERATED WITH A W CONVERSION TARGET AND A 15MEV ENERGY OF THE LINAC.

^{238}Pu	^{239}Pu	^{240}Pu	^{241}Pu	^{242}Pu	^{235}U	^{238}U	^{241}Am	^{242m}Am
10.6	11.1	10.9	10.8	11.1	11.3	11.0	10.9	11.9

The second factor $\overline{\tau_f}(r)$ of equation (2) is a time averaged fission rate while the waste package is rotating in front of the electron/photon beam. With a simple hypothesis of the flux attenuation as a function of the square of the distance, one can write:

$$\overline{\tau_f}(r) = \tau_f(0) \times \left[1 - \left(\frac{r}{d_\gamma + R_P}\right)^2\right] \quad (5)$$

Where the user's input data r, d_γ, R_P stand for the distance of the fissile material from the center of the waste package, the distance between the conversion target and the package surface and finally, the package radius, respectively.

2) Solving the Bateman equations

For a given fissile nucleus amount, we will solve the previously described equation 2. The user has to fix a given mass m_f and the location r in the waste package for this particular isotope. The boundary conditions impose that no fission product are present before the irradiation in the waste package, $N_i(t = 0) = 0$.

We decide to use the DARWIN code package [8] to solve such equations, because of its robustness but also because it contains already preprocessed nuclear data libraries. This code package is able to solve non homogeneous differential equation but the source term has to be associated to neutron induced production and not to photon induced reactions. One can analytically demonstrate that solving the following homogeneous and time-iterative system (see lower case for "homogeneous" number densities $\mathbf{n_i}$):

$$\begin{cases} \frac{dn_i}{dt} = -\lambda_i n_i + \sum_{j \neq i} B_{j \to i} \lambda_j n_j \\ n_i(x.\delta t) = \tilde{Y}_{f \to i}.\overline{\tau_f}(r).m_f.\delta t \end{cases} \quad (6)$$

leads to the following actual number densities which solve equation (2):

$$N_i(\Delta t_{irrad}) = \sum_x n_i(\Delta t_{irrad} - x.\delta t) \quad (7)$$

Our developed DEGAS tool is using a MATLAB® [9] script which reads first the output file from the GEF simulation ($\tilde{Y}_{f \to i}$), produces and submits a DARWIN input file (owing to the user specifications -$m_f, \Delta t_{irrad}$-) in a second step, and then calculates the atomic number densities at the end of the irradiation phase: $N_i(\Delta t_{irrad})$ from equation (6).

B. Acquisition Phase

Once the irradiation beam is stopped, one has to calculate the integrated activity for each fission product during the cooling/acquisition phase. To do so, a similar equation as equation (6) has to be solved, where the number densities obtained at the end of the previous phase are taken as initial conditions:

$$\begin{cases} \frac{dn_i}{dt} = -\lambda_i n_i + \sum_{j \neq i} B_{j \to i} \lambda_j n_j \\ n_i(t = 0) = N_i(\Delta t_{irrad}) \end{cases} \quad (8)$$

The DEGAS script generates a second input file for the DARWIN code package and calculates the delayed photon integrated source, starting after the HPGe deployment (Δt_{Ge}

around 1 to 5 minutes) up to the acquisition time ($\boldsymbol{\Delta t_{acq}}$) as follows:

$$S_\gamma(E_\gamma) = \sum_i \int_{t=\Delta t_{Ge}}^{\Delta t_{acq}+\Delta t_{Ge}} I_{\gamma,i}\{\lambda_i \times n_i(t)\}dt \quad (9)$$

where the continuous sum is approximated by a discrete sum with the same number of time steps as the previous irradiation phase. $I_{\gamma,i}$ stands for the γ or X-rays intensities of the transitions associated to the decay of the i fission product.

IV. Verification of the whole calculation procedure

A first simple verification case consists in producing one atom of ^{135}Xe per second during Δt_{irrad} =1h and calculating the emitted γ-spectrum during Δt_{acq} =1h after the HPGe deployment Δt_{Ge} =1s. The analytical solution of equation (2) gives 3466.88 nuclei at the end of irradiation phase. The DEGAS script gives the following results according to numerous times steps number (Tab. II).

TABLE II
VALIDATION OF THE IRRADIATION PHASE CALCULATION.

Number of integration time steps	End of irradiation atomic number densities of ^{135}Xe
10	3453.75
50	3464.26
100	**3465.57**
1000	3466.75
10000	3466.87

We estimate that 100 time steps for solving equations (6) and (7) are a good compromise versus the calculation time consuming.

The analytical value by solving equations (8) and (9) leads to 227.87 emitted γ for the 250 keV emission. This exact value is obtained with the iterative calculation scheme. See figure 4 for the full γ-spectrum :

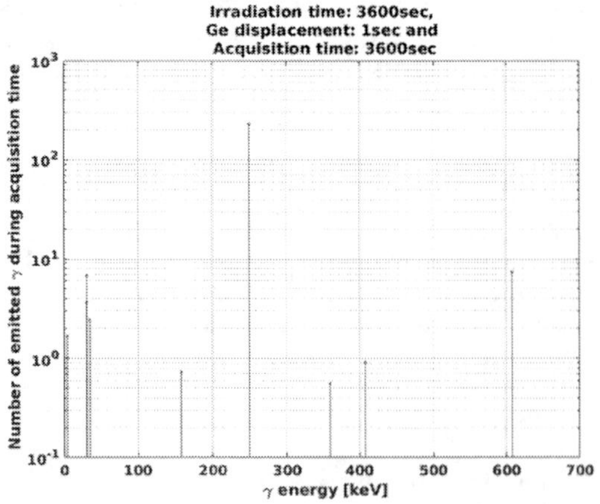

Fig. 4. Delayed γ spectrum for the ^{135}Xe test case.

This validation is also done for two coupled differential equations (2) and (8) and does not show any discrepancies by using 100 time steps.

More of that we compared the results obtained for 10 g of uranium (235 and 238 with various isotopic vectors) to the results of MCNPX/CINDER [10] solving the depletion equation in a stochastic route. The results were fully comparable in spite of the different nuclear data used (GEF version used for the photofission yields, ENDF/B-VI for MCNPX instead of JEFF-3.1.1 Decay data libraries). The advantage of this automated script is to solve Bateman equations in few minutes and without convergence issues instead of the full Monte Carlo process.

V. Delayed γ spectra analysis

We performed the calculation of delayed γ(+X)-spectra for all actinides of interest: ^{235,238}U, ^{238,239,240,241,242}Pu and ^{241,242m}Am photon induced fissions for a 15 MeV LINAC and a W conversion target. Fissile materials were supposed to be located at the center of the waste package. The final spectra is binned (ΔE =0.5 or 1 keV) and analyzed above an energy threshold of E_γ >1.3 MeV (because of their rather high attenuation through the concrete in the waste package). A second filter (fixed by the user) concerns the minimum intensity of γ-lines $\frac{S_{\gamma i}}{\max(S_{\gamma i})} > 1\%$, see figure 5 for the ^{235}U(γ,f) delayed γ-spectrum:

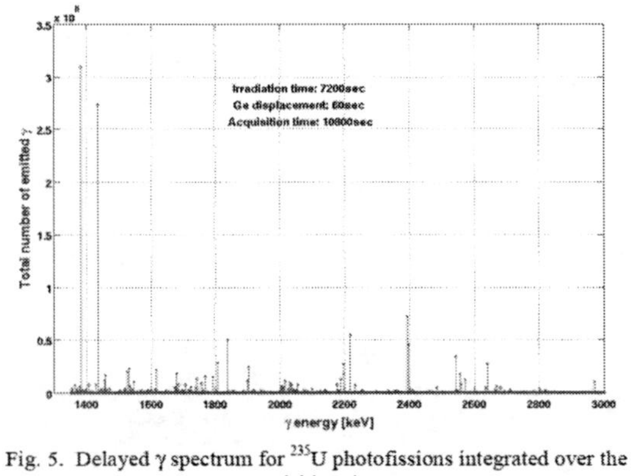

Fig. 5. Delayed γ spectrum for ^{235}U photofissions integrated over the acquisition time.

The goal of the automated script is to distinguish the amount of each isotope in the fissile material. Taking profit that independent fission yields vary from a fissile system to another one, we developed a fully automated procedure to help the user to select specific γ emissions. In order to avoid efficiency corrections in the final HPGe spectrum analyses, we decided to look for γ-line couples labeled 1 and 2 in a rather small energy window. The user has to specify two energy bands limits, for example 5 keV< $E_{\gamma 2} - E_{\gamma 1}$ <60 keV. The script will then look for all the γ line couples that can distinguish uranium from plutonium sliding this energy window on the whole spectra. The final fitting parameter is the discriminant factor which is fixed to the emission line ratio $\frac{S_{\gamma 1}}{S_{\gamma 2}} > 50\%$ or $\frac{S_{\gamma 2}}{S_{\gamma 1}} > 50\%$.

We studied the particular case involving a fissile material made of 1g of depleted uranium mixed with 10 g of plutonium (isotopic vector calculated at 40 GWd-PWR/t_{HM}). The automated script DEGAS found more than 100 γ-lines couples allowing distinguishing uranium from plutonium in the waste package. The figure 6 shows a zoom in the 1590-1670 keV region. The bar histogram represents the plutonium and uranium emissions and the dashed lines symbolize the calculated and selected ratios of intensities for uranium and plutonium:

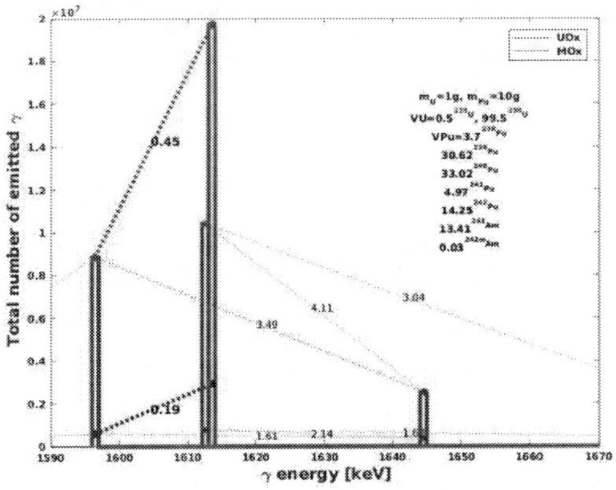

Fig. 6. Superimposed uranium (1g) and plutonium (10g) delayed γ-spectra after 2h for irradiation (15MeV-W conversion target) and 3h for the acquisition time.

In figure 6, the first γ lines allowing to distinguish the uranium to the plutonium materials in this configuration are the following:

- $E_{\gamma 1}$ =1596.7 keV corresponding to the decay emission of the ^{104}Tc,
- $E_{\gamma 2}$ =1613.8 keV corresponding to the decay emission of the ^{134}I.

As shown in the figure 6, the respective emission ratio prediction is:

- $\frac{S_{\gamma 1}}{S_{\gamma 2}} = 0.194$ for uranium,
- $\frac{S_{\gamma 1}}{S_{\gamma 2}} = 0.449$ for plutonium.

The automated selection of γ-ray will help the user to proceed to the second step of the analysis which consists in performing the transport of this few number of γ through MCNPX from the fissile area to the HPGe detectors. Finally the goal will be to optimize the whole facility (irradiation versus acquisition time, on-line spectrum analysis for the characterization of the fissile isotopes involved in the nuclear waste package).

VI. Conclusion

We proposed a simplified but robust and verified tool DEGAS to help the fine conception and the development of a nuclear material characterization facility by using photon interrogation techniques. It is based on the delayed γ-spectra analysis following fission events.

More of that, we proposed an automated procedure to estimate the amount and to give insights on the isotopic vector of fissile materials in a waste package. Results showed more than 100 discriminant γ-ray couples within few minutes of computer calculation time. This script can also be applied to delayed neutron detection in coincidence with the γ lines.

All those calculation results suffer from nuclear data uncertainties (photofission cross sections, independent fission yields, decay data) that can be propagated in a further work, if needed.

To be complete, a necessary validation step versus an experimental actual facility will be set up in the following years at CEA.

References

[1] Bateman, Harry. "The solution of a system of differential equations occurring in the theory of radioactive transformations." In Proc. Cambridge Philos. Soc, vol. 15, no. pt V, pp. 423- 427, 1910.

[2] M.A. Kellett, O. Bersillon, and R.W Mills, "The JEFF-3.1/-3.1.1 radioactive decay data and fission yields sub-libraries, JEFF Report 20", ISBN 978-92-64-99087-6, ©OECD/NEA 2009.

[3] MCNPX 2.7.0, "Monte Carlo N-Particle Transport Code System for Multiparticle and High Energy Applications", RSICC code package C740, (Radiation Safety Information Computational Center, ORNL, Oak Ridge,TN).

[4] M.B. Chadwick, *et al.*, Nuclear Data Sheets 112 (2011) 2887–2996.

[5] https://tendl.web.psi.ch/tendl_2015/tendl2015.html

[6] K. H. Schmidt, B. Jurado, C. Amouroux, and C. Schmitt, Nuclear Data Sheets 131 (2016) 107.

[7] U. Brosa, S. Grossmann, and A. Muller, Phys. Rep. 197, 167 (1990).

[8] A. Tsilanizara, *et al.*, "DARWIN: an Evolution Code System for a Large Range of Applications", J. Nucl. Sci. Technol., Supplement, 1, pp. 845-849, 2000.

[9] MATLAB® 9.2.0.538062 (R2017a). Natick, Massachusetts: The MathWorks Inc., 2010.

[10] Joe W. Durkee, *et al.*, "Delayed-gamma signature calculation for neutron-induced fission and activation using MCNPX, Part I: Theory", Progress in Nuclear Energy 51 (2009) 813–827.

Stochastic approach for radionuclides quantification

A. Clement, N. Saurel, G. Perrin.

***Abstract*— Gamma spectrometry is a passive non-destructive assay used to quantify radionuclides present in more or less complex objects. Basic methods using empirical calibration with a standard in order to quantify the activity of nuclear materials by determining the calibration coefficient are useless on non-reproducible, complex and single nuclear objects such as waste packages. Package specifications as composition or geometry change from one package to another and involve a high variability of objects. Current quantification process uses numerical modelling of the measured scene with few available data such as geometry or composition. These data are density, material, screen, geometric shape, matrix composition, matrix and source distribution. Some of them are strongly dependent on package data knowledge and operator backgrounds. The French Commissariat à l'Energie Atomique (CEA) is developing a new methodology to quantify nuclear materials in waste packages and waste drums without operator adjustment and internal package configuration knowledge. This method suggests combining a global stochastic approach which uses, among others, surrogate models available to simulate the gamma attenuation behaviour, a Bayesian approach which considers conditional probability densities of problem inputs, and Markov Chains Monte Carlo algorithms (MCMC) which solve inverse problems, with gamma ray emission radionuclide spectrum, and outside dimensions of interest objects. The methodology is testing to quantify actinide activity in different kind of matrix, composition, and configuration of sources standard in terms of actinide masses, locations and distributions. Activity uncertainties are taken into account by this adjustment methodology.**

***Index Terms*—Nuclear quantification, Bayes Theorem, MCMC, Monte Carlo sampling**

I. Introduction

THE quantification of nuclear materials is crucial in many branches of nuclear industry as nuclear plants, nuclear criticality safety or nuclear decommissioning. Many nuclear detection techniques like neutron detection methods, gamma spectrometry, and alpha particles detection methods are carried out to accurately quantify radionuclide masses included in a large number of more or less complex objects. Well-known detection systems using gamma spectrometry principles are germanium gamma ray detectors and more precisely the High-Purity Germanium one (HPGe) [1], [2]. Compared to NaI detectors, the dominant feature of HPGe detectors is their excellent energy resolution around 1 keV, very useful for identification and quantification of radionuclides such as ^{239}Pu or ^{241}Am. A numerical method [3] has been developed to propose reliable and accurate characterization of HPGe detectors by combining 3D Monte Carlo particle transport simulation codes such as MCNP [4] with applied mathematical tools. This detector numerical model is built in order to reduce global uncertainties of final quantification results by increasing detection capability knowledge of HPGe detectors.

M. Clement and M. Saurel was with the Nuclear Measurement Laboratory, French Commissariat à l'Energie Atomique, CEA/DAM/VA, F-21120 Is-sur-Tille, e-mails: alois.clement@cea.fr, nicolas.saurel@cea.fr.
M. Perrin was with the French Commissariat à l'Energie Atomique, CEA/DAM/DIF, F-91297 Arpajon CEDEX, France

Actually, to identify and quantify gamma emitter nuclides included in different kind of objects, such as nuclear waste packages, or glove boxes, one needs to calculate the activity A appearing in (1) [5] which leads to the interest radionuclide mass m by (2):

$$A = \frac{S(E)}{\epsilon(E)tI_\gamma(E)} \qquad (1)$$

where:

- E : Energy (MeV)
- A : Activity of the measured object (Bq)
- $S(E)$: Net counting rate of the full energy peak at the energy E (counts)
- $I_\gamma(E)$: Branching ratio of radionuclide at the energy E
- t : Acquisition duration (s)
- $\epsilon(E)$: Absolute efficiency calibration coefficient at the energy E, also called attenuation law

$$m = \frac{A}{A_m} \qquad (2)$$

where:

- m : Interest mass included in the measured object (g)
- A : Activity of the measured object (Bq)
- A_m : Mass activity of the interest radionuclide (Bq/g)

The $\epsilon(E)$ coefficient is related to the capability of a measured object to reduce the gamma signal coming from an interest nuclear material. It depends on object features such as internal layout, screens, density, position of the gamma source relative to the HPGe detector, internal materials and energy. Though the $S(E)$ spectrum extracted areas can be more or less easily determine with accuracy [6], calculation of $\epsilon(E)$ is a hard process for complex objects in terms of internal layout and composition. Current and common calculation methods use Monte Carlo simulation codes as MCNP to model measured scenes and approach as soon as possible the real values of the $\epsilon(\mathbf{E})$ coefficients. For few years, new kind of nuclear quantification numerical methods dealing with applied mathematics and stochastic approaches [3] propose to emulate outcomes of interest as $\epsilon(\mathbf{E})$ to solve the inverse problem of quantification and estimate an interest radionuclide mass distribution.

Section II presents a new approach of the mass quantification problem in considering interest data as random variables in a Bayesian framework. Section III gives some work hypothesis about interest inputs. Sections IV and V propose a method to solve the inverse problem of mass quantification and to estimate an interest mass distribution by sampling techniques. Finally, and before conclusion, sections VI and VII show quantification results, compared with experiments, in considering two different hypotheses about input uncertainties.

II. Statistical approach of quantification

The quantification process which is considered here aims to estimate the probability density function (PDF) of an interest radionuclide mass, so-called m. Let suppose the interest radionuclide is a multi gamma-emitter. Hence, via specific software applied on gamma spectrum analysis, it is possible to obtain extracted areas from the measured gamma spectrum. Let suppose there are N extracted areas gathered in vector $\mathbf{S} = \{S(E_n), n \in [\![1, N]\!]\}$, relatively to $(E_n)_{n\in[\![1,N]\!]}$ energies of the multi gamma-emitter radionuclide, and $\mathbf{X} \in \Xi_D$ represents a D-dimension vector of inputs for the absolute efficiency calibration coefficients vector $\epsilon(\mathbf{X})$ estimation. Hence, (1) and (2) allow us to provide the following equation:

$$\mathbf{S} = A\,\epsilon(\mathbf{E})\,\mathbf{I}_\gamma t = mA_m\,\epsilon(\mathbf{E})\,\mathbf{I}_\gamma t$$

Let suppose that $\mathbf{Y^{obs}}$ is the observation vector defined in (3):

$$\mathbf{Y^{obs}} = \frac{\mathbf{S}}{\mathbf{I}_\gamma t A_m} = m\epsilon(\mathbf{X}) + \xi, \quad \xi \sim \mathcal{N}(0, \sigma_{\mathbf{obs}}) \quad (3)$$

In (3), ξ represents the $\mathbf{Y^{obs}}$ observation uncertainty vector. All of its coefficients are related to normal distributions centered at zero with $\sigma_{\mathbf{obs}} = (\sigma_{obs,n})_{n\in[\![1,N]\!]}$ as standard deviation vector. In considering $\mathbf{Y^{obs}}$, m, $\epsilon(\mathbf{X})$, and $\mathbf{X}$ as random variables, or vectors composed by random variables, the marginal PDF for m mass given $\mathbf{Y^{obs}}$, expressed as $f(m|\mathbf{Y^{obs}})$, is given by Bayes theorem written in terms of PDFs [1]. The probability density $f(m|\mathbf{Y^{obs}})$ depends on joint PDF $f(m, \mathbf{Y^{obs}})$ and $f_{\mathbf{Y^{obs}}}(\mathbf{Y^{obs}})$ the marginal PDF of $\mathbf{Y^{obs}}$ [7], as expressed in following equation:

$$f(m|\mathbf{Y^{obs}}) = \frac{f(m, \mathbf{Y^{obs}})}{f_{\mathbf{Y^{obs}}}(\mathbf{Y^{obs}})} \propto f(\mathbf{Y^{obs}}|m)\pi(m) \quad (4)$$

The distribution $\pi(m)$ is called prior distribution [7] and based on hypothesis, experience, or subjective opinion about m mass. To add $\mathbf{X}$ dependencies lead to obtain (5):

$$f(m|\mathbf{Y^{obs}}) \propto \int_{\mathbf{X}} f(\mathbf{Y^{obs}}|m)\pi(m|\mathbf{X})\pi(\mathbf{X})d\mathbf{X} \quad (5)$$

III. Hypothesis

The objective is to obtain an estimation of $f(m|\mathbf{Y^{obs}})$, the probability density of m mass. Let suppose some hypothesis about necessary PDFs to provide the mass distribution:

- $\mathbf{Y^{obs}}$ is a vector composed by N independent random variables Y_n^{obs} related to normal distributions:

$$\forall n \in [\![1, N]\!],\ Y_n^{obs} \sim \mathcal{N}(\mu_n(m, \mathbf{X}), \sigma_n(m, \mathbf{X})^2) \quad (6)$$

with $\mu_n(m, \mathbf{X})$ and $\sigma_n(m, \mathbf{X})$ respectively the mean and standard variation obtained after the propagation of all uncertainty sources.

- Let suppose there is not any correlation between m and $\mathbf{X}$, and between components of $\mathbf{X}$. Hence, the $\epsilon(\mathbf{X})$ efficiency does not depend on the m radionuclide mass. In this case, with $\mathbf{X}$ of D-dimension composed by $(X_d)_{d\in[\![1,D]\!]}$, prior distributions can be expressed as:

$$\pi(m|\mathbf{X})\pi(\mathbf{X}) = \pi(m)\pi(\mathbf{X}) = \pi(m)\prod_{d=1}^{D}\pi(X_d) \quad (7)$$

- Each prior distribution provides more or less information about its own random variable [7]. A good known variable is associated with tight normal distribution, and so, it brings information about variable behaviour. A bad known variable with no real information about, such as m mass or ρ density of the object, is associated with uniform distribution defining on its own definition domain. This method allows us to control real knowledge about the measured object and test different hypothesis on unknown variables.

IV. MCMC sampling and surrogate models

The calculation of the m mass conditional PDF $f(m|\mathbf{Y^{obs}})$ leads to obtain the integral of $\mathbf{X}$ appearing in (5). Nevertheless, $\mathbf{X}$ is composed by unknown variables such as ρ object density, and to obtain directly the integral, and of course $f(m|\mathbf{Y^{obs}})$, becomes too hard even impossible. A solution to work around the problem is to sample m mass conditional PDF in using Markov Chains Monte Carlo (MCMC) methods. The Metropolis-Hastings algorithm allows us to sample m mass and $\mathbf{X}$ components PDFs in using Markov Chains theory [7]. This method needs several ten thousands of calls to the simulation code which provides the absolute efficiency calibration coefficients vector $\epsilon(\mathbf{X})$. Because of the impossibility to rapidly evaluate $\epsilon(\mathbf{X})$ directly with the particle transport code MCNP, a Kriging surrogate model [8], [9] is proposed to emulate it. Surrogate model [9] requires a limited number of calls to the code, which are gathered in a design of experiments (DoE), in order to first accurately and quickly build an outcome of interest function M, and finally predict new values of interest such as:

$$\forall \mathbf{X} \in \Xi_D, \quad \epsilon(\mathbf{X}) = M(\mathbf{X}) \quad (8)$$

A well-chosen and good-known DoE building technique is the Latin Hypercube Sample one (LHS) [10]. This kind of DoE gives very interesting space filling properties by dealing with different criterion maximization such as the minimax one.

V. Calculation of the $f(m|\mathbf{Y^{obs}})$ probability density

The objective is to obtain m mass conditional PDF $f(m|\mathbf{Y^{obs}})$. Considering (5-7-8) and the given hypothesis exposed below, the calculation of $f(\mathbf{Y^{obs}}|m)$ is due and developed as expressed in the following equation:

$$f(\mathbf{Y^{obs}}|m) = \prod_{n=1}^{N} f(Y_n^{obs}|m) = \prod_{n=1}^{N} \frac{1}{\sqrt{2\pi}\sigma_n} e^{-\frac{(\mu_n^{obs}-\mu_n)^2}{2\sigma_n^2}} \quad (9)$$

where $(\mu_n)_{n\in[\![1,N]\!]}$ and $(\sigma_n)_{n\in[\![1,N]\!]}$ are determined in considering and propagating all uncertainty sources from (3). Here, $(\mu_n^{obs})_{n\in[\![1,N]\!]}$ vector represents observation inputs coming from the extracted gamma spetrum. Let begin in taking (3):

$$\mathbf{Y^{obs}} = \frac{\mathbf{S}}{\mathbf{I}_\gamma t A_m} = m\epsilon(\mathbf{X}) + \xi, \quad \xi \sim \mathcal{N}(0, \sigma_{obs}) \quad (10)$$

Due to its calculation process which uses Kriging emulation method presented below, the $\epsilon(\mathbf{X})$ random variable can be written as following:

$$\epsilon(\mathbf{X}) = \mu^M(\mathbf{X}) + \sigma^M(\mathbf{X}).\delta_\sigma + e_{model}(\mathbf{X}).\delta_e \quad (11)$$
$$\delta_\sigma \sim \mathcal{N}(0, \mathbf{1_N}), \quad \delta_e \sim \mathcal{N}(0, \sigma^2_{\mathbf{model}})$$

with $\mu^M(\mathbf{X}) = (\mu_n^M(\mathbf{X}))_{n\in[\![1,N]\!]}$ the predictive mean vector of $\epsilon(\mathbf{X})$ and $\sigma^M(\mathbf{X}) = (\sigma_n^M(\mathbf{X}))_{n\in[\![1,N]\!]}$ its standard deviation vector coming from the M Kriging regression surrogate model from (8). $e_{model}(\mathbf{X})$ represents the model error. It is noted that this uncertainty is supposed to be insignificant but kept in the calculation process. Please note that uncertainties on necessary MCNP values for Kriging model construction are taken into account during its regression process. Consequently, this error doesn't appear in (11).

Hence, (10-11) allow us to express $\mathbf{Y^{obs}}$ PDF from Y_n^{obs} PDFs:

$$\forall n \in [\![1, N]\!], \quad Y_n^{obs} \sim \mathcal{N}(\mu_n, \sigma_n^2)$$
$$\mu_n = m\mu_n^M(\mathbf{X}) \quad (12)$$
$$\sigma_n^2 = m^2(\sigma_n^M(\mathbf{X})^2 + \sigma^2_{model}) + \sigma^2_{obs,n} \quad (13)$$

To finish, in considering (5-7-11-12-13), the m mass conditional PDF can be expressed as following:

$$f(m|\mathbf{Y^{obs}}) \propto \int_{\mathbf{X}} \prod_{n=1}^{N} \frac{1}{\sqrt{2\pi}\sigma_n} e^{-\frac{(\mu_n^{obs}-\mu_n)^2}{2\sigma_n^2}} \pi(m) \prod_{d=1}^{D} \pi(X_d) d\mathbf{x} \quad (14)$$

VI. About the ξ random variable

Henceforth, the aim is to consider m mass PDF and sample it with MCMC methods as the Metropolis-Hastings algorithm. Nevertheless, a remaining variable has to be approached. Because of different difficulties to accurately quantify the $\sigma_{\mathbf{obs}}$ standard deviation vector appearing in (3) and depending on the spectrum area extraction method, two hypotheses (H_1 and H_2) about it are tested and compared:

$$H_1: \quad \xi \sim \mathcal{N}(0, \sigma_{\mathbf{obs}} = \alpha\mathbf{Y_{obs}}), \quad \alpha \in [0, 1] \quad (15)$$
$$H_2: \quad \xi \sim \mathcal{N}(0, \sigma_{\mathbf{obs}} = \sigma\mathbf{1_N}), \quad \sigma \in \mathbb{R} \quad (16)$$

The first hypothesis H_1 proposes to put σ_{obs} vector proportional to observable values (15). In the second one H_2, σ_{obs} is constant and equal to an arbitrary value σ for all N components (16).

VII. Results

A. Experimental protocol

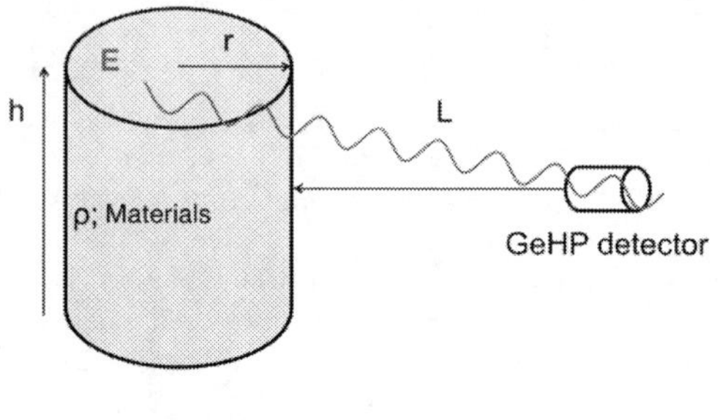

Fig. 1. Model used to build the efficiency emulator

In order to explore the capabilities of the nuclear material quantification method exposed in sections II to VI, a design of experiments related to the measurement of different more or less complex objects has been built. It concerns four gamma spectrometry experiments, so-called E_1, E_2, E_3 and CO. All of them consider the measurement of standard cylindrical objects used in usual processes of nuclear waste package management and contained various quantities of gamma-emitter ^{239}Pu masses. The four objects are filled with different attenuation materials and densities. Experiment features are gathered in Tab. 1:

	L(cm)	*h(cm)*	*r(cm)*	*Materials*	*M(g)*	ρ $(g.cm^{-3})$
E_1	117	80	30	Vinyl	0.37	0.1
E_2	99	80	30	Vinyl/Metal	1.95	0.35
E_3	98.5	80	30	Vinyl	0.28	0.15
CO	100	20	8	X	13	0.4

TABLE I
EXPERIMENTAL DATA

In Tab. 1:

- L : Distance from detector to object (cm)
- h : Cylindrical height (cm)
- r : Cylindrical radius (cm)
- M : Actinide mass (g)
- ρ : Bulk density of materials $(g.cm^{-3})$

The vector $\mathbf{X}$ used for the construction of the detection efficiency surrogate model $\epsilon(\mathbf{X})$ (section IV) consists of the gamma ray energy E, the radius r and the height h of the cylinder, the distance from the detector to the object L, the density ρ and the mixture of three materials: iron, vinyl, and plutonium (Fig. 1). Each of these is defined as follows in the LHS design of experiment used:

- $E \in [90, 450](keV)$
- $r \in [5, 32](cm)$

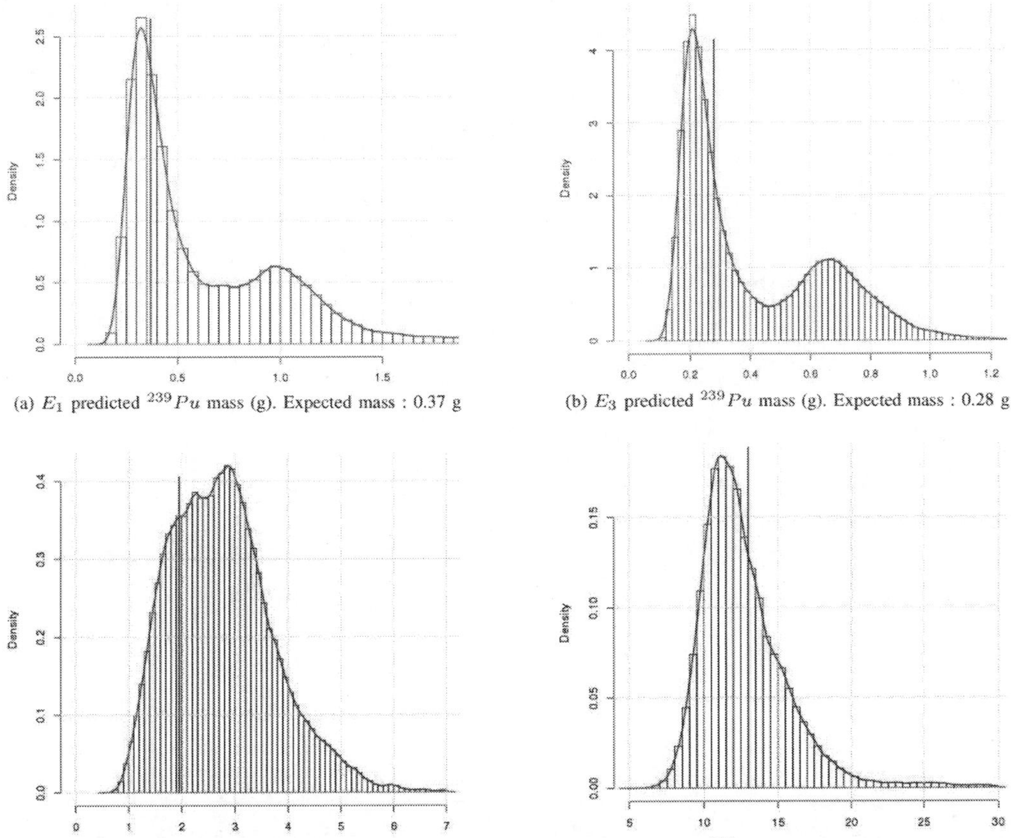

(a) E_1 predicted ^{239}Pu mass (g). Expected mass : 0.37 g

(b) E_3 predicted ^{239}Pu mass (g). Expected mass : 0.28 g

(c) E_2 predicted ^{239}Pu mass (g). Expected mass : 1.95 g

(d) CO predicted ^{239}Pu mass (g). Expected mass : 13 g

Fig. 2. Predicted ^{239}Pu masses (g) related to the four experiments under H_1 hypothesis with α equal to 0.05. Empirical densities appear in red and expected masses in blue.

- $h \in [10, 110](cm)$
- $L \in [80, 140](cm)$
- $\rho \in [0, 1](g.cm^{-3})$
- $Fe, Vin, Pu \in [0, 100](\%)$

Experiments E1 and E3 represent low attenuation cases because of their respective low apparent densities and their vinyl composition. Their difference lies in the definition of the source term. The plutonium source of E1 is punctual and placed in front of the detector and in the center of the object. On the other hand, the plutonium mass of E3 is divided into 16 parts which are distributed homogeneously throughout the cylindrical object. For these two experiments, the plutonium is in the form of a liquid solution. As for them, the experiments E2 and CO are more exotic. They represent greater attenuation cases by their apparent bulk densities and their partially metallic composition. Although their respective source terms are located in a single point and therefore considered as point sources, plutonium is here in metal form and present in larger quantities. The effect of self-abortion is theoretically more expressed for these two experimental cases than for the two previous ones.

A unique characterized HPGe detector is considered for all experiments. Its characterization has been built with a numerical method developed by N. Guillot and N. Saurel [3]. The intrinsic calibration efficiency coefficient of the interest HPGe detector and its uncertainty are taken into account in the proposed quantification method. As explained in section IV, the interest conditional PDF mass $f(m|\mathbf{Y}^{\mathbf{obs}})$ (14) is estimated by Monte Carlo sampling in using the Metropolis-Hastings algorithm.

As explained in Section III, each parameter of the problem has its *a priori* probability density. In the present case, since the radius of the cylinder, its height and the distance from detector to object are available and known, they possess *a priori* Gaussian probability densities centered on their real values with low variances. However, the mass of plutonium to be estimated and the bulk density possess uniform non-informative probability densities. This means that one does not have *a priori* information on these values.

In order to propose ergodic chains [7], thirty-five Markov

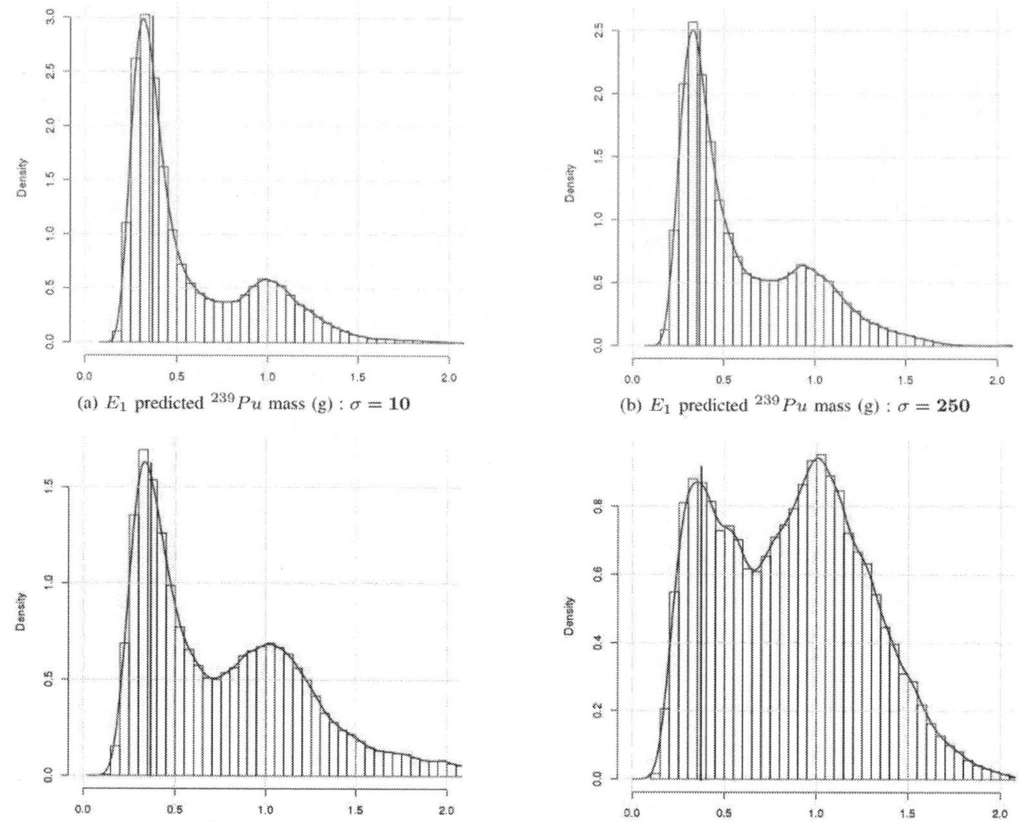

(a) E_1 predicted ^{239}Pu mass (g) : $\sigma = \mathbf{10}$

(b) E_1 predicted ^{239}Pu mass (g) : $\sigma = \mathbf{250}$

(c) E_1 predicted ^{239}Pu mass (g) : $\sigma = \mathbf{500}$

(d) E_1 predicted ^{239}Pu mass (g) : $\sigma = \mathbf{750}$

Fig. 3. Predicted ^{239}Pu masses (g) related to E_1 experiment under H_2 hypothesis. Empirical densities appear in red and expected masses in blue.

chains are generated in parallel. For each of them, initial values of m mass and $\mathbf{X}$ input components are randomly distributed relatively to their prior distributions exposed in section III. Chain burn-ins [7] are not taken into account in mass estimated distribution analysis. Both of the H_1 and H_2 hypothesis presented in section VI have been tested to different values of their parameters.

B. Quantification results

For all of the four experiments, the thirty-five Markov chains are gathered to form the estimated conditional PDF mass. These distributions appear in Fig. 2 with empirical densities. For these figures, the hypothesis H_1 is used with its parameter equal to 0.05, *ie* $\alpha = 0.05$ (15).

First, it is noted that the expected actinide masses of Table.1 are included in each of the distributions. With respect to the E_2 and CO experiments, the mass PDFs are formed from a single distribution whose modal value is close to the magnitude of the expected mass. Gaussian, gamma and lognormal fits were attempted but without success. Moreover, the variance of the distribution of the experiment E_2 is greater the CO one. As for them, the experiments E_1 and E_3 propose a bimodal mass distribution. In both cases, one of the two modal values is close to the expected mass value and has greater probability amplitude. Several mixtures of fits have been tested but none showed any interesting results.

Second, the influence of hypotheses H_1 and H_2 on the form of mass distributions has been tested by varying their parameters. Both hypotheses show similar results for the E_2 and CO experiments. Their variances increase as the uncertainty associated with H_1 and H_2 increases as well. Nevertheless, concerning the experiments E_1 and E_3, it is noted that the ratios of the probability amplitudes of their respective modals decrease as the uncertainty of the observations increases. In other words, the estimated masses from the two modals are more or less probable depending on the value of the uncertainty of the observations as shown in Fig. 3.

C. Discussion

The expected mass values of the four experiments are included in the estimated probability densities and are close to the modal values. This shows the overall consistency of the method. However, there is a significant systematic bias of about 15% between expected and estimated values. This can be explained by a poor ability to predict the surrogate model M of $\epsilon(\mathbf{X})$ efficiency. An optimization of this model is necessary and under progress.

This method of quantification makes it possible to take into account all sources of uncertainty and to easily propagate them. Other sources of uncertainty may also be added. The optimization of the efficiency proposed above may also provide a reduction in its variance of prediction resulting from the technique of emulation by kriging.

Considering the experiment E_1 under the hypothesis H_2 with σ equal to 10 exposed in Fig. 3a, the variance of the estimated probability density of the mass is almost entirely due to the variance of the model of the efficiency. Indeed, given the values of $\mathbf{S^{obs}}$, in this case we have:

$$\forall n \in [\![1, N]\!], \qquad \frac{S_n^{obs}}{\sigma} = \frac{S_n^{obs}}{10} \gg 1 \tag{17}$$

Equation (17) indicates that in the case where σ is equal to 10, it is considered to be negligible with respect to the observation. The increase in the variance visible in Fig. 3 is therefore due to the increase in the uncertainty of the observations under H_2, *ie* σ. In case E_1, a high confidence in the values observed, ie with σ equal to 10, makes it possible to obtain a modal approximately 6 times more likely than the second one. On the other hand, in the absence of precision on the observations, ie with sigma equal to 750, the modals estimated by the method are almost equiprobable. This case reflects the importance of accurately estimating the uncertainty of the observations, and therefore of the surfaces extracted from the gamma spectrum.

VIII. Conclusion

The quantification of nuclear material is essential in the nuclear industry. It allows, among other things, to comply with the legal regulations relating to the management of nuclear waste. The quantification method developed in this study uses a stochastic approach to the problem in a Bayesian framework. This makes it possible to consider all sources of uncertainty and to easily propagate them. It also proposes a resolution of the inverse problem of quantification using an MCMC algorithm. The method was tested on 4 experimental cases of ^{239}Pu measurement. These first results are coherent and allow to glimpse a robust quantification method. The method deals with the emulation of the detection efficiency of the measurement scene. Despite good prediction capabilities, optimization of this model in terms of accuracy and variance reduction is to be expected.

References

[1] G.F. Knoll. *Radiation Detection and Measurement*. John Wiley & Sons, 2010.
[2] K.Debertin and R. G. Helmer. *Gamma and X-ray spectrometry with semiconductor detectors*. North-Holland Amsterdam, 1988.
[3] N. Guillot. *Quantification gamma de radionucléides par modélisation équivalente*. PhD thesis, Universit Blaise Pascal, 2015.
[4] LANL. *MCNP6 Users manual*.
[5] A. Lyoussi. *Détection de rayonnements et instrumentation nucléaire*. EDP Sciences, 2010.
[6] T. Vigineix. *Exploitation de spectre gamma par méthode non paramétrique et indépendante da priori formulés par lopérateur*. PhD thesis, Universit de Caen, 2011.
[7] G. Casella C. P. Robert. *Méthodes de Monte-Carlo avec R*. Springer, 2011.
[8] C. E. Rassmusen. C. K. I. Williams. Gaussian processes for machine learning. *MIT Press*, 2006.
[9] T. Mitchell J. Sacks, W. Welch and H. Wynn. Design and analysis of computer experiments. *Statistical Science*, 4:409-435, 1989.
[10] Jeong-Soo Park. Optimal latin-hypercube designs for computer experiments. *Journal of Statistical Planning and Inference*, 39(1):95–111, April 1994.

Temporal Imaging $CeBr_3$ Compton Camera: A New Concept for Nuclear Decommissioning and Nuclear Waste Management

A. Iltis, H. Snoussi, L. Rodrigues de Magalhaes, M.Z. Hmissi, C. Tata Zafiarifety, G. Zeufack Tadonkeng and C. Morel

***Abstract*—During nuclear decommissioning or waste management operations, a camera that could make an image of the contamination field and identify and quantify the contaminants would be a great progress. Compton cameras have been proposed, but their limited efficiency for high energy gamma rays and their cost have severely limited their application. Our objective is to promote a Compton camera for the energy range (200 keV – 2 MeV) that uses fast scintillating crystals and a new concept for locating scintillation event: Temporal Imaging.**

Temporal Imaging uses monolithic plates of fast scintillators and measures photons time of arrival distribution in order to locate each gamma ray with a high precision in space (*X,Y,Z*), time (*T*) and energy (*E*). This provides a native estimation of the depth of interaction (*Z*) of every detected gamma ray. This also allows a time correction for the propagation time of scintillation photons inside the crystal, therefore resulting in excellent time resolution. The high temporal resolution of the system makes it possible to veto quite efficiently background by using narrow time coincidence (< 300 ps). It is also possible to reconstruct the direction of propagation of the photons inside the detector using timing constraints. The sensitivity of our system is better than 1 nSv/h in a 60 s acquisition with a ^{22}Na source.

The project TEMPORAL is funded by the ANDRA/PAI under the grant No. RTSCNADAA160019.

***Index Terms*—Compton camera, Fast scintillating crystals, Gamma ray imaging, Nuclear decommissioning, Temporal Imaging, Time resolution.**

I. Introduction

THE concept of Compton camera was first investigated to detect gamma rays within the energy range 150 keV – 511 keV in the medical field [1]. It became then more and more used in other fields like astronomy [2] and industry [3], mostly by using semiconductors like CdTe or CZT. However, such detectors are limited by their small active size, detection efficiency and high background noise level. We propose here a new concept called Temporal Imaging to build a Compton camera based on fast monolithic scintillation crystals. In this concept in addition to light sharing, we also use, photon time of flight within the scintillation crystal to accurately calculate the location of every gamma ray event. This acquisition concept allows a good positioning resolution, not only within the plane of detection ($\Delta X, \Delta Y < 2$ mm FWHM), but also a good estimation of the depth of interaction (DOI) ($\Delta Z < 2$ mm FWHM). It is thus possible to use thick monolithic plates (20 mm LYSO or 25 mm $CeBr_3$) without compromising the spatial resolution of the Compton camera. The energy resolution obtained is quite good for large monolithic plates (11.0% FWHM with LYSO and 6.9 % FWHM with $CeBr_3$ at 511 keV).

We will also demonstrate in this paper, that a good DOI measurement allows to reaching very high timing precision (225 ps FWHM with LYSO at 511 keV) for each one of the scintillation plates and will show the benefit that can result for a Compton camera with such a good timing.

We report here first Compton images obtained with LYSO detectors and present first timing measurements with a first generation of windowless $CeBr_3$ detectors.

$CeBr_3$, has been selected against LYSO as the reference crystal for our Compton camera, because of its higher energy resolution, higher number of prompt photons allowing better timing and low radioactive background. Our target is to be able to acquire images of weak and/or distant radioactive sources. Nevertheless, $CeBr_3$ is highly hygroscopic. Therefore, we have developed a new encapsulation technology "Fluoral" that enables direct coupling of the Si-PM to the crystal without a glass window. The results presented here with $CeBr_3$ are obtained with this technology, but are still preliminary. Hence, we present also some results and images obtained using our LYSO demonstrator.

Manuscript submitted June 01, 2017.

A. Iltis, L. Rodrigues de Magalhaes, C. Tata Zafiarifety and G. Zeufack Tadonkeng are with Damavan Imaging, 2, rue Gustave Eiffel, 10430 ROSIERES PRES TROYES, Frnace (e-mail: alain.iltis@damavan-imaging.com, luc.rodrigues@damavan-imaging.com, christian.tata@damavan-imaging.com, gislain.zeufack@damavan-imaging.com).

H. Snoussi and M.Z. Hmissi are with University of Technology of Troyes, UMR CNRS 6281, 12 rue Marie Curie, 10010 TROYES, France (e-mail: hichem.snoussi@utt.fr, mohamed_zied.hmissi@utt.fr).

C. Morel is with Aix-Marseille Univ, CNRS/IN2P3, CPPM, Marseille, France (e-mail : morel@cppm.in2p3.fr).

II. Compton Camera Architecture

A. Principle of the Compton camera

The schematic outline of a Compton camera is described in Fig. 1.

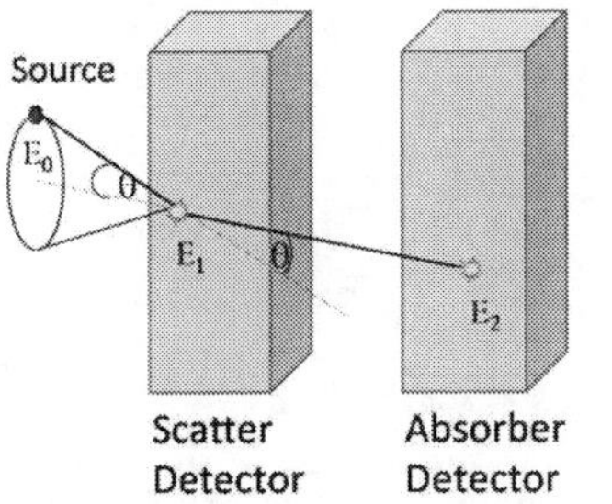

Fig. 1. Schematic outline of a two plate Compton camera

The impinging gamma ray with the initial energy E_0 interacts in the first detector at a well-defined position, resulting in the emission of a scattered photon of energy E_2 by an angle Θ_s. The remaining energy $E_1 = E_0 - E_2$, which is transferred to an electron, is deposited in the first detector almost at the position of the Compton interaction. The scattered photon is then absorbed in the second detector, where it deposits its total energy E_2.

Each Compton event results in a cone shape defined by:

- A vertex, which is the Compton interaction position of the impinging gamma ray in the scatter detector.
- An axis defined by the interaction positions in both the detectors (diffuser and absorber).
- An angle Θ_s determined by the Compton equation:

$$\theta_s = \cos^{-1}\left[1 + m_0c^2\left(\frac{1}{E_0} - \frac{1}{E_2}\right)\right] \qquad (1)$$

- A time delay related to light propagation between E1 and E2

The true path of the impinging gamma ray lies on the mantle of the cone defined by the axis and the angle Θ_s.

B. Experimental set-up

Our camera consists of a LYSO or a $CeBr_3$ monolithic crystal (32 × 32 × 5) mm diffuser plate and a (32 × 32 × 20) mm LYSO or a (32 × 32 x 25) mm $CeBr_3$ monolithic crystal absorber plate. Both are encapsulated in an aluminum housing. For the Compton camera configuration, the 5 mm and the 20 mm detectors are in front of each other. Hence, taking into account the mechanics, the spacing between the two plates is 5 mm.

The photodetector is a digital SiPM Phillips DPC-3200-22. This digital detector consists of 16 independent die sensors arranged in a 4 × 4 array. Each die sensor contains four pixels arranged in 2 × 2 array, and each pixel contains 3200 SPAD cells. A delay-time correction map is applied pixel by pixel to get good timing.

Fig. 2. The Compton camera setup: a ^{22}Na source is located in the red disc 40 mm away from the first detector

The Temporal Imaging Compton camera is designed as shown in Fig. 3. It is subdivided into two parts: a small block on the left including the detectors and a large block on the right where one finds the associated electronics.

Fig. 3. Picture of the TEMPORAL Delta Compton camera

Data acquisitions were performed using a 0.5 MBq ^{22}Na source, which. was located 40 mm away from the diffuser plate. The distance between the diffuser and the absorber plate was set to 5 mm. Images were acquired during 200 s. The Compton camera setup is shown in Fig. 2.

III. Performance of the Temporal detector

Accurate Compton camera reconstruction requires high precision positioning and energy measurements in the diffuser and absorber plates. In this section, we present the spatial and energy resolutions achieved with our setup.

X and Y positions are determined from the spatial distribution of the scintillation photons and the distribution of their time of arrival. The estimated precision on X and Y is 1,98 mm FWHM. DOI (position along Z) is obtained by using a least root mean square method on simulated images [4]. The estimated precision on DOI is 2 mm FWHM.

For energy measurements, a FWHM resolution of 11.0% at 511 keV was observed in the LYSO crystal (see Fig. 4).

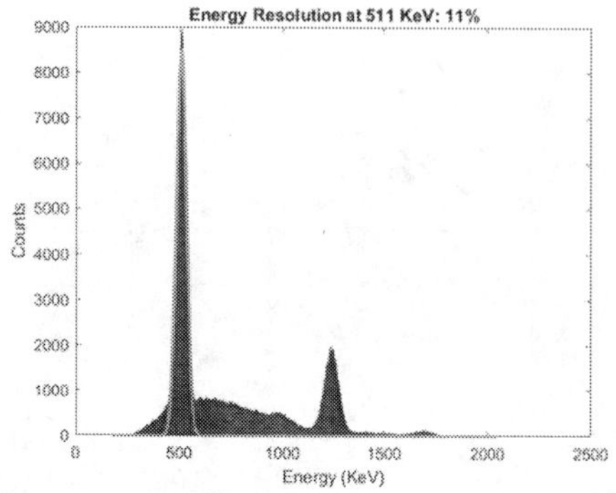

Fig. 4. Energy spectrum of ^{22}Na with the LYSO crystal: we use a Gaussian fit to determine the energy resolution

For the $CeBr_3$ crystal, the energy resolution at 511 keV was 6.9 % FWHM. The discrepancy from published values for $CeBr_3$ comes from the low quantum efficiency of the Phillips digital SiPM at the 370 nm peak emission of $CeBr_3$.

IV. Timing performance

In this section, we present the results of our detectors in terms of timing resolution. In order to calibrate our detectors in time, we use a coincidence setup and measure the coincidence resolving time (CRT) of the detection of electron-positron annihilation pairs.

A. CRT correction for $CeBr_3$ and LYSO detectors

The CRT was measured in another study [5], in which two (32 × 32 × 25) mm $CeBr_3$ crystals were used in coincidence. Each detector was coupled to a Philips Digital Photon Counter tile, DPC-3200-22 sensor with the windowless Fluoral™ technology [6].

We used the DOI measurement in order to correct for the delay time introduced by the scintillation photon travelling along the crystal at a speed of c/n, assuming that the gamma ray travels at the speed of light in vacuum c.

Without corrections, the measured CRT for $CeBr_3$ was 519 ps FWHM. As shown in Fig. 5, the result was obtained by fitting the raw CRT curve with Gaussian distribution.

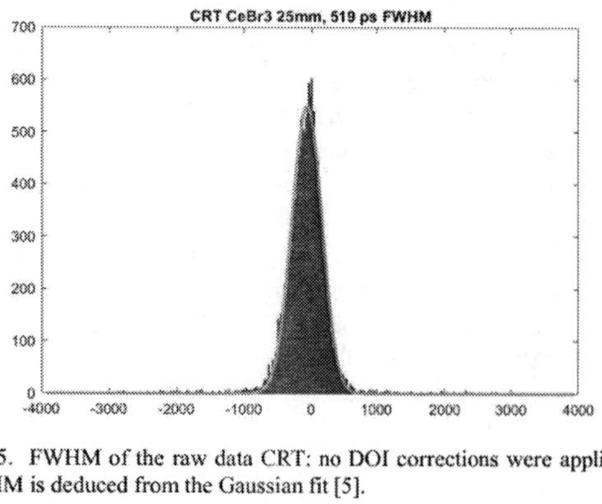

Fig. 5. FWHM of the raw data CRT: no DOI corrections were applied. The FWHM is deduced from the Gaussian fit [5].

When taking DOI into account, the CRT was reduced to 275 ps FWHM [5]. This value is preliminary. We believe it is possible to reach a corrected CRT below 200 ps FWHM with $CeBr_3$. For LYSO detectors using the same setup, the CRT value was 225 ps FWHM when corrected for the 20 mm thick detector plates. The corrected CRT value is part of our Compton image processing.

B. Advantages of corrected CRT

For a Compton camera, a good timing resolution provides three advantages:

- It allows to identifying events with a Compton diffusion in the absorber plate [7].
- It places a stringent coincidence selection of Compton events, which have to be detected on the two plates within a time window smaller than 500 ps. This allows for a very strong reduction of background and thus to acquire high contrast images with few photons, even when using a crystal with a high intrinsic background like LYSO.
- It allows to determining in which crystal has undergone the first interaction, hence recognizing the direction of light propagation between the two plates.

V. Gamma ray source reconstruction

Many gamma ray source reconstruction techniques have been developed based on iterative or analytic algorithms [8], [9]. In our study, we have used the COMPTONREC software package, which consists in a direct method to reconstruct the source position [10].

A. Principle of reconstruction

As presented in [10], the analytic direct method for source reconstruction we have used consists in computing the intersection of the conical surface defined by the Compton event with the plane that contains the source. This plane must be located inside the field of view of the Compton camera.

Then, we have selected the events for which the sum of the energy deposited in both the diffuser and absorber crystals is within 511 keV ± δE, with δE equal to 11% of 511 keV, which corresponds to the energy resolution of the detection system. We have also selected direct Compton events only, which interact at first in the diffuser and then as a photoelectric event in the absorber. Timing information was used to perform this selection.

B. Compton image reconstruction

In the reconstruction process, for each selected Compton event, possible incidences of the impinging gamma ray are all located on a cone. The position of the gamma ray source is then given by the intersection of cones. In Fig. 6, we show the intersection of the conical surfaces with the source plane for some of the selected Compton events.

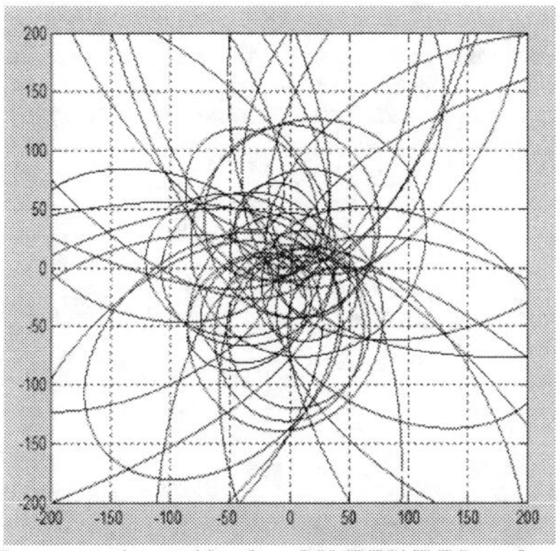

Fig. 6. Reconstruction with the COMPTONREC software package calculating intersections of conical sections with the source plane

All the reconstructed cones intercept at the real source position, which is located in the center of the field of view of the camera.

This kind of representation is difficult to analyze in order to study the performance of the Compton camera such as its spatial resolution. Therefore, in the rest of this work, the reconstructed image will be represented as a density image display where we compute the number of intersections between reconstructed cones per pixel (see Fig 7).

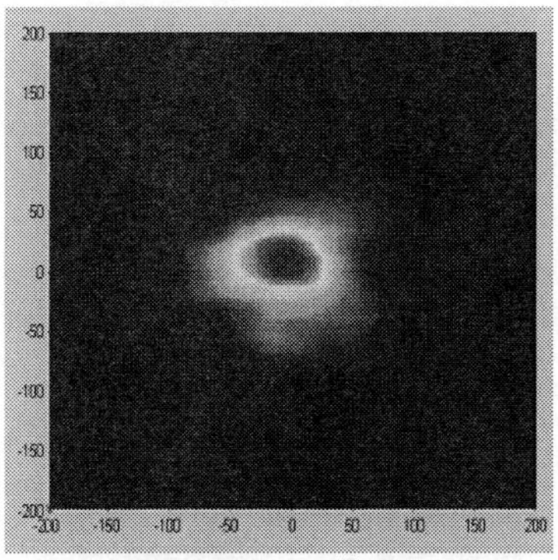

Fig. 7. Density image representation of the ^{22}Na source by selecting 511 keV Compton events: 148 events are used to reconstruct the source.

We can clearly identify the source position as well as the point spread function using this representation.

To determine the source-detector distance, we have used an algorithm described in [11] that is using the solid angle subtended by the reconstructed image as a function of the source-to-detector distance, assuming that an image reconstructed at the correct distance has a smaller solid angle than an image reconstructed at any other distance from the detector. The expected size of the image is minimized when the correct image plane is used.

With the same processing, 1.3 MeV gamma rays were also used to reconstruct the position of the ^{22}Na source. Its density image is shown in Fig. 8. The diffuser plate size is changed from $(32 \times 32 \times 5)$ mm to $(32 \times 32 \times 20)$ mm in order to increase its efficiency. Two 0.06 MBq ^{22}Na sources are used to determine the spatial resolution of our detector at 1.3 MeV. Thus, we have determined the minimum distance after which we cannot resolve anymore the two ^{22}Na sources. The first plate (diffuser) is 50 mm distant from the ^{22}Na sources, which are separated by a 4 mm metallic plate.

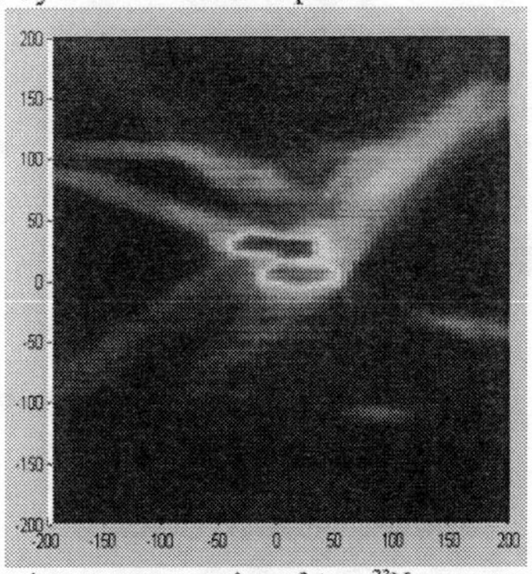

Fig. 8. Density image representation of two ^{22}Na sources using 1.3 MeV Compton events: the two sources are separated by 4 mm metallic plate and distant 50 mm away from the first crystal.

An angular resolution of 4.5° was obtained with our camera for the detection, of 1.3 MeV Compton events. It demonstrates the effectiveness of our use of the time information. Moreover, angular resolution will be improved when $CeBr_3$ crystals will be used.

VI. Conclusion

Reconstruction of Compton events detected by a Compton camera depends on the precision of the gamma ray localization and energy measurement in both the diffuser and the absorber detectors. However, with a standard setup, many of the impinging gamma rays may not be recorded. Here we present a Compton camera based on two high density thick scintillator plates. This setup absorbs more than 50% of the 1.3 MeV impinging gamma rays from a ^{22}Na source. Interaction positions are obtained by using a new concept: Temporal Imaging, which is exploiting both the scintillation photon light distribution and their distribution of time of arrival. This concept allows for good position resolution in three dimensions (< 2 mm FWHM), including DOI. It also allows to recovering an energy resolution close to the intrinsic performance of the scintillators. Timing resolution of this setup is very high (CRT < 225 ps FWHM).

Moreover, the time information in an small detector geometry can be used to discriminate Compton events from LYSO background. This improves the performance of our Compton camera for low energy gamma rays. Thus, we can obtain an image with only few Compton events, resulting in shorter acquisition times needed to localize a gamma ray source.

We have obtained our first images at 511 keV and 1.3 MeV with a functional prototype made of two LYSO crystals: Temporal Delta. The measured angular resolution was at least 4.5° for 1.3 MeV Compton events. We have also presented a first characterization of windowless $CeBr_3$ detector for timing: a CRT of 275 ps FWHM was achieved as a preliminary result after taking into account scintillation light travelling time inside the crystal. A CRT below 200 ps FWHM is expected to be achieved for future developments with $CeBr_3$ crystals.

$CeBr_3$ will be preferred for our future systems to LYSO, because of its better timing and energy resolution potential. Moreover, $CeBr_3$ does not have a radioactive background like LYSO and should thus allow for imaging gamma ray source with very low activities.

References

[1] N. F. Moody, W. Paul, and M. Joy, "A Versatile Gamma-Ray Camera for Dynamic or Static Studies in Medicine," *IEEE Transactions on Nuclear Science*, vol. 16, no. 2, pp. 3-18, Apr. 1969.

[2] V. Schönfelder, A. Hirner, and K. Schneider, "A telescope for soft gamma ray astronomy," *Nuclear Instruments and Methods*, vol. 107, no. 2, pp. 385-394, Mar. 1973.

[3] J. B. Martin, N. Dogan, J. E. Gormley, G. F. Knoll, M. O'Donnell, and D. K.Wehe, "Imaging multi-energy gamma-ray fields with a Compton scatter camera," *IEEE Transactions on Nuclear Science*, vol. 41, no. 4, pp. 1019 1025, Aug. 1994.

[4] A. Iltis and H. Snoussi, The Temporal PET Camera: A New Concept With High Spatial and Timing Resolution for PET Imaging, J.Imaging 2015, 45-59; doi:10.3390/jimaging1010045.

[5] C. Tata Zafiarifety, A. Iltis, H. Snoussi, G. T. Zeufack, M. Z. Hmissi, L. Rodrigues, E. Lamprou, A. Aguilar, V. Ilisie, A. J. Gonzalez and J. M. Benlloch, "Time and energy performance of a temporal imaging LYSO and $CeBr_3$ detectors," submitted in 2017 *IEEE Nuclear Science Symposium and Medical Imaging Conference (NSS/MIC)*, Oct. 2017.

[6] FR n°: 1659397, 30 Sept 2015.

[7] A. Iltis, L. Rodrigues De Magalhaes, H. Snoussi and C. Morel, "Temporal Imaging: Observation and localization of a Compton effect inside a 20 mm monolithic LYSO plate with a Philips digital SiPM," Poster presented at 2016 *IEEE Nuclear Science Symposium and Medical Imaging Conference (NSS/MIC)*, Oct. 2016.

[8] S. J. Wilderman, N. H. Clinthorne, J. A. Fessler, C. h. Hua, and W. L. Rogers, "List mode EM reconstruction of Compton scatter camera images in 3-D," in 2000 *IEEE Nuclear Science Symposium*. Conference Record (Cat.No.00CH37149), vol. 2, 2000, pp. 15/292-15/295 vol.2.

[9] M. J. Cree and P. J. Bones, "Towards direct reconstruction from a gamma camera based on Compton scattering," *IEEE Transactions on Medical Imaging*, vol. 13, no. 2, pp. 398-407, Jun. 1994.

[10] M. Zioga, M. Mikeli, A. Eleftheriou, C. Pafilis, A. N. Rapsomanikis, and E. Stiliaris, "ComptonRec: Mastering conic sections for a direct 3d compton image reconstruction," in 2015 *IEEE Nuclear Science Symposium and Medical Imaging Conference (NSS/MIC)*, Oct. 2015, pp. 1-5.

[11] M. W. Rawool-Sullivan, J. P. Sullivan, S. R. Tornga, and S. P. Brumby, _A simple algorithm for estimation of source-to-detector distance in Compton imaging,_Applied Radiation and Isotopes, vol. 66, no. 12, pp. 1986-1991, Dec. 2008.

A Remote-operated System to Map Radiation Dose in the Fukushima Daiichi Primary Containment Vessel

M. Nancekievill, A. R. Jones, M. J. Joyce, *Member, IEEE,* B. Lennox, *Senior Member IEEE,* S. Watson, *Member, IEEE,* J. Katakura, K. Okumura, S. Kamada, M. Katoh and K. Nishimura

***Abstract*—This paper describes the development of a submersible system based on a remote-operated vehicle coupled with radiation detectors to map the interior of the reactors at the Fukushima Daiichi nuclear power station. It has the aim oflocating fuel debris. The AVEXIS submersible vehicle used in this study has been designed as a low-cost, potentially disposable, inspection platform that is the smallest of its class and is capable of being deployed through a 150 mm diameter access pipe. To map the gamma-ray environment, a cerium bromide scintillator detector with a small form factor has been incorporated into the AVEXIS to identify radioactive isotopes via gamma-ray spectroscopy. This provides the combined system with the potential to map gamma-ray spectra and particle locations throughout submerged, contaminated facilities, such as Units 1, 2 and 3 of the Fukushima Daiichi nuclear power plant. The hypothesis of this research is to determine the sensitivity of the combined system in a submerged environment that replicates the combination of gamma radiation and water submersion but at lower dose rates.**

***Index Terms*— Fukushima Daiichi, Gamma-ray detection, Radiation monitoring, Nuclear Decommissioning**

I. Introduction

After the Fukushima nuclear power plant accident in 2011, many projects have been started with the aim of completing remote inspection and characterization of the inside of the stricken reactors [1]. Due to the evacuation of the surrounding area and flooding of the primary containment vessels (PCV), little is known of the current physical state of the reactor cores inside the PCVs of reactors 1, 2 and 3.

Although large portions of Fukushima Daiichi have been characterized using long-established remote methods, the current state of the environment at the bottom and beneath the submerged pedestal in the reactors is largely uncharacterized [2]. This leads to uncertainties in dose-rate levels and the extent of contaminated materials, and since dose-rate levels as high as 1 kGy h^{-1} are estimated, this could, if not better understood, hamper the progress of decommissioning the plant because of the potential for there to be prohibitive levels of total ionizing dose.

The AVEXIS vehicle was developed in the United Kingdom as part of a project to characterize the Sellafield Legacy ponds [3]. New designs and development cycles have now increased the robustness and adaptability of the vehicle to a multitude of environments.

A live feed camera is included to allow visual inspection of unknown environments with complete manual control of the system from the surface via a neutrally-buoyant, bespoke tether of diameter <10 mm.

The results of experimental testing of the remote-operated vehicle (ROV) with the integrated $CeBr_3$ detector are described in this paper. Testing has been conducted in water tank facilities at both the University of Manchester and Lancaster to determine the capability of the detector to characterize the gamma spectrum from a known source positioned on the exterior of the tank.

Due to the modularity of the AVEXIS design, it is possible to replace the cerium bromide detector with other detectors. In particular the integration of a single crystal diamond chemical vapor deposition neutron detector with a ^{6}Li convertor foil able to map the thermal neutron flux and complementing the gamma-ray sensing functionality.

The remainder of the paper is structured as follows: Section II discusses improvements made to the ROV for suitable operational capacity to be deployed within the Fukushima Daiichi nuclear power plant. Section III introduces the integration of the cerium bromide detector and experimental results of the gamma spectrum analysis with a small ^{137}Cs source. Finally, section IV outlines the work to be completed in the future before the system can be deployed in the field.

Paper submitted May 31st, 2017. This work was funded by the Engineering and Physical Sciences Research Council (EPSRC:EP/N017749/1) and MEXT (Japan). MJJ acknowledges the generous support of the Royal Society as a Wolfson Research Merit Award holder.

M. Nancekievill, B. Lennox and S. Watson were with the Department of Electrical and Electronic Engineering, University of Manchester, United Kingdom, UK e-mail: matthew.nancekievill@manchester.ac.uk

A.R. Jones and M.J. Joyce are with the Department of Engineering, Lancaster University, United Kingdom.

J. Katakura is with the Nagoaka University of Technology, Japan

K. Okumura is with the Japan Atomic Energy Agency, Japan.

S. Kamada, M. Katoh, and K. Nishimura are with the National Maritime Research Institute, Japan.

II. Aqua Vehicle Explorer for in-situ sensing

The Aqua Vehicle Explorer for In-situ Sensing (AVEXIS) is a submersible ROV developed by the University of Manchester and is shown in Figure 1. It was designed for characterization and monitoring of the Sellafield legacy ponds in hard-to-reach areas [3].

The ROV needs to be low-cost allowing easy disposal of the submersible, if required, due to the potential for there to be extensive contamination with radioactive materials. The net material cost for the device was approximately £200, which is consistent with the requirement that multiple ROVs could be manufactured and purchased at relatively easily if required.

Figure 1 AVEXIS design.

A. Overview of the AVEXIS

AVEXIS is cylindrical in shape with an outer diameter of 150 mm. It comprises 5 water pumps on each end-cap that provide movement forwards, backwards, up, down, left and right. Left and right motion occurs through the central axis; therefore the turning radius is half the length of the ROV. This makes the AVEXIS highly maneuverable, consistent with the requirement for characterization of hard-to-access areas.

The end caps are also designed with two O-rings for a watertight seal up to 100 m depth. They are also easily removable for adaptation and maintenance of the internal electronics, for replacement of the actuators or for modifications to the length of the vehicle if necessary.

A Raspberry Pi™ Zero is used for the main control of the ROV. This is connected to a Raspberry Pi™ Camera for a live video feed, displayed on the surface control station for visual inspection. An LED strip on each side of the ROV creates light to allow visual inspection in poorly lit water.

The neutrally-buoyant, custom tether consists of six 26 American Wire Gauge (awg) wires bundled into two sets of three (this will be referred to as a *two-wire tether* for simplicity). Both communication and power are afforded over this two-wire setup.

Powerline technology often found in homes has been adapted for use in the AVEXIS, to constitute a custom Power Over Ethernet (POE) system capable of 200 Mbps transfer speeds. This is more than enough for High-definition (HD) live camera feed and instantaneous control of the ROV. This is very similar to the system used in OpenROV vehicles [4].

A local area network was created by attaching the local Ethernet port of a computer on the surface to a powerline adaptor. This communicates over the two-wire tether to a powerline adaptor inside the AVEXIS that is connected to a mini USB-to-Ethernet adaptor, connected to the Raspberry PiZero. With the Raspberry PiZero acting as a web-server, live video and control can be streamed to any web browser. A commercial gamepad such as a Sony PS4 or Microsoft Xbox controller can be used for intuitive control.

Power is supplied by placing a 48 V d.c. power supply in parallel to the two-wire tether. Communications were unaffected by any bias across the two wires. However, filters were placed on the input of each power supply to avoid interference from the MHz range frequency of Ethernet communication.

The development of a POE tether equated to a reduction in tether size to conventional ROVs as well as an increase in mechanical flexibility. This was especially important for a small ROV as it could be affected by a cumbersome tether reducing the working scope of characterization.

The main ROV board contains a DC-DC converter from 48 V d.c. to 12 Vd.c. for the pumps and 5 Vd.c. for the control electronics. The I/O electronics for the AVEXIS are designed around an ATMEGA 32U4, commonly found on an Arduino Leonardo.

B. $CeBr_3$ Gamma detector integration

For the initial integration of the $CeBr_3$ gamma detector in the ROV, an additional tether was added that consisted of a RG178 coaxial cable. This was used to transmit the signal data from the detector to the surface for analysis. The weight and outer diameter of this additional tether was very small, which allowed it to be wrapped around the AVEXIS custom tether with no reduction in maneuverability or operational flexibility.

Power for the detector was delivered by the main ROV circuitry. This meant no additional power conversion circuit was required allowing quick integration of this detector. It also allows adaptability for future gamma detector/neutron detector integration.

C. Other important specifications

The AVEXIS was designed to carry a payload of approximately 1 kg. This was to accommodate the neutron and gamma detectors and to provide contingency should other sensors be required. For example, a sonar could be attached for understanding the physical topology and dimensions of any debris in an environment where visual inspection is not possible.

Figure 2 shows the layout of one of the AVEXIS endcaps. It shows the tether access point, the coaxial connector for radiation detector signal propagation and the access point for the pump control wires. It also illustrates the position of a depth sensor capable of measuring depth to a tolerance of ±2 mm [5].

The depth sensor allows for semi-autonomous control of the depth of the ROV and better control overall. Visual inspection does not allow for fine control of depth. Therefore, integration of a "depth lock" using this pressure sensor to autonomously maintain a constant depth was applied. This permitted greater security and safety in ROV movement.

Connectors for each individual water pump allow for quick and easy replacement if any pumps became damaged or faulty. This made the ROV modular in design and adaptable to changes in the required thrusters on each ROV.

The use of commercial off-the-shelf components keeps the total manufacturing price low and caters for the prospect of the unit being lost during use.

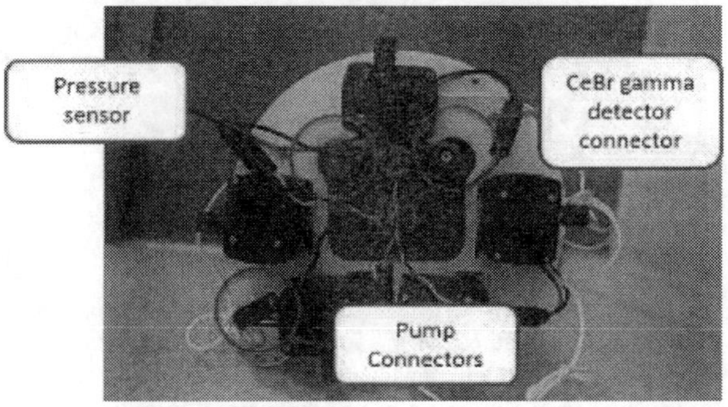

Figure 2 AVEXIS end-cap layout.

III. Cerium Bromide Detector Implementation

A. Detector overview

The detector used for these experiments is a $CeBr_3$ inorganic scintillator that is sensitive to gamma radiation. $CeBr_3$ was considered one of the best alternatives to replace the well-established NaI(Tl) because of its high detection efficiencies and improved energy resolution combined with resilience in high levels of radiation and room temperature operation.

The detector has a 10mm diameter crystal coupled to photomultiplier with a small, integrated, high-voltage supply unit. The integrated HV supply allows the unit to be operated from a 5V signal supplied via the ROV.

The detector unit was manufactured by Scionix [6], Netherlands, and the schematic diagram is shown in Figure 3. Each constituent part is labelled and explained in the schematic.

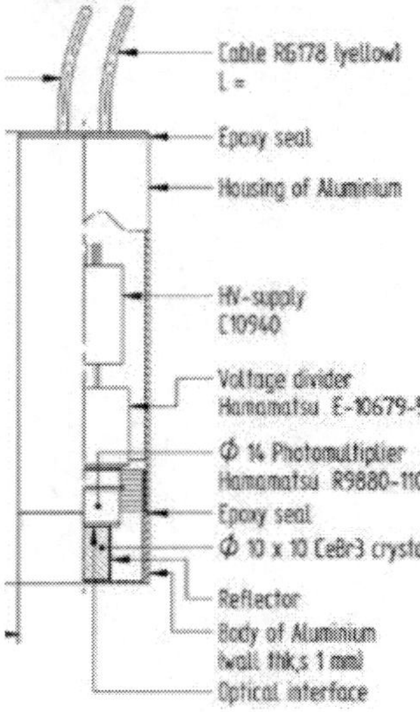

Figure 3 $CeBr_3$ inorganic scintillator detector, model VS-0087-50 supplied by Scionix, Netherlands [6].

For another layer of protection against an unlikely situation of potential leaks within the ROV whilst submerged, the detector was placed within an aluminium casing. The detector can be seen with and without the casing in Figure 4.

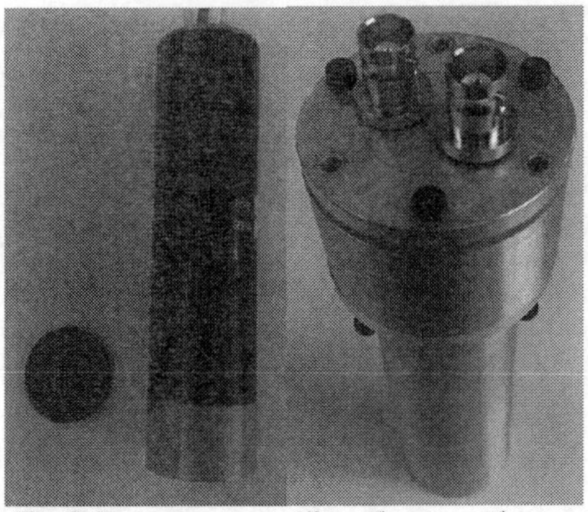

Figure 4 The $CeBr_3$ inorganic scintillator detector without and with aluminium casing.

B. Initial Characterisation

Initial tests were carried out in the laboratory to investigate the impact the aluminium casing itself had on the response of the detector. These initial tests were carried out with both a 330 kBq ^{137}Cs, and a 50 kBq ^{22}Na source. The signal from the detector was passed through and processed by a single-channel, mixed-field analyser (MFA) manufactured by Hybrid Instruments Ltd. [7]. Comparative MCA plots from these tests can be seen in Figure 5 and Figure 6.

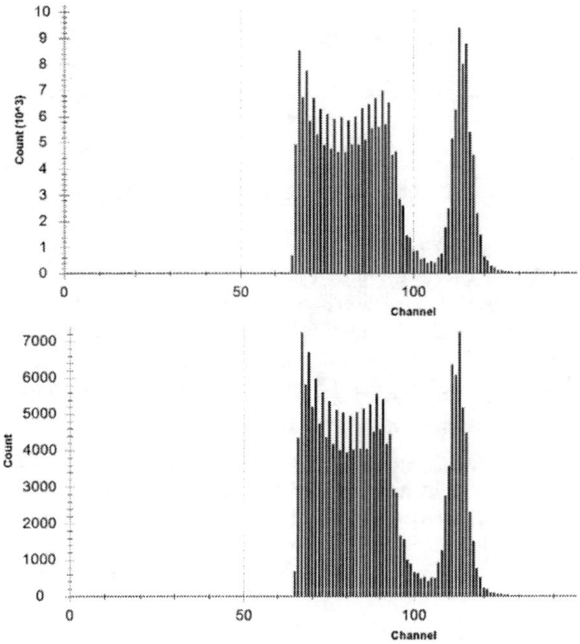

Figure 5 $CeBr_3$ detector exposed to ^{137}Cs source for 10 minutes. Top graph shows results without casing, bottom graph represents detector within casing.

 https://doi.org/10.1051/epjconf/201817006004

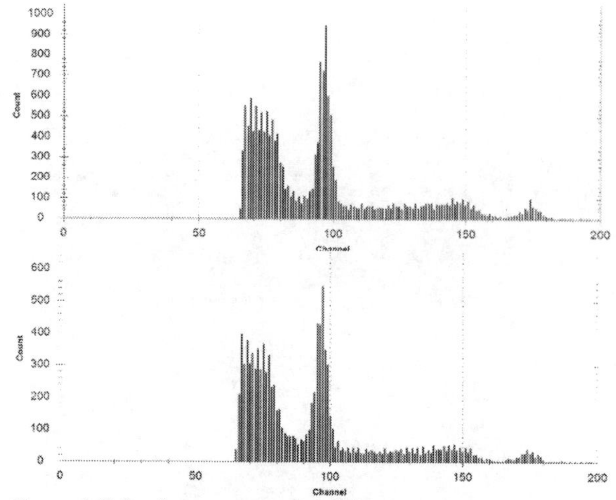

Figure 6 CeBr₃ detector exposed to ^{22}*Na source for 10 minutes. Top graph shows results without casing, bottom graph represents detector within casing.*

As can be seen from the MCA plots in Figure 5 and Figure 6 the spectral peaks are all consistent between the tests with and without the casing. However, the intensity of the peaks relative to each other does vary, as expected.

This intensity change can be observed by the throughput of counts observed by detector during this set time period with and without the casing. The results of this are shown in Table 1.

Table 1 Comparison of Number of Counts With and Without the Aluminium Casing

	10 minutes		Pulses per second	
	No Casing	Casing	No Casing	Casing
Cs source	216291 ±465	174552 ±418	360 ±19	291 ±17
Na Source	19705 ±140	12485 ±112	33 ±6	21 ±5

As can be seen by the results in Table 1, the number of counts was significantly reduced when the casing was used. The difference in attenuation between the two sources is assumed to be associated with the different energy spectrums that are emitted. This is expected, with the lower-energy spectrum of Na, dropping by a higher percentage than the Cs spectrum.

C. Integrated with ROV experiments

The next stage for the experiment was to integrate the detector within the ROV. A power supply from within the ROV was used to power the detector and the additional tether made from a RG178 coaxial cable sent back the readings from the detector. The experimental setup can be seen in Figure 7.

Figure 7 Benchtop test of gamma detector integrated within the ROV

The combined system was tested initially on the bench in the laboratory to ensure that the power supply from the ROV was sufficient for the detector to operate, as well as to confirm that the detector signal was transmitted successfully over a 50 m cable without degradation or noise overpowering the signal. The response of the detector to a 10-minute exposure to a small 330 kBq ^{137}Cs source located firstly inside and subsequently outside the ROV is shown in Figure 8.

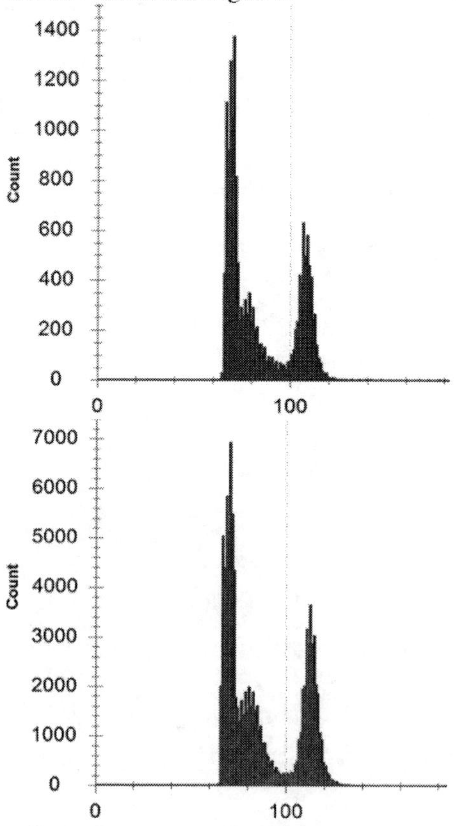

Figure 8 CeBr₃ detector exposed to a ^{137}*Cs source for 10 minutes. Bottom, integrated within the ROV.*

From Figure 8 it can be seen visually that the spectrum is less defined with a lower number of counts when the source was located outside the ROV. This is because of the additional distance and attenuation from the source when the detector was within the ROV unit. However, the results show that the detector is operational and able to function using the power supplied by the ROV.

The signal was transmitted via the 50 m tether and was sufficient to produce a spectrum. It is worthy to note that the signal was attenuated and settings on the data acquisition unit had to be adjusted to account for this in comparison to the previous test results shown in Figure 5 and Figure 6.

The ROV was then tested using a wave tank located at Lancaster University, UK. A ^{137}Cs source was placed on the side of the tank; the same source used in the previously described laboratory tests. A photograph of the experimental facilities is provided in Figure 9.

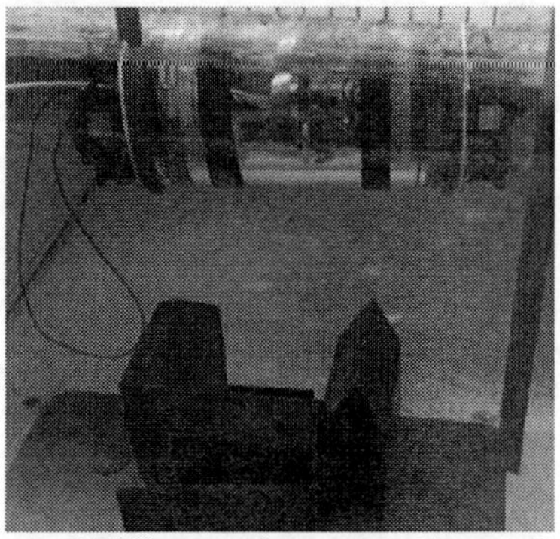

Figure 9 Submerged ROV gamma detector experiments. Cs source was placed up against the edge of the tank with the AVEXIS parallel to the source submerged.

Initial tests involved submerging the ROV with the integrated detector at two specific distances. The first was with the ROV submerged in the tank but as close to the wall as possible, whilst the second was with the ROV at an approximate distance of 0.5 m away from the side of the tank. Two 10-minute exposures were again taken with the spectra provided for comparison purposes in Figure 10.

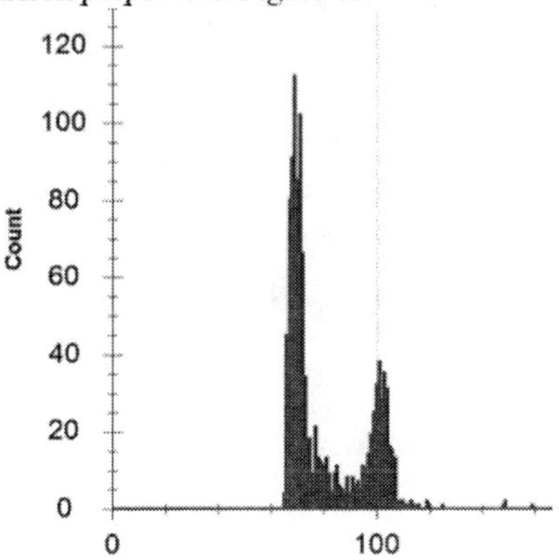

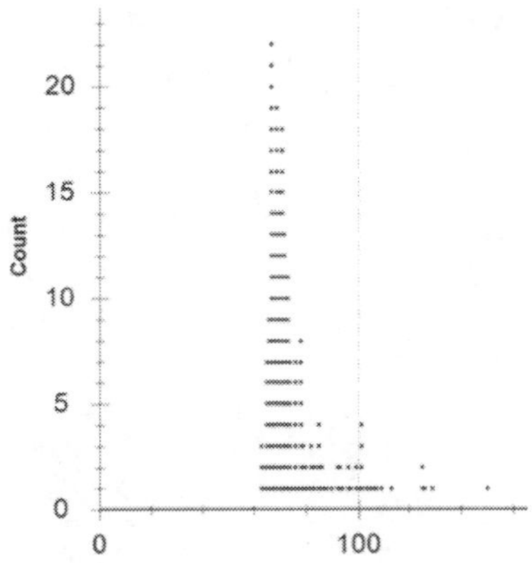

Figure 10 Detector exposed to a ^{137}Cs source for 10 minutes whilst integrated within the ROV and submerged underwater. Top, with the ROV pressed against the side of the tank. Bottom, with the ROV approximately 0.5 m away.

These results show that due to the relatively small activity of the source, it was difficult to obtain a clear gamma-ray spectrum when the ROV was 0.5 m from the tank wall for a 10-minute exposure. However, the outline of the spectrum is starting to become apparent and either a longer exposure or source of higher activity is anticipated to rectify this problem.

These tests demonstrate the detector is able to operate and provide data that can be used to identify a radioactive source whilst submerged and in very close proximity to a weak ^{137}Cs source. However, source identification from a greater distance in a submerged state was not possible.

In theory, whilst in a submerged environment, the $CeBr_3$ gamma detector could take the number of counts over a set time period at different pre-determined positions could then be used to approximate source location; operating in total gamma mode. Using the approximate location of any source, the ROV could then be positioned so that the detector is exposed to the higher counts required to carry out the more detailed spectral analysis required for identification purposes. This two-stage analysis would aid in reducing the radiation damage to the ROV as proximity to the source is not reduced (and thus exposure to the radioactivity is not exacerbated) until necessary.

With a source of greater activity, a reduced time of exposure would be required to identify radioactive isotopes. In this research, an initial duration of 10 minutes was chosen to confirm the operational abilities of the system with a weak source. It is anticipated that isotopic identification will be possible at a much quicker rate, dependent on source activity and ROV distance. In the deployment, the level of activity is not expected to be a problem but, rather, the robustness of the system under high levels of exposure will need to be better understood.

IV. Future Work

It is desirable to remove the extra tether access point for transmitting the detector measurements to reduce the likelihood of entanglement and to reduce weight of the tether. Therefore, methods to integrate the signal into the existing tether are being investigated to reduce the complexity of the tether and also to reduce the possibility of entanglement.

It is also anticipated that a neutron detector will be integrated into the AVEXIS to provide data about the fuel debris within the Fukushima Daiichi pedestal; this is distinct from the 137Cs as the latter is understood to be dispersed widely throughout the water covering the core, the suppression chamber and surrounding facility, thus not being uniquely associated with the debris. A scCVD diamond detector has been acquired and this can be seen in Figure 11.

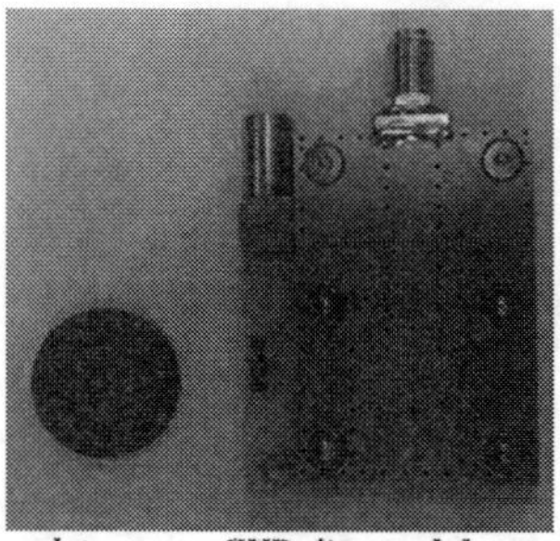

Figure 11 Neutron detector, scCVD diamond detector.

Before this device is installed it will be necessary to calibrate it using a high-energy source of alpha radioactivity. Testing will also need to be completed using a neutron source of significantly flux in order to be comparable with the Fukushima Daiichi environments.

A portable sonar will be added to the bottom of the ROV to aid in detection of debris, particularly suitable for in a murky environments in which the passage of light is impeded by suspended solids. The additional data provided by the sonar could be overlaid onto visual information gathered from the live video feed. A suitable sonar has been identified as the IMAGENEX 831L [8].

A number of options are being investigated to aid in the localisation of the ROV in an uncharted environment. Visual localisation can be achieved through a camera near the surface analysing the *x/y* coordinates of the ROV, whilst the depth sensor adds *z* data. Gamma spectrum analysis can then be overlaid onto this positional data.

Acoustic and RF options are being developed through the use of multiple transducers and receivers positioned along the length of the tether or boundaries of a tank. In water in which vision is impeded by suspended debris, the addition of sonar data will allow the build-up of a 3D point cloud, within which the ROV can localise itself.

V. Conclusion

Integration of a $CeBr_3$ gamma detector into an underwater submersible capable of characterizing the inside of the Fukushima Daiichi nuclear power plant has been achieved.

Experimental validation has been achieved through bench-top and submerged testing. Further characterization of the capabilities of this device will be possible with a stronger gamma source and multiple submerged sources.

Improvements to both the submersible and gamma spectroscopy functions will be conducted with the aim of overlaying positional data of gamma sources in relation to the internal dimensions of the PCV. This will aid in the effort to decommission the nuclear power plant and lead to targeted removal of fuel debris, increasing safety to workers and efficiency of decommissioning.

References

[1] IRID, "R & D on Robots for the Decommissioning of Fukushima Daiichi NPS International Research Institute for Nuclear Decommissioning," Feb. 2016.

[2] IAEA, "Events and highlights on the progress related to recovery operations at Fukushima Daiichi Nuclear Power Station," 2017.

[3] A. Griffiths, A. Dikarev, P. Green, B. Lennox, X. Poteau, and S. Watson, "AVEXIS-Aqua Vehicle Explorer for In-Situ Sensing," *IEEE Robotics and Automation Letters*, vol. 3766, pp. 282–287, Jan-2016.

[4] OpenROV, "OpenROV 2.8 Kit," 2017. [Online]. Available: https://www.openrov.com/products/openrov28/. [Accessed: 14-Apr-2017].

[5] TE connectivity, "MS5837-30BA Ultra Small Gel Filled Pressure Sensor," 2015.

[6] SCIONIX, "SCIONIX," 2017. [Online]. Available: http://scionix.nl/. [Accessed: 02-Jun-2017].

[7] H. I. Ltd., "Single Channel Mixed-Field Analyser," 2016. [Online]. Available: http://hybridinstruments.com/products/MFAX1.3.html. [Accessed: 15-Jun-2017].

[8] IMAGENEX, "IMAGENEX Model 831L Digital Pipe Profiling Sonar," 2011.

The CdZnTe Detector with Slit Collimator for Measure Distribution of the Specific Activity Radionuclide in the Ground

V.E. Stepanov, A.G. Volkovich, V.N. Potapov, I.A. Semin, A.V.Stepanov, Iu.N. Simirskii

National Research Center "Kurchatov Institute" - Moscow (Russian Federation)

Abstract - **From 2011 in the NRC "Kurchatov Institute" carry out the dismantling of the MR multiloop research reactor. Now the reactor and all technological equipment in the premises of the reactor were dismantled. Now the measurements of radioactive contamination in the reactor premises are made. The most contaminated parts of premises - floor and the ground beneath it. To measure the distribution of specific activity in the ground the CdZnTe detector (volume 500мм3) was used. Detector placed in a lead shielding with a slit collimation hole. The upper part of shielding is made movable to close and open the slit of the collimator. At each point two measurements carried out: with open and closed collimator. The software for determination specific activity of radionuclides in ground was developed. The mathematical model of spectrometric system based on the Monte-Carlo method. Measurements of specific activity of ground were made. Using the results of measurements the thickness of the removed layer of ground and the amount of radioactive waste were calculated.**

I. Introduction

From 2011, specialists of the NRC Kurchatov institute carry out the dismantling of MR and RFT research reactors. The RFT 20 MW water-graphite materials-research channel reactor, including the loop facilities, was commissioned in 1952. The reactor consisted of five loop setups for testing fuel assembly and structural materials. After intense operation for 10 years, it was stopped in 1962 and partially dismantled. In the same hall was built a more powerful multiple-loop reactor MP.

The MP reactor was a multipurpose material research nuclear reactor with 9 loops setups. These loops provided an opportunity to model the operating conditions for various types of fuel assemblies from different nuclear power plants, and were equipped with appropriate elements of contour equipment. A pond with reactor internals, a pond for spent fuel assemblies, technological equipment of reactor and loop setups, equipment of first and second cooling circuit pipelines are in the basement rooms of the reactor building. Operating condition of the reactor and its loop setups during the 30-year period of time increased radioactive contamination of equipment, reactor cooling pipelines system, and loops setups.

The reactor MR was finally shutdown in December 1992. In 1996 the fuel was removed from active zone ant reactor was transferred to a nuclear-safe state. In 2013-2014 all loops setup and equipment from 47 basement rooms were dismantled. Dismantling of equipment allowed to reduce the exposure dose rate (EDR) in premises from 5-20 mSv/h to 0.01-0.03 mSv/h. The dismantling of reactor internal constructions was started in 2014. In 2015 the vessel of the MR reactor and the graphite stack of the RFT reactor were dismantled [1-6].

In 2016 the measuring of contamination in basements rooms and removing the radioactive waste (soil, brick and concrete rubble) were begun. The most contaminated parts of the premises are the floor (concrete) and the ground under it. This radioactive contamination is highly inhomogeneous by the area of premises and by the depth of penetration into the ground.

The area under the reactor is about 1300 m^2. If we assume that the thickness of the contaminated layer is, for example, 0.2 m, the volume of radioactive waste will be ~ 260 m^3. Measurement of detailed distribution of specific activity of radionuclides in soil by the area and by depth will allow significantly reduce the amount of the radioactive waste. To measure the distribution of specific activity in ground the collimated spectrometric detector was developed.

II. Collimated Spectrometric Detector

The scheme of the spectrometric collimated detector (SCD) is shown in fig.1. To measure the distribution of specific activity in the soil, the CdZnTe detector (volume 500 mm^3) is used (3). The detector is placed in a lead shield (2.4) with a slit collimation hole. Lead shield is located inside the steel case of the device (1). The upper part of the lead shield is movable (5-rod for moving the upper part of the shield) and provide to close and open the collimator slit. Two measurements are carried out at each point: with an open and closed collimator. The energy resolution of the detector at the 662 keV line is 7.2 keV. The dimensions of the detector CdZnTe: ø24mm, L = 58mm.

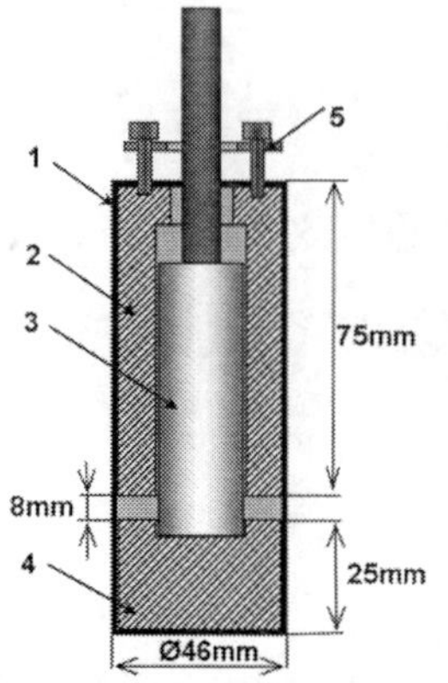

Fig.1. The scheme of collimated spectrometric detector (1-case; 2,4- lead shield; 3 – detector; 5 – movable rod).

SCD move along the hole in the ground with discrete steps and make two measurements in each point. After carrying out measurements, the values of the specific activity of a soil are calculated in specially developed software.

III. Method for Evaluated Specific Activity of Soil

We assume that a specific activity of ^{137}Cs (A_{Cs}) has a uniform distribution in a soil by the depth of the well. The counting rate in the peak of total absorption is formed only by the unscattered emission of ^{137}Cs. The count rate is related to the specific activity by a simple equation:

$$n_{Cs} = \frac{S\varepsilon n_\gamma \vartheta_{\max}}{2\mu(E_0)} A_{Cs} \qquad (1).$$

Where η_γ – is the yield of γ-quanta ^{137}Cs on one decay, S – the effective cross-sectional area of the detector crystal, ε – the detection efficiency of the unscattered emission of ^{137}Cs, $\vartheta_{\max}$ – the effective angle of the detector collimator, $\mu(E_0)$ – the mass attenuation coefficient in medium of the γ-emission with E_0 energy (in this case E_0=662 KeV).
From the equation (1) follows:

$$n_{Cs} = C \times A_{Cs} \qquad (2).$$

Here $C = \frac{S\varepsilon n_\gamma \vartheta_{\max}}{2\mu(E_0)}$ is a coefficient of proportionality.
The coefficient of proportionality is equal to the sensitivity of the detector and determined by calibrating the spectrometric collimated detector. Thus, from equation (2) follows, that the specific activity of ^{137}Cs can be determined by the formula:

$$A_{Cs} = n_{Cs} \times C^{-1}, \qquad (3)$$

C - is the calibration factor, which is equal to the counting rate in the peak of the total absorption of the γ- emission of ^{137}Cs in a medium with a unit specific activity concentration.

According to the equation (2), the calibration factor is universal, i.e. it practically does not depend on chemical composition of the environment, since the mass attenuation coefficient (soil or concrete) for γ-quanta energies from 0.5 to 2 MeV weakly depends on atom number of chemical elements in investigated substance.

The relative error of the measured activity will be determined by the relative error of the counting rate, η_{cs} i.e.

$$\delta_A = \frac{\Delta A_{Cs}}{A_{Cs}} = \frac{\Delta n_{Cs}}{n_{Cs}} \qquad (4).$$

The statistical error $\Delta\eta_{Cs}$, is determined by formula (2). The similar method and the same equations can be used for the other radionuclides specific activity evaluation.

The spatial resolution of the SCD depends on the angle of collimation $\vartheta_{\max}$. Increasing the spatial resolution of the detector (decreasing $\vartheta_{\max}$) leads to sensitivity decrease. Therefore, the optimal choice of the collimator parameters is a compromise between the sensitivity and the spatial resolution. The spatial apparatus feature for the developed spectrometric collimated detector with the geometric parameters shown in fig. 1 was calculated (fig.2).

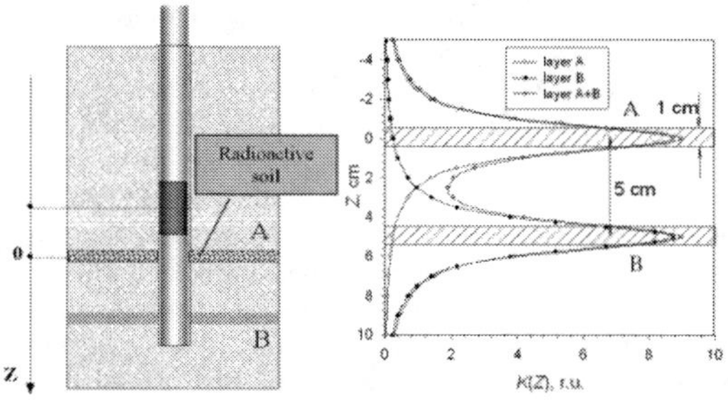

Fig. 2. The spatial hardware feature of the spectrometric collimated detector (A and B layers 1 cm thickness are located at a distance of 5 cm from each other, with the uniform distribution ^{137}Cs activities).

The spatial hardware feature of the spectrometric collimated detector is the dependence of the counting rate in the peak of total absorption on the position of the detector (coordinate Z) relative to the layer of soil (1 cm thickness) with the ^{137}Cs radionuclide activity uniform distributed in it. Two layers with 1 cm thickness (layer A and B) at a distance of 5 cm from each other are shown in left part of fig.2. Dependence of the counting rate on moving of the detector along the Z direction (perpendicular to the layers) is shown in right part of fig.2. From this figure it can be seen that the spatial resolution of the detector with such geometric parameters of the collimator makes it possible to detect two separate thin layers of contaminated soil. This indicates that the spatial resolution of such a spectrometric system is about 3 ÷ 4 cm.

IV. Collibration of Collimated Spectrometric Detector

To carry out the calibration, a point source with activity A_{Cs}^{point} was placed at a distance L from the detector and the counting rate at the peak of the total absorption n_{Cs}^{point} is determined. Then, using the mathematical model of this detector, two calculations are performed: the counting rate n_{Cs}^{calc} for a point source with the same activity and the counting rate $\widetilde{n}_{Cs}^{calc}$ for a uniformly distributed source in a medium with a specific activity of 1Bq/kg . The assumed count rate of the peak of the total absorption of the SCD for a uniform distributed source will be determined by the formula:

$$\widetilde{n}_{Cs} = \widetilde{n}_{Cs}^{calc} \cdot \frac{n_{Cs}^{point}}{n_{Cs}^{calc}} \qquad (5)$$

This count rate determines the calibration coefficient between the soil specific activity and the count rate in a peak of total absorption of the ^{137}Cs. The sensitivity of the detector in the peak total absorption at the 662 keV line (^{137}Cs) was calculated as 1.74 $KBq \cdot kg^{-1} \cdot s^{-1}$. The same method was used to determine calibration coefficients for other radionuclides. This procedure is described in more detail in work [7].

V. Measurments of Radioactive Soil contamination

Exposure dose rate in reactor basements rooms is mainly determined by radiation from the walls and floor. The main γ-emitting radionuclides are ^{137}Cs and ^{60}Co. Before the removal of contaminated soil from the reactor premises, a comprehensive radiation survey was carried out. This survey consisted of:

- EDR distribution measurements in all rooms;
- measurements of radioactive contamination of the surfaces by the β-detector;
- measurements of distribution of a specific activity in soil by the collimated detector [8];
- drilling of technological holes and measuring distributions of EDR in it;
- measurement of a specific activity of a soil along depth by means of spectrometric collimated detector.

The technological holes were drilled in is most radioactive contaminated areas of the rooms. The SCD was moving with a discrete step along the hole. Two measurements (with an open and closed collimator) were carried out at each point of the profile. The resultant spectrum (a difference between spectra with the opened and closed collimator) at a depth of 15 cm in one of the hole is shown in fig.3.

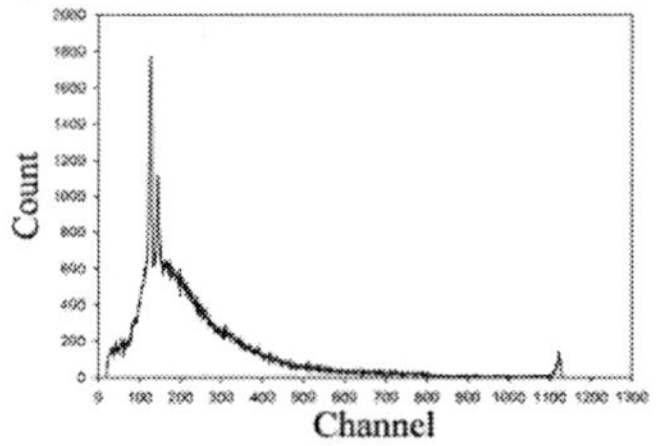

Fig.3. The resultant spectrum (a difference between spectra with the opened and closed collimator)

The distribution of the soil specific activity and the distribution of EDR along the depth of the technological hole in the room №64 are shown in fig.4.

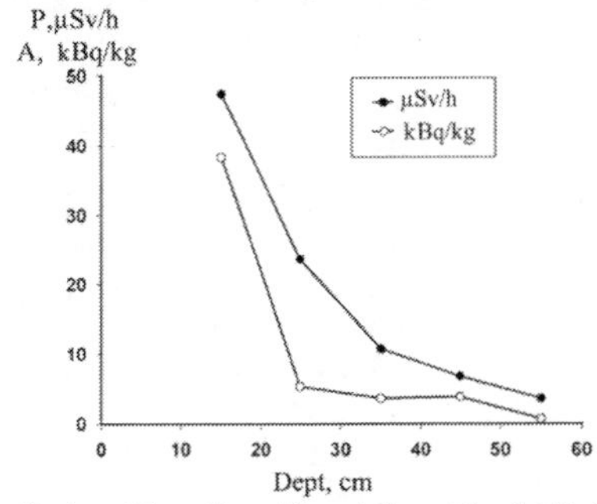

Fig.4. The distribution of the soil specific activity and the distribution of EDR along the depth of the technological hole.

The graph shows, that the change of EDR values along the depth of the hole is more smooth than the change the values of specific activity. This is explained by the fact that the contribution to the EDR at each point gives a greater amount of soil having different specific activity. Measurements made with the spectrometric collimated detector give more detailed information about the distribution of specific activity along the depth in the hole. The results of the measurements with SCD were compared with the results of laboratory measurements of soil samples. These measurements were estimated by the spectrometric complex InSpector-2000 of the Canberra Company that included a semiconductor detector using HP Germanium GC-4018. The analysis of the γ-spectrum was made by the GENIE-2000 software. The value of the specific activity of the soil sample taken from a depth of 15 cm from the hole in room № 64 was – 23 kBq/kg, the SCD measurement at the same depth was 18 kBq/kg. This result shows a good agreement, despite the fact that the measurements by sample method are much less representative compared to measurements by a SCD.

VI. Conclusion

The radiation survey of the reactor basement rooms showed that the radioactive contamination is highly non-uniformed by area and by the depth. Using SCD for measuring the distributions of specific activity in the soil by the depth in the reactor premises showed high efficiency. Based on this measurements the sorting of the soil according to specific activity and estimating the volume of removing ground were made. The most contaminated soils, in accordance with the regulations rules, were packed in containers for disposal. Part of the soil was sent to the soil cleaning unit [9], after which it returned to the backfill.

References

[1] V.G. Volkov, Y.A. Zverkov etc., "Preparation for the decommissioning of research reactor MR in the Kurchatov Institute", Atomic Energy, vol.104, № 5, pp. 259-264, 2008

[2] A.G. Volkovich, O.P. Ivanov, A.V. Lemus etc., "Particularities of the Dismantling of the Intra-Vessel Structures of the RFT Reactor", Atomic Energy, vol.121, № 5, pp. 377-382, 2017
[3] Alexander Chesnokov, Oleg Ivanov, Vyacheslav Kolyadin et al. "HLRW management during MR reactor decommissioning in NRC "Kurchanjv insninute", ICEM'13 Conference Proceedings - 96046, Brussels, Belgium, September 8 – 12, 2013.
[4] Stepanov V.E., Ivanov O.P. et al., "Remote-controlled collimated γ-ray detector for measuring radioactive contamination", Atomic energy, vol.109, №2, pp. 102-105,2010
[5] Ivanov O.P., Danilovich A.S., Stepanov V. E. et al., "Experience of application of new remote controlled instruments for scanning of distribution of radioactive contamination in rooms with high dose rate", ICEM'11 Conference Proceedings - 59354, Reims, France, September 25 – 29, 2011
[6] Alexey Danilovich, Oleg Ivanov, Alexey Lemus, Sergey Smirnov, Vyacheslav Stepanov et al., "Radiological survey of contaminated installations of research reactor before dismantling in high dose conditions with complex for remote measurements of radioactivity." WM2012 Conference Proceedings - 12069, Phoenix, Arizona, USA, February 26 – March 1, 2012
[7] Vyacheslav E.Stepanov, Victor N.Potapov, Sergey V.Smirnov, Oleg P.Ivanov. "Characterizations of the Radioactive Waste by the Remotely-Controlled Collimated Spectrometric System". ANIMMA2015, Book of Proceedings –IQ-45, Lisboa, Portugal, April 20 – 24, 2015.
[8] Alexey Stepanov,Oleg Ivanov, Victor Potapov et al. "Development and Application of Collimated Spectrometric Systems for the Characterization of Radioactive Contamination of Decommissioned Facilities", WM2015 Conference Proceedings - 15030, Phoenix, Arizona, USA, March 1, 2015
[9] Victor Volkov, Alexander Chesnokov, Alexey Danilovich et al., "Techniques of radioactive soil processing at rehabilitation of contamination territories", ICEM'11 Conference Proceedings - 59199, Reims, France, September 25 – 29, 2011

Evaluation of a scattering correction method for high energy tomography

David Tisseur, Navnina Bhatia, Nicolas Estre, Léonie Berge, Daniel Eck, Emmanuel Payan

Abstract*—One of the main drawbacks of Cone Beam Computed Tomography (CBCT) is the contribution of the scattered photons due to the object and the detector. Scattered photons are deflected from their original path after their interaction with the object. This additional contribution of the scattered photons results in increased measured intensities, since the scattered intensity simply adds to the transmitted intensity. This effect is seen as an overestimation in the measured intensity thus corresponding to an underestimation of absorption. This results in artifacts like cupping, shading, streaks etc. on the reconstructed images. Moreover, the scattered radiation provides a bias for the quantitative tomography reconstruction (for example atomic number and volumic mass measurement with dual-energy technique). The effect can be significant and difficult in the range of MeV energy using large objects due to higher Scatter to Primary Ratio (SPR). Additionally, the incident high energy photons which are scattered by the Compton effect are more forward directed and hence more likely to reach the detector. Moreover, for MeV energy range, the contribution of the photons produced by pair production and Bremsstrahlung process also becomes important. We propose an evaluation of a scattering correction technique based on the method named Scatter Kernel Superposition (SKS). The algorithm uses a continuously thickness-adapted kernels method. The analytical parameterizations of the scatter kernels are derived in terms of material thickness, to form continuously thickness-adapted kernel maps in order to correct the projections. This approach has proved to be efficient in producing better sampling of the kernels with respect to the object thickness. This technique offers applicability over a wide range of imaging conditions and gives users an additional advantage. Moreover, since no extra hardware is required by this approach, it forms a major advantage especially in those cases where experimental complexities must be avoided. This approach has been previously tested successfully in the energy range of 100 keV – 6 MeV. In this paper, the kernels are simulated using MCNP in order to take into account both photons and electronic processes in scattering radiation contribution. We present scatter correction results on a large object scanned with a 9 MeV linear accelerator.

***Index Terms*— Computed tomography, scatter correction, Scatter Kernel Superposition**

David Tisseur, Nicolas Estre, Léonie Berge, Daniel Eck and Emmanuel Payan are with the CEA Cadarache DEN/Nuclear Measurement Laboratory, F-13108 Saint-Paul lez Durance, France (e-mail: david.tisseur@cea.fr). Navnina Bhatia works at Imagopole – Citech Institut Pasteur, Paris, France and has performed her Ph.D at CEA List Saclay (2013-2016).

I. Introduction

One of the main drawbacks of MeV energy range CBCT is the scattering of photons inside the object and the detector. Scattered photons intensity is simply added to the primary photons intensity. This effect is seen in the back projection reconstruction algorithms as overestimated reconstructed linear attenuation thus corresponding to an underestimation of absorption. As a result, scatter artifacts can be seen on the reconstruction images like cupping, shading, streaks etc. In case of MeV energy range, forward Compton scattering of photons increases and hence there is a high probability of scattered photons reaching the detector. In addition, these scattered photons are more energetic and therefore more likely to escape from the object. Moreover, when the incident X-ray energy range is increased, the number of photons generated by the bremsstrahlung process also increases. Finally, for the X-ray photons with energies above 1.022 MeV, pair production produces additional 511 keV photons that can contribute to an overall increase in the scatter contribution. Many scatter correction methods for MeV source are listed in the literature. For example, Maltz et al. [1] have used beam stop arrays for scatter correction. Such methods increase the X-ray exposure due to more than one scan per projection and extended scanning time. The classical convolution approaches based on scatter kernel superposition (SKS) method [2] [3] [4] use deconvolution method to calculate the scatter intensity and are based on a discrete set of thickness-dependent kernels. In these techniques, for a range of thickness only one kernel is used. These methods give satisfactory results in many applications. However, these methods are not very efficient in higher energy range due to higher SPR. In this article, we have applied scatter correction by continuously thickness adapted Scatter Kernel Superposition (SKS) method [5] on a data produced by 9 MeV X-ray photon beam generated from a linear accelerator.

II. Method and materials

A. Scatter correction using a continuous SKS approach

In the SKS scatter correction approach based on [5], scatter signal is modeled as the sum of the scatter contributions from

 https://doi.org/10.1051/epjconf/201817006006

a group of pencil beams passing through the object and the detector. The total scatter signal $S(m, n)$ with m and n as the pixel position on the detector, can then be modeled as:

$$S(m,n) = \sum_k \sum_l P(k,l) K_{T(k,l)}(m-k, n-l) \quad (1)$$

Where, P is the primary signal contributed by the photons passing through the object without any attenuation. K_T is the transmission (T) dependent kernel, with amplitude equal to the ratio of the scattered signal at the current pixel to the primary signal P. T is computed with Beer Lambert law:

$$T(k,l) = \frac{P(k,l)}{O(k,l)} \quad (2)$$

Where, O is the full beam intensity (signal without the sample). T is dependent on the attenuation thickness with respect to the classical Beer Lambert law. K_T is non-linearly fitted by an equation formed by a circularly symmetric Lorentzian function describing the shape of the kernel:

$$K_T(m-k, n-l) = \frac{A}{1 + B\,((m-k)^2 + (n-l)^2)} \quad (3)$$

Where parameters A, B are function of transmission T.

B. Kernel generation with MNCP6

In order to generate the kernels, we performed Monte Carlo (MC) simulations with MCNP6 [6]. Fig. 1 presents MC simulation setup and an example of the 2D kernel. Simulated geometry was kept same as the acquisition set up. Pencil beam source corresponding to a 9 MeV bremsstrahlung spectrum was impinged on slabs of the same material as the object under study. Afterwards, a discrete set of point spread 2D kernels was obtained on the detector.

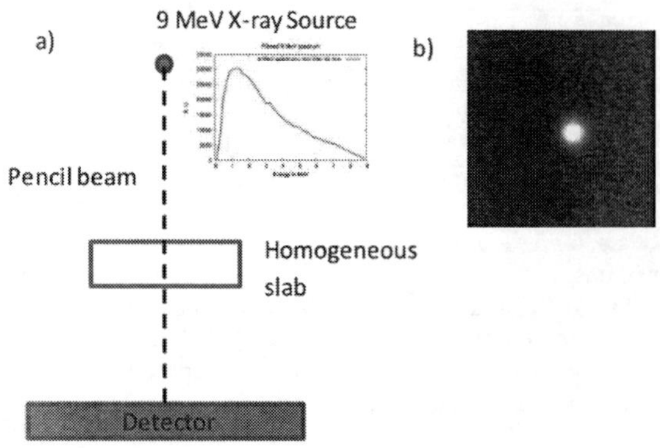

Fig. 1. a) MC kernel simulation setup using pencil beam b) generated 2D kernel

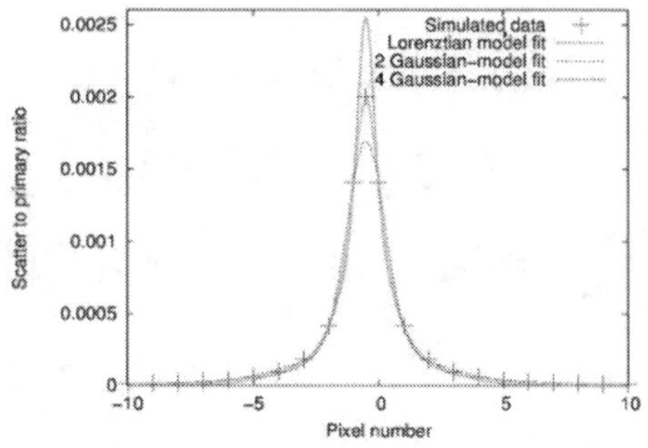

Fig. 2. Simulated and fitted kernel for bitumen slab with a transmission of 0.8.

Eventually, for each kernel, we extracted a radial line profile from the 2D generated kernel image. Then, equation 3 was fit on these 1D kernels using non-linear least square fitting (see Fig. 2 and 3). The values for parameters A and B were calculated for these discrete sets of kernels. Thereupon, in order to obtain the continuous kernel, we analytically computed the expression for the kernels parameters A and B in terms of the transmission of the object [5]. For this case, we fitted a curve with respect to the transmission using a classical non linear least square fitting technique for each parameter. Fig. 3 presents the curve fitted on the

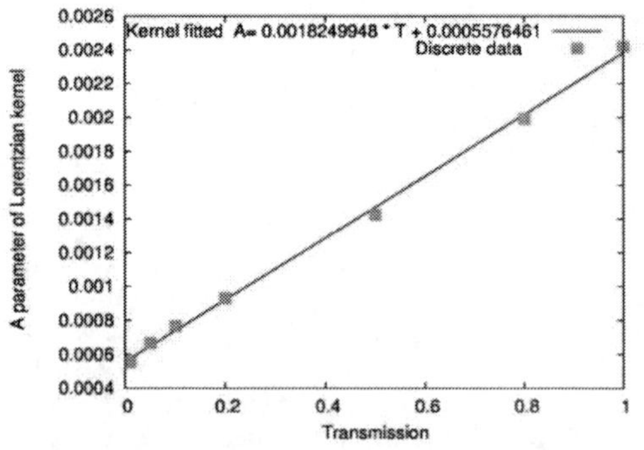

Fig. 3. Curve fitted on the discrete values of parameter A versus transmission

discrete values of parameter A versus transmission.

In a previous work [5], the SKS method has been tested with a 2 and 4 Gaussian-model. Fig.2 shows that with our 9 MeV spectrum the best is obtained with a Lorentzian model. Table II shows a comparison between root mean square error (RMSE) between simulated data and Lorentzian, 2 Gaussian and 4 Gaussian-model fit versus the transmission of the bitumen slab. Lorentzian model gives a better fit along all the 7 kernels. With a high energy spectrum (9 MeV), the Compton effect is less preponderant as compared to pair production effect and Bremsstrahlung effect of the secondary electrons. This could explain the reason of the best fit obtained with Lorentzian model instead of 2-4 Gaussian model.

TABLE I
ROOT-MEAN-SQUARE ERROR (RMSE) BETWEEN THE SIMULATED KERNEL DATA AND LORENTZIAN, 2 GAUSSIAN AND 4 GAUSSIAN-MODEL VERSUS TRANSMISSION

TRANSMISSION	LORENTZIAN	2 GAUSSIAN	4 GAUSSIAN
1	4.06E-06	4.92E-06	4.79E-06
0.8	5.23E-06	6.31E-06	5.94E-06
0.5	5.48E-06	8.65E-06	9.10E-06
0.2	5.31E-06	1.14E-05	1.10E-05
0.1	5.17E-06	1.23E-05	1.19E-05
0.05	2.79E-06	3.78E-06	3.70E-06
0.01	2.64E-06	4.28E-06	4.14E-06

C. Iterative Scatter Correction

We applied an iterative scatter correction scheme described in Fig. 4. It consists of 5 steps:

1) The experimental projection was initialized as the first estimate of the primary *P*.

2) The equivalent transmission was computed for each pixel using eq. (2).

3) The convolution was performed by calculating the suitable kernels for the respective transmission using eq. (3)

4) For each pixel, scatter contribution is estimated using eq. (1).

5) The primary *P* for iteration $i+1$ was updated using eq. (4)

$$P_{i+1}(m,n) = P_0(m,n) \times \frac{P_i(m,n)}{P_i(m,n) + S_i(m,n)} \quad (4)$$

Steps 1 to 5 are repeated until convergence is achieved.

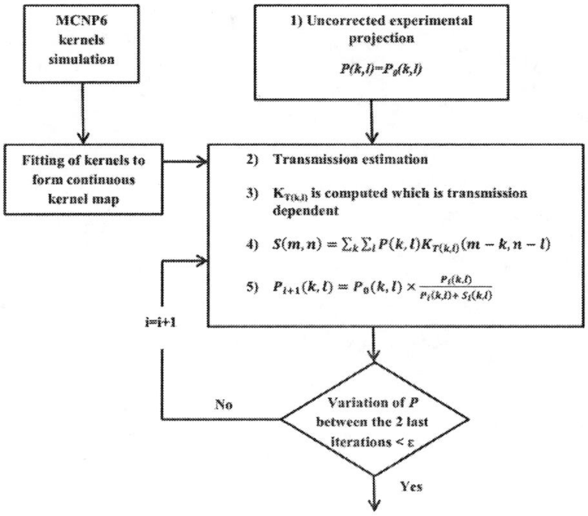

Fig. 4. Scatter correction synopsis

D.Acquisition set up

The acquisitions were performed on a Mini-Linatron VARIAN linear accelerator 9 MeV filtered by 20 mm steel. The source to detector distance was 3.5 m and the source to object distance was 2.85 m. The acquisitions were performed on a cylinder made of bitumen. The detector used was a GADOX scintillator + sCMOS detector. Each of the 1200 projections over 360° has a size of 80 x 60 cm^2 with a pixel size of 334 μm (see Fig. 5).

Fig. 5. 2D scintillator detector. The scintillator screen is imaged by a low noise camera through a 45° tilted mirror. The screen size is 80 x 60 cm^2.

III. CONCLUSIONS

Scatter kernel were simulated accurately with MCNP6 to take into account both photonic and electronic processes in the considered energy range. A comparison study between a two Gaussian and a Lorentzian-model on scatter kernel fit shows best results a single Lorentzian equation. This could be explained by less preponderant influence of Compton scattering and dominant contribution of pair production and Bremsstrahlung effect. In the perspective, we will validate this on the real data.

REFERENCES

[1] J. S. Maltz, B Gangadharan, M. Vidal, A. Paidi, S. Bose, B. A. Faddego, M. Aubi, O. Morin, J. Pouliot, Z. Zheng, M. M. Svatos, and A. R. Bani-Hashemi. "Focused beam-stop array for the measurement of scatter in megavoltage portal and cone beam CT imaging.", Med Phys, 35(6):2452–2462, 2008.

[2] S. F. Petit, W. Elmpt, S. Nijsten, A. Dekker, and P. Lambin. "Correction of cupping artifacts in megavoltage cone beam computed tomography", Department of Radiation Oncology (MAASTRO), Masters thesis.

[3] M. Sun and J. M. Star-Lack. "Improved scatter correction using adaptive scatter kernel superposition", Phys Med. Biol, 55(22):6675 – 6720, 2010.

[4] V. N. Hansen, W. Swindell, and P. M. Evans. "Extraction of primary signal from EPIDs using only forward convolution". Med. Phys., 24(49):1477–1484, 1997.

[5] N. Bhatia, D. Tisseur, F. Buyens, and J. M. Létang, "Scattering correction using continuously thickness-adapted kernels", NDT & E, 78:52–60, 2016.

[6] J.T. Goorley, et al., "Initial MCNP6 Release Overview – MCNP6

Version 1.0", Los Alamos National Laboratory Report, LA-UR-13-22934 (2013).
[7] S. Rit , M. Vila Oliva, S. Brousmiche, R. Labarbe, D. Sarrut, G.C. Sharp, "The Reconstruction Toolkit (RTK), an open-source cone-beam CT reconstruction toolkit based on the Insight Toolkit (ITK)" Journal of Physics: Conference Series, 489, 012079. doi:10.1088/1742-6596/489/1/012079 (2014)

https://doi.org/10.1051/epjconf/201817007001

Modeling Photo-multiplier Gain and Regenerating Pulse Height Data for Application Development

Michael D. Aspinall, *Member, IEEE,* Ashley R. Jones

***Abstract*—Systems that adopt organic scintillation detector arrays often require a calibration process prior to the intended measurement campaign to correct for significant performance variances between detectors within the array. These differences exist because of low tolerances associated with photo-multiplier tube technology and environmental influences. Differences in detector response can be corrected for by adjusting the supplied photo-multiplier tube voltage to control its gain and the effect that this has on the pulse height spectra from a gamma-only calibration source with a defined photo-peak. Automated methods that analyze these spectra and adjust the photo-multiplier tube bias accordingly are emerging for hardware that integrate acquisition electronics and high voltage control. However, development of such algorithms require access to the hardware, multiple detectors and calibration source for prolonged periods, all with associated constraints and risks. In this work, we report on a software function and related models developed to rescale and regenerate pulse height data acquired from a single scintillation detector. Such a function could be used to generate significant and varied pulse height data that can be used to integration-test algorithms that are capable of automatically response matching multiple detectors using pulse height spectra analysis. Furthermore, a function of this sort removes the dependence on multiple detectors, digital analyzers and calibration source. Results show a good match between the real and regenerated pulse height data. The function has also been used successfully to develop auto-calibration algorithms.**

***Index Terms*—Auto-calibration, Detector response matching, Organic scintillation detectors, Pulse height modeling, Response modeling, Synthetic pulse height data production.**

I. Introduction

MANY application systems adopt multiple organic scintillation detectors for fast neutron assay [1-11]. Detector arrays of this sort can consist of tens of detectors that rely on photo-multiplier tube (PMT) technology for scintillation light capture, voltage conversion and signal amplification. Inconsistencies across photo-multiplier gains, even of the same model range, result in discrepancies between detector response functions with a fixed high-voltage supply.

Response matching scintillation detectors by aligning the pulse height spectra (PHS) obtained from a gamma source with a well-defined photo-peak, such as cesium-137, by means of adjusting the bias applied to the PMT is common practice [12]. Recent advances in a real-time embedded control system for fast-neutron scintillation detectors include a procedure that automatically performs detector calibration using this principle [13].

Prior works have modeled and synthesized pulses from organic scintillation detectors in response to neutrons and gammas [14], but do not provide a simulation for pulse height distributions derived from these detectors. Others have more comprehensively simulated response functions and pulse shape discrimination (PSD) for organic scintillation detectors with Geant4 [15], demonstrating excellent agreement, among others, between experimental and simulated energy spectra for a detector exposed to 0.662 MeV gammas from Cs-137. The ability to simulate a detector of varying complexity and configuration within a single Monte Carlo code is impressive but indeed complex and does not encompass features derived from the processing electronics such as pre-amplification, pulse shaping and digitization.

This paper reports on a method for modeling and regenerating pulse height data for testing algorithms for automatically calibrating scintillation detectors. It is proposed that such regenerated data can shorten development by facilitating system testing when access to multiple detectors, digital acquisition electronics or calibration sources are limited.

Pulse height data acquired from an EJ-309 liquid scintillator (supplied by Scionix, The Netherlands) operated at several different high voltage biases and exposed to a Cs-137 source are presented using a multi-channel analyzer (MCA). The MCA channel containing the mode counts in the Compton-edge region is used as a measure of the gain of the detector's PMT and is compared to the supply bias. The relationship between the mode counts in the Compton-edge region on the PHS and the related supply bias is modeled. Different detector PMTs with varying gain parameters have also been modeled. These models have been verified against real data and have been used to regenerate pulse height data that have assisted in the development and validation of several different auto-calibration algorithms.

Date submitted for review. This work was supported in part by Hybrid Instruments Ltd., UK.
M. D. Aspinall is with the Engineering Department at Lancaster University, Lancaster, LA1 4YW, UK (e-mail: m.d.aspinall@lancaster.ac.uk). He was with Aston University, Birmingham, B4 7ET, UK during which the reported work was predominantly developed. He is also the systems engineer responsible for the development of the Hybrid Instruments Ltd. Mixed Field Analyzer suite.

II. Experimental setup

Pulse height data were obtained using a Hybrid Instruments Ltd. Mixed Field Analyzer (model MFAx4.3, serial # 4322) in MCA mode [16]. The MFAx4.3 samples the analogue input signal at 500 MSa/s with 12-bit amplitude resolution. The analogue-to-digital converter (ADC) is operated in bipolar mode, hence zero-baseline is represented as 2047 and the maximum sample amplitude as 4095. The variable gain amplifier (VGA) of the MFAx4.3 was configured with the gain set to low (10 dB) and the attenuation ladder set to -12 dB, providing a net gain of 6.84 dB. The trigger threshold was operated in greater-than mode and set to 100, therefore only capturing pulse height data exceeding 2147 (derived from 2047 zero-baseline + 100 trigger threshold).

Pulses were baseline corrected in real-time by the MFAx4.3. After baseline correction, 0 V is represented as 200 by the 12-bit ADC and the maximum 430 mV signal amplitude (governed by the VGA settings described above) is represented as 2247.

With the MFAx4.3 operating in MCA mode the pulse sample with the largest amplitude, i.e. the pulse height is sent to the host computer via an Ethernet connection and written to a TXT-file. A single Ethernet packet contains 100 pulse heights to increase data transfer efficiency.

Data from an EJ-309 detector (serial # SBM842) exposed to a Cs-137 source were recorded for PMT biases ranging from −1800 V to −1600 V in 50 V increments. The source had an activity of 333 kBq and was positioned on-axis and 50 mm from the front face of the detector. For each PMT bias 100 000 pulse heights were recorded and saved to the TXT-file. These data were imported into MATLAB®, reformatted to represent the pulse heights obtained from each PMT bias and saved as a MAT-file for convenience.

III. Implementation

The developed MATLAB® function utilizes two models to regenerate pulse height data by rescaling real data. The first model replicates the variations in response between different detector-PMTs of the same model equation (2), with predefined values used to model ten different detector-PMTs. The second, models the detector's response with relation to the pulse height spectra (PHS), i.e. the location of the Compton-edge on a Cs-137 spectrum, for different applied PMT biases (5).

The function takes three inputs. The first allows the user to specify the magnitude of the negative high voltage bias, 'hv', applied to the detector's PMT. 'hv' is in positive integer format ranging from 0 to 2000. The second input allows the user to select one of the ten predefined detectors, 'dnum', that correspond to a scalar value D, and is in integer format ranging from 1 to 10. The final user input is used to specify how many pulse height data are generated, 'qty'. 'qty' is in positive integer format. A forced delay has been implemented that is directly proportional to 'qty', this replicates real data collection, i.e. more data improves statistics but takes longer to collect. The delay can be bypassed to speed up development time.

The function outputs pulse heights ranging from 0 to 2047 at roughly the quantity specified by 'qty'. The quantity of pulse heights produced will be slightly different than that specified by 'qty' because of a normally distributed pseudorandom error applied to the regenerated data.

Regenerated pulse height data are produced by the function using the follows steps.

1) Calculate the mode pulse height in the Compton-edge region, M, using (2) for the specified high voltage, 'hv', and detector, 'dnum'. Calculate the bin width, W, necessary to scale the data to the mode pulse height, M, using (5).

2) Read-in 100 000 prerecorded pulse heights and subtract the minimum recorded value to shift the spectrum left to the zero-origin. The data acquired with the EJ-309 detector operated at −1750 V were used in this instance.

3) Partition the data into histogram bins according to the calculated bin width, W. Replace any zero-count bins with the mean count of the two neighbouring bins.

4) Filter the bin counts with a 200-bin moving average filter.

5) Resize the number of bins to 2047 by either padding with zeros if less than 2047 or by deleting the right-most bins if greater than 2047.

6) Normalize the data; bin counts are divided by the total number of counts.

7) Scale the normalized bin counts to the specified number of pulse heights determined by 'qty' and then apply a normally distributed pseudorandom error.

8) Generate pulse heights in the calculated quantities and randomize the order of pulse amplitudes.

9) Delay release of data directly proportional to 'qty'.

IV. Results and discussion

The first model used by the function represents the variations in response between different detector-PMTs of the same type. The model was produced by plotting the PHS acquired from the EJ-309 detector operated at −1800 V, −1750 V, −1700 V, −1650 V and −1600 V whilst exposed to the 333 kBq Cs-137 source, as shown by Fig. 1. The bin counts of the spectra were smoothed using a 100-bin moving average filter.

The mode pulse height in the Compton-edge region of the spectra shown in Fig. 1 were plotted against the applied PMT bias to produce Fig. 2. Using a curve fitting tool these data were fit with an exponential trend (1) which was generalized (2) to serve as a model for the relationship between the mode pulse height, M, and the magnitude of the applied PMT bias, v.

$$M = 0.5424 \cdot e^{0.0045 \cdot v} \quad (1)$$

$$M = D \cdot e^{0.0045 \cdot v} \quad (2)$$

Where D is a scalar used to represent different detectors. Ten detector models were predefined with D values ranging from 0.52 to 0.80.

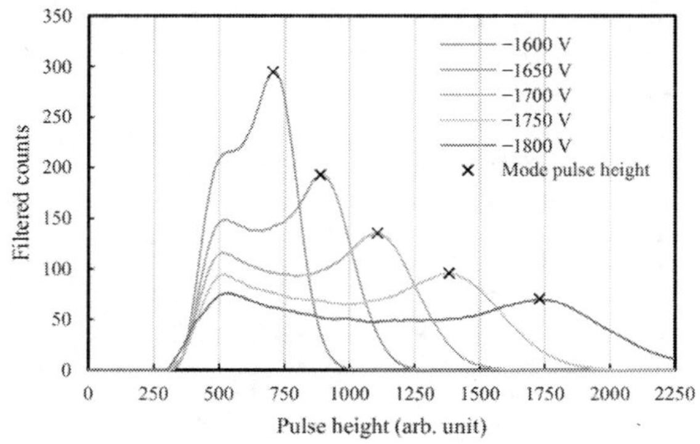

Fig. 1. Pulse height spectra (PHS) acquired from the EJ-309 detector operated at −1800 V, −1750 V, −1700 V, −1650 V and −1600 V whilst exposed to the 333 kBq Cs-137 source. The bin counts of the spectra are filtered using a 100-bin moving average filter.

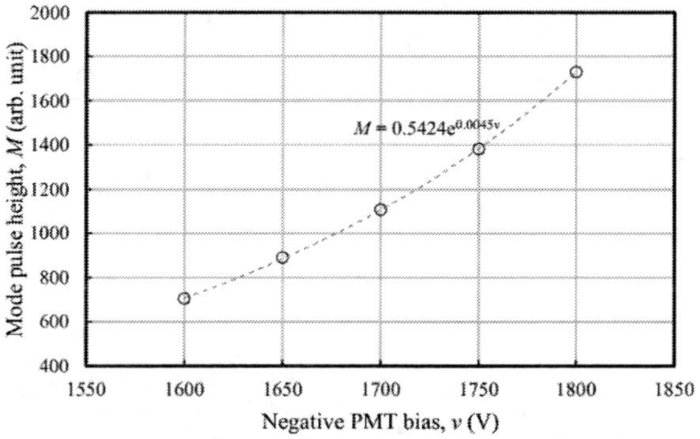

Fig. 2. Nonlinear relationship between mode pulse height, *M*, in the Compton-edge region and the applied PMT bias, *v*. These data were derived from pulse heights acquired from an EJ-309 liquid scintillator exposed to a Cs-137 source and operated at biases ranging from −1800 V to −1600 V in 50 V increments. About 100 000 pulse heights were recorded for each bias.

Fig. 3 was produced by first loading 100 000 prerecorded pulse heights into MATLAB®. These data were acquired from the EJ-309 detector operated at −1750 V whilst exposed to the Cs-137 source. Any events that exceeded a pulse height of 2245 were removed and then the minimum recorded pulse height was subtracted from all pulse height in the data set to shift the spectrum left to the zero-origin. These data were then partitioned into bins and the count in each bin determined using the MATLAB® *histcounts* function. The number of bins were specified by the scalar, *nbins*, expressed by script (3), where W is the bin width.

$$nbins = (0 : W : 2046); \tag{3}$$

Following this partitioning process any bins with zero counts were replaced with the average count of the two neighboring bins. Zero-count bins are more prevalent when the histogram is stretched i.e. when W is less than 1. The bin-counts were then filtered with a 200-bin moving average filter. The number of bins were then resized to 2047 by either padding with zero-count bins (if less than 2047 bins) or by deleting the right-most bins (if greater than 2047 bins). When W is greater than 1 the number of bins produced is less than 2047 and when W is less than 1 the number of bins produced is greater than 2047. The mode pulse height, M, was then recorded for specified bin widths, W, ranging from 0.6 to 1.7 in 0.1 increments (Fig. 3). Using a curve fitting tool a power trend was realized (4).

$$M = 1121.58 \cdot W^{-1.003} \tag{4}$$

Rearranged to make W the subject produces the model used to determine the bin width, W, required to scale the mode pulse height of the real data to the specified pulse height, M (5).

$$W = 1098.2 \cdot M^{-0.997} \tag{5}$$

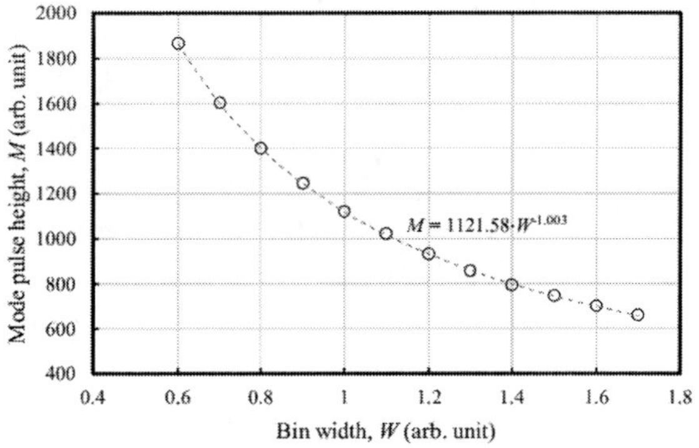

Fig. 3. Nonlinear relationship between mode pulse height, *M*, in the Compton-edge region and the histogram bin width, *W*. These data were derived using different bin widths (0.6-1.7) during the histogram partitioning process and observing the scaling effect this had on the mode pulse height.

Fig. 4 shows the unfiltered PHS employed by the scaling function. These data were acquired from the EJ-309 detector operated at −1750 V whilst exposed to the 333 kBq Cs-137 source. For comparison, Fig. 5 shows the PHS for three modeled detectors produced by the function described in section III and using the real pulse height data shown in Fig. 4. An analysis of variance (ANOVA) was used to compare the real pulse height data (Fig. 4) and the regenerated pulse height data for Detector 2 (Fig. 5). The F ratio (1.71) is smaller than the F *crit* value (3.84) indicating no statistically significant difference and a good resemblance between the real pulse height data and the regenerated pulse height data for the modeled detectors. The effects of the applied normally distributed pseudorandom error are of interest with respect to the likeness to the real data and the challenges that statistical uncertainties present for auto-calibration algorithms. For instance, in this work a moving average filter was used to increase the validity of the measured mode pulse height given the continuum nature of the spectra, however alternative methods could be adopted. For the data shown in Fig. 4 and Fig. 5, both the real detector and the modeled detectors were operated at −1750 V. The D value, see (2), for detector 1 was 0.778, detector 2 was 0.654 and detector 3 was 0.542.

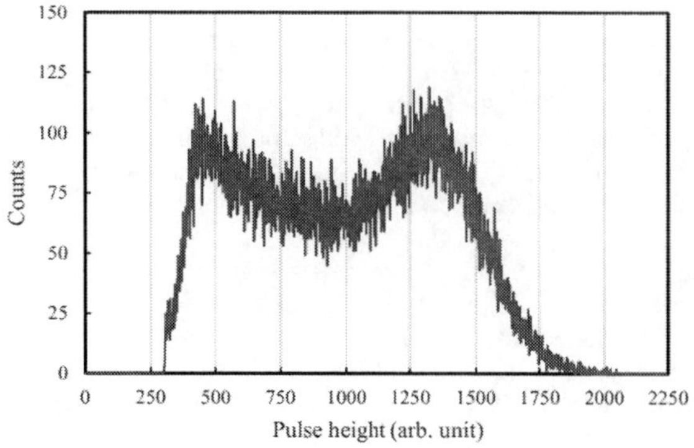

Fig. 4. Unfiltered pulse height spectra (PHS) utilized by the pulse height scaling function. These data were acquired from the EJ-309 detector operated at −1750 V whilst exposed to the 333 kBq Cs-137 source.

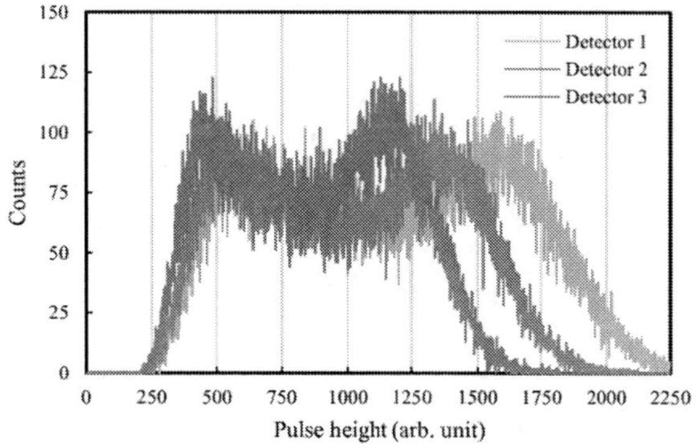

Fig. 5. Pulse height spectra (PHS) produced for three modeled detectors using the pulse height scaling function described in section III and the data shown in Fig. 4. Each spectrum contains approximately 100 000 pulse heights and each detector was operated at −1750 V.

V. Conclusion

These scaled and regenerated pulse height data have proved an acceptable means for validating new auto-calibration algorithms. The reported method has shortened development time for techniques utilizing spectra analysis and has removed the dependence on and access to multiple detectors, digital acquisition electronics and calibration source with associated risks and constraints.

Moreover, an undergraduate Software Engineering project utilized the described function with a brief to design an auto-calibration algorithm for aligning the mode pulse height to a specified pulse height for several model detectors by determining the detector specific PMT bias. Several different auto-calibration algorithms were conceived and extensively validated without any need for the students to access physical detectors, digital analyzers and calibration source.

Further studies are focusing on methods that are capable of automatically determining PSD parameters, including discrimination thresholds, to enable reliable and repeatable neutron/gamma discrimination. Eventually, advances of this sort will enable digital analyzers that have embedded high-voltage supplies that can optimize their entire configuration automatically to suit a range of scintillation detectors. Clearly, developments of this sort will require diverse data sets of the entire digital waveform acquired from a range of scintillation detectors and mixed-field sources or models that can adequately generate equivalent synthetic data.

Acknowledgment

The author thanks the 2016/17 undergraduate first-year students studying Software Engineering at Aston University, UK for application testing the software algorithm described and for enduring a substantially challenging coursework assessment relating to and utilizing scaled and regenerated pulse height data for application development.

References

[1] D. Henzlova, R. Kouzes, R. McElroy, P. Peerani, M. Aspinall, K. Baird, A. Bakel, A. Borella, M. Bourne, L. Bourva, F. Cave, R. Chandra, D. Chernikova, S. Croft, G. Dermody, A. Dougan, J. Ely, E. Fanchini, P. Finocchiaro, V. Gavron, M. Kureta, K. Ianakiev, K. Ishiyama, T. Lee, C. Martin, K. M. Kinny, H. Menlove, C. Orton, A. Pappalardo, B. Pedersen, D. Peranteau, R. Plenteda, S. Pozzi, M. Schear, M. Seya, E. Siciliano, S. Stave, L. Sun, M. Swinhoe, H. Tagziria, J. Takamine, S. Vaccaro, A. L. Weber, T. Yamaguchi, H. Zhu, "Current status of helium-3 alternative technologies for nuclear safeguards," prepared for *National Nuclear Security Administration, U.S. Department of Energy, and The Euratom Atomic Energy Community (Euratom)*, Dec. 2015. DOI: 10.2172/1227248

[2] A. Enqvist, K. J. Weinfurther, M. Flaska and S. A. Pozzi, "Characterization of a mixed multiplicity counter based on liquid organic scintillators," *IEEE Trans. Nucl. Sci.*, vol. 58, pp. 2413–2420, Oct. 2011. DOI: 10.1109/TNS.2011.2164554.

[3] J. L. Dolan, E. C. Miller, A. C. Kaplan, A. Enqvist, M. Flaska, A. Tomanin, P. Peerani, D. L. Chichester and S. A. Pozzi, "Passive measurement of organic-scintillator neutron signatures for nuclear safeguards applications," in *IEEE Nucl. Sci. Symp. and Med. Imag. Conf. (NSS/MIC)*, Anaheim, CA, USA, 2012, pp. 203–206. DOI: 10.1109/NSSMIC.2012.6551093.

[4] M. J. Joyce, K. A. A. Gamage, M. D. Aspinall, F. D. Cave and A. D. Lavietes, "A 4-channel multiplex analyzer for real-time, parallel processing of fast scintillators," in *IEEE Nucl. Sci. Symp. and Med. Imag. Conf. (NSS/MIC)*, Anaheim, CA, USA, 2012, pp. 134–138. DOI: 10.1109/NSSMIC.2012.6551077.

[5] J. K. Polack, A. Poitrasson-Riviere, M. C. Hamel, K. Ito, S. D. Clarke, M. Flaska, S. A. Pozzi, A. Tomanin and P. Peerani, "Image reconstruction of shielded mixed-oxide fuel using a dual-particle imaging system," in *IEEE Nucl. Sci. Symp. and Med. Imag. Conf. (NSS/MIC)*, Seattle, WA, USA, 2014, pp. 1–4. DOI: 10.1109/NSSMIC.2014.7431091.

[6] M. J. Joyce, K. A. A. Gamage, M. D. Aspinall, F. D. Cave and A. D. Lavietes, "Real-time, fast neutron coincidence assay of plutonium with a 4-channel multiplexed analyzer and organic scintillators," *IEEE Trans. Nucl. Sci.*, vol. 61, no. 3, pp. 1340–1348, Jun. 2016.

[7] M. J. Joyce, J. Adamczyk, M. D. Aspinall, F. D. Cave and R. Plenteda, "Real-time, fast neutron detection for stimulated safeguards assay," in *Proc. of 37th ESARDA Symp. on Safeguards and Nucl. Non-Proliferation*, Manchester, UK, 2015, pp. 54–63.

[8] H. M. O. Parker, M. D. Aspinall, A. Couture, F. D. Cave, C. Orr, B. Swinson and M. J. Joyce, "Active fast neutron singles assay of 235U enrichment in small samples of triuranium octoxide," *Progress in Nucl. Energy.*, vol. 93, pp. 59–66, Jul. 2016. DOI: 10.1016/j.pnucene.2016.07.022.

[9] R. Sarwar, M. J. Joyce and C. H. Zimmerman, "Real-time fast-neutron plutonium assay for storage and ageing applications," in *2nd Annual Meeting DISTINCTIVE University Consortium*, Bristol, UK, 2016.

[10] A. Poitrasson-Riviere, M. Flaska, M. C. Hamel, J. K. Polack, M. F. Becchetti, B. M. Wieger, A. Enqvist, S. D. Clarke and S. A. Pozzi, "Digital data acquisition and processing for a neutron-gamma-ray imaging system," in *Nucl. Sci. Symp. and Med. Imag. Conf. (NSS/MIC)*,

Anaheim, CA, USA, 2012, pp. 1597–1599. DOI: 10.1109/NSSMIC.2012.6551381.

[11] M. J. Joyce, S. Agar, M. D. Aspinall, J. S. Beaumont, E. Colley, M. Colling, J. Dykes, P. Kardasopoulos and K. Mitton, "Fast neutron tomography with real-time pulse-shape discrimination in organic scintillation detectors," *Nucl. Instr. Methods Phys. Res. A*, vol. 834, pp. 36–45, Jul. 2016. DOI: 10.1016/j.nima.2016.07.044.

[12] G. R. Gilmore, "Spectrometer Calibration," in *Practical Gamma-ray Spectrometry*, 2nd ed. Chichester, England: Wiley, 2008, ch. 7, pp. 143–147.

[13] M. D. Aspinall, M. J. Joyce, F. D. Cave, R. Plenteda and A. Tomanin, "Automated response matching for organic scintillation detector arrays," *Nucl. Instr. Methods Phys. Res. A.*, vol. 861, pp. 46–54, Jul. 2017. DOI: 10.1016/j.nima.2017.05.004.

[14] S. Marrone, D. Cano-Ott, N. Colonna, C. Domingo, F. Gramegna, E. Gonzalez, F. Gunsing, M. Heil, F. Käppeler, P. Mastinu, P. M. Milazzo, T. Papaevangelou, P. Pavlopoulos, R. Plag, R. Reifarth, G. Tagliente, J. L. Tain and K. Wisshak, "Pulse shape analysis of liquid scintillators for neutron studies," *Nucl. Instr. Methods Phys. Res. A*, vol. 490, no. 1–2, pp. 299–307, Sep. 2002. DOI: 10.1016/S0168-9002(02)01063-X.

[15] Z. S. Hartwig and P. Gumplinger, "Simulating response functions and pulse shape discrimination for organic scintillation detectors with Geant4," *Nucl. Instr. Methods* Phys. Res. A, vol. 737, pp. 155–162, Feb. 2014. DOI: 10.1016/j.nima.2013.11.027.

[16] M. D. Aspinall, M. Joyce, A. Lavietes, R. Plenteda, F. Cave, H. Parker, A. Jones and V. Astromskas, "Real-time capabilities of a digital analyzer for mixed-field assay using scintillation detectors," *IEEE Trans. Nucl. Sci.*, vol. 64, no. 3, pp. 945–950, Mar. 2017. DOI: 10.1109/TNS.2017.2654512.

Pulse Shape Analysis and Discrimination for Silicon-Photomultipliers in Helium-4 Gas Scintillation Neutron Detector

Cathleen Barker, Ting Zhu, Lucas Rolison, Scott Kiff, Kelly Jordan, Andreas Enqvist

***Abstract*— Using natural helium (helium-4), the Arktis 180-bar pressurized gas scintillator is capable of detecting and distinguishing fast neutrons and gammas. The detector has a unique design of three optically separated segments in which 12 silicon-photomultiplier (SiPM) pairs are positioned equilaterally across the detector to allow for them to be fully immersed in the helium-4 gas volume; consequently, no additional optical interfaces are necessary. The SiPM signals were amplified, shaped, and readout by an analog board; a 250 MHz, 14-bit digitizer was used to examine the output pulses from each SiPM-pair channel. The SiPM over-voltage had to be adjusted in order to reduce pulse clipping and negative overshoot, which was observed for events with high scintillation production. Pulse shaped discrimination (PSD) was conducted by evaluating three different parameters: time over threshold (TOT), pulse amplitude, and pulse integral. In order to differentiate high and low energy events, a 30ns gate window was implemented to group pulses from two SiPM channels or more for the calculation of TOT. It was demonstrated that pulses from a single SiPM channel within the 30ns window corresponded to low-energy gamma events while groups of pulses from two-channels or more were most likely neutron events. Due to gamma pulses having lower pulse amplitude, the percentage of measured gamma also depends on the threshold value in TOT calculations. Similarly, the threshold values were varied for the optimal PSD methods of using pulse amplitude and pulse area parameters. Helium-4 detectors equipped with SiPMs are excellent for in-the-field radiation measurement of nuclear spent fuel casks. With optimized PSD methods, the goal of developing a fuel cask content monitoring and inspection system based on these helium-4 detectors will be achieved.**

***Index Terms*—SiPM, helium-4 detector, pulse shaped discrimination, digitizer**

This paper was submitted to conference proceedings on 26 May 2017. This material is in part based upon work supported by the Department of Energy Nuclear Engineering University Program (NEUP) under Award Number DE-NE0008459. Additionally, the United States Army paid for C. Barker's master's degree in Nuclear Engineering from the University of Florida.

C. Barker was with the University of Florida, Gainesville, FL 32606 USA and is now with the United States Military Academy, West Point, NY, 10996 USA (e-mail: cathleen.b.barker.mil@mail.mil).

T. Zhu (email: ting.zhu@ufl.edu), L. Rolison (email: kassodus@ufl.edu), K. Jordan (email: kjordan@mse.ufl.edu), and A. Enqvist (email: enqvist@mse.ufl.edu) are with the University of Florida, Gainesville, FL 32606 USA. S. Kiff is with Sandia National Laboratory (email: skiff@sandia.gov).

I. Introduction

THE Nuclear Engineering Department at the University of Florida has already done a considerable amount of research into neutron spectrometry using helium-4 gas scintillation detectors with PMT readout, developed by Arktis, in order to support research goals from the Department of Energy [1]. In contrast, this research delves into Arktis' new generation of silicon-photomultiplier (SiPM)-based helium-4 detectors in order to analyze their discrimination capabilities and neutron spectrometry for future use in fuel cask content monitoring and inspection systems. Recent studies have noted SiPMs' many useful advantages such as good signal gain, low operation voltage, compactness, and low sensitivity to vibration and magnetic fields [2]. This makes them an ideal detector for the rugged requirements of field radiation measurements.

Using natural helium (^{4}He), the Arktis 180-bar pressurized gas scintillator is capable of detecting and distinguishing fast neutrons and gammas. An extended version of the detector has an additional ^{6}Li-based lining, which allows for detecting thermal neutrons. The most unique feature of the Arktis helium-4 SiPM, shown in Fig. 1, is its 12 SiPM pairs positioned equilaterally across the detector. These pairs are divided into three segments, labeled as segments 1, 2, and 3; all three segments are fully immersed in the helium-4 gas volume. Therefore, no additional optical interfaces are needed. A readout board, shown with the cover removed in Fig. 2., provides signals from individual SiPM-pair channels which are amplified and shaped by internal electronics [3].

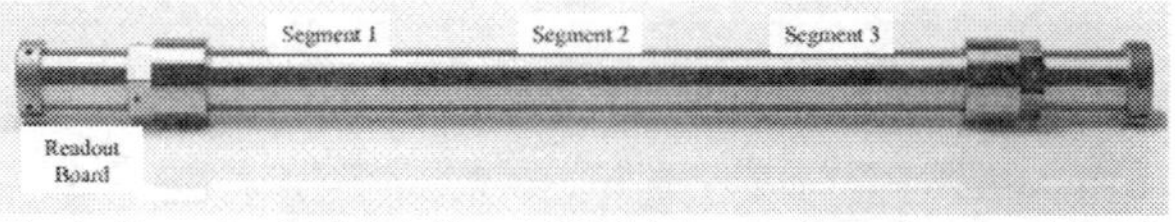

Fig. 1. Example of Arktis helium-4 SiPM detector with readout board (Arktis Radiation Detectors Ltd., 2015).

Fig. 2. Readout (analog) board with the cover removed.

II. Measurement Set-up

A MATLAB script was created to analyze the data measured by the Arktis helium-4 SiPM detector for pulse shaped discrimination. The purpose of PSD is to examine the effectiveness of the Arktis helium-4 SiPM detector's ability to distinguish between gamma and neutron events. Three separate criteria were evaluated in order to determine which method had the best separation spread: time-over-threshold, peak amplitude, and pulse integral.

The Struck SIS3316-DT, 250 MHz, 14-bit digitizer [4], as shown in Fig. 3., consisting of four banks with a total of 16 channels was utilized in all experiments. Each detector segment (i.e. 4 SiPM-pair channels) can be connected to fill one bank of the digitizer.

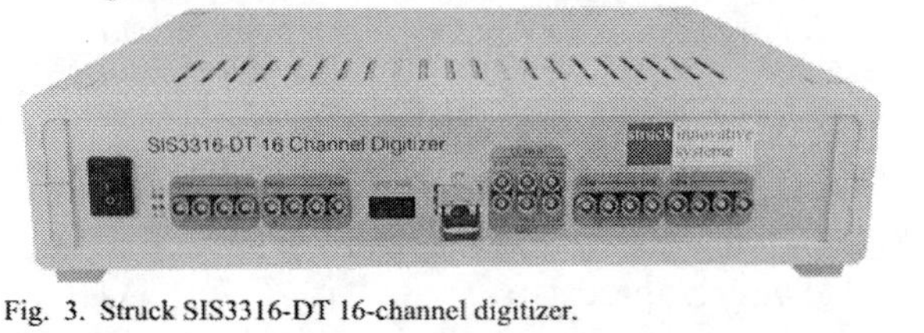

Fig. 3. Struck SIS3316-DT 16-channel digitizer.

Measurements were taken at Sandia National Laboratory using Co-60 , Cf-252, AmBe and AmLi sources. Cf-252 was placed inside HDPE spheres of increasing radii in order to assess the thermal neutron detector capability of the extended-range detector which contains the ^{6}Li-based lining.. AmLi neutron energy spectrum peaks at lower neutron energy levels than Cf-252. In contrast, the AmBe neutron energy spectrum has much higher neutron energies than Cf-252 [5]. Therefore, using three different neutron sources would provide a wider-range of neutron data for PSD analysis.

Finally Arktis' standard applied voltage for SiPM detectors is 30.5V; however, some experiments conducted at the University of Florida examined the effects of lowering SiPM gain for discriminating low-energy gamma events by applying a lower voltage of 29.0V.

A. *Time-over-Threshold PSD*

According to the Arktis manual [6], the SiPM controller uses the time-over-threshold (TOT) method to distinguish gamma, thermal, and fast neutron events. The manual specifies that a 30 ns coincidence window is utilized to group pulses from two or more SiPM channels together. During post-processing, a threshold value is then applied to the grouped pulses. As long as one pulse from the grouping is "active" (above the threshold), the length of time from the second-to-last instance to the second instance is measured. This is demonstrated in Fig. 4.

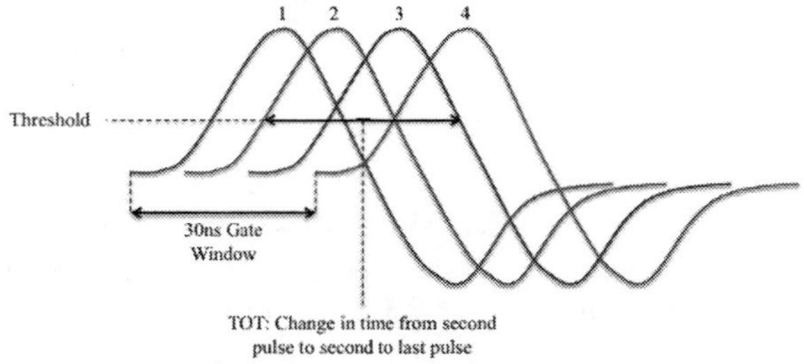

Fig. 4. Time-Over-Threshold (TOT) is defined as the time duration for SiPM pulse signals sustaining a level over a given threshold. A coincident window (default 30ns) can be set to group pulses for a single incident particle event.

In this work, the Time-Over-Threshold parameter is defined as the time duration of consecutive SiPM pulses which sustain a level higher than a given signal threshold. A coincident window of 30 ns (by Arktis' default) can be imposed and for groups of two or more pulses, they are denoted as the "groups of multiples" and the single pulses that do not fit into the 30 ns gate window are "groups of singles". In our work, at least three thresholds were considered: a low, medium, and a high threshold for each set of data.

B. *Peak Amplitude and Pulse Integral PSD*

This research delved into two additional variables: peak amplitude and area in addition to the TOT method by Arktis..

Peak amplitude is the peak amplitude (relative to

the given threshold level) achieved for an individual pulse. Results in this paper are reported in digitizer units (ie. a dynamic range of 2V over 2^{14} bit for the Struck digitizer) unless specified otherwise. The area parameter is defined as the integrated area of the pulse in units of digitizer unit times nanoseconds.

Peak amplitude and area of the pulses were evaluated for with and without a 30 ns gate window applied. Like TOT, when a 30 ns gate window was applied, this research evaluated the peak amplitude and the area for both the groups of "multiples" and the groups of "singles."

A variation on the peak amplitude was evaluated when the 30 ns gate was applied. For groups of two or more pulses, the greatest peak amplitude was selected. Additionally, for groups of two or more pulses, the sum of the peak amplitudes for all the pulses in that grouping was evaluated. Furthermore, the area was summed for all pulses grouped together due to the 30 ns gate window.

III. Results

The focus of the PSD was to evaluate the effectiveness of TOT, Arktis' specified PSD method, and look at other pulse discriminating qualities to distinguish gammas and neutrons.

A. *Effect of Threshold Value for TOT*

Fig. 5 shows a 100-pulse example of Cf-252 measured at the Sandia National Laboratory; various threshold lines have been added to the figure.

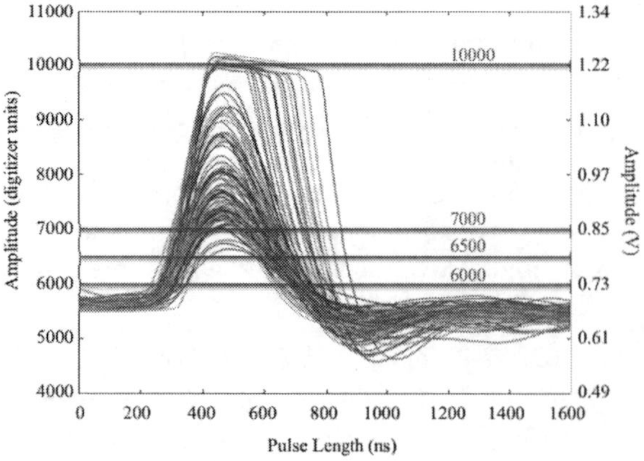

Fig. 5. Depiction of 100 Cf-252 pulses collected. Note the saturation of SiPM pulse above 10000 digitizer unit.

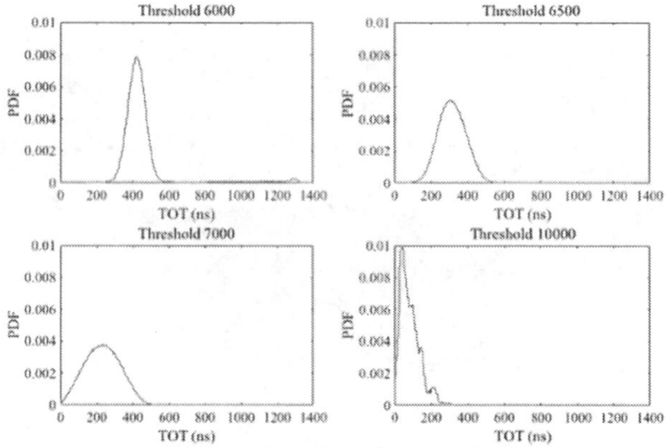

Fig. 6. Quad chart of Cf-252 using data collected from Sandia National Laboratory showing the effects on TOT for different threshold values.

If the threshold is too low, the discrimination ability of TOT for gamma and neutron events is limited. Using an extremely high-value above baseline thresholds, such as at 10000 digitizer units, does two critical things: it removes gamma events almost completely from the data and the TOT distribution will artificially decrease which makes any remaining gamma events difficult to process. This is shown in Fig. 6.

B. *TOT with and without coincident window*

Initially, TOT for all individual pulses without using a 30 ns coincident gate window was evaluated. Fig. 7 shows Cf-252, AmLi, AmBe, Co-60, and a background measurement from Sandia National Laboratory using threshold 6500. The majority of Co-60 has a lower TOT distribution than the neutron sources. This indicates that lower TOTs, if the threshold setting is optimized, are more likely to be gamma events than higher TOTs.

However, there is a secondary Co-60 TOT peak that corresponds to the background TOT distribution. This indicates the helium-4 SiPM detector is sensitive to higher energy events such as cosmic radiation. The SiPM's ability to detect background neutrons and higher energy events was further examined at the University of Florida.

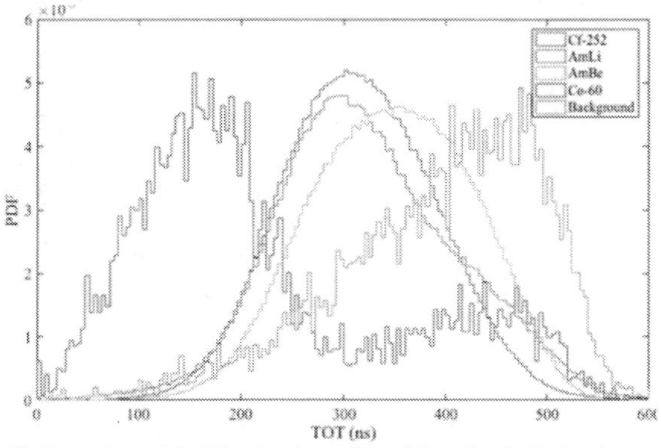

Fig. 7. Quad chart of Cf-252 using data collected from Sandia National Laboratory showing the effects on TOT for different threshold values.

Although there is promising distinction between TOT for individual pulses collected from gamma and neutron sources, Arktis specifies that they use a 30 ns gate window for their TOT evaluation. As long as two or more events are above the threshold value within a 30 ns window, then TOT is measured. When this 30 ns gate window was applied, pulses that were within the 30 ns gate window were grouped together and are referred to as groups of "multiples." Pulses that were not within 30 ns of another pulse are referred to as groups of "singles." Fig. 8 displays the TOT for groups of "singles" while Fig. 9 displays the TOT for groups of "multiples."

Overall, the neutron sources had higher percentages of groups of "multiples," ranging between 37-49% compared to the Co-60 source. In contrast, Co-60, a gamma source, only 12% are groups of "multiples." 88% of Co-60 was found to be in groups of "singles." This shows that the 30 ns gate window discriminates between neutron and gamma sources. Furthermore, 51% of the background pulses are in groups of "multiples" which indicate they are higher energy events. This supports the theory that the majority of the background data were from cosmic rays.

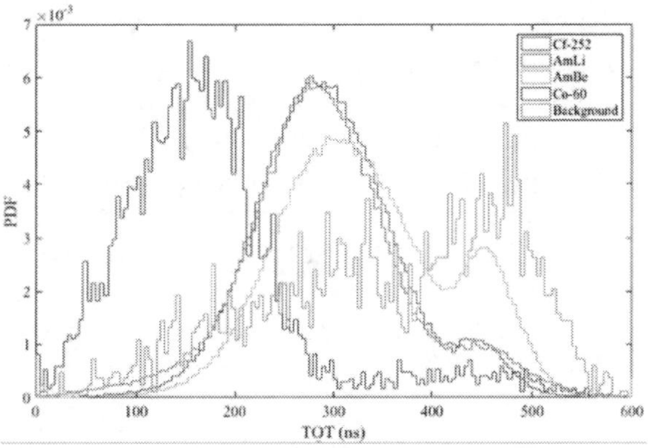

Fig. 8. Quad chart of Cf-252 using data collected from Sandia National Laboratory showing the effects on TOT for different threshold values.

Fig. 8 displays the groups of "singles" which shows a difference in TOT distribution for gamma and neutron sources. There is a secondary TOT distribution peak in the neutron data at higher TOT values. This peak corresponds to the background TOT distribution.

Fig. 9 displays the TOT distribution for groups of "multiples." Here, there are no lower-value TOTs. The Co-60 TOT distribution matches the background TOT distribution for groups of "multiples." Additionally, the rate for groups of "multiples" for Co-60 and background are the within 0.01 groups/sec of each other. This indicates that all of the Co-60 measured within the 30 ns coincident window are actually background. Therefore, the events that make the 30 ns gate window are either neutrons or high-energy background events.

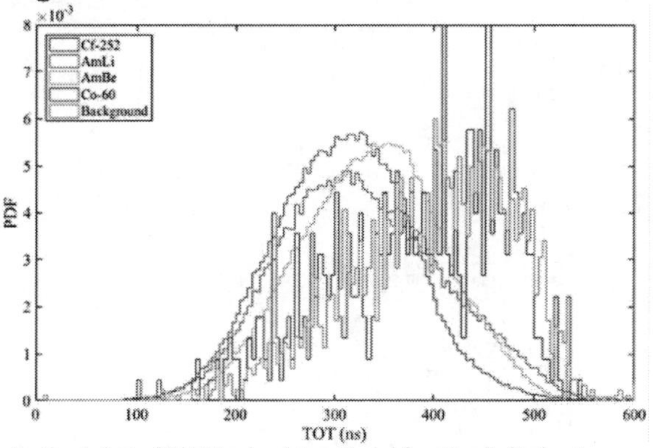

Fig. 9. Quad chart of Cf-252 using data collected from Sandia National Laboratory showing the effects on TOT for different threshold values.

C. PSD for each segment

Analysis of the scintillation pulses in each

segment for TOT, peak amplitude, and integral showed minor discrepancies among the three detector segments. For example, a Cf-252 source was positioned 0.4 m from the center of detector's segment 2. The distributions of the TOT values for SiPM pulses for each segment are shown in Fig. 10. It was expected and confirmed that segment 2 had a greater amount of measured events. However, Fig. 10 shows that segment 1 measured significantly more events than segment 3 and has a lower TOT peak value.

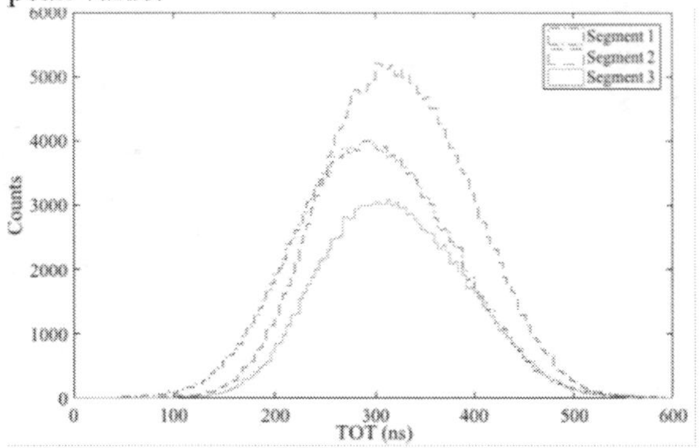

Fig. 10. TOT for individual pulses for all three segments using data collected from Sandia National Laboratory

This discrepancy between segments 1 and 3 is also observed in Figs. 11 and 12, which show the distributions of the values of SiPM pulse peak amplitude and pulse integrated area in each segment.

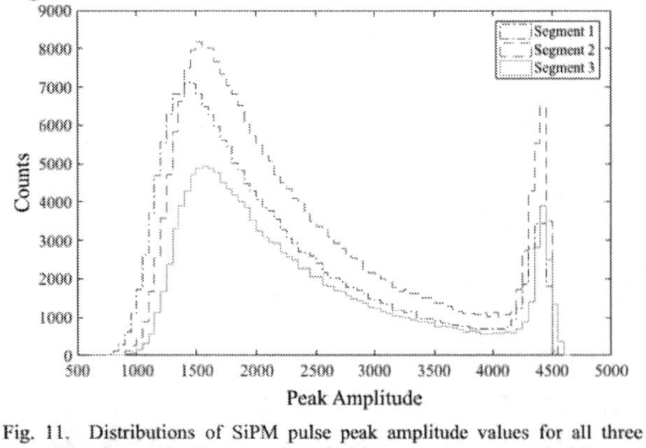

Fig. 11. Distributions of SiPM pulse peak amplitude values for all three segments using a Cf-252 source aligned with segment 2 center and with equal distant to segments 1 and 3.

Overall, Figs. 10-12 indicate that either there may be some inconsistency of scintillation characteristics among the three segments of the detector, which are supposedly optically independent and identical.

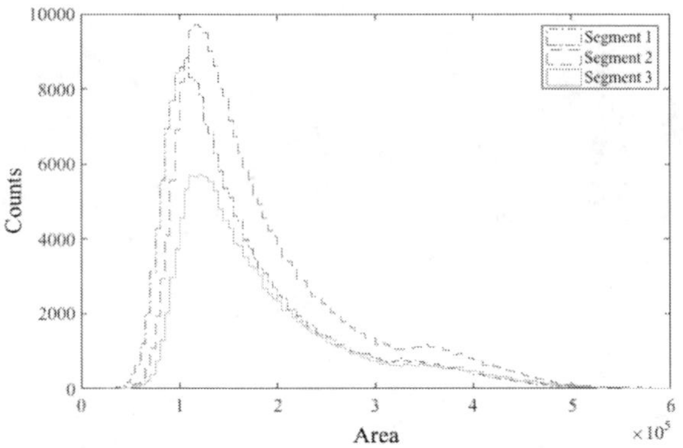

Fig. 12. Distribution of SiPM pulse integrated area values for all three segments using a Cf-252 source aligned with segment 2 center and with equal distant to segments 1 and 3.

D. TOT comparison between 30.5V and 29.0V

Additional research examined the effect of decreasing the applied voltages on the effectiveness of PSD. The nominal supply voltage for the SiPM detector electronics is 30.5V. As shown in Fig. 13, the distributions of TOT for the Cf-252, Co-60 and a background measurement are distinct and demonstrate good PSD capability of the detector. Additionally, there is a small secondary peak in the Co-60 measurement that has a TOT peak around 525 ns. When evaluating the background, it is apparent that the 525 ns -peak corresponds to background. However, at a lower supply voltage of 29.0V, the count rates for Co-60 and the background are nearly the same. Comparing Fig. 13 and Fig. 14, there is a shift of the distributions to lower TOT values in general since the same threshold value was applied, but the SiPM pulses' amplitudes were decreased due to the lower voltage. In Fig. 13, the matching TOT distribution peaks of Co-60 and background implies that the actual gamma events from Co-60 were not measured by the detector; instead the same background contribution was measured. This is a positive finding because helium-4 based SiPM detectors are being considered for spent nuclear fuel detection; lowering the applied voltage can theoretically remove the SiPM detector's capability to measure gamma events.

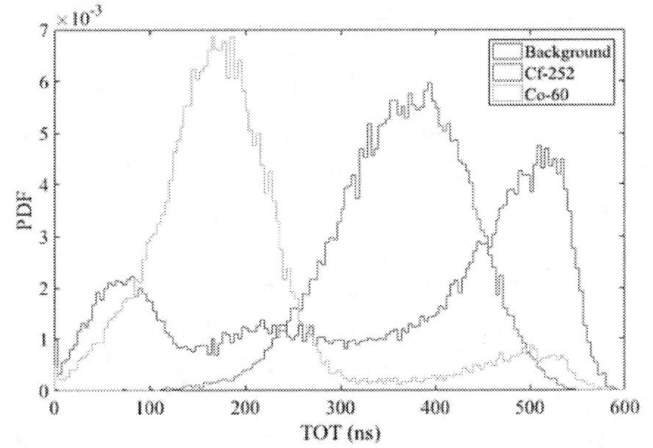

Fig. 13. TOT for individual pulses at 30.5V using data collected from University of Florida. Note the measurement durations were different for the three cases here.

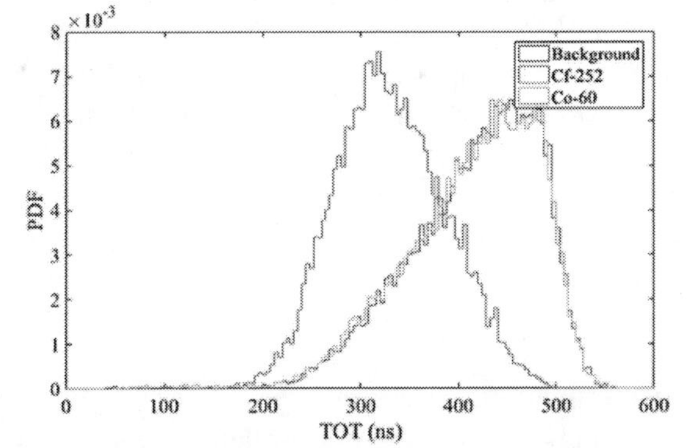

Fig. 14. TOT for individual pulses at 29.0V using data collected from University of Florida.

The effect of lowering supply voltage on the distribution of pulse peak values is shown in Fig. 15. At 30.5V, there is a large background peak at the clipped pulse region (around 4500 digitizer amplitude units) due to SiPM pulse saturation, as noted in Fig. 5. There are similar peaks for both Cf-252 and Co-60. At 29.0V, the Cf-252 peak in that region is almost completely removed. The Co-60 peak matches the background peak in that region. This shows that lowering the applied voltage will remove clipped pulses but also suppress low-energy gamma event measurement.

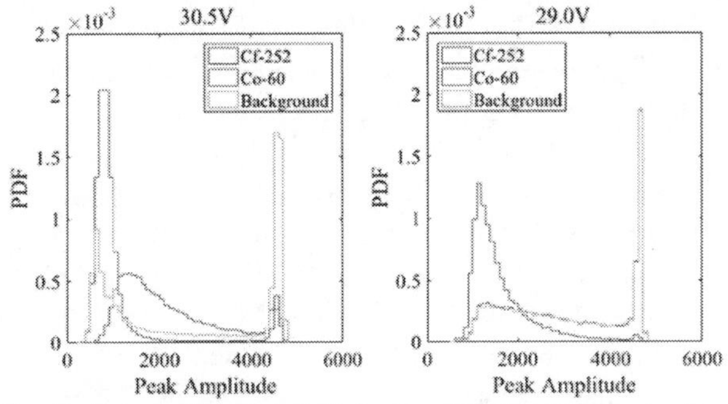

Fig. 15. Peak amplitude for individual pulses at 30.5V and 29.0V using data collected from University of Florida.

E. TOT comparison with longer coincident window

Results above were obtained by using individual SiPM pulse data, without taking into account the fact that one event could induce multiple SiPM pulses. As mentioned, a coincidence window can be applied to group consecutive pulses together. Window lengths of 30 ns, 60 ns, 100 ns, 200 ns, and 400 ns were used and the results are presented in Fig. 16.

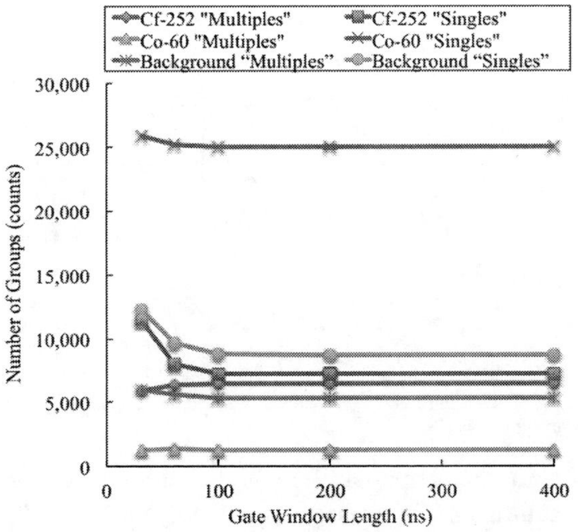

Fig. 16. Number of groups versus coincident window lengths for Cf-252, Co-60, and background.

The groups of "singles' were events where there was only one recorded SiPM pulse in the given coincident window length. The decrease in the number of groups of "singles" in all cases of Cf-252, Co-60, and the background as the window length increase simply means that some single events are grouped together; thus, it is likely that the gate window is "over-cautious" in preventing a false positive (calling an event a neutron when it is really a gamma). Eventually, however, the trend becomes stable for all three measurements, as the gate window length is increased. This shows that the rate of false positives does not increase for gate window lengths greater than 100 ns.

Fig. 17 further evaluates this conclusion by comparing the TOT for 30 ns and 100 ns. The TOT values for groups of "singles" for both Cf-252 and Co-60 stay the same. The TOT for background "singles" looks considerably different at 100 ns than at 30 ns. The peak around 500 ns for background "singles" for the 30 ns data is completely missing from the 100 ns data. Larger TOT tends to indicate that the event is a neutron or other high-energy event but not every high TOT is a neutron event. Therefore, the rate of false positives has likely increased as a result of increasing the gate window length.

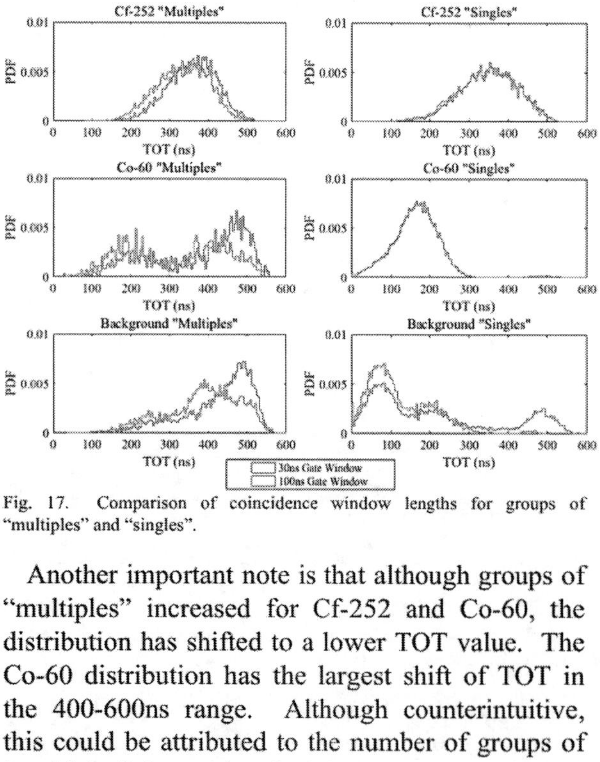

Fig. 17. Comparison of coincidence window lengths for groups of "multiples" and "singles".

Another important note is that although groups of "multiples" increased for Cf-252 and Co-60, the distribution has shifted to a lower TOT value. The Co-60 distribution has the largest shift of TOT in the 400-600ns range. Although counterintuitive, this could be attributed to the number of groups of "multiples" increasing but the average number of pulses in each group decreasing. Fewer pulses in a group will, in general, have a shorter TOT. This a positive effect because it helps distinguish Co-60 group of "multiples" TOT from Cf-252 groups of multiples." Consequently, gate window can be increased slightly with some minimal, but positive, effects on TOT comparison.

IV. Conclusion

Pulse shaped discrimination analysis were conducted on data collected by Arktis helium-4 SiPM detectors. There are clearly many measurement settings that must be adjusted prior to collecting data in order to get the clearest results. Modifications to signal threshold value and supply voltage will impact the ability during post-processing to distinguish gamma and neutron events.

The PSD analysis showed several things. First, TOT has the clearest distinction between gamma and neutron events when compared with peak amplitude and pulse area analysis. Secondly, decreasing the applied voltage improves the helium-4 SiPM detector's ability to suppress gamma events. Furthermore, a 30 ns coincidence window length provides better discrimination between gamma and neutron events than looking at each pulse individually. Finally, the gate window does have some room to be increased from 30ns to 100ns, which can improve gamma and neutron discrimination, except in very high-rate environments where pulse-pileup could become non-negligible with increased gate window.

However, with these positive trends, more research is needed to fully characterize the capabilities of the Arktis helium-4 SiPM detectors. PSD-focused measurements were only conducted using the "non-extended range" version of the detectors. The "extended range" version theoretically should show distinction in the groups of "multiples" between thermal and fast neutrons. Additionally, longer measurements would increase the accuracy of the results and more extensive research needs to be conducted to refine the limits of the coincidence window for TOT. Finally, further research needs to be conducted in order to utilize the PSD method to develop a neutron response matrix.

https://doi.org/10.1051/epjconf/201817007002

REFERENCES

[1] R. P. Kelley, *et al.*, "Neutron response function characterization of 4He scintillation detectors," *Nuclear Instruments and Methods in Physics Research A*, vol. 793, pp. 101-107, 2015.

[2] P. Buzhan, *et al.*, "Silicon photomultiplier and its possible applications," *Nuclear instruments and Methods in Physics Research Section A*, vol. 504, pp. 48-52, May 2003.

[3] Arktis Radiation Detectors Ltd., "Analog Readout Board AA121 Assembly and Operating Manual," Mar. 12, 2015. [Online]. Available: www.arktis-detectors.com.

[4] "SIS3316, 16 Channel VME Digitizer User Manual," Struck Innovative Systeme, Hamburg, Germany, January 2015.

[5] D. Reilly, N. Ensslin, & H. Smith, Jr., "Passive Nondestructive Assay of Nuclear Materials," Los Alamos National Laboratory, Mar. 1991. [Online] http://www.lanl.gov/orgs/n/n1/FMTTD/neut_mc/pdfs/LA_UR_90_0732.pdf

[6] Arktis Radiation Detectors Ltd., "Arktis S-670 Detector Series Operating Manual," Mar. 12, 2015. [Online]. Available: www.arktis-detectors.com. reaction. *Journal of Radioanalytical and Nuclear Chemistry , 256* (1), 163-166.

https://doi.org/10.1051/epjconf/201817007003

Characterization of Low Level Wastes: a new design for calorimetric measurement

Kévin Galliez, Guillaume Jossens, Alain Godot and Christophe Mathonat

Abstract— **Calorimetry is one of the best solutions to estimate the overall quantity of nuclear material on a wide range of masses, from a few milligrams up to kilograms of radionuclides, by measuring the overall thermal power due to the radioactive decay coming from the waste contained in a metallic drum or a different type of container. It has many advantages as it features a non-destructive method which remains independent of matrix effect or the chemical composition. Until now, calorimetry allows to measure at the lowest 0.5 to 1 mW for samples up to 385 liters. But nowadays, thanks to new technological breakthroughs, KEP-Technologies calorimeters are able to measure as low as 50 µW for 40 liters samples. The µLVC is based on a new design with twin cells, a new temperature regulation loop and a heat-flow measurement system inside a vacuum chamber (Patent deposit P005299 LA/VL). The µLVC is a differential heat-flow calorimeter for precise measurement independent of the residual fluctuations caused by environmental changes. The new calorimeter is an industrial product able to work in environmental conditions with wide temperature variations. The first results have shown a great improvement in the detection of very low thermal effect thanks to the thermal noise reduction. The paper presents the developments in Large Volume Calorimetry as a new tool for quantification of nuclear material to characterize Pu-Am samples, i-graphite, and low tritium samples with high precision and reliability.**

Index Terms—**calorimetry, tritium, carbone14, chlorine36, nuclear waste**

I. Introduction

RADIOACTIVE radioactive isotopes within the item. Calorimetry is used as a non-destructive assay technique to determine alpha or beta emitters inventories. It is one of the best solutions to estimate the overall quantity of nuclear material on a wide range of mass, from a few milligrams up to kilograms of radionuclides with a good precision. In the case of drums containing nuclear waste, characterization is very challenging due to complex matrix, volume and low level activities. The advantage of calorimetry towards other techniques is that the measurement is independent of the matrix, only the thermal conductivity of the sample matters, which influences the measurement time. In that case, waste can either be in solid, liquid or gas phase.

The new type of calorimeter developed by KEP Nuclear, called µLVC (Figure 1) meaning microwatts-Large Volume Calorimeter, is able to measure less that one hundred of microwatts for a container volume of 40 liters. This paper presents the conception of this calorimeter and gives some results on the tests currently performed at KEP Nuclear.

Fig. 1. 3D view of the µLVC calorimeter

The µ-LVC is based on a differential heat flow calorimeter designed with twin cells, temperature sensors and heat-flow measurement system inside a vacuum chamber (Patent deposit P005299 LA/VL). Differential heat-flow calorimeters are designed with twin cells for a precise heat flow measurement. Thus, residual fluctuations caused by environmental changes are approximately cancelled out by the signal difference formed between the two cells. The calorimetric signal results from the measurement of the heat flow between the measurement cell containing the radioactive sample and the thermal block. The temperature of the thermal bloc is maintained constant thanks to resistive heater. This thermal heat flow generates a signal (µV) based on time which is recorded by a data acquisition system. When the heat flow signal is stabilized, this one is converted to a thermal activity (W) by means of the linear calibration curve. The Figure 2 shows a schematic view of the calorimeter.

This paper has been submitted on the 1rst of June 2017.
Kévin Galliez, Alain Godot and Christophe Mathonat are from KEP Nuclear, ZA de l'Euze 2, 453 Chemin vieux de Chusclan, 30200 Bagnols sur Cèze, France (contact.nuclear@kep-technologies.com).
Guillaume Jossens is from KTHTP, 7 rue de l'Oratoire, 69300 Caluire, France (contact.htp@kep-technologies.com).

 https://doi.org/10.1051/epjconf/201817007003

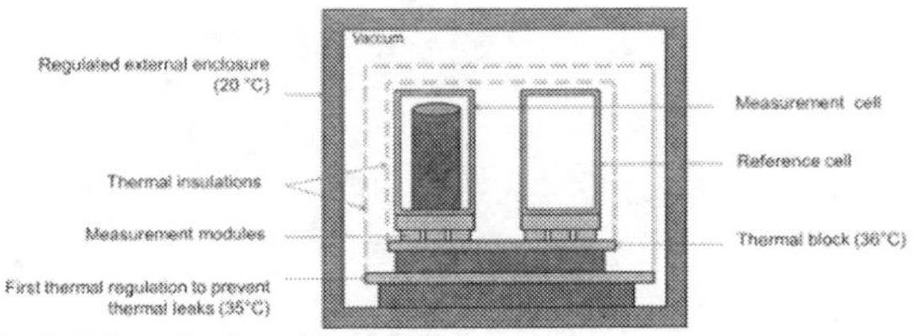

Fig. 2. Schematic view of the μ-LVC

The thermal bloc is regulated using the temperature derivative sensing [1]. The heat flow between the calorimetric bloc and a mass of high thermal inertia - which serves as a reference of temperature - is regulated. The information on the heat flow is given by a thermopile installed between the reference mass and the thermal block. The signal coming from the thermopile is kept constant and close to zero in order to bring the calorimeter at its thermal equilibrium. This principle allows achieving high temperature stability and a very precise regulation. In the case of the calorimeter μLVC, the reference mass is the reference cell and the calorimeter block is the thermal block (see Figure 2). The limitation of thermal leaks of the calorimeter is ensured by the use of a high vacuum (< 1.10-6 mbar) thermally regulated.

The heat flow is measured thanks to a set of Peltier modules arranged in the calorimetric block. The heat flow generated between the sample and the thermal block is thus measured and can be compared with the heat flow measured between the reference cell and the calorimetry block. The heat flow actually released by the sample is thus obtained by subtracting the flow measured on the reference cell from that measured on the measurement cell. The use of several Peltier modules for the same measurement of a heat flow makes it possible to increase the sensitivity of the measurement.

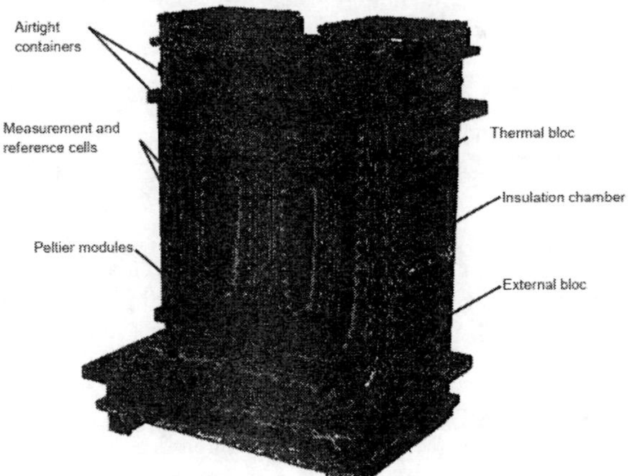

Fig. 3. Illustration of the μLVC

Copper containers have been designed for the measurement of samples. The weight of a container is approximately 60 kg. A guide and guidance arm specifically designed for the containers and cells (Figure 4) are used to place the cells in the calorimeter with an excellent reproducibility. The containers are lifted up thanks to a gantry and hoist in the testing hall.

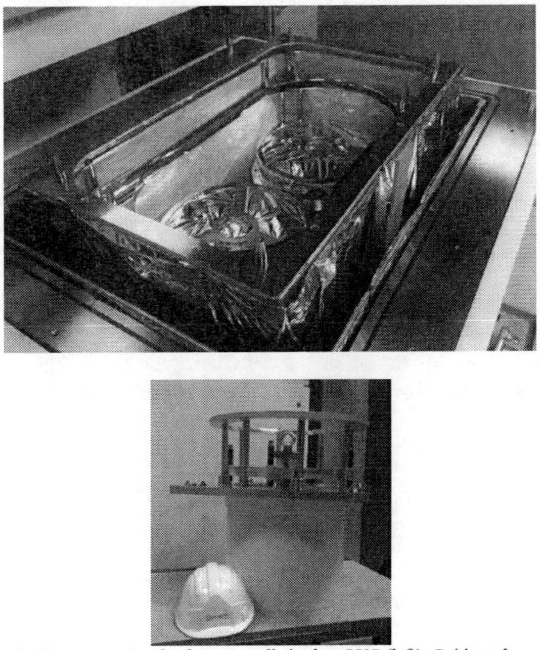
Fig. 4. Measurement and references cells in the μLVC (left). Guide and guidance arm (right)

Fig. 5. closed (left) and opened (right) positions

II. Results and discussion

A. Calibrating the calorimeter

1) Estimating the sensitivity

For estimating the detection limit, the sensitivity and the measurement uncertainty, the injection of a known thermal power using the integrated resistances under the cells are used. For this purpose, a 3-times repetition of a baseline and 100 µW Joule effect measurements have been done. For the repeatability assessment, the containers were withdrawn from the calorimeter between each repetition. The Figure 6 shows the stabilization of the differential heat flow measured in absence of heating (i.e. a baseline). The Figure 7 shows the signal of the differential heat flows when a power of 100 µW is injected in the cells. It also shows the long term variation of the signal, slightly following the ambient temperature variation of the room. Finally the Figure 8 shows the electronic noise measured on the signal.

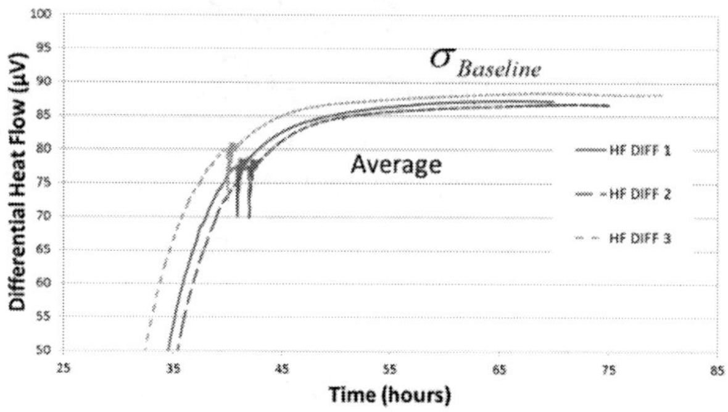

Fig. 6. Baseline of the differential heat flow

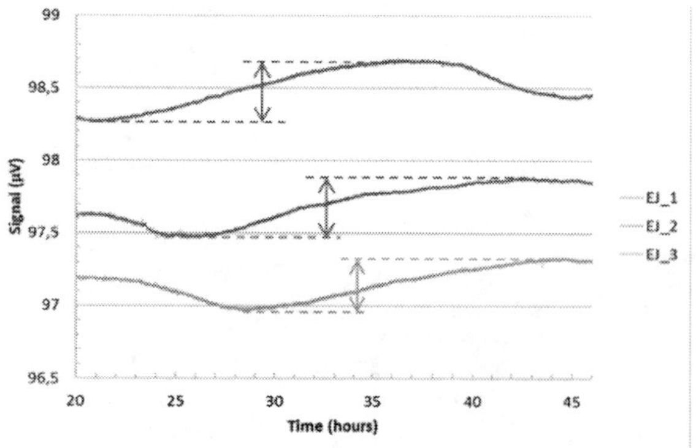

Fig. 7. Long time variations of the signal on three Joule effect measurements

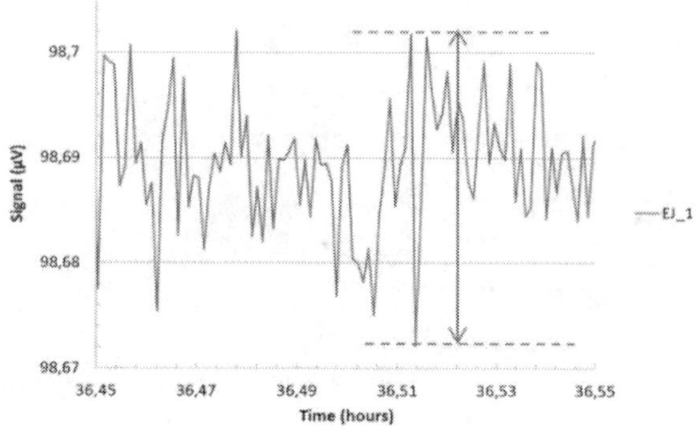

Fig. 8. Short time variation of the heat flow signal

The baselines (Figure 6) and 100 µW (Figure 7) Joule effect measurements allow us to calculate the sensitivity of the calorimeter:

$$\overline{S} = \frac{1}{3}\sum_{i=1}^{3} S_i \quad \text{with} \quad S_i = \frac{\Delta V_{mes} - \Delta V_{baseline}}{P_{injected}}$$

ΔVmes corresponds to the measured differential voltage during a Joule effect in µV and ΔVbaseline corresponds to the measured differential voltage during a baseline in µV. The sensitivity calculated is 105 µV/mW (±2.1%). Uncertainties are calculated according to the GUM [2]. Note that the signal corresponding to the Figure 8 shows a non-predictable variation of the signal corresponding to less than 0.02 % of the signal. According to the GUM [2] this error is negligible compared to the long term variation avec repetitiveness of the measurement. Indeed, the final error estimated for a 100 µW power measurement is 5.6 % (k=1).

2) Estimating the detection and quantification limit

The detection limit is calculated according to Currie [3] using the formula:

$$LD = \frac{\sigma_{LDB}}{\overline{S}} * 3$$

The quantification limit is calculated according to Currie and McDougall [3], [4] using the formula:

$$LQ = \frac{\sigma_{LDB}}{\overline{S}} * 10$$

The standard deviation of the baseline is assessed using the Figure 6. The calculations give a detection limit of 25 µW and a quantification limit of 83 µW.

B. Application to ^{14}C, ^{3}H and ^{36}Cl

Thanks to the Joule effect calibration module of the calorimeter, the µLVC offers the possibility to simulate a heating from tens of µW to several tens of mW. Knowing the specific power of radionuclides, it is easy to assess the amount of radionuclides measureable using our calorimeter. Considering the three radionuclides 3H, 14C and 36Cl, the detection and quantification limits are:

TABLE I
DETECTION LIMITS FOR COMMON RADIONUCLIDES USING THE µLVC

	LD	LQ
3H (324 mW/g)	80 µg – 29 GBq	256 µg – 92 GBq
14C (1.32 mW/g)	19 mg – 3.1 GBq	63 mg – 10 GBq
36Cl (0.06 mW/g)	416 mg – 0.51 GBq	1.40 g – 1.72 GBq

These detection limits are given for samples containing only one type of radionuclide. The specific powers have been calculated using the following formula:

$$SP(W/g) = \frac{2119.3 * Q}{T_{1/2} * A}$$

With:

- 2119.3 : the coefficient converting the period into second, the MeV into J and the mass number into gram.
- Q the average energy of the radionuclide in MeV.
- the period in years.
- A the mass number in g.mol^{-1}.

III. Conclusion

KEP Nuclear has developed a new calorimeter able to measure less than 100 µW with an uncertainty lower than 10 % (k = 1) for a large volume container of 40 liters. The detection limit estimated for tritium, carbon 14 and chlorine 36 are respectively 80 µg, 19 mg and 416 mg. The µLVC truly constitutes a breakthrough in the field of calorimetry and NDA. Further tests are already planned at KEP Nuclear to improve the measurement time and the stability of the signal for improving the daily rate of analysis and the repeatability of the measurement.

References

[1] J. L. Hemmerich, J.-C. Loos, A. Miller, and P. Milverton, "Advances in temperature derivative control and calorimetry," Rev. Sci. Instrum., vol. 67, no. 11, pp. 3877–3884, Nov. 1996.

[2] "Évaluation des données de mesure - Guide pour l'expression de l'incertitude de mesure," JCGM/WG 1, JCGM 100:2008(F), 2008.

[3] L. A. Currie, "Limits for qualitative detection and quantitative determination. Application to radiochemistry," Anal. Chem., vol. 40, no. 3, pp. 586–593, Mar. 1968.

[4] D. MacDougall and W. B. Crummett, "Guidelines for data acquisition and data quality evaluation in environmental chemistry," Anal. Chem., vol. 52, no. 14, pp. 2242–2249, Dec. 1980.

Development of Neutron Energy Spectral Signatures for Passive Monitoring of Spent Nuclear Fuels in Dry Cask Storage

Ira Harkness, Ting Zhu, Yinong Liang, Eric Rauch, Andreas Enqvist, and Kelly A. Jordan

***Abstract*— Demand for spent nuclear fuel dry casks as an interim storage solution has increased globally and the IAEA has expressed a need for robust safeguards and verification technologies for ensuring the continuity of knowledge and the integrity of radioactive materials inside spent fuel casks. Existing research has been focusing on "fingerprinting" casks based on count rate statistics to represent radiation emission signatures. The current research aims to expand to include neutron energy spectral information as part of the fuel characteristics. First, spent fuel composition data are taken from the Next Generation Safeguards Initiative Spent Fuel Libraries, representative for Westinghouse 17x17 PWR assemblies. The ORIGEN-S code then calculates the spontaneous fission and (α,n) emissions for individual fuel rods, followed by detailed MCNP simulations of neutrons transported through the fuel assemblies. A comprehensive database of neutron energy spectral profiles is to be constructed, with different enrichment, burn-up, and cooling time conditions. The end goal is to utilize the computational spent fuel library, predictive algorithm, and a pressurized ^{4}He scintillator to verify the spent fuel assemblies inside a cask. This work identifies neutron spectral signatures that correlate with the cooling time of spent fuel. Both the total and relative contributions from spontaneous fission and (α,n) change noticeably with respect to cooling time, due to the relatively short half-life (18 years) of the major neutron source ^{244}Cm. Identification of this and other neutron spectral signatures allows the characterization of spent nuclear fuels in dry cask storage.**

***Index Terms*—spent fuel, safeguards, neutron measurement**

I. Introduction

THE United States currently has commercial spent fuel accumulating at 78 sites in 34 states. This accounts for over 20,000 significant quantities of plutonium in dry cask storage in the United States. The IAEA has expressed a need for robust safeguards and verification technologies for ensuring the continuity of knowledge and integrity of radioactive materials inside spent fuel casks.

The ultimate objective of safeguards is "the timely detection of diversion of significant quantities of nuclear material from peaceful nuclear activities to the manufacture of nuclear weapons or of other nuclear explosive devices or for purposes unknown and deterrence of such diversion by the risk of early detection" [1]. This is often achieved using convention safeguards measures while the materials are being transported from reactor, to cooling pool, to dry storage casks. Once the spent fuel is inside a dry storage cask, non-destructive assay (NDA) is more difficult.

Existing research has focused on "fingerprinting" casks based on count rate statistics [2] and radiation signatures [3]. The primary method of cask monitoring uses a Bonner sphere spectrometer. Bonner spheres are advantageous due to the wide energy range measured, however the energy resolution is limited due to the smooth fluence response functions of the thermal neutron detectors [4]. Existing cask monitoring methods are unable to restore continuity of knowledge in cases where existing knowledge has been lost.

The current research aims to expand to include neutron energy spectral information as part of the spent fuel characteristics. Computational resources used include the Next Generation Safeguards Initiative (NGSI) Spent Fuel Library, ORIGEN-S and MCNP6. In this paper, we examine the transported neutron source spectra from individual pins from a Westinghouse 17x17 PWR assembly. This work identifies neutron spectral signatures that correlate with the cooling time of the spent fuel. The goal is to utilize a computational spent fuel library, predictive algorithm, and a pressurized ^{4}He scintillator to verify the spent fuel assemblies inside a cask.

II. Computational Methodology

Computational methods in this work include use of the NGSI Spent Fuel Library, ORIGEN-S, and MCNP.

A. Next Generation Safeguards Initiative Spent Fuel Library

The Next Generation Safeguards Initiative began in 2009 with the goals of quantifying plutonium content in spent nuclear fuel through the use of non-destructive assay (NDA) techniques, assessing the capability of the NDA techniques to verify burnup, cooling time, and initial enrichment, and determining the ability to detect pin diversions in spent fuel

This work was supported in part by the U.S. Department of Energy Nuclear Energy University Programs.

I. Harkness, T. Zhu, Y. Liang, A. Enqvist, and K. A. Jordan are with the Department of Materials Science and Engineering, University of Florida, Gainesville, FL 32611 USA (e-mail: ira@mse.ufl.edu).

E. Rauch is with Los Alamos National Laboratory, Los Alamos, NM 87545 assemblies [5]. To achieve these goals, spent fuel assemblies in the form of MCNP input files were developed. NGSI Spent Fuel Library-2a (SFL2a) is a set of 17x17 pressurized water reactor (PWR) fuel assemblies, which were simulated for a broad range of parameters suitable for direct comparison of different NDA techniques [6]. The initial enrichment values used in the library are 2%, 3%, 4% and 5%, burnups range from 15 to 60 GWd/tU incremented by 15 GWd/tU. Cooling times included 1, 5, 20, 40 and 80 years. The NGSI Spent Fuel Library is key to this research as the fuel assembly is modeled pin by pin.

B. ORIGEN-S

The modern ORIGEN-S code was developed from the ORIGEN (Oak Ridge Isotope GENeration) code in the late 1970s as a depletion and decay analysis code for SCALE (Standardized Computer Analyses for Licensing Evaluation). The primary advantage of ORIGEN-S for spent fuel safeguards is the ability to accurately and quickly predict neutron emissions from spent fuel. The neutron source calculation is based on methods from the SOURCES code, using spontaneous fission Watt spectral parameters for 41 actinides and a matrix-dependent (α,n) source method using (α,n) cross sections and yields for α particles on 19 target nuclides, including ^{17}O and ^{18}O which are found in spent fuel. [7]. The ORIGEN-S code and associated nuclear data libraries have been through numerous verification and validation benchmarks. These included validation against more than 100 spent fuel samples from domestic and international programs involving older and modern high-burnup mixed oxide (MOX) and low-enriched uranium (LEU) fuels [8].

C. MCNP6

MCNP6 is a general-purpose Monte Carlo N-Particle code that can be used to calculate continuous energy, three-dimensional, coupled neutron-photon-electron transport. The continuous energy cross section data allow linear interpolation between energy points. MCNP6 includes a general source, criticality source, and surface source along with numerous variance reduction techniques. It also offers a flexible tally structure and an extensive cross-section library. MCNP6 [9] and the ENDF-B/VII library [10] have been validated against numerous transport and shielding scenarios.

III. SIMULATIONS

The Next Generation Safeguards Initiative (NGSI) Spent Fuel Library was used to obtain nuclide masses from individual rods within spent fuel assemblies of varying initial enrichments, cooling times, and burnups. ORIGEN-S used the nuclide mass data from this rod to calculate the neutron spectra for both spontaneous fission and (α,n) reactions. A MCNP model of a single spent fuel rod with cladding was built using the NGSI spent fuel libraries. Three separate models were built for each rod using the following homogeneous source spectra: spontaneous fission, (α,n) reaction neutrons, and total neutrons. MCNP was used to transport one hundred million neutrons per source spectra with a tally on the outside surface of the cladding.

Due to the large amount of nuclides in each spent fuel assembly along with the number of different assemblies to analyze, a manual method was unfeasible. A methodology was developed to link the NGSI spent fuel assemblies in MCNP input files to ORIGEN-S input files for automatically generating both the spontaneous fission and (α,n) reaction neutron spectra. Further development added the ability to write MCNP input files using the data obtained from the ORIGEN-S output files. The total workflow from NGSI spent fuel library to transported neutron source spectra is shown in Figure 1.

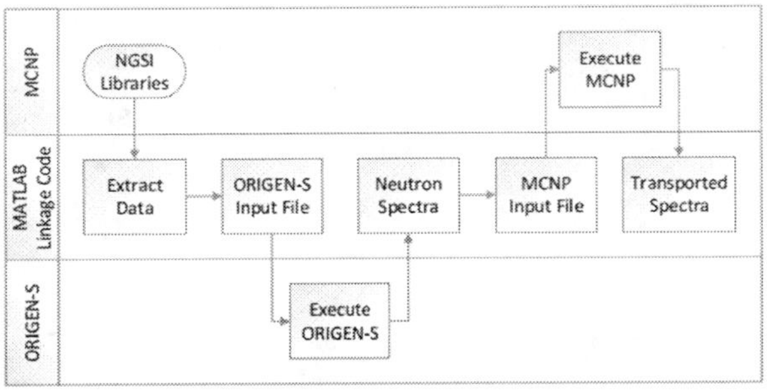

Fig. 1. Total path from NGSI spent fuel library to final transported neutron spectra.

Computational analyses were executed for a single rod from each of the assemblies in the NGSI spent fuel library, which included varying initial enrichments, burnups, and cooling times. The results and discussion below include examination of a rod with an initial enrichment of 4 percent, burnup of 15 GWd/MTU and cooling times of 5, 20, 40, and 80 years.

IV. RESULTS AND DISCUSSION

The simulated total neutron flux on the outside of the assembly is shown in Figure 2. There is a decrease in the total neutron flux with increasing cooling time.

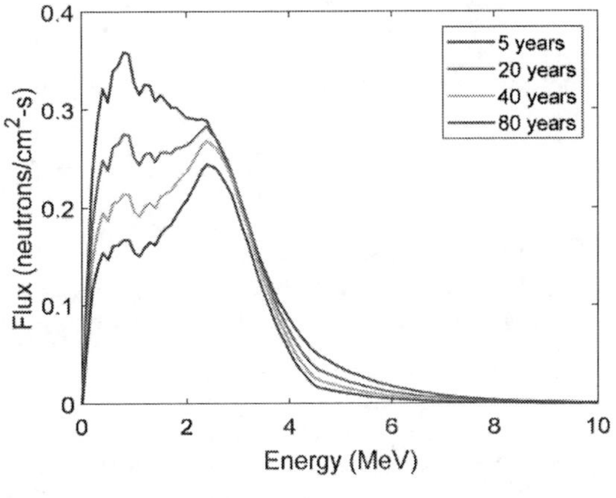

Fig. 2. Change in total neutron spectra as a function of cooling time. The total neutron flux decreases with increasing cooling time, however the peak around 3.0 MeV persists.

Figure 3 shows the neutron source term fractions separated into total neutron flux, spontaneous fission neutron flux and (α,n) reaction neutron flux. The total neutron and spontaneous fission neutron fluxes both decrease, but the neutron flux from the (α,n) reaction increases.

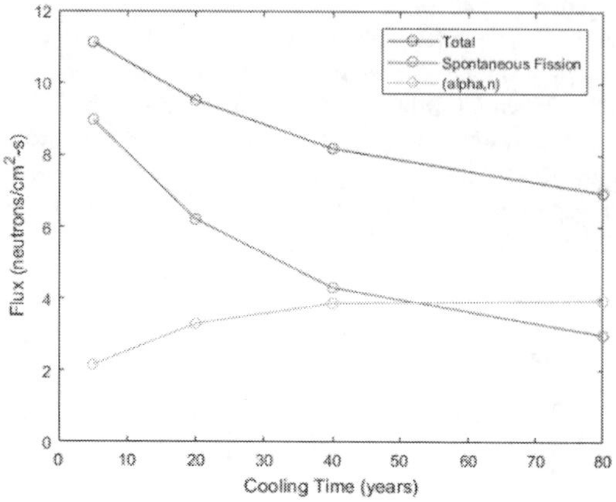

Fig. 3. Change in neutron source strength as a function of cooling time. While the total neutron and spontaneous fission neutron flux both decrease, the (α,n) neutron flux increases.

Analysis of ORIGEN-S data shows that at lower cooling times the majority of neutrons are produced from spontaneous fission of heavy nuclides with ^{244}Cm as the largest contributor. However, the increase in (α,n) reactions with increasing cooling times is due to the increase in ^{241}Am which decays from ^{241}Pu. As shown in Figure 4, ^{241}Am is the major contributor to the (α,n) reaction neutron source term is spent fuel.

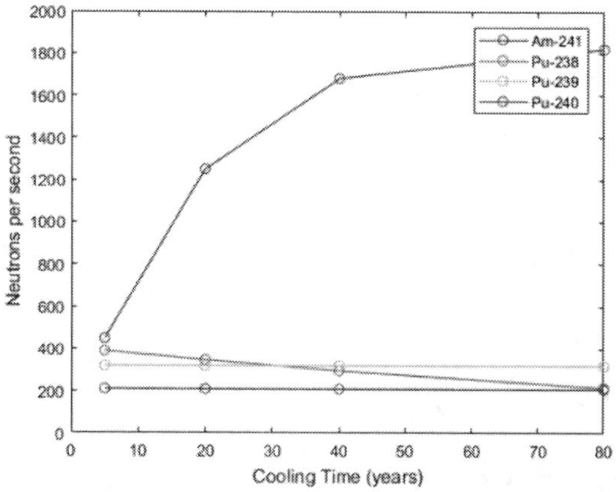

Fig. 4. Change in (α,n) reaction rate by α emitter as a function of cooling time. The increase in alpha emissions from ^{241}Am results in the increase in (α,n) reactions.

Separating spontaneous fission neutrons and (α,n) reaction neutrons in computational models is straightforward. A detector in the field would only be able to measure neutron counts and energies. Therefore, it was necessary to create a method to separate these two neutron sources from a detector measurement.

Based on the characteristics of their spectra, the spontaneous fission neutrons and (α,n) reaction neutrons can be estimated from a total neutron spectrum. The spontaneous fission neutrons peak between 0.8 to 1 MeV and have a long tail extending to 12 to 15 MeV. The spectrum for (α,n) reaction neutrons has a softer curve with a maximum between 2.5 and 3.0 MeV . The (α,n) neutron spectra tail ends at approximately 5 MeV.

Using these characteristics, the results are separated into two integrals. The first is from 0 to 1.6 MeV, covering most of the spontaneous fission neutrons. The second is from 1.6 MeV to 3.6 MeV, spanning most of the (α,n) reaction neutrons. Figure 5 shows the characteristic ranges on a graph of the total neutron spectrum for a spent fuel rod with 4 percent initial enrichment, 15 GWd/MTU burnup and a 40-year cooling time.

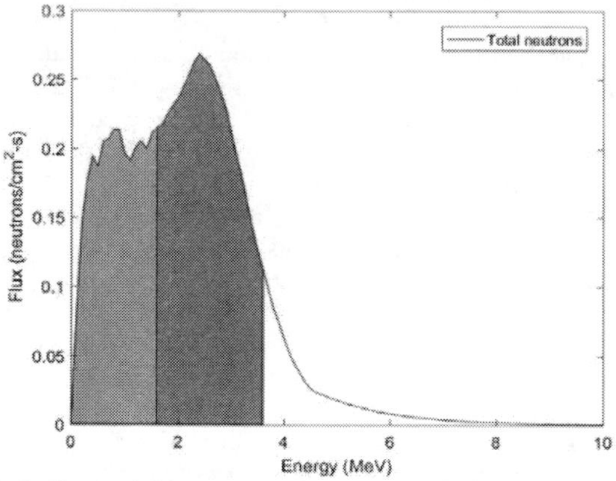

Fig. 5. Characteristic energy ranges for spontaneous fission and (α,n) neutrons. The range on the left from 0 to 1.6 MeV is for spontaneous fission neutrons and the range on the right from 1.6 MeV to 3.6 MeV is for (α,n) neutrons.

Figure 6 shows the energy regions of interest for spontaneous fission and (α,n) reactions plotted versus cooling time. As cooling time increases, the spontaneous fission neutron flux decreases at a faster rate than the (α,n) reaction neutron flux.

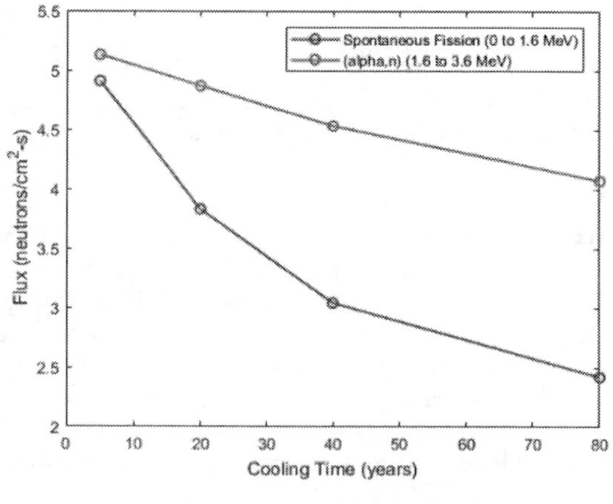

Fig. 6. Change in neutron flux in characteristic energy ranges as a function of cooling time. The spontaneous fission flux decreases at a faster rate, allowing characterizing the cooling time based on the ratio of the two fluxes.

Identification of this trend enables characterizing the cooling time of spent nuclear fuel in dry storage casks. Further refinements to the methodology will be necessary to increase the accuracy of the cooling time prediction.

V. CONCLUSIONS

The linkage code developed for ORIGEN-S and MCNP6 allows analysis of transported neutron source terms for spent fuel rods and assemblies from the NGSI Spent Fuel Libraries. ORIGEN-S calculates the neutron emissions from the spent fuel rods. MCNP6 performs the transport and shielding calculations.

The identification of neutron spectral characteristics in individual rods and 17X17 PWR assemblies are applicable to spent fuel assemblies in dry cask storage. The first characteristic identified is the varying ratio of spontaneous fission neutrons versus (α,n) reactions over time. This is due to the long term increase in ^{241}Am, a significant α source.

This method offers advances over the method of fingerprinting casks. Neutron source spectral characteristics allows restoration of continuity of knowledge, as it is able to determine parameters of spent fuel inside dry storage casks. The next step of this work is examining other trends in the total neutron signal that correlate with spent fuel parameters.

REFERENCES

[1] M. Tarvainen, F. Levai, T. E. Valentine, M. Abhold, and B. Moran, "NDA techniques for spent fuel verification and radiation monitoring," 1997.

[2] E. Rauch, "Developing a Method for Unique Identification of Used Fuel Storage Casks Using Neutrons," 2014.

[3] K. P. Ziock, P. Vanier, L. Forman, G. Caffrey, and A. Lebrun, "The Feasibility of Cask 'Fingerprinting' as a Spent-Fuel, Dry-Storage Cask Safeguards Technique," 2005.

[4] A. Rimpler, "Bonner sphere neutron spectrometry at spent fuel casks," *Nucl. Instruments Methods Phys. Res. Sect. A Accel. Spectrometers, Detect. Assoc. Equip.*, vol. 476, no. 1–2, pp. 468–473, 2002.

[5] R. A. Weldon, M. L. Fensin, and H. R. Trellue, "Total neutron emission generation and characterization for a Next Generation Safeguards Initiative spent fuel library," *Prog. Nucl. Energy*, vol. 80, pp. 45–73, 2015.

[6] J. D. Galloway, H. R. Trellue, M. L. Fensin, and B. L. Broadhead, "Design and Description of the NGSI Spent Fuel Library with Emphasis on the Passive Gamma Signal," *J. Nucl. Mater. Manag.*, vol. XL, no. 3, pp. 25–35, 2012.

[7] I. C. Gauld, S. M. Bowman, B. D. Murphy, and P. Schwalbach, "Applications of ORIGEN to Spent Fuel Safeguards and Non-Proliferation," 2015.

[8] M. D. DeHart and O. W. Hermann, "An Extension of the Validation of SCALE (SAS2H) Isotopic Predictions for PWR Spent Fuel," 1996.

[9] B. C. Kiedrowski, F. B. Brown, L. J. Cox, N. A. Gibson, A. S. Bennett, and M. A. Gonzales, "MCNP6 Shielding Vaidation Suite: Past, Present, and Future," vol. 836, 2011.

[10] M. B. Chadwick, P. Oblozinski, M. Herman, N. M. Greene, and R. D. McKnight, "ENDF/B-VII.0: Next Generation Evaluated Nuclear Data Library for Nuclear Science and Technology," vol. 107, no. 12, pp. 2931–3060, 2006.

Timing Characterization of Helium-4 Fast Neutron Detector with EJ-309 Organic Liquid Scintillator

Yinong Liang[1], Ting Zhu, Andreas Enqvist

***Abstract*— Recently, the Helium-4 gas fast neutron scintillation detectors is being used in time-sensitive measurements, such time-of-flight and multiplicity counting. In this paper, a set of time aligned signals was acquired in a coincidence measurement using the Helium-4 gas detectors and EJ-309 liquid scintillators. The high-speed digitizer system is implanted with a trigger moving average window (MAW) unit combing with its constant fraction discriminator (CFD) feature. It can calculate a "time offset" to the timestamp value to get a higher resolution timestamp (up to 50 ps), which is better than the digitizer's time resolution (4 ns) [1]. The digitized waveforms were saved to the computer hard drive and post processed with digital analysis code to determine the difference of their arrival times. The full-width at half-maximum (FWHM) of the Gaussian fit was used as to examine the resolution. For the cascade decay of Cobalt-60 (1.17 and 1.33 MeV), the first version of the Helium-4 detector with two Hamamatsu R580 photomultipliers (PMT) installed at either end of the cylindrical gas chamber (20 cm in length and 4.4 cm in diameter) has a time resolution which is about 3.139 ns FWHM. With improved knowledge of the timing performance, the Helium-4 scintillation detectors are excellent for neutron energy spectrometry applications requiring high temporal and energy resolutions.**

***Index Terms*—Helium-4 detectors, time resolution, TOF.**

I. Introduction

Time-of-Flight (TOF) technique is based on a pulsed neutron beam, spanning over a wide range of energies, travels a given distance before reaching the detector under study [2]. In this work, the kinetic energy of the neutrons is determined by calculating the difference between the timestamp reported by the analog-digital-converter (ADC) of each neutron event in the Helium-4 detector and the start time of the generator pulse. With a known distance between the target and the detector, the TOF was converted to determine the incident neutron energy. The outputs of the TOF measurement can be used in many neutron spectrometry applications such as detector light response calibration and mathematical spectrum unfolding [3]. The relative energy resolution of a TOF measurement is shown in (1) [2] as a first approximation. The moderation time in the target-moderator assembly, and the detector response, contribute to the fact that even for neutrons with a given kinetic energy, the measured TOF could have a distribution.

$$\frac{\Delta E_n}{E_n} = 2 \times \sqrt{\left(\frac{\Delta t}{t}\right)^2 + \left(\frac{\Delta L}{L}\right)^2} \quad (1)$$

As from (1), a high-accuracy TOF results can be achieved by the use of long flight paths (large L and t) and the relative energy resolution can be estimate by knowing the time resolution of the detection system. Previous studies [3] were mainly focused on the TOF based detector response characterization and the spectrum unfolding algorithm development. The accuracy of the TOF would affect the construction of the detectors response function and therefore the precision of the unfolding results. Therefore, it is crucial to characterize the timing performance of the Helium-4 detectors.

II. Materials and Experimental Methods

In this chapter, two versions of the Helium-4 fast neutron detectors were introduced and compared. For the timing measurement, a Struck SIS3316 16 channel digitizer was used. Its MAW unit is able to characterize the detectors time performance at a higher time resolution (less than 4 ns) through time interpolation.

A. Helium-4 Gas Fast Neutron Scintillation Detector

The Helium-4 gas fast neutron scintillation detector is based on neutron elastic scattering of Helium-4 nuclei. The Helium-4 nucleus gains kinetic energy from neutrons, and it will excite or ionize other helium atoms along its path within the detector volume. The gas pressure of Helium-4 is about 150 bar that the penetration distance of recoil is much smaller than the diameter of the gas chamber (see Fig.1). Scintillation photons are then be generated from the de-excitation process of the helium excimers. The first version of the Helium-4 detector has two Hamamatsu R580 photomultiplier tubes (PMT) on the both ends of the gas chamber are set up in coincidence mode so as to suppress the recording of PMT noise signals [4].

The second version has a series of SensL silicon-photomultipliers (SiPM), also known as the Geiger mode avalanche photodiode (GAPD) or solid state photomultiplier (SSPM), immersed inside a much longer (60 cm) pressurized (180 bars) helium-4 gas chamber [5]. The detector has a unique

[1] Submitted on 6/1/2017. This material is based upon work supported by the Department of Energy under Award Number: DE-NE0008459.

Y. Liang is a PhD student in the University of Florida, Gainesville, FL, 32611, USA (email: yinong@ufl.edu).

design of three optically separated segments in which 12 SiPM pairs (four in each segment) are positioned equilaterally across the detector to allow for them to be fully immersed in the Helium-4 gas volume; consequently, no additional optical interfaces are necessary. A schematic of the SiPM detectors can be seen in Fig. 2.

Typically, PMTs have faster signal rise time and lower dark count rates than SiPMs, but SiPMs have a high gain and quantum efficiency (QE), and are also very compact and rugged, insensitive to magnetic fields, and can be cost efficient when fabricated on a large scale [6]. For SiPM based detectors, they usually present longer tails due to the large output capacitance (C) associated with its relatively large pixel size and the input resistance (R) of the subsequent amplifier [7]. Therefore, the SiPM based detectors are expected having worse time resolution than PMT based ones. Measurements for the SiPM based detectors will be conducted in future work due to their more complex design.

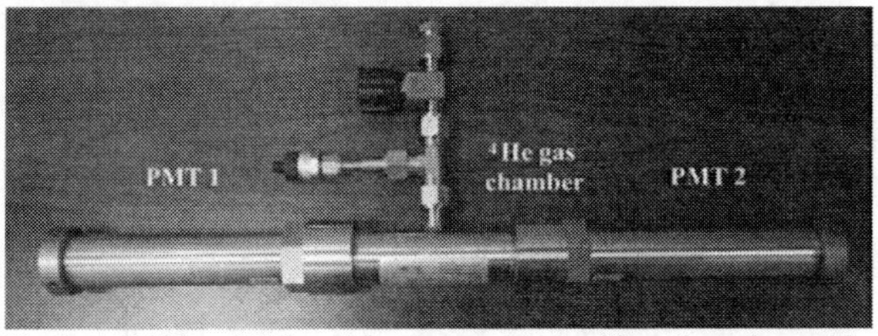

Fig. 1. A PMT based Helium-4 fast neutron detector. The active volume at the center is pressurized to 150 bar.

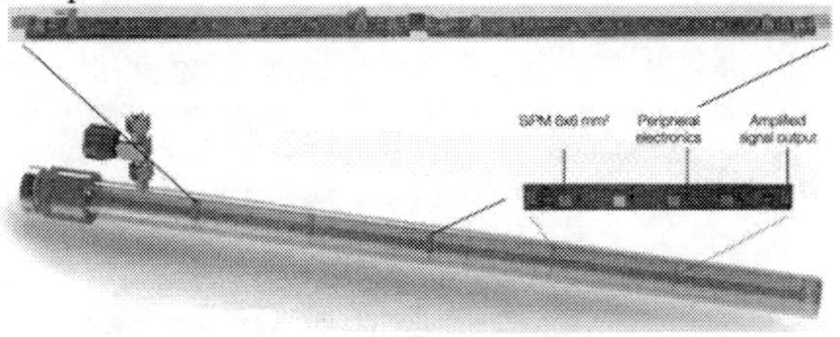

Fig. 2. Schematic of a SiPM based Heliium-4 detector.

The Helium-4 gas fast neutron scintillation detector has been demonstrated to be one of promising candidates to be used in nuclear safeguards and homeland security applications [8]. Energy information of incident neutrons could be retained in measurements based on the Helium-4 detector because no moderating materials are used. They also have fast response time on the order of nanoseconds and low sensitivity to γ-rays. In addition, neutrons and γ-rays can be well separated because of the good pulse shape discrimination capability [4]. These properties make the Helium-4 detector especially attractive for neutron detection in ultra-high γ-ray background. For example, in spent nuclear fuel assay, the γ-ray input count rate from the long-lived fission fragments could easily reach as high as 10^6 counts per second. The low sensitivity and good PSD performance will be greatly beneficial to the improvement of neutron detection throughput at high-rate.

B. SIS3316 16 channel VME digitizer

The data acquisition was performed with a Struck SIS3316 16 channel digitizer card with 250 MHz sampling speed and 14-bit resolution. It combines high channel density per card with the availability of greatly enhanced FPGA resources per channel and a flexible analog input stage followed by high resolution digitizer/ADC chips. The implemented FPGA firmware of the SIS3316 has many advantages which allow the timing characterization measurement. This section will mainly focus on the digitizer's trigger MAW unit and higher "timestamp" resolution ability.

For the internal trigger generation, a trapezoidal finite impulse response (FIR) filter with CFD feature is implemented for each ADC channel to generate a trigger signal. The initial ADC signal goes in to a MAW at first, and a moving average (MA) will be performed over the programmable peaking time. When the CFD feature is enabled, the trigger logic will be armed if the actual trapezoidal value goes above the programmable threshold value. A trigger pulse would be generated only if the trigger logic is armed and the actual trapezoidal value falls below the half of its maximum value. Three MAW values from FIR trigger trapezoidal can be used to get a "timestamp" which is better than the sampling period. With these three MAW values (maximum value, value after trigger, value before trigger), it is possible to make a linear interpolation between the two values (before and after) to calculate a "negative time offset" to the latched timestamp value to get the "real" timestamp [1].

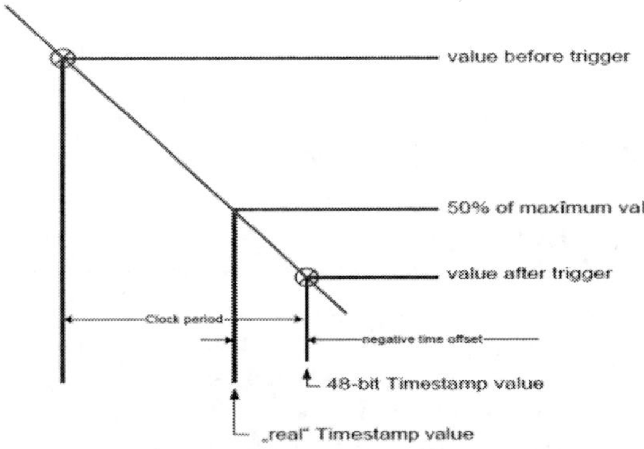

Fig. 3. Linear interpolation by using three MAW values.

C. Time resolution measurement

The hydrogen-based organic liquid scintillator EJ-309 (3" diameter x 3" long) is coupled with a 3" diameter ETEL 9821KB PMT. EJ-309 scintillators are widely used in both neutron and gamma detections and can be used as a reference and comparison with Helium-4 detectors. Time resolution of EJ-309 scintillators was measured at first. Two EJ-309 scintillators with same dimensions were located at an equal distance (5 cm from the Cobalt-60 gamma source to the face of the detectors) at opposite sides. Previous research has proved that the two gamma rays are angularly correlated, with a maximum probability at 180 degrees. Therefore the detectors were placed horizontally with respect to the outgoing gamma rays. Both detectors were set up in the time-coincidence mode (coincidence window length as 40 ns) such that each detector detected one of the two gammas rays emitted from the same Cobalt-60 beta-decay.

The experimental setup for measuring the PMT based Helium-4 detectors was basically the same except for substituting one EJ-309 to one PMT based Helium-4 detector and the employment of a linear fan-in/fan-out units. The anode signal from the two PMTs were connected to a Phillips Scientific 740 quad linear/logic fan-in/fan-out to maintain signal integrity, and was summed into one pulse with twice of the amplitude. The analog output signals were then digitized with a SIS3316-250MHz-14bit ADC chip with a dynamic range of 1.75 V. The programmable trigger threshold was adjust to relatively low (6.2 mV) and the shaping delays (10 ns) were optimized to obtain the best possible time resolution. The digitized data acquired from Struck SIS3316 was processed off-line to obtain the time information of each pulse, and the respective "timestamps" reported by the ADC were compared. One thing to be noticed is the intrinsic time resolution of the analog and digital modules is considered small (115 ps for fan-in/fan-out and 65 ps for Struck digitizer) [1], and is negligible when comparing with the nanosecond range of the time resolution for the whole system.

III. Result and discussion

Time resolution of the EJ-309 scintillators and PMT-based Helium-4 detectors were measured and compared in this chapter. For EJ-309 scintillator, the dependence of the threshold versus time resolution was explored. With increasing threshold, it tends to yield better time resolution. For the Helium-4 detectors, pulses from PMTs were analyzed individually and then summed together for comparison.

A. Time resolution calculation

In a photomultiplier tube, through the two major components inside the tube (photocathode and dynode), the extremely weak light output of a scintillation pulse can be converted to a corresponding usable electrical signal. Because the time required for photoemission in the photocathode or secondary emission from dynodes is very short (0.1 ns or less), the time characteristics of the PMT are determined exclusively by the electron trajectories [9]. The electron transit time of a PMT is defined as the average time difference between the arrival of a photon at the photocathode and the final collection of the subsequent electrons at the anode. In most timing applications, however, the transit time itself is not of the primary interests because if it were always a constant, it would introduce only a fixed delay in the derived signal and can be corrected later in pulse processing. Instead, the spread in transit time is a more important quantity because it determines the time width of the pulse of electrons arriving at the anode of the tube, and can somehow indicate the detection system's time resolution.

Due to various electron trajectories as mentioned above, there will be a slightly difference in the reported "timestamp" for each coincidence gamma rays from the Cobalt-60 cascade decay. The time difference is calculated as:

$$\Delta T = Timestamp_{Det1} - Timestamp_{Det2} \quad (2)$$

The time difference can by fitted by a Gaussian function. For peaks whose shape is Gaussian with standard deviation σ, the full width at half maximum (FWHM) is given by 2.35σ.

$$FWHM = 2.35\sigma \quad (3)$$

The standard deviation of the timing difference is calculated as below based on error propagation.

$$\sigma_{\Delta T}{}^2 = \sigma_{\Delta Timestamp_{Det1}}{}^2 + \sigma_{\Delta Timestamp_{Det2}}{}^2 \quad (4)$$

Combing all the equations above together, the FWHM for each type of detection systems can be obtained.

B. EJ-309 scintillators time resolution

The experimental setup of EJ-309 scintillator time resolution characterization is as shown in Fig. 4. A Cobalt-60 gamma source was placed at the center of the detector's active volume and was 5 cm to the face of both detectors. Within 1 minute, 2400 coincidence events were recorded. The detection pulse-height threshold for all measurements is 100 keVee.

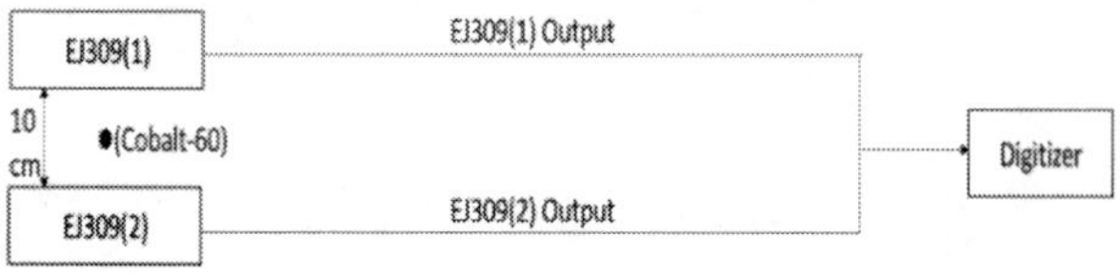

Fig. 4. Experimental setup for EJ-309 scintillator time resolution characterization.

Calibration was conducted as the first step to match the PMT voltages of both detectors. The "MeV-electron-equivalent" conversion was performed as in previous work [10]. Fig. 5 shows the distribution of the difference in the reported "timestamp" for each detector. It is modeled by a Gaussian function at a threshold of 100 keVee. The FWHM was calculated as 1.239 ns. From Fig. 6 and Fig. 7, the detector achieves better time resolution as threshold increases at a sacrifice of statistics since more coincidence events were rejected. Thus 100 keVee was chosen and was kept in later analysis.

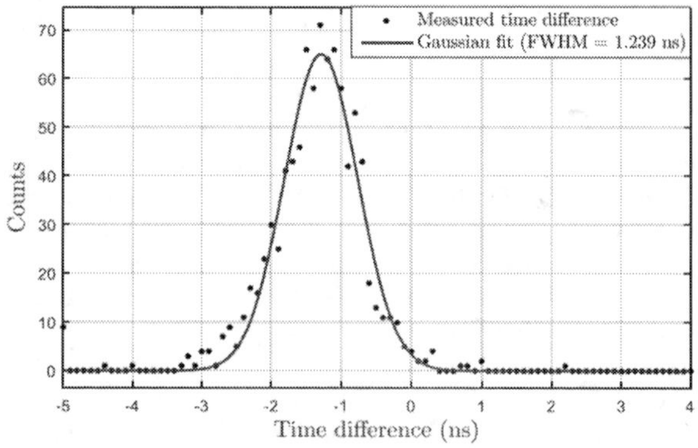

Fig. 5. Timestamp difference distribution of two EJ-309 scintillators as modeled in Gaussian function.

https://doi.org/10.1051/epjconf/201817007005

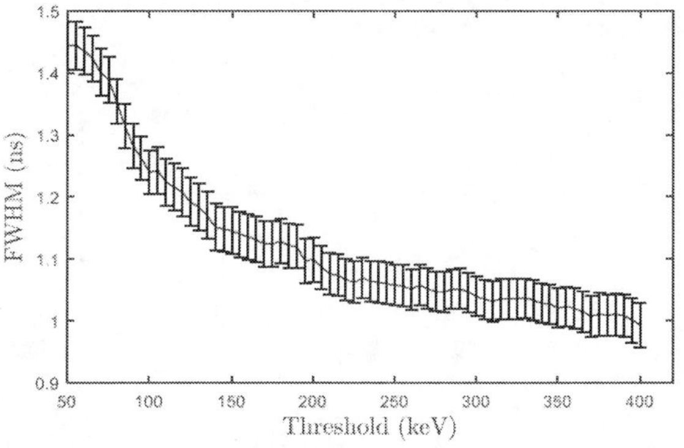

Fig. 6. The FWHM as a function of Threshold.

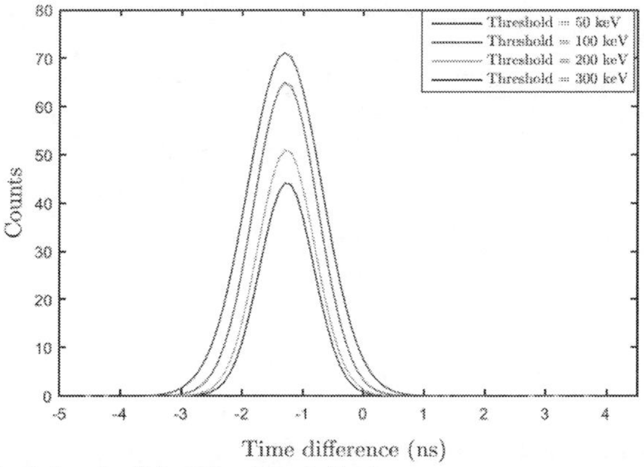

Fig. 7. Gaussian fit for different threshold values.

C. *PMT-based Helium-4 fast neutron detectors*

Fig. 8 shows the experimental setup for measuring the PMT-based Helium-4 fast neutron detectors. The PMT high voltage was set as 1356V (for node A003E) and 1538V (for node A003D) respectively to match the pulse amplitude based on previous calibration [4]. A Cobalt-60 gamma source was placed at the center of the detector's active volume and was 3 cm to the face of both detectors in order to achieve a higher count rate. Two leads were used as collimator. Only the Gamma rays interacted at the center of the detector were counted, thus the FWHM is independent of the interaction location. In 15 minutes, 5086 coincidence events were recorded. The pulses from two PMTs were summed in the fan-in/fan-out module and then measured on coincident mode with the EJ-309 output signals. From the equations in last section, by knowing the FWHM of EJ-309 scintillators, the FWHM for the summed signals was calculated as 3.139 ns, which was shown in Fig. 9. In addition, time resolution characterization for individual PMT was also conducted, and the time difference was calculated and compared separately with the summed PMT outputs as in Fig. 10. The FWHM of the time spread from the two PMTs are calculated as 2.820 ns (for node A003E) and 2.851 ns (for node A003D) respectively. As expected, summing the outputs together brought larger variance thus lead to larger FWHM.

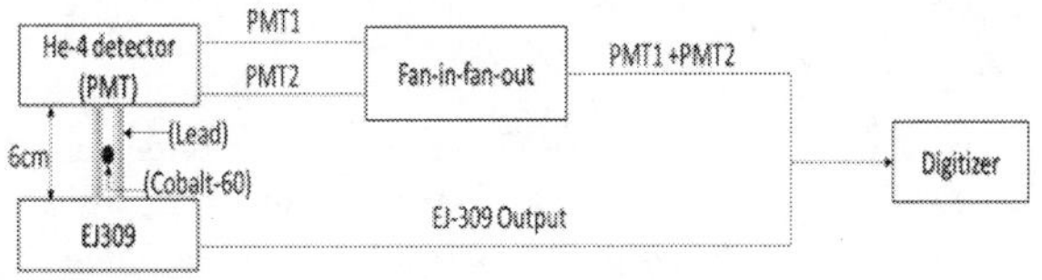

Fig. 8. Experimental setup for the PMT-based Helium-4 fast neutron detectors.

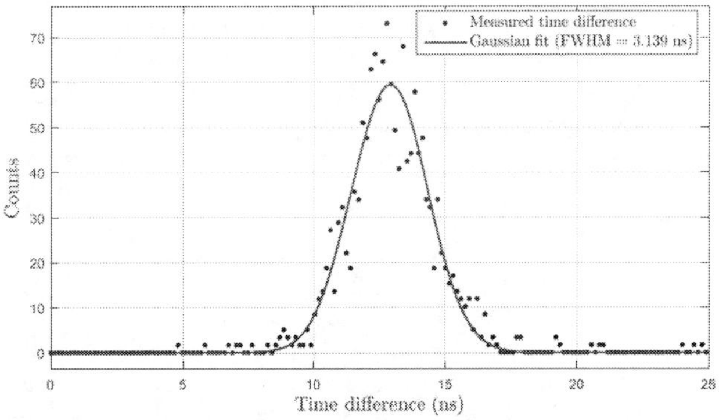

Fig. 9. Timestamp difference (EJ-309 and Helium-4 detectors) distribution as modeled in Gaussian function.

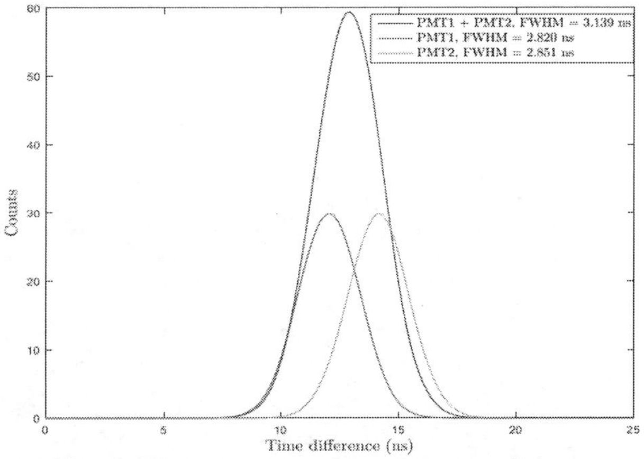

Fig. 10. Gaussian fit for the timestamp difference from summed as well individual PMTs.

By know the time resolution of the Helium-4 detection system, an estimation of the relatively energy resolution of the TOF measurement in previous work [11] can be achieved as in Table 1.

TOF (ns)	Energy (MeV)	ΔE/E (%)
295.2	6	4.25
361.5	4	3.47
511.1	2	2.46
723	1	0.87

Table. 1. Relatively energy resolution of previous TOF measurement (with a distance of 10 m) by knowing the Helium-4 detectors time resolution.

IV. Conclusions

In this paper, the time resolution of PMT-based Helium-4 fast neutron scintillation detectors was measured and compared with a hydrogen-based organic liquid scintillator (EJ-309) for the first time. For the cascade decay of Cobalt-60 (1.17 and 1.33 MeV), the Helium-4 detector was measured and yielded a time resolution as 3.139 ns FWHM. It is slightly worse (2 ns) than EJ-309 scintillators. The time resolution depends mainly on two parameters: the signal rise time and the signal to noise ratio, both of which should be high to give good time resolution [12]. For the Helium-4 detector, its low electron density leads to low gamma interaction probability. Additionally, Gamma rays also have a lower light yield therefore longer scintillation collection time in gaseous helium scintillation when compared to liquid scintillators. Thus, the gamma interaction based characterization measurement could compromise the Helium-4 detector's time resolution. Helium-4 detectors are designed for neutron detection with good gamma rejection and energy discrimination ability. The measured time resolution provides a reference to correct the uncertainties associated with TOF measurement and can be used to examine the detector's energy resolution in later work. Future work will also include the effect of source position, SiPM-based time resolution characterization, PMT operation voltage, and the effects of various digital analysis techniques (leading-edge discrimination [13], constant-fraction-zero-crossing [13], and digital-constant-fraction discrimination [14]) on the measured time resolution."

References

[1] "16 Channel VME Digitizer User Manual," 2015.

[2] C. Guerrero et al., "Performance of the neutron time-of-flight facility n_TOF at CERN," Eur. Phys. J. A, vol. 49, no. 2, pp. 1–15, 2013.

[3] R. P. Kelley et al., "Neutron response function characterization of 4He scintillation detectors," Nucl. Instruments Methods Phys. Res. Sect. A Accel. Spectrometers, Detect. Assoc. Equip., vol. 793, pp. 101–107, 2015.

[4] T. Zhu et al., "Improved fission neutron energy discrimination with 4He detectors through pulse filtering," Nucl. Instruments Methods Phys. Res. Sect. A Accel. Spectrometers, Detect. Assoc. Equip., vol. 848, September 2016, pp. 137–143, 2017.

[5] R. P. Kelley, T. Zhu, S. Gokhale, Y. Liang, H. Chung, S. Kiff, D. Murer, H. Ray, J. E. Baciak, "Neutron Measurements with Extended Range Helium-4 Detectors in High Gamma Environments," in 2016 IEEE Symposium on Radiation Measurements and Applications (SORMA West), 2016.

[6] J. Y. Yeom, R. Vinke, and C. S. Levin, "Optimizing timing performance of silicon photomultiplier based scintillation detectors," IEEE Nucl. Sci. Symp. Conf. Rec., pp. 3119–3121, 2012.

[7] S. Dolinsky, G. Fu, and A. Ivan, "Timing resolution performance comparison for fast and standard outputs of SensL SiPM," IEEE Nucl. Sci. Symp. Conf. Rec., 2013.

[8] H. Chung, R. P. Kelley, W. Lee, Y. H. Chung, and K. A. Jordan, "Spent nuclear fuel cask and storage monitoring with Helium-4 scintillation fast neutron detectors," Proc. Korean Nucl. Soc. 2014 Fall Meeting, 2014.

[9] G. F. Knoll, Radiation detection and measurement. Don Fowley, 2010.

[10] H. Wang, D. Carter, T. N. Massey, and A. Enqvist, "Neutron light output function and resolution investigation of the deuterated organic liquid scintillator EJ-315," Radiat. Meas., vol. 89, pp. 99–106, 2016.

[11] Y. Liang et al., "Neutron spectrum unfolding with 4He fast neutron detectors using an iterative least squares approach." unpublished.

[12] A. Ronzhin et al., "Waveform digitization for high resolution timing detectors with silicon photomultipliers," Nucl. Instruments Methods Phys. Res. Sect. A Accel. Spectrometers, Detect. Assoc. Equip., vol. 668, pp. 94–97, 2012.

[13] J. Nissilä, K. Rytsölä, R. Aavikko, A. Laakso, K. Saarinen, and P. Hautojärvi, "Performance analysis of a digital positron lifetime spectrometer," Nucl. Instruments Methods Phys. Res. Sect. A Accel. Spectrometers, Detect. Assoc. Equip., vol. 538, no. 1–3, pp. 778–789, 2005.

[14] D. G. Cussans and H. F. Heath, "Optimization of the pulse arrival time determination in the ZEUS central tracking detector FADC system," Nucl. Inst. Methods Phys. Res. A, vol. 362, no. 2–3, pp. 277–282, 1995.

A compact equipment for the sulfur content estimation in heavy oil using neutrons and gamma-rays

Pino F.[1,2], Stevanato L.[2], Fabris D.[3], Barros H.[1], Vidal A.[1], Nebbia G.[3], Lunardon M.[2], Moretto S.[2], Sajo-Bohus L.[1,2]

[1]*Universidad Simon Bolivar, Ap.do 89000, 1080A Caracas, Venezuela*
[2]*Dipartimento di Fisica ed Astronomia dell'Università di Padova, I-35131 Padova, Italy*
[3]*INFN Sezione di Padova, Via Marzolo 8, I-35131 Padova, Italy*

A compact and modern equipment for fast neutron and γ-ray transmission (FNGT) technique has been developed. The FNGT is used to perform non-destructive analysis of samples and it makes use of R-ratio that is the ratio of absorption coefficients of fast neutrons and gamma-rays [1].
The FNGT technique has a very good resolving power to determine the average atomic number of compounds with low atomic number (Z<13). Crude oil lays in this range of values, therefore we suggest to employ the mentioned technique for monitoring sulfur content in oil-business. The on-line analysis of fluid provides knowledge of sulfur content that has positive effects on the refinery plant. Nowadays, the used technique requires 7-10 days of delay in getting the information due the offline chemical or nuclear analysis.
The compact system makes use of a Cf-252 source, a single liquid scintillator detector EJ-301 with neutron/gamma discrimination capabilities and a modern nuclear electronics based on fast digitizers.
The use of an optimized Pulse Shape Discrimination algorithm provides a good separation of gammas from neutrons; the method is an alternative to that obtainable by TOF-technique, widely used in the past, that requires almost 1 meter of time of flight and another detector for triggering the fission events. The compactness of this assembly ensures a higher counting efficiency, offering the advantage of a shorter acquisition time to get sufficiently high statistic.
The fast neutron and gamma transmission technique was tested in laboratory for on-line sulfur concentration estimation in a sample of Venezuelan heavy oil. The range of sulfur concentration investigated is between 0.1% - 6.5% in weight. The experimental setup performance and limitations are compared with those predicted by a Monte Carlo model built in GEANT4 v10.01. In particular, we studied with Monte Carlo simulations the radiation buildup effect on the R-values and on the overall sensitivity of the technique. In fact, in the used experimental set-up, the Cf-252 source was not collimated in a "narrow beam", leaving to the detector the possibility to respond either to neutrons and γ-rays directly from the source or from scattered radiation.
The results show that this technique for on-line characterization of the crude oil copes well with the industrial requirements in terms of precision and response time, representing then an interesting alternative to the standard procedures.

The Sulfur-in-Oil Experiment

The experimental set-up was a compact system made of a Cf-252 neutron source, a single liquid scintillator detector with good response for both fast neutrons and γ-rays and a modern nuclear electronic setup based on fast digitizers (for digital processing of the signals). A Monte Carlo model based on the GEANT4 v10.01 code was built in order

to optimize and evaluate the response and the efficiency of the system.
The sample and radioisotope source holder are positioned at 30 cm from the front-face of the scintillator detector. The detector is of a cylindrical EJ-301 liquid scintillator cell of 5 cm in diameter and 5 cm thickness, coupled to an H1949-51 HAMAMATSU linearly focused 12 dynodes photomultiplier (PMT) through an EJ-560 silicone optical interface. The operating voltage was set at -1500 V. The front-end electronics used in this work was composed of the following CAEN VME modules: a V6533 Programmable HV Power Supply (6 Ch., 4 kV, 3 mA, 9 W), a V1730 Digitizer (14 bits, 500 MS/s), and a Bridge USB V1718 connected to a PC with a data acquisition software. Inside the digitizer, Digital Pulse Processing (DPP) algorithms are implemented using an FPGA, providing pulse on-line analysis for each event having: (a) a timestamp, (b) a complete integration of the signal, (c) a partial integration of the signal used for Pulse Shape Discrimination analysis and (d) the possibility of storing a selected part of the digitized signal. The optimization of the DPP parameters was obtained as described in [2]. A 10^5 neutrons/s Cf-252 source was used.
Mono-elemental samples (with thickness from 1 to 3 cm) were used to calibrate the system as reported in [1]. A second group of compounds was used to simulate the composition of the crude oil with different sulfur contents (from 0 to 6 wt%). The sample is made of a plastic cilindric container filled by common engine oil (with a negligible surfur content), and a sulfur disk of known mass attached on the top. An appropriate Monte Carlo simulation was carried out to demonstrate the negligible effect in using two compound's layer instead of a homogenously mixed sample.

Results and discussion

The data analysis was performed using ROOT software. Neutron/gamma discrimination was performed offline by setting a proper discrimination line in the PSD vs. Integrated-charge plot (see ref. [2] for the complete definition of the parameters and the neutron/gamma discrimination procedure). The separation line defines a neutron region with gamma contamination lower than 0.2% at all energies, calculated by measuring a pure ^{22}Na gamma source (see [2, 3] for more details)
Each measurement performed in this work was obtained in a time interval of 15 minutes (time of flight technique requires around 6 h of acquisition time to get equivalent statistics).
Material recognition can be obtained from R-ratio [1, 4, 5, 6] based on absorption coefficients for fast neutrons μ_n, and γ-rays μ_γ, or the logarithmic ratio of the measured transmission factors for neutrons $(I_n/I_{n,0})$ and gamma rays $(I_\gamma/I_{\gamma,0})$, respectively. Therefore the following relationship holds:

$$R = \mu_n/\mu_\gamma = \ln(I_n/I_{n,0})/\ln(I_\gamma/I_{\gamma,0})$$

where: I_n and I_γ ($I_{n,0}$ and $I_{\gamma,0}$) are measured yields for neutrons and γ-rays with (without) the sample. The most important outcome is that the R-ratio characterizes a selected sample independently of its thickness or geometrical shape.
In the R value calculation the same threshold on light output was used both for laboratory measurements and Monte Carlo simulations. The value was set at 50 keVee. The uncertainties associated to the experimental R-values in this work are 2.2%. This value can be calculated by error propagation, but was also verified experimentally.
The mono-elemental and mixtures samples R values are obtained as a function of their effective atomic number and it is shown in Fig. 1, left panel.

 https://doi.org/10.1051/epjconf/201817007006

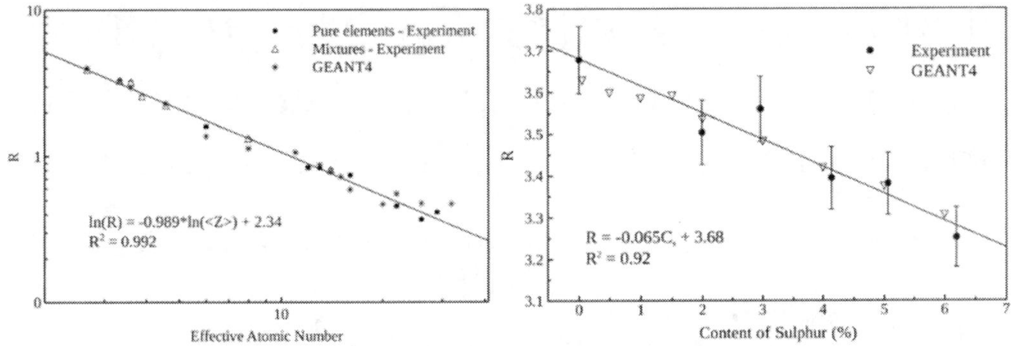

Fig. 1. R-ratio experimental results – see text fo details.

In Fig. 1, right panel, the experimental and Monte Carlo simulation results of the R-values associated to the crude oil samples with different sulfur concentrations are shown. The experimental R-values are well represented by a compound with an effective atomic number around $Z_{eff} \sim 3$, which is very close to the Z_{eff} of crude oil (calculated according to its well-known major composition). Above 2% of sulfur content, a linear trend of the data is observed, with very good agreement with the simulation results.

Monte Carlo results indicate that the determination of the sulfur content for concentrations lower than 1.5% suffers high uncertainty, however, it is more stable and predictable the sulfur concentration in the region corresponding to heavy-sour crude oil, where the technique is intended to be used.

Conclusions

These results indicate the possibility to develop a compact and modern equipment for quality control of heavy-sour oil (sulfur content higher than 2%) with the aim to provide a new technical tool to Venezuelan oil refineries. This non-destructive analysis to determine sulfur content has been demonstrated experimentally and proven by Monte Carlo simulation. The experimental results described in this paper show that this method is fully adequate in the range of average R-value corresponding to sulfur concentration of interest to the oil industrial refineries. This technique showed a lack of precision in the region of sulfur concentration below about 1.5%, nevertheless a useful precision could be obtained increasing the source intensity within reasonable bounds (and keeping the short time intervals for the analysis).

References

[1] Viesti G, et al. (2008). Nucl. Instrum. Meth. A. DOI 10.1016/j.nima.2008.05.024
[2] Stevanato L, Cester D, Nebbia G, Viesti G, (2012). Nucl. Instrum. Meth. A. DOI 10.1016/j.nima.2012.06.047
[3] Stevanato L, et al. (2011). Appl. Radiat. Isotopes. DOI 10.1016/j.apradiso.2010.10.022
[4] Stevanato L, et al. (2013). Appl. Radiat. Isotopes. DOI 10.1016/j.apradiso.2012.11.018
[5] R.J. Rasmussen, et al. (1997). Nucl. Instr. and Meth. B. DOI 10.1016/S0168-583X(97)00120-1
[6] Cywicka-Jakiel T (2003). Applied Energy. DOI 10.1016/S0306-2619(02)00190-3

Instrumentation effects on U and Pu CBNM standards spectra quality measured on a 500 mm^3 CdZnTe and a 2x2 inch $LaBr_3$ detectors

I. Meleshenkovskii, A. Borella, K. Van der Meer, M. Bruggeman, N. Pauly, P.E. Labeau, P. Schillebeeckx

***Abstract*— Nowadays, there is interest in developing gamma-ray measuring devices based on the room temperature operated medium resolution detectors such as semiconductor detectors of the CdZnTe type and scintillators of the $LaBr_3$ type. This is true also for safeguards applications and the International Atomic Energy Agency (IAEA) has launched a project devoted to the assessment of medium resolution gamma-ray spectroscopy for the verification of the isotopic composition of U and Pu bearing samples. This project is carried out within the Non-Destructive Assay Working Group of the European Safeguards Research and Development Association (ESARDA). In this study we analyze medium resolution spectra of U and Pu standards with the aim to develop an isotopic composition determination algorithm, particularly suited for these types of detectors. We show how the peak shape of a CdZnTe detector is influenced by the instrumentation parameters. The experimental setup consisted of a 500 mm^3 CdZnTe detector, a 2x2 inch $LaBr_3$ detector, two types of measurement instrumentation – an analogue one and a digital one, and a set of certified samples – a ^{207}Bi point source and U and Pu CBNM standards. The results of our measurements indicate that the lowest contribution to the peak asymmetry and thus the smallest impact on the resolution of the 500 mm^3 CdZnTe detector was achieved with the digital MCA. Analysis of acquired spectra allowed to reject poor quality measurement runs and produce summed spectra files with the least impact of instrumentation instabilities. This work is preliminary to further studies concerning the development of an isotopic composition determination algorithm particularly suited for CZT and $LaBr_3$ detectors for safeguards applications.**

***Index Terms*—CdZnTe, $LaBr_3$, uranium enrichment, isotopic composition, medium resolution, safeguards.**

I. Introduction

INTEREST in the use of detectors based on cadmium zinc telluride (CZT, room temperature compound semiconductor) and $LaBr_3$ (room temperature scintillator) as X-ray and gamma-ray spectrometers has grown rapidly in recent years[1,2,3,4]. CZT detectors are characterized by a wide energy band gap (E_g~1.6 eV) allowing room temperature operation of these detectors, their high atomic number (Z_{max}=52) yields a high intrinsic efficiency of gamma absorption (and thus a reduced size of the detector[5]. CZT detectors have now been optimized in their design yielding resolution less than 2% at 661 keV (^{137}Cs) for a 10 mm x 10 mm x 5 mm device[6]. $LaBr_3$ detector has a high light output (~60000 photons/MeV), fast response (decay constant <30 ns) and shows good energy resolution (~3% at 662 keV ^{137}Cs). Finally, the spectral resolution of these detectors is significantly better in comparison with other room temperature detectors, as shown in figure 1.

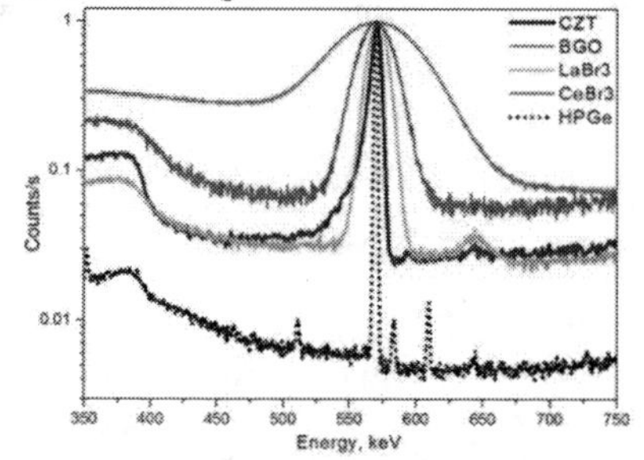

Fig. 1. Comparison of detector resolution for different detectors with a ^{207}Bi source.

Nuclear safeguards is one of the applications for these detectors because these detectors present a viable alternative to HpGe detectors used for quantitative gamma ray spectroscopy of uranium samples, plutonium samples and spent fuel assemblies[7,8,9]. Although neither CZT nor $LaBr_3$ detectors provide the resolution as that of HpGe detectors, they have certain attractive performance features, such as portability, compact design and absence of cryogenics, which make them advantageous for safeguards applications.

Safeguards applications of these detectors require a suitable algorithm for uranium and plutonium isotopic composition determination that would incorporate the spectral particularities of these detectors. Namely, CZT detectors exhibit a peak asymmetry due to incomplete charge collection processes and

I. Meleshenkovskii is with the Nuclear Research Center SCK-CEN, B-2400 Mol, Belgium (e-mail: yaroslav.meleshenkovskii@sckcen.be) and Université libre de Bruxelles, Service de Métrologie Nucléaire (CP/165/84).

A. Borella, K. Van der Meer and M. Bruggeman are with the Nuclear Research Center SCK-CEN, B-2400 Mol, Belgium, (e-mail: alessandro.borella@sckcen.be, klaas.van.der.meer@sckcen.be, michel.bruggeman@sckcen.be).

N. Pauly and P.E. Labeau, P are with Université libre de Bruxelles, Service de Métrologie Nucléaire (CP/165/84), (e-mail: nipauly@ulb.ac.be, pelabeau@ulb.ac.be).

P. Schillebeeckx is with European Commission Joint Research Center, Retieseweg 111, B-2440 Geel, Belgium, (e-mail: Peter.SCHILLEBEECKX@ec.europa.eu).

for both detectors a high degree of peak overlapping in complex spectra, such as that of plutonium, is a common issue.

To develop such an algorithm it is necessary to investigate the spectrometric performance of these detectors in a wide energy range and acquire spectra of certified uranium and plutonium standards in order to further test and validate the algorithm. The former requires a detector characterization procedure with point sources where peaks are well separated, which allows to quantitatively describe the peak shapes at different energies with a mathematical model and parametrize it[10]. The latter requires measurements with certified uranium and plutonium standards the enrichment or isotopic composition of which are known. However, both routines can be subject to bias introduced by the instrumentation, which may affect both the quality of the characterization procedure and the acquired spectra of standards.

In this paper we first investigate the influence of instrumentation parameters on the spectrometric performance of a 500 mm^3 CZT detector, and then perform a routine that aims to acquire high quality spectra of uranium and plutonium standards in different measurement configurations. Specifically, the impact of the shaping time constant on the spectral resolution and low-energy tailing was investigated.

This work is in the framework of the development of a uranium and plutonium isotopic composition determination algorithm for CZT and $LaBr_3$ detectors for safeguards applications.

The current manuscript is organized as follows. In section 2 the methodology, detectors, overview of standards and key instrumentation settings that influence the peak shape are given, section 3 provides the results and section 4 concludes the paper.

II. Materials and Methods

A. Methodology

As discussed in the literature[11], there are three major factors that influence the peak shape of a detector: the statistics of the charge creation processes (1); the completeness of the charge collection processes (2) and electronic noise (3). The first of these contributions is inherent to each detector material. The second contribution to the resulting peak shape is a characteristic of each detector and depends primarily on the mobility-lifetime properties of charge carriers (electrons and holes) and resulting charge collection efficiency. Incomplete charge collection efficiency creates a low-energy asymmetry and may cause serious distortion to the peak shape. The third contribution due to the electronic noise depends on the leakage current of the detector and its capacitance.

Although, there exist different designs of CZT detectors that aim to minimize the second contribution, such as co-planar grid, steering grid, pixelated and strip designs[12], the incomplete charge collection remains a problem for medium to large volume CZT detectors. However, the influence of electronic contribution to the peak asymmetry can be minimized by choosing an appropriate pulse shaping time constant and by operating the detector at the high-voltage bias[13] recommended by the manufacturer.

The resolution of $LaBr_3$ detectors is approaching that of some CZT detectors and its peaks do not exhibit an asymmetrical shape[14]. Due to this reason, in the current paper we focus on choosing an appropriate pulse shaping time constant for the CZT detector.

To quantitatively characterize the spectroscopic performance of the CZT detector with different pulse shaping time constants we evaluated the following parameters: energy resolution and peak shape parameter values. The full-width at half maximum (FWHM) of the full-energy peaks was assumed as a measure of the energy resolution of the detector. To allow for quantitative analysis of the pulse shaping time constant influence on the low energy tail, we evaluated the full width at 1/10th of the maximum (FWTM) and the ratio FWTM/FWHM and compared it to that of an ideal Gaussian peak ($FWTM/FWHM_{gaussian}$ =1.82). We evaluated the mentioned parameters for the ROI representing the 569.7 keV gamma line of the ^{207}Bi source. The spectra were acquired on two types of measurement instrumentation – a digital one and an analogue one.

As suggested in the literature[11], to monitor the impact of instrumentation instabilities on the spectra quality we measured uranium and plutonium standards in runs of 7200 seconds each. The total number of measurement runs was selected based upon a criterion to acquire no less than 10^5 counts in corresponding ROI for each of the standards after summing assuming rejected runs. Such methodology allows to produce a high quality summed spectrum file for each sample. To allow in-depth analysis of the quality of uranium and plutonium spectra measured on the CZT and $LaBr_3$ detectors we evaluated the following parameters for the ROIs representing the 185.7 keV gamma line of ^{235}U and 414.3 keV gamma line of ^{239}Pu for each of the measurement runs: *peak centroid, FWHM, net peak area* and net peak area *uncertainty*.

As indicated in the reference[15]: "The peak centroid is calculated as the sum of the background corrected channel contents multiplied by the channel number divided by the sum of the background corrected channel contents in the range of the channels above the photopeak half maximum":

$$Centroid = \frac{\sum_{i=l}^{h} i * Sp_i}{\sum_{i=l}^{h} Sp_i}$$

Where i is the channel number, l is the lowest channel above the half maximum, h is the highest channel above the half maximum and Sp_i are the net contents of channel i.

As indicated in the reference[15]: "The net peak area (1) is calculated as the number of counts above the background in a specified ROI. The background area (2) is determined by averaging 4 points on both sides of the peak (the ROI limit points and 3 outer points) and linear fit":

$$NPA = Int - BKG \quad (1)$$

$$BKG = \frac{ROI_e - ROI_b + 1}{8}\left(\sum_{i=ROI_b-3}^{ROI_b} Sp_i + \sum_{i=ROI_e}^{ROI_e+3} Sp_i\right) \quad (2)$$

 https://doi.org/10.1051/epjconf/201817007007

$$Int = \sum_{i=ROI_b}^{ROI_e} Sp_i \qquad (3)$$

Where *NPA* is the net peak area, *BKG* is the background, *Int* is the ROI's integral, Sp_i is the absolute number of counts in channels *i*, ROI_b is the ROI's start channel, ROI_e is the ROI's stop channel and *i* is the channel number.

As indicated in the reference[15]: "The Full Width at Half Maximum (FWHM) is calculated as the background corrected photopeak's width at one-half of its maximum amplitude. The algorithm first proceeds down one side of the peak until it finds a channel in which counts fall below the half maximum channel. The same algorithm is applied for the other side of the peak. The left and the right FWHM points are the interpolated channels between the counts of the channel below and above the half maximum value". The same algorithm was used to determine the FWTM.

B. Detectors

The CZT detector used was a 10 mm x 10 mm x 5 mm device of a hemispheric design fabricated by RITEC (Latvia) as shown in figure 2. The high-voltage bias applied to the detector was 1400 V.

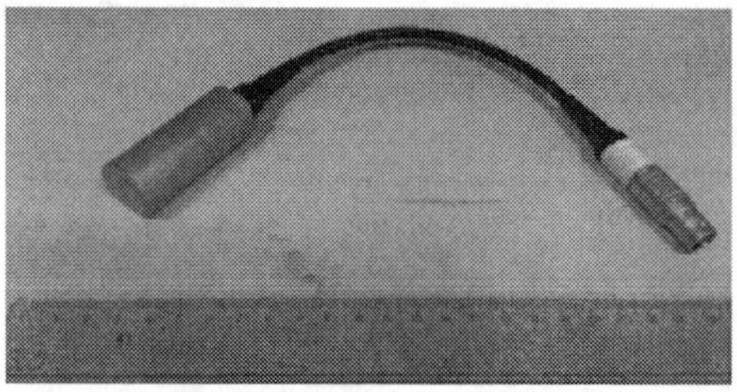

Figure 2: 500 mm^3 CZT detector.

The $LaBr_3$ scintillator used was a 2x2 inch device fabricated by Saint-Gobain Crystals (France) as shown in figure 3. The high-voltage bias applied to the detector was 590 V.

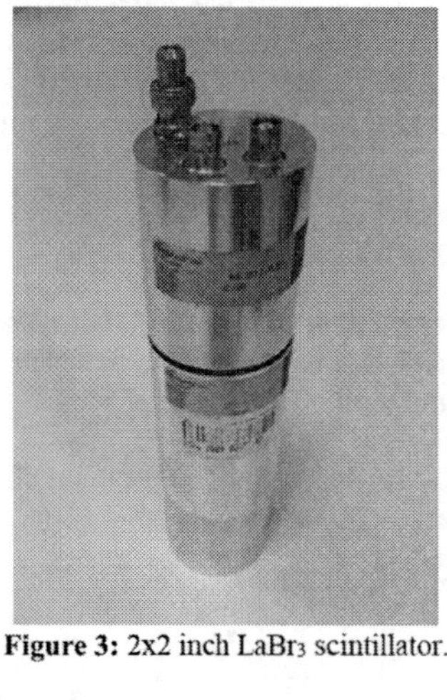

Figure 3: 2x2 inch $LaBr_3$ scintillator.

C. Electronics

The analogue electronics used to measure spectra consisted of standard nuclear spectroscopy signal processing chain (charge sensitive preamplifier, linear amplifier and shaper, analogue-to-digital converter) and a data acquisition software DAQ2000 developed by LabView. The schematic diagram of the analogue instrumentation circuitry is shown in figure 4.

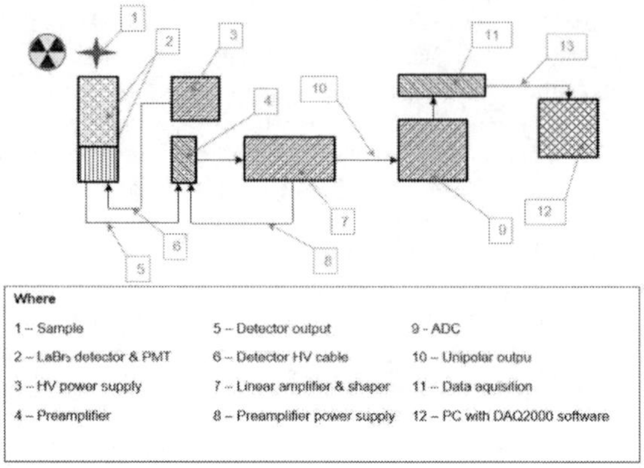

Figure 4: Schematics of the analogue instrumentation circuitry.

The digital instrumentation consisted of a digital multi-channel-analyzer (MCA) module, model 527, produced by GBS-Elektronik[15] (Germany) and a data acquisition software WinSpec[16]. The schematic diagram of the digital instrumentation circuitry is shown in figure 5.

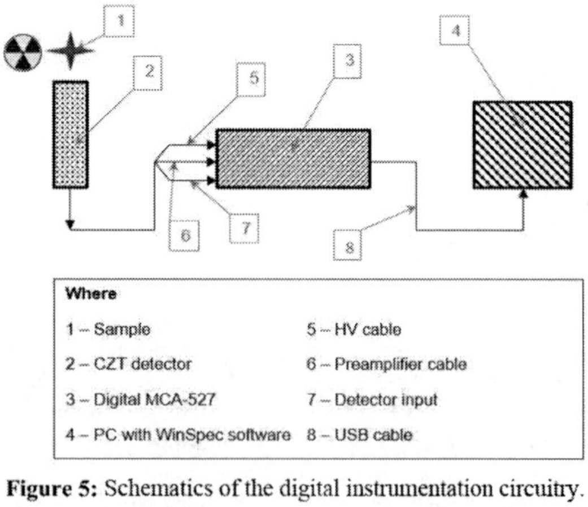

Figure 5: Schematics of the digital instrumentation circuitry.

The analogue instrumentation allowed to vary the pulse shaping time constant in the following range of values: 0.25; 0.5; 1; 2; 4 and 8 μs. The digital instrumentation allowed to vary the pulse shaping time constant in a wide range as well, however in the current paper we investigated the spectrometric performance of the CZT detector at the following values: 0.7 and 1.2 μs.

D. Samples

To characterize the spectroscopic performance of the CZT detector we used a certified ^{207}Bi source (569.7; 1063 and 1770 keV gamma lines), as shown in figure 6.

https://doi.org/10.1051/epjconf/201817007007

Figure 6: ^{207}Bi certified source.

For uranium measurements we used certified uranium CBNM (Central Bureau for Nuclear Measurements) standards, as shown in figure 7, of five different enrichment values: 0.31%, 0.71%, 1.94%, 2.95% and 4.46%.

Figure 7: Uranium certified standards.

For plutonium measurements we used certified plutonium CBNM standards, as shown in figure 8, of four different isotopic compositions as indicated in table 1.

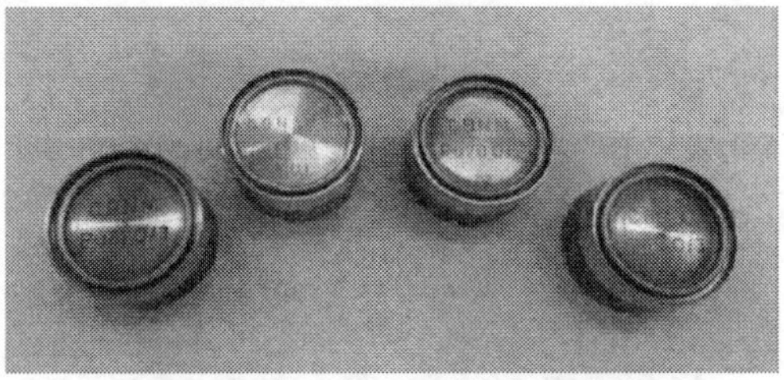

Figure 8: Plutonium certified standards.

CBNM Sample	Mass percent*					
	^{238}Pu	^{239}Pu	^{240}Pu	^{241}Pu	^{242}Pu	^{241}Am/Pu
Pu93	0.0117	93.4123	6.3131	0.2235	0.0395	0.1047
Pu84	0.0703	84.3377	14.2069	1.0275	0.3576	0.2173
Pu70	0.8458	73.3191	18.2945	5.4634	2.0772	1.1705
Pu61	1.1969	62.5255	25.4058	6.6793	4.1925	1.4452

Table 1: Isotopic composition of plutonium standards (*on 1986).

E. Collimators & Attenuators

In order to reduce the impact of scattered photons reaching the detector from the sample and nearby materials, the measurements were conducted with the lead collimators, the exact dimensions of which for CZT and $LaBr_3$ detectors are shown in figures 9 & 10 respectively. The uranium samples were positioned directly on the top surface of the lead collimators for which there were bored seats 81 mm in diameter and 5 mm deep. The sample-to-detector distance was 20 mm.

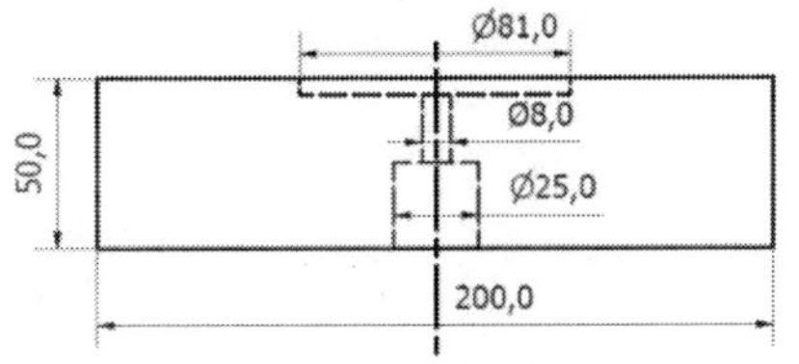

Figure 9: CZT lead collimator.

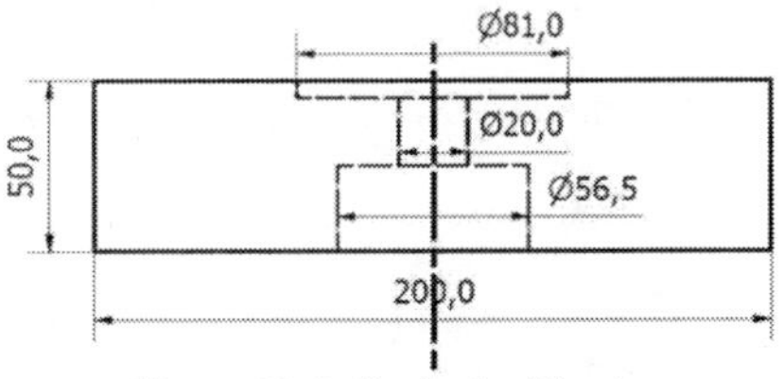

Figure 10: $LaBr_3$ lead collimator.

Measurements of plutonium certified standards were conducted in two configurations – with and without lead collimators on both detectors to investigate their impact on the spectra quality. Plutonium standards were positioned in adjustable height holders above the detectors. To reduce the impact of ^{241}Am 59 keV gamma line on plutonium spectra we used a set of cadmium (1.5 mm thick) and copper (0.5 mm thick) attenuators, as shown in figure 11:

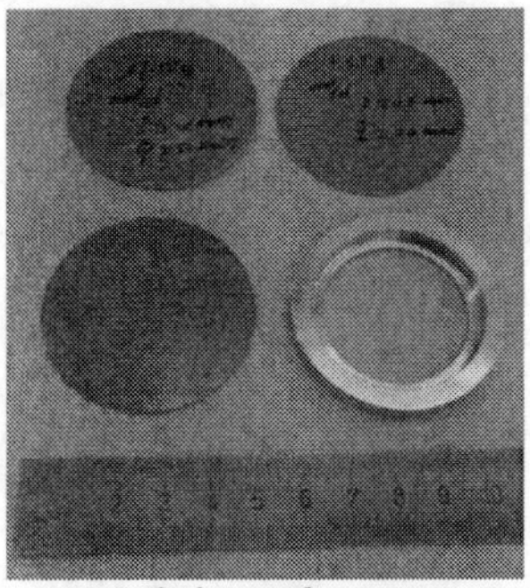

Figure 11: Cadmium and copper attenuators.

III. Results and Discussion

A. Spectroscopic Performance of CZT Detector

We first performed spectroscopic performance assessment of the CZT detector at different pulse shaping time constants on both types of instrumentation for the background corrected ROI corresponding to the 569.7 keV gamma line of the ^{207}Bi source. We measured the ^{207}Bi source for 3600 seconds (real time) on both types of instrumentation. The spectrometric performance of the CZT detector at various pulse shaping time constants for the analogue and digital instrumentation is shown in figures 12 and 13 respectively.

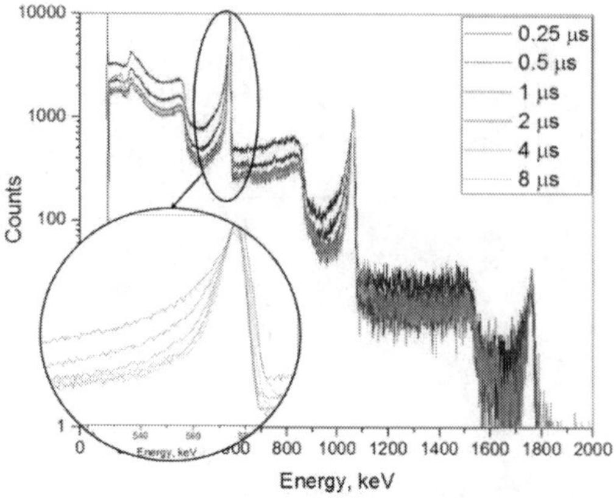

Figure 12: Spectrometric performance of the 500 mm³ CZT detector, analogue instrumentation.

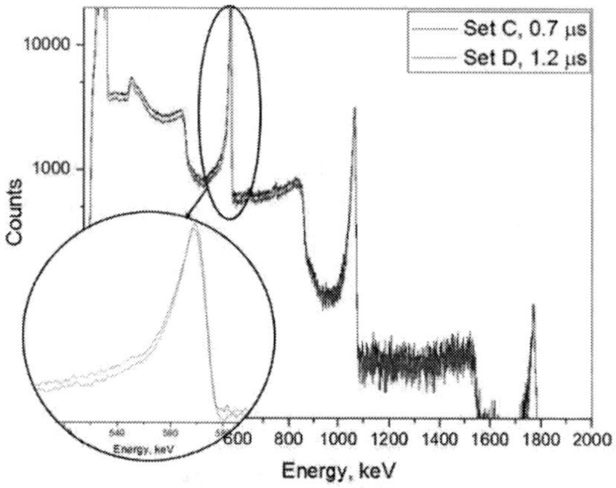

Figure 13: Spectrometric performance of the 500 mm³ CZT detector, digital MCA.

Results indicate that performance improvement for the CZT detector on analogue instrumentation is achieved starting from 1 μs pulse shaping time constant value, however there is no significant improvement in performance for values exceeding 2 μs. Indeed, at 2 μs pulse shaping time constant on analogue instrumentation the impact on the tailing is reduced by 56% given by the peak shape parameter, compared to the results for 0.25 μs pulse shaping time constant, as indicated in table 2.

Pulse shaping constant, μs	FWHM, %	FWTM, %	Peak shape parameter (FWTM/ FWHM)
0.25	2.01	8.04	4
0.5	1.54	5.10	3.3
1	1.59	4.46	2.8
2	1.38	3.11	2.25
4	1.28	2.89	2.25
8	1.26	2.69	2.13

Table 2: Spectrometric performance parameters for the 569.7 keV ROI of the ^{207}Bi source, analogue instrumentation.

Results of the spectrometric performance for the CZT detector on digital MCA indicate that optimal performance is achieved with 1.2 μs pulse shaping time constant value, as indicated in table 3. During the spectrometric performance assessment of the CZT detector with the digital MCA the settings of digital filters were kept constant.

Pulse shaping constant, μs	FWHM, %	FWTM, %	Peak shape parameter (FWTM/ FWHM)
0.7	1.45	3.31	2.28
1.2	1.27	2.71	2.12

Table 3: Spectrometric performance parameters for the 569.7 keV ROI of the ^{207}Bi source, digital MCA.

Comparison of digital and analogue instrumentation performance between each other for the 569.7 keV ROI of the ^{207}Bi source is given in figure 14. Indeed, it is evident that the digital MCA with 1.2 μs pulse shaping time constant exhibits less electronic impact on the resolution and tailing of the CZT detector peak shape.

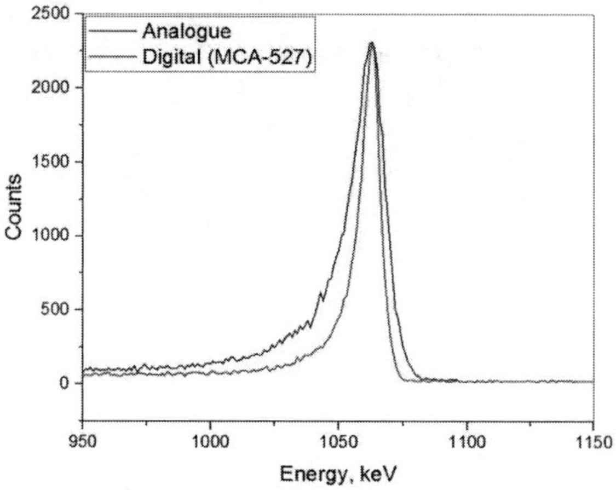

Figure 14: Digital and analogue spectroscopic performance comparison, ROI 569.7 keV of the ^{207}Bi source.

B. Stability of CZT and $LaBr_3$ Detectors

For the measurements of uranium and plutonium standards we used 1.2 μs pulse shaping time constant for the CZT detector on the digital MCA and 0.25 μs pulse shaping time constant for the $LaBr_3$. These settings allowed to achieve good throughput of the analogue pulse processing chain coupled with the $LaBr_3$ detector considering its high efficiency and to minimize the deadtime to less than 2% on the CZT detector coupled with the digital MCA. The sample-to-detector distances used for plutonium measurements are given in table 4.

To investigate the time stability of the CZT and $LaBr_3$ detectors during the measurements of uranium and plutonium standards we evaluated the following parameters for each measurement run: *peak centroid*, *FWHM*, *net peak area* and *net peak area uncertainty*, as shown, for example, for the CBNM-031 uranium sample in figures 15, 16 and 17 respectively. For this purpose, the corresponding Matlab-based programs were developed from scratch; they accept the raw spectra files and the user defined ROI. Identical ROI limits were used for all measurement runs.

CBNM Sample	Sample-to-detector distance, cm			
	$LaBr_3$		CZT	
	Lead	No lead	Lead	No lead
Pu93	25	25	5.4	4
Pu84	25	25	5.4	4
Pu70	25	25	8	8
Pu61	25	25	8	8

Table 4: Sample-to-detector distance for plutonium measurements.

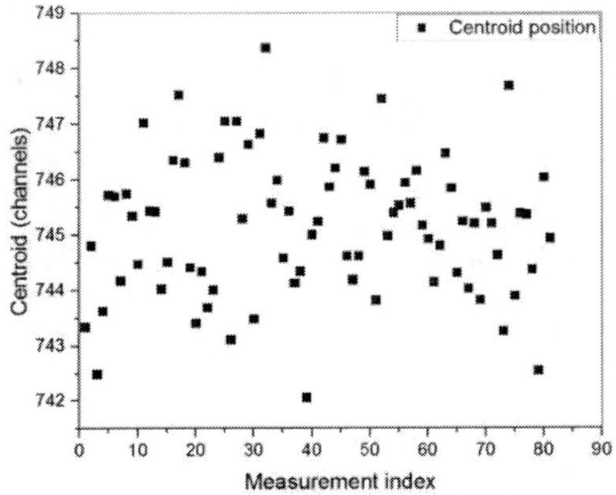

Figure 15: CBNM-031 sample centroid positions.

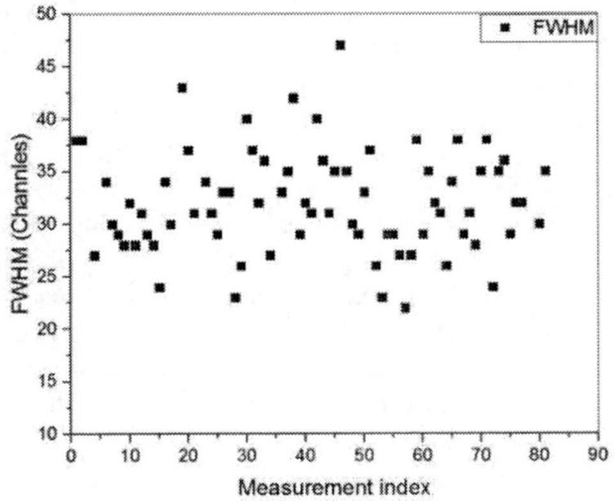

Figure 16: CBNM-031 sample peak widths.

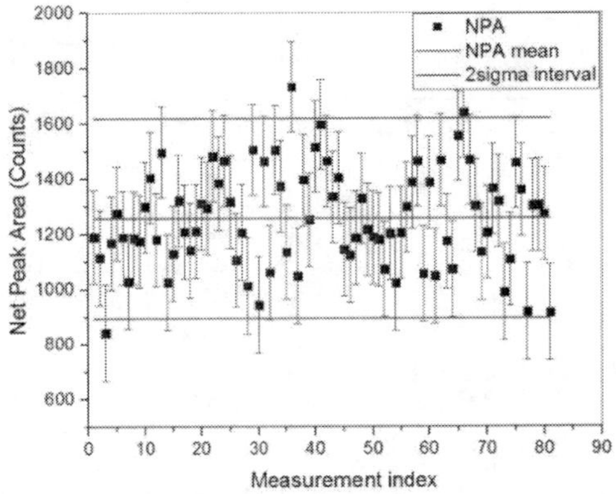

Figure 17: CBNM-031 sample net peak areas and uncertainties.

In order to reject poor quality spectra the rejection criteria based on the 95.5% confidence interval for the net peak area values, ±1 channel centroid drift for the 185.7 keV gamma-line of ^{235}U and ±2 channels centroid position drift for the 413.4 keV gamma-line of ^{239}Pu were adopted. For each standard measured we calculated the mean value of the observed net peak area values for all measurement runs and set the minimum and maximum limits for the net peak area values in corresponding ROIs based on the two sigma interval around that mean value. Those files whose observed net peak area values fall out of this interval, as well as spectra files with significant centroid position drift in corresponding ROIs, were rejected. Spectra that satisfy the criteria were used to produce the summed spectrum file for each given standard. The statistical details of the summed spectra files for all CBNM standards measured are given in tables 5, 6, 7, 8, 9 and 10 respectively.

CBNM Sample	Total m. time, s	N. r. runs	Cor. m. time,s	Σ spectra (ROI 185.7 keV)	
				NPA, counts	NPA unc., δ
U031	583200	36	324000	56585	1.9
U071	410400	9	345600	129620	0.8
U194	345600	4	316800	319203	0.4
U295	223200	2	208800	315350	0.3
U446	151200	1	144000	332037	0.3

Table 5: CZT summed uranium spectra results.

CBNM Sample	Total m. time, s	N. r. runs	Cor. m. time,s	Σ spectra (ROI 185.7 keV)	
				NPA, counts	NPA unc., δ
U031	338400	33	100800	1399278	0.5
U071	165600	4	136800	3510569	0.2
U194	64800	2	50400	3523567	0.1
U295	28800	0	28800	2291632	0.1
U446	21600	0	21600	2311297	0.1

Table 6: $LaBr_3$ summed uranium spectra results.

CBNM Sample	Total m. time, s	N. r. runs	Cor. m. time,s	Σ spectra (ROI 413.4 keV)	
				NPA, counts	NPA unc., δ
Pu93	280800	2	266400	9979283	0.06
Pu84	172800	2	158400	5299094	0.09
Pu70	352800	2	338400	1901123	0.2
Pu61	331200	3	309600	1436089	0.3

Table 7: CZT summed plutonium spectra results, without lead.

CBNM Sample	Total m. time, s	N. r. runs	Cor. m. time,s	Σ spectra (ROI 413.4 keV)	
				NPA, counts	NPA unc., δ
Pu93	172800	1	165600	2174818	0.1
Pu84	172800	0	172800	1982828	0.1
Pu70	172800	1	165600	587762	0.3
Pu61	172800	1	165600	511253	0.4

Table 8: CZT summed plutonium spectra results, with lead.

CBNM Sample	Total m. time, s	N. r. runs	Cor. m. time,s	Σ spectra (ROI 413.4 keV)	
				NPA, counts	NPA unc., δ
Pu93	57600	3	36000	3126014	0.1
Pu84	79200	2	64800	4964420	0.1
Pu70	57600	2	43200	3447786	0.2
Pu61	28800	0	28800	1704848	0.3

Table 9: $LaBr_3$ summed plutonium spectra results, without lead.

CBNM Sample	Total m. time, s	N. r. runs	Cor. m. time,s	Σ spectra (ROI 413.4 keV)	
				NPA, counts	NPA unc., δ
Pu93	72000	4	43200	1079943	0.2
Pu84	21600	0	21600	494498	0.4
Pu70	28800	3	7200	145344	0.9
Pu61	21600	0	21600	407703	0.5

Table 10: $LaBr_3$ summed plutonium spectra results, with lead.

Where *Total m. time* is the total measurement time, *N. r. runs* is the number of rejected runs, *Cor. M. time* is the corrected measurement time and *NPA* is the net peak area.

IV. Conclusion

Measurements of uranium and plutonium certified standards were conducted on two detectors – a 500 mm^3 CZT semiconductor detector and a 2x2 inch $LaBr_3$ scintillator.

Asymmetrical peak shape of a CZT detector can be quite sensitive to the electronic noise introduced by the pulse processing chain. This can negatively influence the quality of acquired spectra. To reduce the impact of the electronic noise on the spectra quality we first investigated the spectrometric performance of a 500 mm^3 CZT detector of hemispheric design at different settings of the pulse shaping time constant on two types of instrumentation – a digital one and an analogue one. The results indicated that asymmetrical peak shape of a CZT detector is very sensitive to the pulse shaping time constant value and best results were achieved with the digital MCA. The $LaBr_3$ detector does not have an asymmetrical peak shape, however its resolution is worse than that of CZT in the whole range of energies.

To analyze the stability of both detectors during the measurements of uranium and plutonium standards we performed measurements in runs of 7200 seconds each. Such methodology allowed us to reject poor quality spectra and to produce the summed spectra files based on the measurement runs that satisfy the rejection criteria. The results of uranium measurements indicate that both detectors showed quite strong gain drift, especially during the long-term measurements of low enriched uranium standards. The observed fluctuations of the net peak area values in the ROI corresponding to the 185.7 keV gamma-line of ^{235}U were smaller.

The results of plutonium measurements indicate strong gain drift on the $LaBr_3$ detector coupled with the analogue instrumentation. Thus, the number of rejected measurement runs was higher for the $LaBr_3$ detector in comparison with the CZT detector. CZT detector coupled with the digital MCA showed better stability in measurements with and without lead collimators during the measurements of plutonium standards.

The results of this study will be further used for the development of a uranium and plutonium isotopic composition determination algorithm particularly suited for CZT and $LaBr_3$ detectors for safeguards applications. The produced spectra of uranium and plutonium standards will be added to the ESARDA database.

References

[1] T.E. Schlesinger, R.B. James (Eds.), Semiconductors for Room Temperature Nuclear Detector Applications, Semiconductors and Semimetals, vol. 43, Academic Press, New York, 1995.

[2] R.B. James, T.E. Schlesinger, P. Siffert, M. Cuzin, M. Squillante, W. Dusi (Eds.), Semiconductors for Room Temperature Nuclear Detector Applications II, in: Materials Research Society Symposium Proceedings, vol. 487, Pittsburgh, MA, 1998.

[3] R.B. James, P. Siffert, Proceedings of the 11th International Workshop on Room Temperature Semiconductor X- and GammaRay detectors and Associated Electronics, Nucl. Instr. and Meth. A 458 (2001) 1.

[4] Raquel Gonzales, "Performance comparison of a large CZT semiconductor detector and a $LaBr_3(Ce)$ scintillator detector", 2005.

[5] T. Takahashi and S. Watanabe, "Recent Progress in CdTe and CdZnTe Detectors", IEEE Transactions on Nuclear Science, Page(s): 950 - 959 (Volume: 48, Issue: 4, Aug 2001.

[6] Etim Iniobong Prosper *et al*, "Characterisation of Cerium-Doped Lanthanum Bromide scintillation detector", Lat. Am. J. Phys. Educ. Vol. 6, No. 1, March 2012.

[7] V. Ivanov, P. Dorogov "Further development of hemispheric CZT detectors for safaguards applications", Proceedings of the 21st ESARDA Annual Symposium, Sevilla, Spain, 4-6 May 1999.

[8] Wayne D. Ruhter "Application of CZT detectors in nuclear materials safeguards", Proc. SPIE 3446, Hard X-Ray and Gamma-Ray Detector Physics and Applications, 204 (July 1, 1998); doi:10.1117/12.312892.

[9] V. Ivanov *et al*, "Performance Evaluation of New Generation CdZnTe Detectors for Safeguards Applications", 2014.

[10] I. Meleshenkovskii *et al*, "Characterization of a 500 mm^3 CdZnTe Detector for U and Pu isotopic Composition Determination tasks in Safeguards", ESARDA proceedings, Dusseldorf, Germany, 2017.

[11] K. Debertin, R. G. Helmer, "Emission-rate determinations" in *Gamma- and X-ray Spectrometry with Semiconductor Detectors*, North-Holland, 1988, ch. 4, pp. 286-287 book.

[12] L. Abbene, S. de Sordo, "Spectroscopic response of a CZT multiple electrode detector".

[13] Gordon R. Gilmore, "Electronics for Gamma-Ray Spectrometry" in *Practical Gamma-ray Spectrometry*, 2nd Edition, Nuclear Training Services Ltd Warrington, UK, John Wiley & Sons, Ltd, 2008, ch. 4, pp. 73-74.

[14] Kromek Coplanar Grid CZT detectors, brochure, http://www.kromek.com/index.php/products/nuclear-technology/czt/cpg

[15] *MCA-527 user manual*, update 2012-08-07, GBS-Elektronik GmbH, Großerkmannsdorf, Germany, 2012.

[16] https://www.gbs-elektronik.de/en/downloads/downloads-nuclear-measurements.php

Modelling of the anti-neutrino production and spectra from a Magnox reactor

Robert W Mills, David J Mountford, Jonathon P. Coleman, Carl Metelko, Matthew Murdoch, and Yan-Jie Schnellbach

***Abstract*—The anti-neutrino source properties of a fission reactor are governed by the production and beta decay of the radionuclides present and the summation of their individual anti-neutrino spectra. The fission product radionuclide production changes during reactor operation and different fissioning species give rise to different product distributions. It is thus possible to determine some details of reactor operation, such as power, from the anti-neutrino emission to confirm safeguards records. Also according to some published calculations, it may be feasible to observe different anti-neutrino spectra depending on the fissile contents of the reactor fuel and thus determine the reactor's fissile material inventory during operation which could considerable improve safeguards. In mid-2014 the University of Liverpool deployed a prototype anti-neutrino detector at the Wylfa R1 station in Anglesey, United Kingdom based upon plastic scintillator technology developed for the T2K project. The deployment was used to develop the detector electronics and software until the reactor was finally shutdown in December 2015. To support the development of this detector technology for reactor monitoring and to understand its capabilities, the National Nuclear Laboratory modelled this graphite moderated and natural uranium fuelled reactor with existing codes used to support Magnox reactor operations and waste management. The 3D multi-physics code PANTHER was used to determine the individual powers of each fuel element (8x6152) during the year and a half period of monitoring based upon reactor records. The WIMS/TRAIL/FISPIN code route was then used to determine the radionuclide inventory of each nuclide on a daily basis in each element. These nuclide inventories were then used with the BTSPEC code to determine the anti-neutrino spectra and source strength using JEFF-3.1.1 data. Finally the anti-neutrino source from the reactor for each day during the year and a half of monitored reactor operation was calculated. The results of the preliminary calculations are shown and limitations in the methods and data discussed.**

***Index Terms*—Antineutrino; Reactor spectrum; Nuclear reactor; Fission Product Yields; Nuclear safeguards**

The work of the UK National Nuclear Laboratory (NNL) authors was supported by the NNL's Innovation Research and Development programme. The University of Liverpool authors acknowledge the support of the STFC and the Royal Society.

R. W. Mills is with the National Nuclear Laboratory, Central Laboratory, Sellafield, Seascale, Cumbria, United Kingdom. (e-mail: robert.w.mills@nnl.co.uk).

D. J. Mountford was with the National Nuclear Laboratory, Central Laboratory, Sellafield, Seascale, Cumbria, United Kingdom during this work.

J. P. Coleman, C. Metelko, M. Murdoch, and Y. Schnellbach are with the University of Liverpool, Liverpool L69 7ZE, United Kingdom.

I. Introduction

THE anti-neutrinos emission from fission has been suggested as a way to monitor nuclear reactors for some time as they have extremely low rates of interaction with matter and thus are considered impossible to shield their emission from reactors. However, this extremely low interaction rate makes the detection of anti-neutrinos very difficult.

There are several steps that are necessary for such reactor monitoring to become a practical technology. Firstly, a detector must be constructed that can uniquely identify reactor anti-neutrinos considering radiations of other types such as reactor neutrons and gamma-rays, and naturally occurring radiations from the environment such cosmic rays, neutrons, gamma-rays and other neutrino sources (including anti-neutrinos from the beta decay of naturally occurring radio-isotopes). Secondly, the detector must be easily used in the area near the nuclear reactor. Nuclear Safety on such sites requires that any equipment must not introduce risks to the reactor or its staff. Specifically, the equipment should not introduce extra chemical, physical, fire or radiation risks, nor should it need to be incorporated into existing safety systems or have significant power requirements. Due to security access issues the equipment should be able to be left as a standalone unit not needing many maintenance visits with most contact being via remote access methods, whilst allowing interference with the equipment to be easily detected remotely. Finally, the detector should easily supply information on the reactor without the need for complex analysis. To ensure that the equipment is useful it must be demonstrated that the detector can function on a nuclear site and that the reported information obtained from the detector be verified against plant records by carrying out experiments on well characterized reactors.

This paper reports work considering the University of Liverpool anti-neutrino detector that was placed near the Wylfa reactor building for 18 months, prior to the last Wylfa reactor core being shut down.

II. The University of Liverpool detector

The detector, now called VIDARR, is about 1 m^3 of plastic scintillator detector using Si photomultipliers. It is based upon the technology of the T2K neutrino experiment, which works by using the inverse beta decay reaction on hydrogen in the

 https://doi.org/10.1051/epjconf/201817007008

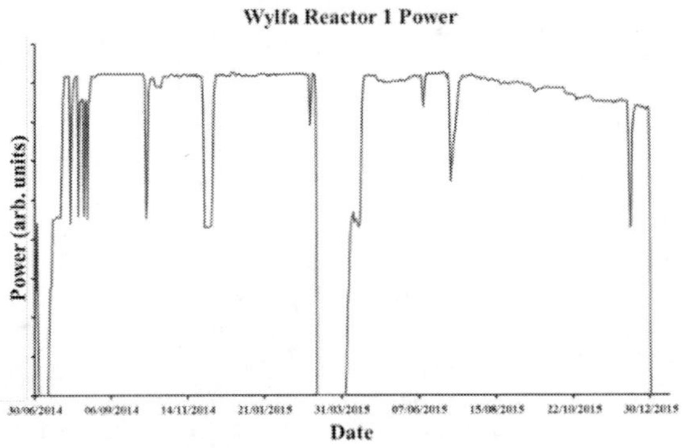

Fig. 1. Wylfa reactor 1 power during detector deployment showing both two periods of start-up and shutdown with periods of low power to replace or remove or shuffle fuel within the core.

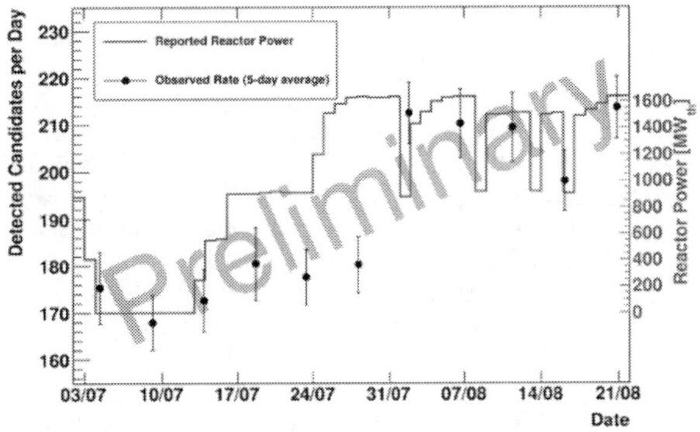

Fig. 2. Wylfa reactor 1 power during the July 2014 start-up in MW thermal and the 5-day averaged event detection rate (per day).

plastic ($\overline{\nu_e} + p \rightarrow e^+ + n$). The positron produces two annihilation photons and the neutron is scattered around the volume until it is captured by the trace of gadolinium introduced into the plastic. The neutron capture reaction results in a delayed radiation signature. From the energies of the two annihilation photons it is possible to trigger on an anti-neutrino capture event and from the radiation produced by the neutron capture reaction to identify the events as an anti-neutrino capture and determine the energy of the anti-neutrino. The detector events from VIDARR were processed and stored on a computer within the instrumentation rack, which could be accessed via a phone connection.

As with all inverse beta decay detectors, the threshold of the anti-neutrino reaction on hydrogen is about 1.8 MeV and then the cross-section rapidly rises. However, the anti-neutrino spectrum above 1.8 MeV rapidly reduces with increasing anti-neutrino energy. Thus the probability of detection is zero below 1.8 MeV and rises rapidly above this, peaking around 4 MeV before rapidly declining at higher energies.

The detector was built at the University of Liverpool and installed in a refrigerated 20ft transport container. The stable, lower than ambient temperature, in the container helped to reduce the electronic noise from the photomultipliers.

In the middle of 2014 the container was transported to the Wylfa reactor site in North Wales and installed in a layby on a site road at about 60m from the centre of the operating Magnox reactor. The instrumentation and cooling equipment was powered by a single cable from a convenient power socket and the equipment was connected via a phone line for remote access to the computer.

The above ground installation of the detector leads to a high cosmic ray background that complicated the identification of anti-neutrino events detected.

III. Reactor Operation during deployment

The Wylfa reactor complex consists of two reactors in a single reactor building linked to an adjacent turbine building which generates the electrical power. Reactors 1 and 2 were built from 1963 and became operational in 1971. They were initially each rated at 570 MW electrical, but during operation this was reduced to 490 MW electrical. The reactor 2 was shut down in 2012, but reactor 1 operated until December 2015, giving about 18 months of operation during which the detector was deployed.

Each of the two Wylfa reactor cores consists of a graphite moderator containing 6156 fuel channels into to each of which can be inserted up to 8 Magnox fuel elements. Each element consists of a uranium metal cylinder surround by a Magnesium alloy cladding (Magnox™) with end-fittings joining the elements in each channel. The uranium is natural or 0.8 wt. % ^{235}U w.r.t. U with each element containing about 12 kg of uranium when installed. The total core loading was approximately 593 metric tons of uranium. Wylfa is unusual for Magnox reactors in that on-power refueling was regularly carried out but this required the reactor power to be significantly reduced during the fuel movements.

The reactor power during the period with the detector installed is shown in figure 1.

IV. Preliminary results from the detector

During the initial period of operation, the detector identified about 170 events per day during shutdown, but this rose to about 210 events during full power operation. Some preliminary results are shown in figure 2. These preliminary results show that the detector can be used to identify reactor power, but that improvements are required in the analysis of the detector response to cosmic rays and other background radiations to improve the identification of anti-neutrinos.

V. Calculations methodology for anti-neutrino emission estimation

The full operational data of the Wylfa reactor 1 was obtained by NNL including the full power history and the fuel element's irradiation (MWd/t), power rating (MW/t), initial enrichment and loading dates for the whole core during the last 18 months of operation. This was calculated using the PANTHER whole core modelling code. The power and shutdown history was then used to estimate the full radionuclide inventory in each of the 49248 fuel elements using the UK spent fuel inventory code FISPIN for each day of the reactor operation and shutdown. This gave the reactor's radionuclide inventory and the beta decay rate of all the isotopes present.

The beta spectra were then calculated using the BTSPEC code with the latest JEFF-3.1.1 decay data for all 670 radionuclides with sufficient information in this library. The code was modified to use the end-point energies of each transition to estimate the corresponding anti-neutrino spectra. It should be noted that these 670 nuclides represent about 90% of the beta decays in the fuel during operation rising to 100% during shutdowns, but that without more information on the short-lived radionuclides and their beta decay properties that it is not possible to estimate their anti-neutrino spectra. In this work the beta particle and anti-neutrino spectra were generated in 10 keV bins between 0 and 15 MeV.

By combining the spectra data and the radionuclide activities for each rod it is possible to estimate the anti-neutrino emission and spectra from the reactor as a 3 dimensional anti-neutron source.

Figure 3 shows the anti-neutrino flux with energy on-power and over several days after shutdown. This shows that the anti-neutrino flux detectable by the inverse beta decay method (above the 1.8 MeV reaction threshold) quickly falls off after shutdown. This suggests that spent fuel storage near to the reactor will not interfere with the detector.

The anti-neutrino flux can also be calculated using FISPIN from fission for the separate fissionable materials present, principally ^{235}U, ^{239}Pu, ^{241}Pu and ^{238}U. These results were compared with the experimental data calculated from experiments on beta decay spectra measured from pulsed irradiations on samples of pure fissile species. The experiment to calculation ratios for ^{235}U, ^{238}U, ^{239}Pu and ^{241}Pu are shown in figure 4. These results show quite good agreement with measured values for ^{235}U and ^{239}Pu above 3 MeV being within the one standard deviation uncertainties on the anti-neutrino spectra calculated from the fission product yield uncertainties (shown in the graphs by the blue lines). The agreement for ^{238}U and ^{241}Pu, although poorer, is still within a range of the ratio between 1.4 and 0.5. This suggests that the result of using this calculation methodology is probably accurate to 10-20 % for thermal reactor fuels where fission of ^{235}U and ^{239}Pu dominate.

VI. Conclusions

The initial result of this work has suggested that the FISPIN calculations used to generate inventories for all 49248 rods in the core during each day of reactor operation with irradiation/cooling history should give an estimate of the anti-neutrino flux and spectra to within 20%.

It is intended to use the results of the Wylfa PANTHER and FISPIN calculations combined with the bespoke anti-neutrino source (number and spectra) calculation code to generate anti-neutrino spectra for each rod in the core during each day of the monitored period. These will then be used to produce an anti-neutrino flux with direction information over the detector location to facilitate GEANT-4 modelling of the detector to help understand its response properties and what the detector can achieve in determining the reactor state.

The study of the underlying nuclear data used in the calculations will also continue to consider if better nuclear data and modelling of the beta decay will improve the estimation of the anti-neutrino source term.

VII. Acknowledgements

The authors wish to thank G. Davies, A. Roberts and A. Tobias of Magnox Ltd, which operated the Wylfa reactor, for information on the reactor and many helpful discussions on its operation.

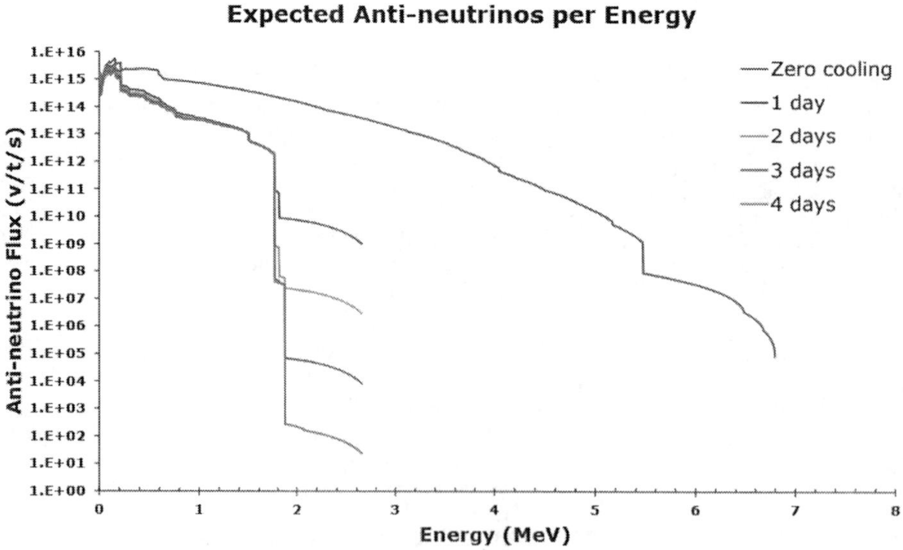

Fig. 3. Estimation of the anti-neutrino spectra per metric ton of uranium in the Wylfa reactor at the moment of shutdown and for the following 4 days using FISPIN.

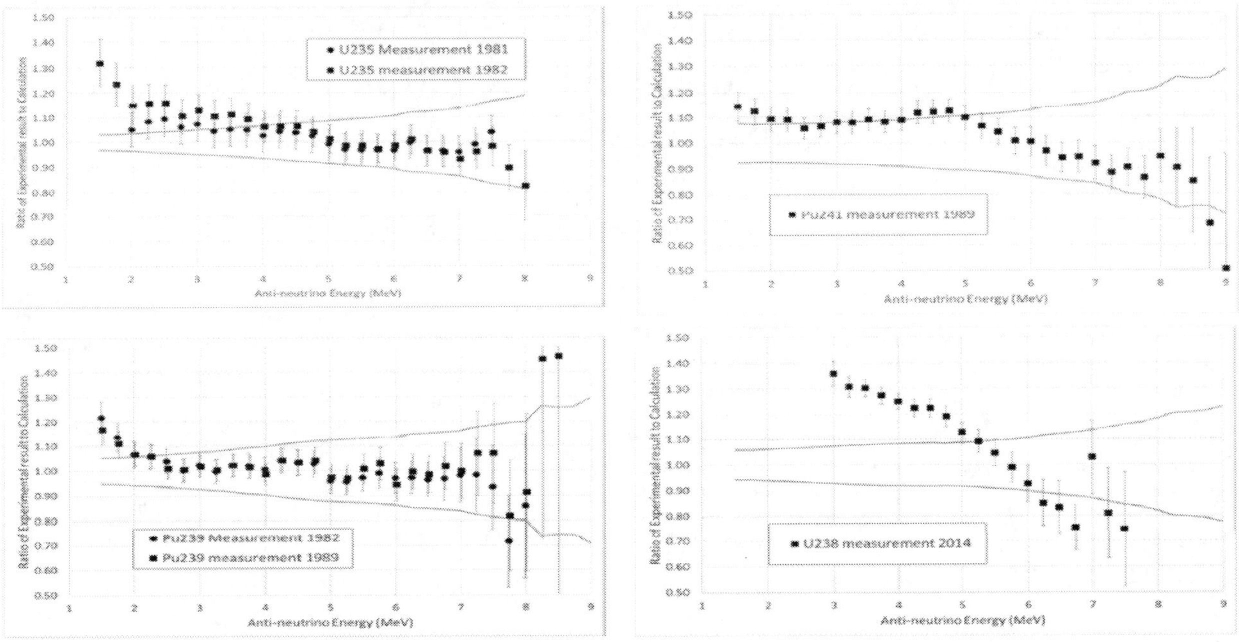

Fig. 4. Comparison of the estimated anti-neutrino flux spectra against the available experimental measurements by plotting experiment over calculation ratios for ^{235}U, ^{238}U, ^{239}Pu and ^{241}Pu. The blue lines show the uncertainties on the calculation from fission product yields.

References

[1] Abe, Y., dos Anjos, J.C. et al. "The Double Chooz collaboration", J. High Energ. Phys. (2014) 86. doi:10.1007/JHEP10(2014)086

[2] An, F.P., Balantekin, A.B., Band, H.R., Bishai, M., Blyth, S., Butorov, I., Cao, D., (...), Zou, J.H. "Measurement of the reactor antineutrino flux and spectrum at Daya Bay" (2016) Physical Review Letters, 116 (6), art. no. 061801

[3] Choi, J.H., Choi, W.Q., Choi, Y., Jang, H.I., Jang, J.S., Jeon, E.J., Joo, K.K., Yu, I. "Observation of Energy and Baseline Dependent Reactor Antineutrino Disappearance in the RENO Experiment " (2016) Physical Review Letters, 116 (21), art. no. 211801

[4] Feilitzsch, F. von, Hahn A.A. and Schreckenbach, K., "Experimental beta-spectra from ^{239}Pu and ^{235}U thermal neutron fission products and their correlated antineutrino spectra" (1982) Physics Letters, Volume 118B, number 1, 2, 3

[5] Haag, N., Gütlein, A., Hofmann, M., Oberauer, L., Potzel, W., Schreckenbach, K., Wagner, F.M. "Experimental determination of the antineutrino spectrum of the fission Products of U 238" (2014) Physical Review Letters, 112 (12)

[6] Hahn A.A., Schreckenbach, K., Gelletly, W., Feilitzsch, F. von, Colvin, G. and Krusche, B. "ANTINEUTRINO SPECTRA FROM ^{241}Pu AND ^{239}Pu THERMAL NEUTRON FISSION PRODUCTS" (1989) Physics Letters B Volume 218, number 3

[7] Huber, P. "NEOS Data and the Origin of the 5 MeV Bump in the Reactor Antineutrino Spectrum" (2017) Physical Review Letters, 118 (4), art. no. 042502.

[8] P.K. Hutt, N. Gaines, M.J. Halsall, M. McEllin, R.J. White, " The UK core performance code package", Nuclear Energy, British Nuclear Energy Society, 30 (1991), p. 291

[9] FISPIN for Nuclide Inventory Calculations: Introductory Guide for Version 10, 2007. Report ANSWERS/FISPIN/REPORT/001 published by Serco Assurance Ltd.

[10] Kellett, M. A., O. Bersillon, and R. W. Mills. "The JEFF-3.1/-3.1.1 radioactive decay data and fission yields sub-libraries." JEFF report 20 (2009)

[11] Mills, R.W. "Uncertainty propagation of fission product yield data in spent fuel inventory calculations" (2014) Nuclear Data Sheets, 118 (1), pp. 484-487

[12] Mougeot, X. "Systematic comparison of beta spectra calculations using improved analytical screening correction with experimental shape factors" (2016) Applied Radiation and Isotopes, 109, pp. 177-182.

[13] Mueller, Th. A., Lhuillier, D., Fallot, M., Letourneau, A., Cormon, S., Fechner, M., Giot, L., Lasserre, T. , Martino, J., Mention, G., Porta, A. and Yermia F. "Improved predictions of reactor antineutrino spectra", (2011) Phys. Rev. C 83, 054615

[14] Oralbaev, A., Skorokhvatov, M., Titov, O. "The inverse beta decay: A study of cross section" (2016) Journal of Physics: Conference Series, 675 (1), art. no. 012003, p. 37.

[15] Schreckenbach, K., Faust, H.R., Feilitzsch, F. von, Hahn, A.A.,Hawrkamp, K. and Vuilleumier J.L. "Absolute measurement of the beta spectrum from 235U fission as a basis for reactor antineutrino experiments" (1981) PHYSICS LETTERS Volume 99B, number 3

[16] Seo, S.-H "New results from RENO and the 5 MeV excess" (2015) AIP Conference Proceedings, 1666, art. no. 080002.ISBN: 978-073541313-9

[17] Tobias, A. "A Retrieval System for Spectral Data from ENDF/B Format Decay Data Files", (1981) CEGB report RD/B/5170N81

Nuclear Fuel Assay through analysis of Uranium L-shell by Hybrid L-edge/XRF Densitometer using a Surrogate Material

Seunghoon Park, Sungyeop Joung, and Jerry Park

***Abstract*—Assay of L-series of nuclear material solution is useful for determination of amount of nuclear materials and ratio of minor actinide in the materials. The hybrid system of energy dispersive X-ray absorption edge spectrometry, i.e. L-edge densitometry, and X-ray fluorescence spectrometry is one of the analysis methods. The hybrid L-edge/XRF densitometer can be a promising candidate for a portable and compact equipment due to advantage of using low energy X-ray beams without heavy shielding systems and liquid nitrogen cooling compared to hybrid K-edge/XRF densitometer.**

A prototype of the equipment was evaluated for feasibility of the nuclear material assay using a surrogate material (lead) to avoid radiation effects from nuclear materials. The uncertainty of L-edge and XRF characteristics of the sample material and volume effects was discussed in the article.

***Index Terms*—L-edge, XRF, Nuclear Material Solution, X-ray absorption, X-ray fluorescence**

I. Introduction

THE low energy X-ray measurement system is intended to be used for on-site analytical measurements of safeguards in various nuclear facilities, such as nuclear fuel fabrication facilities and reprocessing plants. Specially, the hybrid system of X-ray absorption spectrometry, such as L-edge densitometry (LED) [1, 2], and X-ray fluorescence spectrometry (XRF), has an important role of safeguards for nuclear facilities. LED is a technique of determination of uranium concentration as a continuous X-ray energy beams transmit a uranium liquid sample. Compared to K-edge densitometer [3], due to relatively lower energy (L_{III}-edge energy is 17.17 keV) of uranium L-series energy than K-series energy, L-edge densitometer does not require high purity germanium detector with liquid nitrogen cooling and heavy shielding system for shielding of K-series radiation (~ 115 keV) of uranium. Therefore, the L-edge densitometer is appropriate for a portable equipment for on-site nuclear material inspection and safeguards at facility sites. XRF combined with LED is a technique of characterizing the element content of nuclear materials from fluoresced characteristic X-ray photons.

In this study, the hybrid L-edge/XRF densitometer (HLED) was developed for determination of nuclear materials concentration. We verified that the system has feasibility of determination of the concentration of nuclear material from analysis of a surrogate material such as lead nitrate ($Pb(NO_3)_2$) solution prior to the work of actinide-bearing materials.

II. The Hybrid LED/L-XRF Densitometer

The schematic figure of HLED is shown in Fig. 1. The equipment was fabricated based on previous study [4] of Monte Carlo simulation for the instrument design and feasibility study of determination of nuclear material concentration. The prototype of hybrid L-edge/XRF densitometer (HLED) consists of an X-ray tube, shields, a sample container, and two detectors. Both of detectors are silicon drift detectors (SDD) of 25mm^2 × 500 μm which is XR-123SDD by Amptek™ including a digital pulse processor. The X-ray tube of 1.6 mm anode spot diameter is fabricated in Moxtek™ with maximum power, voltage, and current of 4 W, 50 kV, and 200 μA in front of collimator of a hole diameter of 0.5 mm and length of 5-cm-long. The sample container is made of Teflon to protect corrosion by nitric acid with 2 mm-optical path length and 300 μl volume. The windows material is Mylar® (polyester) film of 6 μm to maximize detection efficiency. There are beam lines for detection of transmission and XRF. The XRF detection beam line is aligned with 45 degrees for the sample window normal and the transmission detection line set up in the direction of the X-ray beam as shown in Fig. 1. The X-ray tube is operated by 30 kV and 90 μA with rhodium (Rh) anode.

X-ray photons from the tube transmit or randomly react to the sample by XRF process. The transmission and fluoresced photon energy spectra are recorded by the detectors. The rhodium anode with characteristic peaks of K_α (20.21 keV) and K_β (22.72 keV) can be a role of calibration of energy below 30

This work was supported by the Nuclear Safety Research Program through the Korea Foundation of Nuclear Safety (KOFONS), granted financial resource from the Nuclear Safety and Security Commission (NSSC), Republic of Korea.

Seunghoon Park is with the Korea Institute of Nuclear Nonproliferation and Control, Yusesong-daero 1534, Yuseong-gu, Daejeon, 34054, Republic of Korea (e-mail: shpark@kinac.re.kr). He is the first author and has correspondence.

Sungyeop Joung is with the Korea Institute of Nuclear Nonproliferation and Control, Yusesong-daero 1534, Yuseong-gu, Daejeon, 34054, Republic of Korea (e-mail: jsy1003@kinac.re.kr).

Jerry Park is with ISP Co., Ltd., 422 Palbok-dong 3-ga, Jeonju-si, Deokjin-gu, Jeollabuk-do, Republic of Korea (e-mail: jerry@ispxrf.com).

keV as well as generation of continuous X-ray spectrum, i.e. bremsstrahlung.

The experiments were performed for lead nitrate ($Pb(NO_3)_2$) solution with 0.01, 0.05, 0.1, and 0.2 g/cm^3 to avoid radiation effects for the performance evaluation of the equipment. The blanket material is water in order to obtain transmission.

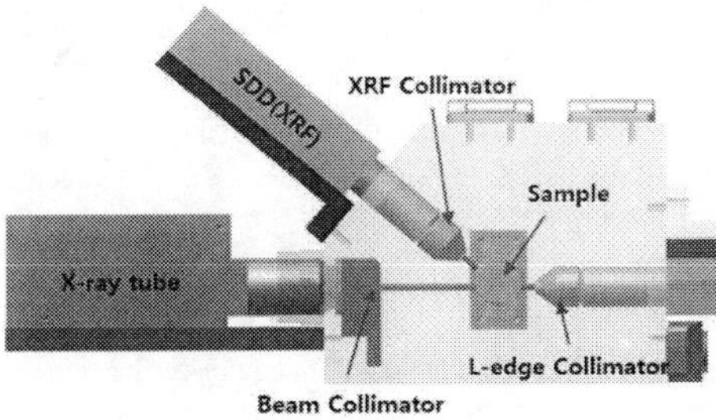

Fig. 1 Design of Hybrid L-edge/XRF Densitometer

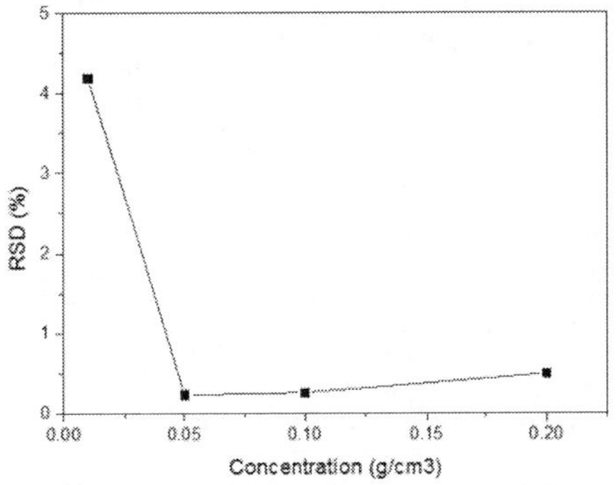

Fig. 2. Relative Standard Deviation (RSD) for each concentration

III. Evaluation of HLED

A. L_{III}-edge Analysis

The discontinuous curves, i.e. jumps, of transmission spectrum are observed. The height of the jump of the spectrum determines the sample concentrations. Material concentrations through transmission spectrum can be calculated by the following equation (1).

$$\rho_{Pb} = \frac{\ln[T(E_-)/T(E_+)]}{\Delta\mu D} \tag{1}$$

$T(E_-)$, $T(E_+)$ are transmission at the energies E_- (lower than L_{III}-edge) and E_+ (upper than L_{III}-edge). D is sample thickness which is exactly called optical path length and $\Delta\mu$ is mass attenuation coefficient difference at upper and lower L_{III}-edge. Extrapolated fitting in linearized a representation of lnln($1/T$)-ln(E) is applied for determining upper and lower transmission for L-edge. Fitting intervals from 11 – 12 keV for E_- and 14 – 15 keV for E_+.

Relative standard deviation (RSD) represents uncertainty of the measurement as shown in Fig. 2. The samples are measured for 3 × 1000s respectively. The RSD reduces at high concentration (> 0.01 g/cm^3) and slightly increases at 0.2 g/cm^3. The total counts at high concentration are relatively low compared to low concentration. The result implies that low transmission above 0.2 g/cm^3 results in low accuracy. As a result, the optimized concentration can be considered above 0.01 g/cm^3 and below 0.2 g/cm^3 of lead. The possible concentration range have consistency with the previous study for L-edge of uranium and plutonium [5].

B. X-ray fluorescence of surrogate material

In order to determine concentrations of minor actinide or impurities in a nuclear material sample, the analysis requires not only L-edge densitometry but also X-ray fluorescence spectrometry.

The typical XRF peaks are L_α (10.4 keV) and L_β (12.6 keV) of lead. The ratio of L_α and L_β for any concentration is shown in Fig. 3. XRF photons created in the sample are absorbed and interact with lead. The ratio decreases as the concentration increases because the lower energy photons can be more affected by the sample. The relation of the peak intensity and the peak ratio between the concentrations of a sample have to be considered of the XRF analysis for mixed solution samples. Unlike a single element analysis, the analysis for mixed solution samples are required calibration data for target materials using various ratio of elements and concentrations.

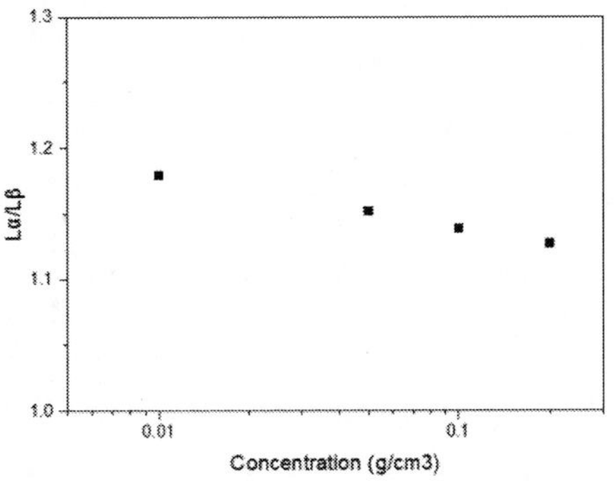

Fig. 3 The peak ratio of L_α and L_β of lead for each concentration

IV. Conclusion

The prototype of a hybrid L-edge/XRF densitometer (HLED) was developed for nuclear material solution assay. The equipment has feasibility due to the uncertainty from L-edge analysis near ITV (International Target Value). The evaluation verified that the concentration and volume effects must be considered for assay of multi-element solution samples.

In future work, based on the study, nuclear material will be analyzed and uncertainty will be reduced by analysis of non-radioactive and radioactive standard samples such as lead nitrate and uranium nitric acid solutions.

References

[1] Analysis Procedure for Timely On-site Uranium Accountancy Verification in LEU FUEL Fabrication Plants", ESARDA BULLETIN No. 43, 18(19), 30-39 2009.

[2] M. L. Brooks, P. A. Russo, J. K. Sprinkle, Jr., "A Compact L-Edge Densitometer for Uranium Concentration Assay", Los Alamos National Laboratory report LA-10306-MS, 1985

[3] H. Ottmar and H. Eberle, "The Hybrid K – Edge/K – XRF Densitometer: Principles – Design – Performance", Kernforschugszenturm Karlsruhe GmbH, Karlsruhe, KfK 4590, Feb. 1991

[4] S. Park, and J.-K. Shin, S.-W. Kwak, "Development of Hybrid L-edge/XRF Densitometer for Determination of Nuclear Material Concentration", 31 October to 7 November, 2015 IEEE Nuclear Science Symposium & Medical Imaging Conference, 2015

[5] D. Reily, N. Ensslin, H. Smih, Jr., and S. Kreiner, Passive Nondestructive Assay of Nuclear Materials, LA-UR-90-732, United States Nuclear Regulatory Commission, 1991

Advanced Multilayer Composite Heavy-Oxide Scintillator Detectors for High Efficiency Fast Neutron Detection

Vladimir D. Ryzhikov, *Member, IEEE*, Sergei V. Naydenov*, Thierry Pochet, Gennadiy M. Onyshchenko, Leonid A. Piven, and Craig F. Smith

***Abstract*—We have developed and evaluated a new approach to fast neutron and neutron-gamma detection based on large-area multilayer composite heterogeneous detection media consisting of dispersed granules of small-crystalline scintillators contained in a transparent organic (plastic) matrix. Layers of the composite material are alternated with layers of transparent plastic scintillator material serving as light guides. The resulting detection medium – designated as ZEBRA – serves as both an active neutron converter and a detection scintillator which is designed to detect both neutrons and gamma-quanta. The composite layers of the ZEBRA detector consist of small heavy-oxide scintillators in the form of granules of crystalline BGO, GSO, ZWO, PWO and other materials. We have produced and tested the ZEBRA detector of sizes 100x100x41 mm and greater, and determined that they have very high efficiency of fast neutron detection (up to 49% or greater), comparable to that which can be achieved by large sized heavy-oxide single crystals of about Ø40x80 cm^3 volume. We have also studied the sensitivity variation to fast neutron detection by using different types of multilayer ZEBRA detectors of 100 cm^2 surface area and 41 mm thickness (with a detector weight of about 1 kg) and found it to be comparable to the sensitivity of a ^{3}He-detector representing a total cross-section of about 2000 cm^2 (with a weight of detector, including its plastic moderator, of about 120 kg). The measured count rate in response to a fast neutron source of ^{252}Cf at 2 m for the ZEBRA-GSO detector of size 100x100x41 mm^3 was 2.84 cps/ng, and this count rate can be doubled by increasing the detector height (and area) up to 200x100 mm^2. In summary, the ZEBRA detectors represent a new type of high efficiency and low cost solid-state neutron detector that can be used for stationary neutron/gamma portals. They may represent an interesting alternative to expensive, bulky gas counters based on ^{3}He or ^{10}B neutron detection technologies.**

***Index Terms*—Fast Neutrons, Neutron/Gamma ray detection, Scintillator detectors, Nuclear Electronics, Nuclear Security**

Manuscript received July 15, 2017. This work was supported in part by the NATO Science for Peace Program (Project SfP-984605)

V. D. Ryzhikov is with Institute for Scintillation Materials (ISMA) of the National Academy of Sciences of Ukraine, 61001 Kharkov, Ukraine (e-mail: ryzhikov@isma.kharkov.ua).

*S. V. Naydenov is with Institute for Single Crystals (ISC) of the National Academy of Sciences of Ukraine, 61001 Kharkov, Ukraine (e-mail: sergei.naydenov@gmail.com) (Corresponding Author).

T. Pochet is with International Atomic Energy Agency (IAEA), 1400 Vienna, Austria (e-mail: tpochet@detecrad.com).

G. M. Onyshchenko, L. A. Piven are with Institute for Scintillation Materials of the National Academy of Sciences of Ukraine, Kharkov, Ukraine.

C. F. Smith is with Naval Postgraduate School, Monterey, CA, USA (email: craigfsmith@yahoo.com).

I. Introduction

One of the main challenges in nuclear security is the creation of detection systems for fast neutrons and mixed neutron/gamma radiation that would be highly efficient, small-sized, inexpensive, easy-to-use, rugged and robust with respect to unfavorable conditions, and reliable in operation. Reliable recognition of illegal nuclear materials, in particular, weapon-grade uranium U-235 and plutonium Pu-239, as well as certain radioactive constituents of "dirty" bombs, requires the creation of detectors with high detection efficiency for fast neutrons (not less than 50%). At the same time, practically all the newly-proposed detectors for the replacement of He-3 counters can efficiently detect only slow/thermal neutrons, relying on the radiative capture reactions of Li-6 or B-10 nuclei with very high thermal absorption cross-sections, but relative insensitivity to fast neutrons. The probability of direct detection of fast neutrons with these Li-6 or B-10 systems without using plastic moderators is orders of magnitude lower and, the efficiency of direct detection therefor does not exceed a few per cent. Even with massive and voluminous moderators to thermalize the fast neutrons, the fast neutron detection efficiency is not higher than 10-20%, and for He-3 detectors about 10%.

At the same time, we developed a new approach to neutron detector design, using single crystals of heavy-oxide scintillators (CWO, ZWO, PWO, BGO, GSO(Ce), etc.) for direct detection of fast neutrons without large scale thermalization [1-8]. This allowed us to realize increased fast neutron detection efficiencies of up to 50-70%. For detection of fast neutrons, the inelastic scattering reaction on heavy nuclei was used, and the scintillator single crystal played a double role (two-in-one) – as an efficient converter of neutrons into secondary gamma-quanta of intermediate energies, and as a scintillator that detects these internal gamma-quanta in either a counting or spectrometric mode.

Our further studies have shown the possibility to also create a comparably efficient detector of fast neutrons on the basis of a novel composite scintillation structure, again relying on direct detection of fast neutrons through their interaction with materials of high effective atomic number. The fabrication costs of such composite scintillator structures can be 5-10 times lower in comparison with single crystals of comparable size. As a basic element, we developed an original heterogeneous scintillation structure (we use the name ZEBRA for it), which is comprised of a series of alternating plates of two types – plates consisting of small particles of heavy crystalline scintillator material (i.e., BGO, ZWO, PWO, GSO(Ce), etc.) embedded in a plastic scintillator matrix, and plates of a clear plastic moderator material, which simultaneously serves as a light guide for output of scintillation photons. In detecting fast neutrons, such a detector medium utilizes, to the maximum extent, not just one, but several channels of neutron interaction with heavy nuclei – inelastic scattering of fast neutrons scattered at different angles into the moderator, as well as resonance scattering and radiative capture of neutrons moderated inside the detector by elastic scattering in hydrogen-containing plastic. For the moderator/light guide layers, one can also use plates of plastic scintillators which serve the additional functions of scintillation and spectrum-shifting activation. (In the latter case, the layer of plastic serves also as scintillator that simultaneously detects both gamma-quanta emitted in the heavy oxide scintillator layer as a result of inelastic and resonance scattering of fast neutrons, and recoil protons from elastic scattering of fast neutrons in the plastic itself). The detection of fast neutrons simultaneously over several channels of their interaction substantially increases total detection efficiency.

II. Experimental Setup for Neutron Measurements

We report on the research and development of fast neutron detectors based on large-size solid-state single crystal as well as multi-layer composite heavy-oxides CWO, ZWO, BGO and GSO(Ce) scintillators. The experimental setup and scheme of the experiments are shown on Figs. 1-2. Detection efficiency measurements for fast neutrons from a ^{239}Pu-Be source by different scintillator detectors were made using a scintillation gamma-spectrometer with an analog-to-digital converter (ADC), which allowed recording of pulses from 1 μs up to 30 μs. The double integration time of the spectrometer amplifier was 30μs, and the voltage on the PMT was 680 V. The spectrometer was equipped with a charge-sensitive preamplifier. To obtain unambiguous correspondence between the occurrence of a neutron interaction and the appearance of the counting pulse in the working window of the detection system, the spectrometric signal integration time constant was chosen to be sufficiently high for all scintillators.

From the amplifier-shaper output, the signals come to the ADC input in the gamma-quanta energy range of 20-300 keV. The measured value was the number of spectrometer pulses in the working spectrometric window (20-300 keV) during the same time interval.

Our method of measuring neutron detection efficiency assumes a "spherical" geometry and isotropic nature of the neutron source. Because of its low ambiguity, repeat measurements did not vary more than 1% with respect to the measured neutron flux.

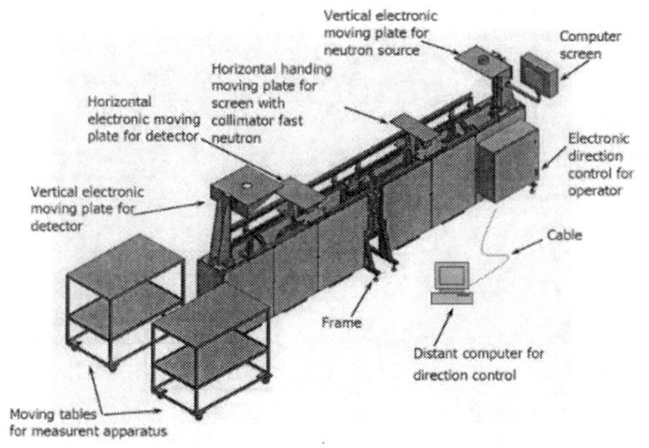

Fig. 1. Design of the experimental setup for fast neutron measurements. Length of hardware system – 3 m. Height – 1.5 m. Position accuracy of electronic-controlled plate – 0.1 cm.

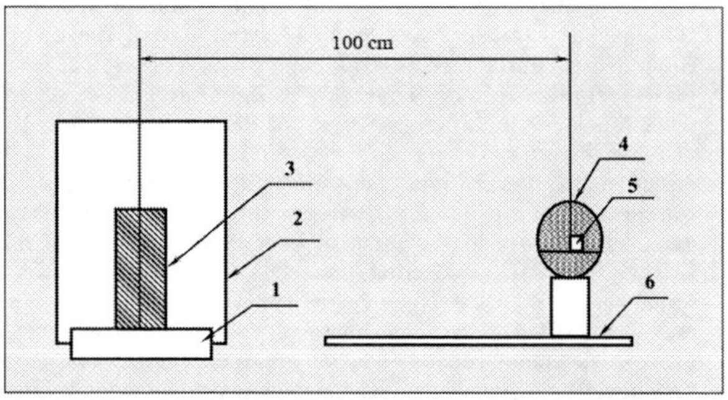

Fig. 2. Scheme of the experiments for fast neutron measurement: 1 – PMT; 2 – enclosure to exclude external light; 3 – detector; 4 – ball (R = 5 cm) with hole for shielding of γ-radiation; 5 – neutron source ^{239}Pu-Be or ^{252}Cf; 6 – foundation base.

III. Multi-Layer Composite Heavy-Oxide Scintillator Detectors

To detect radiation (i.e., neutron/gamma flux), we prepared a new form of composite scintillator fast neutron detectors (brand ZEBRA). These detectors are comprised of layers of a composite of granular scintillator particles embedded in transparent plastic (the dimension of the granules is about 0.2 mm, and the layer thickness of 5 mm). These composite layers are alternated with layers of plastic light guide/ scintillator material of thickness 5mm. In the light channel layers, light is collected and transmitted to the edge of the guide and then registered by a PMT (Hamamatsu model R1307).

The assembly of the 13-layer detectors (100x100x41 mm^3) was performed by optical coupling of the single-layer scintillation panels (1 mm thicknesses) with light guides (5 mm thicknesses) with all surfaces polished. All of these detectors were carefully studied, and their neutron registration effectiveness was preliminarily tested using ^{239}Pu-Be and ^{252}Cf neutron-gamma sources. The sketch on the Fig. 3 and photo on Fig. 4 illustrate the detector medium and its scheme of operation with neutron radiation and its registration.

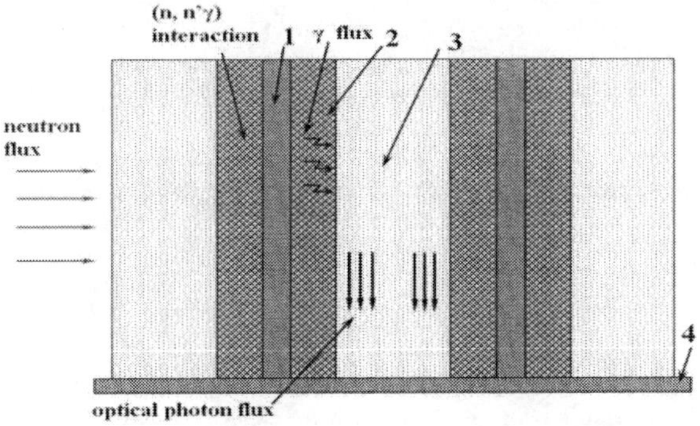

Fig. 3. Sketch of the design and operation of ZEBRA-brand multi-layer composite scintillator detectors.

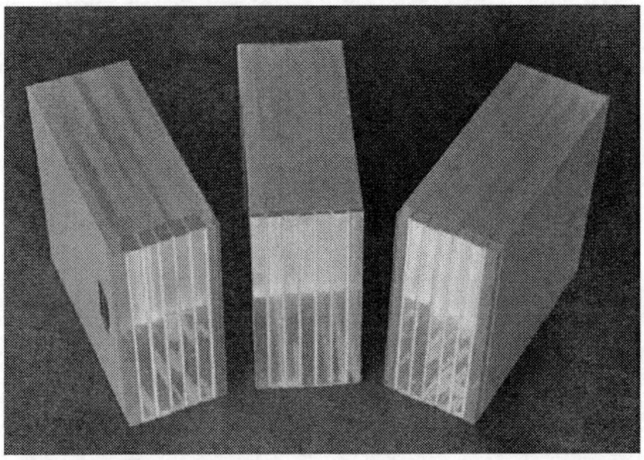

Fig. 4. Photo of ZEBRA-detectors with different composite scintillators GSO(Ce), BGO and ZWO. The size of every detector is 100x100x41 mm^3.

Our measurements of the dependence of detection efficiency of single crystal detectors on crystal thickness show a practically linear relationship up to 40 mm [4]. But for the composite multilayer ZEBRA detectors with 5-6 composite layers of thickness 1-2 mm, i.e., a total thickness of heavy-oxide scintillating layers of about 10 mm, we find that the ZEBRA detector sensitivity is equal to that of single crystals of diameter of 40 mm and height of about 80 mm or more.

We suspect that this is due to the presence of the additional plastic in the composite layers and in the alternate layers of light guide. We believe that the interaction of neutrons with these materials include resonance and radiative capture interactions with much higher interaction cross sections. The process of neutron scattering within the detector can result in a decrease of neutron energy below the inelastic threshold. For this reason, in addition to detecting the signal from inelastic scattering, our method includes measurements in other channels to also detect neutrons of reduced energy which interact by resonance scattering and thermal capture (Fig. 5).

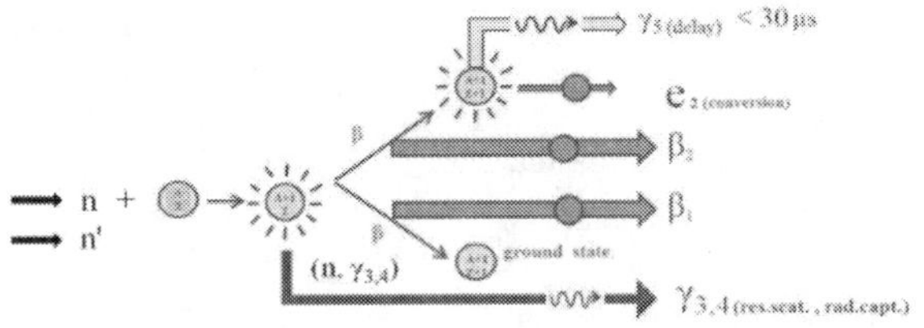

Fig. 5. The principal scheme of sources (physical channels) of registration signals from decelerated neutrons within the scintillator medium including (1) prompt gammas from resonant (non-elastic) neutron scattering and radiative capture as well as (2) beta-particles and internal-conversion electrons following such captures.

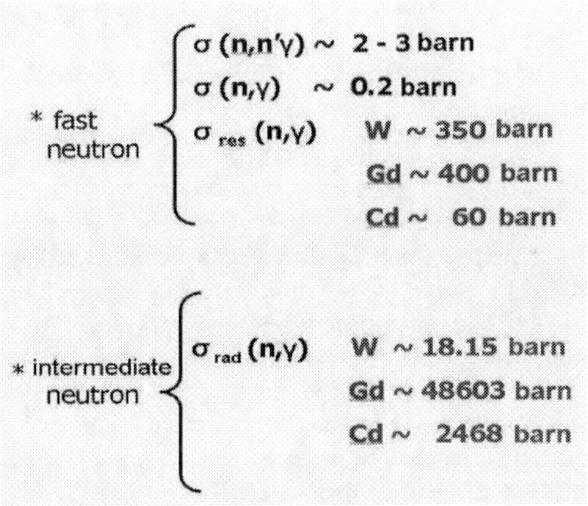

Fig. 6. Specific mechanisms of non-elastic scattering of neutrons with inelastic scattering, resonance and thermal capture cross sections for a Pu-Be source (average fast neutron energy ~4 MeV) for native "heavy" nuclei W, Gd and Cd contained in heavy-oxide scintillators CWO, ZWO, GSO(Ce).

We observe a significant increase (up to 70-80%) in the efficiency of fast neutron detection by single crystal detectors with the use of a surrounding layer of plastic of thickness from 1 to 10 mm. This can be explained as a shift of neutron energy due to moderation in and below the resonance region. Registration of this interaction is enhanced by the fact that the cross-section for radiative capture σ(n, γ) is much greater than that of inelastic scattering σ(n, n'γ) as it is clearly presented on Fig. 6.

IV. Fast Neutron Detectors Efficiency and Performance

Having prepared samples of this multi-layer composite detection medium using different heavy oxide scintillator materials, we tested them and compared their performance with that of a crystalline detector and a ^{3}He systems.

The preliminarily measured detection efficiency for a ZEBRA-GSO(Ce) structure of dimension 100x100x41 mm^3 (containing 6 active layers of the composite scintillator) is 49% and can be increased further by increasing its thickness, that is, by increasing the number of active layers. The detector sensitivity reaches 49 cps / (neutron · s^{-1} · cm^2), which corresponds to a sensitivity of 2.84 cps/ng when the neutron

https://doi.org/10.1051/epjconf/201817007010

flux comes from a Cf-252 source at a distance of 2 m from the detector. Thus, the obtained sensitivity exceeds the generally accepted standard of 2.5 cps/ng. Moreover, the sensitivity of ZEBRA detectors can be made twice as high – up to 5 cps/ng and more – by increasing the area of the plates.

Fig. 7 shows the results of this comparative assessment and also indicates the measured efficiency of the different heavy-oxide scintillators in crystalline and composite forms.

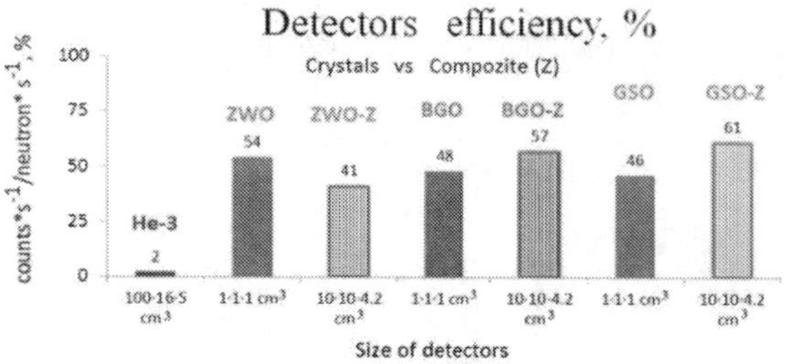

Fig. 7. Comparison of the fast neutron detection efficiency for single crystals and ZEBRA-brand multi-layer composite scintillator detectors.

The efficiency and sensitivity of neutron detectors are improved by increasing the cross sectional area of the crystal across the direction of incident neutron flux. To increase neutron detection sensitivity, theoretical and experimental investigations clearly indicate that the cross-sectional area (as well as the volume) of the detector facing the direction of neutron flux should be increased, with the desired cross-sectional area being up to tens or hundreds of square centimeters. We can accomplish this objective namely by creating of large-volume multi-layer detectors in which a composite of scintillators containing active granules is placed inside a transparent polymer material. Because of the relatively low transparency of the composite inorganic scintillator, better light collection conditions are ensured by incorporation of a light guide between the composite scintillator layers. The design and fabrication technology of this detector allows the creation of neutron-gamma detectors characterized by high sensitivity at relatively low costs (as compared with analogs using oxide scintillators) for portable inspection systems. The sandwich structure can be comprised of any number of plates, with no limitations for thickness or area.

At the same time, to investigate the possibility of suppressing accompanying gamma-radiation, we have also developed a method for detection of fast neutrons against the background of intense gamma-radiation. Neutron-gamma separation was achieved using a specially developed algorithm based on measured data by using the well-known "window" method for different parts of the energy spectrum. This will permit us to achieve a n/γ ratio rejection of 10^4-10^5 and more.

V. Conclusion

We created and characterized a new type of multilayer composite detector (brand ZEBRA). In this detector medium, a composite that is comprised of granules of the scintillators BGO, ZWO, GSO dispersed in transparent plastic is alternated with layers of clear plastic that serve as light guides. The sensitivity of these detectors with a size of 100x100x41 mm^3 compares favorably with detectors based on large size crystals, but the composite structures are much cheaper and capable of manufacture in much larger dimensions.

This work represents a significant advancement from earlier single-crystal detector types as part of our effort to explore alternatives and improvements to the conventional ^3He counter. The current work directly detects fast neutrons while the ^3He counter relies of neutron thermalization prior to detection. This limitation of the ^3He counter restricts its efficiency in fast neutron detection to not more than about 10%, whereas the new method of direct detection of fast neutrons enables efficiencies of up to 50-60%.

The anomalously high efficiency of the composite scintillators can be only qualitatively explained by assuming of interaction of neutrons with the scintillator. More detailed theoretical analysis would be desirable, with quantitative calculations for these mechanisms. Further development of the composite detector design and choice of optimum detector materials need for to ensure the best "price-quality" ratio would be also necessary.

Acknowledgment

This research is supported in part by NATO's Emerging Security Challenges Division in the Science for Peace and Security Program under Project SfP-984605.

References

[1] V.D. Ryzhikov, B.V. Grinyov, G.M. Onyshchenko, L.A. Piven, O.K. Lysetska, L.L. Nagornaya, T. Pochet, "The use of fast and thermal neutron detectors based on oxide scintillators in inspection systems for prevention of illegal transportation of radioactive substances", *IEEE Trans. Nucl. Sci*, vol. 57, No. 5, pp. 2747, 2010.

[2] V.D. Ryzhikov, B.V. Grinyov, G.M. Onyshchenko, L.A. Piven, O.K. Lysetska, O.D. Opolonin, S.A. Kostioukevitch, C.F. Smith, "The neutron detectors based on oxide scintillators for control of fissionable radioactive substances", in *Proc. of SPIE*, Hard X-Ray, Gamma-Ray, and Neutron Detector Physics XVI, vol. 92131B, September 5, 2014; doi:10.1117/12.2058185.

[3] V.D. Ryzhikov, S.V. Naydenov, G.M. Onyshchenko, T. Pochet, C.F. Smith, "High efficiency fast neutron detectors based on inorganic scintillators", in *IEEE 2014 Nuclear Science Symposium*, Seattle, WA, USA, 2014; doi:10.1109/NSSMIC.2014.7431165.

[4] V. Ryzhikov, C. Smith, B. Grinyov, L. Piven, G. Onyshchenko, S. Naydenov, T. Pochet, "Detection of gamma-neutron radiation by novel solid-state scintillation detectors", in *Proceeding of ANIMMA-2015*, Lisbon, Portugal, 2015; doi:10.1109/ANIMMA.2015.7465541.

[5] B. Grinyov, V. Ryzhikov, L. Nagorna, G. Onyshchenko, L. Piven, "Small-sized detector of fast and thermal neutrons", Patent of Ukraine #92193, Oct. 11, 2010.

[6] B. Grinyov, V. Ryzhikov, L. Nagornaa, G. Onyshchenko, L. Piven, "Method of registration of fast neutrons", Patent of Ukraine #96428, Nov. 11, 2011.

[7] B. Grinyov, V. Ryzhikov, L. Nagornaa, G. Onyshchenko, L. Piven "Method of detection of fast neutrons", Patent of USA No. US 8058624 B2, Nov. 15, 2011.

[8] V. Ryzhikov, B. Grinyov, S. Naydenov, G. Onyshchenko, L. Piven, "Method of registration of fast neutrons and combined detector based thereon", Patent Application of Ukraine #a201611057, Nov. 11, 2016.

Gamma signatures of the C-BORD Tagged Neutron Inspection System

A. Sardet, B. Pérot, C. Carasco, G. Sannié, S. Moretto, G. Nebbia, C. Fontana, F. Pino, A. Iovene, C. Tintori

***Abstract*— In the frame of C-BORD project (H2020 program of the EU), a Rapidly relocatable Tagged Neutron Inspection System (RRTNIS) is being developed to non-intrusively detect explosives, chemical threats, and other illicit goods in cargo containers. Material identification is performed through gamma spectroscopy, using twenty NaI detectors and four $LaBr_3$ detectors, to determine the different elements composing the inspected item from their specific gamma signatures induced by fast neutrons. This is performed using an unfolding algorithm to decompose the energy spectrum of a suspect item, selected by X-ray radiography and on which the RRTNIS inspection is focused, on a database of pure element gamma signatures. This paper reports on simulated signatures for the NaI and $LaBr_3$ detectors, constructed using the MCNP6 code. First experimental spectra of a few elements of interest are also presented.**

***Index Terms*—C-BORD project, Rapidly Relocatable Tagged Neutron Inspection System, fast neutron activation analysis, inelastic scattering gamma rays, MCNP6**

I. Introduction

GLOBALIZATION and the increase of trade is accompanied by an increase in the traffic of dangerous substances (such as explosives, illicit drugs, etc.) but also in contraband. Each year, about 4 billion tons of freight is arriving at the European border and the inspection of all these cargo is not compatible with economic constraints. Therefore, the effective Container inspection at BORDer control points (C-BORD) project, launched in the frame of H2020 European projects for the development of "Technologies for inspections of large volume freight", aims at developing a toolbox of five non-intrusive inspection technologies, which, working in combination, would ensure a more efficient screening of cargo containers. Among the developed techniques, the Rapidly Relocatable Tagged Neutron Inspection System (RRTNIS) is designed to be a second-line defense system, which would help identify threats and contraband materials in localized parts of the container, following the identification of a suspect item by X-ray radiography. After a brief presentation of the design of the system and its operation mode, the method used to build simulated elemental gamma spectra is detailed. Finally, the experimental setup used to measure these gamma signatures is presented and first measured spectra are reported.

II. The Rapidly Relocatable Tagged Neutron Inspection System

Fig. 1 presents a schematic view of the system designed in the scope of the C-BORD project [1].

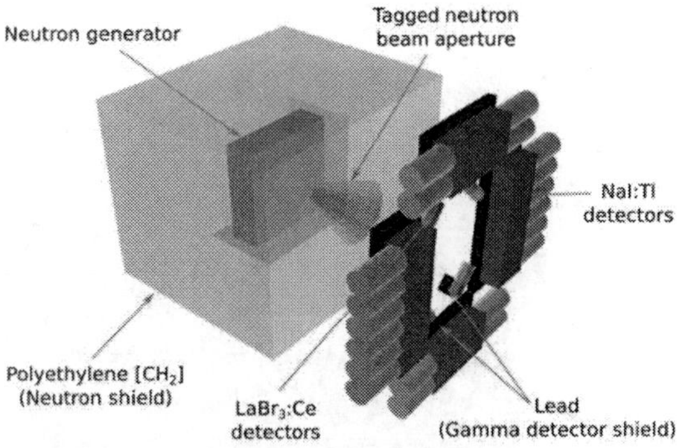

Fig. 1: Schematic view of the RRTNIS of the C-BORD project

A neutron generator, encased in a polyethylene shield to ensure radiation protection, is used to produce 14 MeV neutrons, through the $d + t \rightarrow n + \alpha$ fusion reaction. As the alpha particle is emitted almost back-to-back with the neutron, the use of a position sensitive YAP scintillator enables a spatial selection of the interrogating fast neutrons. In C-BORD RRTNIS, a 20° tagged neutron cone has been chosen, corresponding to the tagged neutron beam aperture shown in Fig. 1. In addition, the coincidence time between the alpha and the neutron-induced gamma detections allows determining the neutron time of flight, and subsequently its distance to interaction. By combining information on the direction of the neutron and its traveled distance, the gamma spectrum of any elementary volume (voxel) in the container can be reconstructed (see further in Fig. 2).

Gamma rays resulting from tagged neutron interactions are detected by twenty 5" × 5" × 10" parallelepiped-shaped NaI

Manuscript received June 30, 2017. This work has received funding from the European Union's Horizon 2020 research and innovation program under grant agreement No 653323. This text reflects only the author's views and the Commission is not liable for any use that may be made of the information contained therein.

A. Sardet, B. Pérot and C. Carasco are with CEA, DEN, Cadarache, Nuclear Measurement Laboratory, F-13108 Saint-Paul-lez-Durance, France (email: alix.sardet@cea.fr).
G. Sannié is with CEA, DRT, LIST, Saclay, F-91191 Gif-Sur-Yvette, France.
S. Moretto, G. Nebbia, C. Fontana and F. Pino are with Dipartimento di Fisica e Astronomia e INFN di Padova, via Marzolo, 8, I-35131 Padova, Italy.
A. Iovene and C. Tintori are with CAEN S.p.A, Via Vetraia, 11, I-55049 Viareggio, Italy.

detectors and four 3" × 3" cylindrical $LaBr_3$ detectors placed in a backscatter position, near the neutron generator. The large NaI scintillators are used to detect gamma rays above 600 keV, which allows identifying most common elements such as carbon, nitrogen, oxygen, silicon, chlorine, calcium, aluminum, iron, chromium, nickel, copper, zinc, lead, etc. The high resolution $LaBr_3$ scintillators are mainly used to detect gamma rays below 500 keV, in view to identify some of the chemical warfare elements that do not show clear gamma signatures at higher energy, such as arsenic, bromine, and iodine. A 5 cm thick lead plate is used to shield all gamma detectors from scattered neutrons and gamma rays coming from the polyethylene shield.

All elements of the system (neutron generator, gamma detectors, shields, electronics, etc.) are placed on the same side of the container in a 270 × 250 × 220 cm^3 box, fixed on a lifting platform so as to be able to inspect the cargo container at different heights. The total weight of the system is estimated to be of about 5 tons.

The gamma spectrum obtained by summing the spectra of the NaI or $LaBr_3$ detectors results from individual gamma signatures of the interrogated elements, weighted by their chemical proportions in the cargo materials. It can be fitted by a linear combination of these individual signatures, using an unfolding algorithm and a database of pure element gamma signatures (see Fig. 2). The retrieved weights give information on the chemical composition of the target, thus allowing its identification. For instance, the target can be categorized as metallic, ceramic or organic. For this last category, additional information is obtained by looking at the proportions of carbon (C), nitrogen (N) and oxygen (O), using a barycentric representation as shown in Fig. 2 (triangle). Unfolding the whole gamma spectrum applies well for the identification of most common elements, but for those having gamma signatures only at low energy, such as arsenic, bromine, or iodine, identification is mainly based on peak detection in the $LaBr_3$ spectrum.

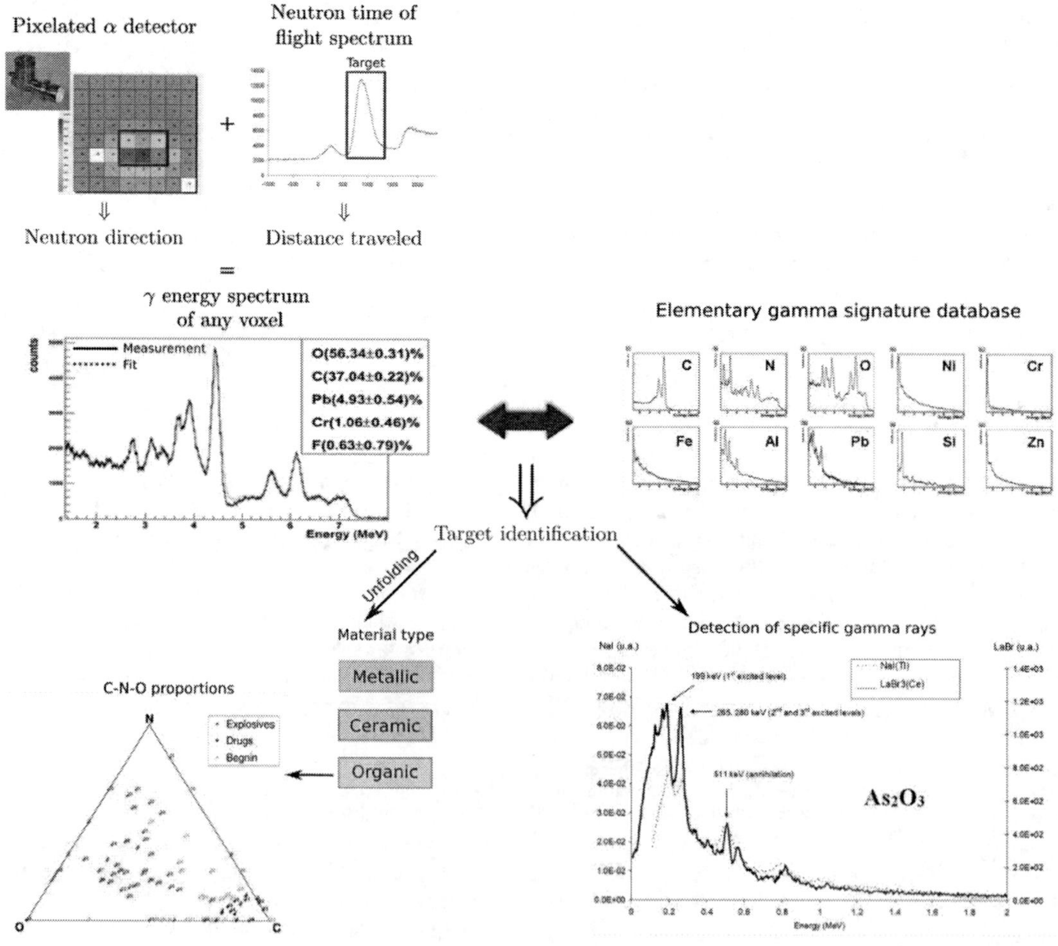

Fig. 2: Data analysis principle of the RRTNIS.

III. Simulation of neutron-induced gamma spectra

In order to assess the performances of different unfolding algorithms developed by C-BORD partners, simulated pure element gamma signatures have been generated for both NaI and $LaBr_3$ detectors, using the MCNP6 code. The interrogation of a pure element target by the RRTNIS was modeled with MCNP6 [2] as presented in Fig. 3.

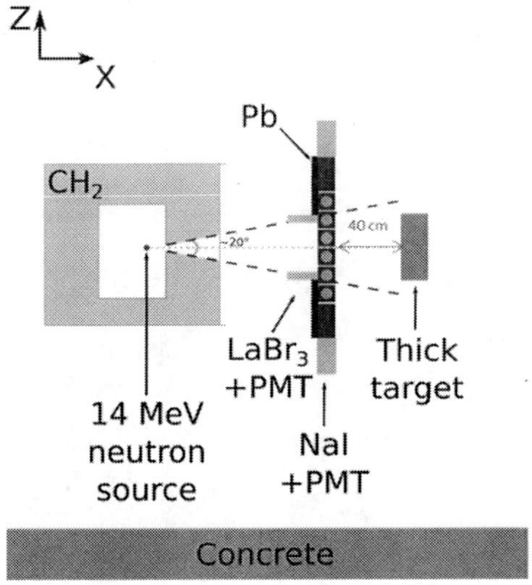

Fig. 3: MCNP6 model of a pure element target inspection, for the construction of the RRTNIS simulated gamma signature database.

The associated particle neutron generator is assimilated to a point-like 14 MeV neutron source with a 20° conical emission. The floor is modeled by a 100 cm thick layer of concrete and the target consists of a $20 \times 50 \times 50$ cm^3 parallelepiped of the pure element of interest, with a density of 1 g/cm^3. The MCNP6 point detector (F5 tally) is used in the front face of each detector to estimate the incoming photon flux. This last is then processed by MODAR software [3] to account for detection efficiency, based on MCNP calculations of the energy deposition in NaI or $LaBr_3$ crystals (F8 pulse height tally), and for time and energy broadening due to the resolution of the detectors taken from [4] [5]. An example of this procedure is presented in Fig. 4 for a thick carbon target.

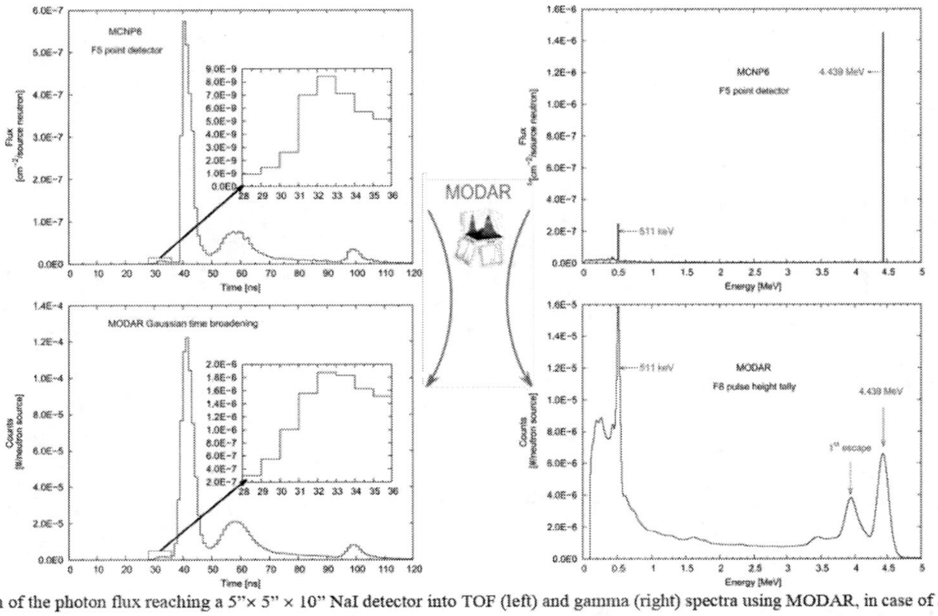

Fig. 4: Conversion of the photon flux reaching a 5"× 5" × 10" NaI detector into TOF (left) and gamma (right) spectra using MODAR, in case of a carbon target.

The time-of-flight (TOF) spectrum of each detector can then be converted into a distance spectrum using the ~ 5 cm.ns^{-1} speed of 14 MeV neutrons, and the spectra of all NaI or $LaBr_3$ detectors can be summed to improve statistics. The gamma spectrum of the region of interest can then be plotted by selecting the corresponding spatial (or TOF) region, as can be seen in Fig. 5 for a carbon target.

Note that MODAR can also be used to process experimental data, in which case time selection allows suppressing random coincidences (see further in Fig. 9), which are not simulated here.

https://doi.org/10.1051/epjconf/201817007011

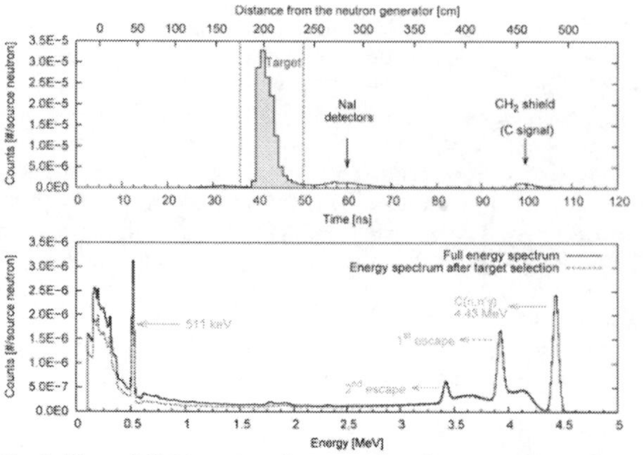

Fig. 5: Time of flight spectrum (upper panel) and energy spectrum (lower panel) of a $LaBr_3$ detector for a thick carbon target. The gamma spectrum presented with a dashed red line was obtained after a time of flight selection of the target. The total spectrum without selection is plotted with the black full line.

Two databases of gamma signatures have been constructed for the NaI and $LaBr_3$ detectors, respectively. An extract of the NaI database is presented in Fig. 6.

As mentioned above, $LaBr_3$ detectors will mainly be used to detect specific gamma rays, below 500 MeV, signing the presence of chemical warfare agents such as arsenic which do not exhibit gamma signatures at higher energy (see Fig. 7).

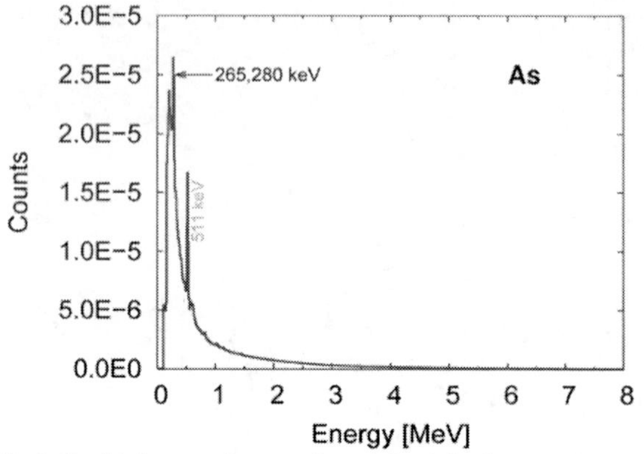

Fig. 7: Simulated gamma signature of arsenic in a $Labr_3$ detector.

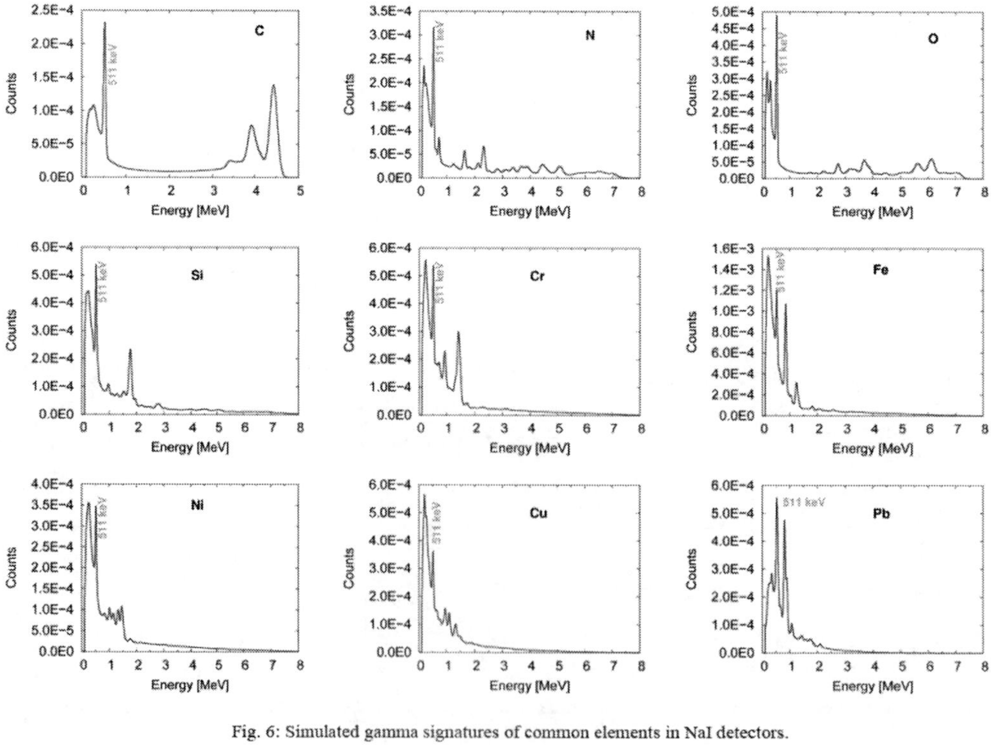

Fig. 6: Simulated gamma signatures of common elements in NaI detectors.

IV. Acquisition of experimental gamma signatures

Experience from past European projects EURITRACK [6] [7], Eritr@ck and UNCOSS [8] [9], showed that simulation does not entirely reproduce measured gamma signatures. For most common elements, a rather good agreement is observed, even if missing peaks or differences in the continuum can be noted. However, for unusual elements like some of those composing chemical warfare agents, unrealistic spectra can be produced by simulation when unrealistic nuclear models are used by MCNP instead of missing nuclear data.
Therefore, experimental gamma spectra are currently being acquired at CEA Cadarache, in DANAIDES irradiation cell of TOTEM nuclear facility. A picture of the experimental setup is presented in Fig. 8.

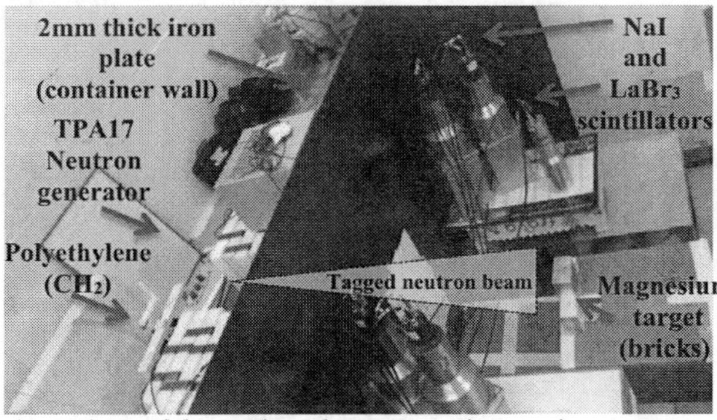

Fig. 8: Picture of the experimental setup to acquire pure element gamma signatures at CEA Cadarache. A 5 cm thick lead screen (not visible in the picture, between polyethylene and the iron wall) is used to shield gamma detectors.

A target of pure element (e.g. magnesium in Fig. 8) is placed in front of the tagged neutron beam produced by a TPA17 neutron generator from SODERN, similar to the one that will be used in C-BORD project. Signals from the embedded YAP detector and of the gamma detectors are processed and recorded using a digital acquisition system and raw data post-processing software, developed in a joint effort by CAEN and UNIPD [10].
From this initial post-processed data, other software developed by CEA Cadarache is used to generate, for each detector, a correlation between the neutron TOF and the deposited gamma energy in the detector. A TOF (or spatial) selection on the target is then performed, as shown in Fig. 9 for a graphite target, to obtain the experimental gamma signature of the interrogated material.

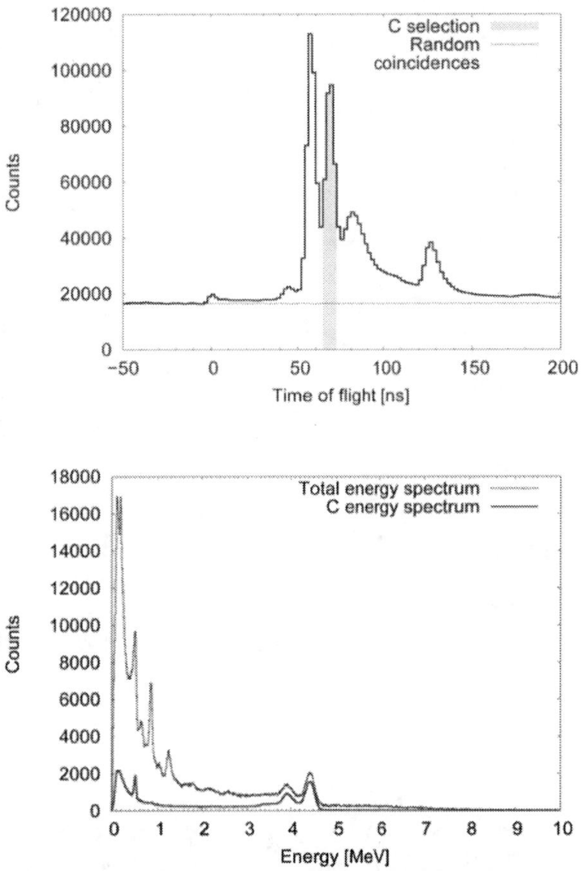

Fig. 9: Time of flight spectrum (upper panel) and energy spectrum (lower panel) on a NaI detector for a graphite target. The gamma spectrum presented with a full red line was obtained after a time of flight selection of the target and after suppression of random coincidences. The total spectrum without time selection is plotted with the black dashed line.

Fig. 10 presents experimental gamma signatures for carbon, nitrogen and oxygen recorded with the NaI detectors.

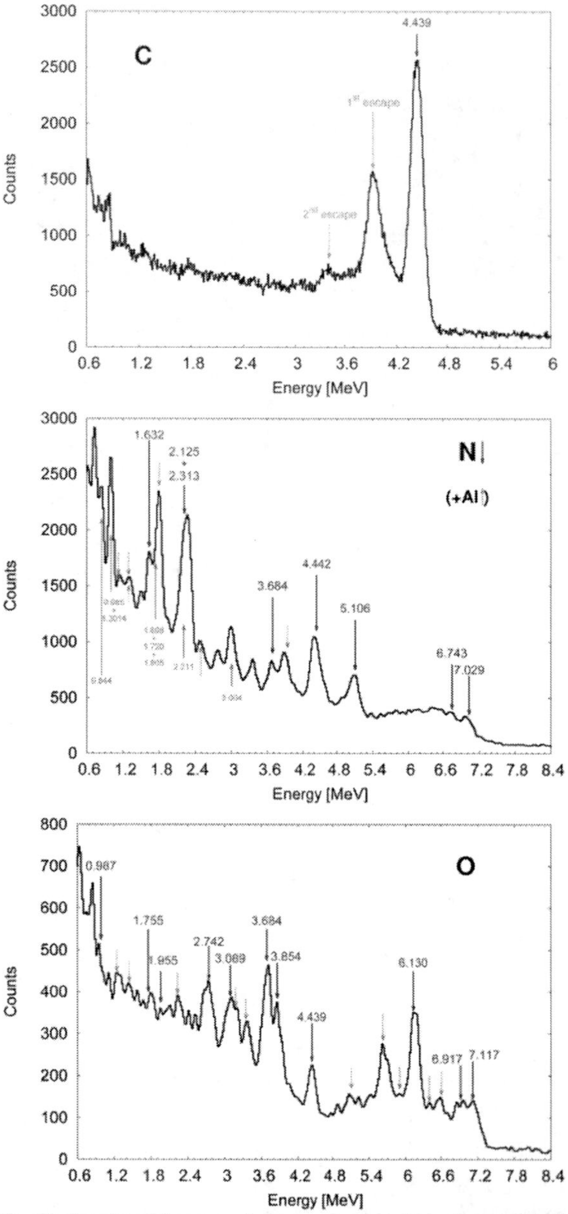

Fig. 10: Experimental gamma signatures of carbon (top), nitrogen (middle) and oxygen (bottom). Dashed arrows are used to point at peaks corresponding to first escapes, while dotted arrows are used for second escapes. On the nitrogen spectrum, the aluminum signature due to the cryogenic storage Dewar has not been subtracted yet (green up arrows).

V. Conclusion

In view to decompose the NaI and $LaBr_3$ gamma spectra of C-BORD Rapidly Relocatable Tagged Neutron Inspection System (RRTNIS) into a weighted sum of elemental spectra, pure element gamma signatures are needed. Therefore, neutron-induced gamma spectra of elements of interest for cargo container inspection have been simulated with MCNP6 Monte Carlo code and post processed with MODAR software. In addition, as differences have been observed in the past between simulation and experimental spectra, such as missing peaks or significant differences in the continuum, the acquisition of neutron-induced gamma signatures is underway at CEA Cadarache. This database of gamma spectra will be used to identify the interrogated materials during RRTNIS tests, first in the laboratory in JRC Ispra (Italy) and CEA Saclay (France) for the commissioning of the system, in 2017, and then for in-situ demonstrations at DCT Gdansk (Poland) and DCA Rotterdam (The Netherlands), in 2018.

References

[1] A. Sardet, B. Perot, C. Carasco, G. Sannié, S. Moretto, G. Nebbia, C. Fontana, M. Moszynski, P. Sibczynski, K. Grodzicki, L. Swiderski, A. Iovene and C. Tintori, Design of the Rapidly Relocatable Tagged Neutron Inspection System of the C-BORD project, in *IEEE Conference Records Nuclear Science Symposium*, Strasbourg, France, 2016.

[2] Los Alamos National Laboratory, Monte Carlo N-Particle eXtended, 2008. Available: https ://mcnpx.lanl.gov/.

[3] C. Carasco, MCNP Output Data Analysis with ROOT (MODAR), *Computer Physics Communications*, vol. 181, pp. 1161-1666, 2010.

[4] M. Gierlik, T. Batsch, M. Moszynski, T. Szczesniak, D. Wolski, W. Klamra, B. Perot et G. Perret, Comparative study of large NaI(Tl) and BGO scintillators for the EURopean Illicit TRAfficking Countermeasures Kit project, *IEEE transactions on nuclear science*, vol. 53, n° 13, pp. 1737-1743, 2006.

[5] C. Eleon, B. Perot et C. Carasco, MCNP simulations for the neutron sensor design and gamma detector selection, UNCOSS project - Deliverable 3.1 & 3.2, 2009.

[6] W. El Kanawati, B. Pérot, C. Carasco, C. Eleon, V. Valkovic, D. Sudac, J. Obhodas and G. Sannié, Acquisition of prompt gamma-ray spectra induced by 14 MeV neutrons and comparison with Monte Carlo simulations, *Applied Radiation and Isotope*, no. 69, pp. 732-743, 2011.

[7] B. Pérot, C. Carasco et al., Measurement of 14MeV neutron-induced prompt gamma-ray spectra from 15 elements found in cargo containers, *Applied Radiation and Isotopes*, n°166, pp. 421-434, 2008.

[8] C. Eleon, B. Perot, C. Carasco, D. Sudac, J. Obhodas et V. Valkovic, Experimental and MCNP simulated gamma-ray spectra for the UNCOSS neutron-based explosive detector, *Nuclear Instruments and Methods in Physics Research Section A*, vol. 629, pp. 220-229, 2011.

[9] B. Pérot, C. Carasco, R. Vogler, E. Cusset, J.-L. Ma, P. Pras, M. Gmar, G. Sannié and S. Normand, Acquisition of neutron-induced gamma signatures of chemical agents, in *Proceedings of the International Topical Meeting on Nuclear Research Applications and Utilization of Accelerators, IAEA, Vienna, Austria, 4-8 May 2009. Proceedings CD Series CN-173*, http://www-pub.iaea.org/MTCD/publications/PDF/P1433_CD/datasets/papers/sm_en-07.pdf, 2009.

[10] C. L. Fontana, A. Carnera, M. Lunardon, F. Pino, C. Sada, F. Soramel, L. Stevanato, G. Nebbia, C. Carasco, B. Pérot, A. Sardet, G. Sannié, A. Iovene, C. Tintori, K. Grodzicki, M. Moszyński, P. Sibczyński, L. Swiderski and S. Moretto, "Detection System of the First Rapidly Relocatable Tagged Neutron Inspection System (RRTNIS), Developed in the Framework of the European H2020 C-BORD Project," in *CAARI proceedings*, Physics Procedia, 2016.

New perspectives for undoped CaF_2 scintillator as a threshold activation neutron detector

Pawel Sibczynski, Andrzej Dziedzic, Krystian Grodzicki, Joanna Iwanowska-Hanke, Marek Moszyński, *Fellow, IEEE,* Lukasz Swiderski, *Member, IEEE,* Agnieszka Syntfeld-Każuch, *Member, IEEE,* Dariusz Wolski, Frédérick Carrel, Amélie Grabowski, Matthieu Hamel, Frederic Laine, Adrien Sari, Alessandro Iovene, Carlo Tintori, Cristiano Fontana, Felix Pino

Abstract— **In this paper we present the prompt photofission neutron detection performance of undoped CaF_2 scintillator using Threshold Activation Detection (TAD). The study is carried out in the frame of C-BORD Horizon 2020 project, during which an efficient toolbox for high volume freight non-intrusive inspection (NII) is under development. Technologies for radiation monitoring are the part of the project. Particularly, detection of various radiological threats on country borders plays an important significant role in Homeland Security applications. Detection of illegal transfer of Special Nuclear Material (SNM) - ^{235}U, ^{233}U and ^{239}Pu - is particular due to the potential use for production of nuclear weapon as well as radiological dispersal device (RDD) – known also as a "dirty bomb". This technique relies on activation of ^{19}F nuclei in the scintillator medium by fast neutrons and registration of high-energy β particles and γ–rays from the decay of reaction products. The radiation from SNM is detected after irradiation in order to avoid detector blinding. Despite the low $^{19}F(n,\alpha)^{16}N$ or $^{19}F(n,p)^{19}O$ reaction cross-section, the method could be a good solution for detection of shielded nuclear material. Results obtained with the CaF_2 detector were compared with the previous study done for BaF_2 and ^3He detector. These experimental results were obtained using ^{252}Cf source and 9 MeV Varian Linatron M9 linear accelerator (LINAC). Finally, performance of the prompt neutron detection system based on CaF_2 will be validated at Rotterdam Seaport during field trails in 2018.**

Index Terms — **photofission, neutron detection, threshold activation detection, TAD, CaF2, BaF2, scintillators, fluorine, border monitoring, SNM, nuclear material, uranium, MCNP**

This project has received funding from the European Union's Horizon 2020 research and innovation programme under grant agreement No 653323. Disclaimer: This text reflects only the author's views and the Commission is not liable for any use that may be made of the information contained therein.

Pawel Sibczynski, Andrzej Dziedzic, Krystian Grodzicki, Joanna Iwanowska-Hanke, Marek Moszyński, Lukasz Swiderski, Agnieszka Syntfeld-Każuch, Dariusz Wolski are with the National Centre for Nuclear Research (NCBJ), Otwock, 05-400 Poland (telephone: +48 22 273 1440, e-mail: pawel.sibczynski@ncbj.gov.pl, pawel.sibczynski@gmail.com)

Frédérick Carrel, Amélie Grabowski, Matthieu Hamel, Frederic Laine, Adrien Sari are with CEA Saclay, LIST, Sensors and Electronic Architectures Laboratory, F-91191 Gif-sur-Yvette, France.

Alessandro Iovene, Carlo Tintori are with the CAEN S.p.A., Via Vetraia 11, 55049 - Viareggio (LU) – Italy.

Cristiano Fontana, Felix Pino are with University of Padua, Department of Physics and Astronomy, Via Marzolo 8, 35131 Padua - Italy

I. Introduction

For several years detection of nuclear materials plays important role in Homeland Security and Border Monitoring. Particularly, detection of Special Nuclear Materials (SNM) is of high importance. Recently, fluorine based scintillation detectors were found to be good candidates for prompt photofission neutrons detection by means of threshold activation (the technique is known as a Threshold Activation Detection – TAD) [1], [2], [3], [4]. Detection of the prompt photofission neutron, emitted several fs after the photofission, is hardly possible directly because of detector blindness during the bremsstrahlung photon pulse from linear accelerator (LINAC). However, prompt photofission neutrons can be registered by the $^{19}F(n,\alpha)^{16}N$ threshold reaction in the scintillator medium using LINAC producing bremsstrahlung photons with endpoint at 9 MeV. By this way, it is possible to detect characteristic β particles (energy endpoint at 4.3 MeV and 10.4 MeV) and γ-rays (6.1 MeV) emitted from the $^{19}F(n,\alpha)^{16}N$ reaction product few seconds after irradiation. Registration of the characteristic radiation can be done after irradiation of nuclear material with high energy bremsstrahlung photons from LINAC due to the fact that the half-life of the ^{16}N nucleus is 7.1 s. Currently, liquid fluorocarbon [1]–[3], pentafluorostyrene-based plastic [5] and BaF_2 [6] scintillators were proposed for the TAD technique. However, some properties of the fluorocarbon liquid scintillators cause the notably limited potential use for Homeland Security. The main disadvantages of the liquid fluorocarbon scintillators, such as EJ-313 or BC-509, are toxicity of hexafluorobenzene as the main compound, high flammability with the flashpoint at 10°C and risk of leakage. On the other hand, BaF_2 scintillator contains rather small mass fraction of fluorine (about 21%). Thus, we decided to investigate the undoped Ø5" × 3" CaF_2 scintillator for the TAD technique. The half of CaF_2:Eu light yield and light emission peak at 280 nm [7], although diminishes energy resolution when a photomultiplier with borosilicate glass would be used, still allows to be applied for the TAD. The energy resolution plays secondary role in the TAD technique. This technique aims on detection of beta continuum with endpoint at 10.4 MeV and high energy delayed γ-rays emitted for variety of photofission products. Both the mass fraction of fluorine about 48.6% and the very low light self-absorption, allowing for production in large volumes, are the main advantages of the undoped CaF_2 crystals. Furthermore, we used an MCNP6 code for simulation of (n,α) reaction rate for the CaF_2 and compared with the BaF_2 of volume of Ø5" × 1", studied in [6].

II. MCNP6 SIMULATIONS

The MCNP6 software package was used for investigation of the CaF_2 and BaF_2 detector sensitivity. The calculations were divided into following processes:

- F4 tally - (n,α) reaction rate for prompt neutrons
- F1 tally - β particle detection efficiency
- F8 tally - 6.1 MeV γ-ray detection efficiency

Visualization of (n,α) interactions is presented in Fig. 1. A ^{252}Cf neutron source was situated 15 cm from the front of the detector. The reactions occur mainly in the first layer of the CaF_2 detector, where about 51% of (n,α) reactions take place. However, the BaF_2 has greater density (ρ = 4.88 g/cm^3), resulting in better detection efficiency of 6.1 MeV γ-rays emitted from ^{16}N. Although the BaF_2 would be more efficient considering the same volume of the detectors, the lower cost

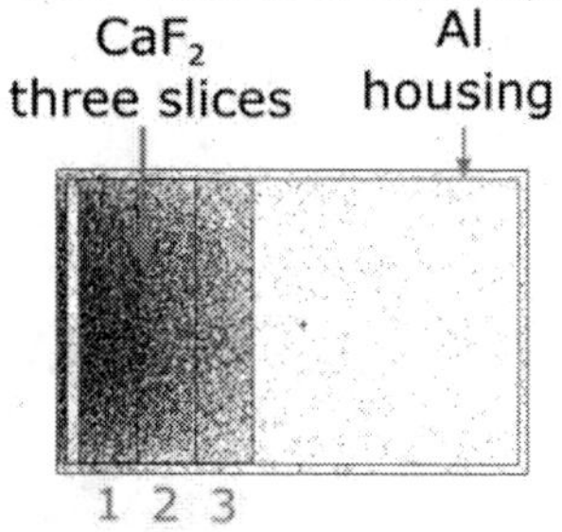

Fig. 1. Visualization of (n,α) reactions in the CaF_2 scintillator medium.

of CaF_2 can play significant role for application in large volume systems. Calculation results were summarized in

TABLE I

THE SIMULATED (N,ALPHA) REACTION RATES, ELECTRON ABSORTPION AND 6.1 MEV GAMMA-RAY DETECTION EFFICIENCY IN DETECTOR SLICES, Ø5" × 1" EACH. VALUES ARE NORMALIZED TO ONE-RADIATION SOURCE.

Detector	Front slice (1)	Central slice(2)	Black slice(3)
CaF_2 (n, α) rate	2.20E-4 (50.9%)	1.34E-4 (30.9%)	7.87E-5 (18.2%)
CaF_2 e$^-$ absorption	0.994%	100%	0.994%
CaF_2 6.1 MeV gamma absorption	1.78E-2	2.30E-2	1.78E-2
CaF_2 efficiency – front slice	2.57E-6	-	-
CaF_2 efficiency – whole detector		5.44E-6	
BaF_2 (n, α) rate	5.88E-5	-	-
BaF_2 e$^-$ absorption	0.997%	-	-
BaF_2 6.1 MeV gamma absorption	3.87E-2	-	-
BaF_2 efficiency – whole detector		1.49E-6	

Table 1. Finally, ø5" × 3" CaF_2 present much greater (n,α) reaction rate and better γ-ray detection efficiency for 6.1 MeV γ-rays than the previously tested BaF_2.

III. MEASUREMENTS WITH A ^{252}CF SOURCE AT NCBJ

Spectra from the ^{252}Cf for the ø5" × 3" CaF_2 and previously tested BaF_2 [6] are presented in Fig. 2. Comparing the relative number of counts, the experimental results and the one obtained with MCNP6 are in reasonable agreement. The number of counts ratio for CaF_2 and BaF_2 in the region of interest (ROI, 6 – 10.5 MeV) from the experiment is 3.1. The total value ratio result from the MCNP simulation for whole detectors, summarized in Table I, is 3.7.

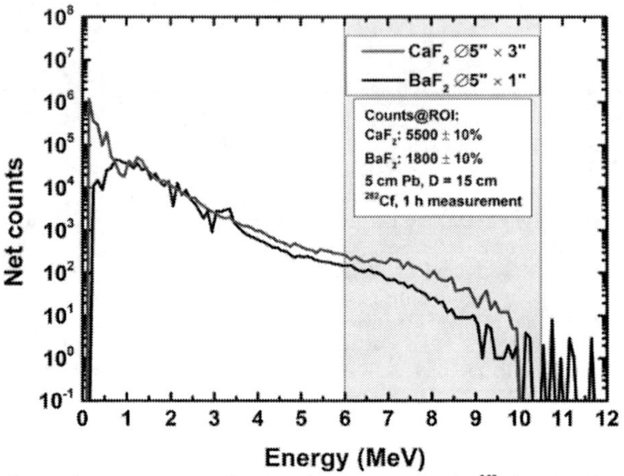

Fig. 2. The energy spectra of BaF_2 and CaF_2 exposed to the ^{252}Cf source. The β continuum up to 10.4 MeV is clearly seen.

IV. TEST SITE AT CEA SACLAY

Fig. 3 presents the test site at CEA Saclay SAPHIR Facility. A Varian LINATRON M9 LINAC working in 9 MeV mode was used to induce photofission in depleted uranium (DU). The DU was placed 1 meter from the conversion target, then, CaF_2 detector was placed 1 meter from the DU. Data from the tested CaF_2 detector were registered with CAEN DT5730 8 channel 500 MS/s desktop digitizer with dedicated FPGA firmware allowing for onboard VETO signal gating using the LINAC trigger.

Fig. 3. Test site at SAPHIR LINAC facility at CEA Saclay, France.

V. Active measurements results at SAPHIR

The single energy spectrum after background subtraction for the CaF_2 for 5 cm Pb shielded DU is showed in Fig. 4. Single measurement was based on 10 s of irradiation, 1 s of cooling time and 20 s of data recording after the global irradiation. Number of net counts for the CaF_2, BaF_2 and ^3He detectors are shown in Table 2. The number of counts between 6 and 10.5 MeV for the CaF_2 and BaF_2 showed in the table were summed from the 3 consecutive measurements.

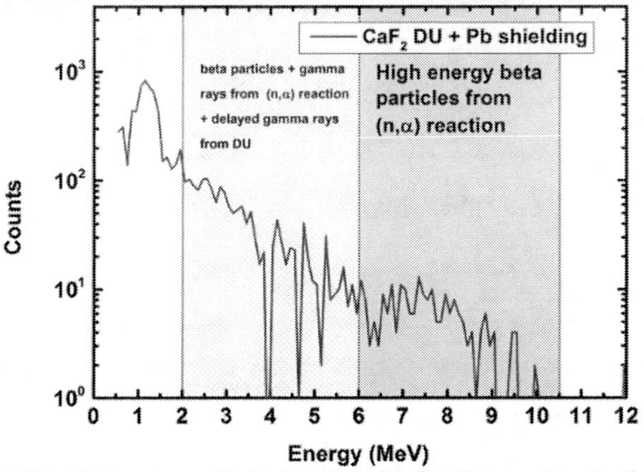

Fig. 4. Net spectrum of depleted uranium after photofission registered with the CaF_2 detector.

VI. Summary

TABLE II
Number of net counts from DU registered with the CaF_2, BaF_2 and single ^3He 150NH100 type detector

Detector	Size	Net counts	Reference
CaF_2 + 5 cm Pb	Ø 5"× 3"	205 ± 14	
CaF_2 without Pb	Ø 5"× 3"	420 ± 20	
BaF_2 + 5 cm Pb	Ø 5"× 1"	189 ± 14	[6]
BaF_2 without Pb	Ø 5"× 1"	275 ± 17	[6]
^3He tube + 5 cm Pb	Ø2.5 cm × 1m	3600 ± 60	[6]
^3He tube without Pb	Ø2.5 cm × 1m	4450 ± 67	[6]

Better performances were achieved with the ø5" × 3" CaF_2 in comparison with the previous solution based on ø5" × 1" BaF_2. Much greater (n,α) reaction rate, resulting in greater high energy β particles emission, was observed for CaF_2 in comparison with the previously tested BaF_2. The ø5" × 3" CaF_2 also has better γ-rays detection efficiency, which is crucial in detection of delayed γ-rays as well as 6.1 MeV γ-rays from (n,α) reaction. The MCNP6 simulations and measurements of relative net counts with ^{252}Cf source are in reasonable agreement. Although measurements at SAPHIR facility confirmed better performance of the CaF_2 over the BaF_2 detector, the dominance is evidently smaller. Despite the fact that the same amount of DU was used in both experiments, the geometry of the DU samples used was different, what could be the reason of the discrepancy. Finally, in the frame of C-BORD project, the prompt neutron detection system based on CaF_2 detectors will be integrated and tested in the field trial at Rotterdam Maasvlakte Seaport in 2018.

Acknowledgment

This project has received funding from the European Union's Horizon 2020 research and innovation programme under grant agreement No 653323. Disclaimer: This text reflects only the author's views and the Commission is not liable for any use that may be made of the information contained therein.

References

[1] T. Gozani, J. Stevenson, and M. J. King, "Neutron threshold activation detectors (TAD) for the detection of fissions," *Nucl. Instrum. Methods Phys. Res. Sect. Accel. Spectrometers Detect. Assoc. Equip.*, vol. 652, no. 1, pp. 334–337, Oct. 2011.
[2] J. Stevenson, T. Gozani, M. Elsalim, C. Condron, and C. Brown, "Linac based photofission inspection system employing novel detection concepts," *Nucl. Instrum. Methods Phys. Res. Sect. Accel. Spectrometers Detect. Assoc. Equip.*, vol. 652, no. 1, pp. 124–128, Oct. 2011.
[3] T. Gozani, M. J. King, and J. Stevenson, "The investigation of fast neutron Threshold Activation Detectors (TAD)," *J. Instrum.*, vol. 7, no. 02, p. C02042, 2012.
[4] P. Sibczynski *et al.*, "Verification of threshold activation detection (TAD) technique in prompt fission neutron detection using scintillators containing 19 F," *J. Instrum.*, vol. 10, no. 09, p. T09005, 2015.
[5] M. Hamel *et al.*, "A fluorocarbon plastic scintillator for neutron detection: Proof of concept," *Nucl. Instrum. Methods Phys. Res. Sect. Accel. Spectrometers Detect. Assoc. Equip.*, vol. 768, pp. 26–31, Dec. 2014.
[6] Pawel Sibczynski *et al.*, "Comparison of prompt and delayed photofission neutron detection techniques using different types of radiation detectors," presented at the IEEE Nuclear Science Symposium / Medical Imaging Conference, Strasbourg, 2016.
[7] R. T. Williams, M. N. Kabler, W. Hayes, and J. P. Stott, "Time-resolved spectroscopy of self-trapped excitons in fluorite crystals," *Phys. Rev. B*, vol. 14, no. 2, pp. 725–740, Jul. 1976.

Data-Fusion for a Vision-Aided Radiological Detection System: Sensor dependence and Source Tracking

Kelsey Stadnikia, Allan Martin, Kristofer Henderson, Sanjeev Koppal, Andreas Enqvist

***Abstract*—The University of Florida is taking a multidisciplinary approach to fuse the data between 3D vision sensors and radiological sensors in hopes of creating a system capable of not only detecting the presence of a radiological threat, but also tracking it. The key to developing such a vision-aided radiological detection system, lies in the count rate being inversely dependent on the square of the distance. Presented in this paper are the results of the calibration algorithm used to predict the location of the radiological detectors based on 3D distance from the source to the detector (vision data) and the detectors count rate (radiological data). Also presented are the results of two correlation methods used to explore source tracking.**

I. Introduction

This project combines 3D vision sensors and radiological detectors in hopes to more efficiently and effectively locate potential nuclear threats in dynamic environments such as airports and package and mailing facilities. Any radioactive material may pose a health threat and could lead to harm if transported undetected and/or used in radiation dispersion devices (RDDs). Special nuclear material (SNM) is also a major concern as it can potentially be used in nuclear weapons [1]. Current systems such as portal monitors and distributed sensor networks approach this issue [2] [3]. However, the system discussed in this paper would not only be able to determine the presence of a nuclear threat, it would be able to *track* it; thus, allowing less disturbance of commerce and more accurate secondary inspector response.

Combining radiological sensors with vision sensors will help to see potential occlusions, suspicious persons or items, and determine if there is a cause for concern. It is possible to combine radiological sensors and vision sensors through data fusion. The ability to correlate a time dependent signal of a moving source is the key to fusing the two systems. Fig. 1 shows the conceptual idea of conjoining the time dependent data sets of the two types of sensors. The concept is based on the radiological count rate being inversely dependent on the square of the distance.

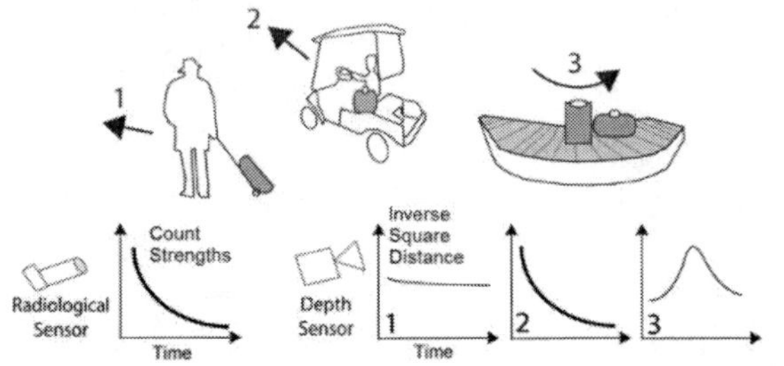

Fig. 1. Tracking scenarios and the principle of the overlapping distance/time domain between the radiological and visual sensor. The scales will be arbitrary but the data trends will be highly correlated.

Ideally, the choice of 3D vision sensors, radiological detectors and fusion algorithms can be flexible for implementation purposes. Due to their long range in air and their ability to travel through some shielding and visually opaque materials, gamma-rays and neutrons are the most useful quanta to detect compared to beta and alpha particles. Detectors with the capability of detecting both Gamma rays and neutrons and then differentiating between the two with pulse shape discrimination methods may be advantageous. A wide range of radiological detectors consisting of EJ-309 liquid scintillators, Na-I, and He-3 have been explored.

Vision sensors such as the Microsoft Kinect and the Velodyne HDL-32E LiDAR have been explored for this research. The Kinect uses an infrared (IR) emitter, which emits infrared light beams, and a depth sensor, which reads the IR beams reflected back to the sensor. The reflected beams are converted into depth information measuring the distance between an object and the sensor [4]. The Velodyne uses Light Detection and Ranging (LiDAR) technology, which measures distances using Time of Flight (TOF) and short laser pulses. Massachusetts Institute of Technology is currently working on depth sensing with polarization cues, which will drastically improve cheap 3D imaging sensors [5]. This paper will only cover the results using the Velodyne HDL-32E and the EJ-309 liquid scintillators. However, two correlation algorithms and their results will be discussed.

Before taking dynamic measurements in a room, the room

June 1, 2017. This material is based upon work supported by the U.S. Department of Homeland Security under Grant Award Number, 2014-DN-077-ARI083-01. The views and conclusions contained in this document are those of the authors and should not be interpreted as necessarily representing the official policies, either expressed or implied, of the U.S. Department of Homeland Security.

Kelsey Stadnikia is with the University of Florida, Gainesville, FL 32606 USA (e-mail: klstad@ufl.edu).

must be calibrated. To do this, a calibration algorithm is used to predict the location of the radiological detectors based on the 3D distance from the source to the detector (vision data) and the detectors count rate (radiological data). Thus, count rate can be correlated to movement in three dimensions.

II. Methods and Measurements

An experimental setup utilizes EJ-309 Liquid Scintillator detectors and an HDL-32E LiDAR. Calibration experiments are used to determine a detector "pseudo-location" and involve several static measurements. The "pseudo-location" is where the data fits best and takes into account room and detector dependencies. Correlation experiments involve the potential use of this discovered "pseudo location" and dynamic measurements.

A. Calibration

For calibration, a Cf-252 source is placed in a camera calibration target resembling a checkerboard and is measured in 27 static locations arranged in a 3x3x3 cube. One EJ-309 detector is setup to be collocated with the HDL-32E while another is setup in an offset location. Fig. 2 shows a schematic of this setup for nine individual experiments.

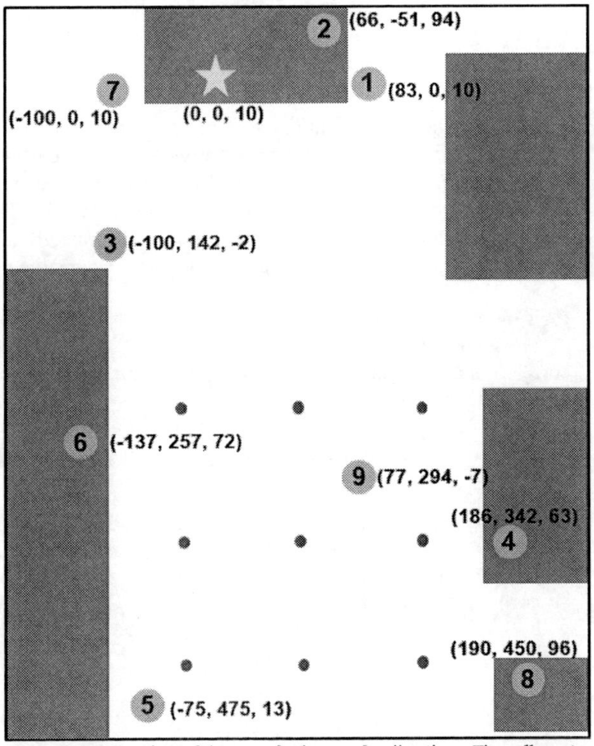

Fig. 2. Top-down view of the room in the x and y directions. The yellow star represents the known location of the co-located detector. Red dots represent the 27 source locations. Orange circles represent the manually measured offset detector locations. Blue blocks represent fixed lab furniture that may cause radiation scattering. Not to scale.

The calibration algorithm is based on (1).

$$R = \lambda_0 \frac{\lambda(R')}{\sqrt{C}} \tag{1}$$

R represents the 3D distance from the source to the offset detector, whereas, R' represents the distance from the source to the collocated detector. C is the count rate collected by the collocated detector and λ_0 is a scaling factor used to account for source strength. The algorithm begins by determining $\lambda(R')$ using the count rate collected by the collocated radiological detector and the 3D distance from the collocated radiological detector to the source as determined by the vision sensor. This function is determined to be a second order polynomial fit to the data which can be seen in Fig. 3.

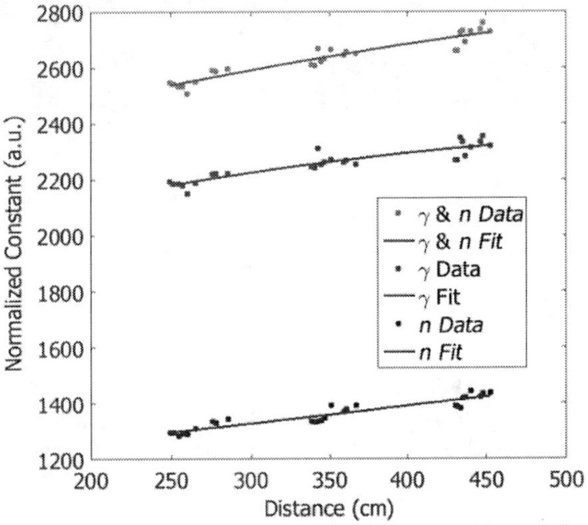

Fig. 3. Second order polynomial fit to combined data, gamma-ray only data, and neutron only data.

The parameter λ_0 is determined by exploring a wide range of values. The algorithm then generates spherical shells of a given thickness and radius R at each individual source location. Where the majority of these spherical shells overlap is said to be the "pseudo-location". The spherical shell generation and progression can be seen in Fig. 4. The point where the most spherical shells overlap then determines the number of voxels overlapped and averages the voxels to output one location with x, y, and z coordinates. The optimal λ_0 value is chosen where the most spherical shells overlap and then where the most voxels are covered by the overlaps.

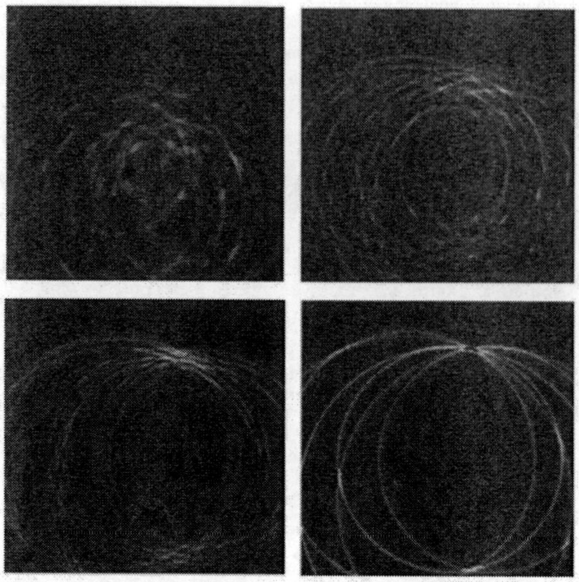

Fig. 4. Horizontal (2D) slices in the x-y plane illustrating the overlapping spherical shells used in the algorithm at various heights in the z directions (top-left: -213 cm, top-right: -129 cm, bottom-left: -72 cm and bottom-right: -12 cm). x, y, z directions correspond to left-right, forward-backward and up-down directions for the vision sensor view respectively.

The location determined by the algorithm now represents the "pseudo-location" where the data works best. Thus, allowing the accuracy of the predicted sensor location to presumably be ignored. Further analysis will determine the eligibility of this approach. Once the room is calibrated, dynamic source tracking can take place. Results of the calibration algorithm are done using hand measured location data, as well as, vision sensor location data.

B. *Correlation*

Numerous measurement scenarios comprising of two to three people walking in a room where one of the persons is carrying a source in a backpack have been taken. Choreographed walking patterns as well as directionless walks were employed. Thus far, all data has been post processed. Fig. 5 is a top-down image of a frame taken from the LiDAR data in two-dimensions (2D). Kalaman filtering is used to obtain the trajectories of the moving targets in the scene [6]. These trajectories are then compared to the radiological trajectory created by taking the inverse square root of the count rate at one second time steps. Two methods of correlating the vision data with the radiological data have been employed.

Fig. 5. Top-Down frame of LiDAR data showing the Kalman filter tracking of three persons. Person 3 is carrying the radioactive material and is given a red track.

1) Cosine Distance Method

The first method investigated was the cosine distance method [7]. This method measures the cosine of the angle between two non-zero vectors as shown in (2).

$$\text{cosine}(\text{A}, \text{B}) = \frac{A \bullet B}{|A| \times |B|} \tag{2}$$

where A and B are vectors. $A \bullet B$ is the dot product between vectors and |A| and |B| are the magnitudes of A and B respectively. In this method values range from negative one to positive one. A value of one means the vectors have the same orientation. Each trajectory created by the Kalman filter is compared to the radiological data using this method. The trajectory with a cosine value closest to one is selected by the algorithm to be the person carrying the radioactive material.

2) Correlation Coefficient Method

This method uses an existing MATLAB function to calculate correlation coefficients, *corrcoef*. The correlation coefficient is defined in terms of (3).

$$\rho(A, B) = \frac{cov(A,B)}{\sigma_A \sigma_B} \tag{3}$$

where A and B are vectors. The *cov(A,B)* is the covariance of A and B and σ_A and σ_B are the standard deviations of A and B respectively. Each trajectory created by the Kalman filter will be correlated with the radiological trajectory. The Trajectory with the highest correlation coefficient is selected by the algorithm to be the person carrying the radioactive material.

III. Results

The background count rate is taken into account in the calibration results. The background gamma-ray count rate is roughly 56 counts per second (cps) for the co-located detector and the offset detector. The neutron count rate is roughly two cps for both the co-located detector and the offset detector. These values vary slightly depending on room location. Background currently remains in the correlation data.

A. Calibration

The count rate used in the calibration, (1), is the net count rate with background subtracted. Measurement times are long enough such that the total rate error as a percentage is kept to around 2% for combined gamma-ray and neutron data.

1) Ground Truth Data

The best-case scenario for the calibration algorithm is examined using manually measured, or ground truth, locations of the offset detector and the 27 source locations. These locations were carefully measured by hand using measuring tapes and laser distance finder. The algorithm predicted offset detector location was compared to the ground truth measured offset detector location. The spherical shell thickness was chosen to be 4 cm, which produces results with the lowest Euclidean distance values for all count rate types and the most experiments. Figure 6 shows the results of recovering the algorithm predicted offset detector location compared to the ground truth measured offset detector location on an aerial schematic of the room where the measurements took place.

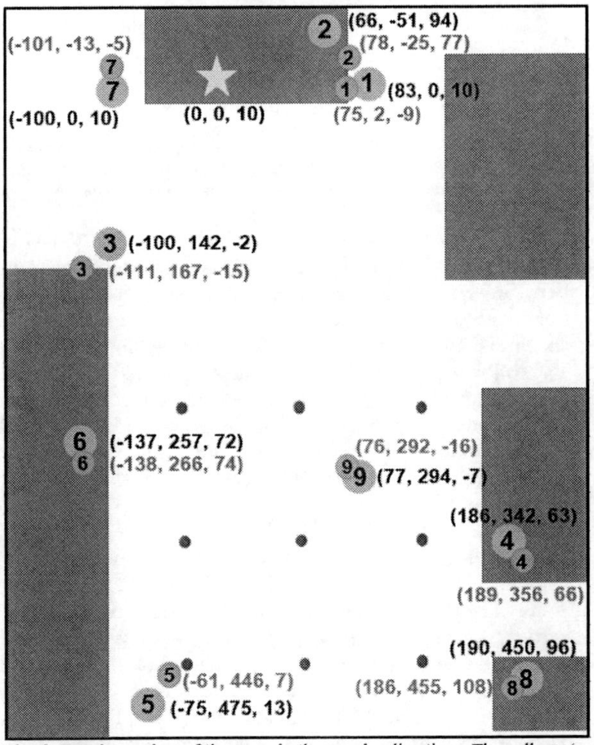

Fig. 6. Top-down view of the room in the x and y directions. The yellow star represents the known location of the co-located detector. Red dots represent the 27 source locations. Orange circles represent the ground truth measured offset detector locations and the green circles represent the calibration predicted offset detector locations. Blue blocks represent fixed lab furniture that may cause radiation scattering. Not to scale.

Table I shows these results. The Euclidean distance is of the predicted offset detector location to the ground truth offset detector location. The average Euclidean distance over the nine experiments using both gamma-ray and neutron data is 22.5 cm. When using only gamma-ray count rate data, the average Euclidean distance is 27.8 cm; and when using neutron only count rate data, the average Euclidean distance was 23.9 cm.

TABLE I
GROUND TRUTH CALIBRATION ALGORITHM RESULTS IN EUCLIDEAN DISTANCES

Exp. No.	Euclidean Distance: Combined Data (cm)	Euclidean Distance: Photon Data (cm)	Euclidean Distance: Neutron Data (cm)
1	31.2	34.9	7.9
2	37.4	55.4	48.1
3	31.3	34.0	31.1
4	15.5	14.9	22.0
5	26.6	35.2	13.2
6	10.9	14.4	14.5
7	29.5	35.5	34.5
8	9.3	12.8	29.3
9	11.1	13.3	14.2

Absolute agreement is not necessarily required for the system to operate accurately. These algorithm-predicted locations are referred to as "pseudo-locations". If there are significant room dependencies that are not easily captured by our allowed quadratic deviation from ideal radiation transport, then this deviation can possibly be incorporated into the model as a "best approximation" for the location of the radiological sensor vs. the radiation source location and the count rate response that will later be tracked in the system. Thus, the pseudo-locations will be utilized along with the non-ideal quadratic deviation. The distance discrepancy is the uncertainty in identifying the radiation detector's location with respect to the overall radiation response from all the source locations compared to an idealized distance dependence of the detectable radiation flux, not the uncertainty in locating the source locations.

2) LiDAR Data

The HDL-32E covers a 360-degree view allowing all nine experiments to be analyzed. The HDL-32E's location within the room allowed a view of all source locations, offset detector locations and the co-located detector. The spherical shell thickness was chosen to be 8 cm, which produces results with the lowest Euclidean distance values for all count rate types and the most experiments. Figure 7 shows the results of recovering the algorithm predicted offset detector location compared to the HDL-32E determined offset detector location on an aerial schematic of the room where the measurements took place.

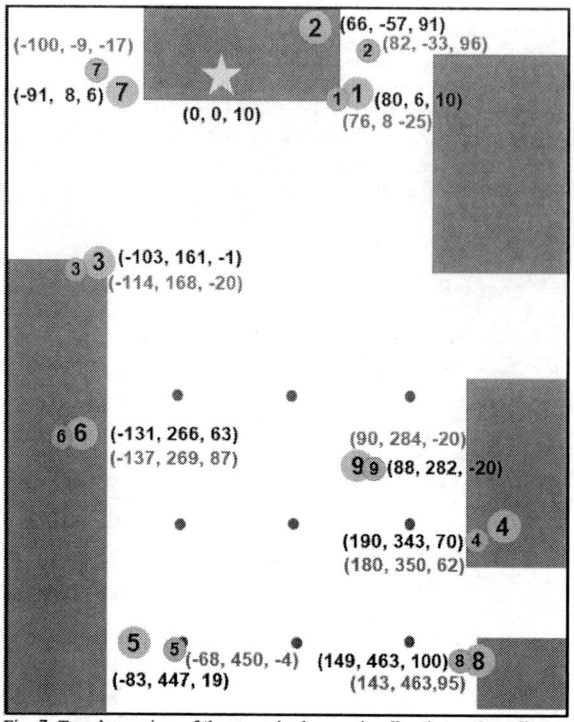

Fig. 7. Top-down view of the room in the x and y directions. The yellow star represents the known location of the co-located detector. Orange circles represent the HDL-32E measured offset detector locations and the green circles represent the calibration predicted offset detector locations. Not to scale.

Table II shows these where the Euclidean distance is from the predicted offset detector location to the HDL-32E determined detector location. The average Euclidean distance over all nine experiments using combined gamma-ray and neutron data is 21.8 cm. When using only gamma-ray count rate data, the average Euclidean distance is 25.5 cm; and when using neutron only count rate data, the average Euclidean distance was 21.9 cm.

TABLE II
GROUND TRUTH CALIBRATION ALGORITHM RESULTS IN EUCLIDEAN DISTANCES

Exp. No.	Euclidean Distance: Combined Data (cm)	Euclidean Distance: Photon Data (cm)	Euclidean Distance: Neutron Data (cm)
1	35.7	43.7	19.6
2	29.3	35.3	37.8
3	23.2	29.5	31.8
4	15.6	14.9	16.6
5	27.0	29.9	12.1
6	25.0	25.0	20.6
7	29.9	38.2	33.5
8	7.7	9.4	18.0
9	2.5	3.7	7.7

The results with the HDL-32E are close to the results obtained using the ground truth data. The HDL-32E has a range of up to 100 m with a typical accuracy of ±2 cm.

B. *Correlation*

The count rate data for correlation was separated into one second time steps. Background was not removed from the data. Fig. 8 shows vision and radiological tracks which are used for correlation.

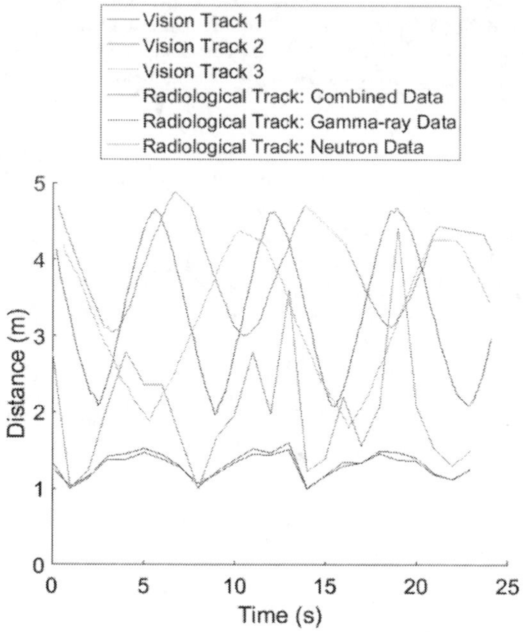

Fig. 8. Vision tracks of three persons to be correlated with the radiological track where the radiological track can be created from combined gamma-ray and neutron data, gamma-ray only data, or neutron only data.

Without subtracting background gamma-ray only data and combined data appear suppressed. This reduces the correlation between the radiological track and vision tracks. It is also important to note that the one over distance squared deviation shown in Fig. 3 can be used to adjust the radiological tracks. The following results, however, show the crude results of the two aforementioned correlation methods without background subtraction and without the use of the "pseudo-location". In the future, results will be calculated with these additions and compared to the results presented here.

1) *Cosine Distance Method*

The cosine distance method predicted 8 out of the 11 experiments correct. Table III shows the results. The "Ground Truth Trajectory" represents the trajectory of the person carrying the radioactive material and therefore is the trajectory that should be most correlated with the radiological data. The "Selected Trajectory" is the trajectory that the algorithm actually chose to be most correlated with the radiological data. The experiments where the algorithm chose the wrong trajectory are highlighted in red.

TABLE III
RESULTS OF THE COSINE DISTANCE METHOD

Exp. No.	Ground Truth Trajectory	Correlation Between Radiation Data and Ground Truth Trajectory	Selected Trajectory	Correlation Between Radiation Data and selected Trajectory
1	1	0.9909	3	0.9942
2	3	0.9909	1	0.9964
3	2	0.9919	2	0.9919
4	1	0.9874	2	0.9939
5	3	0.9940	3	0.9940
6	2	0.9958	2	0.9958
7	2	0.9960	2	0.9960
8	3	0.9930	3	0.9930
9	3	0.9953	3	0.9953
10	2	0.9931	2	0.9931
11	2	0.9987	2	0.9987

2) Correlation Coefficient Method

Out of the 11 experiments, the correlation coefficient method chose the correct Trajectory seven times. Table IV shows the results. The "Ground Truth Trajectory" represents the trajectory of the person carrying the radioactive material and therefore is the trajectory that should be most correlated with the radiological data. The "Selected Trajectory" is the trajectory that the algorithm actually chose to be most correlated with the radiological data. The experiments where the algorithm chose the wrong trajectory are highlighted in red.

TABLE IV
RESULTS OF THE CORRELATION COEFFICIENT METHOD

Exp. No.	Ground Truth Trajectory	Correlation Between Radiation Data and Ground Truth Trajectory	Selected Trajectory	Correlation Between Radiation Data and selected Trajectory
1	1	0.6150	2	0.7186
2	3	0.7240	3	0.7240
3	2	0.6195	2	0.6195
4	1	0.8543	1	0.8543
5	3	0.4152	3	0.4152
6	2	0.1064	1	0.3002
7	2	0.1622	1	0.2070
8	3	0.3718	3	0.3718
9	3	0.2686	3	0.2686
10	2	0.2857	2	0.2857
11	2	0.0948	1	0.8300

IV. Conclusion

Overall, the calibration algorithm for the vision/radiological system predicted the offset detector location well using ground truth and HDL-32E data. The uncertainty seen is within reason for tracking objects of size similar to people, luggage and packages. This calibration algorithm with be a good foundation when transitioning to a larger number of sensors or continuous (real-time) source movement used for calibrating the data fused system. Results of the two correlation methods were less ideal. The cosine distance only slightly more accurate than the correlation coefficient method. Each method proved to have difficulty predicting different experiment numbers only both predicted experiment one incorrectly. Subtracting background is likely to predict the correct track with better accuracy. Using the calibration algorithm determined pseudo location may also improve the correlation algorithms ability to choose the correct track. The use of gamma-ray only data and neutron only data will also be examined. Other more complex and robust correlation algorithms are being looked into as well.

Transitioning to a real-time data processing system is in progress with anticipated challenges such as processing needs for 3D data. A high tracking and source identification accuracy have been achieved in initial stages approaching 90% accuracy in the early test scenarios. Another tracking scenario under investigation involves visual occlusions. In these experiments, a large structure is set in place to obstruct the view of the 3D vision sensor. Distance is then to be determined by the count rate data adjusting for the reduced count rate due to the obstruction. Multi-source tracking is an additional scenario where measurements have been taken are being analyzed. Dynamic scenes with a stationary source as well as scenes without a source entirely are also being explored. These measurements will assist in understanding and optimizing the system threshold. Many tracking scenarios are being studied to continue adjusting and improving the vision-aided detection system.

References and Footnotes

[1] B.D. Geelhood, J.H. Ely, R.R. Hansen, R.T. Kouzes, J.E. Schweppe, and R.A. Warner, "Overview of Portal Monitoring at Border Crossings," *IEEE Conference Record*, 2003.

[2] R. C. Runkle, T. M. Mercier, K. K. Anderson, and D. K. Carlson, "Point Source Detection and Characterization for Vehicle Radiation Portal Monitors," *IEEE Tans. Nucl. Sci., vol. 52, no. 6, pp. 3020-3025*, 2005.

[3] S Brennan, A Mielke and D Torney, "Radiation detection with distributed sensor networks," *IEEE Computer Magazine, vol. 37, no. 8, pp. 57–59*, 2004.

[4] "Kinect for Windows Sensor Components and Specifications," *Microsoft Developer Network*. Microsoft, 2016. Web. 16 May 2016.

[5] Kadambi, Achuta, Vage Taamazyan, Boxin Shi, and Ramesh Raskar, "Polarized 3D: High-Quality Depth Sensing with Polarization Cues,"*2015 IEEE International Conference on Computer Vision (ICCV)* (2015): n. pag. MIT, Dec. 2015. Web. 17 May 2016.

[6] R. E. Kalman. A new approach to linear filtering and prediction problems. Journal of basic Engineering, 82(1):35–45, 1960.

[7] P.-N. Tan et al. Introduction to data mining. Pearson Education. India, 2006.

Remote inspection with multi-copters, radiological sensors and SLAM techniques

Henrique Carvalho, Alberto Vale, Rúben Marques, Rodrigo Ventura, Yoeri Brouwer, Bruno Gonçalves,

***Abstract*—Activated material can be found in different scenarios, such as in nuclear reactor facilities or medical facilities (e.g. in positron emission tomography commonly known as PET scanning). In addition, there are unexpected scenarios resulting from possible accidents, or where dangerous material is hidden for terrorism attacks using nuclear weapons. Thus, a technological solution is important to cope with fast and reliable remote inspection. The multi-copter is a common type of Unmanned Aerial Vehicle (UAV) that provides the ability to perform a first radiological inspection in the described scenarios. The paper proposes a solution with a multi-copter equipped with on-board sensors to perform a 3D reconstruction and a radiological mapping of the scenario. A depth camera and a Geiger-Müler counter are the used sensors. The inspection is performed in two steps: i) a 3D reconstruction of the environment and ii) radiation activity inference to localise and quantify sources of radiation. Experimental results were achieved with real 3D data and simulated radiation activity. Experimental tests with real sources of radiation are planned in the next iteration of the work.**

***Index Terms*—3D reconstruction, radiological inspection, multi-copters**

I. INTRODUCTION

EXPERIMENTAL fusion reactors are aiming for a new and viable source of energy. The normal functioning of these reactors or unforeseen problems will lead to planned and unplanned shutdowns. During this period, Remote Handling Systems (RMS) will proceed with maintenance operations [1]. The RHS require special design for regular operations of inspection and maintenance in hard conditions, such as the high levels of radiation, temperature and residual magnetic fields. The remote maintenance operations are performed during the shutdown of the reactor and shall be previously planned, given the time-consumption and costs.

A Unmanned Aerial Vehicle (UAV) can be deployed in the reactor building to quickly acquire data for a first inspection and insight view. This prompt diagnosis allows to plan ahead the sequence of maintenance operations. However, given the current state-of-the-art of the technologies, the application of a UAV inside reactors for remote inspection during the reactor's shutdown is still an ambitious goal in the near future. Nevertheless, a UAV can be deployed for remote inspection in other scenarios with lower levels of radiation. For instance, in facilities where the material used in nuclear medicine is stored, in locations where possible nuclear weapons are hidden for terrorism attacks or even for search-and-rescue (SaR) like operations in scenarios of nuclear disasters. The outcomes of the research and development performed using a UAV in these scenarios are important to proceed with the remote inspection in nuclear reactors.

A. Vale, H. Carvalho, R. Marques, Y. Brouwer and Bruno Goncalves are with the Instituto de Plasmas e Fusão Nuclear, Instituto Superior Técnico, Universidade de Lisboa, Av. Rovisco Pais 1, 1049-001 Lisboa, Portugal

R. Ventura is with the Institute for Systems and Robotics, Instituto Superior Técnico, Universidade de Lisboa, Av. Rovisco Pais 1, 1049-001 Lisboa, Portugal

The multi-copters are the most common UAV with the powerful capability of a loiter navigation mode, i.e., the vehicle slows to a stop and holds the position during flight while doing data acquisition. This type of UAV can operate autonomously or remotely controlled. In autonomous operation, the predefined trajectory can be optimised according to the dynamics of the vehicle and the target goal of the inspection mission. The main goal of the radiological inspection of a given scenario is to localise and quantify the sources of radiation. For the localisation purpose, a 3D representation of the environment is required.

The proposed solution in this paper provides a methodology for the 3D reconstruction of the scenario as reference for the trajectories and location of the possible sources of radiation. The 3D reconstruction is based on Simultaneous Localisation and Mapping (SLAM) techniques commonly used in mobile robot application [2]. The multi-copter is equipped with a depth camera and a Geiger-Müler (GM) counter to measure the radiological activity along the trajectory and to infer the levels of radiation of the scenario.

This paper is organised as follows. The Section II presents the problem statement of 3D reconstruction using point clouds obtained from depth images and the radiological inference. The Section III proposes a solution for combining different point clouds and the algorithm to estimate how the radiological activity is mapped. Experimental results achieved with real 3D data and simulated radiation activity are presented in Section IV. Conclusions and future work are summarised in Section V.

II. PROBLEM STATEMENT

In activated scenarios, the radiation comes from activated materials or from leaks of containers with activated material inside. During the maintenance operations in nuclear fusion reactors, or during the SaR like operations in nuclear accidents (e.g. Fukushima or Chernobyl), or even during the prevention of terrorism attacks where nuclear weapons may be hidden, the challenge is usually the same: radiological inspection. It is necessary to localise the sources of radiation and then quantify the radiological intensity of the respective sources. As a future development, not covered in this paper, the identification of the type of radiation ($\alpha/\beta/\gamma$) will be addressed, in order to infer the composition of the source (e.g., α emissions from ^{235}U [9], β^- emissions from ^3H, also known as tritium [10]), using other types of sensors, such as the Cadmium Zinc Telluride (CdZnTE) or Scintillation Radiation Detectors.

Since the focus of the radiological inspection with multi-copter system is providing an inspection tool for decision makers, visualisation plays a major role. A possible approach to provide an overview of the scenario and also to support the localisation of the possible points of interest is to build a 3D model, largely used in modern modulation software. Thus, the solution concept towards a 3D reconstruction of the scenario.

Besides the geometrical details, colour and contrast are also essential in the human perception of the world. Therefore, the 3D reconstruction must also include, at least, a grey-scale representation of the scenario.

The 3D of reconstruction of the environment starts with depth images, which are converted to 3D points after calibration using the pin-hole model [3], translated by

$$\lambda \begin{bmatrix} u \\ v \\ 1 \end{bmatrix} = \underbrace{\begin{bmatrix} f_x & f_s & C_x \\ 0 & f_y & C_y \\ 0 & 0 & 1 \end{bmatrix}}_{K} \begin{bmatrix} {}^{\text{cam}}X \\ {}^{\text{cam}}Y \\ {}^{\text{cam}}Z \end{bmatrix}, \qquad (1)$$

where the ${}^{\text{cam}}X$, ${}^{\text{cam}}Y$, and ${}^{\text{cam}}Z$ are the coordinates of the 3D points in the camera's coordinate frame, u and v are the pixel coordinates, and λ is the depth. K is the intrinsic parameters matrix of the camera. This matrix is given by the calibration, which includes the pixels scale factor f_x and f_y, their skewness f_s and the image centre in the camera frame C_x and C_y.

Registering two point clouds, i.e., fitting two point clouds, requires the matching process between points or surfaces, as well as calculating a rigid body transformation that minimises the distance between the points. The (2) is the point-to-plane metric used in this work, where R and T are respectively a rotation and a translation, p_i are the points in the moving point cloud, q_i are projection of the p_i points on the static point cloud planes and n_i are the normals to their respective static plane.

$$E = \sum_i \left[(R\, p_i + T - q_i).n_i \right]^2 \qquad (2)$$

The 3D reconstruction of the scenario provides a point cloud and a set of points describing the path of the camera. In this approach for the radiological inference, it is assumed that each point i of the map is a radioactive source with its own radiological intensity, γ_i. In addition, assuming no interaction with the atmosphere, the radiological intensity decreases at a squared distance to the source [9], [10]. Along a predefined path followed by the multi-copter, the radiological intensity σ_i measured at each point p_i of the path, is given by the sum of the contributions of every source m_j in the map, given by (3). The contribution is proportional to the inverse square distance, "$d(...)$" between p_i and m_j [8].

$$\sigma_i = \sum_{j=1}^{J} \frac{\gamma_j}{d(p_i, m_j)^2} \qquad (3)$$

The path P is defined by a set of I points, p_i. The Σ is the set of measurements σ_i measured at each point of the trajectory. The (4) is the matrix form of (3), where Σ is a $I \times 1$ column vector containing in element i the radiological intensity σ_i measured at point p_i of the path, Δ is the $I \times J$ matrix containing in element ij the inverse square distance between points p_i and m_j of the path and map, respectively, and Γ is a $J \times 1$ column vector containing in element j the radiological activity of the j-th point of the map.

$$\mathbf{\Sigma} = \mathbf{\Delta \Gamma} \qquad (4)$$

In summary, the problems to be addressed in this paper are the 3D reconstruction of the scenario and the radiation activity inference, mathematically described in (2) and (4), respectively. Possible solutions are presented in the next section.

III. Proposed solution

This section introduces the used sensors and describes the 3D reconstruction methodology and the algorithm to infer the radiological activity.

A. Sensors

The proposed solution is based on Kinect like sensors [12], which are sensors that provide RGB or coloured images and depth images. However, the used sensor is a Structure camera from Occipital, which only provides infra-red (IR) and depth images, [13]. Depth images provide, in each pixel of the image, the distance to the closest object in the scenario. The working principle of the sensor is to project an IR laser pattern and capturing its displacement and deformation with an IR camera. From the resulted deformation the camera internally calculates the distance to the objects in the scenario.

To quantify the levels of radiation a GM counter is used, namely the model Sparkfun SEN-09848 with an USB interface. The data acquired by both sensors are synchronized and recorded in a light computer installed on the multi-copter. The data is processed offline, in an external and more powerful computer.

The algorithms for the reconstruction and radiological activity inference are detailed on the next sections.

B. 3D Reconstruction

The 3D reconstructing of the scenario aims for two main objectives. In the first one, the 3D representation provides the means for the scenario visualisation. The reconstructed point clouds colour can create a radiological heat-map, where hotspots are centred on the possible sources of radiation. In the second objective, the position and orientation of the camera relative to a global coordinate frame along the trajectory can be estimated. It is assumed that the GM counter is attached to the camera, hence, the locations and orientations (poses) of the GM counter along the trajectory are also estimated. The

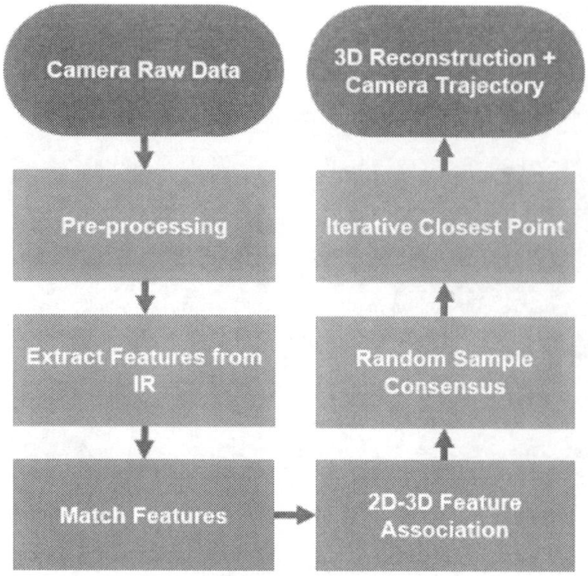

Fig. 1: Reconstruction flowchart.

(a) Colour image.

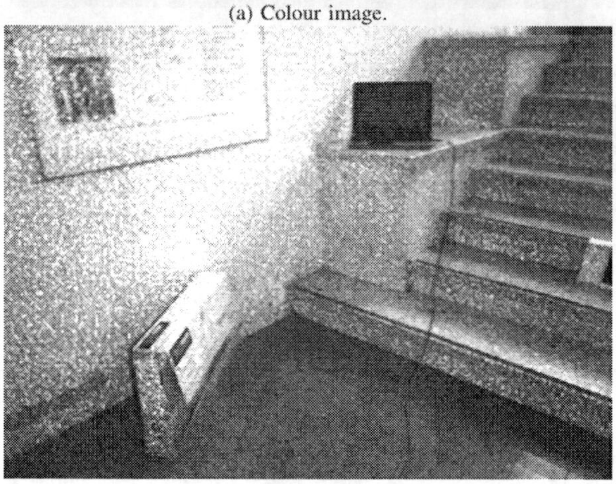

(b) Unfiltered infra-red image.

Fig. 2: The Figure 2a is a photo of part of the scenario and Figure 2b is the respective depth image acquired by the depth sensor.

poses of GM are important to detect and localise the sources of radiation in the map, as suggested in Section II and further detailed in Section III-C.

The algorithm is represented in Figure 1 and detailed in the next sections.

1) Pre-Processing: The 3D reconstruction algorithm receives a set of depth and IR images sorted along the time. The IR images include a projected pattern, as illustrated in Figure 2b, used by the camera to internally calculate depth, i.e., the λ in (1). The projected pattern is similar to salt and pepper noise. The undesired effect of the pattern is reduced using a median filter with a 3×3 kernel, [5].

Point clouds are obtained from the depth images, which are depicted in Figure 3b. In this example, the point cloud coordinates are calculated with (1), where λ is provided by each pixel of the depth image. The grey-scale of the point-cloud comes from the unfiltered IR image.

2) Feature Extraction and Matching: The Scale Invariant Feature Transform (SIFT) was chosen for the proposed solution. The SIFT begins with a difference of Gaussians (DoG). It is a band-pass filter that can be applied to the image and tuned to find sharp transitions of a given scale in the image. The DoG highlights the transitions in the pass-band scale, so that features are then detected in the maxima of the filtered image. The algorithm is detailed in [4]. The output is an array of positions, orientations and scales in the image and their corresponding descriptor in $\mathbb{R}^{128}$.

The euclidean distance criterion is used to match features, where the nearest-neighbour is selected for each feature. To avoid a high number of false-positive matches, e.g., associations of noisy and wrong features, a threshold is used on the nearest-neighbour distance ratio (NNDR), defined by (5), where D_A is the $\mathbb{R}^{128}$ descriptor of the feature being matched, D_B the descriptor of the nearest-neighbour and D_C the second nearest-neighbour. The NNDR is a matching uniqueness factor used to define if the second nearest-neighbour is sufficiently distant in comparison with the first nearest-neighbour [5].

$$\text{NNDR} = \frac{d_{\text{first}}}{d_{\text{second}}} = \frac{||D_A - D_B||}{||D_A - D_C||} \tag{5}$$

3) Pairwise Registration: Using a point-to-plane metric, the Iterative Closest Point (ICP) calculates planes that can be used to minimise the error in (2), as detailed in [7]. The principle of ICP is the following: by iteratively reassigning points to their nearest-neighbour and minimising the distance, the cloud of points will converge to a desired configuration. The main issue of ICP is its high dependence to the initialization. If the point clouds are not "close enough" when the ICP is initialised, the result will converge to an undesired error local minimum.

The SIFT is used to improve the ICP. The SIFT features are associated to the nearest pixel using the coordinates of each feature. Then, the 3D points are obtained with (1). Hence, the

 https://doi.org/10.1051/epjconf/201817007014

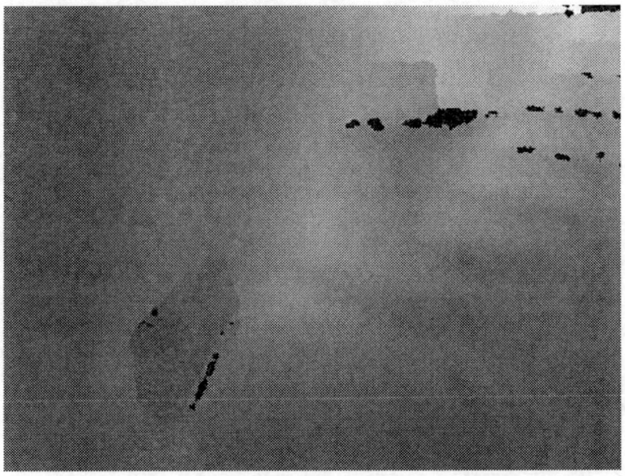

(a) Depth image.

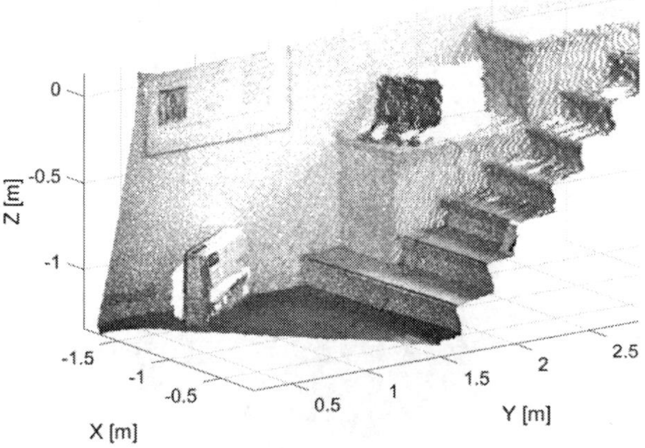

(b) 3D point cloud.

Fig. 3: The Figure 3a is the infra-red image of the scenario and Figure 3b depicts the 3D point cloud.

SIFT matches can be used to associate two points in the 3D space rather than two points in the image.

Afterwards, the Random Sample Consensus (Ransac) algorithm is used to fit mathematical models to data that contains outliers and, thus, to exclude false-positive feature associations. In the feature matching context, the Ransac algorithm fits a rigid body transformation with Procrustes analysis [6] (Singular Value Decomposition), which finds the optimal rotation and translation to a set of matched 3D points.

The rigid body transformation resultant from Ransac provides a good initialisation for ICP. The error evaluated in (2) is then minimised by the point-to-plane ICP, resulting in a fine approximation of the moving points to the planes calculated in the static point cloud.

4) Global Reconstruction Approach: For the proposed solution it is assumed that the images are sorted according to the sequence of acquisition, i.e. there is a time sequence. A simple method is adopted, associating each point cloud with the previous and with the next frame. The global coordinate frame is defined in the first point cloud. The consecutive point clouds are fit to the previous one by the order of acquisition.

C. *Inferring radiological activity*

The goal of the inference is to obtain the matrix Γ from (4) from the measured data as illustrated in Figure 4, i.e., to infer the radiological activity of each point of the map. To estimate the Γ given the measurement data corrupted by noise, a possible solution is given by (6), where Δ^+ is given by (7), the Moore-Penrose pseudoinverse [11].

$$\mathbf{\Gamma} = \mathbf{\Delta}^+\mathbf{\Sigma} \tag{6}$$

$$\mathbf{\Delta}^+ = (\mathbf{\Delta}^\mathrm{T}\mathbf{\Delta})^{-1}\mathbf{\Delta}^\mathrm{T} \tag{7}$$

This paper presents results with real data of depth images, but with simulated data of radiological measurements. The approach to simulate the data is described as follows. The radiation sources are distributed, i.e., each point of the map will contribute with some radiation. The intensity of each point is set with a fall-off 3D Gaussian exponential. The radiation intensity of a point is so much higher as lower is its distance to the centre of the Gaussian.

Let k be the number of sources of radiation, n the index of the n-th source, a_n the "strength", $s_n = (x_n, y_n)$ the position, η_n the width factor and $m_j = (x_j, y_j)$ the j-th point of the map. Then, the radiological activity of a point belonging to the map is given by the sum of the contributions of all radiation sources. The radiological activity, γ_j, is given by (8).

$$\gamma_j = \sum_{n=1}^{k} a_n \exp\Big(-\frac{\|\mathbf{m}_j - \mathbf{s}_n\|^2}{2\eta_n^2}\Big), \tag{8}$$

Figure 5 illustrates a simulated planar scenario, with a rectangular shape (assumed on an horizontal plane) and three Gaussian radiation sources with different intensities. The z-axis represents the radiation intensity in the Figure 5, as well as in the next two figures. According to (8), every point m_j in the map of Figure 5 shall contribute with a given amount of radiation, resulting in σ_i measured at point p_i along a path. The Table I presents the parameters of (8) used to generate the Gaussian sources. The dimensions of the scenario are 20 by 20 meters, with a resolution of 50×50 points.

TABLE I: Parameters of the distributed radioactivity sources.

n (source index)	a_n (strength)	x_n (x coordinate)	y_n (y coordinate)	η_n (width)
1	0.5	7	6	2
2	0.3	13	12	4
3	0.8	3	15	1

Two paths were tested in the described scenario, with 20 waypoints p_i each one. The first path describing a "S" shape along an horizontal plane above the map was tested, as represented in Figure 6. This path is sparse and covers almost the scenario. The second path describing a "L" shape

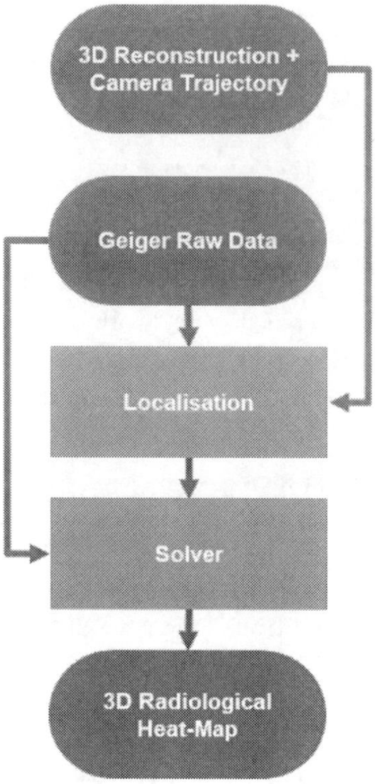

Fig. 4: Radiological inference flowchart.

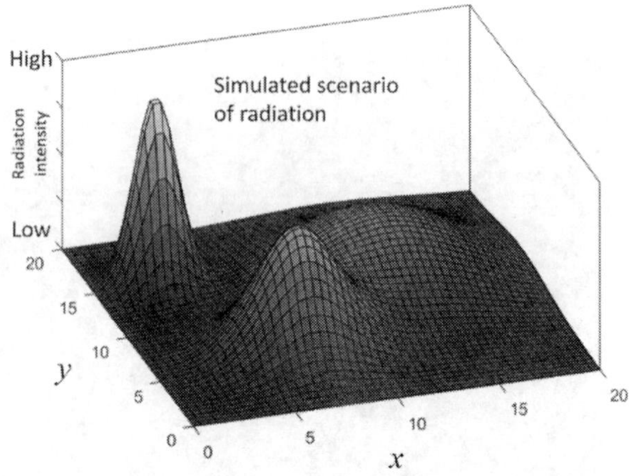

Fig. 5: Representation of the radiological activity of a room where 3 distributed Gaussian radiation sources are positioned in the 2D plane (floor).

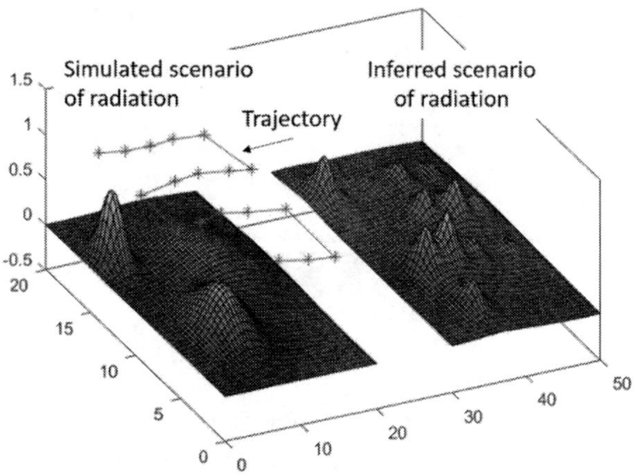

Fig. 6: Representation of the radiological activity of a room with a sparse trajectory on top (left image) and the radiological map estimation obtained after applying 6 (right image).

only covers part of the scenario's periphery, as illustrated in Figure 8.

In both trajectories, it is possible to notice a Gaussian like curve centred in the area corresponding to each point of the path. In addition, the inferred radiological activity of the scenario is higher where the measured radiological activity is higher, and lower where the measured radiological activity is lower. In Figure 6, the first trajectory covers almost the whole map, while the second trajectory in Figure 8 only covers the periphery. The inferred radiological map provides a better approximation of the simulated radiation in the first case. This result was expected, since there are more points (2500) in space than the acquisitions along the trajectory (20). Therefore, the system of equations in (6) is under-determined. The pseudo-inverse matrix Δ^+ found in (7) provides different results, depending on whether the path is spread across all space or just covering part of it.

IV. Experimental and Simulated Results

The results of the proposed solution for radiological inspection are shown in this section. The proposed solution was tested in the entrance of a laboratory building, as illustrated in the panoramic picture of Figure 7. This scenario includes a corridor, walls, doors and stairs.

The experiment was performed using a total of 150 depth images and their corresponding 150 IR images. The sensor was moved at an approximately constant speed and following a path with a spiral shape. The orientation of the sensor is perpendicular to the shape of the path and pointing to the walls or doors. The 3D reconstruction of the scenario is depicted in Figure 9. Then, a source of radiation was simulated in a corner of the scenario as described in Section III-C. The inferred radiation map is illustrated in Figure 10, as well as the estimation of the followed path (represented by a line), where the depth and IR images were acquired. The dots along the line represent the positions and the radiological information was simulated as sensor data acquisition.

V. Conclusion and Future Work

The paper presented a solution for the 3D reconstruction of the scenario as reference for the localisation and quantification of possible sources of radiation. The used sensors are depth cameras and Geiger-Müller counter. The 3D reconstruction is based on techniques for association of images and cloud of

Fig. 7: Panoramic view of the indoor scenario where the experimental test was performed.

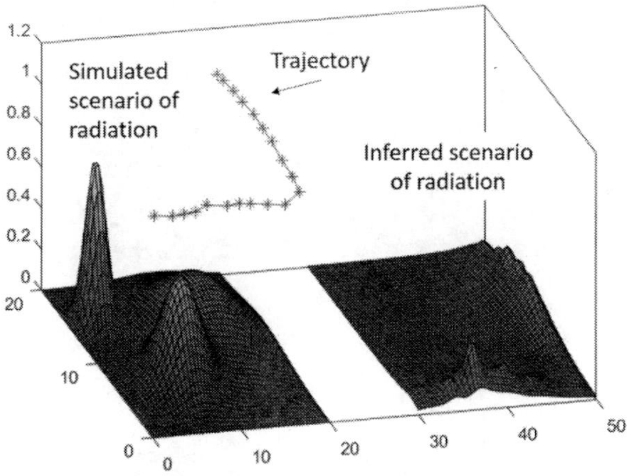

Fig. 8: Representation of the radiological activity of a room with a sparse trajectory on top (left image) and the radiological map estimation obtained after applying 6 (right image).

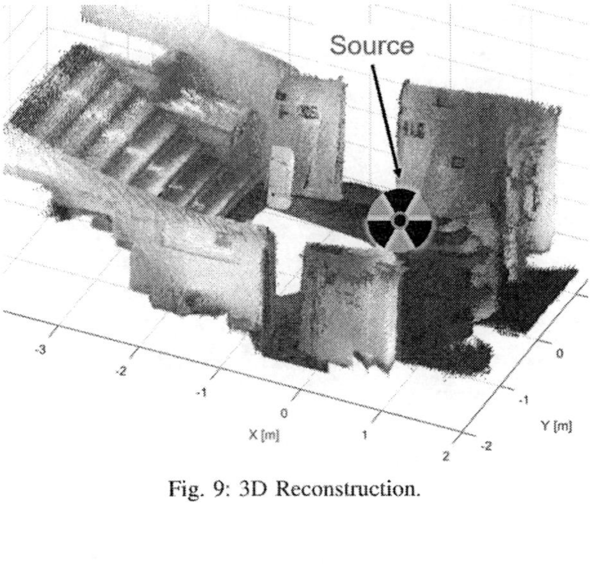

Fig. 9: 3D Reconstruction.

points, commonly used in mobile robot applications. After the 3D reconstruction, the radiation map is inferred given the radiation measurements, with the respective position and orientation of the sensor where the data was acquired.

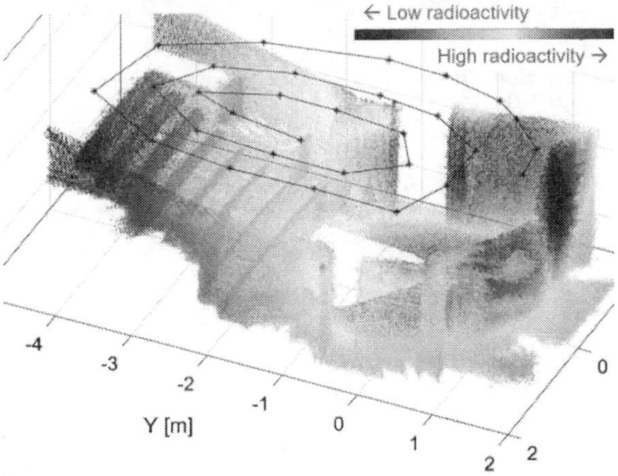

Fig. 10: Radioactivity map estimation.

The solution is dependent to the trajectory, which suggests that the challenge of localise the sources of radiation is also a path planning problem. In addition, the radiological intensity is inversely proportional to the square of the distance between the sensor and the source. In addition, the Geiger-Müller counter is very small, with low efficiency and, thus, the sensor has to move closer to the sources of radiation to acquire any reliable information. For the application with drones, the proximity to the objects in the scenario (below 1 meter) increases the risk of collision and the dynamics interference on the flight.

Therefore, the next work is to test the proposed solution with the drone depicted in Figure 11, flying inside a room with real sources of radiation, namely the Potassium-40. Other sensors, such as Cadmium Zinc Telluride (CdZnTe) or scintillators based sensors are being assessed as alternative radiological sensors, with better efficiency and capability to identify the sources of radiation.

Acknowledgments

IST activities received financial support from "Fundação para a Ciência e Tecnologia" through projects UID/FIS/50010/2013 and UID/EEA/50009/2013.

References

[1] O. Crofts, A. Loving, D. Iglesias, M. Coleman, M. Siuko, M. Mittwollen, V. Queral, A. Vale and E. Villedieu, *Overview of progress on the European*

Fig. 11: Quadcopter equipped with a Sparkfun Geiger counter and a Structure depth camera.

DEMO remote maintenance strategy, Fusion Engineering and Design Vol. 109-111 Part B, 1392 - 1398, 2016.
[2] F. Endres, J. Hess, N. Engelhard, J. Sturm, D. Cremers and W. Burgard, *An evaluation of the RGB-D SLAM system*, IEEE International Conference on Robotics and Automation, Saint Paul, MN, pp. 1691-1696, 2012.
[3] R. Hartley and A. Zisserman, *Multiple View Geometry in Computer Vision*, 2nd Edition. Cambridge University Press, 2004.
[4] D. G. Lowe, *Distinctive image features from scale-invariant keypoints*, International Journal of Computer Vision, vol.2, no.60, pp. 91-110, 2004.
[5] R. Szeliski, *Computer Vision: Algorithms and Applications*. Springer Science & Business Media, pp. 226-232, 2010.
[6] D. Eggert and A. Lorusso and R. Fisher, *Estimating 3-D rigid body transformations: a comparison of four major algorithms*. Machine Vision and Applications 9, Springer-Verlag, pp. 272-290, 1997.
[7] K. Low. *Linear Least-Squares Optimization for Point-to-Plane ICP Surface Registration*. Technical Report TR04-004, Department of Computer Science, University of North Carolina at Chapel Hill, February 2004.
[8] J. Kepler, *Ad Vitellionem paralipomena, quibus astronomiae pars optica traditur*, 1604.
[9] E. Browne and J. K. Tuli, *Nuclear Data Sheets 114, 751*, 2013.
[10] J. E. Purcell and C. G. Sheu, *Nuclear Data Sheets 130, 1*, 2015.
[11] A. Wilansky, *The Row-Sum of the Inverse Matrix*. American Mathematical Monthly, American Mathematical Society, 1951.
[12] P. Zanuttigh, G. Marin, C. Mutto, F. Dominio, L. Minto and G. Cortelazzo, *Time-of-Flight and Structured Light Depth Cameras*. Springer International Publishing, 2016.
[13] 3D scanning, augmented reality, and more for mobile devices. https://structure.io/

Characterization of New-Generation Silicon Photomultipliers for Nuclear Security Applications

Marc A. Wonders, David L. Chichester, *Senior Member, IEEE,* and Marek Flaska, *Member, IEEE*

***Abstract*— Silicon photomultipliers have received a great deal of interest recently for use in applications spanning a wide variety of fields, including nuclear safeguards and nonproliferation. For nuclear-related applications, the ability of silicon photomultipliers to discriminate neutrons from gamma rays using pulse shape discrimination when coupled with certain organic scintillators is a characteristic of utmost importance. This work reports on progress characterizing the performance of twenty different silicon photomultipliers from five manufacturers with an emphasis on pulse shape discrimination performance and timing. Results are presented on pulse shape discrimination performance as a function of overvoltage for 6-mm x 6-mm silicon photomultipliers, and the time response to stilbene is characterized for silicon photomultipliers of three different sizes. Finally, comparison with a photomultiplier tube shows that some new-generation silicon photomultipliers can perform as well as photomultiplier tubes in neutron-gamma ray discrimination.**

I. Introduction

THE ability to detect neutrons and characterize their sources is essential for a variety of nuclear security and safeguards tasks. For decades, the standard method of doing so has been the use of ^{3}He proportional counters, but an increase in its demand coupled with a decrease in production has created a push for the development of alternative methods of detecting neutrons [1]. A promising candidate to replace ^{3}He detectors is the scintillation detector. Organic scintillators are capable of detecting fast neutrons, and neutron capture agents such as ^{6}Li enable the detection of thermal neutrons when incorporated in some scintillating media [2].

All scintillation detectors require some means of converting the optical signal into an electrical signal and amplifying this signal to a measurable amount. A photomultiplier tube (PMT) has been the traditional tool used for this purpose because of its high gain, low noise, and fast response. However, applications requiring low level light sensing in alternate packaging and with other features have stimulated the development of alternative light sensors. A silicon photomultiplier is one such recent development. Silicon photomultiplier devices are called many different names based on the manufacturer, including Geiger-mode avalanche photodiode, multi-pixel photon counter, and solid state photomultiplier, but in this work all these devices will be referred to as silicon photomultipliers (SiPMs). SiPMs are an evolution of the avalanche photodiode and are essentially many avalanche photodiodes miniaturized and connected in parallel [3]. Each avalanche photodiode is generally connected in series with a quenching resistor and this combination of avalanche photodiode and quenching resistor constitutes a microcell of the SiPM.

By operating these avalanche photodiodes in Geiger mode high gain is achieved, and the presence of many different microcells enables signal proportionality. SiPMs have been shown to have excellent time and energy resolution, and comparable gain to many PMTs. They also offer additional advantages characteristic of solid state technology such as mechanical ruggedness, insensitivity to magnetic fields, compactness, and a much lower operating voltage than PMTs [4].

Their main drawback has been higher noise levels than PMTs which is primarily caused by thermal generation of charge carriers in the avalanche region of the SiPM that produce a signal indistinguishable from optically produced charge carriers. Improved performance at low temperatures has consequently been shown, but recent improvements in SiPM technology have also made effective operation at room temperature feasible [4]. Thus, SiPMs are drawing interest in fields as diverse as medical physics, high energy physics, and homeland security. This paper focuses on applications to homeland security and, as such, reports on progress in characterization of a suite of different SiPMs in regard to their ability to discriminate neutrons from gamma rays when coupled to an organic scintillator, as well as their timing properties.

II. SiPMs and Experimental Setup

For this work SiPMs were acquired from five different manufacturers: AdvanSiD, First Sensor, Hamamatsu, Ketek, and SensL [5-9]. Of the packages available from each company, packages were selected that provided direct access to the anode and cathode via conducting pins. These were chosen to both minimize any pulse shaping via added built-in circuitry and provide as uniform testing conditions as possible. The SiPM sizes are 3 mm x 3 mm, 4 mm x 4 mm, and 6 mm x

This material is based upon work supported by the US Department of Energy's National Nuclear Security Administration Office of International Nuclear Safeguards, from the Human Capital Development Subprogram.

M. A. Wonders and M. Flaska are with the Department of Mechanical and Nuclear Engineering, Pennsylvania State University, University Park, PA 16802-4400 (e-mail: mfw5173@psu.edu, mflaska@psu.edu)

D. L. Chichester is with Idaho National Laboratory, Idaho Falls, ID 83402 (e-mail: david.chichester@inl.gov)

 https://doi.org/10.1051/epjconf/201817007015

6 mm, and microcell sizes range from 15 μm to 75 μm. Only one package was selected from each manufacturer except SensL, from which two different package forms were acquired. Only SensL and Ketek provided two different SiPM series; the full list of SiPMs acquired are shown below in Table I. The SiPMs themselves are shown in Fig. 1.

TABLE I
CHARACTERISTICS OF SIPMS ACQUIRED

Manufacturer	Pixel Size (mm)	Microcell Size (μm)	Series	Package	Typical Breakdown Voltage (V)
AdvanSiD	3	40	ASD-NUV	Socket	26
AdvanSiD	4	40	ASD-NUV	Socket	26
First Sensor	3	40	NUV	SMD	26
First Sensor	4	40	NUV	SMD	26
Hamamatsu	3	25	S13360	Cs	53
Hamamatsu	3	50	S13360	Cs	53
Hamamatsu	3	75	S13360	Cs	53
Hamamatsu	6	25	S13360	Cs	53
Hamamatsu	6	50	S13360	Cs	53
Hamamatsu	6	75	S13360	Cs	53
Ketek	3	15	WB	PM	27.5
Ketek	3	25	WB	PM	27.5
Ketek	6	25	EB	PM	26.5
Ketek	6	50	EB	PM	26.5
SensL	3	20	C	SMTPA	24.5
SensL	3	35	C	SMTPA	24.5
SensL	3	35	J	SMTPA	24.5
SensL	3	50	C	SMTPA	24.5
SensL	3	35	C	X13	24.5
SensL	6	35	C	X13	24.5

It should be noted that all the SiPMs studied here have breakdown voltages around 25 V except the Hamamatsu SiPMs, which have breakdown voltages around 53 V. This will have consequences when looking at characteristics based on overvoltage in which the fraction of breakdown voltage will provide more similar results than absolute magnitude of the overvoltage.

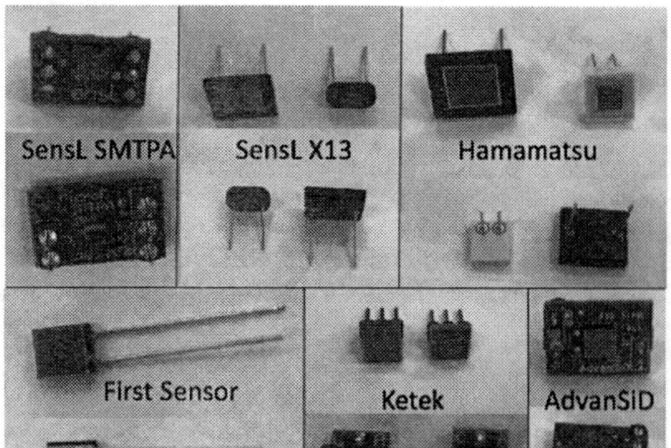

Fig. 1. The SiPMs used in this study.

As seen in Fig. 1, despite all SiPMs having similar readouts, the pins themselves are oriented differently, complicating somewhat the uniform setup of all SiPMs. For this reason, the SiPMs were setup using a simple breadboard with a typical wire layout used to interface with the breadboard as shown in Fig. 2. Measurements were conducted in a light-tight box to minimize wrapping requirements for the scintillator and SiPM combination. The trace from the SiPM was digitized using a 14-bit CAEN DT5730 digitizer with a sampling rate of 500 GHz and analyzed offline using the ROOT data analysis framework [10].

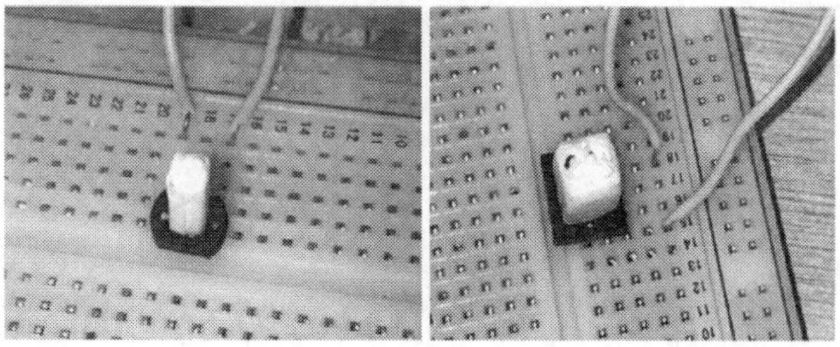

Fig. 2. Biasing and signal readout setup.

Stilbene was chosen as the scintillator for this work because of its excellent ability to perform pulse shape discrimination (PSD). Two different sizes were used: a 3-mm x 3-mm x 10-mm crystal for the 3-mm x 3-mm and 4-mm x 4-mm SiPMs and a 6-mm x 6-mm x 6-mm crystal with the 6-mm x 6-mm SiPMs. The crystals used are shown in Fig. 3. The maximum photon detection efficiencies for the SiPMs used in this study range from 420 nm to 450 nm, yielding effective though imperfect matching of stilbene's emission spectrum, which peaks around 380 nm. The crystals were wrapped in reflective Teflon tape and coupled to the SiPMs using EJ-550 optical grease from Eljen Technology to maximize light collection by the SiPMs [11]. The Teflon wrapping was not changed during these tests.

Fig. 3. The stilbene crystals used in this study.

III. WAVEFORMS

To visualize and quantify the different timing properties of the SiPMs, a ^{137}Cs gamma ray source was used to irradiate the stilbene scintillators. Because stilbene has a very fast decay time on the order of nanoseconds, the different pulse widths observed for each SiPM are largely representative of the SiPM itself and close to the microcell response time for each.

One-hundred pulses from each SiPM were averaged together and then normalized, and these are shown in Figs. 4-7. The full-width tenth-max was determined for each measurement by taking the mean of a Gaussian distribution fit to the distribution of full-width tenth-max values determined over roughly one-hundred thousand events. The rise time of each SiPM was determined in the same manner, where the rise time was calculated as the time between the last sample below 10% of the pulse height and the sample of maximum pulse height. These values are shown in Table II.

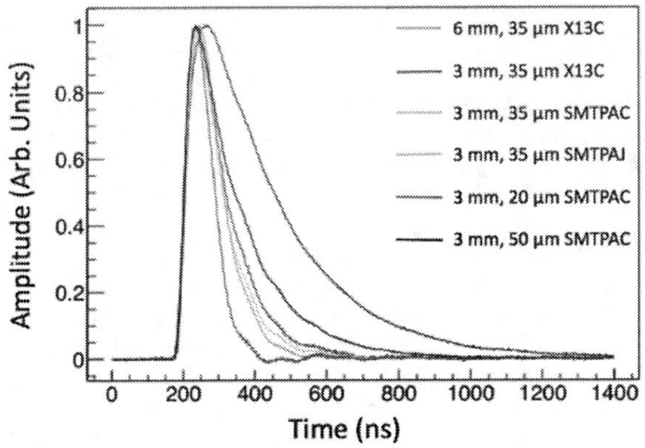

Fig. 4. SensL waveforms.

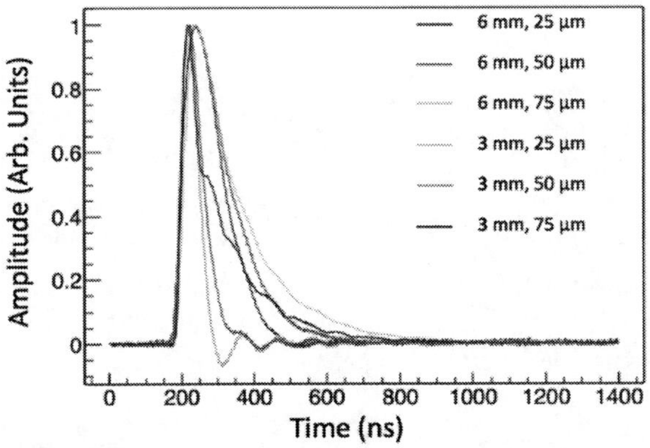

Fig. 5. Hamamatsu waveforms.

From Figs. 4-7 a wide variety of pulse widths are observed ranging from about 100 ns to 600 ns. As expected, larger SiPMs tend to have longer pulses because of their larger capacitances, and it is also observed that for a given SiPM size larger microcells tend to have longer pulses. Despite large differences in pulse width, all SiPMs tested have similar rise times. The undershooting seen in the fastest SiPMs can be removed by adding a small resistance in series after the output of the SiPM but was not included in the setup used here, in order to show unaltered pulse widths for all the devices and provide uniform signal processing for all.

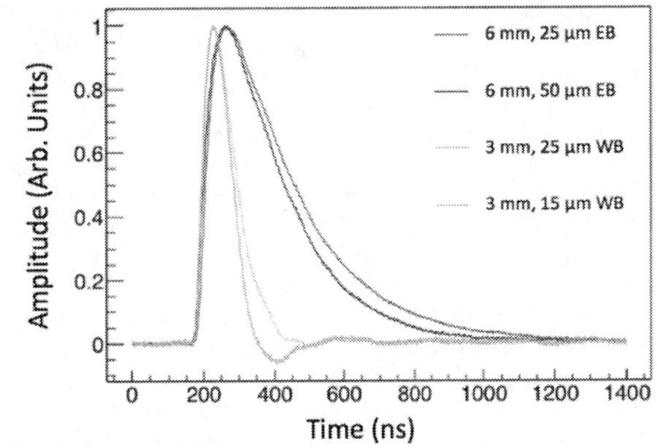

Fig. 6. Ketek waveforms.

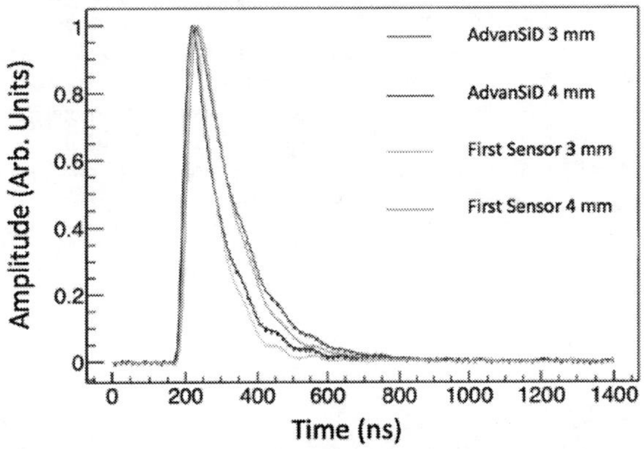

Fig. 7. AdvanSiD and First Sensor waveforms.

TABLE II
TIMING CHARACTERISTICS OF SIPMS TESTED

Manufacturer (Series, Package)	Pixel Size (mm)	Microcell Size (µm)	Rise Time (ns)	Full-Width Tenth-Max (ns)
AdvanSiD	3	40	22.6	240.2
AdvanSiD	4	40	26.4	325.4
First Sensor	3	40	22.6	221.4
First Sensor	4	40	25.8	317
Hamamatsu	3	25	23	94.4
Hamamatsu	3	50	22.2	138.4
Hamamatsu	3	75	20.2	250.6
Hamamatsu	6	25	27	227.6
Hamamatsu	6	50	27	293.8
Hamamatsu	6	75	26.4	358.2
Ketek (WB)	3	15	26.2	139
Ketek (WB)	3	25	25.2	191.2
Ketek (EB)	6	25	27.4	501
Ketek (EB)	6	50	27.6	591
SensL (C, SMTPA)	3	20	25.6	167
SensL (C, SMTPA)	3	35	25.8	250
SensL (C, SMTPA)	3	50	26	372
SensL (J, SMTPA)	3	35	27.4	229.6
SensL (C, X13)	3	35	26.4	276.2
SensL (C, X13)	6	35	27.4	605.8

For manufacturers that provided more than one package or series, this information is also specified in Table II.

IV. Pulse Shape Discrimination Performance

A primary motivating factor for the assessment of SiPMs for nuclear security is their application to neutron detectors, especially organic scintillators that are sensitive to both gamma rays and neutrons. Sensitivity to gamma rays of organic scintillators is generally overcome by analyzing the temporal profiles of pulses using the popular technique of PSD. Because neutrons primarily interact with protons while gamma rays primarily interact with electrons, neutron interactions produce charged particles with higher specific ionization [12]. This creates a higher concentration of triplet states in the scintillator, which enables more frequent triplet-triplet annihilations to occur after the initial light production and, consequently, produces a greater amount of delayed scintillation light. Many ways of exploiting differences in the pulse shape exist to discriminate neutrons and gamma rays including charge comparison, pulse gradient analysis, and frequency gradient analysis [13]. Charge comparison is the most frequently used method because of its effectiveness and simplicity of implementation and is the focus of this paper. In this method, a tail integral and total integral are defined and the ratio of these two values is a pulse shape parameter (PSP) that provides a measure of the amount of delayed light that identifies neutrons and gamma rays. Example neutron and gamma ray pulses are shown in Fig. 8.

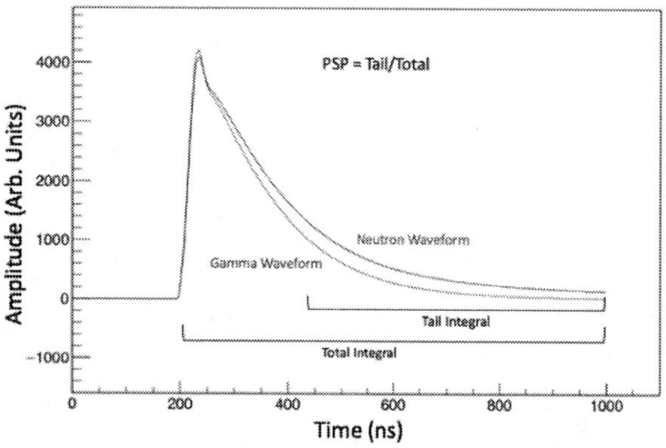

Fig. 8. Depiction of the charge comparison method for neutrons and gamma rays.

A typical one-dimensional histogram of the pulse shape parameter is shown in Fig. 9. The grouping at larger values of the pulse shape parameter corresponds to neutrons. To quantify the effectiveness of PSD between neutrons and gamma rays, Gaussian shapes are fit to each distribution and used to create a figure of merit (FOM, defined in Equation 1).

$$FOM = \frac{Distance\ Between\ Peaks}{FWHM_N + FWHM_G} \quad (1)$$

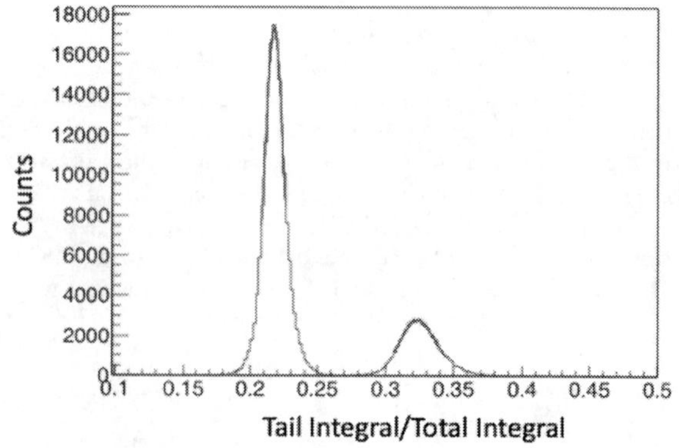

Fig. 9. Example PSD histogram (blue) with Gaussian fits (red) to each distribution.

As seen in Figs. 4-7, the 3-mm x 3-mm and 4-mm x 4-mm SiPMs exhibit undershooting and ringing, and this likely deteriorates the PSD that would be present with an ideal readout. The 6-mm x 6-mm in the setup used produce much smoother pulses optimized for PSD performance likely because of their slower rising edge. Consequently, PSD results will only be shown for the 6-mm x 6-mm SiPMs.

For each SiPM the FOM is determined in the light output window between 200 and 1000 keVee. Calibration is performed using the Compton edge of ^{137}Cs. The PSD performance as a function of overvoltage for the 6-mm x 6-mm SiPMs is shown in Fig. 10.

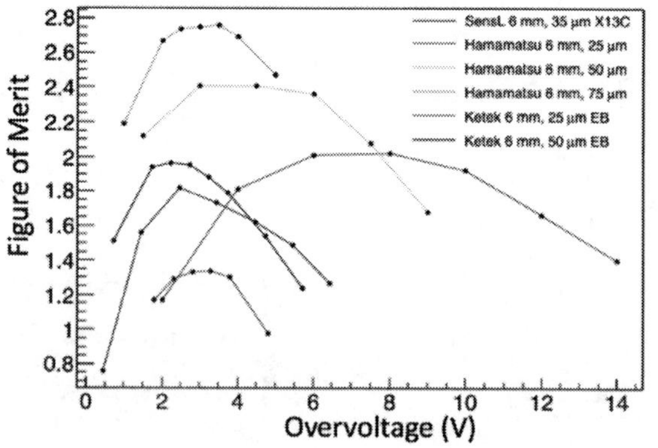

Fig. 10: Pulse shape discrimination performance as a function of overvoltage for 6-mm x 6-mm SiPMs.

All SiPMs initially exhibit improved PSD performance with increasing overvoltage and then reach a plateau with the optimum PSD before the PSD deteriorates. The decrease in PSD performance as overvoltage approaches 0 V is driven at least partly by the presence of digitizer noise that limits the signal-to-noise ratio for smaller pulses. The plateaus occur generally around an overvoltage of 3 V, and it should be recalled that the breakdown voltage of Hamamatsu SiPMs is roughly twice that of the Ketek and SensL SiPMs, explaining the lengthening of and shift to higher overvoltages of their plateaus. Further, there is a greater range of microcell sizes

 https://doi.org/10.1051/epjconf/201817007015

covered by Hamamatsu explaining the larger differences in plateau behavior within that manufacturer's SiPMs. Table III displays the best FOM for each 6-mm x 6-mm SiPM.

TABLE III
PULSE SHAPE DISCRIMINATION PERFORMANCE OF 6-MM X 6-MM SIPMS

Manufacturer (Series, Package)	Microcell Size (μm)	Best Overvoltage (V)	Figure of Merit
Hamamatsu	75	3.5	2.76
Hamamatsu	50	3, 4.5	2.41
Hamamatsu	25	8	2.02
Ketek (EB)	50	2.73	1.96
SensL (C, X13)	35	2.45	1.82
Ketek (EB)	25	2.8	1.34

Table III is ordered by the FOM and there appears to be a correlation between FOM and microcell size. This relationship is shown explicitly in Fig. 11. The impact of the pulse width on the FOM is shown in Fig. 12, and no dependency is observed.

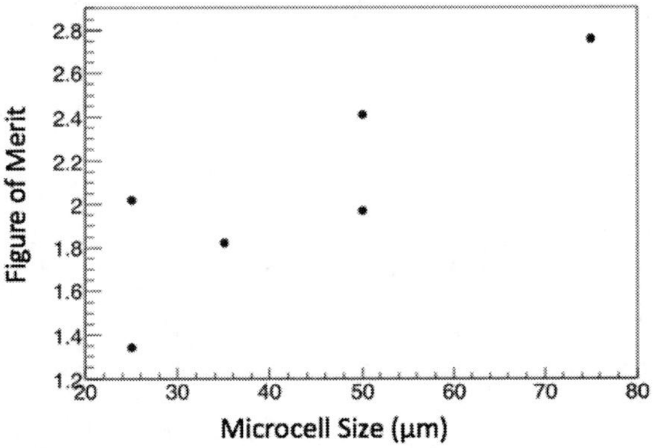

Fig. 11: Relationship between microcell size and FOM for 6-mm x 6-mm SiPMs.

A comparison of the PSD performance of these SiPMs with a fast PMT was also conducted. The 6-mm x 6-mm stilbene crystal was coupled to a Hamamatsu H10580 PMT assembly, and the FOM was calculated to be 1.88 in the 200-1000 keVee window and 2.31 in the 200-300 keVee window. The setup used for these tests is shown in Fig. 13, and the associated PSD plot is shown in Fig. 14. The PSD plot from the SiPM with the highest FOM is shown in Fig. 15.

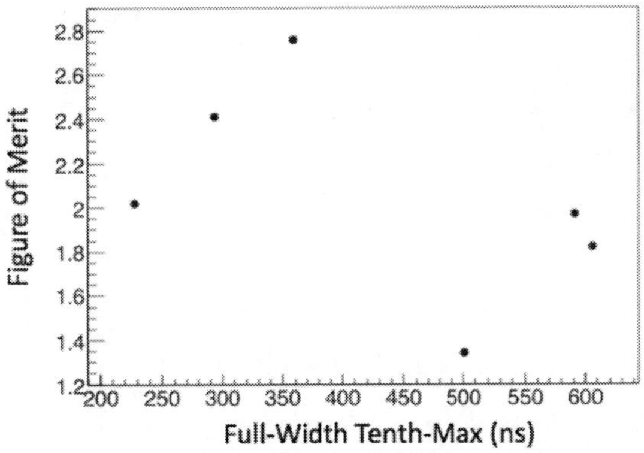

Fig. 12: Relationship between pulse width and FOM for 6-mm x 6-mm SiPMs.

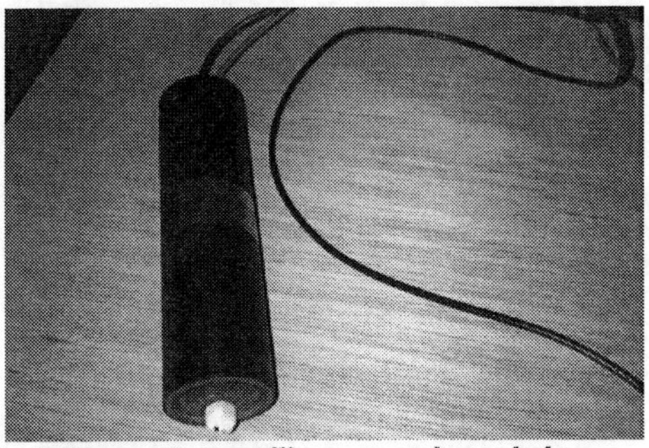

Fig. 13: A 6 mm x 6 mm stilbene crystal coupled to a Hamamatsu H10580 PMT assembly.

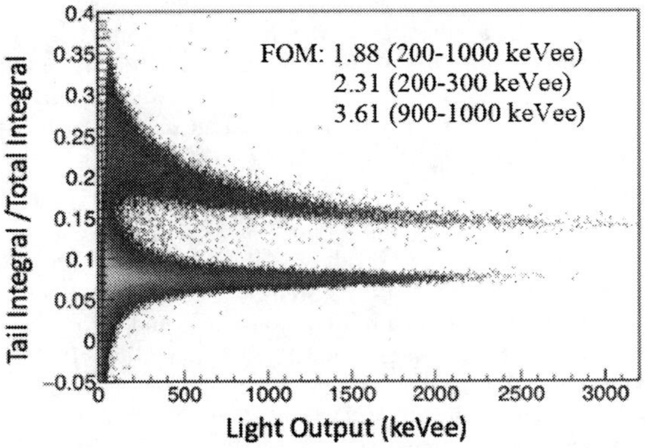

Fig. 14: Pulse shape discrimination plot for the H10580 PMT coupled to stilbene.

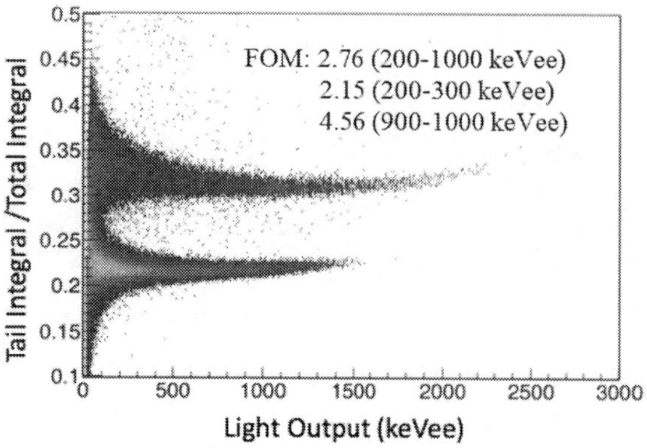

Fig. 15: Pulse shape discrimination plot for the S13360-6075CS SiPM coupled to stilbene.

V. Conclusions

The comparison between a fast PMT capable of effective PSD with SiPMs has shown that new generation SiPMs can perform competitively with PMT technology when coupled with organic scintillators. In terms of effective neutron detection and discrimination, results presented in this paper suggest that SiPMs with the largest microcells tend to perform better. The greater photon detection efficiency caused by a greater geometrical fill factor and the larger gain, defined as the charge produced from a single triggered charge avalanche, likely play a large role in this. As expected, all the SiPMs tested here had similar rise times despite large differences in overall pulse width, and no direct relationship between pulse width and PSD performance was observed between different SiPMs. PSD performance as a function of overvoltage eventually diminishes and this is likely a combination of a sharply increasing noise past a certain overvoltage and other factors such as saturation of the SiPM microcells. Further, the behavior of the PSD plateaus was seen to be affected significantly by the microcell size with smaller microcells showing peak PSD performance at higher overvoltages and over a greater range of overvoltages.

VI. Future Work

In addition to PSD results for the 6-mm x 6-mm SiPMs, only temporal profiles of the SiPMs have been presented here, and, as such, investigation into both noise and signal-to-noise ratios, where signal magnitude is proportional to both the gain and photon detection efficiency, are being carried out as complementary to the PSD results. Further, optimization of SiPM readout for smaller SiPMs will take place so that their PSD performance can be effectively characterized and compared to other SiPMs. This will provide greater support for relationships between different SiPM characteristics and PSD performance. Investigation into the effects of SiPM dynamic range and saturation of the microcells is also necessary, and the acquisition of a picosecond light pulser will allow for tests of wavelength sensitivity and timing response to ultrafast light pulses. Finally, testing of the effects of elevated temperature and radiation damage on the SiPMs will be conducted.

References

[1] A.P. Simpson, et al., "A review of neutron detection technology alternatives to helium-3 for safeguards applications," *INMM 52nd Annual Meeting*, vol. 8, 2011.
[2] P. Peerani, et al., "Testing on novel neutron detectors as alternatives to 3He for security applications," *Nucl. Instr. and Meth. A,* vol. 696, pp. 110-120, Dec. 2012.
[3] P. Buzhan, et al., "Silicon photomultiplier and its possible applications," *Nucl. Instr. and Meth. A* vol. 504, no. 1-3, pp. 48-52, May 2003.
[4] P. K. Lightfoot, "Characterisation of a silicon photomultiplier device for applications in liquid argon based neutrino physics and dark matter searches," *Journal of Instrumentation*, vol. 3, Oct. 2008.
[5] AdvanSiD, Trento, Italy
[6] First Sensor, Berlin, Germany
[7] Hamamatsu, Shizuoka Prefecture, Japan
[8] Ketek, Munich, Germany
[9] SensL, Cork, Ireland
[10] R. Brun and F. Rademakers, "ROOT – An object oriented data analysis framework," *Nucl. Instr. and Meth. A.*, vol. 389, no. 1-2, pp. 81-86, Apr. 1997.
[11] Eljen, Sweetwater, Texas, United States of America
[12] M. L. Roush, "Pulse shape discrimination," *Nuclear Instruments and Methods*, vol. 31, no.1, pp. 112-124, Dec. 1964.
[13] C. Liao and H. Yang, "Pulse shape discrimination using EJ-299-33 plastic scintillator coupled with a Silicon Photomultiplier array," *Nucl. Instr. and Meth. A.,* vol. 789, pp. 150-157, Jul. 2015.

The Self-Powered Detector Simulation 'MATiSSe' Toolbox applied to SPNDs for severe accident monitoring in PWRs

Loïc Barbot, Jean-François Villard, Stéphane Fourrez, Laurent Pichon and Hamid Makil.

Abstract– **In the framework of the French National Research Agency program on nuclear safety and radioprotection, the 'DIstributed Sensing for COrium Monitoring and Safety' project aims at developing innovative instrumentation for corium monitoring in case of severe accident in a Pressurized Water nuclear Reactor. Among others, a new under-vessel instrumentation based on Self-Powered Neutron Detectors is developed using a numerical simulation toolbox, named 'MATiSSe'. The CEA Instrumentation Sensors and Dosimetry Lab developed MATiSSe since 2010 for Self-Powered Neutron Detectors material selection and geometry design, as well as for their respective partial neutron and gamma sensitivity calculations. MATiSSe is based on a comprehensive model of neutron and gamma interactions which take place in Self-powered neutron detector components using the MCNP6 Monte Carlo code. As member of the project consortium, the THERMOCOAX SAS Company is currently manufacturing some instrumented pole prototypes to be tested in 2017. The full severe accident monitoring equipment, including the standalone low current acquisition system, will be tested during a joined CEA-THERMOCOAX experimental campaign in some realistic irradiation conditions, in the Slovenian TRIGA Mark II research reactor.**

Index Terms—Severe accident instrumentation, Corium monitoring, Self-powered neutron detectors, SPND, Monte Carlo codes, Pressurized water reactor, PWR.

I. Introduction

FOLLOWING the 2013 Fukushima accident, the French government launched an important research program on nuclear safety and radioprotection (RSNR), managed by the French National Research Agency (ANR). This strategic initiative is co-funded by the government and run by the General Commissariat of Investment (CGI) through the 'Investments for the Future' Program (PIA).

In this context, the DISCOMS project, which stands for 'DIstributed Sensing for COrium Monitoring and Safety', contributes to develop new under-vessel instrumentation based on both distributed Optical Fiber Sensors and Self-Powered Neutron Detectors (SPNDs) [1].

SPNDs have been widely used in nuclear reactors as on-line IN-CORE neutron flux monitors since 1960 [2]. They are here considered to be part of an EX-CORE innovative instrumentation that would improve nuclear safety through corium monitoring in case of severe accident in Pressurized Water nuclear Reactors (PWR).

Two CEA laboratories (the Instrumentation Sensors and Dosimetry Lab - LDCI and the Sensors and Electronic Architecture Lab - LCAE) and the THERMOCOAX SAS Company co-developed a sensing instrumentation device including SPNDs and thermocouples as well as a standalone low current acquisition system. The paper details the capabilities of this device under mixed neutron and gamma irradiations, both in normal and severe accidental reactor operating conditions.

The LDCI's SPND simulation toolbox, named 'MATiSSe', standing for 'Monte cArlo Tool for SPND Simulation', is used to select SPND materials, to design geometries and to ensure that such detectors remain operational in any reactor situations. Full expected partial neutron and gamma sensitivities are calculated using the MATiSSe toolbox.

The THERMOCOAX SAS Company is currently manufacturing prototypes to be tested in 2017 in a European research reactor.

II. Overview

A. Context

The DISCOMS project intentions are to increase the safety both in existing and future Nuclear Power Plants, inquiring about the status of the third confinement barrier and defining possible mitigation strategies in case of severe accident. Nuclear instrumentation objectives are to detect reactor vessel breach and corium pouring on the concrete reactor pit basemat, to monitor the Molten Core Concrete Interaction, and to follow the corium location (up to the spreading compartment [core catcher] in the EPR) as well as the corium cooling kinetics [1].

Manuscript received April 19th, 2017. This work was supported by the RSNR research program on nuclear safety and radioprotection managed by the French National Research Agency (ANR).

L. Barbot and J-F. Villard are with the CEA, DEN, DER, Instrumentation, Sensors and Dosimetry Laboratory, Cadarache, F-13108 St-Paul-Lez-Durance, France (e-mail: loic.barbot@cea.fr).

S. Fourrez and L. Pichon are with the THERMOCOAX SAS Company, BP26, F-61438 Flers Cedex, France (e-mail: stephane.fourrez@thermocoax.com).

H. Makil is with the CEA, LIST, Sensors and Electronics Laboratory, Saclay, F-91191 Gif sur Yvette, France (e-mail: hamid.makil@cea.fr).

B. Instrumentation implantation

SPNDs are considered within the DISCOMS project for their sensitivity to both neutrons and gamma. Such sensors will be deployed in the reactor core vicinity. Strategically processed, SPND signals will provide key information to the operators in any normal and severe accidental reactor conditions (GEN2&3 PWRs). The sensor locations (see Figure 1) are essential as the severe accident monitoring mainly relies on neutron flux decreasing with the onset of the accident, and gamma flux increasing with the corium pouring (the temperature evolution being monitored with additional thermocouples) [1].

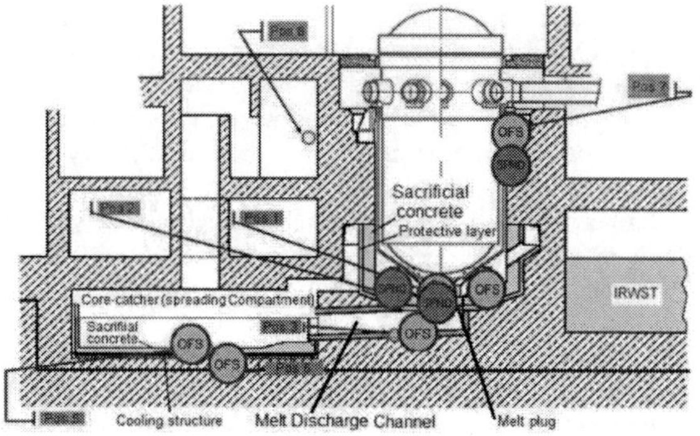

Figure 1: SPND and Optical Fiber Sensors locations for the DISCOMS project corium monitoring instrumentation [1].

GEN 2&3 reactors' normal operations can be monitored with SPNDs located at reactor core mid height, where both neutron and gamma irradiation fields are at relatively high level. At this location, SPND signal is also providing important information at the beginning of the accident as the SPND current will drop drastically with the neutron flux.

To monitor accidental conditions (reactor vessel breach followed by the corium migration), sensors have to be deployed at the reactor vessel bottom. SPNDs could be placed either on the concrete basemat surface (GEN2 reactors) or embedded in the concrete sacrificial layer of the core melt stabilization system (GEN3 reactors). Radiation conditions on the basemat are changing from low level mixed field in normal situations up to high level pure gamma irradiation in accidental conditions.

Another challenge for the DISCOMS project concerns the direct wired connection from SPNDs and thermocouples to the monitoring equipment over a length of typically 200 meters including specifically designed and fully qualified watertight glass-metal connectors [3]. A standalone low current acquisition system is also currently developed with low energy consumption and reliable multichannel low current measurements (bottom limit 1 pA).

C. Neutron and gamma flux calculations

As part of the DISCOMS consortium, the Institute of Radiological Protection and Nuclear Safety (IRSN) and the CEA Reactor Studies and Applied Mathematics Section (SERMA) were in charge of the radiation calculations in order to support the instrumentation implementation in GEN2&3 reactors (normal and accidental conditions).

Particle transport simulations were performed using two Monte Carlo codes (MCNP6 [4] and TRIPOLI4 [5]), the JEFF3.1 [6] and ENDF-B/VII.1 [7] libraries and the Tihange-I (GEN2 - Belgium) and the EPR reactor models. GEN2 simulations results have been validated thanks to open experimental flux distribution data [8]. Only GEN2 data and results are presented in this paper.

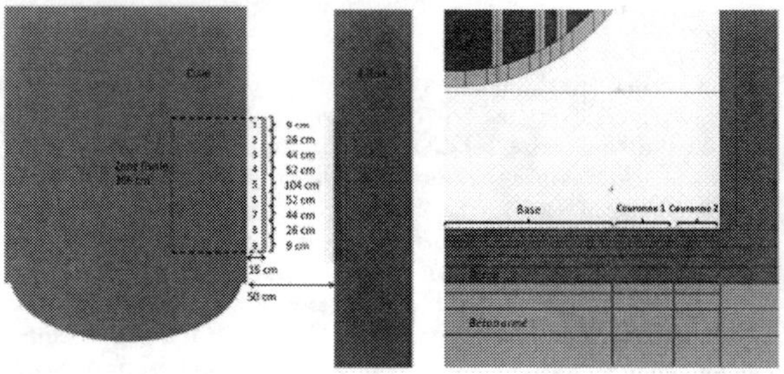

Figure 2: Monte Carlo calculation responses on the side and at the bottom of the vessel with respective segmentations [8].

For both normal and accidental conditions, neutrons and gamma spectra and flux level calculations were performed at SPND locations (see figures 2, 3 and 4). Those are used as entry data for the SPND simulation MATiSSe toolbox. The reactor core height has been divided in segments and the basement shared in a base and rings. The corium definition is approximated as a homogenous mixture of uranium and zirconium oxides and stainless steel. Radiation levels are estimated with a corium relocated on the reactor basemat three hours after the onset of the severe accident.

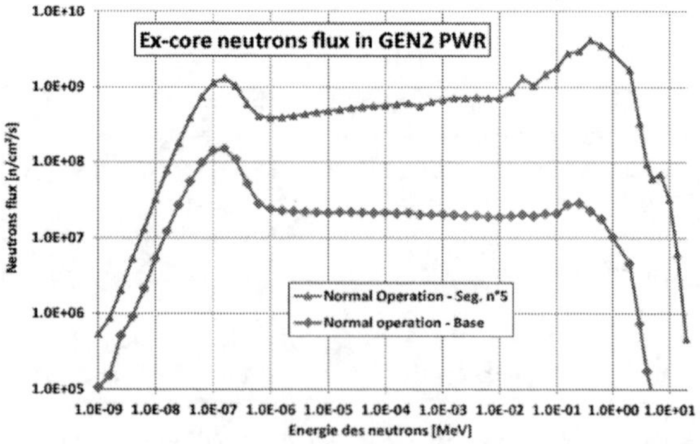

Figure 3: Neutron spectra in GEN2 PWR in Normal Operation; on the side of the vessel (Seg.5) and on the concrete basemat (Base).

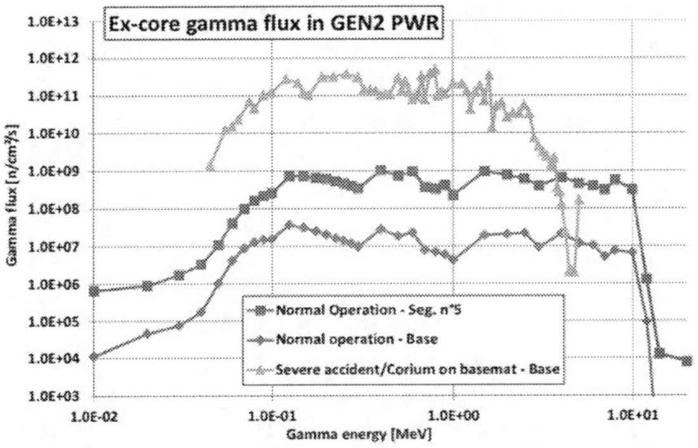

Figure 4: Gamma spectra in GEN2 PWR in Normal Operation (on the side of the vessel (Seg.5) and on the concrete basemat (Base)) and in case of Severe Accident (Corium on basemat – Base).

III. SELF-POWERED DETECTOR SIMULATION

The LDCI MATiSSe toolbox is based on a multistep Monte Carlo calculation model where SPNDs and their immediate environment are finely modelled (geometry and material definition). The SPND model originates from SCK•CEN previous work [9] and has been constantly upgraded at CEA since 2010. The SPND response is predicted from thorough calculations of neutron and gamma partial detector sensitivities, using close neutron and gamma fields (levels and spectra) accurately established beforehand.

A. The SPND Theory

SPNDs are usually designed with a coaxial geometry with a central electrode (emitter) surrounded by an insulator and an external concentric electrode (sheath or collector) (see Figure 5). This sensitive part is connected to the measuring system by an integrated mineral insulated cable. The basic SPND operation principle is based on the collection of electrons created in the emitter, coming from neutron interactions with the detector materials. These moving electrons are generating a measurable direct current.

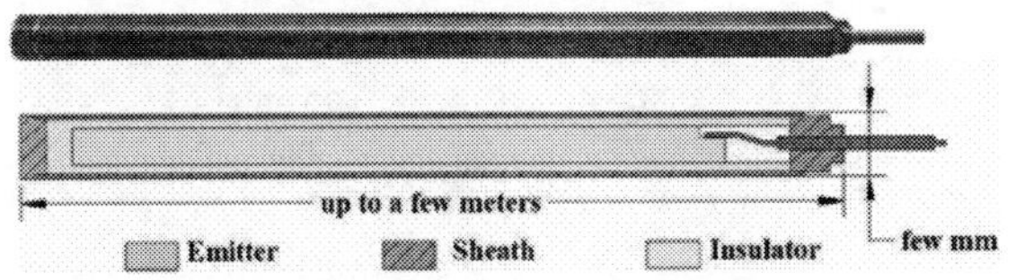

Figure 5: Sketch of a typical self-powered neutron detector

For neutron measurements, two types of SPNDs are considered. First, detectors with dominant (n, β^-) reaction are called 'delayed SPND' since their response time is driven by the decay constant of the created beta emitting isotopes. Second are 'prompt' detectors due to their instantaneous response to predominant (n,γ)(γ,e) reactions. External gamma interactions (fission, activation…) also create some free electrons in detectors denoted as (γ,e) reactions. When the (γ,e) reaction contribution is enhanced through specific material definition, the detector is called Self-Powered Gamma Detector (SPGD) and is dedicated to gamma flux measurements [10].

While moving between the electrodes, created electrons are giving rise to a direct electric current between the emitter and the sheath, which is proportional to the mixed neutron and gamma field at the SPND location. When irradiated in a mixed field, the SPND total output current integrates partial currents coming from the three reactions mentioned above for each materials (emitter, insulator, sheath and close detector surroundings). These partial currents can be either positive or negative.

The mixed neutron-gamma sensitivity of all SPNDs is a key-point in the proposed severe accident monitoring strategy.

B. The SPND Model

The SPND model relies on two fundamental points: the exhaustive study of all possible free electron creation sources within SPND materials and the transport calculation of these different electrons (or beta particles) in the SPND components along with their respective charge depositions profiles. The SPND signal is determined as a combination of the net electron currents at the emitter/insulator and insulator/sheath boundaries. Moreover, under irradiation, electric charges deposited in the insulator are creating an electric field which alters electron paths. The fraction of electron currents reflected by the electric field is introduced in the SPND model as an analytical expression [9].

In SPNDs materials, 76 reactions [(n,β^-), (n,γ)(γ,e) and (γ,e)] [11] have been identified for their potential input to the total current. Only 19 reactions among them are considered to have a significant contribution to the measured signal [12]. Those reaction contributions are evaluated through multistep Monte Carlo calculations.

The 3 main contributions are calculated using neutron and gamma fields (levels and spectra) in the SPND vicinity, a fine model of the SPND and its immediate environment, different steps of Monte Carlo (MCNP6.1) and analytical calculations.

First step is performed using a shell neutron source positioned few millimeters outside the SPND. The neutron reaction rate level and the radial distribution of every β^- emitter precursor ((n, β^-) contribution) are estimated using the track length estimator of neutron flux (F4 Tally in MCNP6). The electron currents at every material boundary ((n, γ)(γ,e) contribution) are calculated with the surface current estimator (F1 Tally in MCNP6) Secondly, (n, β^-) contribution is determined through electron transport calculations using an electron source defined with a β^- radial distribution calculated previously for each β^- emitter. Thirdly, the (γ,e) contribution is computed with a gamma source placed outside the SPND. In addition, the charge deposition is also calculated (+F8 Tally in MCNP6) for the three mentioned contributions.

At the end of the simulation process, the SPND total electric current is derived from the outward net electron currents at both emitter/insulant and insulant/sheath boundaries for all contributions and normalized toward the considered neutrons and gamma flux levels. It is then expressed in Amps.

The multistep Monte Carlo calculations are now performed using the LDCI MATiSSe toolbox.

C. The LDCI MATiSSe Toolbox

The LDCI MATiSSe toolbox was developed to ease operator calculations for self-powered neutron detector design, simulation and operation. It runs the multistep SPND model in adequate order from a user SPND definition (see figure 6). The reaction contributions are evaluated through multistep Monte Carlo calculations using the MCNP6.1 code and the JEFF3.2 neutron data and the ENDF/B-VII.1 gamma data libraries. The JANIS4.0 web nuclear data application [13] provides the β^{-} energy spectra.

In the MATiSSe toolbox, each Monte Carlo and analytical calculation are coded separately in C++ macros. The user defines both the SPND geometry and its close environment, and then launches calculations from a graphical user interface (GUI) developed with the Qt Creator 3.5.0 program (ICS Company).

Two different modes are available: 'automatic' one is running all steps automatically up to the final output current, and 'expert' mode is a step-by-step mode for advanced users.

At the end, the MATiSSe output is a direct electrical current comparable to the current generated in the SPND when irradiated in the same neutron and gamma conditions.

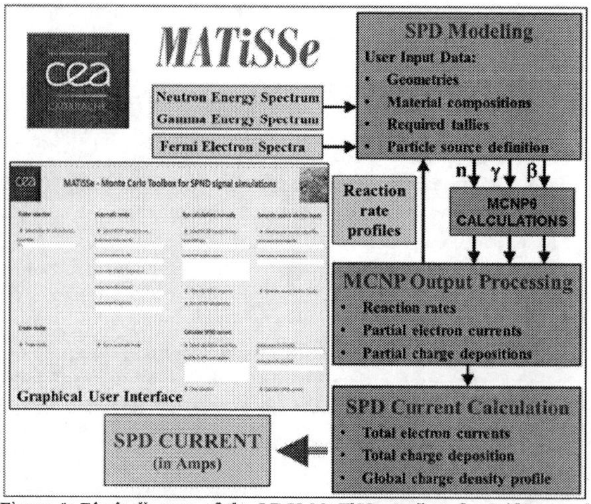

Figure 6: Block diagram of the LDCI MATiSSe toolbox for Self-Powered Neutron Detector simulation.

Experiments dedicated to MATiSSe toolbox qualification have been performed successively in the French CEA Saclay OSIRIS reactor, in the Slovenian TRIGA Mark II reactor (Jozef Stefan Institute - JSI) and in the Polish MARIA reactor (National Centre for Nuclear Research - NCBJ) [14, 15]. Various SPNDs (standards and innovative ones) have been tested in systematically and fully characterized locations and confronted to MATiSSe toolbox simulations.

IV. INSTRUMENTATION DESIGN

The instrumentation proposed in the frame of the DISCOMS project is a specific CEA and THERMOCOAX design. Regarding the relatively low neutron flux conditions, it straightforwardly became mandatory to consider a high thermal neutron sensitivity Rhodium emitter SPND. In addition, Platinum emitter SPND was identified for its mixed response to neutron and gamma irradiations, with rather high gamma sensitivity in favorable conditions. Similar insulation and sheath materials were chosen for the two kinds of SPNDs; alumina and stainless steel 304L (SS304L) due to their common use in nuclear industry and their good behavior under irradiation with time.

Rhodium and Platinum emitter SPNDs show the same cylindrical geometry (see Table 1):

TABLE I. DISCOMS SPND CHARACTERISTICS

DISCOMS SPND	
Emitter	Rh or Pt
- Diameter	1 mm
- Length	up to 6 m
Insulator	Al_2O_3
- Thickness	0.45 mm
Sheath	SS304L
- Outer Diameter	2.8 mm
- Thickness	0.45 mm
Cable	1AcAc10

The two SPNDs, one thermocouple and a background signal cable are embedded in a tubular envelope including spacers (also called followers) as depicted on Figure 8. The sensitive part can be designed up to 6 meters long while the complete device, called an instrumented 'pole' is ranging from 7 to 30 meters (see Figure 7).

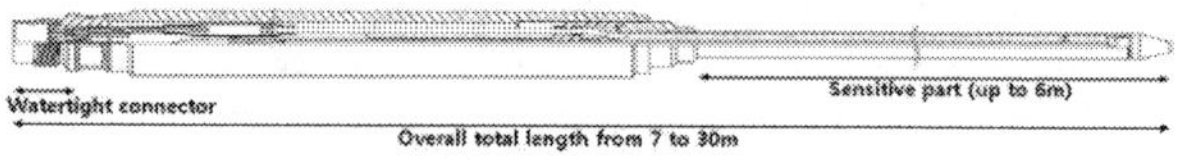

Figure 7: Schematic view of the instrumented DISCOMS pole.

The pole is connected to extension cables (up to 200 meters) using specifically designed and qualified watertight glass-metal connectors [3]. Extension cables are linked to the standalone low current acquisition system placed in a remote location.

V. INSTRUMENTATION PERFORMANCES

A. MATiSSe toolbox results

From the irradiation conditions mentioned in the section II-C, the two bare SPND performances were initially calculated for the different configurations. Then, SPNDs are inserted into the DISCOMS pole and their performances are reevaluated. SPND cable influence on delivered currents is also determined by the MATiSSe toolbox.

All components of the DISCOMS pole and its close environment are then modeled to take into account electrons produced outside SPNDs and possibly depositing their electrical charges in one of the SPND components.

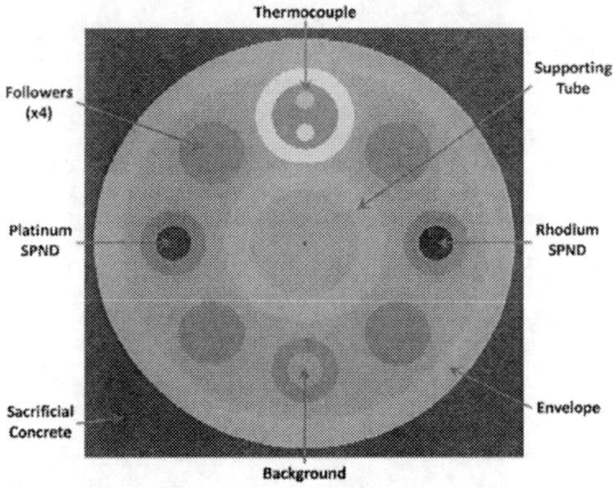

Figure 8: MCNP6 model of the DISCOMS instrumented pole

SPND performances are presented for the GEN2 reactors in 'Normal Operation' (NO) and Severe Accident (SA) when the corium pours on the basemat surface. Two SPND locations are also considered: at reactor core mid height outside the vessel (NO/Seg. 5) and on the concrete basemat surface (NO/Base and SA/Base).

MATiSSe toolbox numerical results are given in Tables II to IV for the Rhodium and the Platinum SPNDs and the Background cable. In these tables, neutron and gamma flux levels are specified for each case using the 'Φ_n' and 'Φ_g' columns respectively. The currents given in A.m^{-1} in the column labeled '*I(A/m)*' represent the final simulation result of the electrical signal delivered by the SPNDs (per unit of emitter length). Such results are directly comparable to the measurement carried out by the operator. Various contributions to the total electric current are expressed as a percentage of the latter.

TABLE II. DISCOMS RHODIUM SPND SIMULATION RESULTS

Rhodium	Φ_n (n/cm^2/s)	Φ_g (γ/cm^2/s)	**I (A/m)**	(n, β$^-$)	(n,γ)(γ,e)	(γ,e)
NO/Seg.5	4.26E+10	1.45E+10	**1.42E-09**	87.9%	10.1%	1.9%
NO/Base	1.34E+09	4.44E+08	**1.17E-10**	89.1%	10.6%	0.3%
SA/Base	N/A	7.27E+12	**2.32E-09**	N/A	N/A	100%

TABLE III. DISCOMS PLATINUM SPND SIMULATION RESULTS

Platinum	Φ_n (n/cm^2/s)	Φ_g (γ/cm^2/s)	**I (A/m)**	(n, β$^-$)	(n,γ)(γ,e)	(γ,e)
NO/Seg.5	4.26E+10	1.45E+10	**4.45E-11**	4.6%	46.8%	48.6%
NO/Base	1.34E+09	4.44E+08	**2.75E-12**	5.3%	73.1%	21.6%
SA/Base	N/A	7.27E+12	**1.42E-08**	N/A	N/A	100%

Background cable currents result from same level but opposite sign contributions. The resultants are very low and poorly reliable and only given as an indication (see TABLE IV).

TABLE IV. DISCOMS BACKGROUND CABLE SIMULATION RESULTS

Background	Φ_n (n/cm^2/s)	Φ_g (γ/cm^2/s)	**I (A/m)**	(n, β$^-$)	(n,γ)(γ,e)	(γ,e)
NO/Seg.5	4.26E+10	1.45E+10	**6.13E-12**	-10.8%	64.2%	46.6%
NO/Base	1.34E+09	4.44E+08	**1.62E-13**	N/A*	N/A*	N/A*
SA/Base	N/A	7.27E+12	**-8.76E-10**	N/A	N/A	100%

B. Discussions

Figure 9 shows that SPND currents delivered during normal reactor operation remain above the lower current measurement limit (> 1 pA). The current levels to be measured are even relatively easy to evaluate with dedicated electronics. The numerical simulations of the instrumented pole developed within the DISCOMS project provide usable signals for all studied configurations. Consequently, MATiSSe toolbox results give DISCOMS partners strong confidence about the use of SPND-based instrumentation for corium monitoring in the event of a severe reactor accident.

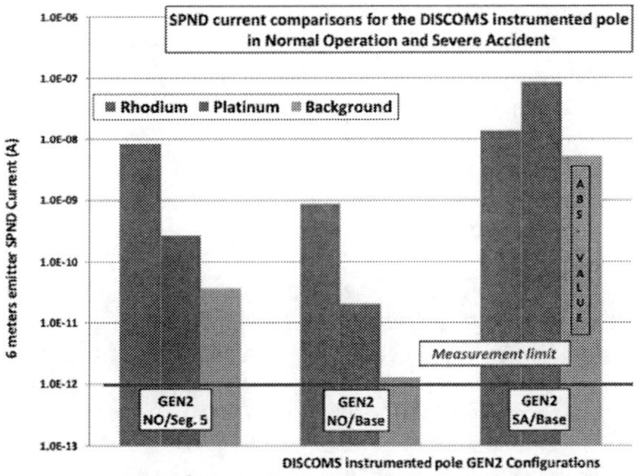

Figure 9: DISCOMS pole SPND current comparisons for Normal Operation and Severe Accident conditions (Negative Background cable SA/Base current is presented in absolute value due to the use of log graphical ordinates).

On-line monitoring and adequate management of SPND current levels in all reactor operation phases should allow the operators to differentiate the reactor conditions: reactor shutdown, normal operation, severe accident without corium migration or severe accident including corium migration towards the concrete basemat.

In addition, the Rhodium and Platinum SPND currents will not evolve the same way at the accident occurrence. Significant current drops following the neutron flux loss and Platinum SPND current large rise during the corium approach will inform the operator about the corium migration towards the concrete basemat.

Calculated Background cable currents are poorly reliable meaning the SPND cable current correction cannot rely on simulations. Hence, correction should be carried out experimentally using a compensation cable. One follower is replaced for this purpose.

Compensation cable end must coincide with the Rhodium and Platinum emitter end positions. Then compensation cable current has to be subtracted from the Rhodium and Platinum SPND measured currents.

VI. Conclusion

In the framework of the ANR DISCOMS project, the feasibility of a remote corium migration monitoring in the unlikely event of a severe nuclear accident is considered and simulated. The corium progression is monitored by means of an instrumented pole equipped with SPNDs and a Thermocouple.

CEA laboratories and the THERMOCOAX SAS Company have designed an instrumented pole according to the considered implementation and irradiation conditions. These conditions for GEN2 reactors have been estimated by DISCOMS partners (IRSN and CEA/SERMA).

A Rhodium and a Platinum emitter SPNDs are encapsulated in this instrumented pole. Their performances have been calculated using the LDCI numerical tool called 'MATiSSe' for different SPND locations.

MATiSSe toolbox simulation results provide SPND currents that demonstrate the proposed instrumented pole capabilities. Its adequate positioning is on the reactor vessel side and on the reactor concrete basemat. These configurations would allow to monitor the corium progression towards the concrete basemat in case of a severe nuclear accident. Coupled temperature measurement would also provide additional information on corium migration.

The full monitoring equipment is planned to be tested during a joined CEA-THERMOCOAX experimental campaign in some realistic irradiation conditions in the Slovenian TRIGA Mark II reactor (Ljubljana, JSI, pool type research reactor 250 kW [16]) during fall 2017. Experiments will be supported by specific neutron and gamma flux and spectra calculations in the instrumented pole irradiation locations (taking into account actual irradiation geometry and materials). Additional measurements will be performed for thermal and fast neutron flux level (with activation dosimetry) and gamma flux level (with miniature ionization chamber [17]) purposes.

Acknowledgment

This project is co-funded by the French *Programme d'Investissements d'Avenir* (PIA) and managed by the French National Research Agency (ANR).

References

[1] P. Ferdinand *et al.*, 'DISCOMS: DIstributed Sensing for COrium Monitoring and Safety', *CANSMART Conf. 2015*, 15-17 July 2015, Vancouver, Canada.

[2] J.W. Hilborn, 'Self-powered neutron detectors for reactor flux monitoring', *Nucleonics*, vol. 22, no 2, pp 69-74, 1964.

[3] THERMOCOAX industrial references n° 304 803 and 304 807.

[4] X-5 Monte Carlo Team, MCNP6 User's Manual, Version 1.0, May 2013, LA-CP-13-00634.

[5] TRIPOLI-4 Monte Carlo Transport Code, http://www.nea.fr/abs/html/nea-1716.html (2013).

[6] J. Kopecky, 'Final verification and validation of JEFF-3.2', NEA document, JEF/DOC-1590, April 2015.

[7] National Nuclear Data Center, Brookhaven National Laboratory. ENDF-6 Formats Manual. Data Formats and Procedures for the Evaluated Nuclear Data File ENDF/B-VI and ENDF/B-VII. CSEWG Document ENDF-102, Report BNL-90365-2009.

[8] M. Brovchenko *et al.*, 'Neutron-gamma flux and dose calculations in a Pressurized Water Reactor (PWR)', *Joint Conf. on Radiation Shielding (ICRS-13) & 19th Topical Meeting of the Radiation Protection & Shielding Division of the American Nuclear Society (RPSD-2016)*, 3-6 October 2016, Paris, France.

[9] L. Vermeeren, R. Van Nieuwenhove, 'Theoretical study of radiation induced electromotive force effects on mineral insulated cables', American Institute of Physics, *Review of Scientific Instruments*, vol. 74 - no 11, 2003.

[10] L. Vermeeren, H. Carcreff, L. Barbot, V. Cloute-Cazalaa, S. Fourrez and L. Pichon, 'Irradiation tests of prototype self-powered gamma and neutron detectors', *2nd ANIMMA Intern. Conf.* publication DOI 10.1109/ANIMMA.2011.6172889.

[11] G.F. Lynch, 'Some theoretical aspects of self-powered detectors', Atomic Energy of Canada Limited, Rapport AECL-5124 paper 1.8, 1974.

[12] C. Blandin, 'Contribution au développement de collectrons pour la mesure instantanée et sélective des différents champs de rayonnements en réacteurs nucléaires', Thèse de l'Institut National Polytechnique de Grenoble, France, 1998.

[13] JANIS4.0 web nuclear data library available at http://www.oecd-nea.org/janis/

[14] L. Barbot *et al.*, 'Calculation to experiment comparison of SPND signals in various nuclear reactor environments', *4th Intern. Conf. ANIMMA 2015*, 20-24 April 2015, Lisbon, Portugal.

[15] L. Barbot *et al.*, 'Experimental validation of a Monte Carlo based toolbox for Self-powered neutron and gamma detector simulation in the OSIRIS MTR', *PHYSOR Conf. 2016*, 1-5 May 2016, Sun Valley, USA.

[16] B. Smodis and L. Snoj, "Utilization and Application of the Slovenian TRIGA Reactor," "Jo˘zef Stefan" Institute, Ljubljana, Slovenia, Tech. Rep., 2011. [Online]. Available: http://www-pub.iaea.org/MTCD/Publications/PDF/ P1575 CD web/datasets/papers/ A29n%20Smodis.pdf.

[17] D. Fourmentel *et al.*, 'Measurement of photon flux with a miniature gas ionization chamber in a Material Testing Reactor', *Nuclear Instruments and Methods in Physics Research Section A*, vol 724, pp. 76–82, 2013.

Malfunctions in radioactivity sensors' networks

Veronika Khalipova, *SCM SA*, Guillaume Damart, *SCM SA*, Bernard Beauzamy, *SCM SA*, Giovanni Bruna, *IRSN*

Abstract— **The capacity to promptly and efficiently detect any source of contamination of the environment (a radioactive cloud) at a local and a country scale is mandatory to a safe and secure exploitation of civil nuclear energy. It must rely upon a robust network of measurement devices, to be optimized vs. several parameters, including the overall reliability, the investment, the operation and maintenance costs. We show that a network can be arranged in different ways, but many of them are inadequate. Through simulations, we test the efficiency of several configurations of sensors, in the same domain. The denser arrangement turns out to be the more efficient, but the efficiency is increased when sensors are non-uniformly distributed over the country, with accumulation at the borders. In the case of France, as radioactive threats are most likely to come from the East, the best solution is densifying the sensors close to the eastern border. Our approach differs from previous work because it is "failure oriented": we determine the laws of probability for all types of failures and deduce in this respect the best organization of the network.**

Index Terms— **Environmental radiation surveillance, network, failure, probabilities, sensor**

I. Introduction

NOWADAYS, nuclear plants are an important source of electricity worldwide and their contribution to the overall sharing is ever increasing, despite the decision of some countries to drop nuclear energy off .

Many countries (including France and Ukraine) perform a constant monitoring of radioactivity in the atmosphere, through a network of sensors which are able to provide with alarms for actions of protection, sheltering, and, in extreme cases, evacuation of populations, see [1], [2].

The topic "Malfunctions in Sensors' Networks" [3] deals with the propagation of radionuclides in the atmosphere and their likelihood of detection by a suitable network of sensors, accounting for the features of the network and its possible malfunctions originating from uncertainties, failures and false alarms.

The objectives of the ongoing research are:

– Checking the effectiveness of a given network of sensors: is it possible to reconstruct the shape of a radioactive cloud relying upon the indications provided by a given network of sensors? And how accurately?

– Optimizing the design of the networks of sensors: is it possible to increase their robustness (vs. uncertainties, failures and false alarms) keeping the investment and exploitation costs at a reasonable level?

To answer these questions, the problem will be addressed through a graduate step-by-step approach, adopting time by time various assumptions upon the type of cloud, its direction, speed, and so on. We start with a complete investigation of the information given by a single sensor and the possible errors, and gradually increase to a whole network.

II. Features of the Modelling

A. Theory

First of all, the cloud is "pixellised". In such an approach, no precise geometrical shape is attributed to the cloud. A radioactivity level, which is a real number, is attributed to each pixel of the domain. Certainly, this assumption leads to a loss in precision, because the cloud has no specific boundaries anymore, moreover the usual difficulties connected with pixels hold: the radioactivity level is homogenized inside a given pixel.

Compared to the presentation of the cloud as a collection of geometrical shapes with different levels of radioactivity, our approach is closer to reality. In fact, there is always a natural level of radioactivity (which may differ from one zone to another) and radioactive clouds do not have precise boundaries. A pixel of 1 x 1 km seems reasonable, in terms of precision.

This work was supported by Société de Calcul Mathématique SA (SCM SA - 111 Faubourg St Honoré, 75008 Paris – France) and Institut de Radioprotection et de Sûreté Nucléaire (IRSN - BP 17 - 92260 Fontenay-aux-Roses – France).

Veronika Khalipova is with SCM SA, Paris, France (e-mail: veronika.khalipova@scmsa.com).
Bernard Beauzamy is the founder and current CEO of SCM SA, Paris, France (e-mail: bernard.beauzamy@scmsa.com).
Guillaume Damart is with SCM SA, Paris, France (e-mail: guillaume.damart@scmsa.eu).
Giovanni Bruna is with IRSN, Fontenay-aux-Roses, France (e-mail: giovanni.bruna@irsn.fr)

Also, the administrator of the network of sensors has no knowledge about the shape of the cloud. He has at his disposal the measurement results provided by the sensors only, plus some meteorological information (such as the speed and direction of the wind).

In this simulation, the radioactive cloud is assumed moving along a straight line from East to West: the entrance point is located in A and the exit point in B. For simulation sake, it is assumed that the cloud has the shape of a disk of the radius r : as we said earlier, the sensors only see pixels.

We denote by V the speed of cloud propagation; in the numerical examples below, we will assume $V = 10$ km/h. Let Δ be the length of the segment AB and C the center of the disk representing the radioactive cloud. The equation of the movement of C is:

$$\overrightarrow{AC} = Vt\frac{\overrightarrow{AB}}{\Delta},$$

assuming the origin of time ($t = 0$) at the moment C is located in position A.

B. *The domain*

Since the shape of the domain does not matter, we take a simplified domain, namely a square, the dimensions of which are comparable to those of continental France, namely 750 x 750 km. Remembering the size of the pixel assumed here below (1 x 1 km), the domain is thus divided into 562 500 pixels. The origin of the axes is located in the square bottom left.

C. *Time units*

The sensor monitors the environment continuously, but transmits the information every ten minutes only. In other words, all motions are discretized and viewed every 10 minutes. The time unit (TU) is denoted by τ.

The network sees the radioactive cloud as a collection of pixels. The coordinates of its center x_C, y_C are determined from the law of uniform linear motion. We denote by n the number of time units; at each TU, n is increased by 1. The time will be written as: $T = \tau \cdot n$

III. Consequences of changes in the cloud's shape

A. *Introduction*

We know from the Fukushima and Chernobyl accidents ([4], [5]) that a radioactive cloud spreads far from the initial emission point and gets larger and less radioactive with time. Therefore, it can be assumed that, on the average and at least in a first approximation, areas which are far from the nuclear power plant are likely to be less contaminated than the nearby ones.

Let's consider situations where the shape of the radioactive cloud changes with time: the radius increases and the radioactivity decreases at each time step. The increase of the radius is likely to allow an earlier detection of the cloud compared to the stationary case (cloud detected after 30 hours). If we assume an increase of 0.2 km per time unit, the sensor will detect the cloud earlier (26.78 hours), and after 24.19 hours if the radius increase is 0.4 km per time unit.

When the radioactivity decreases, the results will be different, depending on the initial radioactivity value of the cloud. Assuming a radioactivity decrease of 0.01 µSv/h per time unit, we can consider three bounding cases:

- A cloud with an initial concentration of contaminants generating 29 µSv/h: even with the decrease, the cloud will remain dangerous, with a minimum value of 25.6 µSv/h at the last time of detection by the sensor;
- A cloud with an initial concentration of contaminants generating 27 µSv/h: with the decrease, the radioactivity will go from 25.4 µSv/h (first time of detection by the sensor) to 23.6 µSv/h (last time of detection);
- A cloud with an initial concentration of contaminants generating 25 µSv/h: with the decrease, the radioactivity will never cross the threshold of 25 µSv/h.

Besides, the false alarms, failures and the uncertainty of measurement can modify the detection, mainly in the last two cases: the sensor can miss the cloud or send a false alarm. However, the increase of the radius, if the cloud is radioactive enough, can improve the likelihood of detection. For instance, if we assume an increase of 0.4 km per time unit, as the cloud increases in radius (average radius 207.4 km at the time of first detection), it becomes almost twice bigger compared to the previous case.

Assuming a repair-time of 24 h in the case with increased radius, part of the cloud can be detected before the failure appears and the last part the sensor has been repaired. This situation cannot happen in the case of constant radius, since the diameter is less than 240 km. In the situation with an increasing radius, we can calculate the true radius of the cloud even if the failure occurs.

B. *General description*

Assuming a linear increase of the cloud radius with (α [km] at each time unit), it can be expressed as follows:

$$r(n) = r(n-1) + \alpha \tag{1}$$

In this situation, the cloud is likely to reach the sensor faster compared to the constant-radius case. Let us denote by the distance spanned by the fixed shape cloud from the entrance point to the point of its first detection. Then the first announced time for the cloud detection is:

$$T_1 = \frac{l}{\frac{\alpha \cdot 60}{10} + V} \quad (2)$$

If n_1 is the number of time units that corresponds to the first detection of the cloud, then the last announced time for the cloud detection is:

$$T_2 = T_1 + \frac{r(n_1) \cdot 2}{V - \frac{\alpha \cdot 60}{10}} \quad (3)$$

The radioactivity of the cloud decreases of an amount [µSv/h] at each time unit. We assume in our simulations that the radioactivity level of the cloud decreases linearly:

$$radioactivity(n) = radioactivity(n-1) - \beta \quad (4)$$

C. Numerical application: parameters

Let us now consider different couples of changes in radius and radioactivity of the cloud. We assume the following couples of values and adopt them in all examples:

x_A=750 km, y_A= 500km, x_B = 0 km, y_B=100 km,

x_U = 400km, y_U = 310 km, V =10 km/h, τ =10 min.

We will keep these values in all examples.

D. First simulation

The initial value of the radioactivity is $radioactivity(1) = 29$ µSv/h and $\beta = 0.01$ µSv/h. The initial value of the radius is $r(1) = 100$ km and $\alpha = 0.2$ km.

The average radius is 150 km. The radius of the cloud, taking into account the uncertainty due to the detection delay, is:

$$r_{b_1} = 149.17 \le r \le r_{b_2} = 150.83 \text{ (km)}$$

The first and last announced times of the cloud detection are $T_1 = 26.83$ (h) and $T_2 = 57.33$ (h). The uncertainty has got a quite low influence so that the sensor actually detects one cloud the size of which is:

$$r_{b_1} = 151.67 \le r \le r_{b_2} = 153.33 \text{ (km)}$$

This size is 2.5 km bigger compared to the original case: We overestimate the size of the cloud.

E. Second simulation

The initial value of the radioactivity is $radioactivity(1) = 27$ µSv/h and $\beta = 0.01$ µSv/h. The initial value of the radius is $r(1) = 100$ km and $\alpha = 0.2$ km. The average radius is 135.9 km. The radius, taking into account the uncertainty due to the delay in detection, is:

$$r_{b_1} = 138.97 \le r \le r_{b_2} = 140.63 \text{ (km)}$$

In this situation, from the information given by the sensor, one may deduce that there are two clouds, since the sensor underestimates the value of radioactivity: the first cloud has times of the detection $T_1 = 26.833$ (h) and $T_2 = 32.33$ (h) and the second $T_1 = 32.666$ (h) and $T_2 = 34.666$ (h). We will simply say that the sensor "sees" two clouds.

Consequently, the sensor can overestimate the value of radioactivity and send a false alarm from the time $T = 33.33$ h. Therefore the sensor sends a false alarm from time 33.33 h to 34.66 h. Then, from the indications given by the sensor, one may deduce that there are five clouds, which is not correct.

The size of the circular region including all small clouds is:

$$r_{b_1} = 151.17 \le r \le r_{b_2} = 153.33 \text{ (km)}$$

That is bigger by 12.2 km compared to the case in the numerical simulation, where the values of the estimated radius are between r_{C_1} =138.97 km, r_{C_2} =140.63 km. We overestimate the size of the cloud because of the false alarms that occur due to the uncertainty of measurements.

F. Third simulation

The initial value of the radioactivity is $radioactivity(1) = 25$ µSv/h and $\beta = 0.01$ µSv/h, the initial value of the radius is $r(1) = 100$ km and $\alpha = 0.2$ km.

Since the radioactivity of the cloud is lower than the threshold, during all the time of possible detection, the sensor is not able to detect the cloud nevertheless it can send a false alarm.

In this case the sensor "sees" three false clouds during the detection time. The size of the circular region including all small false clouds is:

$$r_{b_1} = 41.67 \le r \le r_{b_2} = 43.33 \text{ (km)}$$

Therefore, due to the likelihood of the false alarms, a non-zero radius detection region can show-up.

G. Fourth simulation

The initial value of the radioactivity is $radioactivity(1) = 29$ μSv/h and $\beta = 0.01$ μSv/h. The initial value of the radius is $r(1) = 100$ km and $\alpha = 0.4$ km. The average radius is 207.4 km. The radius, taking into account the uncertainty due to the delay in detection, is:

$$r_{b_1} = 206.57 \leq r \leq r_{b_2} = 208.23 \text{ (km)}$$

In this case, the sensor underestimates the value of radioactivity only once during the detection and it "sees" two clouds.

The size of the circular region including all small false clouds is:

$$r_{b_1} = 209.17 \leq r \leq r_{b_2} = 210.83 \text{ (km)}$$

As a consequence, we overestimate the size of the cloud.

H. Fifth simulation

The sensor undergoes a failure. The repair lasts for 24 h and we can miss 240 km of the cloud. Since the radius increases each time unit and the average radius for this case is 414.8 km, we can see one part of the cloud before the failure and the other after the repair of the sensor. We take the values: $r(1) = 100$ km, $\alpha = 0.4$ km, $radioactivity(1) = 29$ μSv/h, $\beta = 0.01$ μSv/h.

The sensor detects the radioactive cloud at 24.333 h and then, from 38.666 h to 63.333 h, it is not operating. After the repair, the sensor detects the last part of the cloud.

The size of the circular region including all small clouds is:

$$r_{b_1} = 208.335 \leq r \leq r_{b_2} = 209.99 \text{ (km)}$$

That is bigger by 1,765 km compared to the case in the numerical simulation, where the values of the estimated radius are between r_{C_1} =206.57 km, r_{C_2} =208.23 km. We overestimate the size of the cloud. In this situation, the sensor presents a failure but we can calculate the size of the cloud, since we see a part after the repair.

IV. Comparison among several types of networks

A. Introduction

In this paragraph, we assume that the cloud enters a region where it can be detected by several sensors. The information about the radioactive cloud is more reliable when we can rely upon several sensors[1], see [6].

We consider that each sensor provides with information within a circular region of radius 20 km around it. We do not actually know the real shape of the cloud, so that, in a first approximation, we attribute it a circular shape.

Therefore, the area spanned by each sensor and the area where the radioactive cloud is located form two circles, generating different situations:

- The circle of the cloud contains the circle spanned by the sensor;
- The circle of the cloud contains only a part of the circle spanned by the sensor;
- The circle of the cloud does not contain the circle around the sensor.

We define the ratio of "efficient detection" as the area spanned by the network divided by the area of the radioactive cloud. If the ratio is 1, then the network is able to completely detect the cloud. We can find the maximum ratio for a given simulation at the time when the detection is the best. If we have the appropriate position of the sensors in the network and enough sensors, we could detect almost all the radioactive cloud.

An "efficient" network has to:

- Detect the radioactive cloud as soon as possible (time of first detection);
- Detect the cloud as long as possible (detection duration);
- Span the greatest possible area (ratio of detection).

B. General description

Let's consider several cases modifying the location of the sensors as well as their number, and assuming different trajectories for the cloud. We determine the maximum "efficient detection" ratio, the first time of detection and the duration of the detection.

The area spanned by a sensor is delimited by a circle of radius R. It can be in, partially in, or outside the radioactive cloud each time unit. When the sensor starts facing the cloud, its area can be into the cloud or partially in it.

[1] The French network of sensors TELERAY monitors the radioactivity in the environment. There are 400 sensors located on French territory mainly in the vicinity of nuclear sites and major cities, see [7].

https://doi.org/10.1051/epjconf/201817008002

Let's consider different cases of the radioactive cloud area and the area seen by the sensor location:

1. The circles do not intersect, when the distance between them l satisfies the relationship $l > R + r(n)$. The area inside the cloud is $S_{in} = 0$:

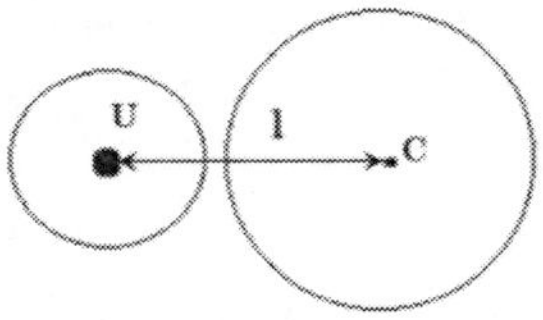

Fig. 1. The cloud and the area spanned by the sensor do not intersect.

2. The area spanned by the sensor is inside the area of the radioactive cloud, when $l \le r(n) - R$. The area inside the cloud is: $S_{in} = \pi R^2$.

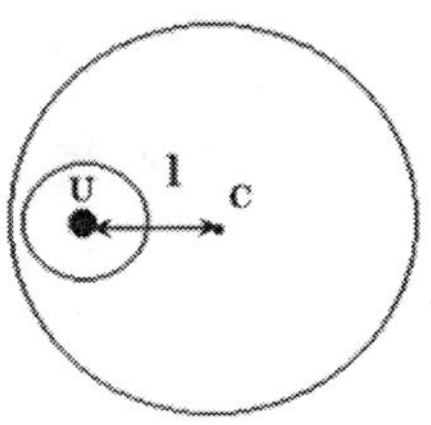

Fig. 2. The area spanned by the sensor is inside the cloud.

3. The area spanned by the sensor is partially inside the radioactive cloud, when $r(n) - R \le l \le r(n)$.

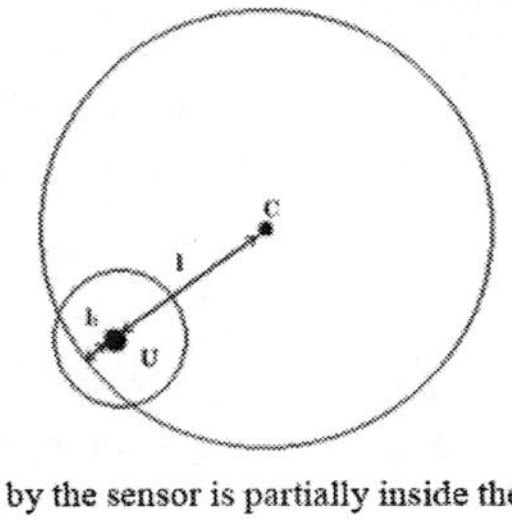

Fig. 3. The area seen by the sensor is partially inside the cloud.

The distances are given by the formulas:

$$l_1 = \frac{r^2(n) - R^2 - l^2}{2 \cdot l}, \quad l_2 = l + l_1$$

The angles are given by the formulas:

$$\alpha_1 = 2 \cdot \arcsin\left(\frac{\sqrt{R^2 - l_1^2}}{R}\right),$$

$$\alpha_2 = 2 \cdot \arcsin\left(\frac{\sqrt{r^2(n) - l_2^2}}{r(n)}\right)$$

The areas are given by the formulas:

$$S_1 = \frac{\alpha_1 \cdot R^2 \cdot \pi}{360} - \frac{R^2 \sin(\alpha_1)}{2},$$

$$S_2 = \frac{\alpha_2 \cdot r^2(n) \cdot \pi}{360} - \frac{r^2(n) \sin(\alpha_2)}{2}$$

Then the area inside the cloud is:

$$S_{in} = \pi R^2 - S_1 + S_2$$

4. The circles intersect but the sensor does not detect the radioactivity level because it is located outside the cloud.

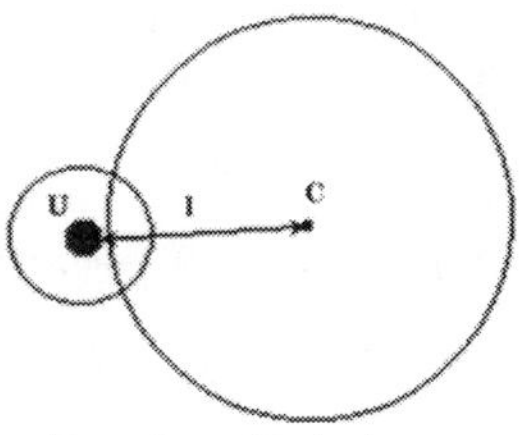

Fig. 4. The sensor is not able to detect the cloud.

Since the sensor does not detect the cloud in this case, the part of its area in the cloud is not considered:

$$S_{in} = 0$$

If several sensors are located in the region facing the cloud, their influence areas are to cover only part of it. It is possible to evaluate the ratio of the cloud the sensors are able to detect over its actual size. If N is the number of sensors, this ratio is:

$$ratio(n) = \frac{\sum_{k=1}^{N} S_{in_k}}{\pi \cdot r^2(n)}$$

C. Numerical simulation

Let's consider different networks with different number of sensors located in different positions. At each time, we record the status of each sensor, whether it detected the cloud or not.

At first, let's assume that the cloud has got constant shape and radioactivity and that the cloud may move along different trajectories, starting from the East border of France. In the simulations, the radioactivity of the cloud is assumed 35 μSv/h, not accounting for the measurement uncertainty.

We simulate the movement of the radioactive cloud and calculate the $ratio(n)$ each time unit. We can find the maximum among the values of the $ratio(n)$ in each simulation ($ratio$), when the radioactive cloud was better detected.

We consider four types of networks:

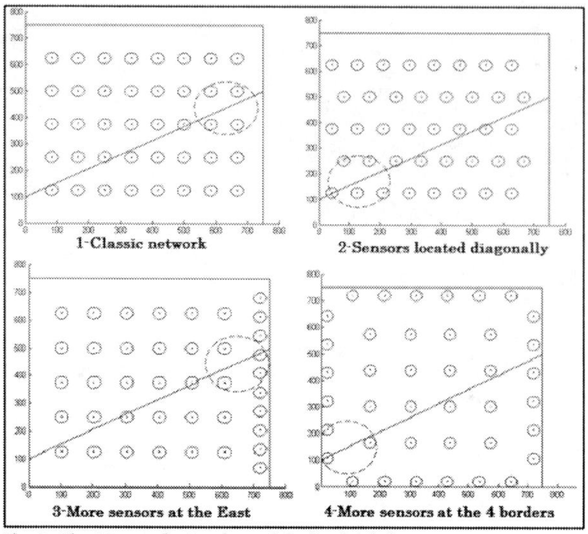

Fig. 5. The 4 types of networks used for the simulations.

The ratio increases with the number of sensors for all types of networks:

- For the networks with 20 sensors, the ratio is bigger for the third network, $ratio = 0.120$. The first time of detection is smaller (0.333 h) for the third network, but the duration of detection is not as long as for the first or second networks.

- For the networks with 40 sensors, the ratio is bigger for the second network, $ratio = 0.1622$. The time of the first detection is 0.333 h for the third network for all the trajectories of the cloud, the duration of detection at some trajectories is less than what the first and second networks require.

- For the networks with 60 sensors, the ratio is bigger for the third network, $ratio = 0.2217$. The time of the first detection is 0.333 for the third and fourth networks independently on the trajectory of the cloud, and the duration of detection is bigger for the fourth network.
- For the networks with 100 sensors, it is bigger for the first network, $ratio = 0.3028$.

- For the networks of 200 sensors, it is bigger for the third network, $ratio = 0.5375$.

- For all the networks with 100, 200 sensors, the time of the first detection is 0.333 h and the duration of detection is maximal for all the trajectories of the cloud.

The network with 400 sensors has a ratio of 0.9523.

The mathematical expectations and standard deviations of the maximum ratio for each network are the following:

- The mathematical expectation is bigger for the third type of network, among the networks with 20 sensors (M = 0.013), but the standard deviation is $\sigma = 0.016$.

- The mathematical expectation is bigger for the third type of network, among the networks with 40 sensors (M=0.158), the standard deviation is $\sigma = 2.585 \cdot 10^{-3}$.

- The mathematical expectation is bigger for the first network, among the networks with 60 sensors (M=0.213), but the standard deviation is $\sigma = 0.016$.

- The mathematical expectation is bigger for the first network, among the networks with 100 sensors (M=0.296), the standard deviation is $\sigma = 6.396 \cdot 10^{-3}$.

- The mathematical expectation is bigger for the third network, among the networks with 200 sensors (M=0.532), the standard deviation is $\sigma = 5.574 \cdot 10^{-3}$.

- The network with 400 sensors covers almost all the area and the mathematical expectation is 0.946.

The third type of network shows good results in achieving maximum ratio and early detection of the cloud. When it has 200 sensors, it shows a maximum duration of radioactive cloud detection and ratio is 0.5375.

D. Average cost of the network per year

It is worth considering not only good ratio and the time characteristics of detection of a network but also its cost. Let's consider the average cost per year of a network with *N* sensors. The network with high number of sensors is good for the detection but has got high cost, too.

Assuming that the probability to provide with a false alarm is $p_{FA} = 0.28 \cdot 10^{-3}$ and $p_{BR} = 0.014$ for the failure for all the sensors at some time unit [2]. The probability for each sensor to have at least one failure per year, without false alarms is:

$$Proba\ (failure\ per\ year) = 0.9804$$

The probability that each sensor provides at least with one false alarm per year, without accounting for failures is:

$$Proba\ (false\ alarm\ per\ year) = 0.01961$$

The average cost of the network per year consists of:

- The casual maintenance cost; the sensors must be inspected all the time to keep them operating. The casual

[2]Data used here are originating from the Teleray Network, [7].

maintenance cost for one sensor per year is K_{CM}, then for N sensors the cost will be $K_{CM} \cdot N$;

- The cost of a false alarm, which mobilizes services. Experts have to the move to the place where the sensor is located and inspect it. The cost of one false alarm is K_{FA}. If we consider that all sensors present malfunctions independently, the cost is likely to increase with the number of false alarms, therefore with the number of sensors. The false alarm occurs with some probability, then for N sensors the cost will be $K_{FA} \cdot N \cdot Proba(false\ alarm\ per\ year)$;
- The cost of the failure, as the device must be repaired or replaced. The cost of one failure is K_{BR}. With the number of failures, therefore with the number of sensors, this cost increases, but failure occurs with some probability, then for N sensors the cost will be $K_{BR} \cdot N \cdot Proba(failure\ per\ year)$.

Therefore, the average cost of the network per year is:

$$K_A = K_{CM} \cdot N + K_{FA} \cdot N \cdot Proba(false\ alarm\ per\ year) + K_{BR} \cdot N \cdot Proba(failure\ per\ year)$$

We take the values: K_{CM}=500 €, K_{FA}=5000 €, K_{BR}=1000 €.

The cost function of the number of sensors increases linearly with the number of sensors.

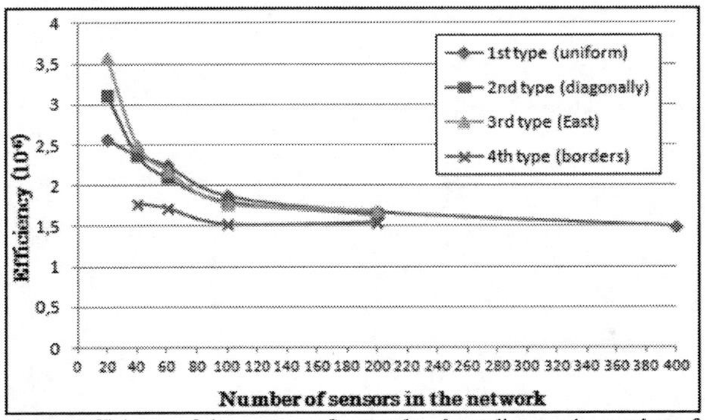

Fig. 6. Efficiency of the 4 types of networks, depending on the number of sensors.

Let us denote the mathematical expectation of the maximum detection ratio for each network as $M(ratio)$, then the efficiency of the network is:

$$Ef = \frac{M(ratio)}{K_A}$$

Let's now investigate the networks' efficiency, which corresponds to the best combination of cost and ratio of detection. The efficiency of each network is shown in Fig. 6.

The greatest efficiency ($3.579 \cdot 10^{-6}$) is met with the third network with 20 sensors (that has more sensors at the East border). We can use the first, the second and the third type of networks with 20, 40, 60 sensors as they have better efficiency than the networks with 100, 200, 400 sensors.

The third type of network shows better results in efficiency almost for all numbers of sensors.

E. Results

The following considerations hold:

- The first time of the cloud detection and the duration of the detection are different for the networks made of 20,40, 60 sensors and equal for the networks with 100, 200 sensors;
- The "efficient detection" ratio increases with the number of sensors. In the case of France, the most interesting network is the one with more sensors at the East border (see footnote 3). In most cases, it has a good ratio and an earlier detection of the cloud;
- If we consider the detection of the cloud only (so that the value of "efficient detection" ratio does not matter) the networks with 60 sensors have got efficiency for detection equivalent to that of the networks with more sensors. The fourth type of network (more sensors on the borders) is able to detect the cloud early (0.333 h) and the duration of the detection is maximum when it has 60 or more sensors;
- If we want to detect and reconstruct the cloud intensity and size, considering the "efficient detection" ratio is mandatory. It is bigger than 0.5 (50%) for almost all types of networks with 200 sensors.

Let's compute the average costs of the networks per year, taking into account the casual maintenance, the false alarms and the failures of the sensors. For each network, we can evaluate its efficiency (i. e. the "efficient detection" ratio divided by the cost of the network).

Our conclusions about the costs of the networks are as follows:

- The average cost of the network per year increases linearly with the number of sensors;
- The third type of network (sensors on the East border) with 20 sensors has the best efficiency;
- The third type of network with 40 sensors has almost the same efficiency as the first type of network with 20 sensors, so we can use more sensors for the detection of the cloud and the efficiency will remain the same;

- The third type of network with 20, 40, 200 sensors has the best efficiency compared to other types of networks with the same number of sensors, therefore it is better to use the network with more sensors at the East border;
- The networks with 20, 40, 60 sensors have better combination of detection ratio and cost than the networks with 100, 200, 400 sensors.

V. Conclusion

The capacity to promptly and efficiently detect any source of contamination to the environment at a local and a country scale is mandatory to a safe and secure exploitation of civil nuclear energy worldwide.

This capacity must rely upon a robust network of measurement devices which is to be optimized vs. several main parameters including its overall reliability, the investment, the operation and maintenance cost.

The present paper investigates the efficiency vs. cost of such network of detectors to several parameters, including their density, lay-out, etc., also considering the major failure modes which may affect them.

Eventually, a modern version of "Archimedes' method" is proposed for optimization. It relies upon systematic comparisons among the results of simulations stored in suitable databases and the actual measurements on-site. In our approach, the best network would be made of three items:

- A few radioactivity sensors, preferably put on the East border;
- Several mobile units, to be sent where detection is supposed to occur;
- A database of simulations, to be used for comparisons.

References

[1] Naosuke Itoigawa, Babette Fahlbruch, Bernhard Wilpert, "Emerging Demands for the Safety of Nuclear Power Operations: Challenge and Response", CRC Press, Boca Raton, FL, 153 pages, 2005.

[2] Claus Grupen, "Introduction to Radiation Protection: Practical Knowledge for Handling Radioactive Sources", Springer, ISBN 978-3-642-02586-0, 415 pages. 2010.

[3] "Malfunctions in Sensors' Networks", a research program by SCM SA. September 2015.

[4] Chernobyl Forum (2005), "Environmental Consequences of the Chernobyl Accident and Their Remediation: Twenty Years of Experience". Report of the UN Chernobyl Forum Expert Group "Environment" (EGE) Working Draft, 280 pages, August 2005.

[5] Sentaro Takahashi, "Radiation Monitoring and Dose Estimation of the Fukushima Nuclear Accident", Springer, 215 pages. 2014.

[6] Bernard Beauzamy, "Archimedes' Modern Works", SCM SA, Paris, France, 224 pages, August 2012.

[7] IRSN, "Report on the Radiological State of the Environment in France in 2010-2011", 308 pages. April 2014.

Fast high-energy X-ray imaging for Severe Accidents experiments on the future PLINIUS-2 platform

L. Berge, N. Estre, D. Tisseur, E. Payan, D. Eck, V. Bouyer, N. Cassiaut-Louis, C. Journeau, R. Le Tellier, and E. Pluyette

***Abstract*— The future PLINIUS-2 platform of CEA Cadarache will be dedicated to the study of corium interactions in severe nuclear accidents, and will host innovative large-scale experiments. The Nuclear Measurement Laboratory of CEA Cadarache is in charge of real-time high-energy X-ray imaging set-ups, for the study of the corium-water and corium-sodium interaction, and of the corium stratification process. Imaging such large and high-density objects requires a 15 MeV linear electron accelerator coupled to a tungsten target creating a high-energy Bremsstrahlung X-ray flux, with corresponding dose rate about 100 Gy/min at 1 m. The signal is detected by phosphor screens coupled to high-framerate scientific CMOS cameras. The imaging set-up is established using an experimentally-validated home-made simulation software (MODHERATO). The code computes quantitative radiographic signals from the description of the source, object geometry and composition, detector, and geometrical configuration (magnification factor, etc.). It accounts for several noise sources (photonic and electronic noises, swank and readout noise), and for image blur due to the source spot-size and to the detector unsharpness. In a view to PLINIUS-2, the simulation has been improved to account for the scattered flux, which is expected to be significant. The paper presents the scattered flux calculation using the MCNP transport code, and its integration into the MODHERATO simulation. Then the validation of the improved simulation is presented, through confrontation to real measurement images taken on a small-scale equivalent set-up on the PLINIUS platform. Excellent agreement is achieved. This improved simulation is therefore being used to design the PLINIUS-2 imaging set-ups (source, detectors, cameras, etc.).**

***Index Terms*— X-ray imaging, MODHERATO, PLINIUS-2, Severe Accidents, Corium**

I. Introduction

The study of severe nuclear accidents is a major research topic at CEA (French atomic energy agency). In this scope, the PLINIUS experimental platform situated in Cadarache studies the behavior of corium (melt fuel-structure mixture), through several experimental set-ups [1]. The tests performed on this facility provide crucial data and validation for the models elaborated at CEA. Among the test facilities of PLINIUS, the KROTOS facility studies the corium-water interaction, through the release of corium into a water tank [2]. It includes high-frequency X-ray imaging through the whole process, which allows a worldwide unique real-time visualization of the reaction [3] (see Fig. 1).

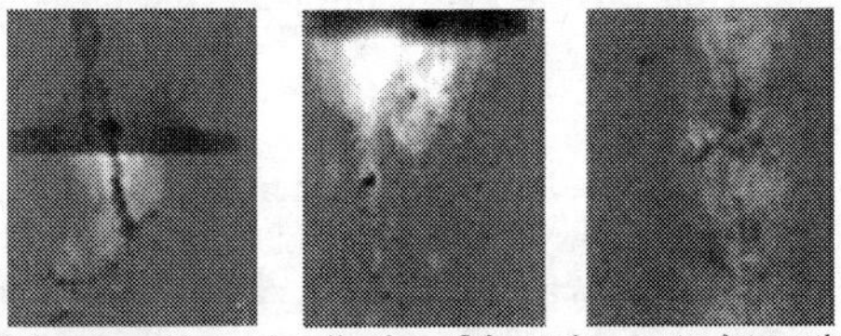

Fig. 1. High-energy X-ray visualization of the corium-water interaction on the KROTOS facility [4]

Imaging the water test column requires an X-ray spectrum of the MeV order of magnitude. A 9 MeV linear electron accelerator (LINAC) coupled to a tungsten target is used to produce an X-ray pulsed cone-beam. The dose rate is about 15 Gy/min at 1 m. The detector is composed of a scintillating Gadox plate coupled to four high-framerate sCMOS cameras. The reaction kinetics (corium fragmentation) imposes a high image frequency in order to decompose the reaction process. Each camera reads up to 50 Hz, and the LINAC pulse frequency is set so that there is one image per pulse (200 Hz at most).

The Fukushima accident in 2011 has launched new effort in the field of severe accidents study. A new platform named PLINIUS-2 is planned in Cadarache, and will be dedicated to larger-scale experiments for corium study (higher corium masses) [5]. In addition to water reactors, the PLINIUS-2 field of study will cover future sodium-cooled fast reactors under design by CEA.

High-energy X-ray imaging will be required on three experiments of PLINIUS-2: the test sections for water-corium interaction and sodium-corium interaction study, and the stratification process study. Higher attenuation than in KROTOS is expected on these bigger test columns. In this scope, the design of imaging set-ups is carefully examined and

All authors work at CEA, DEN, Cadarache, F-13108 Saint-Paul-lez-Durance, France.
L. Berge, N. Estre, D. Tisseur, E. Payan and D. Eck are with Nuclear Measurement Laboratory (LMN).
V. Bouyer, N. Cassiaut-Louis, C. Journeau, R. Le Tellier are with Severe Accidents Physics and Modeling Laboratory (LPMA).
E. Pluyette is with Instrumentation and Technological Assays Laboratory (LIET).

optimized, in order to reach the desired performances in terms of detection limit, noise level, and contrast. The imaging design for PLINIUS-2 is made on a simulation basis, with the KROTOS imaging performances as a target (millimetric detection limit, relative noise level about 1%). The simulation is based on the MODHERATO code, which produces simulated radiographies, quantitatively comparable to the images obtained by a real imaging system. Section II will describe the simulation principle, and validation through confrontation to real images taken on KROTOS. Then, application to the design and imaging performances previsions for the future PLINIUS-2 imaging set-ups will be discussed in section III. Section IV will finally present the conclusion and perspectives of this work.

II. Upgrade of the simulation and validation

The MODHERATO code has been developed by CEA/LMN for high-energy imaging applications in radiography and tomography. In the view of the PLINIUS-2 design studies, the simulation has been improved and validated on an experimental basis, in order to allow fine quantitative prediction of the future imaging system performances. We present in this chapter the simulation principle of MODHERATO applied to point-source radiography, as well as the confrontation with experimental images on the case of KROTOS.

A. Simulation in cone-beam radiography

Computation of the radiographic signal includes two distinct steps. The first step is the exponential attenuation of the X-ray beam through the matter up to the detector. This step requires the description of the angular and energetic distributions of the X-ray source. A Monte Carlo computation is needed using the MCNP-6 code [6], describing the electron beam and conversion target. The calculation is done once for all for a given LINAC. The description of the geometrical objects present in the path of the X radiation to the detector is also needed, with their elementary composition and density (description of the test column, and description of corium and vapor inside the column for example). Finally, the tabulated data for the photon-matter interaction cross-sections [7] and the distances between source, object and detector (magnification factor) are required.

The attenuated flux $I(y,z)$ incident on point (y,z) of the 2D detector is related to the source flux $I_0(y,z)$ emitted in direction of point (y,z) by:

$$\ln\left(\frac{I_0(y,z)}{I(y,z)}\right)=\sum_m \mu_m e_m$$

where, for each material m on the path from the source to point (y,z), μ_m is the attenuation coefficient (in cm^{-1}) and e_m is the thickness crossed by the X-ray in this material.

At this step of the simulation, we get the X-ray flux incident on the detector.

The second part of the simulation is the computation of the detector response. In a KROTOS-like configuration (scintillator and camera coupling), this requires, for a given scintillator material and thickness, a preliminary calculation with MCNP of the detection efficiency ε, and the mean energy deposited by the X-photon as a function of its energy, as well as the associated dispersion ("swank noise"). It also requires the scintillation yield, the de-magnification factor (phosphor screen dimensions over camera sensor dimensions), as well as the camera characteristics: electronic yield, readout noise, numerical conversion.

The detected X-ray flux is randomly sampled in a Poissonian law around the mean value $\varepsilon{\cdot}I$ to account for the statistical noise. It is then converted into visible photon flux, then into electron flux also randomly sampled in a Poissonian law around the mean value, then into numerical signal in ADC[1] units. The simulated signal noise, in addition to the statistical fluctuation of X-photons and visible photons counting, accounts for the swank and readout noise sources.

The simulation also accounts for the source spot-size, and for the detector response FWHM for a point-source, resulting in image blur.

The absolute simulated signal in ADC units is quantitatively comparable to experimental raw signals taken on the corresponding configuration without any background subtraction or normalization.

B. Scattered X-ray flux contribution

The use of a 2D phosphor plate induces high contribution of the scattered X-ray flux, *i.e.* photons that encountered trajectory deviation due to Compton scattering inside the test column. The signal induced by these photons will be misinterpreted as coming directly from the source.

The MCNP code is used to evaluate the bias on the X-ray flux incident on the detector. At each point of the detector, MCNP gives the proportion of X-photons that encountered at least one scattering event on their path from the source to the detector. A map of scattered X-ray flux $I_s(y,z)$ on each point of the detector is then constituted once for all in a given configuration. Scattering due to objects inside the column is considered as high as scattering due to the background (water or sodium for PLINIUS-2). In the presence of corium, the scattering flux will then be slightly overestimated.

The MODHERATO code has been upgraded to take the scattered-photon flux map on the detector surface as an input, and add it to the attenuated direct flux I. The total flux on the detector surface corresponds to both direct and scattered flux contributions: $I_{tot}(y,z) = I(y,z) + I_s(y,z)$. Fig. 2 shows the ratio of direct over scattered flux on the Gadox plate in KROTOS, given by MCNP. This ratio is around 2, *i.e.* the scattered flux represents about 1/3 of the total flux. It is interesting to note that the direct flux I on the detector is peaked on the image center, due to the LINAC emitting preferentially forward, whereas the scattered flux is rather spread on the image. The scattering is therefore less disadvantageous in the image center.

[1] ADC: Analog to Digital Conversion.

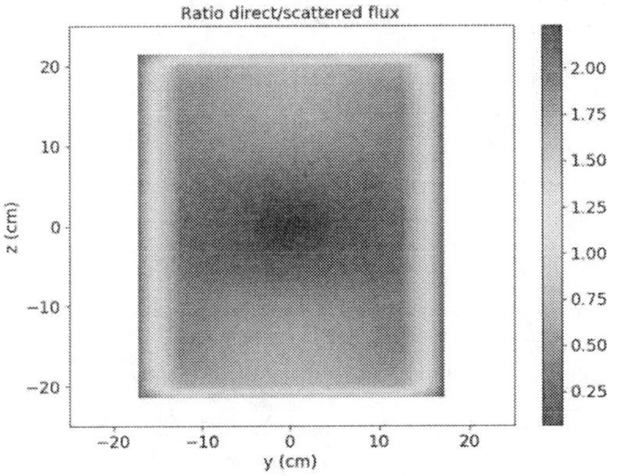

Fig. 2. Ratio of direct flux over scattered flux on the Gadox screen for the KROTOS water test section (MCNP simulation).

C. *Validation on KROTOS images*

KROTOS is a good representation of the future PLINIUS-2 imaging challenges. Especially for the corium-coolant interaction (interaction of corium with water or sodium), the future experiments are extended from KROTOS. Images taken on KROTOS in small-field configuration with a known phantom were used to validate the simulation. The phantom consists in a polyethylene cylinder of density 0.95, representing water (similar density and attenuation power), containing various steel balls of known positions and diameters (see Fig. 3). This phantom was created originally to estimate the detection limit of the set-up.

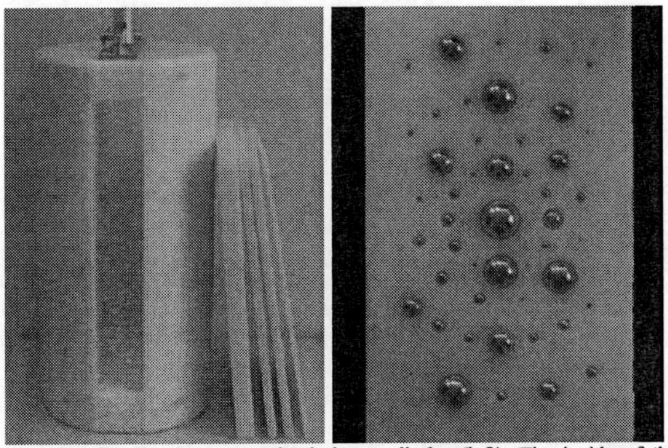

Fig. 3. Steel balls phantom: polyethylene cylinder (left). The inside of the cylinder is filled either with polyethylene or with the plate containing the balls (right).

Fig. 4 presents the real image from one of the four cameras, together with the simulated image using the same grey-scale. Fig. 5 shows the signal on the central horizontal profile (pictured in green on Fig. 4), in the case of real image (raw signal from one camera, in ADC units) and simulated image (simulation signal in ADC units). The simulated profile without scattered flux contribution is also presented to highlight the scattering correction impact.

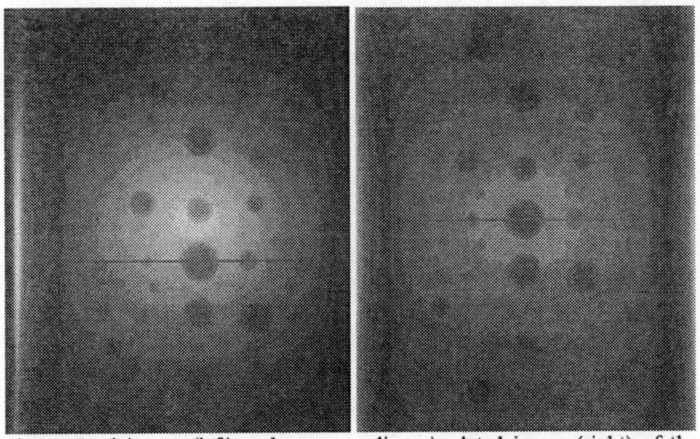

Fig. 4. Real image (left) and corresponding simulated image (right) of the steel balls phantom on KROTOS.

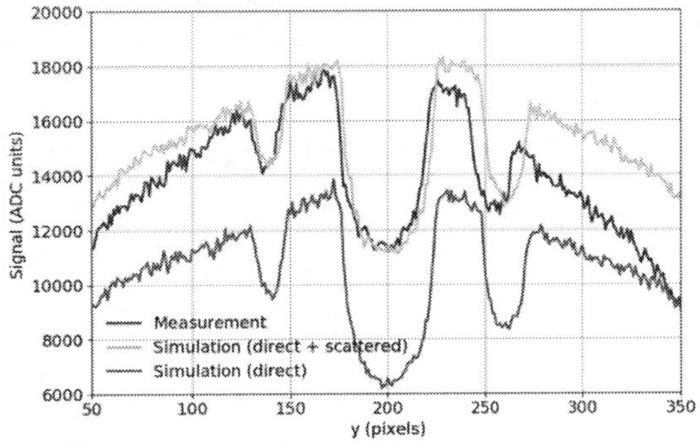

Fig. 5. Comparison between real signal taken on KROTOS and corresponding simulated signal from MODHERATO, with and without scattering, for the green profile of Fig. 4.

The account of scattered flux significantly reduces discrepancy between simulated and measured data as expected. Very good agreement is achieved, especially near the image center. Near the edge of the picture, the measured signal darkens due to a vignetting effect of the camera objective, which is not simulated to date. The noise level is also well reproduced by the simulation: around 1.2% of the total signal.

As it was shown in paragraph A, the simulation only relies on parameters obtained by simulation (MCNP) or known from the cameras documentation. No empirical artefact is added to the calculated signal. Therefore, the good agreement with the measured data is a validation that the simulation is predictive for such systems.

III. Design of the future PLINIUS-2 imaging set-ups and associated predicted performances

The PLINIUS-2 imaging set-ups will rely on the same principle as KROTOS on which the simulation was validated, *i.e.* radiography using a LINAC and phosphor screens coupled to sCMOS cameras. The imaging performances estimation on the different PLINIUS-2 experiments is therefore based on the simulation presented in paragraph II.

A. Corium-coolant interaction study on PLINIUS-2

On the continuation of the KROTOS experiment on PLINIUS, PLINIUS-2 will host bigger test columns for the study of corium-sodium interaction (IC-Na) and corium-water interaction (IC-E). A detailed study of the imaging performances expected on IC-Na has already been performed based on the MODHERATO simulation described in paragraph II. A similar design study will be done for IC-E imaging as well. In the following, we will summarize the results obtained for IC-Na and the main characteristics of the future IC-E imaging set-up.
The principle for IC-Na is the same as on KROTOS: sudden hot corium release in a test column, filled with liquid sodium about 800°C instead of water.

The X-ray attenuation in matter depends for a part on the nature of elements composing the material and on the X-ray energy. But for a given X-ray source, the attenuation is mostly related to the surface density, defined as the product of the material density and thickness $\rho \cdot e$. The surface density of KROTOS is 36 g/cm^2, whereas the IC-Na surface density will be 137 g/cm^2, due to a larger sodium column and to important thermal shielding, and despite the liquid sodium density of 0.75. Therefore, the 9 MeV LINAC in use on KROTOS would not give enough output flux for a good quality image. A 15 MeV LINAC will be used, delivering about 100 Gy/min at 1 m. The optical coupling between the Gadox plate and the sCMOS cameras is also optimized. Instead of having 1 camera per acquisition and 4 cameras alternating to achieve a better image frequency, on IC-Na the 4 cameras will be required for each image. This implies a reduction in the image frequency to about 100 Hz, which is an acceptable frequency for the LINAC. The collected optical signal emitted by the Gadox layer is then 4 times higher than with only one camera. Another option is considered for the scintillating material, by replacing the Gadox 2D plate by a set of LYSO ($Lu_{2(1-x)}Y_{2x}SiO_5$:Ce) scintillating crystals jointed together on a 2D screen. These crystals have high detection efficiency for several MeV photons: about 35% for a 2 cm thick LYSO crystal, for about 4% for the current KROTOS 1.4 mm tick Gadox layer. Due to the achievable size for current LYSO crystals available, imaging the whole column width would require a screen of about 30 crystals, *i.e.* about 45 cm x 50 cm, for a radiographic magnification factor of 1.45. Both options (LYSO assembly or Gadox screen) are currently considered and were evaluated by simulation. Additional MCNP calculations were run prior to the MODHERATO simulation: one to get the characteristics of the 15 MeV source (energy and angular distributions), one to get the scattered X-ray flux associated to the IC-Na configuration, and one to describe the LYSO crystals (deposited energy per X-ray energy and swank noise). Fig. 6 shows the expected signal, obtained with MODHERATO, for the same iron balls as on the phantom of Fig. 3 in the PLINIUS-2 IC-Na test section, for both scintillator options. The KROTOS signal is also shown on the same figure for comparison. With the 15 MeV LINAC and 4 cameras instead of 1, we finally get more output signal than we have on KROTOS, despite the high-attenuating sodium section. Moreover, the use of LYSO crystals instead of the Gadox screen significantly increases the signal, and thus, reduces the statistical noise level. We remind here that the noise is also simulated.

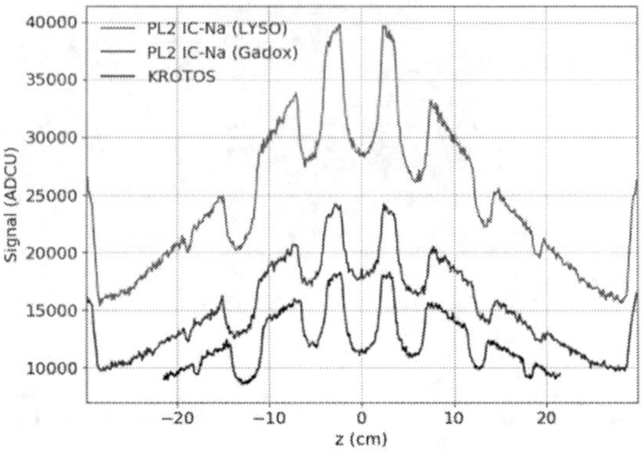

Fig. 6. Expected signal obtained on PLINIUS-2 IC-Na with the same iron balls than on the KROTOS phantom of Fig. 3 (simulation). The KROTOS signal is shown for comparison. The two scintillator options, Gadox and LYSO screens, are shown.

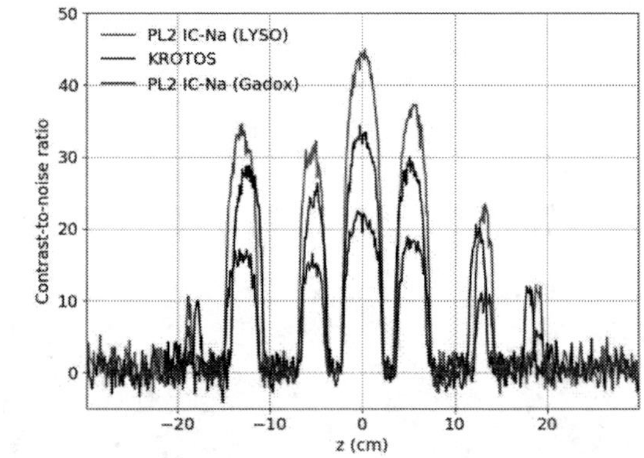

Fig. 7. Contrast-to-noise ratio corresponding to the signals of Fig. 6 for PLINIUS-2 IC-Na (Gadox or LYSO screen) and in the case of KROTOS.

Fig. 7 shows the contrast-to-noise ratio associated to these signals. A contrast of more than 5 times the background noise level (*i.e.* contrast-to-noise ratio superior to 5) is the criterion of object detectability (detection limit). The noise being for a large part related to the X-ray flux, and less dependent on the scintillator-camera coupling, the Gadox configuration leads to a lower contrast-to-noise ratio than on KROTOS where the output signal is lower. In the case of the LYSO screen on the contrary, we get higher contrast-to-noise ratio: around twice the values obtained with Gadox, which leads directly to the ability to detect twice smaller objects. The detection limit is determined using the balls phantom, as the diameter of the smallest detected steel ball. The phantom contains balls of various diameters, especially every 1 mm between 2 mm and 10 mm. The detection limit is about 8 mm for the Gadox solution, which is the same as KROTOS in its large-field configuration. It falls down to 5 mm when using the LYSO crystals, which is the same as KROTOS in small-field configuration. One should note that since the corium

attenuates slightly more than steel, the detection limit for corium will be better.
In both cases (Gadox or LYSO) the use of a 15 MeV LINAC and 4 cameras leads to a detection limit and noise level of the same order than on KROTOS. The design of IC-Na is therefore validated.

The IC-E test section geometry and composition is under design, the associated imaging set-up design will be close to the one for IC-Na. The surface density of the column should not exceed that of the IC-Na test section, around 137 g/cm^2, in order to have a comparable attenuation.

B. Corium stratification experiment

Imaging will also be required for the study of corium stratification. The experiment consists in observing the evolution of different corium phases inside a crucible of diameter about 30 cm. A feasibility study for a real-time X-ray imaging with a framerate of 1 image/s has been performed using MODHERATO, accounting for scattering, based on corium behavior previsions. The corium cylinder is decomposed into 33 slices of different chemical composition and density. The time evolution of these 33 slices is discretized into 17 configurations. Fig. 8 shows images calculated with MODHERATO of 3 typical configurations using a 15 MeV LINAC, and the same detector as on KROTOS (Gadox coupled to 1 camera per image). The interfaces between different type of corium phases (oxide, light metal, heavy metal) are detectable, as shown on Fig. 9. A more detailed study is planned in a close future to establish the ability of high-energy X-ray imaging to study the shape evolution of these interfaces.

Fig. 8. Evolution of the different corium phases (simulation radiography).

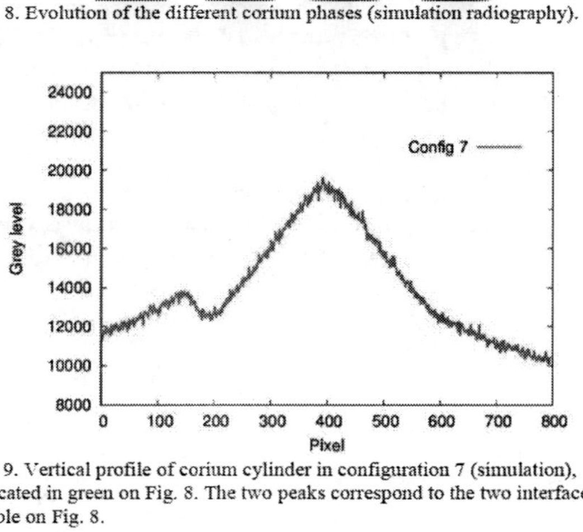

Fig. 9. Vertical profile of corium cylinder in configuration 7 (simulation), indicated in green on Fig. 8. The two peaks correspond to the two interfaces visible on Fig. 8.

IV. Conclusion

In this paper, we have presented real-time X-ray imaging which is a high value-added feature of severe nuclear accidents study performed at CEA in Cadarache. Such an imaging set-up is already used on the PLINIUS platform with success, and will be extended for the future PLINIUS-2 platform large-scale test facilities, on which higher attenuation is expected. These new imaging systems will require the use of a more powerful 15 MeV source, instead of the current 9 MeV LINAC.
We described the simulation principle, using the MODHERATO code developed at CEA Cadarache, from the description of the source spectrum, object geometry and detector characteristics, to obtain a simulated "raw" signal in the same units as the signal from the cameras. The simulation includes calculation of statistical noise, due to X-photons and scintillation photons, swank and readout noises, as well as image blur due to the source spot-size and detector unsharpness. The simulated signal also accounts for the scattered flux contribution, calculated beforehand using the MCNP transport code. We also presented a comparison between measured and simulated "raw" signals, in the case of a phantom used on KROTOS. Since excellent agreement is achieved, the simulation was validated. The imaging performances expected on PLINIUS-2 were then estimated using the simulation. A design study for the future IC-Na imaging set-up was performed using MODHERATO, which confirms performances of the order of the KROTOS imaging quality. A feasibility study has also been performed for high-energy X-ray imaging for corium stratification study, giving satisfactory results. A similar study is planned for the imaging design on the future IC-E test section.

References

1. C. Journeau, "*Severe accident research at the PLINIUS prototypic platform*" in ICAPP'05., Seoul., Korea, 2005, pp. 1074
2. M. Bonnet *et al.*, "*KROTOS FCI Programme at CEA Cadarache*",. ICAPP-05, Seoul, KOREA, May 15-19, 2005, Paper 5381
3. N. Estre *et al.*, "*Fast Megavoltage X-Rays Radioscopy*", proc. of IEEE-NSS/MIC, Strasbourg, 2016
4. S.-W. Hong et al., "*Status of the OECD-SERENA Project for the Resolution of Ex-vessel Steam Explosion Risks*", Journal of Energy and Power Engineering **7** (2013), pp 423-431
5. Ch. Journeau *et al.*, "*Corium-sodium and corium-water fuel-coolant-interaction experimental programs for the PLINIUS-2 prototypic corium platform*", NURETH-17, Xi'an, China, Sept. 3-8, 2017 (to be published)
6. J. Goorley *et al.*, "*Initial MCNP6 Release Overview - MCNP6 version 1.0*", LA-UR-13-22934 NEA report, 2013
7. M.J. Berger and J.H. Hubbell, XCOM: Photon cross-section database: https://www.nist.gov/pml/xcom-photon-cross-sections-database

CEA's Optical Pyrometry Technique for Non-Contact Temperature Measurement in High Temperature Surroundings

B. Bouvry, G. Cheymol, C. Gallou, H. Maskrot, C. Destouches, L. Ferry, and C. Gonnier

The LORELEI (Light-water One-Rod Equipment for Loca Experimental Investigation) device will be integrated into the future experimental Jules Horowitz Reactor [1] of the CEA to provide a feedback on potential nuclear accidents, for example the Loss-Of-Coolant Accident (LOCA) [2]. This device will allow investigating the burst conditions of nuclear fuel claddings when the environment of the cladding, as liquid pressurized water during "normal" nuclear activity, becomes a steam atmosphere. In case of LOCA, the heat produced by the residual power is not evacuated and the cladding surface temperature increases drastically. In the LORELEI device, the thermal contribution of the surrounding rods will be reproduced by using a heating Inconel tube placed around the cladding under test.

To study in detail the temperature profile of the cladding before the burst, the surface temperature of the cladding has to be monitored by a non-intrusive technique avoiding a modification of the thermal environment and, consequently, of the burst conditions. The accuracy of the measurement has to be close to ten degrees.

In preliminary studies, we designed a device and a data treatment procedure able to measure the surface temperature of Zircaloy-4 (Zy-4) claddings, without any contact, under air, up to about 850°C [3, 4]. Two pyrometry methods were investigated: the bichromatic calculation and the Multispectral Radiation Thermometry (MRT) estimation [5]. This latter measurement method revealed an uncertainty < 10°C in the temperature determination in the experimental conditions of the laboratory bench. We selected the spectral ranges 1.0–1.3 μm and 1.45–1.6 μm for the pyrometry measurements according to the future LORELEI conditions [1]. Thermocouples were used to provide a reference temperature and allowed us to validate the constant emissivity profile hypothesis in the spectral ranges of calculation. Emissivity values ranging between 0.85 and about 0.90 were determined, in agreement with the literature [6].

In high temperature surroundings, it appears obvious that a contribution from the surroundings will increase the energy flow collected by the optical fiber, but its impact has not been clearly studied. By using the developed MRT procedure, the objectives of this follow-up study are thus:

(i) To perform a theoretical analysis of the impact of high temperature surroundings on the temperature estimated by MRT, by applying a model of radiative transfers on the conditions of the LORELEI device (i.e., surface properties, temperature range). The aim is to determine potential correction factors to be applied to the retrieved temperature when the pyrometry method will be implemented in the LORELEI device.

(ii) To validate the radiative transfer model by comparing simulated temperature estimations with experimental pyrometry measurements.

Zy-4 is known to present surface modifications under air, for temperatures higher than about 700°C, such as the creation of zirconium nitride [7, 8]. This phenomenon will thus occur in conditions of high temperature surroundings. It was not observed in our previous studies because of the short duration of the experiments and a cladding environment at room temperature. Therefore, measurements will also be performed on an Inconel surface, a material with a more stable surface at high temperature and an emissivity similar to Zy-4 in the temperature range of the experiment.

The simulation of the temperature measurement took into account the major radiative components present in the LORELEI device, with the direct flux from the cladding and all the reflected contributions induced by the presence of a heating Inconel tube to introduce a high temperature surroundings. In the simulations, we noted a temperature difference between the cladding temperature and the temperature estimated by the MRT method as function of the surroundings temperature. These temperature differences will have to be corrected to approach the real temperature of the cladding surface, when the pyrometry method will be implemented in the LORELEI device.

The pyrometry measurements, performed on an Inconel

B. Bouvry, G. Cheymol, C. Gallou, and H. Maskrot are with the Den – Service d'Etudes Analytiques et de Réactivité des Surfaces (SEARS), CEA, Université Paris-Saclay, F-91191, Gif sur Yvette, France (corresponding author: guy.cheymol@cea.fr).
C. Destouches is with the Den – Service de Physique Expérimentale (SPEX/LDCI), CEA, F-13108, St Paul lez Durance, France.
L. Ferry is with the Den – Service Réacteur Jules Horowitz (SRJH/LEDI), F-13108, St Paul lez Durance, France.
C. Gonnier is with the Den – Service Réacteur Jules Horowitz (SRJH), F-13108, St Paul lez Durance, France.

surface, presented the same behavior of temperature identification as the simulation, because of the good stability of the surface properties in high temperature surroundings. This surface stability is similar to the Zy-4 cladding stability in the LORELEI environment. The good agreement found between experimental and simulated results validates the radiative transfer model. This good agreement was not found for Zy-4 cladding under air, because of multiple modifications of its surface properties (nitride and secondary oxide formation) during the experiment. These modifications are not representative of the Zy-4 cladding behavior into the LORELEI device because a neutral gas will be used instead of air.

The optical pyrometry method clearly demonstrated its good potential for the temperature measurement of a stable surface in high temperature surroundings and under air. The outlooks of the development will be focused on:

(i) The validation of the pyrometry method for cladding temperatures up to 1200°C and in steam atmosphere, using the EDGAR-2 test facility available in the CEA research center (Saclay, France).
(ii) The validation of the method under irradiation (neutron and gamma fluxes) taking into account a structural modification of the optical fiber during experiment and the Cherenkov radiation.
(iii) The implementation of the pyrometry device in the LORELEI device, after miniaturization at the millimeter scale.

Acknowledgments

The authors are indebted to Ms. Lezaud-Chaillioux and Vandenberghe, and Mr. Hellouin de Menibus for providing cladding samples.

References

[1] L. Ferry, D. Parrat, C. Gonnier, C. Blandin, Y. Weiss, and A. Sasson, "The LORELEI test device for LOCA experiments in the Jules Horowitz reactor," in *WRFPM 2014 conference*, Japan, 2014.
[2] NEA, "Nuclear fuel behaviour in loss-of-coolant accident (LOCA) conditions - State-of-the-art report," 2009.
[3] B. Bouvry, G. Cheymol, L. Ramiandrisoa, N. Horny, T. Duvaut, C. Gallou, *et al.*, "Optical pyrometry measurement on oxidized Zircaloy-4 cladding," *Journal of Physics: Conference Series,* vol. 745, 2016.
[4] B. Bouvry, G. Cheymol, L. Ramiandrisoa, B. Javaudin, C. Gallou, H. Maskrot, *et al.*, "Multispectral pyrometry for surface temperature measurement of oxidized Zircaloy claddings," *Infrared Physics and Technology,* vol. 83, pp. 78-87, 2017.
[5] H. P. Gavin, "The Levenberg-Marquardt method for nonlinear least squares curve-fitting problems," *Department of Civil and Environmental Engineering,* 2016.
[6] P. M. Mathew, M. Krause, D. M., and M. H. Schankula, "Emittance of Zircaloy-4 sheath at high temperatures in argon and steam atmospheres," in *10. annual conference of the Canadian Nuclear Association*, Ottawa, Canada, 1989.
[7] M. Lasserre, O. Coindreau, M. Pijolat, V. Peres, M. Mermoux, and J.-P. Mardon, "Study of Zircaloy-4 cladding air degradation at high temperature " in *21st International Conference on Nuclear Engineering ICONE21*, Chengdu, China, 2013.
[8] C. Duriez, D. Drouan, and G. Pouzadoux, "Reaction in air and in nitrogen of pre-oxidised Zircaloy-4 and M5™ claddings," *Journal of Nuclear Materials,* vol. 441, pp. 84-95, 2013.

Analytical measurements of fission products during a severe nuclear accident

D. Doizi[1]; S. Reymond la Ruinaz[1] ; I. Haykal[1] ; L. Manceron[2, 3] ; A. Perrin[4] ; V. Boudon[5]; J. Vander Auwera[6] ; F. Kwabia tchana[7] ; M. Faye[2]

Abstract– **The Fukushima accident emphasized the fact that ways to monitor in real time the evolution of a nuclear reactor during a severe accident remain to be developed. No fission products were monitored during twelve days; only dose rates were measured, which is not sufficient to carry out an online diagnosis of the event. The first measurements were announced with little reliability for low volatile fission products. In order to improve the safety of nuclear plants and minimize the industrial, ecological and health consequences of a severe accident, it is necessary to develop new reliable measurement systems, operating at the earliest and closest to the emission source of fission products. Through the French program ANR « Projet d'Investissement d'Avenir », the aim of the DECA-PF project (diagnosis of core degradation from fission products measurements) is to monitor in real time the release of the major fission products (krypton, xenon, gaseous forms of iodine and ruthenium) outside the nuclear reactor containment. These products are released at different times during a nuclear accident and at different states of the nuclear core degradation. Thus, monitoring these fission products gives information on the situation inside the containment and helps to apply the Severe Accident Management procedures. Analytical techniques have been proposed and evaluated. The results are discussed here.**

***Index Terms*—severe accident, fission products, xenon, krypton, iodine, ruthenium, molecular spectroscopy.**

I. Introduction

The Fukushima accident emphasized the need for real time monitoring of fission products [1] to get information on the situation inside the containment in the case of severe accidents. The aim of the DECA-PF project (diagnosis of core degradation from fission products measurements) launched through the French program ANR « Projet d'Investissement d'Avenir », is to propose analytical techniques able to measure major fission products which are released at different time during a nuclear accident and at different states of the nuclear core degradation. Gamma spectrometry is used as the reference technique. A prototype has been designed and will be implemented at the outlet of the U5 filter. Besides Gamma spectrometry, molecular absorption spectrometry has been proposed for the measurement of gaseous forms of iodine and ruthenium. It allows on line chemical speciation of the gaseous forms of fission products and could enable their remote sensing. It has been evaluated and is the subject of this paper.

- For fission gases such as xenon and krypton, micro gas chromatography (micro GC) measurements have been proposed and characterized from the point of view of detection limits and ease of installation on the chimney of the nuclear plant.
- For molecular and organic iodine measurements, optical techniques such as Fourier Transform Infrared (FTIR) spectrometry and Incoherent BroadBand Absorption Spectroscopy (iBBCEAS) have been proposed. To face the low detection limits needed for organic iodine, photoacoustic spectroscopy has also been proposed and evaluated.
- For gaseous ruthenium measurements, Fourier Transform InfraRed (FTIR) spectrometry has been proposed and evaluated.

II. Severe accident

The severity of the accident depends on the degree of damage of the fuel and the degree of confinement loss. During a severe accident, various physical phenomena may occur and lead to the release of fission products from the fuel to the containment [2]:

- emptying of the core ("uncovering" or "dewatering"),
- heating of the fuel rods, and then oxidation of the cladding by the water steam (very exothermic reaction),
- degradation of the fuel rods and rupture of the claddings,

[1]Den-Service d'Etude du Comportement des Radionucléides (SECR), CEA, Université Paris-Saclay, F-91191, Gif-sur -Yvette, France.
[2]Société Civile Synchrotron SOLEIL, L'Orme des Merisiers, St-Aubin BP48, 91192 Gif-sur-Yvette Cedex, France.
[3]MONARIS, UMR 8233, Université Pierre et Marie Curie, 4 Place Jussieu, case 49, F-75252 Paris Cedex 05, France.
[4]LMD UMR 8539, ARA/ABC(t)/LMD/IPSL Ecole Polytechnique RD 36, 91128 Palaiseau Cedex, France.
[5]Laboratoire Interdisciplinaire Carnot de Bourgogne, UMR 6303 CNRS-Université de Bourgogne, 9 avenue Alain Savary, BP 47870, F-21078 Dijon Cedex, France.
[6]Service de Chimie Quantique et Photophysique C.P. 160/09 Université Libre de Bruxelles, 50 avenue F.D. Roosevelt - 1050 Bruxelles - Belgique
[7]Laboratoire Inter-Universitaire des Systèmes Atmosphériques (LISA), UMR 7583, CNRS et Universités Paris Est et Paris Diderot, Institut Paul Simon Laplace, 61 Av. du Général de Gaulle, 94010 Créteil, France.

• fusion of materials (claddings, fuels, structural materials),
• volatilization and release of fission products, which are transported in gaseous or aerosol forms, in the primary circuit, by the hot gases from the core (steam and hydrogen), and then deposited inside the primary circuit and / or transferred in the containment,
• breakthrough of the vessel and erosion of the basemat, due to the corium-concrete interaction,
• transfer of certain fission products, in gaseous or aerosol forms outside of the containment, causing radioactive contamination of the environment (source term). Filtered containment venting systems are implemented in order to avoid a large containment failure and to minimize the release of radioactive materials into the environment.

Another phenomenon can play a key role in the development of a severe accident: an air inlet into the reactor vessel [3]. Studies [4] [5] showed that, just after the breaking of the vessel bottom, air is expelled from the cavity to the surrounding areas due to high flows of gas from the corium-concrete interaction (H_2, CO_2, CO).

The contact between irradiated fuels and air can lead to the production of volatile species of ruthenium especially RuO_4.

III. Xenon and krypton measurements

In the case of the rupture of fuel claddings, fission gases are produced particularly xenon and krypton. The use of molecular spectrometry to measure on line these gases is not easy and would lead to rather complex experimental setups. A robust technique based on gas chromatography has been evaluated. A micro GC Agilent 490 has been chosen to measure xenon and krypton in various gas mixtures containing well known quantities of xenon and krypton. The linearity of the detector response versus concentration has been measured and a detection limit has been estimated for each gas.

A value of ~1ppm has been measured for both gas. This value is in good agreement with the requirements of the source term calculated in the project.

The same technique has also been used to measure CO and CO_2 gases, signatures of an interaction between the corium and the concrete at the ppm level, although molecular spectroscopy could have also been used in this particular case.

IV. FTIR measurements

The Fourier transform infrared spectrometry (FTIR) was chosen to measure gaseous organic iodine and ruthenium tetra-oxide. The species are infrared active and shows absorption bands around 10 microns, a spectral window interesting for atmospheric measurements. This technique is robust, panoramic, and can be used on line. It also allows quantitative measurements when spectral data (positions, absorption cross sections) are available. In our case, such data are unknown and it was necessary to determine them. For that purpose, high resolution infrared measurements were carried out for CH_3I, $^{102}RuO_4$ and natural RuO_4 with Bruker IFS 125 HR Fourier transform spectrometer available at the infrared AILES beamline at the French synchrotron facility SOLEIL, relying on synchrotron radiation, and at LISA Créteil. Theoretical calculations and experimental measurements have been performed to characterize line positions and intensities for each molecule, in collaboration with ICB Dijon, LISA Créteil and UL Bruxelles. The results of this experimental and theoretical work have been published [6-9].

FTIR spectrometry can be used to measure ruthenium in gaseous forms. For organic iodine, the low absorption cross section and the low quantities to be measured would require a very long optical path length to satisfy the specifications of the source term. An alternative way to measure organic iodine was evaluated: photoacoustic spectroscopy.

V. Photoacoustic measurement of organic iodine

Photoacoustic spectroscopy is a robust and sensitive technique. It relies on the absorption of a modulated laser beam by the gas to be studied. In the infrared domain, a vibrational excitation occurs which is followed by a collisional de-excitation creating a transient warming and hence an acoustic wave. This acoustic wave is measured with a microphone.

Our setup uses a CW tunable CO_2 laser, Lasy 4G from Access Laser Company, which is triggered at ~1 kHz.

The CO_2 output wavelength is tuned to an absorption band of CH_3I and sent in the photoacoustic cell made by Aerovia, Reims, France. The acoustic signal is measured using a synchronous detection at the modulation frequency of the laser. A mixture of 5 ppm CH_3I in nitrogen, Air Products, is used. A detection limit of about 30 ppb has been measured. This value meets the requirements of the Source term. This technique may also be used to measure RuO_4 and iodine. Work is in progress to validate these possibilities.

VI. Molecular iodine measurement

Molecular iodine is not infrared active but it exhibits a strong absorption spectrum in the visible range. Due to the low values of the source term, a sensitive technique is necessary to measure iodine.

Incoherent BroadBand Cavity Enhanced Absorption Spectroscopy (iBBCEAS) has been proposed and is under evaluation. This technique has already been successfully used in nuclear [10] and environmental studies [11]. The experimental setup is composed of a green LED (Light Emitting Diode), LuxeonStarLEDs, which is stabilized in current and in temperature using Arroyo instruments drivers. It delivers light between 500 and 600 nm. Light is focused in a high finesse optical cavity. The transmitted light is focused on an Ocean Optics HR 2000 spectrometer and analyzed. The experimental set up is under evaluation at the laboratory.

VII. Conclusion

Source term specifications can be achieved using molecular absorption spectrometry, especially visible and FTIR spectrometries. Results of this work can be used to design new specific sensors to detect and quantify quickly and precisely the amount of fission products at the chimney of a nuclear reactor in case of filtered containment decompression.

The work is in progress in order to determine if one single prototype could measure all gaseous fission products together. This prototype could rely on airborne photoacoustic spectroscopy.

The main advantage of this solution would be to activate the prototype only in the case of severe accidents.

Acknowledgment

The DECA-PF project, launched end of 2013 for 3 years, is sponsored by the French government "Investments for the future" program through the grant ANR-11-RSNR-0003 supervised by the French National Research Agency (ANR) under the "Research in Nuclear Safety and Radioprotection" (RSNR) research initiative.

References

[1] Report of the IAEA expert mission in June 2011

[2] C. Mun, "Etude du comportement du produit de fission ruthénium dans l'enceinte de confinement d'un réacteur nucléaire, en cas d'accident grave", PhD, Université de Paris XI (2007).

[3] Powers D., Kmethyk L.N. and Schmidt R.C., "A review of the technical Issues of air ingression during severe reactor accidents", NUREG/CR-6218, 1994.

[4] Seropian, C., "Analysis of the potential for in-vessel air ingress during a severe accident in a PWR 900 MWe", Note technique IRSN/DPAM/SEMIC/LEPF, 2003.

[5] Freydier, P., Rousset, J.L., "Evaluation of Air Ingress in the Reactor Vessel with the SATURNE Code-SARNET-ST-P19"- EdF n° HI-83/05/006/A, 2006.

[6] H S. Reymond-Laruinaz, M. Faye, V. Boudon, D. Doizi, L. Manceron, "High-resolution infrared spectroscopy and analysis of the v2/v4 bending dyad of ruthenium tetroxide", J. of Mol. Spectroscopy, 336, (2017), 29-35.

[7] A. Perrin, I. Haykal, F. Kwabia Tchana, L. Manceron, D. Doizi, G. Ducros, "New analysis of the v6 and 2v3 bands of methyl iodide (CH_3I)", J. of Mol. Spectroscopy, 324, (2016), 28-35.

[8] I. Haykal, D. Doizi, V. Boudon, A. el Hilali, L. Manceron, G. Ducros, "Line positions in the v6=1band of methyl iodide: validation of the C3v TDS package based on the tensorial formalism", I. Haykal, D. Doizi, V. Boudon, A. el Hilali, L. Manceron, G. Ducros, J. of Quant. Spectros. and Radiat. transfer, 173, (2016), 13-19.

[9] I S. Reymond-Laruinaz, V. Boudon, L. Manceron, L. Lago, D. Doizi,"Infrared Spectrocopy of Ruthenium Tetroxide and High-resolution analysis of the v3 band", J. of Mol. Spectroscopy, 2015, 315 (2015) 46-54.

[10] O. Johansson, H. Mutelle, A. E. Parker, S. Batut, P. Demaux, C. Schoemaecker and C. Fittschen, "Quantitative iBBCEAS measurements of I_2 in the presence of aerosols", Appl. Phys. B, vol 114, pp.421-432, 2014.

[11] S. M. Ball, A. M. Hollingsworth, J. Humbles, C. Leblanc, P. Potin, and G. McFiggans, "Spectroscopic studies of molecular iodine emitted into the gas phase", Atmos. Chem. Phys., 10, 6237–6254, 2010.

EPJ Web of Conferences 170, 08006 (2018) https://doi.org/10.1051/epjconf/201817008006
ANIMMA 2017

Comparison between Silicon-Carbide and diamond for fast neutron detection at room temperature

O.Obraztsova, L.Ottaviani, A.Klix, T. Döring, O. Palais, A. Lyoussi

***Abstract*— Neutron radiation detector for nuclear reactor applications plays an important role in getting information about the actual neutron yield and reactor environment. Such detector must be able to operate at high temperature (up to 600° C) and high neutron flux levels. It is worth nothing that a detector for industrial environment applications must have fast and stable response over considerable long period of use as well as high energy resolution. Silicon Carbide is one of the most attractive materials for neutron detection. Thanks to its outstanding properties, such as high displacement threshold energy (20-35 eV), wide band gap energy (3.27 eV) and high thermal conductivity (4.9 W/cm K), SiC can operate in harsh environment (high temperature, high pressure and high radiation level) without additional cooling system. Our previous analyses reveal that SiC detectors, under irradiation and at elevated temperature, respond to neutrons showing consistent counting rates as function of external reverse bias voltages and radiation intensity. The counting-rate of the thermal neutron-induced peak increases with the area of the detector, and appears to be linear with respect to the reactor power. Diamond is another semi-conductor considered as one of most promising materials for radiation detection. Diamond possesses several advantages in comparison to other semiconductors such as a wider band gap (5.5 eV), higher threshold displacement energy (40-50 eV) and thermal conductivity (22 W/cm K), which leads to low leakage current values and make it more radiation resistant that its competitors. A comparison is proposed between these two semiconductors for the ability and efficiency to detect fast neutrons. For this purpose the deuterium-tritium neutron generator of Technical University of Dresden with 14 MeV neutron output of 10^{10} n s^{-1} is used. In the present work, we interpret the first measurements and results with both 4H-SiC and chemical vapor deposition (CVD) diamond detectors irradiated with 14 MeV neutrons at room temperature.**

***Index Terms*—4H-SiC, neutron detector, SiC neutron detector, Diamond neutron detector, fast neutron detection**

The research work was supported by the CEA Cadarache

O. Obraztsova, L. Ottaviani, O. Palais are with Université Aix-Marseille, Marseille 13397, France (e-mail: olga.obraztsova@im2np.fr; laurent.ottaviani@im2np.fr; olivier.palais@im2np.fr).

A. Klix is with the Karlsruhe Institute of Technology, Institute of Neutron Physics and Reactor Technology, Karlsruhe 76344, Germany (e-mail: Axel.Klix@kit.edu).

T.Döring is with Technical university of Dresden (e-mail: Toralf.Doering@mailbox.tu-dresden.de

A. Lyoussi is with the Laboratoire Dosimétrie Capteurs Instrumentation, CEA, Cadarache 13108, France (e-mail: Abdallah.Lyoussi@cea.fr).

I. Introduction

NEUTRON detector plays a crucial role in many applications where the neutron spectrometry and flux monitoring are important such as the experiments at the material testing reactors (MTR), high energy particle physics experiments or fusion facilities for plasma diagnostics.

These last years the semiconductor-based detectors have received considerable attention thanks to their fast responses, low operating voltage, high energy resolution and radiation resistance. Most suitable semiconductors for harsh environment applications are Silicon-Carbide (SiC) and diamond thanks to their high atomic displacement energy, which is about 20-35 eV [1] for SiC and 40-50 eV [2] for diamond. The high displacement threshold energy makes them more radiation resistant than the conventional semiconductors such as Silicone (Si), Germanium (Ge) and Gallium Arsenide (GaAs) [3]. Moreover, the wide-bandgap energy and low intrinsic carrier concentration over considerable range of temperature make SiC and diamond interesting for use at high temperature applications.

The capability of the diamond-based detector to perform 14 MeV neutrons spectrometry using the $^{12}C(n,\alpha)^{9}Be$ reaction with the energy resolution of 1.5-6% has been demonstrated by several authors [4-7]. The 14 MeV neutron radiation hardness of diamond-based detectors has also been thoroughly studied by some scientific groups. Angelone *et al.* [8] reported that diamond detector show the stable operation under 14 MeV neutrons irradiation up to the neutron fluence of 4×10^{14} n/cm^2. Amosov *et al.* [9] reported that when the 14 MeV neutrons fluence had reached 2×10^{14} n/cm^2, the characteristics of the diamond-based detector started to degrade. The count rate of $^{12}C(n,\alpha)^{9}Be$ reaction was then decreased by 7%, energy resolution was decreased by a factor of 2.3 and the peak shifting to the left on 1 MeV was observed. Concerning the SiC-based detectors the capability to measure the 14 MeV neutron spectrum with the well resolved (~2.13%, full width at half maximum FWHM=192 keV) $^{12}C(n,\alpha)^{9}Be$ peak was demonstrated by F.H. Ruddy *et al.* [10]. The peak due to the $^{28}Si(n,\alpha)^{25}Mg$ reaction was also observed and the energy resolution of about 2.388% (FWHM=271 keV) was obtained. Moreover, the stable operation of the SiC-based detector was demonstrated during long term repeated irradiation tests with 14 MeV neutrons at different temperatures up to 500°C [11]. However, a gradual degradation of the counting rate was observed. This fact was explained by decreasing of the applied bias voltage at high temperature in order to avoid thermal noise effects. The FWHM value of the $^{12}C(n,\alpha_0)^9Be$ peak increased by 11.8 % at 500 °C (-20V) with respect to the estimated FWHM at room temperature (-280V).

II. Background

The fast neutron detection in SiC and diamond is directly provided by the interactions of the impinging neutron with the ^{28}Si and ^{12}C nucleus via the nuclear reactions which result in the production of the ionizing particles. These particles passing through the semiconductor create the electron-hole pairs which drift in a bias electric field inducing the current.

In the case of the diamond, there are five main channels of neutron-induced reactions on carbon that participate in the detector response:

- elastic scattering $^{12}C(n,n)^{12}C$;
- inelasting scattering $^{12}C(n, n')^{12}C^*$. This reaction takes place if the energy of the impinging neutron is higher than 4.8 MeV;
- $^{12}C(n,n')3\alpha$ with the neutron threshold energy of E_n=8.428 MeV;
- $^{12}C(n,\alpha_0)^9Be$ with the neutron threshold energy of E_n=6.419 MeV. This reaction results in the production of the α particle and the ^9Be recoil nuclei with the energy $E_{\alpha+Be}=E_n$ – 5.7 MeV.
- $^{12}C(n,\alpha_1)^9Be$ with the neutron threshold energy of E_n=8.09 MeV.

In the case of the SiC, the detector response contains the contribution from the reactions with both ^{28}Si and ^{12}C nucleus. Additionally to the reactions mentioned earlier the following reactions will contribute to the detector response:

- elastic scattering $^{28}Si(n,n)^{28}Si$;
- inelastic scattering $^{28}Si(n,n')^{28}Si^*$ which takes place if the neutron energy is higher than 1.8 MeV;
- $^{28}Si(n,\alpha)^{25}Mg$ with the neutron threshold energy of E_n= 2.76 MeV;
- $^{28}Si(n,p)^{28}Al$, with the neutron threshold energy of E_n= 4 MeV.

The most important reaction for fast neutron detection is the $^{12}C(n,\alpha_0)^9Be$ reaction which occurs if the neutron energy is higher than 6.4 MeV. If the neutron flux is monoenergetic this reaction produces a well-defined peak which is useful for fast neutron spectrometry and evaluation of the energy resolution of the measurement system (detector+electronics).

III. Devices

In this work we studied two detectors. The first detector is the 4H-SiC-based detector designed by IM2NP laboratory of the Aix-Marseille University. The second detector is the diamond-based detector produced by CIVIDEC Instrumentation Company.

A. 4H-SiC-based neutron detector

The SiC detector is based on a 4H-SiC p^+n diode which has been fabricated using a 4H-SiC n^+ wafer purchased from Ascatron with a substrate of 350 µm thickness, on which a 20-µm thick n-type epitaxial layer with a low nitrogen doping concentration ~2×10^{14} cm^{-3} was grown. To create the pn-junction a 1-µm thick p+ epitaxial layer with a high aluminum doping concentration ~10^{19} cm^{-3} was deposited on the n-type layer. Next the ohmic contact (Ni/Ti/Al/Ni) with 200 nm thickness has been deposited by ultrahigh vacuum electron beam evaporation. In order to reduce the contact resistance a 1 µm Al metallic contact has been deposited under ultrahigh vacuum electron beam evaporation on the top of the ohmic contact. Then a Ti/Ni/Au 555-nm thick overmetallization was deposited. In the present work the area of the investigated diode was 0.33 cm^2. The schematical representation of this diode is illustrated in Fig.1.

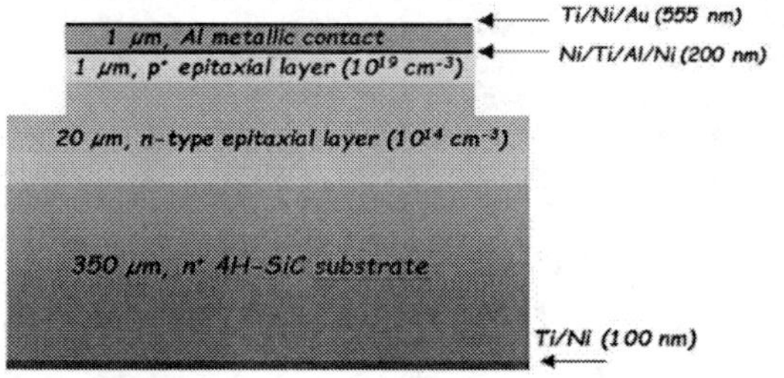

Fig. 1. Schematic representation of the investigated 4H-SiC-based diode.

B. Diamond-based neutron detector

The diamond-based detector has been purchased from CIVIDEC Instrumentation Company. This detector is based on a single crystal CVD (sCVD) diamond of the 500 µm thickness. It has the capacitor-type configuration also named the solid-state ionization chamber [12].

IV. Measurement setup and conditions

The measurements were carried out at the deuterium-tritium (DT) neutron generator at Technical University of Dresden. Two different neutron detectors, one SiC-based and one sCVD diamond-based were used in this study. The experimental conditions were the same for both detectors. The detectors were positioned 13 cm away from the tritium target at an angle of 90° with respect to the deuterium beam. The neutron flux at this point was 9.4×10^6 n/(cm^{2}s) with the neutron energy of 14.12 MeV.

The detectors were connected to the CAEN A1422 charge sensitive preamplifier (CSP) and operated simultaneously. The SiC detector operated at -120V bias voltage while the sCVD diamond at 120V. The bias voltage was provided by CAEN DT 5519M HV bias supply via the charge sensitive preamplifier. The signal output of the CSP was connected to the ORTEC amplifier which amplifies and reshapes the pulses coming from the CPS before digitization in an analog to digital converter (ADC).

https://doi.org/10.1051/epjconf/201817008006

V. Energy calibration

For the energy calibration of the recorded spectrum the peak due to the $^{12}C(n,\alpha_0)^9Be$ reaction was used. The channel numbers of the multichannel analyzer are directly linked to the energy deposited into the active volume of the detector by charged reaction products. The sum of the energies of the α particle and the ^9Be recoil nuclei depends only on the energy of the incoming neutron $E_{\alpha+Be}=E_n - 5.7$ MeV, where E_n is the energy of the incoming neutron, -5.7 MeV corresponds to the Q-value of the $^{12}C(n,\alpha_0)^9Be$ reaction.

In order to calibrate the spectrum the measurements at different angles with respect to the deuterium beam were implemented. The distance from the tritium target was about 13 cm. The energy of the neutrons vary from 13.7 MeV to 14.89 MeV depending on the relative detector position to the deuterium beam. At 90° the neutron energy is 14.12 MeV and the energy distribution is the narrowest. The maximum mean neutron energy of 14.89 MeV is reached at 0°. However at 0° the significant broadening of the neutron energy distribution exist and affect the $^{12}C(n,\alpha_0)^9Be$ peak resolution. The comparative study of the responses of SiC-based and diamond-based detectors to 14 MeV neutrons were then realized at an angle of 90° with respect to the deuterium beam.

The measured $^{12}C(n,\alpha_0)^9Be$ peak positions in the channels and the relevant energies are presented in the Fig.2. for SiC and in the Fig.3. for diamond.

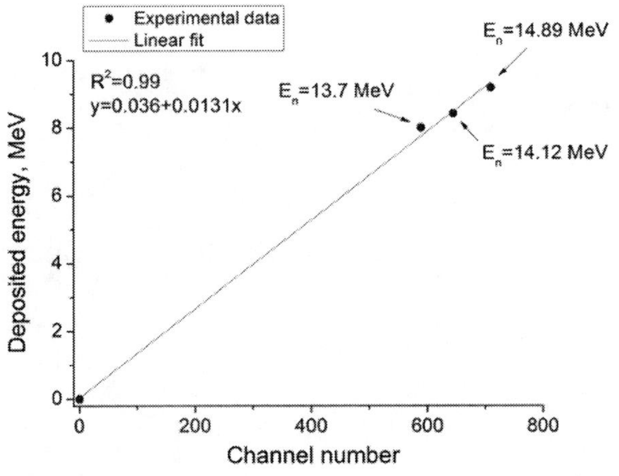

Fig. 2. Energy calibration for the $^{12}C(n,\alpha_0)^9Be$ reaction peak for SiC-based detector.

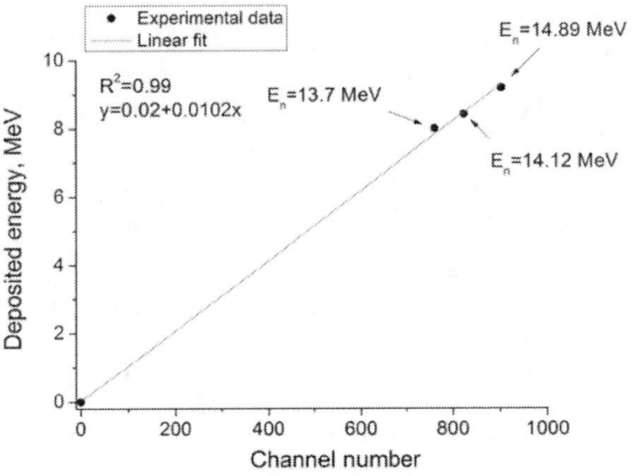

Fig. 3. Energy calibration for the $^{12}C(n,\alpha_0)^9Be$ reaction peak for diamond-based detector.

VI. Results and discussions

The pulse-height spectrum (PHS) of the SiC-based and sCVD diamond-based detectors for 14.12 MeV neutrons are shown in Fig.4. At the low energies below 4 MeV the elastic and inelastic scattering of the neutrons on ^{12}C and ^{28}Si nucleus dominate the response of the SiC detector. Of course, in the case of the diamond only the contribution from the neutron scattering on ^{12}C nucleus exists. At the higher energies the well-defined peak due to the $^{12}C(n,\alpha_0)^9Be$ reaction can be clearly identified for both detectors. For the neutron energy of 14.12 MeV the energy deposited by the α particle and the ^9Be recoil nuclei is about 8.4 MeV. The contribution from the $^{12}C(n,\alpha_1)^9Be$ which results in the production of the charged products with the total energy of 6.72 MeV is obscured by the $^{12}C(n,n')3\alpha$ continuum. In the response of the SiC detector some peaks is observed in the energy range from 7 to 10 MeV which are absent for the diamond and can be explained by the contribution from the multiple branches of the $^{28}Si(n,\alpha)^{25}Mg$ reaction observed by F.H. Ruddy *et al.*[10] for a SiC detector.

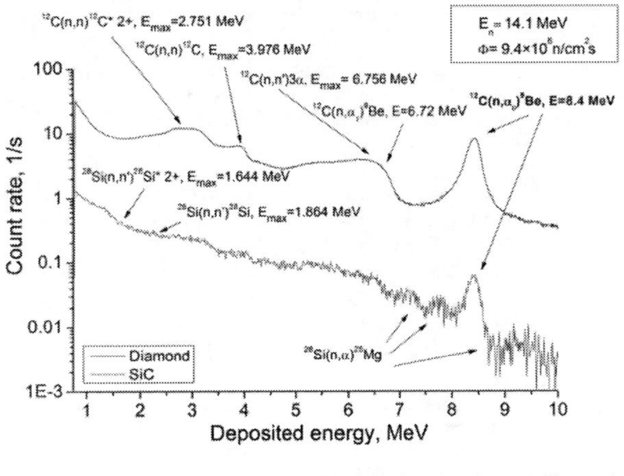

Fig. 4. Comparison of the responses of the SiC-based and sCVD diamond-based detectors to 14.12 MeV neutrons.

The main difference between the two responses is the higher count rate with the diamond-based detector than with the SiC-based one. At the neutron flux of 9.4×10^{6} n/(cm^{2}s) the total count rate of the diamond detector is equal to 4.8×10^{3} c/s and 2.46×10^{2} c/s for the SiC detector. The lower count rate of the SiC can be explained by a smaller thickness of its active detection volume which is limited by the width of the space charge region (SCR). The SCR width depends on the value of the applied reverse-biased voltage. At -120V the SCR is about 21 μm and reaches its maximum value. For the diamond-based detector the thickness of the detection volume is 500 μm, which corresponds to the thickness of the single crystal CVD diamond.

VII. Conclusion

This work has been conducted to compare the responses of the SiC-based and diamond-based detectors of the most commonly used configurations to 14 MeV neutrons irradiation. The linear relation between channel number and energy was obtained for $^{12}C(n,\alpha_0)^{9}Be$ reaction peak for both detectors. Moreover, both detectors showed well-characterized pulse-height spectrum with the well-resolved peak due to the $^{12}C(n,\alpha_0)^{9}Be$ reaction. However, the count rate of the diamond-based detector has been higher than the count rate of the SiC-based detector. This fact is due to the smaller thickness of the active detection volume of the SiC-based detector which is limited to 21 μm.

References

[1] G. Lucas, L. Pizzagalli, "Comparison of threshold displacement energies in β-SiC determined by classical potentials and ab initio calculations", *Nucl. Instr. and Meth. in Phys. Res., B 229*, pp. 359-366, 2005.

[2] J. C. Bourgoin, B. Massarani, "Threshold energy for atomic displacement in diamond", *Phys. Rev. B 14*, pp. 3690-3694, 1976.

[3] F. Nava, G. Bertuccio, A. Cavallini, and E. Vittone, "Silicon carbide and its use as a radiation detector material", *Meas. Sci. Technol.*, vol. 19, no. 10, p. 102001, Oct. 2008.

[4] A. V. Krasilnikov, J. Kaneko, M. Isobe, F. Maekawa, and T. Nishitani, "Fusion neutronic source deuterium-tritium neutron spectrum measurements using natural diamond detectors", *Rev. Sci. Instrum. 68*, pp. 1720–1724, 1997

[5] T.Shimaoka, J.H. Kaneko, K. Ochiai et al.,"A diamond 14 MeV neutron energy spectrometer with high energy resolution", *Rev. Sci. Instrum. 87*, p. 023503, 2016

[6] C. Cazzaniga, M. Nocente, M. Rebai, et al., "A diamond based neutron spectrometer for diagnostics of deuterium-tritium fusion plasmas", *Rev. Sci. Instrum.* 85, p.11E101, 2014

[7] J.H. Kaneko, T. Teraji, Y. Hirai et al.,"Response function measurement of layered type CVD single crystal diamond radiation detectors for 14 MeV neutrons", *Rev. Sci. Instrum.*, vol.75, no. 10, pp. 3581–3584, 2004

[8] M. Angelone, M. Pillon, A. Balducci, «Radiation hardness of a polycrystalline chemical-vapor-deposited diamond detector irradiated with 14 MeV neutrons», *Rev. Sci. Instrum 77,* p. 023505, 2006

[9] V.N. Amosov, S.A. Meshchaninov, N.B. Rodionov, "Тестирование алмазного нейтронного детектора на каскадном сильноточном ускорителе КГ–2,5", *Prikladnaya Fizika*, No. 3, pp. 80-84, 2015 [In Russian]

[10] F.H. Ruddy, A.R. Dulloo, J.G. Seidel et al., "The fast neutron response of 4H Silicon Carbide semiconductor radiation detectors ", *IEEE Trans. Nucl. Sci.*, vol.53, no. 3, pp. 1666-1670, 2006

[11] D. Szalkai, R. Ferone, F. Issa, "Fast Neutron Detection With 4H-SiC Based Diode Detector up to 500 °C Ambient Temperature", *IEEE Trans. Nucl. Sci.*SCIENCE, vol. 63, no. 3, 2016

[12] P. Kavrigin, P. Finocchiaro, E. Griesmayer et al., "Pulse-shape analysis for gamma background rejection in thermal neutron radiation using CVD diamond detectors", *Nuclear Instruments and Methods in Physics Research A 795*, pp. 88–91, 2015

Nuclear Fuel Assay through analysis of Uranium L-shell by Hybrid L-edge/XRF Densitometer using a Surrogate Material

Seunghoon Park, Sungyeop Joung, and Jerry Park

***Abstract*—Assay of L-series of nuclear material solution is useful for determination of amount of nuclear materials and ratio of minor actinide in the materials. The hybrid system of energy dispersive X-ray absorption edge spectrometry, i.e. L-edge densitometry, and X-ray fluorescence spectrometry is one of the analysis methods. The hybrid L-edge/XRF densitometer can be a promising candidate for a portable and compact equipment due to advantage of using low energy X-ray beams without heavy shielding systems and liquid nitrogen cooling compared to hybrid K-edge/XRF densitometer.**

A prototype of the equipment was evaluated for feasibility of the nuclear material assay using a surrogate material (lead) to avoid radiation effects from nuclear materials. The uncertainty of L-edge and XRF characteristics of the sample material and volume effects was discussed in the article.

***Index Terms*—L-edge, XRF, Nuclear Material Solution, X-ray absorption, X-ray fluorescence**

I. Introduction

THE low energy X-ray measurement system is intended to be used for on-site analytical measurements of safeguards in various nuclear facilities, such as nuclear fuel fabrication facilities and reprocessing plants. Specially, the hybrid system of X-ray absorption spectrometry, such as L-edge densitometry (LED) [1, 2], and X-ray fluorescence spectrometry (XRF), has an important role of safeguards for nuclear facilities. LED is a technique of determination of uranium concentration as a continuous X-ray energy beams transmit a uranium liquid sample. Compared to K-edge densitometer [3], due to relatively lower energy (L_{III}-edge energy is 17.17 keV) of uranium L-series energy than K-series energy, L-edge densitometer does not require high purity germanium detector with liquid nitrogen cooling and heavy shielding system for shielding of K-series radiation (~ 115 keV) of uranium. Therefore, the L-edge densitometer is appropriate for a portable equipment for on-site nuclear material inspection and safeguards at facility sites. XRF combined with LED is a technique of characterizing the element content of nuclear materials from fluoresced characteristic X-ray photons.

In this study, the hybrid L-edge/XRF densitometer (HLED) was developed for determination of nuclear materials concentration. We verified that the system has feasibility of determination of the concentration of nuclear material from analysis of a surrogate material such as lead nitrate ($Pb(NO_3)_2$) solution prior to the work of actinide-bearing materials.

II. The Hybrid LED/L-XRF Densitometer

The schematic figure of HLED is shown in Fig. 1. The equipment was fabricated based on previous study [4] of Monte Carlo simulation for the instrument design and feasibility study of determination of nuclear material concentration. The prototype of hybrid L-edge/XRF densitometer (HLED) consists of an X-ray tube, shields, a sample container, and two detectors. Both of detectors are silicon drift detectors (SDD) of 25mm^2 × 500 μm which is XR-123SDD by Amptek™ including a digital pulse processor. The X-ray tube of 1.6 mm anode spot diameter is fabricated in Moxtek™ with maximum power, voltage, and current of 4 W, 50 kV, and 200 μA in front of collimator of a hole diameter of 0.5 mm and length of 5-cm-long. The sample container is made of Teflon to protect corrosion by nitric acid with 2 mm-optical path length and 300 μl volume. The windows material is Mylar® (polyester) film of 6 μm to maximize detection efficiency. There are beam lines for detection of transmission and XRF. The XRF detection beam line is aligned with 45 degrees for the sample window normal and the transmission detection line set up in the direction of the X-ray beam as shown in Fig. 1. The X-ray tube is operated by 30 kV and 90 μA with rhodium (Rh) anode.

X-ray photons from the tube transmit or randomly react to the sample by XRF process. The transmission and fluoresced photon energy spectra are recorded by the detectors. The rhodium anode with characteristic peaks of K_α (20.21 keV) and K_β (22.72 keV) can be a role of calibration of energy below 30

This work was supported by the Nuclear Safety Research Program through the Korea Foundation of Nuclear Safety (KOFONS), granted financial resource from the Nuclear Safety and Security Commission (NSSC), Republic of Korea.

Seunghoon Park is with the Korea Institute of Nuclear Nonproliferation and Control, Yusesong-daero 1534, Yuseong-gu, Daejeon, 34054, Republic of Korea (e-mail: shpark@kinac.re.kr). He is the first author and has correspondence.

Sungyeop Joung is with the Korea Institute of Nuclear Nonproliferation and Control, Yusesong-daero 1534, Yuseong-gu, Daejeon, 34054, Republic of Korea (e-mail: jsy1003@kinac.re.kr).

Jerry Park is with ISP Co., Ltd., 422 Palbok-dong 3-ga, Jeonju-si, Deokjin-gu, Jeollabuk-do, Republic of Korea (e-mail: jerry@ispxrf.com).

 https://doi.org/10.1051/epjconf/201817008007

keV as well as generation of continuous X-ray spectrum, i.e. bremsstrahlung.

The experiments were performed for lead nitrate ($Pb(NO_3)_2$) solution with 0.01, 0.05, 0.1, and 0.2 g/cm^3 to avoid radiation effects for the performance evaluation of the equipment. The blanket material is water in order to obtain transmission.

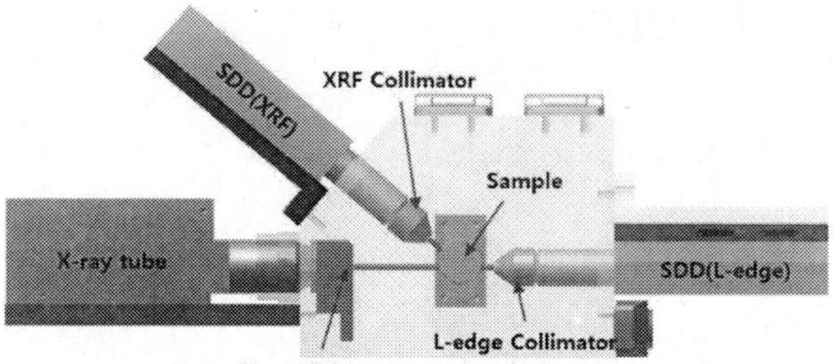

Fig. 1 Design of Hybrid L-edge/XRF Densitometer

III. Evaluation of HLED

A. L_{III}-edge Analysis

The discontinuous curves, i.e. jumps, of transmission spectrum are observed. The height of the jump of the spectrum determines the sample concentrations. Material concentrations through transmission spectrum can be calculated by the following equation (1).

$$\rho_{Pb} = \frac{\ln[T(E_-)/T(E_+)]}{\Delta\mu\Delta} \quad (1)$$

$T(E_-)$, $T(E_+)$ are transmission at the energies E_- (lower than L_{III}-edge) and E_+ (upper than L_{III}-edge). D is sample thickness which is exactly called optical path length and $\Delta\mu$ is mass attenuation coefficient difference at upper and lower L_{III}-edge. Extrapolated fitting in linearized a representation of lnln($1/T$)-ln(E) is applied for determining upper and lower transmission for L-edge. Fitting intervals from 11 – 12 keV for E_- and 14 – 15 keV for E_+.

Relative standard deviation (RSD) represents uncertainty of the measurement as shown in Fig. 2. The samples are measured for 3 × 1000s respectively. The RSD reduces at high concentration (> 0.01 g/cm^3) and slightly increases at 0.2 g/cm^3. The total counts at high concentration are relatively low compared to low concentration. The result implies that low transmission above 0.2 g/cm^3 results in low accuracy. As a result, the optimized concentration can be considered above 0.01 g/cm^3 and below 0.2 g/cm^3 of lead. The possible concentration range have consistency with the previous study for L-edge of uranium and plutonium [5].

B. X-ray fluorescence of surrogate material

In order to determine concentrations of minor actinide or impurities in a nuclear material sample, the analysis requires

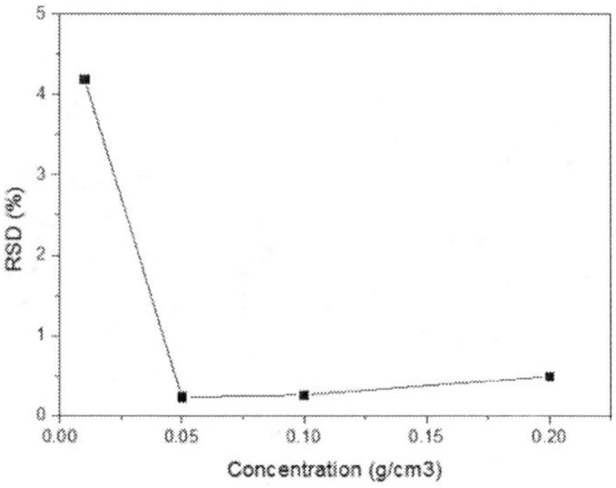

Fig. 2. Relative Standard Deviation (RSD) for each concentration

not only L-edge densitometry but also X-ray fluorescence spectrometry.

The typical XRF peaks are L_α (10.4 keV) and L_β (12.6 keV) of lead. The ratio of L_α and L_β for any concentration is shown in Fig. 3. XRF photons created in the sample are absorbed and interact with lead. The ratio decreases as the concentration increases because the lower energy photons can be more affected by the sample. The relation of the peak intensity and the peak ratio between the concentrations of a sample have to be considered of the XRF analysis for mixed solution samples. Unlike a single element analysis, the analysis for mixed solution samples are required calibration data for target materials using various ratio of elements and concentrations.

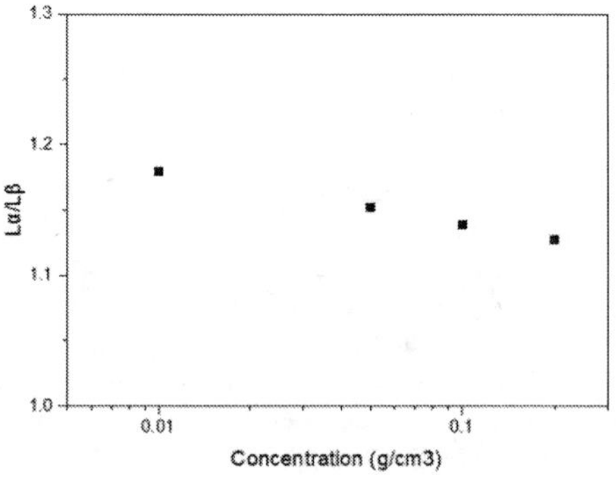

Fig. 3 The peak ratio of L_α and L_β of lead for each concentration

IV. Conclusion

The prototype of a hybrid L-edge/XRF densitometer (HLED) was developed for nuclear material solution assay. The equipment has feasibility due to the uncertainty from L-edge analysis near ITV (International Target Value). The evaluation verified that the concentration and volume effects must be considered for assay of multi-element solution samples.

In future work, based on the study, nuclear material will be analyzed and uncertainty will be reduced by analysis of non-radioactive and radioactive standard samples such as lead nitrate and uranium nitric acid solutions.

References

[1] Analysis Procedure for Timely On-site Uranium Accountancy Verification in LEU FUEL Fabrication Plants", ESARDA BULLETIN No. 43, 18(19), 30-39 2009.

[2] M. L. Brooks, P. A. Russo, J. K. Sprinkle, Jr., "A Compact L-Edge Densitometer for Uranium Concentration Assay", Los Alamos National Laboratory report LA-10306-MS, 1985

[3] H. Ottmar and H. Eberle, "The Hybrid K – Edge/K – XRF Densitometer: Principles – Design – Performance", Kernforschugszenturm Karlsruhe GmbH, Karlsruhe, KfK 4590, Feb. 1991

[4] S. Park, and J.-K. Shin, S.-W. Kwak, "Development of Hybrid L-edge/XRF Densitometer for Determination of Nuclear Material Concentration", 31 October to 7 November, 2015 IEEE Nuclear Science Symposium & Medical Imaging Conference, 2015

[5] D. Reily, N. Ensslin, H. Smih, Jr., and S. Kreiner, Passive Nondestructive Assay of Nuclear Materials, LA-UR-90-732, United States Nuclear Regulatory Commission, 1991

Comparison of CdZnTe neutron detector models using MCNP6 and Geant4

Emma Wilson*, Mike Anderson*†, David Prendergast† and David Cheneler*
*Department of Engineering, Lancaster University, Lancaster, LA1 4YW, UK. E-mail: e.wilson6@lancaster.ac.uk
†Innovative Physics Ltd., Shanklin, Isle of Wight, PO37 7JB, UK.

***Abstract*—The production of accurate detector models is of high importance in the development and use of detectors. Initially, MCNP and Geant were developed to specialise in neutral particle models and accelerator models, respectively; there is now a greater overlap of the capabilities of both, and it is therefore useful to produce comparative models to evaluate detector characteristics. In a collaboration between Lancaster University, UK, and Innovative Physics Ltd., UK, models have been developed in both MCNP6 and Geant4 of Cadmium Zinc Telluride (CdZnTe) detectors developed by Innovative Physics Ltd. Herein, a comparison is made of the relative strengths of MCNP6 and Geant4 for modelling neutron flux and secondary γ-ray emission. Given the increasing overlap of the modelling capabilities of MCNP6 and Geant4, it is worthwhile to comment on differences in results for simulations which have similarities in terms of geometries and source configurations.**

I. Introduction

In order to maintain and improve safety standards within the nuclear sector, more emphasis is placed on novel detector materials and configurations for use in nuclear power, decommissioning, and security, as well as for other applications. It is for this reason that the capability to accurately mathematically model detector response is essential; in the first instance this would be a check of the applicability of the detector for the planned purpose.

Both MCNP6 [1] and Geant4 [2] are widely used nuclear and radiation simulation tools, and both have their specialisations which arise from the background and purpose of their development. An understanding of the difference capabilities and, where appropriate, direct comparisons of these tools is essential, as it give an indication of which is more appropriate in a given circumstance. Previous work in this field has compared the capabilities of MCNP and Geant4 in areas including medicine [3], nuclear power [4], and security [5], although the development of both MCNP and Geant4, and the associated cross-section data, is on-going such that the individual capabilities of the codes vary.

Cadmium zinc telluride (CdZnTe, or CZT) radiation detectors are widely used in γ-ray spectroscopy as a room-temperature, semiconductor-based detector. This is of particular use as CdZnTe detectors possess an improved energy resolution, when compared to scintillation-based radiation detectors, and, unlike high-purity germanium, has a sufficiently wide band-gap to significantly reduce thermal excitation of electrons across the gap, and therefore CdZnTe detectors do not require liquid nitrogen cooling [6].The mechanism for the detection of neutrons with CdZnTe detectors is via thermal neutron capture on ^{113}Cd, resulting in a series of γ-rays which can then be detected in the same manner as with standard γ-ray spectroscopy. CdZnTe neutron detection has been used in a variety of areas, including around medical accelerators [7] and for the detection of special nuclear materials [8].

TABLE I
THE HIGHEST INTENSITY PROMPT SECONDARY γ-RAYS PRODUCED AS A RESULT OF NEUTRON CAPTURE ON A CDZNTE DETECTOR.

Reaction	E_γ (keV)	Reaction	E_γ (keV)
^{112}Cd(n,nγ)	617.5	^{110}Cd(n,nγ)	657.8
^{114}Cd(n,nγ)	558.5	^{111}Cd(n,nγ)	245.4
^{128}Te(n,nγ)	743.2	^{126}Te(n,nγ)	666.3
^{130}Te(n,nγ)	839.4	^{64}Zn(n,nγ)	991.6
^{112}Cd(n,nγ)	606.9	^{113}Cd(n,nγ)	558.3
^{114}Cd(n,nγ)	576.1	^{113}Cd(n,nγ)	651.2
^{128}Te(n,nγ)	753.8	^{111}Cd(n,nγ)	245.3
^{130}Te(n,nγ)	748.8		

II. Materials and Methodology

Both models use a ^{252}Cf neutron source and a CdZnTe detector, in this instance composed of 48% Cd, 2% Zn, and the remainder comprised of Te. Of particular interest is the production of secondary particles within the detector.

The mechanism by which CdZnTe detects neutrons is via neutron capture and the emission of secondary γ-rays from the resulting excited nucleus, which will then result in the production of photoelectrons [9], [10]. ^{113}Cd has an exceptionally high neutron capture cross-section of 20 615 barns at thermal neutron energies [11] [12] and such an event results in characteristic γ-rays of the following energies [13]:

$$^{113}Cd + n \rightarrow ^{114}Cd^{*}$$
$$\rightarrow \gamma_1 \text{ (558.6 keV)} + \gamma_2\text{(651.3 keV)} + \text{further } \gamma$$

Further characteristic γ-rays emitted from CdZnTe from thermal neutron capture are shown in Table I [7], [13], [14]. Most of these γ-ray transitions decay through the first excited state in ^{114}Cd.

The proportions of the constituent materials which make a CdZnTe is important, as this makes a difference to detection efficiency [15].

 https://doi.org/10.1051/epjconf/201817008008

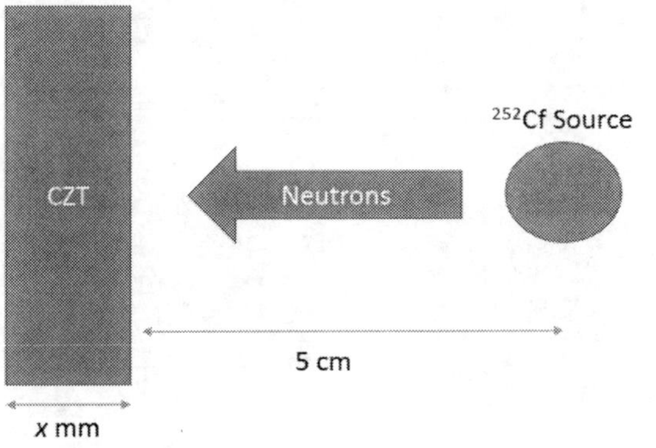

Fig. 1. A simplified sketch of the set-up used in both MCNP and Geant4 simulations.

Figure 1 shows a basic version of the setup used. The neutron source emits either a range of energies from a ^{252}Cf source, which are describes in reference [16]. The ^{252}Cf Lancaster University neutron source may be used in order to best compare theoretical models with experimental results. The thickness of a CdZnTe detector varies, but the value of 1 mm is within the thickness range commonly used [17], [18], [19].

A. MCNP

MCNP6 is a combination of the capabilities of MCNP5 and MCNPX, as well as the further development of both of these codes by groups at Los Alamos National Laboratory, where neutron transport codes have been developed since the 1950s [20]. Later development of photon transport codes, and the merger of the two lead to the release of the MCNP3 (originally Monte Carlo Neutron Photon, now Monte Carlo N-Particle) in 1983.

Figure 2 shows cross-section comparisons of the ENDF/B-VII.1 and JENDL-3.3 cross section files. The cross-sections used in MCNP6 are selected while building the input file, with the cross-section data being selected as the isotopes, or ZAID, of each material are built.

The neutron discrete energy cross section libraries for most nuclei are from the ENDF/B-VII.0 or ENDF/B-VII.1, the latter of which is the most recent. There are various cross-section libraries for reactions taking place at different temperatures [23], [24]. In these simulations, the cross-sections used for each material are listed in table II.

B. Geant4

Geant4 is a Monte Carlo simulation toolkit for the simulation of particles travelling through and interacting with matter. It is a C++ implemented open-source code, developed by a worldwide collaboration of physicists. Geant4 is used within a wide variety of scientific fields, in topics as diverse as high-energy particle physics, nuclear physics, medical science,

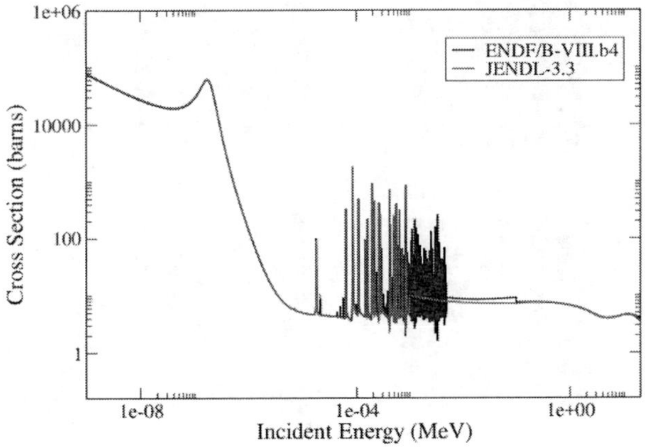

Fig. 2. The cross-section libraries ENDF/B-VII.1 [23] and JENDL-3.3 [25] available for the ^{114}Cd(n,total) reaction. Data extracted from International Atomic Energy Authority database [21].

Fig. 3. A simple model produced in MCNP6, where the detector is in the centre, and on either side are pixellated slices of air, the number of which varied, used to produce the radiation flux on the front and rear of the detector.

TABLE II
THE MCNP6 MATERIALS USED TO MODEL THE DETECTOR, AND THEIR RESPECTIVE CROSS-SECTION INFORMATION. ALSO INCLUDING THE RATIO OF EACH MATERIAL USED IN THESE SIMULATIONS.

Material	Cross-section source	Temperature (K)	Ratio (%)
Zn	ENDF/B-VII.0	293.6	5
Cd	ENDF/B-V	300	45
Te	ENDF/B-VII.1	293.6	50

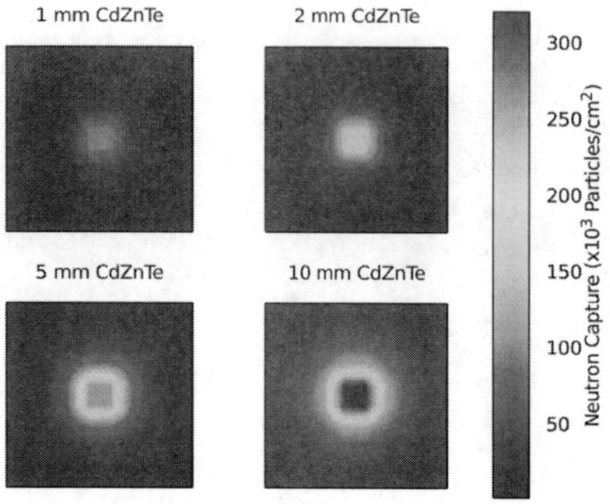

Fig. 4. The simulation of neutron captures using MCNP6, in various thickness of CdZnTe detector. For the purpose of this simulation, the detector was split into 10x10 pixels.

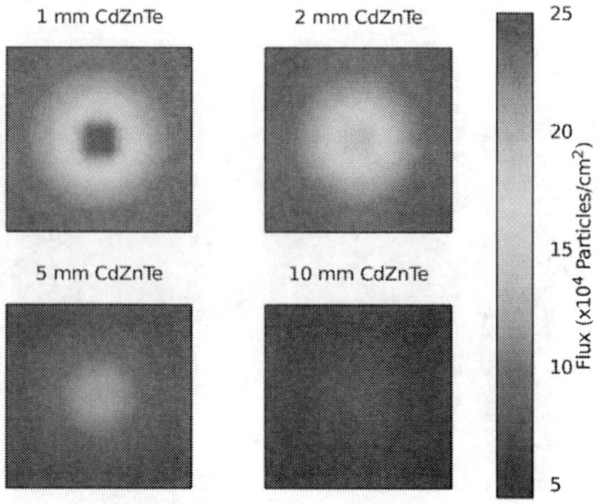

Fig. 5. The simulation of neutron fluxes using MCNP6, on the rear of 1 mm, 2 mm, 5mm, and 10mm CZT blocks. For the purpose of this simulation,the region at the rear of the detector was split into 10x10 pixels.

and accelerator physics. The version used in the simulations described here is Geant4-10.3, released December 2016, patch 1, released February 2017.

Geant4 allows for the modelling of secondary particles. Various physics lists are available for elastic and inelastic hadron scattering; for the relatively low-energy neutrons used here the BERT (Bertini cascade) and BIC (Binary cascade) models are applicable. These both allow for the modelling of protons, neutrons and pions, with BERT also being able to model kaons and hyperons. The energy ranges are 0 - 15 GeV for BERT, and 0 - 5 GeV for BIC [26]. The PhysicsList() used in this instance uses a combination of high precision neutron physics (HP), for the neutron interactions, and electromagnetic physics, for the electrons, and γ-rays within the simulation.

The cross-section data for the simulation comes from a variety of sources, depending on the interaction of interest. For the neutron interactions, this comes from G4NDL4.5, which is mainly based on the ENSDF/B-VII library. Also of use is the TENDL1.3 cross-section data set, which is used in Geant4 for the cross-sections of incident particles. TENDL1.5 cross-sections for neutron capture, and subsequent γ-ray emission, on ^{113}Cd are illustrated in figure 2.

III. Results and Analysis

A. MCNP

Figure 4 shows the capture of neutrons from a ^{252}Cf neutron source in various thickness of CdZnTe detector material. This is the results of the simulation of 10^{10} iniital neutrons. The expectation would be that the thicker a shield or detector is, the more neutron capture events occur. The simulation agrees with this expectation.

Figure 5 shows the fluxes of neutrons from ^{252}Cf source in various thickness of CdZnTe. This is the results of the simulation of 10^{10} iniital neutrons. As would be expected, a greater neutron flux is observed on the rear of the detector when the detector thickness is smaller.

TABLE III
Detailing the major peaks observed in the γ-ray spectrum produced as a result of the neutron capture on a CdZnTe detector. Peak fitting of the theoretical results was performed using ROOT, with the energies and the error measurements on the energies coming from this peak fitting.

$E_\gamma (keV)$	Reaction	Experimental Energy (keV)
245.473(4)	^{111}Cd(n,nγ)	245.395(20) [27]
558.334(3)	^{114}Cd(n,nγ)	558.456(2) [13]
606.931(1)	^{112}Cd(n,nγ)	606.821 (6) [29]
617.651(3)	^{112}Cd(n,nγ)	617.518(3) [29]
650.978(1)	^{114}Cd(n,nγ)	651.256(5) [13]
657.607(8)	^{110}Cd(n,nγ)	657.7600(11) [30]
666.255(30)	^{126}Te(n,nγ)	666.331(12) [31]
742.966(8)	^{128}Te(n,nγ)	743.22(2) [32]
748.968(8)	^{130}Te(n,nγ)	748.76(2) [33]
753.654(9)	^{128}Te(n,nγ)	753.82(2) [32]
839.429(1)	^{130}Te(n,nγ)	839.49(2) [33]
991.853(16)	^{64}Zn(n,nγ)	991.56(5) [34]

B. Geant4

Figure 6 shows the γ-spectrum resulting from the interaction of neutrons from a ^{252}Cf source and a CdZnTe detector. The details of the peaks in the spectrum are described in table III; the γ-rays resulting from the de-excitation of the most common cadmium, zinc, and tellurium have been well-characterised, as described in the relevent Nuclear Data Sheets [13], [27], [29], [30], [31], [32], [33], [34], and the results presented here are in agreement with the energies of the most prominent of these γ-rays.

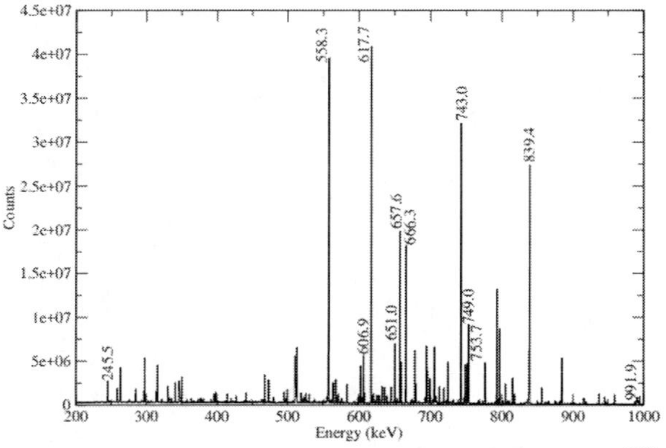

Fig. 6. The truncated γ-ray spectrum produced in Geant4 from a ^{252}Cf neutron source and a 1.5 cm thick CdZnTe detector.

IV. Conclusion

MCNP6 is useful for modelling what reactions are induced by the neutrons, and produces reasonable results for the neutron capture and neutron fluxes. Future work in MCNP may model which other neutron-induced reactions occur in such an experiment. The modelling of secondary radiation is possible with Geant4, and a spectrum which is in line with the experimental data has been produced. In future, the modelling of the electrons produced by the secondary γ-rays, in a realistic electric field, in possible. In addition, a comparison between these theoretical results and experiment may be undertaken, using the ^{252}Cf neutron source at Lancaster University.

Acknowledgment

This work is supported by the Engineering and Physical Sciences Research Council (EPSRC), UK, via the Impact Acceleration Account (IAA), in a collaboration between Lancaster University and Innovative Physics Ltd. The contributions of the Lancaster University technical staff are gratefully acknowledged.

References

[1] T. Goorley, MCNP6.1.1-Beta Release Notes, LA-UR-14-24680, 2014.

[2] S. Agostinelli, J. Allison, K. Amako, J. Apostolakis, H. Araujo, P. Arce, M. Asai, D. Axen, S. Banerjee, G. Barrand, F. Behner, L. Bellagamba, J. Boudreau, L. Broglia, A. Brunengo, H. Burkhardt, S. Chauvie, J. Chuma, R. Chytracek, G. Cooperman, G. Cosmo, P. Degtyarenko, A. Dell'Acqua, G. Depaola, D. Dietrich, R. Enami, A. Feliciello, C. Ferguson, H. Fesefeldt, G. Folger, F. Foppiano, A. Forti, S. Garelli, S. Giani, R. Giannitrapani, D. Gibin, J.J. Gmez Cadenas, I. Gonzlez, G. Gracia Abril, G. Greeniaus, W. Greiner, V. Grichine, A. Grossheim, S. Guatelli, P. Gumplinger, R. Hamatsu, K. Hashimoto, H. Hasui, A. Heikkinen, A. Howard, V. Ivanchenko, A. Johnson, F.W. Jones, J. Kallenbach, N. Kanaya, M. Kawabata, Y. Kawabata, M. Kawaguti, S. Kelner, P. Kent, A. Kimura, T. Kodama, R. Kokoulin, M. Kossov, H. Kurashige, E. Lamanna, T. Lampn, V. Lara, V. Lefebure, F. Lei, M. Liendl, W. Lockman, F. Longo, S. Magni, M. Maire, E. Medernach, K. Minamimoto, P. Mora de Freitas, Y. Morita, K. Murakami, M. Nagamatu, R. Nartallo, P. Nieminen, T. Nishimura, K. Ohtsubo, M. Okamura, S. O'Neale, Y. Oohata, K. Paech, J. Perl, A. Pfeiffer, M.G. Pia, F. Ranjard, A. Rybin, S. Sadilov, E. Di Salvo, G. Santin, T. Sasaki, N. Savvas, Y. Sawada, S. Scherer, S. Sei, V. Sirotenko, D. Smith, N. Starkov, H. Stoecker, J. Sulkimo, M. Takahata, S. Tanaka, E. Tcherniaev, E. Safai Tehrani, M. Tropeano, P. Truscott, H. Uno, L. Urban, P. Urban, M. Verderi, A. Walkden, W. Wander, H. Weber, J.P. Wellisch, T. Wenaus, D.C. Williams, D. Wright, T. Yamada, H. Yoshida, and D. Zschiesche, Geant4 - a simulation toolkit, Nucl. Instru. Meth. A 506, 250-303, 2003.

[3] L. Maigne, Y. Perrot, D. R. Schaart, D. Donnarieix, and V. Breton, Comparison of GATE/GEANT4 with EGSnrc and MCNP for electron dose calculations at energies between 15 keV and 20 MeV, Phys. Med. Biol., 56, 811-827, 2011.

[4] A.A. Hecht, R.E. Blakeley, W.J. Martin, and E. Leonard, Comparison of Geant4 and MCNP6 for use in delayed fission radiation simulation, Ann. Nucl. Energy, 69, 134-138, 2014.

[5] H. Yang, X. Yang, S. Xiao, N. Satvat and T. Jevremovic, Detection of hidden materials using Nuclear Resonance Fluorescence technique: Simulation and measurements, IEEE Nuclear Science Symposuim & Medical Imaging Conference, Knoxville, TN, 2010, pp. 531-533.

[6] T.E. Schlesinger, J.E. Toney, H. Yoon, E.Y. Lee, B.A. Brunett, L. Franks, and R.B. James, Cadmium zinc telluride and its use as a nuclear radiation detector material, Mater. Sci. Eng. R., 32, 103-189, 2001.

[7] A. Martín-Martín, M. P. Iñiguez, P. N. Luke, R. Barquero, A. Lorente, J. Morchón, E. Gallego, G. Quincoces, J. M. Martí-Climent, Evaluation of CdZnTe as neutron detector around medical accelerators. Radiat. Prot. Dosimetry, 133, 193-199, 2009.

[8] M. Streicher, S. Brown, Y. Zhu, D. Goodman and Z. He, "Special Nuclear Material Characterization Using Digital 3-D Position Sensitive CdZnTe Detectors and High Purity Germanium Spectrometers," in IEEE Transactions on Nuclear Science, vol. 63, no. 5, pp. 2649-2656, Oct. 2016.

[9] D.S. McGregor, J.T. Lindsay, R.W. Olsen, Thermal neutron detection with cadmium1x zincx telluride semiconductor detectors, Nuclear Instruments and Methods in Physics Research Section A: Accelerators, Spectrometers, Detectors and Associated Equipment, 381, 498-501, 1996.

[10] S. T. Brown, Time-Encoded Thermal Neutron Imaging Using Large-Volume Pixelated CdZnTe Detectors, PhD Thesis, University of Michigan, USA, 2017.

[11] S.F. Mughabghab, Atlas of Neutron Resonances: Resonance Parameters and Thermal Cross Sections. Z= 1-100, Elsevier, 2006.

[12] K. A. Nelson, N. Edwards, M. J. Harrison, A. Kargar, W. J. McNeil, R. A. Rojeski, and D. S. McGregor, Investigation of CdZnTe and $LiNbO_3$ as electro-optic neutron detectors, Nucl. Instru. Meth. A 620, 363-367, 2010.

[13] J. Blachot, Nuclear Data Sheets for A = 114, Nucl. Data Sheets, 113, 515-714, 2012.

[14] R. C. Reedy, T. H. Prettyman and N. Yamashita, Backgrounds in cadmium zinc telluride (CZT) gamma-ray spectrometers for planetary science, Abstracts for 43rd Lunar & Planetary Science Conference (Houston, 19-23 Mar. 2012), #1284.

[15] M. Chu, S. Terterian, and D. Ting, Role of Zinc in CdZnTe Radiation Detectors, IEEE Transactions on Nuclear Science, 51, 2405-2411, Oct. 2004.

[16] A. B. Smith, P. R. Fields, and J. H. Roberts, Spontaneous Fission Neutron Spectrum of Cf^{252}, Physical Review, 108, 411-413, 1957.

[17] D. Gagnon, G. L. Zeng, J. M. Links, J. J. Griesmer, and F. C. Valentino, Design considerations for a new solid-state gamma-camera: SOLSTICE, IEEE Nuclear Science Symposium and Medical Imaging Conference (Vol. 2, pp. 1156-1160), 2001.

[18] K. B. Parnham, J. Grosholz, R. K. Davis, S. Vydrin, and C. A. Cupec, Development of a CdZnTe-based small field-of-view gamma camera. Proc. SPIE 4508, Penetrating Radiation Systems and Applications III, 134, 2001.

[19] M. Prokesch, D. S. Bale and C. Szeles, Fast High-Flux Response of CdZnTe X-Ray Detectors by Optical Manipulation of Deep Level Defect Occupations, IEEE Transactions on Nuclear Science, 57, 2397-2399, Aug. 2010.

[20] T. Goorley, M. James, T. Booth, F. Brown, J. Bull, L.J. Cox, J. Durkee, J. Elson, M. Fensin, R.A. Forster, J. Hendricks, H.G. Hughes, R. Johns, B. Kiedrowski, R. Martz, S. Mashnik, G. McKinney, D. Pelowitz, R. Prael, J. Sweezy, L. Waters, T. Wilcox, and T. Zukaitis, Initial MCNP6 Release Overview, LA-UR 13-22934, 2013.

[21] V. Zerkin, Evaluated Nuclear Data File (ENDF) Database, Internet: https://www-nds.iaea.org/exfor/endf.htm, Mar. 10th, 2017 [May 23rd, 2017].

[22] M.B. Chadwick, M. Herman, P. Obložinský, M.E. Dunn, Y. Danon, A.C. Kahler, D.L. Smith, B. Pritychenko, G. Arbanas, R. Arcilla, R. Brewer, D.A. Brown, R. Capote, A.D. Carlson, Y.S. Cho, H. Derrien, K. Guber, G.M. Hale, S. Hoblit, S. Holloway, T.D. Johnson, T. Kawano, B.C. Kiedrowski, H. Kim, S. Kunieda, N.M. Larson, L. Leal, J.P. Lestone, R.C. Little, E.A. McCutchan, R.E. MacFarlane, M. MacInnes, C.M. Mattoon, R.D. McKnight, S.F. Mughabghab, G.P.A. Nobre, G. Palmiotti, A. Palumbo, M.T. Pigni, V.G. Pronyaev, R.O. Sayer, A.A. Sonzogni, N.C. Summers, P. Talou, I.J. Thompson, A. Trkov, R.L. Vogt, S.C. van der Marck, A. Wallner, M.C. White, D. Wiarda, P.G. Young, ENDF/B-VII.1 Nuclear Data for Science and Technology: Cross Sections, Covariances, Fission Product Yields and Decay Data, Nuclear Data Sheets, 112, 2011, 2887-2996.

[23] M.B. Chadwick, M. Herman, P. Obloinsk, M.E. Dunn, Y. Danon, A.C. Kahler, D.L. Smith, B. Pritychenko, G. Arbanas, R. Arcilla, R. Brewer, D.A. Brown, R. Capote, A.D. Carlson, Y.S. Cho, H. Derrien, K. Guber, G.M. Hale, S. Hoblit, S. Holloway, T.D. Johnson, T. Kawano, B.C. Kiedrowski, H. Kim, S. Kunieda, N.M. Larson, L. Leal, J.P. Lestone, R.C. Little, E.A. McCutchan, R.E. MacFarlane, M. MacInnes, C.M. Mattoon, R.D. McKnight, S.F. Mughabghab, G.P.A. Nobre, G. Palmiotti, A. Palumbo, M.T. Pigni, V.G. Pronyaev, R.O. Sayer, A.A. Sonzogni, N.C. Summers, P. Talou, I.J. Thompson, A. Trkov, R.L. Vogt, S.C. van der Marck, A. Wallner, M.C. White, D. Wiarda, and P.G. Young, ENDF/B-VII.1 Nuclear Data for Science and Technology: Cross Sections, Covariances, Fission Product Yields and Decay Data, Nuclear Data Sheets, 112, 2887-2996, 2011.

[24] J. L. Conlin (ed.) and Nuclear Data Team XCP-5, Listing of Available ACE Data Tables, LA-UR-17-20709, 2017.

[25] K. Shibata, T. Kawano, T. Nakagawa, O. Iwamoto, J. Katakura, T. Fukahori, S. Chiba, A. Hasegawa, T. Murata, H. Matsunobu, T. Ohsawa, Y. Nakajima, T. Yoshida, A. Zukeran, M. Kawai, M. Baba, M. Ishikawa, T. Asami, T. Watanabe, Y. Watanabe, M. Igashira, N. Yamamuro, H. Kitazawa, N. Yamano and H. Takano, Japanese Evaluated Nuclear Data Library Version 3 Revision-3: JENDL-3.3, Journal of Nuclear Science and Technology, 39, 1125-1136, 2002.

[26] A. V. Ivantchenko, V. N. Ivanchenko, J-M. Quesada Molina and S. L. Incerti, Geant4 hadronic physics for space radiation environment, Int. J. Radiat. Biol., 88, 171-175, 2012.

[27] J. Blachot, Nuclear Data Sheet for A=111, Nuclear Data Sheets 110, 1239-1407, 2009.

[28] J. Blachot, Nuclear Data Sheet for A=114, Nuclear Data Sheets 113, 515-714, 2012.

[29] S. Lalkovski and F. G. Kondev, Nuclear Data Sheet for A=112, Nuclear Data Sheets 124, 157-412, 2015.

[30] G. Gürdal and F. G. Kondev, Nuclear Data Sheet for A=110, Nuclear Data Sheets 113, 1315-1561, 2012.

[31] J. Katakura and K. Kitao, Nuclear Data Sheet for A=126, Nuclear Data Sheets 97, 765-926, 2002.

[32] Z. Elekes and J. Timar, Nuclear Data Sheet for A=1328, Nuclear Data Sheets 129, 191-436, 2015.

[33] B. Singh, Nuclear Data Sheet for A=130, Nuclear Data Sheets 93, 33-242, 2001.

[34] B. Singh, Nuclear Data Sheet for A=64, Nuclear Data Sheets 108, 197-364, 2007.

Conception of a New Recoil Proton Telescope for Real-Time Neutron Spectrometry in Proton-Therapy

Rodolphe Combe, Nicolas Arbor, Ziad el Bitar, Stéphane Higueret, Daniel Husson
Université de Strasbourg
CNRS, IPHC UMR 7178
F-67000 Strasbourg, France
Email: rodolphe.combe@iphc.cnrs.fr

***Abstract*—Neutrons are the main type of secondary particles emitted in proton-therapy. Because of the risk of secondary cancer and other late occurring effects, the neutron dose should be included in the out-of-field dose calculations. A neutron spectrometer has to be used to take into account the energy dependence of the neutron radiological weighting factor. Due to its high dependence on various parameters of the irradiation (beam, accelerator, patient), the neutron spectrum should be measured independently for each treatment.**

The current reference method for the measurement of the neutron energy, the Bonner Sphere System, consists of several homogeneous polyethylene spheres with increasing diameters equipped with a proportional counter. It provides a high-resolution reconstruction of the neutron spectrum but requires a time-consuming work of signal deconvolution. New neutron spectrometers are being developed, but the main experimental limitation remains the high neutron flux in proton therapy treatment rooms. A new model of a real-time neutron spectrometer, based on a Recoil Proton Telescope technology, has been developed at the IPHC. It enables a real-time high-rate reconstruction of the neutron spectrum from the measurement of the recoil proton trajectory and energy. A new fast-readout microelectronic integrated sensor, called FastPixN, has been developed for this specific purpose.

A first prototype, able to detect neutrons between 5 and 20 MeV, has already been validated for metrology with the AMANDE facility at Cadarache. The geometry of the new Recoil Proton Telescope has been optimized via extensive Geant4 Monte Carlo simulations. Uncertainty sources have been carefully studied in order to improve simultaneously efficiency and energy resolution, and solutions have been found to suppress the various expected backgrounds. We are currently upgrading the prototype for secondary neutron detection in proton therapy applications.

I. Introduction

A. Secondary neutrons in proton-therapy

During a proton-therapy treatment, a lot of secondary neutrons are produced by the interaction of the proton beam with the patient (internal neutrons), and with the accelerator (external neutrons) for some treatments, via ${}^{A}_{Z}X(p,n)^{A}_{Z+1}Y$ reactions. Due to their high radiological weighting factor, the neutron dose should be monitored. It can grow up to 500 mGy over a full treatment. It has been shown in [1] that this dose can vary greatly depending on the treatment methods and other parameters. As the biological effect of neutrons is dependent on their energy [2], a neutron spectrometry is mandatory for a precise estimation of the neutron equivalent dose. The fast neutrons being the greatest contributors to the neutron dose, our approach is to focus on this energy range while other systems are developed for measuring thermal neutrons.

The neutron spectrum depends among others on:

- the energy of the protons, which mainly changes the fluence and shifts the endpoint of the spectrum to the energy of the beam [3][4]
- the angle to the proton beam, which changes the shape of the fast neutron spectrum [5][6][7][8][9]
- the distance to isocentre, which changes not only the fluence but also the shape of the spectrum [10][11][12]

Because of that, one spectrum can only be identified to one treatment. Scarce experimental data coupled to extrapolation via Monte Carlo simulation is currently used to estimate the dose, but this leads to large discrepancies between the simulation and experimental data.

B. Neutron spectrometry

The current reference spectrometer for neutrons is the Bonner Sphere System, which has been extensively used and studied in literature. It is composed of several spheres of homogeneous polyethylene of increasing diameters, equipped with a ^{3}He proportional counter in their center. This system allows a precise reconstruction of the spectrum over a large energy range. However, it requires a complex procedure of deconvolution and both the measurement and the deconvolution are time-consuming, which excludes a real-time measurement. More problematic for this application, it is cumbersome and it has been proven in [13] to be limited to a fluence of 4×10^4 n/cm^2/s, which is two orders of magnitudes below the 6×10^6 n/cm^2/s that can be found in treatment rooms [1]. Because of that, the Bonner Sphere System cannot be used in treatment rooms, and is far from being optimal for multi-parameter studies.

Other neutron spectrometers are currently being developed as the MONDO project [14][15] or the capture-gated detector FANS-2 [16]. The first project aims at measuring the double elastic scattering of fast neutrons by the means of arrays of scintillating fibers. It will allow to measure neutrons between 20 and 400 MeV with a size of $10 \times 10 \times 20\,\text{cm}^3$. The second project combines plastic scintillators and ^{3}He counters to measure neutrons between a few MeV and 1 GeV. However, it is 1.15 m long, can only measure 200 n/cm^2/s and requires ^{3}He, which is both expensive and requires high-voltage.

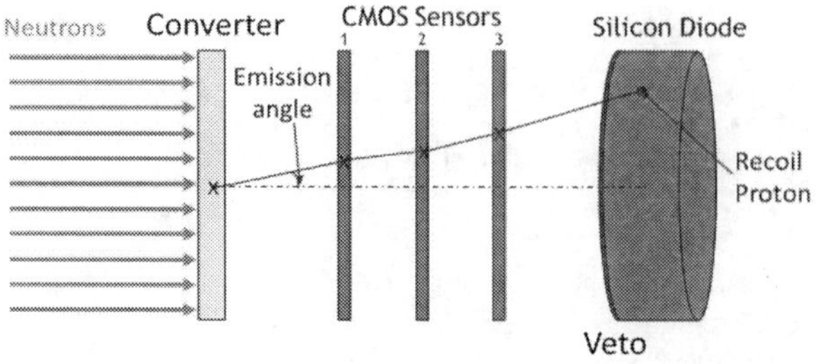

Fig. 1. Scheme of the Recoil Proton Telescope

Our prototype is a new real-time neutron spectrometer based on a Recoil Proton Telescope that has been developed at IPHC. To go beyond existing so-called "zero-degree" telescopes, we choose to reconstruct the full track of recoil protons, inside an opening angle of 41° (95% C.L.). The CMOS pixels Recoil Proton Telescope allows a real-time high-rate (100 kHz) of the fast neutron spectrum within a limited size ($10 \times 15 \times 20\,\text{cm}^3$). Extensive Monte Carlo Geant4 [17] simulations have been carried out to characterize precisely the performances of this device.

II. Recoil Proton Telescope

A. Principle

Fast neutrons ($E_n > 1\,\text{MeV}$) are elastically scattered on a hydrogenated converter (polyethylene). The energy E_p and emission angle θ of each recoil proton is directly related to the incident neutron energy E_n:

$$E_n = \frac{E_p}{\cos^2\theta} \tag{1}$$

The emission angle will be measured using three CMOS sensors and the energy of the proton using a silicon diode. This will allow to measure the neutron energy in real-time with a rate a 100 kHz.

The full scheme of the Recoil Proton Telescope is presented on figure 1.

B. Trajectography

To reconstruct the trajectory, a monolithic active pixels CMOS sensor named FastPixN [18] is used. It is made of 128×128 pixels of $50\,\mu$m pitch. It has been thinned down to $50\,\mu$m to minimize the energy loss inside the Recoil Proton Telescope. With 4-bit column-ADC, the acquisition rate of 100 kHz allows to measure up to 10^9 n/cm^2/s.

The uniformity of these sensors has been tested at the AIFIRA platform (CENBG Bordeaux). Despite the difficulty to calibrate CMOS sensors, due to their charge collection through thermal diffusion, a good calibration curve was achieved up to 20 MeV. This allows to use the energy deposited in the FastPixN as a criterion for the analysis, to constrain the direction of the proton.

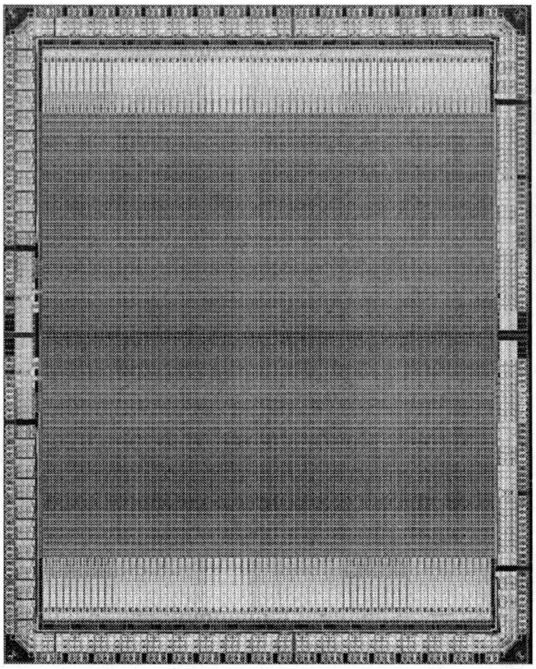

Fig. 2. View of the FastPixN sensor

C. Reconstruction of the neutron energy

To measure the proton energy, the Recoil Proton Telescope uses a silicon diode read by a 13-bit ADC. The thickness of the diode has been set as an optimum between charge collection and proton stopping power. This 3-mm-thick diode can measure protons with energies up to 20 MeV. The proton initial energy is estimated as the sum of the energy deposited in the diode and the estimated mean energy loss in the Recoil Proton Telescope, which has been calculated using a Monte Carlo Geant4 simulation for each deposited energy. This mean energy loss is corrected by the estimated track length, determined by the measurement of the angle between the three sensor hits.

D. First prototype

A first prototype of Recoil Proton Telescope has been designed for the Laboratory of Neutron Metrology and Dosimetry (LMDN) of Institute of Radio-protection and Nuclear safety (IRSN) Cadarache (figure 3). It provides a good reconstruction of mono- and poly-energetic neutrons spectrum between 5 and 20 MeV. The requirements for this device were a 5% resolution on E_n, low background and high efficiency for moderate fluences (< 6000 n/cm^2/s).

III. Measurement of the proton-therapy neutron spectrum

To measure the proton-therapy neutron spectrum, two different sensors will be used. The first one, called AlphaRad [19],

Fig. 3. View of the first prototype

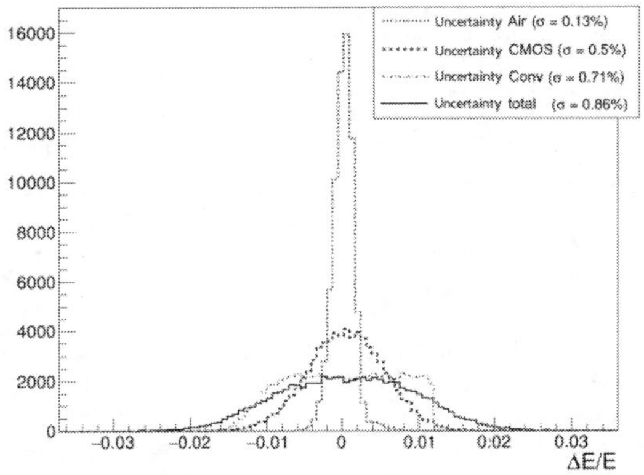

Fig. 4. Contribution of the various uncertainty sources to the total uncertainty (dotted) and total uncertainty (solid) in the reconstruction of the proton energy (Geant4 simulation)

will count thermal and fast neutron under 5 MeV. The Recoil Proton Telescope will be used to reconstruct the fast neutron spectrum over 5 MeV. These experimental measurements will be used, in a first time, to cross-check the various nuclear models used for the dose calculation by the Treatment Planning Softwares and the Monte Carlo simulation codes. Since the telescope has been first optimized for the metrological application at IRSN, we need to use this device over an extended energy range, first up to 30 MeV, to address the survey of the neutron production in proton-therapy. As has been shown in literature, this energy range ($[5, 30]$ MeV) is interesting, especially at high angle and for low energy irradiation such as ocular treatments.

IV. Analysis

A. Uncertainty sources

To improve the reconstruction of the neutron energy, an ideal simulation of the Recoil Proton Telescope, with protons created randomly inside the converter and full knowledge about the trajectory and the energy deposits, was generated. This simulation allowed to study what are the critical parameters for the energy resolution and which ones can eventually be modified.

For example, for 10-MeV protons with a 50-μm-thick converter, the uncertainty (1σ) over the reconstruction of the proton energy is found to be of $0.86 \pm 0.01\%$ (figure 4).

The largest uncertainty source is found to be the inability to determine the position of creation of the protons inside the converter through the elastic H(n,n)p reaction. It creates an uncertainty of $0.71 \pm 0.01\%$ on the proton energy. The other sources of uncertainty are the energetic and angular straggling inside the CMOS, which contribute to $0.50 \pm 0.01\%$, and in the air, which contribute to $0.13 \pm 0.01\%$. Concerning the angular resolution, the uncertainty on $\cos\theta$ is found to be $0.30 \pm 0.01\%$.

B. Background suppression

1) *Coherent background sources:*
As the conversion efficiency of a neutron into a proton is of only 10^{-3} per millimeter of polyethylene, a high background coming from direct neutron hits in the diode is expected. To prevent the contamination of the signal by these hits, a 65-μm veto diode was installed right before the 3-mm-thick diode. A collected track is then validated only if it has one hit in each CMOS plus one hit in the veto.
The high reading frequency of the FastPixN allows a high reduction of the number of neutrons, and therefore protons, per frame, reducing by the way the background. Moreover, an extensive analysis of the geometric and energetic distribution of the events allows to suppress tracks which do not correspond to signal events.
Several other background sources are expected, each of them requesting a dedicated cut, listed thereafter:

- Proton created in the diode: a Si nucleus of the diode scattering inelastically with a neutron can produce one or more protons through Si(n,sp) reactions. A proton flying back through the device would produce a track similar to a signal event. To suppress these events, an increasing energy deposit inside successive CMOS is requested, in accordance with the Bethe-Bloch model.
- Hydrogen isotopes: consecutive to a neutron capture on hydrogen, a deuteron or triton can be ejected, mimicking the signal of a lower energy proton. These events are suppressed by requesting the energy deposited in the veto and in the diode to correspond to those of a proton in accordance with the Bethe-Bloch model.
- Proton created out of the converter: during the inelastic scattering of a neutron, a proton can be produced, which will then generate a track correspond-

ing to a lower energy neutron. To suppress these events, the trajectory of the proton is extrapolated towards the converter.

- Proton escaping the diode: a proton arriving at the periphery of the diode with a certain angle can leave it without depositing its full energy. To suppress these events, it is verified that the proton can't escape the diode by extrapolating its trajectory towards the diode.

2) *Fortuitous coincidence:*
Because of the high neutron flux in proton-therapy treatment rooms and despite the high-rate capability of the Recoil Proton Telescope, fortuitous coincidences are a major background source. To study this source, PCB boards as well as the metallic box, that are a major source of secondaries, were added to the simulation. Moreover, the Recoil Proton Telescope was irradiated in simulation within a solid angle covering the whole box. This allowed to study extensively the effect of fortuitous coincidences.
There are two background sources related to fortuitous coincidences. The first one is the coincidence of a direct neutron hit in the diode with a partial track, corresponding to a proton stopped in the last CMOS or right after. The second one is the coincidence of a neutron hit with a signal event. These two backgrounds are easily suppressed by requesting a hit in the veto for the first one and a correlation between the energy deposited in the veto and the diode for the second one corresponding to the Bethe-Bloch model.

3) *Cuts:*
Hereafter are listed the applied cuts:
   a) Increasing energy deposit and Bragg peak:
      - From the Bethe-Bloch model, (in this energy range) a particle losing its energy has an increasing energy deposit along the track. This means that recoil protons should deposit more and more energy along their trajectory in the Recoil Proton Telescope. Therefore, the energy deposited in the CMOS sensors should be increasing for a signal event. This is softened by the poor energy resolution of the CMOS sensors (4 bits) and by the energetic straggling.
      - At the end of its track, a proton will deposit a lot of energy over a very short distance: this is the so-called "Bragg peak". A proton whose Bragg peak is observed in a CMOS cannot therefore reach the diode.

      The applied criterion is then an increasing energy deposit to the bit level and a lack of the Bragg peak.

$$\Delta E_{FPN} \in [-1\text{bit}; +2\text{bits}] \quad (2)$$

   b) Correlation Veto-Diode:
      The Bethe-Bloch model also allows to establish a correlation between the energy deposited in the veto and in the diode. The energy deposited in the diode E^{diode}_{meas} allows to estimate the energy E^{veto}_{th} that should be deposited in the veto by a proton depositing E^{diode}_{meas} in the diode. By comparing with the energy deposited in the veto E^{veto}_{meas}, a criterion on the validity of the track is obtained. Because of the energetic straggling and after optimization, this criterion is expressed as an upper limit:

$$\frac{E^{veto}_{th} - E^{veto}_{meas}}{E^{veto}_{meas}} < 9\% \quad (3)$$

   c) Extrapolation of the track:
      To verify that the proton emerges from the converter and ends its track inside the diode, its track can be extrapolated in the two directions to cancel background events. With (X_1, Y_1) to (X_3, Y_3) the coordinates of the hits in the CMOS and after optimization, the coordinates of the proton inside the converter and the diode can be extrapolated to:

$$(X_0, Y_0) \approx (2 * X_1 - X_2, 2 * Y_1 - Y_2) \quad (4)$$

$$(X_4, Y_4) \approx (2 * X_3 - X_2, 2 * Y_3 - Y_2) \quad (5)$$

$$R_0 = \sqrt{X_0^2 + Y_0^2} < 5.1\ \mu m \quad (6)$$

$$R_4 = \sqrt{X_4^2 + Y_4^2} < 6.0\ \mu m \quad (7)$$

TABLE I
EFFICIENCY OF THE CUTS

Cut	Efficiency	Purity	Resolution	S/(S+B)
Raw data			$31.3 \pm 0.5\%$	$24.8 \pm 0.7\%$
ΔFPN (2)	$5.7 \pm 0.7\%$	$99.7 \pm 0.1\%$	$30.8 \pm 0.5\%$	$23.8 \pm 0.7\%$
$\Delta Veto$ (3)	$87.1 \pm 1\%$	$75.1 \pm 0.8\%$	$19.3 \pm 0.4\%$	$5.5 \pm 0.5\%$
R0 (6) & R4 (7)	$6.6 \pm 0.8\%$	$97.2 \pm 0.3\%$	$31.0 \pm 0.5\%$	$24.1 \pm 0.7\%$
All cuts	$89.8 \pm 0.9\%$	$73.2 \pm 0.8\%$	$17.9 \pm 0.4\%$	$4.4 \pm 0.4\%$

Table I presents the efficiency, purity, resolution on E_n and background level of the various cuts.

V. STUDY OF THE GEOMETRY

A. Material of the box

According to the literature [20], our aluminum box has a high cross section of secondary production via Al(n,p) and Al(n,n+p) reactions of around 100 mb. Other materials frequently used near accelerators, such as tungsten and lead, have a secondary production cross-section 10 times lower but an overall neutron cross-section 5 times higher. After simulation (Table II), only a small benefit of using tungsten is found. It is therefore chosen to keep aluminum which is easier to process.

B. Distance to the diode

As the third PCB board and the box are two important sources of secondaries, the diode, which lies between them, should be carefully placed. After simulation (figure 5), the efficiency is found to decrease with the distance between the

https://doi.org/10.1051/epjconf/201817009001

TABLE II
EFFECT OF THE BOX MATERIAL ON RPT PERFORMANCES

Material	Al	W
Efficiency ($\times 10^{-5}$)	2.4 ± 0.05	2.22 ± 0.04
B/(S+B)	$7.7 \pm 0.7\%$	$12.8 \pm 0.7\%$
Resolution on E_n	$20.2 \pm 0.5\%$	$18.7 \pm 0.4\%$

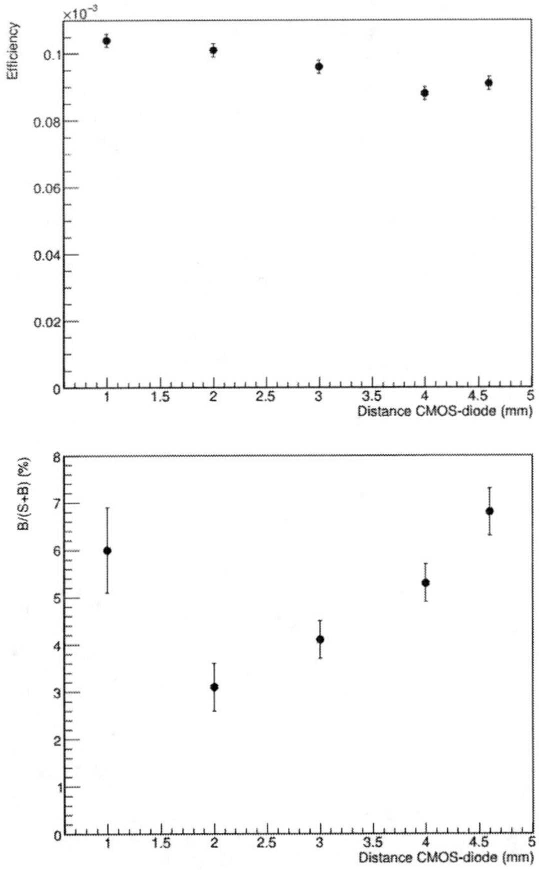

Fig. 5. Fiducial measurement efficiency (right) and background level (left) for a proton therapy spectrum (Geant4 simulation)

third CMOS and the diode. This is due to the decrease of the aperture angle. An optimum for the background level is also found for a distance of 2 mm. If the diode is too close to the third CMOS, it will receive the high-angle proton that should escape the diode. If it is too close to the box, it will detect the secondaries coming from inelastic events in the box. An optimum is then found for a 2 mm distance. The mechanical realization of such a small distance is under investigation.

C. Converter thickness

Because of the high-energy of neutrons in proton-therapy treatment rooms, the H(n,n)p cross-section is low and the

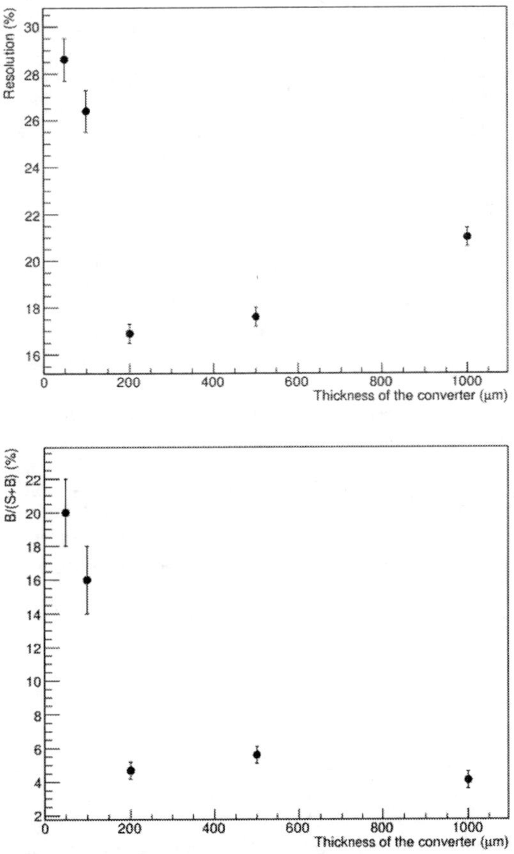

Fig. 6. Resolution on the neutron energy (right) and background level (left) for a proton therapy spectrum as a function of the distance between the diode and the third CMOS (Geant4 simulation)

secondaries production cross-section is high. This creates a high background level that needs to be compensated by a high production efficiency. However, to achieve this high efficiency, the converter needs to be thickened and it has been seen in section IV-A that this will degrade the resolution on E_p and therefore on E_n.

An optimum for the resolution is found between 200 and 500 μm of polyethylene (figure 6). If the converter is too thin, the signal-over-noise ratio is low, which can also be seen through the background level, and if it is too thick, the resolution on E_p starts to be degraded.

VI. FIRST RESULTS UP TO 30 MEV

For a proton-therapy spectrum corresponding to an irradiation of the head with a 178 MeV proton beam [13], a reconstruction of the neutron spectrum is obtained between 5 to 30 MeV with a $16.3 \pm 0.3\%$ resolution. As can be seen in figure 7, the main reason for this resolution is the high level

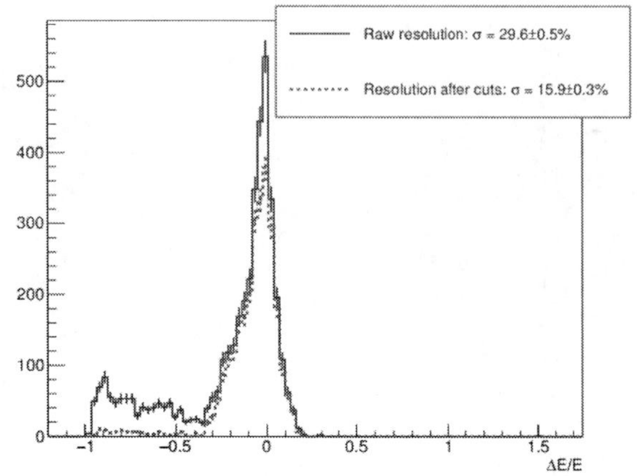

Fig. 7. Raw resolution (solid) and after-cut resolution (dotted) on E_n for a proton therapy spectrum (Geant4 simulation)

of background that can be seen in the tail corresponding to a $\Delta E/E$ smaller than -0.4. This resolution should be sufficient to benchmark nuclear models of neutron production in proton-therapy.

VII. Conclusion and prospective

The possibility of using a new Recoil Proton Telescope as a mean to measure secondary neutrons spectrum in proton-therapy treatment room has been proven. A first prototype is currently being built and should be able to measure neutrons between 5 and 30 MeV.

To measure neutrons of higher energy, an energy degrader is intended to be added right after the converter to slow down the recoil protons. This would allow the Recoil Proton Telescope to measure neutrons up to 70 or even 200 MeV. This requires an extensive work of optimization of the parameters of this new Telescope, namely the converter thickness as well as the degraders material and thicknesses.

References

[1] H. Paganetti, *Proton Therapy Physics*. CRC Press, 2011

[2] ICRP, *The 2007 Recommendations of the International Commission on Radiological Protection*. ICRP Publication 103. Ann. ICRP 37 (2-4), 2007.

[3] S. B. Jia, M. H. Hadizadeha, A. A. Mowlavib and M. E. Loushab, *Evaluation of energy deposition and secondary particle production in proton therapy of brain using a slab head phantom*. Reports of Practical Oncology & Radiotherapy Volume 19, Issue 6, November-December 2014, Pages 376-384

[4] M. R. Islam, Y. Zheng, T. L. Collums, J. M. Monson, S. Ahmad and E. R. Benton, *Measurement ans simulation of secondary neutrons from uniform scanning proton beams in proton therapy*. Radiation Measurements 96 (2017) 8-18

[5] Z. Riazi, H. Afarideh and R. Sadighi-Bonabi, *Influence of ridge filter material on the beam efficiency and secondary neutron production in a proton therapy system*. Z. Med. Phys. 22 (2012) 231-240

[6] Z. Vykydala, M. Králíka, J. Šolca, J. Vilímovskýb and V. Vondráčekb, *Angular distribution of neutron spectral fluence around phantom irradiated with high energy protons*. Radiation Measurements, Volume 92, September 2016, Pages 1-7

[7] X. Yan, U. Titt, A. M. Koehler and W. D. Newhauser, *Measurement of neutron dose equivalent to proton therapy patients outside of the proton radiation field*. Nuclear Instruments and Methods in Physics Research A 476 (2002) 429-434

[8] Y. C. Lin, C. C. Lee, T. C. Chao and H. Y. Tsai, *Ambient neutron dose equivalent during proton therapy using wobbling scanning system : Measurements and calculations*. Radiation Physics and Chemistry (2017)

[9] S. Trinkl, V. Mares, F. S. Englbrecht, J. J. Wilkens, M. Wielunski, K. Parodi, W. Rühm and M. Hillbrand, *Systematic Out-of-field Secondary Neutron Spectrometry and Dosimetry in Pencil Beam Scanning Proton therapy*. Med. Phys. 44 (5), May 2017

[10] M. R. Islam, T. L. Collums, Y. Zheng, J. Monson and E. R. Benton, *Off-axis dose equivalent due to secondary neutrons from uniform scanning proton beams during proton radiotherapy*. Phys. Med. Biol. 58 (2013) 8235-8251

[11] M. R. Islam, *Off-axis Neutron Study from a Uniform Scanning Proton Beam Using Monte Carlo Code FLUKA*. Master thesis, Oklahoma State University, 2013

[12] H. Jiang, B. Wang, X. G. Xu, H. D. Suit and H. Paganetti, *Simulation of organ-specific patient effective dose due to secondary neutrons in proton radiation treatment*. Phys. Med. Biol. 50 (2005) 4337-4353

[13] J. Farah, F. Martinetti, R. Sayah, V. Lacoste, L. Donadille, F. Trompier, C. Nauraye, L. De Marzi, I. Vabre, S. Delacroix, *Monte Carlo modeling of proton therapy installations: a global experimental method to validate secondary neutron dose calculations*. Physics in Medicine and Biology, Volume 59, Number 11 (2014) 2747-2765

[14] S. M. Valle, G. Battistoni, V. Patera, D. Pinci, A. Sarti, A. Sciubba, E. Spiriti and M. Marafini, *The MONDO Project: A secondary neutron tracker detector for particle therapy*. Nuclear Instruments and Methods in Physics Research A (2016)

[15] M. Marafini, L. Gasparini, R. Mirabelli, D. Pinci, V. Patera, A. Sciubba, E. Spiriti, D. Stoppa, G. Traini and A. Sarti, *MONDO: a neutron tracker for particle therapy secondary emission characterisation*. Phys. Med. Biol. 62 (2017) 3299-3312

[16] T. J. Langford, E. J. Beise, H. Breuer, C. R. Heimbach, G. Ji and J. S. Nico, *Development and characterization of a high sensitivity segmented Fast Neutron Spectrometer (FaNS-2)*. Journal of Instrumentation, Volume 11, January 2016

[17] Nuclear Instruments and Methods in Physics Research A 506 (2003) 250-303. IEEE Transactions on Nuclear Science 53 No. 1 (2006) 270-278. Nuclear Instruments and Methods in Physics Research A 835 (2016) 186-225.

[18] M. Kachel, D. Husson, S. Higueret, J. Taforeau and L. Lebreton, *FastPixN, a new integrated pixel chip for a future fast version of the IRSN - Recoil Proton Telescope*. Radiat Prot Dosimetry 214 Oct.; 161 (1-4):249-52.

[19] N. Arbor, S. Higueret, H. Elazhar, R. Combe, P. Meyer, N. Dehaynin, F. Taupin, D. Husson, *Real-time detection of fast and thermal neutrons in radiotherapy with CMOS sensors*. Phys Med Biol. 2017 Mar 7;62(5):1920-1934.

[20] N. Soppera, M. Bossant, E. Dupont, *JANIS 4: An Improved Version of the NEA Java-based Nuclear Data Information System*. Nuclear Data Sheets, Volume 120, June 2014, Pages 294-296.

EPJ Web of Conferences **170**, 09002 (2018) https://doi.org/10.1051/epjconf/201817009002
ANIMMA 2017

Detection limit of a VCO based detection chain dedicated to particles recognition and tracking

K. Coulié, W. Rahajandraibe, H. Aziza, G. Micolau, R. Vauché

Abstract—**A particle detection chain based on CMOS-SOI VCO circuit is presented. The solution is used for the recognition and the tracking of a given particle at circuit level. TCAD simulation of the detector has been performed on a 3x3 matrix of diodes based detector for particles recognition and tracking. The current response of the detector has been used for a case study in order to determine the ability of the chain to recognize an alpha particle crossing a 3x3 detection cell. The detection limit of the proposed solution is investigated and discussed in this paper.**

Index Terms— **VCO, particle detector, TCAD simulation**

I. Introduction

SEMICONDUCTOR radiation detectors can be classified into several categories, depending on the application field and the particles to be detected. These detectors, also called solid-state detectors, are based on charge collection and amplification, with the help of a semiconductor volume such as a *p-n* junction. They exhibit several advantages in comparison with other types of detectors such as gas-filled counters. For instance, they usually offer faster charge-collection times which provide the ability to process higher counting rates. An other advantage is their compactness which allows the measurement of intensity variations over small distances improving the measurement resolution. When a high detector sensitivity is required, a pixel organization including several sensors could be used and will be investigated in this work.

The signal generated by the sensor has to be conditioned in order to allow particles counting and/or recognition. Most of the existing conditioning chains are based on a direct reading of the detected currents created by the crossing of an ionizing particle through the matter. When an embedded reading system is necessary, this solution could not be suitable for the detection of low energy particles where the current signature could be a narrow pulse with a duration of few nanoseconds. Indeed, the readout chain must ensure sufficiently high resolution (optimized Signal-to-Noise-Ratio) for an appropriate post-processing. Moreover, a reading amplifier exhibiting a fast response time is required which means a higher current consumption and complex circuit implementation. Designing high precision, low-power wideband amplifier is a critical issue. As a consequence, most practical detectors are based on direct reading of the detected currents using Charge Sensitive Amplifiers (CSA) [1]. In this case, the circuit delivers a limited amount of information on the actual current shape. Then no information about the electrical signature of the particles is available. In applications such as gas prospection or medical therapy, the characteristics of the electrical signature of the particles must be known. Moreover, the knowledge of the generated current shape at the output of the detector allows easier post processing of the signal

In this context, a new approach based on the use of a ring oscillator was developed with an innovative conditioning method. This new approach is based on the reading of the information related to indirect output parameters of the detection chain signal (*i.e.* phase and frequency), instead of directly measuring the current from the sensor. This solution is interesting since it avoids most of the design problems described upper. Then, the information is extracted by correlating the initial oscillation signal of the system with the oscillating signal after the particle has passed the detector. To be able to make a diagnosis, the requirement is then to link the output information (*i.e.* phase and frequency of the oscillating signal) to the input information (current *stimuli*). The conditioning circuit has been implemented in 130-nm CMOS SOI (Silicon on Insulator) technology. The circuit has been designed at transistor level with transient simulation performed in Cadence Virtuoso platform.

In the second section of this paper, the context of this work is presented. The third section is dedicated to the VCO (Voltage Controlled Oscillator) presentation. Part IV deals with the VCO detection limits followed by a case study of a particle crossing a matrix of detectors (part V). The last section concludes the paper.

Manuscript received June 1st, 2017.
K. Coulié, W. Rahajandraibe, H. Aziza and R. Vauché are with Aix-Marseille University, CNRS and University of Toulon, IM2NP UMR 7334, Marseille France. Gilles Micolau is with Université d'Avignon, UMR 1114 EMMAH (INRA-UAPV), Avignon, France

II. Context of the work

From the late 1970s, high failure rates, due to alpha particles and mainly inducing a financial loss and recall of products, had serious consequences for several semiconductor manufacturers like Xilinx [2]. With the decreasing of the supply voltage and of the node capacitances, the Soft Error Rate (SER) due to alpha particles presents yet a major reliability concern to logic processors because of the critical charge decreasing. Solutions which limit alpha contamination exist but are often costly. For example, packaging alternatives such as a lid coat or flip-chip strongly influence the alpha induced SER. The use of ^{10}B in the process can also greatly influence the SER by enhancing the alpha particle contribution. For a long time, fast neutrons were considered as the main danger for atmospheric applications but nowadays the alpha SER can be comparable to that of fast neutron induced SER [3-4]. To remain reliable, the nanoelectronic devices and systems must take into account upsets due to the transitory errors induced by alpha from radioactive contamination, and ideally, be hardened the best possible. As a consequence, one of the major issues for nanoelectronic companies is today to include/understand the phenomena related to the radioactive contamination in materials in order to prevent, detect, and analyze them.

Concerning medical application, a first clinical study reported in 1998 [5] has shown that cosmic neutron radiation could induce Single Event Upsets (SEU) in implanted devices and has evaluated its incidence in a large population of patients with Implantable Cardioverter Defibrillators, ICD [6]. Note that the SEU is the cell flip of a memory point following the crossing of an ionizing particle. In October 2005, a safety alert has been communicated for a few defibrillator models using a memory chip component, potentially affected by background levels of ionizing radiation [7]. This anomaly could trigger a temporary loss of pacing and permanent loss of defibrillation support. The manufacturer had been able to replicate this condition in a nuclear laboratory. The manufacturer also confirmed independently this anomaly by testing the specific chip component [8]. In 2008, Ferrick et al. reported that cosmic radiation during air travel was the explanation for the electrical reset observed at the follow-up clinic, in three patients with a Medtronic ICD. The exact time of the reset was logged and stored in the devices [9].

Finally, due to dimensions continuously shrinking, more and more effects appear in electronic devices. Among the encountered problems, those created by radioactive impurities and cosmic radiation are significant. The VCO based chain presented here is optimized for the detection and the identification of the particle fluxes lower than 10^9 particles per detection area and per second. Because of the principle of detection, this chain could be particularly suitable for the detection and recognition of low energy particles [10-12]. Then it could be possible to understand how low energy particles can disturb a circuit. Outcomes can benefit the electronic industry and especially electronic devices working in the atmospheric environment as medical application devices.

III. Presentation of the VCO system

The detection chain has been designed and implemented on 130nm CMOS SOI technology, then simulated at circuit level using "Spectre" simulator (SPICE-based) under Cadence DFII ©.

A. Modeling of the input current

At circuit level, the particle crossing the detection cell can be modeled by using a double exponential current pulse at the impacted node (Fig. 3) [7]:

$$I_{input}(t) = \frac{Q}{\tau_\alpha - \tau_\beta}\left(e^{-\frac{t}{\tau_\alpha}} - e^{-\frac{t}{\tau_\beta}}\right) \qquad (1)$$

where, Q is the total deposited charge, τ_α is the collection time constant of the junction and τ_β is the time constant for initially establishing the ion track. This current is injected at the input of the detection chain.

B. The indirect detection chain

The concept itself was presented in [10-11]. The VCO chain is composed of three parts: a CMOS based pixel detector, a shaping circuit based on a Voltage Controlled ring Oscillator (VCO) and a system for frequency and amplitude detection (Fig. 1). This system allows the evaluation of the circuit sensitivity to radiation by measuring the oscillator responses. In the very first versions of the system, the operating frequency was 1MHz. It reaches now 2GHz. This increase of the operating frequency makes possible the development of new methods for signal recognition. Thus if the frequency of the signal to be detected is ten times lower than the operating frequency, the shape of the signal can be directly reproduced at the output of the VCO. Then, the output parameters are now the frequency variation along the time and the average voltage variation (Fig. 5). The advantage of this solution is a direct recognition of the signal shape as an amplitude modulation of the VCO output frequency (Fig. 2).

The original version of the VCO was implemented in bulk technology. As the full integrated circuit (detector + readout electronic) is supposed to work in a radiative environment, the detection chain has to be hardened against radiations. Then, SOI technology seemed to be a good candidate for the integration of the VCO system. As our system is supposed to work under a low flux of particles, dose effects should not be a problem. The main problem could be Single Event Effects. Because of its structure itself, SOI is intrinsically tolerant to Single Events Effects.

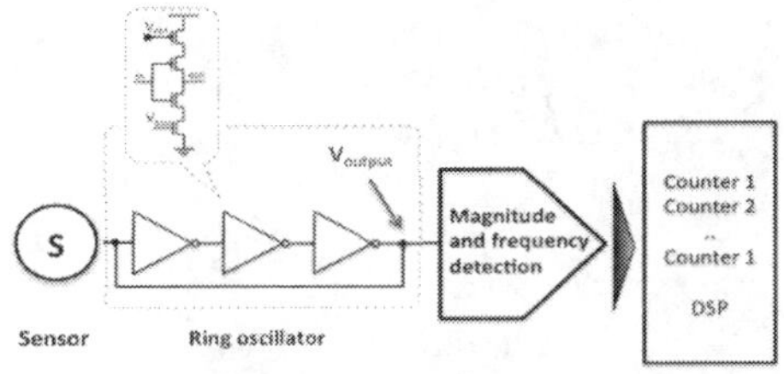

Fig. 1. Presentation of the VCO.

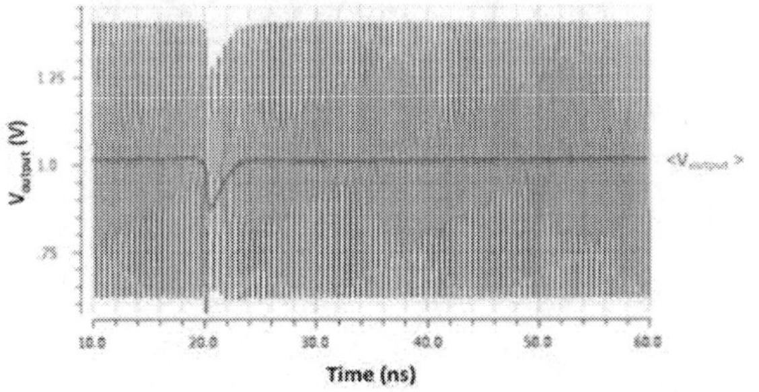

Fig. 2. Variation of the output voltage versus time for the indirect detection chain (VCO based).

SOI technology is suitable for the full integration of the whole system, including both analog and digital functionality necessary to identify the signal at the output of the VCO and to recognize the detected particles.

IV. Exploration of the limit of detection

The key point is how the output parameters of the VCO chain can give information about the input current. This could be done through the analysis of various characteristics extracted from the average voltage variation (Fig. 3). The first parameter is the variation of the average output signal (ΔV_{max}) versus the maximum of the input current (I_{max}) for different particle charges (Fig. 4). As shown in [10], the linearity of Fig. 4 is exactly the same for the different deposited charges. In [12], it has been demonstrated that two different currents with the same maximum value have the same corresponding ΔV_{max}. Then ΔV_{max} does not depend on the shape of the current but only on the current peak. Thus the measurement of ΔV_{max} appeared to be a reliable parameter to recognize the current peak of the input signal. The second parameter is the variation of Δt_{max} (Fig. 3). For the input, Δt_{max_input} is the difference between the time corresponding to the maximum of the current and the beginning of the current variation. For the average voltage at the output, Δt_{max_output} is the difference between the time corresponding to the minimum of the voltage and the beginning of the voltage variation. These two times correspond to peaking times. They are not considered in the paper. The last curve (Fig. 5) is the integral (area of the signal drop) of the average output signal Q_{out} versus Q_{in} (Fig. 3). This curve allows the charge identification through this detection system. Through these curves, the output parameters can be

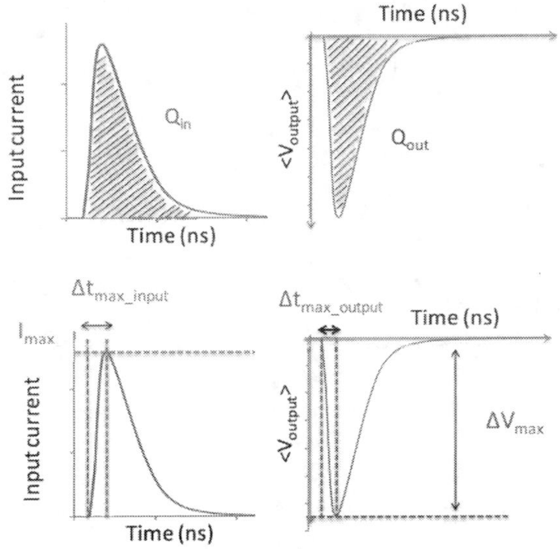

Fig. 3. Definition of the characteristics of the input and output signal .

linked to the input currents, what could allow the incoming particle identification.

To get reliable results, the resolution of the system has to be checked. The degradation of the system response is clearly visible in Fig. 4 and 5. The results are good if the points are included in the "good detection" zone. This zone corresponds to a linear response of the system in the range ±10 %. When results are out of this zone, the resolution is clearly degraded [13] and the pulse characteristics are also not suitable with the resolution of our system.

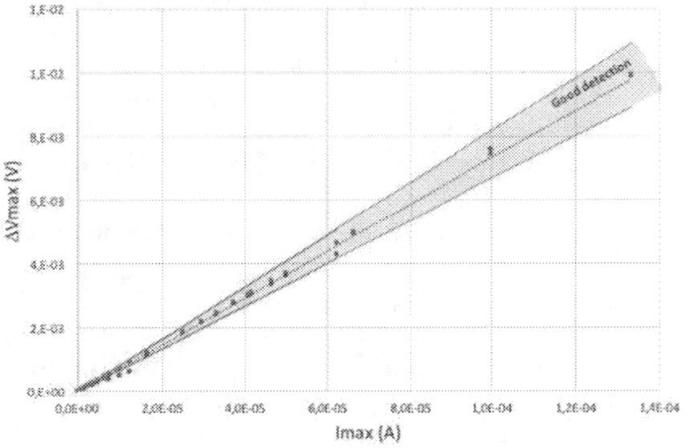

Fig. 4. Variation of the average output signal (ΔV_{max}) versus the maximum of the input current (I_{max}).

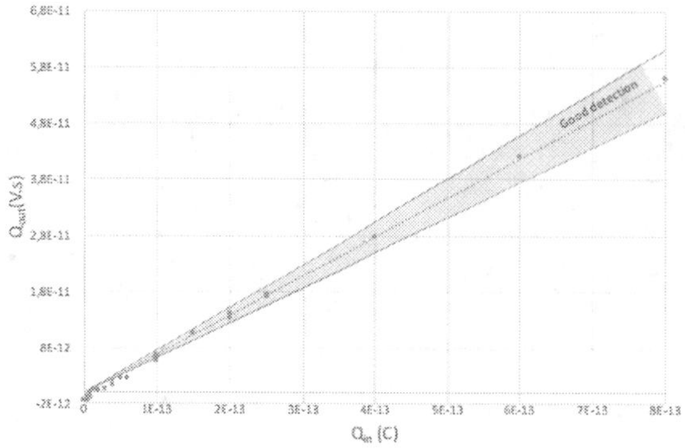

Fig. 5. The integral (area of the signal drop) of the average output signal Q_{out} versus Q_{in}.

V. CASE-STUDY: ALPHA PARTICLE CROSSING A 3X3 DETECTION CELL

A. The 3x3 detection cell

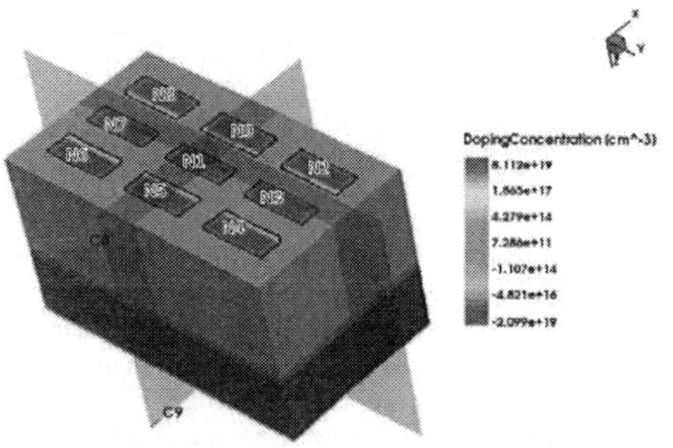

Fig. 6. 3x3 matrix including the name of the contact zones.

In order to improve the efficiency of the detection and the tracking ability, a 3x3 matrix is modeled and analyzed (Fig.6). The input current is no longer the simplified current given in equation (1). Now the current source comes from the realistic simulation of the matrix using TCAD device simulation tools (Synopsis ©) (Fig. 10). The effect of the ion strike is simulated using the Heavy Ion module of Synopsis [14], considering an electron-hole pair column centered on the ion track axis. The Linear Energy Transfer (LET) is defined as the energy lost by the particle, by unit of length. It varies along the track. In TCAD studies, a constant LET is sometimes used for simplification purposes. However, if the approximation of a constant LET is reasonable, it is not realistic. Indeed, this parameter varies along the track and is linked to the initial energy of the particle [15]. Thus a real variation of the LET was integrated in our simulations based on the value given by SRIM tables [16]. As this detector is suitable for low energy particles detection, the simulated ion is an alpha particle crossing the device with an initial energy of 0.6MeV which corresponds to a range of 2.3μm. The generation of such a particle in the whole structure is shown in Fig. 7 (2D heavy

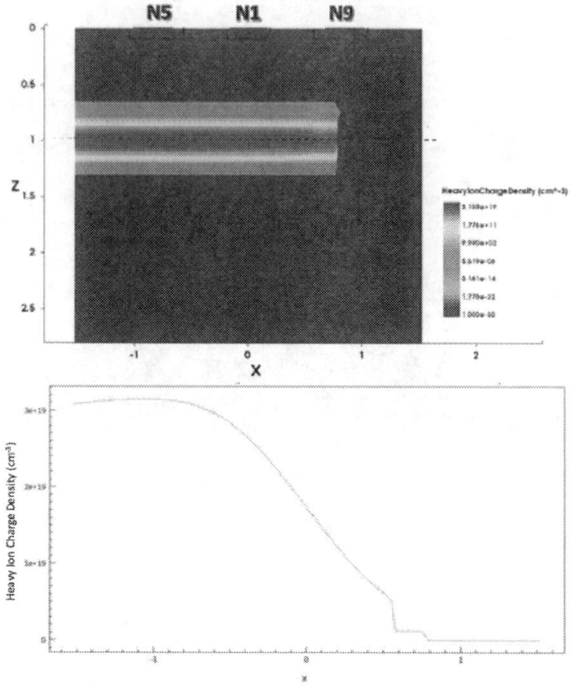

Fig. 7. 2D heavy ion charge density and 1D variation of the heavy ion charge density.

ion charge density and 1D variation of the heavy ion charge density). The evolution of the charge density is shown in Fig. 8 and Fig. 9 for the two cut-planes presented in Fig. 6. The particle is generated at 50ps. The current generated by the particle crossing are shown in Fig. 10 at each electrode.

B. Alpha particle tracking using the VCO

The nine currents generated using Synopis were injected in the nine VCOs in order to test the suitability of the studied VCO to our detection cell. Unfortunately, the resolution of our system was not good enough to allow a good the detection of the currents generated by the detection cell. This was expected because the duration of the signal is near 0.4ns, what corresponds to an input signal frequency of 2.5GHz, too fast to be detected by a 2GHz oscillator. Thus, even if an average voltage is obtained for the zone labeled N5 (Fig. 11), the Imax-ΔVmax couple is clearly out of the detection zone (Fig. 12) what means that the other lower currents are also out of this zone. Moreover, except for the N1 and N5 average voltages, the others are in the noise limit. So we cannot be confident about the obtained curves of Fig. 11. Then we tested the same currents with another VCO developed with a higher frequency (4.3GHz). The detection is better (Fig. 13). However, we cannot conclude about the resolution without the linearity curve.

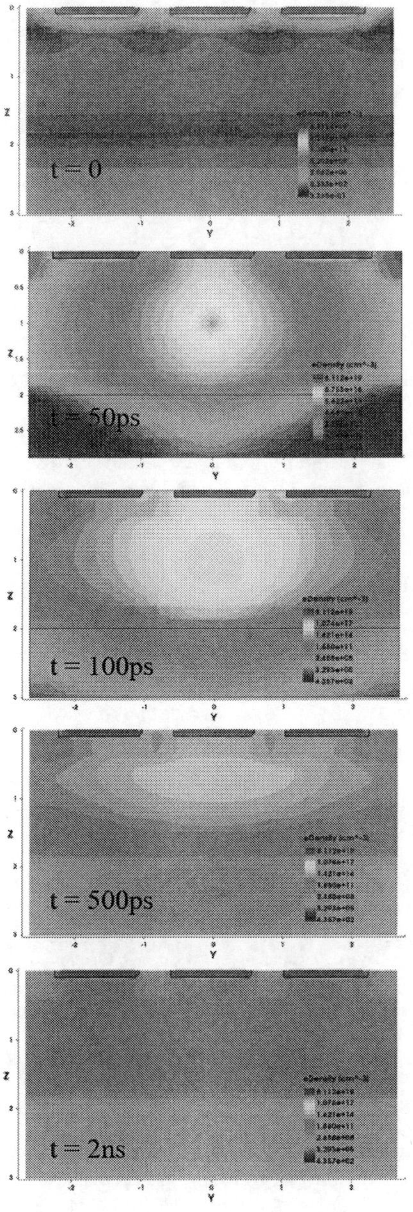

Fig. 8. Evolution of electron density in the Y direction.

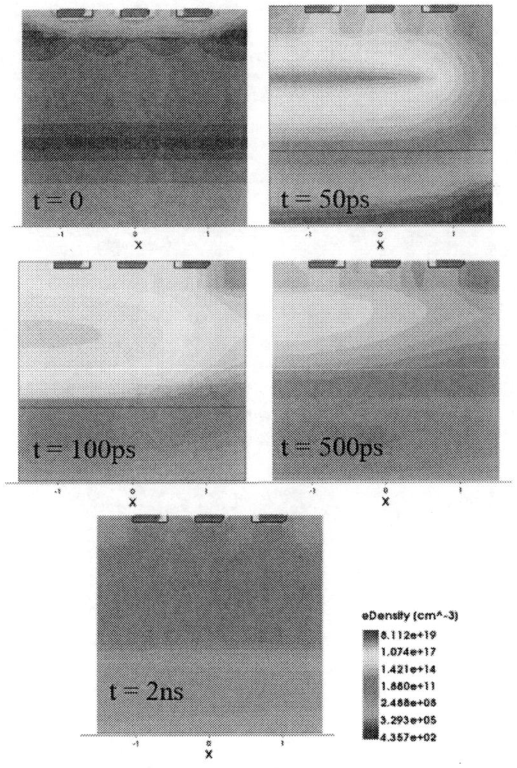

Fig. 9. Evolution of electron density in the X direction.

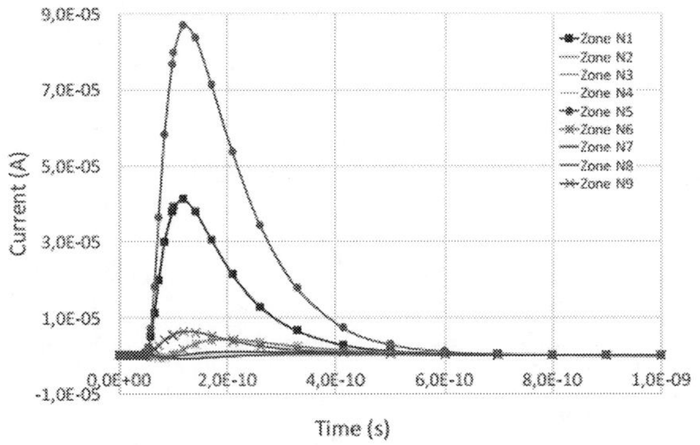

Fig. 10. Currents from the pixels

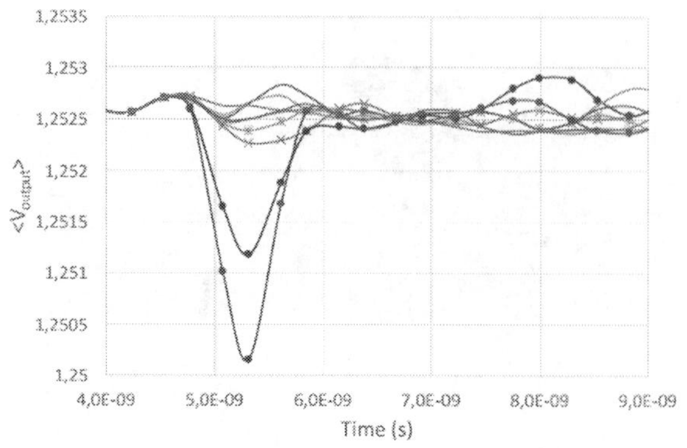

Fig. 11. Average voltages corresponding to the currents of Fig. 10.

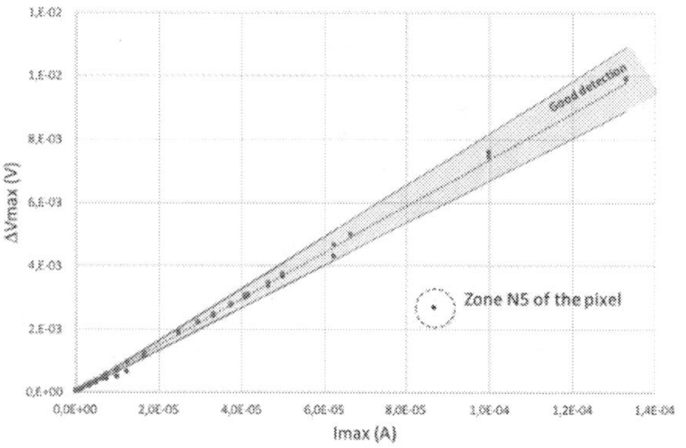

Fig. 12. Variation of the average output signal (ΔV_{max}) versus the maximum of the input current (I_{max}) including zone N5.

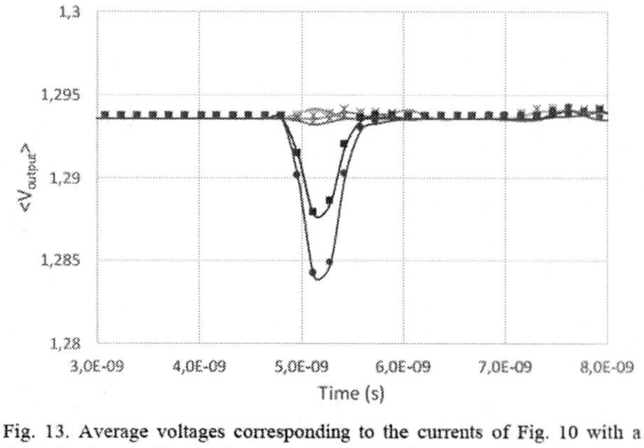

Fig. 13. Average voltages corresponding to the currents of Fig. 10 with a VCO developed with a higher frequency (4.3GHz).

VI. Conclusion

The detection chain presented in this work is based on an innovative concept based on an indirect detection through a voltage controlled oscillator. The analysis of the results extracted from this chain showed the possibility to recognize the input signal by using the average voltage of the output signal. The main advantage of this system is the possibility to get all the information related to the input signal (shape and order of magnitude of the current and duration) with a good sensitivity of detection.

Here we have explored the detection limit of the presented VCO. For the pixel matrix simulated in this paper, a higher frequency has to be used. Indeed, increasing the time resolution of the detected pulse is required and the system resolution could be tuned by changing the oscillation frequency. Calibrating the oscillator frequency is also a solution for spectrometry analysis. In that case, several VCO working at different frequency could be used, giving the opportunity to characterize accurately a given environment.

References

[1] Ying Zhang, Christine Hu-Guo, Daniel Husson, The-Duc Le , Stéphane Higueret, Yann Hu, "A High-Sensitivity Low-Power CMOS Sensor for a Future Neutron Personal Dosimeter", in IEEE Trans. On Nucl. Sc., vol.59, no.4, pp. 1465-1471, August 2012.

[2] Xilinx, Customer Advisory: Flip-Chip Package Substrate Solder Issue, January 19, 2004.

[3] J.F. Ziegler, H. Puchner, "SER – History, Trends and Challenges: A Guide for Designing with Memory ICs", pp. 3-18, 2004.

[4] L.W. Massengill, "Cosmic and Terrestrial Single-Event Radiation Effects in dynamic Random Access Memories", in IEEE Trans. On Nucl. Sc., vol.43, no.2, pp. 576-593, April 1996.

[5] Trigano, A; Hubert, G; Marfaing, J; Castellani-Coulié, K "Experimental study of neutron-induced soft errors in modern cardiac pacemakers", Journal of Interventional Cardiac Electrophysiology (JICE), 2010

[6] Bradley, P., & Normand, E. (1998). implantable cardioverter defibrillators. Institute of Electrical and Electronics Engineers. IEEE Transactions on Nuclear Science, 45 (6), 2929–2940.

[7] Physician Advisory from Saint Jude Medical, Inc, October 6, 2005. Kohanyi, G. and Fain, E.

[8] Saint Jude Medical advisories. Available at: http://www.hc-sc.gc.ca/dhp-mps/medeff/advisories-avis/prof/2005. Accessed 14 April 2011.

[9] Ferrick, A., Bernstein, N., Aizer, A., & Chinitz, L. (2008). Cosmic radiation induced software electrical resets in ICDs during air travel. Heart Rhythm, 5(8), 1201–1203.

[10] K. Castellani-Coulié, H. Aziza, W. Rahajandraibe, G. Micolau, J.-M. Portal, "Development of a CMOS Oscillator Concept for Particle Detection and Tracking", IEEE Trans Nucl. Sci., 2013.

[11] K. Coulié-Castellani, W. Rahajandraibe, G. Micolau, H. Aziza, J.-M. Portal, "Improvement of a detection chain based on a VCO concept for microelectronic reliability under natural radiative environment", IEEE LATW Conference, 2015.

[12] S. Ben Krit, K. Coulié-Castellani, W. Rahajandraibe, L. Ottaviani, G. Micolau, H. Aziza, J.-M. Portal, "Comparison of a Readout Chain Dedicated to the Signal Conditioning of a Particle Detector and an Innovative Chain Based on a VCO Concept", RADECS 2015, Moscou, Russia, September 2015.

[13] Castellani-Coulié K., Aziza H., Micolau G. and Portal J-M., "Optimization of SEU Simulations for SRAM Cells Reliability under Radiation", Journal of Electronic Testing, February 2012

[14] TCAD Sentaurus user's manual (Synopsys).

[15] A. Akkerman, J. Barak, D. Emfietzoglou, "Ion and electron track-structure and its effects in silicon: model and calculation", Nuclear Instruments and Methods in Physics Research, pp. 319-336, 2005.

[16] www.srim.org

Development of High Resolution Gamma-Detectors for PET

A. Kobayashi, E. Yusaku, H. Ito, H. Kawai, T. Mizuno, F. Kento, S. Kimura

I. Introduction

RECENTLY we are developing a new gamma ray detector using wavelength-sifting-fibers (WLSFs) and scintillator plate (34 mm × 34 mm × 4 mm) for Positron Emission Tomography (PET) [1]. It is expected the position resolution and the energy resolution are a few mm [2] and approximately 10% [3], respectively. Mainly the detector has two advantages. First, it can reduce the expense of cutting scintillator into small pieces. Second, it probably achieves the high position resolution by using 0.2 mm phi WLSFs.

In this study, the detector performance for 511-keV gamma ray is estimated by an experiment.

II. Concept of detector

The detector consists of six plate detectors (the effective area is 300 mm × 300 mm) setting up around the human body. The plate detector consists of 8 layers. The structure of a layer is shown in Fig.1.

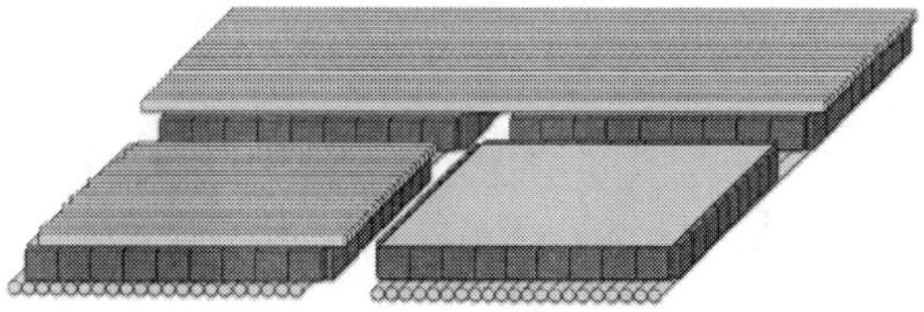

Fig. 1. A schematic of a layer sample which consist of 2 × 2 scintillator plates. WLSFs attached to the surface and SiPMs attached to the side.

Each layer consists of 8 times 8 scintillator plates. The size of each scintillator plate is 34 mm × 34 mm × 4 mm. 160 wavelength-sifting fibers (WLSFs) cover each 34 mm × 34 mm surface of each plate along each axis (x or y). Ends of

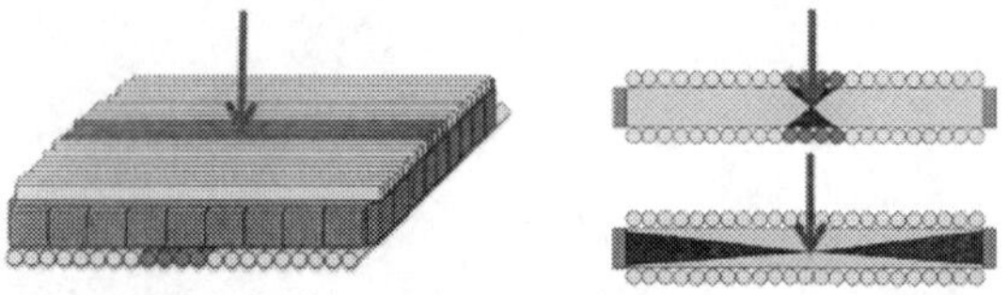

Fig. 2. The method of gamma ray detection.

each fiber are connected to silicon photomultipliers (SiPMs). On each side (34 mm × 3.4 mm surface) of plate 10 SiPMs are glued. (Fig.1)

Fig.2 shows a method of gamma ray detection. When a gamma incidents to the plate and deposit energy, the light emitted. The light satisfied total reflection condition propagates WLSFs, detected by SiPMs connected end of the fiber and the incident position of gamma is measured. On the other hands, the light leaked out from the crystal detected by glued SiPMs on side. Then deposit energy and incident time of gamma are measured.

III. Experiment

The setup is shown in Fig.3.

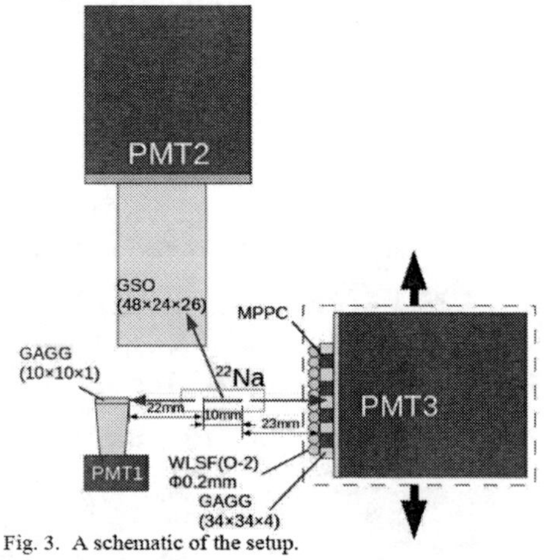

Fig. 3. A schematic of the setup.

of reference consists of a small GAGG scintillator ($Gd_3Al_2Ga_3O_{12}$(Ce), 10 mm × 10mm × 1mm, C&A Co., Ltd.) and PMT1. The other one consists of GSO (48 mm × 24mm × 26mm, Asahi Kasei Co., Ltd.) and PMT2. A main detector consists of a big GAGG scintillator (34 mm × 34 mm × 4 mm, C&A Co., Ltd.), WLSF[O-2] (Kuraray Co., Ltd.), MPPCs

This Fullpaper is submitted June 1, 2017.
Atsushi Kobayashi is with Graduate School of Science and Engineering, Chiba University, Chiba, Japan (e-mail: koba@hepburn.s.chiba-u.ac.jp).
Yusaku Emoto is with Graduate School of Science, Chiba University, Chiba, Japan (e-mail: yusaku@hepburn.s.chiba-u.ac.jp).
Hiroshi Ito is with Graduate School of Science, Chiba University, Chiba, Japan (e-mail: hiroshi@hepburn.s.chiba-u.ac.jp).
Hideyuki Kawai is with Graduate School of Science, Chiba University, Chiba, Japan (e-mail: kawai@hepburn.s.chiba-u.ac.jp).
Takahiro Mizuno is with Graduate School of Science, Chiba University, Chiba, Japan (e-mail: mizuno@hepburn.s.chiba-u.ac.jp).
Shota Kimura is with Graduate School of Science, Chiba University, Chiba, Japan (e-mail: shota@hepburn.s.chiba-u.ac.jp).
Kento Hujihara is with Graduate School of Science, Chiba University, Chiba, Japan (e-mail: kento@hepburn.s.chiba-u.ac.jp).

(Hamamatsu Photonics Co., Ltd.) and a PMT3 (Hamamatsu Photonics Co., Ltd.). 160 WLSFs are used in the main detector and each 20 fibers are tied and connected to MPPCs. At each side of GAGG, 4 MPPCs are glued.

The small GAGG is at the front of the big GAGG and ^{22}Na shielded source is between them. When reference using PMT1 and the main detector observe energy more than 341 keV as well as reference using PMT2 observe 700 keV simultaneously, PMT1~3 and MPPCs are read signals. Therefor narrow 511-keV gamma beam (its width is approximately 1mm) enter to the main detector. We did measurements shifting position of the main detector 2 mm apart.

IV. RESULTS AND DISCUSSION

The relative of energy of two references and distribution of PMT3 is shown in Fig.4.

Events that PMT1 and PMT3 observe 511-keV gamma ray

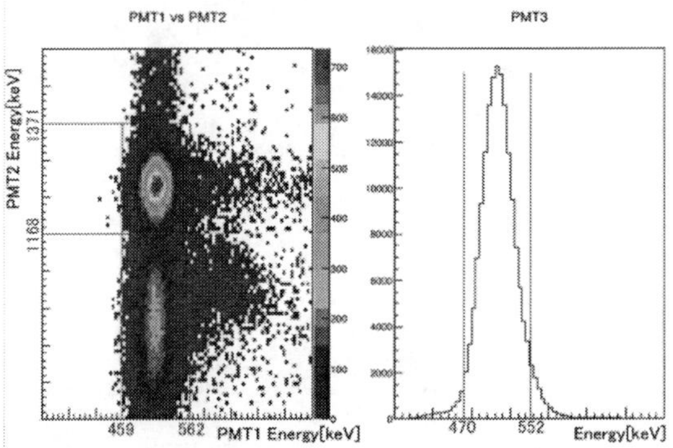

Fig. 4. Distributions of PMTs.

and PMT2 observes 1.27 MeV gamma-ray are selected. Ranges of selection are ± 2 sigma for each peak.

The total number of photoelectrons of MPPCs connected fibers is 4.34 ± 0.02. The incident position is reconstructed by the center of gravity calculation of fiber's position and number of photoelectron of connected MPPC. Fig.5 is distribution of reconstruct position at the incident position of 2mm.

The relative of reconstruct position and the incident position

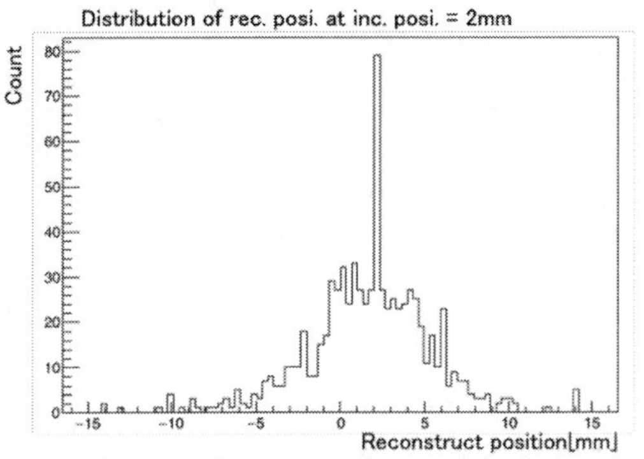

Fig. 5. Distribution of reconstruct position at the incident position of 2mm.

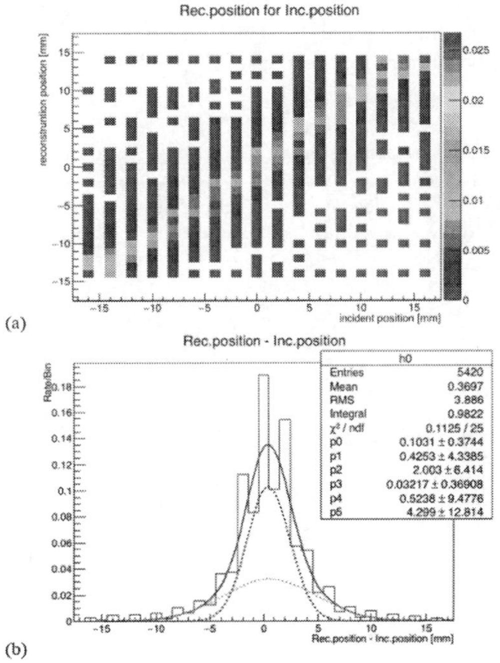

Fig. 6. Those figures show reconstruction of incident position.
(a) A Figure of Reconstruct position vs original position.
(b) A dfference of original position and reconstructed position.

is shown in Fig.6(a). In addition distribution of difference of original incident position and reconstructed position is shown inFig.6(b).

In Fig.6(b), the red line is double Gaussian fitting that is summation of the brue one and green one. The position resolution is calculated 5.4 mm from red one. In this experiment, each 20 WLSFs are tied. If WLSFs are not tied and read one by one, the position resolution probably achieve less than 1mm.

Fig.7 is the distribution of total photoelectron of glued MPPCs at side.

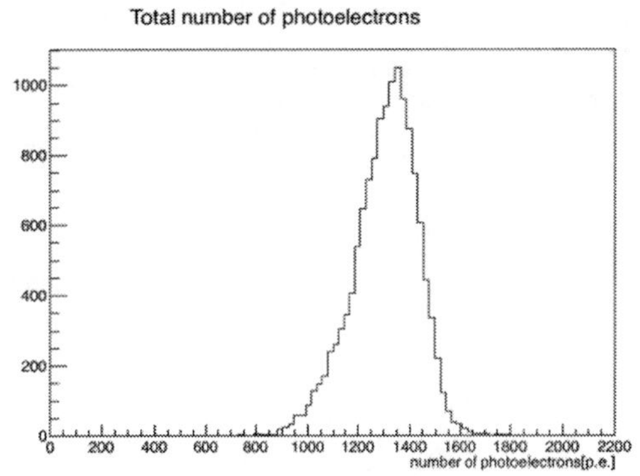

Fig. 7. The distribution of total photoelectron.

The peak of distribution is 1410 p.e.. The energy resolution is also calculated and it is approximately 18.8%(FWHM). In this experiment, 4 MPPCs are glued on a side of GAGG. It mean that effective area of MPPCs cover only 26.5% of surface. If 10 MPPCs are used and cover 78% of surface, a number of photoelectrons become two-and-a-half-times. The energy resolution also become less than 12%.

V. Conclusion

A new PET detector with WLSFs are developed and estimated performance. The position resolution and the energy resolution of demonstration are 5.4 mm and 18.8%, respectively. In addition, we are conducting more accurate experiment and producing a prototyape detector. Its result also will be shown in this conference.

References

[1] H.Ito, *et al.*, IEEE Medical Imaging Conference 2014, 8-15 Nov. 2014, Seattle, Washington, USA
[2] A.Kobayashi, *et al.*, Advancements in Nuclear Instrumentation Measurement Methods and their Applications, 223, Apr. 20-24 2015, Lisbon Convention Center, Portugal
[3] N.naomi *et al.*, Advancements in Nuclear Instrumentation Measurement Methods and their Applications, 220, Apr. 20-24 2015, Lisbon Convention Center, Portugal

Evaluation of Position Resolution for a Prototype Whole-Body PET Detector Based on Suppressing Backgrounds by Compton Scattering

Kento FUJIHARA, Yusaku EMOTO, Hiroshi ITO, Naomi KANEKO, Hideyuki KAWAI, Atsushi KOBAYASHI, and Takahiro MIZUNO

***Abstract*— Existing PET (Positron Emission Tomography) systems make clear images in demonstration (measuring small PET reagent in pure water), however images in real diagnosis become unclear. The authors suspected that this problem was caused by Compton scattering in a detector. When PET systems observe plural photomultiplier tube outputs, an original emission point is regarded as centroid of the outputs. However, even if plural emission in Compton scattering occur, these systems calculate original point in the same way as single emission. Therefore, the authors considered that rejecting Compton scattering events makes PET systems much better, and made prototype counter. Main components of the prototype counter are plate-like high-growth-rate (HGR) La-GPS scintillators and wavelength shifting fibers (WLSF). HGR crystals grow 10 times as fast as a mono-crystal (a normal mono-crystal grows at 2 – 3 mm an hour). Thus, it includes microbubble and its transparency get worth. Consequently, HGR crystals usually are not used in radiation measuring instruments. However, this time they are used on the purpose. Because of their low transparency, scintillation lights come out right above and right under of emission position. Therefore, Compton scattering events is rejected easily. The prototype detector has an effective area of 300 by 300 square mm. The detector consists of 24 layers. One layer consists of HGR La-GPS scintillator of 1 mm thickness. Top and bottom surface of scintillator were covered by dual sheets of WLSF with a diameter of 0.2 mm. Sheets of WLSF on top and bottom of the scintillator make a right angle with each other, and measure X- and Y-components. Z-component is measured by difference of WLSF outputs between top and bottom. If plural layers output signals, this counter regards the event as Compton scattering event, and reject the event. Even if only a layer output signals, the event is rejected when number output signals from WLSF is more than 1.5 times of single emission. Material cost of this system is, 0.2M$ for HGR La-GPS, 0.03M$ for WLSF, 0.03M$ for 600 units of 6 by 6 mm SiPM's, 0.12M$ for 12000 units of 1 by 1 mm SiPM's, and 0.09M$ for 1800 channel of signal readout circuits. Considering total cost, price of this PET will be set 1M$ or less. This idea was confirmed with numerical simulation and experimentation. In experimentation, position resolution in photoelectric absorption was 0.2 mm, and minimum distance that this detector could recognize plural emission in Compton scattering was 1 mm. In parallel, three kinds of model were made: a prototype detector, all the signals readout method, and resistance delay method. Simulation setting was 2 MBq/L in normal tissue and 10 MBq/L in cancer. As a result of simulation, a prototype detector identified 3 mm cancer, however the others made unclear image and was not able to identified cancer. That is to say, the prototype detector is able to reject Compton scattering events and inexpensive. Therefore, whole-body PET system with this detector must diagnose cancer with a diameter of 3 mm or more and be priced 1M$ or less**

This manuscript was submitted on June 1, 2017.

K. FUJIHARA is with the Graduate School of Science and technology, Chiba University, Chiba, 263-8522 Japan (e-mail: kento_fujihara@hepburn.s.chiba-u.ac.jp).

***Index Terms*—Whole-body PET, La-GPS, Wavelength sifting fiber**

I. Introduction

EXISTING PET systems have a problem caused by Compton scattering. In scintillators, scattering occur 4 times as much as photoelectric absorption with 0.511-MeV gamma ray. Even if they have high position resolution, misidentifying plural emissions as a single emission makes reconstructed image unclear. Based on this problem, we are developing a lower cost popular type PET.

II. Method

Main components of the system are wavelength-shifting fibers (WLSF) and platelike high-growth-rate (HGR) La-GPS ($[Ce_{0.01}La_{0.24}Gd_{0.75}]_2Si_2O_7$) scintillators.

WLSF is an optical fiber kneaded fluorescent material into. It can transmit light from the side of a fiber.

HGR crystals are produced at lower cost and have low transparency because of microbubbles. This is useful for reducing cost and identifying scattering. In this system, emission positions are measured by WLSFs' sheets on top and bottom of scintillators. If transparency of scintillator is moderately low, scintillation lights don't spread in scintillator; nevertheless, sufficient quantity of light is ensured.

The size of detection area is 300 mm × 300 mm. The detector consists of some scintillators. Top and bottom surface of scintillator were covered by dual sheets of WLSF with a diameter of 0.2 mm. Sheets of WLSF on top and bottom of the scintillator make a right angle with each other, and measure X- and Y-components. Z-component is measured by the layer number and difference of the number of WLSFs outputting signal between top and bottom. If plural layers output signals, this counter identifies the event as scattering event. Even if only a layer output signals, the event is identified as a scattering event when the number of WLSFs outputting signals is 1.5 times more than that of a single emission. When the event is identified as scattering event, this system regard nearest emission point to body as the first emission point.

 https://doi.org/10.1051/epjconf/201817009004

III. Experiment and simulation

In this study, two preliminary tests were performed for developing the PET detector by using a ^{22}Na sources (Fig. 1-2), where the consistency was demonstrated by using a numerical simulation, Monte Carlo code: GEANT4 (Fig. 3).

In experiment 1, we measured amount of luminescence from La-GPS (1 mm thickness, top surface of scintillator is covered by sheets of WLSF with a diameter of 0.2 mm) with ^{22}Na gamma-ray source. The number of sheets is from 1 to 6. On each number of sheets, we measured 3 runs and calculated the average. (One run includes 10000 events.)

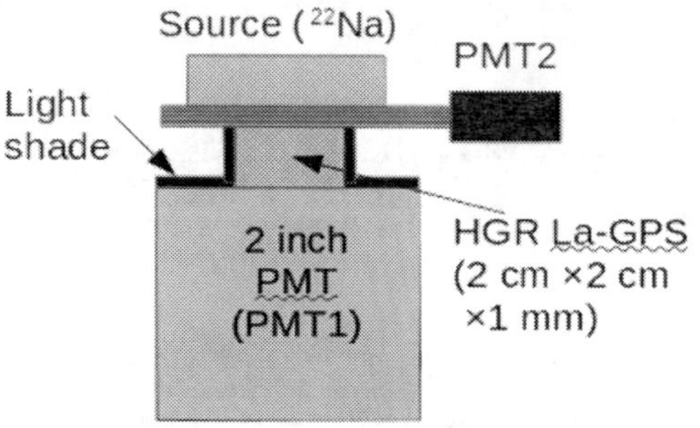

Fig. 1. This is the setting of experiment 1. PMT1 is the trigger detector.

In experiment 2, we measured position resolution of this system. PMT0 and PMT1 are event triggers. Scintillator on PMT1 has small solid angle, therefore gamma-rays from ^{22}Na become narrow. The fiber sheet has 48 fibers and each of them is connected to SiPM. Signals from fibers are analyzed with centroid formula given by

$$\mathrm{pos}_{rec} = \frac{\sum_i n_{p.e.,i} \times \{{}^{10.0}/_{48} \times (i+0.5)\}}{\sum_i n_{p.e.,i}}. \qquad (1)$$

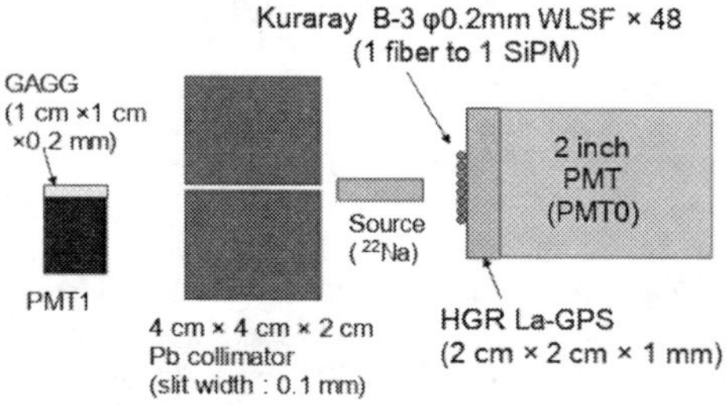

Fig. 2. This is the setting of experiment 2. 511-keV gamma-ray hits center of PMT0.

In the simulation, six detectors surround human (30 cm in diameter and filled with water) like a hexagon. One detector has

TABLE I: Parameters of simulation

Parameter	Quantity
Radioactivity concentration (normal tissue)	2 Mbq/kg
Radioactivity concentration (cancer)	10 Mbq/kg
Width resolution	1 mm
Depth resolution	1 mm
Energy resolution	13.4% (σ)
Time resolution	no error

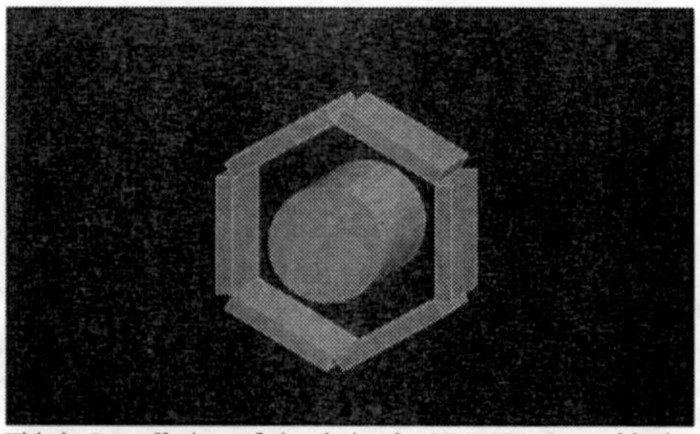

Fig. 3. This is Overall view of simulation in GEANT4. Green blocks are La-GPS scintillators, and a blue tube is human.

300 mm width and 24 mm thickness. If distance between two reaction point is longer than 1 mm, the event is regarded as a scattering event.

IV. Results

In experiment 1, detector with HGR La-GPS and WLSF (B-3, Kuraray) measured 22.3 ± 5.96 photoelectrons (p.e.). in 460-560 keV and 14.1 ± 5.58 p.e. in 290-390 keV from single-face, single-end (Fig. 4-6). This is better value than detector with normal La-GPS. By light output from La-GPS and refractive indexes of La-GPS and WLSFs, efficiency of WLSFs is

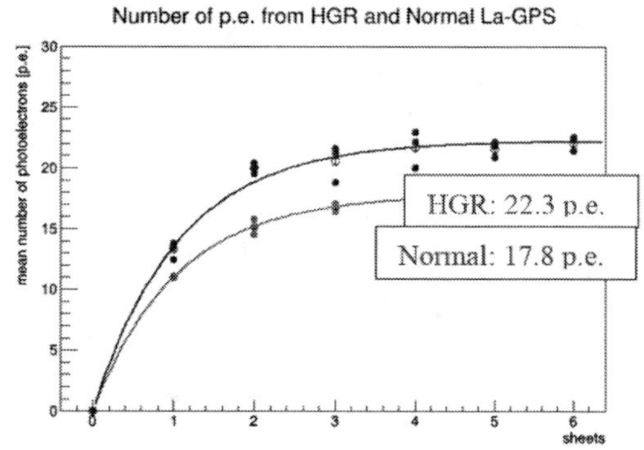

Fig. 4. This is a graph of relationship between the number of WLSF sheets and the number of p.e. in 511-keV gamma rays.

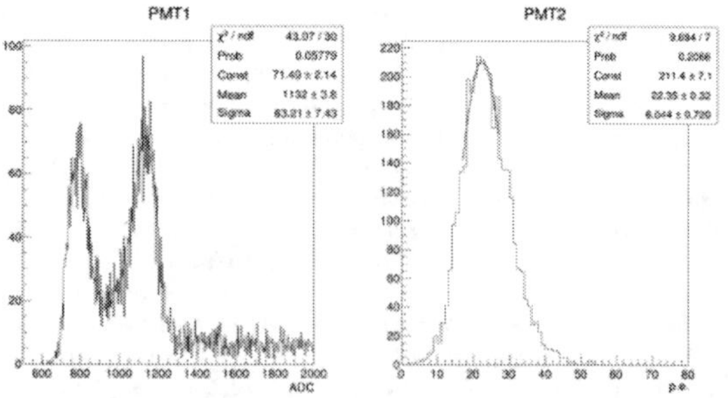

Fig. 5. This is a graph of the number of p.e. in 460-560 keV gamma rays with 6 WLSF sheets.

estimated to be about 4.07%. In addition, we estimated 88 photoelectrons and 36% resolution in 511-keV from both-faces, both-ends readout.

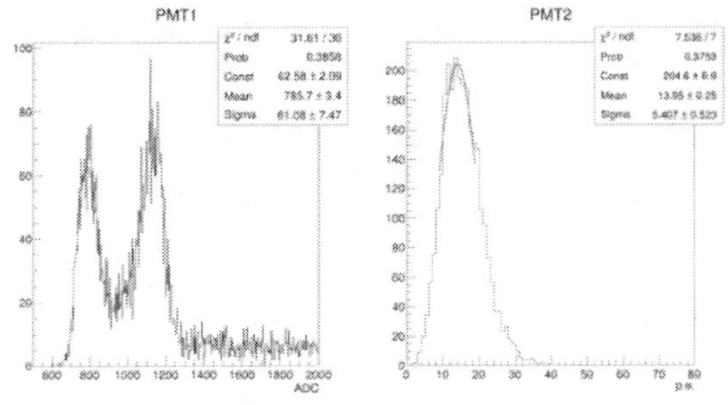

Fig. 6. This is a graph of the number of p.e. in 290-390 keV gamma rays with 6 WLSF sheets.

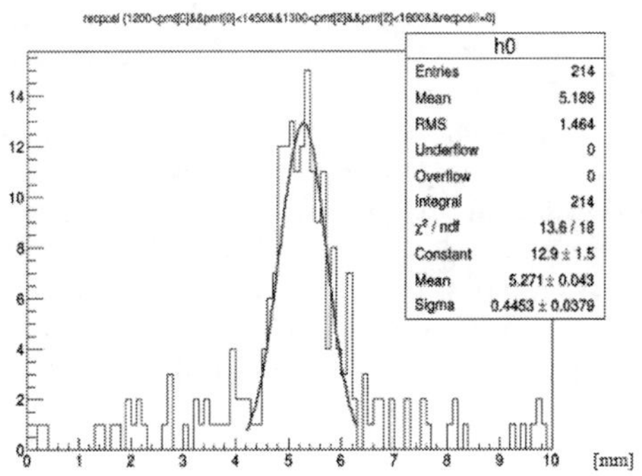

Fig. 7. This is a reconstructed image of the simulation. Central part of histogram is the cancer area.

In experiment 2, position resolution is 1.04 mm (FWHM) with 1 mm thickness La-GPS (Fig. 7.).

In simulation, background level is 25.4 ($\sigma = 5.04$) and cancer level is 17.1. Cancer level is more than 2σ, therefore cancer is visible (Fig. 8-9). If the criterion is set that 21 voxels in 3 mm

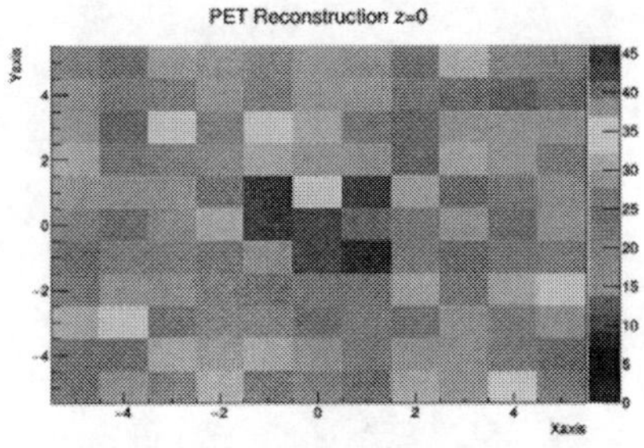

Fig. 8. This is a reconstructed image of the simulation. Central part of histogram is the cancer area.

by 3 mm by 3 mm area (27 voxels) are more than 2σ from background level, misidentifying probability is 1.55×10^{-35}, and missing probability is 1.04×10^{-6}.

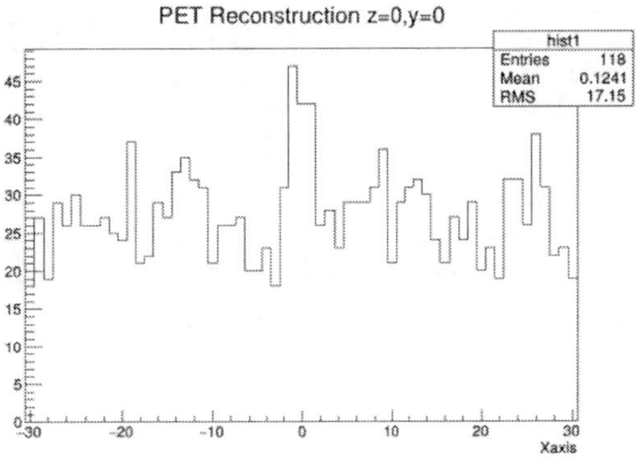

Fig. 9. This is a single dimension reconstructed image of the simulation. Cancer level from background is 17.

V. Discussion

In terms of intensity and position resolution, HGR scintillator has sufficient ability. Linearity of number of p.e. is confirmed; Therefore, the number of p.e. with WLSFs in *460-560* keV is about 1.5 times as it in *290-390* keV. Value of *mean – HWHM in 460-560 keV* is 74.1 and value of *mean + HWHM in 290-390 keV* is 69.5. Therefore, this detector can distinguish between absorption peak and Compton edge. Position resolution is 1.04 mm. In the simulation reflected experiment results, 3 mm cancer is visible. This means that the system is feasible and can reduce the cost by HGR scintillators. However, correlation between growth late of HGR La-GPS and transparency is an important subject. We considered that this is closely related to optimal thickness of scintillators, definitive cost and position resolution.

VI. Conclusion

PET's problem about resolution arise from Compton scattering. Therefore, identifying them makes PET systems much better. According to experiment and simulation, this detector is sufficient for our plan.

References

[1] Berger M J; Hubbell J H; Seltzer S M; Chang J; Coursey J S; Sukumar R; Zucker D S; Olsen K: NIST XCOM: Photon Cross Section Database http://physics.nist.gov/PhysRefData/Xcom/html/xcom1.html (retrieved on the 8th of May 2017)

A large area diamond-based beam tagging hodoscope for ion therapy monitoring

M.-L. Gallin-Martel, L. Abbassi, A. Bes, G. Bosson, J. Collot, T. Crozes, S. Curtoni, D. Dauvergne, W. De Nolf, M. Fontana, L. Gallin-Martel, J.-Y. Hostachy, J. Krimmer, A. Lacoste, S. Marcatili, J. Morse, J.-F. Motte, J.-F. Muraz, F. E. Rarbi, O. Rossetto, M. Salomé, É. Testa, R. Vuiart, M. Yamouni

Abstract—. **The MoniDiam project is part of the French national collaboration CLaRyS (Contrôle en Ligne de l'hAdronthérapie par RaYonnements Secondaires) for on-line monitoring of hadron therapy. It relies on the imaging of nuclear reaction products that is related to the ion range. The goal here is to provide large area beam detectors with a high detection efficiency for carbon or proton beams giving time and position measurement at 100 MHz count rates (beam tagging hodoscope). High radiation hardness and intrinsic electronic properties make diamonds reliable and very fast detectors with a good signal to noise ratio. Commercial Chemical Vapor Deposited (CVD) poly-crystalline, heteroepitaxial and monocrystalline diamonds were studied. Their applicability as a particle detector was investigated using α and β radioactive sources, 95 MeV/u carbon ion beams at GANIL and 8.5 keV X-ray photon bunches from ESRF. This facility offers the unique capability of providing a focused (~1 μm) beam in bunches of 100 ps duration, with an almost uniform energy deposition in the irradiated detector volume, therefore mimicking the interaction of single ions. A signal rise time resolution ranging from 20 to 90 ps rms and an energy resolution of 7 to 9% were measured using diamonds with aluminum disk shaped surface metallization. This enabled us to conclude that polycrystalline CVD diamond detectors are good candidates for our beam tagging hodoscope development. Recently, double-side stripped metallized diamonds were tested using the XBIC (X Rays Beam Induced Current) set-up of the ID21 beamline at ESRF which permits us to evaluate the capability of diamond to be used as position sensitive detector. The final detector will consist in a mosaic arrangement of double-side stripped diamond sensors read out by a dedicated fast-integrated electronics of several hundreds of channels.**

Index Terms— **Synthetic diamond detectors, on-line monitoring of hadron therapy, charge collection efficiency, time resolution, integrated fast read out electronics**

I. Introduction

THE treatment of tumors by a beam of carbon ions or protons is an alternative cancer therapy known as hadron therapy. Ions deposit a large fraction of the dose at the end of their path, in the Bragg peak. Compared to conventional X-ray radiotherapy, hadron therapy allows a more efficient dose delivery in the tumor, with a reduction of the dose deposited in the nearby healthy organs. The conformation of the deposited dose to the tumor volume is provided by distributing Bragg peaks in this volume, either spot by spot, or by means of passive scattering and energy degradation. However, since multiple sources of uncertainty on the ion range may cause deviations from the planned dose distribution [1], online control of the ion range is desired in order to reduce safety margins and optimize the ballistic advantage of ion therapy.

During an irradiation with ion beams, nuclear reactions occur for part of the projectiles along their path in the patient body, and photons in the range 1-10 MeV are emitted almost isotropically within much less than a picosecond after the reactions. It has been shown experimentally that the longitudinal distribution of such prompt-gamma production is highly correlated to the primary ion range [2], [3]. In the CLaRyS collaboration, two systems for prompt gamma detection are currently under development [4], [5], [6]: a

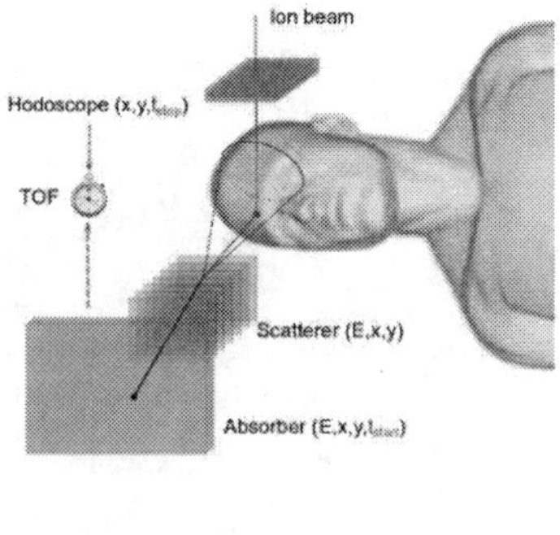

Fig. 1 Scheme of the combination of a Compton camera and a beam hodoscope for time of flight prompt-gamma range verification.

Manuscript written July, 2017. This work was supported by Labex Primes, France Hadron and MONODIAM-HE.

ML. Gallin-Martel, A. Bes, G. Bosson, J. Collot, S. Curtoni, D. Dauvergne, L. Gallin-Martel, JY Hostachy, A. Lacoste, S. Marcatili, JF Muraz, FE. Rarbi, O. Rossetto, R. Vuiart and M. Yamouni are with Laboratoire de Physique Subatomique et de Cosmologie, CNRS/IN2P3, Université Grenoble-Alpes, UMR 5821 38026 Grenoble Cédex France (corresponding author ML. Gallin-Martel e-mail : mlgallin@lpsc.in2p3.fr)

M. Fontana, J. Krimmer and E. Testa are with IPNL, Université de Lyon, Université Lyon 1, CNRS/IN2P3, UMR 5822, France,

W. De Nolf, J. Morse and M. Salomé are with ESRF, Grenoble, France.

L. Abbassi, T. Crozes and JF. Motte are with the Neel Institute, Université Grenoble-Alpes and CNRS UPR2940, Grenoble, France.

collimated gamma camera and a Compton camera (Fig. 1). The originality of the two systems is the use of Time Of Flight (TOF) in order to reduce the background. Depending on the beam time-structure and intensity, a beam tagging hodoscope may be necessary to detect the arrival time of ion bunches or even individual projectiles. This hodoscope will be also used for transverse position measurement, giving access to 3D prompt-gamma imaging. Thus the hodoscope plays a major role in all the detection concepts, and has to overcome the issue of providing position and timing information of the incident hadron beam at about 100 MHz count rates. The collaboration developed a scintillating-fiber based hodoscope, read-out by photomultipliers [6], [7]. In order to improve the timing resolution, and to overcome limitations such as limited count rates of photomultipliers, the collaboration initiated the MoniDiam project which aims to develop a diamond-based hodoscope and its dedicated integrated fast read-out electronics.

II. Material and methods

A. Diamond Sensors

Chemical Vapor Deposition (CVD) synthetic diamonds are grown by the technique of microwave plasma enhanced CVD. Their qualities and detection properties are reproducible by tightly controlling their growth parameters. For the homo-epitaxial ones, diamond is used as a seed for the CVD growth. Diamond nano-crystals are used as substrate for the pCVD (poly-crystal CVD). The resulting diamond bulk contains grain-boundaries corresponding to the junction of crystal growth from different seeds. These structural defects act as charge trapping centers for the free charge carriers. However, the main asset of pCVD diamonds is that they can be grown to large areas of several cm^2 [8]. Besides, sCVD diamond are high quality single-crystal diamonds. The substrate of choice is usually a surface-treated HPHT (High Pressure High Temperature) diamond with a <100> orientation. Unfortunately, the area of these diamonds is restricted up to now to less than 1 cm^2. Last, hetero-epitaxial diamonds are grown using CVD on a substrate with a diamond-like crystal structure, for example iridium (hereafter named as Diamond on Iridium, DOI). They are very promising diamonds with qualities close to the single crystal ones. Moreover, they can be grown up to more than 1 cm^2 area [9], [10].

Compared to other semiconductor detectors, such as silicon ones, detectors based on CVD diamond exhibit several advantages [11], [12], [13], [14]. Their radiation hardness has already been demonstrated. A high resistivity ($>10^{13}$ Ω m) coupled to a large electronic gap (5.48 eV) results in a lower noise level and an almost negligible leakage current even at room temperature. Despite the fact that the energy required to produce electron-hole pairs is 3.6 times larger than in silicon, the signal to noise ratio is higher for diamond. The high charge carrier mobility leads to a very fast detector response allowing excellent time resolution (few tens of picoseconds) and high count rate capability, even for polycrystalline detectors, for which the charge collection efficiency is less than 100%.

Hence, diamond applications are more and more considered in various physics fields. Diamond detectors are commonly used as X-ray synchrotron radiation monitors but also as UV and neutron sensors [15]. In high-energy physics, diamond detectors are used for beam monitoring [16]. In medical physics applications, they are used as dosimeters for radiotherapy [17].

In the present study, we used sCVD and pCVD diamonds from Element6 [8], and heteroepitaxial DOI diamonds from Audiatec [9]. The tested samples were ranging from 300 µm up to 500 µm in thickness, and from 5×5 mm^2 to 2×2 cm^2 area.

B. Detector processing and pulse signal readout

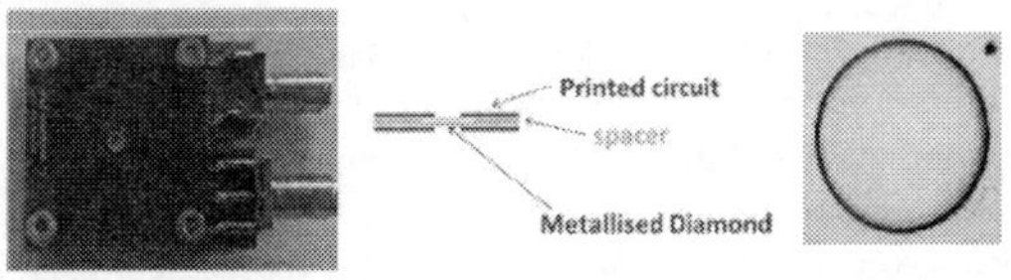

Fig. 2 The diamond sensor with a disk shaped metallization housed in its socket.

Up to 2016, an aluminum disk shaped metallization was locally performed by using the DMW (Distributed Microwave Plasmas), a technology developed by LPSC [18]. The crystal surface preparation and metal deposition were thus performed by a sequential plasma process consisting in two steps of reactive plasma cleaning followed by plasma-assisted sputtering.

The sensor contact for the disk-shaped metallization consists of a 50 nm thick aluminum layer deposited on both sides. Diamonds are mounted on sample holders with 50 Ω adapted impedance and SMA connectors, enabling reversible bias and signal readout from both sides of the diamond as illustrated by Fig. 2. Several preamplifiers were tested: the CIVIDEC C2 low-noise broadband RF amplifier (2 GHz, 40 dB from CIVIDEC Instrumentation company [19]) and the DBA III (Diamond Broadband Amplifier [20]) from GSI (a similar 2 GHz design with a gain remotely controlled from 0 dB to 38 dB). These amplifiers are intended for use with fast detectors. They exhibit a 50 Ω input impedance and are able to work with FWHM pulse widths of less than 1 ns.

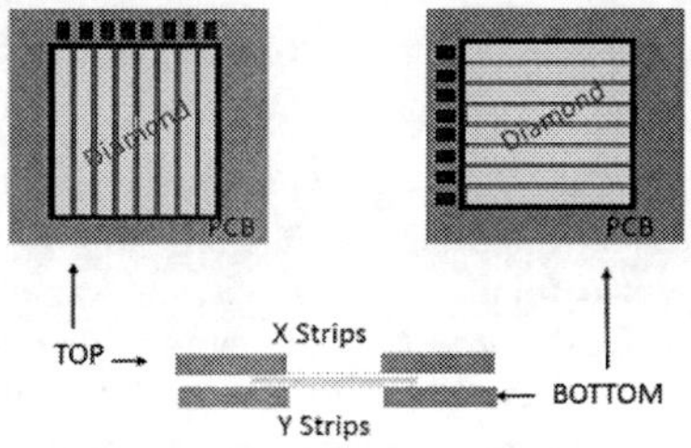

Fig. 3 Scheme of the double-side stripped diamond sensor housed in its dual PCB holder.

In 2017, a double-side strip metallization of the diamond sensors was performed for a position sensitive detector development (one side in X direction and the other in Y direction, see Fig. 3). The lift-off process was used to create the strips. It consists in creating structures (patterning) using a sacrificial material (e.g., photoresist). This was performed at the NANOFAB laboratory from the Néel Institute. Four steps are needed to perform the strip metallization. At first, S1818 photoresist is deposited on the substrate (10×10 mm^2 diamond). Then the pattern is exposed thanks to an extreme ultraviolet lithography (laser wavelength 405 nm). Development will dissolve exposed pattern. The sample is then metallized with a 100 nm aluminum layer. Residual photoresist is finally lifted away in a solvent with the parts of the undesired metallization. The resulting metallization on the diamond sensor consists of 8 strips of 1 mm width separated by a 100 μm gap and surrounded by a guard ring.

Unlike the case of the disk-shaped metallization, wire bonding ensures the electrical connection between each diamond strip and the PCB. In addition, a set of discrete current amplifiers designed at LPSC (1 GHz, 30 dB) are mounted on the circuit board. The whole set-up is placed inside an electromagnetic shielding box surrounded by SMA connectors to enable bias and signal readout.

If such a detector concept with double-sided stripped diamonds satisfies our requirements (in terms of spatial and timing resolution for a hodoscope development purpose) then we will go one step further with a 2 × 2 diamond mosaic arrangement. That time, the sensors read out will be performed by a dedicated, multipurpose fast integrated electronics that is currently under development at LPSC.

C. *Waveform data acquisition*

The pulse shape readout mode was performed with a 500 MHz, 3.2 GS/s digital sampling 'Wavecatcher' [21] system. This system could be configured in a continuous acquisition mode, recording a set of waveforms, and thus enabling large statistics offline analysis.

In addition, a 2 GHz, 20 GS/s analog bandwidth DSO (LeCroy 620Zi) [22] was used. The DSO data acquisition was fully integrated into the ESRF ID21 SPEC software acquisition framework which enabled us to measure the temporal response of the detectors as they were scanned in the beam, thus creating 2D response maps of the detector surfaces.

III. Characterization of CVD Diamonds

A. *Leakage current measurement*

The quality of the diamond was investigated by measuring the current voltage characteristics. For this purpose, a dedicated test bench was developed. This test setup was also used to highlight possible diamond sparking (electric field breakdown) once the diamonds were metallized. Diamonds exhibit very low leakage current (lower than 10pA in current field operation) as illustrated on Fig. 4, where signals were recorded on a heteroepitaxial DOI.

In our application, diamonds are used as solid state ionization chambers. The charge collection properties of CVD diamond

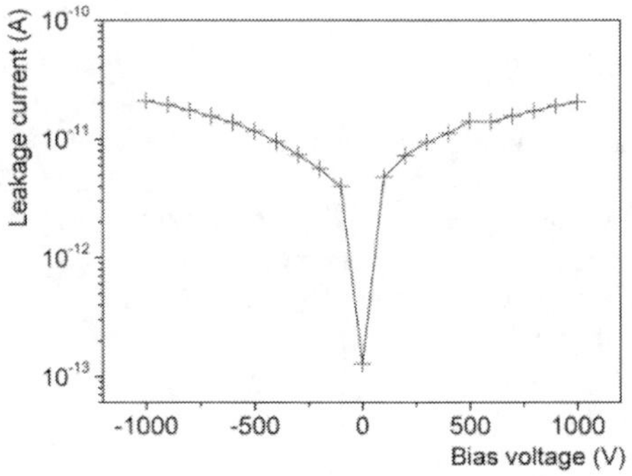

Fig. 4 Measurement on a heteroepitaxial DOI of the leakage current versus the bias voltage.

were investigated with various ionizing particles. We aimed to evaluate the capability of diamond to be used as position sensitive detector.

B. *Measurement with α and β particles*

Radioactive α sources (^{233}U and ^{241}Am) and β source (^{90}Sr) were setup in dedicated test benches. The mean free path of the α particles (about 12 μm) is much smaller than the diamond thickness. All the current due to the migration of charge carriers comes from a single species (either electrons or holes) traversing the whole detector thickness, depending on the bias polarity. This enabled us to explore the diamond sensitivity to polarization effects and to study the difference in the signal shapes between the growth and substrate sides.

On the contrary, the β set-up enabled us to infer the Charge Collection Distance (CCD) since the electrons emitted by the ^{90}Sr source are traversing the diamond (a trigger on transmitted high-energy electrons is used). As an illustration, a measurement done with a 0.45 × 0.45 mm^2 × 518 μm sCVD diamond manufactured by Element 6 is illustrated on Fig. 5. The asymptotic value obtained for the CCD by varying the electric field applied on diamonds electrodes is very close to the diamond thickness. This illustrates the quality of the sCVD diamond.

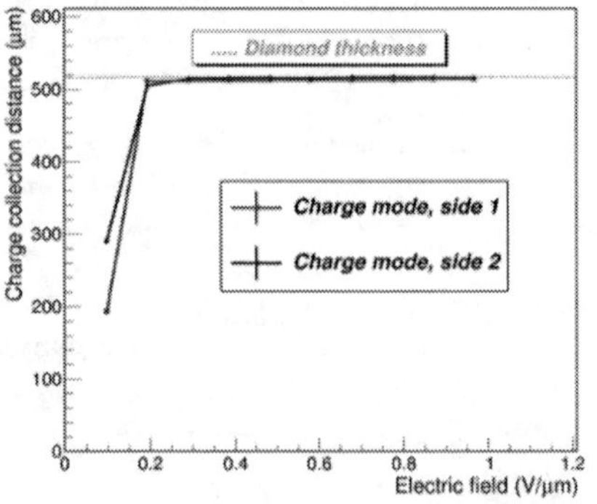

Fig. 5 The charge collection distance measured with the β test bench on a 0.45 × 0.45 mm^2 × 518 μm sCVD diamond manufactured by Element 6.

C. *Measurements with a pulsed photon beam (8.5 keV).*

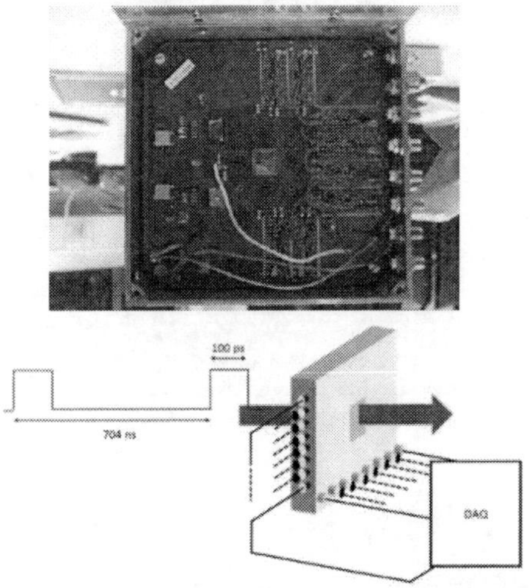

Fig. 6 ESRF experimental test bench (top : the 1 cm^2, double side stripped diamond mounted on the PCB inside an electromagnetic shielding box. 8 discrete in-house preamplifiers are mounted on each side, bottom : the box positioned on the ESRF ID21 beam line).

XBIC (X Beam Induced Current) is of huge interest for our application. In our experiment, a 8.5 keV photon focused micro-beam with a well-defined time structure was used at ESRF. As regards energy deposition in the diamond, 35 % of the beam is absorbed in 300 μm diamond material, which makes the energy deposit almost constant and continuous along the beam path. In the ESRF 4-bunch mode, the ~100 ps duration X-ray pulses, containing a fixed number of photons, adjustable by means of attenuators, with a maximum value of ~1400, mimic the passage of single ionizing particles (maximum energy deposit ~4.1 MeV). Photon bunches are separated by 704 ns intervals, which makes possible individual pulse response measurement, as well as continuous current integration with picoammeters.

The diamond samples were placed inside an electromagnetic shielding box. The box was positioned with micrometric reproducibility at the sample position of the micro-diffraction end station (in air) of the ID21 beamline at ESRF. Fig. 6 illustrates this sample setup for the double side stripped diamond.

D. *Measurements with a 95 MeV/u ^{12}C ion beam*

Finally, to evaluate the diamond performances under carbon ion irradiation, a test with a 95 MeV/u ^{12}C ion beam was carried out at the GANIL facility. The energy deposit by such ions in a 300 μm thick diamond is 25 MeV. This experimental set-up enables us to evaluate single carbon ion detection characteristics.

IV. Results

A. *Disk-shaped diamond metallization*

In the following sections, we report on the charge collection properties of CVD diamonds [23] with disk-shaped metallized sensors (Fig. 2).

1) *Pulse shape analysis*

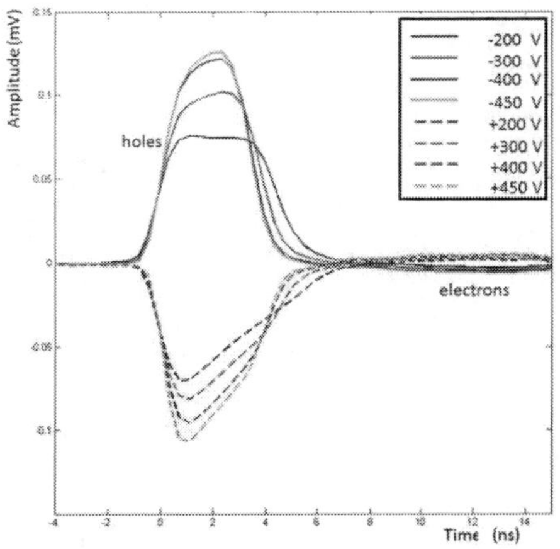

Fig. 7 Signal measurement with an α source on a 0.5 × 0.5 cm^2 × 300 μm DOI detector.

Diamond detectors allow particle identification by pulse-shape analysis. The-pulse shape of the signal collected in diamond material is determined by i) the ionization profile in the diamond sensor, which depends on the particle type and energy, and ii) the transport properties of charge carriers, which depend on the dimensions and properties of the diamond (bias, presence of defects). According to the Shockley-Ramo theoretical description, for very pure crystals such as sCVD diamonds, the shape of the current pulse with incident α particle is rectangular, since it results from drift of a constant amount of charge carriers created close to the surface along the whole crystal thickness. If space charges or trapping defects are present in the diamond bulk, such as for pCVD diamond, the rectangular shape would rather be rhombic. On Fig. 7, the signals recorded on heteroepitaxial diamonds grown on iridium for holes and electrons carriers are plotted by varying the diamond bias voltage using the α test bench. It appears that such diamonds behave as sCVD crystal for the hole carriers and as pCVD for the electron ones. This particular behavior of DOI detector signals was already reported [24].

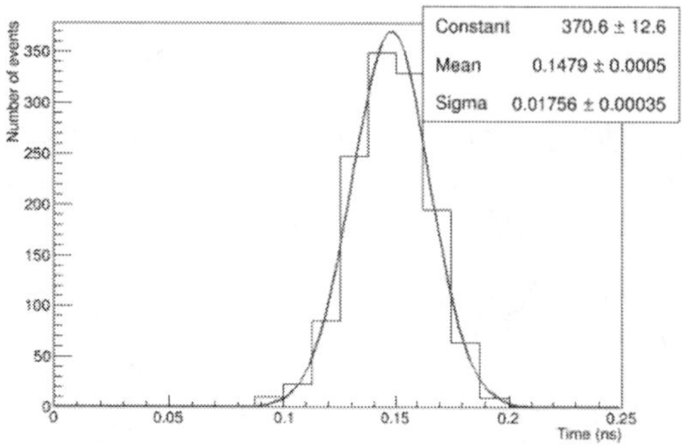

Fig. 8 Time resolution measured between a 0.45 × 0.45 cm² × 518 μm sCVD diamond and a 0.5 × 0.5 cm² × 300 μm DOI detector, traversed by a single carbon ion at 95 MeV/u incidence energy.

2) Time resolution measurements

Diamond detectors are expected to be fast responding detectors, as illustrated in Fig. 7, where rise times smaller than 1 ns are observed. Consequently, an offline procedure was established to determine the time resolution of the detectors, based on the waveform analysis. A numerical Constant Fraction Discrimination (CFD) was performed by averaging the background, determining the maximum of the pulses, and interpolating the 50 % rise time value. At the GANIL facility, the dedicated experimental set-up enables us to measure a time resolution between two diamond samples positioned one behind the other. In Fig. 8, a time resolution lower than 18 ps rms was reached using a 0.45 × 0.45 cm² × 518 μm single-crystal sample provided by Element 6 coupled to a 0.5 × 0.5 cm² × 300 μm DOI provided by Audiatec. The DOI detector was read out with a CIVIDEC preamplifier, and the sCVD with a DBAIII preamplifier, with smaller gain. Thus, this value of timing resolution results from the quadratic sum of both detection device resolutions. Such results with single readout channels are very promising, but, in view of the realization of a multichannel detector prototype, they rely on developments of very fast front-end electronics.

Tests were also carried out at the GANIL facility with a 2 × 2 cm² × 500 μm sample coupled to a 1 × 1 cm² × 500 μm one. Both diamonds were pCVD ones provided by Element 6. That time, the time resolution was at the level of 90 ps. This result is not in contradiction with the one obtained with the DOI coupled to the sCVD, since the active areas are about one order of magnitude larger here, and the timing resolution depends on the detector capacitance [14].

3) Energy resolution measurements

Energy resolution measurements were performed both at the ESRF and GANIL facilities. At GANIL, with a 95 MeV/u carbon ion beam impacting a 0.5 × 0.5 cm² × 300 μm polycrystalline diamond, an energy resolution of ~7 % rms was reached (Fig. 9). Using the same diamond sensor, with the 8.5 keV pulsed photon beam at ESRF an energy resolution of ~9% rms was obtained. Such resolutions are good enough to identify unambiguously the passage of a single ion.

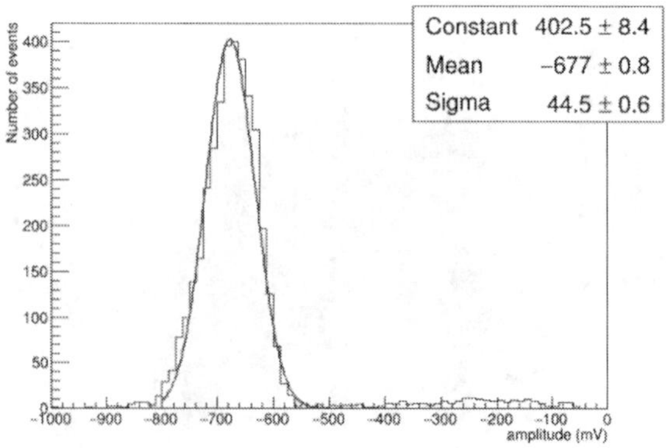

Fig. 9. Energy resolution measured on a 0.5 × 0.5 cm² × 300 μm polycrystalline diamond with 95 MeV/u carbon ion beam.

4) Surface mapping

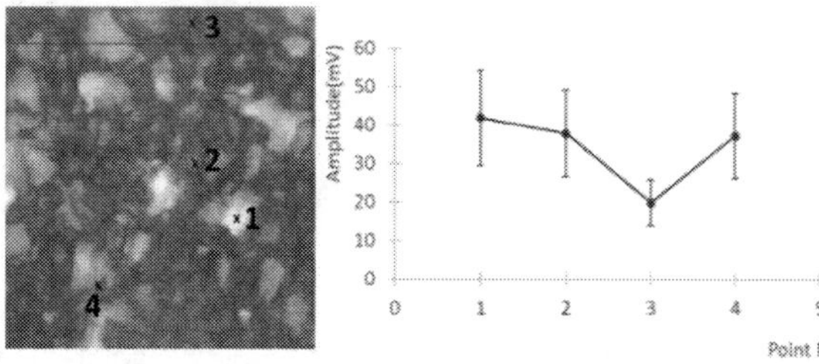

Fig. 10 Amplitude (right) of the signal versus the ESRF-ID21 pencil beam localization from point 1 up to 4 on the current-readout surface map (left) of a 1 mm² area of a 1×1 cm² × 500 μm polycrystalline diamond detector.

At the ESRF facility, we have set up a map of the signal response in current integration mode. Fig. 10 shows an example of a polycrystalline detector map (1 × 1 cm² × 500 μm pCVD detector from Element 6) on a 1 mm² area. The grey scale corresponds to the charge collection efficiency measured by an electrometer. Clearly, the response of the detector reflects the spatial distribution of grain boundaries in the polycrystalline material.

In addition, Fig. 10 shows the diamond response versus the photon pencil beam localization for the same pCVD diamond detector. With such a sample, a difference of a factor two is observed between the clearest point and the darkest one. It has been checked that such an amplitude variation will introduce an acceptable time walk in time measurement. This result could be seen as in contradiction with the result of the previous section, where a resolution of 9 % was obtained with a similar pCVD. Nevertheless, note that the present sample is different (1 cm²) and that we measured the difference between the two extreme response regions of the surface.

B. Double-side stripped diamond

1) Optimization of the strip pitch with a dedicated simulation package

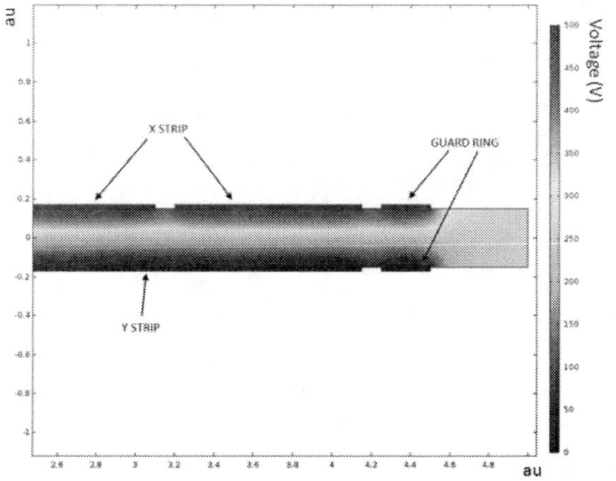

Fig. 11 Transverse view of the voltage distribution in the double side stripped detector width (COMSOL simulation).

Simulations were carried out using COMSOL Multiphysics [25]. The Fig. 11 illustrates the distribution of the voltage while the X strips on the top side of the diamond are biased at +500 V and the Y ones are ground-biased. At first, it demonstrates the importance of the guard ring which permits all the 8 strips on each side to behave in the same way as far as chagre collection is concerned. Then, the 1mm strip width coupled to the 100 µm gap were chosen consequently to these simulations. It was done on one hand to optimize the strip pitch on the 1 × 1 cm^2 diamond detector surface, and, on the other hand, not to lose too many charges in the gaps.

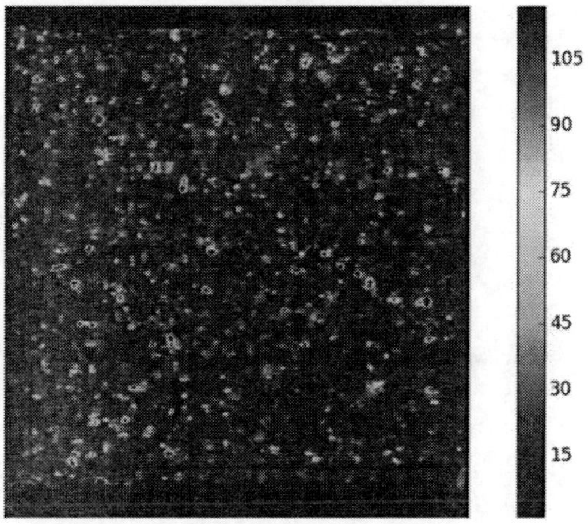

Fig. 12 X-Ray analysis of the surface of a 10 × 10 mm^2 × 300 µm double-side stripped pCVD diamond from Element 6.

2) Signal spatial distribution evaluated at the XBIC ID21 beamline of the ESRF facility

We have performed an X-Ray analysis of a double-side stripped diamond surface in a current integration mode (Fig. 12). The non-uniformity of the diamond response quantified by the color scale on the right reflects the defects distribution in the material like on Fig. 10. However, unlike the previous sample, the defect response appear like hot spots (higher current) rather than low response zones.

A zoom of the previous X-Ray analysis (Fig. 12) on a about 2 × 2 mm^2 area is illustrated by Fig. 13. An ad-hoc adjustment of the colour scale makes visible the inter-strip regions, corresponding to a decrease of the charge collection. This already shows that narrower gaps should be used in order to keep the efficiency constant.

These are preliminary results, a more refined analysis will be performed in a near future, in particular to quantify the cross talk and the reconstructed spatial resolution.

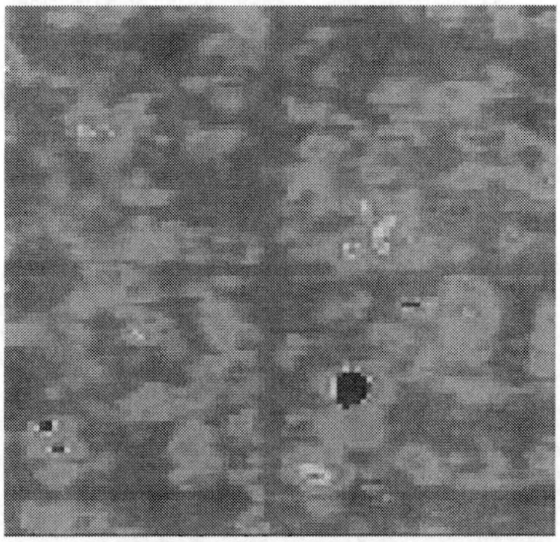

Fig. 13 Zoom of an X-Ray analysis on a about 2 × 2 mm^2 area of a 10 × 10 mm^2 × 300 µm double side stripped pCVD diamond from Element 6.

V. Conclusion

Large-area, synthetic pCVD or DOI diamond detectors are foreseen for on-line hadron therapy beam tagging applications. They will be used as a hodoscope which plays a major role for particle identification in time-of-flight gamma camera projects proposed by the CLaRyS collaboration. The radiation hardness, fast response and good signal to noise ratio make diamonds good candidates.

At first, we have carried out tests with disk-shaped metallized diamonds. A time resolution ranging from 20 up to 90 ps rms and an energy resolution ranging from 7 up to 9 % rms were measured when irradiating the whole surface of CVD diamond using various ionizing particles. These results satisfy our requirement despite the obvious non-uniformity of the crystalline structure observed thanks to XBIC analyses. Recently, we have tested double side stripped detectors (8 strips in X and 8 strips in Y) in a single housing to check signal characteristics.

ACKNOWLEDGMENT

The authors would like to acknowledge the ESRF-ID21 beamline for provision of synchrotron radiation with experiments MI-1243 (2016) and MI-1285 (2017). This work was supported by the Labex PRIMES (ANR-11-LABX-0063), FranceHadron (ANR-11-INBS-0007) and ANR MONODIAM-HE (ANR-089520). The authors are grateful to Matthias Schreck and Martin Fischer from Audiatec_Augsburg for providing the LPSC laboratory with DOI samples, Dominique Breton and Jihanne Maalmi from LAL-Orsay and Eric Delagnes from CEA Saclay are thanked for their implication in dedicated software development and technical support of the Wavecatcher data acquisition system.

REFERENCES

[1] H. Paganetti, "Range uncertainties in proton therapy and the role of Monte Carlo simulations," Phys. Med. Biol., vol. 57, no. 11, pp. R99–R117, Jun. 2012.
[2] C.-H. Min, C. H. Kim, M.-Y. Youn, and J.-W. Kim, "Prompt gamma measurements for locating the dose falloff region in the proton therapy," Appl. Phys. Lett., vol. 89, no. 18, pp. 183517–3, Oct. 2006.
[3] E. Testa *et al.*, "Monitoring the Bragg peak location of 73 MeV/u carbon ions by means of prompt γ-ray measurements," Appl. Phys. Lett., vol. 93, no. 9, p. 93506, 2008.
[4] J. Krimmer *et al.*, "Collimated prompt gamma TOF measurements with multi-slit multi-detector configurations," J. Instrum., vol. 10, no. 1, pp. P01011–P01011, Jan. 2015.
[5] J. Krimmer *et al.*, "Development of a Compton camera for medical applications based on silicon strip and scintillation detectors," Nucl. Instrum. Methods Phys. Res. Sect. Accel. Spectrometers Detect. Assoc. Equip, Elsevier, 787, pp 98-101,2015 <doi:10.1016/j.nima.2014.11.042>.
[6] J. Krimmer et al., "Real-Time Online Monitoring of the Ion Range by Means of Prompt Secondary Radiations", published in 3rd International Conference On Advancements In Nuclear Instrumentation Measurement Methods And Their Applications (ANIMMA 2013), Marseille: France 2013.
[7] S. Deng et al.,"Very fast front end ASIC associated with multi-anode PMTs for a scintillating-fibre beam hodoscope", Journal of Instrumentation 8 (2013)C01047. DOI:10.1088/1748-0221/8/01/C01047.
[8] http://e6cvd.com/
[9] http://www.audiatec.de/
[10] M. Schreck et al., "Large-area high-quality single crystal diamond", MRS Bulletin, vol.39, n°6, pp. 504-510, 2014.
[11] P. Bergonzo et al., "CVD diamond for radiation detection devices", Diamond and Related Materials 10 (2001). 631-638.
[12] E. Berdermann and M. Ciobanu, "CVD-Diamond Detectors for Experiments with Hadrons, Nuclei, and Atoms", *CVD Diamond for Electronic Devices and Sensors* Ed. R. S. Sussmann, 2009 John Wiley & Sons.
[13] M. Ciobanu et al., "In-Beam Diamond Start Detectors" IEEE Trans. Nucl. Sci.58, No. 4 (2011).
[14] F Schirru et al "Development of large area polycrystalline diamond detectors for fast timing application of high-energy heavy-ion beams" JINST No 7 (2012) P05005 doi:10.1088/1748-0221/7/05/P05005.
[15] C. Weiss et al., n-TOF collaboration, "A new CVD diamond mosaic-detector ofr (n,α) cross-section measurements at the n-TOF experiment at CERN", proceedings, 13th Vienna Conference on Instrumentation (VCI 2013), Vienna, Austria, February 11-15, 2013, Nucl. Instrum.Meth A732 pp 190-194, 2013.
[16] E. Griesmayer and B. Dehning, "Diamonds for beam instrumentation", in Proceedings of the 2nd International Conference on Technology and Instrumentation in Particle Physics (TIPP 2011), Chicago U.S.A. (2011) [Phys. Procedia 37 p 1997, 2012].
[17] http://www.ptw.de/2968.html
[18] A. Lacoste et al., "Multi-dipolar plasmas for uniform processphysics", design and performance, SCi. Technol., 11 pp 407-412, 2002..
[19] https://cividec.at/_C2_Broadband_Amplifier,_2_GHz,_40_dB.html
[20] P. Moritz, E. Berdermann, K. Blasche, H. Stelzer, B. Voss, "Broadband electronics for CVD-diamond detectors", Diamond and Related Materials 10 pp 1765-1769, 2001.
[21] D.Breton, E. Delagne, J. Maalmi, "Picosecond time measurement using ultra-fast analog memories", proceedings TWEPP2009, Topical Workshop on Electronics for Particle Physics. Paris 21-25 September 2009.
[22] http://teledynelecroy.com/oscilloscope/
[23] M-L Gallin-Martel et al. "Large Area Polycrystalline Diamond Detectors for Online Hadron Therapy Beam Tagging Applications", IEEE Nuclear Science Symposium & Medical Imaging Conference (2016 IEEE NSS/MIC), Oct 2016, Strasbourg, France. in press, Proceedings of the IEEE NSS MIC 2016.
[24] M. Kis et al.,"TCT characterization of new generation of DoI samples", 4th ADAMAS workshop (2014), Dec. 2014, GSI, Darmstadt.
[25] https://www.comsol.com/release/5.3

Study and Search for Main Reason of Lung Cancers Based on Cherenkov Radiation in Environmental Radiation

Hiroshi Ito, Yusaku Emoto, Kento Fujihara, Hideyuki Kawai, Shota Kimura, Satoshi Kodama and Takahiro Mizuno

***Abstract*—The number of lung-cancer-related death is highest among all cancers in the world, and it is increasing in Japan where population aging in progressing. The main reason for the lung cancer of non-smokers is regarded to be environmental pollution or exposure of the lung to radon in the nature. The risk of lung cancer was estimated to increase by 8 to 13% per every 100 $\mathrm{Bq\,m^{-3}}$ concentration of radon in the air. We observed beta rays with maximum energy of 3.27 MeV emitted from ^{214}Bi as one of the progenies based on a detection of Cherenkov radiation. The surface radioactivity concentration of ^{214}Bi on the sample was measured; the relation between the concentration and exposure time for the sample at the room air is researched. The behavior of the radon progenies in the air is discussed by a research for the progenies attaching on the sample after the radon decay. The inhalation of the radon progenies is not clear. Thus, to understand the behavior of progenies in the air make to clear the causal relation between the radon concentration and lung cancers.**

***Index Terms*—Radon Progenies, Bismuth-214, Cherenkov Radiation**

I. Introduction

THE number of lung-cancer-related death is highest among all cancers in the world, and it is increasing in Japan where population aging in progressing [1]. Although the primal cause of lung cancer is considered to be smoking, non-smokers also suffer. The main reason for the lung cancer of non-smokers is regarded to be environmental pollution or exposure of the lung to radon in the nature [3], [4]. The risk of lung cancer was estimated to increase by 8 to 13% per every 100 $\mathrm{Bq\,m^{-3}}$ concentration in the air, thus showing the strong possibility of decease causing due to inhalation. This result was extracted from the relation between the patient rate and the radon concentration in the North American, European, and Chinese [5]. Thus, the inhalation of radon is a strong candidate for the occurring of the lung cancer in the nature.

The average of annual radioactive dose is estimated as 1.15 mSv from the ^{222}Rn inhalation at the mean concentration of 45 $\mathrm{Bq\,m^{-3}}$ in the air [6]. This value depends on the equilibrium factor after the radon is inhaled [7], which in assumed to be 0.4, for radon progenies of ^{218}Po, ^{214}Pb, ^{214}Bi, and ^{214}Po in radon decaying. However, it is not clear in the papers [6], [8] how the inhalation of theses progenies was treated if they are in the air and inhaled. The inhalation radiative dose can be estimated more in precise if the behavior and concentration of these progenies in the air become to known.

Manuscript received 1 June, 2017.
H. Ito is with the Graduate of School of Science, Chiba University, Chiba, 268-8522, Japan (e-mails: hiroshi@hepburn.s.chiba-u.ac.jp).

The radionuclides of ^{214}Bi, which emit beta rays (max. kinetic energy of 3.27 MeV) are focused in this study. The detection of beta rays in environment has an advantage because a concentration of radioactivity of the radon progenies can be measure mainly, i.e., the behavior of the progenies can be discussed after radon decaying. In this present study, a gamma ray energy spectroscopy in the air and a surface contamination inspection for a sample based on scintillation detection were performed.

II. Experiment

The experimental setup is shown in Fig. 1. The sample (polyethlene) is set under the scintillating fibers after it was exposed in the room air with windless. The scintillating fiber (Kuraray, SCSF-78MJ, ϕ 0.2 mm) is emitted photons when the charged particles deposit energies there. The fibers are assembled to a sheet with an effective area of 10 × 10 cm^2. The photons propagating by a total reflection in the fibers are observed by photomultiplier tubes (PMT): PMT1 and PMT2 (Hamamatsu, R9880U-210), which connect to both ends of the fibers. A Cherenkov detector using silica aerogel was set over the scintillating fibers. The aerogel (a refractive index of 1.041±0.001, the size of 9.6×9.6×1 cm^3) emits Cherenkov photons by a beta ray with kinetic energy over 1.31 MeV. The photons are lead to PMT3 or 4 (Hamamatsu, R1250-03 and H6528) by a reflector with aluminised Mylar and are observed. The black paper sets under the silica aerogel because the Cherenkov light originated from cosmic muons must not be reflect. The scintillation detector was set over the Cherenkov detector. It consists of a plastic scintillator (the size of 10×10×0.5 cm^3), acrylic light guide, and PMT (Hamamatsu, H1161). To suppress events related to cosmic muon, the cosmic-ray-veto counter was set over aluminium plate with thick of 1 cm. The veto counter consists of a plastic scintillator (the size of 10×20×0.5 cm^3), acrylic light guides, and PMT6, 7 (Hamamtsu, H1161). The PMTs connected to both ends of the scintillator via the light guides.

Therefore, the beta rays originated from ^{214}Bi can be observed as events when the coincidence is satisfied followings: hit the scintillating fibers, emitted Cherenkov radiation, hit the scintillator detector, and not hit the veto counter.

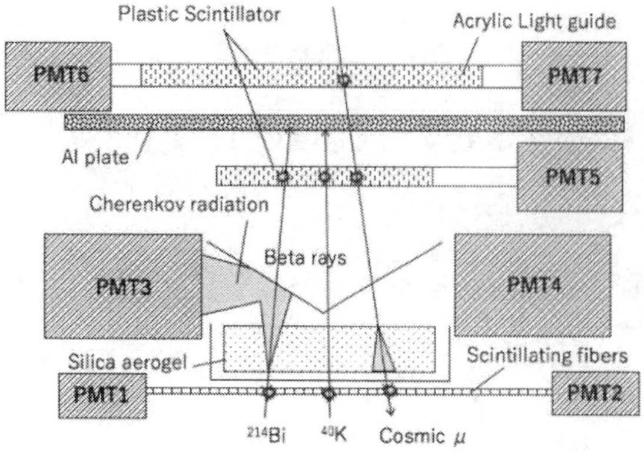

Fig. 1. A schematic of experimental setup.

III. Results and Discussions

The sample was exposed for 6 h in the room air with windless. As a result, the spectra of count rate with 10 min is shown in Fig. 2. The black dot is data. The red solid line is fitting exponential function. The green dashed line represents the background rate. The half-life time constant of the simple exponential function is 39.5±2.2 min. The count rate related to ^{214}Bi becomes to reduce to un-observation amounts after late 4 h 40 min.

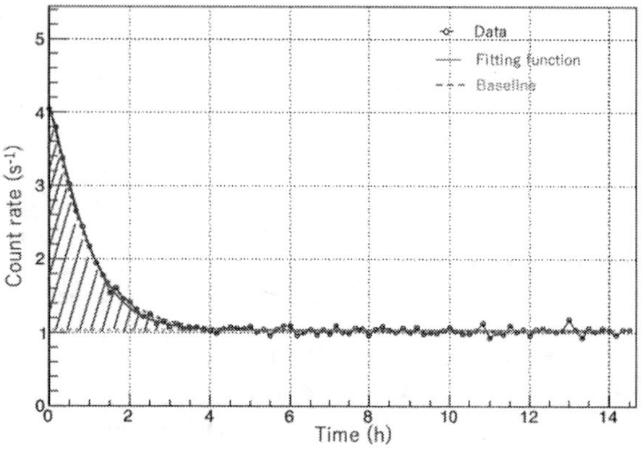

Fig. 2. The time spectrum of count rates for the polyethylene sample.

The integral blue shaded area responds to radioactivity concentration on the sample. This radioactivity is estimated by using corrected efficiency for ^{214}Bi, where the efficiency is calculated from the experimental efficiency for ^{90}Sr and a ratio of efficiency for ^{214}Bi to that for ^{90}Sr by a Monte Carlo simulation, GEANT4. As a result, this radioactivity (4 h 40 min exposes) is 18.0±0.5 Bq/m^2.

The relation between the radioactivity concentration and exposure time is shown in Fig. 3. The black dot is data and the red solid line is fitting function: $(17.9 \pm 0.5)\{1 - \exp(-(t/(1.68 \pm 0.06)))\}$. This result suggests that ^{214}Bi of $17.9 \pm 0.5\ \mathrm{Bq/m^2}$ attaches on our surroundings and the concentration becomes to be saturated just after the radon decay in the air. Furthermore, it also suggests the radon progenies (^{218}Po, ^{214}Pb, ^{214}Bi, 214Po, etc.) do not fall down to ground, and some components behavior in the air. Therefore, the potential by inhaling the progenies should be considered for natural exposure dose.

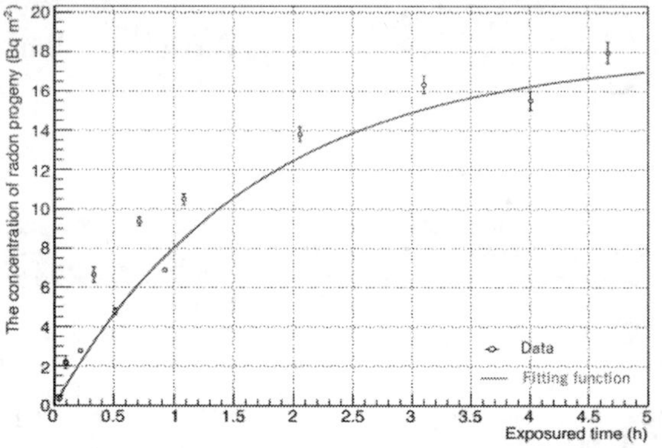

Fig. 3. The relation between the radioactivity concentration and exposure time.

IV. Summary

The radon progenies in the air is focused by the recent study. The behavior of radon progenies is discussed indirectly by measuring the radioactivity concentration of ^{214}Bi on the sample based on Cherenkov detection. The results suggest the potential of inhaling radon progenies. In future, the radioactivity of radon progenies in air will be measured directly.

References

[1] Foundation for Promotion of Cancer Research, 2015. Cancer statistics in Japan 2015.
[2] Funatogawa I., Funatogawa T., Yano E., 2013. Bull World Health Organ. 91 (2013) 332-340.
[3] Subramanianand J., Govindan R., 2007. Lung Cancer in Never Smokers: A Review. J. Clin. Oncol., 25 (2007) 561-570.
[4] Samet J. M., Avila-Tang E., Boffetta P., Hannan L. M., Olivo-Marston S., Thun M. J., Rudin C. M., 2009. Lung Cancer in Never Smokers: Clinical Epidemiology and Environmental Risk Factors. Clin. Cancer. Res. 15 (18) (2009) 5626-5645.
[5] Choi H., Mazzone P., 2014. Radon and lung cancer: Assessing and mitigating the risk. Clev. Cli. J. 81 (9) (2014) 567-575.
[6] UNSCEAR 2008 report vol. 1. Source effects and risks of ionizing radiation.
[7] Martinez J. E., Juste B., Ortiz J., Martorell S., Verdu G., 2017. Air Radon equilibrium factor measurement in a Waste Water Pre-Treatment Plant. Rad. Phys. Chem. 2017. <https://doi.org/10.1016/j.radphyschem.2017.03.011>.
[8] ICRP publication 71. Age-dependent Doses to Members of the Public from Intake of Radionuclides: Part 4 Inhalation Dose Coefficients. A report of a Task Group of Committee 2 of the International Commission on Radiological Protection. Adopted By The Commission In September 1995.

Development of a Beam Trajectory Monitoring System Using e^+/e^- Pair Production Events

Shota KIMURA, Yusaku EMOTO, Kento FUJIHARA, Hiroshi ITO, Hideyuki KAWAI, Atsushi KOBAYASHI, and Takahiro MIZUNO

***Abstract*—In particle therapy, it is important to monitor the Bragg-peak position. It was simulated by GEANT4 Monte Carlo Simulation Code that the distribution of secondary generated gamma rays on the carbon beam therapy and the proton beam therapy. This simulation shows that gamma rays whose energy is 10 MeV or more are intensively generated at the Bragg-peak position. We are developing the system to monitor the Bragg-peak position which can measure pair production events occurred in the detector by gamma rays from irradiation points. The momentum direction of the gamma ray can be determined by measuring passing points and energy of e^+ and e^- generated by pair production. This system has 5 parts. The first is the conversion part. This part consists of several layers. Each layer is composed of a La-GPS ($(Gd_{0.75}La_{0.24}Ce_{0.01})_2Si_2O_7$) scintillator plate and wavelength-shifting fibre (WLSF) sheets. The scintillator plate is sandwiched between sheets, where the directions of the sheets are in orthogonally x and y directions. In this part, gamma rays are converted to e+ e- pairs and the position where the conversion occured is determined. The second is the tracking part. This part consists of 2 layers of scintillating fibre tracker. Each layer has 6 scintillating fibre sheets for x, x', u, u', v, and v'. The third is the energy measurement part. It measures the energy of e^+ and e^- by scintillator array and Silicon Photomultipliers. The fourth is the veto counter for bremsstrahlung gamma rays from e+ and e-. The fifth is the beam monitor. By experiment, the number of photoelectrons of La-GPS with a WLSF (B-3(300)MJ, Kuraray) sheet and scintillating fibre (SCSF-78, Kuraray) when charged particle passed was measured as 9.7 and 7.6 respectively.**

***Index Terms*—particle therapy, pair production, Geant4**

I. Introduction

In particle therapy, cancer cells can be killed by matching the Bragg-peak position and the cancer position. Conversely, if the position of the Bragg-peak deviates from cancer cells even a little, normal cells could be killed. Thus, in particle therapy, it is important to monitor the Bragg-peak position. To monitor the Bragg-peak position, positron emission tomography (PET) detectors [1], Compton cameras [2] have been studied.

We simulated on GEANT4 Monte Carlo Simulation Code that the distribution of secondary generated gamma rays on the carbon beam therapy and the proton beam therapy. The results of proton beam are shown in Fig.1 and parameters of simulations are on Table I. This simulation shows that gamma rays whose energy is 10 MeV or more are intensively generated at the Bragg-peak position. On the other hand, gamma rays whose energy is around 0.511 MeV are widely distributed in the body. Thus, it is clear that using high energy gamma rays is the best way to monitor the Bragg-peak position.

Shota KIMURA is with the Graduate School of Science, Chiba University, Chiba, 263-8522, Japan (e-mail: shota_kimura@hepburn.s.chiba-u.ac.jp).
Manuscript received the 1st of June, 2017.

II. Methods

Most of the reaction of gamma rays which enrgy is 10 MeV or more are pair production [3]. The momentum direction of the gamma ray can be determined by measuring passing points and energy of e^+ and e^- generated by pair production. The Bragg-peak position can be determined by obtaining the intersection of the trajectory of the gamma ray and the trajectory of the particle beam.

We are developing the system to monitor the Bragg-peak position which can measure pair production events occurred in the detector by gamma rays from irradiation points. Fig.2 is a schemetic diagram of the system. This system has 5 parts. The first is the conversion part. This part consists of several layers. Each layer is composed of a La-GPS ($(Gd_{0.75}La_{0.24}Ce_{0.01})_2Si_2O_7$) scintillator plate and wavelength-shifting fibre (WLSF) sheets. The size of the scintillator plate is 300 mm×300 mm×1 mm. The sheets are made by arranging 1500 WLSFs with a diameter of 0.2 mm. The scintillator plate is sandwiched between sheets, where the directions of the sheets are in orthogonally x and y directions. In this part, gamma rays are converted to e^+ e^- pairs and the position where the conversion occured is determined. The second is the tracking part. This part consists of 2 layers of scintillating fibre tracker. Each layer has 6 scintillating fibre sheets for x, x', u, u', v, and v'. Fig.3 shows construction of the layer. The third is the energy measurement part. It measures the energy of e^+ and e^- by plastic scintillator array and SiPM. Each scintillator is the size of 20 mm×20 mm×100 mm. The fourth is the veto counter for bremsstrahlung gamma rays from e^+ and e^-. This part consists of 60 layers of simplyfied detectors of conversion part. WLSF sheet is just on the single surface, and WLSFs are connected to the Photomultiplier tube (PMT) collectively. In this part, 99% efficiency is expected. The fifth is the beam monitor. This part is a small version of the tracking part.

We confirmed the basic performance of the system by experiments. We measured the number of photoelectrons when the charged particle pass through La-GPS plate scintillator of 1 mm thickness with a WLSF (B-3(300)MJ, ϕ0.2 mm, Kuraray) sheet. We also measured the number of photoelectrons when a beta ray from ^{90}Sr source passes through a scintillating fibre (SCSF-78, ϕ1.0 mm, Kuraray). Just one side of the fibres are connected to the PMTs. Setup of the experiment is on Fig.4.

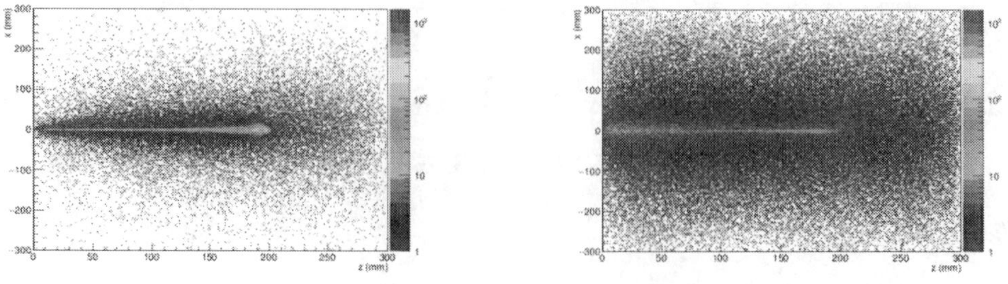

(a) over 10 MeV (b) 0.511 MeV (0.420-0.602 MeV)

Fig. 1. Simulated results of the distribution of secondarily generated gamma rays' final scattered points on the proton beam therapy. Horizontal axis means the depth of the body, and vertical axis means the vertical position of the body.

TABLE I
PARAMETERS OF THE SIMULATION

Particle	Proton	Carbon
Beam Energy	175 MeV	290 MeV/n
Physics List	QGSP_BIC_HP G4RadioactiveDecayPhysics	FTFP_BERT_HP G4RadioactiveDecayPhysics
Material of Phantom	G4_MIX_D_WAX (H: 13.4%, C: 77.8%, O: 3.5%, Mg: 3.9%, Ti: 1.4%)	

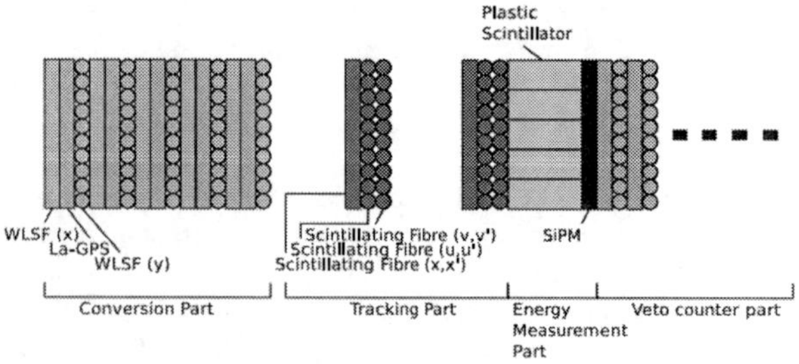

Fig. 2. A schemetic diagram of the system

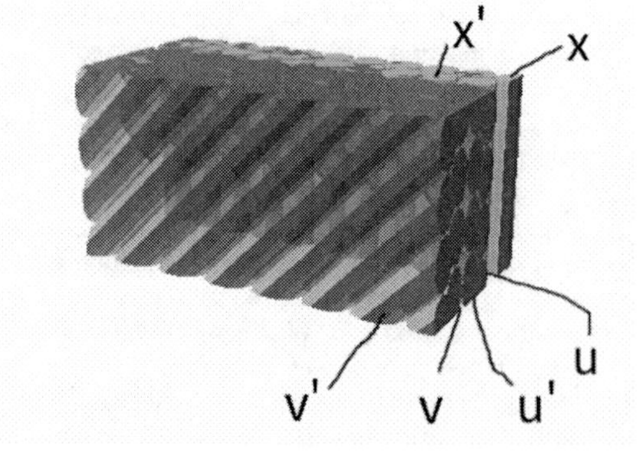

Fig. 3. Construction of scintillating fibre traker.

III. RESULTS

The number of photoelectrons to 0.511 MeV gamma rays was measured as 14.7 in average [4]. We measured 9.7 photoelectrons in average from a WLSF sheet when a beta

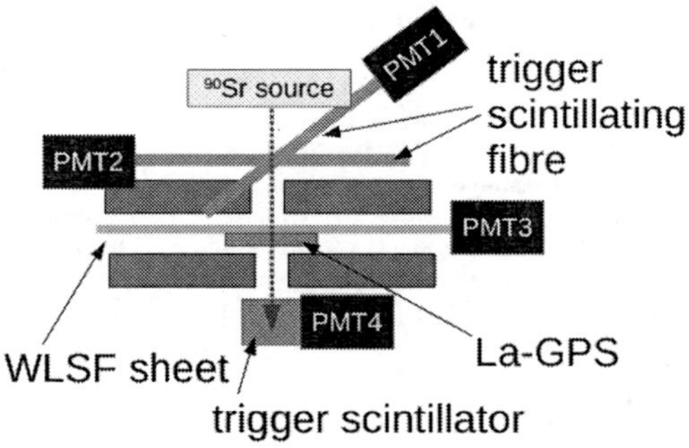

Fig. 4. Setup of the experiment

ray from ^{90}Sr source passes through the La-GPS scintillator plate, and 7.6 photoelectrons in average from a scintillating fibre.

IV. CONCLUSION

The best way to monitor the Bragg-peak position is using gamma ray of several dozen MeV. To monitor with such gamma rays, using pair production events such as our system is the most reasonable. According to the experiment, each component of our system has sufficient amount of luminescence.

REFERENCES

[1] T. Nishio, T. Ogino, K. Nomura, and H. Uchida, "Dose-volume delivery guided proton therapy using beam on-line pet system," *Medical physics*, vol. 33, no. 11, pp. 4190–4197, 2006.
[2] S. Peterson, D. Robertson, and J. Polf, "Optimizing a three-stage compton camera for measuring prompt gamma rays emitted during proton radiotherapy," *Physics in medicine and biology*, vol. 55, no. 22, p. 6841, 2010.

[3] M. Berger, J. Hubbell, S. Seltzer, J. Chang, J. Coursey, R. Sukumar, D. Zucker, and K. Olsen, "Xcom: Photon cross sections database," https://www.nist.gov/pml/xcom-photon-cross-sections-database.
[4] N. KANEKO, "Development of the gamma ray detector with wlsf for pet imaging," Master's thesis, Chiba University, 2016.

 https://doi.org/10.1051/epjconf/201817009008

Construction and tests of an in-beam PET demonstrator (LAPD) for hadrontherapy beam ballistic control

S. Binet, A. Bongrand, E. Busato, P. Force, C. Guicheney, C. Insa, D. Lambert, M. Magne, F. Martin, H. Perrin, F. Podlyski, A. Rozes, G. Montarou

We present the first results obtained with a detector, called Large Area Pixelized Detector (LAPD), dedicated to the study of the ballistic control of the beam delivered to the patient by in-beam and real time detection of secondary particles, emitted during its irradiation in the context of hadrontherapy. These particles are 511 keV gammas from the annihilation of a positron issued from the beta+ emitters induced in the patient tissues along the beam path.

The LAPD basic concepts are similar to a conventional PET camera. The 511 keV gammas are detected and the reconstructed lines of response allow to measure the beta+ activity distribution. Nevertheless, when trying to use gamma from positron annihilation for the ballistic control in hadrontherapy, the large prompt gamma background should be taken into account and properly rejected. This demonstrator should prove the possibility of in line selection of the two gammas annihilation events and highly reject beam induced background noise events.

The presentation is organized in two parts.

The first part starts with a description of the detector geometry and the technical choices made for its construction. The detector electronics will then be described. Some emphasis will be put on the main component of this electronic system which is the ASM board. This board is used for both the generation of trigger signals and the sampling/digitization of physics signals used downstream in the analysis.

The current version of the data acquisition system is then described. One originality of the DAQ software is that it is partly written using the Go language. Reasons for this choice will be shortly described.

The current DAQ system is characterized by a large dead-time due to the use of VME-based technologies. Work is under progress to upgrade this DAQ system and move from VME-based to xTCA-based technologies. This should reduce the dead time by a factor of at least 40. These developments are also briefly described in the presentation.

Then, the monitoring system is described. This system is of crucial importance for this application as results should eventually be served online to clinical staff. It is therefore mandatory to have an online software able to perform the entire analysis (event selection and event reconstruction) in real time and to present the

https://doi.org/10.1051/epjconf/201817009008

reconstruction results in real time too. The online software is briefly discussed as well as the way results are presented by the monitoring system. This discussion will end the first part of the presentation.

The second part is dedicated to the presentation of the tests that were carried recently and the performances that were derived from them.

First, measurements obtained with a phantom filled with a high intensity FDG source at the cancer research center of Clermont-Ferrand are shown. Basic performances obtained from these measurements are presented as well as reconstruction results showing the ability of the LAPD detector to measure the beta+ annihilation signal and reconstruct beta+ activity maps.

Then, measurements performed on low energy proton beam (65 MeV) at the Centre Antoine Lacassagne (CAL) in Nice are reported. After a short presentation of the experimental setup and experimental conditions, first results of online measurements of the beta+ activity profile along the beam axis using PMMA targets are presented.

Finally, some conclusions will be drawn and some perspectives regarding the futur evolution of the system will be given.

Single photon detection and signal analysis for high sensitivity dosimetry based on optically stimulated luminescence with beryllium oxide

J. Radtke, J. Sponner, C. Jakobi, J. Schneider, M. Sommer, T. Teichmann, W. Ullrich, J. Henniger, T. Kormoll

Abstract

Single photon detection applied to optically stimulated luminescence (OSL) dosimetry is a promising approach due to the low level of luminescence light and the known statistical behavior of single photon events. Time resolved detection allows to apply a variety of different and independent data analysis methods. Furthermore, using amplitude modulated stimulation impresses time- and frequency information into the OSL light and therefore allows for additional means of analysis. Considering the impressed frequency information, data analysis by using Fourier transform algorithms or other digital filters can be used for separating the OSL signal from unwanted light or events generated by other phenomena. This potentially lowers the detection limits of low dose measurements and might improve the reproducibility and stability of obtained data. In this work, an OSL system based on a single photon detector, a fast and accurate stimulation unit and an FPGA is presented. Different analysis algorithms which are applied to the single photon data are discussed.

Index Terms
beryllium oxide, dosimetry, luminescence, optical stimulated luminescence, single photon detection

I. Introduction

Optically stimulated luminescence (OSL) is gaining greater importance in the field of personal dosimetry in the last few years. Its principle is based on the release of small amounts of light induced by the prior absorption of ionizing radiation. In comparison to the method of thermoluminescence, which is also a common method in personal dosimetry, the luminophore is stimulated by optical photons and not through heat. This makes the OSL easy to use, more flexible and less sensitive to changing environmental influences.

J. Radtke, J. Sponner, C. Jakobi, J. Schneider, M. Sommer, T. Teichmann, W. Ullrich, J. Henniger and T. Kormoll are with the IKTP (Institute of nuclear and particle physics), Dresden 01069 Germany (e-mail: radtke@asp.tu-dresden.de, sponner@asp.tu-dresden.de, jakobi@asp.tu-dresden.de, schneider@asp.tu-dresden.de, sommer@asp.tu-dresden.de, teichmann@asp.tu-dresden.de, ullrich@asp.tu-dresden.de, henniger@asp.tu-dresden.de, kormoll@asp.tu-dresden.de)

For OSL, there are several possible luminophors, like CaF_2 [1], BeO [2] or Al_2O_3 [3]. One suitable luminophore for OSL in personal dosimetry is beryllium oxide (BeO) because of its near tissue equivalent effective atomic number of 7,1 [4][5]. The luminescence of BeO in the UV region below 370 nm, can be effectively stimulated by broadband blue light. The optical properties of BeO are well-known and were investigated by [6], [7] and [8]. Requirements placed on such systems are high rate of accuracy, a very robust and stable measuring procedure and a very good reproducibility as well as an easy and fast reading process. Therefore a new generation of measurement system based on the OSL of BeO was developed by the radiation physics group at the TU Dresden.

The emission of luminescence photons is a Poisson process. Thus, a time resolved detection of single photons allows access to this statistical character. Moreover, the time resolved detection in a combination with a modulated stimulation should be robust against noise.

II. Material and methods

BeO is most efficiently stimulated by blue light around 435 nm [2]. In this work, a blue LED *CREE XT-E Royal blue* is used as stimulation source.

The peak wavelength of this LED is around 450 nm while the wavelength of the luminescence light is below 370 nm [5]. A combination of different filters is used to allow mainly luminescence light to pass on to the sensor. In this work, a photon counting head by Hamamatsu, *H10682-210*, was used for the detector. This detector has an internal trigger circuit which is able to produce a voltage signal every 20 ns. The schematics of the system are outlined in figure 1. The signal from the photon detector is sampled by an FPGA and a timestamp with 10 ns resolution is assigned to each event.

 https://doi.org/10.1051/epjconf/201817009009

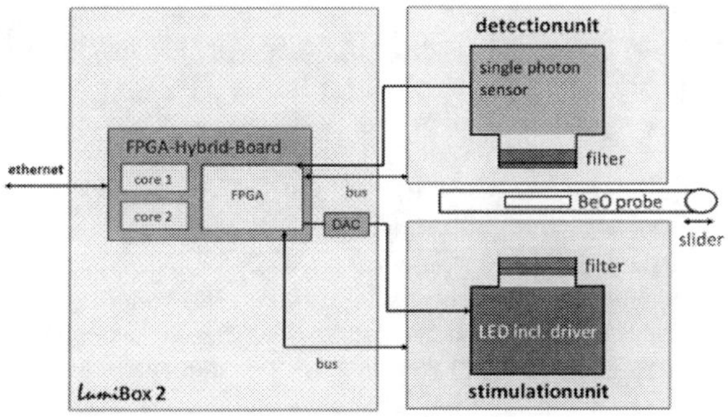

Figure 1 – schematic view of the readout system.

Because of their origin, the signals are independent of each other and obey the Poisson statistics. Consequently the measurement process becomes a statistical experiment. The character of this statistic and the associated uncertainties are well known and allows an accurate determination of the uncertainties. These are only limited by nature. The laws of the Poisson statistics show that multiplication or convolution of the Poisson-distributed events will have no effect to the statistical character of the measurement. Therefore the accuracy of the statistical nature of the measurement remains despite the use of weighting for analysis purposes or dead time correction. An advantage over current measurement is the discrete and character of the single events. These individual events can be separated and resolved, which allows a clear identification. They can then be digitized, which means that inaccuracies of the measurement device have no great influence.

The events are stored in a listmode manner in memory for later analysis. This gives the opportunity to apply a variety of different and independent data analysis methods.

The luminescence process of BeO has a decay constant of around 26 µs at room temperature [9]. The stimulation light is modulated synchronously with the sampling clock. The modulation is characterized by a period of 1 ms and a duty cycle of 0.5 (cf. figure 2).

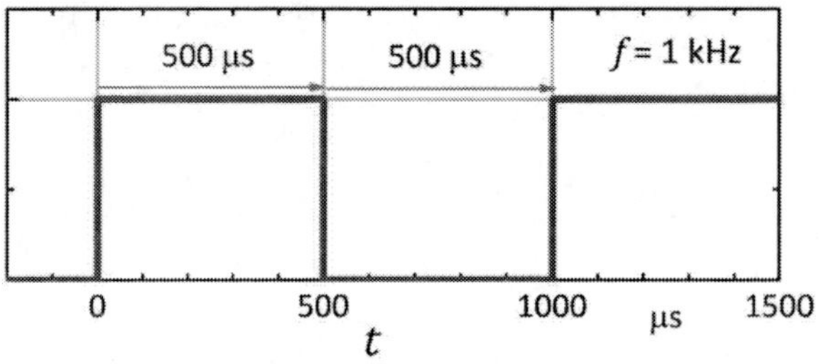

Figure 2 – Periodic modulation of the stimulation light

These parameters of the modulation ensure that the luminescence signal is completely decayed in the stimulation free phase to prevent a superposition of different decay processes of successive periods. With that regime, time and frequency information were imprinted on the signal. This allows the analysis in the time- and frequency domain, which delivers addition options for evaluation versus the continuous wave stimulation. [3]

Different kinds of weighting functions are used as digital filters. The rectangular weighting is adapted to the pulsed stimulation with square wave modulation and the known decay process of BeO. So every counts appearing during the stimulation and the first 180 µs of the decay process is weighted by a factor 1, while signals during the rest of the period are being ignored. Another weighting, that is used is the gate-time weighting. In this case, the signals are collected into two different gates. The first gate is taking all signals during the stimulation process into account while the second gate only considers the signals of the decay process. Thereby it is possible to recognize irregularities in one of these gates and correct them, or use the relation of both gates for further research. Considering the impressed frequency information, data analysis by using Fast Fourier Transform (FFT) algorithms or other digital filters can be used for separating the OSL signal from unwanted light or events generated by other phenomena. This can have significant effect to the lower detection limit. While using this system for dose determination over several orders of magnitude, some properties of the detector must be corrected for. For example there exist no obvious techniques for dead time correction for timestamp data convolved with a filter function. This is a challenge because of the timestamp character of the signal and the modulated stimulation which leads to an inhomogeneous Poisson process and time-dependent event rate.

III. Results and discussion

In figure 3, the count rate stabilized to the stimulation frequency is shown. The measurement without dosimeter shows a steep rise and decay edges which indicates the rise and fall of the stimulation signal in the first 500 µs. Signals with dosimeters, whether irradiated or not, reveal the typical decay curves.

For the signal different kinds of data analysis methods were applied. Therefore especially three different methods are examined are more in detail. As a comparative method it is the simple counting method, which is integration over all signals for the whole measuring time. Using the time impressed information the rectangular weighting will be discussed in the following section. Because of the periodic modulated stimulation there is also frequency information impressed to the measuring signal. This will be investigated by using the Fourier transformation. This data analysis method is implemented by using FFT–algorithm.

These three representative methods will be used to discuss the strength, advantages but also disadvantages of every method.

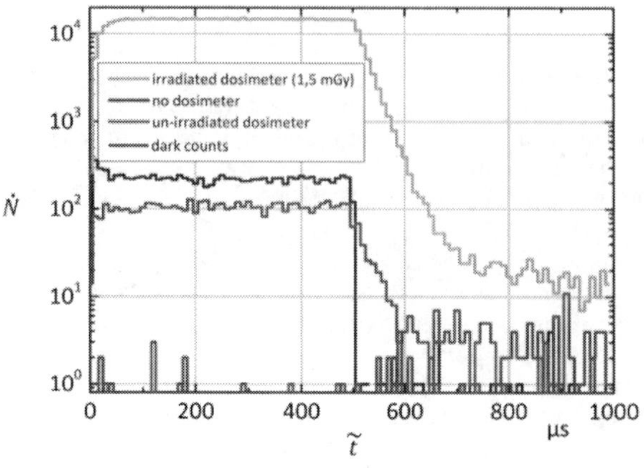

Figure 3 – Single photon count rates stabilized to the stimulation frequency ($\tilde{t} = t\,\%\,\frac{1}{f}$, f: stimulation frequency). The total measurement time was 30 s for each curve.

As a comparison for the different methods, the measurement deviation for a dose range between 2 µGy and 5 mGy was investigated. In this area the dead time correction has not such a big impact on the results. For the analysis a batch of 20 different BeO-detectors was evaluated, to compare the reproducibility of each method. Therefore, the same measurements were analyzed using the various data analysis methods. Figure 4 presents the results of the dose measurements with the different processes. The rectangular weighting has the least uncertainty over all dose ranges, in particular for the low dose. The difference between the counting method and the rectangular weighting is significant.

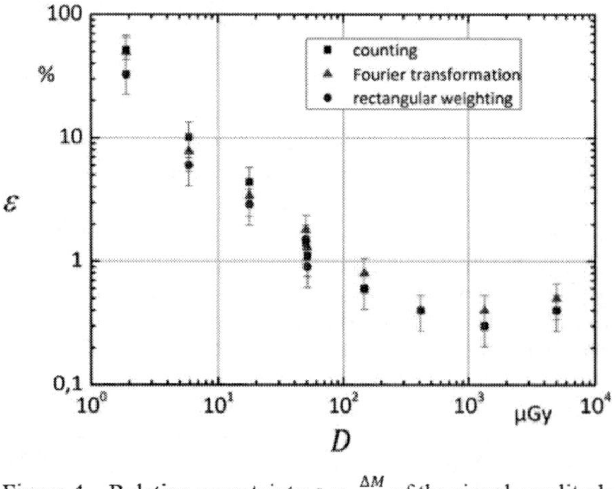

Figure 4 – Relative uncertainty $\varepsilon = \frac{\Delta M}{M}$ of the signal amplitude M based on a batch of 20 dosimeters for different analysis methods as a function of dose.

Using the time impressed information and rectangular weighting can improve the stability of dose measurements by around 20% in the area of the lower detection limit. For the higher dose range, the used method does not have a major impact on uncertainty.

For the measuring process it is of paramount importance to separate the dose induced signal from the background signal. The measurement process can be described by unwanted light detection due to static discharge, scintillation or other uncorrelated events which manifest in the occurrence of a spike in count rate. Empirical studies have revealed the character of such disturbances. It could be observed, that the occurring mean length of such a spike is around 10 µs. Every of them consist of independently occurring events, which can be described by a Poisson process. The number of events per spike has a big variation, depending on the origin and type. On average there are 10-20 detected events for such a disturbance. Using this gained information it was possible to simulate some disturbances. Since the occurrences of such spikes were uncorrelated to the measurement process or event rate, the entrance of this spikes were simulated irrespective of the stimulation function. This has been used for investigate the reproducibility and stability of the system, by combining measuring signals with simulated different kinds of possible occurring disorders. 10 spikes were simulated, each with 10-20 events, spread over a measurement time of 10 seconds with an external simulation software. Subsequently the disturbance signal was combined with the measurement signal, considering the detector reasoned dead time. The impacts of the disorders were investigated for different irradiated doses for the representative data analysis methods (figure 5).

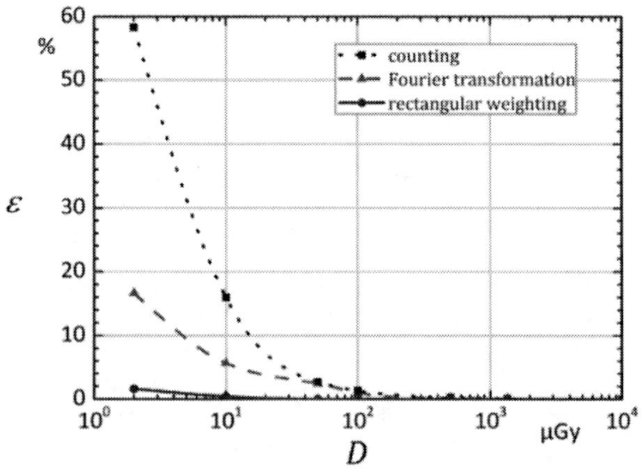

Figure 5 – Simulated impact of disturbances on the relative uncertainty $\varepsilon = \frac{\Delta M}{M}$ for the three analysis methods as a function of dose.

Using a weighting function has a huge impact for the stability of the measuring signal. For the rectangular weighting it is possible to largely suppress this kind of disturbance. It is very stable, even for very low dose. Using the frequency information with the FOURIER transformation, the stability of the signal compared to the comparative method could be improved for lower doses as well. But there are clear differences in comparison to the rectangular function. In the range of higher dose, from around 100 µGy such disorders have not much impact to all measuring methods, because of the very good statistic of the measuring signal and high event rate.

IV. Conclusion

Through the measurement of single events instead of current measurement, this measurement process is based on the statistical nature of counting experiment. This makes it possible on the basis of the well-known character of the Poisson distributed independent events to express the uncertainty by the laws of statistics. Every single event will get an unique time stamp, whereby the maximum amount of accessible information of this measurement process can be retained. This information can be used for any kind of data analysis. Using amplitude-modulated stimulation impresses time- and frequency information into the OSL signal and therefore allows additional means of analysis.

Compared to conventional analytical methods, such has the advantage of simple identification and discrimination of any kind of disturbances, without affecting the actual signal. Thereby the signal-to-noise ratio can be optimized efficiently. This has also some impact to the stability of the whole measurement process. The stability of the measurement especially for small measuring times or small count rates can be improved significantly. Investigation of the background signal shows, that this disorders can have many different sources. Disturbances, which are independent of the modulation or measurement process, and caused by external reasons like cosmic rays, or spontaneous conversions inside the measurement device can be almost completely suppressed, as it was shown in figure 5. The effect of counting rates dependent parasitic, which were identified as detector depended after pulsing is subject of further investigation.

The focus for future development is set on a calibration process, which is independent by the measurement process or used device. Also the time stamp based dead time correction needs further investigation. This is then to implement on an FPGA, in order to allow a parallel evaluation of the measurement process.

References

[1] J. Henniger, B. Horlbeck, K. Hübner, K. Prokert: The evaluation of CaF_2:Mn-Polyethylene Detectors with the Aid of the Optically Stimulated Luminescence (OSL), *Nuclear Instruments and Methods* **204** (1982) 209-212

[2] E. Bulur, H.Y. Göksu, "OSL form BeO ceramics: new observations from an old material", *Radiat. Meas.*, 29 (1998), pp. 639–650

[3] M.S. Akselrod, S.W.S. McKeever: "A Radiation Dosimetry Method using Pulsed Optically Stimulated Luminescence", *Radiat. Prot. Dosimetry* 81 (1999) p.167-175

[4] G. Scarpa, "The dosimetric use of beryllium oxide as a thermoluminescent material: A preliminary study. ", *Phys. Med. Biol.* 15, 1970, 667-672

[5] M. Sommer, R. Freudenberg, J. Henniger, "New aspects of a BeO-based optically stimulated luminescence dosimeter", *Radiat. Meas.*, 42(2007), pp.617-620

[6] C.E. Mandeville, H.O. Albrecht,"Luminescence of Beryllium Oxide", *Phys. Rev.* 94 (1954) p. 494

[7] E. Tochilin, N. Goldstein, W.G. Miller, „BeO as a thermoluminescent dosimeter", *Health Phys.*, 16 (1969), pp. 1–7

[8] K.W. Crase, R.B. Gammage, "Improvements in the use of ceramic BeO for TLD", *Health Phys.*, 29 (1975), p. 739

[9] E. Bulur, B.E. Saraç, "Time-resolved OSL studies on BeO ceramics", *Radiat. Meas.*, 59 (2013), pp. 129–138

Radiation dose response of N channel MOSFET submitted to filtered X-ray photon beam

Luiz C. Gonçalves Filho, David S. Monte, Fabio R. Barros, and Luiz A. P. Santos

Abstract—**MOSFET can operate as a radiation detector mainly in high-energy photon beams, which are normally used in cancer treatments. In general, such an electronic device can work as a dosimeter from threshold voltage shift measurements. The purpose of this article is to show a new way for measuring the dose-response of MOSFETs when they are under X-ray beams generated from 100kV potential range, which is normally used in diagnostic radiology. Basically, the method consists of measuring the MOSFET drain current as a function of the radiation dose. For this the type of device, it has to be biased with a high value resistor aiming to see a substantial change in the drain current after it has been irradiated with an amount of radiation dose. Two types of N channel device were used in the experiment: a signal transistor and a power transistor. The delivered dose to the device was varied and the electrical curves were plotted. Also, a sensitivity analysis of the power MOSFET response was made, by varying the tube potential of about 20%. The results show that both types of devices have responses very similar, the shift in the electrical curve is proportional to the radiation dose. Unlike the power MOSFET, the signal transistor does not provide a linear function between the dose rate and its drain current. We also have observed that the variation in the tube potential of the X-ray equipment produces a very similar dose-response.**

Index Terms—**MOSFET, Radiation, Dosimeter.**

I. Introduction

MOSFET has been used for many years as a radiation detector mainly in high-energy photon beams, which are normally used in cancer treatments. More recently, the MOSFETs are also being used to measure the dose in X-ray beams of the diagnostic radiology energy range. In general, such an electronic device can operate as a dosimeter from the threshold voltage shift measurements. This means that the dose is obtained from the increase in gate voltage value, which causes the transistor to turn on. The purpose of this article is to show a new way for measuring the dose-response of typical MOSFETs when they are under X-ray beam generated from 100kV potential, that is, an energy range which is normally used in diagnostic radiology. The method is based on the analysis of the transistor drain current rather than the threshold voltage. Here, we present how someone can bias the MOSFET so that, by plotting its electrical curve or analyzing the variation of the drain current, it is possible to measure the radiation dose.

Actually, the MOSFET has to be biased with a voltage (V_{DD}) and a high resistance resistor, R_D, connected to the drain (Fig. 1). The device becomes too slow, but it does not matter because it will not operate as switch. To measure the drain current, I_D, an ammeter is in series with V_{DD} source.

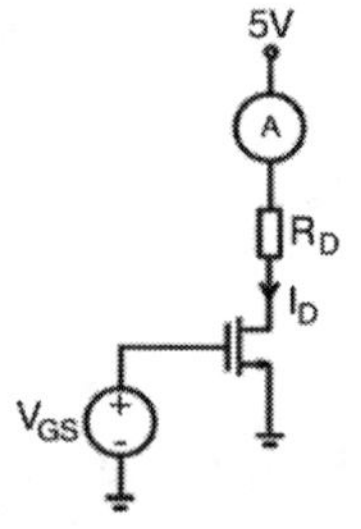

Fig. 1. MOSFET biased with 5V and a resistor R_D connected to the drain. To plot an $I_D \times V_{GS}$ electrical curve, V_{GS} can be varied from 0 up to 3V, for while.

By systematically varying the gate-source voltage, V_{GS}, one can obtain an $I_D \times V_{GS}$ electrical curve as showed in Fig. 2. Notice that the drain current depends on the resistor value, i.e., the saturation point can be changed. For example, if a 100 kΩ resistor is used in the circuit we have I_D=50 μA at V_{GS}=2.1 V. On the other hand, if we choose another resistor with the thousand times lower value, the saturation point naturally goes to I_D=50 mA at V_{GS}=2.9 V. As it will be seen later, there are some considerations that must be taken into account both in the choice of MOSFET bias, type of transistor, and the spectrum of radiation to be measured.

Manuscript received June 1, 2017. This work was supported by the CNPq under Proc. 306681/2015-3, and in part by SCIENTS.
L. C. Gonçalves Filho is with Comissão Nacional de Energia Nuclear (CNEN/CRCN-NE) and with Universidade Federal de Pernambuco (UFPE), Av. Prof. Luiz Freire, 200, Recife, 50740-545 (e-mail: lcar.br@hotmail.com).
D. S. Monte is with Universidade Federal de Pernambuco (UFPE), Av. Prof. Luiz Freire, 1000, Recife, 50740-545, and with SCIENTS, Estrada do Monjope, 5100, Igarassu, 53645-337 (e-mail: dsmonte@scients.com.br).
F. R. Barros and L. A. P. Santos are with Comissão Nacional de Energia Nuclear (CNEN/CRCN-NE), Av. Prof. Luiz Freire, 200, Recife, 50740-545, and with SCIENTS, Estrada do Monjope, 5100, Igarassu, 53645-337 Brazil (e-mail: frbarros12@gmail.com; lasantos@scients.com.br).

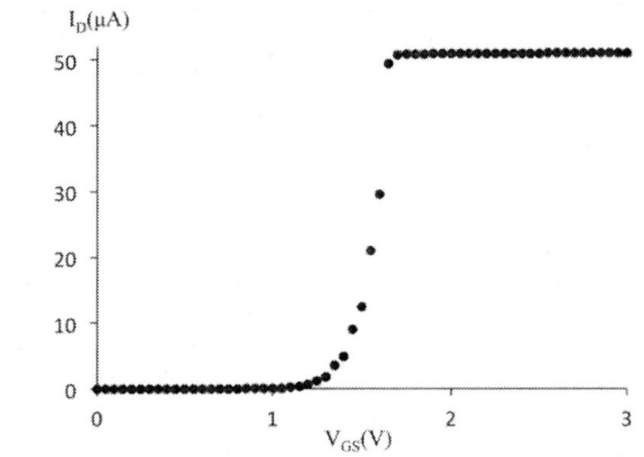

Fig. 2. I_D×V_{GS} curve for a MOSFET biased: V_{DD}=5 V and R_D=100 kΩ. In this case the saturation point is reached at the V_{GS}=1.7 V with I_D=50 μA.

II. Materials And Methods

A. MOSFETs

Three types of N channel MOSFETs and a total of 24 devices were used in the experiments: 6 BSS138 – signal transistor with SOT-23 encapsulation; and 12 FQP12N60C, and 6 BUZ91A – power device with TO-220 package (Fig. 3.) The type of transistor package influences significantly when it is used as a radiation detector. In fact, the epoxy material thickness of the encapsulation depends on the device, and it is a crucial factor for increasing the number of interactions, and consequently the amount of charges generated on the oxide of the MOSFET, as well as the damage in the semiconductor crystal itself. The increase in the photon interactions due to the thickness of the material is known as the build up cap effect.

Fig. 3. MOSFETs with their respective packages in parenthesis: BSS138 (SOT-23); and BUZ91A (TO-220).

B. Instruments and Bias

An EFF1506 SCIENTS system was used to take I_D measurements and simultaneously bias the device with both signals: V_{DD} and V_{GS}. The drain was biased with a R_D=91 kΩ resistor at V_{DD}=4.5 V. The bias circuit is exactly as showed in Fig. 1. The V_{GS} bias was systematically varied from 0 V up to 3 V each 50 mV step to trace the I_D×V_{GS} curves. A W23342 PTW ion chamber connected to the 6517A Keithley system was used to take the dose rate measurements.

C. X-ray unit, radiation beam and experimental set up

An HF-320 Pantak X-ray unit was used to generate the photon beam, which was filtered with an 1 mm aluminum plate. Aiming to test the devices for measuring the radiation dose in the energy range normally used in diagnostic radiology, the tube potential of the X-ray equipment was set to be 100 kV. The photon beam spectrum is fully specified setting the potential on the control panel of the X-ray unit, and selecting a material with a certain thickness to filter the radiation. The experimental set up is illustrated in Fig. 4.

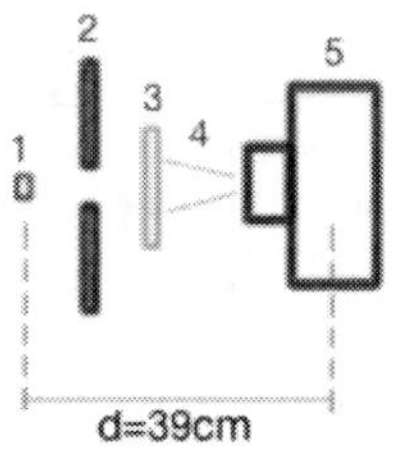

Fig. 4. Experimental set up used to irradiate the MOSFETs with a filtered X-ray photon beam: 80kV, 100kV and 120kV tube potentials. 1) Device under irradiation; 2) Collimator; 3) 1 mm Al plate as a radiation filter; 4) Photon beam without filtration; 5) X-ray generator unit.

D. Experimental procedures

First of all, the technique consisted in evaluating the MOSFET response after the irradiation procedure, which was done delivering 20 Gy radiation dose each time. The I_D×V_{GS} curve was plotted immediately after the irradiation procedure by varying the gate voltage, V_{GS}, in steps of 50 mV up to 3 V.

A second experimental method was based on the I_D×V_{GS} response curves by choosing a V_{GS} value, for each type of MOSFET, and evaluate the drain current as a function of the dose rates, or more specifically as a function of the photon fluence. The dose rates were obtained based on the X-ray tube current, which can be selected in the control panel of the equipment from 5 mA up to 25 mA, for each 5 mA step. The dose rate data are showed in the Table I.

TABLE I
DOSE RATE VALUES USED IN THE EXPERIMENTS

k	X-ray tube current (mA)	Dose rate (mGy/s) at the MOSFET position
1	5.00 ± 0.01	8.07 ± 0.06
2	10.00 ± 0.01	15.90 ± 0.09
3	15.00 ± 0.01	24.20 ± 0.21
4	20.00 ± 0.02	32.00 ± 0.23
5	25.00 ± 0.02	40.40 ± 0.41

The third stage of the experiments was made varying the X-ray tube potential ±20% (80 kV and 120 kV). With this procedure, one can evaluate how the changes in the spectrum can bring about variations in the MOSFET dose-response. All experiments were done with samples of transistors in triplicate aiming to verify reproducibility. The same transistor is not used in another experiment, since if irradiated it is discarded.

III. Results and Discussions

The $I_D \times V_{GS}$ curves for the three types of MOSFETs are showed in Fig. 5. The curves without irradiation (0 Gy) show that the saturation point is not the same. Actually, the structure of the device and its own geometry (e.g. channel dimension, components internally connected) can determine the saturation point. After any irradiation, from 20 Gy onwards, regardless of the encapsulation of each device type, we can observe that the behavior of the $I_D \times V_{GS}$ curves is very similar: as the dose increases, systematically the curves take a shift to the left. Notice that the BUZ91A has a signature as a soft peak, which systematically appears with the same shift to the left. At this point, we may think that the dose-response is practically the same for any type of MOSFET. However, our purpose is to verify how the drain current value varies with the dose rate in order to correlate the present value of I_D with the dose that was received by a transistor. Then, some results for I_D as a function of the dose rate, $\dot{D}$, are in the graphs of Fig. 6.

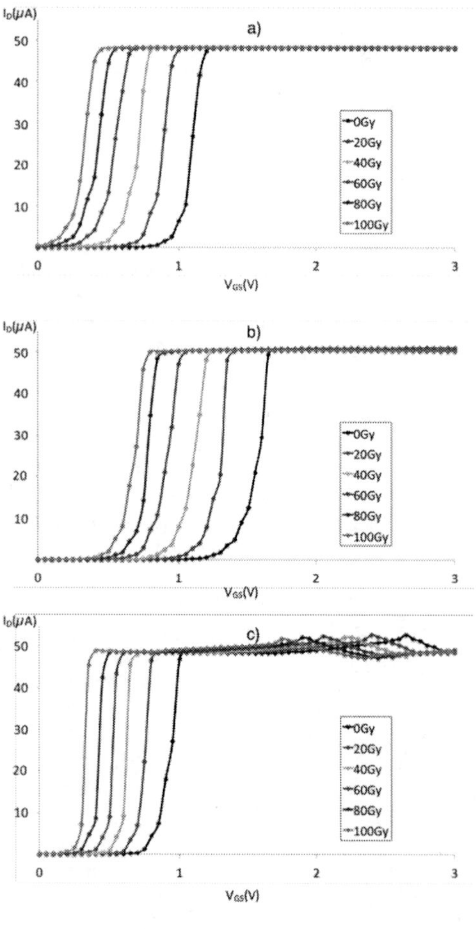

Fig. 5. $I_D \times V_{GS}$ curves after 100 Gy dose in steps of 20 Gy for each Mosfet: a) BSS138; b) FQP12N60C; c) BUZ91A.

To evaluate $I_D \times \dot{D}$, the latest results (Fig. 5) have helped us to choose a V_{GS} value to set the bias for each transistor: 450 mV for BSS138; 950 mV for FQP12N60C; and 500 mV for BUZ91A. In this test, the real-time measurements mode was used. The results show that for the signal transistor the drain current is very low and practically increases continuously with the dose, whereas for the power MOSFETs each step of photon fluence can be seen as practically constant I_D during the irradiation. The I_D variation more pronounced for the BSS138 MOSFET probably means that it could not be used as a dosimeter, since I_D does not hold over time.

The proportionality between I_D and $\dot{D}$ can be seen in Fig. 7, where the results of three samples of FQP12N60 MOSFET are displayed together with the average between them. Notice that T2 transistor has practically the same response of the average.

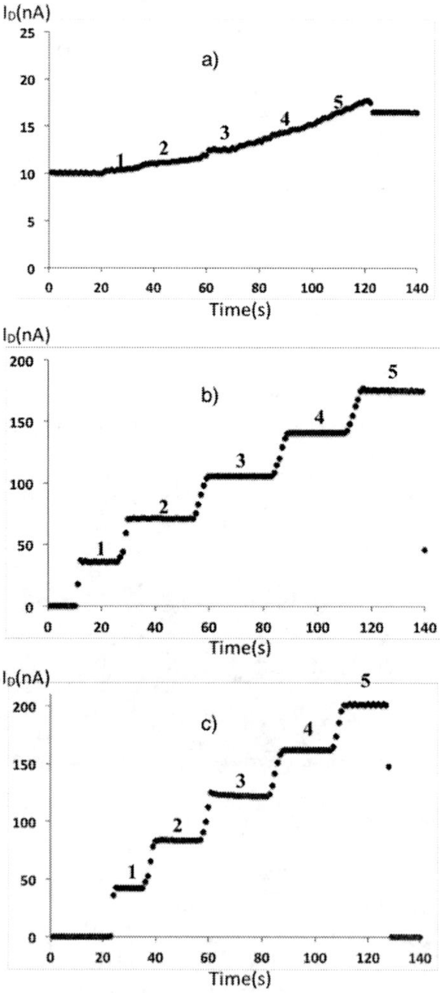

Fig. 6. Real-time measurements of I_D for each $\dot{D}_k$ where k is the index in the Table I. MOSFETs: a) BSS138; b) FQP12N60C; c) BUZ91A.

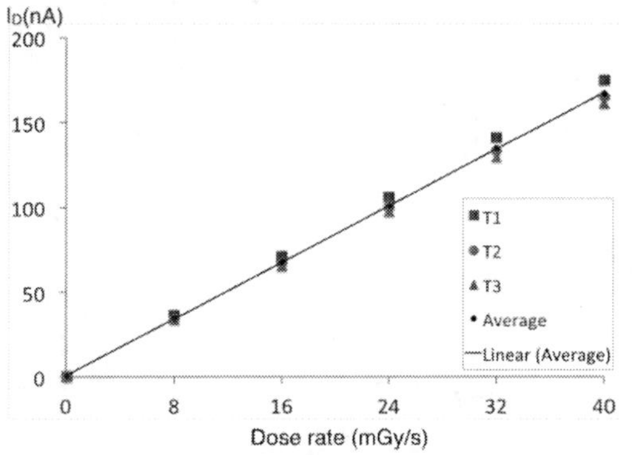

Fig. 7. $I_D \times \dot{D}$ response for three samples of FQP12N60C MOSFET. The average, in black, has the equation $I_D = I_0 + S \cdot \dot{D}$, where $I_0 = 0.7$ (nA) is the initial value of I_D, S=4.176 (nA/mGy·s^{-1}) is the average sensitivity coefficient, and the line has a determination coefficient of R^2=0.99991.

The linear coefficients of each power MOSFET response will depend naturally on the bias and the type of transistor. In fact, every device will have its own sensitivity coefficient, or conversion factor for calibration purposes, as any type of dosimeter has. In order to estimate the useful life of the power MOSFET in diagnostic radiology, consider that the dose of a lung X-ray examination should be about 0.5 mGy, according to the standards guide of the IAEA. In such a case, after 1,000 X-ray examinations, which corresponds to about one month of use in a typical clinic, the device would receive 0.5 Gy dose. Actually, it is the dose during the 3rd irradiation of the device (Fig. 6b) and we can observe that I_D=106±0.2 (nA) practically does not vary in 20 s. Therefore, this result demonstrates that a MOSFET can be used to estimate the dose (in real-time) via the drain current measurements, if it is previously calibrated.

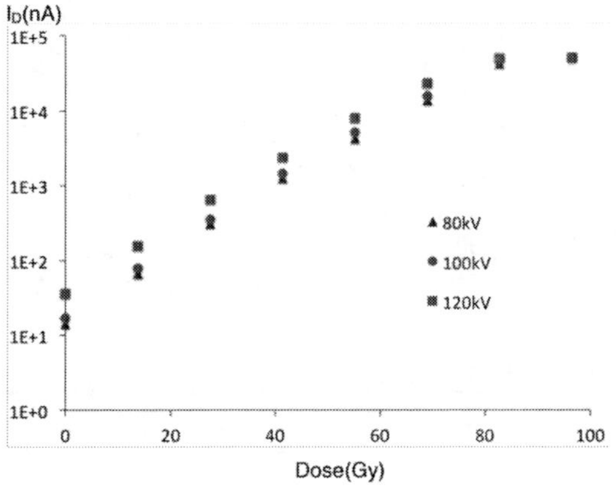

Fig. 8. The drain current I_D as a function of the received dose D for three samples of FPQ12N60 MOSFETs. Each device was submitted to three different spectrum originated from 80 kV, 100kV and 120 kV.

Finally, the radiation dose responses of three N channel MOSFETs whose are submitted to 1 mm Al plate filtered X-ray photon beam can be seen in the Fig. 8. There also is the result of the analysis of how changes in the tube potential (±20%) can modify the response of the MOSFET. We can observe that the drain current is lower if 80 kV tube potential is selected, whereas I_D is higher if the spectrum is originated from 120 kV. This proves that the measurement process has to take into account that a MOSFET must be probably calibrated with respect to the tube potential. Notice that the initial values are different and therefore the curves do not match perfectly, although we know that each device has its own calibration. It can also be observed that after the dose value reaches 70 Gy the dose-response presents saturation. However, for the lung X-ray examination, this may mean that the transistor could operate about more than 100 thousand times.

IV. Conclusions

Unlike the known conventional method for measuring the radiation dose, from the threshold voltage measurements, this paper brings an innovative measurement process, which consists of evaluating the drain current of the MOSFET. Two types of devices were used: a signal transistor and the power transistor. The results found lead us to conclude that with power MOSFET it is easier to measure the radiation dose-response. Actually, the results showed that there are two ways to estimate the dose in X-ray beam: the first is to measure the drain current in real time; and the second one is to correlate the I_D value with the received dose by the power MOSFET. Finally, the dose response indicates that variations in the spectrum lead to variations in the output current of the device. We conclude that the proposed method is feasible to estimate the dose in filtered X-ray photon beam, which is normally used in diagnostic radiology.

References

[1] L. J. Asensio, M. A. Carvajal, J. A. Lopez-Villanueva, M. Vilches, A. M. Lallena, A. J. Palma, "Evaluation of a low-cost commercial mosfet as radiation dosimeter," *Sensors and Actuators A*, vol. 125, pp. 288-295, 2006.

[2] A. B. Rosenfeld, "Electronic dosimetry in radiation therapy," *Radiat. Meas.*, vol. 41, pp. 134-153, 2006.

[3] J. H. Koivisto, J. E. Wolff, T. Kiljunen, D. Schulze, M. Kortesniemi, "Characterization of MOSFET dosimeters for low-dose measurements in maxillofacial anthropomorphic phantoms," *Journal Appl. Clin. Med. Phys.*, vol. 16, n. 4, pp. 266-277, 2015.

[4] A. S. Sedra, K. C. Smith, *Microrelectronic Circuits*. Oxford University Press, 2004.

[5] E. S. Comfort, M. P. Rodgers, W. Allen, S. C. Gausepohl, E. X. Zhang, M.L. Alles, H. L. Hughes, P. J. McMarr, and Ji Ung Lee, "Intrinsic Tolerance to Total Ionizing Dose Radiation in Gate-All-Around MOSFETs," *IEEE Trans. Nucl. Sci.*, vol. 60, pp. 1-5, 2013.

[6] C. M. S. Magalhaes, L. A. P. Santos, D. N. Souza, A. F. Maia, "Effect on the insulation material of a MOSFET device submitted to a standard diagnostic radiation beam," Journal of Physics, 249, IOP Publishing, 2010.

[7] IAEA, "Radiation doses in diagnostic radiology and methods for dose reduction," IAEA-TECDOC-796, 1995.

[8] G. Lutz, *Semiconductor radiation detectors*. Berlin: Springer, 1999, pp. 79-159.

[9] H. Spieler, *Semiconductor detector systems*. Oxford: Oxford University Press, 2008.

EPJ Web of Conferences **170**, 09011 (2018) https://doi.org/10.1051/epjconf/201817009011
ANIMMA 2017

Exploring and Monitoring of Methane Hydrate Deposits

D. Sudac, J. Obhođaš K. Nađ and V. Valković

***Abstract*— Relatively recently, in the last 20 years, it was discovered that methane hydrate (MH) deposits are globally distributed in the permafrost and oceans. Before 1965 when first deposits were discovered in nature, it was believed that MH can occur only in laboratory conditions or in vast parts of the Universe. Presently it is presumed that this solid crystalline compounds in which CH_4 molecules occupies the water ice lattices (nominal chemical formula of MH is $C_4H_{62}O_{23}$) can serve as an energy source favorably to the all of the world remaining conventional hydrocarbon sources. The worldwide estimates of MH deposits range from $2x10^{14}$ m^3 to $3.053x10^{18}$ cubic meters. This uncertainty partly results from our limitations in geological understanding of the MH deposits, which is due to the relatively bad quality of data obtained by presently available seismic and electromagnetic techniques. Moreover, MH deposits can become vulnerable to climate changes, which were already occurring in geological past whit tremendous consequences for the global life on Earth. Thus, further development of advanced techniques is needed to enhance our abilities to better characterize, quantify and monitor the MH deposits. In the work presented 14 MeV neutrons and associated alpha particle imaging (API) where used to quantify the amount of MH in the sample. Samples were prepared from sea sediment, quartz sand and MH simulant.MH simulant with chemical formula $C_4H_{46}O_{23}$ was made from sucrose (25 % by mass) and water. MH quantity was measured by measuring the carbon content in the sample [1-8].**

***Index Terms*—associated alpha particle imaging, fast neutrons, methane hydrate and hydrocarbon energy sources**

I. Results

Measurements were done in the empty basin and in the basin filled with the water. In total six samples were prepared:

a) Samples of sea sediments/ quartz sand with pores filled with air
b) Samples of sea sediments/quartz sand with pores filled with water
c) Samples of sea sediments/quartz sand with pores filled with MH simulant

Detection probability of MH simulant was determined in the sea sediment/quartz sand with respect to the pores filled with air/water in dependence on the measuring time. Figure one shows the gamma ray spectrum of sea sediment which pores are filled with MH simulant.

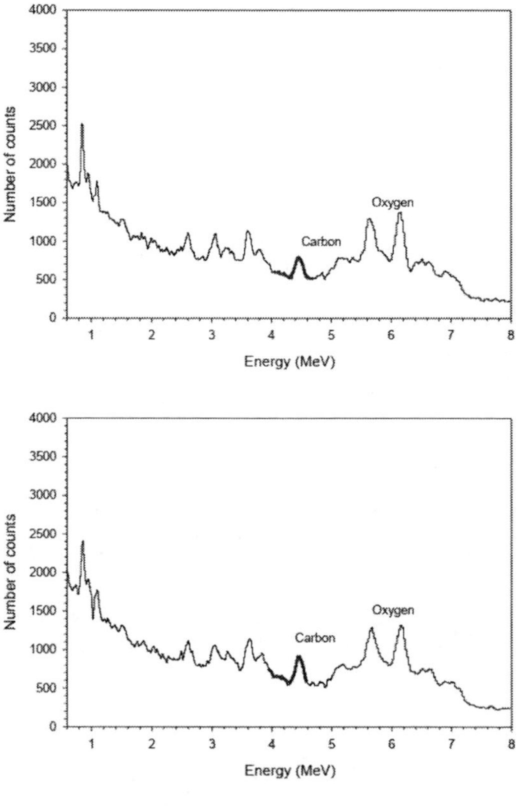

Fig. 1. The gamma ray spectra of sea sediment saturated with water (up) and saturated with MH simulant (down).

According to the data 2.5 h of measurement time is needed for 100% detection probability. Detection probability decreases with decrease of measuring time. For half an hour of

Davorin Sudac, Jasmina Obhodas, Karlo Nad, and Vladivoj Valkovic are with the Rudjer Boskovic Institute, Bijenicka c. 54, 10000 Zagreb Croatia (dsudac@irb.hr)

measurement time detection probability drops to 70%. There are no measurable influence of water on detection probability, at least for a given measurement accuracy. For potential commercial applications at least one order of magnitude more intense neutron beam is needed together with at least four gamma ray detectors placed around the sample. Nevertheless, our investigation shows that MH rich deposits can be detected and investigated by API method.

TABLE I
DETECTION PROBABILITY OF MH SIMULANT IN QUARTZ SAND (WATER IN THE BASIN)

Measurig Time	Detection Probabiliy with respect to pore filled with air (%)	Detection Probabiliy with respect to pore filled with water (%)
2.5 h	100	100
1.25 h	94 ± 4	89 ± 7
50 min	85 ± 5	81 ± 7
37 min	66 ± 8	76 ± 8
30 min	65 ± 7	68 ± 7
25 min	66 ± 5	62 ± 6

TABLE II
DETECTION PROBABILITY OF MH SIMULANT IN SEA SEDIMENT (WATER IN THE BASIN)

Measurig Time	Detection Probabiliy with respect to pore filled with air (%)	Detection Probabiliy with respect to pore filled with water (%)
2.5 h	-	100
1.25 h	-	94 ± 3
50 min	-	89 ± 4
37 min	-	69 ± 8
30 min	-	69 ± 7
25 min	-	64 ± 6

REFERENCES

[1] Kennett JP, Cannariato KG, Hendy IL, Behl RJ.(2000),'' Carbon isotopic evidence for methane hydrate instability during quaternary interstadials'', Science 288, 128-133.

[2] Jahren A.H. Et al. (2005)" A plate tectonic mechanism for methane hydrate relase along subduction zones". Earth and Planetary Science Letters. 236, 691-704.

[3] Mordis GJ, Collet TS, Boswell R, Kurihara M, Reagan MT, Koh C, Sloan ED.(2009),'' Toward production from gas hydrates: current status, assessment of resources, and simulation based evaluation of technology and potential'', SPE Reservoir Evaluation & Engineering,12/5, 745-771

[4] Krey V, Canadell JG, Nakicenovic N, Abe J, Andruleit H, Archer D, Grubler A, Hamilton NTM, Johnson A, Kostov V, Lamarque JF, Langhorne N, Nisbet EG, O'Neill B, Riahi K, Riedel M, Wang W, Yakushev V (2009)'' Gas hydrates: entrance to a methane age or climate Threat'', Env Res Lett 4, 1-6.

[5] Sudac, D., Blagus, S. and Valkovic, V., (2005),''Inspections for contraband in a shipping container using fast neutrons and the associated alpha particle technique: Proof of principle," '', Nucl. Instrum. Methods. Phys. Res. B 241, 798-803.

[6] Valkovic, V., Sudac, D., Obhodas, J., Eleon, C., Perot, B., Carasco, C., Sannié, G., Boudergui, G.K., Kondrasovs, V., Corre, G., Normand, S., Woo, R., Bourbotte, J-M. (2013)," The Use of Alpha Particle Tagged Neutrons for the Inspection of Objects on the Sea Floor for the Presence of Explosives", Nuclear Instruments and Methods A, 703, 133-137.

[7] K.A. Kvenvolden, ''A review of the geochemistry of methane in natural gas hydrate.'', *Org. Geochem. Vol. 23, No. 11/12, pp. 997-1008, (1995).* Published by Elsevier Science Ltd.

[8] M.D. Max (2003),'' Natural Gas Hydrate in Oceanic and Permafrost Environments'', *Kluwer Academic Publishers pp. 4.*

Fiber optical dose rate measurement based on the luminescence of beryllium oxide

Tobias Teichmann, Jan Sponner and Jürgen Henniger, *TU Dresden*

***Abstract*— This work presents a fiber optical dose rate measurement system based on the radioluminescence and optically stimulated luminescence of beryllium oxide. The system consists of a small, radiation sensitive probe which is coupled to a light detection unit with a long and flexible light guide. Exposing the beryllium oxide probe to ionizing radiation results in the emission of light with an intensity which is proportional to the dose rate. Additionally, optically stimulated luminescence can be used to obtain dose and dose rate information during irradiation or retrospectively. The system is capable of real time dose rate measurements in fields of high dose rates and dose rate gradients and in complex, narrow geometries. This enables the application for radiation protection measurements as well as for quality control in radiotherapy. One inherent drawback of fiber optical dosimetry systems is the generation of Cherenkov radiation and luminescence in the light guide itself when it is exposed to ionizing radiation. This so called "stem" effect leads to an additional signal which introduces a deviation in the dose rate measurement and reduces the spatial resolution of the system, hence it has to be removed. The current system uses temporal discrimination of the effect for radioluminescence measurements in pulsed radiation fields and modulated optically stimulated luminescence for continuous irradiation conditions. This work gives an overview of the major results and discusses new-found obstacles of the applied methods of stem discrimination.**

***Index Terms*—Beryllium oxide, Fiber optic dose rate measurements, Optically Stimulated Luminescence, Radioluminescence**

I. Introduction

FIBER OPTICAL DOSIMETERS (FODs) allow real time dose rate measurements with high spatial resolution. FODs consist of a small, radiation sensitive probe which is coupled to a light detection unit with a long and flexible light guide. Exposing the probe to ionizing radiation results in the emission of radioluminescence light (RL) with a characteristic spectrum and an intensity which is, under certain circumstances, proportional to the dose rate. Additional dose and dose rate information can be obtained by in-beam or retrospective optical stimulation of the probe with blue light which leads to optically stimulated luminescence (OSL).
Due to the separation of radiation sensitive volume and light detection and stimulation unit, the system qualifies for measurements in fields of high dose rates and dose rate gradients and in complex and narrow geometries without exposing the operator or the susceptible electronics. Possible applications are radiation protection measurements in facilities for the generation of ionizing radiation and the surveillance and dismantling of nuclear facilities as well as quality control in radiotherapy and nuclear medicine.

One inherent drawback of fiber optical dosimetry systems is the generation of Cherenkov radiation (CR) and luminescence in the light guide itself when it is exposed to ionizing radiation. This "stem" effect results in an additional signal component which is also detected by the PMT. This causes a deviation of the dose rate measurement and reduces the spatial resolution of the system. Therefore the stem effect must be removed. Promising solutions to do so have been investigated in the past [1-3].

In the current system, based on beryllium oxide, two different methods of temporal discrimination of the effect for RL measurements in pulsed radiation fields (already introduced in [4]) have been tested. Furthermore, a combination of optical filtering and modulated OSL, which is an enhancement of [5], has been applied to measurements under continuous irradiation conditions. This work gives an overview of the major results and discusses new-found obstacles of the applied methods of stem discrimination in pulsed radiation fields.

II. Material And Methods

A. Dosimetry System

The presented dosimetry system uses beryllium oxide (BeO) in form of a sintered ceramic cylinder (height 1 mm, diameter 1 mm, Thermalox®995, Materion) as a radiation sensitive probe. The well-known OSL phosphor has been investigated and applied for integrating dose measurements for decades. In the recent past BeO gained greater attention in the field of FODs [4-8]. The material is almost tissue equivalent with an effective atomic number of 7.2. Its luminescence emission spectrum has a broad global maximum at 370 nm. The luminescence life time is 27 μs at room temperature [9].

The BeO detector is coupled to the tip of a PMMA light guide (diameter 1 mm, length 15 m), which is shielded from

ANIMMA 2017 Conference Proceedings, submitted June 27, 2017.
T. Teichmann is with the Radiation Physics Group at the Institute of Nuclear and Particle Physics, TU Dresden, Zellescher Weg 19, 01069, Dresden, Germany (e-mail: teichmann@asp.tu-dresden.de).
J. Sponner was with the Radiation Physics Group at the Institute of Nuclear and Particle Physics, TU Dresden, Zellescher Weg 19, 01069, Dresden, Germany (e-mail: sponner@asp.tu-dresden.de).
J. Henniger is with the Radiation Physics Group at the Institute of Nuclear and Particle Physics, TU Dresden, Zellescher Weg 19, 01069, Dresden, Germany (e-mail: henniger@asp.tu-dresden.de).

ambient light with a black coating. The RL and OSL, generated in the BeO during irradiation and/or optical stimulation, passes the light guide and is detected with a PMT (H10682-210, Hamamatsu) operated in single photon counting mode. In contrast to the previous measurements in [4], an adjustable mechanical aperture was added to reduce the intensity of the luminescence light from the probe in order to measure the influence of the stem signal in the linear detection range of the PMT.

The logical pulses originating the light detector are sampled with 100 MHz by an FPGA. Each incoming pulse is assigned with a time stamp, representing its arrival time since the start of the measurement.

To perform additional OSL measurements, stimulation light from a diode laser (peak wavelength 450 nm, Osram) is coupled into the light guide using a 400 nm dichroic beam splitter (Edmund Optics) and a fiber collimator (Hellma Analytics). The stimulation current can be modulated externally. The laser diode features temperature stabilization. Its maximum optical output power is 1.6 W. However, absorption of stimulation light in the light guide and fiber connector leads to a raise in temperature, which can affect the transmission of the components. In order to prevent the resulting fading effects in the responsivity of the probes, the power is limited to 600 mW in the present measurements.

B. Methods of Gated Discrimination

In contrast to the luminescence life time of BeO (27 µs) the decay time of the stem signal is much smaller, hence the RL signal of BeO is still present shortly after irradiation, when the stem component is already negligible [4]. For measurements in pulsed radiation fields, this is exploited to generate a dose rate proportional signal, which is independent of the stem effect.

Two methods, i.e. “trigger and discriminator” and “artificial dead time”, have been investigated to distinguish irradiation pulses and the time intervals between pulses (the latter providing the dose rate information) on the basis of the difference in event rates.

The first approach uses the distance between sequential time stamps to identify the irradiation pulses. The system stays triggered as long as the time stamp distances are shorter than a preset threshold Δ_{th}. After the end of the irradiation pulse and a following discrimination time T_{d}, the stem component of the signal has decayed and the measured signal is only RL. The procedure is illustrated in Fig. 1.

The second method generates histograms of time stamp distances. Short distances are more likely to occur during an irradiation pulse, while longer distances are assigned to the intervals between irradiation. Introducing an artificial dead time Δ_{d} and neglecting all smaller time stamp distances will result in a signal, which is associated with the part between irradiation pulses. Examples of time stamp distance histograms are displayed in Fig. 4.

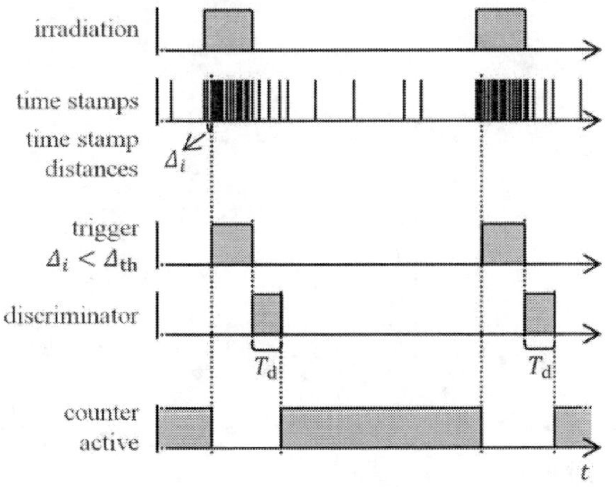

Fig. 1. Trigger and discriminator for gated measurement of the RL signal

C. Irradiation

Pulsed irradiation has been performed in a 30 cm × 30 cm × 20 cm solid water phantom with a 6 MV medical linear accelerator (Elekta). The maximum nominal dose rate in the isocenter was 550 cGy/min. A second pulsed 150 keV X-ray source (Golden Engineering) with 30 µGy per pulse in 30 cm distance, a pulse length of 50 ns, and a repetition rate of 25 Hz was also applied.

III. Results And Discussion

Using the two methods of gated discrimination, profile measurements in the pulsed radiation field of the linac as well as relative depth dose curve measurements in the field of the pulsed X-ray source have been performed. The motivation is to gain further information about the influence of the parameters of both methods on the output of the system at different irradiation conditions.

For the profile measurements the probe was pushed through a 20 cm × 20 cm irradiation field, increasing the length of irradiated light guide with each position. The depth in the phantom was 10 cm for all positions.

Depth dose curves were measured by varying the position of the adapter plate of the probe within the stack of solid water slabs. Another dose depth curve has been measured with a MiniDos OSL dosimeter [10] as a reference.

A. Trigger and Discriminator

The influence of the discrimination time on the profile measurement and the relative dose depth curve is presented in Fig. 2 and Fig. 3.

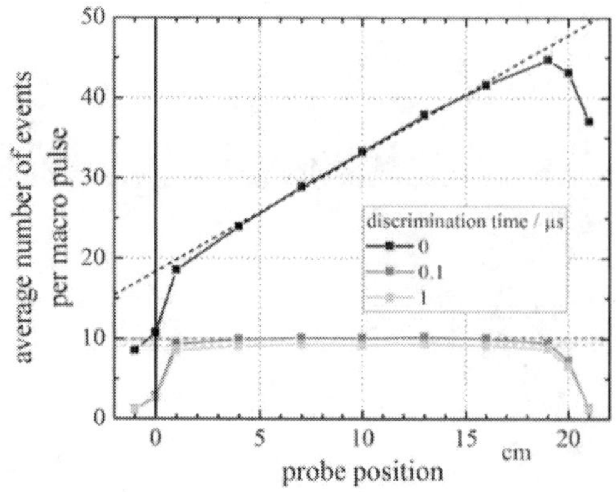

Fig. 2. Measurement of a dose profile in a 20 cm × 20 cm irradiation field of a medical linac. Influence of the discrimination time. Mean of ~ 15000 macro pulses each.

With increasing probe position, i.e. increasing length of irradiated light guide, the unfiltered signal (discrimination time = 0) rises due to the stem effect. The introduction of a dead time of 100 ns leads to a signal with no detectable stem influence left. Further extension of the discrimination time only results in a reduced signal but has no benefit to the stem suppression.

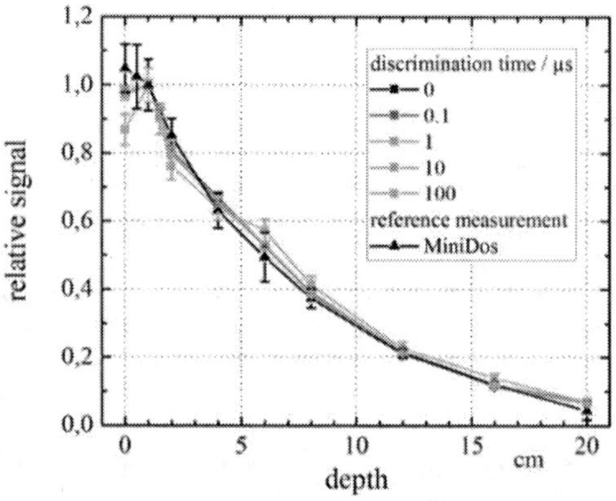

Fig. 3. Measurement of the depth dose curve with the trigger discriminator method in the field of a pulsed X-ray source. Reference measurement with integrating MiniDos OSL dosimeters. Mean of 200 Pulses each.

The dose depth curves are in agreement with the reference for discrimination times up to 10 µs. Larger values result in deviations due to statistical fluctuation because of the small signal. There is a difference between RL probe and reference for the entrance dose at depth 0. The RL probe detects a small build-up effect whereas the reference shows a strictly decreasing behavior. This is due to the different construction of both detectors. The encapsulation of the RL probe is thinner than the one of the reference. So for the RL probe the conditions for secondary particle equilibrium are not met at the surface of the phantom.

To detect all irradiation pulses of the linac, the trigger threshold had to be adjusted with increasing depth. This is caused by the influence of the absorbed dose per irradiation pulse on the measured event rate during the pulse. The problem can be solved in several ways. To obtain a reliable, independent trigger, it is possible to increase the length of irradiated light guide on purpose and use the stem signal for triggering as in [4] or install a second reference fiber with independent light detection which is sensitive on the scattered radiation in the treatment room. A third possibility would be the use of an external trigger signal provided by the linac.

B. Artificial Dead Timer

The same profile measurements as described previously have been analyzed in terms of time stamp distances. Fig. 4 displays the corresponding histograms.

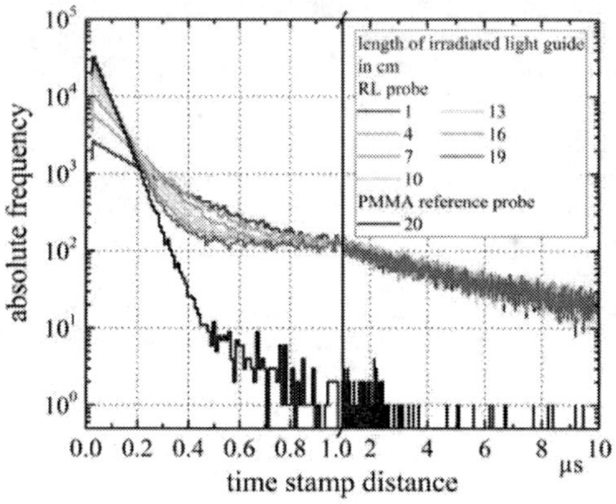

Fig. 4. Histograms of time stamp distances for BeO and PMMA reference probe. Dependency on the length of irradiated light guide. Pulsed irradiation with 550 cGy/min to a total absorbed dose of 2 Gy, each measurement.

The distribution of time stamp distances below 1 µs changes with varying length of irradiated light guide, whereas the distribution for larger distances remains unaffected. An additional measurement with a PMMA reference probe without BeO detector represents the stem only signal whose main contribution is at distances below 1 µs.

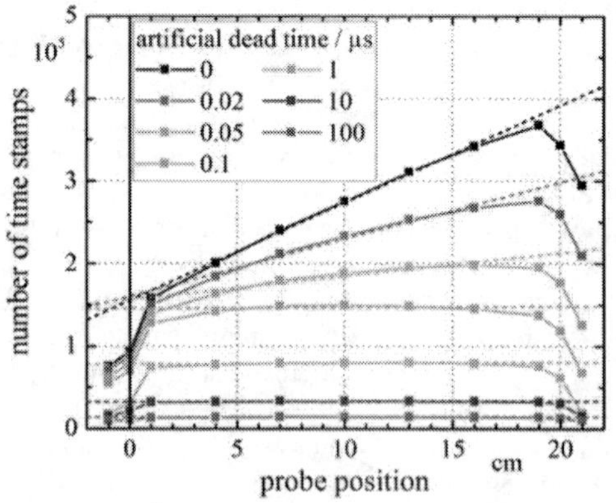

Fig. 5. Influence of the artificial dead time on the measurement of a dose profile in 20 cm × 20 cm irradiation field of a medical linac.

https://doi.org/10.1051/epjconf/201817009012

Applying the method of artificial dead time, the profiles, displayed in Fig. 5 can be obtained. Without artificial dead time the full histograms contribute to the signal and the stem influence is clearly visible. Increasing the dead time leads to a reduction of the effect, but also to a smaller total signal.

The quality of stem discrimination can be quantized, by calculating the ratio between the generated stem per cm irradiated light guide (which is represented by the slope of the profiles in Fig. 5) and the pure RL signal (which is given by the value of the linear approximation at position 0 cm). Since the real beam profile has no perfect rectangular shape, only the central part (4-16 cm) has been used for the linear approximation. A stem to RL ratio of 1%/cm means that, irradiated at the same dose rate, 100 cm of light guide generate the same event rate as the RL originating from the BeO detector at the tip of the probe. The efficiency of the stem discrimination in dependency of the artificial dead time is shown in Fig. 6. The best quality of stem discrimination is achieved with 100 ns artificial dead time. The minimum stem to RL ratio is 0.4%/cm.

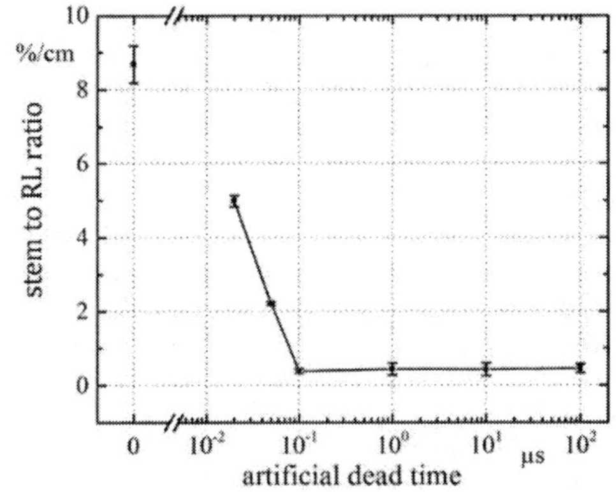

Fig. 6. Efficiency of the stem discrimination in dependency of the artificial dead time.

The results of the depth dose curve measurements with artificial dead time are displayed in Fig. 7. For dead times longer than 0.1 µs there is a strong deviation from the reference. Also the same build-up effect as in Fig. 3 is visible.

For dead times of 0.1 µs and smaller the depth dose curve follow the reference in rough accordance. But there is a noticeable overestimation of the dose curve for larger depths. The explanation is the following. The shape of the histogram is dependent on the dose per pulse. A smaller dose per irradiation pulse results in a smaller event rate during irradiation pulses which causes a right shift of the histogram. With a constant preset value for the artificial dead time, more time stamp distances contribute to the signal. This problem has to be investigated next. A possible solution would be a variable artificial dead time which is adjusted, based on an estimation of the event rate during irradiation pulses.

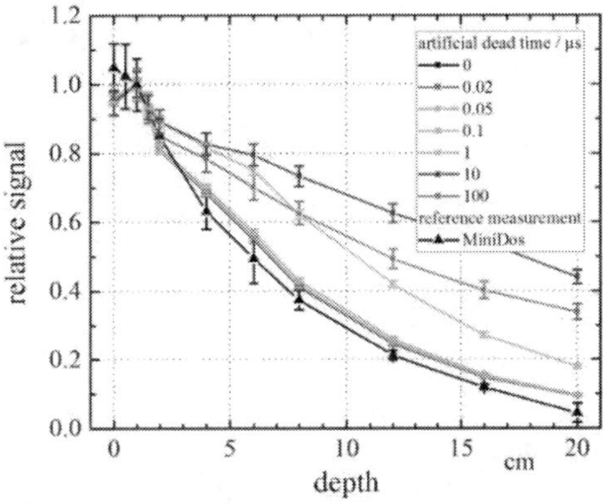

Fig. 7. Depth dose curve measurement with artificial dead time in the field of a pulsed X-ray source. Measurement with integrating MiniDos dosimeters as reference.

IV. Conclusion

Both methods "trigger and discriminator" and "artificial dead time" successfully discriminate the stem effect, which is the main obstacle in fiber optic dose rate measurements. The influence of the parameters, trigger threshold, discrimination time and artificial dead time, has been investigated, leaving open the dependency of the histograms on the dose per pulse.

Acknowledgment

The authors thank H. Blank and the practice for radiotherapy and radiooncology Distler in Pirna, Germany for the possibility to perform measurements at their medical linear accelerator.

References

[1] S. F. de Boer, A. S. Beddar, and J. A. Rawlinson, "Optical filtering and spectral measurements of radiation-induced light in plastic scintillation dosimetry", *Phys. Med. Biol.*, vol. 38, no. 7, pp. 945-958, Apr. 1993.

[2] J. C. Polf, *et al.*, "Real-time luminescence from Al2O3 fiber dosimeters", *Radiat. Meas.*, vol. 38, no. 2, pp. 227–240, Apr. 2004.

[3] R. Gaza *et al.*, "A fiber-dosimetry method based on OSL from Al2O3:C for radiotherapy applications", *Radiat. Meas.*, vol. 38, no. 4–6, pp. 809–812, Aug. 2004.

[4] T. Teichmann *et al.*, "Gated discrimination of the stem signal in pulsed radiation fields for a fiber optic dosimetry system based on the radioluminescence of beryllium oxide", *Radiat. Meas.*, to be published. DOI: 10.1016/j.radmeas.2017.03.046.

[5] T. Teichmann *et al.*, "Real time dose rate measurements with fiber optic probes based on the RL and OSL of beryllium oxide", *Radiat. Meas.*, vol. 90, pp. 201–204, July 2016.

[6] A. M. C. Santos, *et al.*, "Characterization of a real-time fibre-coupled beryllium oxide (BeO) luminescence dosimeter in X-ray beams", *Radiat. Meas.*, vol. 53-54, pp. 1–7, June 2013.

[7] A. M. C. Santos, M. Mohammadi, and S. Afshar V., "Investigation of a fibre-coupled beryllium oxide (BeO) ceramic luminescence dosimetry system", *Radiat. Meas.*, vol. 70, pp. 52–58, Nov. 2014.

[8] A. M. C. Santos, M. Mohammadi, and S. Afshar V., "Evaluation of a real-time BeO ceramic fiber-coupled luminescence dosimetry system for dose verification of high dose rate brachytherapy", *Med. Phys.*, vol. 42, no. 11, pp. 6349-6356, Oct. 2015.

[9] E. G. Yukihara, "Luminescence properties of BeO optically stimulated luminescence (OSL) detectors", *Radiat. Meas.*, vol. 46, no. 6–7, pp. 580–587, June 2011.

[10] A. Jahn *et al.*, "The BeOmax system – Dosimetry using OSL of BeO for several applications", *Radiat. Meas.*, vol. 56, pp. 324–327, Sep. 2013.

ANNETTE Project: Contributing to The Nuclearization of Fusion

W. Ambrosini, L. Cizelj, P. Dieguez Porras, R. Jaspers, J. Noterdaeme, M. Scheffer, and C. Schoenfelder

Abstract— **The ANNETTE Project (Advanced Networking for Nuclear Education and Training and Transfer of Expertise) is well underway, and one of its work packages addresses the design, development and implementation of nuclear fusion training. A systematic approach is used that leads to the development of new training courses, based on identified nuclear competences needs of the work force of (future) fusion reactors and on the current availability of suitable training courses. From interaction with stakeholders involved in the ITER design and construction or the JET D-T campaign, it became clear that the lack of nuclear safety culture awareness already has an impact on current projects. Through the collaboration between the European education networks in fission (ENEN) and fusion (FuseNet) in the ANNETTE project, this project is well positioned to support the development of nuclear competences for ongoing and future fusion projects. Thereby it will make a clear contribution to the realization of fusion energy.**

Index Terms **— ANNETTE, competence, DEMO, education, ENEN, FuseNet, fusion, ITER, nuclear, training.**

I. Introduction

The ANNETTE Project (Advanced Networking for Nuclear Education and Training and Transfer of Expertise, see http://www.enen-assoc.org/en/training/annette.html) is a European Horizon 2020 project that addresses the present situation of nuclear energy in Europe by a continuing effort in the field of education and training. The aim is to assure a qualified work force in the next decades by consolidating and better exploiting the achievements already reached in the past and by tackling the present challenges in preparing the European workforce in the different nuclear areas. Special attention is paid to continuous professional development, life-long learning and cross border mobility.

Keeping in mind the attractiveness of fusion as a possible safe, sustainable and low carbon source of electricity with the potential to contribute to the future mix of different energy sources, the EU created a coherent, ambitious but pragmatic fusion program aiming, via a comprehensive, integrated science, technology and engineering program, to provide electricity to the grid by the middle of the 21st century. As noted in the roadmap to the realization of fusion energy (the Fusion Roadmap, see https://www.euro-fusion.org/wpcms/wp-content/uploads/2013/01/JG12.356-web.pdf), the evolution of the fusion program requires a shift from "*from pure research to designing, building and operating future facilities like ITER and DEMO. This transition requires strengthening the available engineering resources, with a marked change from non-nuclear to nuclear technologies, and has to be facilitated during Horizon 2020 by specific measures in support of training and education*".

At the beginning of 2016, the ENEN (European Nuclear Education Network, see http://www.enen-assoc.org/) association together with 24 partners launched the four-year project ANNETTE. This project is co-funded by the European Commission under the Euratom Research and Training Programme on Nuclear Energy within the H2020 Programme, Call NRFP 2014-2015.

One partner of ENEN in this project is the FuseNet association (European Fusion Education Network, see http://www.fusenet.eu/), responsible for the ANNETTE Work Package 6 (WP6), dealing with Coordinating the Nuclearization of Fusion.

In its activities, FuseNet will rely on the existing initiatives and institutions in Europe that are providing education and training on nuclear (fission) related competences. In the first place, those institutions will be involved that have decided to cooperate within the ANNETTE project.

The human resources involved in development, design and construction of fusion facilities must possess suitable nuclear related competences. Through WP6 the ANNETTE project takes on this challenge. The objective is to contribute to the availability of competent human resources, as needed to enable the transition from non-nuclear to nuclear technologies described in the Fusion Roadmap. Therefore, the inclusion of

"This work was co-funded by the European Commission under the Euratom Research and Training Programme on Nuclear Energy within the H2020 Programme, Grant Agreement Number: 661910."

W. Ambrosini is with the University of Pisa / Italy (e-mail: walter.ambrosini@ing.unipi.it).

L. Cizelj is with the Josef Stefan Institute, Ljubljana / Slovenia, (e-mail: leon.cizelj@ijs.si).

P. Dieguez Porras, is with the ENEN Association, Saclay / France (e-mail: sec.enen@cea.fr).

R. Jaspers is with the Eindhoven University of Technology, Eindhoven / The Netherlands (e-mail: r.j.e.jaspers@tue.nl).

J. Noterdaeme, is with the University of Ghent, Ghent / Belgium (e-mail: noterdaeme@ipp.mpg.de).

M. Scheffer is with the FuseNet Association, Eindhoven / The Netherlands (e-mail: m.scheffer@tue.nl or feo@fusenet.eu).

C. Schoenfelder is with Schoenfelder.Training, Köln / Germany (e-mail: christian@schoenfelder.training)

WP6 within the ANNETTE project is an effective contribution to meeting one important objective of the Fusion Roadmap.

II. The need for nuclear competences in fusion

A. *Why ITER is nuclear*

As already mentioned, the Fusion Roadmap stated in 2012 that the evolution of the fusion program requires "*… a marked change from non-nuclear to nuclear technologies, …*".

ITER (the largest international fusion experiment), currently being built in southern France, is a nuclear facility because of the planned deuterium-tritium plasma in which the fusion reaction will be sustained through internal heating. The ITER reactor will therefore contain an inventory of radioactive material, consisting of the tritium fuel (both to start the fusion reaction and subsequently tritium is bred inside the vacuum vessel) and neutron-activated material, generated when the high-energy neutrons produced during the fusion reaction hit the wall of the reactor.

Consequently, the safety functions of ITER must ensure the confinement of radioactivity and the limitation of radiation exposure, thereby protecting the machine, the ITER staff and the environment from the radiation.

B. *(Nuclear) Licensing of ITER and its implications*

As a nuclear facility, the product life cycle (design, manufacturing, assembly, construction, commissioning, operation and decommissioning) of ITER must follow the applicable national (i.e. French) laws and regulations for licensing a nuclear facility. It implies that the operator (i.e. the ITER Organization / IO) is a nuclear operator that must apply for the licensing of ITER with the French nuclear safety authority (ASN, Autorité de Sûreté Nucléaire). As an example, IO had to prepare the ITER safety case, a detailed document of more than 5,000 pages, demonstrating that fusion technology is not only scientifically viable, but also safe and environmentally responsible.

However, when comparing nuclear fission with fusion reactors, the demands on licensing are less stringent. In fission, reactors must control reactivity, reactor cooling is required for a long time after reactor shutdown, the reactor core will contain a large amount of fuel (tons of Uranium), and during operation it will produce highly activated waste. In contrast, the fusion reactions will terminate intrinsically in abnormal or accidental operation scenarios. Moreover, after termination of the plasma reaction, the residual power will decrease rapidly. There is only a low quantity of fuel in the plasma, and no highly-activated waste will be produced.

In November 2012, a major licensing milestone was achieved when the French government signed a decree authorizing the creation of the ITER nuclear facility. This decree implied also that IO had to establish safety processes that must comply with French regulations. These processes must be regularly verified through audit and inspection by the French nuclear safety authority.

All this has far reaching implications. IO not only must ensure that its own activities, but also that the activities of its sub-contractors will comply with the applicable French nuclear regulation. A real challenge, as sometimes the supply chain can exist of more than 5 levels of sub-contractors.

Examples of implications on design activities include:

1. Systems and components must be safety classified, possibly also qualified for their use in ITER.
2. Relevant codes and standards (e.g. RCC-MR, Règles de conception et de construction des matériels mécaniques des installations nucléaires hautes températures, expérimentales et de fusion, see http://www.afcen.com/en/publications/rcc-mrx) must be applied when designing safety relevant software and equipment, e.g. Protection Important Components.

Most important is that the introduction and continuous practice of a safety culture of all personnel involved must be ensured, particularly during manufacturing, assembly, construction and commissioning activities.

Often the required competences (knowledge, skills and attitudes) will not be available with the personnel employed by the various sub-contractors of IO, as the majority will not have experiences in nuclear (fission) projects.

C. *Relying on fission experiences*

However, these competences must be built up and should therefore exist within companies that are active in the "normal" nuclear (fission) business.

Moreover, current fission new-build projects in Europe are struggling to keep these projects within planned budget and schedule while complying with evolving safety requirements. To some degree this was caused by the insufficient competence and experience transfer between the (fission) new build period in the 1980s and the current time frame. However, by now the fission industry already has conducted appropriate activities to cope with the challenges of modern large international nuclear (new build or modernization) projects.

Consequently, resulting human resources development as well as education and training activities already pursued in the fission area could be taken as a starting point for development and implementation of necessary actions in the fusion area, and then be extended to include fusion-specific aspects and content.

III. Sequence of WP6 Activities

A. *ANNETTE WP6 sub-tasks*

FuseNet agreed with ENEN already during preparation of the ANNETTE grant application to include the topic Nuclearization of Fusion in the ANNETTE work plan. The international cooperation within ANNETTE, and the involvement of different education and training providers specialized in nuclear issues should guarantee the success of WP6.

This work package is structured into four different sub-tasks:

1. Investigate and specify the specific competence needs for the transition of fusion to a nuclear technology
2. Design and development of fusion specific training addressing the competence needs as specified in (1),
3. Implement and evaluate courses or other forms of learning

as designed and developed in (2),

4. Support the participation in nuclear training activities.

This approach reflects a practical implementation of the Systematic Approach to Training (SAT), which was developed in the nuclear (fission) community under the guidance of the IAEA (International Atomic Energy Agency, see [1]). SAT structures a typical training project into the main steps *job and task analysis / training design / training development / training implementation / training evaluation*.

B. Specification of nuclear competences

To identify and specify required nuclear competences, a workshop was organized and held in November 2016 in Eindhoven, where important fusion stakeholders were convened. Participants came from Fusion for Energy (F4E, mainly the European procurement agency for ITER), UKAEA (United Kingdom Atomic Energy Authority, operator of JET / Joint European Torus), University of Ghent, Eindhoven University of Technology, and nuclear (fission) industry, with further input from IO.

The objective was to receive input for a report on competence needs with respect to nuclearization of fusion, and to receive directions for specifying the required training.

In the report, the results had to be documented in terms of expected target groups, their required (nuclear) competences (knowledge, skills and attitudes), learning outcomes of suitable training to develop these competences, and related learning content.

C. From specification to training development

The list of learning outcomes and related content then was further detailed. It was transformed into a questionnaire and distributed to several European education and training institutes, preferably those that are ANNETTE partners. They were asked to provide information about their training that will develop the knowledge, skills and attitudes (i.e. learning content) as listed in the questionnaire.

This work within WP6 is currently on-going, but will be concluded in summer 2017.

Next, one will have to specify suitable criteria for prioritizing the learning content that shall be used for training development and implementation. Criteria could include the potential impact that the learning content will have on the competence development of the fusion work force.

The impact could be assessed in terms of:

- size of target group,
- the need and urgency indicated by stakeholders in the field,
- project phase in which the competences are required,
- to what extent the learning content will provide key competences that are critical for project success or performance,
- to what extent the learning content will provide competences that are indispensable because of external requirements.
- availability of existing training material within the concerned networks.

After this exercise, the fusion stakeholders will be consulted again to review and possibly verify the prioritization conducted by the WP6 team.

If training related to a specific learning content will exist, it may be directly offered and delivered to the target group, or it may have to be possibly adapted. If it does not exist, training must be newly developed. Of course, the adaptation / development actions must be possible within the scope of the ANNETTE WP6, i.e. within its budget and timing constraints.

In all cases, training development will include the activities for development of training material, and possibly also for qualification of trainers. The training material itself, and the effort for its development, will strongly depend on the training method to be applied, which must be in line with the learning outcomes that shall be achieved through the training.

In the case of knowledge competences, in most cases a face-to-face training and related slides, possibly also a training manual will be sufficient. In the case of developing skills, however, more effort must be put into development of further material, e.g. exercises, case studies, possibly also hands-on experiments or even simulations. Development of tests or examinations will further increase the development effort.

In summary, important decisions will have to be made after evaluation of the training survey, and finalization of the specification. The decisions will have an important impact on the further course of the ANNETTE WP6 project, and its relevance for the fusion stakeholders.

D. Development and implementation of nuclear training

Following the specification of the required training, and the selection of the target group based on a set of assessment criteria as explained in section III.C, in the upcoming phase of the ANNETTE project, (if necessary) existing training courses will be adapted, or new training courses developed. In the latter case, all different training methods will be considered for suitable application, e.g. classical presentations in classroom sessions, perhaps enhanced with exercises or case studies, e-learning, hands-on or practical training, or on-the-job training.

For certain types of training, e-learning or online courses offer benefits in terms of costs and ease of access when properly designed and developed. Discussions have started with UNED (Universidad Nacional de Educación a Distancia, Spain), a FuseNet and ENEN member and project partner in ANNETTE. UNED has a broad experience and appropriate tools for implementing online courses. Consequently, pilot implementation of the first training modules may be expected soon.

The impact of pilot courses on the trainees will then be evaluated, e.g. through tests or examinations. The results and other feedback will be used to further increase the quality of the training.

IV. WP6 Results So Far

In the project that will run until the end of 2019, FuseNet has already completed the first sub-task, and issued a report that summarizes what are the Competence Needs with Respect to Nuclearization of Fusion. A significant contribution to this report was given by fusion stakeholders through the stakeholders' workshop held in November 2016.

A. Job positions that require nuclear competences

One important conclusion from this workshop was that nuclear competences will be expected in a variety of different job positions that may be distinguished by

a) educational level (i.e. EQF level or craftsmen / technician / bachelor / master / PhD),

b) position within the product or system life cycle (research and development, design, manufacturing, assembly, construction, commissioning, operation, decommissioning),

c) position within an organization or a project team (e.g. engineer, senior manager, project manager),

d) position within the design, manufacturing or construction process (e.g. component or system design, design integration, chief engineer, industrial architect),

e) technical specialty (e.g. nuclear material, neutronics, electro magnetics, vacuum, tritium),

f) technical discipline (e.g. mechanical / electrical / automation engineering, physicist).

One should also consider that the individuals of the current work force active in the ongoing fusion projects (ITER, research and development for ITER and DEMO, DEMO) can be classified grossly by the two dimensions:

1. the degree to which people have previously been engaged in a nuclear (fission) environment, being either nuclear-aware, nuclearized or nuclear (see, e.g., [2], page 63),

2. the degree to which people have previously been engaged in a nuclear fusion environment, and are aware of fusion specifics (related to main components and systems of a fusion facility).

B. Fusion Nuclear Engineer

Another important statement given by fusion stakeholders was that the Fusion Nuclear Engineer must be a Nuclear Engineer (see, e.g., http://www.enen-assoc.org/en/emsne/information.html) with more specialized knowledge of fusion specific topics such as vacuum technologies, cryogenic technologies, disruption phenomena, electromagnetic fields and electromagnetic loads management, and tritium waste management.

Although the problem of safety is less stringent than in a fission nuclear power plant, a Fusion Nuclear Engineer must be capable to assure the confinement of radioactive material in accidental scenarios. However, he must also assure that the machine is working in normal operation or will restart after an accidental event. These tasks require important knowledge of specialized technologies, see above. Without taking care of these fusion-specific operation aspects, a substantial risk is that the machine is so safe because it never works – but this must not happen!

Another challenge to be met will be alerting the craftsmen and technicians involved in ITER manufacturing, assembly, construction or commissioning on nuclear safety culture. The number of personnel involved, mostly from "normal" (non-nuclear) industry, will finally reach several thousands in the years to come. It will be a challenge to train them effectively on the basics of nuclear safety culture, evaluate the knowledge imparted, and document the successful evaluation.

C. Required nuclear competences in fusion

Based on the results of the workshop, documented in the a.m. report, the current work within WP6 focuses on specifying the required nuclear training for fusion. This will allow subsequently to identify available suitable nuclear training programs as well as existing gaps.

First, nuclear competences were derived and listed that should be considered within the scope of WP6. The competence needs in terms of Knowledge, Skills, and Attitudes were specified. It was concluded that after having attended the training (to be adapted / developed within the scope of WP6), the trainees shall have *knowledge on*

i. Licensing of a nuclear facility and its impact on organizational, project and personal activities,

ii. Basics of a fusion facility (i.e. its design and its operation),

iii. The multi-disciplinary nature of a fusion project,

they shall *be able (or have the skills) to*

i. Implement nuclear regulation in their product life cycle activities,

ii. Perform the necessary processes and activities within the product life cycle,

iii. Evaluate impacts of design changes/modifications on different fields,

iv. Collaborate, communicate, and innovate within a team,

they shall *exhibit a personal behavior* in accordance with

i. Nuclear safety culture,

ii. Social behavior appropriate for a goal oriented team.

D. Training specification - examples

These training objectives (or learning outcomes) and related nuclear competences were taken as a starting point to develop the training specification. For each learning outcome, the associated learning content was identified. To achieve a better overview, the learning content was developed in a hierarchical structure in two levels, the second level providing more details. Thereby it should become easier to identify (in the next step) available training (modules).

As example, the first (knowledge oriented) training objective

Licensing and safety of a nuclear facility and its impact on organizational, project and personal activities

shall be achieved by dealing with the following Learning Content (Level one):

- Nuclear law and regulation, licensing
- Quality Management
- Codes and standards
- Risks, safety requirements, and their impact on licensing and life cycle activities
- Safety analysis and design activities

As example of a second level learning content, the last topic was broken down into the following Learning Content on Level two:

- Introduction to (deterministic and probabilistic) safety

analysis and related tools used by different technical disciplines for simulations in support of licensing

- Development of a safety analysis report
- Thermal hydraulics: analysis by simulation, simulation tools, Computational Fluid Dynamics, risk methodologies
- ALARA (As Low As Reasonably Achievable) principles
- Radiation protection and radiation monitoring: typical simulation tools, shielding and doses etc.

V. Conclusions

Important stakeholders in the design, construction and operation of current and future fusion reactors, such as IO, F4E, UKAEA, and FIIF (Fusion Industry Innovation Forum, see https://www2.euro-fusion.org/fiif/), have acknowledged the need for nuclear competences. So far, WP6 in the ANNETTE project identified an excellent list of competences needs with the help of these stakeholders. This result and the overview of existing training material obtained through the survey, will provide the basis for a systematic specification of the training which needs to be adapted (when existing) or newly developed.

The collaboration between the education networks of ENEN and FuseNet, through their participation in the ANNETTE project, ensures the involvement of the fission *and* the fusion communities (i.e. education and training as well as research institutes plus linked industry). Through the ANNETTE project, we are therefore in a good position to contribute substantially to the realization of fusion energy, as envisioned in the Fusion Roadmap.

References

[1] IAEA Technical Reports Series No. 380: Nuclear Power Plant Personnel Training and its Evaluation, International Atomic Energy Agency, Vienna / Austria, 1996

[2] Simonovska V., Estorff U. von: Putting into Perspective the Supply of and Demand for Nuclear Experts by 2020 within the EU-27 Nuclear Energy Sector. Report EUR 25291 EN, Petten, Netherlands, 2012

AUTHOR INDEX

AUTHOR INDEX

AUTHOR INDEX

AUTHOR INDEX

AUTHOR INDEX

AUTHOR INDEX